McGraw-Hill Benchtop Electronics Handbook

McGraw-Hill Benchtop Electronics Handbook

260 Most Essential Electronics Topics

Victor F. C. Veley

Fourth Edition

McGraw-Hill

New York San Francisco Washington, D.C. Auckland Bogotá
Caracas Lisbon London Madrid Mexico City Milan
Montreal New Delhi San Juan Singapore
Sydney Tokyo Toronto

Library of Congress Cataloging-in-Publication Data

Veley, Victor F. C.
McGraw-Hill benchtop electronics handbook : 260 most essential electronics topics / Victor F. C. Veley. — 4th ed.
p. cm.
Rev. ed. of: The benchtop electronics reference manual
Includes index.
ISBN 0-07-067496- 5
1. Electronics—Handbooks, manuals, etc. I. Veley, Victor F. C. Benchtop electronics reference manual. II. Title.
TK7825.V45 1998
621.381—dc21 97-44033
CIP

McGraw-Hill

A Division of The ***McGraw-Hill*** *Companies*

1 2 3 4 5 6 7 8 9 0 DOC/DOC 9 0 3 2 1 0 9 8

ISBN 0-07-067496-5

The sponsoring editor for this book was Scott Grillo, the editing supervisor was Bernard Onken, and the production supervisor was Pamela Pelton. It was set in Century Schoolbook by McGraw-Hill's Professional Book Group composition unit, Hightstown, N.J.

Printed and bound by R. R. Donnelley & Sons Company.

Contents

Introduction *xix*

PART 1 Direct current principles

1 The international system (SI) of units *2*

2 Unit of charge *5*

3 Unit of electromotive force (EMF) *7*

4 Unit of power *8*

5 Ohm's law, resistance, and conductance *10*

6 Resistance of a cylindrical conductor *13*

7 Temperature coefficient of resistance *15*

8 Composition resistors—the color code *17*

9 Resistors in the series arrangement *19*

10 Ground—voltage reference level *23*

11 Voltage division rule—voltage divider circuit *26*

12 Sources connected in series-aiding and in series-opposing *30*

13 **The potentiometer and the rheostat** *32*

14 **The series voltage-dropping resistor** *34*

15 **Resistors in parallel** *35*

16 **Open and short circuits** *40*

17 **Voltage sources in parallel** *42*

18 **The current division rule** *44*

19 **Series-parallel arrangements of resistors** *46*

20 **The Wheatstone bridge circuit** *51*

21 **The loaded voltage divider circuit** *53*

22 **Internal resistance—voltage regulation** *54*

23 **Maximum power transfer—percentage efficiency** *57*

24 **The constant current source** *60*

25 **Practical sources in series, parallel, and series-parallel** *63*

26 **Kirchhoff's voltage and current laws** *66*

27 **Mesh analysis** *69*

28 **Nodal analysis** *71*

29 **The superposition theorem** *73*

30 **Millman's theorem** *76*

31 **Thévenin's theorem** *78*

32 **Norton's theorem** *81*

33 **Delta-wye and wye-delta transformations** *84*

34 **Magnetic flux—magnetic flux density** *87*

35 The motor effect ***88***

36 Magnetomotive force—magnetic field intensity—permeability of free space ***90***

37 Relative permeability—Rowland's law—reluctance ***92***

38 Electromagnetic induction—Faraday's law—Lenz's law ***94***

39 Self-inductance ***96***

40 Factors determining a coil's self-inductance ***98***

41 Energy stored in the magnetic field of an inductor ***99***

42 Inductors in series ***101***

43 The L/R time constant ***103***

44 Inductors in parallel ***107***

45 Electric flux—Coulomb's law ***110***

46 Charge density—electric field intensity—absolute and relative permittivity ***111***

47 Capacitance and the capacitor ***113***

48 Energy stored in the electric field of a capacitor ***115***

49 Capacitors in series ***116***

50 Capacitors in parallel ***120***

51 The RC time constant ***123***

52 Differentiator circuits ***128***

53 Integrator circuits ***135***

54 Moving-coil (D'Arsonval) meter movement ***139***

55 The milliammeter *141*

56 The loading effect of a voltmeter *143*

57 The ohmmeter *145*

58 The moving-iron meter *147*

59 The wattmeter *148*

60 The cathode-ray oscilloscope *150*

PART 2 Alternating current principles

61 Introduction to alternating current (ac) *154*

62 The root-mean-square (rms) or effective value of an ac voltage or current *158*

63 The average value of a sine wave *161*

64 Phasor representation of an ac voltage or current *162*

65 Phase relationships *164*

66 Addition and subtraction of phasors *166*

67 Resistance in the ac circuit *168*

68 Conductance in the ac circuit *170*

69 Inductive reactance *171*

70 Capacitive reactance *174*

71 The general ac circuit—impedance—power factor *176*

72 Sine-wave input voltage to *R* and *L* in series *178*

73 Sine-wave input voltage to *R* and *C* in series *180*

74 Sine-wave input voltage to *L* and *C* in series *182*

75 Sine-wave input voltage to *R, L,* and *C* in series *185*

76 Sine-wave input voltage to *R* and *L* in parallel *188*

77 Sine-wave input voltage to *R* and *C* in parallel *190*

78 Sine-wave input voltage to *L* and *C* in parallel *193*

79 Sine-wave input voltage to *R, L,* and *C* in parallel *196*

80 Maximum power transfer to the load (ac case) *200*

81 Resonance in a series LCR circuit *202*

82 *Q*—selectivity—bandwidth *205*

83 Parallel resonant LCR circuit *209*

84 The parallel resonant "tank" circuit *213*

85 Free oscillation in an LC circuit *216*

86 Mutually coupled coils *218*

87 Mutually coupled coils in series and parallel *221*

88 The power transformer—transformer efficiency *224*

89 Complex algebra—operator *j*—rectangular/polar conversions *228*

90 Equating real and imaginary parts—addition and subtraction of phasors *230*

91 Multiplication and division of phasors—rationalization *232*

92 Analysis of a series-parallel circuit with the aid of the *j* operator *234*

93 Analysis of a parallel branch circuit with the aid of the *j* operator *236*

94 Kirchhoff's laws for ac circuits *237*

95 Mesh-current analysis for ac circuits *239*

96 The superposition theorem for ac circuits ***240***

97 Nodal analysis for ac circuits ***242***

98 Millman's theorem for ac circuits ***243***

99 Thévenin's theorem for ac circuits ***244***

100 Norton's theorem for ac circuits ***245***

101 Delta ⇄ wye transformations for ac circuits ***247***

102 Reciprocity theorem ***249***

103 Radio frequency transformers ***250***

104 The practical single-phase alternator ***257***

105 The practical two-phase alternator ***258***

106 The three-phase alternator ***260***

107 The line voltage of the three-phase wye connection ***263***

108 The three-phase delta connection ***265***

109 The decibel ***267***

110 The neper ***271***

111 Waveform analysis ***273***

112 Nonsinusoidal voltages and currents in ac circuits ***276***

PART 3 Solid-state devices and their associated circuits

113 The pn junction diode ***280***

114 Half-wave rectifier circuits ***282***

115 Full-wave rectifier circuits ***288***

116 Voltage multiplier circuits ***294***

117 The zener diode ***296***

118 The bipolar transistor ***297***

119 Base bias ***300***

120 Voltage divider bias ***302***

121 Emitter bias ***303***

122 Collector feedback bias ***305***

123 The common-emitter amplifier ***306***

124 The common-base amplifier ***309***

125 The emitter follower ***310***

126 Class-A power amplifier ***312***

127 The class-B push-pull amplifier ***314***

128 The class-C RF power amplifier ***316***

129 Voltage regulator circuits ***318***

130 The junction field-effect transistor ***321***

131 Methods of biasing the JFET ***324***

132 The JFET amplifier and the source follower ***327***

133 The metal-oxide semiconductor field-effect transistor ***330***

134 The DE MOSFET amplifier ***332***

135 The enhancement-only or E MOSFET ***333***

136 The E MOSFET amplifiers ***335***

137 Operational amplifiers ***337***

138 Positive or regenerative feedback ***341***

139 The Wien bridge oscillator ***343***

140 The RC phase-shift oscillator ***345***

141 The multivibrator ***347***

PART 4 Tubes and their associated circuits

142 The triode tube—static characteristics ***350***

143 Ac and dc plate resistances ***351***

144 Transconductance, g_m ***353***

145 Amplification factor ***355***

146 The triode tube as an amplifier ***356***

147 Dynamic characteristics ***359***

148 Types of bias ***362***

149 The development of the amplifier tube ***365***

150 Negative feedback ***371***

151 The grounded-grid triode circuit and the cathode-follower ***374***

PART 5 Principles of radio communications

152 Amplitude modulation—percentage of modulation ***380***

153 AM sidebands and bandwidth ***382***

154 Percentage changes in the antenna current and the sideband power because of amplitude modulation ***385***

155 Plate modulation ***387***

156 The AM transmitter ***390***

157 Frequency and phase modulation ***396***

158 The FM transmitter ***400***

159 AM and FM receivers ***406***

160 Calculation of operating power by the direct and indirect methods ***413***

161 The field strength of an antenna—effective radiated power ***415***

162 The single-sideband system ***417***

163 The piezoelectric crystal ***418***

164 The frequency monitor ***423***

165 Television broadcast frequencies ***424***

166 Distributed constants of transmission lines ***426***

167 The matched line ***429***

168 The unmatched line— reflection coefficient ***432***

169 Voltage standing-wave ratio ***441***

170 The quarter-wave line ***443***

171 The Hertz antenna ***444***

172 Parasitic elements ***447***

173 Electromagnetic (EM) waves in free space ***450***

174 The Marconi antenna ***454***

175 Directional antennas for AM broadcast ***456***

176 Loop antenna ***459***

177 Propagation ***462***

178 Microwave antenna with the parabolic reflector ***472***

179 Propagation in a rectangular waveguide ***473***

180 Circular waveguides ***478***

181 The 50-Ω Smith chart ***482***

182 The universal Smith chart ***485***

183 Determination of the load impedance ***489***

184 Matching stubs ***492***

185 Transmission line losses ***495***

186 Parameters of a pulsed radar set ***498***

187 Radar pulse duration and the discharge line ***501***

188 Bandwidth of the radar receiver—intermediate frequency and video amplifier stages ***502***

189 Radar ranges and their corresponding time intervals ***503***

190 Frequency-modulated and continuous-wave radar systems ***505***

PART 6 Introductory mathematics for electronics

191 Common fractions ***508***

192 Decimal numbers ***512***

193 Scientific notation ***518***

194 Linear equations of the first degree ***524***

195 Simultaneous equations ***530***

196 Determinants ***533***

197 Matrices ***538***

198 Quadratic equations ***543***

199 Graphs ***547***

200 Angles and their measurement ***556***

PART 7 Intermediate mathematics for electronics

201 The theorem of Pythagoras ***562***

202 The sine function ***564***

203 The cosine function ***567***

204 The tangent function ***570***

205 Cosecant, secant, and cotangent ***573***

206 Trigonometric functions for angles of any magnitude ***576***

207 The sine and cosine rules ***579***

208 Multiple angles ***582***

209 The binomial series ***585***

210 The exponential and logarithmic functions ***587***

211 Introduction to differential calculus ***590***

212 Differentiation of a function of a function ***593***

213 Differentiation of a product ***595***

214 Differentiation of a quotient ***597***

215 Differentiation of the trigonometric functions ***598***

216 Differentiation of the exponential and logarithmic functions ***601***

217 The hyperbolic functions and their derivatives ***603***

218 Partial derivatives ***607***

219 Applications of derivatives: maxima, minima, and inflexion conditions ***609***

220 Introduction to integral calculus ***613***

PART 8 Digital principles

221 Number systems ***620***

222 The binary number system ***622***

223 Binary-coded decimal system ***627***

224 The octal number system ***628***

225 The hexadecimal number system ***631***

226 The duodecimal number system ***633***

227 The arithmetic of complements ***636***

228 Introduction to Boolean algebra ***639***

229 The OR operation ***642***

230 The AND operation ***645***

231 The NOT operation ***648***

232 The NOR operation ***650***

233 The NAND operation ***652***

234 The exclusive-OR operation ***654***

235 The exclusive-NOR operation ***656***

236 Combinational logic circuits ***658***

237 Boolean algebra: axioms and fundamental laws ***662***

238 DeMorgan's theorems ***670***

239 Karnaugh maps ***673***

240 Harvard charts ***677***

PART 9 Satellite communications

241 Introduction to satellite communications—the low-earth orbit ***680***

242 The geostationary satellite—angles of elevation and azimuth ***682***

243 Distance between a geostationary satellite and its earth station—spacing of satellites ***686***

244 The satellite signal—frequency-division multiplex ***687***

245 Time-division multiplex ***693***

246 The communications satellite ***696***

247 Launching of geostationary satellites ***699***

248 The earth station transmitter ***703***

249 The earth station receiver ***705***

250 Analysis of the satellite communications system ***707***

PART 10 Fiber-optics technology

251 Introduction to fiber-optics technology ***714***

252 Elements of a fiber-optic communications system ***716***

253 The physics of light ***718***

254 Propagation of light through a cladded optical fiber ***721***

255 Types of optical fiber ***725***

256 Losses in optical fibers ***728***

257 Fiber joins ***730***

258 Light sources ***732***

259 Light detectors ***737***

260 Analysis of fiber-optic networks and systems ***740***

APPENDICES

A Elements of a remote-control AM station ***749***

B AM station license ***750***

C Elements of a directional AM station ***751***

D Directional AM station license ***752***

E Elements of an FM station ***754***

F FM station license ***755***

G FCC emission designations ***756***

H FCC tolerances and standards ***758***

I The phonetic alphabet ***761***

J Answers to practice problems ***762***

Index ***781***

INTRODUCTION

The breadth and depth of this fourth edition have been increased as the result of feedback from instructors and students who have used the first three editions successfully. This new material appears in the following three areas:

1. *Satellite communications* (Part 9) Instructors have pointed out that the third edition did not contain chapters related to the analysis of complete communications systems. This void has been partly filled by ten *new* chapters devoted entirely to modern satellite communications. These chapters refer to such subject areas as low earth orbit (LEO) satellites, geostationary earth orbit (GEO) satellites, the satellite signal, including frequency-division and time-division multiplex, the satellite transponder, launching of satellites, the earth-station transmitter, the earth-station receiver, and the analysis of entire satellite systems. In addition, there is a description of the Global Positioning System (GPS).
2. *Fiber-optics technology* (Part 10) To fill the remainder of the void, a further ten *new* chapters have been added to cover the background of modern fiber-optic systems. These chapters detail the topics of single-mode and multimode fibers, fiber losses, fiber joins, light sources, light detectors, local area networks (LANs), and the analysis of complete fiber-optic systems. The chapters of both Parts 9 and 10 rely considerably on the principles outlined in the chapters of the third edition.
3. The contents of a number of the third edition chapters have been upgraded.

Like the first three editions, the real strength of the fourth edition lies in its unique format. *Two hundred and sixty* of the more common topics in electronics and communications have been selected from the subject areas of dc, ac, solid-state and tube fundamentals, and communications including microwave, applied mathematics, digital principles, satellite communications, and fiber-optics technology. Each topic is treated separately and is explored in four stages:
1. The basic principles are discussed in a "friendly" style, which invites the reader to *think* in terms of electronics and communications.
2. As many relevant equations as possible appear in the mathematical derivations. All of these equations use the SI system of metric units.

3. One or more examples are used to illustrate the application of the various equations. Wherever possible, practical values are used in the examples. It follows that many of the circuits can be constructed in the laboratory. The calculated results can then be compared with the measured values.
4. Reinforcement is provided by the Practice Problems, which appear at the end of each chapter. The anwers to *all* of these problems are included in Appendix J at the back of the book.

The result of using this format is to create an excellent reference guide for the electronics and communications technicians already working in the industry or in the armed services. This book is also a valuable aid to students who are studying electronics in the community colleges, private vocational schools, and senior high schools.

One hundred and twelve of the topics are devoted to the subjects of direct and alternating current. The book can therefore serve as the text for introductory courses in electronics and applied mathematics. When all topics are taken into account, they cover most of the material contained in the qualifying examinations for technicians entering the fields of electronics and communications. The book would also be an important reference handbook for hobbyists and any applicant who wishes to obtain his/her FCC Commercial General Radiotelephone Operator's License (GROL).

Victor F. C. Veley

1 PART

Direct current principles

1
The international system (SI) of units

As far as electrical units are concerned, we are fortunate to live in an age in which one unified system has been adopted. Prior to about 1960, there were really three systems of measurement units. To start with, we had the practical everyday units, some of which you might already know: the ampere, the volt, and the watt. The second system of units was based on magnetism and the third on electrostatics. The last two were referred to as *cgs systems* because they used the centimeter as the unit of length, the gram for mass, and the second for time. Frankly, it was a complete mess! Not only were there three possible units in which an electrical quantity could be measured, but between the three systems, there were horrible conversion factors that were virtually impossible to memorize.

In 1901, Professor Giorgi of Italy proposed a new system founded on the meter (100 centimeters, or slightly greater than 1 yard), the kilogram (1000 grams, or about 2.2 pounds), and the second—as the units of length, mass, and time. For electricity, it was necessary to define a fourth fundamental unit and then build on this foundation to establish other units. In 1948, the *ampere*, which measures electrical current, was internationally adopted as a fourth unit. We therefore have the *MKSA* (meter, kilogram, second, ampere) *system*, which is also referred to as the *international system* or *SI* (*Systéme International d'Unités*). The three previous systems have therefore been replaced by a single system, with the added attraction that the old practical units are part of the new system.

In any study of electricity and electronics, we often need to use mechanical units to measure such quantities as force and energy. Therefore, this chapter is used to establish the mechanical SI units; some of these units are directly transferable to the electrical system. For example, electrical energy and mechanical energy are each measured by the same unit.

UNIT OF FORCE

Isaac Newton stated that when a force is applied to an object or mass (Fig. 1-1A), the object or mass accelerates so that its speed or velocity increases. In the SI system, the velocity is measured in meters per second. The abbreviations are *m* for meter and *s* for second so that meters per second is written as *m/s*. Acceleration is expressed as *meters per second per second*. For example, if a mass starts from rest with

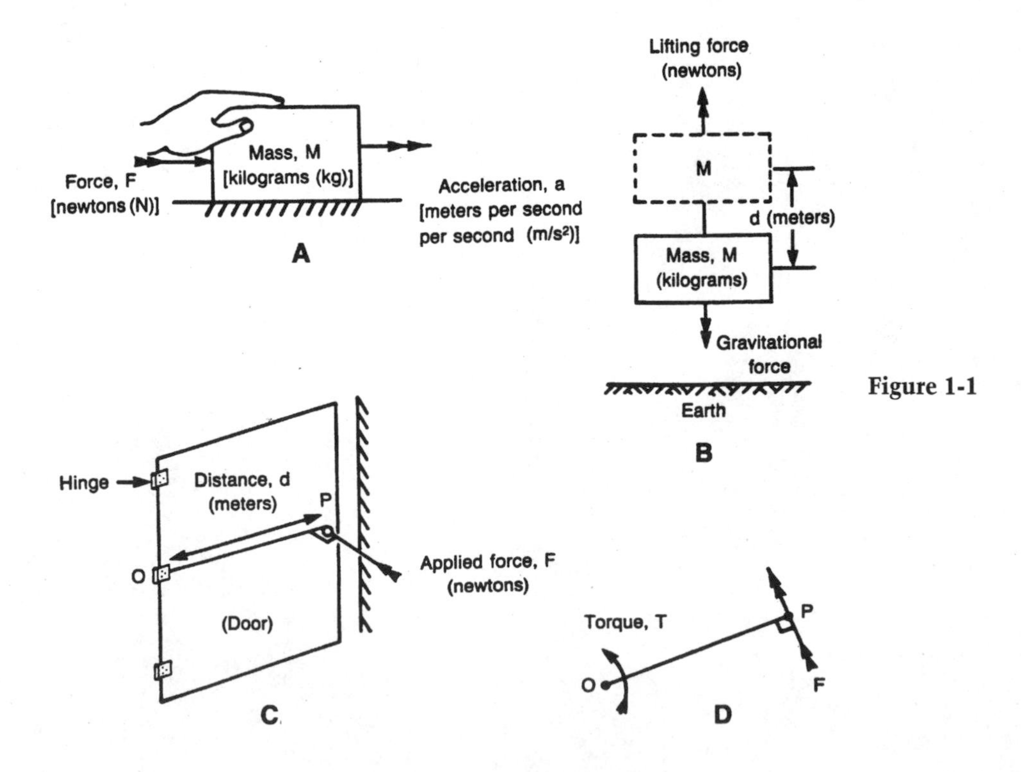

Figure 1-1

zero velocity and is given an acceleration of 3 meters per second per second, its velocity after 1 second is 3 meters per second, after 2 s is 6 m/s, after 3 s is 9 m/s and so on.

If the force applied to a particular mass is increased, the acceleration will be greater. However, if the force is kept the same, but the mass is greater, the acceleration will be less. The unit of force in the SI system is the *newton* (N), which will give a mass of 1 kilogram an acceleration of 1 meter per second per second in the direction of the force.

When a mass is falling under the force of the earth's gravity, its acceleration is 9.81 m/s^2. Therefore, the gravitational force on the mass of 1 kg is 9.81 N. This force is sometimes referred to as *1-kilogram weight*. Of course, on the moon the force of gravity exerted on 1 kg would be less than 9.81 N.

UNIT OF ENERGY OR WORK

Energy is the capacity for doing *work*. Therefore, both of these quantities are measured by the same unit. When a force is applied through a certain distance in the direction of the force, the energy must be supplied so that work can be performed. A good example is lifting a mass at a constant speed against the force of gravity (Fig. 1-1B). When the mass has been raised through a given height or distance, you have expended some energy in performing a certain amount of work. The larger the force and the longer the distance through which the force is applied, the greater is the amount of work that must be done. The value of the work performed is then found by multiplying the force in newtons by the distance in meters. The result of multiplying any two quantities together is called their *product*. Therefore, you can say that work is the product of force and distance. In the SI system, the unit of mechanical energy or work is the *joule (J)*. For example, if a force of 5 N is applied through a distance of 2 m, the work done in joules is 2 m × 5 N = 10 J. One joule can therefore be thought of as 1 meter-newton, as it is the result of multiplying 1 meter by 1 newton.

Notice what is being done. The newton was defined in terms of our fundamental units of mass (kilogram), length (meter), and time (second). The joule is derived from the newton and the meter; in other words, each new unit is defined in terms of its predecessors. This is the logical manner in which a system of units is established.

UNIT OF TORQUE

A *torque* produces a twisting effect, which is used every day when you open a door by either pushing or pulling its handle. The force that we apply is most effective when its direction is at right angles (90°) to the line joining the handle to the door hinge (Fig. 1-1C). This is clearly so because if the force and the line were in the same direction, the door would not open at all. If the handle were positioned close to the hinge, it would be difficult to open the door. We are led to the conclusion that the value of the torque must be equal to the result of multiplying the applied force, F, by the distance, d, which is at right angles (90°) to the force's direction. This distance is the length between the point, P, at which the force is applied and the pivot, O, about which the twisting effect or the rotation occurs (Fig. 1-1D). Torque is equal to the product of force and distance, but with one very important difference. In the case of work, the directions of the force and the distance are the same, but with torque, the directions are 90° apart. Consequently, torque is not measured in joules, but in *newton-meters*.

UNIT OF POWER

There is often much confusion over the distinction between *work* and *power*. *Power* is the rate at which work is performed or energy is expended. As soon as you see the word *rate,* you must realize that time is involved. Here is a mechanical example. Suppose that you have a heavy weight and you ask a powerful adult to lift it up a certain height against the force of gravity. The adult will be able to perform this task quickly (in a short time) because of his or her power. However, you could have a complex pulley system attached to the weight and at the end of the system there might be a wheel. A small child could be capable of turning the handle on the wheel and the weight would slowly rise to the same height achieved by the powerful adult. Neglecting the weight of the pulley system and its friction, the total work performed by the adult and the child is the same, but the child took a much longer time because he or she is much less powerful than the adult. An old British unit is the *horsepower*, which is equivalent to 550 foot-pounds per second. This means that a motor whose mechanical output is 1 horsepower (hp) would be capable of lifting a mass of 550 pounds through a distance of 1 foot against the force of gravity, and do it in a time of 1 second.

The SI unit of power is the *watt,* whose unit symbol is *W*. The power is 1 watt if 1 joule of energy is created or used every second. For example, when you switch on a 60-W electric light bulb, 60 J of energy are released from the bulb every second, mostly in the form of heat, but a small amount as light. Therefore, watts are equivalent to joules per second, or joules are the same as watts × seconds, which can be written as *watt-seconds.* Because the horsepower and the watt both measure the same quantity, the two must be related and, in fact, 1 horsepower is equivalent to 746 watts.

If the 60-W light bulb is left on for 1 h (3600 s), the energy consumed is 60 W × 3600 s = 216,000 J, and this consumption would appear on the electricity bill. It is clear that, for everyday purposes, the joule is too small a unit. In fact, it takes about 8 J of energy to lift a 2-lb book through a distance of 1 yard, and about half a million joules to boil a kettle of water. A larger unit would be the *watthour* (Wh), which is the energy consumed when a power of 1 watt is operated for a time of 1 hour. Because 1 h is the same as 3600 s, 1 watthour = 1 W × 3600 s = 3600 J. The unit on the electricity bill is still larger; it is the *kilowatt-hour* (kWh), which will be equal to 1000 Wh or 3,600,000 J.

MATHEMATICAL DERIVATIONS

Unit of force

Acceleration,

$$a = \frac{F}{m} \text{ meters per second per second} \quad (1\text{-}1)$$

Force,

$$F = m \times a \text{ newtons} \quad (1\text{-}2)$$

Velocity,

$$v = a \times t \text{ meters per second} \quad (1\text{-}3)$$

Acceleration,

$$a = \frac{v}{t} \text{ meters per second per second} \quad (1\text{-}4)$$

where: F = force (newtons, N)
m = mass (kilograms, kg)
a = acceleration (meters per second per second, m/s^2)
v = velocity (meters per second, m/s)

Unit of energy or work

Work,

$$W = d \times F \text{ joules or meter-newtons} \quad (1\text{-}5)$$

where: W = work done (joules, J)
F = force (newtons, N)
d = distance (meters, m)

Unit of torque

Torque,

$$T = F \times d \text{ newton-meters} \quad (1\text{-}6)$$

where: T = torque (newton-meters, N-m)
F = applied force (newtons, N)
d = distance (meters, m)

Unit of power

Power,

$$P = \frac{W}{t} \quad (1\text{-}7)$$

Work,

$$W = P \times t \text{ joules} \quad (1\text{-}8)$$

From Equation 1-5, $W = d \times F$ so that
Power,

$$P = \frac{W}{t} = \frac{d}{t} \times F = v \times F \text{ watts} \quad (1\text{-}9)$$

where: P = power (watts, W)
W = work (joules, J)
d = distance (meters, m)
v = velocity (meters per second, m/s)

Example 1-1

A force of 150 N is continuously applied to a 30 kg mass that is initially at rest. Calculate the values of the acceleration and velocity after 8 s. When the mass has been moved through a distance of 2 km, how much is the total work performed?

Solution

Acceleration,

$$a = \frac{F}{N} \quad (1\text{-}1)$$

$$= \frac{150 \text{ N}}{30 \text{ kg}}$$

$$= 5 \text{ m/s}^2$$

Velocity after 8 s,

$$v = a \times t \quad (1\text{-}3)$$

$$= 5 \text{ m/s}^2 \times 8 \text{ s}$$

$$= 40 \text{ m/s}$$

Work done,

$$W = d \times F \quad (1\text{-}5)$$

$$= 2 \times 10^3 \text{ m} \times 150 \text{ N}$$

$$= 3 \times 10^5 \text{ J}$$

Example 1-2

A metal block whose mass is 250 g is given an acceleration of 15 cm/s^2. Calculate the value of the accelerating force.

Solution

Accelerating force,

$$F = m \times a \quad (1\text{-}2)$$
$$= 250 \times 10^{-3}\ \text{kg} \times 15 \times 10^{-2}\ \text{m/s}^2$$
$$= 3750 \times 10^{-5}\ \text{N}$$
$$= 0.0375\ \text{N}$$

Example 1-3

A mass of 1500 kg is lifted vertically with a velocity of 180 m/min. Calculate the value of the required power in kilowatts.

Solution

Force,

$$F = m \times a \quad (1\text{-}2)$$
$$= 1500 \times 9.81$$
$$= 14715\ \text{N}$$

The distance through which the mass is lifted in 1 s is 180/60 = 3 m. Therefore, the work done in 1 s is:

Work,

$$W = d \times F \quad (1\text{-}5)$$
$$= 3\ \text{m} \times 14715\ \text{N}$$
$$= 44145\ \text{J}$$

Power,

$$P = \frac{W}{t} \quad (1\text{-}7)$$
$$= \frac{44145\ \text{J}}{1\ \text{s}}$$
$$= 44.145\ \text{kW}$$

Example 1-4

A perpendicular force of 250 N is used to create a torque about an axis of rotation. If the distance from the axis to the application point of the force is 8 cm, calculate the value of the torque.

Solution

Torque,

$$T = F \times d \quad (1\text{-}6)$$
$$= 250\ \text{N} \times 8 \times 10^{-2}\ \text{m}$$
$$= 20\ \text{N-m}$$

PRACTICE PROBLEMS

1-1. A mass of 4 kg is lifted through a height of 20 m against gravity. Calculate the amount of work done in joules.

1-2. A mass that is initially at rest, is subjected to an acceleration of 5 m/s^2 for a time of 4 s. Calculate the distance travelled.

1-3. A 2-kg mass, initially at rest, falls under gravity. Calculate the amount of energy acquired by the mass after a time of 3 s has elapsed.

1-4. In 2 min, a motor is capable of lifting a mass of 300 kg at a constant speed through a height of 40 m against gravity. What is the output power of the motor in watts?

1-5. A mass is projected vertically upward with a velocity of 43 m/s. To what height will it rise before it starts to fall? Neglect any effect of air resistance.

2
Unit of charge

The word *charge* means a quantity of electricity and its letter symbol is Q. The SI unit is the *coulomb* (Charles A. Coulomb, 1736–1806) which was originally defined from a series of experiments performed in the 1830s by Michael Faraday (1791–1867). These experiments involved the flow of electron current through a chemical solution and demonstrated the phenomenon of electrolytic conduction.

In one experiment, a bar of silver and a nickel plate (referred to as the *electrodes*) are immersed in a silver nitrate solution (Fig. 2-1), which acts as the electrolyte. The silver bar is called the *anode* and is connected to one terminal of the battery while the nickel plate (*cathode*) is connected to the other terminal. As time passed, Faraday observed that silver was lost from the anode and an equal amount was deposited on the cathode. In addition, the silver was being transferred at a constant rate. At the time of this experiment, the existence of the electron was unknown and therefore Faraday (wrongly) considered that the movement of the elec-

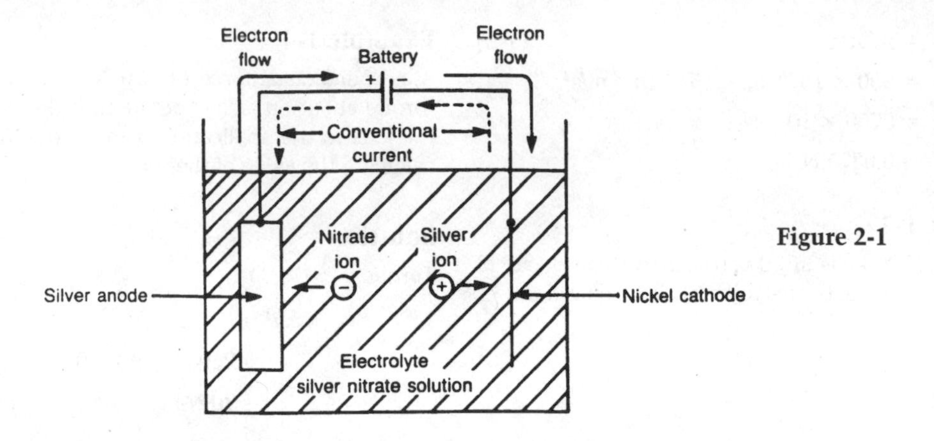

Figure 2-1

tricity through the circuit was carrying the silver from the anode to the cathode. It was then assumed that the anode was connected to the battery's positive terminal while the cathode was joined to the negative terminal so that the direction of the electrical flow or current was from positive to negative. In fact, there is a movement of positive silver ions from the anode to the cathode while negative nitrate ions are traveling in the opposite direction through the electrolyte. The electron flow occurs in the external copper connecting wires with the electrons leaving the battery's negative terminal and entering the positive terminal. Faraday's assumption is often referred to as the *conventional* or *mathematical current flow*, which is considered to flow from the battery's positive terminal to its negative terminal and is, therefore, in the opposite direction to the *actual* physical electron flow.

From his experiments Faraday stated his law of electrolysis. *"The mass of silver leaving the anode and deposited on the cathode is directly proportional to the quantity of electricity or charge passing through the electrolyte."*

For a silver nitrate solution, the coulomb is that quantity of electricity that causes 1.118×10^{-6} kg of silver to be deposited on the cathode. This quantity, 1.118×10^{-6} kg/C, is called silver's *electrochemical equivalent*. Other electrochemical equivalent values are: copper 3.294×10^{-7}, nickel 3.04×10^{-7}, and zinc 3.38×10^{-7} kg/C.

From a study of atomic structure, it follows that the practical charge of 1 coulomb must be equivalent to the negative charge associated with a certain number of electrons. In fact, the coulomb represents the charge carried by 6.24×10^{18} electrons so that the charge, e, possessed by a single electron is only $1/(6.24 \times 10^{18}) = 1.602 \times 10^{-19}$ C.

The *ampere* is the *fundamental* SI unit for measuring the current and has already been referred to in chapter 1. The coulomb can be derived from the ampere by including the unit of time, which is the second. When a current of one ampere flows for a time of one second, a charge of one coulomb passes a particular point in an electrical circuit. In other words, amperes are equivalent to coulombs per second or coulombs are the same as amperes $\times$ seconds.

MATHEMATICAL DERIVATIONS

The equations relating the charge, Q, the current, I, and the time, t, are:

$$Q = I \times t, \qquad I = Q/t, \qquad t = Q/I \qquad (2\text{-}1)$$

where: Q = charge in coulombs (C)
I = current in amperes (A)
t = time in seconds (s)

A larger unit of charge is the ampere-hour (Ah), which is the amount of charge moved past a point when a current of one ampere flows for a time of one hour. Therefore 1 Ah = $1 \times 60 \times 60$ = 3600 C.

Law of electrolysis

Total mass, M, liberated from the anode and deposited on the cathode is given by:

$$M = z\,Q = zIt \text{ kilograms} \qquad (2\text{-}2)$$

where: M = mass liberated (kg)
z = electrochemical equivalent (kg/C)
Q = charge in coulombs (C)
I = current in amperes (A)
t = time in seconds (s)

Example 2-1

If a steady current of 2.5 A exists at a given point for a period of 15 min, calculate the amount of charge flowing past that point in coulombs and ampere-hours.

Solution

Charge,

$$Q = I \times t \quad (2\text{-}1)$$
$$= 2.5 \text{ A} \times 15 \times 60 \text{ s}$$
$$= 2250 \text{ C}$$
$$= \frac{2250}{3600} \text{ Ah}$$
$$= 0.625 \text{ Ah}$$

Example 2-2

A constant current of 4.5 A flows through a copper sulfate solution for a time of 1.7 h. What is the value of the mass liberated from the copper anode?

Solution

Mass liberated,

$$M = zIt \quad (2\text{-}2)$$
$$= 3.294 \times 10^{-7} \text{ kg/C} \times 4.5 \text{ A} \times 1.7 \times 60 \times 60 \text{ s}$$
$$= 9.07 \times 10^{-3} \text{ kg}$$

PRACTICE PROBLEMS

2-1. A current of 8 A flows through a circuit for a time of 4 min. Calculate the amount of charge which has passed a particular point in that circuit.

2-2. In an electrical circuit, 180 C of charge pass a particular point in a time of 5 minutes. What is the value of the current flowing at that point?

2-3. A current of 4 A flows for a time of 2 hr. through a copper sulfate solution. Calculate the mass liberated from the copper anode.

2-4. How many coulombs represent the charge carried by 9.76×10^{19} electrons?

3
Unit of electromotive force (EMF)

In establishing the international system of electrical units, we began in chapter 1 by defining the ampere as the fourth fundamental unit of the MKSA or SI system. By combining the ampere with a unit of time (the second) you were able to derive the coulomb as the unit of charge or quantity of electricity. In chapter 1 we also defined the joule as the SI unit of energy or work. The same unit can be applied to electrical energy, because the various forms of energy (mechanical, electrical, heat, etc.) are interchangeable.

In obtaining the mechanical SI units of chapter 1, we saw that a force was necessary in order to accelerate the mass of an object and cause the object to move. In a similar way, the electrical equivalent of this force imparts a velocity to the free electrons so that a current is able to flow. This "electron moving force" is the *electromotive force* (EMF) whose letter symbol is E. The EMF is the force that gives electricity its motion (Fig. 3-1) and its unit is the *volt,* named after the Italian Count Alessandro Volta (1745–1827) who invented the first chemical cell able to generate electricity. The unit symbol for the volt is the letter V.

In order to drive the charge through a circuit, work must be done, and it is the electrical source (for example, a battery) that must be capable of providing the

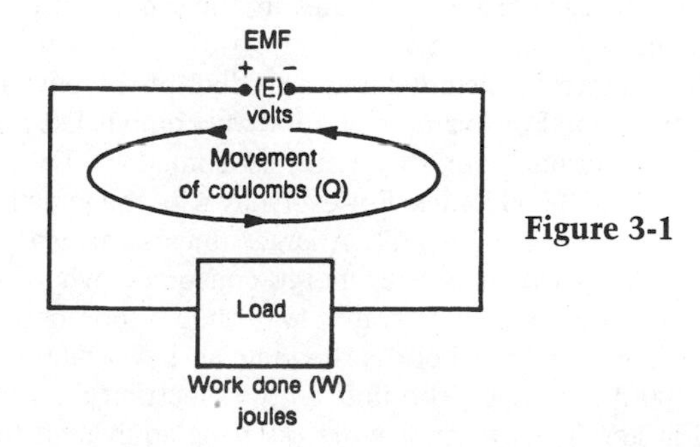

Figure 3-1

necessary energy. The volt may therefore be defined in terms of the coulomb and the joule. The EMF is 1 volt if, when 1 coulomb is driven around an electrical circuit, the work done is 1 joule. If we increase either the EMF and/or the charge, the amount of work done will be greater.

A much smaller unit of work is the *electron-volt* (eV) which is the work done when the charge carried by an electron is driven around a circuit by an EMF of one volt. Because one coulomb is equivalent to the charge carried by 6.24×10^{18} electrons, one joule must be equal to 6.24×10^{18} electron-volts.

MATHEMATICAL DERIVATIONS

The equations relating the work done, the charge and the electromotive force are:

$$E = \frac{W}{Q}, \quad W = Q \times E, \quad Q = \frac{W}{E} \qquad (3\text{-}1)$$

where: W = work done in joules (J)
Q = charge in coulombs (C)
E = EMF in volts (V)

Example 3-1

What is the work done if an EMF of 6 V drives a charge of 3 C around an electrical circuit?

Solution

Work done,

$$\begin{aligned} W &= Q \times E \\ &= 3\text{ C} \times 6\text{ V} \\ &= 18\text{ J} \end{aligned}$$

PRACTICE PROBLEMS

3-1. An electrical source has an EMF of 8 V. If a charge of 6 C is passed between the source terminals, calculate the amount of work done.

3-2. A charge of 25 C is moved through a potential difference of 30 V in a time of 4 s. Calculate the value of the power.

4
Unit of power

In chapter 1, it was decided that power is the rate at which work is performed or the rate at which energy is created. The letter symbol for power is P and its unit is the *watt* (unit symbol, W). For example a 100-watt incandescent light bulb releases 100 joules of energy every second, mostly in the form of heat, but a small amount is changed to light. Therefore, watts are equivalent to joules per second; joules are the same as watts × seconds (watt-seconds).

The energy consumed is the product of the power and the time. For example, if a 100 watt bulb is left on for a time of one hour, the energy consumed is 60 W × 3600 s = 216,000 joules. For everyday use, the joule is obviously too small a unit. A larger unit is the *watt-hour* (Wh) which is the energy consumed when a power of one watt is operated for a time of one hour. Therefore one watt-hour is the same as 1 W × 3600 s = 3600 J. However, the unit on the electricity bill is still larger. The *kilowatt-hour* (kWh) is equivalent to 1000 Wh, or 3,600,000 J.

The power in watts is equal to the same number of joules per second; but the work in joules is the product of the EMF, E, in volts and the charge, Q, in coulombs (equation 3-1). The power is therefore the product of the EMF, E, and the current, I (Fig. 4-1), because the current in amperes is equivalent to the same number of coulombs per second (equation 2-1).

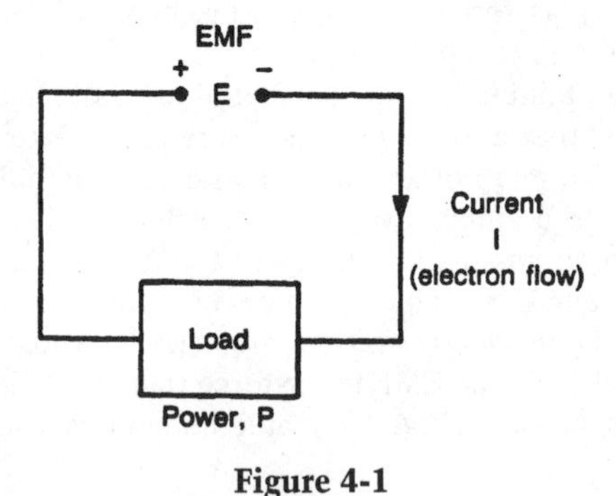

Figure 4-1

MATHEMATICAL DERIVATIONS

Power,

$$P = \frac{W}{t} \text{ watts} \tag{4-1}$$

Work,

$$W = P \times t = E \times Q \text{ joules} \tag{4-2}$$

where: P = power in watts (W)
W = work in joules (J)
E = EMF in volts (E)
Q = charge in coulombs (C)
t = time in seconds (s)

Power,

$$P = \frac{W}{t} \tag{4-3}$$

$$= \frac{E \times Q}{t}$$

$$= E \times \frac{Q}{t}$$

$$= E \times I \text{ watts}$$

where: I = current in amperes (A)

Current,

$$I = \frac{P}{E} \text{ amperes} \tag{4-4}$$

EMF,

$$E = \frac{P}{I} \text{ volts} \tag{4-5}$$

Example 4-1

For a period of 3 h, a current of 4 A is drawn from an electrical source whose EMF is 120 V. Calculate the values of the power and the total energy consumption.

Solution

Power,

$$P = E \times I \tag{4-3}$$

$$= 120 \text{ V} \times 4 \text{ A}$$

$$= 480 \text{ W}$$

Energy consumption,

$$W = P \times t \tag{4-2}$$

$$= 480 \text{ W} \times 3 \text{ h}$$

$$= 1440 \text{ Wh}$$

$$= \frac{1440}{1000} \text{ kWh}$$

$$= 1.44 \text{ kWh}$$

Example 4-2

A 60-W electric light bulb is operated from a 120-V source. What is the value of the current supplied to the bulb?

Solution

Current,

$$I = \frac{P}{E} \tag{4-4}$$

$$= \frac{60 \text{ W}}{120 \text{ V}} = 0.5 \text{ A}$$

PRACTICE PROBLEMS

4-1. A current of 5 A is drawn from a battery whose EMF is 18 V. What is the power delivered from the battery?

4-2. In Practice Problem 4-1, calculate the amount of energy consumed in a time of 3 hours.

4-3. A load has consumed 5 kWh in a time of 8 hours. If the voltage applied across the load is 150 V, what is the value of the load current?

4-4. If the current drawn from a battery is quartered, but the EMF of the battery remains the same, by what factor is the power multiplied?

4-5. Under loaded conditions an electrical motor has an output of 4.5 hp. Over a period of 7 hours, calculate the energy consumed by the load in kWh.

5
Ohm's law, resistance, and conductance

Chapter 3 showed that the electromotive force, E, is responsible for creating the current, I. Consequently, there should be some sort of relationship between E and I. For example, it is only logical to assume that if the EMF applied to an electrical circuit is increased, the current will also increase. However, in 1827 George Simon Ohm stated that the exact relationship between E and I was linear. Under the law which bears Ohm's name, the current flowing through a conductor (such as a very long length of thin wire made from silver, copper, aluminum, etc.) is *directly proportional* to the EMF applied across the conductor. This occurs under constant physical conditions of temperature, humidity, and pressure. This means that if we triple the voltage, the current will also be tripled and if we halve the voltage, the current will also be divided by 2. Whatever we do to the voltage, the same will happen to the current. However, please do not think that Ohm's law is obviously true. Most electronic components *within their operating range* obey Ohm's law, but others do not. For example, if you double the forward voltage across a semiconductor diode, the current will increase, but will not double. In general, solid-state devices and tubes do not obey Ohm's law.

In Fig. 5-1, a voltage source whose EMF can be varied, is connected across a long length of copper wire, which acts as the conductor. An ammeter is placed in the path of the current, I, and will record the current value in amperes. A voltmeter is connected across the battery and will read the value of the EMF, E. To start with, it is obvious that if the applied voltage is zero, the current is also zero. We next assume that when the initial EMF is 12 V, the recorded current is 2 A. If this EMF is now doubled to 24 V, the current will also double to 4 A. When the voltage is again doubled to 48 V, the new current is 8 A. If the EMF is multiplied by 10 to a value of 10×12 V = 120 V, the accompanying current is 10×2 A = 20 A. Finally, if the initial EMF is halved to 12 V/2 = 6 V, the current drops to 2 A/2 = 1 A. These corresponding values of E and I are illustrated in Table 5-1.

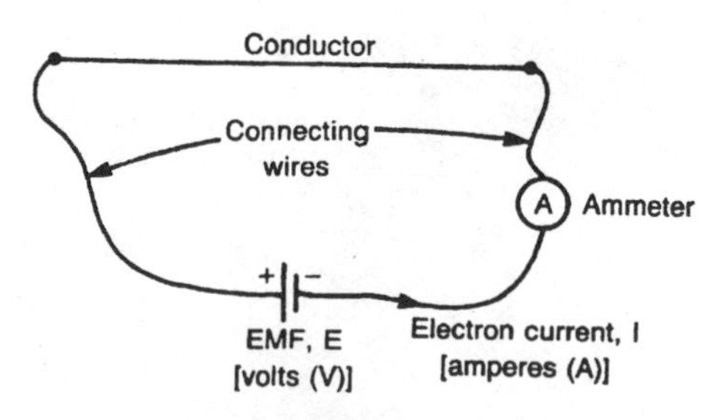

Figure 5-1

Table 5-1. Example of constant ratio of E/I.

EMF (E) volts	Current (I) amperes	Ratio E/I
6 V	1 A	6
12 V	2 A	6
24 V	4 A	6
48 V	8 A	6
120 V	20 A	6

In Table 5-1, the ratio of $E : I$ is calculated for each corresponding voltage and current. In every case the answer is the same (6) which is a constant for the circuit of Fig. 5-1. Ohm's law may therefore be restated as: *"Under constant physical conditions, the ratio of the voltage applied across a conductor to the current flowing through the conductor is a constant."* This constant measures the conductor's opposition to current flow and is called its resistance. The letter symbol for resistance is R and its unit is the *ohm*, which is denoted by the Greek capital letter omega, Ω. Therefore in Fig. 5-1 the conductor's resistance is 6 Ω.

Resistance is that property of an electrical circuit that opposes or limits the flow of current. The component possessing this property is called a resistor whose schematic symbol is ⋀⋀⋀. Practical examples of resistors are described in chapter 8.

If the graph of E versus I is plotted for the values of Table 5-1, the result is the straight line illustrated in Fig. 5-2. This means that there is a linear relationship between E and I, and the straight line is the graphic way of showing that the current is directly proportional to the voltage. By contrast, when a component (such as a transistor) does not obey Ohm's law, its voltage/current relationship is some form of *curve*, not a straight line.

POTENTIAL DIFFERENCE ACROSS A RESISTOR

Figure 5-3 shows a source whose EMF is 12 V is connected across a resistor and the measured (electron

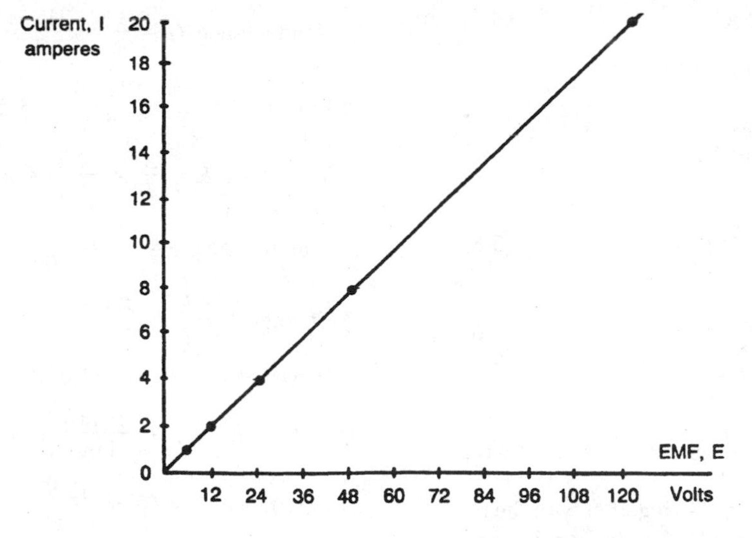

Figure 5-2

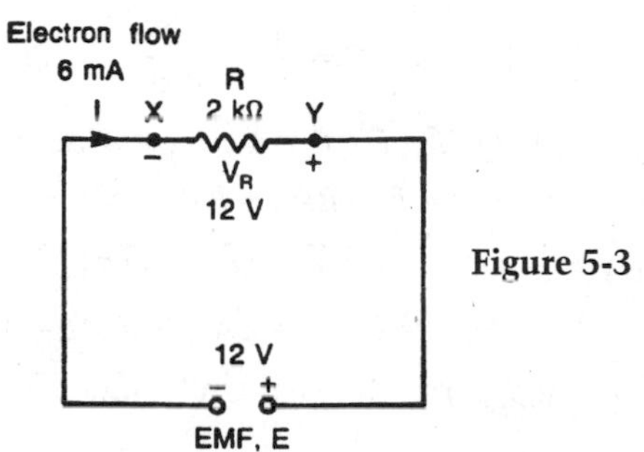

Figure 5-3

flow) current is 6 mA. Therefore, the value of the resistor is $E : I = 12\text{ V} : 6\text{ mA} = 2\text{ k}\Omega$. The electron flow from X to Y develops a voltage, V_R, across the resistor. This voltage is called the *potential difference (PD)*, or *difference of potential (DP)*, and it exactly balances the EMF of the source. Using a water analogy, the back pressure of a pipe always balances the forward pressure of the pump.

When current flows through a resistor, the component becomes hot as the result of friction between the free electrons and the atoms of the material(s) from which the resistor is made. This heat represents an energy *loss* that must be supplied from the voltage source. The resistor dissipates (lost) power in the form of heat, and has a power (or wattage) rating that should not be exceeded.

CONDUCTANCE

You have learned that resistance is a measure of the *opposition* to current flow. However, it is just as valid to introduce an electrical property that measures the *ease* with which current is allowed to flow. Such a property is called the *conductance*, whose letter symbol is G and whose SI unit is the *siemens* (S).

Because a high resistance is obviously equivalent to a low conductance, G and R are reciprocals and are inversely related. It follows that because the ratio $E : I$ is the same as the resistance, the ratio $I : E$ must be equal to the conductance.

MATHEMATICAL DERIVATIONS

Ohm's law

$$\frac{E}{I} = R \text{ ohms} \tag{5-1}$$

$$E = I \times R \text{ volts} \tag{5-2}$$

$$I = \frac{E}{R} \text{ amperes} \tag{5-3}$$

provided the resistance, R, is a constant over the range of operating conditions.

From Equation 4-3:

Power dissipated,

$$\begin{aligned} P &= E \times I \\ &= (I \times R) \times I \\ &= I^2R \text{ watts} \end{aligned} \tag{5-4}$$

It follows that

$$I = \sqrt{\frac{P}{R}} \text{ amperes} \tag{5-5}$$

and

$$R = \frac{P}{I^2} \text{ ohms} \tag{5-6}$$

Also,

Power dissipated,

$$P = E \times I \quad (5\text{-}7)$$

$$= E \times \frac{E}{R}$$

$$= \frac{E^2}{R} \text{ watts}$$

This yields,

$$E = \sqrt{P \times R} \text{ volts} \quad (5\text{-}8)$$

and

$$R = \frac{E^2}{P} \text{ ohms} \quad (5\text{-}9)$$

Conductance,

$$G = \frac{1}{R} = \frac{I}{E} \text{ siemens} \quad (5\text{-}10)$$

The equations 5-1 through 5-9 together with the relationships $P = E \times I$, $I = P/E$, $E = P/I$, enable any two of the four quantities (E, I, R, P) to be found if the other two are given. Of special importance are the two equations $I = \sqrt{P/R}$ and $E = \sqrt{P \times R}$. If you know a resistor's wattage rating, P, and its value, R, you can use these two equations to obtain the highest voltage and current values that will not allow the resistor to overheat.

Example 5-1

All the following questions are related to Fig. 5-1.

(a) If $E = 24$ V and $I = 3$ A, find the values of P, R, and G.

(b) If $E = 80$ V and $P = 320$ W, find the values of I, R, and G.

(c) If $E = 30$ V and $R = 5$ kΩ, find the values of I and P.

(d) If $I = 4$ mA and $P = 48$ mW, find the values of E, R, and G.

(e) If $I = 11$ mA and $R = 3$ kΩ, find the values of E and P.

(f) If $P = 28$ mW and $R = 7$ kΩ, find the values of I and E.

(g) If $P = 12$ μW and $R = 3$ MΩ, find the values of E and I.

(h) If $G = 0.25$ mS and $I = 8$ mA, find the values of E and P.

Solution

(a) Power, $P = E \times I = 24 \text{ V} \times 3 \text{ A} = 72 \text{ W}$.

Resistance, $R = \frac{E}{I} = \frac{24 \text{ V}}{3 \text{ A}} = 8\ \Omega$.

Conductance, $G = \frac{1}{R} = \frac{1}{8\ \Omega} = 0.125$ S.

(b) Current, $I = \frac{P}{E} = \frac{320 \text{ W}}{80 \text{ V}} = 4$ A.

Resistance, $R = \frac{E}{I} = \frac{80 \text{ V}}{4 \text{ A}} = 20\ \Omega$.

Conductance, $G = \frac{1}{R} = \frac{1}{20\ \Omega} = 0.05$ S.

(c) Current, $I = \frac{E}{R} = \frac{30 \text{ V}}{5 \text{ k}\Omega} = 6$ mA.

Power, $P = E \times I = 30 \text{ V} \times 6 \text{ mA} = 180$ mW.

(d) Voltage, $E = \frac{P}{I} = \frac{48 \text{ mW}}{4 \text{ mA}} = 12$ V.

Resistance, $R = \frac{E}{I} = \frac{12 \text{ V}}{4 \text{ mA}} = 3$ kΩ.

Conductance, $G = \frac{1}{R} = \frac{1}{3 \text{ k}\Omega} = 0.33$ mS.

(e) Voltage, $E = I \times R = 11 \text{ mA} \times 3 \text{ k}\Omega = 33$ V.

Power, $P = E \times I = 33 \text{ V} \times 11 \text{ mA} = 363$ mW.

(f) Current, $I = \sqrt{P/R} = \sqrt{28 \text{ mW}/7 \text{ k}\Omega} = 2$ mA.

Voltage, $E = I \times R = 2 \text{ mA} \times 7 \text{ k}\Omega = 14$ V.

(g) Voltage, $E = \sqrt{P \times R} = \sqrt{12\ \mu\text{W} \times 3 \text{ M}\Omega} = 6$ V.

Current, $I = \frac{6 \text{ V}}{3 \text{ M}\Omega} = 2\ \mu$A.

(h) Voltage, $E = \frac{I}{G} = \frac{8 \text{ mA}}{0.25 \text{ mS}} = 32$ V.

Power, $P = E \times I = 32 \text{ V} \times 8 \text{ mA} = 256$ mW.

PRACTICE PROBLEMS

5-1. A current of 1.5 mA flows through a 3300-Ω resistor. Calculate the voltage applied across the resistor and the amount of power dissipated as heat.

5-2. A 75-W light bulb is operated from a 110-Vdc source. What is the current flowing through the bulb's tungsten filament, and what is its resistance?

5-3. When 7 V is applied across a resistor, the power dissipated is 23 mW. Calculate the value of the resistor.

5-4. An EMF of 15 V is applied across a resistor of 1.2 MΩ. Find the energy consumed over a period of 4 hours.

5-5. When a current of 235 mA flows through a wirewound resistor, the power dissipated is 55 W. What is the value of the resistor's conductance?

5-6. If a current flowing through a resistor is doubled, but the resistance remains unchanged, by what factor is the dissipated power multiplied?

5-7. A voltage of 11 V is applied across a 3.9-kΩ resistor. Calculate the values of the current and the power dissipated.

5-8. If the voltage applied across a resistor is tripled, but the resistance remains the same, by what factor is the dissipated power multiplied?

5-9. If the power dissipated in a resistor is halved, but the resistance remains the same, by what factor must you multiply the voltage drop across the resistor?

5-10. If the conductance of a resistor is 0.15 mS and the current flowing through the resistor is 9 mA, calculate the values of the voltage across the resistor and its power dissipation.

6
Resistance of a cylindrical conductor

Normally a load is joined to its voltage by means of copper connecting wires. Ideally these wires should have zero resistance, but in practice their resistance, although small, is not negligible. Consequently, there will be some voltage drop across the connecting wires that will also dissipate power in the form of heat. To keep these effects to a minimum, the wires must have adequate thickness, because the thickness determines the current-carrying capacity.

A length of a metal wire with a circular cross section is an example of a cylindrical conductor. If two identical lengths are joined end-to-end in series, the resistance will be doubled. The conductor's resistance is therefore directly proportional to its length (Fig. 6-1). However, if the conductor is thickened by doubling its cross-sectional area, it is equivalent to connecting two identical lengths in parallel and the resistance is halved. To summarize, the resistance of a cylindrical conductor is directly proportional to its length, inversely proportional to its cross-sectional area and dependent on the material from which the conductor is made.

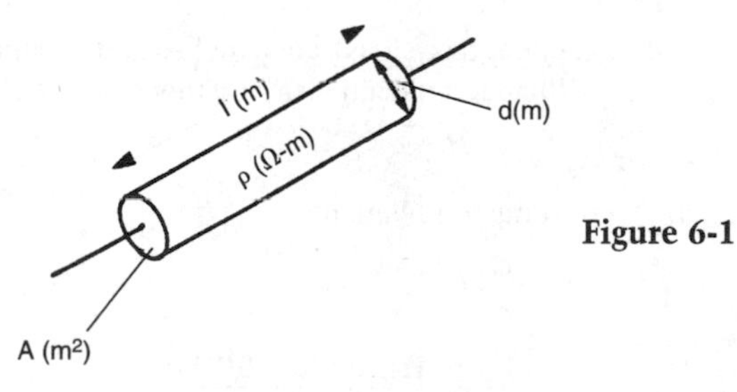

Figure 6-1

The *specific resistance*, or resistivity, is the factor that allows us to compare the resistances of different materials. The letter symbol for the specific resistance is the Greek lowercase letter rho, ρ, while its SI unit is the ohm-meter (Ω-m). A conductor made from a particular material will have a resistance of ρ ohms if its length is one meter and its cross-sectional area is one square meter. The values of specific resistance for various materials are shown in Table 6-1. Because the specific resistance varies with the temperature, the values quoted in the table are for 20° C.

Although copper is excellent for connecting wires, it is far too good of a conductor for a load such as a toaster. For example, the element of a 600-W/120-V toaster has a resistance of E^2/P = (120 V)^2/600 W = 24 Ω, which could not be obtained from any practical length of copper wire with adequate current-carrying capacity. Consequently, the toaster element is made from *nichrome* (an alloy of nickel and chromium) wire, which has a much higher specific resistance than copper (see Table 6-1).

MATHEMATICAL DERIVATIONS

Resistance of a cylindrical conductor,

$$R = \frac{\rho L}{A} \qquad (6\text{-}1)$$

$$= \frac{\rho L}{\pi d^2/4}$$

$$= \frac{4\rho L}{\pi d^2} \text{ ohms}$$

Table 6-1. Specific resistance (ρ) values for various materials.

Material	SI unit (specific resistance in $\Omega \cdot m$ at 20° C)	British unit (specific resistance in $\Omega \cdot$ cmil/ft at 20° C)
Silver	1.46×10^{-8}	9.86
Annealed copper	1.724×10^{-8}	10.37
Aluminum	2.83×10^{-8}	17.02
Tungsten	5.5×10^{-8}	33.08
Nickel	7.8×10^{-8}	46.9
Pure iron	1.2×10^{-7}	72.2
Constantan	4.9×10^{-7}	294.7
Nichrome	1.1×10^{-6}	660.0
Germanium (semiconductor)	0.55	3.3×10^{8}
Silicon (semiconductor)	550	3.3×10^{11}
Mica (insulator)	2×10^{10}	12.0×10^{18}

where: R = conductor's resistance in ohms (Ω)
L = conductor's length in meters (m)
A = cross-sectional area in square meters (m^2)
d = diameter in meters (m)
ρ = specific resistance or resistivity (Ω-m)
π = circular constant, 3.1415926 . . .

Note that the resistance of the cylindrical conductor is inversely proportional to the *square* of the diameter. Consequently, if the diameter is *doubled*, the cross-sectional area is quadrupled, and the resistance is *quartered*.

Unfortunately, it is still more common to quote values in terms of the customary (British) system of units. Here, the chosen length is the foot while the diameter is measured in mils where one linear mil is 0.001 inch. The cross-sectional area is given in *circular mils*, where one circular mil (cmil) is the area of a circle whose diameter is one mil. This simplifies the calculation of the cross-sectional area because, for example, if the diameter is 4 mils, the area is $4^2 = 16$ cmil. In this way the complication of using π is avoided.

In customary units, resistance of cylindrical conductor,

$$R = \frac{\rho L}{A} \qquad (6\text{-}2)$$

$$= \frac{\rho L}{d^2} \text{ ohms}$$

where: R = conductor's resistance in ohms (Ω)
L = length in feet (ft)
A = cross-sectional area in circular mils (cmil)
d = diameter in linear mils (mil)
ρ = specific resistance (Ω-cmil/ft)

The conversion factor between the customary and SI unit system is 1 Ω-m = 6.015×10^8 Ω-cmil/ft.

The American Wire Gage (AWG) number is based on the customary units. For example, household wiring is normally AWG #14, which has a diameter of 64 mils, a cross-sectional area of 4110 cmil, and a resistance of 2.58 ohms per 1000 ft at 25° C. The higher the gauge number, the thinner the wire and the less is its current-carrying capacity.

Example 6-1

What is the resistance of a cylindrical aluminum conductor which is 5 m long and has a cross-sectional area of 6 mm^2?

Solution

Resistance of aluminum conductor,

$$R = \frac{\rho L}{A} \qquad (6\text{-}1)$$

$$= \frac{2.83 \times 10^{-8}\ \Omega\text{-m} \times 5\ \text{m}}{6 \times 10^{-6}\ \text{m}^2}$$

$$= 2.36 \times 10^{-2}\ \Omega$$

Example 6-2

A tungsten filament has a length of 5 inches and a diameter of 2 mil. What is the filament's resistance at 20° C?

Solution

Resistance of tungsten filament,

$$R = \frac{\rho L}{d^2} \qquad (6\text{-}2)$$

$$= \frac{33.08\ \Omega\text{-cmil/ft} \times 5/12\ \text{ft}}{(2\ \text{mil})^2}$$

$$= 3.45\ \Omega$$

PRACTICE PROBLEMS

6-1. A piece of cylindrical wire is 80 cm in length and has a resistance of 120 Ω. What is the resistance of another piece of wire made from the same material if both the length and the cross-sectional area are doubled?

6-2. Calculate the resistance at 20° C of 300 m of nickel wire with a cross-sectional area of 2×10^{-6} m^2.

6-3. A cylindrical conductor, 2.5 m long, has a cross-sectional area of 0.75 mm^2 and a resistance of 0.015 Ω. What is the resistance of 70 m of wire that is made from the same material and has a cross-sectional area of 0.55 mm^2?

6-4. The total voltage drop across a cylindrical annealed copper conductor must not exceed 15 V when the current flowing through the conductor is 18 A. If the conductor's length is 60 m, calculate the minimum value of the cross-sectional area.

6-5. What is the resistance per meter length at 20° C of nickel wire that has a diameter of 4 mm?

6-6. A roll of nichrome wire is manufactured so that the diameter of the circular wire is 2 mm. What length of wire must be cut from the roll to provide a resistance of 50 Ω? Assume that the temperature is 20° C.

6-7. The rectangular copper conductors on a printed circuit board are 0.1 mm thick and 3 mm across. If a current of 0.4 A flows through one such copper conductor, 2 cm in length, calculate the voltage drop across the conductor.

7
Temperature coefficient of resistance

When a conductor is heated, the rise in temperature causes the random motion of the free electrons to increase. If a voltage is then applied to the conductor, it is more difficult to move the electrons along the wire to create the flow of current and the resistance therefore increases. For good conductors such as silver, copper, and aluminum, the increase in the resistance rises in a linear manner with the increase in the temperature so that their resistance versus temperature graphs are straight lines.

By contrast with the good conductors, there are alloys, for example, constantan and manganin, whose resistances are practically independent of temperature. Such alloys are used in the manufacture of precision wirewound resistors.

Semiconductors, such as silicon, germanium, and carbon, have very few free electrons and holes at room temperature. However, if the temperature is increased more electrons are moved from the valence band to the conduction band and the resistance falls.

The change in the resistance with the change in the temperature is measured by the *temperature coefficient* of resistance whose symbol is the Greek lowercase letter alpha, α. If the two changes are in the same direction (for example, the resistance increases when the temperature increases) the coefficient will be positive. Therefore, conductors have positive coefficients and semiconductors have negative coefficients.

The value of a temperature coefficient is normally referred to a temperature of 20° C and is defined as the change in ohms (for every ohm of resistance at 20° C) for every one degree Celsius rise above 20° C. For example, the value of α for silver is + 0.0038 Ω/Ω/° C. Consequently, if a length of silver wire has a resistance of 1 Ω at 20° C, its resistance will *increase* by 0.0038 Ω for every degree Celsius *rise* in temperature above 20° C. The corresponding graph of resistance versus temperature is shown in Fig. 7-1.

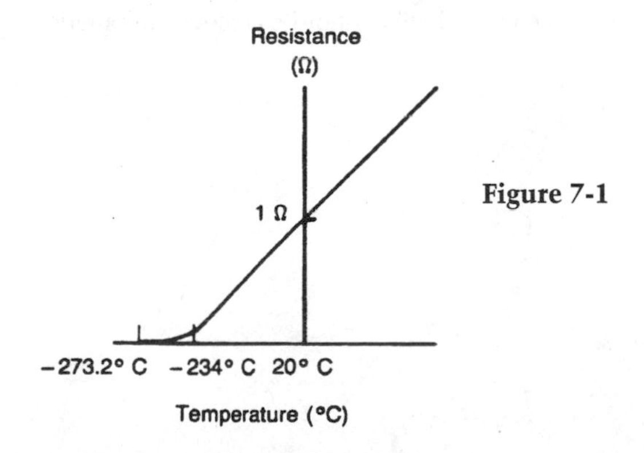

Figure 7-1

Table 7-1 shows the values of the resistance temperature coefficients for various materials.

Table 7-1. Temperature coefficient of resistance for various materials.

Material	Temperature coefficient of resistance, α, at 20°C (Ω/Ω/° C)
Silver	+0.0038
Copper	+0.00393
Aluminum	+0.0039
Tungsten	+0.0045
Iron	+0.0055
Nickel	+0.006
Constantan	+0.0000008
Carbon	−0.0005

MATHEMATICAL DERIVATIONS

If the temperature is increased to $T°$ C,

$$\text{Increase in the resistance} = \alpha \times R_{20°\,C} \times (T°\,C - 20°\,C) \text{ ohms} \qquad (7\text{-}1)$$

Resistance at $T°$C,

$$R_{T°\,C} = R_{20°\,C} + \alpha \times R_{20°\,C} \times (T°\,C - 20°\,C) \qquad (7\text{-}2)$$
$$= R_{20°\,C}\,[1 + \alpha\,(T°\,C - 20°\,C)] \text{ ohms}$$

If the silver conductor is cooled its resistance will decrease by 0.0038 Ω for every 1° C drop in temperature. Consequently when the temperature drops by 1 Ω × 1° C/0.0038 Ω = 263° C, the resistance of the conductor should theoretically be zero. This would occur at a temperature of + 20° C − 263° C = − 243° C. However, the resistance/temperature graph is nonlinear at low temperatures (Fig. 7-1) so that the resistance does not virtually disappear until we approach a temperature which is close to − 273.2° C. This temperature is the ultimate limit of absolute zero, or 0 K (kelvin scale), which cannot be achieved in practice.

Example 7-1

At 20° C a length of silver wire has a resistance of 12 Ω. What is its resistance at (a) 0° C and (b) 100° C?

Solution

(a) Resistance at 0° C,

$$R_{0°\,C} = R_{20°\,C}[1 + \alpha(0°\,C - 20°\,C)] \qquad (7\text{-}1)$$
$$= 12\,\Omega\,[1 - 20 \times 0.0038]$$
$$= 11.09\,\Omega$$

(b) Resistance at 100° C,

$$R_{100°\,C} = R_{20°\,C}\,[1 + \alpha(100°\,C - 20°\,C)] \qquad (7\text{-}1)$$
$$= 12\,\Omega\,[1 + 80 \times 0.0038]$$
$$= 15.65\,\Omega$$

PRACTICE PROBLEMS

7-1. A piece of cylindrical wire has a resistance of 3 Ω at room temperature. If the temperature coefficient of resistance for the material used is + 0.004 Ω/Ω/° C and the temperature is raised by 35° C, what is the value of the new resistance?

7-2. When a 100-W bulb is operated from a 110-V source, the temperature of the tungsten filament is 2700° C. What is the filament's resistance at room temperature (20° C)?

7-3. A certain length of wire has a measured resistance of 255 Ω at 20° C and 320 Ω at 70° C. Calculate the value of the wire's temperature coefficient of resistance.

7-4. At a temperature of 0° C, a coil of wire is found to have a resistance of 12 Ω. When the temperature is raised to 100° C, the new resistance measurement is 16.7 Ω. What is the value of the temperature coefficient for the material from which the wire is made?

7-5. The resistance of a coil of aluminum wire is 960 Ω at 20° C. When the coil is submerged in a liquid for an appreciable time, its resistance drops to 936 Ω. What is the temperature of the liquid?

8
Composition resistors—the color code

A composition resistor is the most common type of resistor used in electronics. These resistors are inexpensive and are manufactured in standard resistance values which normally range from 2.7 Ω to 22 MΩ. The resistance values remain reasonably constant over a limited temperature range so that composition resistors are linear components and obey Ohm's law. However, such resistors can only dissipate a limited amount of power so that their normal wattage ratings are only ⅛ W, ¼ W, ½ W, 1 W, and 2 W. These ratings are judged from the resistor's physical size (Fig. 8-1A). Knowing the values of the power rating (P) and the resistance (R), the formulas $I = \sqrt{P/R}$ amperes and $E = \sqrt{P \times R}$ volts will enable you to calculate the maximum values of the voltage and the current that will not allow the resistor's power rating to be exceeded.

Composition resistors are normally manufactured from powdered carbon with its specific resistance of more than 3.325×10^{-5} Ω-m. The carbon is mixed with an insulating substance (for example, talc) and a binding material such as resin. The proportions of carbon and talc are then varied to produce a wide range of resistances. The resistor is finally coated with an insulating material that is baked to a hard finish. This coating provides protection against moisture and mechanical damage (Fig. 8-1B). At either end of the resistor, a tinned copper "pigtail" wire is deeply embedded and provides an adequate contact area for making a good connection with the carbon/talc mixture.

The main disadvantage of composition resistors is that they are not manufactured to precise values, but are normally sorted into three groups. The first group has a 5% tolerance, which means that their measured resistances are within ±5% of their rated values. The second group has a tolerance between ±5% and ±10%, while the tolerance of the third and final group lies between ±10% and ±20%. Any resistor whose tolerance is greater than ±20%, is discarded.

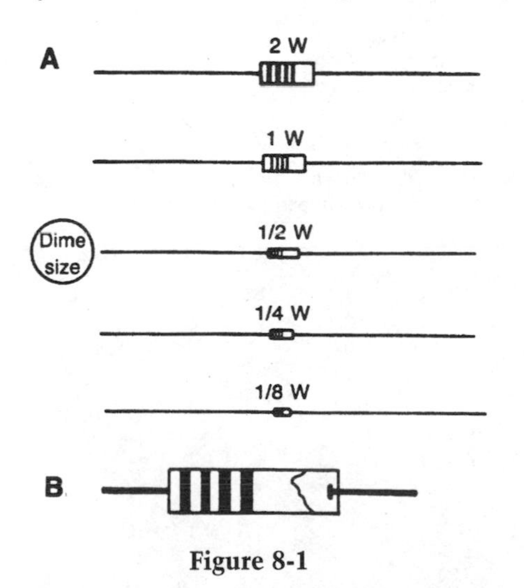

Figure 8-1

As an example, let us consider a 10%-tolerance resistor with a rated value of 10 kΩ. The actual resistance value must lie between 10 kΩ + (10/100) × 10 kΩ = 11 kΩ and 10 kΩ − (10/100) × 10 kΩ = 9 kΩ. Consequently there would be no point in manufacturing 10.5-kΩ or 9.5-kΩ resistors since both these values lie within the tolerance of the rated 10 kΩ resistor. It follows that composition resistors are only made in certain standard values. Between 10 Ω and 100 Ω in the 20%-range, the only values manufactured are 10 Ω, 15 Ω, 22 Ω, 33 Ω, 47 Ω, 68 Ω, and 100 Ω. These values are chosen so that the upper tolerance limit of one resistor equals (or slightly overlaps) the lower limit of the next highest value resistor. For example, the upper limit of a 22-Ω resistor is 22 + (20/100) × 22 = 26.4 Ω, while the lower limit of a 33-Ω resistor is 33 − (20/100) × 33 = 26.4 Ω. The ±10% and ±5% ranges are then formed by including additional standard values in the gaps of the 20% range (Table 8-1).

The higher resistances are formed by adding zeros to the values shown in Table 8-1. The least costly method of indicating the rated resistance value is to use a four-band color code. The first two bands represent the significant figures of the rated value, the third band is the multiplier, which indicates the *number* of zeros, while the fourth band (if present) indicates the tolerance (Fig. 8-2). For example:

33 ± 5%	Orange, orange, black (no zeros) gold.
560 ± 10%	Green, blue, brown, silver.
2200 ± 20%	Red, red, red (no fourth band for ±20% tolerance).
39000 ± 10%	Orange, white, orange, silver.
150000 ± 20%	Brown, green, yellow.

Black is not allowed in the first band so that if the rated value is less than 10 Ω, the four band system cannot be used. Gold and silver are then placed in the third band as multipliers of 0.1 and 0.01 respectively. For example, a 5.6-Ω resistor would be color-coded green, blue, and gold.

Table 8-1. Standard resistor values.

5% tolerance	10% tolerance	20% tolerance
10	10	10
11		
12	12	
13		
15	15	15
16		
18	18	
20		
22	22	22
24		
27	27	
30		
33	33	33
36		
39	39	
43		
47	47	47
51		
56	56	
62		
68	68	68
75		
82	82	
91		

Wirewound resistors are manufactured to more precise values than the composition type and are available in high wattage ratings (2 W and upwards). They are commonly made from constantan wire with its low temperature coefficient.

Example 8-1

A ½-W composition resistor is color-coded brown, green, brown, gold. What are the permitted upper and lower limits of its resistance value?

Solution

The rated value of the resistor is 150 Ω with a ±5% tolerance. The permitted upper and lower limits are therefore:

$$150 \pm \frac{5}{100} \times 150\ \Omega = 150 \pm 7.5\ \Omega = 157.5\ \Omega \text{ and } 142.5\ \Omega$$

Example 8-2

The color bands of a 2-W composition resistor are red, violet, brown, silver. What is the maximum value of the current that can flow through the resistor without exceeding its power rating?

Solution

$$\text{Maximum current, } I = \sqrt{\frac{P}{R}}$$

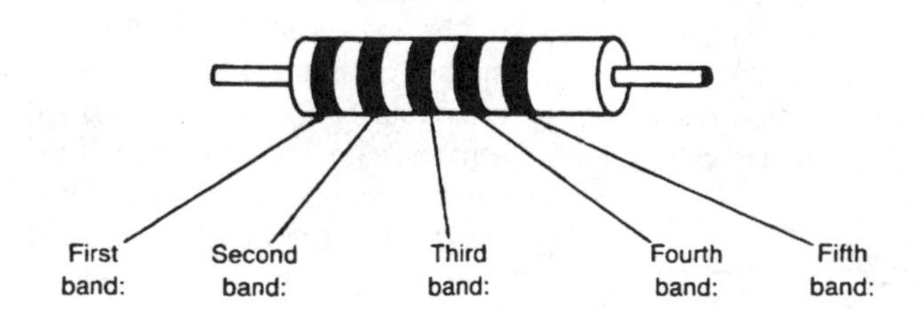

Color code	First significant figure	Second significant figure	Multiplier	Tolerance (%)	Fail-rate percent per 1000 hr
Black		0	$\times 1 = \times 10^0$		
Brown	1	1	$\times 10 = \times 10^1$		1.0
Red	2	2	$\times 100 = \times 10^2$		0.1
Orange	3	3	$\times 1000 = \times 10^3$		0.01
Yellow	4	4	$\times 10000 = \times 10^4$		0.001
Green	5	5	$\times 100000 = \times 10^5$		
Blue	6	6	$\times 1000000 = \times 10^6$		
Violet	7	7	$\times 10000000 = \times 10^7$		
Gray	8	8			
White	9	9			
Gold			$\times 0.1$	±5	
Silver			$\times 0.01$	±10	
No color				±20	

Figure 8-2

$$= \sqrt{\frac{2\text{ W}}{270\ \Omega}} = 0.086\text{ A}$$

$$= 86\text{ mA}$$

Note:

$$\text{Maximum voltage, } E = I \times R$$

$$= 0.086\text{ A} \times 270\ \Omega$$

$$= 23.2\text{ V}$$

PRACTICE PROBLEMS

8-1. A resistor is color-coded brown, red, red, silver. What is the permitted upper limit of its resistance value?

8-2. A thin film resistor has a nominal value of 3300 Ω with a ± 1% tolerance. What are the permitted upper and lower limits of its resistance value?

8-3. The first three bands of a ½-W composition resistor are red, violet, and red. Assuming that the actual resistance is the same as the nominal value, what is the maximum current in mA that can flow through the resistor without its power rating being exceeded?

8-4. The first three bands of a 2-W composition resistor are brown, green, brown. Assuming that the actual resistance is the same as the nominal value, what is the highest voltage that can be applied across the resistor without exceeding its power rating?

8-5. A resistor is color-coded green, blue, gold. Calculate the conductance value of this resistor.

8-6. When a current of 11.15 mA flows through a resistor with a 5% tolerance, its dissipated power is ¼ W. What is the nominal value of the resistance?

8-7. A resistor is color-coded brown, black, red, silver. What is the rated value of the resistance and its tolerance?

8-8. A resistor is color-coded brown, black, black gold. What is the permitted upper limit of its resistance value?

9
Resistors in the series arrangement

Resistors in series are joined end-to-end so that there is only a single path for the current. Starting at the negative battery terminal (where a surplus of electrons exists) there is an electron flow through the connecting wires (which are assumed to possess zero resistance) and the three resistors. Finally, this flow reaches the positive terminal, where there is a deficit of electrons. The battery, through its chemical energy, is then responsible for maintaining the surplus of electrons at the negative terminal and the deficit at the positive terminal. It follows from this description that the current is the same throughout the circuit and that the ammeters (Fig. 9-1A) *A*1, *A*2, *A*3, and *A*4 all have the same reading.

Figure 9-1B represents the top and bottom views of a printed circuit board. With the ammeters removed, the series resistors (R1, R2, and R3) are solder-mounted on the top surface while the conducting paths are etched on the bottom surface (refer to "Soldering Techniques" in appendix H).

Across each resistor is developed a difference of potential (DP) [sometimes referred to as the potential difference (PD)]; in this sense "potential" is another word for "voltage." For example, V_1 is the amount of voltage that is required to drive the current, I, through the resistor, R1. Because $V_1 = IR_1$, the voltage is often called the "*IR* drop," where the meaning of the word drop is illustrated in Example 9-2.

The sum of the voltages across the resistors must exactly balance the source voltage, E; this is an example of Kirchhoff's voltage law (KVL) which you will fully explore in chapter 26. Because the current through each resistor is the same, the highest value resistor will develop the greatest voltage drop, and the lowest value resistor will have the smallest voltage drop. In the extreme case of the connecting wires, which theoretically have zero resistance, there will be no voltage drop so that if, for example, a voltmeter were connected between points *A* and *B* (Fig. 9-1A), its reading would be zero.

Notice that other voltages exist in the circuit apart from V_1, V_2, V_3, and E. If a voltmeter is connected between points *B* and *E*, its reading will be the sum of the voltages V_1 and V_2, but between the points *D* and *G* the

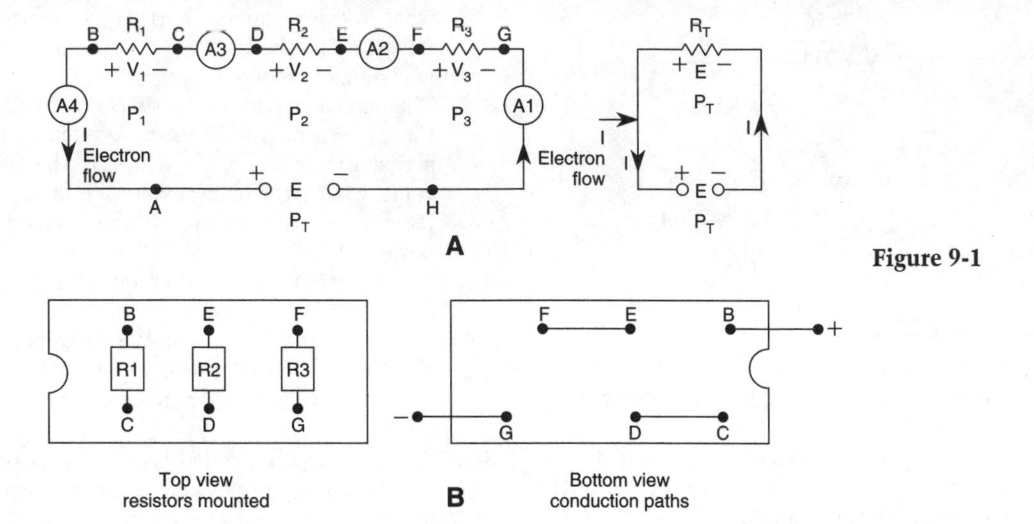

Figure 9-1

voltage is equal to $V_2 + V_3$. It is often said that the order in which series resistors are connected is immaterial. This is true as far as the current and the individual voltage drops are concerned, but not in terms of the other voltages that exist in the circuit (see Example 9-1).

In each resistor a certain amount of power is dissipated in the form of heat. (Care must be taken that the amount of power dissipated does not exceed the resistor's wattage rating.) The sum of the individual powers dissipated is equal to the total power derived from the source.

Finally, what is the purpose of connecting resistors in series? One obvious reason is to increase the total equivalent resistance, R_T, and thereby limit the current to a safe value; however this could also be achieved by using a higher value single resistor. More importantly, adding resistors enables you to increase the overall wattage rating, and to obtain nonstandard resistance values by using standard components (chapter 8). In addition, series resistors can be connected across a source voltage to provide a voltage divider circuit (chapter 11).

MATHEMATICAL DERIVATIONS

In the circuit of Fig. 9-1A:

Voltage drop across the resistor R1,

$$V_1 = V_{BC} = I \times R_1, \qquad (9\text{-}1)$$

$$I = \frac{V_1}{R_1}, R_1 = \frac{V_1}{I}$$

Voltage drop across the resistor R2,

$$V_2 = V_{DE} = I \times R_2, \qquad (9\text{-}2)$$

$$I = \frac{V_2}{R_2}, R_2 = \frac{V_2}{I}$$

Voltage drop across the resistor R3,

$$V_3 = V_{FG} = I \times R_3, \qquad (9\text{-}3)$$

$$I = \frac{V_3}{R_3}, R_3 = \frac{V_3}{I}$$

Source voltage,

$$E = V_1 + V_2 + V_3 \qquad (9\text{-}4)$$

$$= I \times R_1 + I \times R_2 + I \times R_3$$

$$= I \times (R_1 + R_2 + R_3) \text{ volts} \qquad (9\text{-}5)$$

If the total equivalent resistance is R_T,

Source voltage,

$$E = I \times R_T \text{ volts} \qquad (9\text{-}6)$$

Comparing equations 9-4 and 9-5,

Total equivalent resistance,

$$R_T = R_1 + R_2 + R_3 \text{ ohms} \qquad (9\text{-}7)$$

If N resistors are connected in series,

Total equivalent resistance,

$$R_T = R_1 + R_2 + R_3 \ldots + R_N \text{ ohms} \qquad (9\text{-}8)$$

If all N resistors are of equal value R,

Total equivalent resistance,

$$R_T = NR \text{ ohms} \qquad (9\text{-}9)$$

Power dissipated in the R1 resistor,

$$P_1 = I \times V_1 = I^2 R_1 = V_1^2/R_1 \text{ watts} \qquad (9\text{-}10)$$

Power dissipated in the R2 resistor,

$$P_2 = I \times V_2 = I^2 R_2 = V_2^2/R_2 \text{ watts} \qquad (9\text{-}11)$$

Power dissipated in the R3 resistor,

$$P_3 = I \times V_3 = I^2R_3 = V_3^2/R_3 \text{ watts} \quad (9\text{-}12)$$

Total power dissipated,

$$P_T = P_1 + P_2 + P_3 \text{ watts} \quad (9\text{-}13)$$

Power derived from the source voltage,

$$P_T = I \times E = I^2R_T = E^2/R_T \text{ watts} \quad (9\text{-}14)$$

Example 9-1

In Fig. 9-2, calculate the values of R_T, I_1, V_1, V_2, V_3, V_4, P_1, P_2, P_3, P_4, P_T. What are the voltages between the points *PA, PB, PC, PD, PE, PF, PG, PH, PN, AB, AC, AD, AE, AF, AG, AH, AN, BC, BD, BE, BF, BG, BH, BN, CD, CE, CF, CG, CH, CN, DE, DF, DG, DH, DN, EF, EG, EH, EN, FG, FH, FN, GH, GN, HN?*

Solution

You must first convert the 1.8 kΩ into ohms so that 1.8 kΩ = 1800 Ω

Total equivalent resistance,

$$R_T = R_1 + R_2 + R_3 + R_4 \quad (9\text{-}7)$$
$$= 1800 + 200 + 680 + 820 = 3500\ \Omega = 3.5\ \text{k}\Omega$$

$$\text{Current, } I = \frac{E}{R_T} = \frac{70\text{ V}}{3.5\text{ k}\Omega} = 20\text{ mA} \quad (9\text{-}7)$$

Voltage drop,

$$V_1 = V_{AB} = I \times R_1 = 20\text{ mA} \times 1.8\text{ k}\Omega = 36\text{ V} \quad (9\text{-}1)$$

Voltage drop,

$$V_2 = V_{CD} = I \times R_2 = 20\text{ mA} \times 200\ \Omega = 4\text{ V} \quad (9\text{-}1)$$

Voltage drop,

$$V_3 = V_{EF} = I \times R_3 = 20\text{ mA} \times 680\ \Omega = 13.6\text{ V} \quad (9\text{-}1)$$

Voltage drop,

$$V_4 = V_{GH} = I \times R_4 = 20\text{ mA} \times 820\ \Omega = 16.4\text{ V} \quad (9\text{-}1)$$

Notice that the highest value resistor (R_1 = 1800 Ω) carries the greatest voltage drop (V_1 = 36 V) while the lowest value resistor (R_2 = 200 Ω) has the least voltage drop (V_2 = 4 V).

Voltage check:

Source voltage,

$$E = V_1 + V_2 + V_3 + V_4 \quad (9\text{-}4)$$
$$= 36 + 4 + 13.6 + 16.4 = 70\text{ V}$$

Power dissipated in the R1 resistor,

$$P_1 = I \times V_1 = 20\text{ mA} \times 36\text{ V} \quad (9\text{-}10)$$
$$= 720\text{ mW}$$

Power dissipated in the R2 resistor,

$$P_2 = I \times V_2 = 20\text{ mA} \times 4\text{ V} \quad (9\text{-}11)$$
$$= 80\text{ mW}$$

Power dissipated in the R3 resistor,

$$P_3 = I \times V_3 = 20\text{ mA} \times 13.6\text{ V} \quad (9\text{-}12)$$
$$= 272\text{ mW}$$

Power dissipated in the R4 resistor,

$$P_4 = I \times V_4 = 20\text{ mA} \times 16.4\text{ V} \quad (9\text{-}13)$$
$$= 328\text{ mW}$$

Total power dissipated,

$$P_T = P_1 + P_2 + P_3 + P_4 \quad (9\text{-}14)$$
$$= 720 + 80 + 272 + 328$$
$$= 1400\text{ mW}$$

The highest value resistance (R_1 = 1800 Ω) dissipates the greatest power (P_1 = 720 mW) while the lowest value resistance (R_2 = 200 Ω) has the least power dissipation (P_2 = 80 mW).

Notice that ½-watt resistors would be adequate for R_2, R_3 and R_4. By contrast, R_1 would need to be a 1-watt resistor.

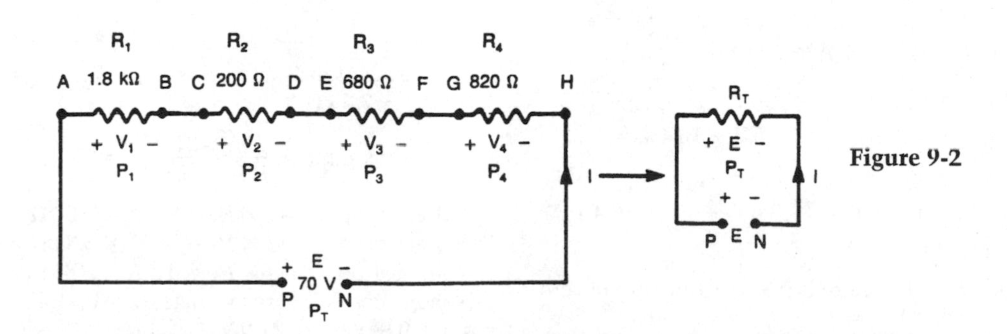

Figure 9-2

Total power derived from the source,

$$P_T = I \times E = 20 \text{ mA} \times 70 \text{ V} \quad (9\text{-}15)$$
$$= 1400 \text{ mW}$$

Voltage between the points PA, $V_{PA} = 0$ V because PA is a connecting wire of zero resistance.

$V_{PB} = V_{PA} + V_{AB} = 0 \text{ V} + 36 \text{ V} = 36$ V.

$V_{PC} = V_{PA} + V_{AB} + V_{BC} = 0 \text{ V} + 36 \text{ V} + 0 \text{ V} = 36$ V.

$V_{PD} = V_{PA} + V_{AB} + V_{BC} + V_{CD} = 0 \text{ V} + 36 \text{ V} + 0 \text{ V} + 4 \text{ V} = 40$ V.

$V_{PE} = V_{PA} + V_{AB} + V_{BC} + V_{CD} + V_{DE} = 0 \text{ V} + 36 \text{ V} + 0 \text{ V} + 4 \text{ V} + 0 \text{ V} = 40$ V.

$V_{PF} = V_{PA} + V_{AB} + V_{BC} + V_{CD} + V_{DE} + V_{EF} = 0 \text{ V} + 36 \text{ V} + 0 \text{ V} + 4 \text{ V} + 0 \text{ V} + 13.6 \text{ V} = 53.6$ V.

$V_{PG} = V_{PA} + V_{AB} + V_{BC} + V_{CD} + V_{DE} + V_{EF} + V_{FG} = 0 \text{ V} + 36 \text{ V} + 0 \text{ V} + 4 \text{ V} + 0 \text{ V} + 13.6 \text{ V} + 0 \text{ V} = 53.6$ V.

$V_{PH} = V_{PA} + V_{AB} + V_{BC} + V_{CD} + V_{DE} + V_{EF} + V_{FG} + V_{GH}$
$= 0 \text{ V} + 36 \text{ V} + 0 \text{ V} + 4 \text{ V} + 0 \text{ V} + 13.6 \text{ V} + 0 \text{ V} + 16.4 \text{ V} = 70$ V.

$V_{PN} = V_{PA} + V_{AB} + V_{BC} + V_{CD} + V_{DE} + V_{EF} + V_{FG} + V_{GH} + V_{HN}$
$= 0 \text{ V} + 36 \text{ V} + 0 \text{ V} + 4 \text{ V} + 13.6 \text{ V} + 0 \text{ V} + 16.4 \text{ V} + 0 \text{ V} = 70$ V.

Notice that if for example, the resistors R2 and R3 were interchanged, V_{PD} would become 36 V + 13.6 V = 49.6 V. Because P and A are only joined by a connecting wire of zero resistance, $V_{AB} = V_{PB}$, $V_{AC} = V_{PC}$, etc.

$V_{BC} = 0$ V.

$V_{BD} = V_{BC} + V_{CD} = 0 \text{ V} + 4 \text{ V} = 4$ V.

$V_{BE} = V_{BC} + V_{CD} + V_{DE} = 0 \text{ V} + 4 \text{ V} + 0 \text{ V} = 4$ V.

$V_{BF} = V_{BC} + V_{CD} + V_{DE} + V_{EF} = 0 \text{ V} + 4 \text{ V} + 0 \text{ V} + 13.6 \text{ V} = 17.6$ V.

$V_{BG} = V_{BC} + V_{CD} + V_{DE} + V_{EF} + V_{FG} = 0 \text{ V} + 4 \text{ V} + 0 \text{ V} + 13.6 \text{ V} + 0 \text{ V} = 17.6$ V.

$V_{BH} = V_{BC} + V_{CD} + V_{DE} + V_{EF} + V_{FG} + V_{GH}$
$= 0 \text{ V} + 4 \text{ V} + 0 \text{ V} + 13.6 \text{ V} + 16.4 \text{ V} = 34.0$ V.

$V_{BN} = V_{BC} + V_{CD} + V_{DE} + V_{EF} + V_{FG} + V_{GH} + V_{HN}$
$= 0 \text{ V} + 4 \text{ V} + 0 \text{ V} + 13.6 \text{ V} + 0 \text{ V} + 16.4 \text{ V} + 0 \text{ V} = 34.0$ V.

Between B and C there is a connecting wire of zero resistance so that $V_{BD} = V_{CD}$, $V_{BE} = V_{CE}$, etc.

$V_{DE} = 0$ V.

$V_{DF} = V_{DE} + V_{EF} = 0 \text{ V} + 13.6 \text{ V} = 13.6$ V.

$V_{DG} = V_{DE} + V_{EF} + V_{FG} = 0 \text{ V} + 13.6 \text{ V} + 0 \text{ V} = 13.6$ V.

$V_{DH} = V_{DE} + V_{EF} + V_{FG} + V_{GH} = 0 \text{ V} + 13.6 \text{ V} + 16.4 \text{ V} = 30$ V.

$V_{DN} = V_{DE} + V_{EF} + V_{FG} + V_{GH} + V_{HN} = 0 \text{ V} + 13.6 \text{ V} + 16.4 \text{ V} + 0 \text{ V} = 30$ V.

Between D and E there is a connecting wire of zero resistance so that $V_{DF} = V_{EF}$, $V_{DG} = V_{EG}$, etc.

$V_{FG} = 0$ V.

$V_{FH} = V_{FG} + V_{GH} = 0 \text{ V} + 16.4 \text{ V} = 16.4$ V.

$V_{FN} = V_{FG} + V_{GH} + V_{HN} = 0 \text{ V} + 16.4 \text{ V} + 0 \text{ V} = 16.4$ V.

Between F and G there is a connecting wire of zero resistance so that $V_{FH} = V_{GH}$, $V_{FN} = V_{GN}$.

$V_{HN} = 0$ V.

Example 9-2

In Fig. 9-3 the negative probe of the voltmeter is connected permanently to the point G. What are the readings of the voltmeter when the positive probe is (in turn) connected to the points D, E, F, and G?

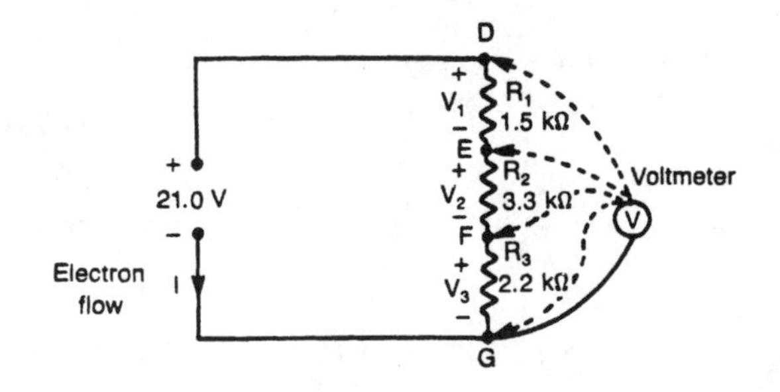

Figure 9-3

Solution

Total equivalent resistance, $R_T = R_1 + R_2 + R_3 = 1.5 + 3.3 + 2.2 = 7.0$ kΩ

$$\text{Current, } I = \frac{E}{R_T} = \frac{21.0 \text{ V}}{7 \text{ k}\Omega} = 3 \text{ mA}.$$

Voltage drop, $V_1 = I \times R_1 = 3 \text{ mA} \times 1.5 \text{ k}\Omega = 4.5$ V.
Voltage drop, $V_2 = I \times R_2 = 3 \text{ mA} \times 3.3 \text{ k}\Omega = 9.9$ V.
Voltage drop, $V_3 = I \times R_3 = 3 \text{ mA} \times 2.2 \text{ k}\Omega = 6.6$ V.
Voltage check: Source voltage, $E = V_1 + V_2 + V_3 = 4.5 + 9.9 + 6.6 = 21.0$ V.

Positive probe connected to the point D: Reading of the voltmeter = source voltage = 21.0 V.

Positive probe connected to the point E: Reading of the voltmeter = $V_2 + V_3 = 9.9 + 6.6 = 16.5$ V.

We can therefore say that there has been a *voltage drop* of 21.0 − 16.5 = 4.5 V between the points D and E. This voltage drop is, of course, equal to the value of V_1.

Positive probe connected to the point F: Reading of the voltmeter = $V_3 = 6.6$ V.

There has been a voltage drop of 16.5 − 6.6 = 9.9 V between points E, F and a voltage drop of 21.0 − 6.6 = 14.4 V between the points D, F.

Positive probe connected to the point G: The two probes are joined together so that the voltmeter reading is zero. The voltage drops between points D and G, E and G, F and G are respectively 21.0 V, 16.5 V, and 6.6 V.

Example 9-3

A series circuit consists of four 1-W resistors whose values are 120 Ω, 150 Ω, 270 Ω, and 330 Ω. Calculate their maximum power dissipation, without exceeding the wattage rating of any of the resistors.

Solution

The current in each resistor to produce its 1-W rating is different; using the relationship $I = \sqrt{P/R}$, these currents are:

$$I_{120\,\Omega} = \sqrt{\frac{1\text{ W}}{120\,\Omega}} = 0.0913\text{ A.}$$

$$I_{150\,\Omega} = \sqrt{\frac{1\text{ W}}{150\,\Omega}} = 0.0816\text{ A.}$$

$$I_{270\,\Omega} = \sqrt{\frac{1\text{ W}}{270\,\Omega}} = 0.0609\text{ A.}$$

$$I_{330\,\Omega} = \sqrt{\frac{1\text{ W}}{330\,\Omega}} = 0.0550\text{ A.}$$

You must select the smallest of these currents, 0.0550 A, because if the current were allowed to exceed this value, the power dissipation in the 330-Ω resistor would be exceeded. Using the relationship $P = I^2R_T$ the maximum power dissipation of this series string is:

$$(0.0550\text{ A})^2 \times (120\,\Omega + 150\,\Omega + 270\,\Omega + 330\,\Omega) = 2.27\text{ W.}$$

The maximum power dissipation is therefore only 2.27 W, and *not* 4 W.

PRACTICE PROBLEMS

9-1. Three resistors are connected in series across a voltage source. The color-coding of these resistors is brown-red-red-silver, yellow-violet-brown-silver, blue-gray-black. What is the total rated resistance of this series string?

9-2. Three resistors whose values are 1.8 kΩ, 2.7 kΩ, and 820 Ω are connected in series across a 12-Vdc source. What is the value of the current that is flowing in the series circuit?

9-3. A 33-kΩ resistor and a 56-kΩ resistor are connected in series across a 30-Vdc source. What is the voltage drop across the 33-kΩ resistor?

9-4. A 20,000-Ω, 200-W resistor, a 40,000-Ω 100-W resistor, and a 10,000-Ω 20-W resistor are connected in series. What is the maximum value of the total applied voltage that will not cause the wattage rating of any of the resistors to be exceeded?

10
Ground—voltage reference level

Ground can be regarded as any large mass of conducting material in which there is essentially *zero* resistance between any two ground points. Examples of ground are the metal chassis of a transmitter, the aluminum chassis of a receiver, or a wide strip of copper plating on a printed-circuit board.

The main reason for using a ground system is to simplify the circuitry by saving on the amount of wiring. This is done by using ground as the return paths for many circuits so that at any time there are a number of currents flowing through the ground system.

Figures 10-1A and B show two circuits that contain four connecting wires. However, if a common ground (alternative symbols ⏚ or ⏚, usually to sig-

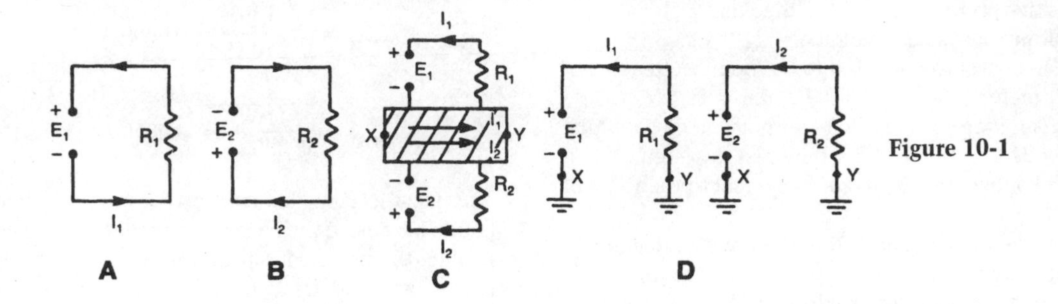

Figure 10-1

nify earth ground; and ⏊ to signify chassis ground) is used for the return path as in Fig. 10-1C, only two connecting wires are required; the schematic equivalent of Fig. 10-1C is shown in Fig. 10-1D. Here is where the concept of zero resistance is vital. If the ground resistance were not zero, the current flow in one circuit would develop a voltage between the ground points X and Y, and this would cause a current to flow in the other circuit. In other words the two circuits would interfere with each other. It is therefore important to regard all ground points in a schematic as being electrically joined together as a single path. This zero resistance property of ground is essential in order to achieve isolation between all those circuits that use the same ground.

The previous descriptions have regarded a voltage drop, potential difference, or difference of potential as existing between *two* points; to have referred to the voltage at a single point would have been meaningless. However, if you create a common reference level to represent the voltage at one point, you can quote the voltages at other points in relation to the reference level. Ground is therefore chosen as a reference level of zero volts and the voltage or *potential* at any point is then measured relative to ground. This is the manner in which the expected potentials at various points are indicated on a schematic.

In Fig. 10-2A, you could either say that the point X is 6-V positive with respect to the point Y or that the point Y is 6-V negative with respect to the point X. However, if the negative battery terminal is grounded at zero volts (Fig. 10-2B), point X carries a potential of positive 6 V (+6 V) with respect to ground. By contrast, if the battery is reversed and its positive terminal is grounded (Fig. 10-2C), the potential at X is now negative 6 V (−6 V) with respect to ground.

When troubleshooting a circuit, it is normal practice to connect the black or common probe of a voltmeter to ground and to measure the potentials at various points using the red probe. The voltmeter's function switch has two positions marked "+DC" and "− DC"; these enable either positive or negative potentials to be measured without removing the common probe from ground. Remember that a potential with respect to ground must always be indicated as either "+" or " − ".

In any particular circuit the potentials at the various points will depend on which point is grounded. However, the potential differences between any two points must always remain the same.

It is worth mentioning that both positive and negative potentials are used in the operation of electronic circuits. For example, pnp transistor amplifiers normally need negative potentials, but npn amplifiers require positive potentials.

Example 10-1

In Fig. 10-3, ground the points A, B, C, and D in turn. In each case, calculate the values of the potentials at the various points in the circuit.

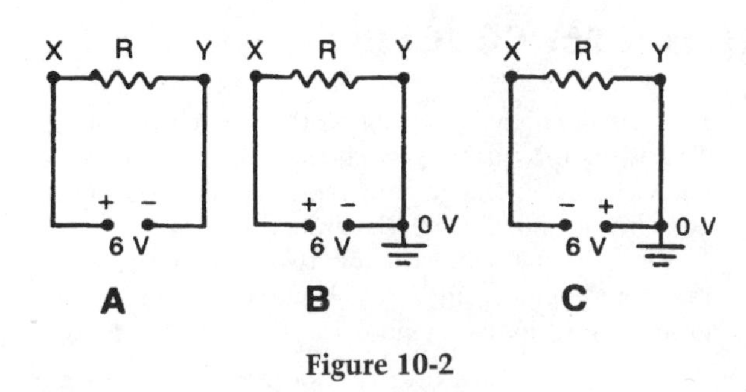

Figure 10-2

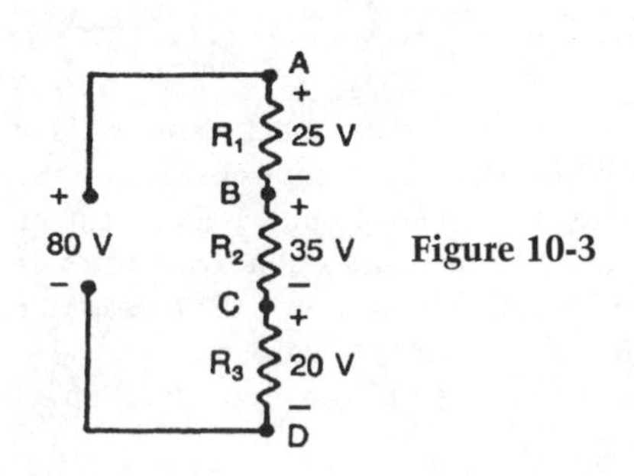

Figure 10-3

Solution

Point A Grounded

Potential at $A = 0$ V

Potential at $B = -25$ V

Potential at $C = -(25 + 35) = -60$ V

Potential at $D = -(25 + 35 + 20) = -80$ V

Point B Grounded

Potential at $A = +25$ V

Potential at $B = 0$ V

Potential at $C = -35$ V

Potential at $D = -(35 + 20) = -55$ V

Point C Grounded

Potential at $A = +(25 + 35) = +60$ V

Potential at $B = +35$ V

Potential at $C = 0$ V

Potential at $D = -20$ V

Grounding at a point such as B or C would be regarded as an *intermediate* ground.

Point D Grounded

Potential at $A = +(25 + 35 + 20) = +80$ V

Potential at $B = +(35 + 20) = +55$ V

Potential at $C = +20$ V

Potential at $D = 0$ V

Note that in this circuit, you must not ground two separate points simultaneously. For example, if points B and C were grounded at the same time, it would be equivalent to placing a short circuit of zero resistance across the resistor R2.

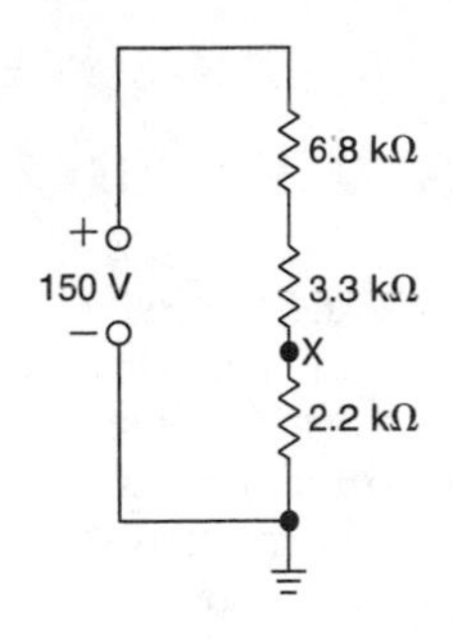

Figure 10-4

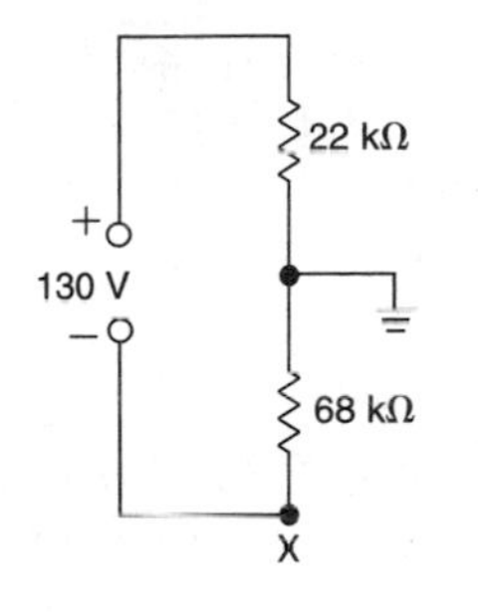

Figure 10-5

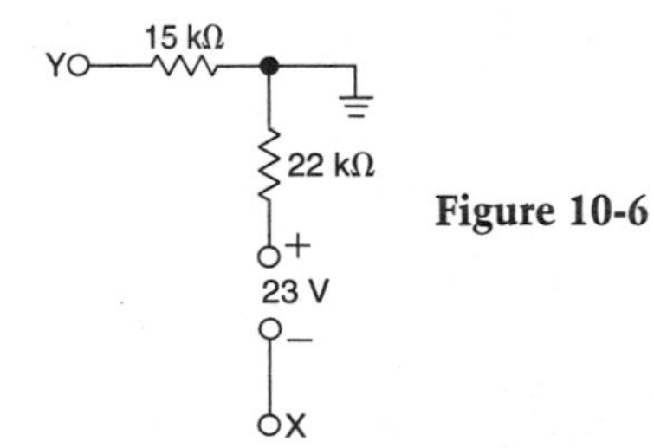

Figure 10-6

PRACTICE PROBLEMS

10-1. In Fig. 10-4, what is the potential at the point X?

10-2. In Fig. 10-5, what is the potential at the point X?

10-3. In Fig. 10-6, the points X and Y are joined by a path of zero resistance. Calculate the potential at the point Y.

11
Voltage division rule—voltage divider circuit

In a series string of resistors, the highest voltage drop is developed across the largest value resistor, while the lowest voltage drop appears across the smallest value resistor. In fact, the voltage drops across all of the resistors are directly proportional to the resistor values. You may use the method of proportion to solve a problem in which, for example, you are given the values of R_1, R_2, V_1, and then asked to find the value of V_2 (Fig. 11-1). The same rule of proportion can be applied to the powers dissipated in the individual resistors.

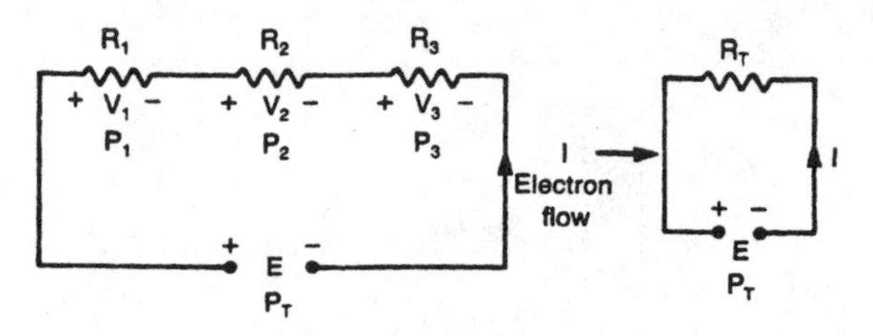

Figure 11-1

The source voltage is divided between the series resistors in a proportional manner that is determined by the resistor values. The fraction of the source voltage developed across a particular resistor is equal to the ratio of the resistor's value to the circuit's total resistance. This result is known as the *voltage division rule* (VDR). In the same manner, the total power derived from the source divides between the resistors to produce the individual powers dissipated.

MATHEMATICAL DERIVATIONS

Voltage drop,

$$V_1 = IR_1 \text{ volts} \tag{11-1}$$

Voltage drop,

$$V_2 = IR_2 \text{ volts} \tag{11-2}$$

Therefore: Current,

$$I = \frac{V_1}{R_1} = \frac{V_2}{R_2} \text{ amperes} \tag{11-3}$$

and

$$\frac{V_1}{V_2} = \frac{R_1}{R_2} \tag{11-4}$$

This yields

Voltage drop,

$$V_1 = V_2 \times \frac{R_1}{R_2}, V_2 = V_1 \times \frac{R_2}{R_1} \text{ volts} \tag{11-5}$$

Power dissipated,

$$P_1 = I^2R_1 \text{ watts} \tag{11-6}$$

Power dissipated,

$$P_2 = I^2R_2 \text{ watts} \tag{11-7}$$

Then,

$$I^2 = \frac{P_1}{R_1} = \frac{P_2}{R_2} \tag{11-8}$$

and

$$\frac{P_1}{P_2} = \frac{R_1}{R_2} \tag{11-9}$$

This yields:

Power dissipated,

$$P_1 = P_2 \times \frac{R_1}{R_2}, P_2 = P_1 \times \frac{R_2}{R_1} \text{ watts} \tag{11-10}$$

These results apply to any number of resistors in series.

Voltage Division Rule (VDR): Current,

$$I = \frac{E_T}{R_T} \text{ amperes} \tag{11-11}$$

Voltage drop, $V_1 = IR_1$ volts

It follows that: Voltage drop,

$$V_1 = \frac{E_T}{R_T} \times R_1 = E_T \times \frac{R_1}{R_T} \text{ volts} \tag{11-12}$$

(Voltage Division Rule)

Then

Voltage drop,

$$V_2 = E_T \times \frac{R_2}{R_T} \text{ and } V_3 = E_T \times \frac{R_3}{R_T} \text{ volts} \tag{11-13}$$

Also,

$$\frac{V_1}{E_T} = \frac{R_1}{R_T} \text{ and} \tag{11-14}$$

$$E_T = V_1 \times \frac{R_T}{R_1} = V_2 \times \frac{R_T}{R_2} = V_3 \times \frac{R_T}{R_3} \text{ volts}$$

For *two* resistors in series:

Voltage drop,

$$V_1 = E_T \times \frac{R_1}{R_T} = E_T \times \frac{R_1}{R_1 + R_2} \text{ volts} \quad (11\text{-}15)$$

Voltage drop,

$$V_2 = E_T \times \frac{R_2}{R_1 + R_2} \text{ volts} \quad (11\text{-}16)$$

For N *equal* value resistors:
Voltage drops,

$$V_1 = V_2 = V_3 \ldots = V_N = \frac{E}{N} \text{ volts} \quad (11\text{-}17)$$

Power division: Total power delivered from the source,

$$P_T = I^2 R_T \text{ watts} \quad (11\text{-}18)$$

Power dissipated in the R1 resistor,

$$P_1 = I^2 R_1 \frac{E_2}{R_T} \text{ watts} \quad (11\text{-}19)$$

Therefore:

$$I^2 = \frac{P_T}{R_T} = \frac{P_1}{R_1} \quad (11\text{-}20)$$

This yields:

$$P_1 = P_T \times \frac{R_1}{R_T}, P_2 = P_T \times \frac{R_2}{R_T}, \quad (11\text{-}21)$$

$$P_3 = P_T \times \frac{R_3}{R_T} \text{ watts}$$

The voltage divider circuit is the practical outcome of the voltage division rule. By its use you can obtain a number of different voltages from a single voltage source. These voltages will then be available as the supplies for various loads.

Referring to Fig. 11-2, V_1 is the voltage drop across the series combination of R_1, R_2, R_3, R_4, and is therefore equal to the source voltage, $+E$. V_2 is dropped across R_2, R_3, and R_4 in series.

Therefore, using the voltage division rule,

$$V_2 = +E \times \frac{(R_2 + R_3 + R_4)}{R_1 + R_2 + R_3 + R_4} \text{ volts} \quad (11\text{-}22)$$

Similarly,

$$V_3 = +E \times \frac{(R_3 + R_4)}{R_1 + R_2 + R_3 + R_4} \text{ volts} \quad (11\text{-}23)$$

and

$$V_4 = +E \times \frac{R_4}{R_1 + R_2 + R_3 + R_4} \text{ volts} \quad (11\text{-}24)$$

Notice that V_1, V_2, V_3, and V_4 are all positive potentials.

To develop the voltages V_1, V_2, V_3, V_4 it is necessary for a so-called "bleeder current", I_B, to flow through the resistors. The provision of the different output voltages is therefore achieved at the expense of the power dissipated in the series string.

The voltage divider described so far would be regarded as operating under no load conditions. As soon as loads are connected across the voltages V_1, V_2, V_3, V_4, the circuit becomes a more complex series-parallel arrangement and the equations 11-22, 11-23, 11-24 no longer apply.

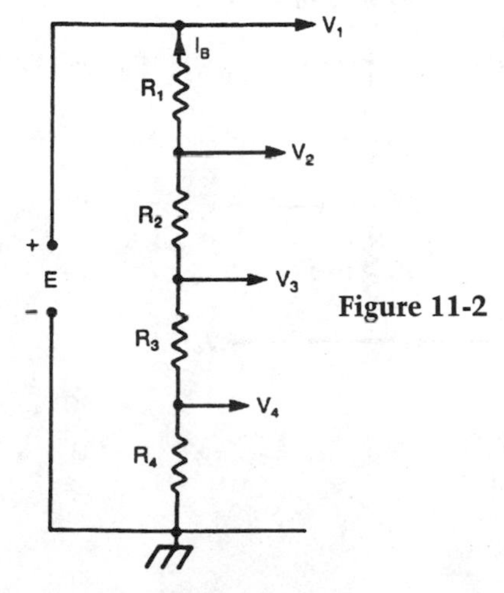

Figure 11-2

Example 11-1

In Fig. 11-3, calculate the values of V_1, V_3, and the potential at the point X.

Solution

Voltage drop,

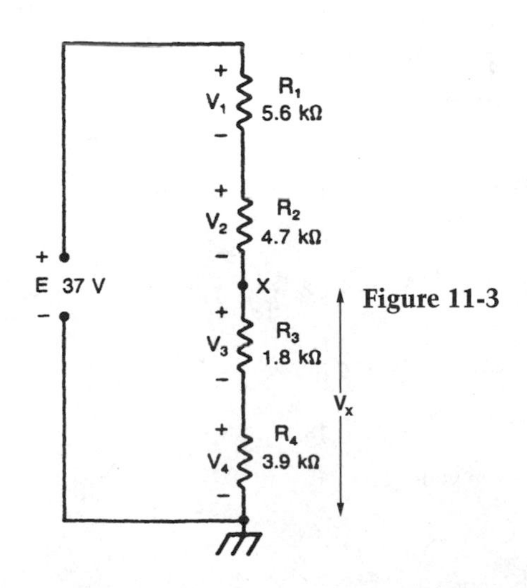

Figure 11-3

$$V_1 = E \times \frac{R_1}{R_T} \quad (11\text{-}12)$$

$$= 37\text{ V} \times \frac{5.6\text{ k}\Omega}{5.6\text{ k}\Omega + 4.7\text{ k}\Omega + 1.8\text{ k}\Omega + 3.9\text{ k}\Omega}$$

$$= 37\text{ V} \times \frac{5.6\text{ k}\Omega}{16.0\text{ k}\Omega}$$

$$= 12.95\text{ V}.$$

Voltage drop,

$$V_3 = V_1 \times \frac{R_3}{R_1} \quad (11\text{-}5)$$

$$= 12.95\text{ V} \times \frac{1.8\text{ k}\Omega}{5.6\text{ k}\Omega} = 4.16\text{ V}.$$

Voltage check:

Voltage drop,

$$V_3 = E \times \frac{R_3}{R_T} = 37\text{ V} \times \frac{1.8\text{ k}\Omega}{16.0\text{ k}\Omega} = 4.16\text{ V} \quad (11\text{-}13)$$

Potential at X,

$$V_X = +37\text{ V} \times \frac{(1.8\text{ k}\Omega + 3.9\text{ k}\Omega)}{5.6\text{ k}\Omega + 4.7\text{ k}\Omega + 1.8\text{ k}\Omega + 3.9\text{ k}\Omega}$$

$$= +37\text{ V} \times \frac{5.7\text{ k}\Omega}{16.0\text{ k}\Omega} = +13.2\text{ V}.$$

Example 11-2

In Fig. 11-4, calculate the values of V_1, E, and P_3.

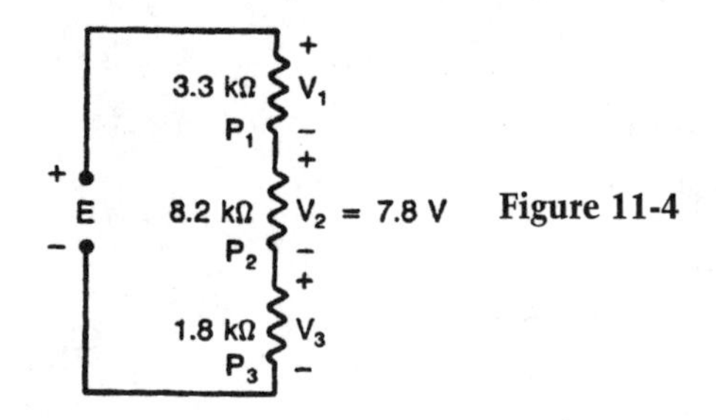

Figure 11-4

Solution

Voltage drop,

$$V_1 = V_2 \times \frac{R_1}{R_2} \quad (11\text{-}5)$$

$$= 7.8\text{ V} \times \frac{3.3\text{ k}\Omega}{8.2\text{ k}\Omega}$$

$$= 3.14\text{ V}$$

Source voltage,

$$E = V_2 \times \frac{R_T}{R_2} \quad (11\text{-}14)$$

$$= 7.8\text{ V} \times \frac{(3.3\text{ k}\Omega + 8.2\text{ k}\Omega + 1.8\text{ k}\Omega)}{8.2\text{ k}\Omega}$$

$$= 7.8\text{ V} \times \frac{13.3\text{ k}\Omega}{8.2\text{ k}\Omega}$$

$$= 12.65\text{ V}.$$

Total power,

$$P_T = \frac{E^2}{R_T} \quad (11\text{-}18)$$

$$= \frac{(12.65\text{ V})^2}{13.3\text{ k}\Omega}$$

$$= 12.03\text{ mW}$$

Power dissipated,

$$P_3 = P_T \times \frac{R_3}{R_T} \quad (11\text{-}21)$$

$$= 12.03\text{ mW} \times \frac{1.8\text{ k}\Omega}{13.3\text{ k}\Omega}$$

$$= 1.63\text{ mW}$$

Example 11-3

In Fig. 11-5, calculate the values of the potentials V_1, V_2, V_3, and V_4.

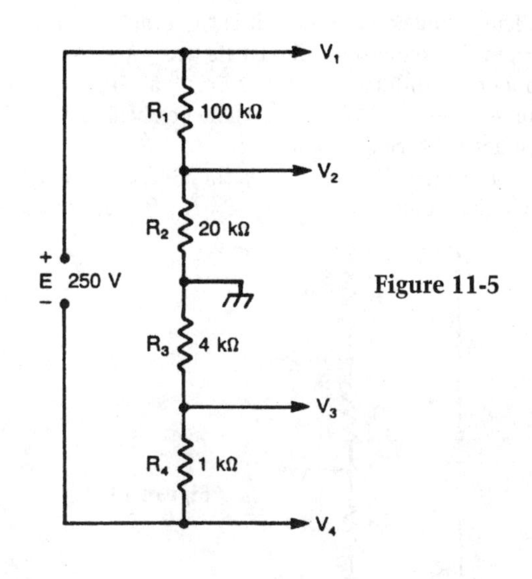

Figure 11-5

Solution

Potential,

$$V_1 = +E \times \frac{(R_1 + R_2)}{R_1 + R_2 + R_3 + R_4} \qquad (11\text{-}23)$$

$$= +250\text{ V} \times \frac{(100\text{ k}\Omega + 20\text{ k}\Omega)}{100\text{ k}\Omega + 20\text{ k}\Omega + 4\text{ k}\Omega + 1\text{ k}\Omega}$$

$$= +250\text{ V} \times \frac{120\text{ k}\Omega}{125\text{ k}\Omega}$$

$$= +240\text{ V}$$

Potential,

$$V_2 = +E \times \frac{R_2}{R_1 + R_2 + R_3 + R_4} \qquad (11\text{-}24)$$

$$= +250\text{ V} \times \frac{20\text{ k}\Omega}{100\text{ k}\Omega + 20\text{ k}\Omega + 4\text{ k}\Omega + 1\text{ k}\Omega}$$

$$= +250\text{ V} \times \frac{20\text{ k}\Omega}{125\text{ k}\Omega}$$

$$= +40\text{ V}$$

Potential,

$$V_3 = -E \times \frac{R_3}{R_1 + R_2 + R_3 + R_4} \qquad (11\text{-}24)$$

$$= -250\text{ V} \times \frac{4\text{ k}\Omega}{100\text{ k}\Omega + 20\text{ k}\Omega + 4\text{ k}\Omega + 1\text{ k}\Omega}$$

$$= -250\text{ V} \times \frac{4\text{ k}\Omega}{125\text{ k}\Omega}$$

$$= -8\text{ V}$$

Potential,

$$V_4 = -E \times \frac{(R_4 + R_4)}{R_1 + R_2 + R_3 + R_4} \qquad (11\text{-}23)$$

$$= -250\text{ V} \times \frac{(4\text{ k}\Omega + 1\text{ k}\Omega)}{100\text{ k}\Omega + 20\text{ k}\Omega + 4\text{ k}\Omega + 1\text{ k}\Omega}$$

$$= -250\text{ V} \times \frac{5\text{ k}\Omega}{125\text{ k}\Omega}$$

$$= -10\text{ V.}$$

PRACTICE PROBLEMS

11-1. A voltage divider consists of two resistors connected across a 30-V supply. The output voltage, 12 V, is developed across a 2.2-kΩ resistor. Calculate the total power dissipated in the divider circuit.

11-2. In Fig. 11-6, what are the values of the voltage potentials at the points X and Y?

11-3. A voltage divider consists of a 470-Ω resistor and a second resistor which are connected in series across a 25-V supply. If the output voltage drop across the second resistor is 6 V, how much power is dissipated in the 470-Ω resistor?

11-4. In the voltage divider circuit of Fig. 11-7, calculate the voltage values of the potentials at the points X, Y, and Z.

11-5. In Practice Problem 11-4, a short circuit occurs between the points Y and Z. Calculate the new voltage values of the potentials at the points X, Y, and Z.

11-6. In Practice Problem 11-4, the 3.3-kΩ resistor opens circuits. Calculate the new voltage values of the potentials at the points X, Y, and Z.

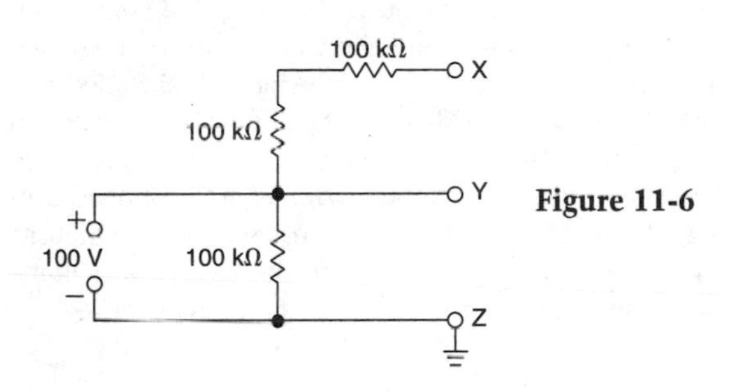

Figure 11-6

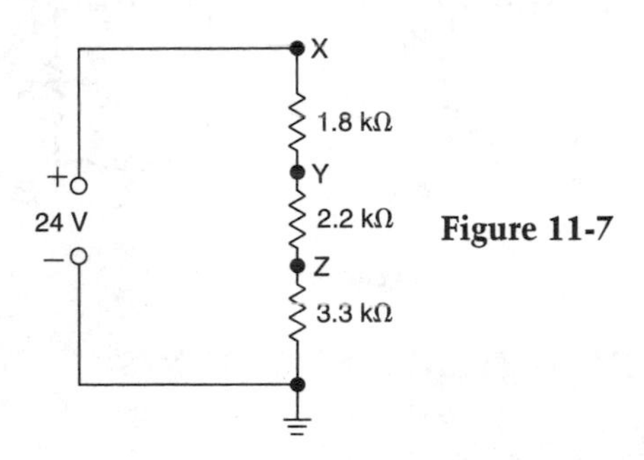

Figure 11-7

12
Sources connected in series-aiding and in series-opposing

Two sources are in series-aiding when their polarities are such as to drive the current in the same direction around the circuit. This is illustrated in Fig. 12-1A, where the negative terminal of E_1 is directly connected to the positive terminal of E_2. Both voltages will then drive the electrons around the circuit in the counterclockwise direction. If both voltage sources are reversed, they will still be connected in series-aiding but the electron flow will now be in the clockwise direction.

The normal purpose of connecting sources in series-aiding is to increase the amount of voltage applied to a circuit. A good example is the insertion of two 1½-V "D" cells into a flashlight to create a total of $2 \times 1\frac{1}{2}$ V = 3 V. In the case of the flashlight, the positive center terminal of one cell must be in contact with the negative casing of the other cell to provide the series-aiding connection.

If the E_2 cell is reversed as in Fig. 12-1B, the polarities of the sources will be such as to drive currents in opposite direction around the circuit. The connection is, therefore, series-opposing and the total voltage available is the *difference* of the individual EMFs. The greater of the two EMFs will then determine the actual direction of the current flow. In the particular case when two identical voltage sources are connected in series-opposing, the total voltage is zero. Consequently, the current is also zero, as are the individual voltage drops and the powers dissipated in the resistors.

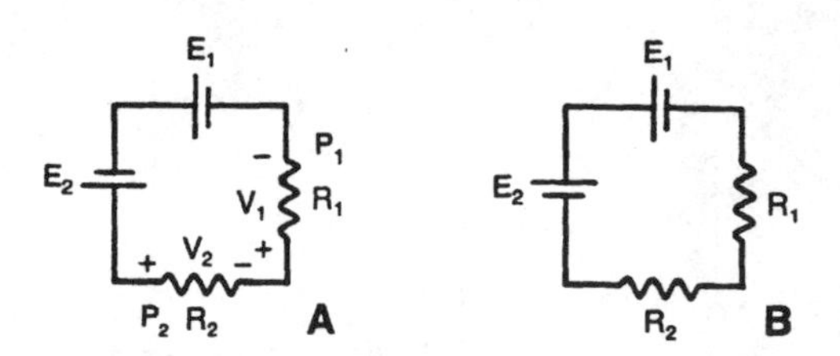

Figure 12-1

MATHEMATICAL DERIVATIONS

In Fig. 12-1A:

Total equivalent EMF,

$$E_T = E_1 + E_2 \text{ volts} \tag{12-1}$$

Total equivalent resistance,

$$R_T = R_1 + R_2 \text{ ohms} \tag{12-2}$$

Current,

$$I = \frac{E_1 + E_2}{R_1 + R_2} \text{ amperes} \tag{12-3}$$

Voltage drop,

$$V_1 = I \times R_1 \text{ volts} \tag{12-4}$$

Voltage drop,

$$V_2 = I \times R_2 \text{ volts} \tag{12-5}$$

Power dissipated,

$$P_1 = I^2 R_1 = V_1^2/R_1 = I \times V_1 \text{ watts} \tag{12-6}$$

Power dissipated,

$$P_2 = I^2 R_2 = V_2^2/R_2 = I \times V_2 \text{ watts} \tag{12-7}$$

Total power delivered from the source,

$$\begin{aligned} P_T &= P_1 + P_2 \\ &= I \times E_T \\ &= I \times (E_1 + E_2) \\ &= I \times (V_1 + V_2) \\ &= I^2 \times (R_1 + R_2) \\ &= \frac{(E_1 + E_2)^2}{R_1 + R_2} \text{ watts} \end{aligned} \tag{12-8}$$

In Fig. 12-1B:

Total equivalent EMF,

$$E_T = E_1 - E_2 \text{ volts} \tag{12-9}$$

This assumes that $E_1 > E_2$, and that the resultant electron flow is in the clockwise direction.

Current,

$$I = \frac{E_1 - E_2}{R_1 + R_2} \text{ amperes} \tag{12-10}$$

Total power,

$$P_T = I \times (E_1 - E_2) = \frac{(E_1 - E_2)^2}{R_1 + R_2} \text{ watts} \tag{12-11}$$

The other equations are the same as those for the series-aiding connection.

Example 12-1

In Fig. 12-2, calculate the values of E_T, R_T, I, V_1, V_2, P_1, P_2, and P_T. What is the value of the potential at the

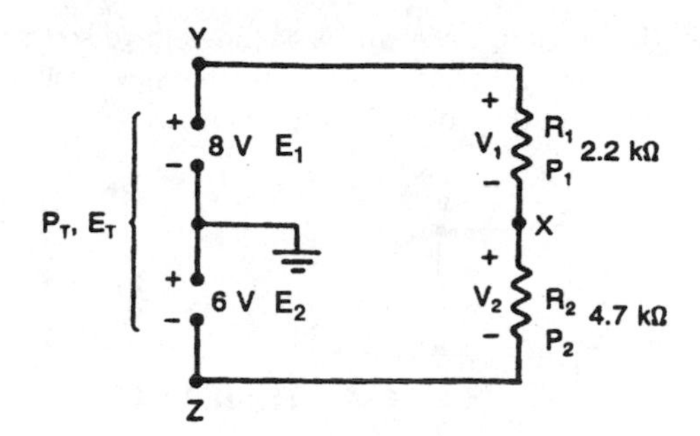

Figure 12-2

point X? If the E_2 source is reversed, recalculate the values of E_T, I, V_1, V_2, P_1, P_2, and P_T.

Solution

Total equivalent EMF,

$$E_T = E_1 + E_2 = 8 + 6 = 14 \text{ V} \qquad (12\text{-}1)$$

Total equivalent resistance,

$$R_T = R_1 + R_2 = 2.2 + 4.7 = 6.9 \text{ k}\Omega \qquad (12\text{-}2)$$

$$\text{Current, } I = \frac{E_T}{R_T} = \frac{14 \text{ V}}{6.9 \text{ k}\Omega} = 2.03 \text{ mA} \qquad (12\text{-}3)$$

and is in the counterclockwise direction (electron flow).

Voltage drop,

$$V_1 = I \times R_1 = 2.03 \text{ mA} \times 2.2 \text{ k}\Omega = 4.46 \text{ V} \qquad (12\text{-}4)$$

Voltage drop,

$$V_2 = I \times R_2 = 2.03 \text{ mA} \times 4.7 \text{ k}\Omega = 9.54 \text{ V} \qquad (12\text{-}5)$$

Voltage check:

$$E_T = V_1 + V_2 = 4.46 + 9.54 = 14 \text{ V}$$

Power dissipated,

$$P_1 = I \times V_1 = 2.03 \text{ mA} \times 4.46 \text{ V} = 9.01 \text{ mW} \qquad (12\text{-}6)$$

Power dissipated,

$$P_2 = I \times V_2 = 2.03 \text{ mA} \times 9.54 \text{ V} = 19.4 \text{ mW} \qquad (12\text{-}7)$$

Total power,

$$P_T = I \times E_T = 2.03 \text{ mA} \times 14 \text{ V} = 28.4 \text{ mW} \qquad (12\text{-}8)$$

Power check:

$$P_T = P_1 + P_2 = 9.01 + 19.4 = 28.41 \text{ mW}.$$

Potential at $Y = +8$ V.

Potential at $X = (+8 \text{ V}) + (-4.46 \text{ V}) = +3.54$ V.

Potential check:

Potential at $Z = -6$ V.

Potential at $X = (-6 \text{ V}) + (+9.54 \text{ V}) = +3.54$ V.

E_2 source reversed:

Total equivalent EMF,

$$E_T = E_1 - E_2 = 8 \text{ V} - 6 \text{ V} = 2 \text{ V}. \qquad (12\text{-}9)$$

$$\text{Current, } I = \frac{E_T}{R_T} = \frac{2 \text{ V}}{6.9 \text{ k}\Omega} = 0.29 \text{ mA} \qquad (12\text{-}10)$$

and is in the counterclockwise direction (electron flow).

Voltage drop, $V_1 = I \times R_1 = 0.29 \text{ mA} \times 2.2 \text{ k}\Omega = 0.64$ V.

Voltage drop, $V_2 = I \times R_2 = 0.29 \text{ mA} \times 4.7 \text{ k}\Omega = 1.36$ V.

Voltage check:

$$E_T = V_1 + V_2 - 0.64 + 1.36 = 2.0 \text{ V}$$

Power dissipated,

$$P_1 = I \times V_1 = 0.29 \text{ mA} \times 0.64 \text{ V} = 0.186 \text{ mW}.$$

Power dissipated,

$$P_2 = I \times V_2 = 0.29 \text{ mA} \times 1.36 \text{ V} = 0.394 \text{ mW}.$$

Total power,

$$P_T = I \times E_T = 0.29 \text{ mA} \times 2 \text{ V} = 0.58 \text{ mW}.$$

Power check:

$$P_T = P_1 + P_2 = 0.186 + 0.394 = 0.58 \text{ mW}.$$

Example 12-2

In Fig. 12-3, calculate the values of I and the voltmeter readings V_1 and V_2.

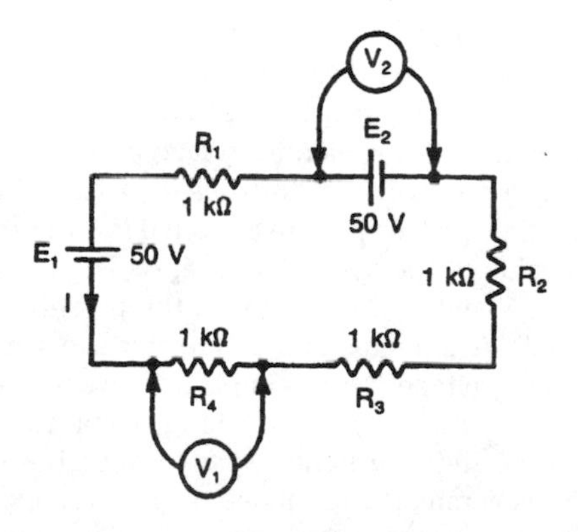

Figure 12-3

Solution

The two sources are connected in series-opposing.

Total equivalent EMF,

$$E_T = E_1 - E_2 = 50 - 50 = 0 \text{ V}. \quad (12\text{-}9)$$

Current, $I = 0$ A.

Voltage drop, $V_1 = 0$ V.

With respect to the voltmeter V_2, one probe is effectively connected to both positive terminals while the other probe is joined to the two negative terminals. Therefore, the reading of V_2 is 50 V.

PRACTICE PROBLEMS

12-1. A number of cells, each of EMF 1.5 V, is connected in series-aiding to produce an output voltage of 12 V. If two of the cells are reversed, what is the new value of the total output voltage?

12-2. In Fig. 12-4, calculate the values of the potentials at the points W, X, Y, and Z.

12-3. In Fig. 12-4, the terminals of the 4-V source are reversed. Calculate the new values of the potentials at the points W, X, Y, and Z.

12-4. In the circuit of Fig. 12-5, what is the voltage value of the potential at the point X?

12-5. In Practice Problem 12-4, the voltage sources are interchanged. What is the new voltage value of the potential at the point X?

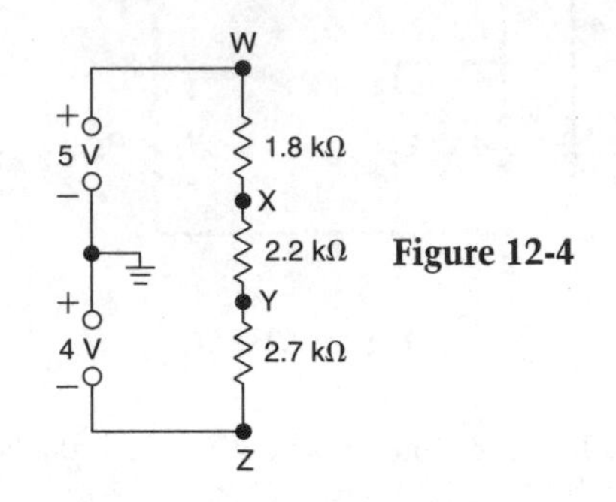

Figure 12-4

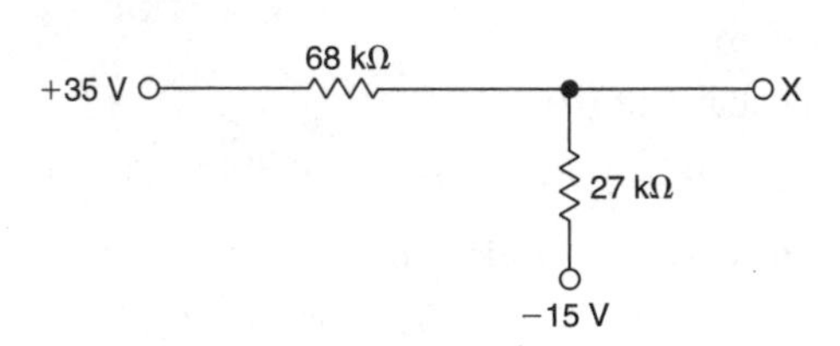

Figure 12-5

13
The potentiometer and the rheostat

The potentiometer (or *pot*) is another practical application of the voltage division rule. Essentially it consists of a length of resistance wire (for example, nichrome wire), or a thin carbon track, along which a moving contact or slider can be set at any point (Fig. 13-1A). The wire is wound on an insulating base that can either be straight or, more conveniently, formed into a circle.

The purpose of the *potentiometer* is to obtain an output voltage, V_o, which can be varied from zero up to the full value of the source voltage, E. There are two terminals, X and Y, at the ends of the potentiometer, and the slider is connected to the third terminal, Z. The source voltage is then applied between the end terminals, X and Y, while the output voltage appears between the slider terminal, Z, and the end terminal, Y, which is commonly grounded. A practical example of the potentiometer is a receiver's volume control. In this case the source is an audio signal but the principles of the potentiometer apply equally well to both dc and ac. In either case care must be taken not to exceed the potentiometer's power rating.

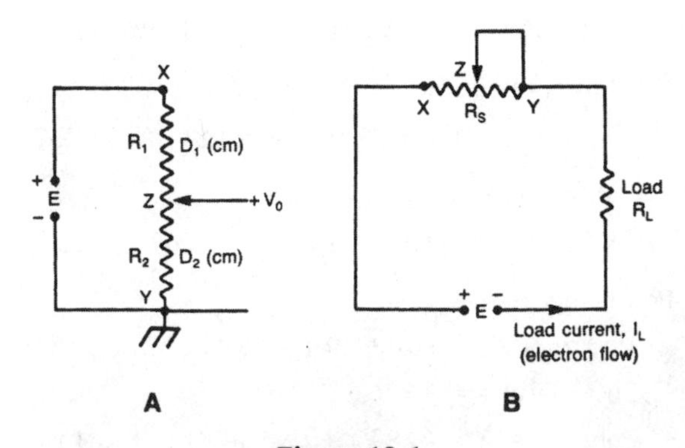

Figure 13-1

Whereas a potentiometer varies an output *voltage*, the rheostat is used for controlling the *current* in a series circuit (Fig. 13-1B). The construction of a rheostat is similar to that of a potentiometer except that the rheostat needs only one end terminal, X, and the terminal, Z, which is connected to the moving contact. The third terminal, Y, is then directly joined to the terminal, Z. A practical example of a rheostat is the dimmer control for the lights on a car's dashboard.

MATHEMATICAL DERIVATIONS

In Fig. 13-1A, the total resistance of the potentiometer is $R_1 + R_2$, and the resistance between Z and Y is R_2. Then by the voltage division rule (VDR),

Output voltage,

$$V_o = E \times \frac{R_2}{R_1 + R_2} \text{ volts} \qquad (13\text{-}1)$$

If the slider is moved to the terminal, Y,

$$R_2 = 0\ \Omega \text{ and } V_o = 0\text{ V} \qquad (13\text{-}2)$$

When the slider is moved to the terminal, X,

$$R_1 = 0\ \Omega \text{ and } V_o = E \text{ volts} \qquad (13\text{-}3)$$

If the terminal, Z, is at the center position,

$$R_1 = R_2 \text{ and } V_o = E/2 \text{ volts} \qquad (13\text{-}4)$$

The results of Equations 13-1, 13-2, 13-3, and 13-4 are only true if no load is connected across the output voltage. Once a load is connected, there would be an additional current flow through R_1, increasing its voltage drop so that the output voltage would fall.

If the ratio of R_2 to the total resistance of the potentiometer is equal to the ratio of the slider's travel, D_2 (from its starting position), to the total travel available, $D_1 + D_2$, the potentiometer is said to be "linear." However, if the resistance and the travel ratios are connected by a *logarithmic* relationship, we have a "log pot." Whether a potentiometer is linear or logarithmic depends on the winding of the resistance wire or the construction of the carbon track.

For a linear potentiometer,

Output voltage,

$$V_o = E \times \frac{D_2}{D_1 + D_2} \text{ volts} \qquad (13\text{-}5)$$

where D_1 and D_2 can be conveniently measured in centimeters.

In Fig. 13-1B, the slider is moved to terminal X. None of the rheostat's resistance is now included in the circuit; the current has its maximum value given by:

Maximum current,

$$I_{max} = \frac{E}{R_L} \text{ amperes} \qquad (13\text{-}6)$$

When the slider is subsequently moved to the terminal, Y, the whole of the rheostat's resistance, R_S, is in series with the load, R_L, so that a minimum current will flow.

Minimum current,

$$I_{min} = \frac{E}{R_S + R_L} \text{ amperes} \qquad (13\text{-}7)$$

The rheostat can therefore control any level of current between the maximum and minimum values.

Example 13-1

In Fig. 13-1A, $E = 20$ V, $D_1 = 35$ cm, $D_2 = 25$ cm. If the potentiometer is linear and has a total resistance of 500 Ω, calculate the values of the output voltage and the power dissipation.

Solution

Output voltage,

$$V_o = +E \times \frac{D_2}{D_1 + D_2} \qquad (13\text{-}5)$$

$$= +20\text{ V} \times \frac{25\text{ cm}}{25\text{ cm} + 35\text{ cm}}$$

$$= +20\text{ V} \times \frac{25\text{ cm}}{60\text{ cm}}$$

$$= +8.33\text{ V.}$$

Power dissipation,

$$P = \frac{E^2}{R_1 + R_2} = \frac{(20\text{ V})^2}{500\ \Omega} = 0.8\text{ W.}$$

Example 13-2

In Fig. 13-1B, $E = 9$ V, $R_S = 6\ \Omega$ and $R_L = 4\ \Omega$. Calculate the maximum and minimum values of the circuit current. Assuming that the rheostat is linear, what is the value of the load current when the slider is in its center position?

Solution

Maximum load current,

$$I_{max} = \frac{E}{R_L} \qquad (13\text{-}6)$$

$$= \frac{9\text{ V}}{4\ \Omega} = 2.25\text{ A}$$

Minimum load current,

$$I_{min} = \frac{E}{R_L + R_S} \quad (13\text{-}7)$$

$$= \frac{9\text{ V}}{4\ \Omega + 6\ \Omega} = 0.9\text{ A}$$

With the slider in the center position,

$$R_S = \frac{6\ \Omega}{2} = 3\ \Omega.$$

The load current is then,

$$I_L = \frac{E}{R_L + R_S} = \frac{9\text{ V}}{4\ \Omega + 3\ \Omega} = 1.29\text{ A}.$$

PRACTICE PROBLEMS

13-1. A 10-kΩ linear potentiometer is connected across a 24-V supply. If the shaft is rotated through two thirds of its entire movement, what are the values of the output voltage, and the power dissipated, in the potentiometer?

13-2. A 100-kΩ potentiometer is connected in series with an unknown resistor across a 20-Vdc source. Calculate the value of the unknown resistor if the maximum output voltage is 15 V.

13-3. In the circuit of Fig. 13-2, calculate the maximum and minimum values of the output voltage, V_o.

13-4. In the circuit of Fig. 13-2, a short circuit develops across the 330-Ω resistor. Find the new maximum and minimum values of the output voltage, V_o.

13-5. In the circuit of Fig. 13-3, calculate the values of the maximum and minimum values of the power dissipated in the lamp, L_1.

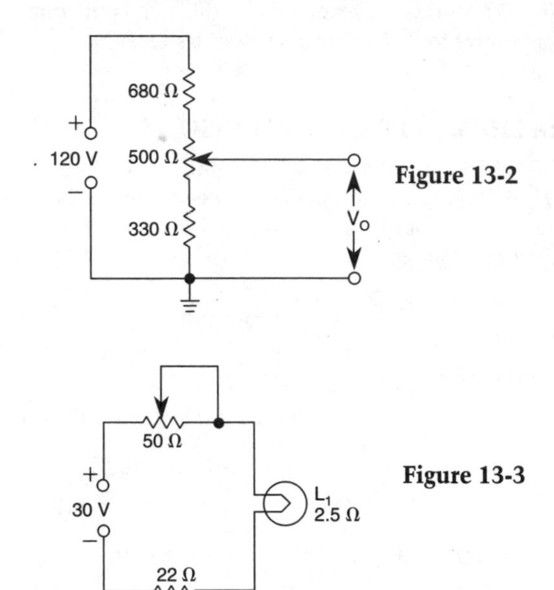

Figure 13-2

Figure 13-3

14 The series voltage-dropping resistor

Every load normally has a voltage/current rating, as well as a wattage rating. For example, a 75-W bulb is also rated at 120 V. If the available source voltage is greater than the required load voltage, it is possible to insert a series dropping resistor so that the load is still operated correctly. Figure 14-1 illustrates a load such as a transistor amplifier, which needs a supply voltage of 9 V with a corresponding current of 8 mA. If the available source voltage is 15 V, the series resistor is inserted to drop the voltage down to the 9 V required by the load. The voltage across R_D is therefore 15 V − 9 V = 6 V. This advantage of using a dropping resistor is achieved at the expense of its cost and the power dissipation.

Care must be taken to calculate the required ohmic value of the dropping resistor, R_D, as well as its required power dissipation because the resistor would overheat if its power rating were too low (see Example 14-1).

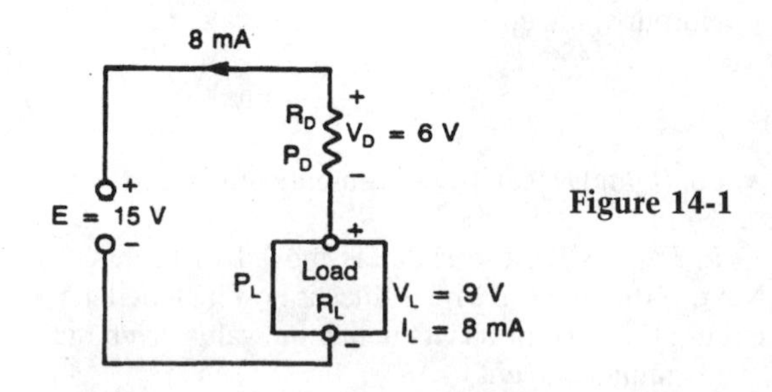

Figure 14-1

MATHEMATICAL DERIVATIONS

Voltage across the dropping resistor,

$$V_D = E - V_L \qquad (14\text{-}1)$$

$$= E - I_L R_L \text{ volts}$$

Ohmic value of the dropping resistor,

$$R_D = \frac{V_D}{I_L} \text{ ohms} \qquad (14\text{-}2)$$

Load resistance,

$$R_L = \frac{V_L}{I_L} \text{ ohms} \qquad (14\text{-}3)$$

Power dissipation of the dropping resistor,

$$P_D = I_L \times V_D \text{ watts} \qquad (14\text{-}4)$$

Load power,

$$P_L = I_L \times V_L \text{ watts} \qquad (14\text{-}5)$$

Total power delivered from the source,

$$P_T = P_L + P_D \qquad (14\text{-}6)$$

$$= I_L \times (V_L + V_D)$$

$$= I_L^2(R_L + R_D)$$

$$= \frac{E^2}{(R_L + R_D)} \text{ watts}$$

Example 14-1

A relay whose coil resistance is 400 Ω is designed to operate when the voltage across the relay is 80 V. If the relay is operated from a 120-V supply, calculate the ohmic value of the required series dropping resistor and its power dissipation.

Solution

Voltage across the dropping resistor,

$$V_D = E - V_L \qquad (14\text{-}1)$$

$$= 120 \text{ V} - 80 \text{ V}$$

$$= 40 \text{ V}$$

Relay current,

$$I_L = \frac{V_L}{R_L} = \frac{80 \text{ V}}{400\ \Omega} = 200 \text{ mA} \qquad (14\text{-}2)$$

Ohmic value of the dropping resistor,

$$\frac{40 \text{ V}}{200 \text{ mA}} \qquad R_D \equiv \frac{V_D}{I_L} \qquad (14\text{-}3)$$

$$= 200\ \Omega$$

Power dissipation of the dropping resistor,

$$P_D = I_L \times V_D \qquad (14\text{-}4)$$

$$= 200 \text{ mA} \times 40 \text{ V}$$

$$= 8 \text{ W}$$

A 200-Ω, 10-W resistor would be a suitable choice.

PRACTICE PROBLEMS

14-1. An 11.0-V battery is to be charged through a series resistor from a 12.5-Vdc source. If the charging current is 4 A, what is the required resistance value for the series resistor?

14-2. A 500-Ω resistive load has a power rating of 30 W and is connected in series with a dropping resistor across a 440-Vdc source. What is the power dissipated in the dropping resistor?

14-3. Ten 6-V, 5-W lamps are connected in series with a dropping resistor across a 90-Vdc source. Calculate the values of the resistance of the dropping resistor, and its power dissipation.

14-4. It is required to operate a resistive load rated at 12 V, 100 W from a 100-Vdc source. What is the value of the dropping resistor that must be connected in series with the load?

15
Resistors in parallel

All loads (lights, vacuum cleaner, toaster, etc.) in your home are normally stamped with a 120-V rating. It follows that all these loads must be connected directly across the household supply because, although each load has its own power rating, its required voltage value in every case is the same. This type of arrangement is known as a *parallel circuit* and is illustrated in Figs. 15-1A and B in which the three resistors are connected between two common points X, Y (Fig. 15-1A) or two common lines (Fig. 15-1B). The points, or lines, are then directly connected to the voltage source. It follows that the voltage drop across each parallel resistor is the same and equal to the source voltage. This is in contrast with the series arrangement, where voltage division occurred.

Each of the resistors in a parallel circuit is a path for current flow, and each path is called a branch. Because the voltage across each branch is the same, each branch will carry its own current, and the individual branch currents will all be different (unless the branch resistances are the same, in which case their currents would also be equal). Notice that the branch currents are independent of one another. In other words, if one branch current is switched off, the other branch currents are unaffected. This would be equivalent to saying that, if you switch off the light in the kitchen, the TV set in the lounge continues to operate.

The total source current, I_T, splits at the point X into individual branch currents. The current in a particular branch is inversely proportional to the value of the resistance in that branch. Consequently, the lowest-value resistor will carry the greatest current, while the smallest current will flow through the highest-value resistor. In terms of electron flow, the branch currents (I_1, I_2, and I_3) are all leaving the junction point, X, while the total current I_T is entering the same junction point. Recombination of the branch currents then occurs at the other junction point, Y. It follows that the total current is the *sum* of the individual branch currents. This is an expression of *Kirchhoff's Current Law* (KCL), which is further explored in chapter 26.

The total equivalent resistance, R_T, is defined as the ratio of the source voltage, E, to the source current, I_T (Fig. 15-1C). If an additional branch is added to a parallel arrangement of resistors, the total current must *increase* by the amount of the new branch current, and therefore, the total equivalent resistance must *decrease*. For parallel resistors, the value of R_T must al-

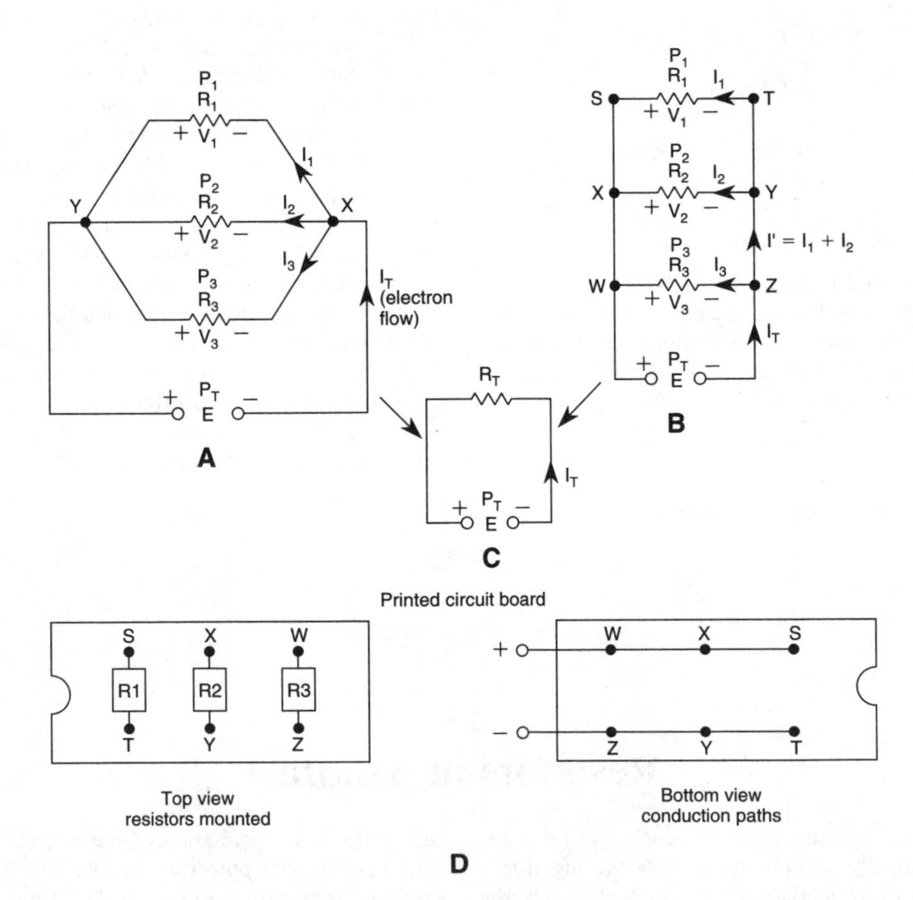

Figure 15-1

ways be less than the lowest-value resistor in the arrangement.

Each parallel resistor dissipates its own power and the sum of the individual powers dissipated is equal to the total power delivered from the source. The greatest power is dissipated by the lowest-value resistor, which carries the greatest branch current. For example, the resistance of a nickel-iron 1000-W heater element is only one-tenth that of the tungsten filament for a 100-W electric light bulb.

MATHEMATICAL DERIVATIONS

Source voltage,

$$E = V_1 = V_2 = V_3 \text{ volts} \quad (15\text{-}1)$$

Branch currents,

$$I_1 = \frac{V_1}{R_1} = \frac{E}{R_1}, I_2 = \frac{V_2}{R_2} = \frac{E}{R_2}, \quad (15\text{-}2)$$

$$I_3 = \frac{V_3}{R_3} = \frac{E}{R_3}$$

Total source current,

$$I_T = I_1 + I_2 + I_3 \quad (15\text{-}3)$$

$$= \frac{E}{R_1} + \frac{E}{R_2} + \frac{E}{R_3}$$

$$= E \times \left(\frac{1}{R_1} + \frac{1}{R_2} + \frac{1}{R_3}\right) \text{ amperes}$$

Notice that the current I', which exists in the circuit of Fig. 15-1B, cannot be measured in the circuit of Fig. 15-1A.

Total source current is also given by,

$$I_T = \frac{E}{R_T} = E \times \frac{1}{R_T} \text{ ohms} \quad (15\text{-}4)$$

Comparing equations 15-3, 15-4:

$$\frac{1}{R_T} = \frac{1}{R_1} + \frac{1}{R_2} + \frac{1}{R_3} \quad (15\text{-}5)$$

For N resistors in parallel,

$$I_T = I_1 + I_2 + I_3 \ldots + I_N \quad (15\text{-}6)$$

$$\frac{1}{R_T} = \frac{1}{R_1} + \frac{1}{R_2} + \frac{1}{R_3} \ldots + \frac{1}{R_N} \quad (15\text{-}7)$$

This is known as the "reciprocal formula" because $1/R_T$, $1/R_1$, etc., are the reciprocals of R_T, R_1, etc. Then

$$R_T = \frac{1}{\frac{1}{R_1} + \frac{1}{R_2} + \frac{1}{R_3} \ldots + \frac{1}{R_N}} \text{ ohms} \quad (15\text{-}8)$$

If all N resistors are of equal value R,

$$R_T = \frac{R}{N} \text{ ohms} \quad (15\text{-}9)$$

Conductance,

$$G_1 = \frac{1}{R_1}, G_2 = \frac{1}{R_2}, G_3 = \frac{1}{R_3} \quad (15\text{-}10)$$

$$\ldots G_N = \frac{1}{R_N} \text{ siemens}$$

Combining equations 15-7 and 15-8,
Total conductance,

$$G_T = \frac{1}{R_T} = G_1 + G_2 + G_3 \ldots + G_N \text{ siemens} \quad (15\text{-}11)$$

For *two* resistors in parallel:

$$R_T = \frac{R_1 \times R_2}{R_1 + R_2}, R_1 = \frac{R_2 \times R_T}{R_2 - R_T}, \quad (15\text{-}12)$$

$$R_2 = \frac{R_1 \times R_T}{R_1 - R_T} \text{ ohms}$$

The equation $R_T = (R_1 \times R_2)/(R_1 + R_2)$ is commonly referred to as the *product-over-sum* formula. If more than two parallel resistors are involved, the formula may be repeated indefinitely by considering pairs of resistors, but the method then becomes far more cumbersome than the use of the reciprocal formula to which the electronic calculator is particularly suited.

Power dissipated in the R1 resistor,

$$P_1 = I_1 \times E = I_1^2 R_1 = E^2/R_1 \text{ watts} \quad (15\text{-}13)$$

Power dissipated in the R2 resistor,

$$P_2 = I_2 \times E = I_2^2 R_2 = E^2/R_2 \text{ watts} \quad (15\text{-}14)$$

Power dissipated in the R3 resistor,

$$P_3 = I_3 \times E = I_3^2 R_3 = E^2/R_3 \text{ watts} \quad (15\text{-}15)$$

Total power delivered from the source,

$$P_T = P_1 + P_2 + P_3 \quad (15\text{-}16)$$

$$= I_T \times E$$

$$= I_T^2 \times R_T$$

$$= E^2/R_T \text{ watts}$$

Example 15-1

In Fig. 15-2, calculate the values of I_1, I_2, I_3, I_4, I', I'', I_T, R_T, P_1, P_2, P_3, P_4, and P_T.

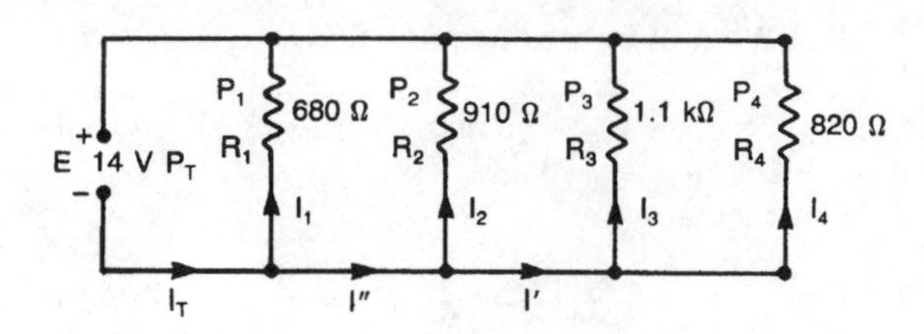

Figure 15-2

Solution

$$\text{Branch current, } I_1 = \frac{E}{R_1} \quad (15\text{-}2)$$

$$= \frac{14 \text{ V}}{680\ \Omega}$$

$$= 20.6 \text{ mA}$$

$$\text{Branch current, } I_2 = \frac{E}{R_2} \quad (15\text{-}2)$$

$$= \frac{14 \text{ V}}{910\ \Omega}$$

$$= 15.4 \text{ mA}$$

$$\text{Branch current, } I_3 = \frac{E}{R_3} \quad (15\text{-}2)$$

$$= \frac{14 \text{ V}}{1.1 \text{ k}\Omega}$$

$$= 12.7 \text{ mA}$$

$$\text{Branch current, } I_4 = \frac{E}{R_4} \quad (15\text{-}2)$$

$$= \frac{14 \text{ V}}{820\ \Omega}$$

$$= 17.1 \text{ mA}$$

$$\text{Current } I' = I_3 + I_4$$

$$= 12.7 + 17.1$$

$$= 29.8 \text{ mA.}$$

$$\text{Current } I'' = I_2 + I'$$

$$= 15.4 + 29.8$$

$$= 45.2 \text{ mA.}$$

$$\text{Total current, } I_T = I_1 + I_2 + I_3 + I_4 \quad (15\text{-}6)$$

$$= 20.6 + 15.4 + 12.7 + 17.1$$

$$= 65.8 \text{ mA}$$

$$\text{Total resistance, } R_T = \frac{E}{I} \quad (15\text{-}4)$$

$$= \frac{14 \text{ V}}{65.8 \text{ mA}}$$

$$= 213\ \Omega.$$

Total resistance check:
Total resistance,

$$R_T = \frac{1}{\frac{1}{R_1} + \frac{1}{R_2} + \frac{1}{R_3} + \frac{1}{R_4}} \quad (15\text{-}8)$$

$$= \frac{1}{\frac{1}{680} + \frac{1}{910} + \frac{1}{1100} + \frac{1}{820}}$$

$$= 213\ \Omega.$$

Note that the value of R_T (213 Ω) is less than the lowest-value resistor (680 Ω) in the parallel arrangement.

Power dissipated in the R1 resistor,

$$P_1 = I_1 \times E \quad (15\text{-}13)$$

$$= 20.6 \text{ mA} \times 14 \text{ V}$$

$$= 288 \text{ mW}$$

Power dissipated in the R2 resistor,

$$P_2 = I_2 \times E \quad (15\text{-}14)$$

$$= 15.4 \text{ mA} \times 14 \text{ V}$$

$$= 216 \text{ mW}$$

Power dissipated in the R3 resistor,

$$P_3 = I_3 \times E \quad (15\text{-}15)$$

$$= 12.7 \text{ mA} \times 14 \text{ V}$$

$$= 178 \text{ mW}$$

Power dissipated in the R4 resistor,

$$P_4 = I_4 \times E$$

$$= 17.1 \text{ mA} \times 14 \text{ V}$$

$$= 239 \text{ mW}$$

Total power dissipated,

$$P_T = P_1 + P_2 + P_3 + P_4 \quad (15\text{-}16)$$

$$= 288 + 216 + 178 + 239$$

$$= 921 \text{ mW}$$

Power check:
Total power delivered from the source,

$$P_T = I_T \times E \quad (15\text{-}16)$$

$$= 65.8 \text{ mA} \times 14 \text{ V}$$

$$= 921 \text{ mW}$$

Example 15-2

A parallel circuit consists of four 1-watt resistors whose values are 120 Ω, 150 Ω, 270 Ω, and 330 Ω. Calculate their maximum power dissipation without exceeding the wattage rating of any of the resistors.

Solution

The voltage drop across each resistor to produce its 1-W rating is different; using the relationship $E = \sqrt{P \times R}$, these voltages are

$$E_{120\,\Omega} = \sqrt{1\text{W} \times 120\,\Omega} = 10.95\text{ V}$$
$$E_{150\,\Omega} = \sqrt{1\text{W} \times 150\,\Omega} = 12.25\text{ V}$$
$$E_{270\,\Omega} = \sqrt{1\text{W} \times 270\,\Omega} = 16.43\text{ V}$$
$$E_{330\,\Omega} = \sqrt{1\text{W} \times 330\,\Omega} = 18.17\text{ V}$$

You must select the smallest of these voltages, 10.95 V, because if the current were allowed to exceed this value, the maximum allowed dissipation in the 120-Ω resistor would be exceeded. Using the relationship $P_T = E^2G_T$, the maximum power dissipation of this network is:

$$(10.95\text{ V})^2 \times \left(\frac{1}{120\,\Omega} + \frac{1}{150\,\Omega} + \frac{1}{270\,\Omega} + \frac{1}{330\,\Omega}\right)$$
$$= 2.61\text{ W}$$

This result should be compared with the solution to Example 9-3.

Example 15-3

Two equal value resistors are connected in series across a dc voltage source. These two resistors are then reconnected in parallel across the same voltage source. What is the ratio of the power dissipated in the parallel circuit to the power dissipated in the series string?

Solution

Let the value of each resistor be R ohms. If the source's output is E volts, the power dissipated in the series arrangement,

$$P_s = \frac{E^2}{R_T} = \frac{E^2}{2R}\text{ watts} \qquad (9\text{-}14)$$

Power dissipated in the parallel formation,

$$P_p = \frac{E^2}{R_T} = \frac{E^2}{R/2}\text{ watts} \qquad (15\text{-}16)$$

Value of ratio,

$$\frac{P_p}{P_s} = \frac{E^2/2R}{E^2/(R/2)} = 4{:}1$$

Note that this result is only true if the two resistors have equal values. For example, if $E = 10$ V, $R_1 = 5\,\Omega$, $R_2 = 10\,\Omega$,

$$P_s = \frac{(10\text{ V})^2}{5\,\Omega + 10\,\Omega} = \frac{100}{15}\text{ W}$$

$$P_p = \frac{(10\text{ V})^2}{5\,\Omega \times 10\,\Omega\,(5\,\Omega + 10\,\Omega)} = \frac{100 \times 15}{50}\text{ W}$$

Ratio,

$$\frac{P_p}{P_s} = \frac{100 \times 15}{50 \times 100/15} = \frac{225}{50} = 4.5{:}1$$

PRACTICE PROBLEMS

15-1. Three resistors are connected in parallel across a dc voltage source. The color-coding of these resistors is brown-red-red-silver, gray-red-brown-silver, and blue-gray-brown-silver. What are the total equivalent rated resistance and conductance values of this parallel bank?

15-2. A 680-Ω resistor and a 220-Ω resistor are connected in parallel across a dc source. Calculate the value of their total equivalent resistance. If a current of 17 mA flows through the 680-Ω resistor, what is the value of the current flowing through the 220-Ω resistor?

15-3. One thousand 1-MΩ, ½-W resistors are connected in parallel across a 100 Vdc source. Calculate the value of the total equivalent resistance presented to the source. Find the values of the current flowing in each branch and the total current drawn from the source. Determine the total power dissipated in the circuit.

15-4. The values of three parallel resistors are 270 Ω, 330 Ω, and 470 Ω. If the current flowing through the 470-Ω resistor is 6 mA, what is the value of the total current drawn from the source?

15-5. Three resistors in parallel have a total equivalent resistance of 1.86 kΩ. If two of the resistors have values of 5.6 kΩ and 6.8 kΩ, what is the value of the third resistor?

15-6. Two resistors in parallel possess a total conductance of 48 μS. If the value of one resistor is 33 kΩ, what is the value of the other resistor?

15-7. A 120-Vdc source is connected through a 20-A fuse to a number of parallel connected 120-V, 25-W lamps. What is the maximum possible number of lamps that can be safely connected?

15-8. A 20,000-Ω, 100-W resistor, a 40,000-Ω, 50-W resistor, and a 5,000-Ω, 20-W resistor are connected in parallel. What is the maximum value of the source voltage that will not cause the wattage rating of any of the resistors to be exceeded?

16
Open and short circuits

An open circuit is regarded as a break that occurs in a circuit. For example a resistor might burn out (Fig. 16-1A) as a result of excessive power dissipation or become disconnected from an adjacent resistor. In either case, the break would have theoretically infinite resistance so that no current can exist in an open circuit.

THE OPEN CIRCUIT

In the series string of Fig. 16-1A, resistor R2 has burned out (points *D*, *E*) so that the total equivalent resistance is theoretically infinite. There is zero current throughout the circuit, and no voltage drops across the resistors R1, R3. Consequently, the source voltage appears across the open circuit and it could be measured by a voltmeter connected between the points *D* and *E*.

To locate the position of the break, the switch, S, is opened to remove power from the circuit. Using an ohmmeter (on a high resistance range), the common probe is connected to the point *H*. With the red probe at point *A*, the ohmmeter reading is virtually infinite because the measured resistance contains the open circuit. The reading will not change as the red probe is moved in turn to the points *B*, *C*, and *D*. However, as soon as the red probe is moved to the point *E*, the resistance falls to the relatively low value of R3. This change in the reading indicates that the open circuit exists between the points *D* and *E*. The ohmmeter has been used to conduct a *continuity test,* which has located a *discontinuity*—the open circuit.

If resistor R1 burns out in the parallel circuit of Fig. 16-1B, current I_1 is zero, but currents I_2 and I_3 remain the same. Total current I_T is reduced and the total equivalent resistance is greater. However, if a break occurs at point *X*, all current ceases and the total equivalent resistance is theoretically infinite.

THE SHORT CIRCUIT

A short circuit has, theoretically, zero resistance so that a current can flow through a short circuit without developing any voltage drop. Although resistors rarely become shorted, a short circuit can occur as the result

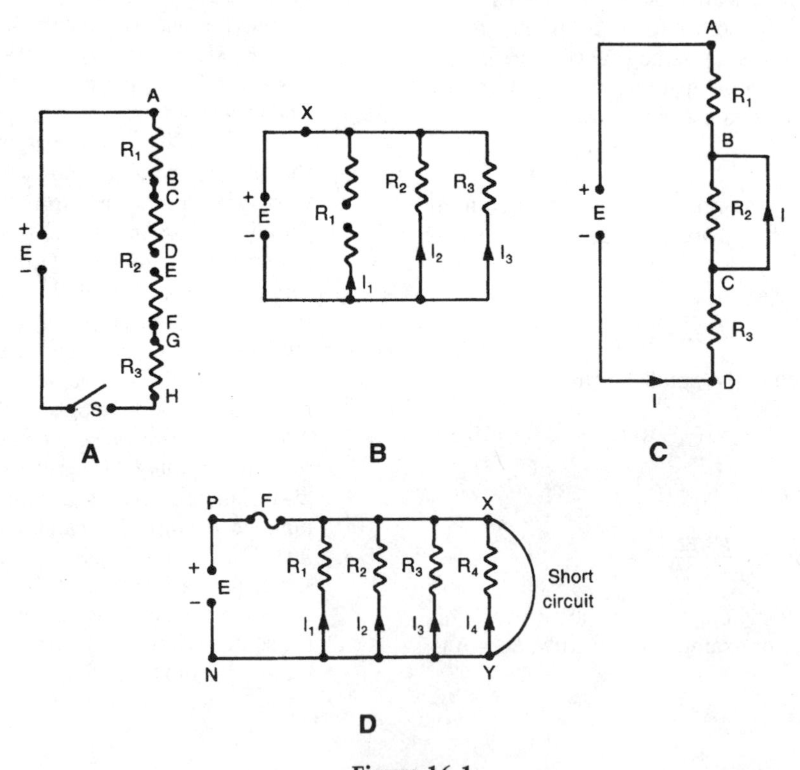

Figure 16-1

of the bare connecting wires coming into contact or being joined together by a stray drop of solder. The short circuit is therefore a zero-resistance path in parallel with (shunting across) the resistor.

An example of a short circuit is shown in the series circuit of Fig. 16-1C. No current will flow through the resistor R2, and there will be no voltage drop between the points B and C. The total equivalent resistance will be given by $R_T = R_1 + R_3$, and current I will increase as a result of the short circuit.

In the parallel circuit of Fig. 16-1D, a short circuit has developed across the resistor R4. The same short circuit also exists in parallel with the resistors R1, R2, R3 so that the currents I_1, I_2, I_3, and I_4 are all zero. However, the path *NYXP* has an extremely low resistance so that a high current would flow and the protection fuse, F, would melt. This would prevent possible damage to the connecting wires and the source.

Example 16-1

In Fig. 16-2, an open circuit occurs between the points B and C. What are the potential values at the points A, B, C, D?

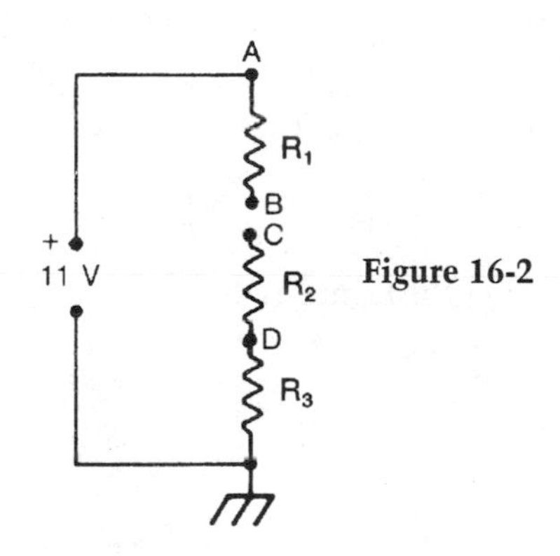

Figure 16-2

Solution

The current throughout the circuit is zero and there are zero voltage drops across the resistors. Consequently, the potentials at the points A, B are both +11 V, and the potentials at the points C, D are *zero*.

Example 16-2

In Fig. 16-3 the 180-Ω resistor burns out. What are the new values of I_1, I_2, I_3, and I_T?

Solution

$$\text{Current, } I_1 = \frac{15\text{ V}}{220\ \Omega}$$

$$= 68.2\text{ mA.}$$

$$\text{Current, } I_2 = 0.$$

$$\text{Current, } I_3 = \frac{15\text{ V}}{470\ \Omega}$$

$$= 31.9\text{ mA.}$$

$$\text{Current, } I_T = I_1 + I_2 + I_3$$

$$= 68.2 + 0 + 31.9$$

$$= 100.1\text{ mA.}$$

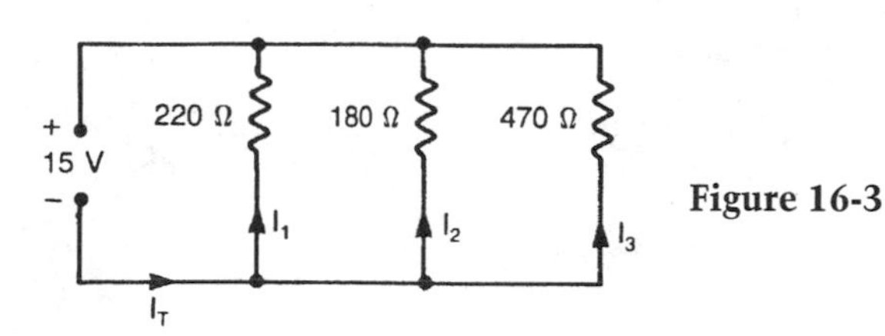

Figure 16-3

Example 16-3

In Fig. 16-4, a short develops across resistor R2. Calculate the new potential values at points A, B, C, and D.

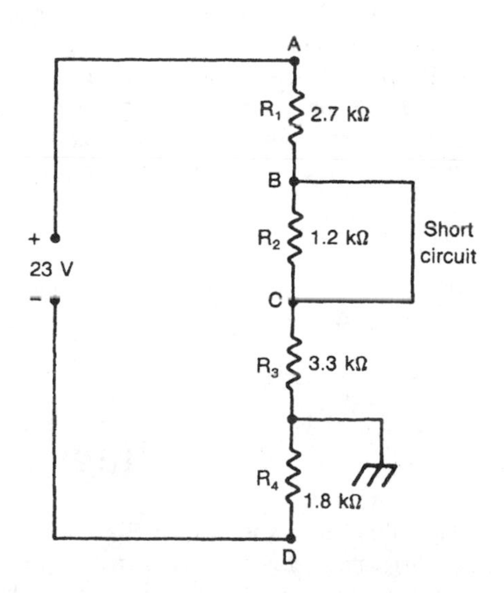

Figure 16-4

Solution

By the voltage division rule (VDR),
Potential at A,

$$V_A = +23\text{ V} \times \frac{(2.7\text{ k}\Omega + 3.3\text{ k}\Omega)}{2.7\text{ k}\Omega + 3.3\text{ k}\Omega + 1.8\text{ k}\Omega}$$

$$= +23\text{ V} \times \frac{6.0\text{ k}\Omega}{7.8\text{ k}\Omega}$$

$$= +17.69\text{ V}.$$

Potentials at B, C,

$$V_B = V_C$$

$$= +23\text{ V} \times \frac{3.3\text{ k}\Omega}{2.7\text{ k}\Omega + 3.3\text{ k}\Omega + 1.8\text{ k}\Omega}$$

$$= +23\text{ V} \times \frac{3.3\text{ k}\Omega}{7.8\text{ k}\Omega}$$

$$= +9.73\text{ V}.$$

Potential at D,

$$V_D = -23\text{ V} \times \frac{1.8\text{ k}\Omega}{2.7\text{ k}\Omega + 3.3\text{ k}\Omega + 1.8\text{ k}\Omega}$$

$$= -23\text{ V} \times \frac{1.8\text{ k}\Omega}{7.8\text{ k}\Omega}$$

$$= -5.31\text{ V}.$$

PRACTICE PROBLEMS

16-1. A 470-Ω resistor and a 330-Ω resistor are connected in parallel across a 60-V dc source. If a short circuit develops across the 470-Ω resistor, what is the value of the current that flows through the 330-Ω resistor?

16-2. In Practice Problem 16-1, the short across the 470-Ω resistor is removed. If, subsequently, the 330-Ω resistor open-circuits, calculate the values of the total current drawn from the source.

16-3. The values of three parallel resistors are 220 Ω, 270 Ω and 390 Ω; the current flowing through the 390-Ω resistor is 15 mA. If the 270-Ω resistor open-circuits, calculate the amount of the total power dissipated in the circuit.

16-4. The total resistance of four resistors in parallel is 278 Ω. If the values of three of the resistors are 910 Ω, 1.2 kΩ, and 1.5 kΩ, what is the value of the fourth resistor? If a short circuit develops across the 1.2-kΩ resistor, what is the value of the new total equivalent resistance?

16-5. In Fig. 16-5, find the value of the total equivalent resistance presented to the source. If an open circuit occurs between the points X and Y, calculate the new value of the total equivalent resistance presented to the source.

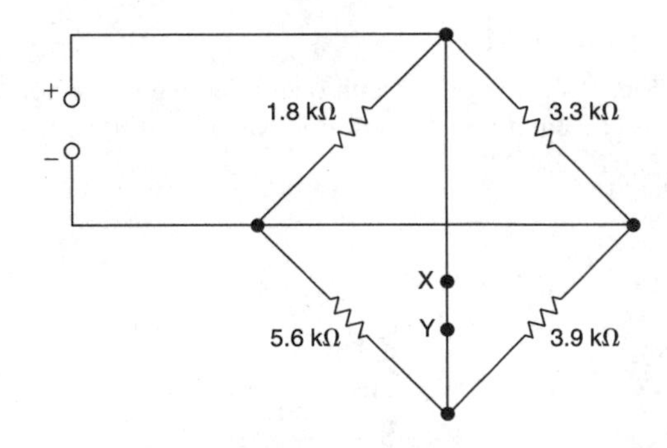

Figure 16-5

17
Voltage sources in parallel

Figure 17-1 illustrates N identical cells, each of EMF E, which are parallel-connected across a number (M) of resistive loads. All positive terminals are joined together so that any one of these terminals may be chosen as the positive output terminal; the negative terminals are likewise connected. If a voltmeter (V) is connected across any one of the cells, it must also be in parallel with all the other cells, so that the voltmeter reading is only E volts (the EMF of one cell). The purpose of connecting cells in parallel is, therefore, *not* to increase the total voltage. However, the current capability has been increased, because the total load current, I_T, will be shared between the batteries.

An everyday example of parallel-connected batteries occurs when you give someone a "jump start" for his/her car. You use the cables to join the positive terminals of the two 12-V batteries together and make the same type of connection with the negative terminals. However, with this parallel arrangement the total voltage available is still only equal to 12 V (the EMF of one battery).

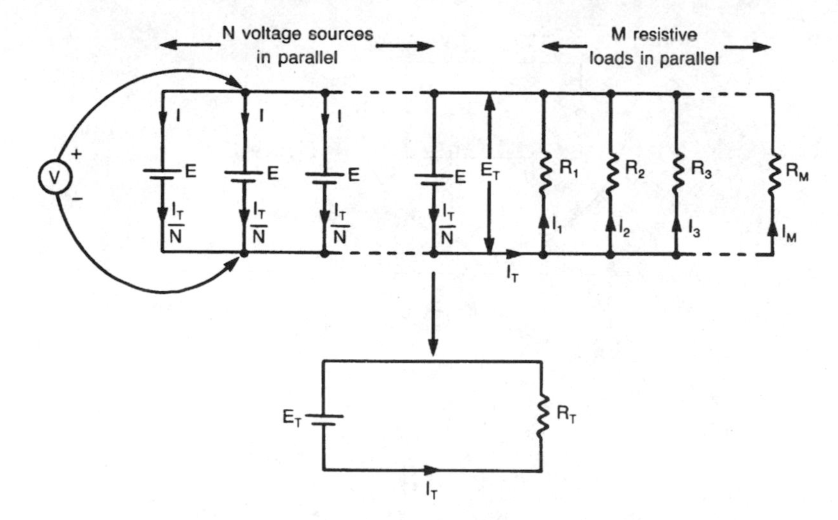

Figure 17-1

Notice that, in the loops containing the cells, the arrangement is series-opposing, so that there is no loading effect of one cell on another. With identical cells, no parallel-opposing arrangement is possible with respect to the loads, because the cells would be in series-aiding around the loops. A very high circulating current would flow between the cells, but no current would be supplied to the loads.

MATHEMATICAL DERIVATIONS

For N identical cells in parallel:

Total voltage of the parallel combination,

$$E_T = E \text{ volts} \tag{17-1}$$

Total equivalent resistance of the parallel loads,

$$R_T = \frac{1}{\frac{1}{R_1} + \frac{1}{R_2} + \frac{1}{R_3} \ldots + \frac{1}{R_N}} \tag{17-2}$$

Total load current,

$$I_T = \frac{E_T}{R_T} \text{ amperes} \tag{17-3}$$

Current supplied by each cell,

$$I = \frac{I_T}{N} \text{ amperes} \tag{17-4}$$

Example 17-1

Four identical 6-V cells are connected in parallel across two loads whose resistances are 20 Ω and 5 Ω. Calculate the value of the current supplied by each cell.

Solution

Total voltage applied to the parallel loads,

$$E_T = 6 \text{ V (EMF of one cell)} \tag{17-1}$$

Total load,

$$R_T = \frac{20 \times 5}{20 + 5}$$

$$= 4\ \Omega$$

Total load current,

$$I_T = \frac{E_T}{R_T} = \frac{6 \text{ V}}{4\ \Omega} \tag{17-3}$$

$$= 1.5 \text{ A}$$

Current supplied by each cell,

$$I = \frac{I_T}{N} = \frac{1.5 \text{ A}}{4} \tag{17-4}$$

$$= 0.375 \text{ A}$$

PRACTICE PROBLEMS

17-1. For the configuration of cells shown in Fig. 17-2, what is the reading of the voltmeter, V?

17-2. Two identical 60-V batteries are connected in parallel across a bank of resistive loads whose values are 20 Ω, 12 Ω, 6 Ω, and 5 Ω. Determine the value of the current supplied by each battery.

17-3. In Practice Problem 17-2, the 6-Ω load open-circuits. Determine the new value of the current supplied by each battery.

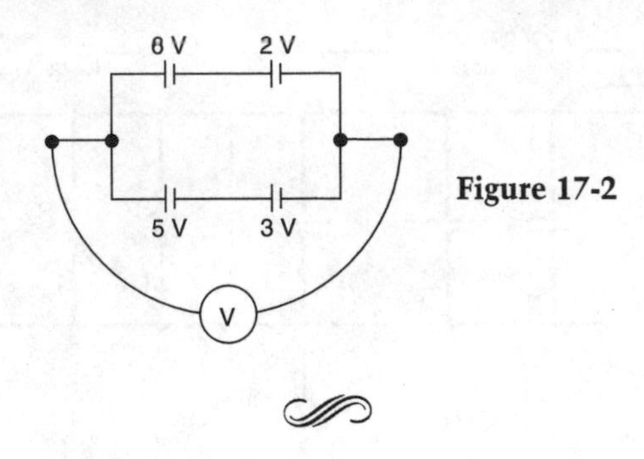

Figure 17-2

18
The current division rule

The *current division rule* is used if you are given the values of the parallel resistors R1, R2, R3 . . . R_N (Fig. 18-1) and the value of the total current, I_T, and are then asked to determine the individual branch currents I_1, I_2, I_3 . . . I_N. Because the voltage across each branch is the same, the individual branch currents are in inverse proportion to the values of the resistors. In other words, the branch with the lowest resistance will carry the highest current, and the smallest branch current will flow in the branch with the greatest resistance.

Because the circuit's total resistance (R_T) is equal to the source voltage (E_T) divided by the total current (I_T) the fraction of the source current flowing through a particular branch is equal to the ratio of the total equivalent resistance, to the branch resistance. This relationship is called the current division rule.

MATHEMATICAL DERIVATIONS

Source voltage,

$$E_T = I_1R_1 = I_2R_2 = I_3R_3 \ldots = I_NR_N = I_TR_T \text{ volts}$$

Then,

$$\frac{I_1}{I_2} = \frac{R_2}{R_1} \text{ etc.}$$

Therefore,

$$I_1 = I_2 \times \frac{R_2}{R_1}, I_2 = I_1 \times \frac{R_1}{R_2} \ldots \text{ etc., amperes} \quad (18\text{-}1)$$

Also,

$$\frac{I_1}{I_T} = \frac{R_T}{R_1}$$

This yields,

$$I_1 = I_T \times \frac{R_T}{R_1}, I_2 = I_T \times \frac{R_T}{R_2} \text{ etc., amperes} \quad (18\text{-}2)$$

and,

$$I_T = I_1 \times \frac{R_1}{R_T} = I_2 \times \frac{R_2}{R_T} \text{ amperes} \quad (18\text{-}3)$$

Equation 18-2 represents the *current division rule.*

If there are only two resistors in parallel,

$$R_T = \frac{R_1 R_2}{R_1 + R_2} \text{ ohms}$$

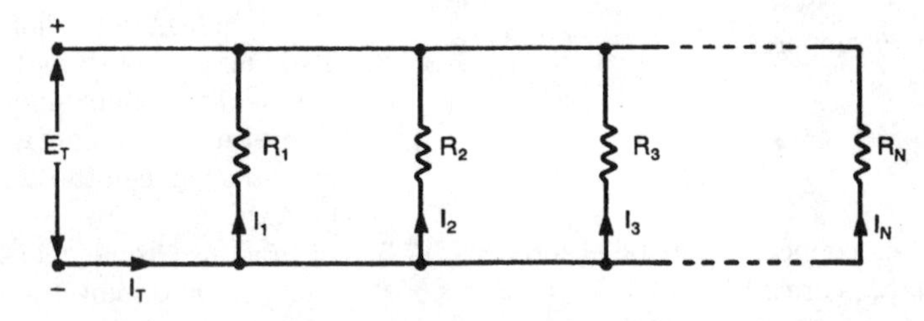

Figure 18-1

Then,

$$I_1 = I_T \times \frac{R_1 R_2}{R_1(R_1 + R_2)} \qquad (18\text{-}4)$$

$$= I_T \times \frac{R_2}{R_1 + R_2} \text{ amperes}$$

and,

$$I_2 = I_T \times \frac{R_1 R_2}{R_2(R_1 + R_2)} \qquad (18\text{-}5)$$

$$= I_T \times \frac{R_1}{R_1 + R_2} \text{ amperes}$$

To calculate one of the branch currents for two resistors in parallel, you must first multiply the source current by the resistance *in the other branch* and then divide by the sum of the two resistances.

Example 18-1

In Fig. 18-2, find the value of the current, I_2.

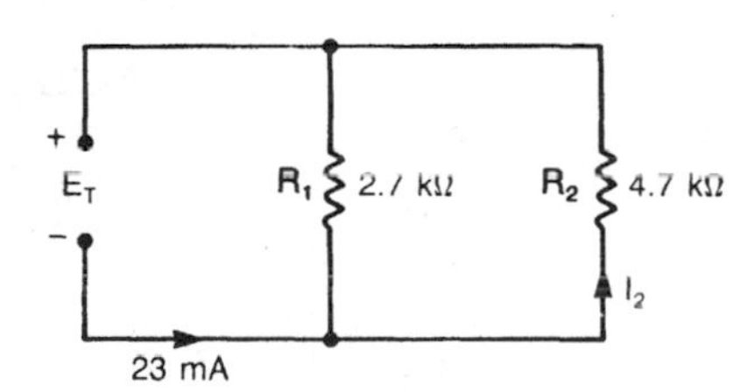

Figure 18-2

Solution

Current,

$$I_2 = I_T \times \frac{R_1}{R_1 + R_2} \qquad (18\text{-}5)$$

$$= 23 \text{ mA} \times \frac{2.7 \text{ k}\Omega}{2.7 \text{ k}\Omega + 4.7 \text{ k}\Omega}$$

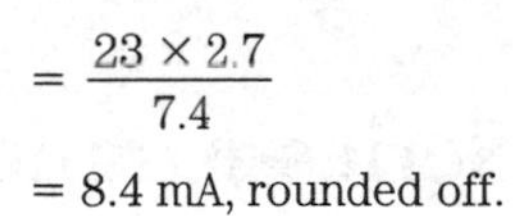

$$= \frac{23 \times 2.7}{7.4}$$

$= 8.4$ mA, rounded off.

Example 18-2

In Fig. 18-3, find the value of the total current, I_T.

Solution

Total equivalent resistance,

$$R_T = \frac{1}{\frac{1}{8.2} + \frac{1}{6.8} + \frac{1}{10} + \frac{1}{5.6}} \qquad (15\text{-}8)$$

$= 1.83$ kΩ.

Total current,

$$I_T = I_4 \times \frac{R_4}{R_T} \qquad (18\text{-}3)$$

$$= 7.3 \text{ mA} \times \frac{5.6 \text{ k}\Omega}{1.83 \text{ k}\Omega}$$

$= 22.3$ mA, rounded off.

PRACTICE PROBLEMS

18-1. A 6.8-kΩ resistor and a 5.6-kΩ resistor are connected in parallel across a dc voltage source. If the total current drawn from the source is 9.4 mA, calculate the value of the current flowing through the 5.6-kΩ resistor.

18-2. Three resistors, whose values are 330 Ω, 470 Ω, and 680 Ω, are connected in parallel across a dc source. If the current through the 470-Ω resistor is 23 mA, determine the value of the total current drawn from the source.

18-3. Four resistors, whose values are 910 Ω, 1.5 kΩ, 2.2 kΩ, and 2.7 kΩ, are connected in parallel across a dc source. If the total current drawn from the source is 56 mA, find the value of the current that flows through the 910-Ω resistor.

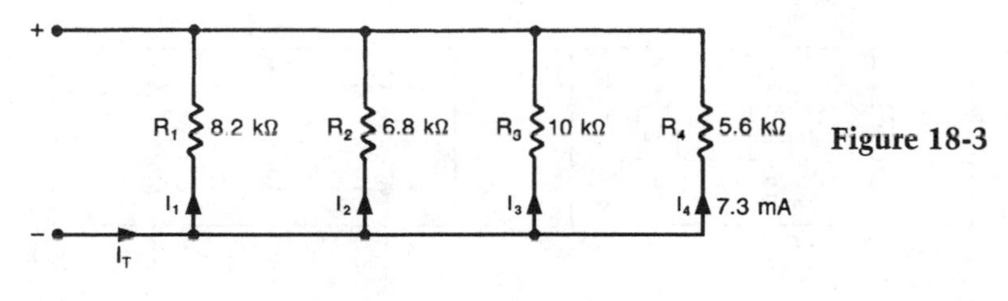

Figure 18-3

19
Series-parallel arrangements of resistors

Chapter 9 showed how the properties related to series strings of resistors, and chapter 15 covered the results of connecting resistors in parallel banks. Briefly summarizing, the single current was the same throughout the series circuit and the source voltage was divided among the resistors in proportion to their resistances. By contrast, the (same) source voltage was applied across each resistor in parallel and the (total) source current was divided between the various branches in inverse proportion to their resistances. However, series strings and parallel banks may be combined to create so-called *series-parallel* circuits of which the two simplest are shown in Figs. 19-1A and B.

In Fig. 19-1A, the resistors R2 and R3 are joined end-to-end so that the same current must flow through each of these resistors. Consequently, the resistors R2 and R3 are in series. However, the (electron flow) current splits at the point *Y* and recombines at the point *X*. Therefore, R1 is in parallel with the series combination of R2 and R3. The equivalent printed circuit is shown in Fig. 19-1C.

The (electron flow) current splits at the point, *E*, in Fig. 19-1B and it recombines at the point, *F*, so that R2 and R3 are in parallel. However, this parallel combination is joined to one end of R1 so that this resistor is in series with R2,3. ($R_{2,3}$ is the equivalent resistance of the R2, R3 parallel combination); this may also be abbreviated to $R_2//R_3$. The equivalent printed circuit is shown in Fig. 19-1D

The equations for the circuits of Figs. 19-1A, B are fully developed in the mathematical derivations. However, there is an infinite variety of complex series-parallel networks of resistors. For example, you

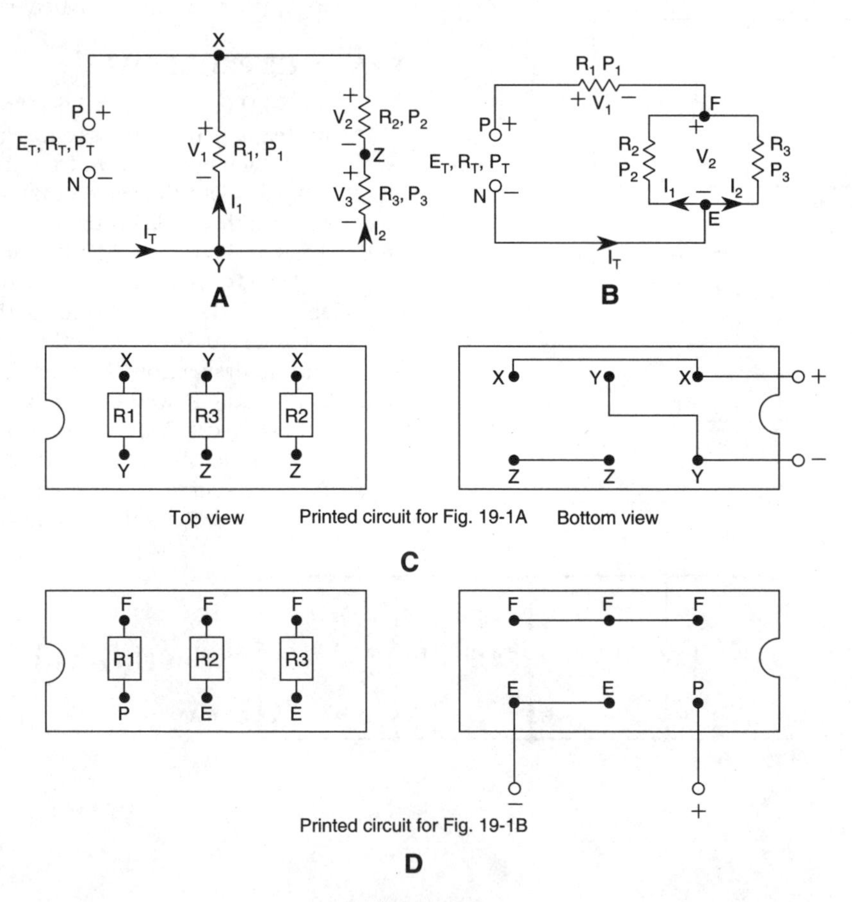

Figure 19-1

might be faced with the circuit of Fig. 19-2A and be asked to find the total equivalent resistance presented to the source. This problem is solved in a number of steps.

Step 1

Identify all points that are electrically different in the circuit. These are the points *P*, *N*, *X*, *Y*, and *Z*. The circuit can then be redrawn in the more conventional manner of Fig. 19-2B.

Corresponding points are labeled in the two drawings and you must make certain that the same resistors are connected between any two such points.

Step 2

Combine all obvious series strings and parallel banks. Redraw the resulting simplified circuit. The general practice is to start furthest away from the source, then work your way toward the source.

In this example, R4 and R5 are in parallel, and their equivalent resistance,

$$R_{4,5} = \frac{R_4 \times R_5}{R_4 + R_5} = \frac{4.7 \times 2.7}{4.7 + 2.7}$$

$$= 1.715 \text{ k}\Omega$$

The circuit can then be redrawn as shown in Fig. 19-2C.

Step 3

Add together the series equivalent resistances, and again draw the circuit. In Fig. 19-2C, R3 is in series with R4,5 so that the equivalent resistance of the combination is $R_{3,4,5} = 1.2 + 1.715 = 2.915$ kΩ. The circuit, when redrawn, is shown in Fig. 19-2D.

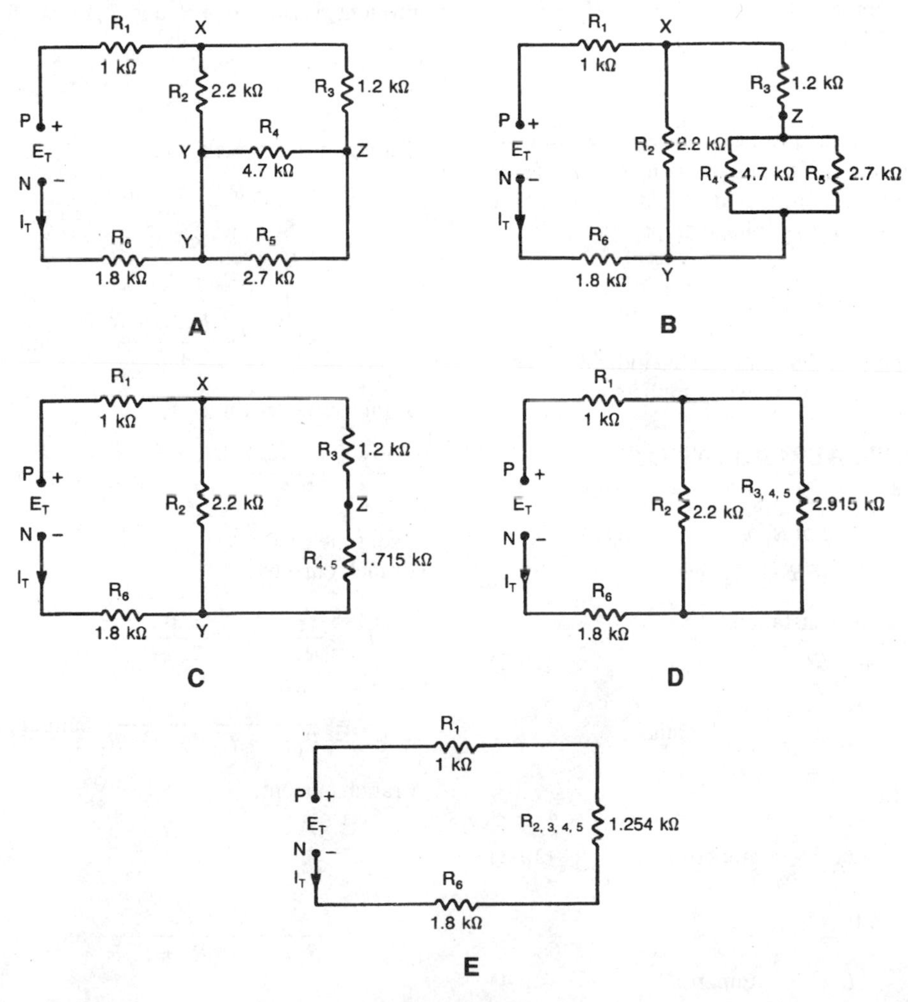

Figure 19-2

Step 4

Use the product-over-sum (or reciprocal) formula to combine the equivalent parallel resistances. In our example, R2 is in parallel with R3,4,5 and the equivalent resistance of this combination is,

$$R_{2,3,4,5} = \frac{2.915 \times 2.2}{2.915 + 2.2} = 1.254\ \text{k}\Omega.$$

When the circuit is again redrawn, its appearance is as in Fig. 19-2E.

Step 5

Repeat steps 3 and 4 alternatively and redraw the circuit as many times as necessary (note that, with experience, it is possible to eliminate most, if not all, of the intermediate drawings).

The redrawn circuit of Fig. 19-2E is clearly a simple series arrangement, and therefore the total equivalent resistance presented to the source is $R_T = R_1 + R_{2,3,4,5} + R_6 = 1 + 1.254 + 1.8 = 4.054\ \text{k}\Omega$.

Step 6

Knowing the value of the total equivalent resistance, you can determine the total source current, I_T. Then, work away from the source and calculate, in turn, the branch currents, and the voltage drops, across the individual resistors.

Step 7

Calculate the powers dissipated in the individual resistors and the total power drawn from the source.

MATHEMATICAL DERIVATIONS

In Fig. 19-1A

Equivalent resistance of R_1 and R_2 in series,

$$R_{1,2} = R_1 + R_2 \text{ ohms} \qquad (19\text{-}1)$$

Total equivalent resistance,

$$R_T = R_{1,2}//R_3 \qquad (19\text{-}2)$$

$$= \frac{(R_1 + R_2) \times R_3}{R_1 + R_2 + R_3} \text{ ohms}$$

Total current,

$$I_T = \frac{E_T}{R_T} \text{ amperes} \qquad (19\text{-}3)$$

Branch current,

$$I_1 = \frac{E_T}{R_1} \text{ amperes} \qquad (19\text{-}4)$$

By Kirchhoff's current law (KCL):

$$I_T = I_1 + I_2, I_1 = I_T - I_2, I_2 = I_T - I_1 \text{ amperes} \qquad (19\text{-}5)$$

By the voltage division rule (VDR)

$$V_2 = I_2 R_2 = E_T \times \frac{R_2}{R_2 + R_3} \qquad (19\text{-}6)$$

$$V_3 = I_2 R_3 = E_T \times \frac{R_3}{R_2 + R_3} \text{ volts}$$

$$E_T = V_1 = V_2 + V_3 \text{ volts} \qquad (19\text{-}7)$$

Total power delivered from the source,

$$P_T = E_T I_T \qquad (19\text{-}8)$$

$$= P_1 + P_2 + P_3 \text{ watts}$$

In Fig. 19-1B

Equivalent resistance of R_2 and R_3 in parallel,

$$R_{2,3} = \frac{R_2 \times R_3}{R_2 + R_3} \text{ ohms}$$

Total equivalent resistance,

$$R_T = R_1 + R_{2,3} \qquad (19\text{-}9)$$

$$= R_1 + \frac{R_2 \times R_3}{R_2 + R_3}$$

$$= \frac{R_1 R_2 + R_2 R_3 + R_3 R_1}{R_2 + R_3} \text{ ohms}$$

Total source current,

$$I_T = \frac{E_T}{R_T} = \frac{E_T \times (R_2 + R_3)}{R_1 R_2 + R_2 R_3 + R_3 R_1} \text{ amperes} \qquad (19\text{-}10)$$

Using the CDR,
Branch current,

$$I_1 = \frac{V_2}{R_2} = I_T \times \frac{R_3}{R_2 + R_3} \qquad (19\text{-}11)$$

$$= \frac{E_T \times R_3}{R_1 R_2 + R_2 R_3 + R_3 R_1} \text{ amperes}$$

Branch current,

$$I_2 = \frac{V_2}{R_3} = I_T \times \frac{R_2}{R_2 + R_3} \qquad (19\text{-}12)$$

$$= \frac{E_T \times R_2}{R_1 R_2 + R_2 R_3 + R_3 R_1} \text{ amperes}$$

Also,

$$I_T = I_1 + I_2 = \frac{V_1}{R_1}, I_1 = \frac{V_2}{R_2} = I_T - I_2, \quad (19\text{-}13)$$

$$I_2 = \frac{V_2}{R_3} = I_T - I_1 \text{ amperes}$$

Voltage drop across resistor R1,

$$V_1 = I_T R_1 \text{ volts} \quad (19\text{-}14)$$

By KVL,

$$E_T = V_1 + V_2, V_1 = E_T - V_2, \quad (19\text{-}15)$$

$$V_2 = E_T - V_1 \text{ volts}$$

Total power drawn from the source,

$$P_T = E_T I_T$$

$$= P_1 + P_2 + P_3 \text{ watts}$$

Example 19-1

In Fig. 19-1A, $R_1 = 3.9$ kΩ, $R_2 = 1$ kΩ, $R_3 = 1.5$ kΩ. If $E_T = 9$ V, calculate the values of $R_T, I_T, I_1, I_2, V_1, V_2$.

Solution

Equivalent resistance of R_2 and R_3 in series,

$$R_{2,3} = R_2 + R_3 \quad (19\text{-}1)$$

$$= 1 + 1.5$$

$$= 2.5 \text{ k}\Omega.$$

Total equivalent resistance,

$$R_T = \frac{R_1 \times R_{2,3}}{R_1 + R_{2,3}} \quad (19\text{-}2)$$

$$= \frac{3.9 \times 2.5}{3.9 + 2.5}$$

$$= 1.52 \text{ k}\Omega.$$

Total source current,

$$I_T = \frac{E_T}{R_T} \quad (19\text{-}3)$$

$$= \frac{9 \text{ V}}{1.52 \text{ k}\Omega}$$

$$= 5.91 \text{ mA}.$$

Branch current,

$$I_1 = \frac{E_T}{R_1} \quad (19\text{-}4)$$

$$= \frac{9 \text{ V}}{3.9 \text{ k}\Omega}$$

$$= 2.31 \text{ mA}.$$

Branch current,

$$I_2 = I_T - I_1 \quad (19\text{-}5)$$

$$= 5.91 - 2.31$$

$$= 3.6 \text{ mA}.$$

Voltage drop,

$$V_1 = I_2 R_2 \quad (19\text{-}6)$$

$$= 3.6 \text{ mA} \times 1 \text{ k}\Omega$$

$$= 3.6 \text{ V}.$$

Voltage drop,

$$V_2 = I_2 R_3 \quad (19\text{-}6)$$

$$= 3.6 \text{ mA} \times 1.5 \text{ k}\Omega$$

$$= 5.4 \text{ V}.$$

Voltage check:

Source voltage,

$$E_T = V_1 + V_2 = 3.6 \text{ V} + 5.4 \text{ V} = 9 \text{ V} \quad (19\text{-}7)$$

Example 19-2

In Fig. 19-1B, $R_1 = 1.5$ kΩ, $R_2 = 3.3$ kΩ, $R_3 = 5.6$ kΩ and $E_T = 7$ V. Calculate the values of $R_T, I_T, I_1, I_2, V_1, V_2$.

Solution

Equivalent resistance of R_2 and R_3 in parallel,

$$R_{2,3} = \frac{R_2 \times R_3}{R_2 + R_3}$$

$$= \frac{3.3 \times 5.6}{3.3 + 5.6}$$

$$= 2.076 \text{ k}\Omega.$$

Total equivalent resistance,

$$R_T = R_1 + R_{2,3} \quad (19\text{-}9)$$

$$= 1.5 + 2.076$$

$$= 3.576 \text{ k}\Omega.$$

Total source current,

$$I_T = \frac{E_T}{R_T} \quad (19\text{-}10)$$

$$= \frac{7 \text{ V}}{3.576 \text{ k}\Omega}$$

$$= 1.96 \text{ mA}.$$

Voltage drop,

$$V_1 = I_T \times R_1 \quad (19\text{-}14)$$

$$= 1.96 \text{ mA} \times 1.5 \text{ k}\Omega$$

$$= 2.94 \text{ V}.$$

Voltage drop,

$$V_2 = E - V_1 \quad (19\text{-}15)$$
$$= 7 - 2.94$$
$$= 4.06 \text{ V}.$$

Branch current,

$$I_1 = \frac{V_2}{R_2} \quad (19\text{-}11)$$
$$= \frac{4.06 \text{ V}}{3.3 \text{ k}\Omega}$$
$$= 1.23 \text{ mA}.$$

Branch current,

$$I_2 = \frac{V_2}{R_3} \quad (19\text{-}12)$$
$$= \frac{4.06 \text{ V}}{5.6 \text{ k}\Omega}$$
$$= 0.73 \text{ mA}.$$

Current check:
Total current,

$$I_T = I_1 + I_2 = 1.23 + 0.73 = 1.96 \text{ mA} \quad (19\text{-}13)$$

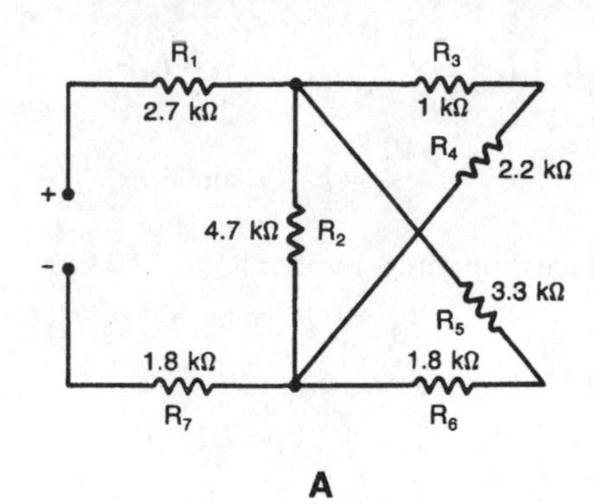

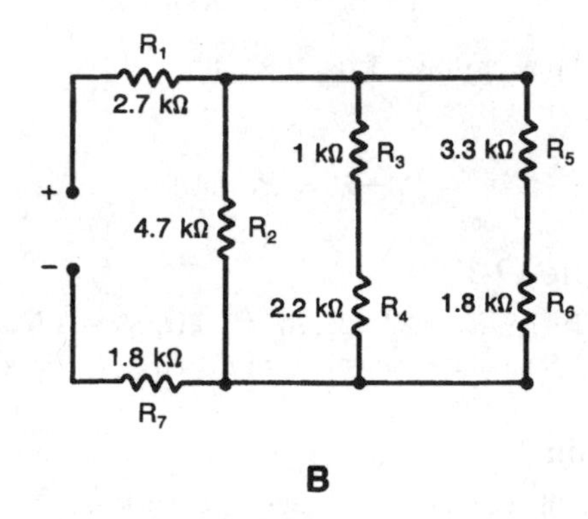

Figure 19-3

Example 19-3

In Fig. 19-3A, calculate the value of the total equivalent resistance presented to the source.

Solution

The circuit is redrawn in Fig. 19-3B.

Total resistance of R_3 and R_4 in series,

$$R_{3,4} = 1 + 2.2$$
$$= 3.2 \text{ k}\Omega.$$

Total resistance of R_5 and R_6 in series,

$$R_{5,6} = 3.3 + 1.8$$
$$= 5.1 \text{ k}\Omega.$$

Total resistance of $R_2//R_{3,4}//R_{5,6}$ in parallel,

$$R_{2,3,4,5,6} = \frac{1}{\frac{1}{4.7} + \frac{1}{3.2} + \frac{1}{5.1}}$$
$$= 1.386 \text{ k}\Omega.$$

Total equivalent resistance,

$$R_T = R_1 + R_{2,3,4,5,6} + R_7$$
$$= 2.7 + 1.386 + 1.8$$
$$= 5.886 \text{ k}\Omega.$$

PRACTICE PROBLEMS

19-1. In the circuit of Fig. 19-4, what is the value of the total current drawn from the 70-V source?

19-2. In the circuit of Fig. 19-4, points X and Z are joined by a connecting wire. What is the new value of the total equivalent resistance presented to the source?

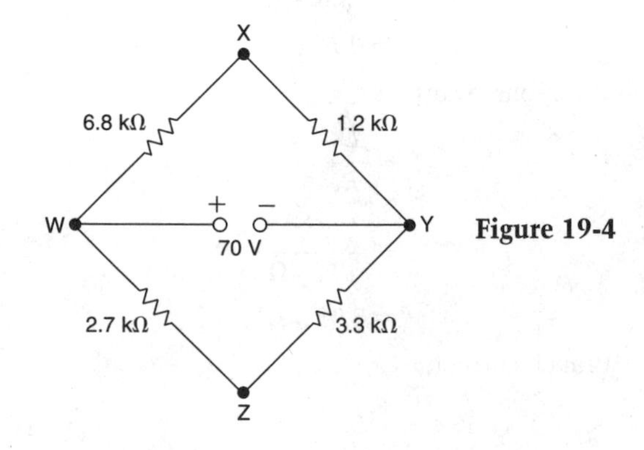

Figure 19-4

19-3. A circuit consists of two parallel resistors, 120 Ω and 680 Ω respectively, connected in series with a 220-Ω resistor. If the current through the 680-Ω resistor is 7.6 mA, calculate the amount of the total power dissipated in the circuit.

19-4. A series combination of two resistors whose values are 220 Ω and 680 Ω, is connected in parallel with an 820-Ω resistor across a dc source. If the current through the 220-Ω resistor is 16 mA, calculate the amount of the total power dissipated in the circuit.

19-5. Three resistors whose values are 180 Ω, 270 Ω, and 330 Ω are connected in parallel. Two more resistors whose values are 120 Ω and 150 Ω are also connected in parallel. The two parallel combinations are then joined in series across a dc source. If the current through the 120-Ω resistor is 74 mA, calculate the value of the current flowing through the 270-Ω resistor.

19-6. Derive the corresponding schematic to the printed circuit board configuration illustrated in Fig. 19-5.

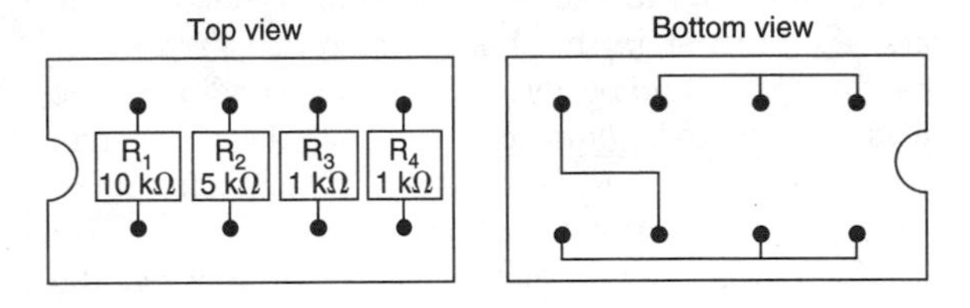

Figure 19-5

20
The Wheatstone bridge circuit

The Wheatstone bridge circuit is used to obtain an accurate measurement of an unknown resistance, R_x. Conventionally, the bridge circuit consists of four resistor arms R1, R2, R3, and Rx (Fig. 20-1A) with a center link ("bridge") that contains a sensitive current indicating device, or galvanometer, G. However, the circuit can be redrawn as in Fig. 20-1B. Then, you can observe that the bridge circuit is really an example of a series-parallel resistor arrangement.

The whole of the source voltage must be dropped across R1, R2, and also, R3, Rx. However, unless there is a special relationship between the four resistance values, the current through the resistor R1 will not be equal to the current through the resistor R2, and the same applies to the currents through the resistors R3, Rx.

As an example, consider the circuit of Fig. 20-1B. If the low resistance of the sensitive galvanometer is neglected, the resistor R1 is in parallel with the resistor Rx, and the resistor R3 is in parallel with the resistor R2. The two parallel combinations are then in series. The total equivalent resistance presented to the source is 1 kΩ//1 kΩ + 8 kΩ//10 kΩ = 500 Ω + 4444 Ω = 4944 Ω. The total current, I_T = 220 V/4944 Ω = 44.5 mA, which divides equally between the two 1-kΩ resistors. The voltage drop across each of these resistors is 44.5/2 mA × 1 kΩ = 22.25 V. The voltage drop across each of the 8-kΩ and 10-kΩ resistors is 220 V − 22.25 V = 197.75 V. Consequently, the current in the 8-kΩ resistor is 197.75 V/8 kΩ = 24.72 mA, while the 10-kΩ

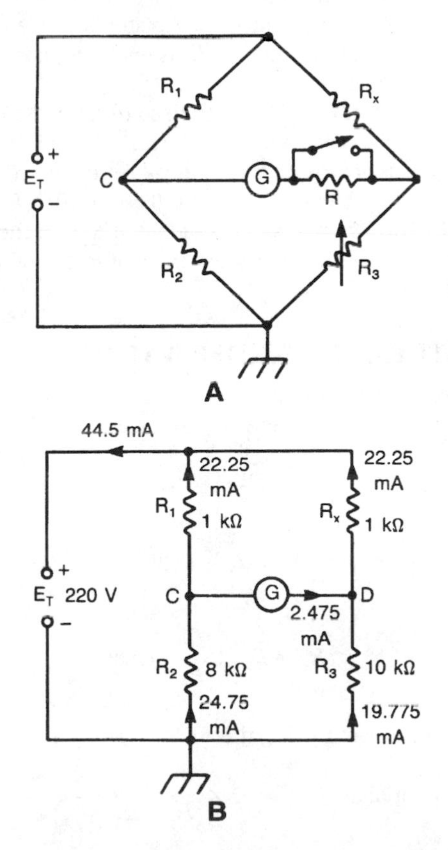

Figure 20-1

resistor carries a current of 197.75 V/10 kΩ = 19.775 mA. The current in the link is then 22.25 − 19.775 (or 24.72 − 22.25) = 2.475 mA, with the electron flow in the direction from left to right (Fig. 20-1B). However, if the 8-kΩ resistor was replaced by a 10-kΩ resistor, the currents in the four resistors would be all the same and the current in the link (as recorded by the galvanometer) would be zero. Under such conditions, the bridge is said to be *balanced*, although the situation in which $R_1 = R_x$ and $R_2 = R_3$ is a special case, and is not the general condition for a balance to occur.

There will be no voltage difference between the points C and D if the ratios $R_1{:}R_2$ and $R_x{:}R_3$ are equal. The reading of the galvanometer is zero, and the bridge is balanced. The value of the unknown resistor, R_x is then equal to $R_3 \times R_1/R_2$ (see mathematical derivations). Notice that in this condition, products, R_2R_x and R_1R_3, of the opposing resistors are equal in the conventional bridge circuit of Fig. 20-1A.

To measure the value of R_x, R_3 is an accurately calibrated variable resistor, which is used to balance the bridge. A protection resistor (R) is commonly included to prevent the galvanometer from being subjected to excessive current. With R included, a rough balance is first obtained. Afterwards the resistor is shorted out by the switch and the value of R_3 is finally adjusted for an accurate balance.

Normally the high-quality fixed resistors R1 and R2 have possible values of 1 Ω, 10 Ω, 100 Ω or 1000 Ω. By using switches to change the values of the resistors, the ratio $R_1{:}R_2$ can be set for 1000, 100, 10, 1, 0.1, 0.01 and 0.001. The measured reading of R_x can therefore range from 1/1000 of the lowest value of R_3 to 1000 times the highest value of R_3.

MATHEMATICAL DERIVATIONS

For a balanced bridge,

$$\frac{R_x}{R_3} = \frac{R_1}{R_2} \qquad (20\text{-}1)$$

or

$$R_x R_2 = R_1 R_3 \qquad (20\text{-}2)$$

or

$$R_x = R_3 \times \frac{R_1}{R_2} \text{ ohms} \qquad (20\text{-}3)$$

Alternatively, using the VDR:

$$\text{Potential at } C,\ V_C = E \times \frac{R_2}{R_1 + R_2}$$

$$\text{Potential at } D,\ V_D = E \times \frac{R_3}{R_3 + R_x}$$

If the bridge is balanced, $V_C = V_D$. Therefore,

$$E \times \frac{R_2}{R_1 + R_2} = E \times \frac{R_3}{R_3 + R_x} \qquad (20\text{-}4)$$

$$R_2 R_3 + R_2 R_x = R_1 R_3 + R_2 R_3$$

$$R_2 R_x = R_1 R_3$$

$$R_x = R_3 \times \frac{R_1}{R_2} \text{ ohms}$$

Example 20-1

In Fig. 20-1A, $R_1 = 10\ \Omega$, $R_2 = 1000\ \Omega$. When the bridge is balanced, the value of R_3 is adjusted to 5.7 Ω. What is the value of the unknown resistor, R_x?

Solution

For a balanced bridge,

$$R_x = R_3 \times \frac{R_1}{R_2} \qquad (20\text{-}3)$$

$$= 5.7\ \Omega \times \frac{10\ \Omega}{1000\ \Omega}$$

$$= 0.057\ \Omega.$$

PRACTICE PROBLEMS

20-1. In Fig. 20-1A, $R_1 = 100\ \Omega$ and $R_2 = 1\ \Omega$. For the bridge to be balanced so that the reading of G is zero, the value of R_3 is adjusted to be 7.46 Ω. What is the value of the unknown resistance, R_X?

20-2. In Fig. 20-1A, $E_T = 24$ V, $R_1 = 1.5$ kΩ, $R_2 = 2.7$ kΩ, $R_3 = 4.7$ kΩ, $R_X = 2.2$ kΩ. What is the current reading of the meter, G? Ignore any resistance associated with the meter.

20-3. In Practice Problem 20-2, the resistor R_X open-circuits. What is the new value of the current flowing through the meter, G?

20-4. In Practice Problem 20-2 a short circuit develops across the resistor R_X. What is the new value of the current flowing through the meter, G?

20-5. In Fig. 20-2, what is the value of the current flowing through the 10-kΩ resistor?

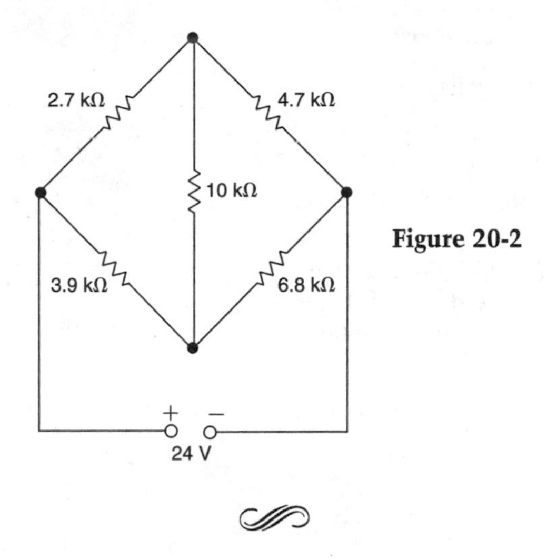

Figure 20-2

21
The loaded voltage divider circuit

Chapter 11 described the unloaded voltage divider circuit, which was a series string of resistors, capable of producing a number of different voltages from a single voltage source. In such a circuit, only the single bleeder current flowed through the series resistors. The values of those resistors were then calculated to provide the necessary voltages. However, as soon as you try to make use of the divider circuit by connecting loads across the various output points *X*, *Y*, and *Z*, the arrangement becomes an application of a series-parallel network (Fig. 21-1). Knowing the values of the load currents, it is necessary to recalculate the required values of R_1, R_2, and R_3, in order to provide the correct output voltages, V_{L1}, V_{L2} and V_{L3}.

MATHEMATICAL DERIVATIONS

The bleeder current, I_B, is normally about 10% of the total load current, I_{LT}, which is the sum of I_{L1}, I_{L2}, and I_{L3}. Therefore,

$$I_{LT} = I_{L1} + I_{L2} + I_{L3} \text{ amperes} \qquad (21\text{-}1)$$

The voltage drop across R3 must equal the load voltage, V_{L3}. It follows that:

$$V_{L3} = I_B R_3, \; R_3 = \frac{V_{L3}}{I_B} \qquad (21\text{-}2)$$

The sum of the currents I_{L3} and I_B must flow through R2. When V_{L3} is added to the voltage drop across R2,

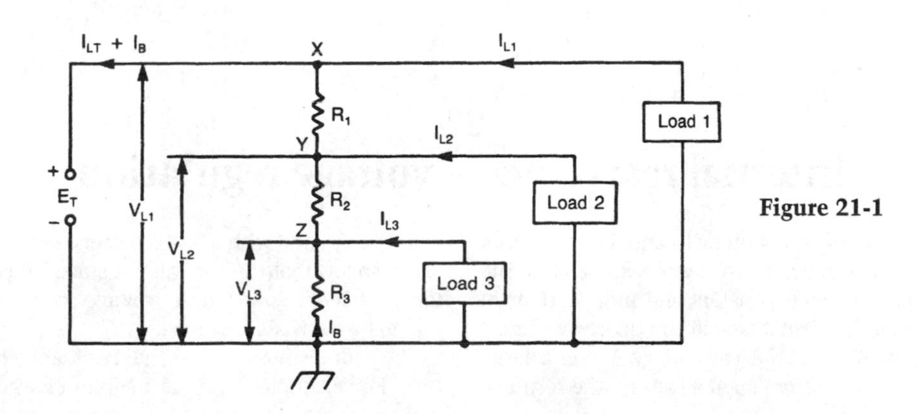

Figure 21-1

the result must equal the value of V_{L2}. In equation form:

$$V_{L2} = V_{L3} + (I_{L3} + I_B) R_2 \text{ volts} \quad (21\text{-}3)$$

This yields,

$$R_2 = \frac{V_{L2} - V_{L3}}{I_{L3} + I_B} \text{ ohms} \quad (21\text{-}4)$$

The sum of the currents I_{L2}, I_{L3} and I_B flows through R1. The sum of this resistor's voltage drop and V_{L2} is equal to V_{L1}, which is the same as the source voltage, E_T. Then,

$$V_{L1} = E_T = R_1(I_{L2} + I_{L3} + I_B) + V_{L2} \text{ volts} \quad (21\text{-}5)$$

so that,

$$R_1 = \frac{V_{L1} - V_{L2}}{I_{L2} + I_{L3} + I_B} \text{ ohms} \quad (21\text{-}6)$$

Example 21-1

In Fig. 21-1, $E = 60$ V, $V_{L2} = 35$ V, $V_{L3} = 10$ V, $I_{L1} = 170$ mA, $I_{L2} = 70$ mA, $I_{L3} = 40$ mA. Suggest suitable values for R_1, R_2, and R_3 (assume that the bleeder current is 10% of the total load current).

Solution

Total load current,

$$I_{LT} = I_{L1} + I_{L2} + I_{L3} \quad (21\text{-}1)$$

$$= 170 + 70 + 40$$

$$= 280 \text{ mA.}$$

Bleeder current, $I_B = 280 \times 10/100 = 28$ mA.
The required voltage divider resistances are:

$$R_3 = \frac{V_{L3}}{I_B} \quad (21\text{-}2)$$

$$= \frac{10 \text{ V}}{28 \text{ mA}}$$

$$= 357 \ \Omega.$$

$$R_2 = \frac{V_{L2} - V_{L3}}{I_{L3} + I_B} \quad (21\text{-}4)$$

$$= \frac{35 \text{ V} - 10 \text{ V}}{40 \text{ mA} + 28 \text{ mA}} = \frac{25 \text{ V}}{68 \text{ mA}}$$

$$= 368 \ \Omega$$

$$R_1 = \frac{V_{L1} - V_{L2}}{I_{L2} + I_{L3} + I_B} \quad (21\text{-}6)$$

$$= \frac{60 \text{ V} - 35 \text{ V}}{70 \text{ mA} + 40 \text{ mA} + 28 \text{ mA}}$$

$$= \frac{25 \text{ V}}{138 \text{ mA}}$$

$$= 181 \ \Omega$$

The power dissipations of these resistors are:

$$P_{R1} = 25 \text{ V} \times 138 \text{ mA} = 3450 \text{ mW.}$$

$$P_{R2} = 25 \text{ V} \times 68 \text{ mA} = 1700 \text{ mW.}$$

$$P_{R3} = 10 \text{ V} \times 28 \text{ mA} = 280 \text{ mW.}$$

Suitable resistors would be R1: 180 Ω, 5 W, R2: 390 Ω, 2 W, and R3: 330 Ω, ½ W.

PRACTICE PROBLEMS

21-1. In Fig. 21-1, $V_{L2} = 300$ V, $V_{L3} = 150$ V, $I_{L3} = 40$ mA, and $I_{L2} = 60$ mA. If $I_B = 15$ mA, what are the values of R_2 and its power dissipation?

21-2. In Fig. 21-1, $R_1 = 470\ \Omega$, $R_2 = 270\ \Omega$, $R_3 = 150\ \Omega$, $I_{L1} = 170$ mA, $I_{L2} = 160$ mA, $I_{L3} = 120$ mA. Calculate the values of V_{L1}, V_{L2}, and V_{L3}.

22
Internal resistance—voltage regulation

The sources covered in chapters 12 and 17 were (in a certain sense) idealized because it was assumed that the source voltage remained constant and independent of the load current. This is not true in practice, as we know from our experience with a car battery. When a large load current is drawn from such a battery, the terminal voltage falls, and with a "bad" battery, the terminal voltage is so low that the car will not start. The explanation for this effect lies within the voltage source itself, and is due to its internal resistance.

All sources possess internal resistance to some extent. For example, a primary (nonrechargeable) 1½-V

"D" cell has an internal resistance on the order of 1 ohm. This value of resistance depends on the size of the electrodes, their separation, and the nature of the electrolyte. By contrast, the basic lead acid secondary (rechargeable) cell has a negative electrode of sponge lead, a positive electrode of lead peroxide, and an electrolyte of sulfuric acid.

During the cell's discharge, both the sponge lead and the lead peroxide are converted to lead sulphate while the specific gravity of the electrolyte falls; these conditions are reversed when the cell is recharged. If it is required to store the lead-acid battery for an appreciable time, the electrolyte is drained out and replaced by distilled water; this will prevent the lead sulphate from hardening on the electrodes. The large size of the electrodes results in a very low internal resistance, which is typically less than 10 mΩ. This is the major difference between a 12-V car battery, and a 12-V cell for operating a transistor radio. The cell for the transistor radio has a much higher internal resistance and is, therefore, quite incapable of starting a car (whose initial current requirement is about 100 A or more!).

The capacity of a lead-acid battery is measured in ampere-hours (1 Ah = 3600 C). The product of the discharge current and the rating in hours over which the discharge occurs, is equal to the value of the capacity. For example, a battery with a capacity of 200 Ah and an 8-hour rating permits a discharge current of 200 Ah/8 h = 25 A.

Another secondary cell is the nickel-cadmium type, which has a cadmium anode, a nickel-hydroxide cathode, and a potassium-hydroxide electrolyte. Such a cell has an operating voltage of approximately 1.25 V which is accompanied by a very low internal resistance.

When a load current is drawn from a voltage source, the terminal voltage, V_L, falls as the load current, I_L, is increased. A model that explains this effect is illustrated in Fig. 22-1. This model assumes that the source contains a constant voltage (EMF, or E), which is in series with the internal resistance (R_i). Then, if the load resistance (R_L) is decreased, the load current (I_L) will rise. This will cause an increase in the voltage drop (V_i) across the internal resistance (R_i) and therefore the terminal voltage (V_L) will drop.

Having established the model, you must have some means of finding the values of E and R_i for any electrical source. The value of E is best measured by disconnecting the load (R_L) and then placing a voltmeter across the terminals. Assuming that the voltmeter draws virtually no current from the source, the voltage drop across the internal resistance is negligible. The voltmeter reading will then be equal to the value of E, which is often referred to as the open-circuit, or no-load, terminal voltage.

Having determined the value of E, then connect an ammeter of negligible resistance between the terminals so that only the internal resistance will now limit the current. The value of this current is therefore E/R_i amperes. Because there is practically zero resistance between the terminals, this level is normally referred to as the *short-circuit terminal* current. The internal resistance can then be calculated from the ratio of the open-circuit terminal voltage to the short circuit terminal current.

It might not be practical to measure the short-circuit current because such a large current can possibly damage the electrical source. A superior method of finding the internal resistance is to vary the value of the load resistance until the terminal load voltage (V_L) is equal to $E/2$. V_i will also be $E/2$ and the measured value of R_L will then equal the internal resistance, R_i.

The degree to which the load voltage (V_L) varies with changes in the load current is measured by the voltage regulation percentage. Ideal regulation means that the load voltage remains constant and independent of any changes in the load current.

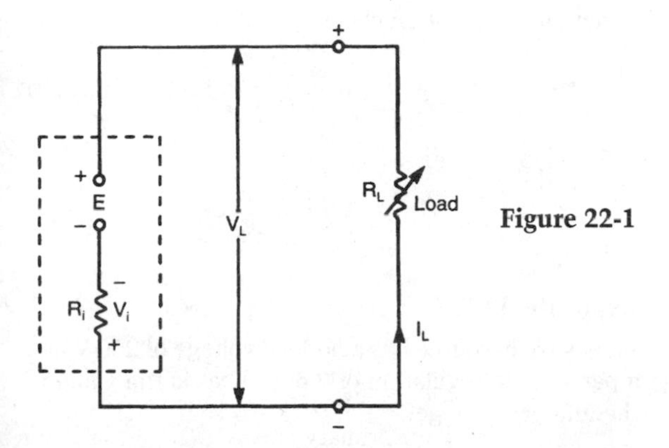

Figure 22-1

MATHEMATICAL DERIVATIONS

In Fig. 22-1,

$$\text{Open-circuit or no-load terminal voltage} = E \text{ volts} \qquad (22\text{-}1)$$

Load current,

$$I_L = \frac{V_L}{R_L} = \frac{V_i}{R_i} = \frac{E}{R_i + R_L} \text{ amperes} \qquad (22\text{-}2)$$

Terminal load voltage,

$$V_L = I_L R_L \qquad (22\text{-}3)$$

$$= E - I_L R_i$$

$$= E - \frac{E\,R_i}{R_i + R_L} \text{ volts}$$

This yields

Load voltage,

$$V_L = E \times \frac{R_L}{R_i + R_L} \text{ volts} \qquad (22\text{-}4)$$

Typical graphs of V_L vs. R_L and I_L vs. R_L are illustrated in Fig. 22-2.

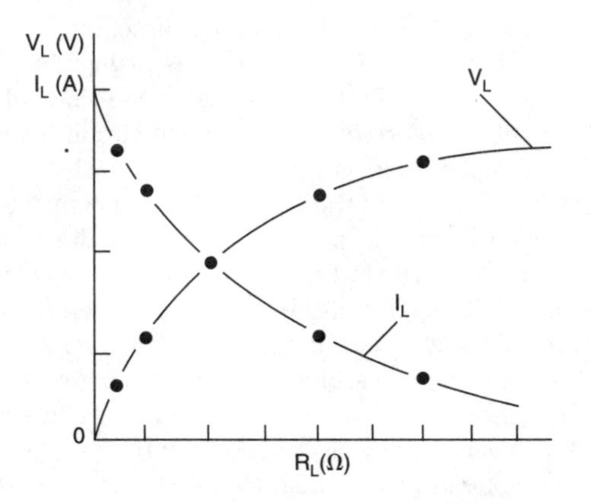

Figure 22-2

Equation 22-4 is an example of the voltage division rule (VDR) because the source voltage (E) is divided between the load resistance (R_L) and the internal resistance, (R_i). To obtain a high voltage across the load, R_L must typically be at least 5 to 10 times the value of R_i.

$$\text{Short-circuit terminal current} = \frac{E}{R_i} \text{ amperes} \qquad (22\text{-}5)$$

Internal resistance,

$$R_i = \frac{\text{open-circuit terminal voltage, } E}{\text{short-circuit terminal current, } E/R_i} \text{ ohms} \qquad (22\text{-}6)$$

Voltage regulation percentage (22-7)

$$= \frac{V_{NL} - V_{FL}}{V_{FL}} \times 100\%$$

where: V_{NL} = terminal voltage under no-load (or minimum load) conditions

V_{FL} = terminal voltage under full-load (current) conditions when the load resistance is R_{FL}

No-load voltage,

$$V_{NL} = V_{FL}\left(1 + \frac{\%\ \text{Regulation}}{100}\right) \text{ volts} \qquad (22\text{-}8)$$

Full-load voltage,

$$V_{FL} = \frac{V_{NL}}{\left(1 + \frac{\%\ \text{Regulation}}{100}\right)} \text{ volts} \qquad (22\text{-}9)$$

For an ideal source, $V_{FL} = E$ and the percentage regulation is zero.

If the no-load voltage is E volts,

$$\text{Percentage regulation} = \frac{R_i}{R_{FL}} \times 100\% \qquad (22\text{-}10)$$

Example 22-1

An electrical source has an open-circuit terminal voltage of 48 V and a short-circuit current of 32 A. If the full-load resistance is 12 Ω, calculate the values of the full-load voltage and the percentage of regulation.

Solution

Internal resistance,

$$R_i = \frac{\text{open-circuit voltage, } E}{\text{short-circuit current, } E/R_i} \qquad (22\text{-}6)$$

$$= \frac{48\text{ V}}{32\text{ A}}$$

$$= 1.5\ \Omega$$

Full-load voltage,

$$V_{FL} = \frac{E \times R_{FL}}{R_{FL} + R_i} \qquad (22\text{-}4)$$

$$= \frac{48\text{ V} \times 12\ \Omega}{12\ \Omega + 1.5\ \Omega}$$

$$= 42.7\text{ V}$$

$$\text{Percentage regulation} = \frac{V_{NL} - V_{FL}}{V_{FL}} \times 100\% \qquad (22\text{-}7)$$

$$= \frac{48 - 42.7}{42.7} \times 100$$

$$= 12.5\%$$

Percentage regulation check:

$$\text{Percentage regulation} = \frac{R_i}{R_{FL}} \times 100\% \qquad (22\text{-}10)$$

$$= \frac{1.5}{12} \times 100$$

$$= 12.5\%$$

Example 22-2

An electrical source has a no-load voltage of 250 V and a percentage regulation of 3.5%. What is the value of the full-load voltage?

Solution

Full-load voltage,

$$V_{FL} = \frac{V_{NL}}{\left(1 + \frac{\%\ \text{Regulation}}{100}\right)} \quad (22\text{-}9)$$

$$= \frac{250}{\left(1 + \frac{3.5}{100}\right)}$$

$$= \frac{250}{1.035}$$

$$= 241.5\ \text{V}$$

Example 22-3

A 12.6-V, 55-Ah battery continuously supplies 325 W to a transmitter and 50 W to a receiver. For how many hours can the battery supply full power to both the transmitter and the receiver?

Solution

Total power,

$$P_T = 325 + 50$$

$$= 375\ \text{W}$$

Total current,

$$I_T = \frac{375\ \text{W}}{12.6\ \text{V}}$$

$$= 29.76\ \text{A}$$

Period of time,

$$t = \frac{55\ \text{Ah}}{29.76\ \text{A}}$$

$$= 1.8\ \text{h}$$

PRACTICE PROBLEMS

22-1. A source has an open-circuit voltage of 48 V. When a load current of 8 A is drawn from the source, the terminal voltage falls to 44 V. Calculate the values of the source's internal resistance, the load resistance, and the short-circuit terminal current. What new value of the load resistance will reduce the load voltage to half the value of the open-circuit voltage?

22-2. A source's terminal voltage drops from its 40-V no-load value down to 30 V when it is connected to its full-load resistance of 0.2 Ω. Calculate the values of the load current, source's internal resistance, and percentage regulation.

22-3. A dc source has a full-load voltage of 45 V and a percentage regulation of 8%. What is the value of the no-load voltage under open-circuit conditions?

22-4. A dc source has a terminal voltage of 80 V when a 5-Ω load is connected to its terminals. If the load is changed to 4 Ω, the terminal voltage falls to 76 V. Calculate the values of the source's internal resistance and its open-circuit voltage.

22-5. A dc source whose open-circuit voltage is 120 V delivers 40 W to a resistive load when the load current is 0.4 A. What is the percentage regulation of the source?

23
Maximum power transfer—percentage efficiency

Chapter 22 explored the condition for a high voltage across the load. It was decided that the load resistance (R_L) should be many times greater than the internal resistance (R_i) so that when the constant EMF (E) divided between these two resistances, very little voltage was dropped across R_i and most appeared across the load.

Because there is both a load voltage (V_L) and a load current (I_L) it follows that there is a power (P_L) developed in the load (Fig. 23-1A). However, the condition for maximum power in the load cannot be the same as the condition for a high load voltage because for open- and short-circuit loads, the load power in both cases is zero. Consequently, the maximum power must occur between these two extremes and this is well illustrated in the circuit of Fig. 23-1B. Here, choose particular load resistances (R_L) and calculate the corresponding values of I_L, V_L, and P_L. The results are shown in Table 23-1.

As an example, adjust the value of R_L to be 8 Ω. The total resistance is $R_i + R_L = 4\ \Omega + 8\ \Omega = 12\ \Omega$. The

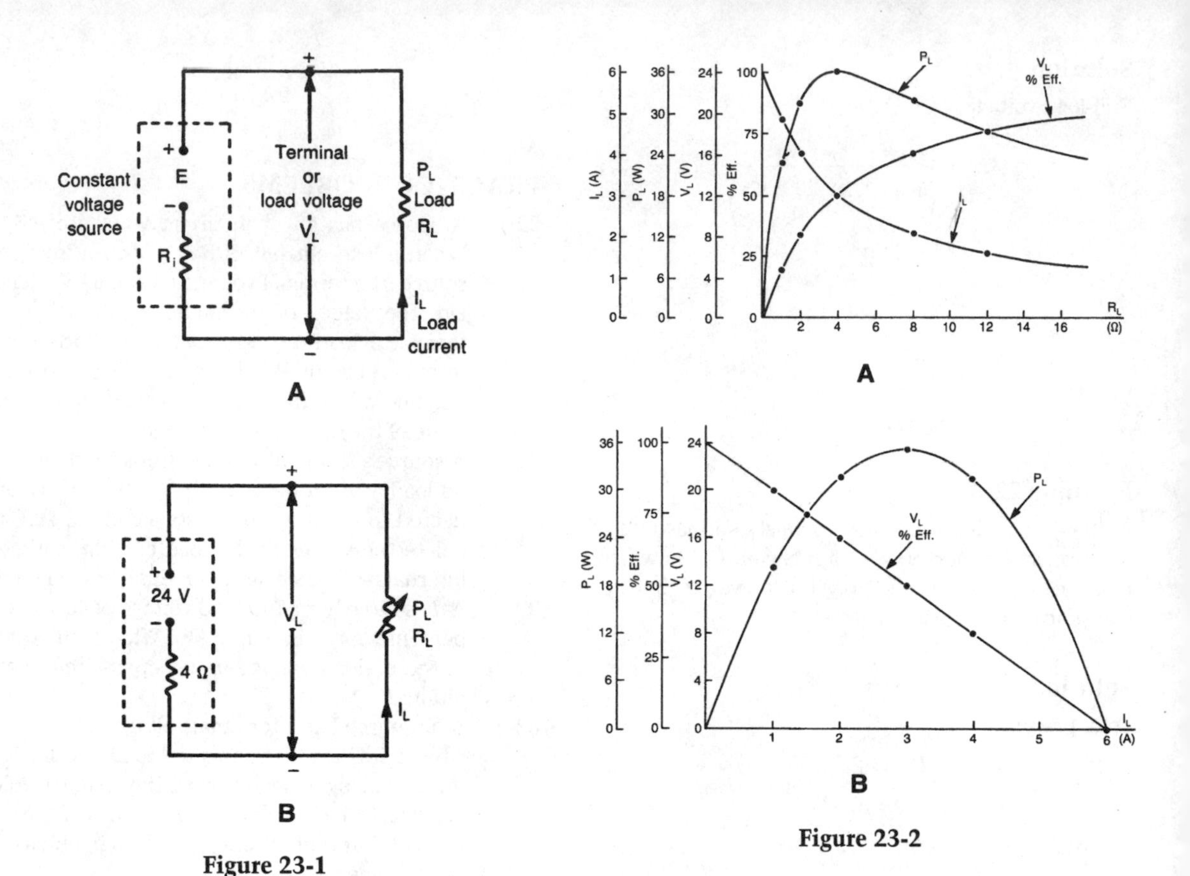

Figure 23-1

Figure 23-2

load current is $E/(R_i + R_L)$ = 24 V/12 Ω = 2 A, and the load voltage is $V_L = I_L \times R_L$ = 2 A × 8 Ω = 16 V. The corresponding load power is $P_L = I_L \times V_L$ = 2 A × 16 V = 32 W. The graphs of V_L, P_L, and percentage efficiency versus R_L and I_L are illustrated in Figs. 23-2A and B.

From the graph of P_L versus R_L, it is clear that the load power reaches its maximum value of 36 W when the load resistance is 4 Ω. It is no numerical coincidence that this value of load resistance is the same as the internal resistance of the source.

In the mathematical derivations, it will be proved that there is maximum power transfer to the load when the load resistance is *matched (made equal)* to the internal resistance. This result is of more than academic significance. In the case of a radio transmitter, the antenna represents a load whose ohmic value must be matched to the load required by the final stage of

Table 23-1. Load Resistance R_L vs. I_L, V_L, and P_L.

Load Resistance (Ω) R_L	Load Current (A) $I_L = E/(R_i + R_L)$	Load Voltage (V) $V_L = I_L \times R_L$	Load Power (W) $P_L = I_L \times V_L$	% Efficiency $P_L/P_T \times 100\%$
Zero (short-circuit)	6	0	0	0
2	4	8	32	33⅓
4	3	12	36	50
8	2	16	32	66⅔
20	1	20	20	83⅓
Infinite (open-circuit)	0	24	0	100

the transmitter; only then can RF (radio frequency) power be effectively transferred from the transmitter to the antenna. In another example, the load of a loudspeaker must be matched to the load required by its receiver's output audio stage.

The percentage efficiency is determined from the ratio of the *load* power, to the *total* power drawn from the constant voltage EMF. When there is maximum power transfer to the load, the percentage efficiency is only 50%; it might therefore be necessary to compromise between the values of load power and percentage efficiency.

MATHEMATICAL DERIVATIONS

Load current,

$$I_L = \frac{V_L}{R_L} = \frac{E}{R_i + R_L} \text{ amperes} \qquad (23\text{-}1)$$

Load voltage,

$$V_L = I_L \times R_L = \frac{ER_L}{R_i + R_L} \text{ volts} \qquad (23\text{-}2)$$

Load power,

$$P_L = I_L \times V_L = \frac{E}{R_i + R_L} \times \frac{ER_L}{R_i + R_L} \qquad (23\text{-}3)$$

$$= \frac{E^2R_L}{(R_i + R_L)^2} \text{ watts}$$

The load power reaches its maximum value when its reciprocal ($1/P_L$) is at its minimum level.

$$\frac{1}{P_L} = \frac{(R_i + R_L)^2}{E^2R_L} = \frac{1}{E^2}\left[\frac{(R_i - R_L)^2}{R_L} + 4R_i\right] \qquad (23\text{-}4)$$

Because the lowest value of a square term [such as $(R_i - R_L)^2$] is zero, $1/P_L$ will reach its minimum level when $(R_i - R_L)^2 = 0$ or $R_L = R_i$. This proves that there is maximum power transfer to the load when the load and the internal resistances are matched.

Alternatively:

Load power,

$$P_L = \frac{E^2R_L}{(R_i + R_L)^2}$$

Differentiating P_L with respect to R_L,

$$\frac{dP_L}{dR_L} = E^2 \times \left[\frac{(R_i + R_L)^2 - 2R_L(R_i + R_L)}{(R_i + R_L)^4}\right]$$

$$= E^2 \times \left[\frac{(R_i + R_L) - 2R_L}{(R_i + R_L)^3}\right]$$

$$= E^2\left[\frac{R_i - R_L}{(R_i + R_L)^3}\right]$$

For the maximum value of P_L, $dP_L/dR_L = 0$. Therefore,

$$R_i = R_L \text{ ohms} \qquad (23\text{-}5)$$

Maximum power developed in the load,

$$P_{L\max} = \frac{E^2R_i}{(R_i + R_L)^2} \qquad (23\text{-}6)$$

$$= \frac{E^2}{4R_i}$$

$$= \frac{E^2}{4R_L} \text{ watts}$$

$$\text{Percentage efficiency} = \frac{P_L}{P_T} \times 100 \qquad (23\text{-}7)$$

$$= \frac{I_L^2R_L}{I_L^2(R_i + R_L)} \times 100$$

$$= \frac{R_L}{R_i + R_L} \times 100\%$$

Because $V_L = E \times R_L/(R_i + R_L)$, the graphs of percentage efficiency, and V_L versus R_L, will be similar in appearance (Fig. 23-2A).

Example 23-1

In Fig. 23-3, what is the value of R_L that will permit maximum power transfer to the load? Calculate the value of the maximum load power involved. What would be the new values of the load power if the chosen value of R_L were (a) halved and (b) doubled? In each case, calculate the value of the percentage efficiency.

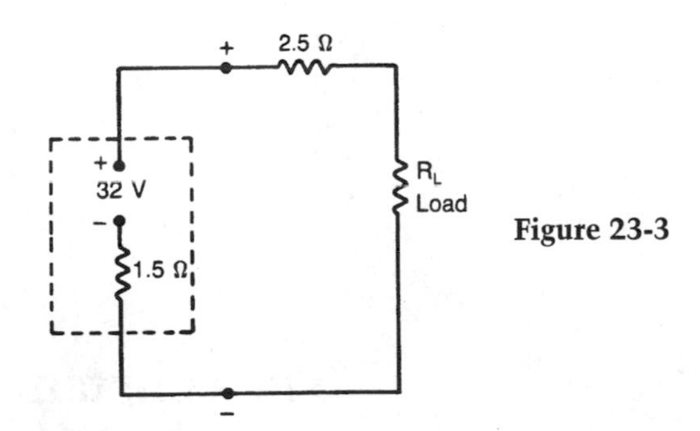

Figure 23-3

Solution

For the maximum power transfer to the load, the value of R_L must be matched to the total of all the resistances that are not associated with the load. Therefore, the required value for R_L is 1.5 + 2.5 = 4 Ω. Then:

$$\text{Maximum load power} = \frac{E^2}{4R_L} \qquad (23\text{-}6)$$

$$= \frac{(32\text{ V})^2}{4 \times 4\ \Omega}$$

$$= 64\text{ W}$$

$$\text{Percentage efficiency} = \frac{R_L}{R_i + R_L} \times 100 \qquad (23\text{-}7)$$

$$= \frac{R_L}{2R_L} \times 100$$

$$= 50\%$$

(a) When R_L is doubled to 2 × 4 Ω = 8 Ω,

$$\text{Load power} = \frac{E^2 R_L}{(R_i + R_L)^2} \qquad (23\text{-}3)$$

$$= \frac{(32\text{ V})^2 \times 8\ \Omega}{(4\ \Omega + 8\ \Omega)^2}$$

$$= 56.9\text{ W}$$

$$\text{Percentage efficiency} = \frac{R_L}{R_i + R_L} \times 100 \qquad (23\text{-}7)$$

$$= \frac{8\ \Omega \times 100}{4\ \Omega + 8\ \Omega}$$

$$= 66.7\%$$

(b) When R_L is halved to 4 Ω/2 = 2 Ω,

$$\text{Load power} = \frac{E^2 R_L}{(R_i + R_L)^2} \qquad (23\text{-}3)$$

$$= \frac{(32\text{ V})^2 \times 2\ \Omega}{(4\ \Omega + 2\ \Omega)^2}$$

$$= 56.9\text{ W}$$

$$\text{Percentage efficiency} = \frac{R_L}{R_i + R_L} \times 100 \qquad (23\text{-}7)$$

$$= \frac{2\ \Omega \times 100}{4\ \Omega + 2\ \Omega}$$

$$= 33.3\%$$

It is worth noting that doubling and halving the matched value of R_L always provides equal load powers.

PRACTICE PROBLEMS

23-1. A dc source whose open-circuit voltage is 30 V, delivers a maximum power of 100 W to a resistive load. Calculate the value of the source's internal resistance.

23-2. A variable resistive load is connected across a dc source. When the load resistance is adjusted to 4 Ω, the power developed in the load is its maximum value of 60 W. Calculate the values of the source's open-circuit voltage and its internal resistance.

23-3. A dc source whose open-circuit voltage is 225 V, has an internal resistance of 50 Ω. If the circuit efficiency is 60%, what is the required value for the load resistance?

23-4. A source has an open-load terminal voltage of 50 V, which drops to 40 V when a 4-Ω load is connected. Calculate the maximum possible value of the load power.

23-5. A dc source has an open-circuit voltage of 35 V and an internal resistance of 5 Ω. What are the two values of load resistance that will only provide 50% of the maximum possible power? In each case, what is the value of the circuit efficiency?

24
The constant current source

In chapter 22 a *constant voltage* model was proposed to represent an electrical source. This consisted of the constant EMF (*E*) in series with the internal resistance (R_i) (Fig. 24-1A). But is this the only model that can be devised? As far as the load is concerned, it is also possible to have a *constant current* source that is capable of generating a variable EMF. The value of the constant current is the short-circuit terminal value

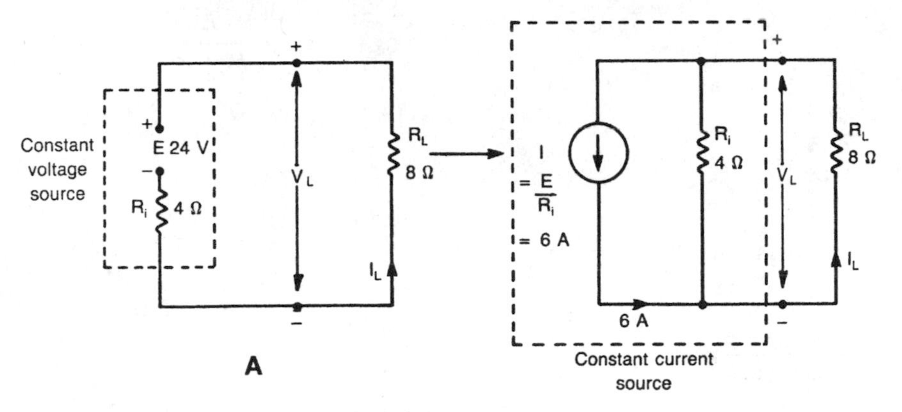

Figure 24-1

previously covered in chapter 22. Across this current generator is a parallel internal resistance that has the same value as the series internal resistance of the constant voltage model (Fig. 24-1B).

Compare the two models by an example in which an electrical source has an open-circuit voltage of 24 V and a short-circuit current of 6 A. Then the internal resistance is 24 V/6 A = 4 Ω. The two comparable models are illustrated in Figs. 24-1A and B, and the arrow convention in the constant current source indicates the same direction of electron flow as produced by the constant voltage source.

If we now connect an 8-ohm load across the terminals of each source, the load current for the constant voltage generator is:

$$I_L = \frac{24\ \text{V}}{4\ \Omega + 8\ \Omega} = 2\ \text{A}.$$

For the constant current model the load current by the CDR rule is:

$$I_L = 6\ \text{A} \times \frac{4\ \Omega}{4\ \Omega + 8\ \Omega} = 2\ \text{A}.$$

Consequently, as far as the load is concerned, both models are equally valid. However, the generators themselves are not exact equivalents because under open-circuit load conditions, there is power dissipated in the internal resistance of the current generator, but not in the internal resistance of the voltage generator.

One question has not been answered. Why do we need the concept of the constant current source when we already have the constant voltage generator? You will see that in the analytical methods of chapters 26 through 32, that one model is sometimes preferable to the other. Moreover, some active devices approximate more to the constant voltage model (Fig. 24-2A), but others tend to behave like constant current sources (Fig. 24-2B).

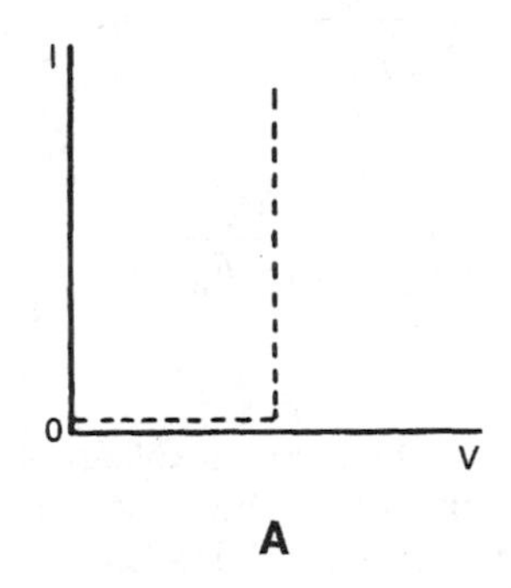

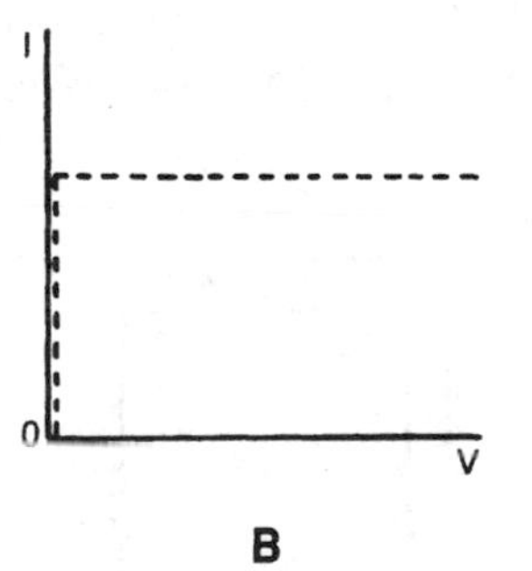

Figure 24-2

MATHEMATICAL DERIVATIONS

Constant voltage source

Constant voltage EMF,

$$E = \text{open-circuit terminal voltage (volts)} \qquad (24\text{-}1)$$

Series internal resistance,

$$R_i = \frac{\text{open-circuit voltage, } E}{\text{short-circuit current, } I}\ \text{ohms} \qquad (24\text{-}2)$$

Load current,

$$I_L = \frac{E}{R_i + R_L}\ \text{amperes} \qquad (24\text{-}3)$$

Load voltagc,

$$V_L = I_L R_L = \frac{ER_L}{R_i + R_L} \text{ volts} \tag{24-4}$$

Constant current source

Constant current,

I = short-circuit terminal current (amperes) (24-5)

Parallel internal resistance,

$$R_i = \frac{\text{open-circuit voltage, } E}{\text{short-circuit current, } I} \text{ ohms} \tag{24-6}$$

Load current,

$$I_L = I \times \frac{R_i}{R_i + R_L} = \frac{E}{R_i + R_L} \text{ amperes} \tag{24-7}$$

Load voltage,

$$V_L = I_L R_L = I \times \frac{R_i R_L}{R_i + R_L} = \frac{ER_L}{R_i + R_L} \text{ volts} \tag{24-8}$$

Example 24-1

An electrical source has an open-circuit voltage of 32 V and a short-circuit current of 16 A. Draw the schematic of its constant current source model. If a load of 2.5 Ω is connected across the terminals of the source, calculate the values of the load current and the load voltage. See Fig. 24-3.

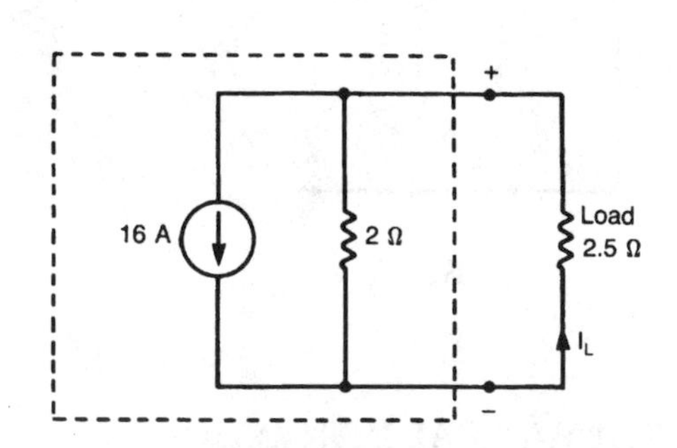

Figure 24-3

Solution

Internal resistance,

$$R_i = \frac{E}{I} \tag{24-6}$$

$$= \frac{32 \text{ V}}{16 \text{ A}}$$

$$= 2\ \Omega$$

Load current,

$$I_L = I \times \frac{R_i}{R_i + R_L} \tag{24-7}$$

$$= 16 \text{ A} \times \frac{2\ \Omega}{2\ \Omega + 2.5\ \Omega}$$

$$= 7.11 \text{ A}$$

Load voltage,

$$V_L = I_L R_L \tag{24-8}$$

$$= 7.11 \text{ A} \times 2.5\ \Omega$$

$$= 17.8 \text{ V}$$

Example 24-2

An electrical source has an open-circuit voltage of 15 V and a short-circuit current of 25 A. Draw the schematic of its constant voltage source model. If a load of 4 Ω is connected across the terminals of the source, calculate the values of the load current and the load voltage. See Fig. 24-4.

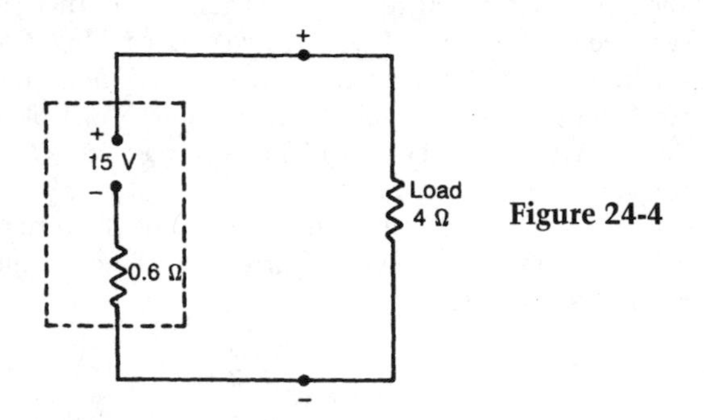

Figure 24-4

Solution

Internal resistance,

$$R_i = \frac{E}{I} = \frac{15 \text{ V}}{25 \text{ A}} = 0.6\ \Omega \tag{24-2}$$

Load current,

$$I_L = \frac{E}{R_i + R_L} \tag{24-3}$$

$$= \frac{15 \text{ V}}{0.6\ \Omega + 4\ \Omega} = \frac{15 \text{ V}}{4.6\ \Omega}$$

$$= 3.26 \text{ A}$$

Load voltage,

$$V_L = I_L R_L$$

$$= 3.26 \text{ A} \times 4\ \Omega$$

$$= 13.0 \text{ V}$$

PRACTICE PROBLEMS

24-1. A constant current source has a short-circuit current of 20 A and an internal resistance of 4 Ω. If this source is connected to a 6-Ω load, calculate the value of the load current. What is the maximum value of power that can be delivered to a load?

24-2. A constant voltage source has an open-circuit voltage of 25 V. When a 2-A current is drawn from the source, the terminal voltage drops to 20 V. For the equivalent constant current source, determine the values of the short-circuit current and the parallel internal resistance.

25
Practical sources in series, parallel, and series-parallel

When the cells of chapter 12 were connected in series-aiding, the total EMF available was the sum of the individual EMFs. However, these cells were idealized in the sense that their internal resistances were ignored.

PRACTICAL SOURCES IN SERIES

Figure 25-1A illustrates *N* practical sources connected in series-aiding. The total no-load terminal voltage will still be the sum of the individual EMFs, but the internal resistances are also in series and must be added to obtain the total equivalent internal resistance. As a result, a series-aiding combination of sources will increase the total EMF available, but this advantage is only achieved at the expense of the higher total internal resistance.

PRACTICAL SOURCES IN PARALLEL

If practical parallel sources are expressed in terms of their constant voltage equivalents, the resulting circuit cannot be analyzed without the need for simultaneous equations (Fig. 25-1B). However, if each source is replaced by its constant current generator, the individual currents can be added to produce the total current of the final *equivalent generator*. At the same time, the individual internal resistances are directly in parallel and, therefore, the reciprocal formula can be used to obtain the relatively low total equivalent resistance. It follows that the main purpose of connecting cells in parallel is to reduce the effective internal resistance, as opposed to increasing the available voltage. If necessary, the final constant current generator can be converted to its equivalent voltage model.

IDENTICAL (PRACTICAL) SOURCES IN SERIES-PARALLEL

A series connection increases the available voltage at the expense of a greater internal resistance. By contrast, a parallel connection reduces the internal resistance, but does not raise the voltage. It is obviously possible to have the best of both worlds by using a series-parallel arrangement of identical cells. This is illustrated in Fig. 25-1C, where *N* cells in series are connected to form a single "bank" and raise the voltage while there are *M* such banks in parallel to lower the internal resistance. However, these advantages are achieved at the expense of a large, cumbersome, and costly dc source.

MATHEMATICAL DERIVATIONS

Cells in series (Fig. 25-1A)

Total EMF,

$$E_T = E_1 + E_2 + E_3 \ldots + E_N \text{ volts} \tag{25-1}$$

Total internal resistance,

$$R_{iT} = R_{i1} + R_{i2} + R_{i3} \ldots + R_{iN} \text{ ohms} \tag{25-2}$$

If the sources are identical,
Total EMF,

$$E_T = NE \text{ volts} \tag{25-3}$$

Total internal resistance,

$$R_{iT} = NR_i \text{ ohms} \tag{25-4}$$

Cells in parallel (Fig. 25-1B)

Equivalent constant currents,

$$I_1 = \frac{E_1}{R_{i1}}, I_2 = \frac{E_2}{R_{i2}}, \tag{25-5}$$

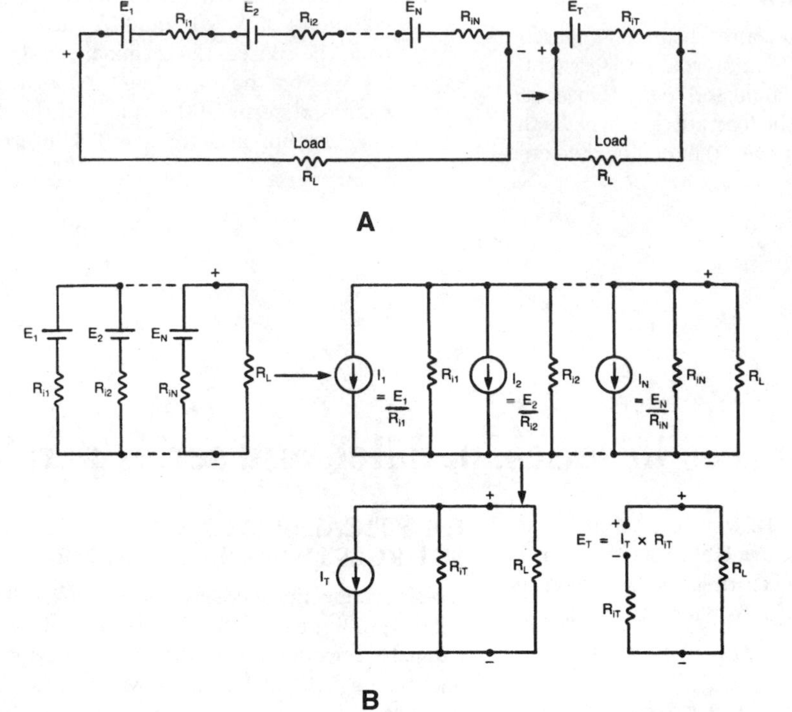

Figure 25-1

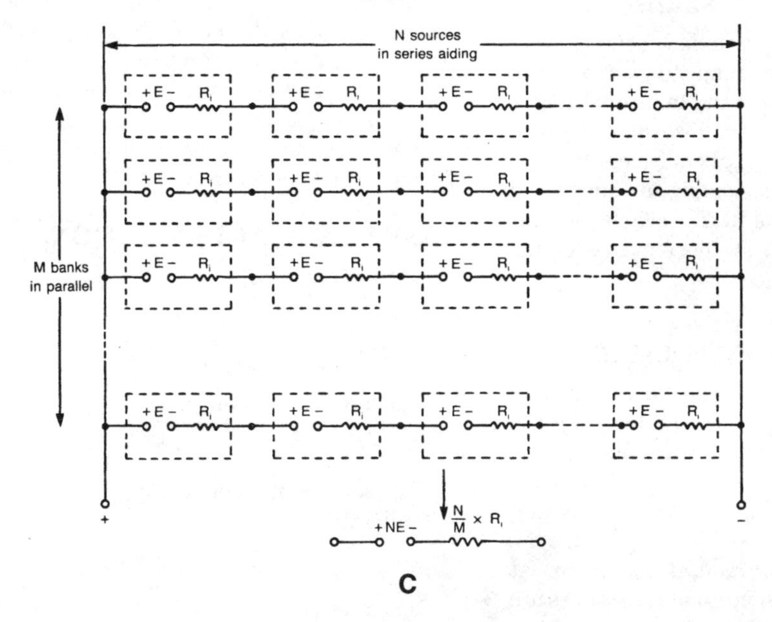

$$I_3 = \frac{E_3}{R_{i3}} \ldots I_N = \frac{E_N}{R_{iN}} \text{ amperes}$$

Total equivalent constant current,

$$I_T = I_1 + I_2 + I_3 \ldots + I_N \text{ amperes} \qquad (25\text{-}6)$$

Total equivalent internal resistance,

$$R_{iT} = \frac{1}{\dfrac{1}{R_{i1}} + \dfrac{1}{R_{i2}} + \dfrac{1}{R_{i3}} \ldots + \dfrac{1}{R_{iN}}} \text{ ohms} \qquad (25\text{-}7)$$

Total equivalent EMF,

$$E_T = I_T \times R_{iT} \text{ volts} \qquad (25\text{-}8)$$

If the sources are identical,

Total equivalent constant current,

$$I_T = NI \text{ amperes} \tag{25-9}$$

Total equivalent internal resistance,

$$R_{iT} = \frac{R_i}{N} \text{ ohms} \tag{25-10}$$

$$\text{Total equivalent EMF} = E \text{ volts} \tag{25-11}$$

Cells in series-parallel (Fig. 25-1C)

$$\text{Total EMF of one bank} = NE \text{ volts} \tag{25-12}$$

$$\text{Total internal resistance of one bank} = NR_i \text{ ohms} \tag{25-13}$$

$$\text{Total EMF of } M \text{ banks} = NE \text{ volts} \tag{25-14}$$

$$\text{Total internal resistance of } M \text{ banks} = \frac{NR_i}{M} \text{ ohms} \tag{25-15}$$

Example 25-1

A voltage source has an EMF of 1.8 V and an internal resistance of 0.3 Ω. If 12 such sources are connected in series-aiding across a 25-Ω load, what is the value of the load voltage?

Solution

Total constant voltage EMF,

$$E_T = 12 \times 1.8 \text{ V} = 21.6 \text{ V} \tag{25-3}$$

Total internal resistance,

$$R_{iT} = 12 \times 0.3\ \Omega = 3.6\ \Omega \tag{25-4}$$

Using the voltage division rule,
Load voltage,

$$V_L = E_T \times \frac{R_L}{R_{iT} + R_L}$$

$$= 21.6 \text{ V} \times \frac{25\ \Omega}{3.6\ \Omega + 25\ \Omega}$$

$$= 21.6 \text{ V} \times \frac{25\ \Omega}{28.6\ \Omega}$$

$$= 18.9 \text{ V}$$

Example 25-2

In Fig. 25-2A, calculate the values of I_L and V_L.

Solution

Convert the voltage sources of Fig. 25-2A into the current sources of Fig. 25-2B. The values of the equivalent current generators are 18 V/6 Ω = 3 A, 24 V/4 Ω = 6 A, and 20 V/5 Ω = 4 A. The total equivalent current is therefore,

$$I_T = 3 \text{ A} + 6 \text{ A} + 4 \text{ A} \tag{25-6}$$

$$= 13 \text{ A}$$

Total equivalent resistance,

$$R_{iT} = \frac{1}{\frac{1}{6} + \frac{1}{4} + \frac{1}{5}} \tag{25-7}$$

$$= 1.62\ \Omega$$

Load current,

$$I_L = 13 \text{ A} \times \frac{1.62\ \Omega}{1.62\ \Omega + 22\ \Omega}$$

$$= 0.93 \text{ A}$$

Load voltage,

$$V_L = 0.93 \text{ A} \times 22\ \Omega$$

$$= 20.5 \text{ V}$$

Example 25-3

Twelve cells, each with an EMF of 1.5 V and an internal resistance of 0.1 Ω, are connected in series-aiding. Eight such banks are joined in parallel so that the total number of individual cells is 12 × 8 = 96 cells. Calculate the open-circuit terminal voltage available and the corresponding value of the total internal resistance.

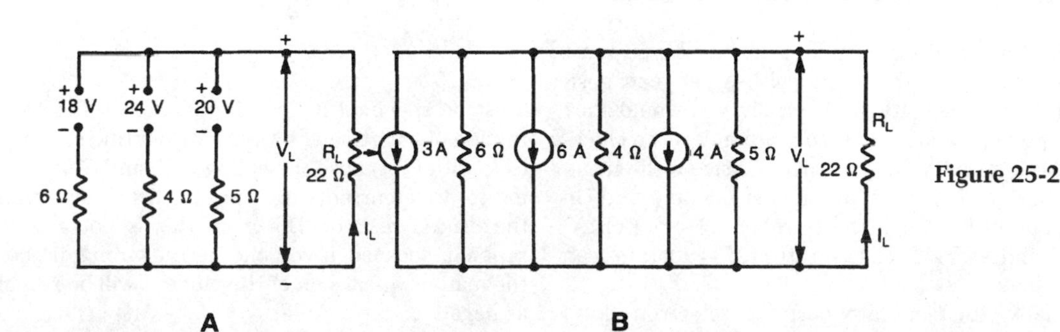

Figure 25-2

Solution

Total EMF,

$$E_T = NE \qquad (25\text{-}12)$$
$$= 12 \times 1.5\text{ V}$$
$$= 18\text{ V}$$

Total internal resistance,

$$R_{iT} = \frac{NR_i}{M} \qquad (25\text{-}15)$$
$$= \frac{12 \times 0.1\ \Omega}{8}$$
$$= 0.15\ \Omega$$

PRACTICE PROBLEMS

25-1. Twenty 2-V cells, each with an internal resistance of 0.1 Ω, are connected in series-aiding across a 4.6-Ω load. Calculate the value of the load current.

25-2. Twenty 2-V cells, each with an internal resistance of 1.0 Ω, are parallel connected to a 1.2-Ω load. Find the value of the load current.

25-3. One hundred and twenty 3-V cells, each with an internal resistance of 0.6 Ω, are connected in 12 parallel banks, with each bank containing 10 cells in series-aiding. If this series-parallel arrangement is connected to a 2-Ω load, what is the value of the load power?

25-4. Six 2-V cells, each with an internal resistance of 0.8 Ω, are connected in series-aiding. What value of load resistance will allow maximum power transfer to the load? Calculate the value of this maximum load power.

25-5. Six 2-V cells, each with an internal resistance of 0.8 Ω, are correctly connected in parallel. What value of load resistance will allow maximum transfer to the load? Calculate the value of this maximum load power.

26
Kirchhoff's voltage and current laws

The previous sections have loosely referred to *Kirchhoff's voltage* law (KVL) as the requirement for a voltage balance around any closed circuit or loop, and *Kirchhoff's current law* (KCL) fulfilled the need for the current balance that must exist at any junction point. The laws are commonly used to analyze circuits that are too difficult to be solved by Ohm's law alone.

A more formal KVL statement would be: "The *algebraic* sum of the constant voltage EMFs and the voltage drops around *any closed* electrical loop is *always* zero."

Because the algebraic sum is involved, you must have a convention that distinguishes between positive and negative voltages. Normally, you would start at any point in a loop and move around in the clockwise direction. A voltage is then positive if the negative polarity of that voltage is first encountered. In the circuit of Fig. 26-1, there are three loops (*XYZX*, *XWYX*, and *XWYZX*) so that three KVL equations can be obtained.

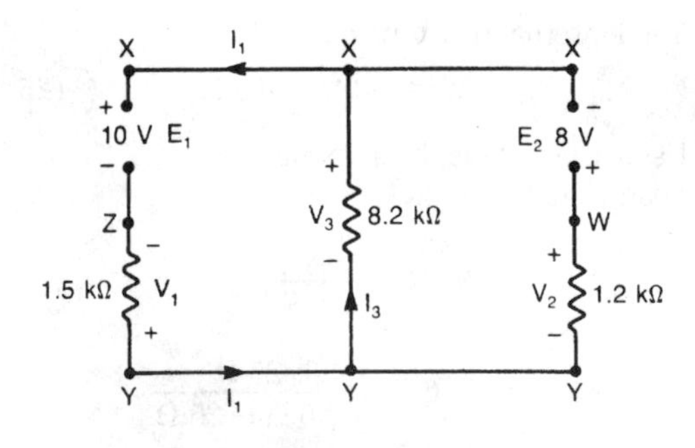

Figure 26-1

Initially, the separate currents (electron flow) must be specified in the circuit. Sometimes the direction of a particular current is doubtful; such is the case with I_3 because the voltages E_1 and E_2 are tending to drive currents in opposite directions through the 8.2-kΩ resistor. However, this is not a serious problem because, if you choose the wrong direction, the mathematical sign of the current will be revealed as negative.

MATHEMATICAL DERIVATIONS

If the currents are measured in milliamperes:

Loop *XYZX* (starting at *X*)

The first KVL equation is:

$$+(-V_3) + (-V_1) + (+E_1) = 0 \qquad (26\text{-}1)$$

$$-I_3 \times 8.2\text{ k}\Omega - I_1 \times 1.5\text{ k}\Omega + 12\text{ V} = 0$$

$$1.5\,I_1 + 8.2\,I_3 = 12$$

Loop *XWYX* (starting at *X*)

The second KVL equation is:

$$+(+E_2) + (-V_2) + (+V_3) = 0 \qquad (26\text{-}2)$$

$$+8\text{ V} - I_2 \times 1.2\text{ k}\Omega + I_3 \times 8.2\text{ k}\Omega = 0$$

$$+1.2\,I_2 - 8.2\,I_3 = 8$$

Loop *XWYZX* (starting at *X*)

The third KVL equation is:

$$+(+E_2) + (-V_2) + (-V_1) + (+E_1) = 0 \qquad (26\text{-}3)$$

$$+8\text{ V} - I_2 \times 1.2\text{ k}\Omega - I_1 \times 1.5\text{ k}\Omega + 12\text{ V} = 0$$

$$1.5\,I_1 + 1.2\,I_2 = 8 + 12 = 20$$

Although you have three equations, they are not independent because the third KVL equation is the result of adding together the first and second equations. Consequently, there are three unknowns (I_1, I_2 and I_3) and you still need a third equation, which is of the KCL type.

Formally stated, Kirchhoff's current law is: "The *algebraic* sum of the currents existing at any junction point is zero."

Conventionally, you can assume that the electron flow currents entering a junction point are positive and those leaving are negative. Therefore at the junction point, *Y*:

$$+(+I_1) + (-I_2) + (-I_3) = 0 \qquad (26\text{-}4)$$

$$I_1 = I_2 + I_3$$

Combining equations 26-1 and 26-4:

$$1.5\,(I_2 + I_3) + 8.2\,I_3 = 12 \qquad (26\text{-}5)$$

$$1.5\,I_2 + 9.7\,I_3 = 12$$

Multiplying equation 26-2 by 1.5 and equation 26-5 by 1.2:

$$1.8\,I_2 - 12.3\,I_3 = 12 \qquad (26\text{-}6)$$

$$1.8\,I_2 + 11.64\,I_3 = 14.4 \qquad (26\text{-}7)$$

Subtracting equation 26-6 from equation 26-7:

$$23.9\,I_3 = 2.4$$

$$I_3 \approx +0.1\text{ mA.}$$

The positive sign for I_3 indicates that the chosen direction for this current was correct.

Substituting this value of I_3 in equation 26-1,

$$1.5\,I_1 + 0.823 = 12$$

$$I_1 = \frac{11.177}{1.5} = 7.45\text{ mA.}$$

From equation 26-4,

$$I_2 = I_1 - I_3 = 7.35\text{ mA.}$$

Checking for loop *XYZX*:

$$-0.1\text{ mA} \times 8.2\text{ k}\Omega - 7.45\text{ mA} \times 1.5\text{ k}\Omega + 12\text{ V}$$

$$= -0.82 - 11.18 + 12 = 0\text{ V.}$$

The individual voltage drops and powers dissipated can be obtained from the normal Ohm's law equations.

Example 26-1

In Fig. 26-2, calculate the value of the load current (I_L) and the potential at the junction *X*.

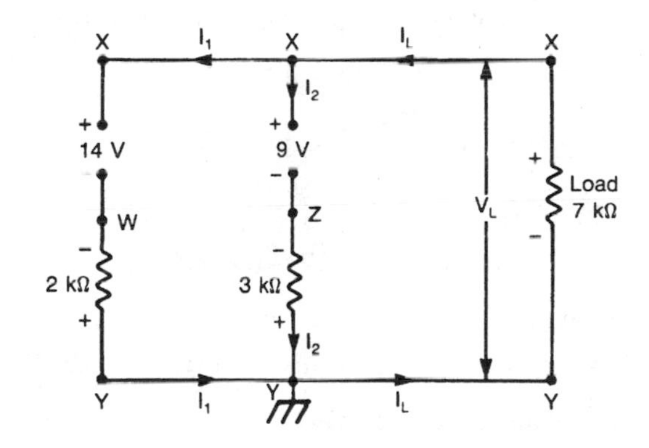

Figure 26-2

Solution

The (electron flow) currents in milliamperes are as assigned.

The KVL equation for the loop *XYWX* (starting at *X*) is:

$$+(-I_L \times 7\text{ k}\Omega) + (-I_1 \times 2\text{ k}\Omega) + 14\text{ V} = 0$$

$$2\,I_1 + 7\,I_L = 14.$$

For loop *XYZX* (starting at *X*):

$$+(-I_L \times 7\text{ k}\Omega) + (-I_2 \times 3\text{ k}\Omega) + 9\text{ V} = 0$$

$$3\,I_2 + 7\,I_L = 9$$

The KCL equation for the X junction is:

$$I_1 + I_2 = I_L$$

Therefore,

$$2\,I_1 + 7\,(I_1 + I_2) = 9\,I_1 + 7\,I_2 = 14 \qquad (26\text{-}1\text{-}1)$$

and,

$$3\,I_2 + 7\,(I_1 + I_2) = 7\,I_1 + 10\,I_2 = 9 \qquad (26\text{-}1\text{-}2)$$

Multiplying the equation 26-1-1 by 7 and the equation 26-1-2 by 9,

$$63\,I_1 + 49\,I_2 = 98 \qquad (26\text{-}1\text{-}3)$$

$$63\,I_1 + 90\,I_2 = 81 \qquad (26\text{-}1\text{-}4)$$

Subtracting the equation 26-1-3 from the equation 26-1-4,

$$(90 - 49)\,I_2 = 41\,I_2 = 81 - 98 = -17$$

$$I_2 = -\frac{17}{41} = -0.417 \text{ mA.}$$

The negative sign indicates that the electron flow direction is from the junction Y to the junction X (and not from X to Y as indicated). This means that the 14-V source is, in fact, charging the 9-V source.

Substituting the value of I_2 in the equation 26-1-3,

$$63\,I_1 = 98 + (49 \times 0.417) = 118.4$$

$$I_1 = \frac{118.4}{63} = 1.88 \text{ mA.}$$

Therefore,

$$I_L = 1.88 - 0.417 = 1.463 \text{ mA.}$$

Potential at the junction X,

$$V_L = +I_L \times R_L = +(1.463 \text{ mA} \times 7 \text{ k}\Omega) = +10.24 \text{ V.}$$

Checking the loop $XYWX$:

$$2 \text{ k}\Omega \times 1.88 \text{ mA} + 7 \text{ k}\Omega \times 1.463 \text{ mA} - 14 \text{ V}$$

$$= 3.76 + 10.24 - 14 = 0.$$

PRACTICE PROBLEMS

26-1. In the circuit of Fig. 26-3, use Kirchhoff's laws to determine the value of the current flowing through the 2-kΩ resistor.

26-2. In the circuit of Fig. 26-4, calculate the value of the current flowing through the 4-kΩ resistor.

26-3. In the circuit of Fig. 26-5, calculate the voltage potential at the point N.

26-4. In the current of Fig. 26-6, what are the values of currents I_1 and I_2?

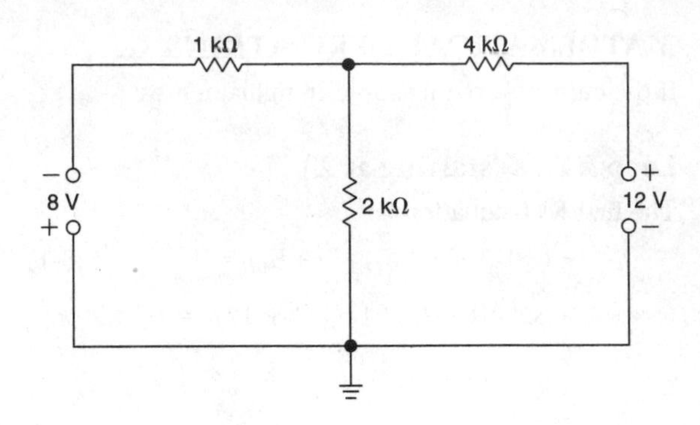

Figure 26-3

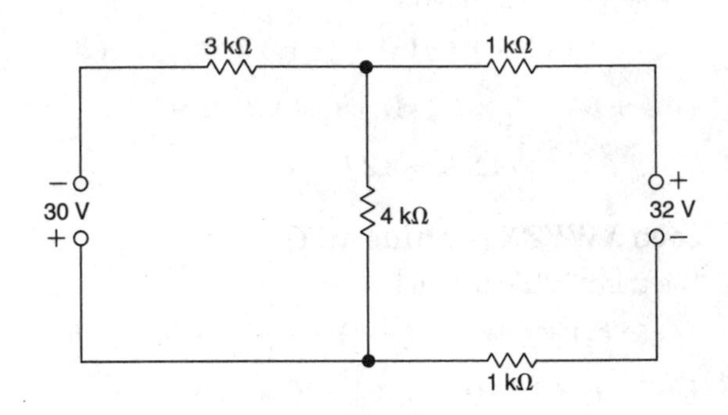

Figure 26-4

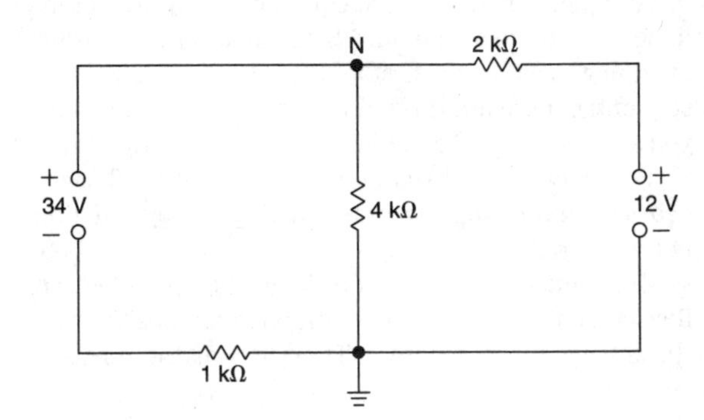

Figure 26-5

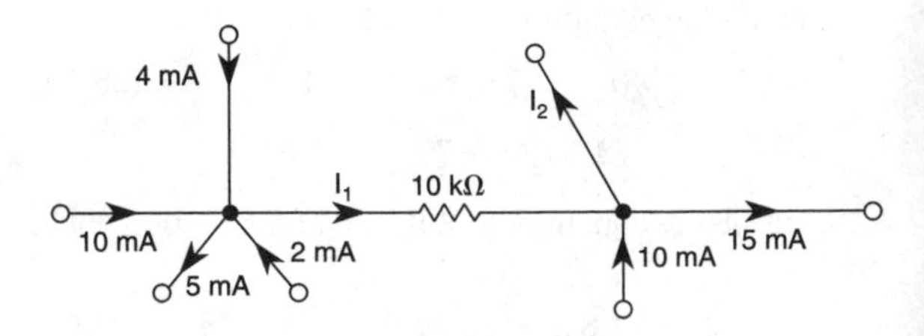

Figure 26-6

27
Mesh analysis

A closed voltage loop can also be referred to as a *mesh*. In mesh analysis, each of the mesh currents i_1, i_2, i_3 (Fig. 27-1) flow around a complete loop, although an individual resistor can carry one or more mesh currents. For example, the mesh currents i_1 and i_2 flow in opposite directions through resistor R1.

The normal convention is to assign clockwise mesh (electron flow) currents in each of the loops, then write down the KVL equation for each loop. The main advantages of this method are: (1) no KCL equations are required, and (2) the KVL equations can be written by inspection and do not require precise voltage conventions.

MATHEMATICAL DERIVATIONS

In Fig. 27-1

KVL equation for the mesh *PWXQP* (starting at *P*):

$$i_1R_3 - E_1 + i_1R_1 + i_1R_2 - i_2R_1 = 0 \qquad (27\text{-}1)$$

$$-i_2R_1 + i_1(R_1 + R_2 + R_3) = E_1$$

KVL equation for the mesh *QXYSQ* (starting at *Q*):

$$i_2R_1 - i_1R_1 + i_2R_4 + i_2R_5 - i_3R_5 + i_2R_6 = 0 \qquad (27\text{-}2)$$

$$-i_1R_1 - i_3R_5 + i_2(R_1 + R_4 + R_5 + R_6) = 0$$

KVL equation for the mesh *SYZTS* (starting at *S*):

$$i_3R_5 - i_2R_5 + E_2 + i_3R_7 + i_3R_8 = 0 \qquad (27\text{-}3)$$

$$-i_2R_5 + i_3(R_5 + R_7 + R_8) = -E_2$$

From the pattern of these results, it follows that the KVL equations can be written down from inspection, if we observe the following rules:

1. Add together all the resistances in the loop, and multiply the results by that loop's mesh current. This applies to terms $i_1(R_1 + R_2 + R_3)$, $i_2\,(R_1 + R_4 + R_5 + R_6)$, and $i_3\,(R_5 + R_7 + R_8)$.
2. If a resistor in a particular mesh carries a current assigned to another mesh, the corresponding *IR* drop is given a negative sign. This refers to terms $-i_2R_1$, $-i_1R_1$, $-i_3R_5$, and $-i_2R_5$.
3. The signs of the voltage sources are determined by the normal KVL convention.

Example 27-1

In Fig. 27-2, calculate the value of the potential at the point *X*. Note: This is the same problem that was previously solved by Kirchhoff's laws in Example 26-1.

Solution

For mesh *XZGYX*:

$$i_1\,(2\text{ k}\Omega + 3\text{ k}\Omega) - i_2 \times 3\text{ k}\Omega - 9\text{ V} + 14\text{ V} = 0$$

$$5\,i_1 - 3\,i_2 = -5 \qquad (27\text{-}1\text{-}1)$$

For mesh *XGZX*:

$$-i_1 \times 3\text{ k}\Omega + i_2(3\text{ k}\Omega + 7\text{ k}\Omega) + 9\text{ V} = 0 \qquad (27\text{-}1\text{-}2)$$

$$-3\,i_1 + 10\,i_2 = -9$$

Multiply Equation 27-1-1 by 3 and Equation 27-1-2 by 5,

$$15\,i_1 - 9\,i_2 = -15$$

$$-15\,i_1 + 50\,i_2 = -45$$

Therefore,

$$41\,i_2 = -60$$

$$i_2 = \frac{-60}{41}\text{ mA.}$$

Potential at point *X* is $i_2R_L = -(-60/41\text{ mA}) \times 7\text{ k}\Omega = +10.24\text{ V}$.

Notice that the mesh analysis solution is shorter than that obtained from Kirchhoff's laws.

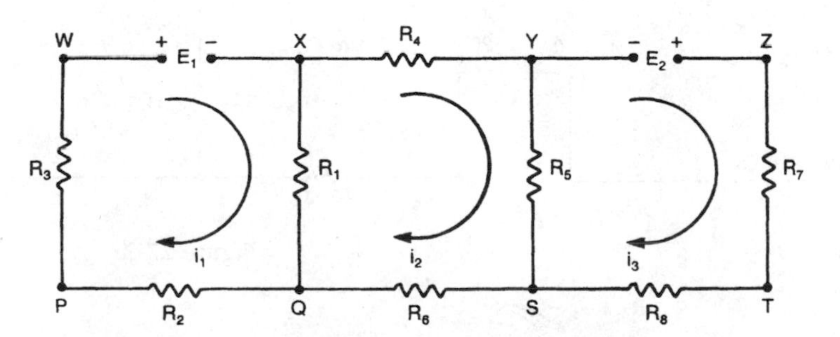

Figure 27-1

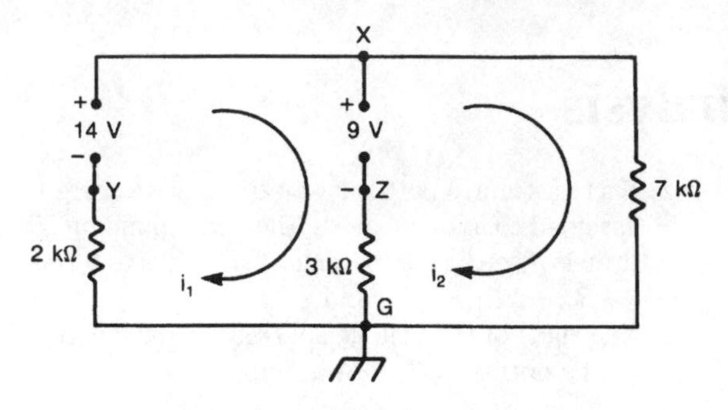

Figure 27-2

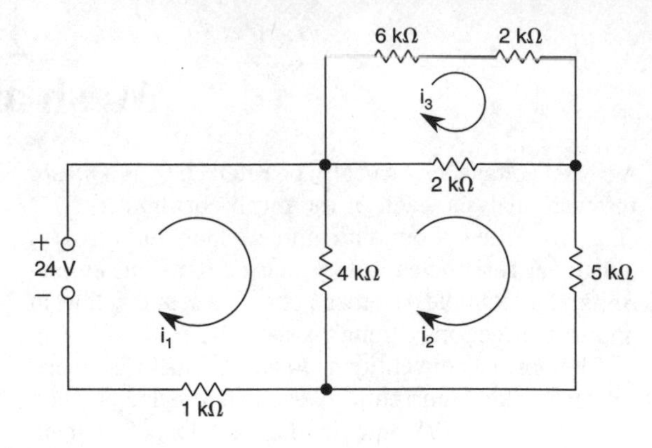

Figure 27-5

PRACTICE PROBLEMS

27-1. In Fig. 27-3, use mesh analysis to determine the voltage between points *A* and *B*.

27-2. In Fig. 27-4, use mesh analysis to find the current flowing through the 7-kΩ resistor.

27-3. In Fig. 27-5, find the values of mesh currents i_1, i_2, and i_3.

27-4. In Fig. 27-6, use mesh analysis to find the voltage between points *A* and *B*.

27-5. In Fig. 27-7, use mesh analysis to find the value of the current flowing through the 3.3-kΩ resistor.

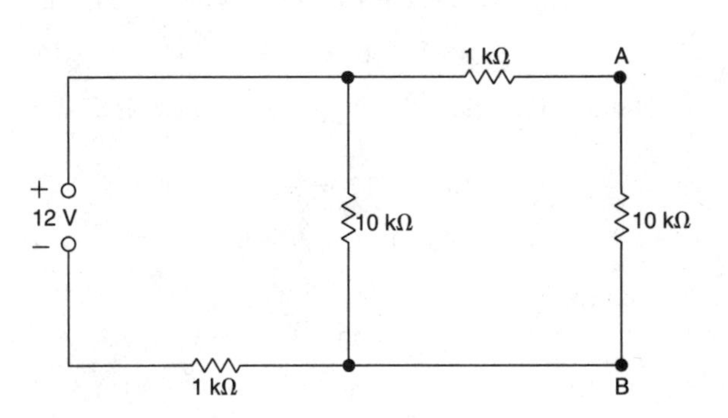

Figure 27-3

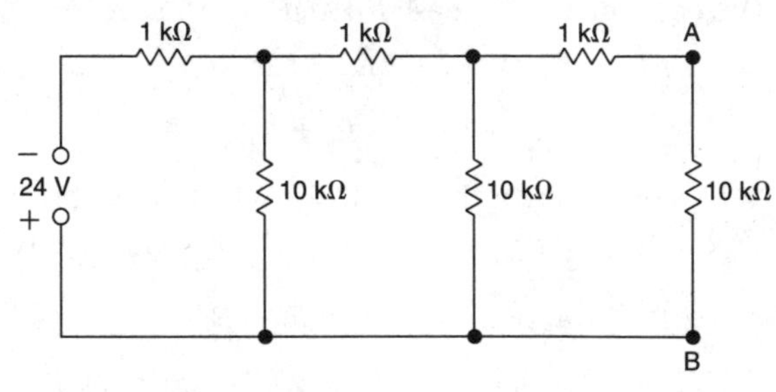

Figure 27-6

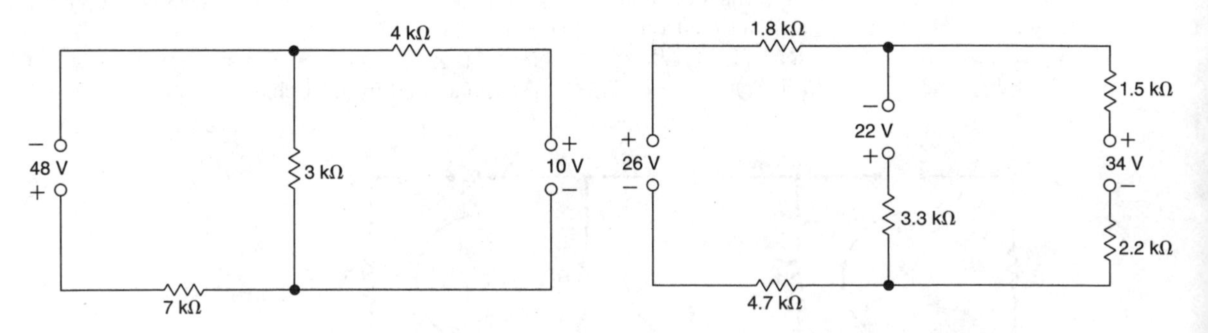

Figure 27-4

Figure 27-7

28
Nodal analysis

Nodal analysis is another method for solving circuits, but this time, only current sources are involved. By contrast, those circuits that were solved by mesh analysis or Kirchhoff's laws contained voltage sources and required the use of simultaneous equations. However, on many occasions it is possible to convert the voltage sources into their equivalent current sources and subsequently solve the problem with a single nodal equation, rather than with two or more simultaneous equations.

A node is another term for a junction point at which two or more electrical currents exist. Therefore, the junction N is a node (Fig. 28-1), and the purpose of the analysis is to find the value of the potential at N. As a convention, assume that the potential at N is negative. The true polarity will be shown by the sign of the value for V_L. If the potential at N is negative, the electrons through the resistors must flow from the nodal point to ground. The generators that force (electron flow) currents into the node are then positive, and those driving currents out of the node are negative.

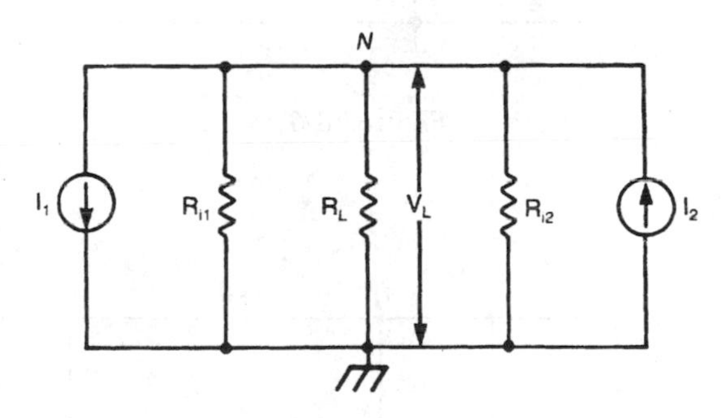

Figure 28-1

MATHEMATICAL DERIVATIONS

In Fig. 28-1

Total electron flow currents leaving node N through the resistors is:

$$\frac{V_L}{R_{i1}} + \frac{V_L}{R_L} + \frac{V_L}{R_{i2}} \text{ amperes} \tag{28-1}$$

Total of the generator currents is:

$$(-I_1) + (+I_2) \text{ amperes} \tag{28-2}$$

Consequently, the nodal equation is:

$$(-I_1) + (+I_2) = \frac{V_L}{R_{i1}} + \frac{V_L}{R_L} + \frac{V_L}{R_{i2}} \tag{28-3}$$

Example 28-1

In Fig. 28-2A, calculate the value of the potential at the point N. Note: This is the same problem that was previously solved by three simultaneous equations in Example 26-1 (Kirchhoff's laws), and by two simultaneous equations in Example 27-1 (Mesh analysis).

Solution

Convert the voltage sources of Fig. 28-2A into their constant current equivalents (Fig. 28-2B).

Total electron flow currents leaving node N through the resistors is:

$$\frac{V_N}{2\text{ k}\Omega} + \frac{V_N}{3\text{ k}\Omega} + \frac{V_N}{7\text{ k}\Omega} \tag{28-1}$$

$$= 0.9762\ V_N \text{ mA}$$

Total of generator currents is:

$$(-7) + (-3) = -10 \text{ mA} \tag{28-2}$$

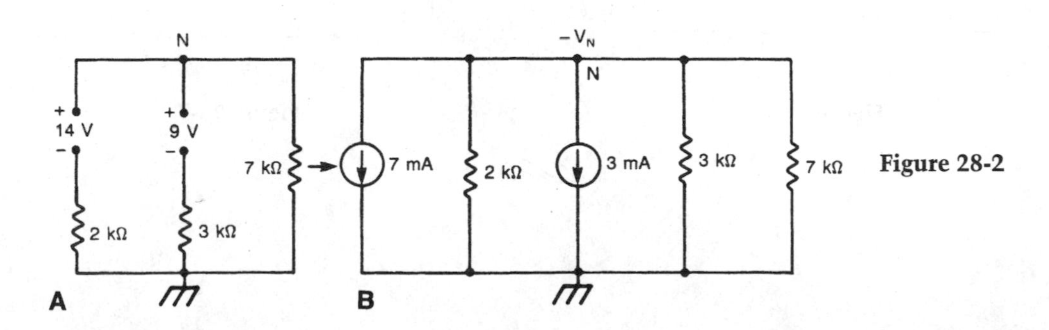

Figure 28-2

Therefore,

$$0.9762\ V_N = -10\ \text{mA}$$

$$V_N = -\frac{10}{0.9762}$$

$$= -10.24\ \text{V}.$$

The potential at N is therefore +10.24 V.

Note that the nodal solution involved only a *single* equation.

PRACTICE PROBLEMS

28-1. In the circuit of Fig. 28-3, use nodal analysis to determine the potential at node N.

28-2. In the circuit of Fig. 28-4, use nodal analysis to determine the potential at point N.

28-3. In the circuit of Fig. 28-5, use nodal analysis to determine the potential at point N.

28-4. In the circuit of Fig. 28-6, use nodal analysis to determine the potential at point N.

28-5. In the circuit of Fig. 28-7, use nodal analysis to determine the potential at point N.

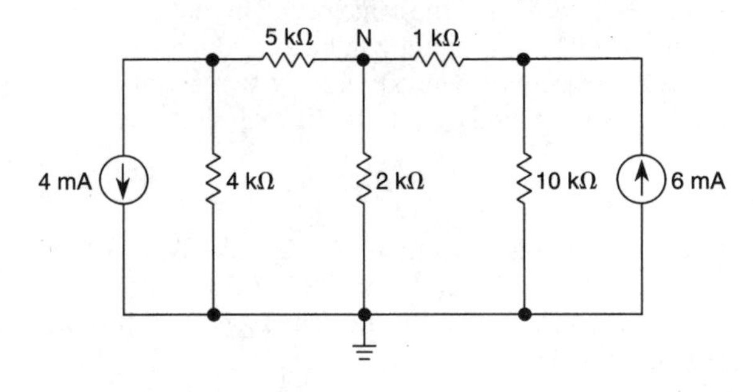

Figure 28-3

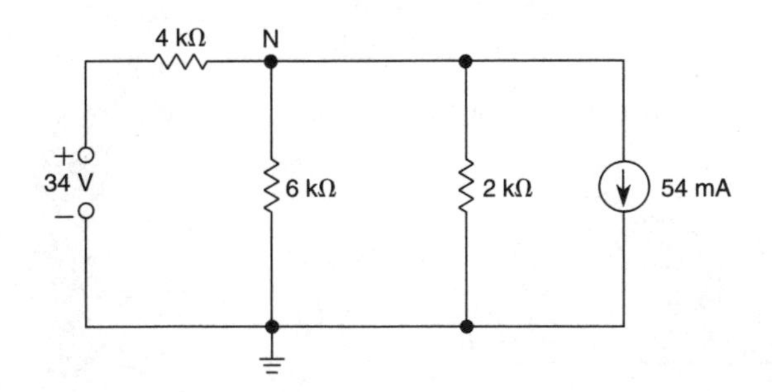

Figure 28-4

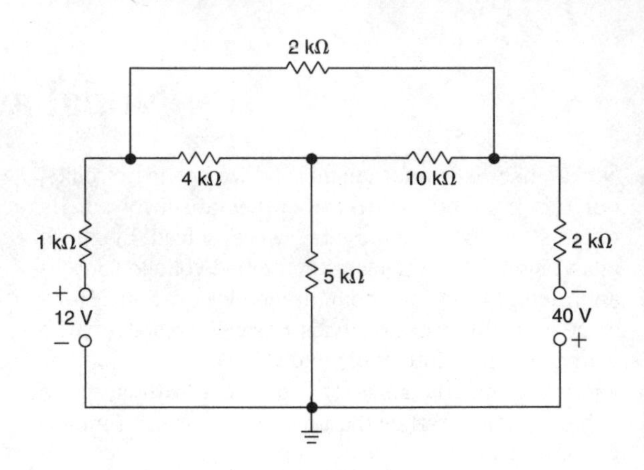

Figure 28-5

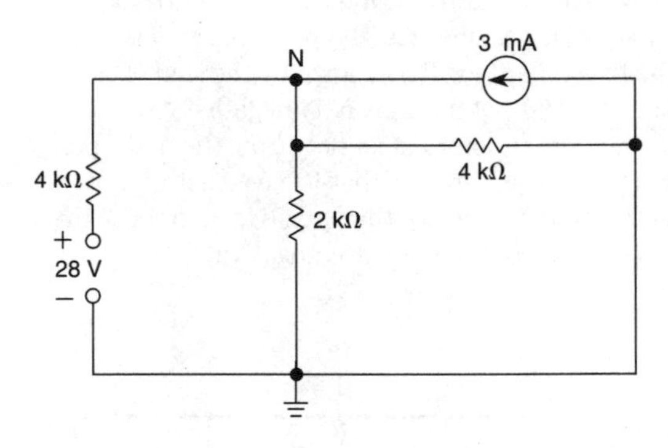

Figure 28-6

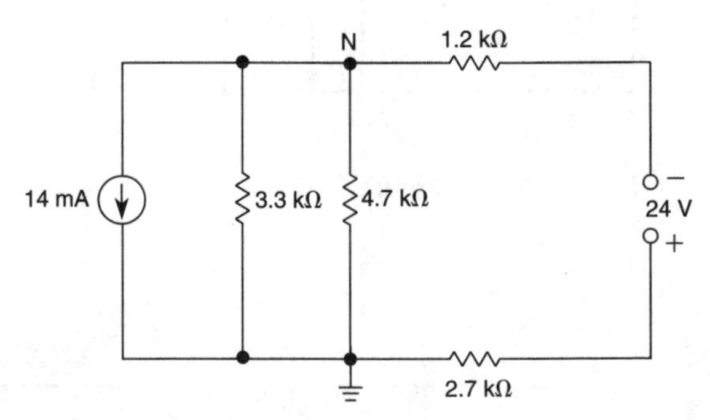

Figure 28-7

29
The superposition theorem

The principle of superposition can be used to solve a number of problems that contain more than one source and only linear resistances. The method requires you to consider the currents and voltages created by each source, in turn, and then finally combine (superimpose) the results from all the sources. A formal statement of this theorem follows.

In a network of ***linear*** *resistances, containing more than one source, the resultant current flowing at any one point is the algebraic sum of the currents that would flow at that point if each source is considered separately, and all other sources are replaced by their equivalent internal resistances. This last step is carried out by short-circuiting all sources of constant voltage and open-circuiting all sources of constant current.*

This theorem has the advantage of allowing each source to be considered separately so that only Ohm's law equations are required in the solution. However, each time one of the sources is applied to the circuit, a different voltage will normally appear across a particular resistor. The superposition theorem, therefore, requires the resistance to be linear so that its resistance does not change with the amount of the voltage drop, but remains at a constant value. In other words, the voltage drop is always directly proportional to the current.

If a large number of sources are involved, use of the superposition theorem is not recommended because the circuit has to be solved separately for each source before combining the results. The final analysis would probably be far more tedious than if we used either the mesh or nodal methods.

MATHEMATICAL DERIVATIONS

In Fig. 29-1A, assume that our purpose is to find the value of the current, I_3.

Short out the constant voltage source (E_1) to produce the circuit of Fig. 29-1B.

Total equivalent resistance (R_T) presented to the E_2 source is

$$R_T = R_2 + R_4 \,||\, (R_1 + R_3) \text{ ohms} \qquad (29\text{-}1)$$

Then,

$$I_2' = \frac{E_2}{R_T} \text{ and } I_3' = I_2' \times \frac{R_1 + R_3}{R_1 + R_3 + R_4} \text{ amperes} \qquad (29\text{-}2)$$

Short out the constant voltage source (E_2) to produce the circuit of Fig. 29-1C.

Total equivalent resistance (R_T) presented to the E_1 source is

$$R_T = R_1 + R_3 + R_2 \,||\, R_4 \text{ ohms} \qquad (29\text{-}3)$$

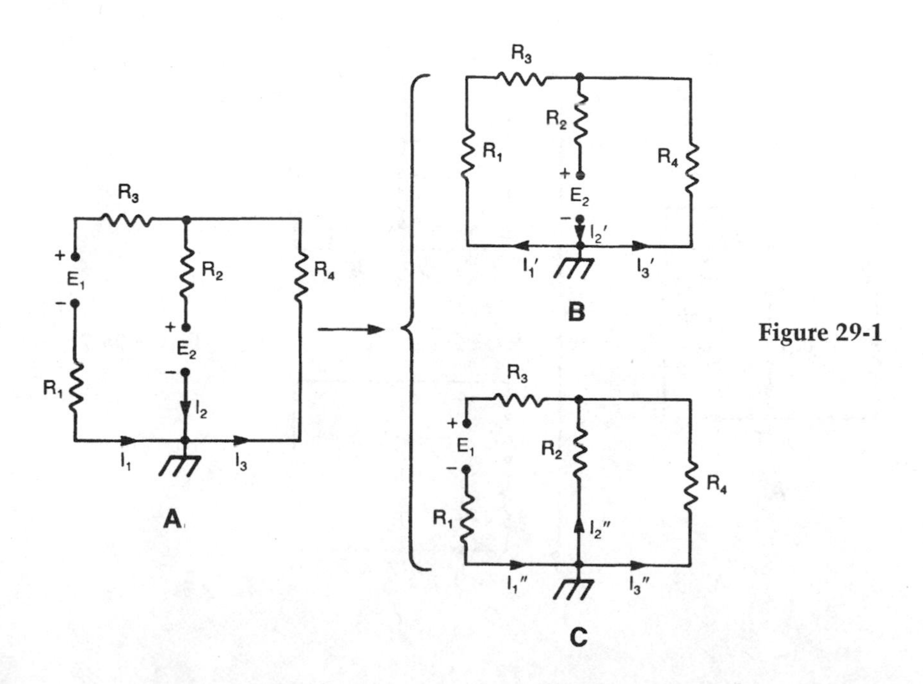

Figure 29-1

Then,

$$I_1'' = \frac{E_1}{R_T} \text{ and } I_3'' = I_1 \times \frac{R_2}{R_2 + R_4} \text{ amperes} \quad (29\text{-}4)$$

Superimposing these results,

$$I_3 = I_3' + I_3'' \text{ amperes} \quad (29\text{-}5)$$

Similarly,

$$I_1 = I_1' - I_1'' \text{ and } I_2 = I_2' - I_2'' \text{ amperes} \quad (29\text{-}6)$$

The signs of the answers for I_1 and I_2 will indicate the actual directions of the electron flows.

Example 29-1

In Fig. 29-2A, calculate the value of the potential at the point X. Note: This is the same problem that was previously solved by the methods of Kirchhoff's laws, mesh analysis, and nodal analysis.

Solution

Short out the 14-V constant voltage source to produce the circuit of Fig. 29-2B. Total resistance presented to the 9-V source,

$$R_T = 3 \text{ k}\Omega + 2 \text{ k}\Omega \parallel 7 \text{ k}\Omega$$

$$= 3 + \frac{2 \times 7}{2 + 7} = 3 + \frac{14}{9} = \frac{41}{9}$$

$$= 4.56 \text{ k}\Omega.$$

$$\text{Current, } I_2' = \frac{9 \text{ V}}{4.56 \text{ k}\Omega}$$

$$= 1.97 \text{ mA}.$$

$$\text{Current, } I_3' = 1.97 \text{ mA} \times \frac{2 \text{ k}\Omega}{2 \text{ k}\Omega + 7 \text{ k}\Omega}$$

$$= 0.44 \text{ mA (CDR rule)}.$$

Short out the 9-V constant source to produce the circuit of Fig. 29-2C.

Total resistance presented to the 14-V source,

$$R_T = 2 \text{ k}\Omega + 3 \text{ k}\Omega \parallel 7 \text{ k}\Omega = 2 + \frac{3 \times 7}{3 + 7}$$

$$= 4.1 \text{ k}\Omega.$$

$$\text{Current } I_1'' = \frac{14 \text{ V}}{4.1 \text{ k}\Omega}$$

$$= 3.41 \text{ mA}.$$

$$\text{Current } I_3'' = 3.41 \text{ mA} \times \frac{3 \text{ k}\Omega}{3 \text{ k}\Omega + 7 \text{ k}\Omega}$$

$$= 1.023 \text{ mA}.$$

When the results are superimposed,

$$I_3 = I_3' + I_3'' = 0.440 + 1.023 = 1.463 \text{ mA} \quad (29\text{-}5)$$

$$\text{Potential at } X = +(1.463 \text{ mA} \times 7 \text{ k}\Omega)$$

$$= +10.24 \text{ V}.$$

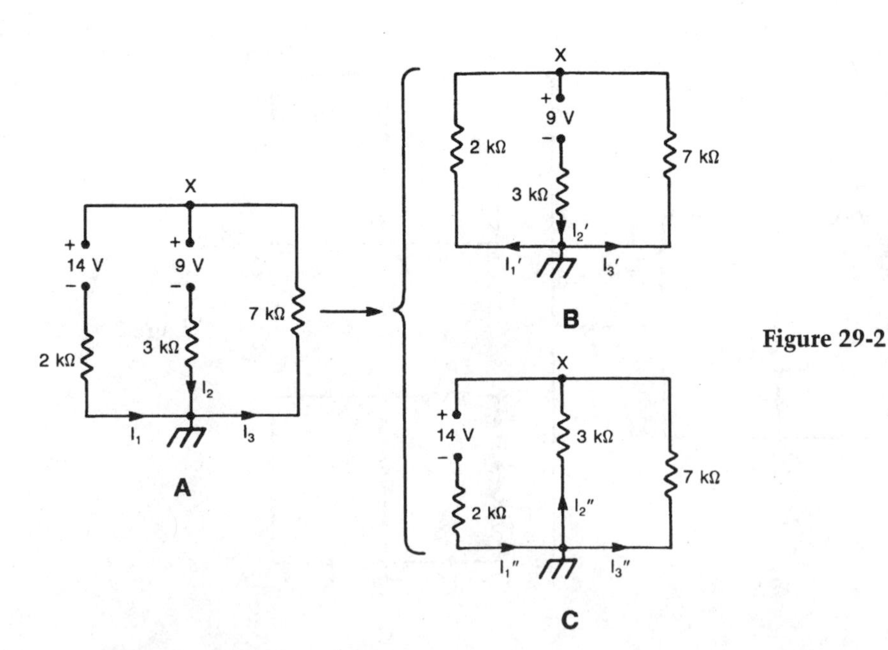

Figure 29-2

PRACTICE PROBLEMS

29-1. In the circuit of Fig. 29-3, use the superposition theorem to calculate the value of the current flowing through the 1-kΩ resistor.

29-2. In the circuit of Fig. 29-4, use the superposition theorem to determine the value of the current flowing through the 230-kΩ resistor.

29-3. In the circuit of Fig. 29-5, use the superposition theorem to find the current flowing through the 5-kΩ resistor.

29-4. In the circuit of Fig. 29-6, use the superposition theorem to find the potential at point *N*.

29-5. In the circuit of Fig. 29-7, use the superposition theorem to find the value of current *I*.

29-6. In the circuit of Fig. 29-8, use the superposition theorem to find the value of the current flowing through the 4.7-kΩ resistor.

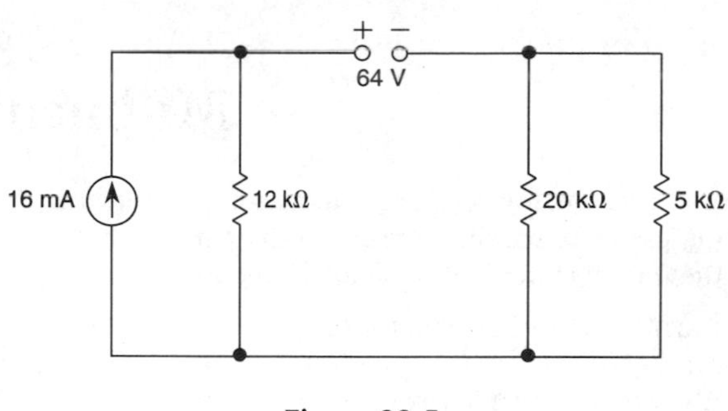

Figure 29-5

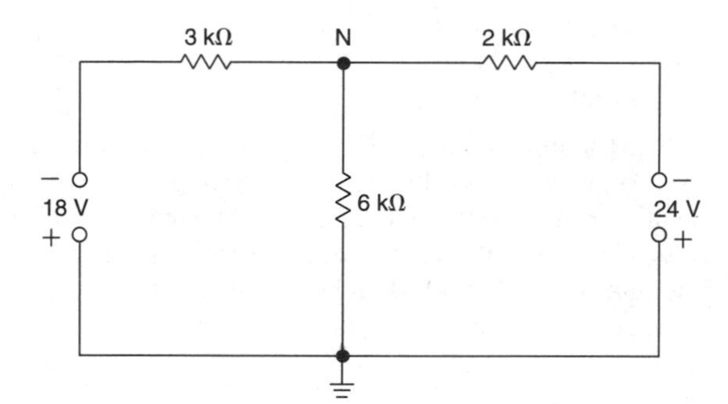

Figure 29-6

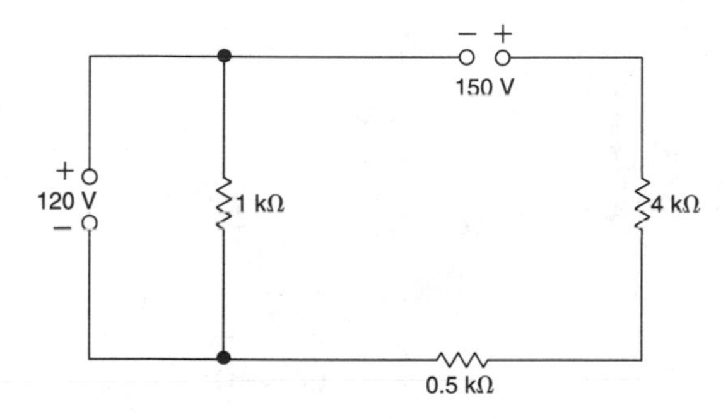

Figure 29-3

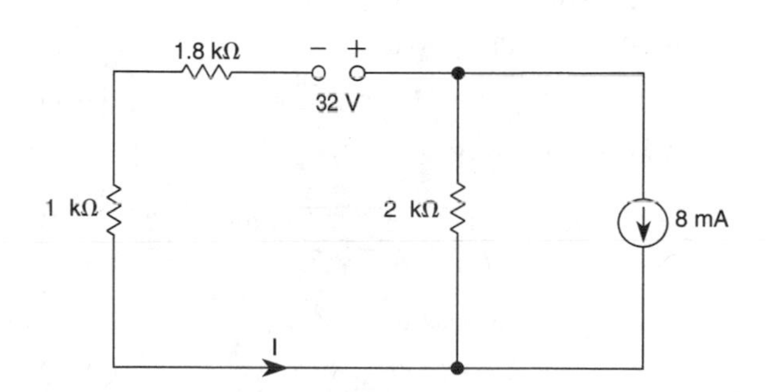

Figure 29-7

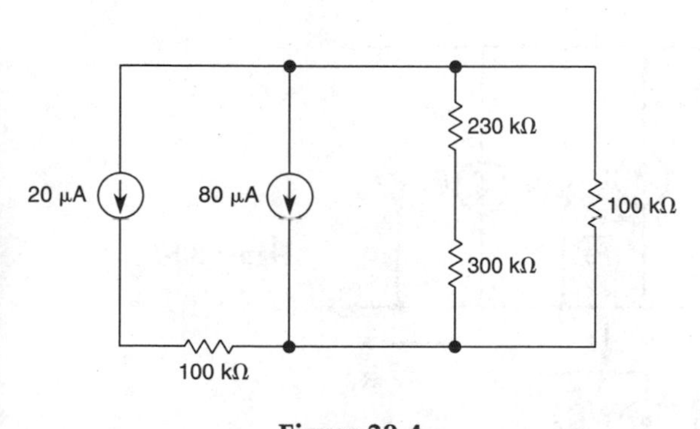

Figure 29-4

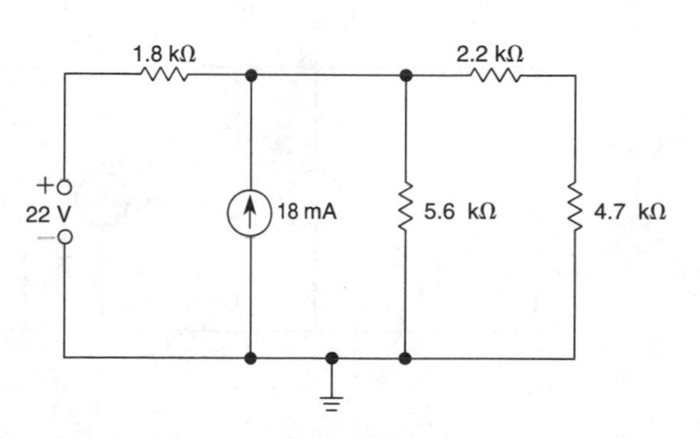

Figure 29-8

30
Millman's theorem

You have already used Millman's theorem when analyzing practical parallel sources in chapter 25. However, the formal statement of Millman's theorem is:

Any number of constant current sources that are directly connected in parallel can be converted into a single current source whose total generator current is the algebraic sum of the individual source currents, and whose total internal resistance is the result of combining the individual source resistances in parallel.

The theorem actually refers only to current sources, but it may also be applied to a mixture of parallel voltage and current sources (Fig. 30-1A) by converting all voltage sources into their constant current equivalents. The final generator can then be shown either as a constant current, or as a constant voltage, source.

MATHEMATICAL DERIVATIONS

In Fig. 30-1B

Total generator current,

$$I_T = +(+I_1) + (+I_2) + (-I_3) \text{ amperes} \qquad (30\text{-}1)$$

Total internal resistance,

$$R_{iT} = \frac{1}{\frac{1}{R_{i1}} + \frac{1}{R_{i2}} + \frac{1}{R_{i3}}} \text{ ohms} \qquad (30\text{-}2)$$

Total generator voltage,

$$E_T = I_T \times R_{iT} \text{ volts} \qquad (30\text{-}3)$$

Example 30-1

In Fig. 30-2A, calculate the value of the potential at the point N. Note: This is the same problem that was previously solved by three simultaneous equations in Example 26-1 (Kirchhoff's laws), by two simultaneous equations in Example 27-1 (mesh analysis), and by one equation in Example 28-1 (nodal analysis).

Solution

Convert the two constant voltage sources into their equivalent constant current sources (see Fig. 30-2B).

Total generator current,

$$I_T = +(+7 \text{ mA}) + (+3 \text{ mA}) \qquad (30\text{-}1)$$
$$= 10 \text{ mA}$$

Total internal resistance,

$$R_{iT} = \frac{1}{\frac{1}{2} + \frac{1}{3}} \qquad (30\text{-}2)$$
$$= 1.2 \text{ k}\Omega$$

Total resistance of the circuit,

$$1.2 \text{ k}\Omega \parallel 7 \text{ k}\Omega = \frac{1.2 \times 7}{1.2 + 7}$$
$$= \frac{8.4}{8.2}$$
$$= 1.024 \text{ k}\Omega$$

Potential at the point N,

$$= +1.024 \text{ k}\Omega \times 10 \text{ mA}$$
$$= +10.24 \text{ V.}$$

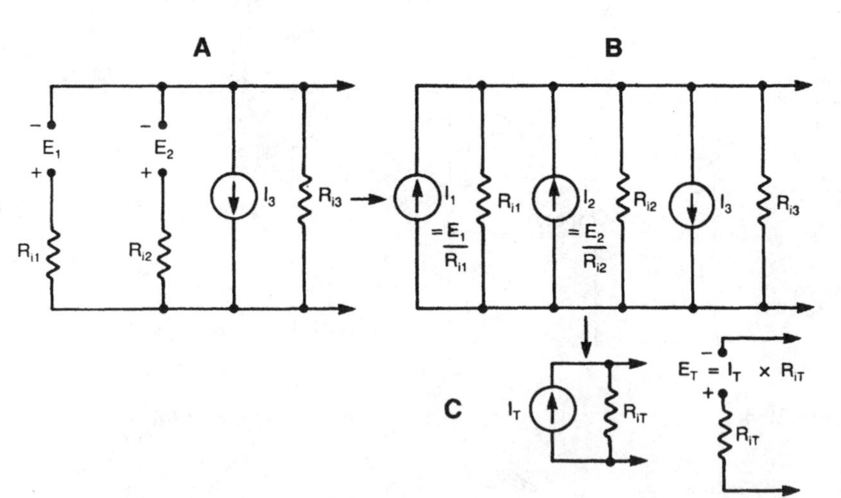

Figure 30-1

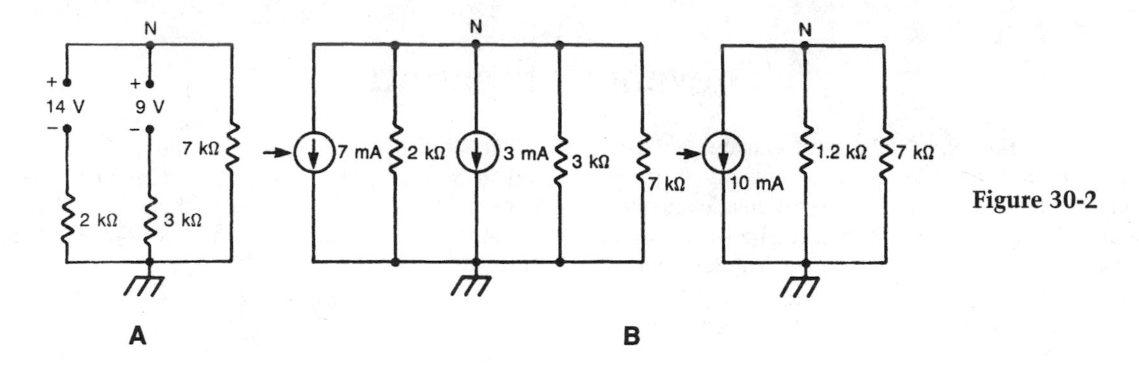

Figure 30-2

Notice that the use of Millman's theorem does not even require the solution of one linear equation.

PRACTICE PROBLEMS

30-1. In the circuit of Fig. 30-3, use Millman's theorem to obtain the constant current generator to the left of points X and Y.

30-2. In the circuit of Fig. 30-4, use Millman's theorem to combine the sources to the left of points X and Y into a single constant current generator.

30-3. In the circuit of Fig. 30-5, use Millman's theorem to reduce, to a single constant current generator, the circuit to the left of points X and Y.

30-4. In the circuit of Fig. 30-6, use Millman's theorem to reduce the circuit to a single constant voltage generator that is connected between points A and B.

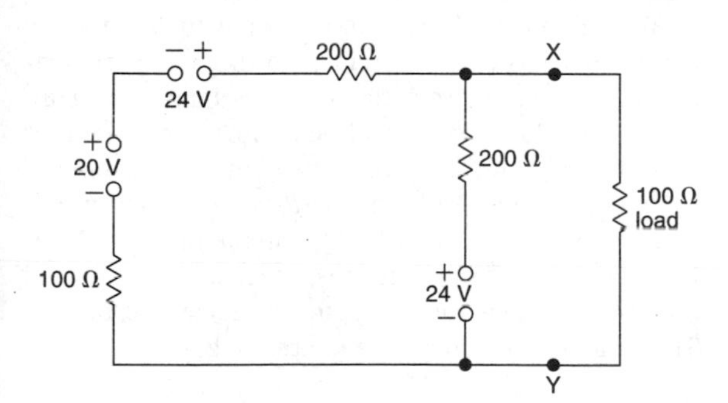

Figure 30-3

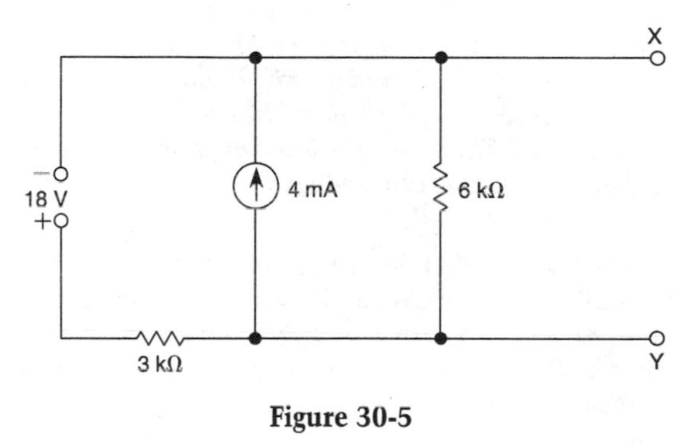

Figure 30-5

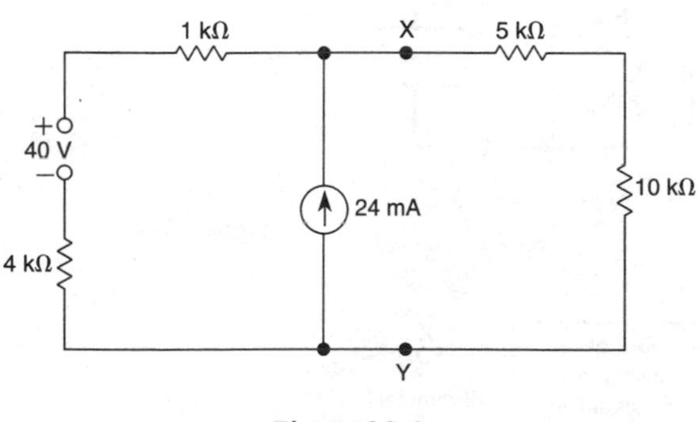

Figure 30-4

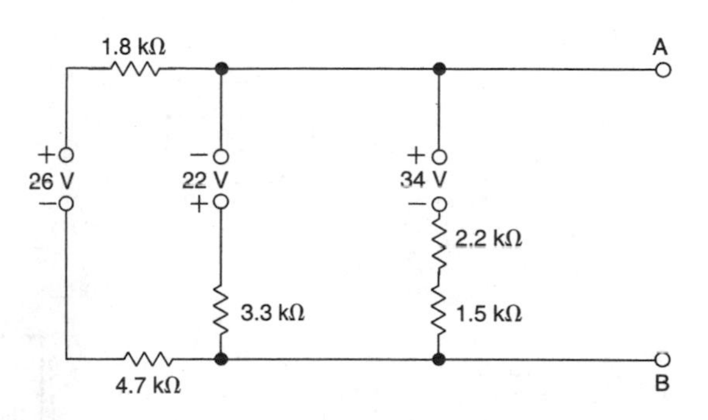

Figure 30-6

31
Thévenin's theorem

Thévenin's theorem is the most valuable analytical tool when dealing with complex networks. It enables you to focus your attention on a particular component, or any circuit part that is connected between two terminals, and is regarded as the load. The remainder of the circuit is then represented by a single generator with a constant voltage (E_{TH}) which is in series with an internal resistance, R_{TH}. The purpose of Thévenin's theorem is then to obtain the values of E_{TH} and R_{TH} from the components and sources in the original circuit. Stated formally, Thévenin's theorem is:

The current in a load connected between two output terminals (X and Y) of a network of resistors and electrical sources (Fig. 31-1A), is no different than if that same load were connected across a simple constant voltage generator (Fig. 31-1B) whose EMF (E_{TH}) is the open circuit voltage measured between X and Y (Fig. 31-1C), and whose series internal resistance (R_{TH}), is the resistance of the network looking back into terminals X and Y with all sources replaced by resistances equal to their internal resistances (Fig. 31-1D). This last step involves short-circuiting all sources of constant voltage and open-circuiting all sources of constant current.

The process of Thévenizing the circuit is illustrated in Fig. 31-1. As an example, Thévenize the same circuit (Fig. 31-2A), which has already been explored with Kirchhoff's laws, mesh analysis, nodal analysis, and Millman's theorem.

The 7-kΩ resistor is regarded as the load. When this load is removed, consider that a voltmeter is connected between X and Y to record the value of E_{TH}. In loop $ABCDA$, the sources are in series-opposing and the current (I) is

$$\frac{14\text{ V} - 9\text{ V}}{2\text{ k}\Omega + 3\text{ k}\Omega} = 1\text{ mA}$$

in the direction shown.

The value of E_{TH} can be thought of as either 14 V − (1 mA × 2 kΩ) = 12 V or 9 V + (1 mA × 3 kΩ) = 12 V (Fig. 31-2B).

To find the value of R_{TH}, the 14-V and 9-V constant voltage sources are shorted out and we visualize that an ohmmeter is connected between X and Y (Fig. 31-2C). The ohmmeter reading will then be 2 kΩ || 3 kΩ = (2 × 3)/(2 + 3) = 1.2 kΩ. The load is now replaced across the Thévenin equivalent generator (Fig. 31-2D), and by the VDR, the potential at X will be +12 V × 7 kΩ/(1.2 kΩ + 7 kΩ) = +10.24 V.

You are probably not impressed with the performance of Thévenin's theorem in solving this particular problem. The solution certainly seemed much more involved than the Millman treatment. However, with other types of problems, Thévenin's theorem has a number of advantages; nowhere is this better shown than in the case of the unbalanced bridge circuit of Fig. 31-3A. Using mesh or nodal analysis, such a circuit would require three algebraic simultaneous equations, but the Thévenin method does not require algebra at all.

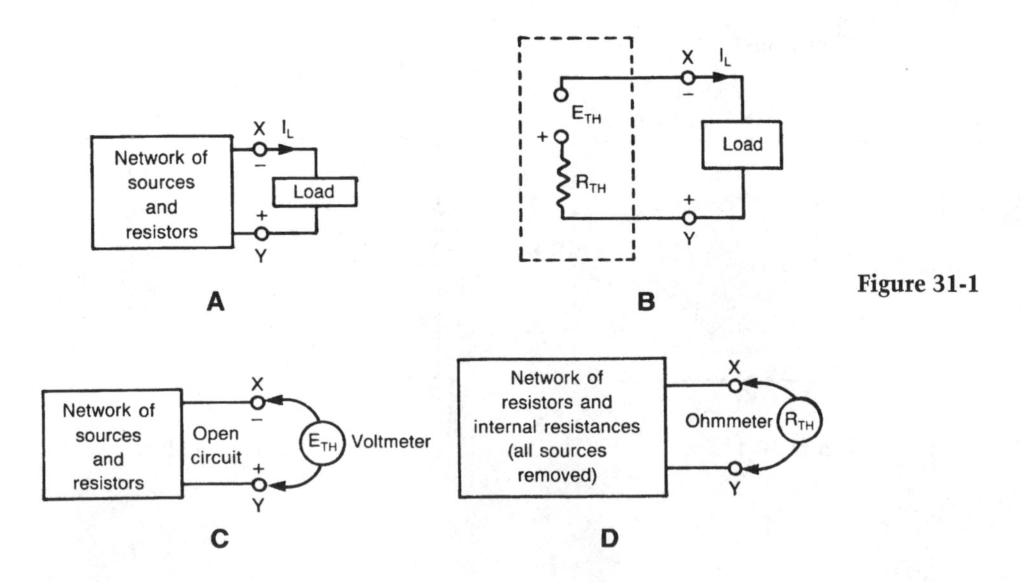

Figure 31-1

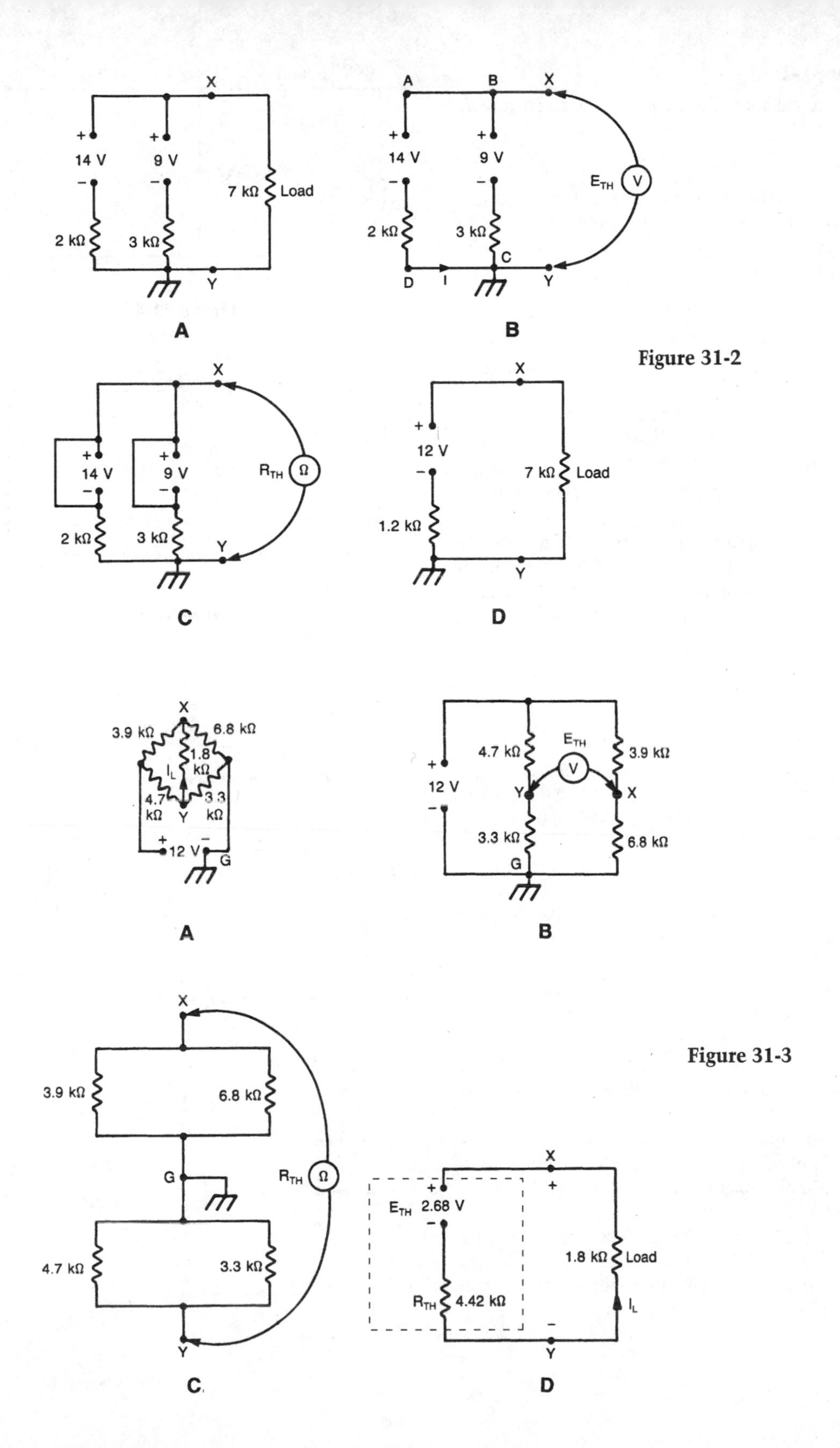

Figure 31-2

Figure 31-3

Example 31-1

In Fig. 31-3A, calculate the value of the load current, I_L.

Solution

Regard the 1.8-kΩ resistor as the load. When this load is removed and replaced by a voltmeter, the remaining circuit is as shown in Fig. 31-3B. By the voltage division rule: The potential at Y is:

$$+12\text{ V} \times \frac{3.3\text{ k}\Omega}{3.3\text{ k}\Omega + 4.7\text{ k}\Omega} = +4.95\text{ V}.$$

The potential at X is:

$$+12\text{ V} \times \frac{6.8\text{ k}\Omega}{6.8\text{ k}\Omega + 3.9\text{ k}\Omega} = +7.63\text{ V}.$$

Consequently, the value of E_{TH} is $+(7.63 - 4.95) = +2.68$ V (terminal X is positive with respect to the terminal Y).

When a short is placed across the 12-V source, the equivalent circuit is shown in Fig. 31-3C. The value of R_{TH} is the series combination of 3.9 kΩ || 6.8 kΩ and 3.3 kΩ || 4.7 kΩ. Using the "product-over-sum" formula,

$$R_{TH} = \frac{3.9 \times 6.8}{3.9 + 6.8} + \frac{3.3 \times 4.7}{3.3 + 4.7} = 2.48 + 1.94$$

$$= 4.42\text{ k}\Omega$$

The 1.8-kΩ load resistor is now replaced across the equivalent Thévenin generator (Fig. 31-3D). The load current, $I_L = 2.68\text{ V}/(4.42\text{ k}\Omega + 1.8\text{ k}\Omega) = 0.43$ mA.

PRACTICE PROBLEMS

31-1. Thévenize the circuit of Fig. 31-4 between points X and Y by regarding the 10-kΩ resistance as the load.

31-2. Thévenize the circuit of Fig. 31-5 between points X and Y by regarding the 40-kΩ resistance as the load.

31-3. Thévenize the circuit of Fig. 31-6 between points X and Y by regarding the 10-kΩ resistance as the load.

31-4. In the common emitter amplifier circuit of Fig. 31-7, obtain the equivalent Thévenin generator for the bias circuit to the *left* of points X and Y.

31-5. In Fig. 31-8, regard the 4.7-kΩ resistor as the load, and obtain the values of E_{TH} and R_{TH} for the equivalent Thévenin generator that is connected between points X and Y.

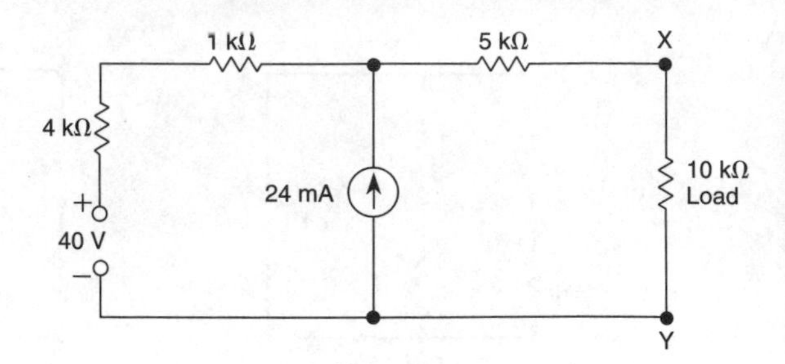

Figure 31-4

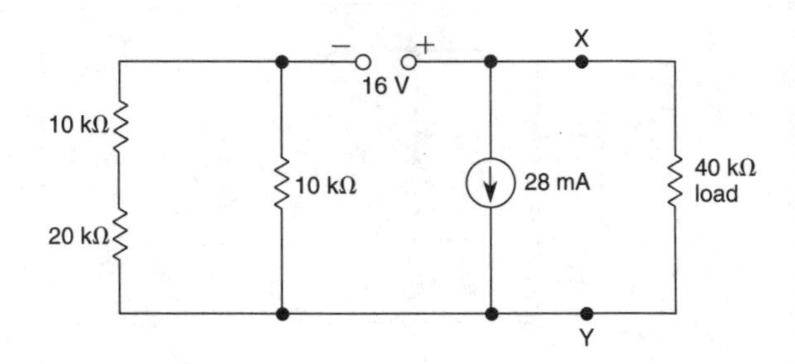

Figure 31-5

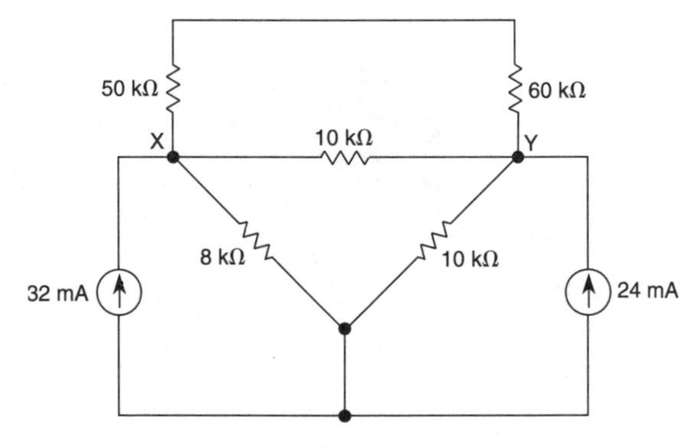

Figure 31-6

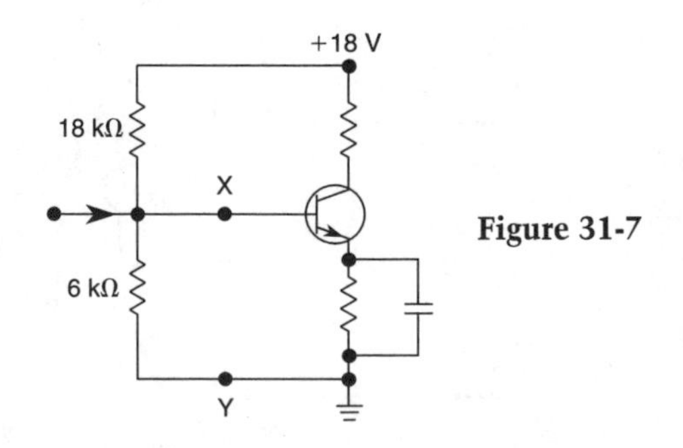

Figure 31-7

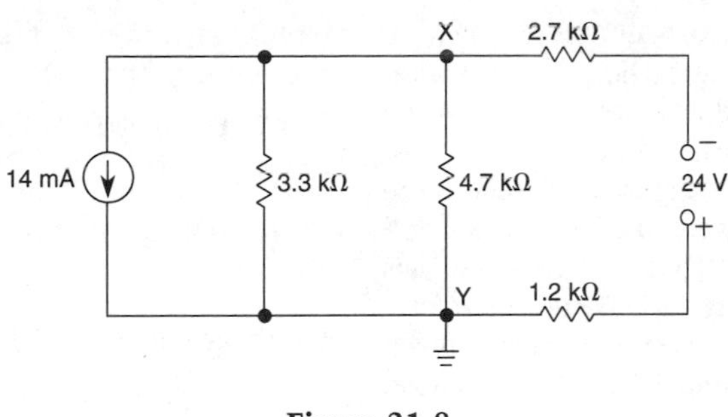

Figure 31-8

32
Norton's theorem

A formal statement of Norton's theorem is: *The current in a load connected between two output terminals, X and Y, of a complex network containing electrical sources and resistors, is no different than if that same load were connected to a* ***constant current*** *source, whose generator current (I_N) is equal to the* ***short-circuit*** *current measured between X and Y. This constant current generator is placed in parallel with a resistance (R_N), which is equal to the resistance of the network looking back into the terminals X and Y with all sources replaced by resistances equal to their internal resistances.*

This last step would involve short-circuiting all sources of constant voltage, and open-circuiting all sources of constant current. Figures 32-1A, B, C, and D illustrate the various steps that are taken in Nortonizing a circuit.

It is clear that Norton's Theorem involves a constant current generator, but Thévenin's theorem reduces to a constant voltage source. You can readily convert between the two generators by using the results of chapter 24. For example, the series Thévenin resistance and the parallel Norton resistance have the same value.

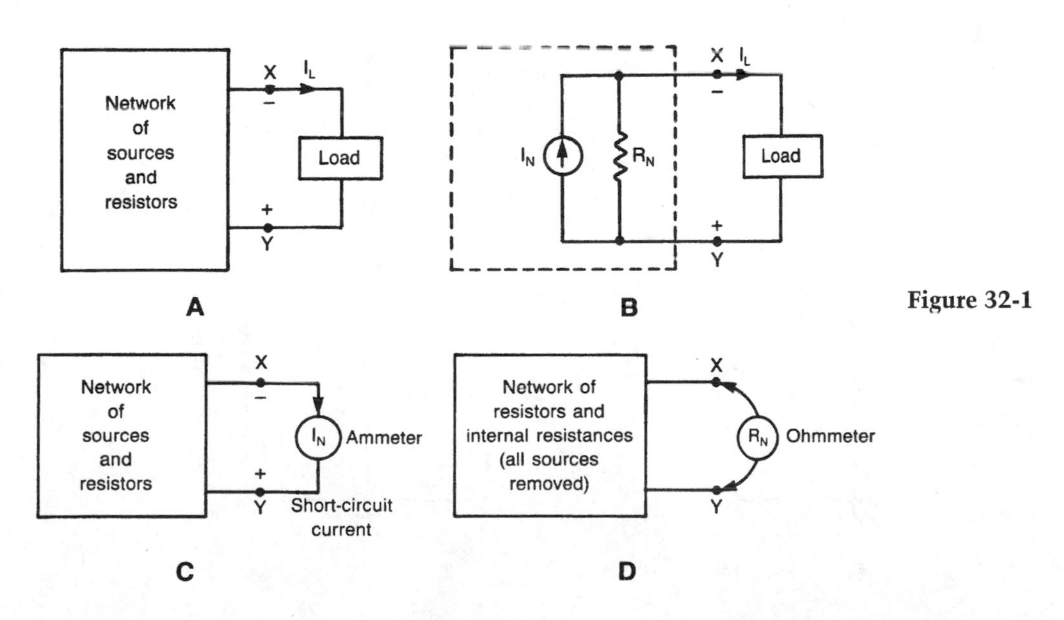

Figure 32-1

We will now Nortonize the same circuit to which we have applied all other analytical methods. The 7-kΩ resistor is regarded as the load (Fig. 32-2A). When the load is removed, we consider that an ammeter is connected between X and Y. This ammeter will record the Norton current (I_N) which is equal to 14 V/2 kΩ + 9 V/3 kΩ = 10 mA (Fig. 32-2B). The Norton resistance is (2 × 3)/(2 + 3) = 1.2 kΩ (Fig. 32-2C) and, therefore, the equivalent constant current generator is as shown in Fig. 32-2D. Using the current division rule, the load current:

$$I_L = 10 \text{ mA} \times \frac{1.2 \text{ k}\Omega}{1.2 \text{ k}\Omega + 7 \text{ k}\Omega} = 1.4634 \text{ mA}$$

and the potential at X is +(1.4634 mA × 7 kΩ) = +10.24 V.

For this type of problem there is little to choose between the Thévenin and Norton methods of analysis.

Example 32-1

In Fig. 32-3, calculate the value of the load current, I_L.

Solution

Regard the 1.8-kΩ resistor as the load. When this load is removed and replaced by an ammeter, the remaining circuit as shown in Fig. 32-3B. The total resistance presented to the 12-V source is:

$$4.7 \text{ k}\Omega \parallel 3.9 \text{ k}\Omega + 3.3 \text{ k}\Omega \parallel 6.8 \text{ k}\Omega = 2.13 + 2.22 = 4.35 \text{ k}\Omega$$

so that the current I is:

$$12 \text{ V}/4.35 \text{ k}\Omega = 2.76 \text{ mA}$$

Using the current division rule,

$$I_1 = 2.76 \text{ mA} \times 3.3 \text{ k}\Omega/(3.3 \text{ k}\Omega + 6.8 \text{ k}\Omega) = 0.90 \text{ mA}$$

while,

$$I_2 = 2.76 \text{ mA} \times 4.7 \text{ k}\Omega/(4.7 \text{ k}\Omega + 3.9 \text{ k}\Omega) = 1.51 \text{ mA}.$$

Therefore, I_N = 1.51 − 0.90 = 0.61 mA with the electron flow in the direction of the terminal Y, to the terminal X.

The Norton resistance (R_N) is the same as the Thévenin resistance (R_{TH}) and is equal to 4.42 kΩ. With the equivalent Norton generator as shown in Fig. 32-3D, the load current

$$I_L = 0.61 \text{ mA} \times \frac{4.42 \text{ k}\Omega}{1.8 \text{ k}\Omega + 4.42 \text{ k}\Omega} = 0.43 \text{ mA}$$

This is the same result as was obtained with Thévenin's theorem in chapter 31.

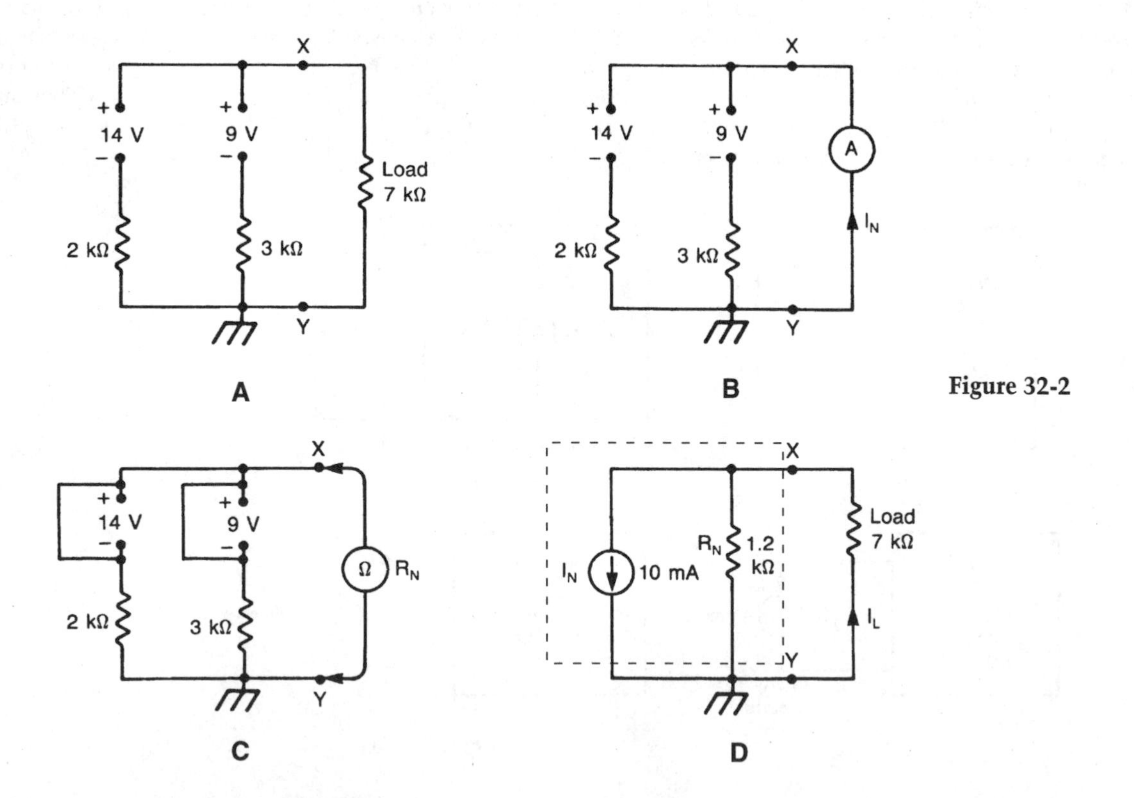

Figure 32-2

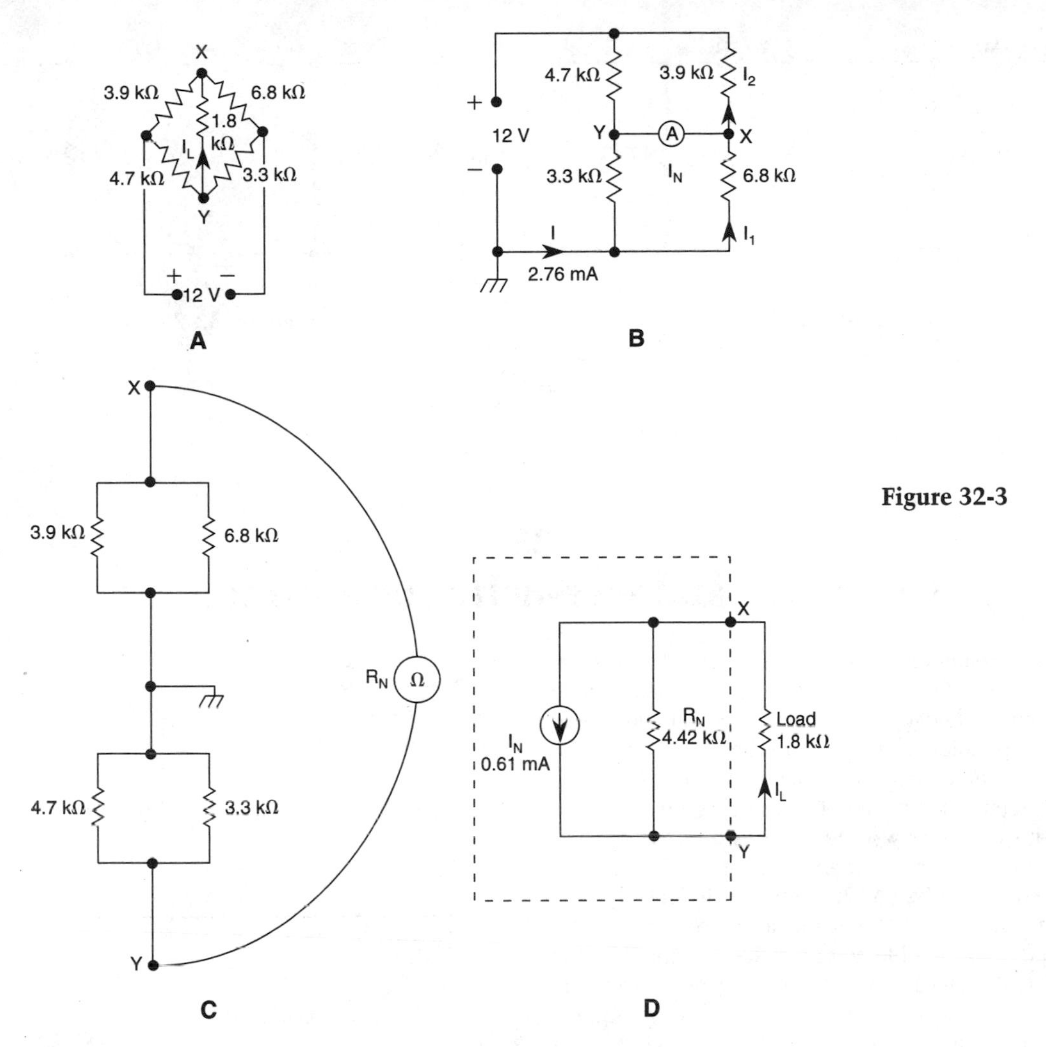

Figure 32-3

PRACTICE PROBLEMS

32-1. Nortonize the circuit of Fig. 32-4 between points X and Y.

32-2. In the circuit of Fig. 32-5, obtain the equivalent Norton generator to the left of points X and Y.

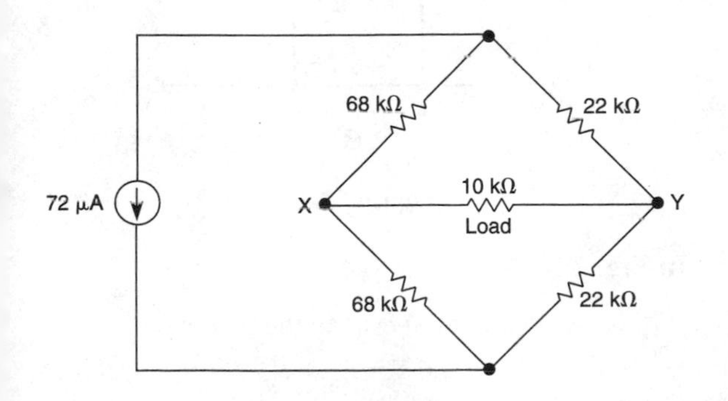

Figure 32-4

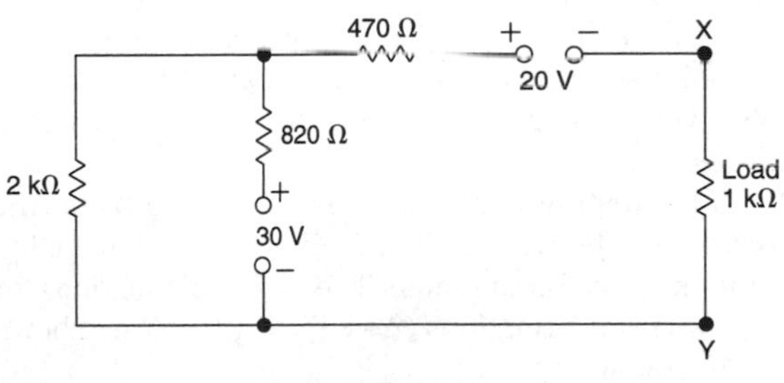

Figure 32-5

32-3. In the circuit of Fig. 32-6, calculate the values of I_N and R_N for the equivalent Norton generator between points X and Y.

32-4. In Fig. 32-7, calculate the values of I_N and R_N for the equivalent Norton generator between points X and Y.

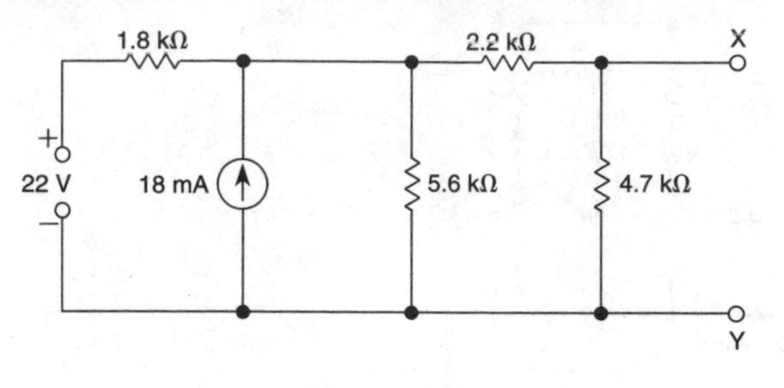

Figure 32-6

−   +   500 Ω
X
8 V
2 mA
1 kΩ
200 Ω
Y

Figure 32-7

33
Delta-wye and wye-delta transformations

We have already examined the unbalanced bridge circuit, which cannot be regarded as a series-parallel arrangement, because the center arm bears no simple series or parallel relationship to the other four resistors. However, on many occasions, it is possible to convert such circuits into simpler arrangements by means of a delta-wye (or a wye-delta) transformation.

A triangular arrangement of resistors (Fig. 33-1A) is known as a *delta* (Δ, the Greek capital letter) connection. However, the same arrangement can be changed to resemble another Greek capital letter (pi) Π (Fig. 33-1B) so that the two names are interchangeable. The same applies to a second configuration, which is referred to as a *wye* (letter *Y*) *connection* (Fig. 33-2A). This can also be modified to look like the letter tee (*T*) so that either tee (Fig. 33-2B) or wye is used to describe identical networks of resistors.

If the delta-wye networks are to be equivalent, the resistances between points, *X*, *Y*, and *Z* in both configurations must be the same. The required equations for the two transformations ($\Delta \rightarrow Y$ and $Y \rightarrow \Delta$) are shown in the following:

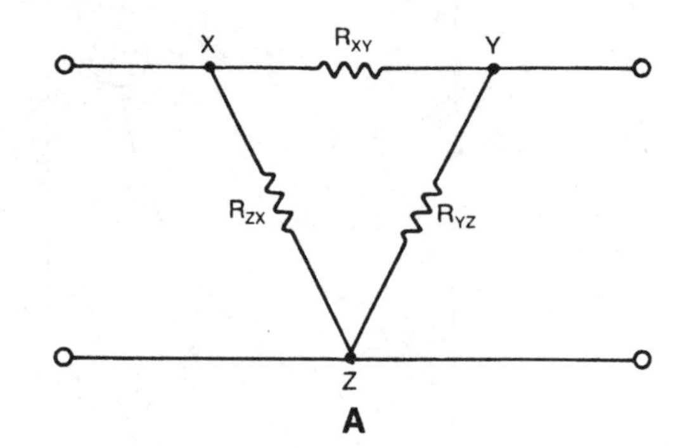

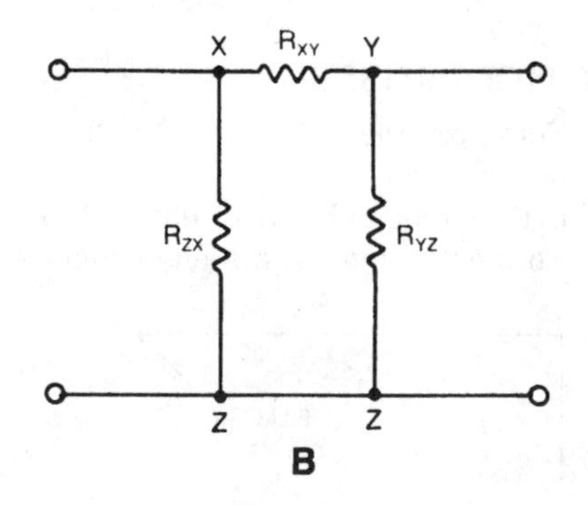

Figure 33-1

MATHEMATICAL DERIVATIONS
In Fig. 33-1A

Resistance between the points *X* and *Y*,

$$\frac{R_{XY} \times (R_{YZ} + R_{ZX})}{R_{XY} + (R_{YZ} + R_{ZX})} \qquad (33\text{-}1)$$

In Fig. 33-2A

Resistance between the points

$$X \text{ and } Y = R_X + R_Y \qquad (33\text{-}2)$$

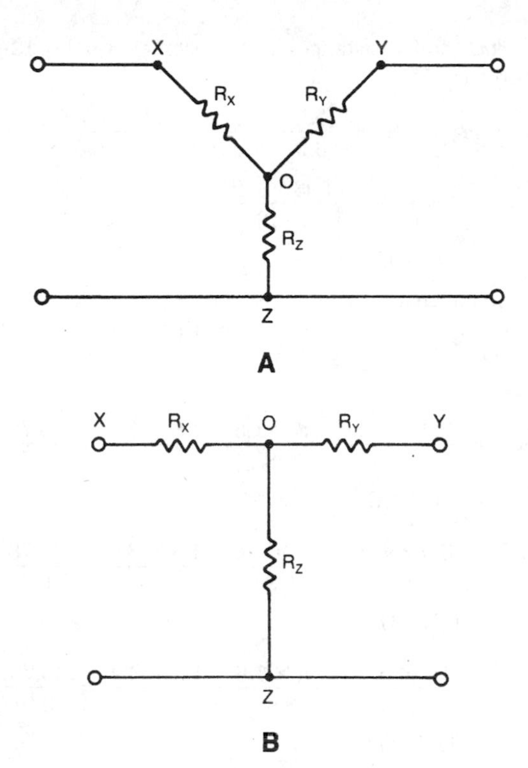

Figure 33-2

Therefore,

$$R_X + R_Y = \frac{R_{XY}(R_{YZ} + R_{ZX})}{R_{XY} + R_{YZ} + R_{ZX}} \qquad (33\text{-}3)$$

By symmetry,

$$R_Y + R_Z = \frac{R_{YZ}(R_{ZX} + R_{XY})}{R_{XY} + R_{YZ} + R_{ZX}} \qquad (33\text{-}4)$$

$$R_Z + R_X = \frac{R_{ZX}(R_{XY} + R_{YZ})}{R_{XY} + R_{YZ} + R_{ZX}} \qquad (33\text{-}5)$$

Add Equations 33-3 and 33-5. From their sum, subtract Equation 33-4 and divide the result by 2. This yields:

$$R_X = \frac{R_{XY}R_{ZX}}{R_{XY} + R_{YZ} + R_{ZX}} \qquad (33\text{-}6)$$

Similarly,

$$R_Y = \frac{R_{YZ}R_{XY}}{R_{XY} + R_{YZ} + R_{ZX}} \qquad (33\text{-}7)$$

and,

$$R_Z = \frac{R_{ZX}R_{YZ}}{R_{XY} + R_{YZ} + R_{ZX}} \qquad (33\text{-}8)$$

Equations 33-6, 33-7, 33-8 are used to achieve a $\Delta \rightarrow Y$ transformation. For the $Y \rightarrow \Delta$ transformation, the following three equations can be derived:

$$R_{XY} = \frac{R_XR_Y + R_YR_Z + R_ZR_X}{R_Z} \qquad (33\text{-}9)$$

$$R_{YZ} = \frac{R_XR_Y + R_YR_Z + R_ZR_X}{R_X} \qquad (33\text{-}10)$$

$$R_{ZX} = \frac{R_XR_Y + R_YR_Z + R_ZR_X}{R_Y} \qquad (33\text{-}11)$$

These transformations can be used to advantage with the analysis of the unbalanced bridge circuit. In the following analysis, choose the same circuit, which has already been solved with the aid of Thévenin's and Norton's theorems.

Example 33-1

In Fig. 33-3A, calculate the value of the total equivalent resistance presented to the 12-V source.

Solution

In Fig. 33-3A, regard X, Y, Z as the corners of a delta formation of resistors, then convert this arrangement into its wye equivalent. Then,

$$R_X = \frac{R_{XY}R_{ZX}}{R_{XY} + R_{YZ} + R_{ZX}} \qquad (33\text{-}6)$$

$$= \frac{1.8 \times 6.8}{3.3 + 6.8 + 1.8}$$

$$= 1.03\ \text{k}\Omega$$

$$R_Y = \frac{R_{XY}R_{YZ}}{R_{XY} + R_{YZ} + R_{ZY}} \qquad (33\text{-}7)$$

$$= \frac{1.8 \times 3.3}{3.3 + 6.8 + 1.8}$$

$$= 0.50\ \text{k}\Omega$$

$$R_Z = \frac{R_{XZ}R_{YZ}}{R_{XY} + R_{YZ} + R_{ZX}} \qquad (33\text{-}8)$$

$$= \frac{6.8 \times 3.3}{3.3 + 6.8 + 1.8}$$

$$= 1.89\ \text{k}\Omega$$

The circuit has now been transformed to the relatively simple series-parallel arrangement of Fig. 33-3B.

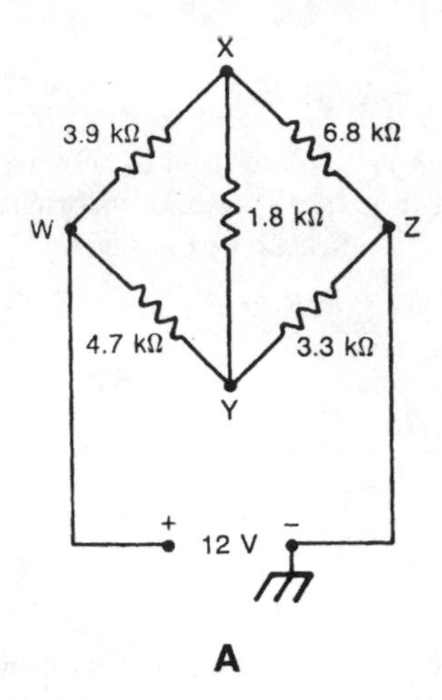

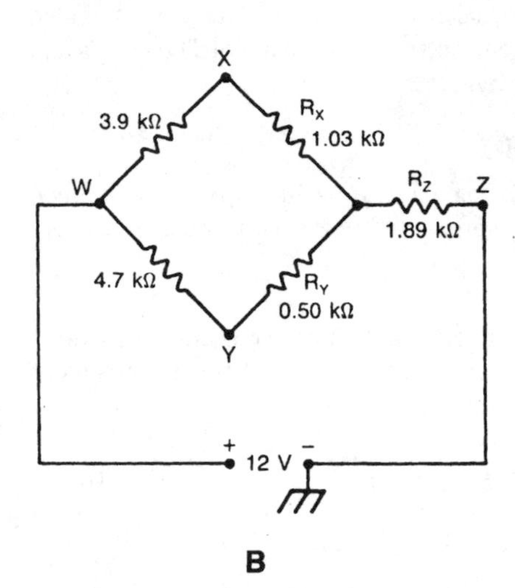

Figure 33-3

The total resistance (R_T) presented to the 12-V source is:

$$R_T = 1.89 + \frac{(3.9 + 1.03) \times (4.7 + 0.5)}{(3.9 + 1.03) + (4.7 + 0.5)}$$

$$= 1.89 + \frac{4.93 \times 5.2}{10.13}$$

$$= 4.42 \text{ k}\Omega$$

An alternative solution is to regard X as the center of a wye formation and convert this formation to its delta equivalent (Fig. 33-4A).
Then,

$$R_{WZ} = \frac{3.9 \times 6.8 + 6.8 \times 1.8 + 1.8 \times 3.9}{1.8} = \frac{45.78}{1.8}$$

$$= 25.43 \text{ k}\Omega$$

$$R_{WY} = \frac{3.9 \times 6.8 + 6.8 \times 1.8 + 1.8 \times 3.9}{6.8} = \frac{45.78}{6.8}$$

$$= 6.73 \text{ k}\Omega$$

$$R_{YZ} = \frac{3.9 \times 6.8 + 6.8 \times 1.8 + 1.8 \times 3.9}{3.9} = \frac{45.78}{3.9}$$

$$= 11.74 \text{ k}\Omega$$

The circuit then reduces to the schematic of Fig. 33-4B.

Total equivalent resistance (R_T) presented to the 12-V source is:

$$R_T = \frac{25.43 \times (2.76 + 2.57)}{25.43 + (2.76 + 2.57)}$$

$$= 4.41 \text{ k}\Omega$$

PRACTICE PROBLEMS

33-1. Convert the delta network (Fig. 33-5) into its equivalent wye circuit.

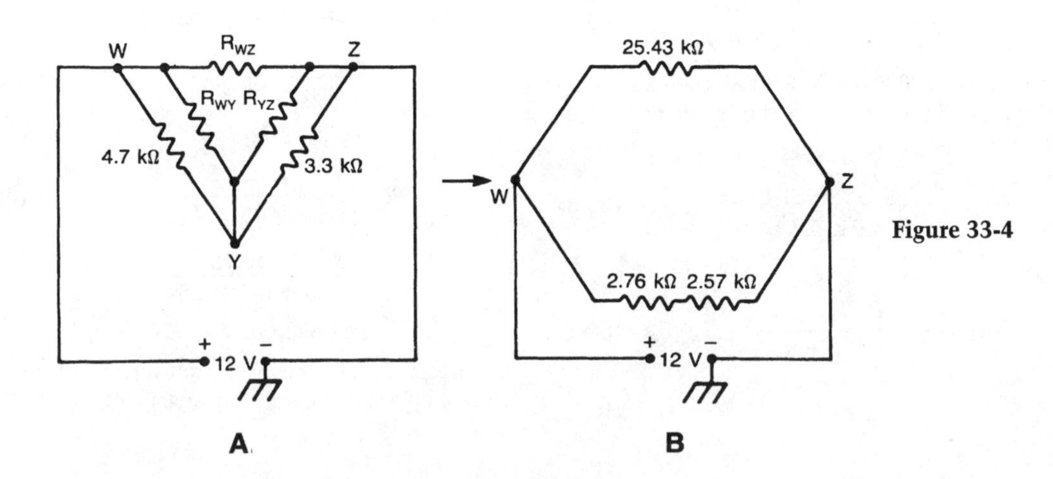

Figure 33-4

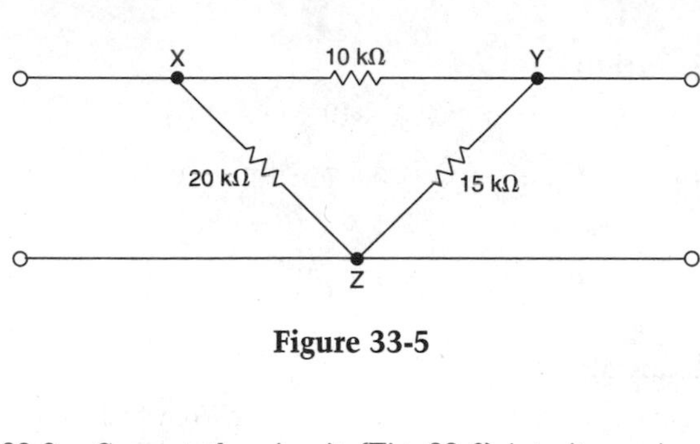

Figure 33-5

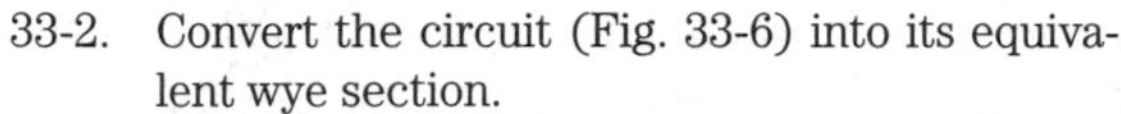

33-2. Convert the circuit (Fig. 33-6) into its equivalent wye section.

33-3. In Fig. 33-7, convert the circuit between the points X, Y, and Z into its wye equivalent.

33-4. Reduce the circuit of Fig. 33-8 to a single delta formation, and calculate the equivalent resistance values of R_{XY}, R_{YZ}, and R_{ZX}.

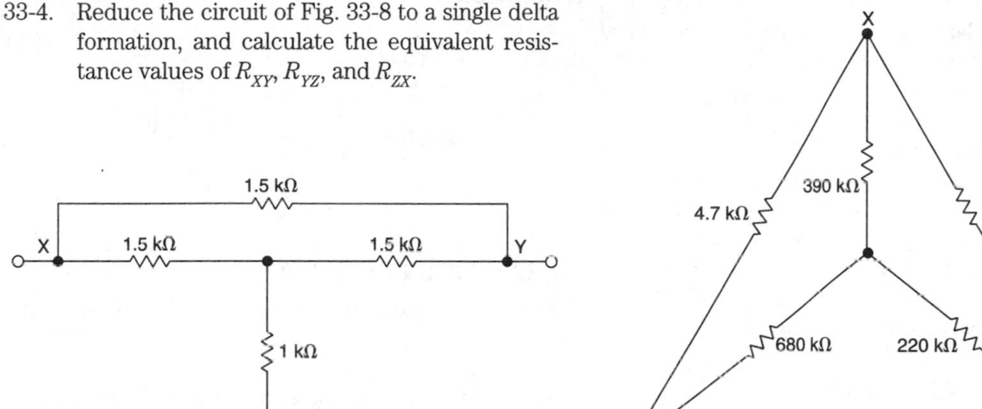

Figure 33-6

X
10 kΩ
Y
15 kΩ
10 kΩ
15 kΩ
10 kΩ
Z
Z

Figure 33-7

X
390 kΩ
4.7 kΩ
3.3 kΩ
680 kΩ
220 kΩ
Y
Z
2.7 kΩ

Figure 33-8

34
Magnetic flux—magnetic flux density

Figure 34-1 shows a toroidal soft iron ring, on which is wound a coil of N turns. A current of I amperes flows through the coil so that the ring is magnetized and a *magnetic flux* is established in the iron.

The flux is represented by magnetic lines of force. These are the directions in which isolated north poles would travel in a magnetic field. Such lines form closed loops, cannot intersect, and mutually repel one another. The degree of concentration of the lines of force indicates the strength of the magnetic field.

In a solid magnetized ring, the flux lines would be continuous and no magnetic poles would exist. However, if a radial cut is made in the ring to create an air gap (Fig. 34-1), a north pole exists on one side of the gap and a south pole exists the other side.

In a magnetic field, the total number of lines is referred to as the magnetic flux, whose letter symbol is the Greek lower case letter phi (ϕ). In the SI system, the unit of magnetic flux is the *weber*, whose unit symbol is Wb. This unit is defined through the phenome-

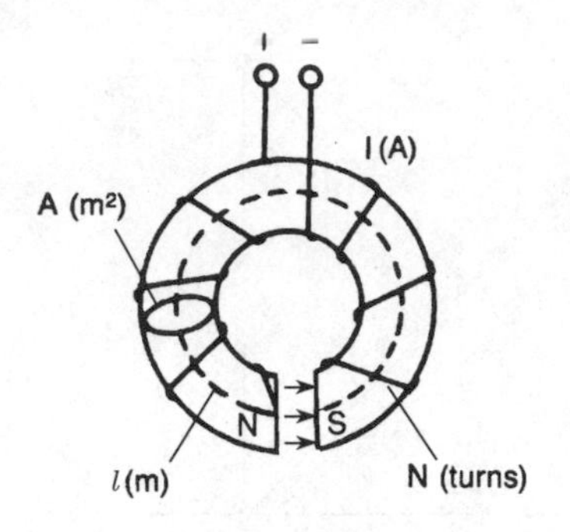

Figure 34-1

non of electromagnetic induction (chapter 38). If a conductor cuts a flux of 1 weber in one second, the voltage induced in the conductor is 1 volt. In practice, the weber is too large a unit for measuring a typical magnetic flux so that the more common units are the microweber (μWb) and the milliweber (mWb).

To indicate the actual strength of a magnetic field, we must have a unit to measure the concentration of the lines of force. Referring to Fig. 34-1, the concentration in the air gap will be directly proportional to the magnetic flux, ϕ, but inversely proportional to the circular cross-sectional area (A) of the radial cut. The ratio of ϕ (Wb) to A (m^2) is called the *flux density*, whose letter symbol is B and whose unit is the tesla (T).

MATHEMATICAL DERIVATIONS

$$\text{Flux density, } B = \frac{\phi}{A} \text{ teslas} \qquad (34\text{-}1)$$

$$\text{Flux, } \phi = BA \text{ webers} \qquad (34\text{-}2)$$

Example 34-1

A flux density of 0.6 T is constant over a total area of 2.5 cm². Calculate the value of the total flux.

Solution

Area, $A = 2.5 \text{ cm}^2 = 2.5 \times 10^{-4} \text{ m}^2$

$$\begin{aligned}\text{Total flux, } \phi = B \times A &= 0.6 \text{ T} \times 2.5 \times 10^{-4} \text{ m}^2 \quad (34\text{-}2)\\ &= 1.5 \times 10^{-4} \text{ Wb}\\ &= 150\ \mu\text{Wb.}\end{aligned}$$

Example 34-2

A magnetic field has a flux of 0.37 mWb, which is uniformly spread over an area of 4.3 cm². Calculate the value of the flux density.

Solution

$$\begin{aligned}\text{Area, } A &= 4.3 \text{ cm}^2 \qquad (34\text{-}1)\\ &= 4.3 \times 10^{-4} \text{ m}^2\\ \text{Flux density, } B &= \frac{\phi}{A} = \frac{0.37 \times 10^{-3} \text{ Wb}}{4.3 \times 10^{-4} \text{ m}^2}\\ &= 0.86 \text{ T.}\end{aligned}$$

PRACTICE PROBLEMS

34-1. A magnetic flux of 800 μWb is uniformly distributed over an area of 10 cm². Calculate the value of the flux density.

34-2. When a magnetic flux of 84 mWb is uniformly distributed over an area, the flux density is 1.2 T. Calculate the value of the area in square meters.

34-3. When a magnetic flux is uniformly distributed over an area of 15 cm², the flux density is 0.65 T. Calculate the value of the magnetic flux.

34-4. The cylindrical core of a coil has a flux of 18.5 mWb, and a flux density of 1.5 T. Calculate the length of the core's diameter.

35
The motor effect

Let an external voltage (E) drive an electron flow (I) through a conductor that is situated in a uniform magnetic field. In other words, electrical energy is being supplied so that a force can be exerted on the conductor. The result is the so-called *motor effect*, whereby electrical energy is converted into mechanical energy. The reverse effect occurs in a generator that changes mechanical energy into electrical energy.

Because it is carrying a direct current, the conductor will have its own surrounding magnetic field whose flux pattern consists of concentric circles, with the conductor at the center. The two fields will therefore interact to produce a resultant flux pattern. Below the conductor (Fig. 35-1A), the magnetic fields have the same direction and will therefore combine to produce a concentration of the flux lines. However, the fields above the conductor have opposite directions and, therefore, the flux lines are spread far apart. The resultant flux pattern is illustrated in Fig. 35-1B and because the flux lines mutually repel one another, there will be an upward force exerted on the conductor. If the conductor is free to accelerate, the result would be to equalize the distribution of the flux lines in the resultant field.

The direction of the force can be found by the right hand or motor rule, which is illustrated in Fig. 35-1C. The *first* finger indicates the direction of the uniform magnetic *field*, whose flux density is B teslas. The second finger is the direction of the *c*urrent (electron flow) while the thu*m*b shows the direction of the *mo*tion due to the force.

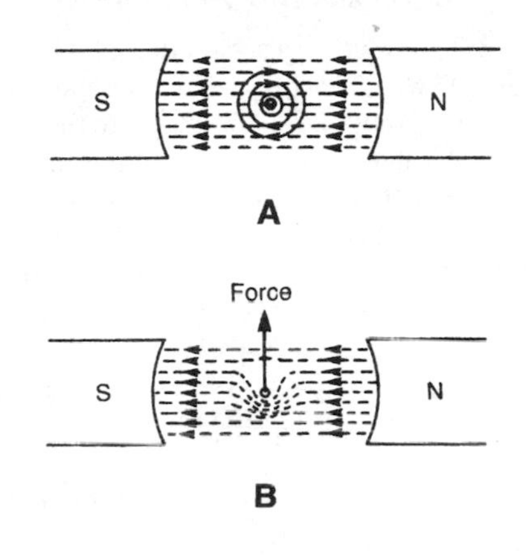

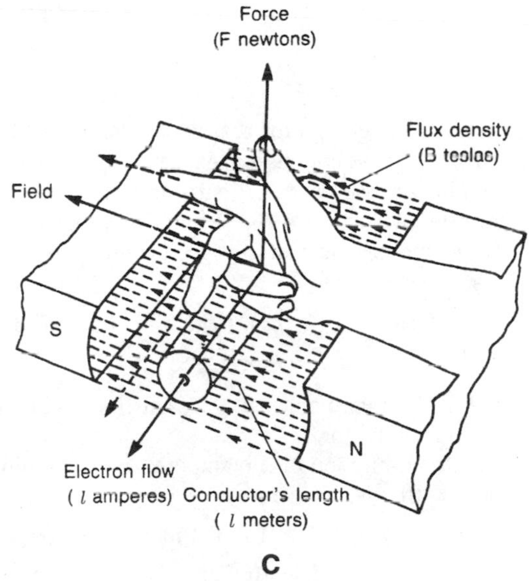

Figure 35-1

The principle, that we have illustrated, is used in electrical motors and the moving coil instrument; it is also the means by which an electrical beam is deflected in one type of cathode-ray tube (CRT).

MATHEMATICAL DERIVATIONS

The magnitude on the force (F) exerted on the conductor is directly proportional to the flux density (B teslas) of the uniform magnetic field, the current (I amperes) flowing through the conductor and its length, l meters. The equation is therefore:

$$\text{Force, } F = B \times I \times l \text{ newtons} \qquad (35\text{-}1)$$

EXAMPLE 35-1

A conductor of 20 cm length is placed at 90° to the direction of a uniform magnetic field whose flux density is 0.9 T. If the conductor is carrying a current of 4 A, what is the magnitude of the force exerted on the conductor?

Solution

$B = 0.9\ T, I = 4\text{ A}, l = 20\text{ cm} = 0.2\text{ m}.$

Force exerted on the conductor,

$$F = B \times I \times l \qquad (35\text{-}1)$$

$$= 0.9 \times 4 \times 0.2$$

$$= 0.72\text{ N}$$

PRACTICE PROBLEMS

35-1. A coil of wire consists of 20 rectangular turns whose dimensions are 25 cm long and 20 cm wide. The long dimension is at right angles to the direction of a magnetic field whose flux density is 0.6 T. If the current through the coil is 7 A, calculate the value in N-m of the torque which can be exerted on the coil about its axis of rotation.

35-2. A single copper loop has a length of 20 cm and a width of 10 cm. The loop carries a current of 8 A, and is situated in a uniform magnetic field whose flux density is 1.5 T. Calculate the maximum value of the torque exerted on the loop.

35-3. A current-carrying conductor of length 15 cm is placed at 90° to the direction of a uniform magnetic field whose flux density is 0.85 T. If the force exerted on the conductor is 1.1 N, calculate the value of the current flowing through the conductor.

36 Magnetomotive force—magnetic field intensity—permeability of free space

Consider a toroidal soft-iron ring that is wound with an exciting coil of N turns (Fig. 36-1). The ring is being magnetized by the flow of the current through the coil, and the strength of the magnetic field will therefore depend on the number of turns and the value of the current. Their product (IN in ampere-turns or amperes) is called the *magnetomotive force* (MMF) whose letter symbol is $\mathcal{F}$. The magnetomotive force is the magnetic equivalent of the electromotive force (EMF). The EMF *creates* a current I, the MMF *establishes* a flux ϕ. Flux in magnetism is, therefore, the equivalent of current in electricity, but with one important difference. A current flows around an electrical network, but a flux is merely established in a magnetic circuit and does not flow.

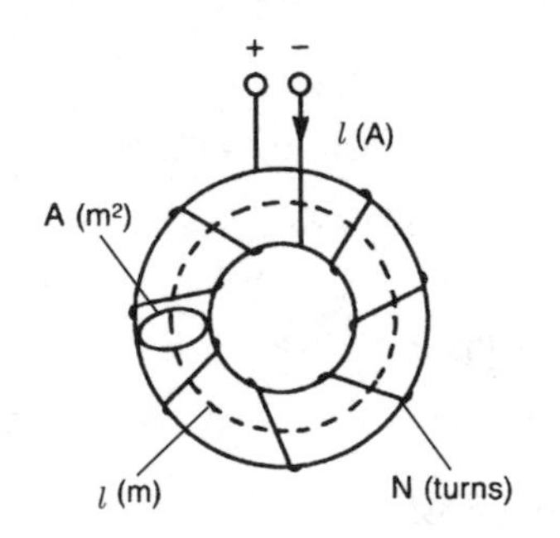

Figure 36-1

Let us assume that we keep the same values of the current (I), the number of turns (N) and the cross-sectional area (A), but the ring is enlarged by increasing its circumference (the average of the inner and outer circumferences). Clearly the turns will be spread further apart, and the flux density (B teslas) in the iron ring will be reduced. Consequently, the value of the flux density depends not only on the current (I) and the number of turns (N), but also on the length (l) of the magnetic path over which the flux is established. These three factors are contained in the *magnetic field intensity, H* (sometimes called the magnetizing force) which is equivalent to the MMF per unit length, and is therefore measured in ampere-turns per meter, or amperes per meter.

It follows that the magnetic field intensity is directly responsible for producing the flux density and that some relationship must exist between H and B.

MATHEMATICAL DERIVATIONS

Magnetomotive force (MMF),

$$\mathcal{F} = IN = H \times l \text{ ampere-turns (AT) or amperes (A)} \qquad (36\text{-}1)$$

Magnetizing force,

$$H = \frac{IN}{l} = \frac{\mathcal{F}}{l} \text{ ampere-turns per meter (AT/m) or amperes per meter (A/m)} \qquad (36\text{-}2)$$

Figure 36-2 shows the cross section of a long straight conductor X, which is situated in a vacuum and carries an electron current of 1 A, exiting from the paper. One of the circular lines of force has a radius of 1 m so that the length of its magnetic path is 2π meters. The MMF associated with this path is 1 A and the corresponding field intensity (H) is $1/(2\pi)$ ampere per meter.

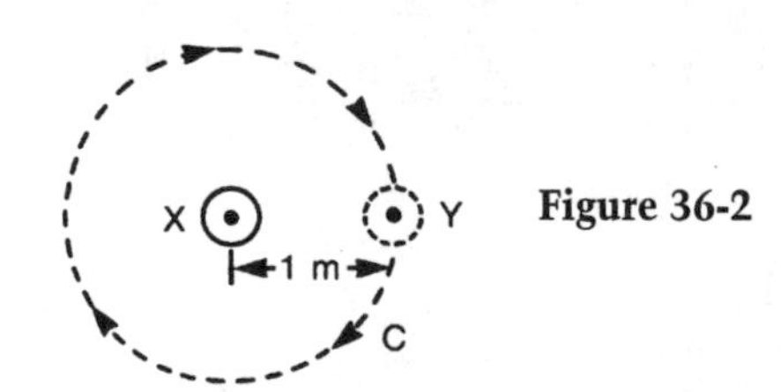

Figure 36-2

Y is a second conductor that is parallel to and 1 m away from the first conductor. Assuming that the second conductor also carries a current of 1 A, use the "motor rule" of Equation 35-1 to find the force exerted on the one-meter length of the second conductor. This force is:

$$F \text{ (newton)} = B \times I \times l \qquad (36\text{-}3)$$
$$= B \times 1 \text{ A} \times 1 \text{ m}$$

where B is the flux density in teslas at the position of the second conductor.

From the definition of the ampere, this force must also be 2×10^{-7} N. Therefore,

$$F = B \times 1 \text{ A} \times 1 \text{ m}$$
$$= 2 \times 10^{-7} \text{ N}$$

This yields,

$$B = 2 \times 10^{-7} \text{ T} \qquad (36\text{-}4)$$

Consequently, a field intensity (H) of $1/(2\pi)$ ampere per meter is responsible for a flux density of 2×10^{-7} T. Therefore, in a vacuum (free space), the ratio of B to H is:

$$\mu_o = \frac{B}{H} = \frac{2 \times 10^{-7}\ \text{T}}{1/(2\pi)\ \text{A/m}} = 4\pi \times 10^{-7} \qquad (36\text{-}5)$$

$$= 12.57 \times 10^{-7}\ \text{H/m}$$

In a vacuum the ratio of B to H is called the *permeability of free space*, whose letter symbol is μ_o and whose unit is the henry per meter. The term *permeability* measures the ability of a medium to permit the establishment of a flux density (B) by applying a magnetic field intensity, H. For free space only,

$$\mu_o = \frac{B}{H} \qquad (36\text{-}6)$$

$$B = \mu_o H$$

$$H = \frac{B}{\mu_o}$$

Most materials are nonmagnetic and have virtually the same permeability as that of free space.

Example 36-1

An air gap has a length of 0.75 mm and a cross-sectional area of 2.5 cm². What is the value of the MMF required to establish a magnetic flux of 0.45 mWb in the air gap?

Solution

Because air is nonmagnetic, its permeability is virtually the same as in free space.

$$A = 2.5 \times 10^{-4}\ \text{m}^2$$

$$l = 0.75 \times 10^{-3}\ \text{m}$$

$$\phi = 0.45 \times 10^{-3}\ \text{Wb.}$$

Flux density,

$$B = \frac{\phi}{A} = \frac{0.45 \times 10^{-3}\ \text{Wb}}{2.5 \times 10^{-4}\ \text{m}^2}$$

$$= 1.8\ \text{T.}$$

Magnetic field intensity,

$$H = \frac{B}{\mu_o} = \frac{1.8\ \text{T}}{12.57 \times 10^{-7}\ \text{H/m}} \qquad (36\text{-}6)$$

$$= 1.43 \times 10^6\ \text{AT/m.}$$

Magnetomotive force,

$$\mathcal{F} = H \times l \qquad (36\text{-}1)$$

$$= 1.43 \times 10^6\ \text{AT/m} \times 0.75 \times 10^{-3}\ \text{m}$$

$$= 1072.5\ \text{AT.}$$

PRACTICE PROBLEMS

36-1. The air gap in a certain magnetic circuit is 0.12 cm long and has a cross-sectional area of 25 cm². Calculate the field intensity in the gap and the number of ampere-turns required to establish a flux of 800 μWb across the gap.

36-2. An air gap is 0.1524 cm long and has a cross-sectional area of 32.26 cm². If there is an MMF of 600 AT available to establish the flux across the air gap, determine the magnetic flux intensity and the total flux.

36-3. A long straight conductor is situated in air and is carrying a current of 400 A. If the return conductor is far away, calculate the values of the magnetic field intensity and the flux density at a radius of 10 cm.

36-4. Two straight parallel conductors are situated in air with a spacing of 10 cm between their centers. If these conductors carry currents of 500 A each, in opposite directions, calculate the values of the magnetic field intensity at a position midway on a straight line joining, and at right angles to, the two conductors.

36-5. In Practice Problem 36-4, determine the value of the flux density at a position which is 3 cm from one conductor on a straight line joining, and at right angles to, the two conductors.

36-6. In Practice Problem 36-4 the direction of the current in one of the conductors is reversed. Calculate the value of the magnetic field intensity at a position which is 3 cm from one conductor on a straight line joining, and at right angles to, the two conductors.

36-7. An air-cored inductor has a length of 0.18 m and a diameter of 6 cm. If this coil has 325 turns, what value of current is required to provide a magnetic field intensity of 4500 AT/m.

36-8. A coil with an air core consists of 1200 turns and is 35 cm long. If the flux density in the air core is 0.36 T, what is the value of the current flowing through the coil?

36-9. The dimensions of an air gap with a rectangular cross-section, are 2.4 cm long, 5.3 cm high, and 3.3 cm wide. If the flux across the gap is 0.34 Wb, calculate the required value of the magnetic field intensity.

37
Relative permeability—Rowland's law—reluctance

A circular coil that consists of N turns and carries a current of I amperes, is placed in a vacuum (free space). As the result of the magnetic field intensity (H), a certain flux density B ($B = \mu_0 H$ teslas) will be established. If the same coil is now wound on a ferromagnetic toroidal ring (Fig. 37-1), the flux density for the same magnetic field intensity will be greatly increased. The factor by which the flux density is increased, is called the *relative permeability* (μ_r), which has no units and is just a number. However, the value of μ_r is not a constant, but depends on the magnetic field intensity (Fig. 37-2). This is a result of the ferromagnetic material's atomic structure. Note that if the magnetic field intensity is reduced to zero, there is a small amount of residual (or remanent) magnetism so that the flux density is *not* zero.

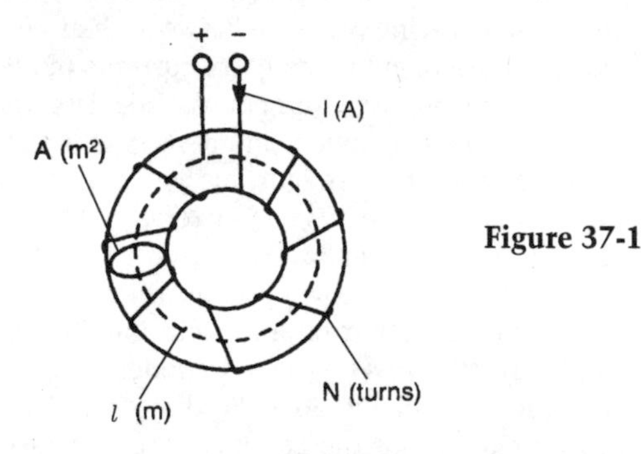

Figure 37-1

Because the MMF of a magnetic circuit can be compared with the EMF of an electrical circuit, and the flux is regarded as similar to the current, it follows that the ratio of MMF: flux must be the magnetic equivalent of the resistance. This ratio is equal to a factor called the *reluctance* so that MMF, $\mathcal{F}$ / Flux, ϕ = Reluctance, $\mathfrak{R}$; this equation is sometimes referred to as *Rowland's law*. The reluctance is measured in ampere-turns per weber, and is directly proportional to the length of the magnetic circuit, but is inversely proportional to the relative permeability and the cross-sectional area. This relationship recalls the equation for the resistance, R, of a cylindrical conductor, namely

$$R = \frac{\text{length, } \ell}{\text{Conductivity, } \sigma \times \text{cross-sectional area, } A} \text{ ohms.}$$

The principles outlined in the analyses of series and parallel resistor circuits can also be applied to magnetic circuits. For example, if a radial cut is made in the iron ring and you assume that the same flux is es-

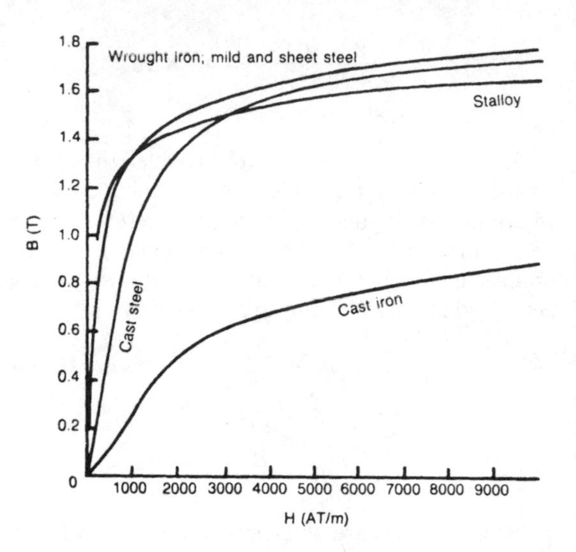

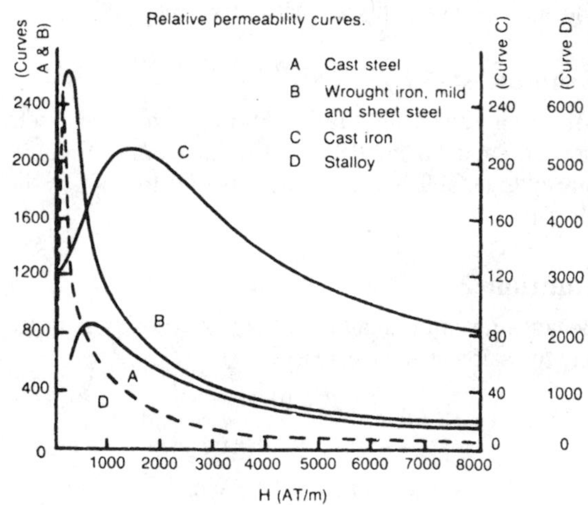

Figure 37-2

tablished in both the air gap and the iron, this is equivalent to saying that the current is the same throughout a series circuit. A certain number of ampere turns would be necessary to establish the flux in the air gap and a different number would create the same flux in the iron. The total required MMF would then be obtained by adding together the individual numbers of ampere turns. This is the same as finding the source voltage by adding together the individual voltage drops across two resistors in series.

Finally, the concept of power is another important difference between magnetic circuits and electrical circuits. For an electrical circuit containing resistors, energy must be continuously supplied from the source, but when a flux is established in a magnetic circuit, no further energy is required.

MATHEMATICAL DERIVATIONS

Relative permeability,

$$\mu_r = \frac{\text{flux density with a magnetic core}}{\text{flux density with a free space core}} \quad (37\text{-}1)$$

The flux density for a free space core is $\mu_o H$.

It follows that:

Flux density with a magnetic core,

$$B = \mu_o \mu_r H = \mu H \text{ teslas} \quad (37\text{-}2)$$

where $\mu = \mu_o \mu_r$ and is the absolute permeability which is measured in henrys per meter.

Remembering that the total flux, $\phi = B \times A$ Wb and the magnetomotive force, $\mathcal{F} = H \times \ell$ ampere turns,

$$\frac{\text{MMF}, \mathcal{F}}{\text{Flux}, \phi} = \frac{H \times \ell}{B \times A} = \frac{H \times \ell}{\mu_o \mu_r H \times A} \quad (37\text{-}3)$$

$$= \frac{\ell}{\mu_o \mu_r A} = \frac{\ell}{\mu A}$$

Therefore Rowland's law is,

$$\text{Reluctance}, \mathfrak{R} = \frac{\mathcal{F}}{\phi} \quad (37\text{-}4)$$

$$= \frac{\ell}{\mu A} \text{ ampere-turns per weber}$$

Example 37-1

An air gap has a length of 2.5 mm and a cross-sectional area of 7 cm^2. What are the values of the gap's reluctance and the MMF required to establish a flux of 0.9 mWb across the gap?

Solution

$\mu_r = 1$, A = 7×10^{-4} m^2, $\ell = 2.5 \times 10^{-3}$ m.

Reluctance of the air gap,

$$\mathfrak{R} = \frac{\ell}{\mu A} = \frac{\ell}{\mu_o \mu_r A} \quad (37\text{-}4)$$

$$= \frac{2.5 \times 10^{-3}}{12.57 \times 10^{-7} \times 7 \times 10^{-4}}$$

$$= 2.841 \times 10^6 \text{ AT/Wb}$$

$$\text{MMF}, \mathcal{F} = \mathfrak{R} \times \phi$$

$$= 2.841 \times 10^6 \times 0.9 \times 10^{-3}$$

$$= 2557 \text{ AT}$$

Alternatively:

Flux density,

$$B = \frac{\phi}{A} = \frac{0.9 \times 10^{-3}}{7 \times 10^{-4}}$$

$$= 1.286 \text{ T}$$

Magnetic field intensity,

$$H = \frac{B}{\mu_o} = \frac{1.286}{12.57 \times 10^{-7}}$$

$$= 1.023 \times 10^6 \text{ AT/m}$$

$$\text{MMF}, \mathcal{F} = H \times \ell = 1.023 \times 10^6 \times 2.5 \times 10^{-3}$$

$$= 2557 \text{ AT}$$

Example 37-2

A toroidal ring of cast steel has a mean circumference of 45 cm, and a cross-sectional area of 4 cm^2. A radial cut creates an air gap with a thickness of 1.6 mm. Neglecting any flux leakage, what is the total MMF required to establish a flux density of 1.32 T in the air gap?

Solution

The air gap and the cast steel ring must be regarded as in series. From Fig. 37-2, a flux density of 1.32 T corresponds to a magnetic field intensity of 2000 AT/m. MMF required to establish the flux in the steel ring is

$$\mathcal{F}_1 = H_1 \times \ell_1 = 2000 \times 45 \times 10^{-2}$$

$$= 900 \text{ AT}$$

For the air gap, the magnetic field intensity is

$$H_2 = \frac{B}{\mu_o} = \frac{1.32}{12.57 \times 10^{-7}}$$

$$= 1.05 \times 10^6 \text{ AT/m}$$

MMF required to establish the flux in the air gap is

$$\mathcal{F}_2 = H_2 \times \ell_2 = 1.05 \times 10^6 \times 1.6 \times 10^{-3}$$

$$= 2520 \text{ AT}$$

Total MMF required,

$$\mathcal{F}_T = \mathcal{F}_1 + \mathcal{F}_2 = 900 + 2520$$

$$= 3420 \text{ AT}$$

PRACTICE PROBLEMS

37-1. A magnetic circuit, with a relative permeability of 5000, has a cross-sectional area of 4 cm^2 and a length of 70 cm. If the flux density is 1.5 T, calculate the values of the circuit's reluctance, and the total MMF required.

37-2. An iron magnetic circuit has a cross-sectional area of 7.5 cm^2 and a length of 20 cm. A coil of 100 turns is wound uniformly over the circuit. When the current flowing through the coil is 1 A, the total flux is 450 μWb, and when the current is 3 A, the flux is 900 μWb. Calculate the values of the magnetic field intensity and the relative permeability of the iron for each value of the current.

37-3. A cast-steel ring (refer to Fig. 37-2) has a cross-sectional area of 4 cm^2 and a mean diameter of 15 cm. If a coil of 120 turns is uniformly wound on the ring, determine the current required to establish a flux of 500 μWb.

37-4. A cast-iron ring (refer to Fig. 37-2) has a mean diameter of 25.4 cm and a cross-sectional area of 12.9 cm^2. If a coil of 40 turns is uniformly wound on a ring, calculate the value of the total flux if the current flowing through the coil is 6 A.

37-5. A metal toroidal ring with a circular cross-sectional area has an internal diameter of 10 cm and an external diameter of 14 cm. A coil of 250 turns is wrapped around the ring and carries a current of 1.6 A. If the metal's absolute permeability is 7.0×10^{-4} H/m, determine the values of the ring's reluctance and the flux established in the ring.

38
Electromagnetic induction—Faraday's law—Lenz's law

Electromagnetic induction is the link between mechanical energy, magnetism, and electrical energy. In other words, the principles behind the electrical generator. In 1831, Faraday discovered that when a conductor cuts (mechanical energy), or is cut by a magnetic field (magnetism), an EMF (electricity) is *induced* in the conductor. Notice that it is only necessary for there to be relative motion between the conductor and the flux lines of the magnetic field.

Faraday's law is illustrated in Fig. 38-1A. This shows a conductor that is mechanically moved to cut the flux lines of a magnetic field at right angles (90°). The ends of the conductor are connected to a sensitive voltmeter that records the value of the EMF. When the conductor is stationary, the free electrons are moving in random directions. However, when the conductor cuts the magnetic flux, the electrons are given a movement in a particular direction. Through the motor effect, forces are then exerted on the moving charges so that the electrons are driven along the conductor. One end is then negatively charged and the other end is positively charged so that the EMF is induced in the conductor. The magnitude of this EMF is directly proportional to the flux density, the length of the conductor, and its velocity at right angles to the lines of force. These are the results of Faraday's law, which states that the magnitude of the EMF depends on the rate of the cutting of the magnetic flux.

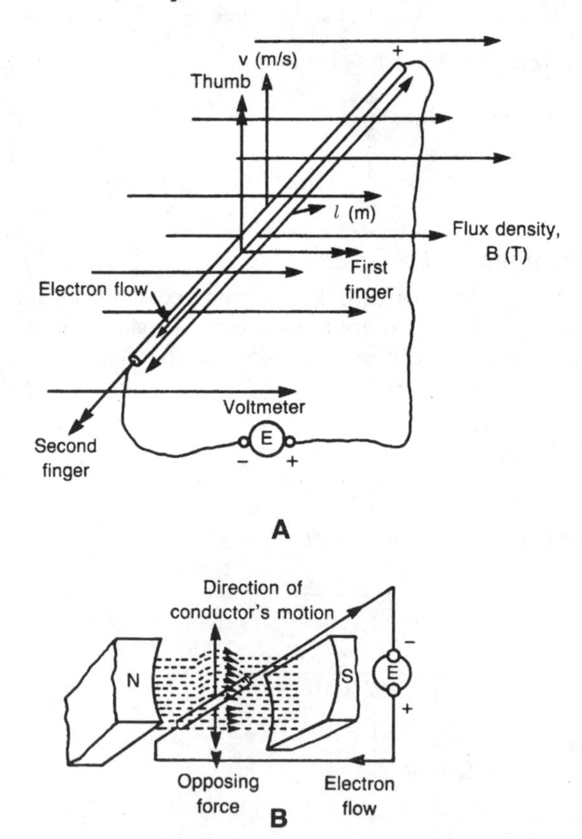

Figure 38-1

The directions of the current (electron flow), the magnetic flux and the conductor's motion are mutually perpendicular. These may be found by the *left-hand* "generator" rule (as opposed to the right-hand "motor" rule of chapter 35). The *first* finger indicates the direction of the *flux*, the second finger is the direction of the electron flow through the conductor and the thu*m*b shows the direction of the conductor's *m*ovement (Fig. 38-1A).

The output electrical energy is created from the work done in moving the conductor. The conductor itself carries a current, and is therefore surrounded by its own flux, which reacts with the uniform magnetic field. Consequently, the conductor experiences a force whose direction opposes the motion of the conductor. This may be verified by applying the right-hand "motor" rule. Mechanical work must, therefore, be done in order to move the conductor against the opposing force (Fig. 38-1B). This is all summarized in *Lenz's law*, which states that *the direction of the induced EMF is such as to oppose the change that originally created the EMF.* Such a law sounds more like a riddle, so examine the statement in some detail. The "original change" is the conductor's motion that creates the induced EMF by cutting the uniform flux. This EMF produces a current flow in a particular direction so that a second magnetic field appears around the conductor. The reaction between the two fields causes a force to be exerted on the conductor and the direction of this force opposes the conductor's original motion. Lenz's law is really saying that you cannot obtain the electrical energy without supplying the mechanical work.

MATHEMATICAL DERIVATIONS

Faraday's law states that the magnitude of the EMF depends on the amount of the magnetic flux, ϕ, cut by the conductor in one second. Therefore,

$$\text{Induced EMF, } E = -\frac{\Delta\phi}{\Delta T} \text{ volts} \qquad (38\text{-}1)$$

The minus sign illustrates the meaning of Lenz's law.

If a flux of one weber is cut in a time of one second, the induced EMF is one volt. This enables the weber unit to be defined.

Assume the uniform flux density is B teslas, the conductor's length is ℓ meters, and the conductor is moving upwards at 90° to the magnetic field with a velocity of v m/s. In 1 second (ΔT), the conductor covers a vertical distance of v meters and sweeps out an area, $\Delta A = v \times \ell$ m². Then:

$$\text{Flux cut by the conductor in 1 s} = B \times \Delta A = B\ell v \text{ webers} \qquad (38\text{-}2)$$

Value of the induced EMF,

$$E = \frac{\Delta\phi}{\Delta T} \qquad (38\text{-}3)$$
$$= \frac{Blv \text{ Wb}}{1 \text{ s}}$$
$$= Blv \text{ volts}$$

Example 38-1

A conductor cuts a flux of 650 μWb in a time of 0.4 ms. What is the value of the induced EMF?

Solution

Flux change, $\Delta\phi = 650 \times 10^{-6}$ Wb.
Time change, $\Delta T = 0.4 \times 10^{-3}$ s.

Value of the induced EMF,

$$E = \frac{\Delta\phi}{\Delta T} \qquad (38\text{-}1)$$
$$= \frac{650 \times 10^{-6} \text{ Wb}}{0.4 \times 10^{-3} \text{ s}}$$
$$= 1.625 \text{ V.}$$

Example 38-2

A conductor of length 30 cm is moving at 90° to the direction of a uniform magnetic field with a flux density of 0.7 T. If the conductor's velocity is 130 m/s, what is the value of the induced EMF?

Solution

$B = 0.7$ T, $l = 30 \times 10^{-2}$ m, $v = 130$ m/s

Value of the induced EMF,

$$E = Blv \qquad (38\text{-}3)$$
$$= 0.7 \text{ T} \times 30 \times 10^{-2} \text{ m} \times 130 \text{ m/s}$$
$$= 27.3 \text{ V.}$$

PRACTICE PROBLEMS

38-1. A coil of 600 turns cuts a flux of 400 μWb in a time of 2 ms. Calculate the value of the total EMF induced in the coil.

38-2. A conductor of length 40 cm cuts the flux lines of a magnetic field at 90°. If the flux density is 0.9 T and the voltage induced in the conductor is 36 V, calculate the value of the conductor's velocity.

38-3. A conductor cuts a certain magnetic flux in a time of 0.2 ms. If the EMF induced in the conductor is 40 V, calculate the value of the magnetic flux.

39
Self-inductance

An electrical circuit can only possess three passive quantities—resistance, inductance, and capacitance. The previous chapters have already covered that resistance opposes and, therefore, limits the current. By contrast, inductance is defined as that electrical property that prevents any sudden, or abrupt, *change* of current and also limits the rate of the *change* in the current. This means that, if an electrical circuit contains inductance, the current of that circuit can neither rise nor fall instantaneously.

Although a straight length of wire possesses some inductance, the property is most marked in a coil, which is called an *inductor* and whose circuit symbol is ⌒⌒⌒. When a current increases or decreases in a coil, the surrounding magnetic field will expand or collapse and cut the turns of the coil itself. On the principle of electromagnetic induction (chapter 38), a voltage or counter EMF will be induced into the coil but this voltage will depend on the *rate* at which the current is *changing*, rather than on the value of the current itself. Because of the *change* in the current, the coil creates its own *moving* flux that cuts its own turns, and induces the voltage across the coil. For these reasons, the property of the coil is referred to as its self-inductance, whose letter symbol is L.

Figure 39-1A compares the properties of resistance and inductance. When the switch is closed in position 1, the current in the resistor immediately rises from zero to a constant value of 10 V/5 Ω = 2 A (Fig. 39-1B). The voltage drop across the resistor (10 V) then exactly balances (or opposes) the source voltage (10 V).

If the switch is now moved to position 2, the resistor current immediately drops to zero, but the inductor current starts to grow in the coil. This creates a magnetic flux, which, as it expands outwards, cuts the turns of the coil and induces the counter EMF, which must exactly balance the source voltage (assuming that the coil only possesses inductance, and has zero resistance). Because the source voltage is fixed, it follows that the current must start at zero and must then grow at a constant rate that is measured in amperes per second (Fig. 39-1C).

Although the counter EMF is proportional to the rate of the change of the current, its value must also depend on some factor of the coil. This factor is called the *self-inductance* (L), which is determined by such physical quantities as the number of turns, the length, the cross-sectional area and the nature of the core on which the coil is wound. The unit of inductance is the henry (H), which is defined as follows. The inductance is one henry if, when the current is changing at the rate of one ampere per second, the induced EMF is 1 V. Consequently, in Fig. 39-1A, where the source voltage is 10 V and the inductance is 5 H, the rate of the growth of the current is 10 V/5 H = 2 A/s.

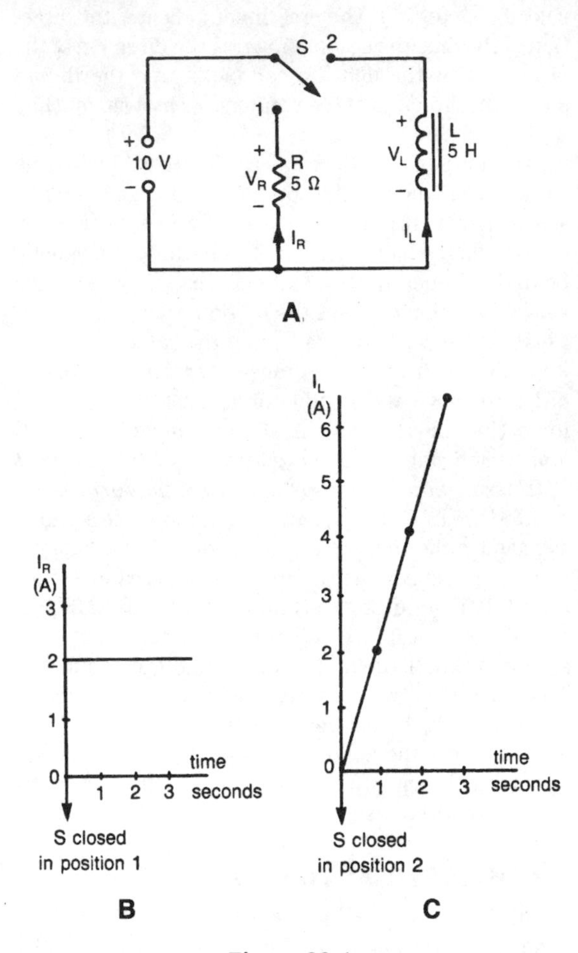

Figure 39-1

MATHEMATICAL DERIVATIONS

Switch S in position 1

Voltage drop across the resistor,

$$V_R = -I_R \times R \text{ volts} \tag{39-1}$$

Using Kirchhoff's voltage law (KVL):

$$E + V_R = 0 \tag{39-2}$$

Equations 39-1, 39-2 yield

$$E = I_R \times R \text{ volts} \quad (39\text{-}3)$$

Switch S in position 2

Counter EMF,

$$V_L = -L \times \frac{\Delta I_L}{\Delta T} \text{ volts} \quad (39\text{-}4)$$

where: $\Delta I_L/\Delta T$ = rate of the change of the current in amperes per second

Using Kirchhoff's voltage law (KVL),

$$E + V_L = 0 \quad (39\text{-}5)$$

From Equations 39-4, 39-5
Source voltage,

$$E = L \times \frac{\Delta I_L}{\Delta T} \text{ volts} \quad (39\text{-}6)$$

Therefore,
Rate of the change of the current,

$$\frac{\Delta I_L}{\Delta T} = \frac{E}{L} \text{ amperes per second} \quad (39\text{-}7)$$

Self-inductance,

$$L = \frac{E}{\Delta I_L/\Delta T} \text{ henrys} \quad (39\text{-}8)$$

Current after t seconds,

$$I_L = \frac{\Delta I_L}{\Delta T} \times t \text{ amperes} \quad (39\text{-}9)$$

Example 39-1

A 12-V dc source is suddenly switched across a 3-H inductor of negligible resistance. What are the values of the initial current, the rate of the growth of the current, the counter EMF, and the current after 0.6 second?

Solution

Initial current is *zero*.

Rate of the growth of the current,

$$\frac{\Delta I_L}{\Delta T} = \frac{E}{L} \quad (39\text{-}7)$$

$$= \frac{12 \text{ V}}{3 \text{ H}}$$

$$= 4 \text{ A/s.}$$

Counter EMF,

$$V_L = (-)E = 12 \text{ V.} \quad (39\text{-}5)$$

Current after 0.6 s,

$$I_L = \frac{\Delta I_L}{\Delta T} \times t \quad (39\text{-}9)$$

$$= 4 \text{ A/s} \times 0.6 \text{ s} = 2.4 \text{ A.}$$

Example 39-2

The counter EMF induced in a coil is 30 V, when the current changes by 12 mA in a time of 5 μs. What is the value of the coil's self-inductance?

Solution

Rate of the change of the current,

$$\frac{\Delta I_L}{\Delta T} = \frac{12 \times 10^{-3} \text{ A}}{5 \times 10^{-6} \text{ s}}$$

$$= 2400 \text{ A/s}$$

Self-inductance,

$$L = \frac{E}{\Delta I_L/\Delta T} = \frac{30 \text{ V}}{2400 \text{ A/s}} \quad (39\text{-}8)$$

$$= 0.00125 \text{ H}$$

$$= 1.25 \text{ mH.}$$

PRACTICE PROBLEMS

39-1. The current flowing through a 200-mH inductor decays from 9 A to 6 A in 12 ms. What is the average value of the induced counter EMF?

39-2. A 9-V source is switched across a 3-mH inductor. If the inductor's resistance is neglected, what is the value of the inductor's current after 2.5 ms?

39-3. The average EMF induced in a 500-μH coil is 10 V, when the current changes by 20 mA. Over what interval of time did the current change?

40
Factors determining a coil's self-inductance

For a coil, a high value of inductance requires that a large counter EMF is induced for a given rate of change in the current. Clearly the number of turns is one of the factors involved because each turn can be regarded as a conductor and the magnetic field intensity is directly proportional to the number of turns (Fig. 40-1). In fact, the value of the inductance is directly proportional to the *square* of the number of turns (as will be proved in the mathematical derivations that follow).

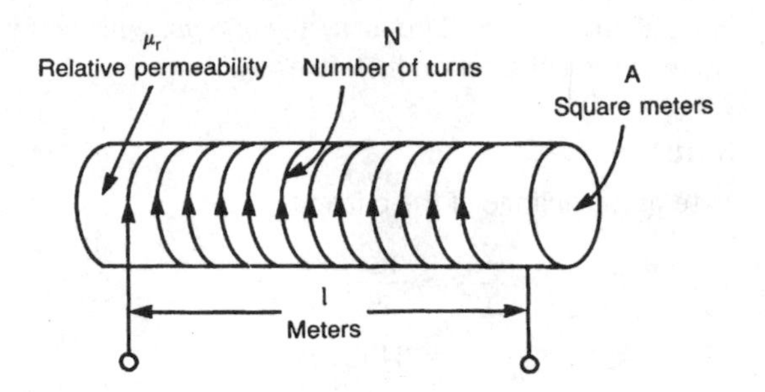

Figure 40-1

The size of each turn is another factor that determines the value of the inductance. If the cross-sectional area of each turn is increased, the coil's reluctance [$\Re = l/(\mu_o \mu_r A)$] will be less and the flux density will be greater. Looking at it another way, if we continue to reduce the cross-sectional area, the coil would ultimately vanish and the inductance would obviously be zero.

If the number of turns, and their size, are kept constant, but the coil's length is increased, the turns are spread farther apart and there is a greater reluctance. Consequently, the flux density and the inductance are reduced.

Finally, you can replace a nonmagnetic core with one that is made from ferromagnetic material with a high relative permeability. This greatly lowers the amount of the reluctance, and therefore increases the flux density. The result is a much higher inductance value.

Summarizing, the inductance value is directly proportional to the square of the number of turns, their cross-sectional area, and the coil's permeability, but is inversely proportional to the coil's length.

The values of practical inductors range from henrys to microhenrys. Those with a value of several henrys are referred to as chokes (circuit symbol) which are made with a large number of copper turns wound on an iron core. The presence of the core is indicated by the two lines drawn to one side of the circuit symbol. Such a core is laminated (cut into slices that are insulated from each other) to reduce the loss owing to eddy currents. Because the relative permeability varies with the amount of the direct current flowing through the coil, the inductance value is normally quoted for a particular dc current level.

A coil with a value of only a few microhenrys consists of a few copper turns that are wound on a nonmagnetic core (circuit symbol). Between these two extremes are millihenry inductors with cores of either iron dust (circuit symbol) or a ferrite material. Both of these cores serve to *increase* the inductance; however, the ferrite material, although possessing magnetic properties comparable with iron, is an insulator which will not suffer from the loss due to eddy currents.

MATHEMATICAL DERIVATIONS

Regarding each turn as a conductor and using Faraday's law,

$$\text{EMF}, E = N\frac{\Delta\phi}{\Delta T} \tag{40-1}$$

$$= L\frac{\Delta I}{\Delta T} \text{ volts}$$

This yields

Self-inductance,

$$L = N\frac{\Delta\phi}{\Delta I} \tag{40-2}$$

$$= \frac{\text{change in flux linkages}}{\text{change in current}} \text{ henrys}$$

The "flux linkages" is the product of the flux and the number of turns with which the flux is linked. The inductance is one henry if a current change of one ampere creates a change in the flux linkages of 1-Wb turn.

Self-inductance,

$$L = N\frac{\Delta\phi}{\Delta I} \tag{40-3}$$

$$= N\frac{\Delta(BA)}{\Delta I}$$

$$= NA\frac{\Delta B}{\Delta I} \text{ henrys}$$

From $B = \mu_o \mu_r H$ teslas (equation 37-2),

$$L = N \times \frac{\Delta B}{\Delta I} \quad (40\text{-}4)$$

$$= NA \times \mu_o \mu_r \frac{\Delta H}{\Delta I} \text{ henrys}$$

Using $H = IN/l$ amperes per meter,

$$L = NA \times \mu_o \mu_r \frac{\Delta(IN/l)}{\Delta I}$$

Self-inductance,

$$L = \frac{\mu_o \mu_r N^2 A}{l} \text{ henrys} \quad (40\text{-}5)$$

where the permeability (μ_o) of free space $= 4\pi \times 10^{-7}$ henry per meter. Equation 40-5 demonstrates mathematically that the inductance value is directly proportional to the square of the number of turns, the cross-sectional area, and the core's relative permeability, but is inversely proportional to the coil's length.

For the nonmagnetic core, $\mu_r = 1$, and therefore, Self-inductance,

$$L = \frac{\mu_o N^2 A}{l} \text{ henrys} \quad (40\text{-}6)$$

Example 40-1

A coil is wound with 800 turns on a ferromagnetic core with a relative permeability of 1100. If the coil's length is 5.5 cm and its cross-sectional area is 3.75 cm^2, calculate the value of the coil's self-inductance.

Solution

Self-inductance,

$$L = \frac{\mu_o \mu_r N^2 A}{l} \text{ henrys} \quad (40\text{-}5)$$

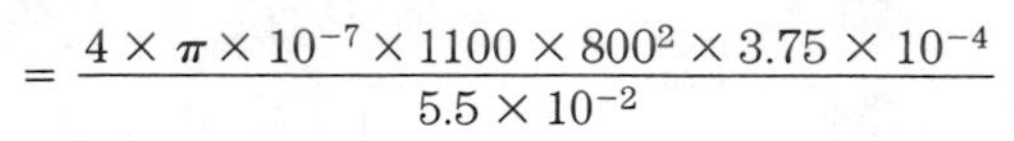

$$= \frac{4 \times \pi \times 10^{-7} \times 1100 \times 800^2 \times 3.75 \times 10^{-4}}{5.5 \times 10^{-2}}$$

$$= 6.03 \text{ H.}$$

If the same coil were wound on a nonmagnetic core, the self-inductance would only be 6.03/1100 H = 5.48 mH.

PRACTICE PROBLEMS

40-1. A coil is wound with 600 turns on an air core, and has a self-inductance of 4 mH. If the number of turns is increased to 900, without changing either the coil's cross-sectional area or its length, what is the new value of the self-inductance?

40-2. An average voltage of 15 V is induced in a coil of 120 turns as a result of a change in the magnetic flux. If this event occurs in the total time of 90 ms, what is the amount of the total flux change?

40-3. A coil of 1500 turns surrounds a magnetic circuit whose reluctance is 3×10^6 ampere-turns per weber. What is the value of the self-inductance?

40-4. An inductor has a length of 30 cm and a diameter of 1 cm. How many turns are required to produce a self-inductance of 135 μH?

40-5. An 8-H inductor is manufactured by wrapping a number of turns of copper wire around an iron ring with a circular cross-sectional area. The inner and outer diameters are 8 cm and 10 cm respectively, and the relative permeability of the iron is 1280. Calculate the number of turns required.

40-6. A current of 8 A establishes a flux of 60 mWb in an iron-cored inductor that is wound with 1000 turns. What is the value of the self-inductance?

41
Energy stored in the magnetic field of an inductor

The inductor is assumed to be ideal, in the sense that there are no resistance losses associated with the component. When S is closed, the growth rate of the current is E/L amperes per second so that the counter EMF induced in the inductor exactly balances the source voltage (Fig. 41-1A). After a period of t seconds the current is equal to I amperes, and because the growth of the current is linear with respect to time, the average current during the period is $I/2$ amperes (Fig. 41-1B). A magnetic field is established

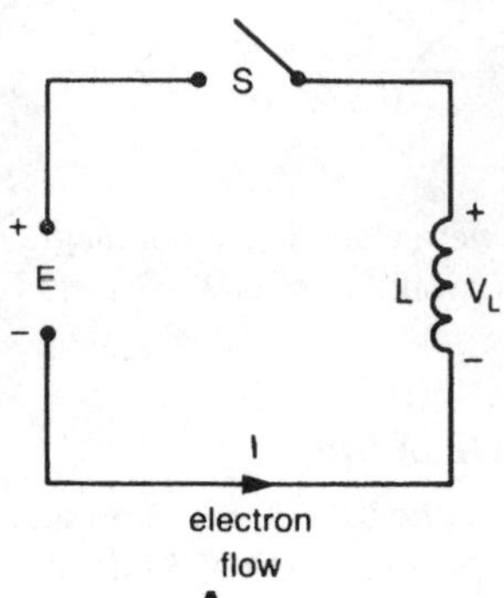

Figure 41-1

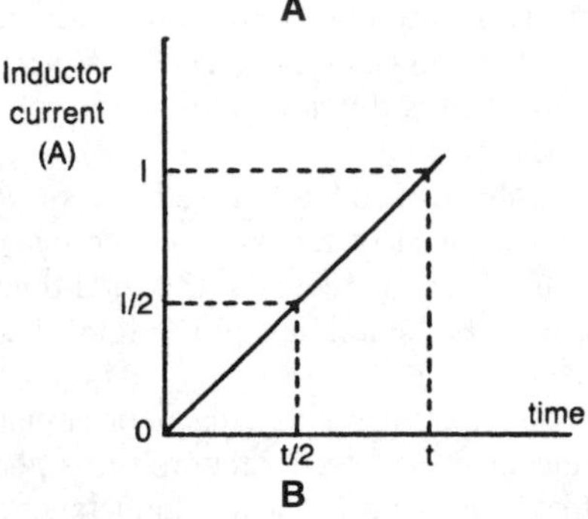

around the inductor that represents the energy derived from the source. If the switch is then opened, the magnetic field collapses and the energy is dissipated in the form of an arc, which occurs between the contacts of the switch.

MATHEMATICAL DERIVATIONS

Source voltage,

$$E = L \times \text{rate of change of current} \tag{41-1}$$

$$= \frac{L \times I}{t} \text{ volts}$$

Average power at the source over the period of t seconds,

$$P = E \times \frac{I}{2} = L \times \frac{I}{t} \times \frac{I}{2} \tag{41-2}$$

$$= \frac{1}{2} \times \frac{LI^2}{t} \text{ watts}$$

Total energy stored in the magnetic field = average power × time,

$$= \frac{1}{t} \times \frac{LI^2}{2} \times t \tag{41-3}$$

$$= \frac{1}{2} \times LI^2 \text{ joules}$$

Alternatively

Let current I be flowing through an inductor, L. If the current increases by an amount (δI) in a time of δT, the induced voltage is $L \times \delta I/\delta t$, and the additional energy delivered to the inductor is $(L \times \delta I)/\delta t \times I \times \delta t = LI\delta I$ joules. Integrating over the period of t seconds, during which the current increased from zero to I amperes:

Total energy stored in the magnetic field,

$$W = \int_o^I LI\, dI \tag{41-4}$$

$$= \frac{1}{2} LI^2 \text{ joules}$$

Example 41-1

In Fig. 41-1, $E = 60$ V and $L = 8$ H. When the switch is closed, what is the rate of growth of the current? Calculate the amount of energy stored in the magnetic field after a time of 5 seconds.

Solution

Rate of growth of the current = E/L = 60 V/8 H = 7.5 amperes per second.

After 5 seconds, current I = 7.5 A/s × 5 s = 37.5 amperes.

Energy stored in magnetic field,

$$W = \frac{1}{2} LI^2 = \frac{1}{2} \times 8\text{ H} \times (37.5\text{ A})^2 \tag{41-3}$$

$$= 5625\text{ J}$$

PRACTICE PROBLEMS

41-1. A 3-H inductor is suddenly switched across a 12-Vdc source. If the inductor's resistance is neglected, what is the energy stored in the inductor after a time interval of 0.25 s?

41-2. A coil with 500 turns is wrapped around a magnetic core, and has a diameter of 2 cm and a length of 12 cm. If the relative permeability of the core is 450 and the current through the coil is 13.0 A, what is the amount of energy stored in the magnetic field surrounding the coil?

41-3. A coil has an inductance of 5 H and a resistance of 50 Ω. If a steady current of 4 A flows through the coil, determine the voltage across the inductor and the amount of energy stored in its magnetic field.

41-4. A coil has an inductance of 4 H. If the current flowing through the coil is reduced from 2 A to 1 A, calculate the amount of energy returned to the source.

42
Inductors in series

It is assumed that all three inductors of Fig. 42-1 are ideal in the sense that they have no losses and they only possess the property of inductance. Because these components are joined end-to-end, the same *rate of change* in the current will be associated with all of the inductors. In each inductor the expanding magnetic field will induce a counter EMF (V_{L1}, V_{L2}, V_{L3}) and the sum of these EMFs must exactly balance the source voltage.

When the switch (S) is closed, there will be a constant rate of growth in the current, and the magnetic field surrounding each inductor will expand. After a period of time, there will be a certain amount of energy stored in each of the magnetic fields, and the sum of these energies will equal the total energy derived from the source voltage. Because the inductors are regarded as ideal components, there is no power *dissipated* in the circuit as heat.

From the mathematical derivations, you can see that the total inductance is the sum of the individual inductances. Consequently, one use of connecting inductors in series is to create a nonstandard total inductance value from standard components; in the same way, a series arrangement of small inductors can avoid the use of a physically larger inductor. However, in such an arrangement, it is essential that the current ratings of the series inductors are compatible.

MATHEMATICAL DERIVATIONS

Rate of growth in the current is:

$$-\frac{\Delta I}{\Delta T} \text{ amperes per second} \qquad (42\text{-}1)$$

ΔI represents "a small change in the current", and ΔT is the corresponding "small change in the time".

Counter EMF,

$$V_{L1} = -L_1 \frac{\Delta I}{\Delta T}, V_{L2} = -L_2 \frac{\Delta I}{\Delta T}, \qquad (42\text{-}2)$$

$$V_{L3} = -L_3 \frac{\Delta I}{\Delta T} \text{ volts}$$

The negative sign for each counter EMF and the rate of growth in the current indicates that the counter EMF opposes (balances) the source voltage. It would be equally correct to write $V_R = -I \times R$ for the voltage drop across the resistor although this is rarely done.

Kirchhoff's voltage law (KVL):

$$E + V_{L1} + V_{L2} + V_{L3} = 0 \qquad (42\text{-}3)$$

Combining Equations 42-2, 42-3:

$$E = L_1 \frac{\Delta I}{\Delta T} + L_2 \frac{\Delta I}{\Delta T} + L_3 \frac{\Delta I}{\Delta T} \qquad (42\text{-}4)$$

$$= \frac{\Delta I}{\Delta T} \times (L_1 + L_2 + L_3) \text{ volts}$$

If L_T is the total equivalent inductance,

$$E = L_T \times \frac{\Delta I}{\Delta T} = \frac{\Delta I}{\Delta T} \times L_T \text{ volts} \qquad (42\text{-}5)$$

Comparing Equations 42-4, 42-5:

$$L_T = L_1 + L_2 + L_3 \text{ henrys} \qquad (42\text{-}6)$$

For N inductors connected in series,

$$L_T = L_1 + L_2 + L_3 \ldots + L_N \text{ henrys} \qquad (42\text{-}7)$$

If the N inductors are all of equal value, L,

$$L_T = NL \text{ henrys} \qquad (42\text{-}8)$$

Voltage division:

$$V_{L1} = -L_1 \times \frac{\Delta I}{\Delta T} = L_1 \times \frac{E}{L_T} = E \times \frac{L_1}{L_T} \text{ volts} \qquad (42\text{-}9)$$

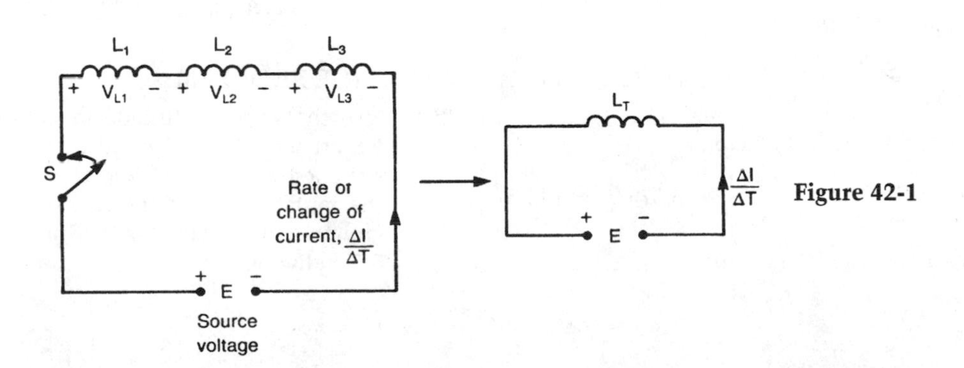

Figure 42-1

$$V_{L2} = E \times \frac{L_2}{L_T} \text{ volts} \quad (42\text{-}10)$$

$$V_{L3} = E \times \frac{L_3}{L_T} \text{ volts} \quad (42\text{-}11)$$

Consequently, the equations for inductors in *series* are comparable with the equations for resistors in *series*.

Energy stored in the magnetic field surrounding the L1 inductor,

$$W_1 = \frac{1}{2} L_1 I^2 \text{ joules} \quad (42\text{-}12)$$

Energy stored in the magnetic field surrounding the L2 inductor,

$$W_2 = \frac{1}{2} L_2 I^2 \text{ joules} \quad (42\text{-}13)$$

Energy stored in the magnetic field surrounding the L3 inductor,

$$W_3 = \frac{1}{2} L_3 I^2 \text{ joules} \quad (42\text{-}14)$$

Total energy derived from the source,

$$W_T = W_1 + W_2 + W_3 \quad (42\text{-}15)$$

$$= \frac{1}{2} L_T I^2 \text{ joules}$$

Example 42-1

In Fig. 42-1, $L_1 = 4$ H, $L_2 = 8$ H, $L_3 = 20$ H, and $E = 43$ V. Calculate the values of L_T, and the rate of growth of the current, V_1, V_2, and V_3. If the switch is closed for a period of four seconds, what is the energy stored in the magnetic field surrounding each inductor and what is the total energy derived from the source voltage?

Solution

Total inductance,

$$L_T = L_1 + L_2 + L_3 = 4 + 8 + 20 \quad (42\text{-}7)$$

$$= 32\ H$$

Rate of growth in the current,

$$\frac{\Delta I}{\Delta T} = \frac{E}{L_T} = \frac{43 \text{ V}}{32 \text{ H}} = 1.344 \text{ A/s} \quad (42\text{-}5)$$

Voltage across the $L1$ inductor,

$$V_{L1} = L_1 \times \frac{E}{L_T} = 4 \text{ H} \times \frac{43 \text{ V}}{32 \text{ H}} = 5.375 \text{ V} \quad (42\text{-}9)$$

Voltage across the $L2$ inductor,

$$V_{L2} = L_2 \times \frac{E}{L_T} = 8 \text{ H} \times \frac{43 \text{ V}}{32 \text{ H}} = 10.75 \text{ V} \quad (42\text{-}10)$$

Voltage across the $L3$ inductor,

$$V_{L3} = L_3 \times \frac{E}{L_T} = 20 \text{ H} \times \frac{43 \text{ V}}{32 \text{ H}} = 26.875 \text{ V} \quad (42\text{-}11)$$

Voltage check:

Source voltage,

$$E = V_{L1} + V_{L2} + V_{L3}$$

$$= 5.375 + 10.75 + 26.875 = 43 \text{ V}$$

After a period of four seconds,

Current,

$$I = 1.344 \text{ A/s} \times 4 \text{ s} = 5.376 \text{ A}$$

Energy stored in the L1 inductor,

$$W_1 = \frac{1}{2} L_1 I^2 = \frac{1}{2} \times 4 \text{ H} \times (5.376 \text{ A})^2 \quad (42\text{-}12)$$

$$= 57.8 \text{ joules}$$

Energy stored in the L2 inductor,

$$W_2 = \frac{1}{2} L_2 I^2 = \frac{1}{2} \times 8 \text{ H} \times (5.376 \text{ A})^2 \quad (42\text{-}13)$$

$$= 115.6 \text{ joules}$$

Energy stored in the L3 inductor,

$$W_3 = \frac{1}{2} L_3 I^2 = \frac{1}{2} \times 20 \text{ H} \times (5.376 \text{ A})^2 \quad (42\text{-}14)$$

$$= 289.0 \text{ joules}$$

Total energy stored,

$$W_T = 57.8 + 115.6 + 289.0 = 462.4 \text{ joules} \quad (42\text{-}15)$$

Energy check:

Total energy derived from the source,

$$W_T = \frac{1}{2} L_T I^2 \quad (42\text{-}15)$$

$$= \frac{1}{2} \times 32 \text{ H} \times (5.376 \text{ A})^2$$

$$= 462.4 \text{ joules}$$

PRACTICE PROBLEM

42-1. Two chokes with self-inductances of 3 H and 5 H are joined in series and are then suddenly switched across a 24-Vdc source. What are the initial values of the rate of the current growth and the counter EMF induced in the 5-H inductor?

43
The L/R time constant

From chapter 39, you know that the property of self-inductance prevents any sudden change of the current, and also limits the rate at which the current can change. Consequently, when the switch S, is closed in position 1 (Fig. 43-1A), the current must take a certain time before reaching a final steady level, which is determined by the value of the resistance, R. We are therefore going to discuss the factors that control the duration of the so-called *transient*, or *changing*, state. This is the interval between the time at which the switch (S) is closed in position 1, and the time when the final, or steady-state, conditions are reached.

Immediately after the switch is closed in position 1, the current must initially be zero so that the voltage drop across the resistor (R) must also be zero. Consequently, to satisfy Kirchhoff's voltage law, the counter EMF created in the inductor (L) must equal the source voltage, E. The initial rate of the current growth is therefore at its maximum value of E/L amperes per second.

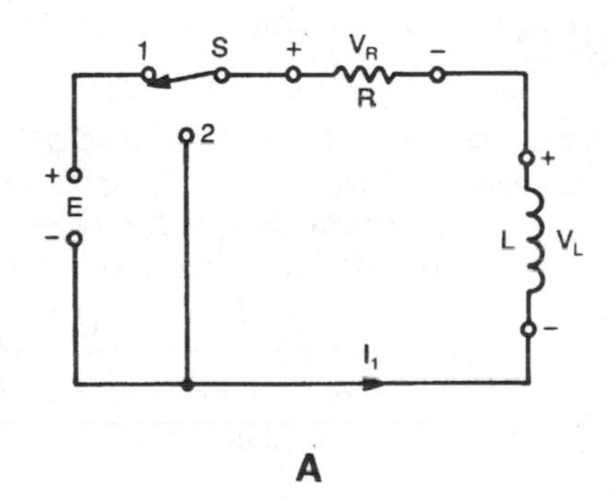

A

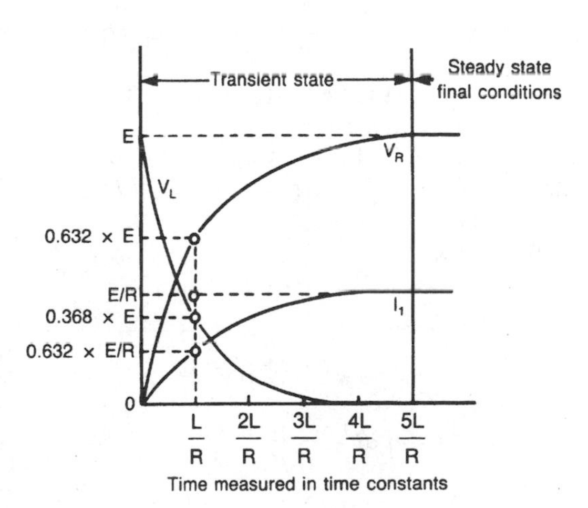

B

Figure 43-1

As the current increases and the magnetic field expands outwards, the voltage drop across the resistor rises, and the counter EMF in the inductor falls (because at all times the sum of these two voltages must exactly balance the source voltage, E). When the counter EMF falls, the rate of the current's growth is correspondingly less, so there is a situation in which the greater the value of the current, the less is its rate of growth. Theoretically, it would take infinite time for the current to reach its final steady value of E/R amperes, but (as you will see in the mathematical derivations) the current reaches its final value (to within 1%) after a finite time interval that is determined by the values of the inductance and the resistance.

Remember that the initial rate of current growth is E/L amperes per second, but the final current value is E/R amperes. Consequently, *if* the initial rate of current growth were maintained, the current would reach its final value after an interval of

$$\frac{E/R \text{ A}}{E/L \text{ A/s}} = \frac{L}{R} \text{ seconds}$$

which is referred to as the *time constant* of the circuit. However, because the rate of current growth falls off, the current only reaches 63.2% of its final value after one time constant. It requires *five* time constants before the current can rise to within 1% of E/R amperes (Fig. 43-1B), and we can assume that the transient growth period has been completed.

Notice that the conditions at the start of the transient state can be predicted by regarding the inductor as an open circuit, but the final steady-state values can be obtained by considering the inductor to be a short circuit.

Assuming that the steady-state conditions have been reached, now move the switch (S) to position 2. The current cannot change instantaneously so that I_2 must still be E/R amperes, and the voltage across the resistor must remain at E volts. By the KVL rule, the inductor's counter EMF will also be E volts, but its polarity is reversed because the magnetic field is starting to collapse rather than expand (Fig. 43-2A).

As the current decays, V_R and V_L must fall together in order to maintain the voltage balance around the closed loop. When V_L decreases, the rate of current

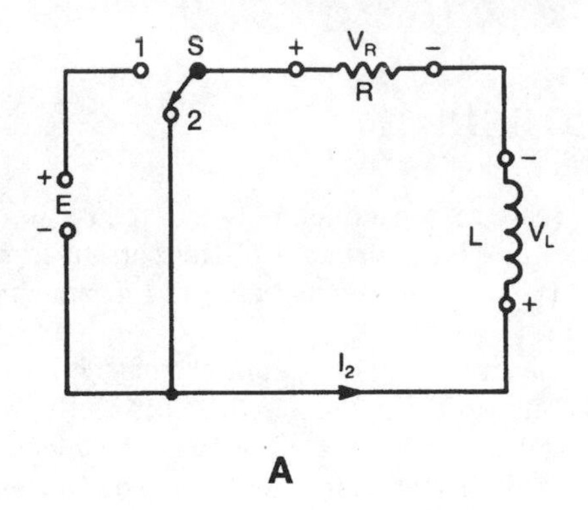

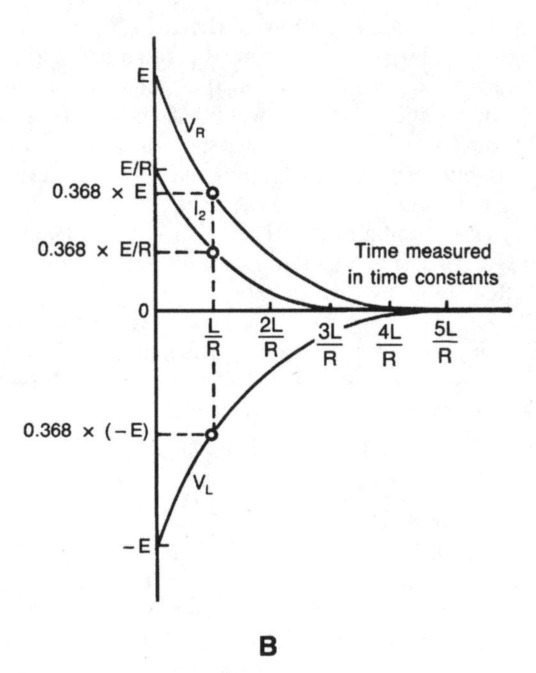

Figure 43-2

decay is less; so that, theoretically, it would take infinite time for the current to reach zero. However, after one time constant, the current has *lost* 63.2% of its initial value and has, therefore, dropped to 36.8% of E/R amperes. After five time constants, assume that the transient decay period has been concluded in the sense that I_2, V_R, and V_L have all fallen to less than 1% of their original values (Fig. 43-2B).

MATHEMATICAL DERIVATIONS

In Fig. 43-1A, the switch, S, is closed in position 1. The initial conditions are:

$I_1 = 0$ amperes, $V_R = 0$ volts, $V_L = E$ volts, initial rate of current growth $= E/L$ amperes per second.

At all times:

$$V_R + V_L = E \tag{43-1}$$

$$I_1 \times R + L\frac{dI_1}{dt} = E$$

Integrating the equation throughout,

$$\int dt = \int \frac{L}{E - I_1R}\, dI_1$$

$$t + k = -\frac{L}{R}\ln (E - I_1R)$$

When $t = 0$, $I_1 = 0$ and $k = -L/R \ln E$. Therefore,

$$t = -\frac{L}{R}\ln\left(\frac{E - I_1R}{E}\right) \tag{43-2}$$

$$I_1 = \frac{E}{R}\,(1 - e^{-Rt/L}) \text{ amperes}$$

where $e = 2.7183\ldots$ and is the base of the natural logarithms (chapter 210).

Voltage drop across the resistor,

$$V_R = I_1 \times R = E(1 - e^{-Rt/L}) \text{ volts} \tag{43-3}$$

The expressions for I_1 and V_R represent the equations of exponential growth.

When $t = L/R$ seconds, $I_1 = E/R\ (1 - e^{-1}) = 0.632$ E/R amperes (63.2% of the final steady-state current). After $t = 5L/R$ seconds, $I_1 = E/R\ (1 - e^{-5}) = 0.993$ E/R amperes so that the current has reached a level which is within 1% of its ultimate value.

Counter EMF induced in the coil,

$$V_L = E - V_R = E\, e^{-Rt/L} \tag{43-4}$$

The expression for V_L represents an equation for exponential decay.

In Fig. 43-2A, the switch (S) is moved to position 2. The initial conditions are:

$I_2 = E/R$ amperes, $V_R = E$ volts, $V_L = -E$ volts and the initial rate of the current's *decay*, $dI_2/dt = -E/L$ amperes per second.

At all times:

$$V_R + V_L = 0.$$

$$I_2R + L\frac{dI_2}{dt} = 0.$$

$$\int dt = -\frac{L}{R}\int \frac{1}{I_2}\, dI_2$$

$$t + k = -\frac{L}{R}\ln I_2$$

When $t = 0$, $I_2 = E/R$ amperes and $k = -L/R \ln E/R$.

Therefore,

$$t = -\frac{L}{R}\ln\left(\frac{R I_2}{E}\right) \quad (43\text{-}5)$$

$$I_2 = \frac{E}{R}\, e^{-Rt/L} \text{ amperes}$$

$$V_R = I_2 \times R \quad (43\text{-}6)$$

$$= E\, e^{-Rt/L} \text{ volts}$$

$$V_L = -\, E\, e^{-Rt/L} \text{ volts} \quad (43\text{-}7)$$

The expressions for I_2, V_R, and V_L all represent equations of exponential decay (Fig. 43-2B).

When $t = L/R$ seconds, $I_2 = 0.368\, E/R$ amperes, $V_R = 0.368\, E$ volts and $V_L = -0.368\, E$ volts. After $t = 5\, L/R$ seconds, the values of I_2, V_R, and V_L have all dropped to 0.7% of their original levels. The transient decay period is therefore virtually finished.

Example 43-1

In Fig. 43-1A, $R = 3.9$ kΩ, $L = 25$ mH and $E = 80$ V. When S is closed in position 1, what are the initial values of I_1, V_R, V_L, and their values after time intervals of 4.49, 6.41, 12.82, 19.23, 25.64, 32.05, and 64.10 microseconds?

Solution

Time constant, $L/R = 25$ mH/3.9 kΩ $= 6.41\ \mu$s. Initial values of I_1, V_R, and V_L are, respectively, zero amperes, zero volts, and 80 volts.

The time interval of 4.49 μs is equal to $4.49/6.41 = 0.7 \times$ time constant. Final current, E/R, is 80 V/3.9 kΩ $= 20.5$ mA. Then,

$I_1 = E/R\ (1 - e^{-Rt/L}) = 20.5\ (1 - e^{-0.7}) = 10.25$ mA.

Note that the current has reached 50% of its final value in the interval of $0.7 \times$ time constant.

$$V_R = 80 \times 0.5 = 40 \text{ V.}$$

$$V_L = 80 - 40 = 40 \text{ V.}$$

The time interval of 6.41 μs is equal to one time constant. Therefore,

$I_1 = 20.5\ (1 - e^{-1}) = 20.5 \times 0.632 = 12.96$ mA.

The current has reached 63.2% of its final value in the interval of one time constant.

$$V_R = 80 \times 0.632 = 50.56 \text{ V.}$$

$$V_L = 80 - 50.56 = 29.44 \text{ V.}$$

The time interval of 12.82 μs is equal to two time constants. Therefore,

$I_1 = 20.5\ (1 - e^{-2}) = 20.5 \times 0.865 = 17.72$ mA.

$V_R = 80 \times 0.865 = 74.8$ V.

$V_L = 80 - 74.8 = 5.2$ V.

The interval of 19.23 μs is equivalent to three time constants. Then:

$I_1 = 20.5\ (1 - e^{-3}) = 20.5 \times 0.95 = 19.48$ mA.

$V_R = 80 \times 0.95 = 76.02$ V.

$V_L = 80 - 76.02 = 3.98$ V.

The time interval of 25.64 μs is the same as four time constants. Therefore,

$I_1 = 20.5\ (1 - e^{-4}) = 20.5 \times 0.982 = 20.13$ mA.

$V_R = 80 \times 0.982 = 78.56$ V.

$V_L = 80 - 78.56 = 1.44$ V.

The time interval of 32.05 μs is equal to five time constants. This virtually completes the transient state in the growth of the current I_1.

$I_1 = 20.5\ (1 - e^{-5}) = 20.5 \times 0.993 = 20.36$ mA.

$V_R = 80 \times 0.993 = 79.46$ V.

$V_L = 80 - 79.46 = 0.54$ V.

The time interval of 64.10 μs is equal to ten time constants. Therefore,

$I_1 = 20.5\ (1 - e^{-10}) = 20.5 \times 0.99995 = 20.499$ mA.

$V_R = 80 \times 0.99995 = 79.996$ V.

$V_L = 80 - 79.996 = 0.004$ V.

Example 43-2

In Example 43-1, it is assumed that the growth transient state has been concluded. If S is now closed in position 2, what are the initial values of I_2, V_R, and V_L, and their values after time intervals of 6.41 μs and 32.05 μs?

Solution

Initial values of I_2, V_R, and V_L are, respectively, 20.5 mA, 80 V, and −80 V.

After the time interval of 6.41 μs or one time constant:

$$I_2 = 20.5\, e^{-1} = 20.5 \times 0.368 = 7.36 \text{ mA.}$$

$$V_R = 80 \times 0.368 = 29.43 \text{ V.}$$

$$V_L = -29.43 \text{ V.}$$

After the time interval of 32.05 μs, or five time constants,

$$I_2 = 20.5\,e^{-5} = 0.00674 \times 20.5 = 0.138 \text{ mA}.$$

$$V_R = 80 \times 0.00674 = 0.539 \text{ V}.$$

$$V_L = -0.539 \text{ V}.$$

The transient decay time is virtually finished.

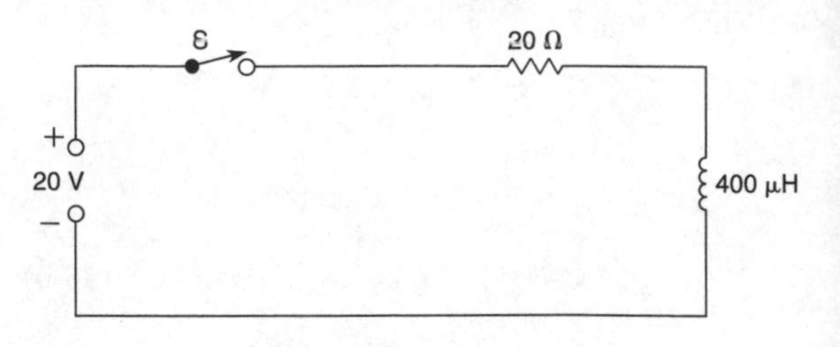

Figure 43-3

PRACTICE PROBLEMS

43-1. In Fig. 43-3, calculate the values of the time constant, and the initial rate of current growth after the switch (S) is closed. What are the values of the circuit current after intervals of 20 μs and 60 μs, from the time the switch is closed? What is the total duration of the transient state?

43-2. When a 12-V battery is switched across a coil the current increases from zero to a final level of 240 mA in a time of 80 ms. Calculate the values of the coil's resistance, its self-inductance, and the time constant.

43-3. In Fig. 43-4, the switch (S) is closed in position 1. Find the values of the time constant, and the values of the inductor's current after time intervals of 50, 80, 120, 200 and 300 ms.

43-4. In Practice Problem 43-3, the circuit reaches its final steady-state condition with the switch in position 1. S is now switched to position 2. Calculate the initial values of the induced EMF and the rate of the current's decay. What is the new value of the circuit's time constant?

43-5. In Fig. 43-5, the switch S1 is closed, but the switch S2 is left open. How long does it take for the inductor's current to reach 1 A?

43-6. In Practice Problem 43-5, the switch S1 has been closed for so long that the current through the inductor is limited only by the circuit's resistance. Switch S1 is then opened and switch S2 is closed. How long does it take for the inductor current to decay to 1 A?

43-7. In Fig. 43-6, what are the voltage drops across the 4.7-kΩ resistor after time intervals of 3.75 μs and 18.75 μs have elapsed from the time the switch S is closed?

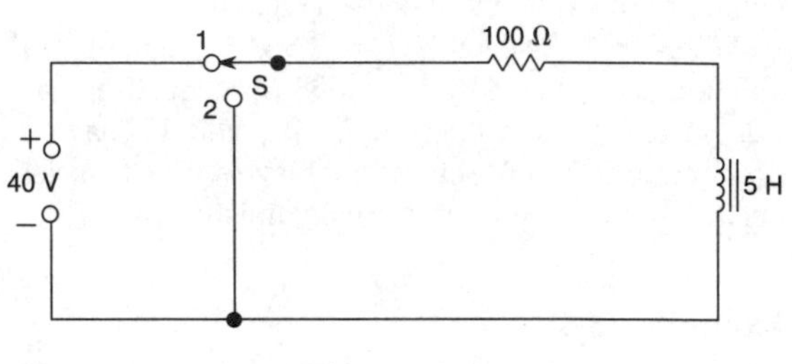

Figure 43-4

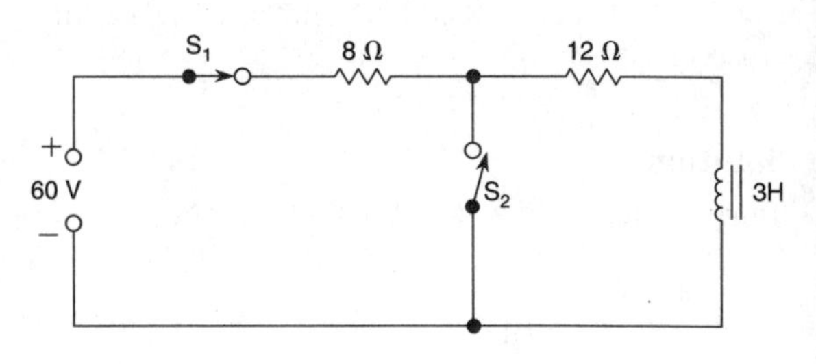

Figure 43-5

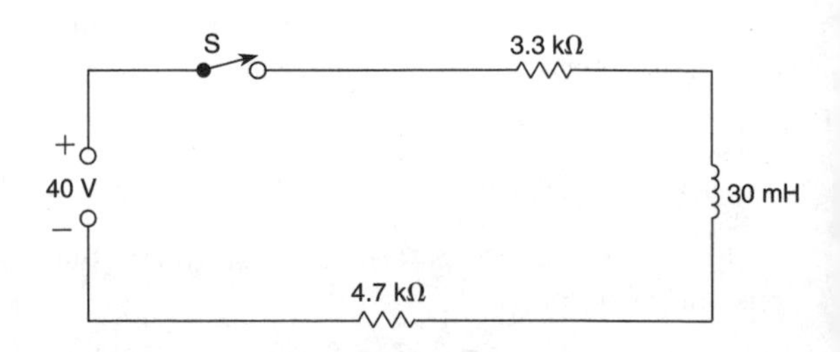

Figure 43-6

44
Inductors in parallel

It is assumed that all three inductors in Fig. 44-1 are ideal, in the sense that they have no losses and they only possess the property of inductance. Because these components are joined between two common points A and B, the source voltage will be across every inductor. Each inductor will then carry a different rate of growth of current so that its magnetic field will induce the same counter EMF, which exactly balances the source voltage. The sum of the individual rates of current growth will equal the total rate of growth associated with the source voltage.

When the switch (S) is closed, there will be a constant rate of current growth in each inductor and the associated magnetic fields will continue to expand. After a period of time, there will be a certain amount of energy stored in each of the three magnetic fields, and the sum of these energies will equal the total energy derived from the source voltage. Because the inductors are regarded as ideal components, there is no power *dissipated* in the circuit as heat.

From the mathematical derivations you can see that the parallel arrangement of inductors reduces the value of the total inductance. The reasons for connecting inductors in parallel are to increase the current capability and to obtain nonstandard values from standard components.

The principles of inductors in series (chapter 43) and parallel may be extended to series-parallel arrangements (see Example 44-2).

MATHEMATICAL DERIVATIONS

Rate of growth of current in the inductor L1,

$$\frac{\Delta I_1}{\Delta T} = \frac{E}{L_1} \text{ amperes per second} \tag{44-1}$$

Rate of growth of current in the inductor L2,

$$\frac{\Delta I_2}{\Delta T} = \frac{E}{L_2} \text{ amperes per second} \tag{44-2}$$

Rate of growth of current in the inductor L3,

$$\frac{\Delta I_3}{\Delta T} = \frac{E}{L_3} \text{ amperes per second} \tag{44-3}$$

For the meaning of the "Δ" sign, see chapter 42.

Total rate of growth of the source current,

$$\frac{\Delta I_T}{\Delta T} = \frac{\Delta I_1}{\Delta T} + \frac{\Delta I_2}{\Delta T} + \frac{\Delta I_3}{\Delta T} \tag{44-4}$$

$$= \frac{E}{L_1} + \frac{E}{L_2} + \frac{E}{L_3}$$

$$= E \times \left(\frac{1}{L_1} + \frac{1}{L_2} + \frac{1}{L_3}\right) \text{ amperes per second}$$

If L_T is the total inductance,

$$\frac{\Delta I_T}{\Delta T} = \frac{E}{L_T} = E \times \frac{1}{L_T} \text{ amperes per second} \tag{44-5}$$

Comparing Equations 44-4, 44-5:

$$\frac{1}{L_T} = \frac{1}{L_1} + \frac{1}{L_2} + \frac{1}{L_3}$$

$$\text{or } L_T = \frac{1}{\frac{1}{L_1} + \frac{1}{L_2} + \frac{1}{L_3}} \text{ henrys} \tag{44-6}$$

For N inductors in parallel:

$$L_T = \frac{1}{\frac{1}{L_1} + \frac{1}{L_2} + \frac{1}{L_3} \ldots + \frac{1}{L_N}} \text{ henrys} \tag{44-7}$$

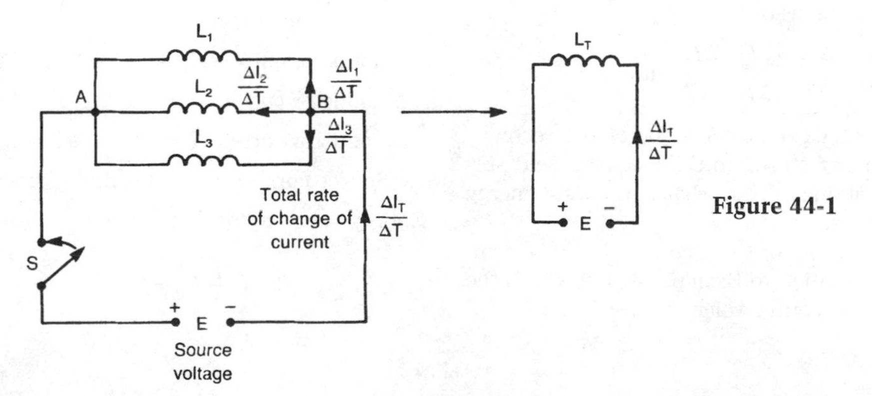

Figure 44-1

If all N inductors are of equal value, which is L henrys,

$$L_T = L/N \text{ henrys} \tag{44-8}$$

For two inductors in parallel,

$$L_T = \frac{1}{\dfrac{1}{L_1} + \dfrac{1}{L_2}} \tag{44-9}$$

$$= \frac{L_1 \times L_2}{L_1 + L_2} \text{ ("product-over-sum" formula)}$$

This yields,

$$L_1 = \frac{L_2 \times L_T}{L_2 - L_T} \text{ and } L_2 = \frac{L_1 \times L_T}{L_1 - L_T} \tag{44-10}$$

Consequently, the equations for inductors in parallel are comparable with those for resistors in *parallel*.

Energy stored in the magnetic field surrounding the L1 inductor,

$$W_1 = \frac{1}{2} L_1 I_1^2 \text{ joules} \tag{44-11}$$

Energy stored in the magnetic field surrounding the L2 inductor,

$$W_2 = \frac{1}{2} L_2 I_2^2 \text{ joules} \tag{44-12}$$

Energy stored in the magnetic field surrounding the L3 inductor,

$$W_3 = \frac{1}{2} L_3 I_3^2 \text{ joules} \tag{44-13}$$

Total energy derived from the source,

$$W_T = W_1 + W_2 + W_3 \tag{44-14}$$

$$= \frac{1}{2} L_T I_T^2 \text{ joules}$$

Example 44-1

In Fig. 44-1, $L_1 = 4$ H, $L_2 = 8$ H, $L_3 = 20$ H, and $E = 43$ V.

Calculate the values of:

$$\frac{\Delta I_1}{\Delta T}, \frac{\Delta I_2}{\Delta T}, \frac{\Delta I_3}{\Delta T}, \frac{\Delta I_T}{\Delta T}, \text{ and } L_T.$$

If the switch is closed for a period of four seconds, what is the energy stored in the magnetic field surrounding each inductor and what is the total energy derived from the source?

Note: The solution for this example should be compared with the solution to Example 43-1, in which the same inductor and source values are used.

Solution

Rate of the current growth in the L1 inductor,

$$\frac{\Delta I_1}{\Delta T} = \frac{43 \text{ V}}{4 \text{ H}} \tag{44-1}$$

$$= 10.75 \text{ A/s}$$

Rate of the current growth in the L2 inductor,

$$\frac{\Delta I_2}{\Delta T} = \frac{43 \text{ V}}{8 \text{ H}} \tag{44-2}$$

$$= 5.375 \text{ A/s}$$

Rate of the current growth in the L3 inductor,

$$\frac{\Delta I_3}{\Delta T} = \frac{43 \text{ V}}{20 \text{ H}} \tag{44-3}$$

$$= 2.15 \text{ A/s}$$

Total rate of growth of the source current,

$$\frac{\Delta I_T}{\Delta T} = 10.75 + 5.375 + 2.15 \tag{44-4}$$

$$= 18.275 \text{ A/s}$$

Total equivalent inductance,

$$L_T = \frac{E}{\Delta I_T/\Delta T} \tag{44-5}$$

$$= \frac{43 \text{ V}}{18.275 \text{ A/s}}$$

$$= 2.35 \text{ H}$$

Total inductance check:

$$L_T = \frac{1}{\dfrac{1}{L_1} + \dfrac{1}{L_2} + \dfrac{1}{L_3}} \tag{44-6}$$

$$= \frac{1}{\dfrac{1}{4} + \dfrac{1}{8} + \dfrac{1}{20}}$$

$$= 2.353 \text{ H}$$

After a period of four seconds,

L_1 inductor current, $I_1 = 10.75$ A/s $\times$ 4 s = 43 A.

L_2 inductor current, $I_2 = 5.375$ A/s $\times$ 4 s = 21.5 A.

L_3 inductor current, $I_3 = 2.15$ A/s $\times$ 4 s = 8.6 A.

Source current, $I_T = 43 + 21.5 + 4.6 = 73.1$ A.

Energy stored in the L1 inductor,

$$W_1 = \frac{1}{2} L_1 I_1^2 \tag{44-11}$$

$$= \frac{1}{2} \times 4\text{ H} \times (43\text{ A})^2$$

$$= 3698\text{ joules.}$$

Energy stored in the L2 inductor,

$$W_2 = \frac{1}{2} L_2 I_2^2 \qquad (44\text{-}12)$$

$$= \frac{1}{2} \times 8\text{ H} \times (21.5\text{ A})^2$$

$$= 1849\text{ joules.}$$

Energy stored in the L3 inductor,

$$W_3 = \frac{1}{2} L_3 I_3^2 \qquad (44\text{-}13)$$

$$= \frac{1}{2} \times 20\text{ H} \times (8.6\text{ A})^2$$

$$= 739.6\text{ joules.}$$

Total energy stored,

$$W_T = W_1 + W_2 + W_3 \qquad (44\text{-}14)$$

$$= 3698 + 1849 + 739.6$$

$$= 6286.6\text{ joules.}$$

Energy check:

Total energy derived from the source,

$$W_T = \frac{1}{2} L_T I_T^2 \qquad (44\text{-}14)$$

$$= \frac{1}{2} \times 2.353\text{ H} \times (73.1\text{ A})^2$$

$$= 6286.8\text{ joules.}$$

Example 44-2

In Fig. 44-2, calculate the total value of the inductance between points A and B.

Solution

Inductance of

$$5\text{ H} // 20\text{ H} = \frac{5 \times 20}{5 + 20} = 4\text{ H} \qquad (44\text{-}9)$$

Combined inductance in the right-hand branch between points A and B,

$$4 + 8 = 12\text{ H} \qquad (44\text{-}6)$$

Inductance in the left-hand branch between points A and B = 6 H.

Total inductance between points A and B,

$$L_T = \frac{12 \times 6}{12 + 6} \qquad (44\text{-}9)$$

$$= \frac{72}{18} = 4.0\text{ H.}$$

PRACTICE PROBLEMS

44-1. In Fig. 44-3, calculate the value of the equivalent total inductance between points A and B.

44-2. A 20-H inductor is connected in parallel with a 5-H inductor. This parallel combination is then joined in series with a 4-H inductor. If the complete circuit is switched across a 12-Vdc source, calculate the value of the initial EMF induced in the 5-H inductor.

44-3. A 30-mH inductor is connected in parallel with a 60-mH inductor. If the rate of change of current through the 30-mH inductor is 10 A/s, determine the total rate of change of current for both inductors.

44-4. Two chokes with self-inductances of 20 H and 5 H are connected in parallel. They are then suddenly switched across a dc source. If the total rate of the current growth drawn from the source is 4 A/s, what is the value of the source voltage?

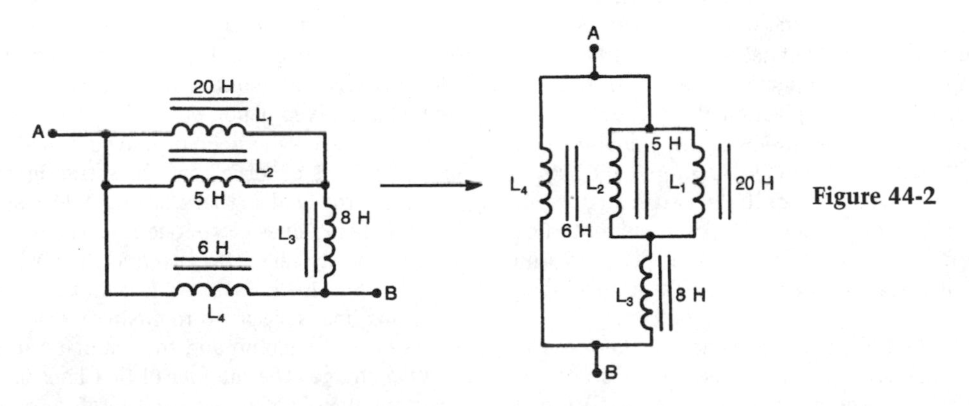

Figure 44-2

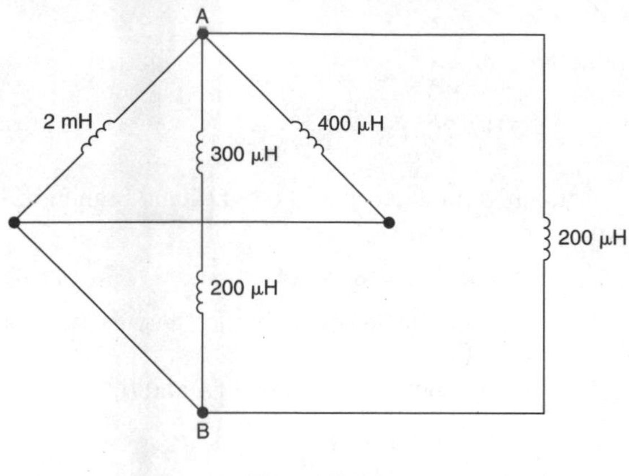

Figure 44-3

44-5. Three coils have self-inductances of 20 mH, 40 mH, and 60 mH. How can these be connected to provide a total self-inductance of 44 mH?

44-6. Two coils have a total self-inductance of 50 H when connected in series, but only 8 H when connected in parallel. What are the values of the individual self-inductances?

44-7. In Fig. 44-4, what are the initial values of I_1, I_2, I_T, V_1, V_2, and V_3 *immediately* after the switch (S) is closed.

44-8. In Practice Problem 44-7, assume that the transient state has been concluded. What are the steady-state values of I_1, I_2, I_T, V_1, V_2, and V_3?

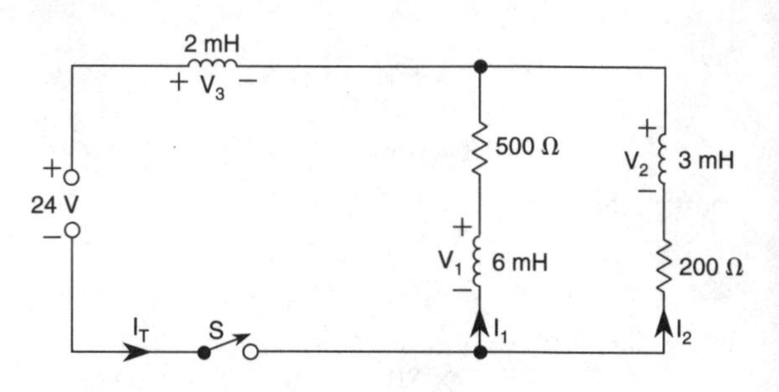

Figure 44-4

45
Electric flux—Coulomb's law

When two neutral insulators (such as a rubber rod and a piece of fur) are rubbed together, electrons will literally be wiped off from one insulator to the other. This is referred to as the *triboelectric effect,* or *friction effect.* Electrons are transferred from the fur to the rod; the fur has a deficit of electrons and is positively charged, but the rod has a surplus of electrons and carries an equal negative charge. If the two charged insulators are positioned near to one another, there is no electron flow and the charges cannot move. This phenomenon is therefore called *static electricity*, or *electrostatics.* If the two insulators are then brought into contact, equalizing currents flow in the form of sparks (which produce a crackling sound) that continue until the insulators are again electrically neutral.

In magnetism, like poles repel and unlike poles attract. A similar law applies to electrostatics because charges with the same polarity (for example, two negative charges) repel each other, but charges with the opposite polarity (a positive charge and a negative charge) are subjected to a force of attraction. The force of attraction, or repulsion, is caused by the influence that a charged body exerts on its surroundings. This influence is called the *electric* (or *electrostatic*) *field*, which terminates on a charged body, and extends between positive and negative charges. Such a field can exist in a vacuum (free space) or in insulators of air, glass, waxed paper, etc.

As with the magnetic field, an electric field is represented by lines of force that show the direction and strength of the field so that a strong field is shown as a large number of lines close together. The total number of force lines is called the *electric flux*, whose letter symbol is the Greek letter psi (ψ). In the SI system, a single flux line is assumed to emanate from a positive charge of one coulomb and to terminate on an equal negative charge. The number of flux lines is the same as the number of coulombs and, therefore, can be measured directly in coulombs.

Figures 45-1A and B show the electric flux patterns for the forces of attraction and repulsion. Coulomb discovered that the magnitude of the force between two charged bodies is directly proportional to the product of the charges, but is inversely proportional to the square of their separation.

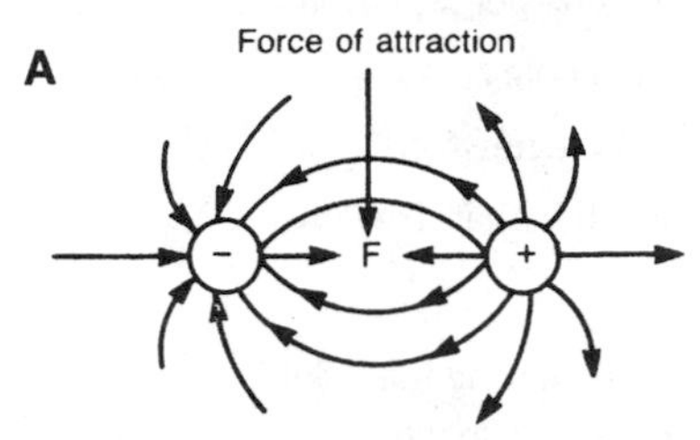

Figure 45-1

MATHEMATICAL DERIVATIONS

Coulomb's law

Force of attraction or repulsion,

$$F = \frac{8.99 \times 10^9 \times Q_1 \times Q_2}{d^2} \text{ newtons} \qquad (45\text{-}1)$$

where: Q_1, Q_2 = charges (C)
d = separation (m).

Example 45-1

In free space a negative charge of 8 μC is positioned 5 cm away from another negative charge of 8 μC. What are the values of the total electrical flux and the force exerted on the charges? Is the force one of attraction or repulsion?

Solution

Total electrical flux, $\psi = 8\ \mu$C

$$\text{Force, } F = \frac{8.99 \times 10^9 \times Q_1 \times Q_2}{d^2} \qquad (45\text{-}1)$$

$$= \frac{8.99 \times 10^9 \times (8 \times 10^{-6})^2}{(5 \times 10^{-2})^2}$$

$$= \frac{8.99 \times 64 \times 10}{25}$$

$$= 230 \text{ N.}$$

Because the charges are of the same polarity, the force will be one of *repulsion*.

PRACTICE PROBLEM

45-1. In free space, a positive charge of 12 μC is positioned 8 cm away from a negative charge of 12 μC. What are the values of the force of attraction exerted on the charges, and the total electrical flux between the charges?

46 Charge density—electric field intensity—absolute and relative permittivity

Figure 46-1 illustrates two identical rectangular plates, which are made from a conducting material, such as silver, copper, aluminum, etc. The area of one side of one plate is A square meters, and the two plates are separated by a distance of d meters in free space. These plates are charged to a potential difference of V_C volts so that one plate carries a negative charge of Q coulombs and the other plate has an equal positive charge. The total electric flux (ψ) is therefore Q coulombs and consists of Q lines of force, which are spread over an area of A square meters. The flux density, whose letter symbol is D, is found by dividing ψ by A and is measured in coulombs per square meter.

The flux density (D) is the result of the electric field intensity (ε) which exists between the two plates. Increasing the value of V_C and/or reducing the separation (d) between the plates, causes a stronger electric field; so that the electric field intensity (ε) is expressed in terms of the voltage gradient (V_C/d) which is measured in volts per meter. The ratio of D:ε is

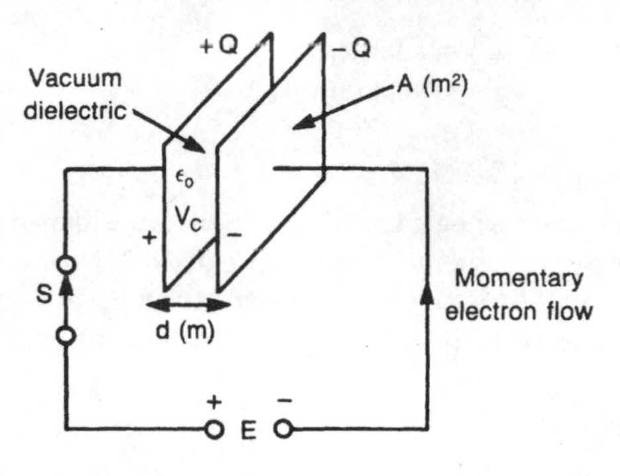

Figure 46-1

termed the *permittivity of free space* whose letter symbol is ϵ_o (also known as the *absolute permittivity of a vacuum*), and whose value is 8.85×10^{-12} farad per meter in SI units.

If a negative charge (coulombs) is placed in the vacuum, it will be repelled from the negative plate and will be accelerated towards the positive plate. The accelerating force in newtons is directly proportional to the magnitude of the charge and the value of the electric field intensity.

If the free space between the plates is replaced by an insulating material, such as mica, waxed paper, etc., the flux density for a given value of electric field intensity, is increased by a factor called the *relative permittivity*, ϵ_r. This factor has no units and is defined by:

Relative permittivity, ϵ_r =

$$\frac{\text{electric flux density in the insulating material for a given field intensity}}{\text{electric flux density in free space for the same field intensity}}$$

Values for ϵ_r are 1.0006 for air, 3 to 6 for mica and 3.5 for waxed paper.

It follows that, for an insulating material, the ratio of D:ε is $\varepsilon_o\epsilon_r = \epsilon$, which is the insulator's absolute permittivity, measured in farads per meter.

MATHEMATICAL DERIVATIONS

Flux density,

$$D = \frac{\psi}{A} \text{ coulombs per square meter} \quad (46\text{-}1)$$

Total electric flux,

$$\psi = D \times A \text{ coulombs} \quad (46\text{-}2)$$

Electric field intensity,

$$\varepsilon = \frac{V}{d} \text{ volts per meter} \quad (46\text{-}3)$$

In a vacuum,

$$\frac{\text{Electric flux density, } D}{\text{Electric field intensity, } \varepsilon} = \epsilon_o \quad (46\text{-}4)$$

$$= 8.85 \times 10^{-12} \text{ farad per meter}$$

Electric flux density,

$$D = \varepsilon\,\epsilon_o \text{ coulombs per square meter} \quad (46\text{-}5)$$

Force exerted on a charge,

$$F = Q\varepsilon \text{ newtons} \quad (46\text{-}6)$$

For an insulator other than free space,

Relative permittivity, ϵ_r =

$$\frac{D \text{ in the insulating material}}{D \text{ in free space}} \quad (46\text{-}7)$$

For a given electric field intensity,

$$\frac{\text{Electric flux density, } D}{\text{Electric field intensity, } \varepsilon} = \epsilon_o\epsilon_r \quad (46\text{-}8)$$

$$= \epsilon \text{ farads per meter}$$

Example 46-1

In Fig. 46-1, the area of one side of each plate is 4 cm^2 and their separation in free space is 0.8 cm. If the voltage between the plates is 50 V, what are the values of the electric field intensity, the electric flux density and the charge on each plate? If an insulator whose relative permittivity is 3.5 is inserted between the plates, what are the new values of the electric flux density and the charge on each plate?

Solution

Free space

Electric field intensity,

$$\varepsilon = \frac{V}{d} \quad (46\text{-}3)$$

$$= \frac{50 \text{ V}}{0.8 \times 10^{-2} \text{ m}}$$

$$= 6250 \text{ V/m.}$$

Flux density,

$$D = \varepsilon \times \epsilon_o \quad (46\text{-}5)$$

$$= 6250 \times 8.85 \times 10^{-12}$$

$$= 5.53 \times 10^{-8} \text{ C/m}^2.$$

Charge on each plate,

$$Q = \text{flux, } \psi = D \times A \quad (46\text{-}2)$$

$$= 5.53 \times 10^{-8} \text{ C/m}^2 \times 4 \times 10^{-4} \text{ m}^2$$

$$= 22.1 \text{ pC.}$$

Insulator

There is no change in the value of the electric field intensity.

Flux density,

$$D = \varepsilon \times \epsilon_o \epsilon_r \quad (46\text{-}8)$$
$$= 5.53 \times 10^{-8} \times 3.5$$
$$= 1.935 \times 10^{-7}\ \text{C/m}^2.$$

Charge on each plate,

$$Q = 1.935 \times 10^{-7}\ \text{C/m}^2 \times 4 \times 10^{-4}\ \text{m}^2$$
$$= 77.4\ \text{pC}.$$

PRACTICE PROBLEMS

46-1. Two aluminum foil surfaces are separated by a paper ($\epsilon_r = 3.5$) dielectric whose thickness is 0.15 mm. The area of one side of each surface is 80 cm². If the voltage between the surfaces is 120 V, calculate the values of the electric flux, the electric field intensity, and the electric flux density.

46-2. In Practice Problem 46-1, the dielectric is removed and replaced by free space. What are the new values of the electric flux density, and the charge on each plate?

47
Capacitance and the capacitor

Capacitance is that property of an electrical circuit that prevents any sudden change in *voltage* and limits the rate of change in the voltage. In other words, the voltage across a capacitance, whose letter symbol is C, cannot change instantaneously.

The property of capacitance is possessed by a capacitor, which is a device for storing charge, and consists of two conducting surfaces (copper, silver, aluminum, tin foil, etc.), which are separated by an insulator or dielectric (air, mica, ceramic, etc.).

When a dc voltage (E) is switched across the capacitor (C) in Fig. 47-1, electrons are drawn off the right-hand plate, flow through the voltage source and are then deposited on the left-hand plate. This electron flow is only momentary and ceases when the voltage (V_C) between the plates exactly balances the source voltage, E. The charge stored by the capacitor directly depends on the value of the applied voltage and also on a constant of the capacitor, called the *capacitance*, which is determined by the component's physical construction. The unit for the capacitance is the *farad* (F). The capacitance is one farad if, when the voltage applied across the capacitor's plates is one volt, the charge stored is one coulomb. Unfortunately, the farad is far too large a unit for practical purposes so that capacitances are normally measured in either microfarads (μF) or picofarads (pF). Notice that a picofarad can alternatively be referred to as a micromicrofarad ($\mu\mu$F).

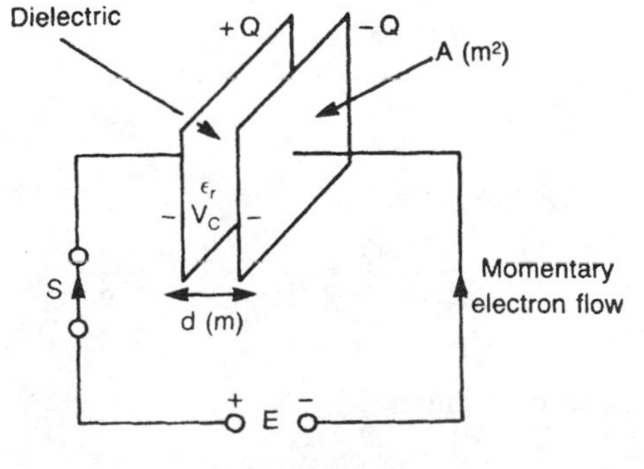

Figure 47-1

We already know that a particular electric field intensity (which is inversely proportional to the separation between the plates) produces a certain flux (which is directly proportional to the area of one of the plates). This means that a capacitor's capacitance is directly proportional to the area of the plates, but is inversely proportional to their distance apart. It is true that the capacitance is increased if the conductor plates are brought closer together. However, if the dielectric is made too thin, the voltage between the plates can cause arcing to occur and the capacitor will be damaged. Each type of insulator has a certain dielectric strength, which is a measure of the insulator's ability to withstand a high electric field intensity. For example, the dielectric strength of mica is typically 50 kV/mm so that, in order to avoid breakdown, no more than 50 kV should be applied across a 1-mm thickness of mica. It follows that each capacitor is rated for a particular dc working voltage (WVdc).

MATHEMATICAL DERIVATIONS

For the capacitor in Fig. 47-1,

$$V_C \text{ (volts)} = -\frac{Q \text{ (coulombs)}}{C \text{ (farads)}} \qquad (47\text{-}1)$$

Using KVL,

$$E + V_C = 0.$$

Therefore,

$$E = \frac{Q}{C}, Q = CE, C = \frac{Q}{E} \qquad (47\text{-}2)$$

Flux density,

$$D = \frac{Q}{A} \text{ coulombs per square meter} \qquad (47\text{-}3)$$

Electric field intensity,

$$\varepsilon = \frac{E}{d} \text{ volts per meter} \qquad (47\text{-}4)$$

For a capacitor with a vacuum dielectric,
Permittivity of free space,

$$\epsilon_o = \frac{D}{\varepsilon} \qquad (47\text{-}5)$$

$$= \frac{Q/A}{E/d}$$

$$= \frac{Q}{E} \times \frac{d}{A}$$

$$= \frac{C \times d}{A}$$

Equation 47-5 yields,
Capacitance,

$$C = \epsilon_o \times \frac{A}{d} \text{ farads} \qquad (47\text{-}6)$$

If the dielectric has a relative permittivity of ϵ_r,
Capacitance,

$$C = \frac{\epsilon_o \epsilon_r A}{d} = \frac{\epsilon A}{d} \text{ farads} \qquad (47\text{-}7)$$

where ϵ is the dielectric's absolute permittivity, which is measured in farads per meter.

Since $Q = CE$,

$$I = \frac{dQ}{dt} = C\frac{dE}{dt}$$

or,

$$\frac{dE}{dt} = \frac{I}{C} \text{ volts per second} \qquad (47\text{-}8)$$

Because dE/dt is inversely proportional to C, the capacitance will limit the rate of change in the voltage.

Example 47-1

Two parallel aluminum plates are separated by 0.8 mm in air. If the area of one side of each plate is 50 cm^2, what is the value of the capacitance?

Solution

Capacitance,

$$C = \frac{\epsilon_o \epsilon_r A}{d} \qquad (47\text{-}7)$$

$$= \frac{8.85 \times 10^{-12} \times 1.0006 \times 50 \times 10^{-4}}{0.8 \times 10^{-3}} \text{ F}$$

$$= 55 \text{ pF.}$$

Example 47-2

A paper capacitor consists of two sheets of tin foil, each side of which has an area of 1500 cm^2. The sheets are separated by a 0.15 mm thickness of a waxed paper dielectric with a relative permittivity of 2.7. What is the value of the capacitance? If the capacitor is charged to 120 V, what are the values of the electric field intensity, the electric flux density, and the charge stored?

Solution

Capacitance,

$$C = \frac{\epsilon_o \epsilon_r A}{d} \qquad (47\text{-}7)$$

$$= \frac{8.85 \times 10^{-12} \times 2.7 \times 1500 \times 10^{-4}}{0.15 \times 10^{-3}} \text{ F}$$

$$= 0.0239\ \mu\text{F.}$$

Charge stored,

$$Q = CE$$

$$= 0.0239\ \mu\text{F} \times 120 \text{ V}$$

$$= 2.87\ \mu\text{C.}$$

Electric flux density,

$$D = \frac{\psi}{A} = \frac{Q}{A} = \frac{2.87\ \mu\text{C}}{1500 \times 10^{-4} \text{ m}^2}$$

$$= 19.1\ \mu\text{C/m}^2.$$

Electric field intensity,

$$\varepsilon = \frac{V_C}{d}$$

$$= \frac{120 \text{ V}}{0.15 \times 10^{-3} \text{ m}} = 8 \times 10^5 \text{ V/m}.$$

PRACTICE PROBLEMS

47-1. A capacitor has two identical conducting surfaces which are separated by a 0.1-mm thickness of mica ($\epsilon_r = 5.4$). If the area of one side of each conducting surface is 0.2 cm², and the voltage between the surfaces is 100 V, calculate the values of the capacitance, the flux density, and the electric field intensity.

47-2. In Practice Problem 47-1, determine the force of attraction between the two surfaces.

47-3. The capacitance of a parallel plate capacitor with a free space dielectric is 250 pF. If the area of one side of each plate is doubled and the separation between the plates is also doubled, determine the new value of the capacitance when an insulator ($\epsilon_r = 2$) is substituted for the free-space dielectric.

47-4. The voltage to which a capacitor is charged rises from 60 V to 80 V in a time of 2.5 μs. If the average value of the charging current is 0.8 A, calculate the value of the capacitance.

47-5. The initial voltage across a 0.03-μF capacitor is 60 V, but after a time interval of 5 ms, this voltage has risen to 105 V. What is the average value of the charging current? If the capacitor is then discharged by a constant current of 1.8 mA, how long would it take for the capacitor to discharge completely?

48
Energy stored in the electric field of a capacitor

Consider that a capacitor is being charged from a constant current source of I amperes over a period of t seconds (Fig. 48-1A). The voltage across the capacitor then increases in a linear manner, as shown in the voltage-versus-time graph of Fig. 48-1B. At any time after the switch is closed, one of the plates has acquired a positive charge and the other is negatively charged; the charge stored by the capacitor is given by Q (coulombs) = C (capacitance in farads) × V_C (volts); see equation 47-2.

When the capacitor is charged, an electric field (ψ), also measured in coulombs, exists between the plates. This represents the stored energy that has been derived from the constant current source. In another sense, the same energy is related to the force of attraction between the positive and negative plates.

MATHEMATICAL DERIVATIONS

Final charge,

$$Q = I \times t \text{ coulombs} \quad (48\text{-}1)$$

Voltage across the capacitor, $V_C = \dfrac{Q}{C}$ volts (48-2)

Average voltage across the capacitor,

$$V_{av} = \frac{V_C}{2} \text{ volts} \quad (48\text{-}3)$$

Average power at the source

$$= I \times \frac{V_C}{2} \text{ watts} \quad (48\text{-}4)$$

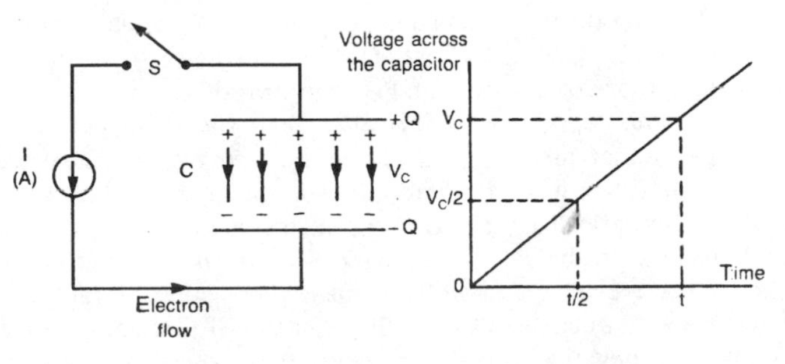

Figure 48-1

Total energy supplied from the source during the charging process is:

$$\text{average power} \times \text{time} = I \times \frac{V_C}{2} \times t$$

$$= \frac{1}{2} \times V_C Q = \frac{1}{2} \times \frac{Q^2}{C} = \frac{1}{2} \times CV_C^2 \text{ joules} \qquad (48\text{-}5)$$

Alternatively:

The work done in moving a charge through a potential difference (δV_C) is $Q \times \delta V_C$ (joules = coulombs × volts) = $CV_C \delta V_C$ (48-6)

Now integrate over a total period of t seconds. Total work done or total energy stored,

$$W = \int_o^{V_C} CV_C \, dV_C \qquad (48\text{-}7)$$

$$= \frac{1}{2} CV_C^2 \text{ joules}$$

Example 48-1

In Fig. 48-1, the switch (S) is closed and an 8-μF capacitor is charged for a period of 7 seconds from a 5 μA constant current source. At the end of the charging period, what is the amount of charge acquired by the capacitor, and what is the corresponding amount of energy stored in the electric field?

Solution

Final charge,

$$Q = I \times t = 5\ \mu\text{A} \times 7\ \text{s} = 35\ \mu\text{C} \qquad (48\text{-}1)$$

Final voltage across the capacitor,

$$V_C = \frac{Q}{C} = \frac{35\ \mu\text{C}}{8\ \mu\text{F}} = 4.75\ \text{V} \qquad (48\text{-}2)$$

Total energy stored in the capacitor,

$$W = \frac{1}{2} \times V_C \times Q = \frac{1}{2} \times 4.75\ \text{V} \times 35\ \mu C \qquad (48\text{-}5)$$

$$= 83.125\ \mu\text{J}$$

Notice that the amount of energy stored in a capacitor is normally very small.

PRACTICE PROBLEMS

48-1. A variable capacitor with an initial capacitance of 800 pF is charged to 200 V. The plates of the capacitor are then separated until the capacitance is reduced to 200 pF. What is the value of the new voltage across the capacitor? Calculate the values of the energy stored before, and after, the plates are separated. Explain the difference between these two energy values.

48-2. To what voltage must a 0.02-μF capacitor be charged, in order that the energy stored is 100 μJ?

48-3. When a capacitor is connected across a 400-Vdc source, the energy stored is 0.3 J. Calculate the values of the capacitance, and the charge stored by the capacitor.

49
Capacitors in series

Because the capacitors are joined end-to-end in Fig. 49-1, there is only one path for current flow. Consequently, when S is closed in position 1, the charging current for each capacitor is the same and, therefore, each capacitor must store the same charge, Q (coulombs). But there are two problems that are worth considering. First, since the electron flow cannot pass through the dielectrics, which theoretically are open circuits, how does the middle capacitor become charged? The explanation lies in *electrostatic induction*. When the electron flow causes the plate (P1) to acquire a positive charge, an equal negative charge is induced on the plate, P2. This causes another positive charge of the same size to appear on the plate (P3) and so on. As the result of this process, the total charge stored between the plates P1 and P6 is Q (*not* $3Q$). Secondly, a current can only flow in a closed circuit but clearly, the dielectrics represent discontinuities in the circuit of Fig. 49-1. James Clerk Maxwell resolved this difficulty by proposing the so-called *displacement current,* which is not an electron flow, but is the result of a rate of change in the electric flux. On the SI system, the electric flux in a dielectric consists of lines where one line is assumed

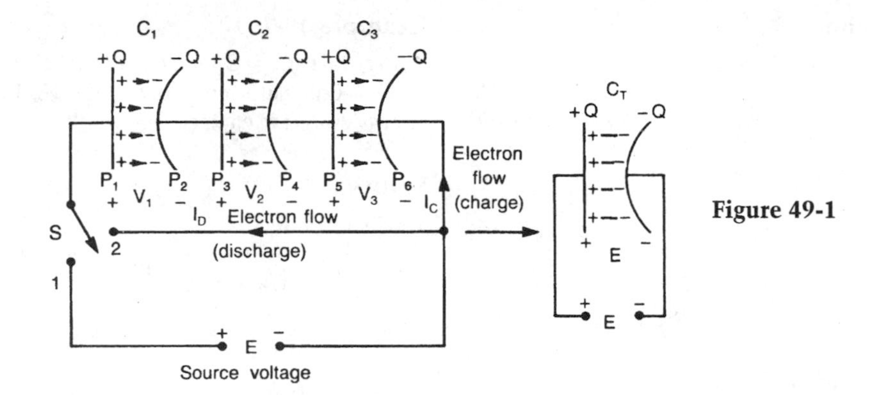

Figure 49-1

to emanate from a positive charge of one coulomb, and terminate on an equal negative charge. Consequently, the number of lines is the same as the number of coulombs, and the electric flux can therefore be measured in coulombs. When the capacitor is being charged, there is a change in the electric flux, and its rate of change will be equivalent to coulombs per second or amperes. This naturally results in the displacement current being measured in the same units as the electron flow.

After the capacitors are charged, there is a voltage across each capacitor, and the sum of these voltages must exactly balance the source voltage. Because $Q = VC$ (equation 47-2) and each capacitor carries the same charge, it follows that the *highest* voltage is across the *lowest* value of capacitance, and vice-versa.

Each capacitor stores a certain amount of energy, in the form of the electric field, between its plates. When the switch is moved to position 2, the capacitors discharge, and the total energy stored is released in the form of the spark at the contacts of the switch. Incidentally, the spark creates an electromagnetic wave and this was the principle behind the early spark-gap transmitter.

Finally, what is the purpose of connecting capacitors in series? As you shall see in the mathematical derivations, connecting capacitors in series *reduces* the total capacitance, so that the total capacitance is less than the value of the smallest capacitance in series. Basically, this is because the series arrangement effectively increases the distance between the end plates, P1 and P6, connected to the source voltage, and the capacitance is inversely proportional to this distance. Because the capacitor has a dc working voltage (WVdc), the series arrangement may be used to distribute the source voltage between the capacitors so that the voltage across an individual capacitor does not exceed its rating.

MATHEMATICAL DERIVATIONS

$$V_1 = \frac{Q}{C_1},\ Q = C_1V_1,\ C_1 = \frac{Q}{V_1} \qquad (49\text{-}1)$$

$$V_2 = \frac{Q}{C_2},\ Q = C_2V_2,\ C_2 = \frac{Q}{V_2} \qquad (49\text{-}2)$$

$$V_3 = \frac{Q}{C_3},\ Q = C_3V_3,\ C_3 = \frac{Q}{V_3} \qquad (49\text{-}3)$$

Source voltage,

$$E = V_1 + V_2 + V_3 \qquad (49\text{-}4)$$

$$= \frac{Q}{C_1} + \frac{Q}{C_2} + \frac{Q}{C_3}$$

$$= Q \times \left(\frac{1}{C_1} + \frac{1}{C_2} + \frac{1}{C_3}\right) \text{ volts}$$

If C_T is the total equivalent capacitance,

$$E = \frac{Q}{C_T} = E \times \frac{1}{C_T} \text{ volts} \qquad (49\text{-}5)$$

Comparing Equations 49-4 and 49-5,

$$\frac{1}{C_T} = \frac{1}{C_1} + \frac{1}{C_2} + \frac{1}{C_3} \qquad (49\text{-}6)$$

For N capacitors in series,

$$C_T = \frac{1}{\frac{1}{C_1} + \frac{1}{C_2} + \frac{1}{C_3} \cdots + \frac{1}{C_N}} \text{ farads} \qquad (49\text{-}7)$$

If the N capacitors are all of equal value C,

$$C_T = \frac{C}{N} \text{ farads} \qquad (49\text{-}8)$$

The equations for capacitors in series can therefore be compared with those for resistors in parallel. Consequently connecting capacitors in series *reduces* the total capacitance.

Voltage division

Source voltage,

$$E = Q \times \frac{1}{C_T} = C_1 V_1 \times \frac{1}{C_T} \text{ volts} \quad (49\text{-}9)$$

Therefore,

$$V_1 = E \times \frac{C_T}{C_1}, V_2 = E \times \frac{C_T}{C_2}, \quad (49\text{-}10)$$

$$V_3 = E \times \frac{C_T}{C_3} \text{ volts}$$

Energy stored in the C1 capacitor,

$$W_1 = \frac{1}{2} C_1 V_1^2 \quad (49\text{-}11)$$

$$= \frac{1}{2} \frac{Q^2}{C_1} = \frac{1}{2} QV_1 \text{ joules}$$

Energy stored in the C2 capacitor,

$$W_2 = \frac{1}{2} C_2 V_2^2 \quad (49\text{-}12)$$

$$= \frac{1}{2} \frac{Q^2}{C_2} = \frac{1}{2} QV_2 \text{ joules}$$

Energy stored in the C3 capacitor,

$$W_3 = \frac{1}{2} C_3 V_3^2 \quad (49\text{-}13)$$

$$= \frac{1}{2} \frac{Q^2}{C_3} = \frac{1}{2} QV_3 \text{ joules}$$

Total energy stored,

$$W_T = \frac{1}{2} QE = \frac{1}{2} C_T E^2 \quad (49\text{-}14)$$

$$= \frac{1}{2} \frac{Q^2}{C_T} \text{ joules}$$

For two capacitors in series,
Total capacitance,

$$C_T = \frac{1}{\frac{1}{C_1} + \frac{1}{C_2}} = \frac{C_1 \times C_2}{C_1 + C_2} \text{ farads} \quad (49\text{-}15)$$

This yields

$$C_1 = \frac{C_2 \times C_T}{C_2 - C_T}, C_2 = \frac{C_1 \times C_T}{C_1 - C_T} \text{ farads} \quad (49\text{-}16)$$

$$V_1 = E \times \frac{C_T}{C_1} = E \times \frac{C_2}{C_1 + C_2},$$

$$V_2 = E \times \frac{C_1}{C_1 + C_2} \text{ volts} \quad (49\text{-}17)$$

Example 49-1

In Fig. 49-1, $C_1 = 20\ \mu F$, $C_2 = 5\ \mu F$, $C_3 = 4\ \mu F$, and $E = 30$ V. Calculate the values of C_T, V_1, V_2, V_3, and the energy stored in each capacitor.

Solution

Total capacitance,

$$C_T = \frac{1}{\frac{1}{C_1} + \frac{1}{C_2} + \frac{1}{C_3}} \quad (49\text{-}7)$$

$$= \frac{1}{\frac{1}{20} + \frac{1}{5} + \frac{1}{4}} = 2\ \mu F$$

Voltage across the C1 capacitor,

$$V_1 = E \times \frac{C_T}{C_1} = 30 \text{ V} \times \frac{2\ \mu F}{20\ \mu F} \quad (49\text{-}10)$$

$$= 3 \text{ V.}$$

Voltage across the C2 capacitor,

$$V_2 = E \times \frac{C_T}{C_2} = 30 \text{ V} \times \frac{2\ \mu F}{5\ \mu F} \quad (49\text{-}10)$$

$$= 12 \text{ V}$$

Voltage across the C3 capacitor,

$$V_3 = E \times \frac{C_T}{C_3} \quad (49\text{-}10)$$

$$= 30 \text{ V} \times \frac{2\ \mu F}{4\ \mu F}$$

$$= 15 \text{ V}$$

Voltage check:
Source voltage,

$$E = V_1 + V_2 + V_3 \quad (49\text{-}4)$$

$$= 3 + 12 + 15$$

$$= 30 \text{ V}$$

Charge stored by each capacitor,

$$Q = \text{total charge stored} \quad (49\text{-}5)$$

$$= E \times C_T$$

$$= 30 \text{ V} \times 2\ \mu F$$

$$= 60\ \mu C$$

Energy stored by the C1 capacitor,

$$W_1 = \frac{1}{2} QV_1 \quad (49\text{-}11)$$

$$= \frac{1}{2} \times 60\ \mu C \times 3\ V$$

$$= 90\ \mu J$$

Energy stored by the C2 capacitor,

$$W_2 = \frac{1}{2} QV_2 \qquad (49\text{-}12)$$

$$= \frac{1}{2} \times 60\ \mu C \times 12\ V$$

$$= 360\ \mu J$$

Energy stored by the C3 capacitor,

$$W_3 = \frac{1}{2} QV_3 \qquad (49\text{-}13)$$

$$= \frac{1}{2} \times 60\ \mu C \times 15\ V$$

$$= 450\ \mu J$$

Total energy stored = 90 + 360 + 450 = 900 μJ.

Energy check:

$$\text{Total energy stored} = \frac{1}{2} QE \qquad (49\text{-}14)$$

$$= \frac{1}{2} \times 60\ \mu C \times 30\ V$$

$$= 900\ \mu J$$

Example 49-2

In Fig. 49-2, calculate the values of V_1 and V_2.

Solution

Voltage,

$$V_1 = E \times \frac{C_2}{C_1 + C_2} \qquad (49\text{-}17)$$

$$= 50\ V \times \frac{330\ pF}{220\ pF + 330\ pF}$$

$$= 30\ V.$$

$$V_2 = E \times \frac{C_1}{C_1 + C_2}$$

$$= 50\ V \times \frac{220\ pF}{220\ pF + 330\ pF}$$

$$= 20\ V.$$

Voltage check:

Source voltage, $E = V_1 + V_2 = 30 + 20 = 50$ V.

PRACTICE PROBLEMS

49-1. In Fig. 49-3, calculate the amount of energy stored in the 8-μF capacitor.

49-2. Three capacitors of identical dimensions possess dielectrices whose relative permittivities are respectively one, three, and five. If these capacitors are connected in series across a 180-Vdc source, calculate the value of the voltage across each capacitor.

49-3. A 6-μF and a 3-μF capacitor are joined in series across a 90-Vdc source. What are the values of the charge, and the energy stored in each capacitor? What is the value of the voltage across each capacitor?

49-4. The total equivalent capacitance of two capacitors in series is 8 μF. If the value of one capacitor is 20 μF, what is the value of the other capacitor?

49-5. Three capacitors whose values are 40 μF/100 WVdc, 20 μF/150 WVdc and 25 μF/125 WVdc are connected in series. What is the working voltage of the combination?

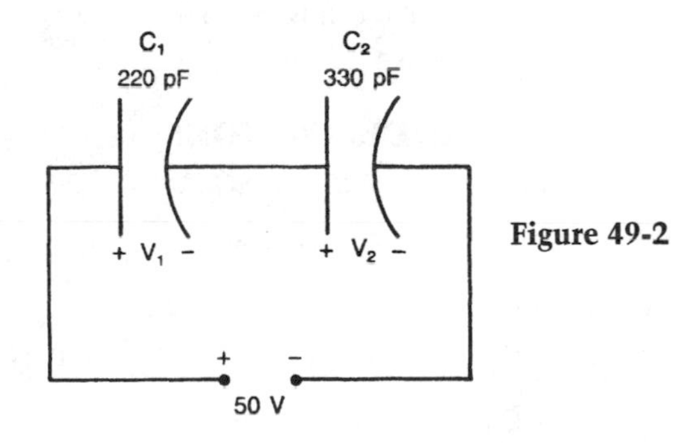

Figure 49-2

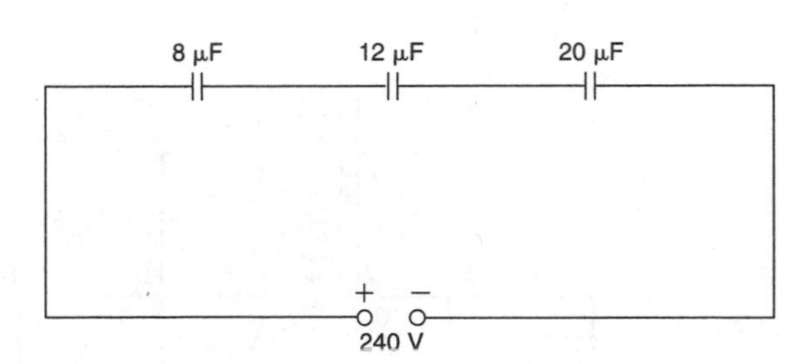

Figure 49-3

50
Capacitors in parallel

Capacitors in parallel are connected across two common points (A, B) so that the source voltage is applied across each of the capacitors (Fig. 50-1). Because $Q = CV$, it follows that a different charge must be stored by each capacitor. This is due to the different *momentary* charging currents that exist at the instant the switch (S) is closed. The total charge is then equal to the sum of the individual charges stored by each capacitor.

As we shall see in the mathematical derivations, the purpose of connecting capacitors in parallel is to increase the total capacitance. In Fig. 50-1, we have connected together on one side the plates P1, P3, P5, and, on the other side, the plates P2, P4, P6. The result is to increase the effective surface area to which the capacitance is directly proportional.

Each of the three capacitors in Fig. 50-1, stores energy in the form of the electric field between its plates. The total energy stored is the sum of the capacitor's individual energies.

The principles of capacitors in series (chapter 49) and capacitors in parallel can be extended to series-parallel arrangements (see Example 50-3).

MATHEMATICAL DERIVATIONS

$$\text{Charge } Q_1 = C_1E,\ Q_2 = C_2E,\ Q_3 = C_3E \text{ coulombs} \quad (50\text{-}1)$$

Total charge stored,

$$Q_T = Q_1 + Q_2 + Q_3 \quad (50\text{-}2)$$

$$= C_1E + C_2E + C_3E$$

$$= E \times (C_1 + C_2 + C_3) \text{ coulombs}$$

If C_T is the total equivalent capacitance,

$$\text{Total charge, } Q_T = E \times C_T \text{ coulombs} \quad (50\text{-}3)$$

Comparing Equations 50-2 and 50-3, it follows that,

$$\text{Total capacitance, } C_T = C_1 + C_2 + C_3 \text{ farads} \quad (50\text{-}4)$$

Consequently, adding capacitors in parallel increases the total capacitance. If a multiplate capacitor is made up of N parallel plates, with alternate plates connected as in Fig. 50-2, the total capacitance is,

$$C_T = \frac{\epsilon_o \epsilon_r (N-1)A}{d} \text{ farads} \quad (50\text{-}5)$$

where: $\epsilon_o = 8.85 \times 10^{-12}$ F/m,
ϵ_r = relative permittivity
A = area of *one* side of *one* plate (m^2)
d = distance between adjacent plates (m)

For N capacitors in parallel,

$$C_T = C_1 + C_2 + C_3 + \cdots + C_N \text{ farads} \quad (50\text{-}6)$$

If the N capacitors are all of equal value, C,

$$C_T = NC \text{ farads} \quad (50\text{-}7)$$

Energy stored by the C1 capacitor,

$$W_1 = \frac{1}{2}Q_1E \quad (50\text{-}8)$$

$$= \frac{1}{2}C_1E^2$$

$$= \frac{1}{2}Q_1^2/C_1 \text{ joules}$$

Energy stored by the C2 capacitor,

$$W_2 = \frac{1}{2}Q_2E \quad (50\text{-}9)$$

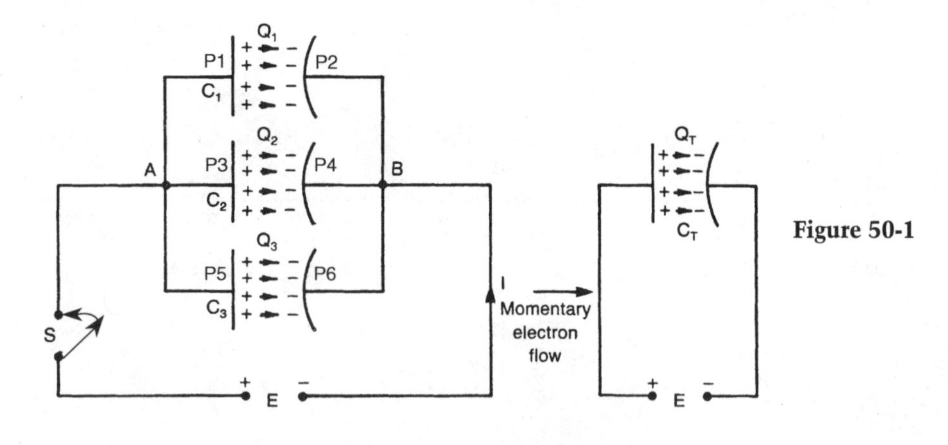

Figure 50-1

$$= \frac{1}{2} C_2 E^2$$

$$= \frac{1}{2} Q_2^2/C_2 \text{ joules}$$

Energy stored by the C3 capacitor,

$$W_3 = \frac{1}{2} Q_3 E \quad (50\text{-}10)$$

$$= \frac{1}{2} C_3 E^2$$

$$= \frac{1}{2} Q_3^2/C_3 \text{ joules}$$

Total energy stored,

$$W_T = \frac{1}{2} Q_T E = \frac{1}{2} C_T E^2 = \frac{1}{2} Q_T^2/C_T \text{ joules} \quad (50\text{-}11)$$

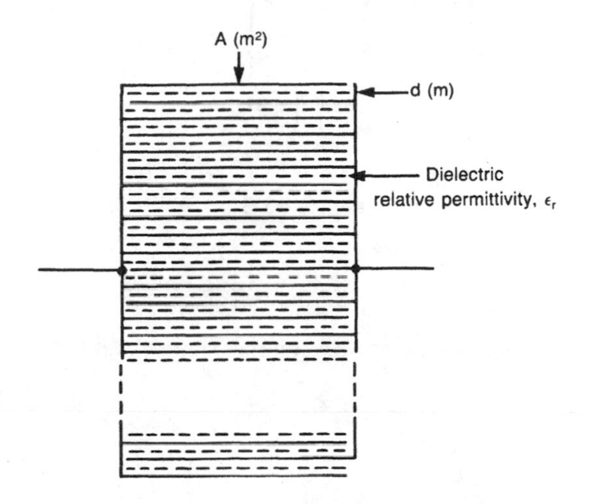

Figure 50-2

Example 50-1

In Fig. 50-1, $C_1 = 20\ \mu F$, $C_2 = 5\ \mu F$, $C_3 = 4\ \mu F$, and $E = 30$ V. Calculate the values of Q_1, Q_2, Q_3, Q_T, C_T, the energy stored by each capacitor, and the total energy derived from the source voltage. Note: The solution to this example should be compared with the solution to example 49-1, in which the series circuit contains the same values for the capacitors and the source voltage.

Solution

Charge, $Q_1 = C_1 E = 20\ \mu F \times 30\ V = 600\ \mu C$. (50-1)

Charge, $Q_2 = C_2 E = 5\ \mu F \times 30\ V = 150\ \mu C$. (50-1)

Charge, $Q_3 = C_3 E = 4\ \mu F \times 30\ V = 120\ \mu C$. (50-1)

Total charge stored,

$$Q_T = Q_1 + Q_2 + Q_3 \quad (50\text{-}2)$$

$$= 600 + 150 + 120$$

$$= 870\ \mu C.$$

Total capacitance,

$$C_T = C_1 + C_2 + C_3 \quad (50\text{-}4)$$

$$= 20 + 5 + 4$$

$$= 29\ \mu F.$$

Charge check:

Total charge,

$$Q_T = C_T \times E = 29\ \mu F \times 30\ V = 870\ \mu C \quad (50\text{-}3)$$

Energy stored in the C1 capacitor,

$$W_1 = \frac{1}{2} Q_1 \times E \quad (50\text{-}8)$$

$$= \frac{1}{2} \times 600\ \mu C \times 30\ V$$

$$= 9000\ \mu J$$

Energy stored in the C2 capacitor,

$$W_2 = \frac{1}{2} Q_2 \times E \quad (50\text{-}9)$$

$$= \frac{1}{2} \times 150\ \mu C \times 30\ V$$

$$= 2250\ \mu J$$

Energy stored in the C3 capacitor,

$$W_3 = \frac{1}{2} Q_3 \times E \quad (50\text{-}10)$$

$$= \frac{1}{2} \times 120\ \mu C \times 30\ V$$

$$= 1800\ \mu J$$

Total energy stored,

$$W_T = W_1 + W_2 + W_3$$

$$= 9000 + 2250 + 1800$$

$$= 13050\ \mu J.$$

Energy check:

Total energy stored and, therefore, derived from the source,

$$\frac{1}{2} Q_T E = \frac{1}{2} \times 875\ \mu C \times 30\ V \quad (50\text{-}11)$$

$$= 13050\ \mu J$$

Example 50-2

A 40-μF capacitor is charged from a 150-V source. Another 20-μF capacitor is charged from a 100-V source.

These capacitors are correctly paralleled immediately after they are disconnected from the sources. What is (a) the voltage across the combination and (b) the total energy stored before and after the parallel connection?

Solution

(a) Charge stored in the 40-μF (C1) capacitor,

$$Q_1 = C_1E_1 \qquad (50\text{-}1)$$

$$= 40\ \mu\text{F} \times 150\ \text{V}$$

$$= 6000\ \mu\text{C}$$

Charge stored in the 20-μF (C2) capacitor,

$$Q_2 = C_2E_2 \qquad (50\text{-}1)$$

$$= 20\ \mu\text{F} \times 100\ \text{V}$$

$$= 2000\ \mu\text{C}$$

Total charge,

$$Q_T = Q_1 + Q_2 = 6000 + 2000 = 8000\ \mu\text{C} \qquad (50\text{-}2)$$

Total capacitance of the parallel combination,

$$C_T = C_1 + C_2 \qquad (50\text{-}6)$$

$$= 40 + 20$$

$$= 60\ \mu\text{F}$$

Voltage across the parallel combination,

$$E = \frac{Q_T}{C_T} \qquad (50\text{-}3)$$

$$= \frac{8000\ \mu\text{C}}{60\ \mu\text{F}}$$

$$= 133.3\ \text{V}.$$

(b) Energy stored in the 40-μF capacitor,

$$W_1 = \frac{1}{2}Q_1E_1 \qquad (50\text{-}8)$$

$$= \frac{1}{2} \times 6000\ \mu\text{C} \times 150\ \text{V}$$

$$= 0.45\ \text{J}$$

Energy stored in the 20-μF capacitor,

$$W_2 = \frac{1}{2}Q_2E_2 \qquad (50\text{-}8)$$

$$= \frac{1}{2} \times 2000\ \mu\text{C} \times 100\ \text{V}$$

$$= 0.10\ \text{J}$$

Total energy stored *before* the parallel combination,

$$0.45 + 0.10 = 0.550\ \text{J}.$$

Total energy stored *after* the parallel combination,

$$1/2(Q_TE) = 1/2 \times 8000\ \mu\text{C} \times 133.3\ \text{V}$$

$$= 0.533\ \text{J}.$$

Notice that energy has been lost as the result of the parallel connection. Neglecting the resistance of any connecting wires, the lost energy appears in the form of the spark that occurs when the parallel connection is made.

Example 50-3

In Fig. 50-3, calculate the value of the total capacitance between the points A and B.

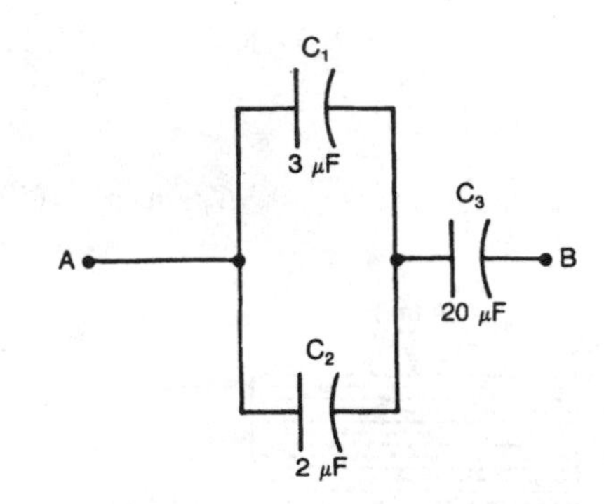

Figure 50-3

Solution

Capacitance of 2 μF || 3 μF = 2 + 3 = 5 μF (50-6)

Total capacitance between the points A and B,

$$C_T = \frac{5 \times 20}{5 + 20} \qquad (49\text{-}15)$$

$$= 4\ \mu\text{F}$$

PRACTICE PROBLEMS

50-1. In Fig. 50-4, calculate the value of the total equivalent capacitance presented to the source. To what voltage is the 10-μF capacitor charged?

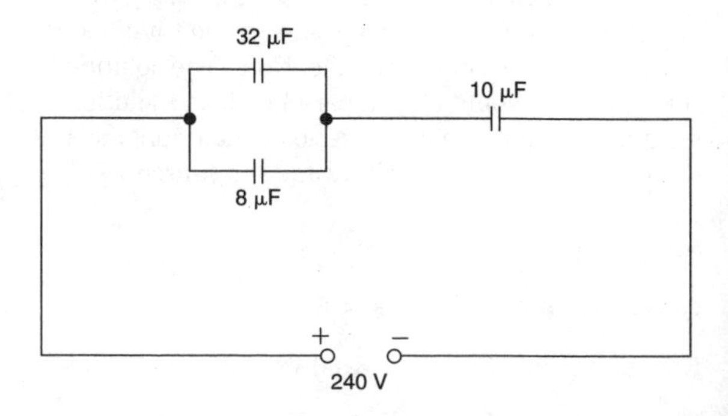

Figure 50-4

50-2. In Fig. 50-5, what is the value of the total equivalent capacitance presented to the source? Calculate the amount of energy stored in the C1 capacitor.

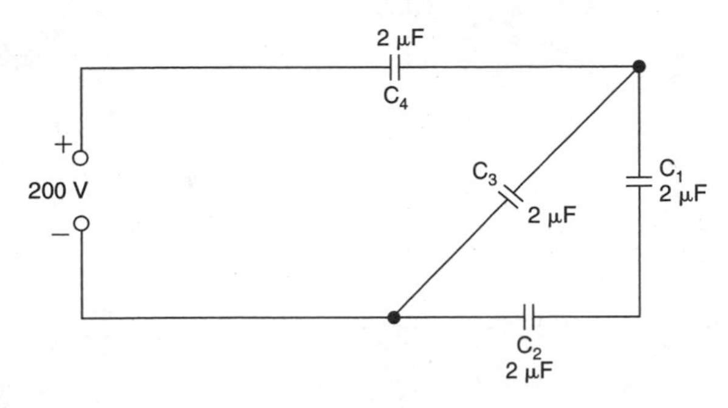

Figure 50-5

50-3. Three capacitors whose capacitances, and voltage ratings, are respectively 40 μF/50 WVdc, 25 μF/40 WVdc, and 50 μF/20 WVdc are connected in parallel. What is the maximum charge that can be stored by this combination without exceeding the voltage rating of any of the capacitors?

50-4. Two 0.1-μF capacitors are charged in series from a 180-V source. When the charging is completed, the source is carefully removed and the two free leads are connected together. What is the value of the subsequent voltage across each of the capacitors?

50-5. In Fig. 50-6, points A and B are connected to the terminals of a 200-Vdc source. What is the amount of the total energy stored?

50-6. Two capacitors, whose values are 0.2 and 0.05 μF, are connected in series across a 240-Vdc source. After the capacitors have been charged, the source is disconnected and replaced by a third, uncharged 0.1-μF capacitor. What is the final voltage across the third capacitor?

50-7. A variable capacitor with an air dielectric has a rotor and a stator with a total of 21 aluminum plates. The effective area of one side of each plate is 8 cm^2. If the separation between adjacent plates is 0.5 mm, calculate the maximum capacitance that the capacitor can provide.

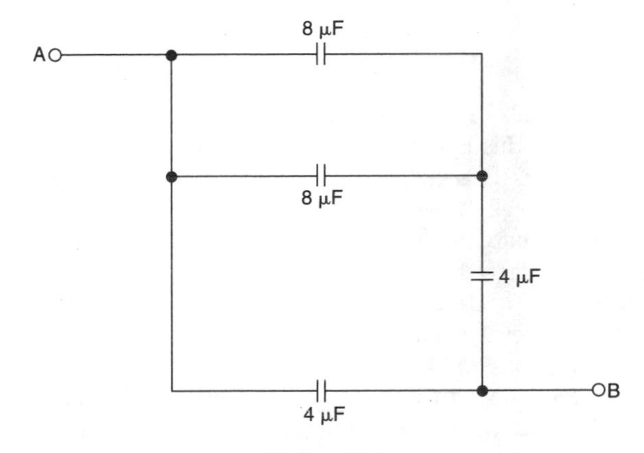

Figure 50-6

51
The RC time constant

From chapter 47, the property of capacitance prevents any sudden change of voltage, and also limits the rate at which the voltage can change. Consequently, when the switch (S) is closed in position 1 (Fig. 51-1A), the capacitor must take a certain time before it acquires its full charge. Therefore, the factors that determine the duration of the so-called *transient* or *changing* state are covered next. This is the interval between the time at which the switch (S) is closed in position 1, and the time when the final (or steady-state) conditions are reached.

Immediately after the switch is closed in position 1, the capacitor cannot charge instantaneously, so that the initial value of V_C must be zero. Consequently, to satisfy KVL, the voltage drop across the resistor (V_R) must balance the source voltage E. The initial current must be at its maximum level of E/R amperes.

As the capacitor charges, its voltage (V_C) rises and, therefore, the resistor's voltage (V_R) falls because at all times the sum of these voltages exactly balances the source voltage, E. When the voltage drop (V_R) falls, the current is correspondingly less and there is a decrease

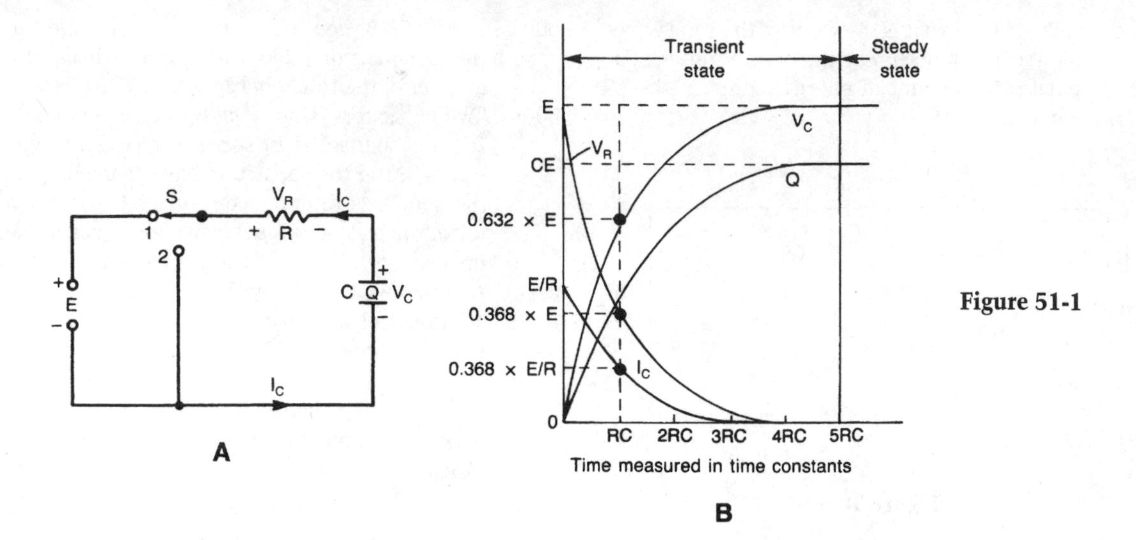

Figure 51-1

in the rate at which the capacitor charges. Consequently, this is a situation in which the more the capacitor is charged, the less is its rate of charging. Theoretically, it would take infinite time for the capacitor to charge fully; but as you can see in the mathematical derivations, the capacitor acquires its ultimate charge of *CE* coulombs (to within 1%) after a finite time interval, which is determined by the values of the capacitance and the resistance.

Remember that the initial current is *E*/*R* amperes (coulombs per second), but the capacitor's final charge is *CE* coulombs. If the initial current had been maintained, the capacitor would have been fully charged after a time of *CE* coulombs /(*E*/*R*) coulombs per second = *RC* seconds, which is the *time constant* of the circuit. However, because the rate at which the capacitor charges falls off, the capacitor only acquires 63.2% of its final charge after one time constant, and it requires *five* time constants before the capacitor is charged to within 1% of *CE* coulombs (Fig. 51-1B). Only then can you assume that the transient charging period has been concluded.

Note that the conditions, at the start of the transient interval, can be predicted by regarding the capacitor as a short circuit, but the final (steady-state) values can be obtained by considering the capacitor to be an open circuit.

Assuming that the steady-state conditions have been reached, now move the switch (S) to position 2 (Fig. 51-2A). The capacitor cannot discharge instantaneously so that V_C must still be equal to *E* volts. By the KVL rule, the voltage drop across the resistor must abruptly rise to *E* volts, but its polarity is reversed be-

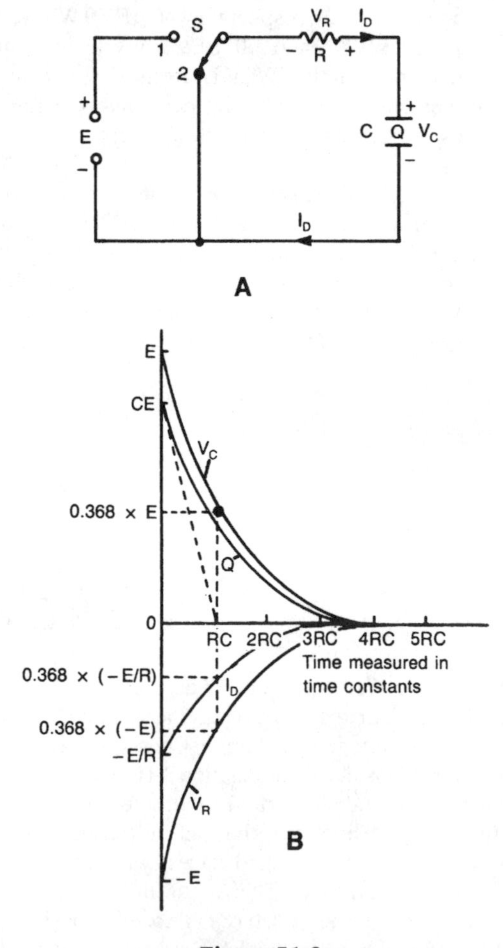

Figure 51-2

cause the discharge current (I_D) is in the opposite direction to the original charging current, (I_C).

As the capacitor discharges, V_C and V_R fall together in order to maintain the voltage balance around the closed loop. When V_R decreases, the current is reduced so that the capacitor discharges more slowly. Theoretically, it would take infinite time for the capacitor to discharge fully; but after one time constant, the capacitor has *lost* 63.2% of its initial charge and its charge has therefore dropped to 36.8% of CE coulombs. After five time constants, assume that the transient discharge period has been concluded in a sense that I_D, V_C, and V_R have all fallen to less than 1% of their original values (Fig. 51-2B).

MATHEMATICAL DERIVATIONS

In Fig. 51-1A the switch (S) is closed in position 1. The initial conditions are,

$$Q = 0 \text{ coulombs}, V_C = 0 \text{ volts},$$

$$V_R = E \text{ volts}, I_C = \frac{E}{R} \text{ amperes.}$$

At all times,

$$V_R + V_C = E \tag{51-1}$$

$$I_C \times R + \frac{Q}{C} = E$$

Differentiating the equation throughout with respect to t,

$$R\frac{dI_C}{dt} + \frac{1}{C} \times \frac{dQ}{dt} = 0$$

$$R\frac{dI_C}{dt} + \frac{I_C}{C} = 0$$

$$\int dt = -CR \int \frac{1}{I_C}\, dI_C$$

$$t + k = -CR \ln I_C$$

When $t = 0$, $I_C = \frac{E}{R}$ and $k = -RC \ln \frac{E}{R}$.

Therefore,

$$t = -CR \ln \frac{E}{RI_C}$$

$$I_C = \frac{E}{R} e^{-t/RC} \text{ amperes} \tag{51-2}$$

$$V_R = I_C \times R = E\, e^{-t/RC} \text{ volts} \tag{51-3}$$

$$V_C = E - V_R = E(1 - e^{-t/RC}) \text{ volts} \tag{51-4}$$

where $e = 2.7183 \ldots$ and is the base of the natural logarithms.

Equation 51-4 is an example of exponential growth and Equations 51-2 and 51-3 are expressions of exponential decay.

When $t = RC$ seconds,

$$I_C = \frac{E}{R} \times e^{-1} = 0.368 \frac{E}{R} \text{ amperes,}$$

$V_R = 0.368\,E$ volts and $V_C = 0.632\,E$ volts.

After $t = 5\,RC$ seconds, $I_C = 0.007 \frac{E}{R}$ amperes,

$V_R = 0.007\,E$ volts and $V_C = 0.993\,E$ volts.

In Fig. 51-2A, the capacitor is assumed to be fully charged and the switch, S, is then moved to position 2. The initial conditions are:

$$I_D = -\frac{E}{R} \text{ amperes}, V_C = E \text{ volts, and } V_R = -E \text{ volts.}$$

At all times,

$$V_R + V_C = 0.$$

$$I_D \times R + \frac{Q}{C} = 0.$$

$$R\frac{dI_D}{dt} + \frac{I_D}{C} = 0.$$

This leads to,

$$I_D = -\frac{E}{R} e^{-t/RC} \text{ amperes} \tag{51-5}$$

$$V_R = -Ee^{-t/RC} \text{ volts} \tag{51-6}$$

$$V_C = Ee^{-t/RC} \text{ volts} \tag{51-7}$$

When $t = RC$ seconds, $I_D = -0.368\,E/R$, $V_R = -0.368\,E$, and $V_C = 0.368\,E$. After $t = 5\,RC$ seconds, the values of I_D, V_R, and V_C have all dropped to 0.7% of their original levels; the transient discharge period is, therefore, virtually finished.

Example 51-1

In Fig. 51-1A, $C = 100$ pF, $R = 150$ kΩ, and $E = 80$ V. From the instant that S is closed in position 1, what are the values of Q, V_R, V_C, and I_C after time intervals of t = 0, 10.5, 15, 30, 45, 60, 75, and 150 μs?

Solution

Time constant, $RC = 150$ kΩ × 100 pF = 15 μs.
When $t = 0$, the initial conditions are:

$$Q = 0 \text{ C}, V_R = 80 \text{ V}, V_C = 0 \text{ V},$$

$$I_C = \frac{80 \text{ V}}{150 \text{ k}\Omega} = 0.533 \text{ mA.}$$

When,

$t = 10.5\ \mu s = 10.5/15 = 0.7 \times RC,$

$$Q = CE(1 - e^{-t/RC}) = 100\text{ pF} \times 80\text{ V}(1 - e^{-0.7})$$
$$= 8000 \times 0.503\text{ pC}$$
$$= 4024\text{ pC}.$$

$$V_C = \frac{Q}{C} = \frac{4024\text{ pC}}{100\text{ pF}} = 40.24\text{ V}.$$

$$V_R = 80 - 40.24 = 39.76\text{ V}.$$

$$I_C = \frac{V_R}{R} = \frac{39.76\text{ V}}{150\text{ k}\Omega} = 0.265\text{ mA}.$$

When,

$t = 15.0\ \mu s = 1$ time constant,

$$Q = 8000\ (1 - e^{-1}) = 8000 \times 0.632 = 5056\text{ pC}.$$

$$V_C = \frac{5056\text{ pC}}{100\text{ pF}} = 50.56\text{ V}.$$

$$V_R = 80 - 50.56 = 29.44\text{ V}.$$

$$I_C = \frac{29.44\text{ V}}{150\text{ k}\Omega} = 0.196\text{ mA}.$$

When,

$t = 30\ \mu s = 2 \times$ time constants,

$$Q = 8000\ (1 - e^{-2}) = 8000 \times 0.865 = 6917\text{ pC}.$$

$$V_C = \frac{6917\text{ pC}}{100\text{ pF}} = 69.17\text{ V}.$$

$$V_R = 80 - 69.17 = 10.83\text{ V}.$$

$$I_C = \frac{10.83\text{ V}}{150\text{ k}\Omega} = 0.072\text{ mA}.$$

When,

$t = 45\ \mu s = 3 \times$ time constants,

$$Q = 8000\ (1 - e^{-3}) = 8000 \times 0.950 = 7602\text{ pC}.$$

$$V_C = \frac{7602\text{ pC}}{100\text{ pF}} = 76.02\text{ V}.$$

$$V_R = 80 - 76.02 = 3.98\text{ V}.$$

$$I_C = \frac{3.98\text{ V}}{150\text{ k}\Omega} = 0.0265\text{ mA}.$$

When,

$t = 60\ \mu s = 4 \times$ time constants,

$$Q = 8000\ (1 - e^{-4}) = 8000 \times 0.982 = 7853\text{ pC}.$$

$$V_C = \frac{7853\text{ pC}}{100\text{ pF}} = 78.53\text{ V}.$$

$$V_R = 80 - 78.53 = 1.47\text{ V}.$$

$$I_C = \frac{1.47\text{ V}}{150\text{ k}\Omega} = 0.00977\text{ mA}.$$

When,

$t = 75\ \mu s = 5 \times$ time constants,

$$Q = 8000\ (1 - e^{-5}) = 8000 \times 0.993 = 7944\text{ pC}.$$

$$V_C = \frac{7944\text{ pC}}{100\text{ pF}} = 79.44\text{ V}.$$

$$V_R = 80 - 79.44\text{ V} = 0.56\text{ V}.$$

$$I_C = \frac{0.56\text{ V}}{150\text{ k}\Omega} = 0.00373\text{ mA}.$$

The transient charging state is now assumed to be completed.

When,

$t = 150\ \mu s = 10 \times$ time constants,

$$Q = 8000(1 - e^{-10}) =$$
$$8000 \times 0.9999546 = 7999.6368\text{ pC}.$$

$$V_C = \frac{7999.6368\text{ pC}}{100\text{ pF}} = 79.9963668\text{ V}.$$

$$V_R = 80 - 79.996368\text{ V} = 0.003632\text{ V}.$$

$$I_C = \frac{0.003632\text{ V}}{150\text{ k}\Omega} = 0.0242\ \mu\text{A}.$$

Example 51-2

In Fig. 51-2A, $C = 100$ pF, $R = 150$ kΩ and $E = 80$ V. It is assumed that the capacitor is fully charged to 80 V, and that the switch (S) is then closed in position 2. What are the values of Q, V_C, V_R, and I_D after time intervals of $t = 0$, 10.5, 15, 30, 45, 60, 75 and 150 μs?

Solution

From Example 51-1, the time constant is 15 μs. When $t = 0$, the initial conditions are:

$$Q = 8000\text{ pC}, V_C = 80\text{ V}, V_R = -80\text{ V},$$

$$\text{and } I_D = -\frac{80\text{ V}}{150\text{ k}\Omega} = -0.533\text{ mA}$$

When,

$t = 10.5\ \mu s = 0.7 \times$ time constant,

$$Q = CE\,e^{-t/RC} = 8000\,e^{-0.7} = 3973\text{ pC}.$$

$$V_C = \frac{3973\text{ pC}}{100\text{ pF}} = 39.73\text{ V}.$$

$$V_R = -39.73\text{ V}.$$

$$I_D = \frac{-39.73\text{ V}}{150\text{ k}\Omega} = -0.265\text{ mA}.$$

When,

$t = 15\ \mu s = 1 \times$ time constant,

$$Q = 8000\,e^{-1} = 2943\text{ pC}.$$

$$V_C = \frac{2943\text{ pC}}{100\text{ pF}} = 29.43\text{ V}.$$

$V_R = -29.43$ V.

$$I_D = \frac{-29.43 \text{ V}}{150 \text{ k}\Omega} = -0.196 \text{ mA.}$$

When,

$t = 30\ \mu s = 2 \times$ time constants,

$Q = 8000\, e^{-2} = 1083$ pC.

$$V_C = \frac{1083 \text{ pC}}{100 \text{ pF}} = 10.83 \text{ V.}$$

$V_R = -10.83$ V.

$$I_D = \frac{-10.83 \text{ V}}{150 \text{ k}\Omega} = -0.072 \text{ mA.}$$

When,

$t = 45\ \mu s = 3 \times$ time constants,

$Q = 8000\, e^{-3} = 398.3$ pC.

$$V_C = \frac{398.3 \text{ pC}}{100 \text{ pF}} = 3.983 \text{ V.}$$

$V_R - -3.983$ V.

$$I_D = \frac{-3.983 \text{ V}}{150 \text{ k}\Omega} = -0.2655 \text{ mA.}$$

When,

$t = 60\ \mu s = 4 \times$ time constants,

$Q = 8000\, e^{-4} = 146.5$ pC.

$$V_C = \frac{146.5 \text{ pC}}{100 \text{ pF}} = 1.465 \text{ V.}$$

$V_R = -1.465$ V.

$$I_D = \frac{-1.465 \text{ V}}{150 \text{ k}\Omega} = -0.00977 \text{ mA.}$$

When,

$t = 75\ \mu s = 5 \times$ time constants,

$Q = 8000\, e^{-5} = 53.9$ pC.

$$V_C = \frac{53.9 \text{ pC}}{100 \text{ pF}} = 0.539 \text{ V.}$$

$V_R = -0.539$ V.

$$I_D = \frac{-0.539 \text{ V}}{150 \text{ k}\Omega} = -3.59\ \mu\text{A.}$$

The transient decay state is now assumed to have ended.
When,

$t = 150\ \mu s = 10 \times$ time constants,

$Q = 8000 \times e^{-10} = 0.363$ pC.

$$V_C = \frac{0.363 \text{ pC}}{100 \text{ pF}} = 0.00363 \text{ V.}$$

$V_R = -0.00363$ V.

$$I_D = \frac{-0.00363 \text{ V}}{150 \text{ k}\Omega} = -0.0242\ \mu\text{A.}$$

PRACTICE PROBLEMS

51-1. In the circuit of Fig. 51-3, determine the value of the voltage across the 10-μF capacitor after an interval of 1.2 s elapses from the time the switch (S) is closed.

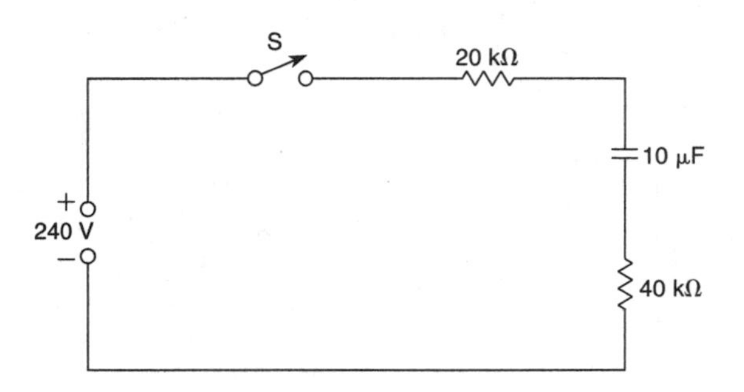

Figure 51-3

51-2. In the circuit of Fig. 51-4, calculate the values of the potentials at points X, Y, and Z immediately after the switch is closed. Assuming that the transient state has concluded, determine the steady-state potentials at points X, Y, and Z.

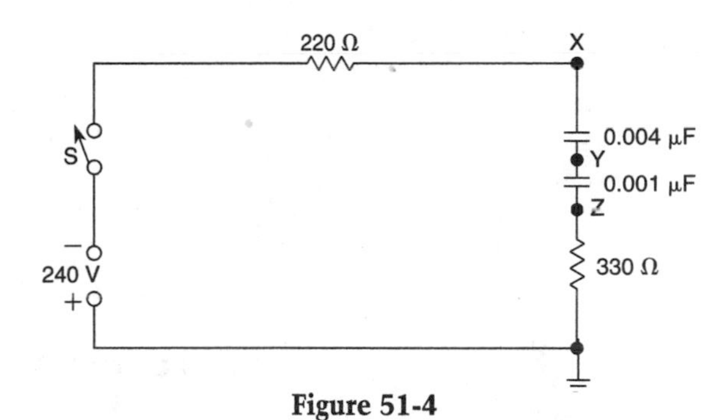

Figure 51-4

51-3. If the switch (S) is closed in the circuit of Fig. 51-5, what are the initial, and final, values of the potential at point X? How long will it take for the circuit to reach its steady-state condition?

51-4. If the switch (S) is closed in the circuit of Fig. 51-6, what are the initial, and final, values of the potentials at points X and Y? What is the value of the circuit's time constant.

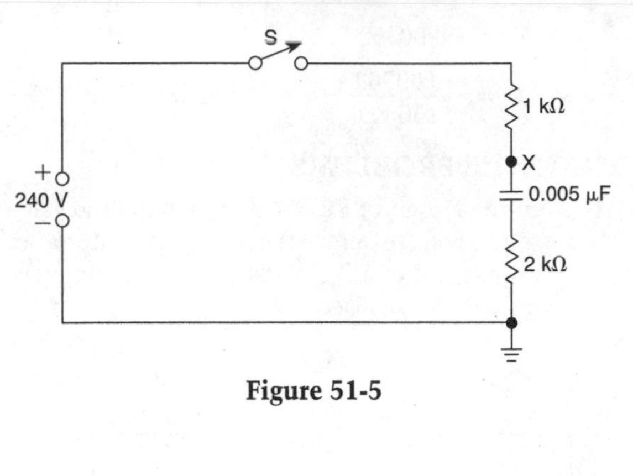

Figure 51-5

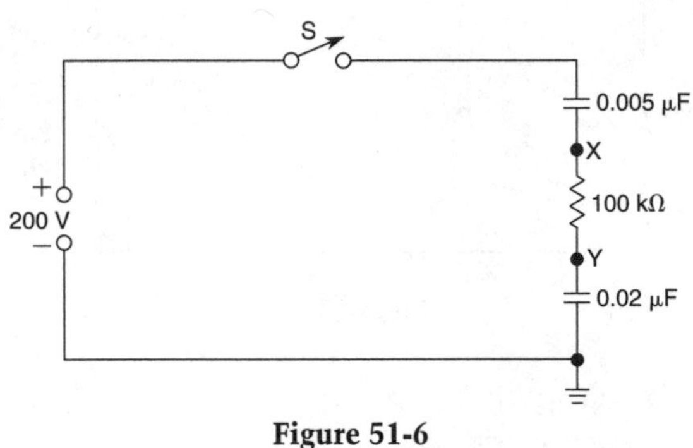

Figure 51-6

51-5. If the switch (S) is closed in the circuit of Fig. 51-7, what are the initial, and final, values of the potential at point X? What is the duration of the transient state?

51-6. If the switch (S) is closed in the circuit of Fig. 51-8, what are the values of I_T, I_1, and I_2 at the beginning, and end, of the transient state? Calculate the value of the circuit's time constant (Hint: Thévenize the circuit to the left of points X and Y.)

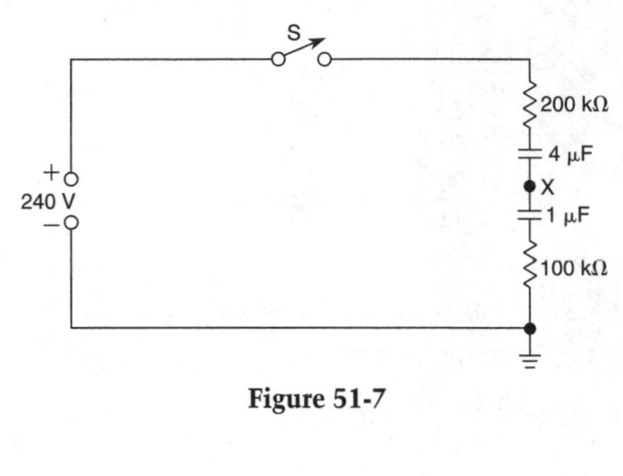

Figure 51-7

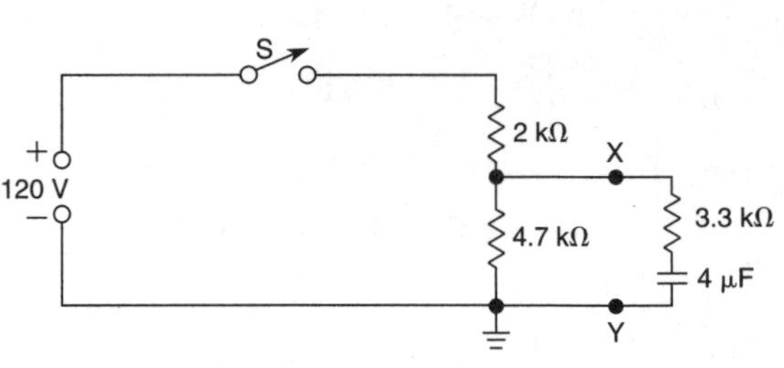

Figure 51-8

52
Differentiator circuits

A differentiator circuit uses the concept of the time constant to achieve its objective. The process of *differentiation* is related to the subject of calculus in mathematics (chapters 211 through 220). Assume that you have two waveforms, which are identified as waveform A and waveform B. Then, if waveform B is the result of differentiating waveform A, the instantaneous value of any point in waveform B is equal to the value of the *slope* at the corresponding point in waveform A. In this case, the word *corresponding* means "occurring at the same time."

As an example, consider a mass that is falling as the result of gravity. In chapter 1, you learned that the acceleration under the force of the earth's gravitational pull is 9.81 m/s^2. Consequently, if the mass is dropped from rest, with an initial velocity of zero, its velocity after one second will be 9.81 m/s so that its average velocity during this interval is $9.81/2 = 4.905$ m/s and the distance through which the mass has fallen is equal to $4.905 \text{ m/s} \times 1 \text{ s} = 4.905$ m.

After two seconds, the velocity of the mass is $2 \times 9.81 = 19.62$ m/s, the average velocity over the *two* second interval is 19.62/2 = 9.81 m/s, and the total distance through which the mass has fallen is 9.81 m/s × 2 s = 19.62 m. Notice that when the time is doubled (for instance, from one second to two seconds), the velocity is doubled, but the distance fallen has quadrupled (from 4.905 m to 19.62 m). This suggests that the distance fallen is directly proportional to the *square* of the elapsed time.

Table 52-1 shows corresponding values of time, velocity, and distance fallen during the first four seconds. The symbol, v, represents the velocity acquired by the mass at the *end* of each time interval, and s indicates the corresponding total distance through which the mass has fallen.

The equations relating these results are:
Acceleration,

$$a = 9.81 \text{ m/s}^2 \qquad (52\text{-}1)$$

Velocity,

$$v = 9.81t \text{ m/s} \qquad (52\text{-}2)$$

This equation states that velocity = acceleration × time (Equation 1-1).
Total distance fallen,

$$s = \text{average velocity} \times \text{time} \qquad (52\text{-}3)$$

$$= \frac{v \times t}{2}$$

$$= \frac{9.81t}{2} \times t = 4.905t^2 \text{ m}$$

Equation 52-3 indicates that the distance fallen is directly proportional to the square of the elapsed time.

The three graphs of distance, velocity, and acceleration versus time are illustrated in Figs. 52-1A, B, and C. The "distance versus time" graph is a parabolic curve, which is the result of the "square-law" relationship between the two variables (distance, $s = 4.905t^2$ m). By contrast, there is a "linear" equation connecting the velocity and the time (velocity, $v = 9.81t$ m/s), so that this graph is a straight line (Fig. 51-2B). The gravitational acceleration remains constant (a = 9.81 m/s^2) and, therefore, the graph of Fig. 52-1C is a *horizontal* straight line.

At the point specified by t = 1 sec., s = 4.905 m, draw (by estimation) a tangent that touches the parabolic curve, as shown in Fig. 52-1A. A right-angled triangle is then constructed with the tangent line as the hypotenuse. For convenience, the horizontal side of the triangle is chosen to represent one second and the corresponding vertical side is equivalent to 9.8 m (approximately). The slope of the tangent can then be found by dividing the vertical side of 9.8 m by the horizontal side of 1 sec. As a result, the slope is equal to 9.8 m/1 sec. = 9.8 m/s, which is the velocity of the mass corresponding to t = 1 sec. Repeating the process for the point t = 3 sec., s = 44.14 m, the slope of the tangent is approximately 29.4 m/1 sec. = 29.4 m/s. In other words, the slope of the tangent measures the ratio of the variation in the distance fallen, to the variation in the corresponding time interval; the value of this ratio is equal to the *instantaneous velocity of the mass*. In calculus terminology, the result of *differentiating* the distance with respect to the time, is equal to the velocity; the differentiation of the "distance versus time" graph produces the "velocity versus time" graph.

Turn your attention to the "velocity versus time" graph. This is an inclined straight line, so that its slope is the same at all points. In Fig. 52-1B, triangles have been constructed for the two points specified by t = 2 sec., s = 19.62 m/s, and t = 3 sec., s = 29.43 m/s; in each case the slope in 9.8 m/s/1 sec. = 9.8 m/s^2 which is the value of the gravitational acceleration. It follows that, the result of differentiating the "velocity versus time" graph is the "acceleration versus time" graph.

Table 52-1. Values of velocity and distance for an object falling under gravity.

Time, t seconds	Velocity, v meters per second	Distance, s meters	Acceleration, a meters per second per second
0	0	0	9.81
1	9.81	4.905	9.81
2	19.62	19.62	9.81
3	29.43	44.45	9.81
4	39.24	78.48	9.81

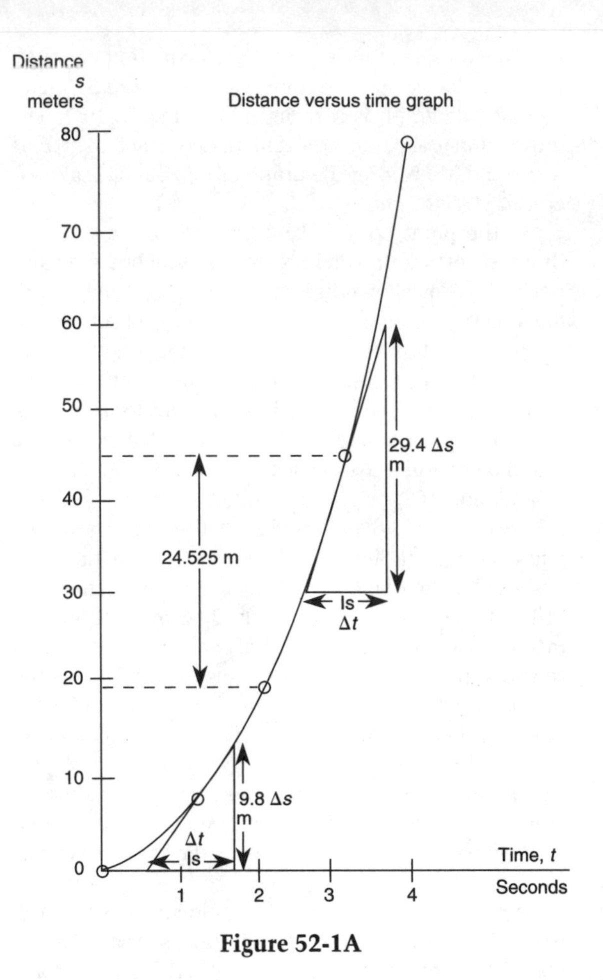

Figure 52-1A

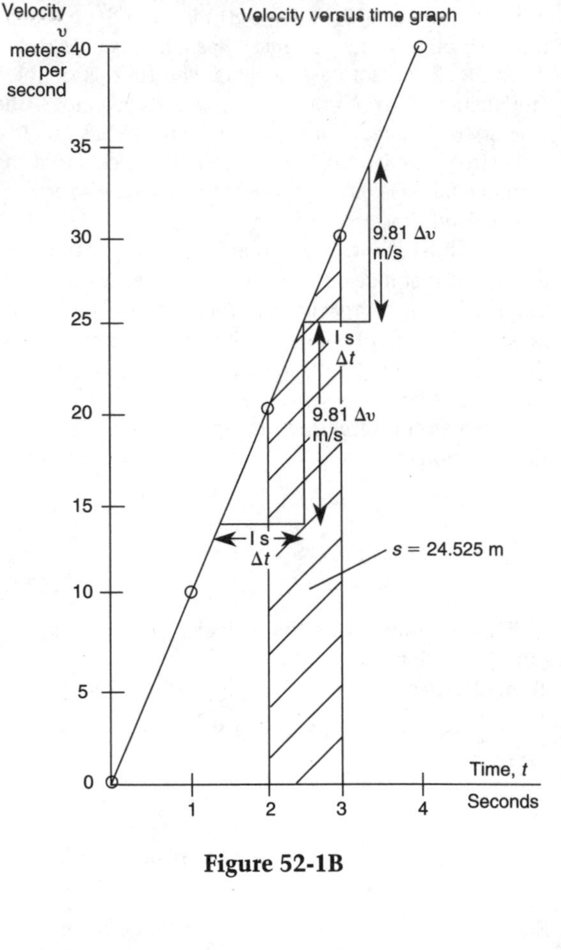

Figure 52-1B

Parabolic curves are not commonly encountered in electronics, which is more concerned with square waves, sine waves, and "ramp" or "sawtooth" waveforms. For example, what would you expect the resultant waveform to look like if you differentiated a square wave? But to begin with, how would you generate a square wave, such as shown in Fig. 52-2B? When you look at the circuit of Fig. 52-2A, assume that the switch (S) is closed in position 1; as a result, E volts are applied across the series combination of the capacitor (C) and the resistor, R. After an appreciable time (T), S is switched *instantaneously* to position 2. The source voltage is now replaced by a short circuit, which is equivalent to applying zero volts. After the same time interval (T), the switch is instantaneously returned to position 1 and the sequence is repeated. The applied voltage is then represented by the symmetrical square wave of Fig. 52-2B. Of course, this square wave is idealized because, in practice, the switch cannot move instantaneously from position 1 to position 2, or vice-versa. There must be a certain "rise" time in the movement of the voltage from point A to point B, and also a "fall" time in its decrease from point C to point D.

In order to differentiate the square wave, you must consider the slope of the waveform at every point during its entire sequence. When the voltage rises instantaneously from zero volts to E volts, the time involved is (ideally) zero, so that the slope is theoretically infinite. Moreover, this slope is *positive* because the waveform is increasing from the low level of zero volts to the high level of E volts. Consequently, the mathematical result of differentiating the waveform from point A to point B is an infinite positive "spike" as shown in Fig. 52-2C.

As you move from point B to point C, the voltage remains unchanged so that this section of the square wave is a horizontal line whose slope is zero. Between point C and point D, the level of the square wave instantaneously falls from E volts to zero volts, so that the differentiated result is an infinite *negative* spike.

Summarizing, when you differentiate a square wave, you produce alternate positive and negative spikes, which are separated from one another by a zero level.

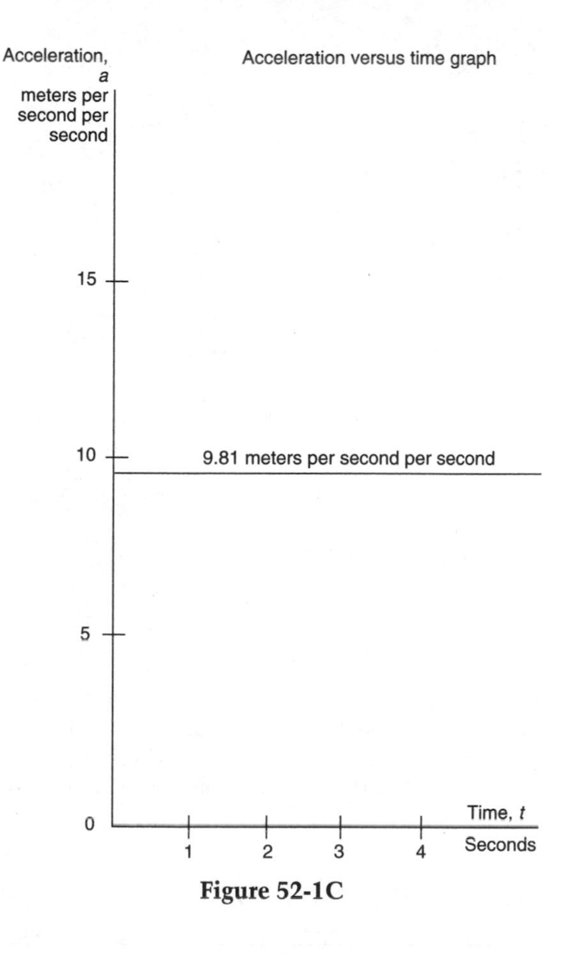

Figure 52-1C

C/R DIFFERENTIATOR CIRCUIT

Electronically, you cannot create the precise waveform of Fig. 52-2C. How can you obtain a close approximation? Here, you must refer back to the curves of Figs. 51-1B and 51-2B. The V_R curve of Fig. 51-1B looks something like a positive spike, and the V_R curve of Fig. 51-2B resembles a negative "spike." Therefore, it appears that you could apply a symmetrical square wave to a series C,R combination, and obtain the differentiated output from across the resistor (Fig. 52-3A). But when you join the two V_R curves together (Fig. 52-3B), it is clear that the "spikes" are far too broad and must be *shortened* to improve the degree of approximation.

In the waveform of Fig. 52-3B, the decay time of $5CR$ seconds, associated with each V_R curve, is equal to the interval of T seconds for the symmetrical square wave. A practical differentiator circuit normally uses a time constant that is much less than one tenth of T seconds; this means that the decay time of $5CR$ seconds is less than $T/2$ so that the capacitor has either charged, or discharged, completely before one half of each T interval has lapsed.

As an example, assume that the square wave is repeating at the rate of 500 times per second, so that one complete sequence involves a time of 1/500 s = 2000 μs. The time interval (T) is 1000 μs and, therefore, the basic differentiator circuit would require a time constant of less than 1000/10 = 100 μs. In the circuit of Fig. 52-4A, $R = 500\ \text{k}\Omega$ and $C = 100$ pF, so that the time constant is $100\ \text{pF} \times 0.5\ \text{M}\Omega = 50\ \mu\text{s}$. Therefore, we have fulfilled the condition that $CR << T/10$, and the

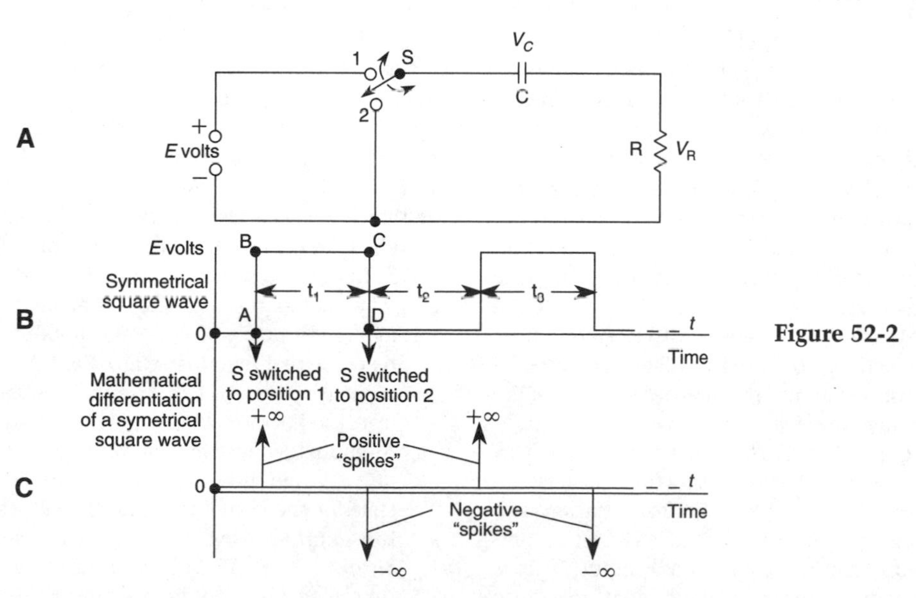

Figure 52-2

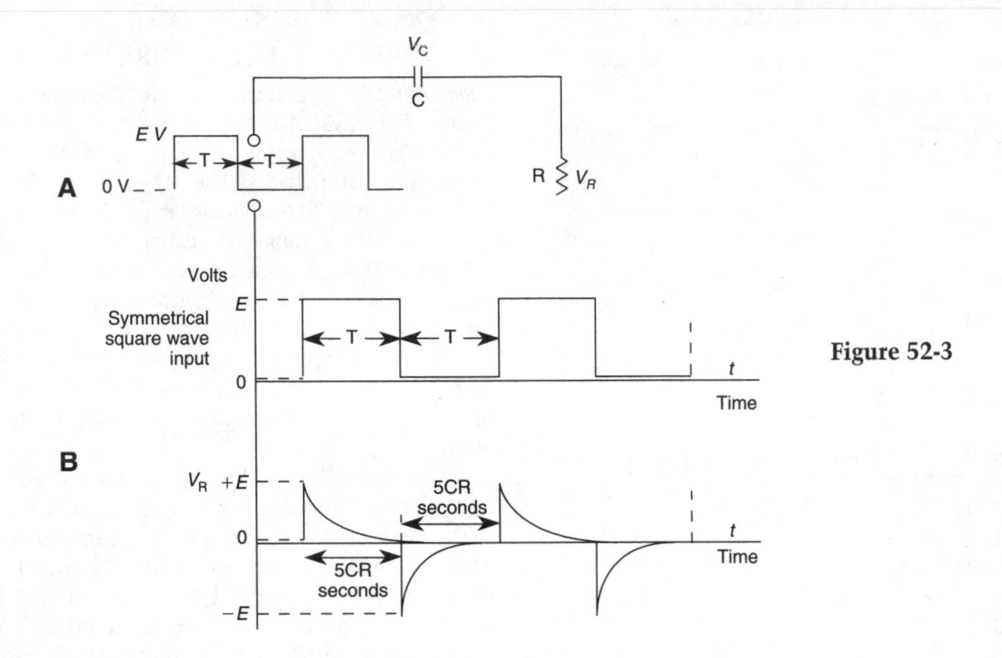

Figure 52-3

capacitor will either change or discharge completely after ¼ of each T interval has elapsed. The differentiated output voltage across the resistor then consists of narrow positive and negative "spikes," and it is a reasonable approximation to the ideal waveform of Fig. 52-2C.

For the circuit of Fig. 52-4A, the waveforms of V_C and V_R are shown in Figs. 52-4B and C. Notice that the mean level of the V_R waveform is zero, but the mean level of the V_C waveform is $E/2$ volts (which is the same as the mean level of the source voltage). To obey Kirchhoff's Voltage Law at all times, the sum of the instantaneous values of the voltages V_R and V_C must equal the instantaneous value of the source voltage.

You might well ask, "why can't I reduce the values of C and R still further, for an even better approximation to the ideal differentiated waveform?" First, consider the effect of reducing the value of the resistance. The source of the square wave must possess a certain internal resistance, so that the abrupt rise from zero to E volts is divided between the internal resistance and the differentiating resistor. If the value of R is reduced to the point where it becomes comparable with the internal resistance, the output positive "spike" will be decreased by the amount dropped across the internal resistance; the differentiated waveform will then be distorted. Similarly, if the value of C is reduced to the same order as the stray capacitance in the circuit, futher distortion will occur.

To summarize, one purpose of differentiation is to create short interval "spikes" (or pulses) from a comparatively long duration, square wave input. This is achieved by using a series C,R combination whose time constant is small, compared with the duration of the square wave. The differentiated output is taken from across the resistor, and then one set of pulses (either the positive group or the negative group) can be used to "trigger" (control) the operation of the following circuitry.

THE L,R DIFFERENTIATOR CIRCUIT

Refer back to the graphs displayed in Figs. 43-1B and 43-2B. These show the growth and decay of current in the series L,R circuit of Figs. 43-1A and 43-2A. The two graphs of "V_L versus time" have the same appearance as the two "V_R versus time" graphs for the series C,R circuit (Figs. 51-1B, 51-2B).

Therefore, it seems that differentiation can be provided by either a C,R circuit, or an L,R circuit; in the latter case the time constant L/R seconds would be less than $T/10$ seconds, and the differentiated output would be taken from across the inductor (Fig. 52-5).

The L,R differentiator circuit has two disadvantages when compared with the C,R arrangement:

1. Inductors normally possess appreciable resistance, so that their losses are greater than those of capacitors. As a result, the differentiated output from the L,R circuit is lower and is more distorted.
2. An inductor must have a distributed self-capacitance. When the square wave is applied, it is possible for the inductance and the self-capacitance to set up an oscillation which is referred to as *ringing*. The effect of ringing is to introduce further distortion into the waveform of the differentiated output voltage.

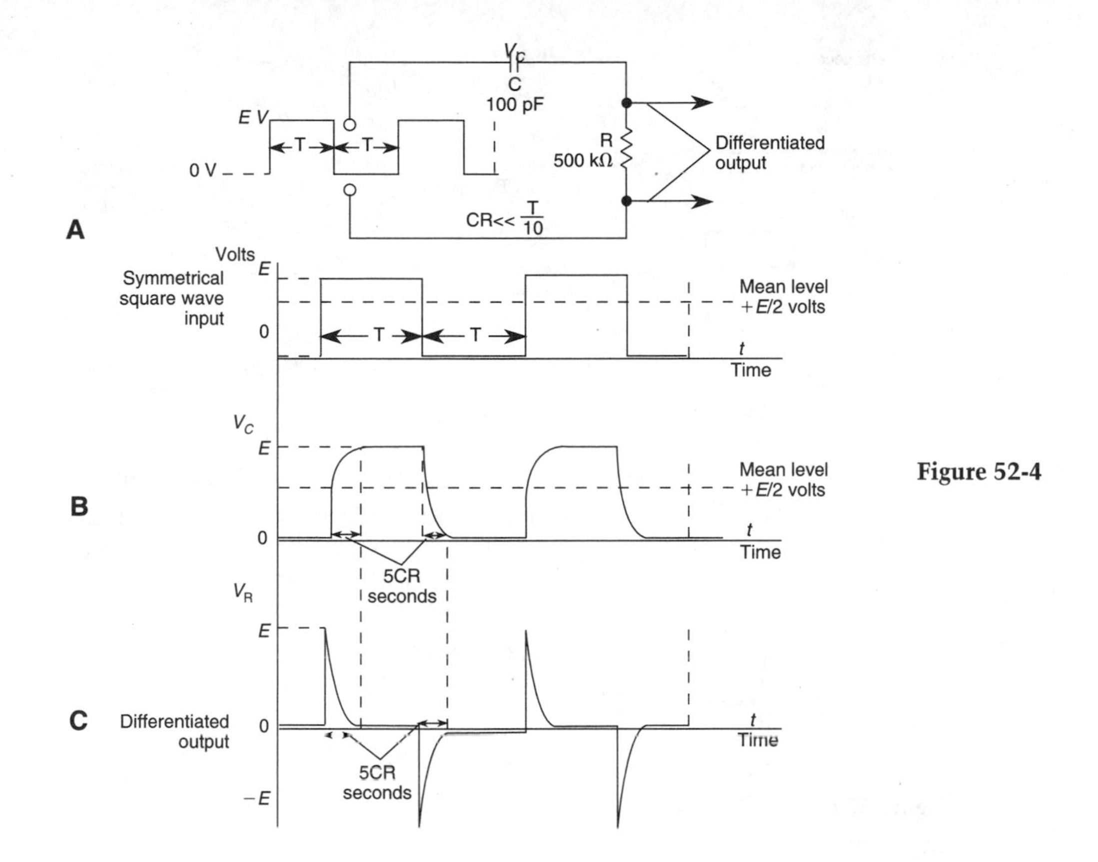

Figure 52-4

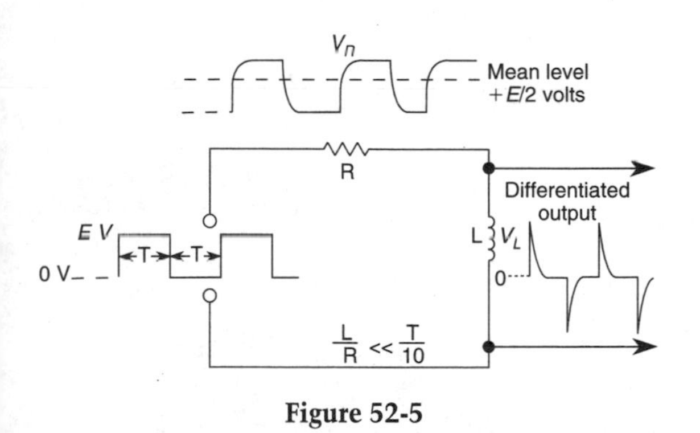

Figure 52-5

Because of these disadvantages the L,R differentiator is not commonly found in electronic circuits.

DIFFERENTIATION OF A SAWTOOTH WAVEFORM

The symmetrical sawtooth waveform of Fig. 52-6A is frequently encountered in electronics and communications. For example, a sawtooth voltage is applied between the X plates of an electrostatic cathode ray tube in order to move the spot of light at uniform speed across the face of the tube (chapter 60).

Between points A and B, the slope of the sawtooth is m volts per second so that the mathematically differentiated waveform has a constant value over this interval, and will, therefore, be represented by a horizontal line (Fig. 52-6B). As the sawtooth drops from point B to point C, in theoretically zero time, the slope is infinitely steep, and the differentiated result is an instantaneous negative spike. Over the interval between points C and D, the value of the sawtooth is always zero so that the differentiated waveform is a horizontal line which is at the zero level.

When the sequence is repeated, the mathematically differentiated waveform (Fig. 52-6B) has the appearance of a symmetrical square wave. It follows that the result of differentiating a sawtooth waveform is to create a square wave.

The basic C,R differentiator circuit is shown in Fig. 52-6C. Once again, assume that the repetition rate of the sawtooth waveform is 500 times per second; consequently,

$$T = \frac{1}{2 \times 500} \text{ s} = 1000\ \mu\text{s}.$$

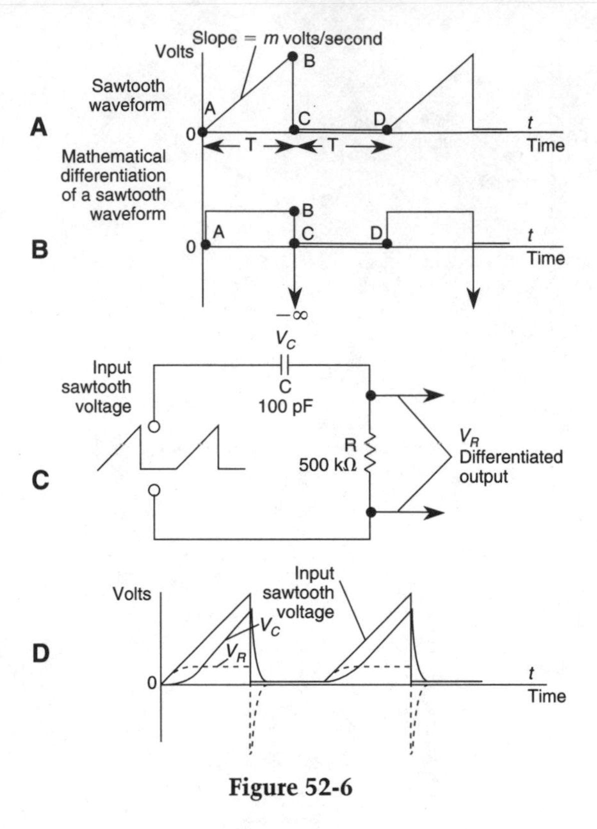

Figure 52-6

The value of $C \times R$ is 100 pF × 0.5 MΩ = 50 μs, and the requirement for a short time constant ($CR << T/10$) is fulfilled.

After a brief transient time, (see mathematical derivations) the circuit has reached its steady state condition in which the rate at which the voltage across the capacitor is increasing, must be the same as the rate at which the sawtooth voltage is rising, namely m volts per second. Because the charge (q) = voltage (V_C) × capacitance (C), it follows that the capacitor is being charged at the rate of $m \times C$ coulombs per second. Therefore, the charging current is mC amperes, and the voltage across the resistor has a constant value of mCR volts. The waveforms of V_C and V_R are illustrated in Fig. 52-6D.

If you assume that $m = 0.1$ V/μs, the peak value of the sawtooth is 0.1 V/μs × 1000 μs = 100 V, and the steady voltage across the resistor is mCR = 0.1 V/μs × 100 pF × 0.5 MΩ = 5 V.

MATHEMATICAL DERIVATIONS

L,R differentiator circuit

For a series L,R circuit,

$$e = v_R + v_L = iR + L\frac{di}{dt} \qquad (52\text{-}4)$$

Then,

$$\frac{de}{dt} = R\frac{di}{dt} + L\frac{d^2i}{dt^2} \qquad (52\text{-}5)$$

$$\text{Because, } \frac{di}{dt} = \frac{v_L}{L} \text{ and } \frac{d^2i}{dt^2} = \frac{1}{R} \times \frac{d^2v_R}{dt^2}$$

$$\frac{de}{dt} = \frac{Rv_L}{L} + \frac{L}{R} \times \frac{d^2v_R}{dt^2} \qquad (52\text{-}6)$$

Because L/R is a *short* time constant, the term,

$$\frac{L}{R} \times \frac{d^2v_R}{dt^2}$$

is involved with the initial transient condition which is rapidly concluded. The waveform of the voltage, (v_L), then directly represents the result of differentiating the input voltage (e) with respect to time.

C,R differentiator circuit

For a series C,R circuit,

$$e = v_C + v_R = \frac{q}{C} + Ri \qquad (52\text{-}7)$$

Then,

$$\frac{de}{dt} = \frac{i}{C} + R\frac{di}{dt} \qquad (52\text{-}8)$$

$$\text{Because, } i = \frac{v_R}{R} \text{ and } \frac{di}{dt} = C\frac{d^2v_C}{dt^2},$$

$$\frac{de}{dt} = \frac{v_R}{CR} + CR\frac{d^2v_C}{dt^2} \qquad (52\text{-}9)$$

Because CR is a short time constant, the term,

$$CR\frac{d^2v_C}{dt^2}$$

is involved with the initial transient condition, which is rapidly concluded. The waveform of the voltage (v_C) then directly represents the result of differentiating the input voltage (e) with respect to time.

For the "ramp" waveform, $e = mt$, where m volts per second is equal to the value of the slope. When the short duration transient state is ended,

$$\frac{de}{dt} = \frac{d\,(mt)}{dt} = m = \frac{v_R}{CR}$$

Then,

Differentiated output, $v_R = mCR$ volts.

Example 52-1

A symmetrical square wave with a frequency of 2000 Hz is applied to the C,R differentiator circuit. If the value of the capacitor is 100 pF, calculate the resistor's maximum value, which will allow adequate differentiation to occur.

Solution

Square wave's period,

$$2T = \frac{1}{2000}\ \text{s} = 500\ \mu\text{s}$$

Maximum time constant for the differentiator circuit,

$$\frac{T}{10} = \frac{500}{2 \times 10} = 25\ \mu\text{s}$$

Maximum resistor value,

$$\frac{25\ \mu\text{s}}{100\ \text{pF}} = 250\ \text{k}\Omega$$

Example 52-2

A "ramp" voltage, whose slope is 10^5 volts per second, is applied to a C,R differentiator circuit that consists of a 250-pF capacitor and a 200-kΩ resistor. Describe the waveform of the differentiated output across the resistor.

Solution

The waveform of the differentiated output is a square wave with a peak value of $mCR = 10^5 \times 250 \times 10^{-12} \times 200 \times 10^3 = 5$ V.

PRACTICE PROBLEMS

52-1. A symmetrical 2000-Hz square wave has upper and lower levels of +10 V and −10 V respectively. This square wave is applied to an C,R differentiator circuit in which the value of the capacitor is 100 pF. Calculate the resistor's highest value, which must not be exceeded if adequate differentiation is to occur.

52-2. An asymmetrical 2500-Hz square wave has an initial time interval of 100 μs, followed by a second time interval of 300 μs. If this waveform is applied to an C,R differentiator circuit in which the value of the capacitor is 100 pF, determine the resistor's highest value, which must not be exceeded if adequate differentiation is to occur.

52-3. The complete sequence of a symmetrical square wave repeats at a rate of 500 times per second. This square wave is applied to an C,R differentiator circuit in which the value of the resistor is 100 kΩ. What is the capacitor's highest value, which will allow adequate differentiated to occur.

52-4. A C,R circuit consists of a 250-pF capacitor in series with a 200-kΩ resistor. If a symmetrical square wave is applied to the circuit, what is the square wave's highest repetition rate, which will allow an adequately differentiated output to appear across the resistor?

52-5. In Practice Problem 52-4, the square wave is replaced by a "ramp" voltage with a slope of 3 × 10^5 V/s. Calculate the value of the voltage across the resistor.

53 Integrator circuits

As far as calculus is concerned, *integration* is the reverse process of differentiation. In terms of the falling mass, you discovered in chapter 52 that the velocity was the result of differentiating the distance, with respect to the time. Therefore, it follows that the distance is the result of integrating the velocity, with respect to the time. With regard to waveforms, if waveform X is the result of *differentiating* waveform Y, with respect to time, then waveform Y is the result of *integrating* waveform X. For example, when you differentiated a sawtooth waveform, a square wave was obtained. Reversing the process means that if you integrate a square wave, a sawtooth waveform will be produced.

C,R INTEGRATOR CIRCUIT

Looking back to Fig. 52-2A, you have already observed the creation of a square wave by the rapid movement of the switch between positions 1 and 2. But which of the waveforms in Fig. 53-1 represents a reasonable approximation to the integrated sawtooth output? Only a graph of "V_C versus time" shows an increasing voltage, but this graph is a curve of exponential growth; rather,

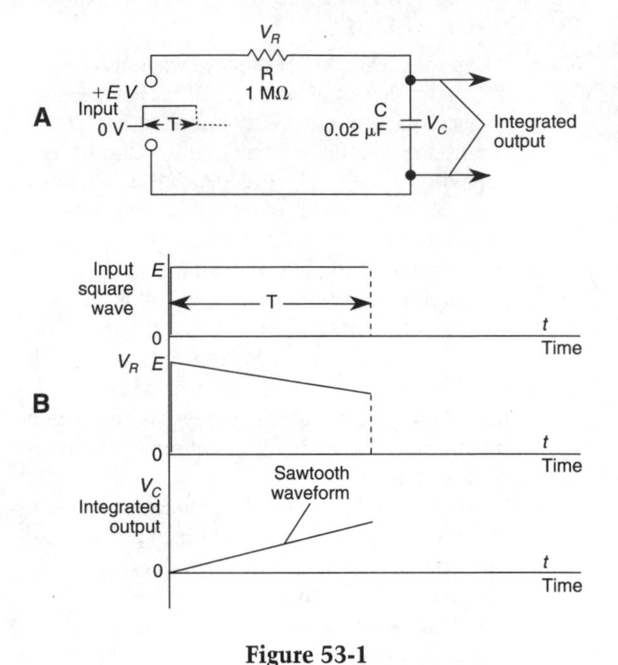

Figure 53-1

we need a linear ("straight line") rise of constant slope. However, you will observe that the curve shows good linearity between the times of zero and $CR/10$ seconds. Consequently, if values of C and R are chosen such that $CR >> 10T$, the V_C waveform will be a close approximation to the required integrated output. The basic integrator circuit is shown in Fig. 53-1A. Assume that the input square wave is repeating at the rate of 500 times per second; the time interval (T) is 1000 μs and, therefore, the circuit will require a time constant which is greater than $10 \times 1000 = 10000$ μs. Because $C = 0.02$ μF and $R = 1$ MΩ, the time constant is 0.02 μF $\times$ 1 MΩ = 20000 μs, and the necessary condition is fulfilled. The integrated sawtooth output is taken from across the capacitor, as indicated by the waveforms of Fig. 53-1B.

Integration in calculus is also a method of determining the area beneath the curve. Referring back to Figs. 52-1A, B, integrate the "velocity versus time" graph between the times $t = 2$ s and $t = 3$ s. The result is the shaded area, which has a value of $1/2 \times (19.62 + 29.43)$ m/s $\times$ (3 − 2) s = 24.525 m. Remembering that the distance/time graph is the integral of the velocity /time graph, this area should have the same value as the distance covered between $t = 2$ s and $t = 3$ s; from Table 52-1, this distance is 44.145 − 19.62 = 24.525 m and, therefore, the anticipated result has been verified.

Assume that a repeating symmetrical square wave is applied to the integrating circuit of Fig. 53-2A. When steady-state conditions have been reached, the capacitor will charge during the time that the square wave is at the E volt level. However, there will only be a slow "charging" rate because of the long time constant. In a similar way, the capacitor will only discharge slowly when the square wave is at the zero volt level. The V_C waveform of Fig. 53-2B fluctuates slightly about the +

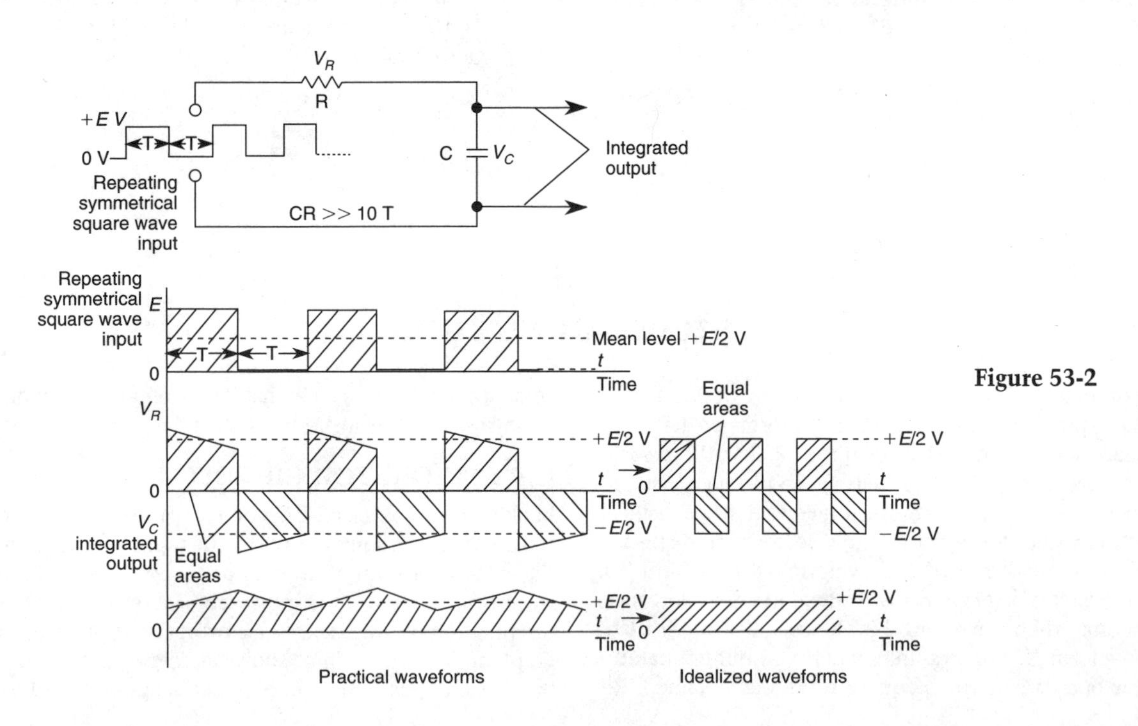

Figure 53-2

$E/2$ volt level, but the average value of the V_R waveform is zero. At all times, the sum of the instantaneous V_R, V_C values must equal the instantaneous value of the input square wave. The total of the shaded areas for the square wave is equal to the shaded area shown in the V_C waveform; therefore, the circuit satisfies the area concept in integration.

If you idealize the V_C waveform, it becomes a steady level of $+ E/2$ volts; by contrast, the corresponding V_R waveform is a square wave with levels between $+ E/2$ and $- E/2$ volts. The capacitor has, in fact, charged to the mean level of the input square wave, and the square wave itself appears across the resistor. These results are used when coupling a signal from one circuit to another (Fig. 53-3). The mean level of the input signal is "blocked" by the capacitor, and the voltage across the resistor has an average value of zero, but the shape of the waveform is virtually undistorted. This will only be true, provided that the time constant of the coupling components is long compared with the time taken for one complete sequence of the signal. To summarize, a C,R integrator circuit requires a high time constant, and the integrated output is developed across the capacitor.

If the symmetrical square wave of Fig. 53-2A is followed by the asymmetrical square wave of Fig. 53-4A, the mean level is no longer equal to $+ E/2$ volts. Assuming that $T_1 = 3T_2$, the new mean level is,

$$+ \frac{T_1}{T_1 + T_2} \times E$$

$$= + \frac{3\,T_2 \times E}{4\,T_2} = +3E/4 \text{ volts}$$

The integrated output (V_C) will therefore climb from $+ E/2$ to $+ 3E/4$, and the new waveforms are as shown in Fig. 53-4B. You have deduced that the value of the integrated output depends on the degree of asymmetry in the square wave input. Once again, the sum of the areas under the input square wave is equal to the area under the integrated output. The V_R waveform has equal areas above and below the zero line so that its average value is again zero.

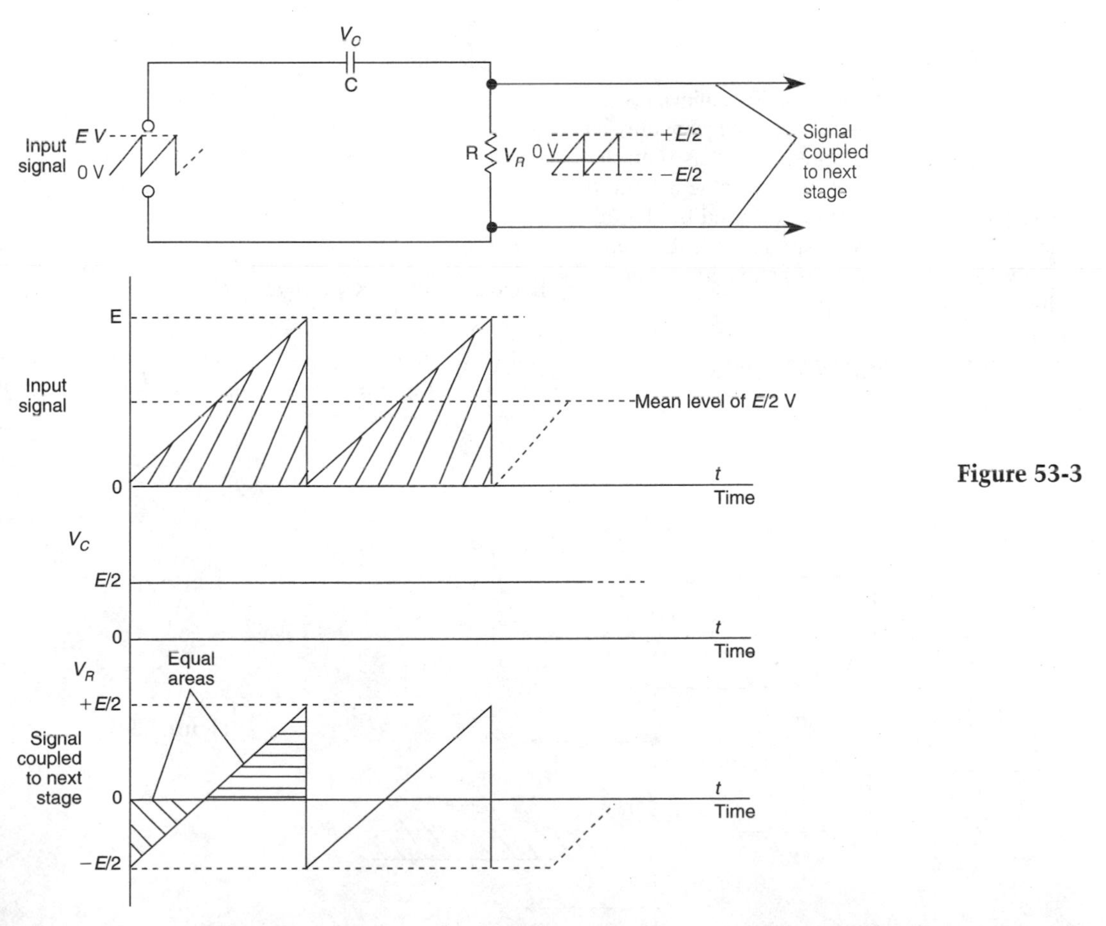

Figure 53-3

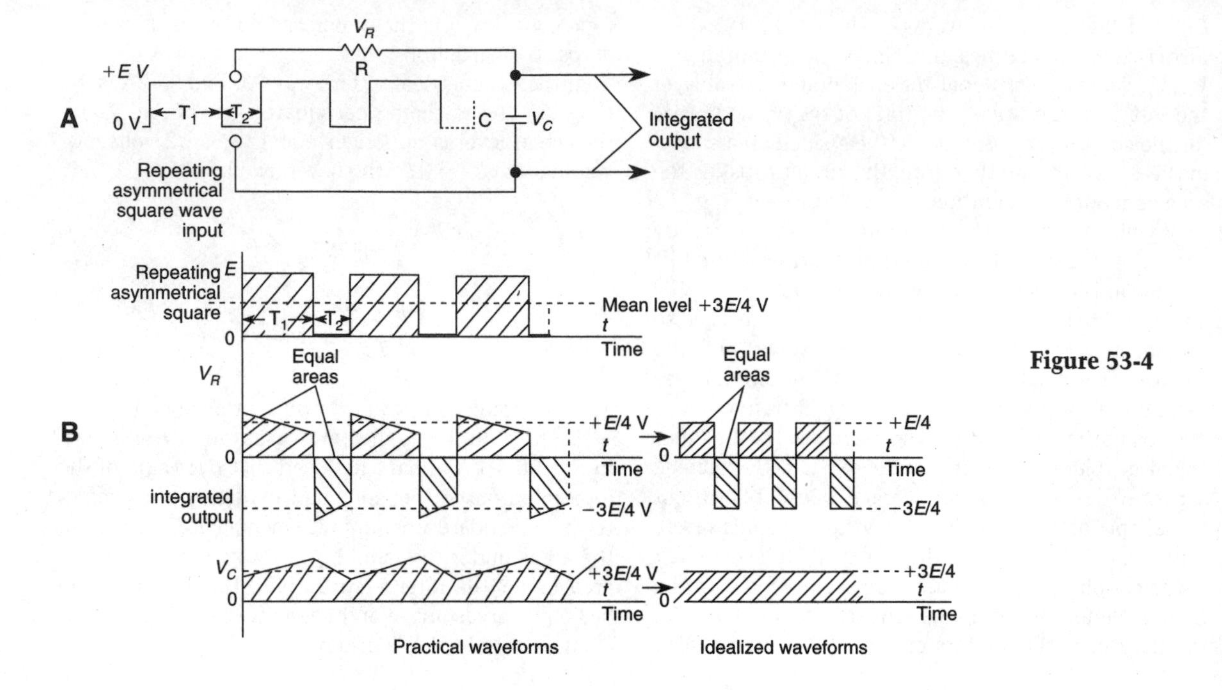

Figure 53-4

L,R INTEGRATOR CIRCUIT

This circuit is illustrated in Fig. 53-5. To achieve the integrating action, it requires a long time constant so that $L/R >> 10\ T$. The integrated output is then developed across the resistor (R) whose value is limited by the effective load, which is represented by the following circuitry; this load is in parallel with R. Consequently, the total equivalent resistance can never exceed the value of the load.

MATHEMATICAL DERIVATIONS

L,R integrator circuit

For a series L,R circuit,

$$e = v_R + v_L = iR + L\frac{di}{dt}$$

Then,

$$\int e\,dt = Rq + Li \tag{53-1}$$

Because, $q = \frac{1}{L} \times \int \left[\int v_L dt\right] dt$ and $i = \frac{v_R}{R}$,

$$\int e\,dt = \frac{R}{L}\int\left[\int v_L dt\right] dt + \frac{L}{R}\,v_R \tag{53-2}$$

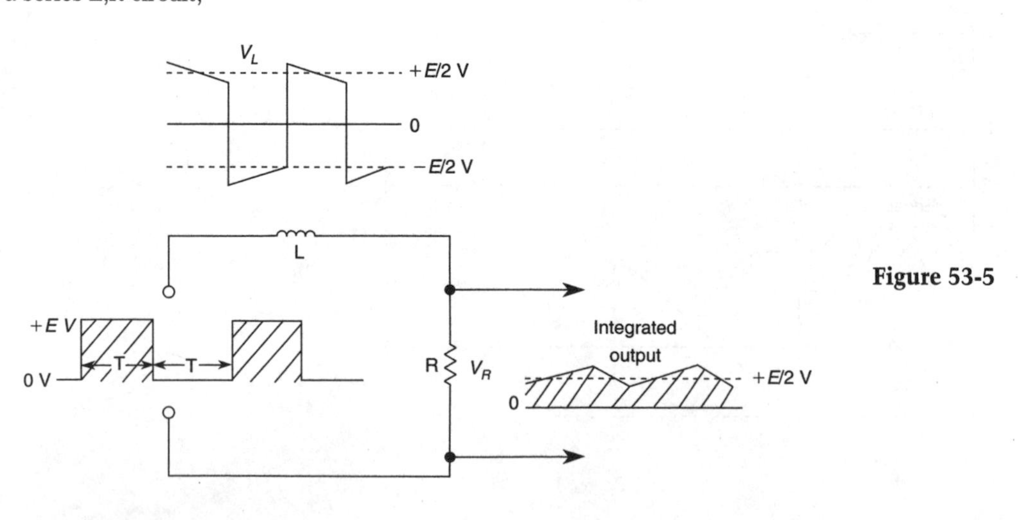

Figure 53-5

Because L/R is a *long* time constant, the value of R/L is small and the term

$$\frac{R}{L}\int\left[\int v_L dt\right]dt$$

is involved with the initial transient condition which is soon concluded. The waveform of the voltage (v_R) then directly represents the result of integrating the input voltage (e) with respect to time.

C,R integrator circuit

For a series C,R circuit,

$$e = v_C + v_R = \frac{q}{C} + iR$$

Then,

$$\int e\,dt = \frac{1}{C}\int q\,dt + Rq \qquad (53\text{-}3)$$

Because $\int q\,dt = \frac{1}{R}\left[\int v_R dt\right]$ and $q = C\,v_C$,

$$\int e\,dt = \frac{1}{CR}\left[\int v_R dt\right]dt + CR\,v_C \qquad (53\text{-}4)$$

Because CR is a *long* time constant, the term

$$\frac{1}{CR}\int\left[\int v_R dt\right]dt$$

is involved with the initial transient condition which is rapidly concluded. The waveform of the voltage (v_C) then directly represents the result of integrating the input voltage (e) with respect to time.

If e is a step voltage of magnitude E,

$$\int e\,dt = \int E\,dt = Et = CRv_C.$$

Then,

$$v_C = \frac{E}{CR} \times t$$

and the v_C waveform is that of a "ramp" voltage.

Example 53-1

In Fig. 53-2, assume that the symmetrical square-wave input has a repetition rate of 1000 times per second. If the value of R is 1 MΩ, and the time constant is $20T$ seconds, what is the required value of the capacitor for integration to occur?

Solution

$$\text{Time interval, } T = \frac{1}{2 \times 1000}\text{ s} = 500\ \mu\text{s}$$

$$\text{Time constant} = 20 \times T = 10000\ \mu\text{s}$$

$$\text{Capacitance, } C = \frac{\text{time constant}}{R} = \frac{10000\ \mu\text{s}}{1\ \text{M}\Omega} = 0.01\ \mu\text{F}.$$

PRACTICE PROBLEMS

53-1. The sequence of an 2500-Hz asymmetrical square wave contains time intervals of 100 μs and 300 μs. If this square wave is applied to a C,R integrator circuit in which R = 1 MΩ, calculate the minimum allowed value for the capacitor.

53-2. A step voltage which extends from 0 V to 10 V is applied to a C,R integrator circuit in which R = 400 kΩ and C = 0.005 μF. Calculate the initial slope of the integrated output.

53-3 A square wave which extends from 0 to 5 V is applied to an integrator circuit in which R = 1 MΩ and C = 0.002 μF. Calculate the slope of the integrated output across the capacitor.

54 Moving-coil (D'Arsonval) meter movement

The moving coil meter movement is an electromechanical device whose action depends on the motor effect (chapter 35). One of its main applications is in the voltmeter—ohmmeter—milliammeter, or VOM, which is a multimeter capable of measuring voltage, current, and resistance. Such an instrument does not require an external power supply for its operation.

The meter movement itself essentially consists of a

rectangular coil that is made from fine insulated copper wire wound on a light aluminum frame (Fig. 54-1A). This frame is carried by a spindle that pivots in jeweled bearings. The current that creates the deflection is lead into and out of the coil by the spiral hair springs (*HH'*), which behave as the controlling device and provide the restoring torque. The whole assembly is mounted between the poles of a permanent magnet so that the amount of current flowing through the coil, creates the deflecting torque. The coil is then free to move in the gaps between the permanent magnetic pole pieces (*P*) and the soft-iron cylinder (*A*), which is normally carried by a nonmagnetic bridge attached to *P* (Fig. 54-1A). The purpose of the soft-iron cylinder is to concentrate the magnetic flux and to produce a radial magnetic field with a uniform flux density. For a given current, the deflecting torque will then be constant over a wide arc (Fig. 54-1B).

The motor effect (Equation 35-1) shows that the deflecting torque is directly proportional to the current so that the divisions on the scale of the VOM are equally spaced.

When the coil rotates towards the equilibrium position, where the deflecting and restoring torques are balanced, there could be an overswing and the coil could oscillate before finally coming to rest. This unwanted oscillation is prevented by eddy current damping of the aluminum frame (Lenz's Law).

The sensitivity of the moving meter movement is inversely related to the amount of the current required to produce full-scale deflection. A high sensitivity will require many turns of fine copper wire to be mounted on a light aluminum frame and then attached to delicate hair springs. In particular, the sensitivity will depend on the strength of the meter's permanent magnet; with a high flux density, even a weak current passing through the coil will be able to produce an appreciable deflection torque.

Any flow of current through the deflecting copper coil will raise its temperature and increase its resistance. A swamping resistor with a negative temperature coefficient is therefore added in series with the coil to provide a suitable total resistance (for example, 50 Ω), which is virtually independent of temperature.

The accuracy of the meter is given as the percentage error related to the full-scale current. For example, if the error of a 0–1 mA meter movement is ±2%, the accuracy for a reading of 1 mA is $\pm 2/100 \times 1$ mA = ±0.02 mA, but if the reading is only 0.1 mA, the accuracy is still ±0.02 mA and the percentage error rises to $(\pm 0.02/0.1) \times 100 = \pm 20\%$.

MATHEMATICAL DERIVATIONS

In Fig. 54-1B, the coil consists of N turns, each with a length of l meters, and a width of d meters. If the radial magnetic field has a uniform flux of B teslas,

Deflection torque,

$$T = BINld = BINA \text{ newton-meters} \qquad (54\text{-}1)$$

where: I = current flowing through the coil (A)
A = area of the coil (m^2).

The restoring torque (T') provided by the hair springs is directly proportional to the deflection angle ($\theta°$) and therefore $T' = k\theta°$, where k is the constant for the hairspring system. In the final rest position of the needle, the two torques exactly balance.

Therefore,

$$T = T' \qquad (54\text{-}2)$$

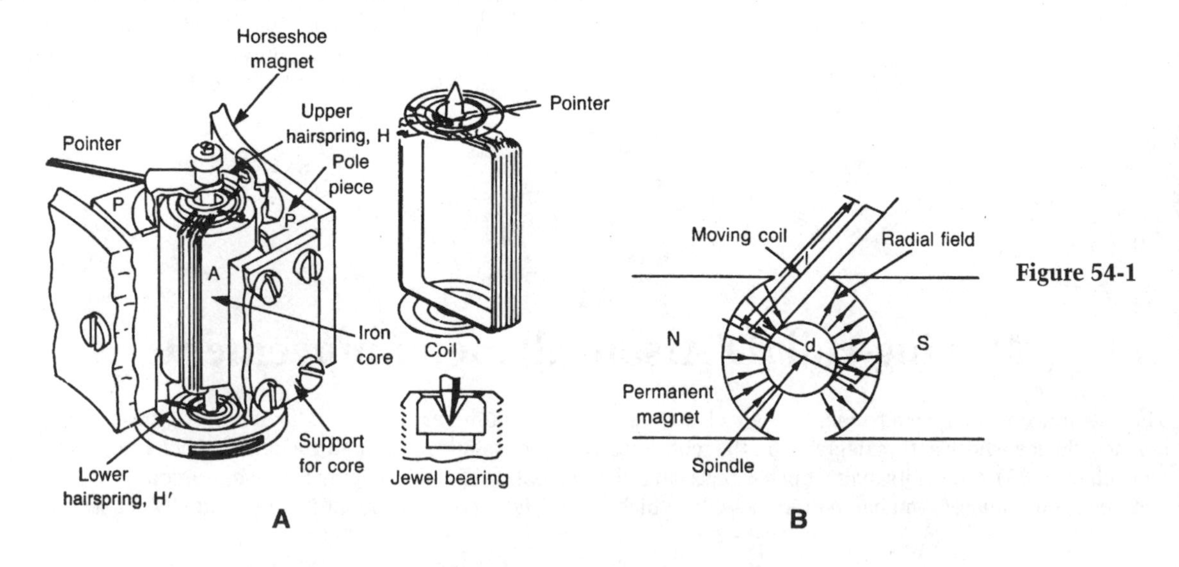

Figure 54-1

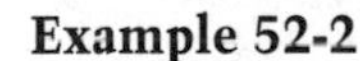

$$BINA = k\theta^{\circ}$$

$$I = \frac{k}{BAN} \times \theta^{\circ} = K\theta^{\circ}$$

where K is the constant of proportionality for the entire meter movement.

If the full-scale deflection current is I amperes.

$$\text{Sensitivity} = \frac{1}{\text{full-scale deflection current}} \quad (54\text{-}3)$$

$$= \frac{1}{I \text{ amperes or volts per ohm}}$$

$$= \frac{1}{I} \text{ ohms-per-volt}$$

Full-scale deflection current in amperes

$$= \frac{1}{\text{sensitivity in ohms-per-volt}} \quad (54\text{-}4)$$

Example 52-1

A moving coil meter movement has a sensitivity of 20,000 Ω/V. What is the full-scale deflection current in μA?

Solution

$$\text{Full-scale deflection current} = \frac{1}{\text{sensitivity}} \quad (54\text{-}4)$$

$$= \frac{1}{20{,}000\ \Omega/\text{V}}$$

$$= \frac{1{,}000{,}000}{20{,}000}\ \mu\text{A}$$

$$= 50\ \mu\text{A}.$$

Example 52-2

A moving coil meter movement has a full-scale deflection current of 1 mA. What is the value of its sensitivity?

Solution

$$\text{Sensitivity} = \frac{1}{\text{full-scale deflection current}} \quad (54\text{-}3)$$

$$= \frac{1}{1\ \text{mA}}$$

$$= \frac{1}{1/1000\ \text{A}}$$

$$= 1000\ \Omega/\text{V}.$$

PRACTICE PROBLEMS

54-1. A moving coil meter movement has a full-scale deflection current of 10 μA. What is the value of its sensitivity?

54-2. If the sensitivity of a moving coil meter movement is 2000 Ω/V, what is the value in microamperes of its full-scale deflection current?

54-3. The torque provided by the hairsprings is 2×10^{-5} N-m when the needle indicates full-scale deflection. The flux density in the air gap is $5.0 \times 10^{-2}\ T$, and there are 100 turns of wire wound on a frame whose cross-sectional area is 4 cm^2. Determine the amount of current that will provide 50% of the full-scale deflection.

54-4. The flux density for a moving coil meter movement is doubled, the number of turns is doubled, but the area of each turn remains the same. If the diameter of the wire is halved, by what factor is the torque multiplied?

55 The milliammeter

Assume that you have a basic moving coil meter movement whose total resistance (coil resistance and swamping resistance) is 50 Ω and whose full-scale deflection is 50 μA (sensitivity = 1/50 μA = 20000 Ω/V). Clearly, such an instrument, as it stands, would not be capable of measuring a current of more than 50 μA; whereas, you might be required to measure currents of hundreds of milliamperes or even amperes. To solve this problem, a shunt resistor (R_{sh}) is connected across the series combination of the meter movement and the swamp resistor (Figure 55-1A). This shunt resistor is a low-value, precision type (which is usually made from constantan wire with its negligible temperature coefficient). As a result of the shunt's low resistance, most of the current to be measured will be diverted through R_{sh} and only a small part of the current will pass through the meter movement.

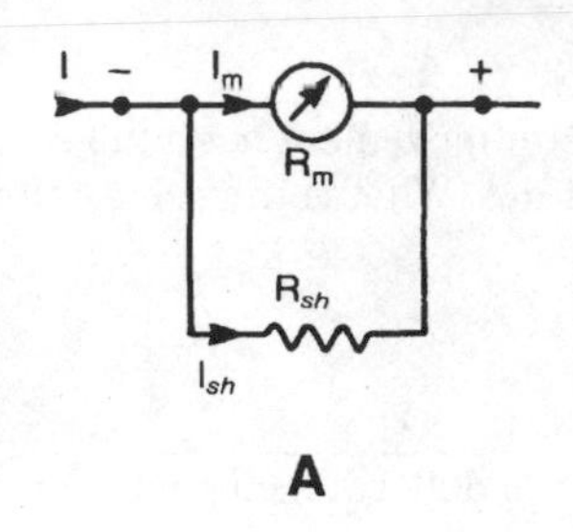

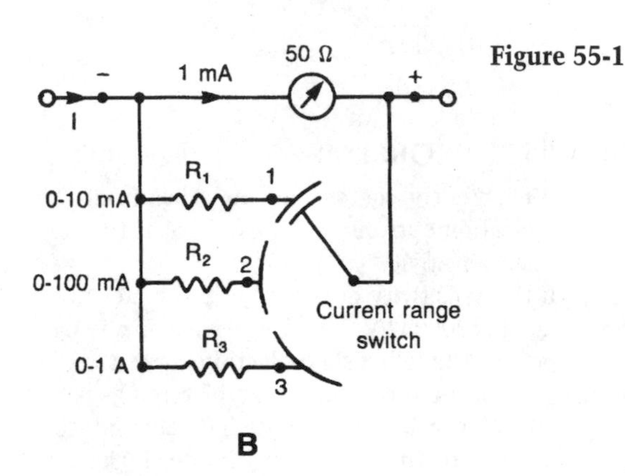

Figure 55-1

Figure 55-1B shows an arrangement that has a number of shunts for different current ranges. However, in order to take a current reading, the circuit must first be broken. The instrument is then inserted in the break so that it is directly in the path of the current to be measured. When switching from one current range to another, the meter movement could momentarily be placed in the circuit without the protection of the shunt; as a result, the meter movement might be damaged. This could be avoided by using a switch of the *make-before-break* type.

Because of its very low resistance, a current meter must never be placed directly across a voltage source. After your current measurements have been completed, you should, for safety's sake, switch to a high-voltage range. Moreover if you have no idea of the value of the current to be measured, you should start with the highest current range and then, if necessary, move to lower ranges until an appropriate deflection is obtained.

MATHEMATICAL DERIVATIONS

In Fig. 55-1A, using KCL,

$$I = I_m + I_{sh} \text{ milliamperes} \qquad (55\text{-}1)$$

Because the shunt resistance and the total resistance of the meter movement are in parallel,

$$I_{sh} \times R_{sh} = I_m \times R_m \text{ volts} \qquad (55\text{-}2)$$

Combining Equations 55-1 and 55-2,

$$R_{sh} = \frac{I_m \times R_m}{I_{sh}} = \frac{I_m \times R_m}{I - I_m} \text{ ohms} \qquad (55\text{-}3)$$

where: I = current to be measured (mA or A)
I_m = current flow through the meter movement (mA or μA)
R_{sh} = shunt resistance (Ω)
R_m = total resistance of the meter movement (Ω).

Example 55-1

In Fig. 55-1B the sensitivity of the meter movement is 1000 Ω/V and its total resistance is 50 Ω. Calculate the values of the required shunt resistors for the ranges: (a) 0 – 10 mA, (b) 0 – 100 mA, and (c) 0 – 1 A.

Solution

(a) Shunt resistance,

$$R_{sh} = \frac{I_m \times R_m}{I - I_m} \qquad (55\text{-}3)$$

$$= \frac{1 \text{ mA} \times 50\ \Omega}{10 \text{ mA} - 1 \text{ mA}}$$

$$= \frac{50}{9}$$

$$= 5.555\ \Omega.$$

(b) Shunt resistance,

$$R_{sh} = \frac{I_m \times R_m}{I - I_m} \qquad (55\text{-}3)$$

$$= \frac{1 \text{ mA} \times 50\ \Omega}{100 \text{ mA} - 1 \text{ mA}}$$

$$= \frac{50}{99}$$

$$= 0.505\ \Omega.$$

(c) Shunt resistance,

$$R_{sh} = \frac{I_m \times R_m}{I - I_m} \qquad (55\text{-}3)$$

$$= \frac{1 \text{ mA} \times 50\ \Omega}{1000 \text{ mA} - 1 \text{ mA}}$$

$$= \frac{50}{999}$$

$$= 0.05005\ \Omega.$$

Example 55-2

If the meter movement of Example 55-1 is used to measure the current in Fig. 55-2, what will be the meter reading?

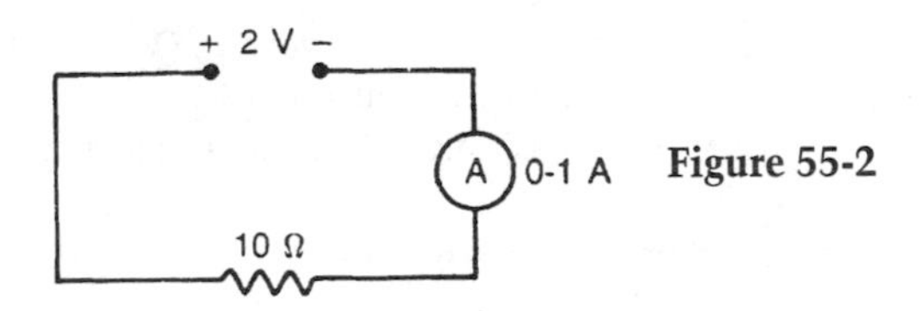

Figure 55-2

Solution

Because the theoretical current is 2 V/10 Ω = 0.2 A, the meter must be switched to the 0- to 1-A range. The total resistance of the meter movement and the shunt is 1 mA × 50 Ω/1000 mA = 0.05 Ω. Therefore, the meter reading will be 2 V/(10 Ω + 0.05 Ω) = 0.199 A. This indicates the loading effect of the ammeter on the circuit being monitored. Such a loading effect is negligible with high-resistance circuits, which are normally encountered in electronics.

PRACTICE PROBLEMS

55-1. A VOM contains a meter movement whose resistance is 50 Ω. On the 0- to 1-A range the total resistance of the instrument is 0.05 Ω. Determine the sensitivity of the VOM.

55-2. A moving coil meter movement has a resistance of 20 Ω, and has a full-scale deflection current of 1 mA. After shunting the meter movement with a 3.2-Ω resistor, a reading of 0.8 mA is recorded. What is the actual value of the current being measured?

55-3. A moving coil meter movement has a resistance of 15 Ω, and requires a current of 0.5 mA to provide full-scale deflection. A shunt resistor of 0.075 Ω is connected across the meter movement. Determine the maximum value of the new current range.

55-4. An ammeter, whose resistance is 0.5 Ω, is connected in series with a 12-Vdc source, and a load which normally requires a current of 12 A from the source. What is the reading of the current indicated by the ammeter?

56
The loading effect of a voltmeter

A voltmeter is placed in parallel with (shunting across) the component whose potential difference is to be measured. This means that the voltmeter must possess as high a resistance as possible so that it has minimum loading effect on the circuit being monitored. By itself, the moving-coil meter movement is basically a millivoltmeter (or microvoltmeter) so that, in order to adapt the movement for higher voltages, it is necessary to connect a series "multiplier" resistor, R_S (Fig. 56-1A). Most of the voltage to be measured is then dropped across this high-value series resistor, and only a small amount appears across the meter movement to provide the necessary deflection. For a multirange voltmeter, a separate multiplier resistor can be used for each range.

THE VOLTMETER

On a particular range, the total resistance of a VOM (including the multiplier resistor) can be found from the product of the sensitivity and the range's full-scale deflection voltage. For example, if the sensitivity is 20,000 Ω/V and the full-scale deflection voltage is 10 V, the voltmeter's resistance is 20,000 Ω/V × 10 V = 200,000 Ω = 200 kΩ. On the 0- to 100-V range, the voltmeter's resistance would be 20,000 Ω/V × 100 V = 2,000,000 Ω = 2 MΩ.

LOADING EFFECT OF A VOLTMETER

An ideal voltmeter would have infinite resistance so that the instrument does not load the circuit being monitored. However, we have already seen that the resistance of a moving-coil voltmeter is far from infinite, and moreover, its resistance changes with the particular voltage range which you select. The worst loading effect occurs with a low-voltage, high-resistance circuit such as shown in Fig. 56-1B.

With no voltmeter present, the voltage existing between points A and B is 6 V so that we would switch to the 0- to 10-V range. Assuming that the voltmeter's sensitivity is 20,000 Ω/V, the voltmeter's resistance is

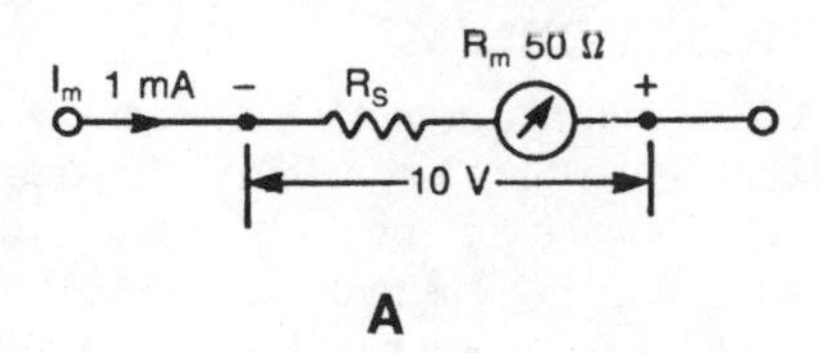

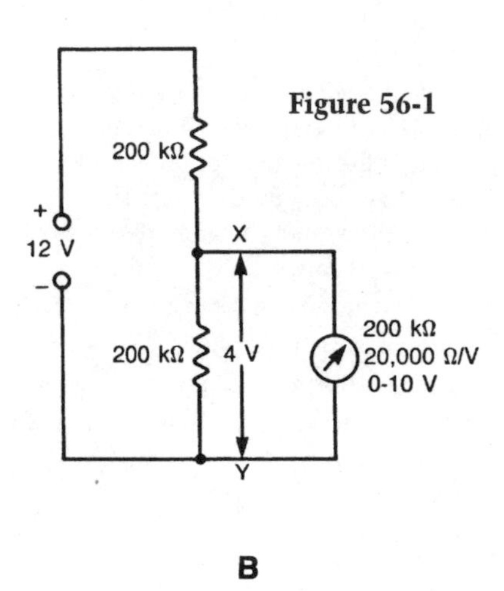

Figure 56-1

200 kΩ. To read the voltage between points A and B, the voltmeter is connected across the lower 200-kΩ resistor. However, the parallel combination of this resistor and the voltmeter only has a total resistance of 200 kΩ/2 = 100 kΩ and, therefore, by using the VDR, the voltmeter reading will be 12 V × 100 kΩ/(100 kΩ + 200 kΩ) = 4 V. The loading effect of the voltmeter has, therefore, reduced the voltage between A and B from 6 V down to 4 V.

Most solid-state voltmeters have very high resistances (several megohms) that are independent of the voltage range chosen; such voltmeters have a minimum loading effect on the circuit being monitored.

MATHEMATICAL DERIVATIONS

In Fig. 56-1A,

Total resistance of the voltmeter,

$$R_V = \frac{V}{I_m} \text{ ohms} \qquad (56\text{-}1)$$

$$= V \times (\text{sensitivity in ohms-per-volt}) \qquad (56\text{-}2)$$

Resistance of the series "multiplier" resistor,

$$R_S = \frac{V}{I_m} - R_m \text{ ohms} \qquad (56\text{-}3)$$

Also,

$$\text{Resistance, } R_S = V \times (\text{sensitivity in ohms per volt}) - R_m \text{ ohms} \qquad (56\text{-}4)$$

where: R_S = resistance of series "multiplier" resistor (Ω)
V = full-scale deflection voltage (V)
I_m = total deflection current (A)
R_m = total resistance of the meter movement (Ω)
R_V = total resistance of the voltmeter (Ω).

Example 56-1

In Fig. 56-2, the potential difference between the points A and B is measured by a 1000 Ω/V voltmeter, which is switched to the 0- to 10-V range. What is the reading of the voltmeter?

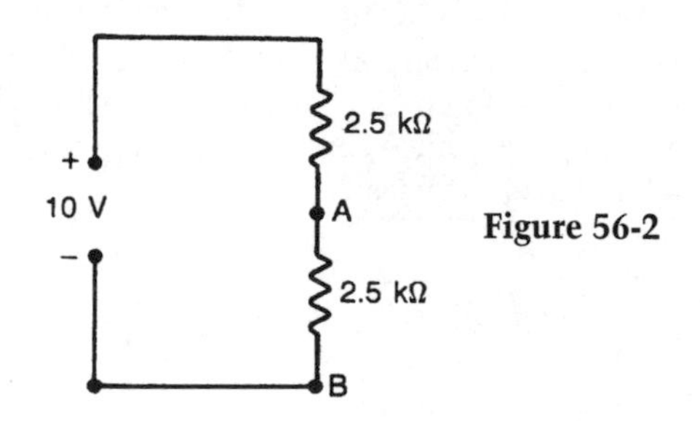

Figure 56-2

Solution

Total resistance of the voltmeter,

$$R_V = V \times (\text{sensitivity in } \Omega/\text{V}) \qquad (56\text{-}2)$$
$$= 10 \text{ V} \times 1000\ \Omega/\text{V}$$
$$= 10000\ \Omega$$
$$= 10 \text{ k}\Omega.$$

Total equivalent resistance between points A and B, with the voltmeter included, is (10 × 2.5)/(10 + 2.5) = 2 kΩ.

By the VDR, voltmeter reading = 10 V × 2 kΩ/(2 kΩ + 2.5 kΩ) = 4.444 V. This reading compares with 5 V, which exists between points A and B in the absence of the voltmeter.

PRACTICE PROBLEMS

56-1. Two identical voltmeters, each with a sensitivity of 1000 Ω/V, are connected in series across a 200-Vdc source. The first voltmeter is switched to the 0- to 250-V scale, and the second one is operated on the 0- to 100-V scale. Determine the reading of the second voltmeter.

56-2. A dc source has an open-circuit voltage of 12 V. When a 20000-Ω/V VOM on its 0- to 10-V range is connected across the terminals of the source, its reading is 11.5 V. Calculate the value of the source's internal resistance.

56-3. When a series combination of a voltmeter and a 47-kΩ resistor is connected across a 120-Vdc source, the voltmeter reading is 110 V. Another series resistor is now inserted into the circuit and the voltmeter reading falls to 95 V. What is the value of the second resistor?

56-4. In the circuit of Fig. 56-3, the 2750-Ω resistor is a series multiplier which has been used to convert the milliammeter (M) to a voltmeter. With the switch (S) in position 1, M shows a full-scale deflection when 100 V is applied between points X and Y. With S in position 2, 125 V is applied between X and Y, and the meter then reads 70 V. Neglecting the internal resistance of M, calculate the value of R.

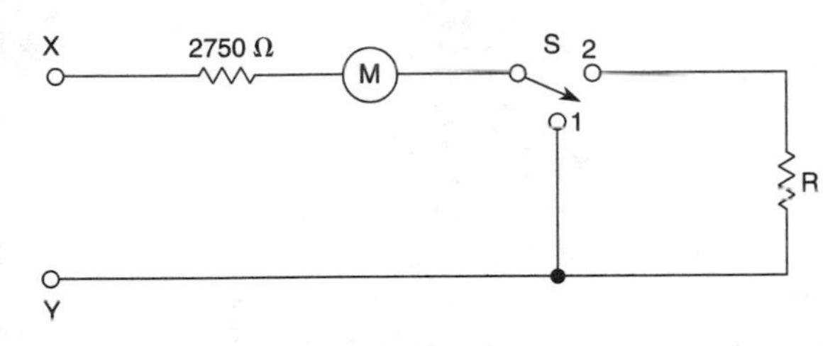

Figure 56-3

56-5. In the circuit of Fig. 56-4, the voltmeter (V) is switched to the 0- to 10-V range, and shows a reading of 1.92 V. What is the value of the voltmeter's sensitivity?

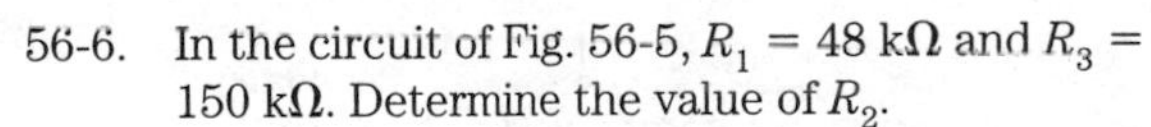

56-6. In the circuit of Fig. 56-5, R_1 = 48 kΩ and R_3 = 150 kΩ. Determine the value of R_2.

56-7. Two resistors whose values are 1 MΩ and 2 MΩ are connected in series across a 12-Vdc source. The voltage across the 2-MΩ resistor is measured by a 0- to 100-μA meter movement which has been adapted for use as a voltmeter. What are the readings of the voltmeter on the 0- to 10-V and 0- to 100-V scales?

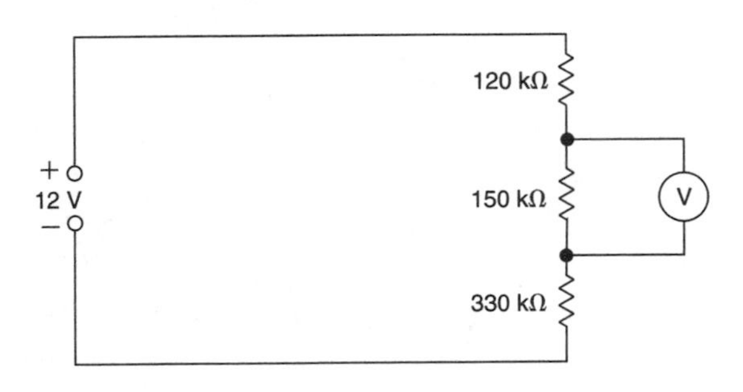

Figure 56-4

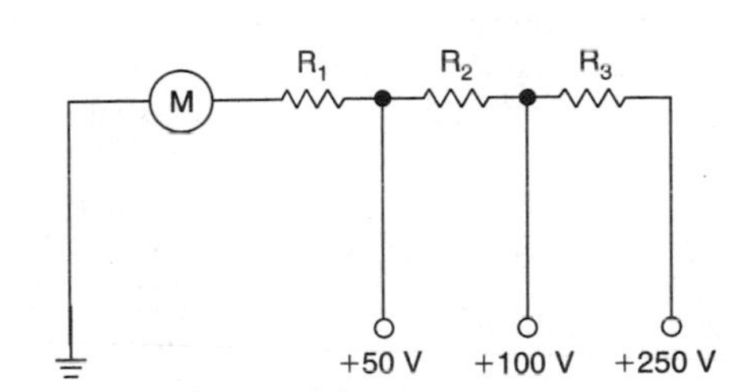

Figure 56-5

57
The ohmmeter

Because all resistance measurements must be made with power removed from the circuit, it is necessary to provide one (or more) primary cells to create the necessary deflection. These cells are normally included within the instrument itself.

In its simplest form (Fig. 57-1A), an ohmmeter would consist of a 1.5-V primary cell, a rheostat, and a 0- to 1-mA moving-coil meter movement with its 50-Ω total resistance. The unknown resistance (R_x) is then connected between the probes of the ohmmeter.

The steps involved with taking a resistance reading are:

1. Place the probes apart so that an open circuit (infinite resistance) exists, and there is zero current. The needle will then be on the far left-hand side of the scale (Fig. 57-1A) so that the reading is zero current, but infinite ohms. The symbol for

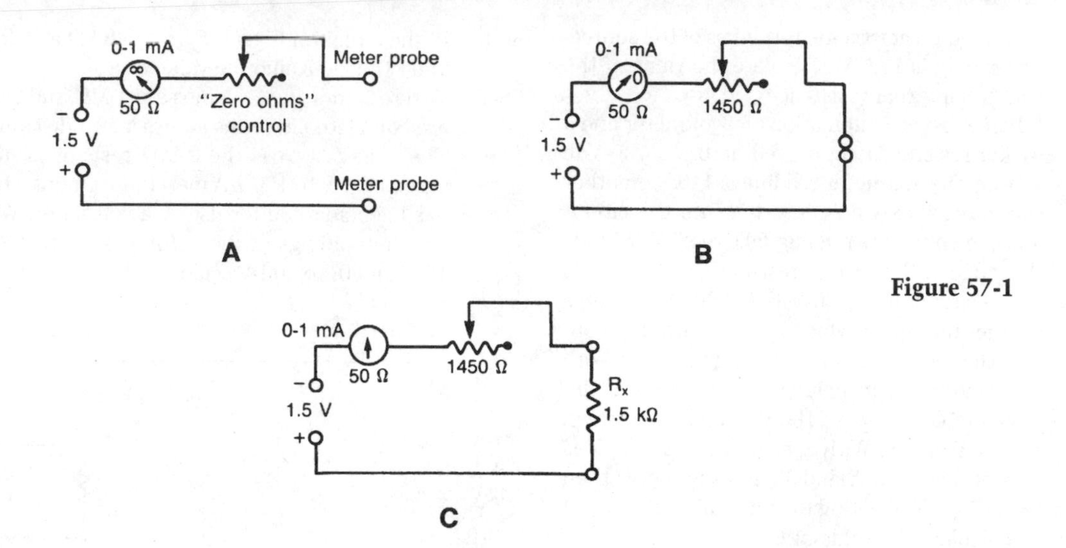

Figure 57-1

infinity is ∞, which is normally marked on the ohms scale. If the needle is not correctly on the zero current mark, the torque provided by the hairsprings can be mechanically adjusted.

2. Short the probes together (Fig. 57-1B) to create zero resistance, then adjust the rheostat until the full-scale deflection current of 1 mA flows, and the needle is on the far right of the scale. The circuit's total resistance is now 1.5 V/1 mA = 1500 Ω and the value of the rheostat (R_{rh}) is then 1500 Ω − 50 Ω = 1450 Ω. As the cell ages, its terminal voltage decreases, and the rheostat can be adjusted to bring the current back up to 1 mA.
3. Assume that a 1500-Ω resistor (R_x) is now connected between the probes (Fig. 57-1C). The current would then be 1.5 V/(1450 Ω + 50 Ω + 1500 Ω) = 0.5 mA, and the needle would be in the center of the scale. Consequently, infinite ohms are on the left of the scale, zero ohms on the right, and 1500 ohms in the center. Therefore, the scale is nonlinear because the deflection current is inversely dependent on the value of the unknown resistance. The deflection marks are expanded on the low-resistance side, but crowded on the high-resistance end. Clearly such a scale could not be used for accurate measurements of high resistances (greater than 10 kΩ).

With the VOM, it is common practice to have three (or more) resistance ranges: $R \times 1$, $R \times 100$, and $R \times 10000$. This is achieved by including the multiplier resistors (R_S), as shown in Fig. 57-2. On the $R \times 1$ range the multiplier resistor (R_{S1}) has a high resistance so that if R_x has a value of 150 kΩ (for example), the needle would be close to the infinite mark and an accurate reading would be impossible. However, if you now

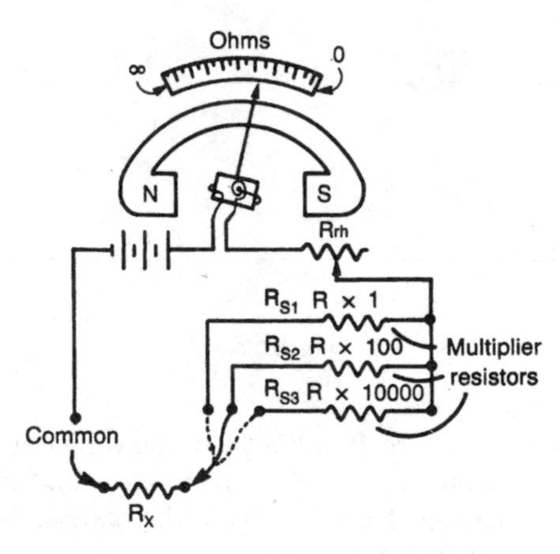

Figure 57-2

switch to the $R \times 100$ range, with its lower multiplier resistor (R_{S2}), the needle is deflected to the middle of the scale and the reading would be 1500 Ω × 100 = 150000 Ω = 150 kΩ.

Note that because the multiplier resistors have different values, you must always readjust the "zero ohms" rheostat when switching from one resistance range to another.

Example 57-1

From Fig. 57-1A, what is the deflecting current if the unknown resistance is: (a) 500 Ω, and (b) 4.5 kΩ?

Solution

(a) Total circuit resistance = 1500 Ω + 500 Ω = 2000 Ω.

$$\text{Deflection current, } I_m = \frac{1.5\text{ V}}{2000\ \Omega} = 0.75\text{ mA.}$$

(b) Total circuit resistance = 1500 Ω + 4.5 kΩ = 6 kΩ.

$$\text{Deflection current, } I_m = \frac{1.5\text{ V}}{6\text{ k}\Omega} = 0.25\text{ mA.}$$

PRACTICE PROBLEMS

57-1. A basic ohmmeter uses a 0- to 1-mA moving coil meter movement in conjunction with a 1.5-V cell. When an unknown resistor is connected between the ohmmeter probes, the deflection is exactly one fifth of the way across the linear current scale. Determine the value of the unknown resistor.

57-2. A basic ohmmeter contains a 200,000-Ω/V meter movement and a 1.5-V battery. What is the resistance reading for 30% of full-scale deflection as measured from the "∞" ohms mark.

57-3. A basic ohmmeter circuit consists of a 1.5-V cell, a rheostat, and a 0- to 100-μA meter movement whose resistance is 50 Ω. When the measured resistance is 10 kΩ, what is the current flowing through the meter movement?

57-4. An ohmmeter uses a 0- to 100-μA meter movement whose resistance is 30 Ω. This meter movement is connected in series with a 4-V cell and a rheostat. Across the meter movement is a variable shunt resistor whose value is 30 Ω. To what value is the rheostat set? Find the resistances measured when the deflection of the needle is one third, one half, and two thirds of its full-scale value. If the cell's voltage falls to 3 V, calculate the new resistance value to which the variable shunt must be adjusted.

58
The moving-iron meter

There are two types of moving-iron instrument:

1. The repulsion type, in which two parallel soft-iron vanes are equally magnetized inside a solenoid, and, therefore, repel one another.
2. The attraction type, in which a solenoid is used to attract a piece of soft iron.

REPULSION MOVING-IRON METER

Figure 58-1A shows two iron vanes which are separated axially in a short solenoid. One vane is moveable and attached to a pivot that also carries the needle, while the other vane is fixed. When a current flows through the coil, the vanes are equally magnetized to create a force of repulsion. The pointer is then deflected across the scale, which is calibrated for a direct reading. However, the force of repulsion is dependent on the flux of each iron vane and directly proportional to the *square* of the current. This scale is therefore nonlinear, being cramped at the low end, but open at the high end.

The restoring torque of the controlling device is provided by either a spring or by gravity. An air piston is commonly used as a damping device.

Because the amount of the deflection torque depends on the number of ampere-turns for the coil, it is possible to achieve various ranges by winding different numbers of turns on the coil. In addition, you can vary the type of wire used so that the meter's resistance can be changed and there is no need for swamping and shunt resistances.

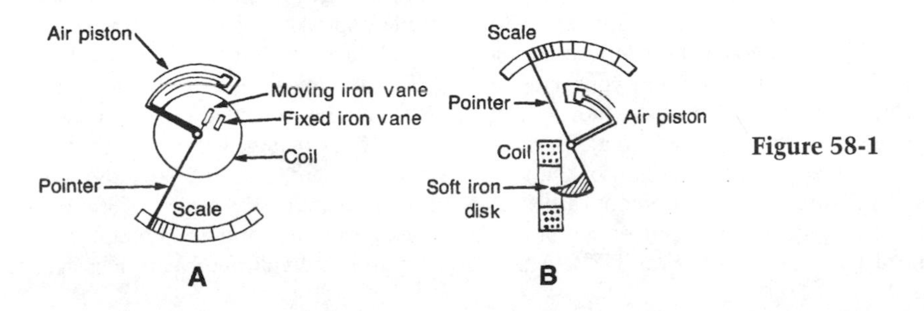

Figure 58-1

ATTRACTION MOVING-IRON METER

This type of meter is illustrated in Fig. 58-1B. The current to be measured flows through the coil and establishes a flux, which magnetizes the soft-iron disk. As a result, the soft-iron disk is drawn into the center of the coil and the pointer moves across the scale. However, the force acting on the iron disk depends on the flux density and the coil's magnetic field intensity, which are both proportional to the current. Therefore, the deflection is not proportional to the current alone (as in the moving-coil meter movement) but to the *square* of the current. Consequently, the meter has a nonlinear scale that is cramped at the low end, but is open at the high end. It is possible to improve the linearity by accurate shaping of the iron disk.

The moving-iron milliammeter can be converted to a voltmeter by adding a suitable noninductive series resistor. When compared with moving-coil instruments, moving-iron meters are relatively inexpensive and robust, but have low sensitivity. The accuracy of moving-iron meters is normally ±5% of the full-scale deflection value. Unlike the moving-coil meter, the moving-iron instruments can be used for direct *ac measurements*.

Example 58-1

A moving-iron meter requires 550 ampere-turns to provide full-scale deflection. How many turns on the coil will be required for the 0- to 10-A range? For a voltage scale of 0 to 250 V, with a full-scale deflection current of 25 mA, how many turns are required, and what is the total resistance of the instrument?

Solution

Number of turns required for the

$$\text{0- to 10-A range} = \frac{550\text{ AT}}{10\text{ A}}$$

$$= 55\text{ turns.}$$

Number of turns for the

$$\text{0- to 250-V range} = \frac{550}{0.025}$$

$$= 22000\text{ turns.}$$

$$\text{Total resistance} = \frac{250\text{ V}}{25\text{ mA}} = 10\text{ k}\Omega.$$

PRACTICE PROBLEMS

58-1. When a moving-iron meter is wound with 60 turns, its full-scale deflection current is 5 A. How many turns would be required to provide a full-scale deflection current of 20 A?

58-2. A current-squared moving-iron meter has a full-scale deflection current of 10 A. If the measured current increases from 5 A to 7 A, determine the rise in the percentage of the full-scale deflection.

58-3. A current-squared moving-iron meter has a full-scale deflection current of 5 A. What percentage of the full-scale deflection corresponds to a current of 2 A? What value of current will correspond to 40% of the full-scale deflection?

59
The wattmeter

Because the moving-coil meter movement works in conjunction with a permanent magnet, this type of meter can only be used for direct dc measurements (to measure an ac voltage, for example, the alternating voltage must first be rectified and thereby converted to a dc voltage, which is then applied to the moving-coil instrument). However, if the permanent magnet is replaced by an electromagnet, the direction of the flux may be reversed and direct ac measurements are then possible. This *electrodynamometer* movement of Fig. 59-1A consists of two fixed coils (F and F'), which form the electromagnet. The third moving coil (M) is carried by a spindle and the controlling torque is exerted by spiral hairsprings (H and H'), which also serve to lead the current into and out of the moving coil.

For dc measurements, the electrodynamometer movement has no advantage over the D'Arsonval type. It is better to measure the source voltage, and the source current, separately and then calculate their product to obtain the dc power. Compared with the ac moving-iron instruments of chapter 58, dynamometer ammeters and voltmeters are less sensitive and more

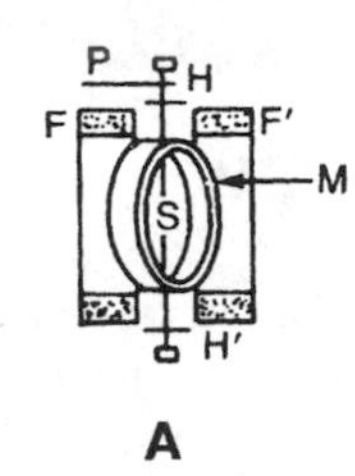

A

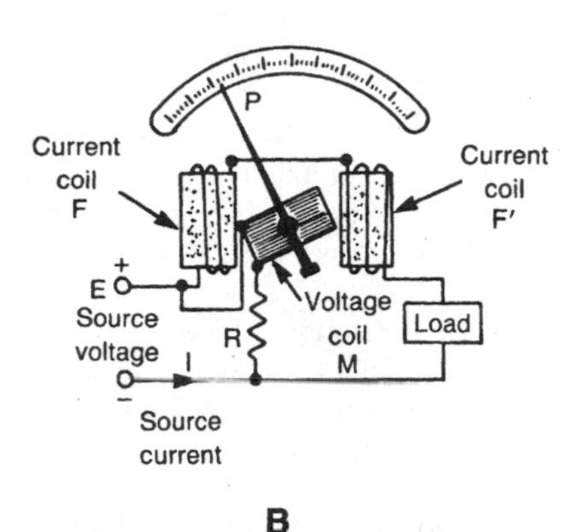

B

Figure 59-1

expensive so that they are rarely used. The most important application is the *dynamometer wattmeter*, which is the common way of measuring power directly in ac circuits.

Figure 59-1B illustrates the manner in which the wattmeter is connected into the circuit. The two fixed coils (F and F') are joined in series with the load and, therefore, they carry the load current, I. The third moving coil is connected in series with a high-value multiplier resistor, R. This combination is connected across the source so that the current through the moving coil is directly proportional to the source voltage, E. Consequently, the torque on the moving coil is directly proportional to the product of the current through the fixed (current) coils, and the current through the moving (voltage) coil. It then follows that this torque, and its resulting deflection, are directly proportional to the power absorbed by the load. Because the controlling torque provided by the hairsprings is also directly proportional to the deflection, electrodynamometer wattmeters have linear scales with the markings evenly spaced.

When used for ac measurements, the moving coil cannot follow the rapid fluctuations in the level of the instantaneous power so that it assumes a mean (or average) position that corresponds to the true power in the circuit. In other words, the electrodynamometer movement automatically takes into account the circuit's *power factor.*

MATHEMATICAL DERIVATIONS

Power measurement: dc

$$\text{Wattmeter reading} = E \times I \text{ watts} \qquad (59\text{-}1)$$

where: E = source voltage (V)
I = source current (A)

Power measurement: ac

$$\text{Wattmeter reading} = E \times I \times \text{power factor} \qquad (59\text{-}2)$$

where: E = rms source voltage (V)
I = rms source current (A).

Example 59-1

An electrodynamometer wattmeter has a full-scale deflection of 100 W. What value of power will correspond to 65% of the full-scale deflection?

Solution

Because the electrodynamometer wattmeter has a linear scale, the power corresponding to a 65% deflection is 65/100 × 100 W = 65 W.

PRACTICE PROBLEMS

59-1. A wattmeter is switched to its 0- to 100-W range and is connected to a 50-Vdc source. When an 80-Ω resistive load is connected to the source, what is the percentage of the full-scale deflection on the wattmeter's scale?

59-2. A wattmeter is connected to a 120-Vdc source and is operated on the 0- to 100-W range. If the source current is 400 mA, what percentage of full-scale deflection is the reading on the wattmeter?

60
The cathode-ray oscilloscope

The basic triggered sweep cathode ray oscilloscope (CRO) is probably the most flexible, general-purpose measuring instrument in use today and, perhaps, for some time in the future. The key element of the CRO is the electrostatic cathode-ray tube, which operates with a high degree of vacuum and in which a stream of electrons is produced by a heated cathode. This acts together with other metallic elements to form the *electron gun*. The stream of electrons is accelerated by high voltages towards the inside surface of the tube face. Light is produced when the electrons strike the fluorescent screen that coats this inside surface.

Between the beginning of the electron gun and the fluorescent screen are a number of electrodes held at positive potentials with respect to the cathode. Some of these serve the purpose of focusing the electron beam to a point at the screen. A control on the front panel can be used to improve the focus and affect the size and appearance of the spot seen by the operator. Other electrodes cause the electron beam to be deflected. One pair of electrodes, called the X plates, causes a horizontal deflection, which is proportional to the impressed voltage. A second pair, the Y plates, causes vertical deflection (Fig. 60-1A) whose amount depends on the length of the vertical plates and their separation, the values of the vertical deflection voltage and the horizontal accelerating voltage, as well as the distance between the vertical plates and the screen.

An electrostatic CRT commonly has a sensitivity that is about 1 mm of vertical deflection for each volt applied to the Y plates, with 1 kV of horizontal accelerating voltage. If $V_Y = 100$ V and $V_X = 10$ kV, the vertical deflection is $1 \times (100/10)$ mm = 1 cm. In other words, the sensitivity is 100 V/cm for a horizontal accelerating voltage of 10 kV.

In the most commonly used mode of operation, the horizontal deflection is linear and is produced by a sawtooth voltage waveform applied to the X plates. The beam, and the spot that it produces, move equal distances in equal intervals of time. For slow movements, the spot can be seen moving to the right, leaving a rapidly disappearing trail. Upon reaching the end of its travel on the right, the spot disappears and quickly reappears on the left to repeat its linear journey. With fast movements of the beam, required for most electronic measurements, the beam cannot be seen, but its trace, repeated hundreds, thousands, or even millions of times per second, appears as a solid horizontal line.

The vertical deflection of the beam is usually caused by an amplified signal to be observed and measured. The amount of amplification, and hence the deflection, is controlled by a knob on the CRO panel.

The combined horizontal sweep motion and vertical signal deflection give a graphical picture of the signal waveform. If a dc voltage is applied to the vertical plates, the appearance on the screen is a horizontal line, which is deflected upwards or downwards according to the polarity of the deflecting voltage.

MATHEMATICAL DERIVATIONS
In Fig. 60-1B,

Vertical deflection,

$$y = \frac{Ll\,V_Y}{2d\,V_X} = K \times V_Y \text{ millimeters} \qquad (60\text{-}1)$$

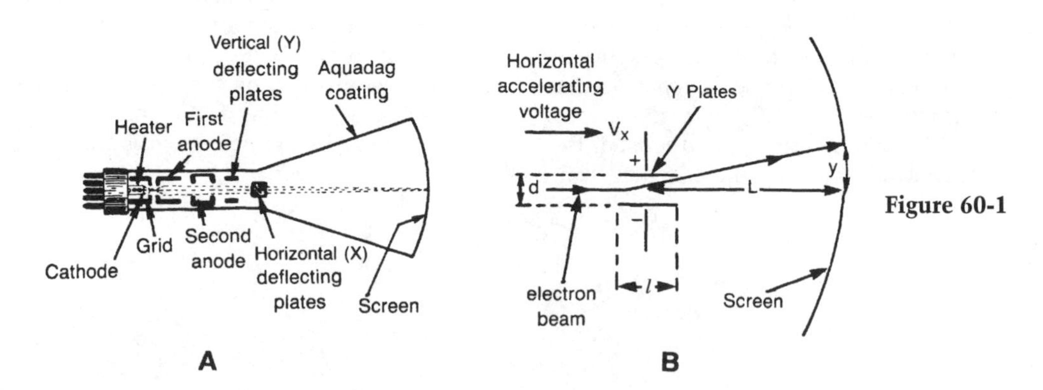

Figure 60-1

Sensitivity of the CRT,

$$K = \frac{Ll}{2d\,V_X} \text{ millimeters per deflection volt, for each 1 kV of horizontal accelerating voltage} \quad (60\text{-}2)$$

where: y = vertical deflection (mm)
L = distance between the vertical plates and the screen (mm)
l = length of the vertical plates (mm)
d = separation between the plates (mm)
V_Y = vertical deflection voltage (V)
V_X = horizontal accelerating voltage provided by the "gun" (kV).

Note that the amount of vertical deflection depends neither on the charge nor on the mass of the particle being deflected. Therefore, electrons and negative residual gas ions are subjected to the same deflection.

Example 60-1

The sensitivity of a CRT is 0.6 mm per deflection volt for each 1 kV of accelerating voltage. If the accelerating voltage is 8 kV, what value of dc voltage must be applied to the Y plates in order to produce a deflection of 2.7 cm?

Solution

Required deflection voltage,

$$V_Y = \frac{y}{K} \quad (60\text{-}1)$$

$$= \frac{27 \text{ mm} \times 10 \text{ kV} \times 1 \text{ V}}{0.6 \text{ mm} \times 1 \text{ kV}}$$

$$= 450 \text{ V.}$$

PRACTICE PROBLEMS

60-1. The vertical amplifier of a CRT is switched to the 10-V/cm position. When the dc voltage is measured, the amount of deflection is 2.5 cm. Calculate the value of the dc voltage.

60-2. The sensitivity of a CRT is 1.5 mm per deflection volt, for each 1 kV of accelerating voltage. If the accelerating voltage is 10 kV, what amount of deflection will be produced by applying 240 Vdc to the Y plates?

2
PART
Alternating current principles

61
Introduction to alternating current (ac)

Previous chapters have been primarily concerned with direct current in which the electron flow has always been in the same direction, although the magnitude of the flow has not necessarily been constant.

With alternating current, the current reverses its direction periodically and has an average value of zero. If the average value of the waveform is not zero, it is regarded as a combination of a dc component and an ac component. The alternating current can assume a variety of waveforms, some of which are shown in Figs. 61-1A, B, C, and D. In every case, a single complete waveform is known as a *cycle* and the number of cycles occurring in one second is the *frequency*, which is therefore the rate at which the waveform repeats.

The letter symbol for the frequency is f and its SI unit is the hertz (Hz), which is equivalent to one cycle per second.

The time interval taken by a complete cycle is the *period*, which is equal to the reciprocal of the frequency. The letter symbol for the period is T and its basic unit is the second (s). As an example, the commercial ac line voltage has a frequency of 60 Hz and, therefore, its period is 1/60 of a second.

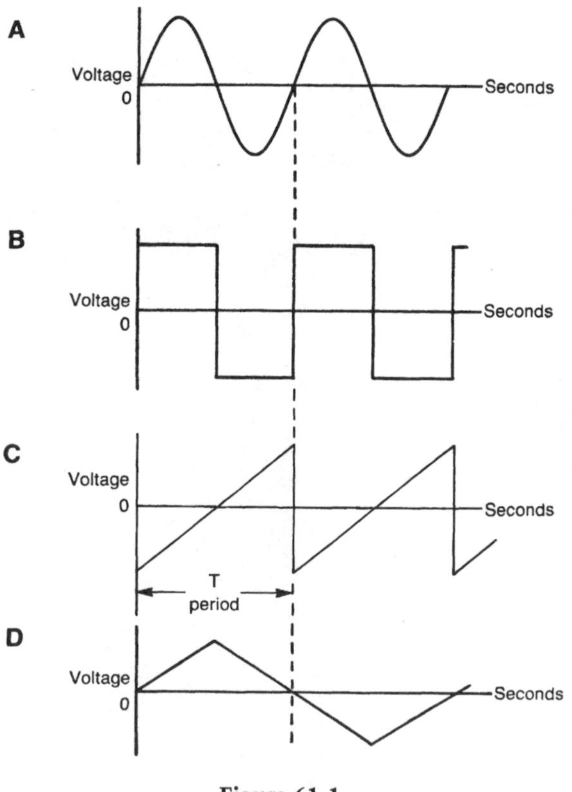

A Voltage 0 Seconds
B Voltage 0 Seconds
C Voltage 0 Seconds T period
D Voltage 0 Seconds

Figure 61-1

The *sine wave* is not only associated with the commercial line voltage, but it is also widely used for communications. The curve itself is the result of plotting the mathematical sine function (sin θ) versus the angle θ, measured in either degrees or radians (Fig. 61-2).

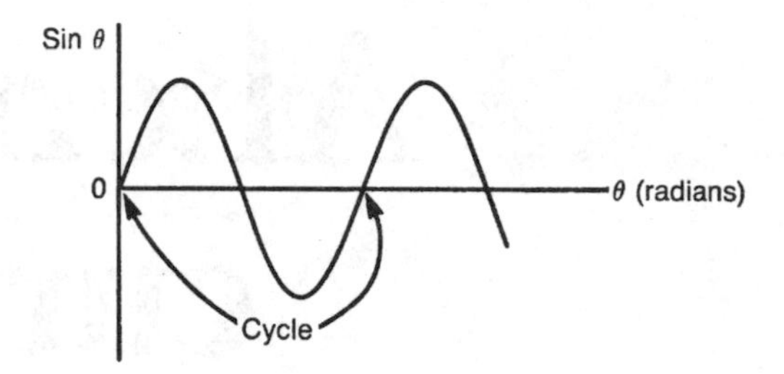

Figure 61-2

The vertical axis measures only one of a number of electrical parameters such as voltage, current, or power, while the horizontal axis is a time scale. It follows that, for every sine wave, a certain number of radians on the horizontal scale must correspond to a particular time interval. The angular frequency whose letter symbol is the Greek lowercase omega (ω) is found by dividing the number of radians by the equivalent time interval and is, therefore, measured in radians per second (rad/s).

One cycle of a sine wave consists of a "positive" alternation and a "negative" alternation (Fig. 61-3). The maximum excursion from its average (or zero) line is called the *peak* or *maximum* (E_{max}) *value*, and the distance between the wave's crest and trough is a measure of the *peak-to-peak* ($E_{p\text{-}p}$) *value*.

Sine waves are of particular importance, because Fourier analysis allows the waveforms of Figs. 61-1B, C, and D to be broken down into a series of sine waves, that consist of a fundamental component together with its harmonics (Fig. 61-4). The *harmonic* of a sine wave is another sine wave whose frequency is a whole number multiple of the original (or fundamental) frequency. For example, if the fundamental frequency is 1 kHz (f), the second harmonic has a frequency of 2 kHz ($2f$), the third harmonic frequency is 3 kHz ($3f$), etc. The sine wave syntheses of a square wave and a sawtooth wave are, respectively, illustrated in Figs. 61-5A, B.

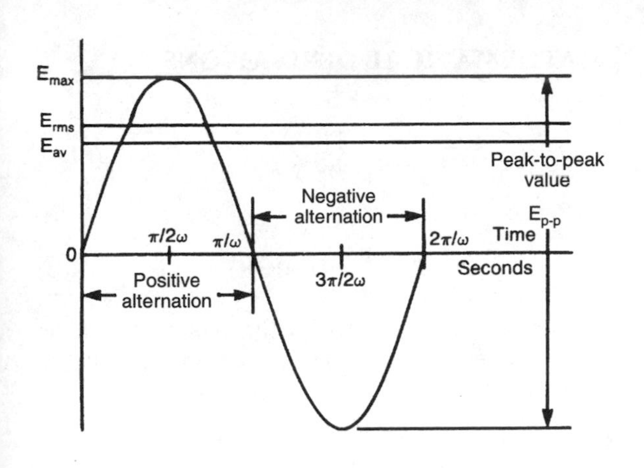

Figure 61-3

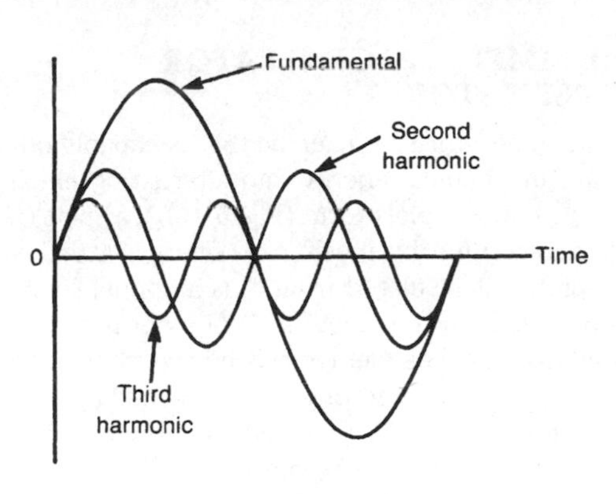

Figure 61-4

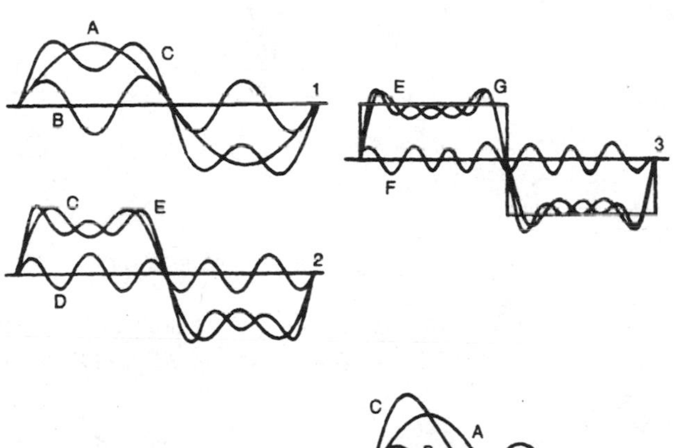

A

A Fundamental
B 3rd Harmonic
C Fundamental plus 3rd harmonic
D 5th harmonic
E Fundamental plus 3rd & 5th harmonics
F 7th harmonic
G Fundamental plus 3rd, 5th, & 7th harmonics

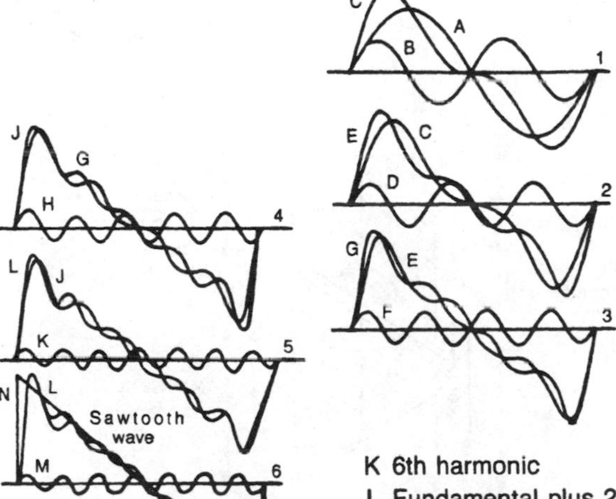

B

A Fundamental
B 2nd harmonic
C Fundamental plus 2nd harmonic
D 3rd harmonic
E Fundamental plus 2nd and 3rd harmonics
F 4th harmonic
G Fundamental plus 2nd, 3rd, & 4th harmonics
H 5th harmonic
J Fundamental plus 2nd, 3rd, 4th, & 5th harmonics
K 6th harmonic
L Fundamental plus 2nd, 3rd, 4th 5th & 6th harmonics
M 7th harmonic
N Fundamental plus 2nd, 3rd, 4th, 5th, 6th & 7th harmonics.

Figure 61-5

THE SIMPLE AC GENERATOR (ALTERNATOR)

The ac generator is a machine that is capable of converting mechanical energy into alternating electrical energy. In its simplest form (Fig. 61-6A), a shaft (DD') is driven around by mechanical energy. Attached to the shaft, but insulated from it, is a copper loop that rotates between the poles (PP') of a permanent magnet or an electromagnet (energized by direct current).

The ends of the rotating loop are joined to two separate slip rings (SS') that make continuous contact with the stationary carbon brushes (CC'); these brushes are then connected to the electrical load.

The poles' pieces (PP') are specially shaped to provide a constant flux density (B) in which the loop rotates (Fig. 61-6B). One complete rotation of the loop will generate one cycle of alternating voltage, which is applied across the load. If the output frequency is 60 Hz, the loop's speed of rotation is $60 \times 60 = 3600$ revolutions per min. However, if the alternator has four poles, each complete rotation of the loop generates two cycles of alternating voltage; therefore, the speed required for a 60-Hz output is only 1800 rpm.

When a sine-wave voltage is displayed on an oscilloscope (chapter 60), the vertical setting indicates the number of volts (or millivolts) for one graticule division (normally 1 cm). The horizontal setting represents the number of milliseconds or microseconds for one graticule division. For example, Fig. 61-7 illustrates a sine-wave voltage for which the vertical, and horizontal, settings are respectively 10 mV/div. and 50 μs/div. Because the peak value is equal to 2 divs., the amplitude of the sine wave is $2 \times 10 = 20$ mV. The period (T) occupies 4 divs. horizontally so that $T = 4 \times 50 = 200$ μs, and the frequency, $f = 1/T = 1/200\ \mu\text{s} = 5000$ Hz.

MATHEMATICAL DERIVATIONS

Frequency,

$$f = \frac{1}{T}\ \text{Hz} \tag{61-1}$$

Period,

$$T = \frac{1}{f}\ \text{seconds} \tag{61-2}$$

One radian is the angle at the center of a circle, which subtends (is opposite to) an arc (l) equal in length to the radius (r).

Angle,

$$\theta = \frac{l}{r}\ \text{radians} \tag{61-3}$$

Because the total length of the circle's circumference is $2\pi r$ radians, the angle of 360° must be equivalent to $2\pi r/r = 2\pi$ radians. Therefore,

$$\theta\ (\text{degrees}) = \frac{360}{2\pi} \times \theta\ \text{radians}$$

$$= \frac{180}{\pi} \times \theta\ \text{radians} \tag{61-4}$$

$$\theta\ (\text{radians}) = \frac{\pi}{180} \times \theta\ \text{degrees} \tag{61-5}$$

It follows that,

$$1\ \text{radian} = \frac{180}{\pi}\ \text{degrees} = 57.296° = 57°17'45'' \tag{61-6}$$

Because the angular velocity, or angular frequency (ω) is measured in radians per second, angular frequency,

$$\omega = \frac{\theta}{t} = \frac{2\pi}{T} = 2\pi f\ \text{radians per second} \tag{61-7}$$

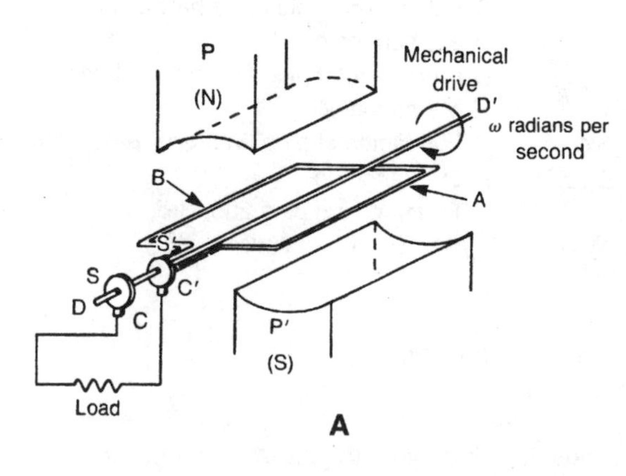

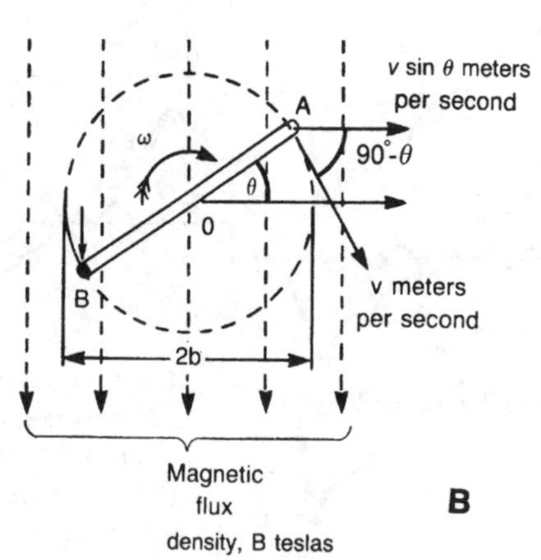

Figure 61-6

Angle, $\theta = 2\pi ft$ radians (61-8)

where t is the time in seconds.

Peak-to-peak voltage,

$$E_{p\text{-}p} = 2 \times E_{max} \text{ volts} \qquad (61\text{-}9)$$

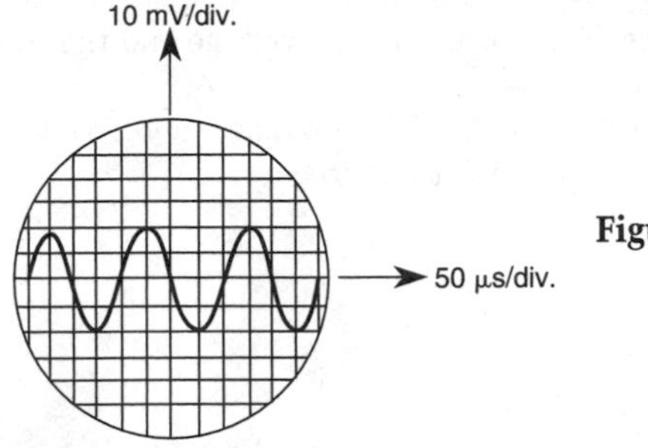

Figure 61-7

Assume that the loop of N turns in Fig. 61-6B moves with an angular velocity of ω radians per second, so that the linear velocity of each of the conductors, A and B, is $v = \omega b$ (where $2b$ is the width of the loop in meters). The component of the velocity at right angles to the magnetic flux is $v \sin \theta = v \sin \omega t$. From Faraday's law (Equation 36-3), the instantaneous total voltage induced in the loop is:

Instantaneous voltage,

$$e = 2BlNv \sin \theta \qquad (61\text{-}10)$$

$$= 2BlbN\omega \sin \omega t$$

$$= BAN\omega \sin \omega t$$

$$= E_{max} \sin \omega t = E_{max} \sin 2\pi ft \text{ volts}$$

where: B = uniform flux density (T)
$2b$ = width of loop (m)
N = number of turns
l = length of each conductor (m)
A = area of each turn (m^2)
t = time (s)
$E_{max} = BAN\omega$ = peak or maximum voltage value (V)

$$\text{Generated frequency, } f = \frac{np}{60} \text{ hertz} \qquad (61\text{-}11)$$

Speed of rotation,

$$n = \frac{60f}{p} \text{ revolutions per minute} \qquad (61\text{-}12)$$

where: p = number of pairs of poles
n = speed of rotation in rpm.

Example 61-1

A coil of 2000 turns is rotated at 3600 rpm in a uniform magnetic field with a flux density of 0.05 tesla. The average area of each turn is 25 cm^2 and the axis of rotation is at right angles to the direction of the flux. Calculate the values of: (*a*) the frequency, (*b*) the period, (*c*) the angular frequency, and (*d*) the peak-to-peak voltage. Write down the trigonometric expression for the instantaneous voltage and calculate its value when the coil has rotated through nine radians from the position of zero voltage.

Solution

(*a*) Frequency,

$$f = \frac{np}{60} \qquad (61\text{-}11)$$

$$= \frac{3600 \times 1}{60}$$

$$= 60 \text{ Hz.}$$

(*b*) Period,

$$T = \frac{1}{f} \qquad (61\text{-}2)$$

$$= 1/60 \text{ s}$$

$$= 16.67 \text{ ms}$$

(*c*) Angular frequency,

$$\omega = 2\pi f \qquad (61\text{-}7)$$

$$= 377 \text{ rad/s}$$

(*d*) Peak voltage,

$$E_{max} = BAN\omega \qquad (61\text{-}10)$$

$$= 0.05 \times 25 \times 10^{-4} \times 2000 \times 377$$

$$= 94.25 \text{ V}$$

Peak-to-peak voltage,

$$E_{p\text{-}p} = 2 \times 94.25 \text{ V} = 188.5 \text{ V}$$

Instantaneous output voltage,

$$e = E_{max} \sin \omega t \qquad (61\text{-}10)$$

$$= E_{max} \sin \theta$$

$$= 94.25 \sin 377t \text{ volts}$$

If $\theta = 9$ radians, instantaneous voltage,

$$e = E_{max} \sin \theta \qquad (61\text{-}10)$$

$$= 94.25 \sin (9 \text{ rad})$$

$$= 38.84 \text{ V}$$

PRACTICE PROBLEMS

61-1. Five cycles of a sine-wave voltage occur in a time interval of 0.02 s. Calculate the value of the frequency.

61-2. The frequency of a sine-wave voltage is 400 kHz. Calculate the value of the period.

61-3. A 6-pole alternator is rotating at a speed of 1200 rpm. What is the value of the generated frequency?

61-4. A symmetrical square-wave ac voltage (Fig. 61-5 A) has a period of 40 μs. List all frequencies (below 150 kHz) that are contained in the square waveform.

61-5. A negative sawtooth ac voltage (Fig. 61-5 B) has a period of 20 μs. List all frequencies (below 250 kHz) that are contained in the sawtooth waveform.

61-6. If a sine-wave voltage has a frequency of 175 kHz, determine the value of its angular velocity (frequency).

61-7. The equation of an instantaneous sine-wave voltage is $e = 11 \sin(3000t)$ V. Calculate the values of the peak-to-peak voltage and the frequency.

61-8. Convert (a) 111.63° into radians, and (b) 3.62 radians into decimal degrees.

62 The root-mean-square (rms) or effective value of an ac voltage or current

In chapter 61 we observed that the magnitude of a sine-wave voltage or current could be measured in terms of its peak or peak-to-peak value. In fact it is common practice to use a cathode-ray oscilloscope (CRO) to obtain the peak-to-peak value of an alternating voltage. Because the peaks of a sine wave only occur instantaneously twice during a cycle, you would normally prefer to use a value that reflects the conditions during the whole of the cycle. However, you cannot use the average over the complete cycle because this value is always zero. Instead we turn to the *root-mean-square* (rms), or effective, value. Nonmathematically, this is defined as that ac value which will provide the same heating (or power effect) as its equivalent dc value. In other words, a 100-W, 100-V bulb that is connected to a 100-V rms 60-Hz supply will appear to have the same brightness as an identical bulb joined to a 100-Vdc source.

The bulb's tungsten filament, which is operated from the dc source, will glow continuously at a constant brightness. However, when the identical bulb is connected to the ac supply, the filament with its low thermal capacity, alternately glows very brightly, and is extinguished at the rate of 120 times per second. Because of its persistence of vision, the eye cannot directly observe a flicker effect, whose rate exceeds 30 times per second; consequently, the brightness is "averaged out" and appears to be the same as that of the bulb operated from the dc source. The flicker effect may be indirectly perceived with the aid of a stroboscope.

In Fig. 62-1A an ac sine-wave source, with a peak value of 10 V, is connected across a 1-Ω resistor, but in Fig. 62-1B another 1-Ω resistor is joined to a dc source with a terminal voltage, E. If the average power in the ac circuit (over the complete cycle) is the same as the constant power of the dc circuit, E must equal the rms value of the ac voltage. When the ac voltage is instanta-

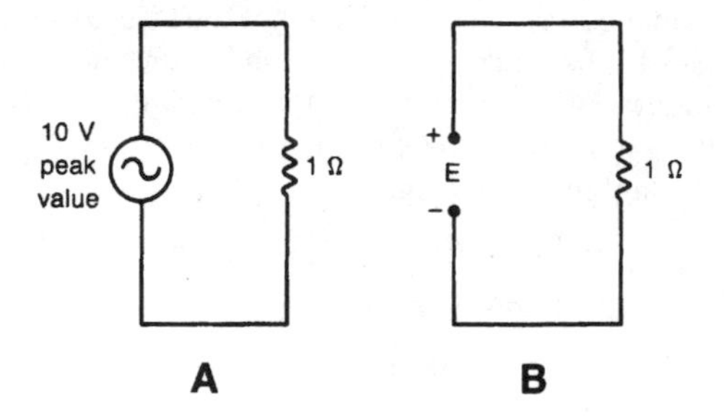

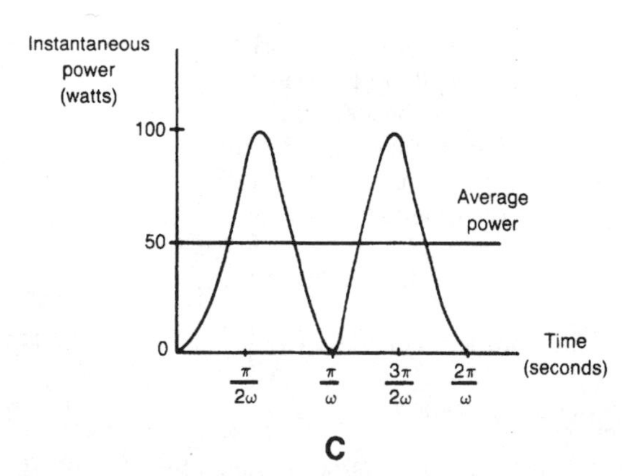

Figure 62-1

neously zero, the instantaneous power is also zero; but when the voltage is 10 V, the instantaneous power rises to its peak value of $(10\text{ V})^2/1\ \Omega = 100$ W (this result is independent of the voltage's polarity, which changes with each alternation). Therefore, the instantaneous power fluctuates between zero and 100 W, and has an average value of 50 W. Notice that the frequency of the power curve is twice that of the ac voltage.

If the power in the dc circuit is 50 W, the terminal voltage $E = \sqrt{P \times R} = \sqrt{50\text{ W} \times 1\ \Omega} = 7.07$ V. Consequently, an ac voltage with a peak value of 10 V has an effective (or rms) value of 7.07 V. This value is independent of the numbers chosen, as will be shown in the mathematical derivations. Therefore, for a *sine* wave, the *effective value* = 0.707 × the peak value; or, the peak value = (1/0.707) × the effective value = 1.414 × the effective value. These relationships are only true for the sine wave. For example, the effective value of the square wave in Fig. 61-1B must be the same as its peak value, but for the triangular wave of Fig. 61-1D, the effective value is only 0.577 × the peak value.

Finally, what is the meaning of the term *root mean square*, abbreviated rms? The value of 7.07 V was obtained from the expression

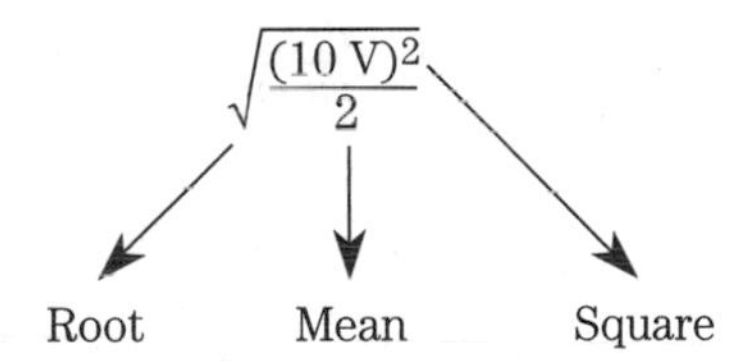

In other words, square the peak value to obtain a value that is proportional to the peak power. Then divide the result by 2 in order to obtain the average (or mean) power and finally take the square root to obtain the effective value.

Most alternating voltage and current measurements are in terms of effective values. However, the insulation in ac circuits must normally be able to withstand the peak voltage. For example, the commercial 110 V rms, 60-Hz line voltage has a peak value of 110-V × 1.414 = 155.5 V and a peak-to-peak value of 2 × 155.5 V = 311 V.

MATHEMATICAL DERIVATIONS

Let an ac voltage, whose instantaneous value is $e = E_{max} \sin \omega t$, be applied across a resistor, R. The instantaneous power dissipated is,

$$p = \frac{e^2}{R} = \frac{E^2_{max} \sin^2 \omega t}{R} \qquad (62\text{-}1)$$

$$= \frac{E^2_{max}(1 - \cos 2\,\omega t)}{2R} \text{ watts}$$

The term "cos 2 ωt" indicates that the frequency of the power curve is twice that of the voltage waveform. The instantaneous power ranges from zero to a peak value of E^2_{max}/R, while the average value of "cos 2 ωt" over a complete cycle is zero.

Therefore,

Average power,

$$P_{AV} = \frac{E^2_{max}}{2R} \qquad (62\text{-}2)$$

$$= \frac{(E_{max}/\sqrt{2})^2}{R}$$

$$= \frac{(E_{max}/1.414)^2}{R}$$

$$= \frac{(0.707\,E_{max})^2}{R}$$

$$= \frac{E^2_{rms}}{R} \text{ watts}$$

Effective value,

$$E_{rms} = 0.707 \times E_{max} \qquad (62\text{-}3)$$

$$= \frac{E_{max}}{1.414}$$

$$= \frac{E_{max}}{\sqrt{2}} \text{ volts}$$

Peak value,

$$E_{max} = \sqrt{2} \times E_{rms} \qquad (62\text{-}4)$$

$$= 1.414 \times E_{rms} \text{ volts}$$

Similarly,

Effective current,

$$I_{rms} = E_{rms}/R \qquad (62\text{-}5)$$

$$= 0.707 \times I_{max} \text{ amperes}$$

Peak current,

$$I_{max} = E_{max}/R \qquad (62\text{-}6)$$

$$= 1.414 \times I_{rms} \text{ amperes}$$

Alternatively,

rms value,

$$E_{rms} = \sqrt{\frac{\int_0^{2\pi/\omega} (E_{max} \sin \omega t)^2\, dt}{\int_0^{2\pi/\omega} dt}}$$

$$= \sqrt{\frac{\int_0^{2\pi/\omega} (E^2_{max}/2)(1 - \cos 2\omega t)\, dt}{\int_0^{2\pi/\omega} dt}}$$

$$= \sqrt{\frac{(E^2_{max}/2) \times 2\pi/\omega}{2\pi/\omega}}$$

$$= \frac{E_{max}}{\sqrt{2}} = \frac{E_{max}}{1.414} = 0.707 \times E_{max} \text{ volts}$$

If the waveform of the current flowing through a resistor contains both dc and ac components (Fig. 62-2), the effective value of the total current is $\sqrt{I^2_{dc} + I^2_{rms}}$, where I_{dc} is the dc component and I_{rms} is the effective value of the ac component.

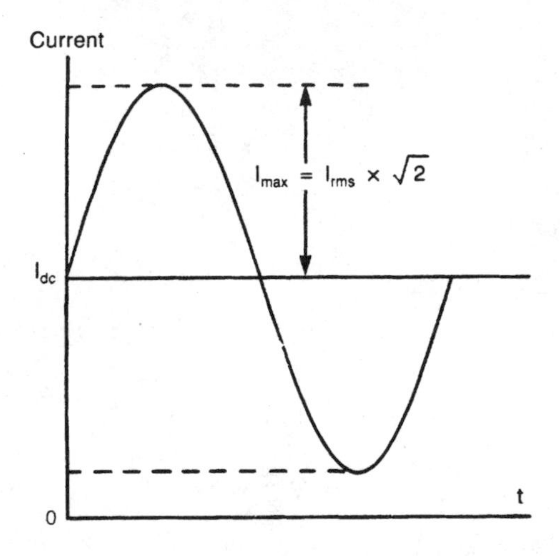

Figure 62-2

Example 62-1

A sine-wave voltage, whose peak-to-peak value is 32 V, is applied across a 68-Ω resistor. Calculate the values of the peak voltage, the effective voltage, the peak current, the effective current, the peak power, and the average power dissipated over the cycle.

Solution

Peak voltage,

$$E_{max} = \frac{E_{p\text{-}p}}{2} = \frac{32 \text{ V}}{2} = 16 \text{ V}$$

Effective voltage,

$$E_{rms} = 0.707 \times E_{max} \qquad (62\text{-}3)$$

$$= 0.707 \times 16 \text{ V} = 11.3 \text{ V}$$

Peak current,

$$I_{max} = \frac{E_{max}}{R} \qquad (62\text{-}6)$$

$$= \frac{16 \text{ V}}{68\ \Omega} = 235 \text{ mA}$$

Effective current,

$$I_{rms} = \frac{E_{rms}}{R} \qquad (62\text{-}5)$$

$$= \frac{11.3 \text{ V}}{68\ \Omega} = 166 \text{ mA}$$

Peak power,

$$P_{max} = \frac{E^2_{max}}{R} \qquad (62\text{-}1)$$

$$= \frac{(16 \text{ V})^2}{68\ \Omega} = 3.76 \text{ W}$$

Average power,

$$P_{AV} = \frac{P_{max}}{2} \qquad (62\text{-}2)$$

$$= \frac{3.76}{2} = 1.88 \text{ W}$$

PRACTICE PROBLEMS

62-1. A sine-wave current has a peak-to-peak value of 17 mA. What is the rms value of the current?

62-2. A sine-wave voltage has an rms value of 37 V. Calculate the peak value of this voltage.

62-3. The current flowing through a 18-kΩ resistor has a dc component of 5 mA, and a sine-wave ac component of 7 mA peak. Does the current through the resistor ever reverse its direction? What are the values of the effective current and what is the average power dissipated in the resistor?

62-4. The upper and lower limits of a symmetrical square wave ac voltage are +8 V and –8 V. What is the effective value of this voltage waveform?

62-5. If the peak power dissipated in a 120-Ω resistor is 2 W, calculate the rms value of the sine-wave voltage across the resistor.

62-6. If the average power dissipated in a 1.8-kΩ resistor is ½ W, calculate the peak value of the sine-wave current flowing through the resistor.

63
The average value of a sine wave

The full cycle of a sine wave is composed of two alternations (Fig. 63-1). One "positive" alternation extends from 0° to 180°, and the other "negative" alternation stretches from 180° to 360°. Although the *average value* of the sine wave over the complete cycle is zero, the average value over an alternation can be found by the use of calculus (see mathematical derivations) and is important in any discussion of half-wave and full-wave rectifier circuits

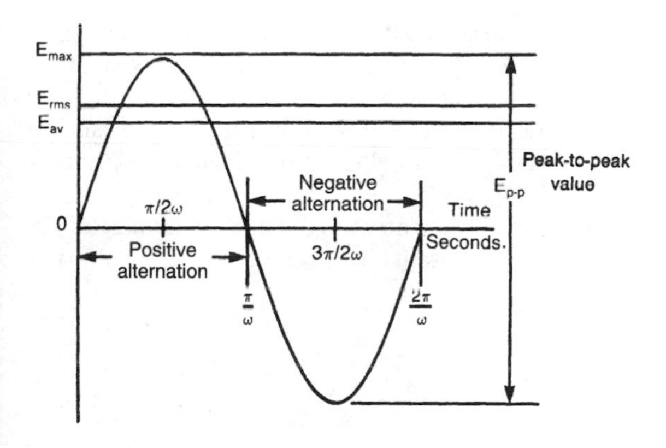

Figure 63-1

Because the "positive" alternation is symmetrical about the 90° mark, it is possible to find the approximate relationship between the average and peak values by computing the mean of the sine values between 0° and 90°. Table 63-1 shows the values for 5° steps, and the computation of the average value. In fact, the average value = 0.63662 × peak value ≈ 0.637 × peak value.

Table 63-1. Average value of a sine-wave alternation.

Angle θ	Sin θ
0°	0.000000
5°	0.087156
10°	0.173648
15°	0.258819
20°	0.342020
25°	0.422618
30°	0.500000
35°	0.573576
40°	0.642788
45°	0.707107
50°	0.766044
55°	0.819152
60°	0.866025
65°	0.906308
70°	0.939693
75°	0.965926
80°	0.984808
85°	0.996195
90°	1.000000
Total	11.951883

The 19 values total 11.951883. Average value = 11.951883/19 = 0.63 × Peak Value.

The ratio of the effective value to the average value of an alternating current (or voltage) is called the "form factor" of the waveform. For the sine wave,

$$\text{Form factor} = \frac{\text{rms value}}{\text{average value}} = \frac{0.707 \times \text{peak value}}{0.637 \times \text{peak value}}$$

$$= 1.11$$

The square wave has a form factor of 1.0 and the form factor of a triangular wave is

$$\frac{0.557}{0.5} = 1.12.$$

Compared with other ac waveforms, the sine wave is relatively easy to generate and has a high rms value with a good form factor.

MATHEMATICAL DERIVATIONS

Consider a sine-wave voltage whose instantaneous value is $e = E_{max} \sin \omega t$. Average value of the sine-wave voltage over the "positive" alternation,

$$E_{av} = \frac{\int_0^{\pi/\omega} E_{max} \sin \omega t \, dt}{\int_0^{\pi/\omega} dt} \quad (63\text{-}1)$$

$$= \frac{\left[-\frac{E_{max}}{\omega} \cos \omega t\right]_0^{\pi/\omega}}{\frac{\pi}{\omega}}$$

$$= \frac{\frac{2}{\omega} \times E_{max}}{\frac{\pi}{\omega}}$$

$$= \frac{2}{\pi} \times E_{max} = 0.63662 \times E_{max} \text{ volts}$$

Peak value,

$$E_{max} = \frac{E_{av}}{0.63662} = 1.5708 \times E_{av} \text{ volts} \quad (63\text{-}2)$$

Similarly,

$$I_{av} = 0.63662 \times I_{max} \text{ amperes} \quad (63\text{-}3)$$

$$I_{max} = 1.5708 \times I_{av} \text{ amperes} \quad (63\text{-}4)$$

Example 63-1

A sine-wave alternating current has a peak-to-peak value of 37 mA. What is its average value over an alternation?

Solution

Peak value,

$$I_{max} = \frac{I_{p\text{-}p}}{2} = \frac{37 \text{ mA}}{2} = 18.5 \text{ mA}.$$

Average value,

$$I_{av} = 0.63662 \times I_{max} \quad (63\text{-}3)$$

$$= 0.63662 \times 18.5$$

$$= 11.78 \text{ mA}.$$

PRACTICE PROBLEMS

63-1. A sine-wave ac voltage has a peak value of 47 V. What is its average value over an alternation?

63-2. A sine-wave alternating current has an average value of 7.4 mA over an alternation. What is the current's peak value?

63-3. A sine-wave ac voltage has an average value of 5.3 V over an alternation. What is its peak-to-peak value?

63-4. A sine-wave ac voltage has an rms value of 7.8 V. Calculate its average value over an alternation.

64
Phasor representation of an ac voltage or current

If two components are in series across an ac source there will be an alternating voltage across each component. If these two voltages are then added, their resultant exactly balances the source voltage (Kirchhoff's voltage law). In the case of a parallel circuit, the individual branch currents must be added to obtain the total current drawn from the source. Previous sections have expressed a sine wave either as a trigonometric equation ($e = E_{max} \sin \omega t$), or graphically by a waveform. When you need to add or subtract ac quantities, the trigonometric equations become difficult to manipulate and the combination of waveforms tends to be tedious.

A third method involves the representation of ac quantities by means of *phasors*. If the equation of an ac voltage is $e = E_{max} \sin \omega t$, its phasor is a line whose length is a measure of E_{max}. By convention, the line phasor is assumed to rotate in the positive (or counterclockwise) direction with an angular velocity of $\omega = 2\pi f$ radians per second, where f is the voltage's frequency in hertz. The vertical projection (PN) of the phasor on the horizontal reference line is then equal to the instantaneous value (e) of the ac voltage (Fig. 64-1). Therefore, as the phasor rotates, the extremity of the line can be said to trace out the voltage's sinusoidal waveform, with a frequency that is equal to the phasor's speed of rotation (in revolutions per second). One complete rotation of the phasor then traces out one cycle of the sine wave. The phasor diagram, therefore, contains the same information as the waveform presentation, but it is obviously easier to work with lines, rather than sine waves.

Because there is a constant relationship between effective and peak values (effective value = 0.707 × peak value), the phasor's length can also be used to indicate the rms value.

Prior to about 1960, the word *vector* was used, rather than *phasor.* A vector is a quantity that possesses both magnitude and direction so that a mechanical force is an example of a vector. By contrast a *scalar* quantity, such as mass, possesses magnitude only. The rules for adding (or subtracting) mechanical vectors are the same as those for adding or subtracting ac phasors. However, the vector rules for multiplication and division are totally different from the corresponding phasor rules. For this reason, *vectors* were no longer used to represent ac quantities and the word *phasor* was introduced instead.

MATHEMATICAL DERIVATIONS

In Fig. 64-1

Instantaneous voltage,

$$e = PN = OP \sin \theta$$

$$= E_{max} \sin \theta \qquad (64\text{-}1)$$

$$= E_{max} \sin \omega t$$

$$= E_{max} \sin 2\pi ft \text{ volts}$$

where: E_{max} = peak voltage (V)
θ = angle measured relative to the horizontal reference line (rad)
ω = angular velocity or frequency (rad/s)
f = frequency (Hz)
t = time (s)

Example 64-1

An ac voltage has an rms value of 6.7 V and a frequency of 400 kHz. What are the values of the peak voltage and the angular velocity? What is the instantaneous voltage after an interval of 2.73 μs, as measured from the horizontal reference line?

Solution

Peak voltage,

$$E_{max} = 1.414 \times E_{rms}$$

$$= 1.414 \times 6.7 \text{ V}$$

$$= 9.47 \text{ V}$$

Angular velocity,

$$\omega = 2\pi f$$

$$= 2 \times \pi \times 400 \times 10^3$$

$$= 2.51 \times 10^6 \text{ rad/s}$$

Instantaneous voltage,

$$e = E_{max} \sin 2\pi ft \qquad (64\text{-}1)$$

$$= 9.47 \sin (2 \times \pi \times 400 \times 10^3 \times 2.73 \times 10^{-6})$$

$$= 5.17 \text{ V}$$

PRACTICE PROBLEMS

64-1. The equation of a sine-wave ac voltage is $e = 13 \text{ Sin } 3750t$ V. Calculate the frequency of its phasor and the rms value of the voltage.

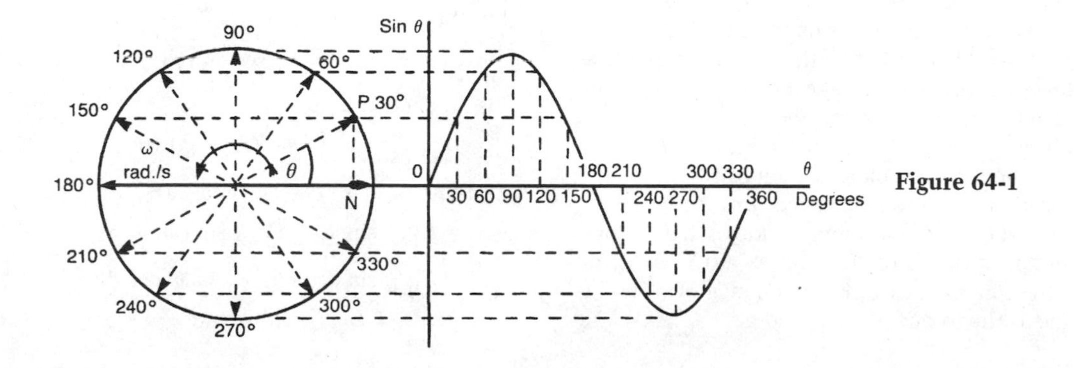

Figure 64-1

64-2. A sine-wave ac current has a peak-to-peak value of 17 A and a frequency of 440 Hz. Determine the angular velocity of its phasor and the rms value of the current. Write down the expression for the instantaneous value of the current.

64-3. A voltage phasor rotates at a rate of 17,380 revolutions per second. Determine the value of its angular frequency.

65 Phase relationships

When two dc voltages, E_1 and E_2, are in series, they are either in series-aiding, or in series-opposing, so that only addition or subtraction is necessary in order to obtain their combined voltage (chapter 12). However, two ac sine-wave voltages of the same frequency might not reach similar points in their cycles at the same time. As an example (chapter 72), when an inductor and a resistor are in series across an ac source, the voltage across the inductor reaches its peak value when the voltage across the resistor is zero. In other words, the two voltages are a quarter of a cycle apart. The amount by which the two sine waves are out of step is referred to as their *phase difference*, which is either measured in degrees or radians. In our example, the shift of one quarter of a cycle is equivalent to a phase difference of 90°, or $\pi/2$ radians.

In Fig. 65-1, the e_2 waveform reaches its peak (X) earlier in time than the e_1 waveform with its corresponding peak, Y. The difference between the two waveforms is ϕ radians, with e_2 leading e_1 (or e_1 lagging e_2). Phase differences can, therefore, be either leading or lagging with their angles usually extending up to 180°. In the particular case, where ϕ is 180° (or π radians), the terms "leading" or "lagging" are not used, because 180° leading has the same meaning as 180° lagging. In most cases, there is little point in using angles greater than 180° because, for example, a phase difference of 270° leading, is equivalent to 90° lagging.

It is impossible to add together two or more series ac voltages (or currents, if some form of parallel circuit is involved) without knowing their peak values and their phase relationships. Figure 65-2 shows the results of adding two ac voltages, each of 10-V peak, but with phase differences that are in turn 0°, 90°, 120°, and 180°; the resultant (sum) voltages have corresponding peak values of 20 V, 14.4 V, 10 V, and zero. Notice that, in each case, the frequency of $e_1 + e_2$ is the same as the frequency of e_1 and e_2.

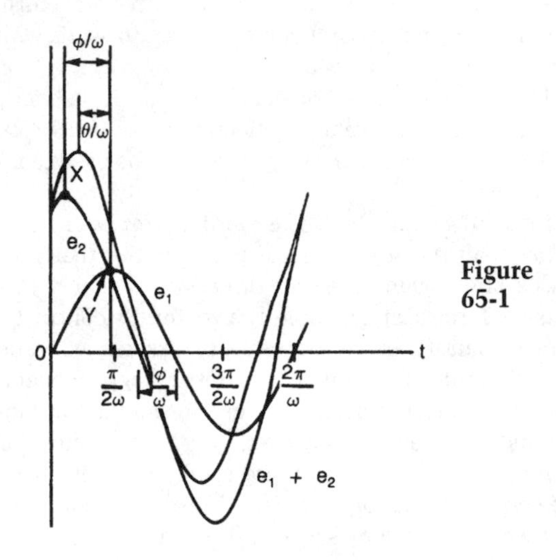

Figure 65-1

As the mathematical derivations will show, the frequency of $e_1 + e_2$ is unchanged. The peak value of $e_1 + e_2$ is obtained from the peak values of e_1, e_2 and their phase relationship. The same factors determine the phase relationships between $e_1 + e_2$ and e_1, e_2.

MATHEMATICAL DERIVATIONS

In Fig. 65-1, e_2 leads e_1 by an angle of ϕ radians. Therefore,

Instantaneous voltage,

$$e_1 = E_{1max} \sin \omega t \text{ volts} \tag{65-1}$$

Instantaneous voltage,

$$e_2 = E_{2max} \sin (\omega t + \phi) \text{ volts} \tag{65-2}$$

The sum of these two ac voltages is,

$$e_1 + e_2 = E_{1max} \sin \omega t + E_{2max} \sin (\omega t + \phi) \tag{65-3}$$

$$= E_{1max} \sin \omega t + E_{2max} \sin \omega t \cos \phi + E_{2max} \cos \omega t \sin \phi$$

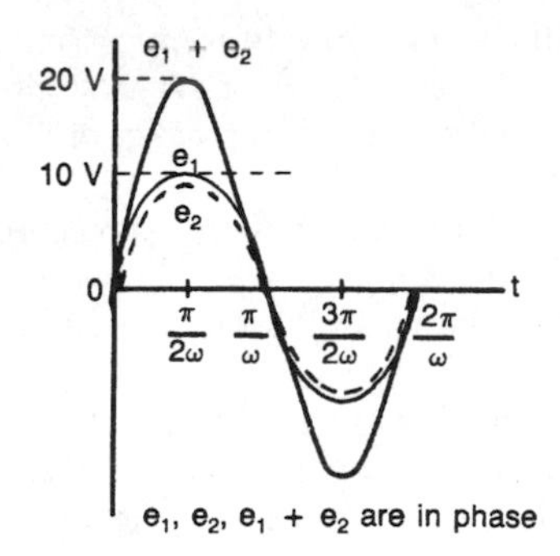

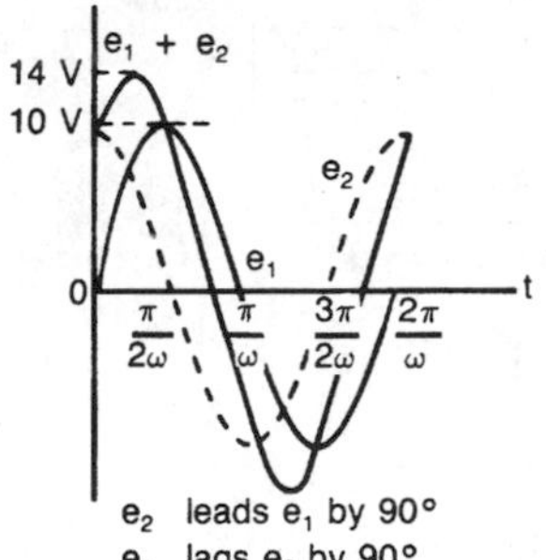

Figure 65-2

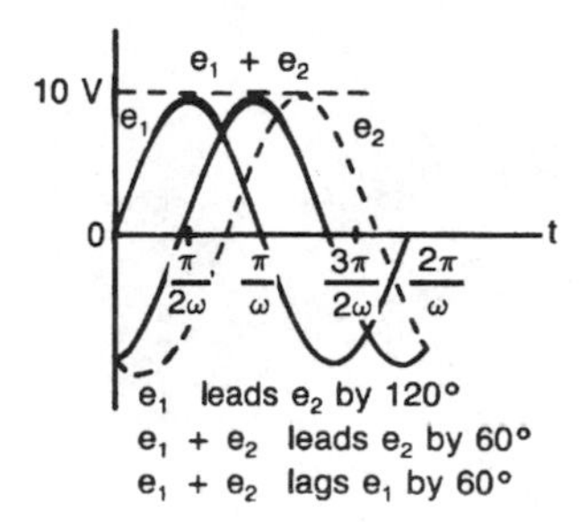

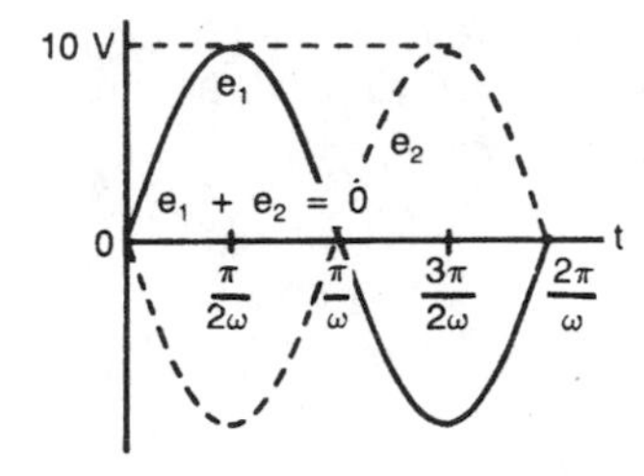

$$= (E_{1max} + E_{2max} \cos \phi) \sin \omega t$$
$$+ E_{2max} \sin \phi \cos \omega t \text{ volts}$$

Let $E_{1max} + E_{2max} \cos \phi = E_{max} \cos \theta$ volts (65-4)

and,

$$E_{2max} \sin \phi = E_{max} \sin \theta \text{ volts} \quad (65\text{-}5)$$

Then,

$$e_1 + e_2 = E_{max} \sin \omega t \cos \theta + E_{max} \cos \omega t \sin \theta \quad (65\text{-}6)$$
$$= E_{max} \sin (\omega t + \theta) \text{ volts}$$

Therefore, E_{max} is the peak value of $e_1 + e_2$, and $e_1 + e_2$ leads e_1 by θ radians, but lags e_2 by $\phi - \theta$ radians. From Equations 65-4 and 65-5,

$$E_{max} = \sqrt{(E_{1max} + E_{2max} \cos \phi)^2 + (E_{2max} \sin \phi)^2} \quad (65\text{-}7)$$
$$= \sqrt{E^2_{1max} + E^2_{2max} + 2\,E_{1max}\,E_{2max} \cos \phi} \text{ volts}$$

and,

$$\theta = \text{inv tan} \left[\frac{E_{2max} \sin \phi}{E_{1max} + E_{2max} \cos \phi} \right] \text{radians} \quad (65\text{-}8)$$

The sine wave representing $e_1 + e_2$ is shown in Fig. 65-1.

By substituting $-E_2$ for $+E_2$, you can obtain the results of subtracting e_2 from e_1, namely $e_1 - e_2$. The results for $e_1 - e_2$ are:

$$E_{max} = \sqrt{E^2_{1max} + E^2_{2max} - 2E_{1max}\,E_{2max} \cos \phi} \text{ volts} \quad (65\text{-}9)$$

and,

$$\theta = \text{inv tan} \left[\frac{-E_{2max} \sin \phi}{E_{1max} - E_{2max} \cos \phi} \right] \text{radians} \quad (65\text{-}10)$$

Example 65-1

Two alternating voltages are represented by $e_1 = 8 \sin \omega t$ V and $e_2 = 12 \sin (\omega t - \pi/3)$ V. What is the phase relationship between e_1 and e_2? Obtain the trigonometric results for $e_1 + e_2$.

Solution

e_2 lags e_1 by $\pi/3$ radians, or 60°. Alternatively, e_1 leads e_2 by 60°.

Peak value of $e_1 + e_2$, (65-7)

$$E_{max} = \sqrt{(E^2_{1max} + E^2_{2max} + 2E_{1max}\,E_{2max} \cos \phi)}$$
$$= \sqrt{8^2 + 12^2 + 2 \times 8 \times 12 \times \cos 60^\circ}$$
$$= 17.4 \text{ V}$$

Phase angle of $e_1 + e_2$, (65-8)

$$\theta = \text{inv tan}\left[\frac{E_{2\max}\sin\phi}{E_{1\max} + E_{2\max}\cos\phi}\right]$$

$$= \text{inv tan}\left(\frac{12\sin(-60°)}{8 + 12\cos(-60°)}\right)$$

$$= \text{inv tan}\left(-\frac{10.4}{14}\right)$$

$$= -36.6°$$

Therefore, $e_1 + e_2$ lags e_1 by 36.6°

PRACTICE PROBLEMS

65-1. Two alternating voltages are represented by $v_1 = 12.6 \sin \omega t$, and $v_2 = 12.6 \sin \omega t$. What are the rms values of $v_1 + v_2$ and $v_1 - v_2$?

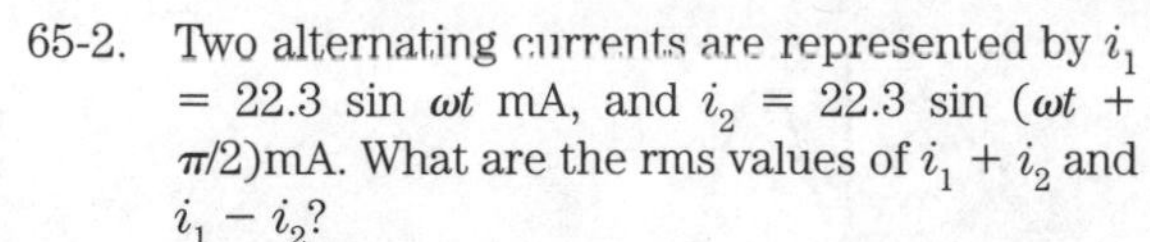

65-2. Two alternating currents are represented by $i_1 = 22.3 \sin \omega t$ mA, and $i_2 = 22.3 \sin(\omega t + \pi/2)$mA. What are the rms values of $i_1 + i_2$ and $i_1 - i_2$?

65-3. Two alternating voltages are represented by $v_1 = 8.7 \sin \omega t$ V, and $v_2 = 8.7 \sin(\omega t + 2\pi/3)$V. What are the rms values of $v_1 + v_2$ and $v_1 - v_2$?

65-4. Two alternating currents represented by $i_1 = 17.8 \sin(\omega t + \pi)$ mA and $i_2 = 17.8 \sin \omega t$ mA. What are the rms values of $i_1 + i_2$ and $i_1 - i_2$?

66
Addition and subtraction of phasors

In the mathematical derivations of chapter 63, you added together two voltage sine waves that were represented by the equations $e_1 = E_{1\max} \sin \omega t$, and $e_2 = E_{2\max} \sin(\omega t + \phi)$. This chapter shows how these same voltages are added together in a phasor diagram. Because e_2 leads e_1 by ϕ radians, you can use e_1 as the reference phasor; the e_2 phasor will appear in the first quadrant, with its direction inclined at the angle ϕ to the horizontal line (Fig. 66-1).

The sine waves of e_1 and e_2 are added together to produce the $e_1 + e_2$ sine wave which has a peak value of $E_{\max}$ and leads e_1 by the angle θ. The corresponding $e_1 + e_2$ phasor is found by completing the parallelogram whose sides are OP and OQ; the diagonal OR is then the $e_1 + e_2$ phasor and the other diagonal QP is the $e_2 - e_1$ phasor.

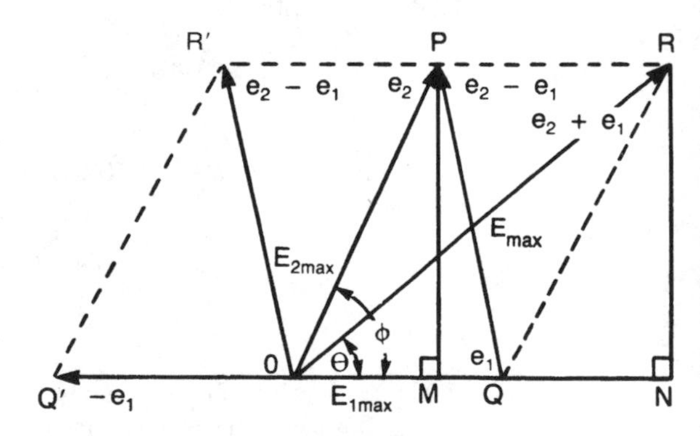

Figure 66-1

MATHEMATICAL DERIVATIONS

In Fig. 66-1

$$RN = E_{2\max} \sin\phi \tag{66-1}$$

$$OM = E_{2\max} \cos\phi \tag{66-2}$$

$$ON = OM + MN = E_{2\max}\cos\phi + E_{1\max} \tag{66-3}$$

Using the Phythagorean theorem,

$$OR^2 = RN^2 + ON^2$$

$$E^2_{\max} = (E_{2\max}\sin\phi)^2 + (E_{1\max} + E_{2\max}\cos\phi)^2$$

$$= E^2_{1\max} + E^2_{2\max}\sin^2\phi + 2E_{1\max}E_{2\max}\cos\phi + E^2_{2\max}\cos^2\phi$$

$$= E^2_{1\max} + E^2_{2\max} + 2E_{1\max}E_{2\max}\cos\phi$$

because,

$$\sin^2\phi + \cos^2\phi = 1$$

Therefore,

$$E_{max} = \sqrt{E^2_{1max} + E^2_{2max} + 2E_{1max}E_{2max} \cos \phi} \text{ volts} \quad (66\text{-}4)$$

$$\tan \theta = \frac{RN}{ON} = \frac{E_{2max} \sin \phi}{E_1 + E_{2max} \cos \phi}$$

and,

$$\theta = \text{inv tan}\left[\frac{E_{2max} \sin \phi}{E_1 + E_{2max} \cos \phi}\right] \text{radians} \quad (66\text{-}5)$$

These equations match those that appeared in the mathematical derivations of chapter 65. This establishes the validity of the geometric construction used in the parallelogram of phasors.

A similar analysis leads to the following results:

Peak value of $e_2 - e_1$ phasor (66-6)

$$= \sqrt{E^2_{1max} + E^2_{max} - 2E_{1max}E_{2max} \cos \phi} \text{ volts}$$

Phase angle of $e_2 - e_1$ phasor (66-7)

$$= \text{inv tan}\left[\frac{E_{2max} \sin \phi}{-E_{1max} + E_{2max} \cos \phi}\right] \text{radians}$$

Example 66-1

If $e_1 = 7 \sin (\omega t + \pi/6)$ V and $e_2 = 11 \sin (\omega t + \pi/4)$ V, what is the phase relationship between e_1 and e_2? Find the values of $e_1 + e_2$ and $e_2 - e_1$.

Solution

e_2 leads e_1 by the angle $\phi = \frac{\pi}{4} - \frac{\pi}{6} = \frac{\pi}{12}$ radian or 15°

The peak value of $e_1 + e_2$, (66-4)

$$E = \sqrt{E^2_{1max} + E^2_{2max} + 2E_{1max}E_{2max} \cos \phi}$$

$$= \sqrt{7^2 + 11^2 + 2 \times 7 \times 11 \times \cos 15°}$$

$$= 17.85 \text{ V}$$

Phase angle of $e_1 + e_2$, (66-5)

$$\theta = \text{inv tan}\left[\frac{E_{2max} \sin \phi}{E_{1max} + E_{2max} \cos \phi}\right]$$

$$= \text{inv tan}\left[\frac{11 \sin 15°}{7 + 11 \cos 15°}\right]$$

$$= 9.17°$$

$$= 0.16 \text{ radian}$$

Therefore,

$$e_1 + e_2 = 17.85 \sin\left[\omega t + \left(\frac{\pi}{6} + 0.16\right) \text{radian}\right]$$

$$= 17.85 \sin (\omega t + 0.684 \text{ radian})$$

Peak value of

$$e_2 - e_1 = \sqrt{7^2 + 11^2 - 2 \times 7 \times 11 \times \cos 15°}$$

$$= 4.61 \text{ V.}$$

$$\text{Phase angle of } e_2 - e_1 = \text{inv tan}\left[\frac{+11 \sin 15}{-7 + 11 \cos 15}\right]$$

$$= +38.1°$$

$$= 0.665 \text{ radian}$$

Therefore,

$$e_2 - e_1 = 4.61 \sin\left[\omega t + \left(\frac{\pi}{6} + 0.665\right) \text{radian}\right]$$

$$= 4.61 \sin (\omega t + 1.189 \text{ radian})$$

PRACTICE PROBLEMS

66-1. Two alternating currents are represented by $i_1 = 9.4 \sin (\omega t + \pi/6)$ mA and $i_2 = 7.3 \sin \omega t$ mA. What is the phase relationship between i_1 and i_2? Determine the peak values, and the phase angles, of $i_1 + i_2$ and $i_1 + i_2$.

66-2. Two alternating voltages are represented by $v_1 = 7.8 \sin (\omega t + \pi/3)$ V and $v_2 = 5.6 \sin (\omega t - \pi/6)$ V. What is the phase relationship between v_1 and v_2? Determine the rms values of $v_1 + v_2$, and $v_1 - v_2$. What is the phase relationship between v_1, and $v_1 + v_2$?

66-3. Two alternating voltages are represented by $v_1 = 33.7 \cos \omega t$ V, and $v_2 = 33.7 \sin (\omega t - \pi/6)$ V. Determine the rms value of $v_1 + v_2$ and $v_1 - v_2$. What is the phase relationship between v_2 and $v_1 + v_2$?

67
Resistance in the ac circuit

If a sine wave is connected across a resistor, R (Fig. 67-1A), Ohm's law applies throughout the cycle of the alternating voltage, e. When e is instantaneously zero, there is zero current flowing in the circuit, but when the applied voltage reaches one of its peaks (X) the current is also a maximum in one direction. When you reach the peak, Y of the other alternation, the current is again at its maximum value, but its direction is reversed. At all times $e/i = R$, and therefore, it follows that the sine waves of e and i are in phase (Fig. 67-1B) with a phase difference of zero degrees. This is indicated by the e and i phasors, which lie along the same horizontal line (Fig. 67-1C).

The instantaneous power (p) in the circuit is the product of the instantaneous voltage (e) and the instantaneous current, i. When e and i are simultaneously zero, the instantaneous power is also zero but, when both e and i reach their peaks together, the power reaches its peak value of $E_{max} \times I_{max}$ watts. When the voltage reverses polarity, the current reverses direction, but the resistor continues to dissipate (lost) power in the form of heat. The whole of the instantaneous power curve must be drawn above the zero line and its frequency is twice that of the applied voltage. The mean value of the power curve is a measure of the average power dissipated over the voltage cycle.

MATHEMATICAL DERIVATIONS

Let the equation of the applied ac voltage be $e = E_{max} \sin \omega t$. Because Ohm's law applies throughout the cycle,

Instantaneous current,

$$i = \frac{e}{R} \tag{67-1}$$

$$= \frac{E_{max} \sin \omega t}{R}$$

$$= I_{max} \sin \omega t \text{ amperes}$$

where $I_{max} = E_{max}/R$ and is the peak value of the alternating current. Therefore,

$$\frac{E_{max}}{I_{max}} = \frac{\sqrt{2} \times E_{rms}}{\sqrt{2} \times I_{rms}} \tag{67-2}$$

$$= \frac{E_{rms}}{I_{rms}}$$

$$= R \text{ ohms}$$

The instantaneous power (p) is given by:
Instantaneous power,

$$p = e \times i \tag{67-3}$$

$$= E_{max} \sin \omega t \times I_{max} \sin \omega t$$

$$= E_{max} \times I_{max} \sin^2 \omega t$$

$$= \frac{E_{max} \times I_{max}}{2} (1 - \cos 2\omega t) \text{ watts}$$

The term "cos $2\omega t$" indicates that the instantaneous power curve is a second harmonic of the voltage wave-

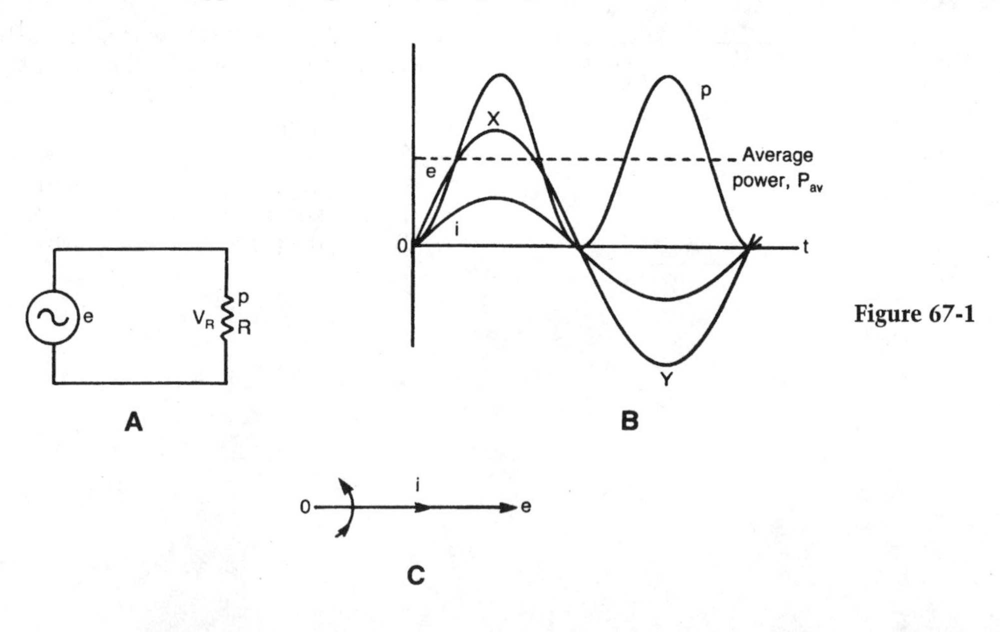

Figure 67-1

form. Because the mean value of cos $2\omega t$ over the complete cycle is zero, the average power is:

Average power,

$$P_{av} = \frac{E_{max} \times I_{max}}{2} \qquad (67\text{-}4)$$

$$= \frac{E_{max}}{\sqrt{2}} \times \frac{I_{max}}{\sqrt{2}}$$

$$= E_{rms} \times I_{rms}$$

$$= I^2_{rms} \times R$$

$$= \frac{E^2_{rms}}{R} \text{ watts}$$

By using the *effective values*, the ac equations for resistors in series and parallel are the same as those for dc.

Example 67-1

In Fig. 67-1, let $e = 18 \sin(\omega t - \pi/3)$ V and $R = 2.7$ kΩ. Calculate the values of the effective current, the peak power and the average power over the cycle. What is the trigonometric expression for the instantaneous current?

Solution

Effective voltage,

$$E_{rms} = \frac{E_{max}}{\sqrt{2}}$$

$$= 0.707 \times E_{max}$$

$$= 0.707 \times 18$$

$$= 12.7 \text{ V}.$$

Effective current,

$$I_{rms} = \frac{E_{rms}}{R} \qquad (67\text{-}2)$$

$$= \frac{12.7 \text{ V}}{2.7 \text{ k}\Omega}$$

$$= 4.71 \text{ mA}.$$

Peak power,

$$P_{max} = \frac{E^2_{max}}{R}$$

$$= \frac{18 \text{ V}^2}{2.7 \text{ k}\Omega}$$

$$= 120 \text{ mW}.$$

Average power,

$$P_{av} = E_{rms} \times I_{rms} \qquad (67\text{-}4)$$

$$= 12.7 \text{ V} \times 4.71 \text{ mA}$$

$$= 60 \text{ mW}.$$

Because e and i are in phase, the instantaneous current,

$$i = \frac{18 \text{ V}}{2.7 \text{ k}\Omega} \sin\left(\omega t - \frac{\pi}{3}\right)$$

$$= 6.67 \sin\left(\omega t - \frac{\pi}{3}\right) \text{ mA}$$

PRACTICE PROBLEMS

67-1. An alternating current is represented by $i = 17.8 \sin(\omega t + \pi/6)$ mA. If this current flows through a 2.7-kΩ resistor, calculate the values of the effective voltage drop across the resistor, as well as the peak and average powers dissipated in the resistor. Write down the trigonometric expression for the instantaneous voltage drop across the resistor.

67-2. In Practice Problem 67-1, write down the trigonometric expression for the instantaneous power dissipated in the resistor.

67-3. A 270-Ω resistor is dissipating an average power of 700 mW. What are the peak values of the voltage drop across the resistor and the current flowing through the resistor? What is the phase difference between the voltage and the current?

67-4. A sine-wave ac voltage drop across a resistor has a peak value of 13.8 V. If the rms value of the current flowing through the resistor is 5.42 mA, calculate the value of the resistor.

68
Conductance in the ac circuit

In chapter 67 it was decided that the same resistor formulas applied to both dc and ac circuits, provided the formulas used effective values of voltage and current. This is also true for the property of *conductance*, which is defined as the reciprocal of the resistance, and is the measure of the "ease" with which current flows in an electrical circuit. In the series circuit of Fig. 68-1A, the reciprocal formula will determine the total conductance, G_T. This will enable you to calculate the effective current, I_{rms} and the effective voltages V_{1rms}, V_{2rms}, V_{3rms}, . . . V_{Nrms}.

The main advantage of using conductance (as opposed to resistance) lies with parallel circuits (Fig. 68-1B), because the total conductance is simply found by adding the individual conductances. You can then easily calculate the effective value of the total current, I_{Trms}.

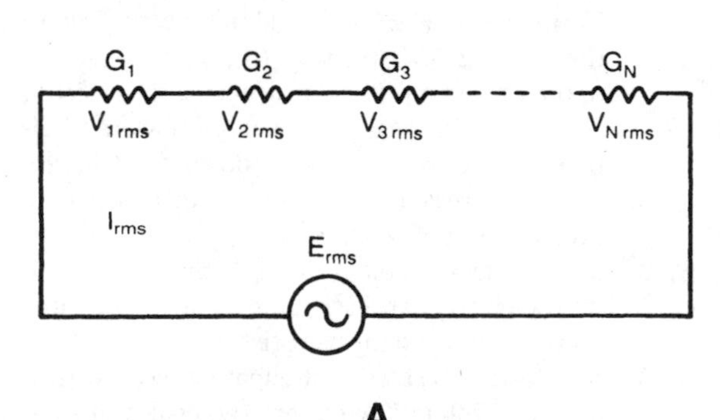

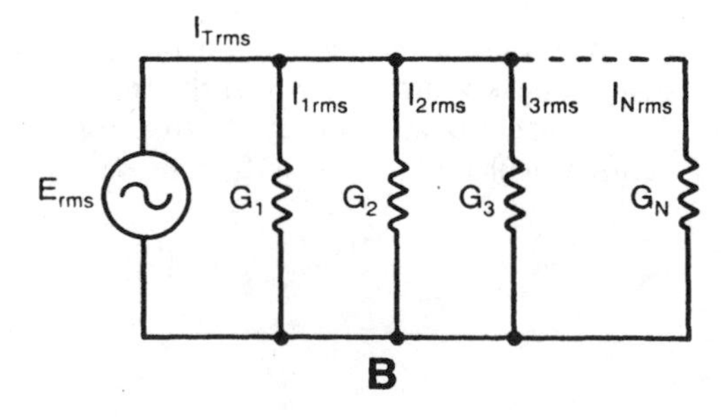

Figure 68-1

MATHEMATICAL DERIVATIONS

In Fig. 68-1A

Total conductance,

$$G_T = \frac{1}{\frac{1}{G_1} + \frac{1}{G_2} + \frac{1}{G_2} \dots + \frac{1}{G_N}} \text{ siemens} \tag{68-1}$$

Circuit current,

$$I_{rms} = E_{rms} \times G_T \text{ amperes} \tag{68-2}$$

Effective voltage across the R_N resistor,

$$V_{Nrms} = I_{rms} \times G_N \text{ volts} \tag{68-3}$$

Power dissipated in the R_N resistor,

$$\begin{aligned} P_N &= V_{Nrms} \times I_{rms} \\ &= I^2_{rms} \times R_N \\ &= I^2_{rms}/G_N \\ &= V^2_{Nrms}/R_N \\ &= V^2_{Nrms} \times G_N \text{ watts} \end{aligned} \tag{68-4}$$

In Fig. 68-1B

Total conductance,

$$G_T = G_1 + G_2 + G_3 \dots + G_N \text{ siemens} \tag{68-5}$$

Total current,

$$I_{Trms} = E_{rms} \times G_T \text{ amperes} \tag{68-6}$$

Effective current flowing through the R_N resistor,

$$I_{Nrms} = E_{rms} \times G_N \text{ amperes} \tag{68-7}$$

Power dissipated in the R_N resistor,

$$\begin{aligned} P_N &= E^2_{rms} \times G_N \\ &= E_{rms} \times I_{Nrms} \\ &= I^2_{Nrms}/G_N \\ &= E^2_{Nrms}/R_N \\ &= I^2_{Nrms} \times R_N \text{ watts} \end{aligned} \tag{68-8}$$

Example 68-1

In Fig. 68-2 calculate the value of I_{rms}.

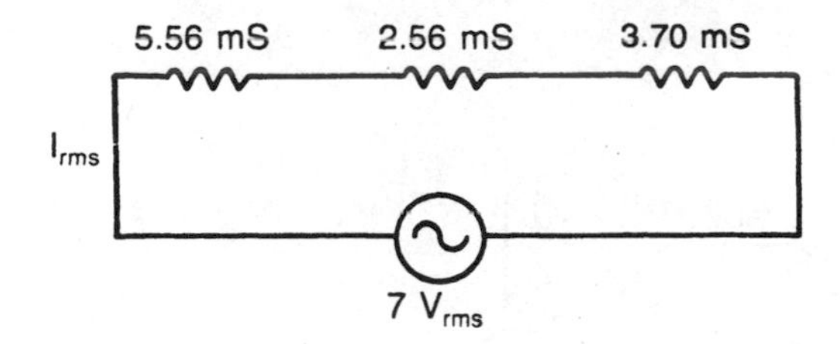

Figure 68-2

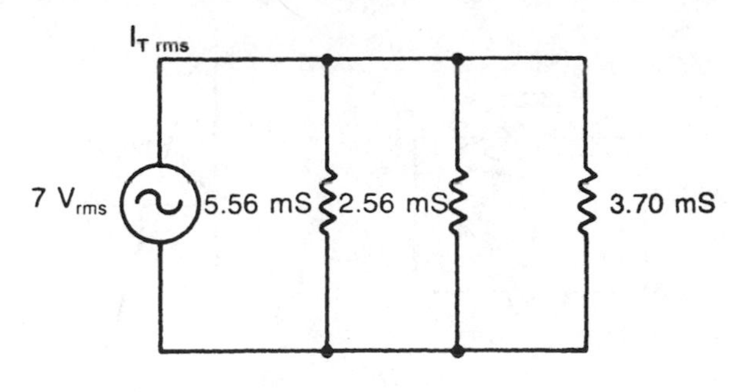

Figure 68-3

Solution

Total conductance,

$$G_T = \frac{1}{\frac{1}{G_1} + \frac{1}{G_2} + \frac{1}{G_3}} \quad (68\text{-}1)$$

$$= \frac{1}{\frac{1}{5.56 \text{ mS}} + \frac{1}{2.56 \text{ mS}} + \frac{1}{3.70 \text{ mS}}}$$

$$= 1.19 \text{ mS}$$

Effective current,

$$I_{\text{rms}} = E_{\text{rms}} \times G_T \quad (68\text{-}6)$$

$$= 7 \text{ V} \times 1.19 \text{ mS}$$

$$= 8.33 \text{ mA}.$$

Example 68-2

In Fig. 68-3, calculate the value of $I_{T\text{rms}}$.

Solution

Total conductance,

$$G_T = G_1 + G_2 + G_3 \quad (68\text{-}5)$$

$$= 5.56 + 2.56 + 3.70$$

$$= 11.82 \text{ mS}$$

Total effective current,

$$I_{T\text{rms}} = E_{\text{rms}} \times G_T \quad (68\text{-}6)$$

$$= 7 \text{ V} \times 11.82 \text{ mS}$$

$$= 82.74 \text{ mA}.$$

PRACTICE PROBLEMS

68-1. Four resistors whose conductances are 3.03 ms, 3.70 mS, 6.67 mS, and 2.13 mS, are connected in series across a 12-V rms source. Determine the total power dissipated in the circuit.

68-2. In Practice Problem 68-1, calculate the values of the voltage drop and power dissipated in the resistor whose conductance is 3.70 mS.

68-3. Four resistors whose conductances are 3.03 mS, 3.70 mS, 6.67 mS, and 2.13 mS, are connected in parallel across a 12-V rms source. Determine the total power dissipated in the circuit.

68-4. In Practice Problem 68-3, calculate the values of the current and the power dissipated in the resistor whose conductance is 3.03 mS.

69
Inductive reactance

In Fig. 69-1A, assume that the inductor only possesses the property of inductance, and that you can ignore its resistance. When an alternating current is flowing through the coil, the surrounding magnetic field is continually expanding and collapsing so that there is an induced voltage (v_L), whose value must at all times exactly balance the instantaneous value of the source voltage, e. Therefore, when e is instantaneously at its peak value, the value of v_L is also at its peak and because:

$$v_L = -L \frac{di}{dt} \quad (39\text{-}4)$$

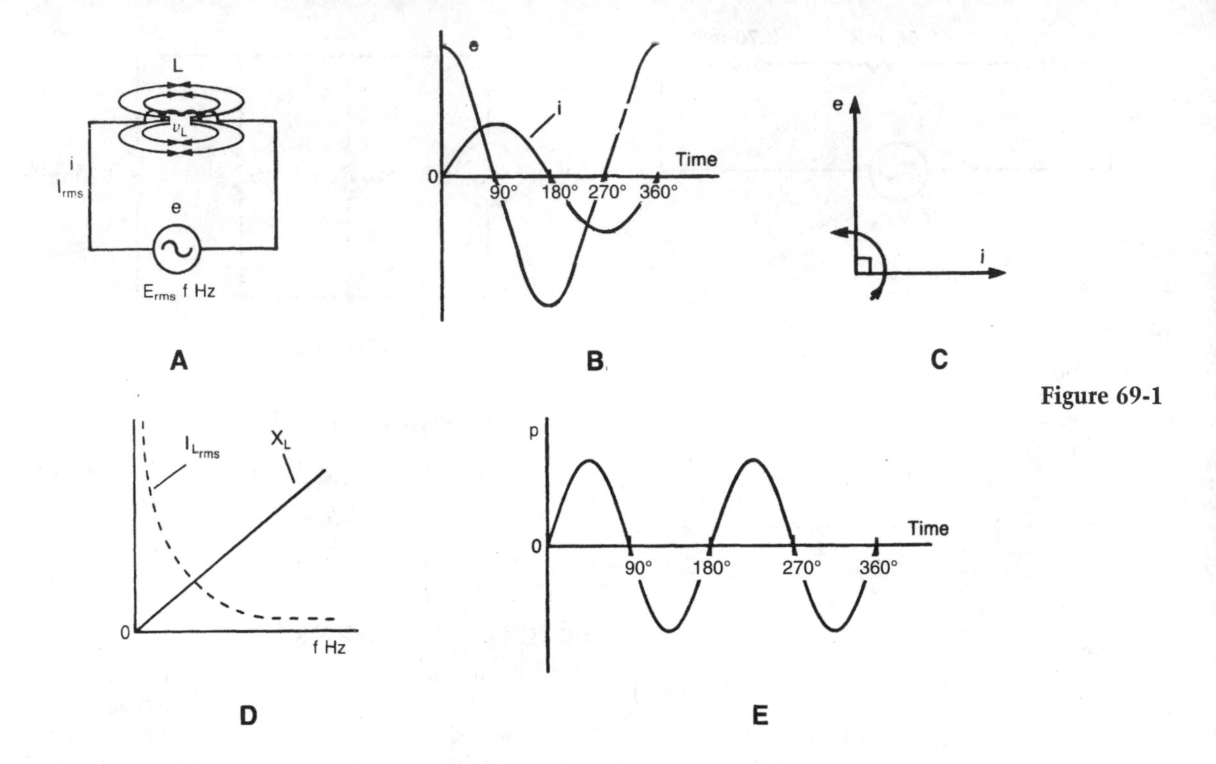

Figure 69-1

the *rate of change of the current* is at its highest level. This occurs when the slope of the current sine wave is steepest (at the point of zero time and zero current). Consequently, when e is at its peak value, i is instantaneously zero and, therefore, e and i are 90° out of phase (Figs. 69-1B and C). This can be remembered by the word "**eLi**" because for an inductor (L) the instantaneous voltage (e) leads the instantaneous current (i) by 90°.

You cannot obtain the opposition to the current flow through dividing the instantaneous voltage by the instantaneous current. For example, at the zero degree point, the value of e/i is infinite but at the 90° mark, e/i is zero. However, you can attempt to find the value of E_{rms}/I_{rms}, because the effective values are derived from the complete cycles of the voltage and the current.

Assume that the source voltage and the inductance are kept constant, but the frequency is raised. The voltage induced in the coil must remain the same, but the magnetic field is expanding and collapsing more rapidly so that the required magnetic flux is less. As a result, the current is reduced, and is inversely proportional to the frequency. For example, if the frequency is doubled, the value of the effective current is halved.

Now, keep the source voltage and the frequency constant, but replace the inductor with another coil having a higher inductance. Because the induced voltage, $v_L = -L\, di/dt$ is the same, the value of di/dt must be less and, therefore, the effective current is again reduced. Finally, by raising the value of the source voltage while the frequency and the inductance are unchanged, the induced voltage must be increased and, consequently, the effective current is greater.

Summarizing, the effective current is *directly* proportional to the effective voltage, but is *inversely* proportional to the frequency and the inductance. The opposition to the alternating current is, therefore, determined by the product of the frequency and the inductance. This opposition is called the *inductive reactance* ($X_L = E_{rms}/I_{rms}$) and it is measured in ohms. As shown in the mathematical derivations, $X_L = 2\pi fL$ Ω and because X_L is directly proportional to the frequency, the graph of "X_L versus frequency" is a straight line (Fig. 69-1D).

The instantaneous power (p) is equal to the product of e and i. The graph of the "instantaneous power (p) versus time" is a second harmonic sine wave (Fig. 69-1E) whose average value is zero over the source voltage's cycle. During the first quarter cycle, the magnetic field is established around the inductor, and the energy is drawn from the source. However, during the second quarter cycle, the magnetic field collapses and the energy is returned to the source. This action is repeated during the third and fourth quarter cycles so that the average power over the complete cycle is zero.

This highlights the difference between resistance and reactance. Both resistance and reactance limit the value of the alternating current, but while resistance dissipates (lost) power in the form of heat, reactance does not lose any power at all.

MATHEMATICAL DERIVATIONS

Let a sine-wave current, $i = I_{max} \sin 2\pi ft$, flow through the inductor, L.

Then, $v_L = -L\, di/dt$ and, by Kirchhoff's voltage law, $e + v_L = 0$.

This yields,

$$e = L\frac{di}{dt} = 2\pi fL\, I_{max} \cos 2\pi ft \qquad (69\text{-}1)$$

$$= E_{max} \sin\left(2\pi ft + \frac{\pi}{2}\right)$$

Therefore, e leads i by 90° and,

$$E_{max} = 2\pi fL\, I_{max}$$

$$\frac{E_{max}}{I_{max}} = \frac{E_{rms}}{I_{rms}} = 2\pi fL = X_L\ \Omega \qquad (69\text{-}2)$$

Then,

$$f = \frac{X_L}{2\pi L}\ \text{Hz} \quad \text{and} \quad L = \frac{X_L}{2\pi f}\ \text{H} \qquad (69\text{-}3)$$

Instantaneous power,

$$p = e \times i$$

$$= E_{max} \cos 2\pi ft \times I_{max} \sin 2\pi ft$$

$$= \frac{E_{max} I_{max}}{2} \sin (2 \times 2\pi ft)$$

$$= E_{rms} I_{rms} \sin 4\pi ft.$$

The average value of the instantaneous power over the complete cycle is zero.

Example 69-1

In Fig. 69-1A, L = 150 μH, f = 820 kHz and E_{rms} = 8 V. Find the value of I_{rms}.

Solution

The inductive reactance is given by,

$$X_L = 2 \times \pi \times 820 \times 10^3 \times 150 \times 10^{-6} = 773\ \Omega \qquad (69\text{-}2)$$

The current is given by,

$$I_{rms} = \frac{E_{rms}}{X_L} = \frac{8\ \text{V}}{773\ \Omega} = 10.4\ \text{mA}. \qquad (69\text{-}2)$$

Example 69-2

In Fig. 69-1A, f = 15 kHz, E_{rms} = 20 V, I = 12.4 mA. Calculate the value of the inductance, L.

Solution

The inductive reactance is given by,

$$X_L = \frac{E_{rms}}{I_{rms}} = \frac{20\ \text{V}}{12.4\ \text{mA}} = 1613\ \Omega \qquad (69\text{-}2)$$

The equation for the inductance is,

$$L = \frac{X_L}{2\pi f} = \frac{1613}{2 \times \pi \times 15 \times 10^3}\ \text{H} = 17.1\ \text{mH} \qquad (69\text{-}3)$$

PRACTICE PROBLEMS

69-1. A 125-μH inductor has a reactance of 535 Ω. Calculate the value of the frequency.

69-2. At a frequency of 11.5 kHz, an inductor has a reactance of 673 Ω. Calculate the value of the inductance.

69-3. An 8-H inductor is connected across an 110-V rms, 60-Hz source. Determine the value of the current.

69-4. A 225-μH inductor is connected across a sine-wave voltage source. If the equation of the current flowing through the inductor is $i = 27.2 \sin (1.2 \times 10^6 \times t)$ mA, determine the equation of the voltage source.

69-5. Calculate the reactances of a 85-μH inductor at frequencies of 350 kHz, 700 kHz, and 1050 kHz.

69-6. In Practice Problem 69-5, the inductor is connected across a 1-V rms source. Determine the value of the current for each of the same frequencies.

70
Capacitive reactance

When an alternating current (I) is flowing in the circuit of Fig. 70-1A, the capacitor is continuously charging and discharging so that the voltage (v_C) across the capacitor must at all times exactly balance the instantaneous value of the source voltage, e. Therefore, when e is momentarily at its peak value, the value of v_C is also at its peak and the capacitor is fully charged. The current is then instantaneously zero and, therefore, e and i are 90° out of phase (Figs. 70-1B, C). This can be remembered by the word *iCe;* for an ideal capacitor (C) the instantaneous current (i) leads the instantaneous voltage (e) by 90°. Then, combine **"iCe"** with the word **"eLi"** for the inductor and create **"eLi, the iCe man!"**

Now, derive the factors that determine the capacitor's opposition to the flow of alternating current. Assume that the source voltage and the capacitance are kept constant, but the frequency is raised. This reduces the period so that the capacitor must acquire or lose the same amount of charge in a shorter time. The charging (or discharging) current must, therefore, be greater and, in fact, the effective current is directly proportional to the frequency so that, if the frequency is doubled, the effective current is also doubled. By contrast, when the frequency is doubled for an *inductor,* the effective current is halved (chapter 69).

Now, keep the source voltage and the frequency constant, but raise the capacitance. The new capacitor will have to store (or lose) more charge in the same time so that the effective current is again increased. The values of the capacitance, and the effective current, are *directly* proportional so that if the capacitance is halved, the effective current is also halved. Finally, by raising the source voltage, while the frequency and the capacitance are unchanged, the voltage across the capacitor must be greater, the capacitor must store more charge in the same period of time and, consequently, the effective current is once more increased.

Summarizing, the value of the effective current is *directly* proportional to the voltage, the capacitance, and the frequency. The opposition to the alternating current flow is, therefore, *inversely* determined by

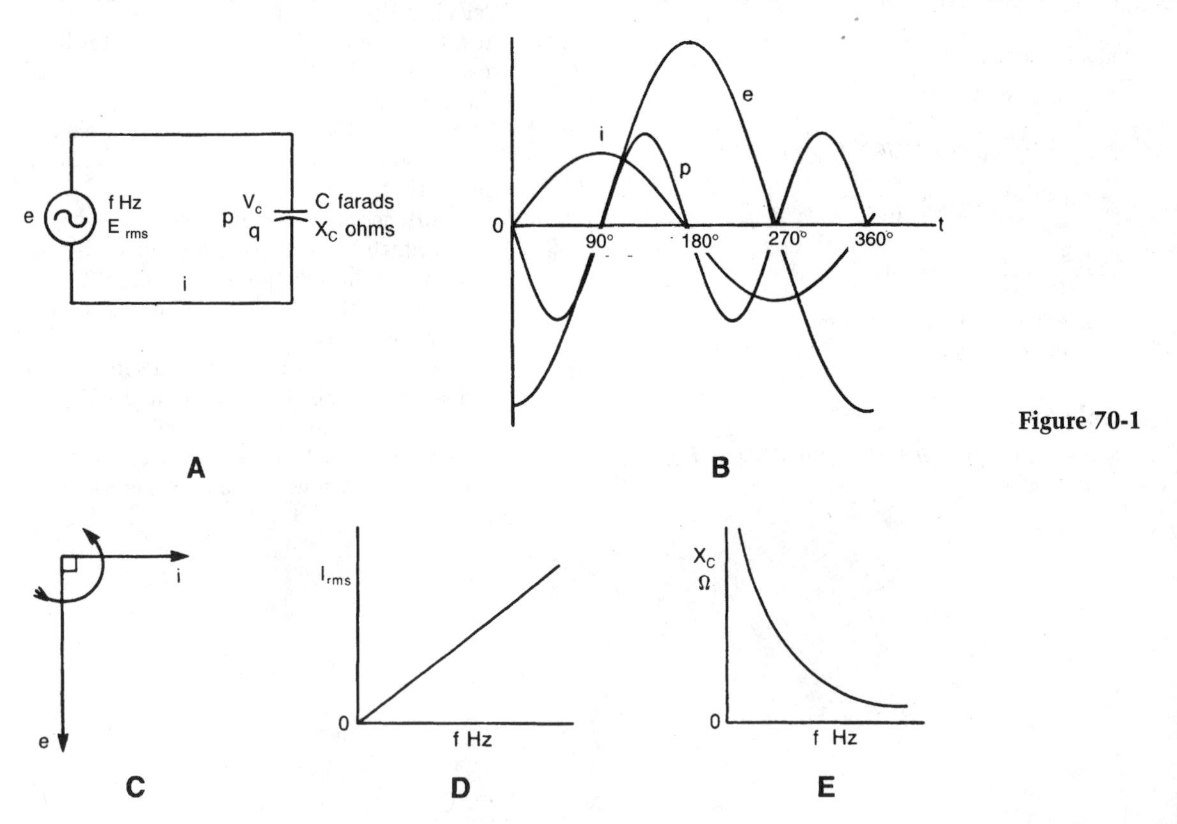

Figure 70-1

the product of the frequency and the capacitance. This opposition is called the *capacitive reactance*,

$$X_C = \frac{E_{rms}}{I_{rms}}$$

and is measured in ohms. As shown in the mathematical derivations,

$$X_C = \frac{1}{2\pi fC}\ \Omega$$

and, because X_C is inversely proportional to the frequency, the graph of "X_C versus frequency" is the rectangular hyperbolic curve of Fig. 70-1E. By contrast, the effective current is directly proportional to the frequency, so that the graph of "I_{rms} versus frequency" is a straight line (Fig. 70-1D).

The instantaneous power (p) is equal to the product of e and i. The graph of the "instantaneous power (p) versus time" is a second harmonic sine wave (Fig. 70-1B) whose average value is zero over the source voltage's cycle. During one quarter cycle, the capacitor charges; energy is drawn from the source and it appears in the form of the electric field between the capacitor's plates. However, during the following quarter cycle, the capacitor discharges and the energy is returned to the source.

Like the inductor, the importance of the capacitor lies in the fact that its opposition to alternating current depends on frequency; as the frequency is raised, the effective current increases from zero to infinity (Fig. 69-1D), while the capacitive reactance decreases from infinity towards zero (Fig. 70-1E). Consequently, for given values of L and C, there must always be a (resonant) frequency for which $X_L = X_C$; this fact is used in the LC tuning circuit, which, for example, is capable of selecting a particular station through its ability to distinguish between one frequency and another (chapters 82 and 84).

MATHEMATICAL DERIVATIONS

Let a sine-wave voltage, $e = E_{max} \sin 2\pi ft$, be applied to the capacitor, C.

Then $v_c = -q/C$ and, by Kirchhoff's voltage law, $e + v_c = 0$. This yields,

$$e = \frac{q}{C} \text{ or } q = Ce.$$

Then,

$$\frac{dq}{dt} = i = C\frac{de}{dt} = 2\pi fCE_{max} \cos 2\pi ft \qquad (70\text{-}1)$$

$$= 2\pi fCE_{max} \sin\left(2\pi ft + \frac{\pi}{2}\right)$$

$$= I_{max} \sin\left(2\pi ft + \frac{\pi}{2}\right) \text{ amperes}$$

Therefore i leads e by 90° and,

$$I_{max} = 2\pi fCE_{max} \qquad (70\text{-}2)$$

or,

$$\frac{E_{max}}{I_{max}} = \frac{E_{rms}}{I_{rms}} = \frac{1}{2\pi fC} = X_C\ \Omega$$

Then,

$$f = \frac{1}{2\pi C X_C} = \frac{0.159}{CX_C}\ \text{Hz} \qquad (70\text{-}3)$$

$$C = \frac{1}{2\pi fX_C} = \frac{0.159}{fX_C}\ \text{F}$$

Instantaneous power,

$$p = e \times i \qquad (70\text{-}4)$$

$$= E_{max} \sin 2\pi ft \times I_{max} \cos 2\pi ft$$

$$= \frac{E_{max} \times I_{max}}{2} \sin 4\pi ft$$

$$= E_{rms} I_{rms} \sin 4\pi ft$$

The average value of the instantaneous power over the complete cycle is zero.

Example 70-1

In Fig. 70-1A, $C = 250$ pF, $f = 822$ kHz, and $E_{rms} = 8$ V. Find the value of I_{rms}.

Solution

Capacitive reactance,

$$X_C = \frac{1}{2 \times \pi \times 822 \times 10^3 \times 250 \times 10^{-12}} \qquad (70\text{-}2)$$

$$= 774\ \Omega$$

Current,

$$I = \frac{8\text{ V}}{774\ \Omega} \qquad (70\text{-}2)$$

$$= 10.3\text{ mA}$$

These results should be compared with the results of Example 69-1.

Example 70-2

In Fig. 70-1A, $f = 15$ kHz, $E_{rms} = 20$ V, $I_{rms} = 12.4$ mA. Calculate the value of the capacitance.

Solution

Capacitive reactance,

$$X_C = \frac{E_{rms}}{I_{rms}} = \frac{20\text{ V}}{12.4\text{ mA}} = 1613\ \Omega \qquad (70\text{-}2)$$

Capacitance,

$$C = \frac{1}{2 \times \pi \times f \times X_C} \qquad (70\text{-}3)$$

$$= \frac{1}{2 \times \pi \times 15 \times 10^3 \times 1613}\text{ F}$$

$$= 0.00658\ \mu\text{F}$$

PRACTICE PROBLEMS

70-1. A 215-pF capacitor has a reactance of 637 Ω. Calculate the value of the frequency.

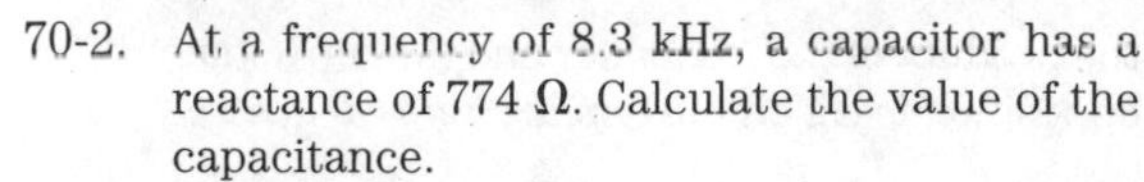

70-2. At a frequency of 8.3 kHz, a capacitor has a reactance of 774 Ω. Calculate the value of the capacitance.

70-3. A 25-μF capacitor is connected across a 110-V, 60-Hz source. Calculate the value of the current.

70-4. A 275-pF capacitor is connected across a sine-wave voltage source. If the equation of the current associated with the capacitor is $i = 17.4 \sin(7.5 \times 10^6 \times t)$ mA, determine the equation of the voltage source.

70-5. Calculate the reactances of a 125-pF capacitor at frequencies of 545 kHz, 1.09 MHz, and 1.635 MHz.

70-6. In Practice Problem 70-5, the capacitor is connected across a 1-V rms source. Determine the value of the current for each of the same frequencies.

71
The general ac circuit—impedance—power factor

Provided there is only a single sine-wave source, all ac circuits, however complex, can ultimately be analyzed into a resistance in series with a reactance (which can be either inductive or capacitive). The total opposition to the flow of alternating current is, therefore, a resistance/reactance combination that is called the *impedance*, *z*. See Fig. 71-1.

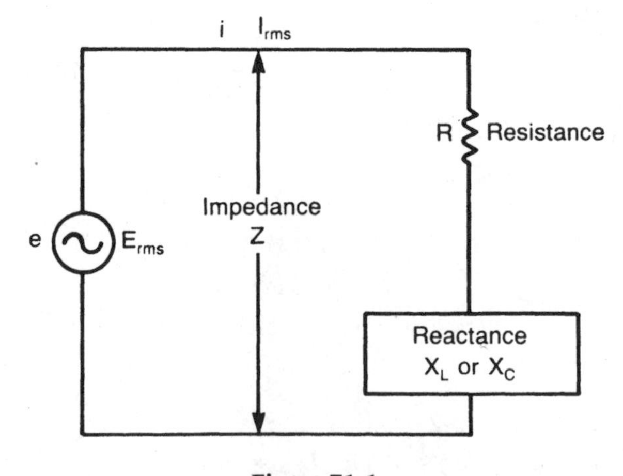

Figure 71-1

The impedance (z) is a phasor that is defined as the ratio of the source voltage phasor to the source current phasor. The magnitude (Z)of the impedance phasor is measured in ohms and is equal to the ratio of E_{rms} : I_{rms}, where E_{rms} and I_{rms} are the effective values of the source voltage and the source current.

The phase angle (ϕ) of the impedance is the angle between the direction of z and the horizontal reference line. The value of this angle normally ranges from $-90°$ (capacitive reactance only) through $0°$ (resistance only) to $+90°$ (inductive reactance only).

In the general ac circuit, the resistive component dissipates (lost) power in the form of heat and the reactive component stores power in the form of either an electric field (capacitive reactance) or a magnetic field (inductive reactance). Therefore, the product of $E_{rms} \times I_{rms}$ does *not* represent the dissipated power, but is instead referred to as the *apparent* power, as measured at the source in volt-amperes (VA). By contrast, the dissipated power is called the *true* power, and the ratio of true power to apparent power is the *power factor*.

If an ac circuit is *inductive,* the phase angle is positive and the power factor is *lagging* (the source current lags the source voltage). With a *capacitive* circuit, the phase angle is negative and the power factor is *leading* (the source current leads the source voltage). The value of the power factor extends from zero (reactance only) to unity (resistance only).

MATHEMATICAL DERIVATIONS

Let the instantaneous current be represented by,

$$i = I_{max} \sin \omega t \text{ amperes.}$$

Then the instantaneous source voltage,

$$e = E_{max} \sin (\omega t \pm \phi) \text{ volts.}$$

Magnitude of the impedance,

$$Z = \frac{E_{max}}{I_{max}} = \frac{E_{rms}}{I_{rms}} \text{ ohms} \quad (71\text{-}1)$$

This yields,

$$E_{rms} = I_{rms} \times Z, I_{rms} = E/Z \quad (71\text{-}2)$$

Instantaneous power,

$$p = e \times i \quad (71\text{-}3)$$

$$= E_{max} I_{max} \sin \omega t \sin (\omega t \pm \phi)$$

$$= \frac{E_{max} I_{max}}{2} [\cos \phi - \cos (2\omega t \pm \phi)]$$

The mean value of $\cos (2\omega t \pm \phi)$ is zero over the complete cycle of the source voltage. Therefore, the average power over the complete cycle is:

True power,

$$P_{av} = \frac{E_{max} \times I_{max} \cos \phi}{2} \quad (71\text{-}4)$$

$$= \frac{E_{max}}{\sqrt{2}} \times \frac{I_{max}}{\sqrt{2}} \times \cos \phi$$

$$= E_{rms} \times I_{rms} \times \cos \phi$$

$$= \text{Apparent power} \times \cos \phi \text{ watts}$$

The power factor

$$= \frac{\text{True power (watts)}}{\text{Apparent power (volt-amperes)}} = \cos \phi \quad (71\text{-}5)$$

Phase angle,

$$\phi = \text{inv. cos (power factor)} \quad (71\text{-}6)$$

The power factor is, therefore, the cosine of the phase angle.

Example 71-1

An alternating source has an instantaneous voltage of $e = 19 \sin (\omega t + \pi/3)$ V. If the instantaneous source current is $i = 3.6 \sin (\omega t - \pi/12)$ mA, calculate the values of the impedance, the phase angle and the power factor. Is the circuit inductive or capacitive? What are the values of the true power, and the apparent power?

Solution

The voltage (e) leads the current (i) by an angle of $\pi/4 - (-\pi/12) = +\pi/3$ radians. The circuit is, therefore, inductive.

Magnitude of the impedance,

$$Z = \frac{E_{rms}}{I_{rms}} = \frac{E_{max}}{I_{max}} = \frac{19 \text{ V}}{3.6 \text{ mA}} = 5.28 \text{ k}\Omega$$

The power factor is $\cos (+\pi/3) = 0.5$, lagging.

Apparent power = 19.4 V × 3.6 mA = 69.84 mVA.

True power (71-5)

= Apparent power × power factor

= 69.84 × 0.5

= 34.92 mW.

PRACTICE PROBLEMS

71-1. In an ac circuit the source (E) leads to the source current (I) by 37°. If $E = 110$ V and $I = 6.4$ A, calculate the values of the apparent power and the true power.

71-2. An alternating sinewave source has an apparent power of 337 mVA and a true power of 285 mW. Determine the phase difference between the source voltage and the source current.

71-3. An alternating source has an instantaneous voltage of $e = 13.7 \sin (\omega t - \pi/6)$ V. If the instantaneous source current is $5.2 \sin (\omega t + \pi/12)$ mA, calculate the values of the impedance, the phase angle, the power factor, the true power, and the apparent power. Is the circuit inductive or capacitive?

71-4. For an ac source, $E = 53$ V rms, and $I = 3.7$ A rms. If the value of the power factor is 0.82, determine the value of the true power.

72
Sine-wave input voltage to R and L in series

Assume that the value of $X_L (= 2\pi fL\ \Omega)$ is greater than the value of R. Because R and L are in series, the same alternating current must flow through each component (Fig. 72-1A). The instantaneous voltage drop (v_R) across the resistor and the instantaneous current (i) through the resistor are in phase (Fig. 72-1B) so that their phasors lie along the same horizontal line (Fig. 72-1C). By contrast, the instantaneous voltage (v_L) across the inductor leads the instantaneous current (i) by 90° and, consequently, their phasors are perpendicular. The phasor sum of v_R and v_L is the supply voltage (e); the current (i) then lags the source voltage (e) by the phase angle, ϕ. This inductive circuit is then considered to have a lagging power factor (resistance, R/impedance, Z), and the phase angle is *positive.*

The total opposition to the alternating current flow is measured by the *impedance phasor* (z), which is defined as the ratio of the e phasor to the i phasor and is equal to the phasor sum of R and X_L (Fig. 72-1D).

The *true power* (TP in watts) is the power dissipated (or lost) as heat in the resistor; it is also the average value of the instantaneous power curve (Fig. 72-1B). The reactive, idle, or wattless power (RP) in volt-amperes reactive (VAr) is the power stored by the inductor as a magnetic field during one quarter of the ac cycle; this power is subsequently returned to the source during the next quarter cycle. The *apparent power* (AP) in volt-amperes (VA) is the product of the source voltage and the source current, and is the phasor sum of the true power and the reactive power (Fig. 72-1E). The *power factor* is the ratio of the true power to the apparent power and is equal to the cosine of the phase angle.

MATHEMATICAL DERIVATIONS

Impedance phasor,

$$z = Z \angle \phi = R + jX_L = R + j \times 2\pi fL \text{ ohms} \qquad (72\text{-}1)$$

In terms of the effective values (capital letters):

$$\frac{E}{I} = Z, E = I \times Z, I = \frac{E}{Z} \qquad (72\text{-}2)$$

$$\phi = \text{inv tan } \frac{X_L}{R} \text{ and is a positive angle} \qquad (72\text{-}3)$$

When $R = X_L$, $\phi = +45°$,

$$Z = \sqrt{R^2 + X_L^2}, X_L = \sqrt{Z^2 - R^2}, \qquad (72\text{-}4)$$

$$R = \sqrt{Z^2 - X_L^2} \text{ ohms}$$

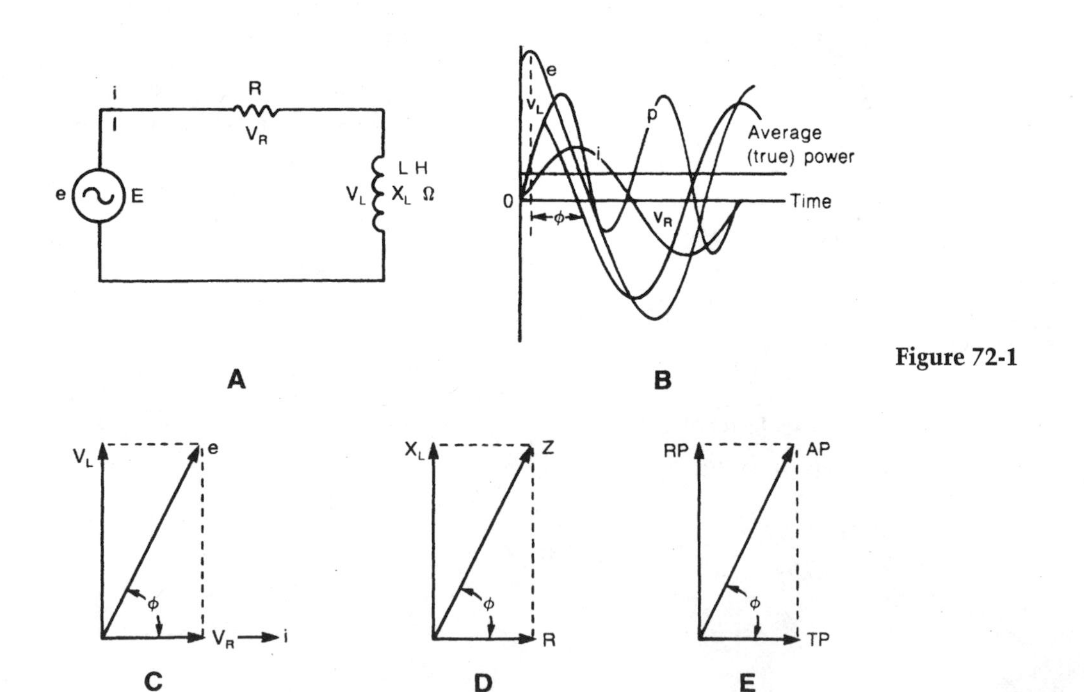

Figure 72-1

$$V_R = I \times R, R = \frac{V_R}{I}, I = \frac{V_R}{R} \quad (72\text{-}5)$$

$$V_L = I \times X_L, X_L = \frac{V_L}{I}, I = \frac{V_L}{X_L} \quad (72\text{-}6)$$

$$E = \sqrt{V_R^2 + V_L^2}, \quad (72\text{-}7)$$

$$V_L = \sqrt{E^2 - V_R^2}, V_R = \sqrt{E^2 - V_L^2} \text{ volts}$$

True power (TP) $= I \times V_R$ (72-8)

$= I^2R = V_R^2/R$ watts

Reactive power (RP) $= I \times V_L$ (72-9)

$= I^2X_L = V_L^2/X_L$ VArs

Apparent power (AP) $= I \times E$ (72-10)

$= I^2Z = E^2/Z$ VA

$$AP = \sqrt{TP^2 + RP^2}, \quad (72\text{-}11)$$

$$RP = \sqrt{AP^2 - TP^2}, TP = \sqrt{AP^2 - RP^2}$$

$$\text{Power factor} = \frac{TP}{AP} = \frac{I^2R}{I^2Z} \quad (72\text{-}12)$$

$$= \frac{R}{Z} = \cos \phi, \text{ and is lagging}$$

Phase angle, ϕ = inv cos (power factor) (72-13)
and is a positive angle

True power (72-14)
= apparent power × power factor
= $EI \cos \phi$ watts

Example 72-1

In Fig. 72-2, $R = 680\ \Omega$, $L = 150\ \mu$H and the supply voltage is 2 V at a frequency of 800 kHz. Calculate the values of Z, I, V_R, V_L, true power, reactive power, apparent power, the power factor, and the phase angle, ϕ.

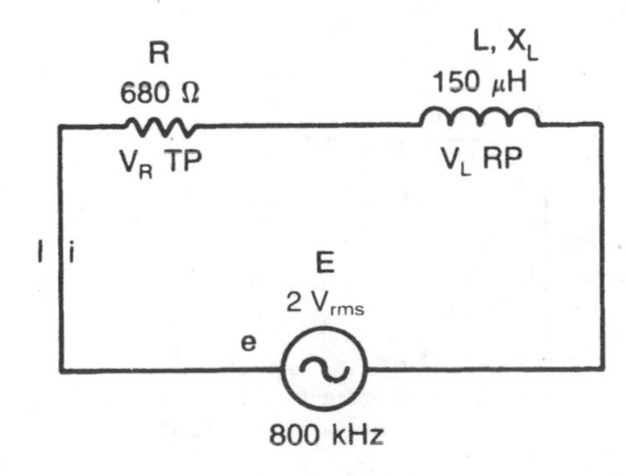

Figure 72-2

Solution

Inductive reactance,

$$X_L = 2\pi fL \quad (72\text{-}1)$$
$$= 2 \times \pi \times 800 \times 10^3 \times 150 \times 10^{-6}$$
$$= 754\ \Omega$$

Impedance,

$$Z = \sqrt{R^2 + X_L^2} = \sqrt{680^2 + 754^2} = 1015\ \Omega \quad (72\text{-}4)$$

Current,

$$I = \frac{E}{Z} = \frac{2 \text{ V}}{1015\ \Omega} = 1.97 \text{ mA} \quad (72\text{-}2)$$

Voltage across the resistor, (72-5)

$$V_R = I \times R = 1.97 \text{ mA} \times 680\ \Omega = 1.34 \text{ V}.$$

Voltage across the inductor, (72-6)

$$V_L = I \times X_L = 1.97 \text{ mA} \times 754\ \Omega = 1.49 \text{ V}.$$

Voltage check:

$$E = \sqrt{V_R^2 + V_L^2} = \sqrt{1.34^2 + 1.49^2} = 2.0 \text{ V} \quad (72\text{-}7)$$

$$TP = I \times V_R = 1.97 \text{ mA} \times 1.34 \text{ V} = 2.64 \text{ mW} \quad (72\text{-}8)$$

$$RP = I \times V_L = 1.97 \text{ mA} \times 1.49 \text{ V} = 2.93 \text{ mVAr} \quad (72\text{-}9)$$

$$AP = I \times E = 1.97 \text{ mA} \times 2 \text{ V} = 3.94 \text{ mVA} \quad (72\text{-}10)$$

Power check:

$$AP = \sqrt{TP^2 + RP^2} = \sqrt{2.64^2 + 2.93^2} \quad (72\text{-}11)$$
$$= 3.94 \text{ mVA}$$

$$\text{Power factor} = \frac{TP}{AP} = \frac{2.64 \text{ mW}}{3.94 \text{ mVA}} \quad (72\text{-}12)$$
$$= 0.67, \text{ lagging}$$

Phase angle, ϕ = inv cos (power factor) (72-13)
= inv cos 0.67 = +47.9°

PRACTICE PROBLEMS

72-1. In Fig. 72-1A, $E = 110$ V rms, $I_L = 0.1$ A rms, and $X_L = 600\ \Omega$. Determine the value of the true power dissipated in the resistor.

72-2. In Fig. 72-1A, $X_L = 470\ \Omega$ and $R = 470\ \Omega$. Calculate the value of the phase angle, and the power factor.

72-3. In Fig. 72-1A, $E = 110$ V rms, $f = 60$ Hz, $L = 4$ H, $R = 1.2$ kΩ. Determine the values of Z, I, V_R, V_L, TP, RP, AP, the power factor, and the phase angle.

72-4. In Fig. 72-1A, $E = 1$ V rms, $f = 11$ kHz, $L = 40$ mH, $R = 2.2$ kΩ. Determine the values of Z, I, V_R, V_L, TP, RP, AP, the power factor, and the phase angle.

72-5. In Practice Problem 72-4, the frequency is doubled. Calculate the new value of the impedance.

72-6. In Fig. 72-1A, $E = 3$ V rms, $f = 735$ kHz, $I = 2.2$ mA, and $L = 175$ μH. Calculate the value of R.

73
Sine-wave input voltage to *R* and *C* in series

Assume that the value of $X_C = 1/(2\pi fC)$ Ω is greater than the value of R. Because R and C are in series, the same alternating current must flow through each component (Fig. 73-1A). The instantaneous voltage drop (v_R) across the resistor, and the instantaneous current (i) through the resistor, are in phase (Fig. 73-1B) so that their phasors lie along the same horizontal reference line (Fig. 73-1C). By contrast, the instantaneous voltage (v_C) across the capacitor lags the instantaneous current (i) by 90°; consequently, their phasors are perpendicular. The phasor sum of v_R and v_C is the supply voltage (e); the current (i) then leads the source voltage (e) by the phase angle, ϕ. This capacitive circuit is considered to have a *leading* power fac-

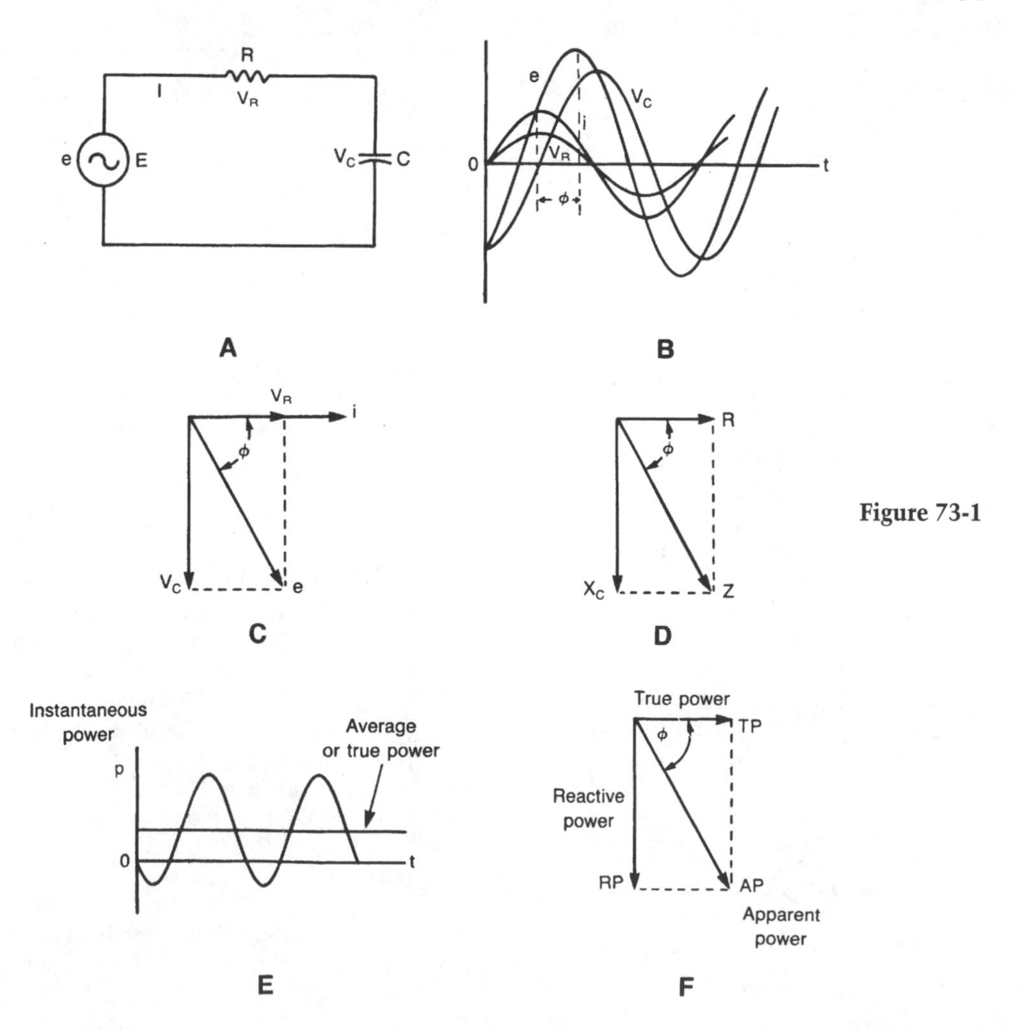

Figure 73-1

tor (resistance, R/impedance, Z), and the phase angle is *negative.*

The total opposition to the alternating current flow is measured by the *impedance phasor* (z)which is defined as the ratio of the e phasor to the i phasor, and is equal to the phasor sum of R and X_C (Fig. 73-1D).

The *true power* (TP in watts) is the power dissipated (or lost) as heat in the resistor; it is also the average value of the instantaneous power curve (Fig. 73-1E). The reactive, idle, or wattless power (RP) in volt-amperes, reactive (VAr) is the power stored by the capacitor as an electric field during one quarter of the ac cycle; this power is subsequently returned to the source during the next quarter cycle. The *apparent power* (AP) in volt-amperes (VA) is the product of the source voltage and the source current, and is the phasor sum of the true power and the reactive power (Fig. 73-1F). The *power factor* is the ratio of the true power to the apparent power and is equal to the cosine of the phase angle.

MATHEMATICAL DERIVATIONS

Impedance phasor,

$$z = Z \angle -\phi = R - jX_C = R - \frac{j \times 1}{2\pi fC} \text{ ohms} \quad (73\text{-}1)$$

In terms of the effective values (capital letters):

$$\frac{E}{I} = Z, E = I \times Z, I = \frac{E}{Z} \quad (73\text{-}2)$$

$$\phi = \text{inv tan}\left(\frac{-X_C}{R}\right) \text{ and is a negative angle} \quad (73\text{-}3)$$

When $R = X_C$, $\phi = -45°$.

$$Z = \sqrt{R^2 + X_C^2}, X_C = \sqrt{Z^2 - R^2}, \quad (73\text{-}4)$$

$$R = \sqrt{Z^2 - X_C^2} \text{ ohms}$$

$$V_R = I \times R, R = \frac{V_R}{I}, I = \frac{V_R}{R} \quad (73\text{-}5)$$

$$V_C = I \times X_C, X_C = \frac{V_C}{I}, I = \frac{V_C}{X_C} \quad (73\text{-}6)$$

$$E = \sqrt{V_R^2 + V_C^2}, V_C = \sqrt{E^2 - V_R^2}, \quad (73\text{-}7)$$

$$V_R = \sqrt{E^2 - V_C^2} \text{ volts}$$

$$\text{True power } (TP) = I \times V_R = I^2R \quad (73\text{-}8)$$

$$= V_R^2/R \text{ watts}$$

$$\text{Reactive power } (RP) = I \times V_C = I^2X_C \quad (73\text{-}9)$$

$$= V_C^2/X_C \text{ VArs}$$

$$\text{Apparent power } (AP) = I \times E = I^2Z \quad (73\text{-}10)$$

$$= E^2/Z \text{ VA}$$

$$AP = \sqrt{TP^2 + RP^2}, RP = \sqrt{AP^2 - TP^2}, \quad (73\text{-}11)$$

$$TP = \sqrt{AP^2 - RP^2}$$

$$\text{Power factor} = \frac{TP}{AP} = \frac{I^2R}{I^2Z} = \frac{R}{Z} \quad (73\text{-}12)$$

$$= \cos\phi \text{ and is leading}$$

$$\text{Phase angle, } \phi = \text{inv cos (power factor)} \quad (73\text{-}13)$$
and is a negative angle

$$\text{True power} = \text{apparent power} \quad (73\text{-}14)$$
$$\times \text{ power factor} = EI\cos\phi \text{ watts}$$

Example 73-1

In Fig. 73-2, $R = 820\ \Omega$, $C = 0.068\ \mu$F, and the supply voltage is 30 V at the frequency of 4.5 kHz. Calculate the values of Z, I, V_R, V_C, true power, reactive power, apparent power, the power factor, and the phase angle ϕ.

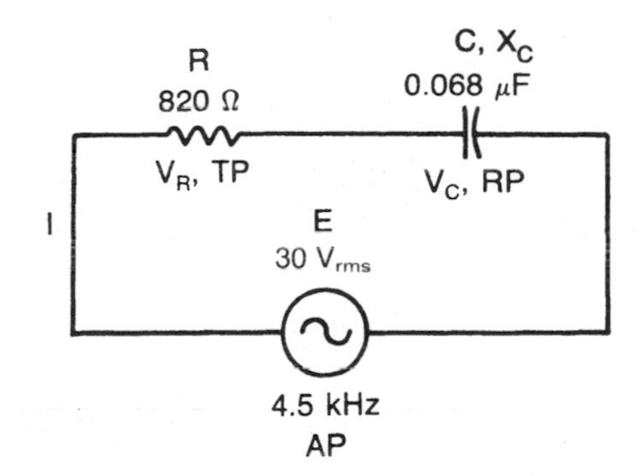

Figure 73-2

Solution

Capacitive reactance,

$$X_C = \frac{1}{2\pi fC} \quad (73\text{-}1)$$

$$= \frac{1}{2 \times \pi \times 4.5 \times 10^3 \times 0.068 \times 10^{-6}}$$

$$= 520\ \Omega$$

Impedance,

$$Z = \sqrt{R^2 + X_C^2} = \sqrt{820^2 + 520^2} \quad (73\text{-}4)$$

$$= 971\ \Omega$$

Current,

$$I = \frac{E}{Z} = \frac{30 \text{ V}}{971\ \Omega} \quad (73\text{-}2)$$

$$= 30.9 \text{ mA}$$

Voltage across the resistor,

$$V_R = I \times R = 30.9 \text{ mA} \times 820 \,\Omega \qquad (73\text{-}5)$$

$$= 25.34 \text{ V}$$

Voltage across the capacitor,

$$V_C = I \times X_C = 30.9 \text{ mA} \times 520 \,\Omega \qquad (73\text{-}6)$$

$$= 16.09 \text{ V}$$

Voltage check:

Source voltage,

$$E = \sqrt{V_R^2 + V_C^2} = \sqrt{25.34^2 + 16.09^2} \qquad (73\text{-}7)$$

$$= 30.0 \text{ V}$$

$$\text{True Power } (TP) = I \times V_R \qquad (73\text{-}8)$$

$$= 30.9 \text{ mA} \times 25.34 \text{ V}$$

$$= 783.0 \text{ mW}$$

$$\text{Reactive Power } (RP) = I \times V_C \qquad (73\text{-}9)$$

$$= 30.9 \text{ mA} \times 16.09 \text{ V}$$

$$= 497.2 \text{ mVAr}$$

$$\text{Apparent Power } (AP) = I \times E \qquad (73\text{-}10)$$

$$= 30.9 \text{ mA} \times 30 \text{ V}$$

$$= 927.0 \text{ mVA}$$

Power check:

$$AP = \sqrt{TP^2 + RP^2} = \sqrt{783.0^2 + 497.2^2} \qquad (73\text{-}11)$$

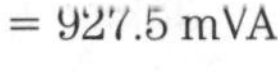

$$= 927.5 \text{ mVA}$$

$$\text{Power factor} = \frac{TP}{AP} = \frac{783.0 \text{ mW}}{927.0 \text{ mVA}} \qquad (73\text{-}12)$$

$$= 0.845, \text{ leading}$$

$$\text{Phase angle, } \phi = \text{inv cos } (0.845) = -32.4° \qquad (73\text{-}13)$$

PRACTICE PROBLEMS

73-1. In Fig. 73-1A, X_C = 560 Ω, and R = 560 Ω. What is the value of the power factor and the phase angle? If the frequency is doubled, calculate the new values of the power factor and the phase angle.

73-2. In Fig. 73-1A, E = 2 V rms, f = 735 kHz, C = 275 pF, R = 680 Ω. Determine the values of Z, I, V_R, V_C, TP, RP, AP, the power factor, and the phase angle.

73-3. In Practice Problem 73-2, the frequency is halved. Calculate the new values of Z, I, V_R, V_C, TP, RP, AP, the power factor, and the phase angle.

73-4. In Fig. 73-1A, E = 110 V rms, f = 60 Hz, C = 20 μF, R = 200 Ω. Determine the values of Z, I, V_R, V_C, TP, RP, AP, the power factor, and the phase angle.

73-5. In Fig. 73-1A, E = 1 V rms, f = 11 kHz, C = 0.004 μF, R = 2.7 kΩ. Determine the values of Z, I, V_R, V_C, TP, RP, AP, the power factor, and the phase angle.

74
Sine-wave input voltage to *L* and *C* in series

Assume that the value of $X_L (= 2\pi fL \,\Omega)$ is greater than the value of $X_C = 1/(2\pi fC) \,\Omega$. Because L and C are in series, the same alternating current must be associated with both the inductor and the capacitor (Fig. 74-1A). The instantaneous voltage (v_L) across the inductor leads the instantaneous current (i) by 90°, and the instantaneous voltage across the capacitor (v_C) lags i by 90° (Fig. 74-1B). Consequently, v_L and v_C are 180° out of phase, and their phasors are pointing in opposite directions (Fig. 74-1C). Because it has been assumed that $X_L > X_C$ and $v_L > v_C$, the supply voltage (e) (which is the phasor sum of v_L and v_C) is in phase with v_L, but has a magnitude of $V_L - V_C$. Because i lags e by 90°, the circuit behaves inductively, and has a lagging power factor with a phase angle of + 90°. The magnitude of the impedance (Z) is found by combining X_L and X_C so that $Z = X_L - X_C \,\Omega$ (Fig. 74-1D).

Because the inductor and the capacitor are considered to be ideal components, there is no resistance present in the circuit and no true power is dissipated. The inductive reactive power (IRP) is greater than the capacitive reactive power (CRP) and the apparent

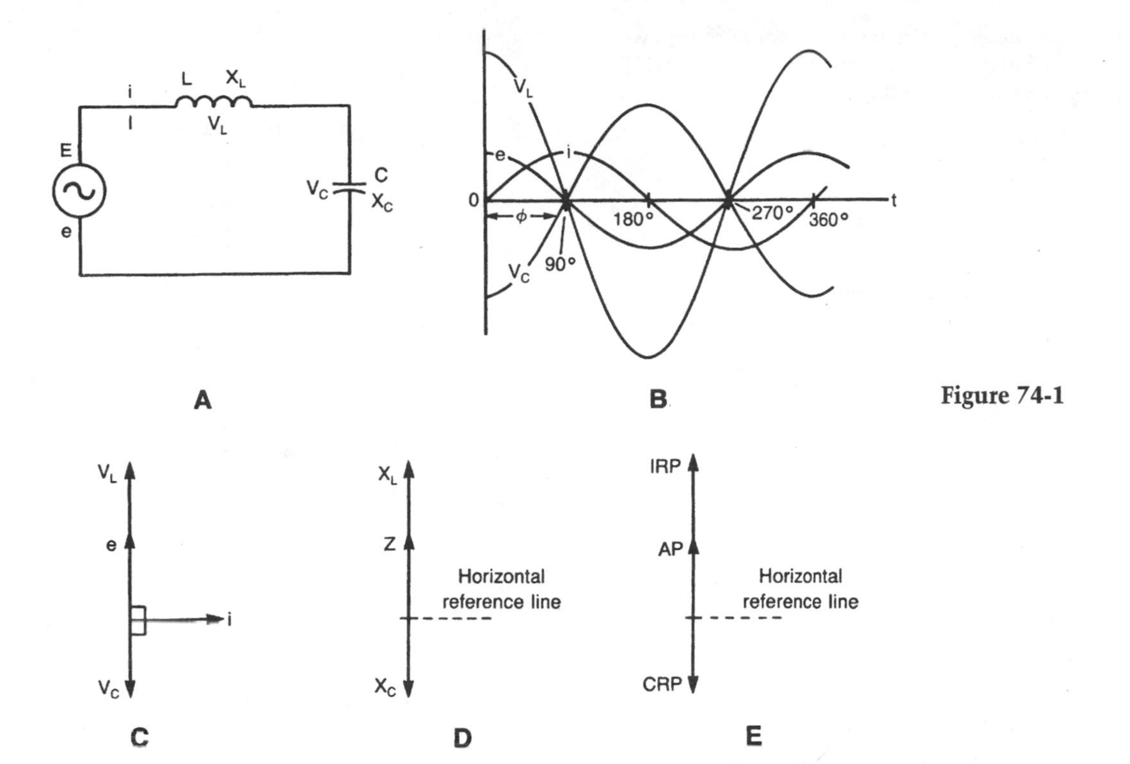

Figure 74-1

power (AP) is entirely reactive with a value equal to *IRP* − *CRP* (Fig. 74-1E).

If, by contrast, you lowered the frequency to the point where X_C became greater than X_L, E would be equal to $V_C - V_L$, and Z would be $X_C - X_L$. The current (i) would then lead e by 90°, the circuit would behave capacitively with a leading power factor and a phase angle of − 90°. The apparent power would equal *CRP* − *IRP*.

In the particular case where the frequency is chosen so that X_L is equal to X_C, V_L is equal to V_C and the impedance (Z) is zero. The current would then be theoretically infinite.

MATHEMATICAL DERIVATIONS

Impedance phasor,

$$z = j\,(X_L - X_C) = j\left(2\pi fL - \frac{1}{2\pi fC}\right) \text{ ohms} \qquad (74\text{-}1)$$

In terms of the effective values (capital letters):

$$\frac{E}{I} = Z, E = I \times Z, I = \frac{E}{Z} \qquad (74\text{-}2)$$

$$\phi = +\,90° \text{ if } X_L > X_C \qquad (74\text{-}3)$$

and

$$\phi = -\,90° \text{ if } X_C > X_L$$

$$\text{Impedance, } Z = X_L \sim X_C \text{ ohms} \qquad (74\text{-}4)$$

The sign "∼" means the "difference"; you are required to subtract the smaller quantity from the larger quantity so that the result is always positive. This takes into account the two cases, $X_L > X_C$ and $X_C > X_L$.

$$V_L = I \times X_L, X_L = \frac{V_L}{I}, I = \frac{V_L}{X_L} \qquad (74\text{-}5)$$

$$V_C = I \times X_C, X_C = \frac{V_C}{I}, I = \frac{V_C}{X_C} \qquad (74\text{-}6)$$

$$\text{Source voltage, } E = V_L \sim V_C \text{ volts} \qquad (74\text{-}7)$$

True power,

$$TP = 0 \text{ watts} \qquad (74\text{-}8)$$

Inductive reactive power,

$$IRP = I \times V_L = I^2 \times X_L = V_L^2/X_L \text{ VArs} \qquad (74\text{-}9)$$

Capacitive reactive power,

$$CRP = I \times V_C = I^2 \times X_C \qquad (74\text{-}10)$$

$$= V_C^2/X_C \text{ VArs}$$

Apparent power,

$$AP = I \times E = I^2 \times Z = E^2/Z \text{ VA(r)s} \qquad (74\text{-}11)$$

It is permissible to measure the apparent power in VArs because the power is entirely reactive and there is no true power in the circuit.

Apparent power,

$$AP = IRP \sim CRP \text{ VA(r)s} \qquad (74\text{-}12)$$

$$\text{Power factor} = 0 \text{ and is lagging if } X_L > X_C\text{, but is leading if } X_C > X_L. \qquad (74\text{-}13)$$

At the particular frequency (f) for which $X_L = X_C$, $V_L = V_C$, $Z = 0$, and I is theoretically infinite. Then,

$$2\pi fL = \frac{1}{2\pi fC}$$

This yields,

$$f = \frac{1}{2\pi\sqrt{LC}} \text{ Hz}, \quad L = \frac{1}{4\pi^2 f^2 C} \text{ H}, \qquad (74\text{-}14)$$

$$C = \frac{1}{4\pi^2 f^2 L} \text{ F}$$

Example 74-1

In Fig. 74-2, $L = 125\ \mu\text{H}$, $C = 275$ pF, and the source voltage is 50 mV at 750 kHz. Calculate the values of Z, I, V_L, TP, IRP, CRP, AP, the power factor, and the phase angle, ϕ. At which frequency will the inductive and capacitive reactances be equal?

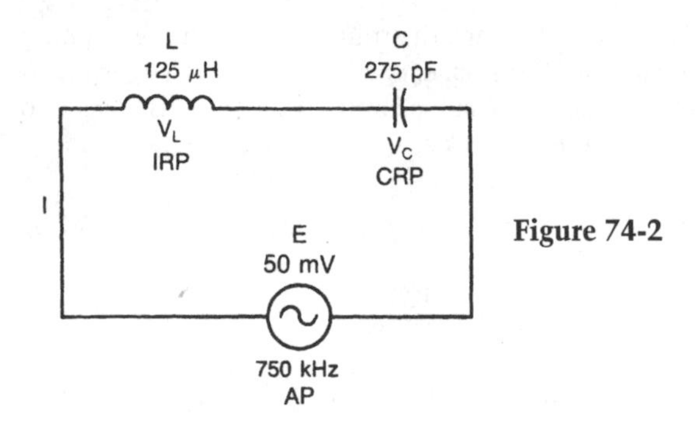

Figure 74-2

Solution

Inductive reactance,

$$X_L = 2\pi fL = 2 \times \pi \times 750 \times 10^3 \times 125 \times 10^{-6} \qquad (74\text{-}1)$$

$$= 589\ \Omega$$

Capacitive reactance,

$$X_C = \frac{1}{2\pi fC} \qquad (74\text{-}1)$$

$$= \frac{1}{2 \times \pi \times 750 \times 10^3 \times 275 \times 10^{-12}}$$

$$= 772\ \Omega$$

Impedance,

$$Z = X_C - X_L = 772 - 589 \qquad (74\text{-}4)$$

$$= 183\ \Omega \text{ and is capacitive}$$

Current,

$$I = \frac{E}{Z} = \frac{50 \text{ mV}}{183\ \Omega} \qquad (74\text{-}2)$$

$$= 0.273 \text{ mA}$$

Voltage across the inductor,

$$V_L = I \times X_L = 0.273 \text{ mA} \times 589\ \Omega \qquad (74\text{-}5)$$

$$= 161 \text{ mV}$$

Voltage across the capacitor,

$$V_C = I \times X_C = 0.273 \text{ mA} \times 772\ \Omega \qquad (74\text{-}6)$$

$$= 211 \text{ mV}$$

Voltage check:

Source voltage,

$$E = V_C - V_L = 211 - 161 \qquad (74\text{-}7)$$

$$= 50 \text{ mV}$$

Notice that both V_L and V_C are greater than E. This is allowed because the power in the circuit is entirely reactive and there is no true power dissipated. It is always true that either V_L, or V_C, or both V_L and V_C are greater than E.

$$\text{True power, } TP = 0 \qquad (74\text{-}8)$$

Inductive reactive power,

$$IRP = I \times V_L = 0.273 \text{ mA} \times 161 \text{ mV} \qquad (74\text{-}9)$$

$$= 44.0\ \mu\text{VAr}$$

Capacitive reactive power,

$$CRP = I \times V_C = 0.273 \text{ mA} \times 211 \text{ mV} \qquad (74\text{-}10)$$

$$= 57.6\ \mu\text{VAr}$$

Apparent power

$$AP = I \times E = 0.273 \times 50 \text{ mV} \qquad (74\text{-}11)$$

$$= 13.6\ \mu\text{VAr}$$

Power check:

$$AP = CRP - IRP = 57.6 - 44.0 \qquad (74\text{-}12)$$

$$= 13.6\ \mu\text{VAr}$$

$$\text{Power factor} = 0\text{, leading} \qquad (74\text{-}13)$$

$$\text{Phase angle, } \phi = -90°$$

Frequency for which the reactances are equal,

$$f = \frac{1}{2\pi\sqrt{LC}} \qquad (74\text{-}14)$$

$$= \frac{1}{2\pi\sqrt{125 \times 10^{-6} \times 275 \times 10^{-12}}} \text{ Hz}$$

$$= 858 \text{ kHz}$$

Notice that the Pythagorean theorem is never used in the solution of this type of problem.

PRACTICE PROBLEMS

74-1. In Fig. 74-1A, $X_L = 280\ \Omega$ and $X_C = 370\ \Omega$. Determine the values of the impedance, the power factor, and the phase angle.

74-2. In Practice Problem 74-1, the frequency is doubled. Determine the new values of the impedance, the power factor, and the phase angle.

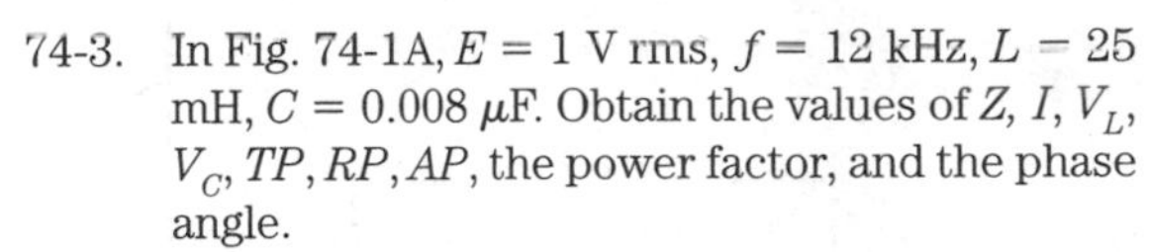

74-3. In Fig. 74-1A, $E = 1$ V rms, $f = 12$ kHz, $L = 25$ mH, $C = 0.008\ \mu$F. Obtain the values of Z, I, V_L, V_C, TP, RP, AP, the power factor, and the phase angle.

74-4. In Practice Problem 74-3, the frequency is halved. Determine the new values of Z, I, V_L, V_C, TP, RP, AP, the power factor, and the phase angle.

74-5. In Practice Problem 74-3, calculate the value of the frequency for which the reactances are equal.

74-6. In Fig. 74-1A, $E = 110$ V rms, $f = 60$ Hz, $C = 4\ \mu$F, $L = 3$ H. Determine the values of Z, I, V_L, V_C, TP, RP, AP, the power factor, and the phase angle.

75
Sine-wave input voltage to *R*, *L*, and *C* in series

This circuit must contain all of the information contained in the previous three chapters (72, 73, and 74). For example, if the capacitor was eliminated, the results would then be the same as for R and L in series.

Assume that X_L is greater than X_C, and that R is less than $X_L - X_C$. Since all three components are in series, the instantaneous current (i) is the same throughout the circuit. In terms of phase relationships, v_R is in phase with i, v_L leads i by 90°, and v_C lags i by 90° (Figs. 75-1B and C). The combined voltage across the inductor and the capacitor is represented by the phasor v_X, which is in phase with v_L and leads i by 90°. Current i lags source voltage e, which is the phasor sum of v_X and v_R. The circuit is, therefore, overall inductive, the power factor is lagging, and the phase angle, ϕ, is positive.

The net reactance (X) is equal to $X_L - X_C$, which is then combined with R to produce the impedance, Z (Fig. 75-1D). The total reactive power is the phasor sum of the inductive reactive power, and the capacitive reactive power. When the total reactive power is combined with the true power, the result is the apparent power (Fig. 75-1E).

If X_C was greater than X_L, v_X would be in phase with v_C, i would lead e by the phase angle, ϕ. The power factor would then be leading, and ϕ would be a negative angle.

At the particular frequency, $f = 1/(2\pi\sqrt{LC})$ Hz, the reactances are equal and, therefore, cancel each other. The impedance is then equal to the resistance and the phase angle is zero. As a result, the values of the true power and the apparent power are the same.

MATHEMATICAL DERIVATIONS

Impedance phasor,

$$z = R + jX = R + j(X_L - X_C) \qquad (75\text{-}1)$$

$$= R + j\left(2\pi fL - \frac{1}{2\pi fC}\right) \text{ ohms}$$

In terms of the effective values (capital letters):

$$\frac{E}{I} = Z, E = I \times Z, I = \frac{E}{Z} \qquad (75\text{-}2)$$

$$\phi = \text{inv tan}\left(\frac{X_L - X_C}{R}\right) \text{ degrees} \qquad (75\text{-}3)$$

ϕ is a positive angle when $X_L > X_C$, but is a negative angle if $X_C > X_L$. When $R = X_L - X_C$, $\phi = +\ 45°$.

Impedance,

$$Z = \sqrt{R^2 + (X_L \sim X_C)^2} = \sqrt{R^2 + X^2} \text{ ohms} \qquad (75\text{-}4)$$

For the meaning of the difference " ~ " sign, see chapter 74.

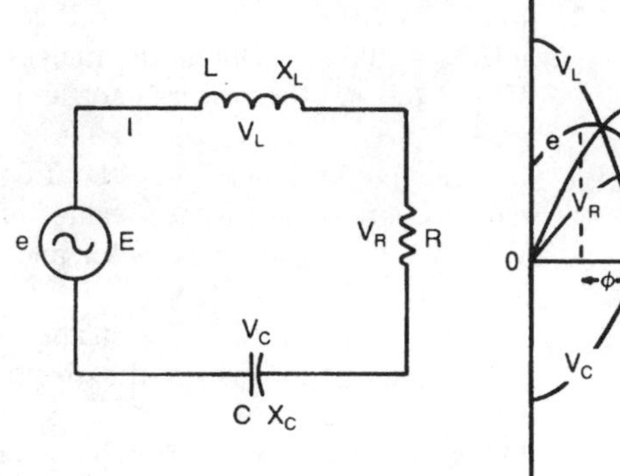

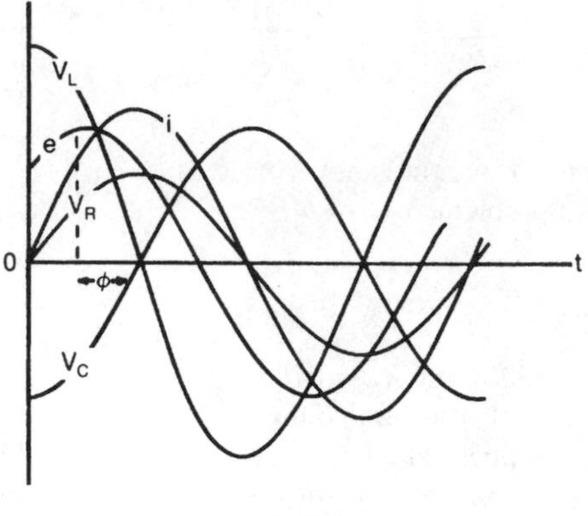

Figure 75-1

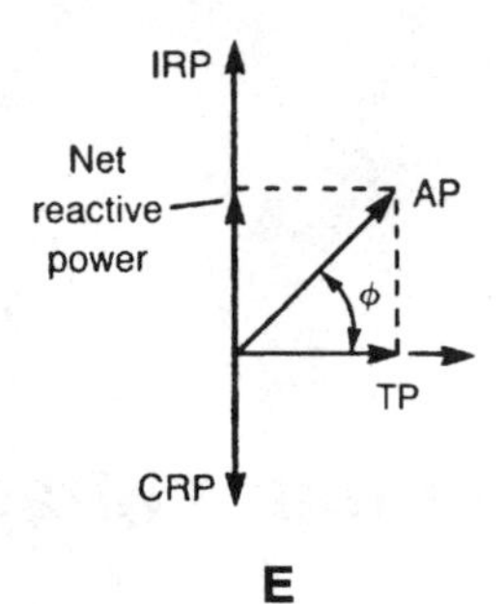

$$V_R = I \times R, R = \frac{V_R}{I}, I = \frac{V_R}{R} \quad (75\text{-}5)$$

$$V_L = I \times X_L, X_L = \frac{V_L}{I}, I = \frac{V_L}{X_L} \quad (75\text{-}6)$$

$$V_C = I \times X_C, X_C = \frac{V_C}{I}, I = \frac{V_C}{X_C} \quad (75\text{-}7)$$

Source voltage,

$$E = \sqrt{V_R^2 + (V_L \sim V_C)^2} = \sqrt{V_R^2 + V_X^2} \text{ volts} \quad (75\text{-}8)$$

True power,

$$TP = I \times V_R = I^2R = V_R^2/R \text{ watts} \quad (75\text{-}9)$$

Inductive reactive power,

$$IRP = I \times V_L \quad (75\text{-}10)$$

$$= I^2 \times X_L = V_L^2/X_L \text{ VArs}$$

Capacitive reactive power,

$$CRP = I \times V_C \quad (75\text{-}11)$$

$$= I^2 \times X_C = V_C^2/X_C \text{ VArs}$$

Apparent power,

$$AP = I \times E = I^2Z = E^2/Z \text{ VA} \quad (75\text{-}12)$$

Apparent power,

$$AP = \sqrt{TP^2 + (IRP \sim CRP)^2} \quad (75\text{-}13)$$

Power factor,

$$PF = \frac{TP}{AP} = \frac{I^2R}{I^2Z} = \frac{R}{Z} = \cos\phi \text{ and is lagging} \quad (75\text{-}14)$$

Phase angle,

ϕ = inv cos (PF) and is a *positive* angle (75-15)

True power,

$$TP = E \times I \times \cos\phi \text{ watts} \quad (75\text{-}16)$$

Example 75-1

In Fig. 75-2, $R = 350\ \Omega$, $L = 2.5$ H, $C = 5\ \mu$F, and the source voltage is 55 V at 60 Hz. Calculate the values of Z, I, V_R, V_L, V_C, true power, inductive reactive power, capacitive reactive power, apparent power, the power factor, and the phase angle, ϕ.

Solution

Inductive reactance,

$$X_L = 2\pi fL = 2 \times \pi \times 60 \times 2.5 \quad (75\text{-}1)$$

$$= 942.5\ \Omega$$

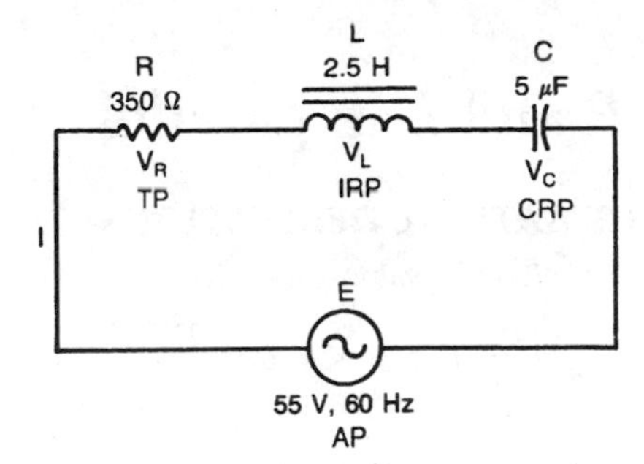

Figure 75-2

Capacitive reactance,

$$X_C = \frac{1}{2\pi fC} \qquad (75\text{-}1)$$

$$= \frac{1}{2 \times \pi \times 60 \times 5 \times 10^{-6}}$$

$$= 530.5\ \Omega$$

Total reactance,

$$X = X_L - X_C = 942.5 - 530.5 \qquad (75\text{-}1)$$

$$= 412.0\ \Omega$$

Impedance,

$$Z = \sqrt{R^2 + X^2} = \sqrt{350^2 + 412^2} \qquad (75\text{-}4)$$

$$= 514\ \Omega$$

Current,

$$I = \frac{E}{Z} = \frac{55\text{ V}}{541\ \Omega}$$

$$= 0.1017\text{ A.}$$

Voltage across the resistor,

$$V_R = I \times R = 0.1017\text{ A} \times 350\ \Omega \qquad (75\text{-}5)$$

$$= 35.6\text{ V}$$

Voltage across the inductor,

$$V_L = I \times X_L = 0.1017\text{ A} \times 942.5\ \Omega \qquad (75\text{-}6)$$

$$= 95.85\text{ V}$$

Voltage across the capacitor,

$$V_C = I \times X_C = 0.1017\text{ A} \times 530.5\ \Omega \qquad (75\text{-}7)$$

$$= 54.0\text{ V}$$

Combined voltage across the inductor and the capacitor, $V_X = 95.85 - 54.0 = 41.85$ V. Voltage check:

$$E = \sqrt{V_R^2 + V_X^2} = \sqrt{35.6^2 + 41.85^2} \qquad (75\text{-}8)$$

$$= 54.9\text{ V}$$

True power,

$$TP = I \times V_R = 0.1017\text{ A} \times 35.6\text{ V} \qquad (75\text{-}9)$$

$$= 3.62\text{ W}$$

Inductive reactive power,

$$IRP = I \times V_L = 0.1017\text{ A} \times 95.85\text{ V} \qquad (75\text{-}10)$$

$$= 9.75\text{ VAr}$$

Capacitive reactive power,

$$CRP = I \times V_C = 0.1017\text{ A} \times 54.0\text{ V} \qquad (75\text{-}11)$$

$$= 5.49\text{ VAr}$$

Apparent power,

$$AP = I \times E = 0.1017\text{ A} \times 55\text{ V} \qquad (75\text{-}12)$$

$$= 5.59\text{ VA}$$

Power check:

$$AP = \sqrt{TP^2 + (IRP - CRP)^2} \qquad (75\text{-}13)$$

$$= \sqrt{3.62^2 + (9.75 - 5.49)^2}$$

$$= 5.59\text{ VA}$$

$$\text{Power factor} = \frac{TP}{AP} = \frac{3.62\text{ W}}{5.59\text{ VA}} \qquad (75\text{-}14)$$

$$= 0.647\text{, lagging}$$

$$\text{Phase angle, } \phi = \text{inv cos } 0.647 = +49.6° \qquad (75\text{-}15)$$

PRACTICE PROBLEMS

75-1. In Fig. 75-1A, $R = 680\ \Omega$, $X_L = 420\ \Omega$, $X_C = 1100\ \Omega$. Calculate the values of the impedance, the power factor, and the phase angle.

75-2. In Practice Problem 75-1, the frequency is doubled. Calculate the new values of the impedance, the power factor, and the phase angle.

75-3. In Fig. 75-1A, $E = 1$ V rms, $f = 12$ kHz, $R = 270\ \Omega$, $L = 40$ mH, $C = 0.004\ \mu$F. Determine the values of $Z, I, V_R, V_L, V_C, TP, RP, AP$, the power factor, and the phase angle.

75-4. In Fig. 75-1A, $E = 2$ V rms, $f = 680$ kHz, $R = 270\ \Omega$, $L = 115\ \mu$H, $C = 275$ pF. Obtain the values of $Z, I, V_R, V_L, V_C, TP, RP, AP$, the power factor, and the phase angle.

75-5. In Practice Problem 75-4, the frequency is doubled. Determine the new values of $Z, I, V_R, V_L, V_C, TP, RP, AP$, the power factor, and the phase angle.

75-6. In Practice Problem 75-4, calculate the value of the frequency for which the reactances are equal. At this frequency, what is the value of the circuit's impedance?

76
Sine-wave input voltage to *R* and *L* in parallel

Because R and L are in parallel, the source voltage, (e) is across each of the components and the source or supply current (i_S) will be the phasor sum of the two branch currents, i_R and i_L (Figs. 76-1A and B). If you assume that the value of R is greater than the value of X_L, the resistor current ($I_R = E/R$) will be less than the inductor current ($I_L = E/X_L$). Because i_R is in phase with e, and i_L lags e by 90°, the source current will lag the source voltage by the phase angle, ϕ (Figs. 76-1B and C). The power factor will be lagging and the phase angle is positive (note that in both the series, and parallel, combinations of R and L, the source current lags the source voltage so that, in each case, the phase angle is positive).

The "ease" with which the source current flows, is measured by the *admittance*, $Y_T(= I_S/E)$, which is the reciprocal of the total impedance (Z_T), and is measured in siemens. Because $I_R = E \times G$, $I_L = E \times B_L$ and $I_S = E \times Y_T$, the admittance (Y_T) is the phasor sum of the conductance (G) and the inductive susceptance, B_L (Fig. 76-1D).

The true power (TP) is equal to E^2/R watts and is independent of the frequency. The reactive power (RP) is associated with the inductor and the apparent power (AP) at the source is the phasor sum of the true power and the reactive power (Fig. 76-1E).

MATHEMATICAL DERIVATIONS

In terms of effective values:

$$I_R = \frac{E}{R}, E = I_R \times R, R = \frac{E}{I_R} \tag{76-1}$$

$$I_L = \frac{E}{X_L}, E = I_L \times X_L, X_L = \frac{E}{I_L} \tag{76-2}$$

$$I_S = \frac{E}{Z_T}, E = I_S \times Z_T, Z_T = \frac{E}{I_S} \tag{76-3}$$

Source current,

$$I_S = \sqrt{I_R^2 + I_L^2} \tag{76-4}$$

$$I_R = \sqrt{I_S^2 - I_L^2}$$

$$I_L = \sqrt{I_S^2 - I_R^2} \text{ amperes}$$

Because $I_S^2 = I_R^2 + I_L^2$,

$$\left(\frac{E}{Z_T}\right)^2 = \left(\frac{E}{R}\right)^2 + \left(\frac{E}{X_L}\right)^2$$

This yields,

Total impedance,

$$Z_T = \frac{R \times X_L}{\sqrt{R^2 + X_L^2}} \text{ ohms} \tag{76-5}$$

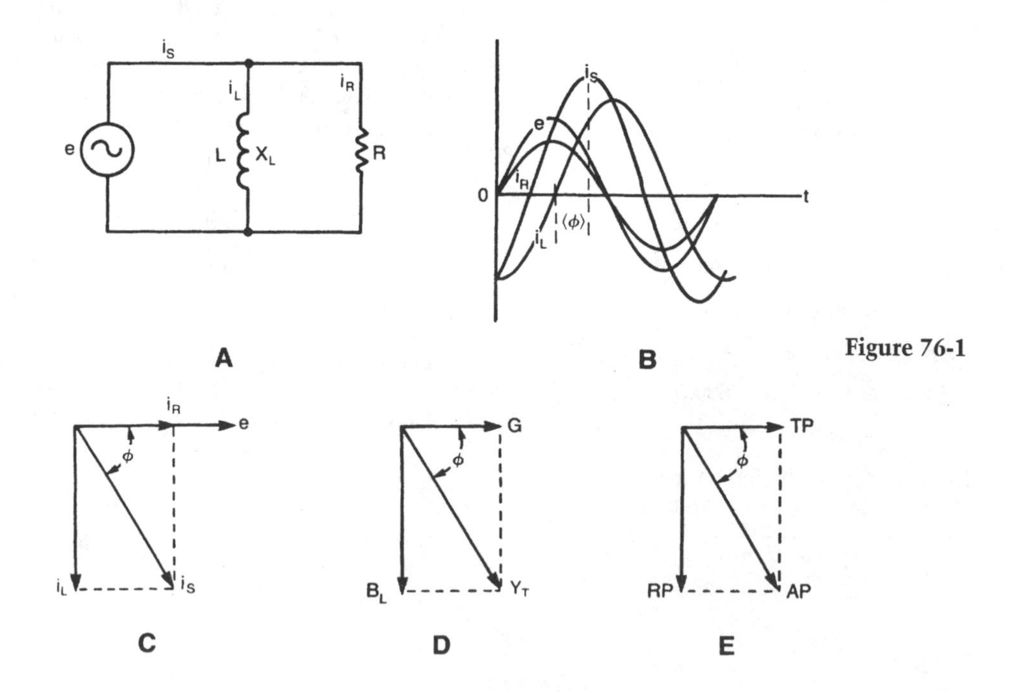

Figure 76-1

Equation 76-5 represents the "product-over-sum" formula when the "Pythagorean sum" is involved.

Because the conductance $G = 1/R$, the inductive susceptance $B_L = 1/X_L$ and the admittance $Y_T = 1/Z_T$,

$$\text{Admittance, } Y_T = \sqrt{G^2 + B_L^2} \quad (76\text{-}6)$$

$$G = \sqrt{Y_T^2 - B_L^2}$$

$$B_L = \sqrt{Y_T^2 - G^2} \text{ siemens}$$

True power,

$$TP = E \times I_R = I_R^2 \times R = E^2/R = E^2G \text{ watts} \quad (76\text{-}7)$$

Reactive power,

$$RP = E \times I_L = I_L^2 \times X_L \quad (76\text{-}8)$$

$$= E^2B_L \text{ volt-amperes reactive}$$

Apparent power,

$$AP = E \times I_S = I_S^2 \times Z_T \quad (76\text{-}9)$$

$$= E^2/Z_T = E^2Y_T \text{ volt-amperes}$$

$$AP = \sqrt{TP^2 + RP^2} \quad (76\text{-}10)$$

$$TP = \sqrt{AP^2 - RP^2}$$

$$RP = \sqrt{AP^2 - TP^2}$$

Power factor,

$$PF = \frac{TP}{AP} = \cos\phi = \frac{E^2/R}{E^2/Z_T} \quad (76\text{-}11)$$

$$= \frac{Z_T}{R}\left(not\ \frac{R}{Z_T}\right), \text{ lagging}$$

Phase angle,

$$\phi = \text{inv cos (PF) and is a positive angle} \quad (76\text{-}12)$$

Note that if $R = X_L$, $\phi = +45°$, and the power factor is 0.707, lagging.

Example 76-1

In Fig. 76-2, calculate the values of I_R, I_L, I_S, Z_T, Y_T, true power, reactive power, apparent power, the power factor, and the phase angle, ϕ.

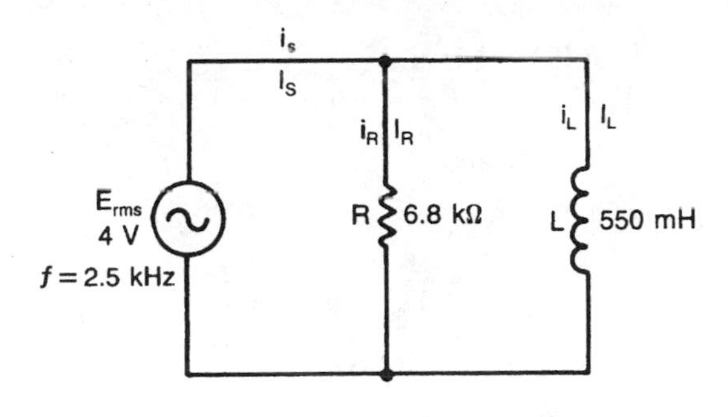

Figure 76-2

Solution

Inductive reactance,

$$X_L = 2\pi fL = 2 \times \pi \times 2.5 \times 10^3 \times 550 \times 10^{-3}\ \Omega$$

$$= 8.64\ \text{k}\Omega$$

Resistor current,

$$I_R = \frac{E}{R} = \frac{4}{6.8\ \text{k}\Omega} \quad (76\text{-}1)$$

$$= 0.588\ \text{mA}$$

Inductor current,

$$I_L = \frac{E}{X_L} = \frac{4\ \text{V}}{8.64\ \text{k}\Omega} \quad (76\text{-}2)$$

$$= 0.463\ \text{mA}$$

Source current,

$$I_S = \sqrt{I_R^2 + I_L^2} = \sqrt{0.588^2 + 0.463^2} \quad (76\text{-}4)$$

$$= 0.7484\ \text{mA}$$

Total impedance,

$$Z_T = \frac{E}{I_S} = \frac{4\ \text{V}}{0.7484\ \text{mA}} \quad (76\text{-}3)$$

$$= 5.34\ \text{k}\Omega$$

Impedance check #1:
Total impedance,

$$Z_T = \frac{R \times X_L}{\sqrt{R^2 + X_L^2}} \quad (76\text{-}5)$$

$$= \frac{6.8 \times 8.64}{\sqrt{6.8^2 + 8.64^2}}$$

$$= 5.34\ \text{k}\Omega$$

Impedance check #2:
Conductance,

$$G = \frac{1}{R} = \frac{1}{6.8\ \text{k}\Omega}$$

$$= 0.147\ \text{mS}$$

Inductive susceptance,

$$B_L = \frac{1}{X_L} = \frac{1}{8.64\ \text{k}\Omega}$$

$$= 0.116\ \text{mS}$$

Admittance,

$$Y_T = \sqrt{G^2 + B_L{}^2} = \sqrt{0.147^2 + 0.116^2} \qquad (76\text{-}6)$$

$$= 0.187 \text{ mS}$$

Total impedance,

$$Z_T = \frac{1}{Y_T} = \frac{1}{0.187 \text{ mS}}$$

$$= 5.34 \text{ k}\Omega$$

True power,

$$TP = E \times I_R = 4 \text{ V} \times 0.588 \text{ mA} \qquad (76\text{-}7)$$

$$= 2.35 \text{ mW}$$

Reactive power,

$$RP = E \times I_L = 4 \text{ V} \times 0.463 \text{ mA} \qquad (76\text{-}8)$$

$$= 1.85 \text{ mVAr}$$

Apparent power,

$$AP = E \times I_S = 4 \text{ V} \times 0.7484 \text{ mA} \qquad (76\text{-}9)$$

$$= 2.99 \text{ mVA}$$

Power check:

$$AP = \sqrt{TP^2 + RP^2} = \sqrt{2.35^2 + 1.85^2} \qquad (76\text{-}10)$$

$$= 2.99 \text{ mVA}$$

Power factor,

$$PF = \frac{TP}{AP} = \frac{2.35 \text{ W}}{2.99 \text{ VA}} \qquad (76\text{-}11)$$

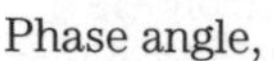

$$= 0.786, \text{ lagging}$$

Phase angle,

$$\phi = \text{inv cos } 0.786 = +\ 38.2° \qquad (76\text{-}12)$$

PRACTICE PROBLEMS

76-1. In Fig. 76-1A, E = 3 V rms, f = 735 kHz, L = 175 μH, R = 1.2 kΩ. Obtain the values of I_R, I_L, I_S, Z_T, TP, RP, AP, the power factor, and the phase angle.

76-2. In Practice Problem 76-1, the frequency is doubled. Determine the new values of I_R, I_L, I_S, Z_T, TP, RP, AP, the power factor, and the phase angle.

76-3. In Fig. 76-1A, calculate the value of the frequency for which the phase angle of the circuit is 45°.

76-4. In Fig. 76-1A, E = 110 V rms, f = 60 Hz, R = 1.2 kΩ, L = 4 H. Determine the values of I_R, I_L, I_S, Z_T, TP, RP, AP, the power factor, and the phase angle. Note: Compare these results with the answers to Practice Problem 72-3. What is the relationship between the phase angles in Practice Problems 72-3 and 76-4?

76-5. In Practice Problem 76-4, calculate the values of the conductance, susceptance, and admittance.

76-6. In Practice Problem 76-4, what is the frequency for which the power factor is 0.707, lagging?

77
Sine-wave input voltage to *R* and *C* in parallel

Because R and C are in parallel, the source voltage (E) is across each of the components, and the source (or supply) current (i_S) will be the phasor sum of the two branch currents, i_R and i_C (Figs. 77-1A and B). If you assume that the value of R is greater than the value of X_C, the resistor current ($I_R = E/R$) will be less than the capacitor current ($I_C = E/X_C$). Because i_R is in phase with e while i_C leads e by 90°, the source current will *lead* the source voltage by the phase angle (ϕ) (Figs. 77-1B and C). The power factor will then be *leading*, and the phase angle is negative.

The "ease" with which the source current flows is measured by the *admittance* $Y_T (= I_S/E)$, which is the reciprocal of the total impedance (Z_T) and is measured in siemens. Because $I_R = E \times G$, $I_C = E \times B_C$ and $I_S = E \times Y_T$, the admittance (Y_T) is the phasor sum of the conductance (G) and the capacitive susceptance, B_C (Fig. 77-1D).

The true power (TP) is equal to E^2/R watts, and is independent of the frequency. The reactive power (RP) is associated with the capacitor and the apparent power (AP) at the source is the phasor sum of the true power and the reactive power (Fig. 77-1E).

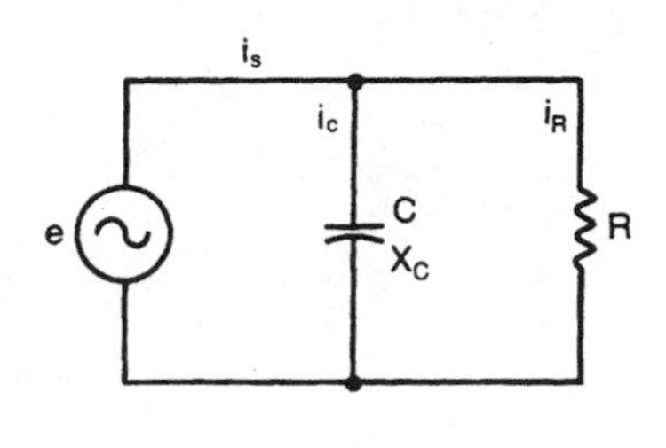

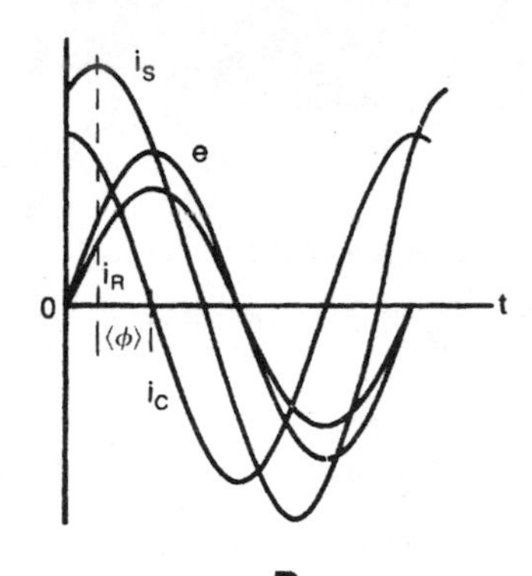

Figure 77-1

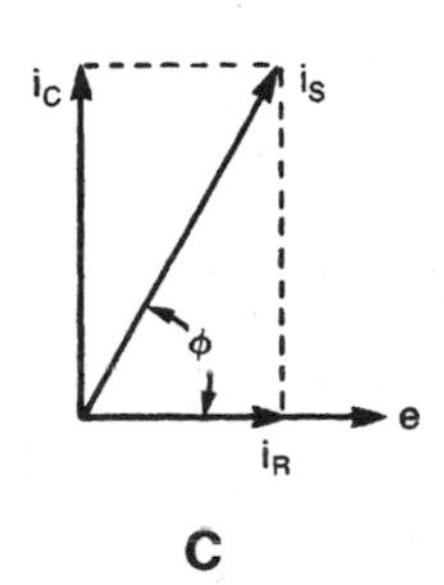

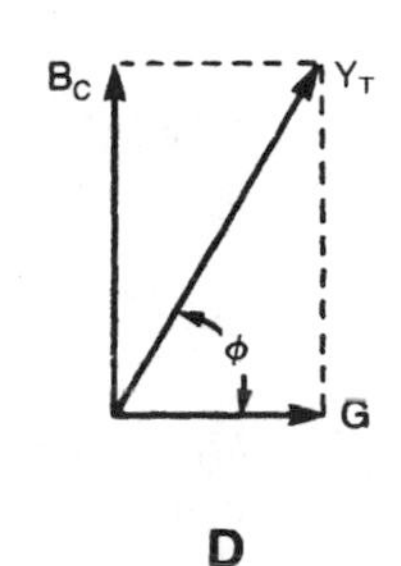

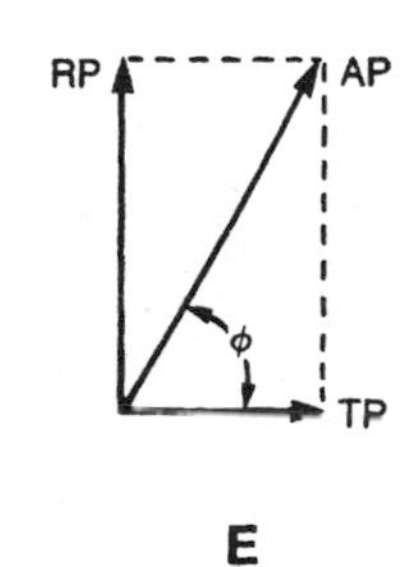

MATHEMATICAL DERIVATIONS

In terms of effective values:

$$I_R = \frac{E}{R}, E = I_R \times R, R = \frac{E}{I_R} \tag{77-1}$$

$$I_C = \frac{E}{X_C}, E = I_C \times X_C, X_C = \frac{E}{I_C} \tag{77-2}$$

$$I_S = \frac{E}{Z_T}, E = I_S \times Z_T, Z_T = \frac{E}{I_S} \tag{77-3}$$

Source current,

$$I_S = \sqrt{I_R^2 + I_C^2}, I_R = \sqrt{I_S^2 - I_C^2}, \tag{77-4}$$

$$I_C = \sqrt{I_S^2 - I_R^2} \text{ amperes}$$

Because $I_S^2 = I_R^2 + I_C^2$,

$$\left(\frac{E}{Z_T}\right)^2 = \left(\frac{E}{R}\right)^2 + \left(\frac{E}{X_C}\right)^2$$

This yields,

Total impedance,

$$Z_T = \frac{R \times X_C}{\sqrt{R^2 + X_C^2}} \text{ ohms} \tag{77-5}$$

Equation 77-5 represents the "product-over-sum" formula, when the "Pythagorean sum" is involved.

Because the conductance, $G = 1/R$, the capacitive susceptance, $B_C = 1/X_C$, and the admittance $Y_T = 1/Z_T$,

$$Y_T = \sqrt{G^2 + B_C^2}, G = \sqrt{Y_T^2 - B_C^2} \tag{77-6}$$

$$B_C = \sqrt{Y_T^2 - G^2} \text{ siemens}$$

True Power,

$$TP = E \times I_R = I_R^2 \times R = E^2/R = E^2G \text{ watts} \tag{77-7}$$

Reactive Power,

$$RP = E \times I_C = I_C^2 \times X_C = E^2/X_C \tag{77-8}$$

$$= E^2B_C \text{ volt-amperes reactive}$$

Apparent Power,

$$AP = E \times I_S = I_S^2 \times Z_T = E^2/Z_T \tag{77-9}$$

$$= E^2Y_T \text{ volt-amperes}$$

$$AP = \sqrt{TP^2 + RP^2}, TP = \sqrt{AP^2 - RP^2} \tag{77-10}$$

$$RP = \sqrt{AP^2 - TP^2}$$

Power factor,

$$PF = \frac{TP}{AP} = \cos\phi = \frac{E^2/R}{E^2/Z_T} \tag{77-11}$$

$$= \frac{Z_T}{R} \left(not\ \frac{R}{Z_T} \right), \text{ leading}$$

Phase angle,

ϕ = inv cos (PF) and is a negative angle (77-12)

Note that if $R = X_C$, $\phi = -45°$, and the power factor = 0.707, leading.

Example 77-1

In Fig. 77-2, calculate the values of I_R, I_C, I_S, Z_T, Y_T, true power, reactive power, apparent power, the power factor, and the phase angle ϕ.

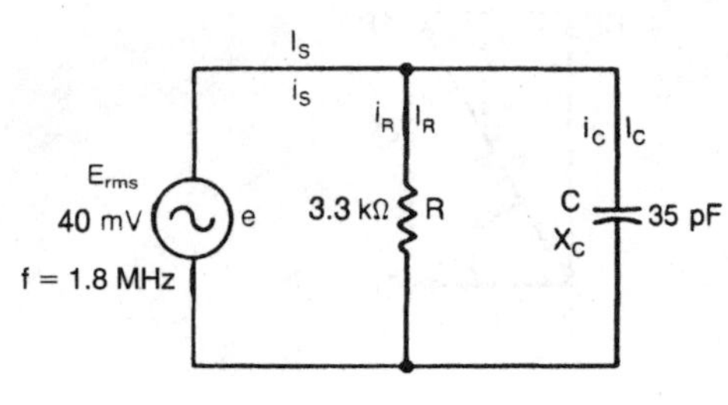

Figure 77-2

Solution

Capacitive reactance,

$$X_C = \frac{1}{2\pi fC}$$

$$= \frac{1}{2 \times \pi \times 1.8 \times 10^6 \times 35 \times 10^{-12}}\ \Omega$$

$$= 2.53\ \text{k}\Omega$$

Resistor current,

$$I_R = \frac{E}{R} = \frac{40\ \text{mV}}{3.3\ \text{k}\Omega} \qquad (77\text{-}1)$$

$$= 12.1\ \mu\text{A}$$

Capacitor current,

$$I_C = \frac{E}{X_C} = \frac{40\ \text{mV}}{2.53\ \text{k}\Omega} \qquad (77\text{-}2)$$

$$= 15.8\ \mu\text{A}$$

Source current,

$$I_S = \sqrt{I_R^2 + I_C^2} = \sqrt{12.1^2 + 15.8^2} \qquad (77\text{-}4)$$

$$= 19.9\ \mu\text{A}$$

Total impedance,

$$Z_T = \frac{E}{I_S} = \frac{40\ \text{mV}}{19.9\ \mu\text{A}} \qquad (77\text{-}3)$$

$$= 2.01\ \text{k}\Omega$$

Impedance check #1:

Total impedance,

$$Z_T = \frac{R \times X_C}{\sqrt{R^2 + X_C^2}} \qquad (77\text{-}5)$$

$$= \frac{3.3 \times 2.53}{\sqrt{3.3^2 + 2.53^2}}$$

$$= 2.01\ \text{k}\Omega$$

Impedance check #2:

Conductance,

$$G = \frac{1}{R} = \frac{1}{3.3\ \text{k}\Omega}$$

$$= 0.303\ \text{mS}$$

Capacitive susceptance,

$$B_C = \frac{1}{X_C} = \frac{1}{2.53\ \text{k}\Omega}$$

$$= 0.395\ \text{mS}$$

Admittance,

$$Y_T = \sqrt{G^2 + B_C^2} = \sqrt{0.303^2 + 0.395^2} \qquad (77\text{-}6)$$

$$= 0.498\ \text{mS}$$

Total impedance,

$$Z_T = \frac{1}{Y_T} = \frac{1}{0.498\ \text{mS}}$$

$$= 2.01\ \text{k}\Omega.$$

True Power,

$$TP = E \times I_R = 40\ \text{mV} \times 12.1\ \mu\text{A} \qquad (77\text{-}7)$$

$$= 0.484\ \mu\text{W}$$

Reactive power,

$$RP = E \times I_C = 40\ \text{mV} \times 15.8\ \mu\text{A} \qquad (77\text{-}8)$$

$$= 0.632\ \mu\text{VAr}$$

Apparent power,

$$AP = E \times I_S = 40\ \text{mV} \times 19.9\ \mu\text{A} \qquad (77\text{-}9)$$

$$= 0.796\ \mu\text{VA}$$

Power check:

$$AP = \sqrt{TP^2 + RP^2} = \sqrt{0.484^2 + 0.632^2} \qquad (77\text{-}10)$$

$$= 0.796\ \mu\text{VA}$$

$$\text{Power factor} = \frac{TP}{AP} = \frac{0.484\ \mu\text{W}}{0.796\ \mu\text{VA}} \qquad (77\text{-}11)$$

$$= 0.61, \text{ leading}$$

Phase angle,

$$\phi = \text{inv cos } 0.61 = -52.5° \qquad (77\text{-}12)$$

PRACTICE PROBLEMS

77-1. In Fig. 77-1A, $E = 10$ V rms, $f = 11$ kHz, $R = 2.7$ kΩ, $C = 0.004$ μF. Determine the values of I_R, I_C, I_S, Z_T, TP, RP, AP, the power factor, and the phase angle.

77-2. In Practice Problem 77-1, the frequency is halved. Calculate the new values of I_R, I_C, I_S, Z_T, G, B_C, Y_T, TP, RP, AP, the power factor, and the phase angle.

77-3. In Practice Problem 77-1, what is the frequency for which the power factor is 0.707, leading?

77-4. In Fig. 77-1A, $E = 110$ V rms, $f = 60$ Hz, $R = 470$ Ω, $C = 4$ μF. Obtain the values of I_R, I_C, I_S, Z_T, TP, RP, AP, the power factor, and the phase angle.

78
Sine-wave input voltage to *L* and *C* in parallel

Because L and C are in parallel, the source voltage (E) is across each of the components and the source (or supply) current (i_S) will be the phasor sum of the two branch currents, i_L and i_C (Figs. 78-1A and B). If you assume that the value of X_C is *greater* than the value of X_L, the capacitor current ($I_C = E/X_C$) will be *less* than the inductor current ($I_L = E/X_L$). Because i_C *leads* the source voltage by 90°, and i_L *lags* the same source voltage by 90°, i_C and i_L are 180° out-of-phase (Fig. 78-1B) and their phasors are pointing in opposite directions (Fig. 78-1C). The phasor sum of i_L and i_C is the source current (i_S), which will be in phase with i_L and will *lag* the source voltage by 90°. Therefore, the circuit behaves inductively, the power factor is *lagging* and the phase angle is *positive*. Notice that this is opposite to the result obtained from the series LC circuit of chapter 74 where if X_C was greater than X_L, the power factor was leading. This is because, in the

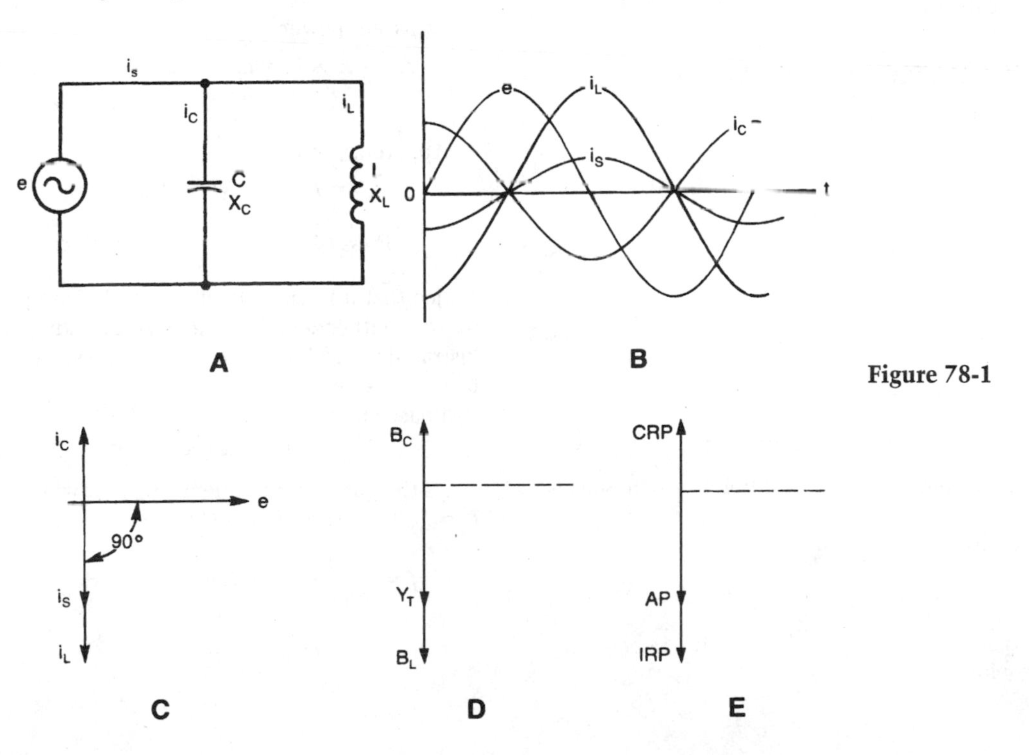

Figure 78-1

parallel circuit, you are concerned with the branch currents, which are *inversely* proportional to the reactances, but in the series circuit, you considered the component voltages, which were *directly* proportional to the reactances.

If X_L is greater than X_C, the capacitor current is greater than the inductor current; i_S is in phase with i_C and leads the source voltage by 90°. The power factor is then leading and the phase angle is negative.

The admittance, $Y_T (= I_S/E)$ is the reciprocal of the total impedance (Z_T) and is measured in siemens. Because $I_L = E \times B_L$, $I_C = E \times B_C$ and $I_S = E \times Y_T$, the admittance (Y_T) is the phasor sum of the inductive susceptance (B_L) and the capacitive susceptance, B_C (Fig. 78-1D).

Because the components are assumed to be ideal, there are no resistance losses associated with the circuit, and the true power (*TP*) is zero. The apparent power (*AP*) is the phasor sum of the inductive reactive power (*IRP*) and the capacitive reactive power (*CRP*) (Fig. 78-1E).

Notice at the particular frequency (f) for which the reactances are equal, the branch currents are also equal and, therefore, the supply current is zero. The total impedance is then infinite and the parallel combination behaves as an *open* circuit. This is theoretically possible because the power in the circuit is entirely reactive and there is no true power dissipated as heat.

MATHEMATICAL DERIVATIONS

In terms of effective values:

$$I_L = \frac{E}{X_L}, E = I_L \times X_L, X_L = \frac{E}{I_L} \quad (78\text{-}1)$$

$$I_C = \frac{E}{X_C}, E = I_C \times X_C, X_C = \frac{E}{I_C} \quad (78\text{-}2)$$

$$I_S = \frac{E}{Z_T}, E = I_S \times Z_T, Z_T = \frac{E}{I_S} \quad (78\text{-}3)$$

Source current,

$$I_S = I_L \sim I_C \text{ amperes} \quad (78\text{-}4)$$

For the meaning of the " ~ " difference sign, see chapter 74.

Because $I_S = I_L \sim I_C$,

$$\frac{E}{Z_T} = \frac{E}{X_L} \sim \frac{E}{X_C}$$

This yields:

Total impedance,

$$Z_T = \frac{X_L \times X_C}{X_L \sim X_C} \text{ ohms} \quad (78\text{-}5)$$

Equation 78-5 represents the "product-over-sum" formula when the "difference" sign is involved. From this formula, it follows that Z_T must be greater in value than either X_L, or X_C, or both X_L and X_C.

The inductive susceptance, $B_L = 1/X_L$, the capacitive susceptance, $B_C = 1/X_C$, and the admittance $Y_T = 1/Z_T$; then:

Admittance,

$$Y_T = B_L \sim B_C \text{ siemens} \quad (78\text{-}6)$$

True power,

$$TP = 0 \text{ watts} \quad (78\text{-}7)$$

Inductive reactive power,

$$IRP = E \times I_L = I_L^2 \times X_L \quad (78\text{-}8)$$
$$= E^2/X_L$$
$$= E^2 B_L \text{ volt-amperes reactive}$$

Capacitive reactive power,

$$CRP = E \times I_C = I_C^2 \times X_C \quad (78\text{-}9)$$
$$= E^2/X_C$$
$$= E^2 B_C \text{ volt-amperes reactive}$$

Apparent power,

$$AP = E \times I_S = I_S^2 Z_T = E^2/Z_T \quad (78\text{-}10)$$
$$= E^2 Y_T \text{ volt-amperes}$$

Apparent power,

$$AP = IRP \sim CRP \text{ volt-amperes} \quad (78\text{-}11)$$

$$\text{Power factor} = \frac{TP}{AP} = 0, \text{ lagging} \quad (78\text{-}12)$$

Notice that it is theoretically possible for a power factor to be either zero, lagging, or zero, leading. This is illustrated in Fig. 78-2, which shows the scale of a power factor meter.

Phase angle,

$$\phi = \text{inv cos } 0 = +90° \quad (78\text{-}13)$$

At the particular frequency (f) for which $X_L = X_C$, $I_L = I_C$, $I_S = 0$ and $Z_T = \infty\ \Omega$. Then from chapter 74,

$$f = \frac{1}{2\pi\sqrt{LC}} \text{ Hz}, L = \frac{1}{4\pi^2 f^2 C} \text{ H} \quad (78\text{-}14)$$

$$C = \frac{1}{4\pi^2 f^2 L} \text{ F}$$

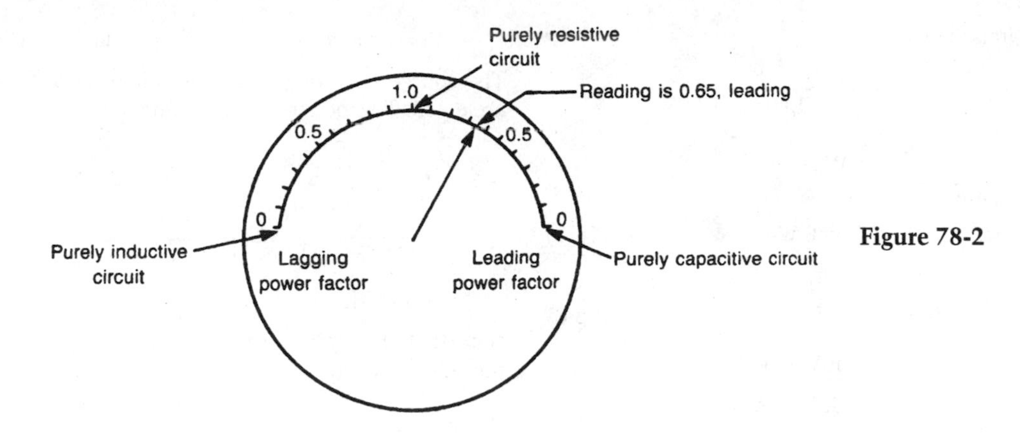

Figure 78-2

Example 78-1

In Fig. 78-3, calculate the values of I_L, I_C, I_S, Z_T, Y_T, TP, IRP, CRP, AP, the power factor, and the phase angle, ϕ. What is the particular frequency for which $I_S = 0$?

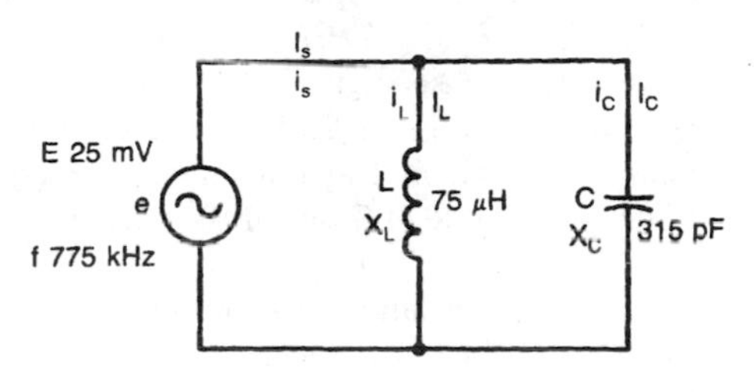

Figure 78-3

Solution

Inductive reactance,

$$X_L = 2\pi fL = 2 \times \pi \times 775 \times 10^3 \times 75 \times 10^{-6}$$

$$= 365\ \Omega.$$

Capacitive reactance,

$$X_C = \frac{1}{2\pi fC}$$

$$= \frac{1}{2 \times \pi \times 775 \times 10^3 \times 315 \times 10^{-12}}$$

$$= 652\ \Omega.$$

Inductor current,

$$I_L = \frac{E}{X_L} = \frac{25\text{ mV}}{365\ \Omega} \qquad (78\text{-}1)$$

$$= 68.50\ \mu\text{A}$$

Capacitor current,

$$I_C = \frac{E}{X_C} = \frac{25\text{ mV}}{652\ \Omega} \qquad (78\text{-}2)$$

$$= 38.34\ \mu\text{A}$$

Source current,

$$I_S = I_L - I_C - 68.50 - 38.34 \qquad (78\text{-}4)$$

$$= 30.16\ \mu\text{A}$$

Total impedance,

$$Z_T = \frac{E}{I_S} = \frac{25\text{ mV}}{30.16\ \mu\text{A}} \qquad (78\text{-}3)$$

$$= 829\ \Omega$$

To find the total impedance by the branch current method, you can assume any convenient value for the source voltage if none is given.

Impedance check #1:

Total impedance,

$$Z_T = \frac{X_L \times X_C}{X_L - X_C} = \frac{652 \times 365}{652 - 365} \qquad (78\text{-}5)$$

$$= 829\ \Omega$$

Impedance check #2:

Inductive susceptance,

$$B_L = \frac{1}{X_L} = \frac{1}{365\ \Omega}$$

$$= 2.740\text{ mS}$$

Capacitive susceptance,

$$B_C = \frac{1}{X_C} = \frac{1}{652\ \Omega}$$

$$= 1.534\text{ mS}$$

Admittance,

$$Y_T = B_L - B_C = 2.740 - 1.534 \qquad (78\text{-}6)$$

$$= 1.206\text{ mS}$$

Total impedance,

$$Z_T = \frac{1}{Y_T} = \frac{1}{1.206 \text{ mS}}$$

$$= 829\ \Omega$$

True power,

$$TP = 0 \text{ W} \tag{78-7}$$

Inductive reactive power,

$$IRP = E \times I_L \tag{78-8}$$

$$= 25 \text{ mV} \times 68.50\ \mu\text{A}$$

$$= 1.7125\ \mu\text{VAr}$$

Capacitive reactive power,

$$CRP = E \times I_C \tag{78-9}$$

$$= 25 \text{ mV} \times 38.34\ \mu\text{A}$$

$$= 0.9585\ \mu\text{VAr}$$

Apparent power,

$$AP = E \times I_S = 25 \text{ mV} \times 30.16\ \mu\text{A} \tag{78-10}$$

$$= 0.7540\ \mu\text{VAr}$$

Power check:

$$AP = IRP - CRP = 1.7125 - 0.9585 \tag{78-11}$$

$$= 0.7540\ \mu\text{VAr}$$

$$\text{Power factor} = \frac{TP}{AP} = 0, \text{ lagging} \tag{78-12}$$

$$\text{Phase angle, } \phi = \text{inv cos } 0 = +90° \tag{78-13}$$

The source current (I_S) is zero when the reactances are equal. This occurs at the frequency,

$$f = \frac{1}{2\pi\sqrt{LC}} \tag{78-14}$$

$$= \frac{1}{2 \times \pi \times \sqrt{75 \times 10^{-6} \times 315 \times 10^{-12}}} \text{ Hz}$$

$$= 1.03 \text{ MHz}$$

Note that the Pythagorean theorem was not used in the solution of this example.

PRACTICE PROBLEMS

78-1. In Fig. 78-1A, $E = 10$ V rms, $f = 11$ kHz, $L = 40$ mH, $C = 0.004\ \mu$F. Determine the values of I_L, I_C, I_S, Z_T, TP, RP, AP, the power factor, and the phase angle.

78-2. In Practice Problem 78-1, the frequency is increased to 17 kHz. Obtain the new values of I_L, I_C, I_S, Z_T, B_L, B_C, Y_T, TP, RP, AP, the power factor, and the phase angle.

78-3. In Practice Problem 78-1, calculate the frequency for which the reactances are equal. What is the value of Z_T at this frequency?

78-4. In Fig. 78-1A, $E = 110$ V rms, $f = 60$ Hz, $L = 4$ H, $C = 2\ \mu$F. Determine the values of I_L, I_C, I_S, Z_T, TP, RP, AP, the power factor, and the phase angle.

79
Sine-wave input voltage to *R*, *L*, and *C* in parallel

This circuit must contain all the information of chapters 76, 77, and 78 because if (for example) the capacitor were eliminated, you would be left with R and L in parallel (chapter 76). Because all three components are in parallel across the source voltage, the source current (i_S) will be the phasor sum of the three branch currents, i_R, i_L, and i_C (Figs. 79-1A and B). If you assume that the inductive reactance is greater than the capacitive reactance, and that the value of the resistance is relatively large, the capacitor current ($I_C = E/X_C$) will be greater than the inductor current ($I_L = E/X_L$), and the resistor current ($I_R = E/R$) will be small. Because i_C *leads* the source voltage by 90°, and i_L *lags* the same source voltage by 90°, i_C and i_L are 180° out-of-phase (Fig. 79-1B) and their phasors are pointing in opposite directions (Fig. 79-1C). The phasor sum of i_L and i_C is the total reactive current (i_X), which will be in phase with i_C. The resistor current (i_R) is then combined with i_X to produce the source current (i_S), which *leads* the source voltage so that the power factor is *leading* and the phase angle (ϕ) is negative.

If X_C is greater than X_L, the inductor current is greater than the capacitor current; i_X is in phase with i_L and i_S *lags* the source voltage by the *positive* phase angle, ϕ. The power factor is then *lagging*.

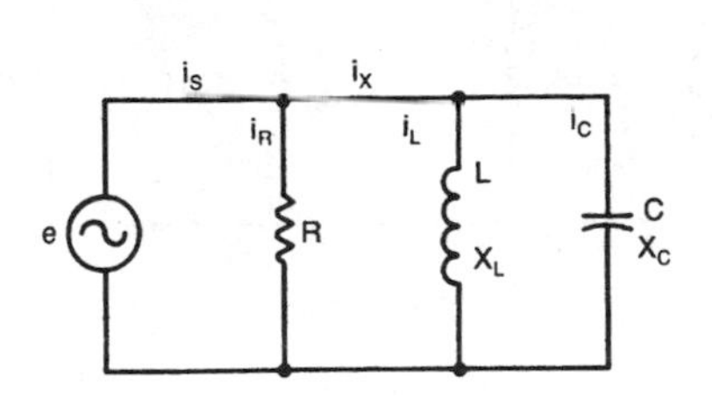

A

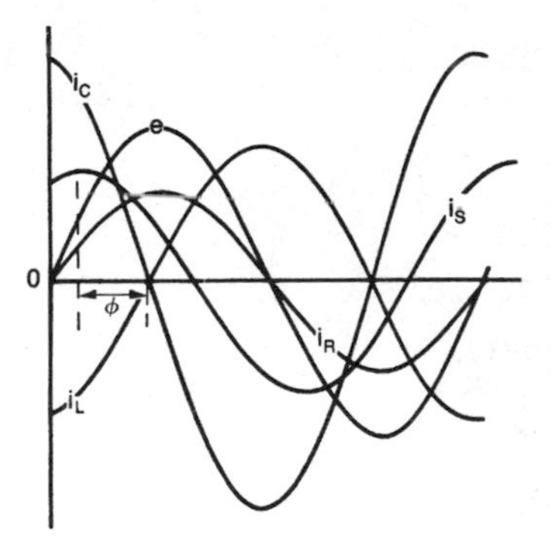

B

Figure 79-1

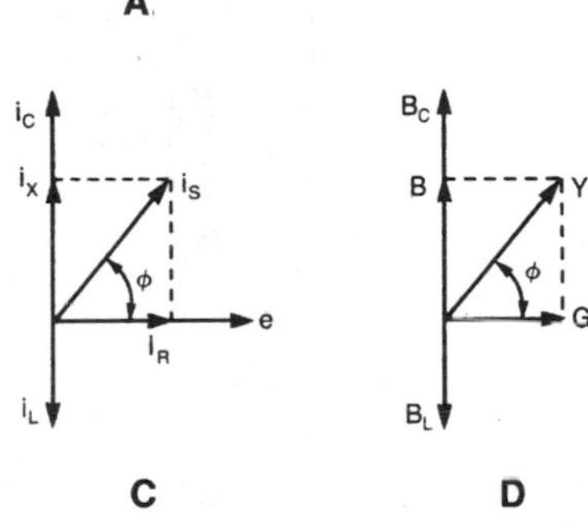

C

D

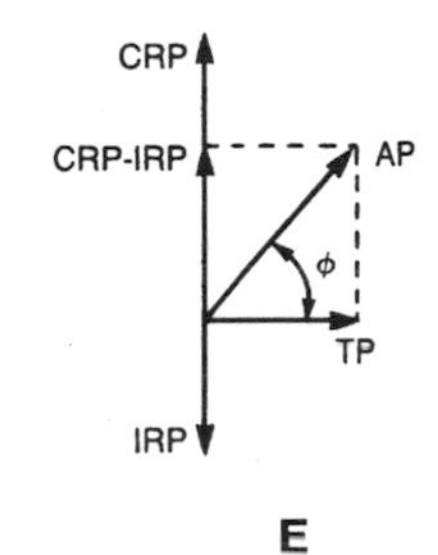

E

The admittance $Y_T (= I_S/E)$ is the reciprocal of the total impedance (Z_T)and is measured in siemens. Because $I_R = E \times G$, $I_L = E \times B_L$, $I_C = E \times B_C$ and $I_S = E \times Y_T$, the admittance (Y_T) is the phasor sum of the conductance (G), the inductive susceptance (B_L), and the capacitive susceptance, B_C (Fig. 79-1D). In a similar way, the apparent power (AP) is the phasor sum of the true power (TP), the inductive reactive power (IRP), and the capacitive reactive power (CRP) (Fig. 79-1E).

At the particular frequency (f) for which the reactances are equal, the inductor and the capacitor currents cancel out (the parallel LC combination behaves as an open circuit), and the source current is the same as the resistor current. The total impedance is then equal in value to the resistance. At frequencies above and below f, the parallel LC combination behaves as a certain reactance value which, when placed in parallel with the resistance, produces a total impedance that is less the value of the resistance. When these two statements are combined, it means that the total impedance cannot be higher than the value of the resistance.

MATHEMATICAL DERIVATIONS

In terms of effective values:

$$I_R = \frac{E}{R}, E = I_R \times R, R = \frac{E}{I_R} \quad (79\text{-}1)$$

$$I_L = \frac{E}{X_L}, E = I_L \times X_L, X_L = \frac{E}{I_L} \quad (79\text{-}2)$$

$$I_C = \frac{E}{X_C}, E = I_C \times X_C, X_C = \frac{E}{I_C} \quad (79\text{-}3)$$

$$I_S = \frac{E}{Z_T}, E = I_S \times Z_T, Z_T = \frac{E}{I_S} \quad (79\text{-}4)$$

Total reactive current, $I_X = I_L \sim I_C$ amperes (79-5)

For the meaning of the " ~" difference sign, see chapter 74.

Source current,

$$I_S = \sqrt{I_R^2 + (I_L \sim I_C)^2} \text{ amperes} \quad (79\text{-}6)$$

Total reactance,

$$X_T = \frac{X_L \times X_C}{X_L \sim X_C} \text{ ohms} \quad (78\text{-}5, 79\text{-}7)$$

Total impedance,

$$Z_T = \frac{R \times X_T}{\sqrt{R^2 + X_T^2}} \text{ ohms} \quad (76\text{-}5, 77\text{-}5, 79\text{-}8)$$

Because $I_R = EG$, $I_L = EB_L$, $I_C = EB_C$ and $I_S = EY_T$,

Admittance,

$$Y_T = \sqrt{G^2 + (B_L \sim B_C)^2} \text{ siemens} \quad (79\text{-}9)$$

True power,

$$TP = E \times I_R \tag{79-10}$$
$$= I_R^2 \times R$$
$$= E^2/R$$
$$= E^2G \text{ watts}$$

Inductive reactive power,

$$IRP = E \times I_L \tag{79-11}$$
$$= I_L^2 \times X_L$$
$$= E^2/X_L$$
$$= E^2B_L \text{ volt-amperes reactive}$$

Capacitive reactive power,

$$CRP = E \times I_C \tag{79-12}$$
$$= I_C^2 \times X_C$$
$$= E^2/X_C$$
$$= E^2B_C \text{ volt-amperes reactive}$$

Apparent power,

$$AP = E \times I_S \tag{79-13}$$
$$= I_S^2 \times Z_T$$
$$= E^2/Z_T$$
$$= E^2Y_T \text{ volt-amperes}$$

Apparent power,

$$AP = \sqrt{TP^2 + (IRP \sim CRP)^2} \text{ volt-amperes} \tag{79-14}$$

Power factor,

$$PF = \frac{TP}{AP} = \cos\phi = \frac{Z_T}{R}, \text{ leading} \tag{79-15}$$

Phase angle,

$$\phi = \text{inv cos (power factor)} \tag{79-16}$$
and is a negative angle

Example 79-1

In Fig. 79-2, calculate the values of I_R, I_L, I_C, I_S, Z_T, Y_T, TP, IRP, CRP, AP, the power factor, and the phase angle.

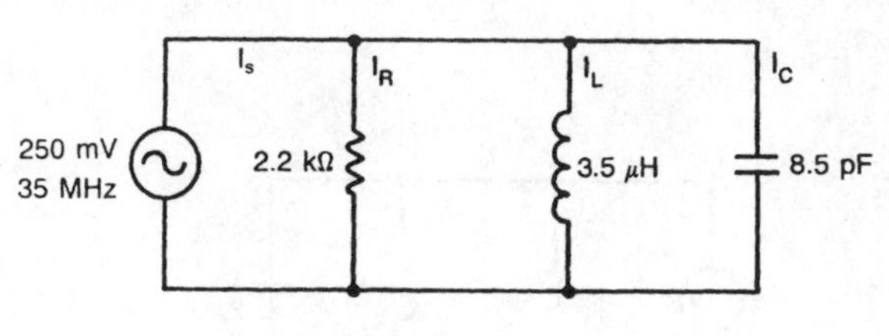

Figure 79-2

Solution

Inductive reactance,

$$X_L = 2\pi fL = 2 \times \pi \times 35 \times 10^6 \times 3.5 \times 10^{-6}$$
$$= 770\ \Omega.$$

Capacitive reactance,

$$X_C = \frac{1}{2\pi fC}$$
$$= \frac{1}{2 \times \pi \times 35 \times 10^6 \times 8.5 \times 10^{-12}}$$
$$= 535\ \Omega.$$

Resistor current,

$$I_R = \frac{E}{R} = \frac{250 \text{ mV}}{2.2 \text{ k}\Omega} \tag{79-1}$$
$$= 113.6\ \mu\text{A}$$

Inductor current,

$$I_L = \frac{E}{X_L} = \frac{250 \text{ mV}}{770\ \Omega} \tag{79-2}$$
$$= 324.7\ \mu\text{A}$$

Capacitor current,

$$I_C = \frac{E}{X_C} = \frac{250 \text{ mV}}{535\ \Omega} \tag{79-3}$$
$$= 467.3\ \mu\text{A}$$

Total reactive current,

$$I_X = I_C - I_L\ 467.3 - 324.7 \tag{79-5}$$
$$= 142.6\ \mu\text{A}$$

Source current,

$$I_S = \sqrt{I_R^2 + I_X^2} = \sqrt{113.6^2 + 142.6^2} \tag{79-6}$$
$$= 182.3\ \mu\text{A}$$

Total impedance,

$$Z_T = \frac{E}{I_S} = \frac{250 \text{ mV}}{182.3\ \mu\text{A}} \tag{79-4}$$
$$= 1.37 \text{ k}\Omega$$

Impedance check #1:

Total reactance,

$$X_T = \frac{X_L \times X_C}{X_L - X_C} \tag{79-7}$$
$$= \frac{770 \times 535}{770 - 535}\ \Omega$$
$$= 1.753 \text{ k}\Omega$$

Total impedance,

$$Z_T = \frac{R \times X_T}{\sqrt{R^2 + X_T^2}} \quad (79\text{-}8)$$

$$= \frac{2.2 \times 1.753}{\sqrt{2.2^2 + 1.753^2}}$$

$$= 1.37 \text{ k}\Omega$$

Impedance check #2:

Conductance,

$$G = \frac{1}{2.2 \text{ k}\Omega} = 0.4545 \text{ mS}.$$

Inductive susceptance,

$$B_L = \frac{1}{770 \ \Omega} = 1.299 \text{ mS}.$$

Capacitive susceptance,

$$B_C = \frac{1}{535 \ \Omega} = 1.869 \text{ mS}.$$

Admittance,

$$Y_T = \sqrt{G^2 + (B_C - B_L)^2} \quad (79\text{-}9)$$

$$= \sqrt{0.4545^2 + (1.869 - 1.299)^2}$$

$$= 0.729 \text{ mS}.$$

Impedance,

$$Z_T = \frac{1}{Y_T} = \frac{1}{0.729 \text{ mS}}$$

$$= 1.37 \text{ k}\Omega.$$

True power,

$$TP = E \times I_R = 250 \text{ mV} \times 113.6 \ \mu\text{A} \quad (79\text{-}10)$$

$$= 28.4 \ \mu\text{W}.$$

Inductive reactive power,

$$IRP = E \times I_L = 250 \text{ mV} \times 324.7 \ \mu\text{A} \quad (79\text{-}11)$$

$$= 81.2 \ \mu\text{VAr}.$$

Capacitive reactive power,

$$CRP = E \times I_C = 250 \text{ mV} \times 467.3 \ \mu\text{A} \quad (79\text{-}12)$$

$$= 116.8 \ \mu\text{VAr}.$$

Apparent power,

$$AP = E \times I_S = 250 \text{ mV} \times 182.3 \ \mu\text{A} \quad (79\text{-}13)$$

$$= 45.58 \ \mu\text{VA}.$$

Power check:

$$AP = \sqrt{TP^2 + (CRP - IRP)^2} \quad (79\text{-}14)$$

$$= \sqrt{28.4^2 + (116.8 - 81.2)^2}$$

$$= 45.54 \ \mu\text{VA}.$$

Power factor,

$$PF = \frac{TP}{AP} = \frac{28.4}{45.58} \quad (79\text{-}15)$$

$$= 0.62, \text{ leading}.$$

Phase angle,

$$\phi = \text{inv cos } 0.62 = -51.5°. \quad (79\text{-}16)$$

PRACTICE PROBLEMS

79-1. In Fig. 79-1A, $E = 10$ V rms, $f = 15$ kHz, $R = 8.2$ kΩ, $L = 40$ mH, $C = 0.004$ μF. Obtain the values of I_R, I_L, I_C, I_X, I_S, Z_T, TP, RP, AP, the power factor, and the phase angle.

79-2. In Practice Problem 79-1, the frequency is increased to 17 kHz. Calculate the new values of I_R, I_L, I_C, I_X, I_S, Z_T, TP, RP, AP, the power factor, and the phase angle.

79-3. In Practice Problem 79-1, determine the frequency for which the inductive reactance is equal to the capacitive reactance. What is the value of Z_T at this frequency?

79-4. In Fig. 79-1A, $E = 110$ V rms, $f = 60$ Hz, $R = 8.2$ kΩ, $L = 4$ H, $C = 2$ μF. Calculate the values of I_R, I_L, I_C, I_S, Z_T, TP, RP, AP, the power factor, and the phase angle.

80
Maximum power transfer to the load (ac case)

Chapter 23 showed that, in a dc circuit, there is maximum power transfer to the load when its resistance is matched (made equal to) to the sum of all resistances not associated with the load. In the ac circuit, there is the added complication of reactance, which limits the flow of current, but does not dissipate any true power.

There are two separate ac cases to consider. In Fig. 80-1A, the ac voltage source has an internal impedance with a resistive component, and a reactive component that may be either inductive or capacitive. The mathematical derivations then show that the load power is at maximum when the *load resistance* is varied to match the *magnitude* of the source's internal impedance.

The second case is illustrated in Fig. 80-1B. Both the internal impedance and the load impedance have resistive and reactive components that are either inductive or capacitive. Maximum power to the load is then achieved by a two-stage procedure.

1. The load reactance is varied until its *magnitude* is equal to the *magnitude* of the source's internal reactance. However, if the source reactance is inductive, the load reactance must be capacitive, and vice-versa. As a result, the net reactance of the complete circuit is reduced to zero.
2. Once the load reactance cancels the internal reactance, the remaining load resistance is matched (made equal) to the source resistance.

MATHEMATICAL DERIVATIONS

In Fig. 80-1A

In the following equations, all the voltages and currents are effective values,

Magnitude of the total impedance of the circuit,

$$Z_T = \sqrt{(R_S + R_L)^2 + X_S^2} \text{ ohms} \qquad (80\text{-}1)$$

Load current,

$$I_L = \frac{E}{Z_T} = \frac{E}{\sqrt{(R_S + R_L)^2 + X_S^2}} \text{ amperes} \qquad (80\text{-}2)$$

Load power,

$$P_L = I_L^2 \times R_L = \frac{E^2 R_L}{(R_S + R_L)^2 + X_S^2} \text{ volts} \qquad (80\text{-}3)$$

For the load power to reach its maximum value,

$$\frac{dP_L}{dR_L} = E^2\left[\frac{(R_S + R_L)^2 + X_S^2 - R_L[2(R_S + R_L)]}{[(R_S + R_L)^2 + X_S^2]^2}\right] = 0$$

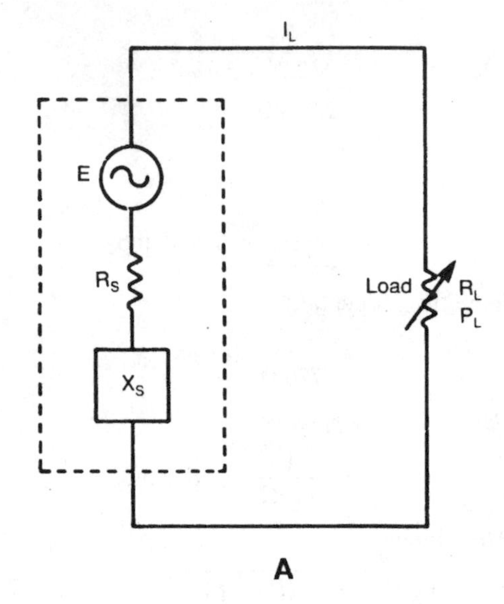

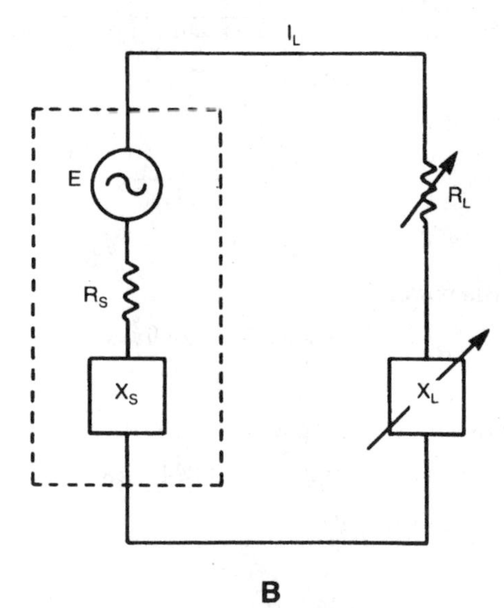

Figure 80-1

It follows that,

$$(R_S + R_L)^2 + X_S^2 = 2R_L R_S + 2R_L^2 \qquad (80\text{-}4)$$

$$R_L = \sqrt{R_S^2 + X_S^2} \text{ ohms}$$

Therefore, the load resistance is matched to the magnitude of the internal impedance.

From Equations 80-3, 80-4,

Value of the maximum power,

$$P_{L\max} = \frac{E^2R_L}{(R_S + R_L)^2 + X_S^2} \tag{80-5}$$

$$= \frac{E^2R_L}{(R_S + R_L)^2 + (R_L^2 - R_S^2)}$$

$$= \frac{E^2R_L}{2(R_L^2 + R_LR_S)}$$

$$= \frac{E^2}{2(R_L + R_S)} \text{ watts}$$

In Fig. 80-1B

Magnitude of the total circuit impedance,

$$Z_T = \sqrt{(R_L + R_S)^2 + (X_L + X_S)^2} \text{ ohms} \tag{80-6}$$

where X_L and X_S can be either inductive or capacitive. The load current is,

$$I_L = \frac{E}{Z_T} \tag{80-7}$$

$$= \frac{E}{\sqrt{(R_L + R_S)^2 + (X_L + X_S)^2}} \text{ amperes}$$

The power in the load is,

$$P_L = I_L^2R_L = \frac{E^2R_L}{(R_L + R_S)^2 + (X_L + X_S)^2} \text{ watts} \tag{80-8}$$

If X_L is varied, P_L will reach its maximum value when $X_L + X_S = 0$ or $X_L = -X_S$. The load reactance (X_L) will then have the same magnitude as X_S, but the two reactances are required to cancel so that, if, for example, X_S is inductive, X_L is made capacitive, and vice-versa.

After the reactance in the circuit has been reduced to zero,

$$P_L = \frac{E^2R_L}{(R_L + R_S)^2} \text{ watts} \tag{80-9}$$

The load power (P_L) will then have its maximum value when R_L is matched to R_S:

$$P_{L\max} = \frac{E^2}{4R_S} = \frac{E^2}{4R_L} \text{ watts} \tag{80-10}$$

and the corresponding load current is:

$$I_L = \frac{E}{2R_S} \text{ amperes} \tag{80-11}$$

When $R_L = R_S$ and $X_L = -X_S$, the load and internal impedances are said to be conjugates (chapter 91).

Example 80-1

In Fig. 80-1A, $R_S = 2$ kΩ, $X_S = 3$ kΩ and $E = 10$ V rms. Determine the load power when R_L is (a) 2 kΩ, and (b) 3 kΩ. What is the value of R_S for which the load power is a maximum? Calculate the amount of the maximum load power.

Solution

(a) Total impedance,

$$Z_T = \sqrt{(R_S + R_L)^2 + X_S^2} \tag{80-1}$$

$$= \sqrt{(2 + 2)^2 + 3^2}$$

$$= 5 \text{ k}\Omega.$$

Load current,

$$I_L = \frac{E}{Z_T} = \frac{10 \text{ V}}{5 \text{ k}\Omega} \tag{80-2}$$

$$= 2 \text{ mA}.$$

Load power,

$$P_L = I_L^2 \times R_L = (2 \text{ mA})^2 \times 2 \text{ k}\Omega \tag{80-3}$$

$$= 8 \text{ mW}.$$

(b) Total impedance,

$$Z_T = \sqrt{(R_L + R_S)^2 + X_S^2} \tag{80-1}$$

$$= \sqrt{(2 + 3)^2 + 3^2}$$

$$= 5.83 \text{ k}\Omega.$$

Load current,

$$I_L = \frac{10 \text{ V}}{5.83 \text{ k}\Omega} \tag{80-2}$$

$$= 1.71 \text{ mA}.$$

Load power,

$$P_L = I_L^2 \times R_L = (1.71 \text{ mA})^2 \times 3 \text{ k}\Omega \tag{80-3}$$

$$= 8.8 \text{ mW}.$$

The load power is a maximum when:

$$R_L = \sqrt{R_S^2 + X_S^2} = \sqrt{2^2 + 3^2} \tag{80-4}$$

$$= 3.61 \text{ k}\Omega.$$

Total load impedance,

$$Z_T = \sqrt{(R_L + R_S)^2 + X_S^2} \tag{80-1}$$

$$= \sqrt{(2 + 3.61)^2 + 3^2}$$

$$= 6.36 \text{ k}\Omega.$$

Load current,

$$I_L = \frac{10 \text{ V}}{6.36 \text{ k}\Omega} \tag{80-2}$$

$$= 1.57 \text{ mA}.$$

Maximum load power,

$$P_{L\max} = I_L^2 \times R_L \qquad (80\text{-}3)$$

$$= (1.57 \text{ mA})^2 \times 3.61 \text{ k}\Omega$$

$$= 8.89 \text{ mW}.$$

Example 80-2

In Fig. 80-1B, E = 18 V rms and R_S = 4 kΩ. The source reactance (X_S) is capacitive and equal to 5 kΩ. Determine the values of R_L and X_L that will allow maximum power transfer to the load and calculate the amount of the maximum load power.

Solution

For maximum power transfer to the load, the load impedance is the conjugate of the source impedance. Therefore, R_L = 4 kΩ and X_L is an inductive reactance of 5 kΩ.

Maximum load power,

$$P_{L\max} = \frac{E^2}{4R_L} \text{ watts} \qquad (80\text{-}10)$$

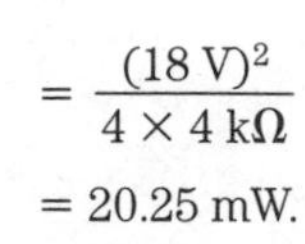

$$= \frac{(18 \text{ V})^2}{4 \times 4 \text{ k}\Omega}$$

$$= 20.25 \text{ mW}.$$

PRACTICE PROBLEMS

80-1. In Fig. 80-1A, E = 12 V rms, R_S = 1 kΩ, X_S = 2 kΩ (capacitive reactance). What is the value of R_L for which the power is a maximum? Calculate the amount of the maximum load power.

80-2. In Practice Problem 80-1, the frequency of the source is doubled. Recalculate the required value of R_L, and the corresponding amount of the maximum load power.

80-3. In Fig. 80-1B, E = 15 V rms, R_S = 3 kΩ. The source reactance (X_S) is inductive and equal to 2 kΩ. Determine the values of R_L and X_{load}, which will allow maximum power transfer to the load, and calculate the amount of the maximum load power.

80-4. In Practice Problem 80-3 the frequency of the source is doubled. Recalculate the required values of R_L and X_{load}, and the corresponding amount of the maximum load power.

81 Resonance in a series LCR circuit

The condition of electrical resonance in *all* circuits is defined as follows: *any two-terminal* (single source) network containing resistance and reactance is said to be in *resonance* when the source voltage and the current drawn from the source are in phase.

It follows from this definition that a resonant circuit has *a phase angle of zero* and *a power factor of unity*.

If the series LCR circuit of Fig. 81-1A is at resonance, the values of the inductive reactance and the capacitive reactance must be equal. Therefore, the phasor sum of v_L and v_C is zero (Figs. 81-1B and C) so that $e = v_R$ and the circuit is purely resistive. The total impedance of the circuit is equal to the resistance, and is at its minimum value (Fig. 81-1D); for this reason the series resonant LCR combination is sometimes referred to as an *acceptor circuit*. It follows that, at resonance, the circuit current is at its maximum value, which is equal to E/R amperes.

Because the values of the inductive reactance and the capacitive reactance are both dependent on the frequency, there must be a particular resonant frequency for which the two reactances are equal. The manner in which the behavior of the series LCR circuit varies with frequency, is illustrated by means of response curves (Figs. 81-2A and B). These are the graphs of certain variables (such as impedance, current, voltages across inductor, capacitor, etc.) versus frequency. Such response curves are important because they show the circuit's ability to distinguish between one frequency and another.

Tuning a series LCR circuit means adjusting the value of the inductor or the capacitor, until the resonant frequency is the same as the desired signal frequency. Take the amplitude modulation (AM) broadcast band as an example. Each station is assigned a particular operating frequency, while the frequencies of the nearest stations on either side are 10 kHz away. Literally, hun-

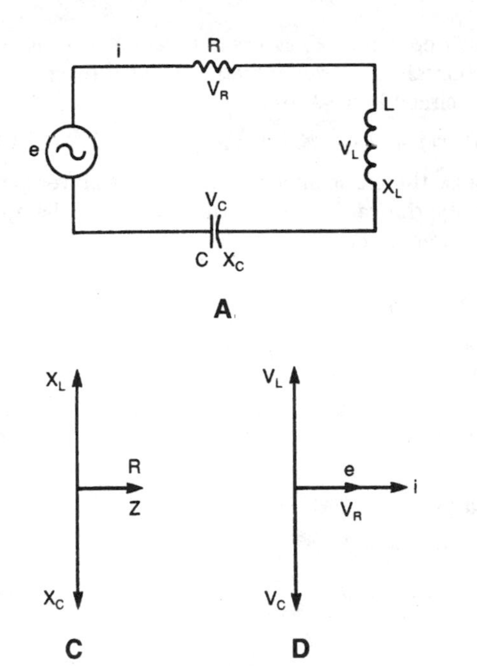

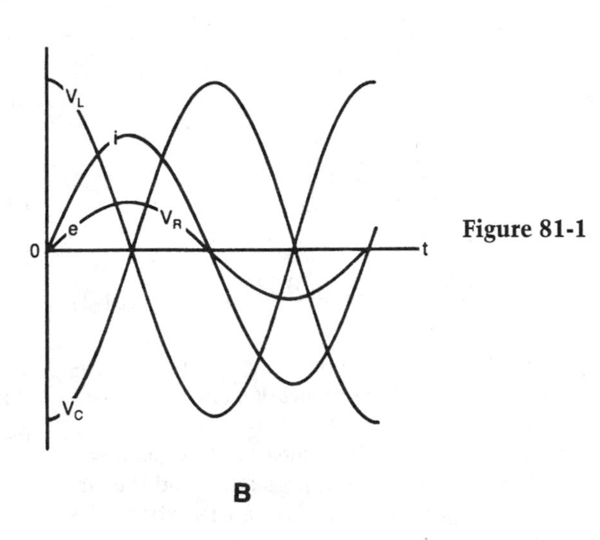

Figure 81-1

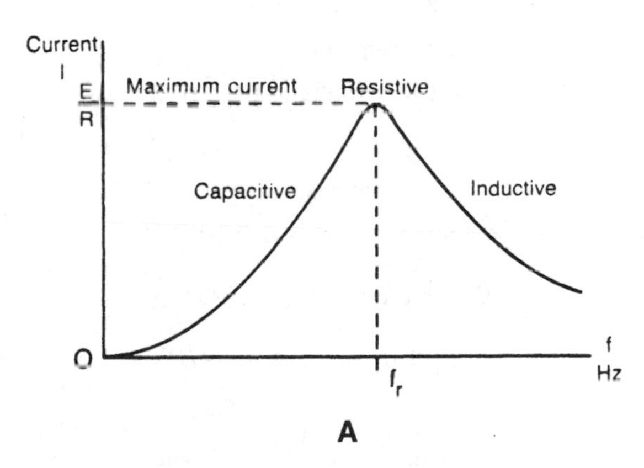

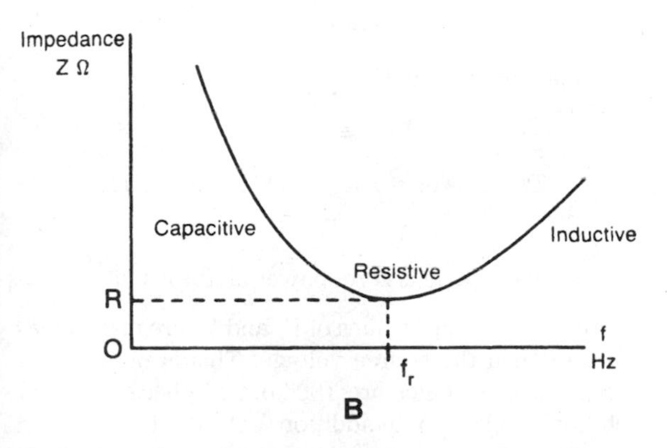

Figure 81-2

dreds of signals from all the radio waves in the vicinity are induced in the antenna (Figs. 81-3A and B) of an AM receiver. The purpose of the tuned circuit is to provide maximum response at the frequency of the wanted signal, but much smaller responses at the other unwanted signals. This is achieved by adjusting the capacitor until the resonant frequency is equal to the assigned frequency of the desired station.

Note that the circuit of Fig. 81-3B is a series arrangement because the wanted signal induced in the antenna drives a current through the coil L1. The alternating magnetic flux surrounding L1 cuts the other coil so that an RF (radio frequency) voltage is induced in L2. This voltage is within the loop containing the coil L2 and the capacitor C so that these components and their source voltage are in series.

MATHEMATICAL DERIVATIONS

At resonance,

Phase angle, ϕ, is zero.

Phasor $e = v_R$.

Phasor sum of v_L and v_C is zero.

Inductive reactance, X_L = capacitive reactance, X_C.

Inductor voltage, V_L = capacitor voltage, V_C.

Impedance (Z) is equal in value to the resistance (R) and is at its *minimum* level.

Current (I) is equal to E/R and is at its *maximum* level.

Because,

$$X_L = X_C,$$

$$2\pi f_r L = \frac{1}{2\pi f_r C}$$

This yields:

Resonant frequency,

$$f_r = \frac{1}{2\pi\sqrt{LC}} = \frac{0.159}{\sqrt{LC}} \text{ Hz} \qquad (81\text{-}1)$$

Then,

$$L = \frac{0.0253}{f_r^2 C} \text{ H and } C = \frac{0.0253}{f_r^2 L} \text{ F} \qquad (81\text{-}2)$$

Notice that the value of the resonant frequency is *inversely* dependent on the product of L and C, but is independent of R.

At frequencies below the value of the resonant frequency (X_C is greater than X_L), i leads e and the circuit behaves capacitively. At frequencies above the resonant frequency (X_L is greater than X_C), i lags e and the circuit is overall inductive. Only at resonance does the circuit behave resistively.

$$\text{True Power} = I^2R = E \times I = E^2/R \text{ watts} \qquad (81\text{-}3)$$

Because the phase angle is zero and the power factor is unity, the values of the true power and the apparent power are equal.

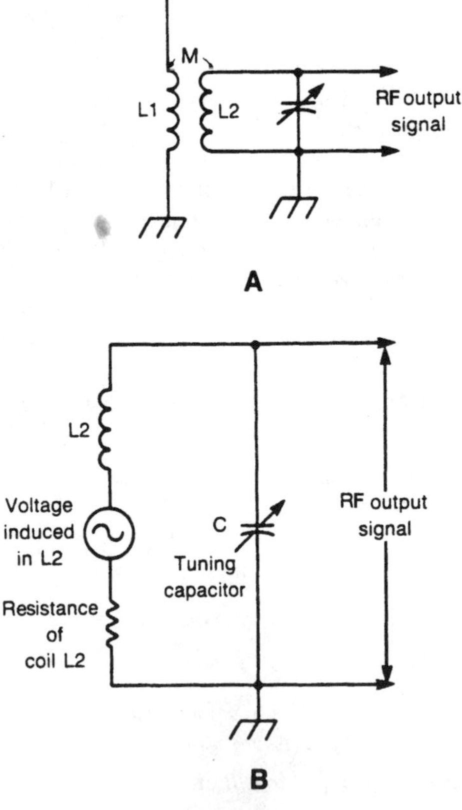

Figure 81-3

Example 81-1

In Fig. 81-1, $R = 10\ \Omega$, $L = 150\ \mu\text{H}$, $C = 250$ pF, and $E = 2$ V rms. Determine the value of the resonant frequency and calculate the resonant values of Z, I, V_R, V_L, and V_C.

Solution

Resonant frequency,

$$f_r = \frac{1}{2\pi\sqrt{LC}} \text{ Hz} \qquad (81\text{-}1)$$

$$= \frac{1}{2 \times \pi \times \sqrt{150 \times 10^{-6} \times 250 \times 10^{-12}}} \text{ Hz}$$

$$= 822 \text{ kHz.}$$

At the resonant frequency,

$$X_L = X_C = 2 \times \pi \times 822 \times 10^3 \times 150 \times 10^{-6}$$

$$= 775\ \Omega.$$

When the circuit is at resonance,

$$\text{Impedance, } Z = R = 10\ \Omega.$$

$$\text{Current, } I = \frac{2 \text{ V}}{10\ \Omega} = 200 \text{ mA.}$$

$$\text{Resistor voltage, } V_R = E$$

$$= 2 \text{ V.}$$

$$\text{Inductor voltage, } V_L = I \times X_L = 200 \text{ mA} \times 775\ \Omega$$

$$= 155 \text{ V.}$$

$$\text{Capacitor voltage, } V_C = V_L$$

$$= 155 \text{ V.}$$

$$\text{True Power} = I^2R = (200 \text{ mA})^2 \times 10\ \Omega$$

$$= 0.4 \text{ W.}$$

Phase angle is zero, power factor is unity.

Note that the equal values of V_L and V_C are many times greater than the source voltage. This is possible because these voltages are 180° out of phase and their phasor sum is zero. In addition, both the inductor and the capacitor are reactive components that only store power and do not dissipate power.

PRACTICE PROBLEMS

81-1. In Fig. 81-1A, $E = 3$ V rms, $f = 715$ kHz, $R = 15$ Ω, $L = 120$ μH. What value of C is required to make the circuit resonant?

81-2. In Practice Problem 81-1, assume that the circuit is at resonance. Calculate the resonant values of X_L, X_C, Z, I, V_R, V_L, V_C, TP, RP, AP, the power factor, and the phase angle.

81-3. In a series LCR circuit $R = 12$ Ω, $C = 24$ pF. What value of inductance is required to bring the circuit to resonance at a frequency of 1.37 MHz?

81-4. In Practice Problem 81-3, the value of the capacitance is changed to 37 pF. If the circuit is to resonate at the same frequency, determine the new value of the required inductance.

82
Q—selectivity—bandwidth

When a series LCR circuit is at resonance, the current has its maximum possible value for a given source voltage. Developed across the inductor and the capacitor are equal but 180° out-of-phase voltages. These voltages can each be many times greater than the source voltage (refer to Example 81-1). The number of times greater is called the *voltage magnification factor*, which is referred to as Q. The Q factor is just a number and has no units. However, it is a measure of the inductor's merit, in the sense that a "good" coil will have a high value of inductive reactance compared with its resistance.

The main importance of Q is its indication of a series tuned circuit's *selectivity*; this is defined as its ability to distinguish between the signal frequency to which it is resonant, and other unwanted signals on nearby frequencies. It follows that the higher the selectivity, the greater is the freedom from adjacent channel interference. The degree of the selectivity is related to the sharpness of the current response curve (the sharper the curve, the greater the selectivity), and may be measured by the frequency separation between two specific points on the curve (Fig. 82-1). The arbitrarily chosen points are those for which the true power in the circuit is one half of the maximum true power, which occurs when the circuit is resonant. These positions on the response curve are also referred to as the *3-decibel (dB) points* because a loss of 3 dB is equivalent to a power ratio of one half. The frequency separation between these points is called the *bandwidth* (or *bandpass*) of the tuned circuit.

At the 3-dB points, the rms circuit current will be $1/\sqrt{2}$ or 0.707 times the maximum rms value of the current at resonance (do not confuse this result with the relationship between the rms and the peak values of a sinewave alternating current). In addition, the overall reactance at the 3-dB points is equal to the circuit's resistance, so that the phase angle is 45°, and the power factor is cos 45° = 0.707.

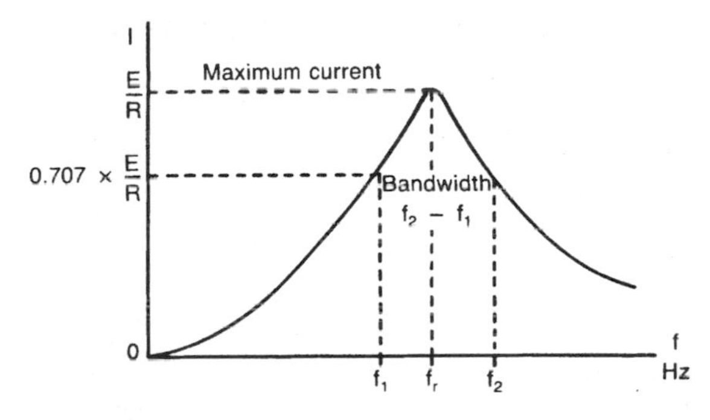

Figure 82-1

The narrower the bandwidth, and the higher the resonant frequency, the sharper is the response curve, and the greater is the selectivity. The mathematical derivations will show that the Q value is a direct measure of the selectivity.

MATHEMATICAL DERIVATIONS

Voltage magnification factor

At resonance,

$$Q = \frac{V_L}{E} = \frac{V_C}{E} \tag{82-1}$$

$$Q = \frac{I \times X_L}{I \times R} = \frac{X_L}{R} = \frac{2\pi f_r L}{R} \tag{82-2}$$

$$= \frac{\text{inductor's reactance}}{\text{inductor's resistance}}$$

The resonant frequency,

$$f_r = \frac{1}{2\pi\sqrt{LC}} \text{ Hz}$$

It follows that

$$Q = \frac{2\pi L}{2\pi\sqrt{LC} \times R} = \frac{1}{R} \times \sqrt{\frac{L}{C}} \qquad (82\text{-}3)$$

Because the resonant frequency depends on the product of L and C, it follows that, for the given resonant frequency, there are an infinite number of possible pairs of values for L and C. However, because the Q value is determined by the ratio of L and C, there are (for particular values of f_r and Q) only a limited range of values for L and C (assuming that the value of R does not alter appreciably). For audio frequency (AF) circuits involving a few kHz, Q is on the order of 10 to 20, but with radio frequency (RF) circuits, Q can exceed 100.

Inductor's power factor and figure of merit

$$\text{Inductor's power factor} = \frac{R}{\sqrt{R^2 + X_L^{\,2}}} \qquad (82\text{-}4)$$

$$= \frac{R}{X_L \times \sqrt{\left(1 + \frac{R^2}{X_L^{\,2}}\right)}}$$

$$= \frac{1}{Q \times \sqrt{1 + \frac{1}{Q^2}}}$$

The inductor's power factor equals $1/Q$ to within 1%, provided Q is greater than 10. Therefore, the values of Q and the power factor for radio frequency inductors are reciprocals. The higher the Q, the lower is the power factor and a low power factor is the feature of a "good" coil. Therefore, Q directly measures the inductor's merit.

Bandwidth

At any point on the current response curve,

$$\text{True power} = EI \cos\phi \qquad (82\text{-}5)$$

$$= E \times \frac{E}{Z} \times \frac{R}{Z} = \frac{E^2 R}{Z^2} \text{ watts}$$

At the peak of the response curve,

$$\text{Maximum power} = \frac{E^2}{R} \text{ watts}$$

At the half-power points,

$$\frac{E^2 R}{Z^2} = \frac{1}{2} \times \frac{E^2}{R} \qquad (82\text{-}6)$$

$$2R^2 = Z^2 = R^2 + (X_L - X_C)^2$$

$$R^2 = (X_L - X_C)^2$$

$$X_L - X_C = \pm R \text{ ohms}$$

At frequency f_1 (Fig. 82-1), X_C is greater than X_L. Therefore,

$$X_C - X_L = R$$

$$\frac{1}{2\pi f_1 C} - 2\pi f_1 L = R$$

$$4\pi^2 f_1^{\,2} LC + 2\pi f_1 CR - 1 = 0.$$

This yields,

$$f_1 = -\frac{R}{4\pi L} + \frac{1}{2\pi}\sqrt{\frac{1}{LC} + \frac{R^2}{4L^2}} \text{ Hz} \qquad (82\text{-}7)$$

At frequency f_2, X_L is greater than X_C. Therefore,

$$X_C - X_L = -R$$

This yields,

$$f_2 = +\frac{R}{4\pi L} + \frac{1}{2\pi}\sqrt{\frac{1}{LC} + \frac{R^2}{4L^2}} \qquad (82\text{-}8)$$

$$\text{Bandwidth, } f_2 - f_1 = \frac{R}{4\pi L} - \left(\frac{-R}{4\pi L}\right) \qquad (82\text{-}9)$$

$$= \frac{R}{2\pi L} \text{ Hz}$$

Therefore,

$$\frac{\text{Bandwidth}}{\text{Resonant Frequency}} = \frac{R}{2\pi f_r L} = \frac{R}{X_L} = \frac{1}{Q} \qquad (82\text{-}10)$$

$$\text{Bandwidth} = \frac{\text{resonant frequency}}{Q} \qquad (82\text{-}11)$$

$$Q = \frac{\text{resonant frequency}}{\text{bandwidth}} \qquad (82\text{-}12)$$

Q is a direct measure of the degree of selectivity. The higher the value of Q, the sharper is the current response curve, the greater is the degree of selectivity, and the narrower is the bandwidth.

Sharpness of the impedance response curve

Impedance,

$$Z = \sqrt{R^2 + (X_L - X_C)^2} \text{ ohms}$$

Except near the trough of the response curve, the value of R can normally be neglected in comparison with the overall reactance.

Therefore,

$$Z \approx X_L - X_C = 2\pi fL - \frac{1}{2\pi fC} \qquad (82\text{-}13)$$

$$\frac{dZ}{df} = 2\pi L + \frac{1}{2\pi C f^2}$$

$$= \sqrt{\frac{L}{C}}\left(\frac{1}{f_r} + \frac{f_r}{f^2}\right)$$

The higher the value of R, the flatter is the impedance response curve, and the lower is the selectivity. Therefore, the slope of the impedance response is *inversely* proportional to R, but *directly* proportional to $\sqrt{L/C}$. It follows that the sharpness of the impedance response is *directly* proportional to Q. These results are illustrated in Figs. 82-2A and B.

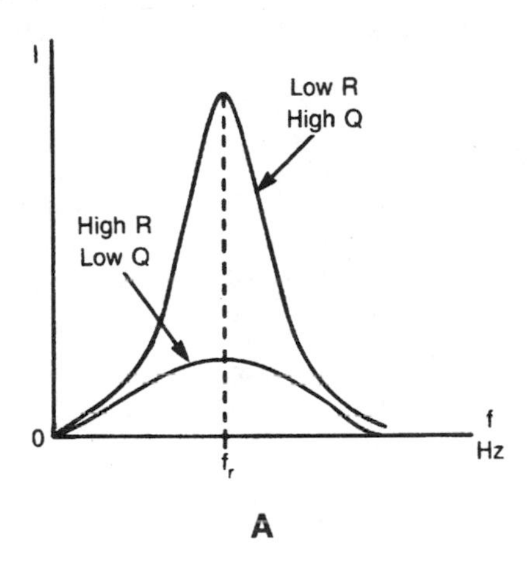

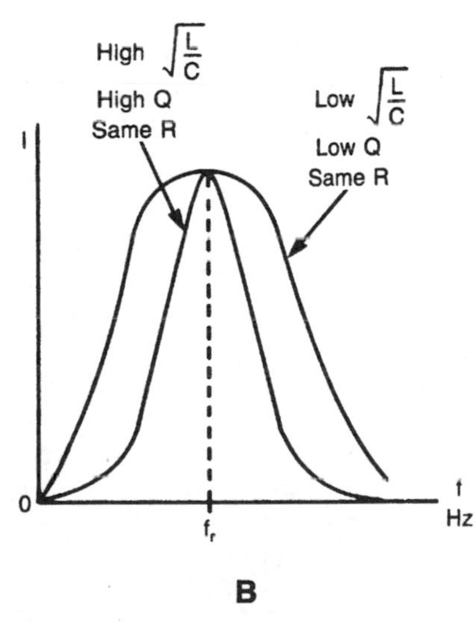

Figure 82-2

Energy relationships

The value of Q is equal to 2π times the ratio of the maximum energy stored during the cycle to the mean energy dissipated over the period.

The maximum energy stored during the cycle is in the form of the magnetic field surrounding the inductor; it is also in the form of the electric field associated with the capacitor.

$$\text{Maximum energy stored in the inductor} = \frac{1}{2} L (I_{peak})^2 \qquad (82\text{-}14)$$

$$= \frac{1}{2} L (\sqrt{2} \times I_{rms})^2$$

$$= L I_{rms}^{\ 2} \text{ joules}$$

where I_{peak} and I_{rms} respectively are the peak and the effective values of the current at resonance.

The mean energy dissipated over the period, T seconds, is equal to the product of the power and the time. Therefore,

$$\text{Mean energy dissipated} = I_{rms}^{\ 2} \times R \times T \qquad (82\text{-}15)$$

$$= \frac{I_{rms}^{\ 2} \times R}{f_r} \text{ joules}$$

Then,

$$2\pi \times \frac{\text{maximum energy stored during the cycle}}{\text{mean energy dissipated over the period}} \qquad (82\text{-}16)$$

$$= 2\pi \frac{L I_{rms}^{\ 2}}{I_{rms}^{\ 2} \times R/f_r}$$

$$= \frac{2\pi f_r L}{R} = Q$$

Note that if two identical series LCR circuits are joined end-to-end, with zero mutual coupling between the coils, the total impedance of the series combination at resonance is doubled ($Z = 2R$), but the Q and the resonant frequency are unchanged. However, if a single additional damping resistor (R_d) is connected to the series LCR circuit, the new total impedance at resonance is greater, and equal to $R + R_d$. Therefore, the new Q is reduced and is given by:

$$Q_{new} = \frac{1}{R + R_d} \times \sqrt{\frac{L}{C}} = \frac{R}{R + R_d} \times Q_{old} \qquad (82\text{-}17)$$

$$\text{New bandwidth} = \frac{R + R_d}{R} \times \text{old bandwidth} \qquad (82\text{-}18)$$

The new bandwidth is greater but the resonant frequency remains the same.

Skin effect

Because $Q = 2\pi f_r L/R$ it would appear that the value of Q would increase indefinitely as the resonant frequency is raised, and that the graph of "Q versus f_r" would be a straight line. However, if the frequency is raised, the equivalent inductive changes (due to the effect of the coil's distributed capacitance), and the coil's resistance increases for a number of reasons:

- *Proximity effect* This is the effect of winding one turn on top of another. The greater mutual inductance results in an increased resistance.
- *Dielectric loss* This is caused by leakage currents associated with the insulation between the turns.
- *Core loss* If an iron dust core is present, the hysteresis and eddy current losses increase with frequency.
- *Mutual inductance* There will be losses associated with the currents induced in neighboring conductors (such as screens), which act as secondary circuits of virtually zero resistance.
- *Skin effect* Of these five reasons, skin effect is the most important. If a steady dc current is passing through a length of copper wire, the current density (A/m^2) is the same throughout the wire's cross-sectional area. This is also approximately true for the line frequency (60 Hz) and for audio frequencies (up to 15 kHz).

To consider the case of an RF current, assume that the conductor consists of a number of separate conductors, which are welded together (Fig. 82-3). The RF current flowing through the conductor (A) at the center of the wire produces an alternating magnetic field, which induces voltages in the surrounding conductors. By Lenz's Law, these voltages are such as to oppose the change which was responsible for their creation, namely the RF current. Consequently, there will be an increase in the impedance to the flow of RF current at the center of the wire. However, fewer conductors surround the conductor (B) at the wire's perimeter and, therefore, the impedance at that position is less. As the frequency is raised, the current retreats more and more to the circumference, and is then confined to the wire's "skin." Consequently, the cross-sectional area available for the flow of current is reduced, and the resistance is increased. The effect is quite dramatic; if a length of 2-mm diameter copper wire has a dc resistance of 1 Ω, its resistance at 10 MHz exceeds 800 Ω! At frequencies above 30 MHz, it is practical to use hollow, rather than solid conductors. Note that the skin effect is less for conductors with a large surface area so that, for example, a flat sheet of copper will have less impedance than a wire of comparable length.

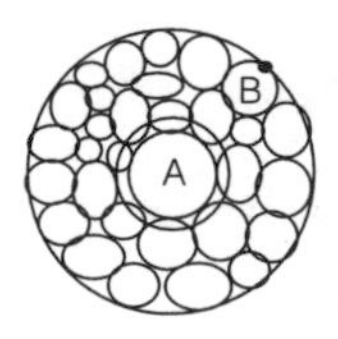

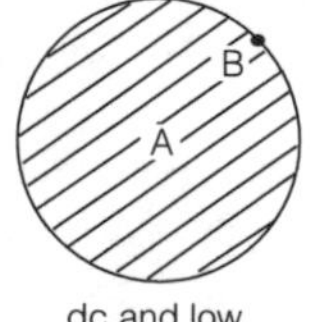

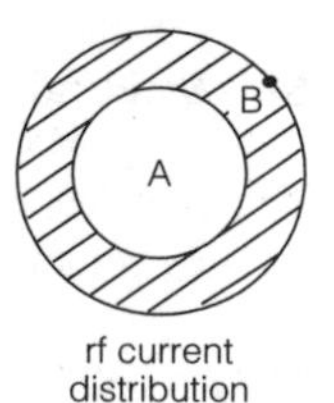

Figure 82-3

It is clear that if the resistance increases rapidly as the frequency is raised, the Q of a coil will not increase indefinitely, but it will reach a shallow peak and subsequently decrease in value (Fig. 82-4). A coil is therefore designed to be used over a limited frequency range, with the maximum value of Q occurring near the midfrequency of the band. However, at either end of the band, the Q value is still sufficient to provide an adequate degree of selectivity.

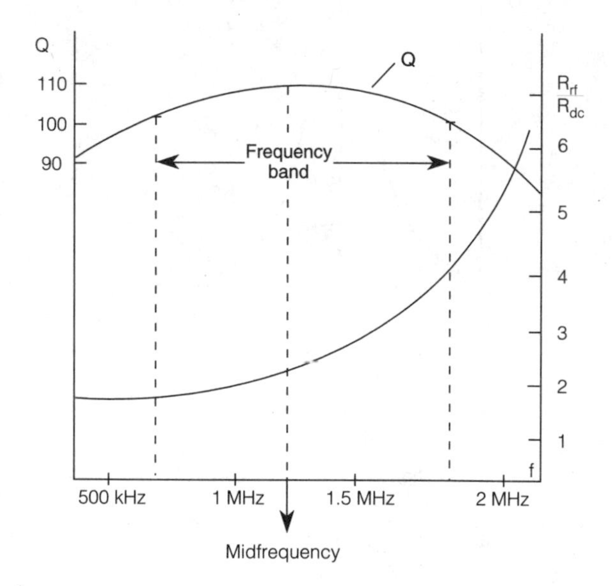

Figure 82-4

Example 82-1

In Fig. 82-1A, $R = 8\ \Omega$, $L = 150\ \mu$H, $C = 250$ pF, and $E = 2$ V rms. What are the values of Q and the circuit's bandwidth? If an additional 10-Ω resistor is inserted in series, calculate the new values of the resonant frequency, Q, and the bandwidth.

Solution

$$Q = \frac{1}{R} \times \sqrt{\frac{L}{C}} \qquad (82\text{-}3)$$

$$= \frac{1}{8} \times \sqrt{\frac{150 \times 10^{-6}}{250 \times 10^{-12}}}$$

$$= 97$$

Bandwidth,

$$BW = \frac{R}{2\pi L} \quad (82\text{-}9)$$

$$= \frac{8}{2 \times \pi \times 150 \times 10^{-6}} \text{ Hz}$$

$$= 8.5 \text{ kHz}$$

Resonant frequency,

$$f_r = \frac{1}{2\pi\sqrt{LC}}$$

$$= \frac{1}{2 \times \pi \times \sqrt{150 \times 10^{-6} \times 250 \times 10^{-12}}} \text{ Hz}$$

$$= 822 \text{ kHz}$$

Check

$$Q = \frac{\text{resonant frequency}}{\text{bandwidth}} = \frac{822 \text{ kHz}}{8.5 \text{ kHz}}$$

$$= 97$$

When an additional 10-Ω resistor is connected in series, the resonant frequency is unchanged at 822 kHz.

The new Q is given by

$$Q = \frac{R}{R + R_d} \times \text{old } Q \quad (82\text{-}17)$$

$$= \frac{8\ \Omega}{8\ \Omega + 10\ \Omega} \times 97$$

$$= 43$$

$$\text{New bandwidth} = \frac{\text{resonant frequency}}{\text{new } Q} \quad (82\text{-}11)$$

$$= \frac{822 \text{ kHz}}{43}$$

$$= 19.1 \text{ kHz.}$$

PRACTICE PROBLEMS

82-1. In Fig. 82-1A, E = 3 V rms, R = 11 Ω, L = 122 μH, C = 235 pF. Determine the values of the resonant frequency, Q, and bandwidth.

82-2. In Practice Problem 82-1, what is the rms value of the current at the half-power points? What are the upper and lower frequency limits of the bandwidth?

82-3. In Practice Problem 82-1, an additional 15-Ω resistor is in series with the L,C,R circuit. Calculate the new values of Q, bandwidth, and the current at the half-power points.

82-4. In Practice Problem 82-1, the value of C is quartered. What are the new values of the resonant frequency, Q, and bandwidth?

83 Parallel resonant LCR circuit

From the definition of resonance, the supply or source current (i_S) must be in phase with the source voltage, e (Fig. 83-1A). This will occur if the values of the inductor and capacitor currents are equal; only then will the phasor sum (i_x) of their currents be zero (remember that i_L and i_C are 180° out of phase). Then, the resistor current (which is independent of the frequency) will be the same as the supply current (whose value will be at its minimum level of E/R amperes). It follows that the total impedance (Z_T) of the complete circuit will be entirely resistive, and have its maximum value of R ohms.

At resonance the inductive and capacitive reactances are equal, and the parallel LC combination theoretically behaves as an open circuit; the total impedance of the circuit is then equal to the resistance. At frequencies other than the resonant frequency, the LC combination behaves as a certain value of reactance, which when placed in parallel with the resistance, produces a total impedance that is less than the value of the resistance. Combining these two statements, it means that the total impedance can never exceed the value of the resistance.

At frequencies below the resonant frequency, the capacitive reactance is greater than the inductive reactance, the capacitor current is less than the inductor current, and the circuit behaves inductively. At frequencies above the resonant frequency, the inductive reactance is greater than the capacitive reactance, the inductor current is less than the capacitor current, and the circuit behaves capacitively. These results are the reverse of those for the series LCR circuit, which be-

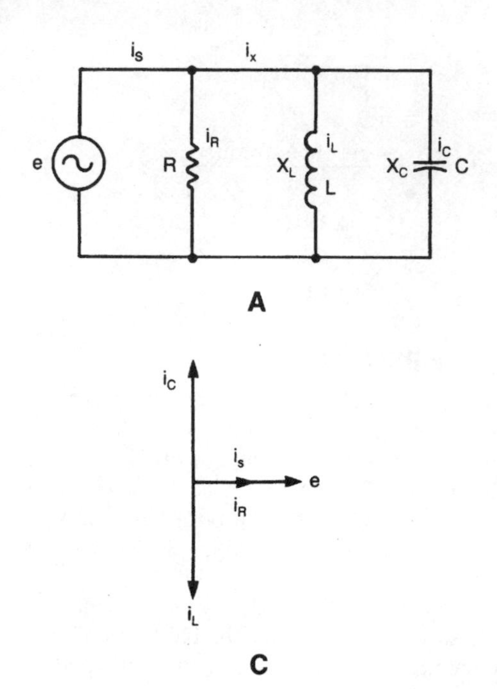

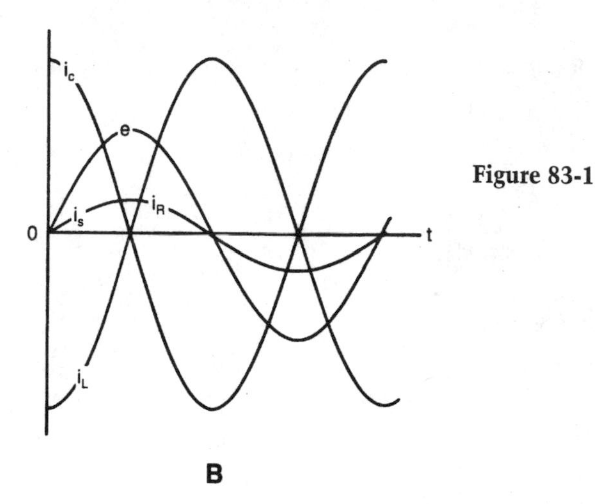

Figure 83-1

haved capacitively for frequencies below f_r, and inductively for frequencies above f_r. The current and the impedance response curves are illustrated in Figs. 83-2A and B.

In the resonant phasor diagram of Fig. 82-1C, currents I_L and I_C are equal in magnitude, and each might be many times greater than the supply current. The number of times greater is equal to the circuit's Q factor, which is a direct measure of the selectivity (chapter 82).

MATHEMATICAL DERIVATIONS

At resonance,

Phase angle of the entire circuit, $\phi = 0°$.

Inductor current, $I_L (= E/X_L)$ is equal in value to the capacitor current, $I_C (= E/X_C)$ amperes.

Supply current (I_S) is equal to the resistor current, $I_R = E/R$ amperes, which is at its minimum value.

The impedance (Z_T) of the circuit is at its maximum level, and equal to the value of the resistance, R ohms.

Inductive reactance, X_L = capacitive reactance, X_C.

Therefore,

$$2\pi f_r L = \frac{1}{2\pi f_r C}$$

This yields,

Resonant frequency,

$$f_r = \frac{1}{2\pi\sqrt{LC}} \text{ Hz} \tag{83-1}$$

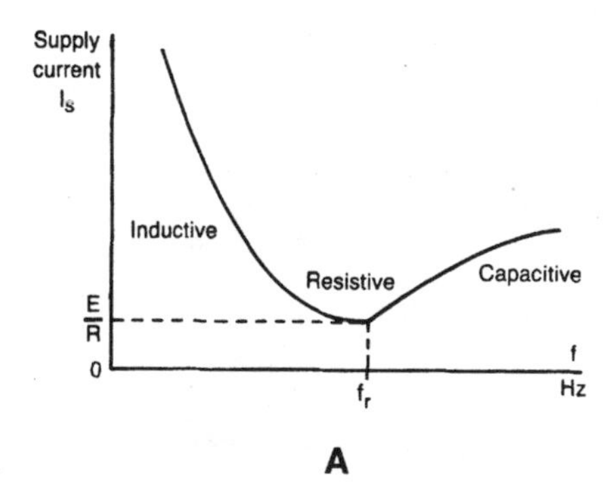

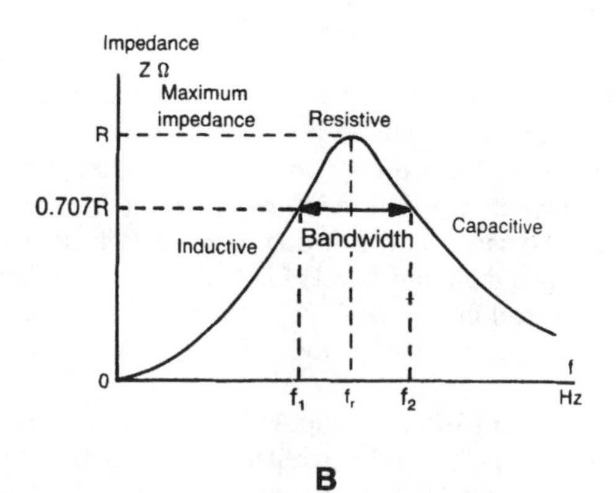

Figure 83-2

Then,

$$L = \frac{1}{4\pi^2 f_r^2 C} = \frac{0.0253}{f_r^2 C}\ \text{H} \qquad (83\text{-}2)$$

and,

$$C = \frac{1}{4\pi^2 f_r L} = \frac{0.0253}{f_r^2 L}\ \text{F} \qquad (83\text{-}3)$$

These last three equations are the same as the corresponding expressions for series resonance.

Current magnification

At resonance, the currents I_L and I_C are equal in magnitude, and each is Q times the supply current, I_S.

Then,

$$I_L = \frac{E}{X_L}, I_C = \frac{E}{X_C}, I_S = I_R = \frac{E}{R} \text{ amperes.}$$

and,

$$Q = \frac{I_L}{I_S} = \frac{E/X_L}{E/R} = \frac{R}{X_L} \qquad (83\text{-}4)$$

$$= \frac{R}{X_C}\left(not\ \frac{X_L}{R} \text{ or } \frac{X_C}{R}\right)$$

Because,

$$Q = \frac{R}{X_L} = \frac{R}{2\pi f_r L} \qquad (83\text{-}5)$$

and,

$$f_r = \frac{1}{2\pi\sqrt{LC}}\ \text{Hz},$$

$$Q = \frac{R \times 2\pi\sqrt{LC}}{2\pi L} = R \times \sqrt{\frac{C}{L}}$$

Impedance magnification

From Equation 83-4,

$$R = QX_L = QX_C \qquad (83\text{-}6)$$

The impedance at resonance is Q times the reactance of either the inductor or the capacitor.

Selectivity

As in the case of the series LCR circuit, Q determines the sharpness of the response curve, and is therefore a direct measure of the selectivity.

$$Q = \frac{\text{resonant frequency}}{\text{bandwidth}} = \frac{f_r}{f_2 - f_1} \qquad (83\text{-}7)$$

The *bandwidth* is defined from the impedance response curve as the frequency separation between the points where the total impedance of the circuit is 0.707 times the maximum impedance, R. From this definition,

$$\text{Bandwidth} = f_2 - f_1 = \frac{1}{2\pi RC}\ \text{Hz} \qquad (83\text{-}8)$$

where C and R are, respectively, measured in farads and ohms. The concept of the half-power points cannot be used in the case of the parallel LCR circuit because the true power is always E^2/R watts, and it is independent of the frequency.

Energy relationships

As in the case of the series LCR circuit, Q is equal to,

$$2\pi \times \frac{\text{maximum energy stored during the cycle}}{\text{mean energy dissipated during the period, } T}$$

The maximum energy stored in the capacitor is,

$$\frac{1}{2}C(E_{peak})^2 = \frac{1}{2}C(\sqrt{2}E_{rms})^2 = CE^2_{rms} \text{ joules} \qquad (83\text{-}9)$$

The mean energy dissipated during the period is,

$$E^2_{rms} \times T/R \text{ joules}$$

Then,

$$Q = 2\pi \times \frac{CE^2_{rms}}{E^2_{rms} \times T/R} = 2\pi f_r C \times R = R/X_C \qquad (83\text{-}10)$$

Note that, if two identical parallel resonant LCR circuits are shunted across each other, with zero mutual coupling between the coils, the equivalent resistance is halved. Therefore, the impedance at resonance is halved, but the Q, the resonant frequency, and the bandwidth remain the same.

If a single damping resistor (R_d) is connected across the parallel *LCR* circuit, the new impedance at resonance is $R_d || R = R \times R_d/(R + R_d)$ ohms and is reduced in value.

Then, the new value of Q is given by:

$$Q_{new} = \frac{R \times R_d}{R + R_d} \times \sqrt{\frac{C}{L}} = \frac{R_d}{R + R_d} \times Q_{old} \qquad (83\text{-}11)$$

Therefore, the value of Q has been decreased.

$$\text{New bandwidth} = \frac{R + R_d}{R_d} \times \text{old bandwidth} \qquad (83\text{-}12)$$

The bandwidth is consequently increased, although the resonant frequency remains the same.

Example 83-1

In Fig. 83-1A, $R = 8.2\ \text{k}\Omega$, $L = 3.3\ \mu\text{H}$, $C = 6.5$ pF, and $E = 15$ mV. Calculate the resonant frequency, the resonant values of I_S, I_R, I_L, I_C, and the true power. What are the values of the Q factor and the circuit's band width? If an additional 10-kΩ resistor is added in parallel, what are the new values of the resonant frequency, Q, and bandwidth?

Solution

Resonant frequency,

$$f_r = \frac{1}{2\pi\sqrt{LC}} \quad (83\text{-}1)$$

$$= \frac{1}{2 \times \pi \times \sqrt{3.3 \times 10^{-6} \times 6.5 \times 10^{-12}}} \text{ Hz}$$

$= 34.4$ MHz.

The inductive reactance at resonance,

$$X_L = 2 \times \pi \times f_r \times L$$

$$= 2 \times \pi \times 34.4 \times 10^6 \times 3.3 \times 10^{-6}$$

$$= 712.5\ \Omega$$

The capacitive reactance at resonance,

$$X_C = 712.5\ \Omega.$$

Supply current,

$$I_S = I_R = \frac{E}{R} = \frac{15 \text{ mV}}{8.2 \text{ k}\Omega}$$

$$= 1.83\ \mu\text{A}.$$

Inductor current,

$$I_L = \frac{E}{X_L} = \frac{15 \text{ mV}}{712.5\ \Omega}$$

$$= 21.05\ \mu\text{A}.$$

Capacitor current,

$$I_C = 21.05\ \mu\text{A}.$$

$$\text{True power} = E^2/R = (15 \text{ mV})^2/8.2 \text{ k}\Omega$$

$$= 27.4 \text{ nW}.$$

$$Q = R \times \sqrt{\frac{C}{L}} \quad (83\text{-}5)$$

$$= 8.2 \times 10^3 \times \sqrt{\frac{6.5 \times 10^{-12}}{3.3 \times 10^{-6}}}$$

$$= 11.5$$

Check:

$$Q = \frac{I_C}{I_S} = \frac{21.05\ \mu\text{A}}{1.83\ \mu\text{A}}$$

$$= 11.5$$

$$\text{Bandwidth} = \frac{1}{2\pi RC} \quad (83\text{-}8)$$

$$= \frac{1}{2 \times \pi \times 8.2 \times 10^3 \times 6.5 \times 10^{-12}} \text{ Hz}$$

$= 2.99$ MHz.

Check:

$$\text{Bandwidth} = \frac{f_r}{Q} = \frac{34.4 \text{ MHz}}{11.5} \quad (83\text{-}7)$$

$= 2.99$ MHz.

When the additional 10-kΩ resistor is added in parallel, the new total equivalent resistance at resonance is 10 kΩ||8.2 kΩ = 10 × 8.2/(10 + 8.2) = 4.5 kΩ. The resonant frequency is unchanged, but the new value of Q is 11.5 × 4.5 kΩ/8.2 kΩ = 6.3, and the new bandwidth is 2.99 MHz × 8.2 kΩ/4.5 kΩ = 5.45 MHz.

PRACTICE PROBLEMS

83-1. In Fig. 83-1A, E = 34 mV rms, R = 5.6 kΩ, L = 47 μH, C = 83 pF. Calculate the resonant frequency and determine the resonant values of I_R, I_L, I_C, I_S, and the true power.

83-2. In Practice Problem 83-1, what are the values of the Q factor, the bandwidth, and the impedance at the bandwidth points?

83-3. In Practice Problem 83-1, an additional 12-kΩ resistor is added in parallel. Calculate the new values of the resonant frequency, Q, and bandwidth.

83-4. In Practice Problem 83-1 the value of the inductance is doubled, but the other values remain unchanged. What are the new values of the resonant frequency, Q, and bandwidth?

84
The parallel resonant "tank" circuit

The parallel resonant *tank* circuit consists of a practical inductor with its series resistance, in parallel with a capacitor whose losses are assumed to be negligible (Fig. 84-1A). Such a circuit is commonly used as the collector load of certain transistor RF (radio frequency) amplifiers; in such stages, a high value of load impedance is required at the frequency to which the circuit is tuned.

Because the inductor branch contains both inductive reactance and resistance, the current (i_L) lags the source voltage (e) by the phase angle, ϕ. If the frequency is raised, the impedance of the inductor branch increases, and the angle (ϕ) moves closer to 90°.

The capacitor current (i_C) leads the source voltage (e) by 90°, and the supply current (i_S) is the phasor sum of i_C and i_L. Figure 84-1B shows the changes that occur in the circuit's behavior as the frequency is varied. At all frequencies below the resonant frequency, the supply current lags the source voltage by the *total* phase angle (θ) and the circuit behaves inductively. When the frequency exceeds its resonant value, the supply current leads the source voltage, and the circuit is capacitive. These results are similar to those obtained for the parallel resonant *LCR* circuit of chapter 83, but are opposite to those for the series *LCR* circuit of chapter 81.

At resonance, the supply voltage and the supply current are in phase, so that the angle (θ) is zero. As suggested by the phasor diagrams of Fig. 84-1B, the supply current at resonance is at its minimum level and, consequently, the total impedance at resonance is at its maximum value. This last statement is not mathematically exact, but it can be regarded as true, provided that the value of the Q factor is sufficiently high.

The supply current and impedance response curves are illustrated in Figs. 84-2A and B.

MATHEMATICAL DERIVATIONS

In the resonant phasor diagram of Fig. 84-1B,

$$I_C = I_L \sin\phi \text{ amperes} \tag{84-1}$$

Capacitor current,

$$I_C = \frac{E}{X_C} = 2\pi f_r CE \text{ amperes}$$

Inductor current,

$$I_L = \frac{E}{\sqrt{R^2 + X_L^2}} = \frac{E}{\sqrt{R^2 + 4\pi^2 f_r^2 L^2}} \text{ amperes}$$

$$\text{Sin } \phi = \frac{X_L}{\sqrt{R^2 + X_L^2}} = \frac{2\pi f_r L}{\sqrt{R^2 + 4\pi^2 f_r^2 L^2}}$$

Substituting these results in Equation 84-1,

$$2\pi f_r CE = \frac{E}{\sqrt{R^2 + 4\pi^2 f_r^2 L^2}} \times \frac{2\pi f_r L}{\sqrt{R^2 + 4\pi^2 f_r^2 L^2}}$$

$$R^2 + 4\pi^2 f_r^2 L^2 = \frac{L}{C} \tag{84-2}$$

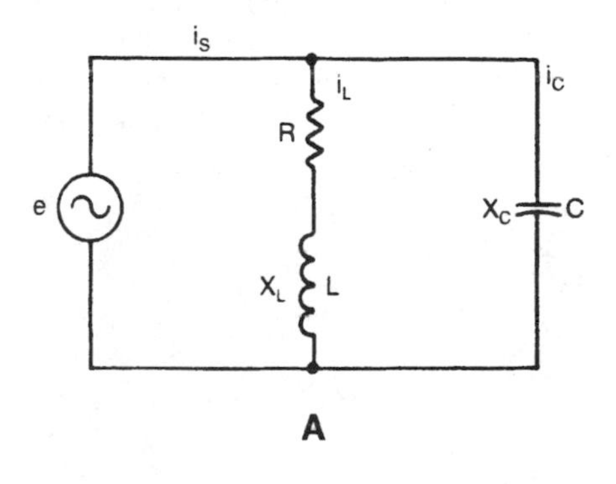

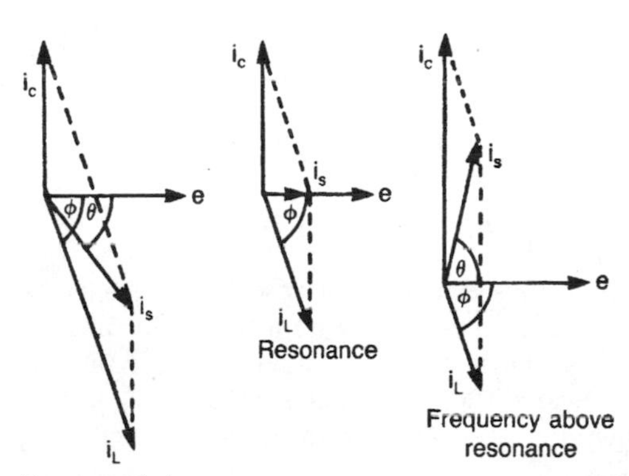

Figure 84-1

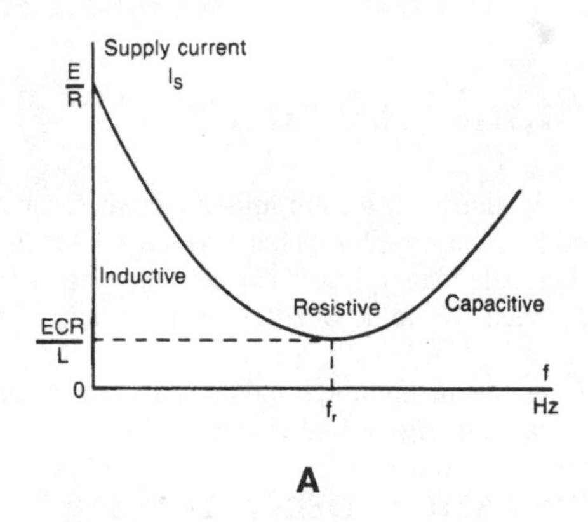

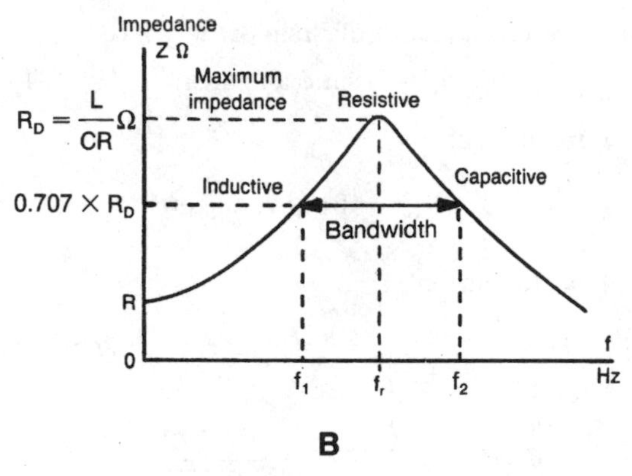

Figure 84-2

This yields:

Resonant frequency,

$$f_r = \frac{1}{2\pi}\sqrt{\frac{1}{LC} - \frac{R^2}{L^2}} \qquad (84\text{-}3)$$

$$= \frac{1}{2\pi\sqrt{LC}}\sqrt{\frac{1 - R^2C}{L}} \qquad (84\text{-}4)$$

$$= \frac{1}{2\pi\sqrt{LC}}\sqrt{1 - \frac{1}{Q^2}} \text{ Hz}$$

where Q is the merit factor of the coil, and is equal to X_L/R or $(\sqrt{L/C})/R$.

Notice that the formula for the resonant frequency of a tank circuit differs slightly from the comparable expression for the series and parallel LCR circuits. However, provided the value of Q is sufficiently high, the formula for the tank circuit reduces to $f_r = 1/(2\pi\sqrt{LC})$.

The total impedance at resonance is given by:

Total impedance,

$$Z_T = \frac{E}{I_S} = \frac{E}{I_L \cos\phi}$$

Because,

$$I_L = \frac{E}{\sqrt{R^2 + X_L^2}}$$

and,

$$\cos\phi = \frac{R}{\sqrt{R^2 + X_L^2}}$$

$$Z_T = \frac{E \times (R^2 + X_L^2)}{E \times R} = \frac{R^2 + X_L^2}{R}$$

But from Equation 84-2,

$$R^2 + X_L^2 = \frac{L}{C}$$

Therefore,

$$Z_T = R_D = \frac{L}{CR} \text{ ohms} \qquad (84\text{-}5)$$

The quantity L/CR Ω is called the *dynamic resistance* (R_D) because it only appears under operating conditions. The value of the dynamic resistance at resonance is virtually equal to the maximum level of the impedance.

The true power in the tank circuit can be expressed either as $I_S^2 \times L/CR$ watts, or $I_L^2 \times R$ watts. Therefore,

$$\frac{I_L^2}{I_S^2} = \frac{L/CR}{R} = \frac{L}{R^2C} = Q^2$$

or,

$$I_L\,(\approx I_C) = Q \times I_S \text{ amperes} \qquad (84\text{-}6)$$

The capacitor current is nearly the same as the inductor current. Consequently, at resonance there is a large circulating (or "flywheel") current associated with the inductor and the capacitor, and this current is Q times the supply or line current. It follows that the value of the dynamic resistance is approximately equal to Q times the inductive reactance, or Q times the capacitive reactance. Moreover because,

$$Z_T \approx Q \times X_L$$

then,

$$X_L = Q \times R,$$

$$Z_T \approx Q^2 \times R \text{ ohms.}$$

Selectivity

The Q factor is a measure of the tank circuit's selectivity, and is equal to the resonant frequency/bandwidth. From the impedance response curve, the *bandwidth* is defined as the frequency separation between those points where the circuit's total impedance is equal to

0.707 times the dynamic resistance (the total impedance at resonance).

Energy relationships

Q factor =

$$2\pi \times \frac{\text{maximum energy stored during the cycle}}{\text{mean energy dissipated over the period}}$$

Note that, if two identical tank circuits are paralleled with zero mutual coupling between the coils, the total dynamic resistance at resonance is halved, but the Q and the resonant frequency remain unchanged.

If a single damping resistor (R_d) is connected across the tank circuit, the new total impedance at resonance is:

New total impedance,

$$Z_{\text{new}} = \frac{R_d \times L/CR}{R_d + L/CR} \text{ ohms} \qquad (84\text{-}7)$$

and,

$$Q_{\text{new}} = \frac{R_d}{R_d + L/CR} \times Q_{\text{old}} \qquad (84\text{-}8)$$

Therefore, the values of Q and the total impedance are both reduced. The bandwidth is greater, but the resonant frequency remains the same (assuming that the value of Q is greater than 10).

Example 84-1

In Fig. 84-1A, $R = 12\ \Omega$, $L = 25\ \mu\text{H}$, $C = 15$ pF, and $E = 80$ mV rms. Find the value of the resonant frequency, and calculate the resonant values of I_S, I_L, and I_C. What is the value of Q, and the tank circuit's bandwidth?

Solution

The exact formula for the resonant frequency is:

$$f_r = \frac{1}{2\pi} \times \sqrt{\frac{1}{LC} - \frac{R^2}{L^2}} \text{ Hz} \qquad (84\text{-}3)$$

However,

$$\frac{1}{LC} = \frac{1}{25 \times 10^{-6} \times 15 \times 10^{-12}}$$

$$= 2.667 \times 10^{15}$$

and,

$$\frac{R^2}{L^2} = \frac{144}{(25 \times 10^{-6})^2}$$

$$= 2.304 \times 10^{11}$$

Clearly R^2/L^2 is negligible when compared with $1/LC$.

Therefore, the formula reduces to Resonant frequency,

$$f_r = \frac{1}{2\pi\sqrt{LC}}$$

$$= \frac{1}{2 \times \pi \times \sqrt{25 \times 10^{-6} \times 15 \times 10^{-12}}} \text{ Hz}$$

$$= 8.22 \text{ MHz.}$$

At the resonant frequency,

$$X_C = \frac{1}{2\pi f_r C}$$

$$= \frac{1}{2 \times \pi \times 8.22 \times 10^6 \times 15 \times 10^{-12}}$$

$$= 1291\ \Omega.$$

The impedance of the inductor branch is

$$Z_L = \sqrt{R^2 + X_L{}^2} = \sqrt{18^2 + 1291^2}$$

$$\approx 1291\ \Omega.$$

Circulating current,

$$I_L = I_C = \frac{80 \text{ mV}}{1291\ \Omega}$$

$$= 62.00\ \mu\text{A.}$$

Dynamic resistance,

$$R_D = \frac{L}{CR}\ \Omega \qquad (84\text{-}5)$$

$$= \frac{25 \times 10^{-6}}{15 \times 10^{-12} \times 12}\ \Omega$$

$$= 138.9 \text{ k}\Omega.$$

Supply, line, or "make-up" current,

$$I_S = \frac{E}{R_D} = \frac{80 \text{ mV}}{138.9 \text{ k}\Omega}$$

$$= 0.576\ \mu\text{A.}$$

$$Q = \frac{X_L}{R} = \frac{1291\ \Omega}{12\ \Omega} = 107.6$$

Check:

$$Q = \frac{I_C}{I_S} = \frac{62.00\ \mu\text{A}}{0.576\ \mu\text{A}} \qquad (84\text{-}6)$$

$$= 107.6$$

Bandwidth,

$$BW = \frac{f_r}{Q} = \frac{8.22 \text{ MHz}}{107.6}$$

$$= 76.4 \text{ kHz.}$$

PRACTICE PROBLEMS

84-1. In Fig. 84-1A, $R = 15\ \Omega$, $L = 85\ \mu H$, $C = 55$ pF and $E = 0.1$ V rms. Calculate the value of the resonant frequency and the resonant values of I_S, I_L, and I_C.

84-2. In Practice Problem 84-1, calculate the values of the dynamic resistance, Q, and the bandwidth.

84-3. In Practice Problem 84-1, the value of R is doubled, but the other given values remain the same. Calculate the new values of the dynamic resistance, Q, and bandwidth.

84-4. In Practice Problem 84-1, the value of C is doubled, but the other given values remain the same. Calculate the new values of the resonant frequency, the dynamic resistance, Q, and the bandwidth.

85
Free oscillation in an LC circuit

The resonant circuits that were covered in previous chapters were connected to an ac source that generated the resonant frequency; the resultant oscillation therefore, was "forced." By contrast, it is possible to create a "free," or "natural," oscillation by removing the source and connecting a coil across a charged capacitor (Figs. 85-1A, B, C, and D).

Figure 85-1A shows the tank circuit immediately after the inductor has been connected. The capacitor then produces a current (electron flow) in the direction shown. The inductor provides a counter EMF that balances the voltage across the capacitor. As the capacitor discharges through the inductor, the inductor is storing energy in the form of a magnetic field. At the time when

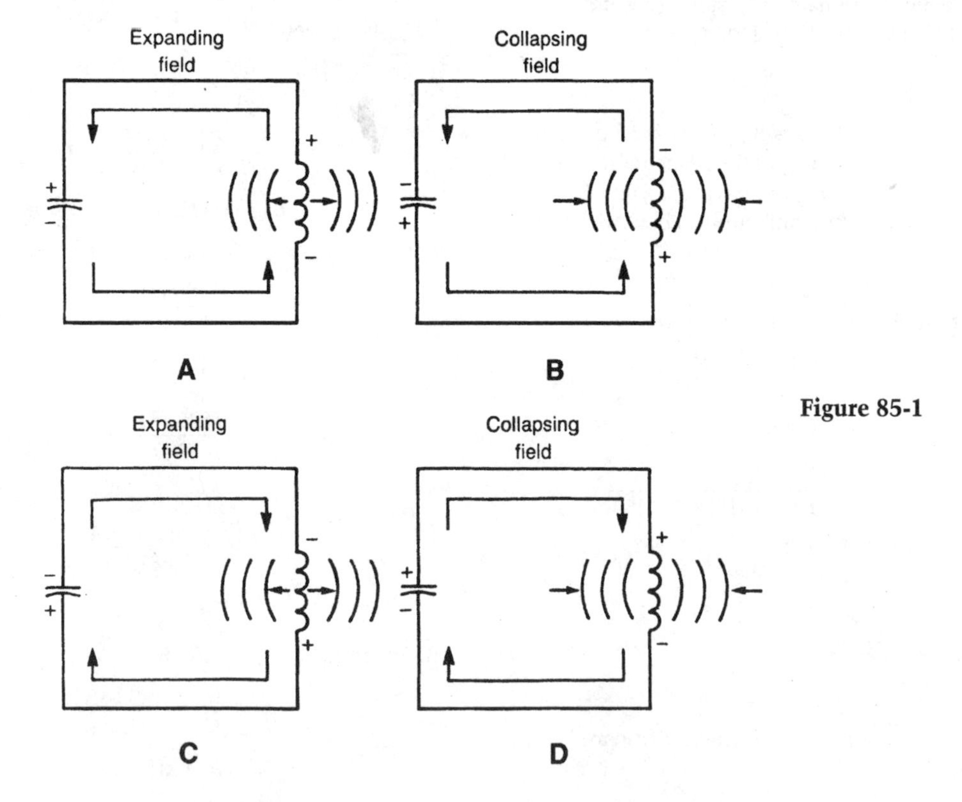

Figure 85-1

the capacitor has completed its discharge, there is zero voltage across the inductor, and all of the circuit's energy is stored in the inductor's magnetic field.

Figure 85-1B shows the second quarter cycle of the circuit's operation. The magnetic field of the inductor is collapsing in order to maintain the inductor current. The collapsing magnetic field produces a current in the same direction as that previously established. This current now charges the capacitor with the polarity opposite to that which it originally had. When the magnetic field of the inductor is exhausted, the capacitor is fully charged, and all of the circuit's energy is stored in the form of an electrostatic field across the capacitor.

Figure 85-1C shows the third quarter cycle of operation. The capacitor is now discharging, and once again the inductor is building up its magnetic field. The direction of the electron flow is in the direction shown.

In Fig. 85-1D, the capacitor is seen to be fully discharged, so the voltage across the capacitor is again zero. The collapsing magnetic field of the inductor will continue the circuit current, and recharge the capacitor back to the original condition shown in Fig. 85-1A. Further cycles of operation will be identical to the one just described.

In this electrical circuit, there exists some resistance that is mainly associated with the coil. This resistance will constitute a power loss that will eventually kill the oscillations in the tank circuit. The action of a tank circuit in which the amplitude of oscillations decreases to zero is called a *damped wave.* Damping can only be overcome by resupplying energy to the tuned circuit at a rate comparable to that at which it is being used. This is usually accomplished through the use of a dc source and a transistor, which acts as an amplifying device.

The inductance and capacitance values in the LC circuit determine the natural frequency of oscillation:

$$f_0 = \frac{1}{2\pi}\sqrt{\frac{1}{LC} - \frac{R^2}{4L^2}} \text{ Hz}$$

where: L = coil's inductance (H)
C = capacitance (F)
R = coil's resistance (Ω).

Note that this frequency differs slightly from the "forced" resonant frequencies of the series LCR circuit, the parallel LCR circuit, and the tank circuit. However, for high Q circuits, the differences are negligible.

MATHEMATICAL DERIVATIONS

In Fig. 85-2, using Kirchhoff's voltage law,

$$V_L + V_R + V_C = 0.$$

$$L\frac{di}{dt} + iR + \frac{Q}{C} = 0.$$

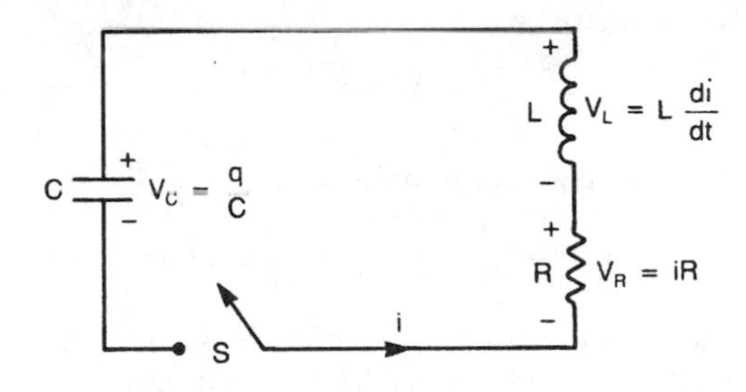

Figure 85-2

Differentiating with respect to t,

$$L\frac{d^2i}{dt^2} + R\frac{di}{dt} + \frac{1}{C}\frac{dq}{dt} = 0. \qquad (85\text{-}1)$$

$$\frac{d^2i}{dt^2} + \frac{R}{L}\times\frac{di}{dt} + \frac{i}{LC} = 0.$$

The solution to this differential equation is:

$$i = Ae^{m_1t} + Be^{m_2t}$$

where m_1 and m_2 are the roots of the quadratic equation,

$$m^2 + \frac{R}{L}m + \frac{1}{LC} = 0$$

The roots are:

$$m = \pm\sqrt{\frac{R^2}{4L^2} - \frac{1}{LC}} - \frac{R}{2L}$$

In the practical case, $1/LC >> R^2/4L^2$. Therefore,

$$i = e^{-\alpha t}(F\cos\omega_0 t + G\sin\omega_0 t)$$

where: $\alpha = \dfrac{R}{2L}$ sec^{-1}

$$\omega_0 = \sqrt{\frac{1}{LC} - \frac{R^2}{4L^2}} \text{ rad/s.}$$

When,

$$t = 0, i = 0, \frac{di}{dt} = \frac{E}{L} \text{ amperes per second}$$

Therefore,

$$F = 0, G = \frac{E}{\omega_0 L} \text{ amperes.}$$

This yields
"Free" oscillation current,

$$i = \frac{E}{2\pi f_0 L}e^{-Rt/L}\sin 2\pi f_0 t \text{ amperes} \qquad (85\text{-}2)$$

This is the equation of the damped oscillation (Fig. 84-3) where:

$$\text{Initial peak current value} = \frac{E}{2\pi f_0 L} \text{ amperes} \qquad (85\text{-}3)$$

Decay factor,

$$\alpha = \frac{R}{L}\ \text{s}^{-1} \tag{85-4}$$

Natural frequency of oscillation,

$$f_0 = \frac{1}{2\pi}\sqrt{\frac{1}{LC} - \frac{R^2}{4L^2}}\ \text{Hz} \tag{85-5}$$

The damped oscillation principle is used in the "ringing" coil which is capable of generating a short duration pulse.

Example 85-1

A 150-μH coil has a resistance of 8 Ω, and is connected across a charged 500-pF capacitor. Calculate the values of the decay factor, and the natural frequency of oscillation. See Fig. 85-3.

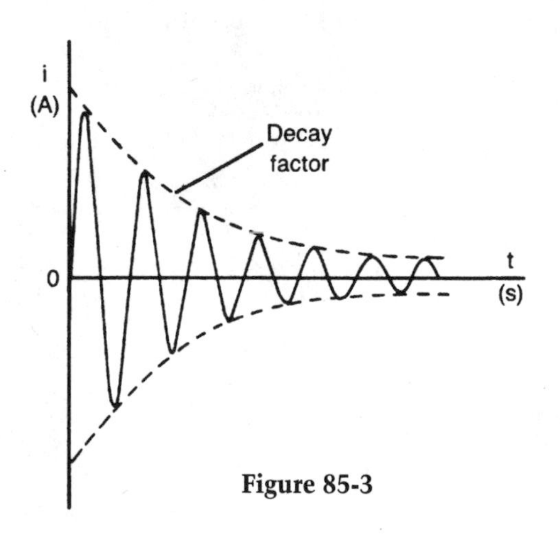

Figure 85-3

Solution

Natural frequency of oscillation,

$$f_0 = \frac{1}{2\pi}\sqrt{\frac{1}{LC} - \frac{R^2}{4L^2}} \tag{85-5}$$

The value of $\frac{1}{LC}$ is

$$\frac{1}{150 \times 10^{-6} \times 500 \times 10^{-12}} = 1.33 \times 10^{13}$$

The value of $\frac{R^2}{4L^2}$ is $\frac{8^2}{4 \times (50 \times 10^{-6})^2} = 7.11 \times 10^8$

Therefore, $1/LC >> R^2/4L^2$ and the expression for the natural frequency reduces to:

$$f_0 = \frac{\sqrt{1.33 \times 10^{13}}}{2 \times \pi}\ \text{Hz} = 580\ \text{kHz}.$$

Decay factor,

$$\alpha = \frac{R}{L} = \frac{8}{150 \times 10^{-6}} \tag{85-4}$$

$$= 5.33 \times 10^4\ \text{s}^{-1}$$

PRACTICE PROBLEMS

85-1. An 80-μH coil has a resistance of 5 Ω, and is connected across a 350-pF capacitor which is initially charged to 50 V. Determine the values of the natural frequency of oscillation and the decay factor.

85-2. In Practice Problem 85-1, what is the initial peak value of the instantaneous current? Write down the trigonometric expression for the instantaneous current.

85-3. In Practice Problem 85-1, how long will it take for the current to decay to less than 1% of its initial value?

86 Mutually coupled coils

In previous chapters you learned about inductors that have a magnetic flux linked with their *own* turns, and therefore possess the property of *self*-inductance. Now, consider the situation where the magnetic flux associated with one coil links with the turns of another coil. Such inductors are said to be magnetically, inductively, mutually, or transformer *coupled*; they possess the property of *mutual inductance*, whose letter symbol is M and whose unit is the henry (H). In most of the circuits of the previous chapters, there was a pair of input terminals (*2-port network*). You now have a pair of input, and another pair of output terminals; this is referred to as a *4-port network*.

The alternating current (i_1) in the coil #1 (Fig. 86-1) creates an alternating flux, only part of which links with coil #1 and the remainder links with coil #2. This causes an induced voltage (v_2) whose size depends on the rate of change of the current (i_1) and on the mutual inductance (M) between the coils. The mutual inductance is 1 H if, when the current (i_1) is instantaneously changing at the rate of one ampere per second, the induced voltage (v_2) is 1 volt. The factors which determine the mutual inductance include the number of turns (N_1 and N_2), the cross-sectional area of the coils, their separation, the orientation of their axes, and the nature of their cores. Consequently, the induced voltage $v_2 = M \times$ rate of change of i_1.

If the alternating current (i_1) has a sinusoidal waveform with a frequency of f hertz, $V_{2rms} = 2\pi f M I_{rms}$ (compare $V_{Lrms} = 2\pi f L I_{rms}$ for the property of self-inductance). When the two coils are wound in the same sense, v_2 lags i_1 by 90°; but if the two coils are wound in the opposite sense, v_2 leads i_1 by 90°. Note that the property of the mutual inductance is reversible in the sense that, if the same rate of change in the current (i_1) is flowing in coil #2, then the voltage induced in the coil #1 is $v_1 = M \times$ rate of change in the current, i_1.

The mathematical derivations will show that, in the extreme case, where the coils are tightly wound, one on top of the other, with a common soft iron core, the leakage flux is extremely small and can therefore be neglected. Assuming perfect flux linkage between the coils (corresponding to zero flux leakage), the mutual inductance is given by:

Mutual inductance,

$$M = \frac{\mu_0 \mu_r N_1 N_2 A}{l} \text{ H}$$

where: A = cross-sectional area for each coil (m^2).
μ_r = relative permeability of the soft iron.
l = the coil's length (m).
μ_0 = permeability of free space ($4\pi \times 10^{-7}$ H/m).

The self-inductance of coil #1 is:

$$\text{Self-inductance, } L_1 = \frac{\mu_0 \mu_r N_1^2 A}{l} \text{ H}$$

and the self-inductance of coil #2 is:

$$\text{Self-inductance, } L_2 = \frac{\mu_0 \mu_r N_2^2 A}{l} \text{ H}$$

This yields,

$$M^2 = L_1 L_2$$

and,

$$\text{Mutual inductance, } M = \sqrt{L_1 L_2} \text{ H.}$$

If the leakage flux is not negligible, then only a fraction (k) of the total flux links with the two coils. This fraction (k) whose value cannot exceed unity, is called the *coefficient of coupling*, or coupling factor. Its value can be close to unity if a common soft iron core is used for the two coils, but can be very small (less than 0.01) with a nonmagnetic core and the two coils widely separated (loose coupling).

MATHEMATICAL DERIVATIONS

In Fig. 86-1,

Induced voltage,

$$v_2 = -M\frac{\Delta i_1}{\Delta t} \text{ volts} \tag{86-1}$$

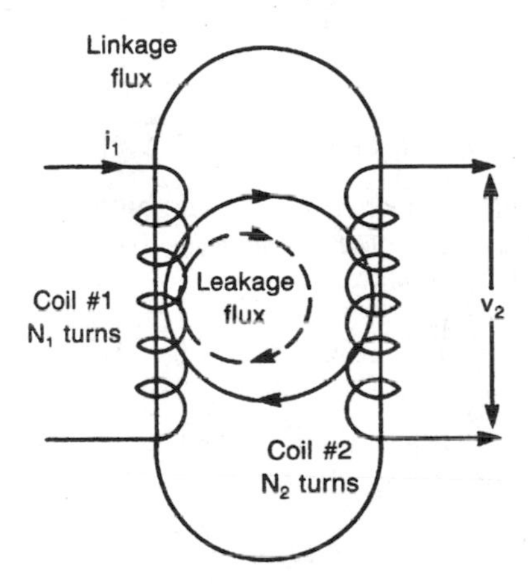

Figure 86-1

where: M = mutual inductance (H)

$\frac{\Delta i_1}{\Delta t}$ = rate of change in the current, i_1 (A/s)

If the two coils are *tightly* wound,

$$v_2 = -N_2\frac{\Delta \phi}{\Delta t} \tag{86-2}$$

where: ϕ = linkage flux (Wb).

The minus sign is the direct result of Lenz's law.

But,

$$\phi = BA \tag{86-3}$$

$$= \mu_0 \mu_r HA$$

$$= \frac{\mu_0 \mu_r i_1 N_1 A}{l}$$

where: B = flux density in the common iron core (T).
N_1 = number of turns in coil #1.
A = cross-sectional area of each coil (m^2).
l = length of each coil (m).
μ_r = relative permeability of the iron core.
μ_0 = permeability of free space ($4\pi \times 10^{-7}$ H/m).

Combining Equations 86-2 and 86-3,

$$v_2 = -\frac{\mu_0 \mu_r N_1 N_2 A}{l} \times \frac{\Delta i_1}{\Delta t} \qquad (86\text{-}4)$$

Combining Equations 86-1 and 86-4,
Mutual inductance,

$$M = \frac{\mu_0 \mu_r N_1 N_2 A}{l} \text{H} \qquad (86\text{-}5)$$

Because,

$$L_1 = \frac{\mu_0 \mu_r N_1^2 A}{l} \text{H and } L_2 = \frac{\mu_0 \mu_r N_2^2 A}{l} \text{ H},$$

it follows that:

$$M^2 = L_1 L_2 \text{ and } M = \sqrt{L_1 L_2} \text{ H} \qquad (86\text{-}6)$$

The result of Equation 86-6 would be equivalent to a coupling factor, k, of unity. However, in the general case,
Coupling factor,

$$k = \frac{M}{\sqrt{L_1 L_2}} \text{ or } M = k \times \sqrt{L_1 L_2} \text{ H} \qquad (86\text{-}7)$$

Notice that k is just a number and has no units.
Equation 86-7 yields

$$L_1 = \frac{M^2}{k^2 L_2} \text{ H and } L_2 = \frac{M^2}{k^2 L_1} \text{ H} \qquad (86\text{-}8)$$

and,

$$M = k\,L_1 \times \frac{N_2}{N_1} = k\,L_2 \times \frac{N_1}{N_2} \text{H} \qquad (86\text{-}9)$$

If L_1 and L_2 are each equal to L,

$$k = \frac{M}{L} \text{ and } M = kL \qquad (86\text{-}10)$$

Note that if a steady direct current is flowing through coil #1, the linkage flux will be constant in magnitude and direction, so the voltage induced in coil #2 will be zero.

Example 86-1

Two inductors whose coupling factor is unity, have self-inductances of 1.7 H and 2.5 H. What is the value of their mutual inductance? If the rate of change in the current flowing through the 1.7-H inductor is 8 A/s, what is the voltage induced in the 2.5-H inductor?

Solution

Mutual inductance,

$$M = \sqrt{L_1 \times L_2} \qquad (86\text{-}6)$$

$$= \sqrt{1.7 \times 2.5}$$

$$= 2.06 \text{ H.}$$

Voltage induced in the 2.5-H inductor,

$$v_2 = (-)M\frac{\Delta i_1}{\Delta t} \qquad (86\text{-}1)$$

$$= 2.06 \text{ H} \times 8 \text{ A/s}$$

$$= 16.5 \text{ V.}$$

Example 86-2

Two coils with self-inductances of 85 μH and 115 μH have a mutual inductance of 14 μH. What is the value of their coupling factor?

Solution

Coupling factor,

$$k = \frac{M}{\sqrt{L_1 L_2}} \qquad (86\text{-}7)$$

$$= \frac{14}{\sqrt{84 \times 115}}$$

$$= 0.14.$$

PRACTICE PROBLEMS

86-1. Two inductors, whose self-inductances are 1.52 H and 2.36 H, have a coupling factor of 0.85. What is the value of their mutual inductance?

86-2. In Practice Problem 86-1, the voltage induced in the 1.52-H inductor is 20.2 V. What is the value of the rate of change of current in the 2.36-H inductor?

86-3. Two identical inductors have a mutual inductance of 22 μH and a coupling factor of 0.18. What is the value of the self-inductance of each inductor?

86-4. The mutual inductance between two coils L1 and L2 is 0.75 H and the coupling factor is 0.85. If the inductance (L_1) of the first coil is 1.35 H, determine the value of the turns ratio, N_1/N_2, for the two inductors.

86-5. In Practice Problem 86-4, obtain the value of the self-inductance, L_2.

86-6. In Practice Problem 86-4, the current in the first coil (L1) grows at a constant rate from 0.5 A to 2.5 A in 0.04 s. What is the voltage induced in the second coil as a result of the mutual coupling between the coils?

86-7. In Practice Problem 86-4, the current flowing through the first coil is 2 Adc. What is the value of the voltage induced in the second coil?

87
Mutually coupled coils in series and parallel

If two mutually coupled coils have a common axis and are connected in series, their individual fluxes may aid or oppose, depending on the sense in which the coils are wound. In Fig. 87-1A, the directions of the arrows are indicative of the electron current flow, and the individual fluxes surrounding L1 and L2 will be aiding.

MUTUALLY COUPLED COILS IN SERIES

In the series-aiding connection, the rate of change in the current ($\Delta I/\Delta T$) is the same throughout the circuit (Fig. 87-1B). The dots at the ends of the coil symbols are called *polarity marks*, or *sense dots*; when the current enters the dotted end of L1 and leaves, it must

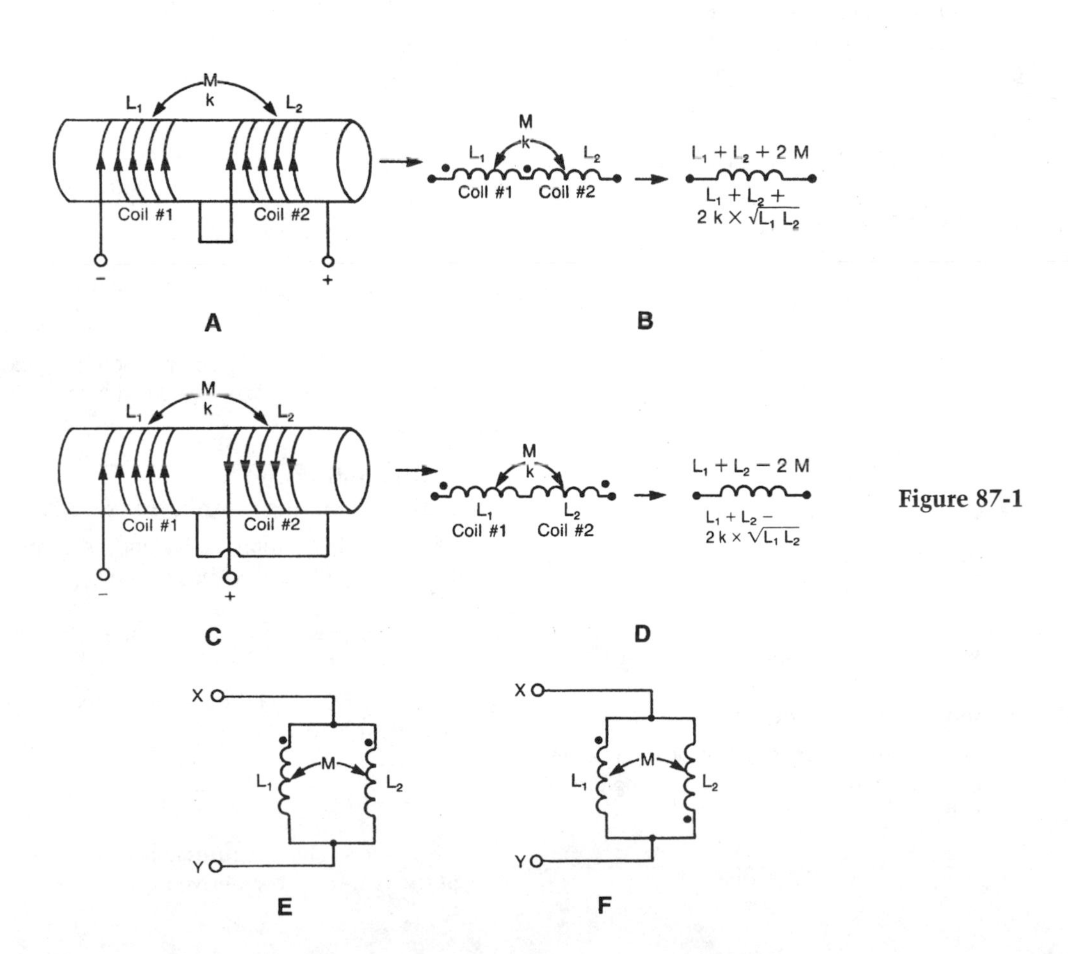

Figure 87-1

enter L2 at its dotted end, in order for the two fluxes to aid. The voltage induced in the coil #1 will be a combination of $L_1(\Delta I/\Delta T)$ volts because of self-inductance, and $M(\Delta I/\Delta T)$ volts because of the flux that surrounds coil #2 and links partially with coil #1. In a similar way, the total voltage induced in coil #2 will be the addition of $L_2(\Delta I/\Delta T)$ volts because of self-inductance, and $M(\Delta I/\Delta T)$ volts as the result of mutual inductance. The effect of this mutual coupling is to increase the total equivalent inductance.

In Fig. 87-1C, the fluxes surrounding L1 and L2 will subtract; consequently, the coils are connected in series-opposing (as indicated by the positions of the polarity marks). The connections to L2 are reversed so that the current enters its undotted end (Fig. 87-1D). The mutually induced voltages then oppose the voltages created by self-induction, and the total equivalent inductance is reduced.

MUTUALLY COUPLED COILS IN PARALLEL

Figure 87-1E shows two mutually coupled coils in parallel. The connection is parallel-aiding as shown by the positions of the polarity dots. Compared with zero mutual coupling, the effect of the parallel-aiding connection is to increase the total equivalent inductance.

In Fig. 87-1F, the connection is parallel-opposing, and the result is to reduce the total equivalent inductance.

MATHEMATICAL DERIVATIONS

Series-aiding connection

The total voltage across the series-aiding circuit of Fig. 87-1B is:

$$L_1\frac{\Delta I}{\Delta T} + M\frac{\Delta I}{\Delta T} + L_2\frac{\Delta I}{\Delta T} + M\frac{\Delta I}{\Delta T} \qquad (87\text{-}1)$$

$$= \left(L_1 + L_2 + 2\,M\right)\frac{\Delta I}{\Delta T} \text{ volts}$$

Therefore, the total equivalent inductance is:

$$L_T^{+} = L_1 + L_2 + 2\,M \text{ henrys} \qquad (87\text{-}2)$$

$$L_T^{+} = L_1 + L_2 + 2\,k\sqrt{L_1L_2} \text{ henrys} \qquad (87\text{-}3)$$

The positive sign associated with L_T means "aiding."

Series-opposing connection

If the fluxes are opposing, the sign of M is reversed. The total voltage across the series-opposing circuit of Fig. 87-1C is:

$$L_1\frac{\Delta I}{\Delta T} - M\frac{\Delta I}{\Delta T} + L_2\frac{\Delta I}{\Delta T} - M\frac{\Delta I}{\Delta T} \qquad (87\text{-}4)$$

$$= \left(L_1 + L_2 - 2\,M\right)\frac{\Delta I}{\Delta T} \text{ volts}$$

Therefore, the total equivalent inductance is:

$$L_T^{-} = L_1 + L_2 - 2\,M \text{ henrys} \qquad (87\text{-}5)$$

$$L_T^{-} = L_1 + L_2 - 2\,k\sqrt{L_1L_2} \text{ henrys} \qquad (87\text{-}6)$$

The negative sign associated with L_T means "opposing."

From Equations 87-2, 87-5,

$$L_1 + L_2 = \frac{L_T^{+} + L_T^{-}}{2} \text{ henrys} \qquad (87\text{-}7)$$

and,

$$M = \frac{L_T^{+} - L_T^{-}}{2} \text{ henrys} \qquad (87\text{-}8)$$

If L_1 and L_2 are each equal to L, Equations 87-2 and 87-5 become

$$L_T^{+} = 2(L + M) = 2L(1 + k) \text{ henrys} \qquad (87\text{-}9)$$

and,

$$L_T^{-} = 2(L - M) = 2L(1 - k) \text{ henrys} \qquad (87\text{-}10)$$

Then,

$$L = \frac{L_T^{+} + L_T^{-}}{4} \text{ henrys} \qquad (87\text{-}11)$$

$$M = \frac{L_T^{+} - L_T^{-}}{4} \text{ henrys} \qquad (87\text{-}12)$$

and,

$$k = \frac{L_T^{+} - L_T^{-}}{L_T^{+} + L_T^{-}} \qquad (87\text{-}13)$$

Equations 87-12 and 87-13 are used in an experimental determination of the M and k values.

Parallel-aiding and parallel-opposing connections

In Fig. 87-1E, the total equivalent inductance L_T^{+} between the points X and Y is given by:

$$L_T^{+} = \frac{L_1L_2 - M^2}{L_1 + L_2 - 2M} \text{ henrys} \qquad (87\text{-}14)$$

In Fig. 87-1F, the total equivalent inductance L_T^{-} between the points X and Y is:

$$L_T^{-} = \frac{L_1L_2 - M^2}{L_1 + L_2 + 2M} \text{ henrys} \qquad (87\text{-}15)$$

Note: The above expressions for L_T^{+} and L_T^{-} are *not* the same as those derived from

$$\frac{1}{L_T^{\pm}} = \frac{1}{L_1 \pm M} + \frac{1}{L_2 \pm M}$$

which is erroneously quoted in some textbooks.

If there is zero mutual coupling ($M = 0$) between the coils, Equations 87-14 and 87-15 both reduce to:

$$L_T = \frac{L_1 L_2}{L_1 + L_2} \text{ henrys}$$

which is the familiar "product-over-sum" formula for two self-inductances in parallel.

If L_1 and L_2 are each equal to L, Equations 87-14 and 87-15 become:

$$L_T^{+} = \frac{L^2 - M^2}{2(L - M)} = \frac{L + M}{2} \text{ henrys} \qquad (87\text{-}16)$$

and,

$$L_T^{-} = \frac{L - M}{2} \text{ henrys} \qquad (87\text{-}17)$$

Example 87-1

Two mutually coupled coils are joined in a series-aiding arrangement. If the self-inductances are 0.65 H and 0.45 H, and the coupling factor is 0.85, what is the value of the total equivalent inductance? If one of the coils is reversed, without changing the magnitude of the coupling factor, what is the new value of the total equivalent inductance?

Solution

Mutual inductance,

$$M = k \times \sqrt{L_1 L_2}$$

$$= 0.85 \times \sqrt{0.65 \times 0.45}$$

$$= 0.46 \text{ H.}$$

Total equivalent inductance,

$$L_T^{+} = L_1 + L_2 + 2M \qquad (87\text{-}2)$$

$$= 0.65 + 0.45 + 2 \times 0.46$$

$$= 2.02 \text{ H.}$$

If one of the coils is reversed, the new connection is series-opposing,

Total equivalent inductance,

$$L_T^{-} = L_1 + L_2 - 2M \qquad (87\text{-}5)$$

$$= 0.65 + 0.45 - 2 \times 0.46$$

$$= 0.18 \text{ H.}$$

Example 87-2

Two coils, whose self-inductances are 75 mH and 55 mH, are connected in parallel-aiding. If the coupling factor is 0.35, what is the total equivalent inductance of the parallel combination? If one of the coils is now reversed, without changing the value of the coupling factor, what is the new value of the total equivalent inductance?

Solution

Mutual inductance,

$$M = k \times \sqrt{L_1 L_2}$$

$$= 0.35 \times \sqrt{75 \times 55}$$

$$= 22.5 \text{ mH.}$$

Total equivalent inductance,

$$L_T^{+} = \frac{L_1 L_2 - M^2}{L_1 + L_2 - 2M} \qquad (87\text{-}14)$$

$$= \frac{75 \times 55 - 22.5^2}{75 + 55 - 2 \times 22.5}$$

$$= 42.6 \text{ mH.}$$

Total equivalent inductance,

$$L_T^{-} = \frac{L_1 L_2 - M^2}{L_1 + L_2 + 2M} \qquad (87\text{-}15)$$

$$= \frac{75 \times 55 - 22.5^2}{75 + 55 + 2 \times 22.5}$$

$$= 20.7 \text{ mH.}$$

PRACTICE PROBLEMS

87-1. Two 1.36-H inductors are joined in a series-opposing arrangement. If the coupling factor is 0.76, determine the value of the total equivalent inductance.

87-2. Two 500-mH inductors are joined in series. Their total inductance is measured and found to be 1.5 H. Determine the value of the coupling factor between the inductors.

87-3. A 75-mH inductor and a 100-mH inductor are connected in series. If their total inductance is 125 mH, calculate the value of the coupling factor between the coils.

87-4. Two identical coils are connected in series. If the total inductance is half the value of the self-inductance of one of the coils, determine the value of the coupling factor between the coils.

87-5. When two chokes are connected in series-aiding, their total inductance is 1.6 H. If the chokes are connected in series-opposing, the total inductance falls to 0.4 H. Determine the value of the mutual inductance between the chokes.

87-6. In Practice Problem 87-5, calculate the value of the total self-inductance possessed by the two chokes.

87-7. Two 1.5-H chokes are connected in parallel-aiding. If the coupling factor is 0.8, calculate the total equivalent inductance of the parallel combination.

88
The power transformer—transformer efficiency

The purpose of a power transformer is to increase (or decrease) the value of the ac line (or supply) voltage without altering the frequency. This operation is achieved with a high level of efficiency.

A power transformer consists of a primary coil and a secondary coil whose number of turns are, respectively, N_p and N_s (Figs. 88-1A and B). The alternating source voltage (E_P) is applied across the primary coil while the load is connected across the secondary coil. These two coils might each consist of thousands of turns, which are wound on a common soft-iron core so that the leakage flux is reduced to a low value.

In the case of the *ideal* transformer, the leakage flux is zero so that the mutual inductance $M = \sqrt{L_p L_s}$, and the coupling factor (k) is unity. When an alternating current (I_p) flows in the primary coil, it creates a magnetic flux that links with the secondary coil and induces the secondary voltage, E_s. With zero flux leakage, there are the *same* volts per turn associated with both the primary and secondary coils. It follows that, if the number of secondary turns exceeds the number of primary turns, the secondary voltage is greater than the primary voltage, and you have a "step-up" transformer. Similarly, if the number of secondary turns is

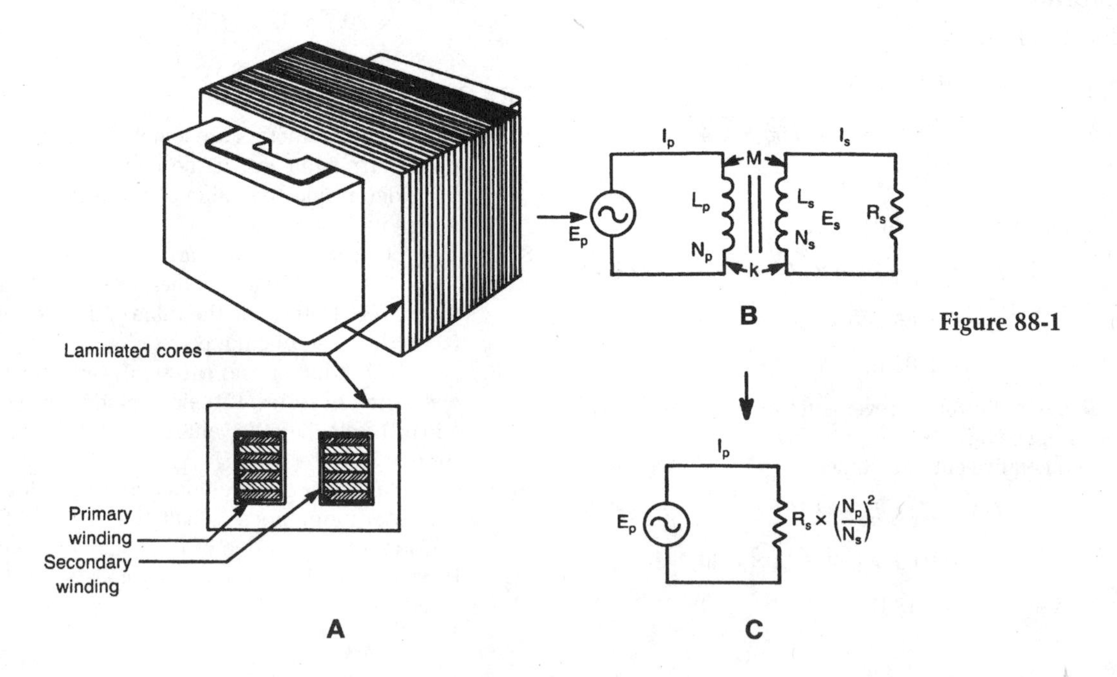

Figure 88-1

less than the number of primary turns, the secondary voltage is smaller than the primary voltage, and the transformer is of the "step-down" type. The terms *step-up* and *step-down* normally refer to the voltage, and not to the current. From the phase point of view, the primary and secondary voltages are 180° out of phase, provided that the two coils are wound in the same sense.

An *ideal* power transformer has zero power losses, and therefore, the power input to the primary circuit equals the power output from the secondary circuit. It follows that a step-up of the voltage level from the primary circuit to the secondary circuit must be accompanied by a corresponding reduction in the current level.

PRACTICAL POWER TRANSFORMER

The *practical* power transformer suffers from the following losses:

1. The copper loss, which is the power dissipated in the resistances of the primary and secondary windings.
2. The iron loss, which is dissipated in the core. This lost energy is subdivided into:
   (a) The eddy-current loss, that is caused by the moving magnetic flux, which cuts the core and induces circulating currents within the metal. This loss is reduced by laminating the core into thin slices, with each slice insulated from its neighbor.
   (b) The hysteresis loss, which is caused by the rapid magnetizing, demagnetizing, and remagnetizing of the core during the cycle of the primary current.

For practical power transformers, the total losses are less than 10% of the primary power.

When a load resistance is connected across the secondary coil, the secondary current creates an alternating magnetic flux which, by Lenz's Law, opposes and partially cancels the primary flux. As a result, the primary current increases and an effective resistance of value E_p/I_p ohms is presented to the primary source; this ohmic value is called the *resistance reflected* from the secondary circuit into the primary circuit, owing to the presence of the secondary load.

Note that, if a steady direct current flows in the primary coil, the leakage flux will be constant in magnitude and direction, so that the voltage induced in the secondary coil is zero. However, if the steady dc voltage applied to the primary coil is "chopped" to produce a square wave, the transformer responds to this type of input, and a form of alternating voltage (not however, a simple sine wave) will be induced in the secondary coil; the secondary voltage is then rectified to provide a final output dc voltage which is larger (or smaller) than the dc input voltage.

MATHEMATICAL DERIVATIONS

Ideal transformer

$$\text{Turns ratio, } \frac{N_p}{N_s} = \frac{E_p}{E_s} \qquad (88\text{-}1)$$

Because an ideal transformer has no losses,

$$\text{Primary power, } P_p = \text{secondary power, } P_s \qquad (88\text{-}2)$$

$$E_p \times I_p = E_s \times I_s \text{ watts}$$

Combining Equations 88-1 and 88-2,

$$\frac{E_p}{E_s} = \frac{I_s}{I_p} = \frac{N_p}{N_s} \qquad (88\text{-}3)$$

Primary voltage,

$$E_p = \frac{E_s I_s}{I_p} = \frac{E_s N_p}{N_s} \text{ volts} \qquad (88\text{-}4)$$

Primary current,

$$I_p = \frac{E_s I_s}{E_p} = \frac{I_s N_s}{N_p} \text{ amperes} \qquad (88\text{-}5)$$

Secondary voltage,

$$E_s = \frac{E_p I_p}{I_s} = \frac{E_p N_s}{N_p} \text{ volts} \qquad (88\text{-}6)$$

Secondary current,

$$I_s = \frac{E_p I_p}{E_s} = \frac{I_p N_p}{N_s} \text{ amperes} \qquad (88\text{-}7)$$

Primary turns,

$$N_p = \frac{E_p N_s}{E_s} = \frac{I_s N_s}{I_p} \qquad (88\text{-}8)$$

Secondary turns,

$$N_s = \frac{E_s N_p}{E_p} = \frac{I_p N_p}{I_s} \qquad (88\text{-}9)$$

Practical transformer

Transformer efficiency,

$$\eta = \frac{\text{secondary output power}}{\text{primary input power}} \times 100\% \qquad (88\text{-}10)$$

$$= \frac{E_s I_s}{E_p I_p} \times 100\%$$

Secondary power output,

$$P_s = E_s I_s \quad (88\text{-}11)$$

$$= P_p \times \frac{\eta}{100}$$

$$= E_p I_p \times \frac{\eta}{100} \text{ watts}$$

Then:

Secondary voltage,

$$E_s = \frac{E_p I_p \eta}{I_s \times 100} \text{ volts} \quad (88\text{-}12)$$

Secondary current,

$$I_s = \frac{E_p I_p \eta}{E_s \times 100} \text{ amperes} \quad (88\text{-}13)$$

Primary voltage,

$$E_p = \frac{E_s I_s \times 100}{\eta I_p} \text{ volts} \quad (88\text{-}14)$$

Primary current,

$$I_p = \frac{E_s I_s \times 100}{\eta E_p} \text{ amperes} \quad (88\text{-}15)$$

The total power loss in the transformer is given by:

$$P_{loss} = \text{primary power} - \text{secondary power} \quad (88\text{-}16)$$

$$= E_p \times I_p - E_s \times I_s$$

$$= \left(\frac{100 - \eta}{100}\right) \times P_p$$

$$= \left(\frac{100 - \eta}{\eta}\right) \times P_s \text{ watts}$$

Reflected resistance

If the secondary is loaded with a resistance (R_s),

$$R_s = \frac{E_s}{I_s} \text{ ohms} \quad (88\text{-}17)$$

Because,

$$E_p = \frac{E_s N_p}{N_s} \text{ (Equation 88-4) and}$$

$$I_p = \frac{I_s N_s}{N_p} \text{ (Equation 88-5)}$$

$$R_p = \frac{E_p}{I_p} \quad (88\text{-}18)$$

$$= \frac{E_s N_p / N_s}{I_s N_s / N_p}$$

$$= \frac{E_s}{I_s} \times \left(\frac{N_p}{N_s}\right)^2$$

$$= R_s \times \left(\frac{N_p}{N_s}\right)^2 \text{ ohms}$$

where R_p = effective resistance presented to the primary source.

The equation

$$\frac{E_p}{I_p} = R_s \times \left(\frac{N_p}{N_s}\right)^2$$

is represented by the equivalent circuit of Fig. 88-1C. The expression $R_s \times (N_p/N_s)^2$ is the effective resistive load presented to the primary source, and is referred to as the value of the resistance reflected from the secondary circuit into the primary circuit, owing to the introduction of the secondary load, R_s. If you choose the turns ratio so that the value of the reflected resistance ($R_s \times (N_p/N_s)^2$) is equal to the internal resistance of the primary source, the secondary load is then matched to the primary source for maximum power transfer to the secondary load.

If R_p is the resistance associated with the primary source, the matched condition is

$$R_p = R_s \times \left(\frac{N_p}{N_s}\right)^2 \text{ and } R_s = R_p \times \left(\frac{N_s}{N_p}\right)^2 \text{ ohms} \quad (88\text{-}19)$$

or,

$$\left(\frac{N_p}{N_s}\right)^2 = \frac{R_p}{R_s} \text{ and turns ratio, } \frac{N_p}{N_s} = \sqrt{\frac{R_p}{R_s}} \quad (88\text{-}20)$$

Example 88-1

In Fig. 88-1B, E_p = 110 V, 60 Hz, N_p = 1200 turns, N_s = 4800 turns, R_s = 220 Ω. Assuming that the transformer is 100% efficient, and has a coupling factor of unity, what are the values of E_s, I_s, I_p, primary power, secondary power, and the reflected resistance?

Solution

$$\text{Turns ratio, } \frac{N_p}{N_s} = \frac{1200}{4800} = 1{:}4$$

The transformer is of the "step-up" variety.

Secondary voltage,

$$E_s = E_p \times \frac{N_s}{N_p} \quad (88\text{-}6)$$

$$= 110 \text{ V} \times 4$$
$$= 440 \text{ V}.$$

Secondary current,

$$I_s = \frac{E_s}{R_s}$$
$$= \frac{440 \text{ V}}{220 \ \Omega}$$
$$= 2 \text{ A}.$$

Primary current,

$$I_p = I_s \times \frac{N_s}{N_p} \quad (88\text{-}5)$$
$$= 2 \text{ A} \times 4$$
$$= 8 \text{ A}.$$

Primary power,

$$P_p = \text{secondary power, } P_s \quad (88\text{-}2)$$
$$= E_p \times I_p$$
$$= 110 \text{ V} \times 8 \text{ A}$$
$$= 880 \text{ W}.$$

Check:

Secondary power,

$$P_s = E_s \times I_s$$
$$= 440 \text{ V} \times 2 \text{ A}$$
$$= 880 \text{ W}.$$

Reflected resistance,

$$\frac{E_p}{I_p} = \frac{110 \text{ V}}{8 \text{ A}} = 13.75 \ \Omega.$$

Check:

Reflected resistance,

$$R_p = R_s \times \left(\frac{N_p}{N_s}\right)^2 \quad (88\text{-}18)$$
$$= 220 \ \Omega \times \left(\frac{1}{4}\right)^2$$
$$= 13.75 \ \Omega.$$

Example 88-2

In Fig. 88-1B, $E_p = 220$ V, 60 Hz, $I_p = 2.3$ A, $E_s = 55$ V, $R_s = 6.5\ \Omega$. What are the values of the primary power, the secondary power, and the transformer efficiency?

Solution

The transformer of the "step-down" variety with a turns ratio of 220 V:55 V = 4:1.

Primary power,

$$P_p = E_p \times I_p \quad (88\text{-}2)$$
$$= 220 \text{ V} \times 2.3 \text{ A}$$
$$= 506 \text{ W}.$$

Secondary power,

$$P_s = \frac{E_s^2}{R_s} = \frac{(55 \text{ V})^2}{6.5 \ \Omega}$$
$$= 465 \text{ W}.$$

Transformer efficiency,

$$\eta = \frac{P_s}{P_p} \times 100 \quad (88\text{-}10)$$
$$= \frac{465 \text{ W}}{506 \text{ W}} \times 100$$
$$= 92\%.$$

PRACTICE PROBLEMS

88-1. In a power transformer, the primary voltage is 120 V, 60 Hz, and the turns ratio is 6:1 step down; the total secondary load resistance is 4 Ω. Assuming a coupling factor of unity and a transformer that is 100% efficient, calculate the values of the primary current and the effective load presented to the primary source.

88-2. In Practice Problem 88-1, what must be the value of the resistance associated with the primary source, in order for maximum power to be transferred to the secondary load?

88-3. The audio output amplifier of a receiver requires a load of 3.6 kΩ. The loudspeaker is 4 Ω. Calculate the required turns ratio of the transformer.

88-4. In a power transformer. operating at 60 Hz, the primary voltage and the primary current are 120 V and 5.5 A. If the secondary voltage and current are, respectively, 550 V and 1.1 A, calculate the value of the transformer efficiency.

88-5. In Practice Problem 88-4, calculate the amount of the power losses in the transformer.

89
Complex algebra—operator *j*—rectangular/polar conversions

In previous chapters, an alternating voltage or current has been represented either by a sine wave or a phasor or a trigonometric expression. Although these representations are adequate for simple series and parallel arrangements, they are too cumbersome for the analysis of more complicated circuits, such as series-parallel combinations and those circuits that require the use of the network theorems. What is clearly needed is a form of algebra that can be applied directly to the solution of general ac circuits. Such an algebra must be capable of accounting for the circuit's phase relationships by distinguishing between the resistive and reactive elements. This is achieved, in complex algebra, by the introduction of the operator j. See Fig. 89-1.

The definition of the *operator j* is as follows: A phasor, when multiplied by the operator j, is rotated through 90° (or $\pi/2$ radians) in the positive (or counterclockwise) direction, but the magnitude of the phasor is unchanged. In a similar way, a phasor, when multiplied by the operator $-j$, is rotated through 90° in the clockwise direction.

Consider a simple case in which a resistance of 3 Ω is connected in series with an *inductive* reactance of 4 Ω. The corresponding phasor diagram is shown in Fig. 89-1A. Because the inductive reactance phasor is pointing vertically upwards, it can be said to lie along the "$+j$" axis (assuming the reference line to be horizontal). The phasor equation for the series combination is therefore:

$$\text{Total impedance phasor, } z = 3 + j4\ \Omega$$

This expression for the impedance phasor is known as *rectangular notation* because the 3 Ω and the $+j4$ Ω phasors are perpendicular. Alternatively, you can say that the magnitude of the impedance phasor is $\sqrt{3^2 + 4^2} = 5\ \Omega$, and its phase angle is inv cos (3/5) = + 53.1°. These results may be combined by stating that the impedance phasor is $5 \angle +53.1°$ (a magnitude of 5 Ω with a phase angle of + 53.1°): this expression for the impedance is referred to as *polar notation*.

If the resistance of 3 Ω were connected in series with a *capacitive* reactance of 4 Ω, the phasor of the total impedance would be $3 - j4$ (rectangular notation), or $5\ \Omega \angle -53.1°$ (polar notation).

An impedance phasor of $3 + j4\ \Omega$ (rectangular notation) can be alternatively expressed as $5\ \Omega \angle +53.1°$ (polar notation). Now, the means by which you can convert between rectangular and polar notations (R→P key and P→R key, on a scientific calculator) are examined. But, first of all, why are both notations required?

In the analysis of ac circuits, you need to add, subtract, multiply and divide phasors. These operations are involved with the rules of complex algebra (chapters 90 and 91) which require both rectangular and polar notations. For example, if two impedances (z_1 and z_2) are in parallel, the total impedance phasor can be found from the product-over-sum formula, $z_T = (z_1 \times z_2)/(z_1 + z_2)$. Clearly, this formula involves the addition of the phasors z_1 and z_2, their multiplication, and finally the division of the numerator phasor by the denominator phasor. If z_2 and z_T are given, the formula is then rearranged as $z_1 = (z_2 \times z_T)/(z_2 - z_T)$ so that subtraction of phasors is required in the calculation of z_1.

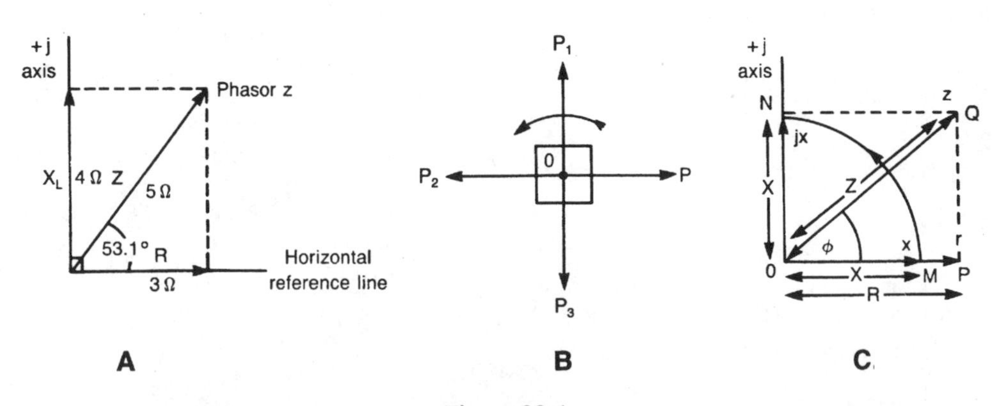

Figure 89-1

MATHEMATICAL DERIVATIONS

In Fig. 89-1B,

$$\text{Phasor } OP_1 = j \times \text{phasor } OP$$
$$\text{Phasor } OP_2 = j \times \text{phasor } OP_1$$
$$= j^2 \times \text{phasor } OP.$$

Because phasors OP and OP_2 are 180° apart,

$$\text{Phasor } OP_2 = -\text{ phasor } OP.$$

Multiplication by -1 rotates a phasor through 180°; of course, the direction of the rotation is immaterial.

But phasor $OP_2 = j^2 \times$ phasor OP.

$$\text{Therefore, } j^2 = -1 \tag{89-1}$$

It follows that $+j$ and $-j$ are reciprocals.

Because the square of a "real" positive or negative number is always positive, j cannot be evaluated in terms of "real" numbers and is, therefore, known as an "imaginary" quantity. The term "imaginary" is used as the opposite of "real," and is not to be thought of as something that does not exist. Also, a complex number, such as $3 + j4$, is composed of a real number and an imaginary number. The word *complex* indicates a combination of real and imaginary quantities, and must not be confused with "complicated."

Figure 89-1C represents a phasor diagram in which phasor OP = phasor r, phasor OM = phasor x and phasor OQ = phasor z. Then, phasor ON = phasor jx and because phasor OQ = phasor ON + phasor OP,

$$\text{Phasor } z = r + jx \tag{89-2}$$

This is known as the *rectangular notation* because the phasors r and jx are separated by 90°. When the phasor z is specified in terms of two phasors, whose *order* is important (phasor $2 + j3$ is *not* the same as phasor $3 + j2$), the representation of the phasor z is referred to as *complex algebra*.

The polar method of denoting a phasor is in terms of its magnitude and its direction. The magnitude of z is indicated by the length of the line OQ, and its direction is measured by the angle, ϕ, between OQ and the horizontal reference line. In polar notation,

$$z = Z \angle \phi \tag{89-3}$$

where Z is the magnitude of the phasor z, and the value of ϕ lies between 0° and ±180°.

When the phasor lies in the lower two quadrants, the angle (ϕ) is negative so that $z = Z \angle -\phi$. The phasor $Z \angle -\phi$ should not be confused with $-Z \angle \phi$, which equals $Z \angle \phi \pm \pi$.

If R and X are the magnitudes of the phasors r and x, the rectangular to polar (R→P) conversion is achieved by the equations:

$$Z = +\sqrt{R^2 + X^2} \text{ ohms} \tag{89-4}$$

and,

$$\phi = \text{inv tan} \frac{X}{R} \tag{89-5}$$

Note that Z is always considered to be positive.

In the polar to rectangular conversion (P→R),

$$R = Z \cos \phi \text{ ohms} \tag{89-6}$$

and,

$$X = Z \sin \phi \text{ ohms} \tag{89-7}$$

Many electronic calculators have the capability of rapid conversions between rectangular and polar notations. You should familiarize yourself with the methods used in your own calculator.

Example 89-1

A series circuit contains a 6-kΩ resistor, an inductor whose reactance is 7 kΩ, and a capacitor with a reactance of 4 kΩ. Write down the expression for the total impedance phasor of the series combination.

Solution

Total impedance,

$$z = 6 + j7 - j4$$
$$= 6 + j3 \text{ k}\Omega \text{ (rectangular notation)}$$

The result that $j7 - j4 = j3$, agrees with the rules of complex algebra, as covered in Chapter 90.

The total impedance has a magnitude of $\sqrt{6^2 + 3^2}$ = 6.71 kΩ and a phase angle of inv. cos (6/6.71) = +26.6°. Therefore, in polar notation, the impedance phasor is 6.71 kΩ $\angle +26.6°$.

Example 89-2

Convert (a) $z = 2 - j3$, $z = 5 + j2$, $z = 4 + j7$, $z = -3 - j$ into polar notation and (b) $z = 7 \angle 32°$, $z = 4 \angle 83°$, $z = 5 \angle -17°$, $z = 3 \angle -127°$ into rectangular notation.

Solution

(a) From Equations 89-4 and 89-5,

$$z = 2 - j3 = +\sqrt{2^2 + 3^2} \angle \text{inv tan } (-3/2)$$
$$= +3.61 \angle -56.3°$$

$$z = 5 + j2 = +\sqrt{5^2 + 2^2} \angle \text{inv tan } (2/5)$$
$$= +5.39 \angle 21.8°$$

$$z = 4 + j7 = +\sqrt{4^2 + 7^2} \angle \text{inv tan } (7/4)$$
$$= +8.06 \angle 60.3°$$

$$z = -3 - j = -3 - j1$$
$$= +\sqrt{3^2 + 1^2} \angle \text{inv tan } (-1/-3)$$
$$= +3.16 \angle -161.6^\circ$$

Caution should be exercised in the determination of the angle. It is sometimes advisable to use a rough sketch in order to establish the quadrant in which the phasor lies. It should also be remembered that the angle, derived mathematically from the calculator, is always measured with respect to the horizontal line, and is an acute angle.

Note the following special cases of rectangular to polar conversions:

$$2 + j0 = 2 = 2 \angle 0^\circ$$
$$0 + j5 = +j5 = 5 \angle 90^\circ$$
$$0 - j3 = -j3 = 3 \angle -90^\circ$$
$$1 + j = +\sqrt{1^2 + 1^2} \angle \text{inv tan } (1/1) = \sqrt{2} \angle 45^\circ$$
$$= 1.414 \angle 45^\circ$$

(b) From Equations 89-6 and 89-7,

$$z = 7 \angle 32^\circ = 7 \cos 32^\circ + j7 \sin 32^\circ$$
$$= 5.94 + j\,3.71.$$

$$z = 4 \angle 83^\circ = 4 \cos 83^\circ + j4 \sin 83^\circ$$
$$= 0.49 + j\,3.97$$

$$z = 5 \angle -17^\circ = 5 \cos (-17^\circ) + j\,5 \sin (-17^\circ)$$
$$= 4.78 - j\,1.46$$

$$z = 3 \angle -127^\circ = 3 \cos (-127^\circ) + j\,3 \sin (-127^\circ)$$
$$= -1.81 - j\,2.40$$

PRACTICE PROBLEMS

89-1. Convert the following rectangular expressions into polar notation: $3 + j11$, $12 - j1$, $7 + j2$, $1 - j8$, $5 + j6$, $-j5$, $-2 + j7$, $-4 - j9$.

89-2. Convert the following polar expressions into rectangular notation: $5.4\angle 37^\circ$, $7.2\angle -43^\circ$, $4\angle 0^\circ$, $8\angle 90^\circ$, $11\angle -90^\circ$, $3.4\angle -117^\circ$, $4.7\angle 147^\circ$.

89-3. Determine the rectangular and polar forms of the impedance phasors for the following series combinations:
(a) 5 Ω of resistance and 6 Ω of inductive reactance.
(b) 7 Ω of resistance and 8 Ω of capacitive reactance.
(c) 11 Ω of inductive reactance and 3 Ω of capacitive reactance.

90 Equating real and imaginary parts—addition and subtraction of phasors

In electronics analysis, you are often faced with an equation in which both the left-hand side and the right-hand side of the equation contain "real" terms and "imaginary" terms. Because "real" numbers and "imaginary" numbers lie on axes that are mutually perpendicular, it follows that the "real" terms on the left-hand side must equal the "real" terms on the right-hand side. Similarly, the "imaginary" terms on the left-hand side can be equated with the "imaginary" terms on the right-hand side. Therefore, the process derives two equations from the original single equation, and is known as "equating real and imaginary parts."

So far, we have added ac quantities together by means of sine waves and phasor diagrams. When complex algebra is used as a means of adding or subtracting phasors, the rectangular notation is much to be preferred. As examples, $(4 - j2) + (3 + j5) = (7 + j3)$ and $(4 - j2) - (3 + j5) = 1 - j7$; however $3 \angle 20^\circ + 2 \angle 35^\circ$ is *not* equal to $5 \angle 55^\circ$. Consequently, if you are asked to add together two phasors which are expressed in polar notation, the best procedure is to convert the two phasors into rectangular notation, and then carry out the addition; the answer can afterwards be converted back into polar notation (if necessary).

MATHEMATICAL DERIVATIONS

Rules of complex algebra
Equating "real" and "imaginary" parts

Let phasor $z_1 = R_1 + jX_1$ and phasor $z_2 = R_2 + jX_2$. If phasor z_1 = phasor z_2, then,

$$R_1 = R_2 \text{ and } X_1 = X_2 \quad (90\text{-}1)$$

This is the process of equating the "real" and the "imaginary" parts of an equation.

Addition and subtraction of phasors

If $z_1 = R_1 + jX_1$ and $z_2 = R_2 + jX_2$, then the sum of z_1 and z_2 is

$$\text{Phasor } z_1 + z_2 = (R_1 + jX_1) + (R_2 + jX_2) \quad (90\text{-}2)$$
$$= (R_1 + R_2) + j(X_1 + X_2)$$

When subtracting phasor z_2 from phasor z_1, the result is

$$\text{Phasor } z_1 - z_2 = (R_1 + jX_1) - (R_2 + jX_2) \quad (90\text{-}3)$$
$$= (R_1 - R_2) + j(X_1 - X_2)$$

Example 90-1

If phasor $z_1 = 4 - j2$ and phasor $z_2 = 3 + j5$, find the values of the phasors $z_1 + z_2$ and $z_1 - z_2$. Illustrate the results on a phasor diagram.

Solution

$$\text{Phasor } z_1 = 4 - j2 = \sqrt{4^2 + 2^2} \text{ inv tan } (-2/4)$$
$$= 4.47 \angle -26.6°$$

$$\text{Phasor } z_2 = 3 + j5 = \sqrt{3^2 + 5^2} \text{ inv tan } (5/3)$$
$$= 5.83 \angle 59.0°$$

$$\text{Phasor } z_1 + z_2 = (4 - j2) + (3 + j5) \quad (90\text{-}2)$$
$$= 7 + j3$$
$$= \sqrt{7^2 + 3^2} \angle \text{inv tan } (3/7)$$
$$= 7.62 \angle 23.2°$$

$$\text{Phasor } z_1 - z_2 = (4 - j2) - (3 + j5) \quad (90\text{-}3)$$
$$= 1 - j7 = \sqrt{1^2 + 7^2} \angle \text{inv tan } (-7/1)$$
$$= 7.07 \angle -81.9°$$

These results are illustrated to scale in Fig. 90-1A. This phasor diagram confirms the parallelogram method of obtaining the sum and difference phasors.

Example 90-2

If phasor $z_1 = 3 \angle 20°$ and phasor $z_2 = 2 \angle 35°$, find the value of phasor $z_1 + z_2$.

Solution

$$\text{Phasor } z_1 = 3 \angle 20° = 3 \cos 20° + j3 \sin 20° \quad (90\text{-}2)$$
$$= 2.82 + j1.03$$

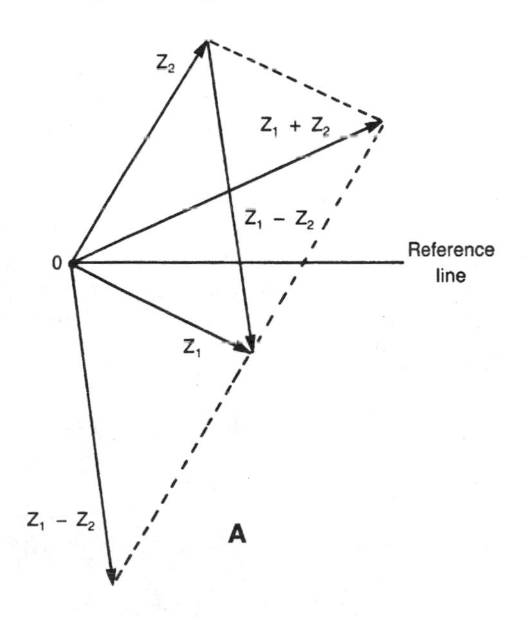

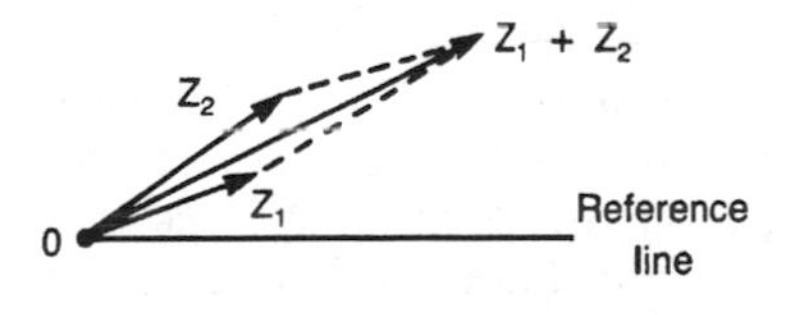

Figure 90-1

$$\text{Phasor } z_2 = 2\angle 35° = 2\cos 35° + j2\sin 35°$$
$$= 1.64 + j1.15$$

$$\text{Phasor } z_1 + z_2 = (2.82 + j1.03) + (1.64 + j1.15)$$
$$= 4.46 + j2.18$$
$$= \sqrt{4.46^2 + 2.18^2}\ \angle \text{inv tan } (2.18/4.46)$$
$$= 4.96\ \angle 26.04°$$

This result is illustrated in Fig. 90-1B.

PRACTICE PROBLEMS

90-1. If $x + jy = 3.7\angle 74°$, determine the values of x and y.

90-2. If $z_1 = 7 - j3$ and $z_2 = 9 + j4$, find the values of the phasors $z_1 + z_2$ and $z_1 - z_2$. Express the answers in both rectangular and polar notations.

90-3. If $z_1 = 5.6\angle -117°$ and $z_2 = 3.4\angle 156°$, find the values of the phasors $z_1 + z_2$ and $z_2 - z_1$. Express the answers in both rectangular and polar notations.

91 Multiplication and division of phasors—rationalization

If an ac circuit consists of a general impedance (z_1) in parallel with another general impedance (z_2), the total impedance (z_T) is equal to $(z_1 \times z_2)/(z_1 + z_2)$. In order to obtain the value of z_T, you need to have both the ability to multiply phasors and to divide phasors.

Although it is possible to multiply phasors in their rectangular notation, it is generally preferable to use polar notation. As you can see from the mathematical derivations, $2\angle -35° \times 3\angle 40° = 2 \times 3\angle[(-35°) + (+40°)] = 6\angle 5°$; consequently, the magnitudes of the phasors are multiplied, but their angles are added algebraically. Knowing the rule for multiplication, you can also square a phasor or take its square root.

Division is also preferably carried out with polar notation. From the mathematical derivations, $3\angle 40°/2\angle -35° = 1.5\ \angle[(+40°) - (-35°)] = 1.5\ \angle 75°$. The magnitude of the numerator is divided by the magnitude of the denominator, but the angle of the denominator is algebraically subtracted from the angle of the numerator. From the rule of division, you can also obtain the reciprocal of a phasor.

MATHEMATICAL DERIVATIONS

Multiplication of phasors

If phasor $z_1 = R_1 + jX_1$, and phasor $z_2 = R_2 + jX_2$, then by rectangular notation,

Product phasor,

$$z_1 z_2 = (R_1 + jX_1)(R_2 + jX_2) \tag{91-1}$$
$$= R_1R_2 + jR_1X_2 + jX_1R_2 + j^2X_1X_2$$
$$= (R_1R_2 - X_1X_2) + j(R_1X_2 + X_1R_2)$$

because $j^2 = -1$.

Using polar notation, $z_1 = Z_1\angle\phi_1$ and $z_2 = Z_2\angle\phi_2$.

Product phasor,

$$z_1z_2 = Z_1(\cos\phi_1 + j\sin\phi_1) \times \tag{91-2}$$
$$Z_2(\cos\phi_2 + j\sin\phi_2)$$
$$= Z_1Z_2[(\cos\phi_1\cos\phi_2 - \sin\phi_1\sin\phi_2) + j(\sin\phi_1\cos\phi_2 + \cos\phi_1\sin\phi_2)]$$
$$= Z_1Z_2[\cos(\phi_1 + \phi_2) + j\sin(\phi_1 + \phi_2)]$$
$$= Z_1Z_2\angle(\phi_1 + \phi_2)$$

When multiplying phasors, the magnitudes are multiplied, but the angles are added algebraically.

If a phasor, $z = Z\angle\phi$, the square of the phasor is:

$$z^2 = Z\angle\phi \times Z\angle\phi = Z^2\angle 2\phi \tag{91-3}$$

Therefore, when squaring a phasor, the magnitude is squared, but the angle is doubled. It follows that by reversing the process to obtain the square root of a phasor, you must take the square root of the magnitude but halve the angle.

The square root of the phasor,

$$\sqrt{z} = \sqrt{Z\angle\phi} = \sqrt{Z}\angle(\phi/2) = Z^{1/2}\angle(\phi/2) \tag{91-4}$$

Division of phasors

If phasor $z_1 = R_1 + jX_1$ and phasor $z_2 = R_2 + jX_2$,

$$\text{Phasor } \frac{z_1}{z_2} = \frac{R_1 + jX_1}{R_2 + jX_2}$$

In order to separate out the "real" and "imaginary" parts of z_1/z_2, it is necessary to eliminate j from the denominator. This is achieved by the process of *rationalization,* which means multiplying both the numerator and the denominator by the conjugate of the denominator. The *conjugate* of a phasor is another phasor that has the same magnitude, but whose angle (although equal in size) has the opposite sign. Therefore, the conjugate of $Z\angle\phi$ is $Z\angle-\phi$, and the conjugate of $R+jX$ is $R-jX$. It is important to recognize that the product of a phasor and its conjugate is always a totally "real" quantity with no j component.

Rationalizing z_1/z_2 yields

$$\text{Phasor } \frac{z_1}{z_2} = \frac{(R_1+jX_1)(R_2-jX_2)}{(R_2+jX_2)(R_2-jX_2)} \tag{91-5}$$

$$= \frac{R_1R_2+X_1X_2}{R_2^2+X_2^2} + j\frac{(X_1R_2-X_2R_1)}{R_2^2+X_2^2}$$

In polar form, let $z_1 = Z_1\angle\phi_1$ and $z_2 = Z_2\angle\phi_2$. Then

$$\frac{z_1}{z_2} = \frac{Z_1(\cos\phi_1 + j\sin\phi_1)}{Z_2(\cos\phi_2 + j\sin\phi_2)}$$

$$= \frac{Z_1}{Z_2}\times\left[\frac{(\cos\phi_1 + j\sin\phi_1)(\cos\phi_2 - j\sin\phi_2)}{(\cos\phi_2 + j\sin\phi_2)(\cos\phi_2 - j\sin\phi_2)}\right]$$

$$= \frac{Z_1}{Z_2}\times\left[\frac{(\cos\phi_1\cos\phi_2 + \sin\phi_1\sin\phi_2)}{\cos^2\phi_2+\sin^2\phi_2} + \frac{j(\sin\phi_1\cos\phi_2 - \cos\phi_1\sin\phi_2)}{\cos^2\phi_2+\sin^2\phi_2}\right]$$

$$= \frac{Z_1}{Z_2}\times\left[\frac{\cos(\phi_1-\phi_2) + j\sin(\phi_1-\phi_2)}{1}\right]$$

because $\cos^2\phi_2 + \sin^2\phi_1 = 1$.

Therefore,

$$\frac{z_1}{z_2} = \frac{Z_1}{Z_2}\angle(\phi_1-\phi_2) \tag{91-6}$$

The magnitude of the numerator is divided by the magnitude of the denominator, but the angle of the denominator is algebraically subtracted from the angle of the numerator.

In the special case that involves the reciprocal of the phasor, $z = Z\angle\phi$,

$$\frac{1}{z} = \frac{1\angle 0°}{Z\angle\phi} = \frac{1}{Z}\angle-\phi \tag{91-7}$$

Example 91-1

If phasor $z_1 = 1+j2$ and $z_2 = 3-j2$, find the values of the phasors z_1z_2, z_1/z_2, z_1^2, $\sqrt{z_2}$, $1/z_1$, and the conjugate of z_1. Illustrate your results on phasor diagrams.

Solution

$$\text{Phasor } z_1 = 1+j2 = \sqrt{1^2+2^2}\angle\text{inv tan }(2/1)$$
$$= 2.24\angle 63.4°$$

$$\text{Phasor } z_2 = 3-j2 = \sqrt{3^2+2^2}\angle\text{inv tan }(-2/3)$$
$$= 3.61\angle-33.7°$$

$$\text{Phasor } z_1z_2 = 2.24\angle 63.4° \times 3.61\angle-33.7°$$
$$= 8.09\angle 29.7°$$

$$\text{Phasor } \frac{z_1}{z_2} = \frac{2.24\angle 63.4°}{3.61\angle-33.7°}$$
$$= 0.62\angle[(+63.4°)-(-33.7°)]$$
$$= 0.62\angle 97.1°$$

$$\text{Phasor } z_1^2 = (2.24\angle 63.4°)^2$$
$$= 5.02\angle 126.8°$$

$$\text{Phasor } \sqrt{z_2} = \sqrt{3.61\angle-33.7°}$$
$$= 1.90\angle-16.85°$$

$$\text{Phasor } \frac{1}{z_1} = \frac{1}{2.24\angle 63.4°}$$
$$= 0.45\angle-63.4°$$

$$\text{Phasor conjugate of } z_1 = 2.24\angle-63.4°$$

These results are illustrated in the phasor diagrams of Figs. 91-1A and B.

PRACTICE PROBLEMS

91-1. Rationalize the expression $(14+j5)/(3+j2)$. Express the answer in both rectangular and polar notations.

91-2. If $z_1 = 4\angle 30°$ and $z_2 = 5\angle-70°$, find the values in polar form of z_1z_2, z_1/z_2, z_2^2, $\sqrt{z_1}$ and $1/z_2$.

91-3. Write down the conjugates of $-3+j2$ and $7\angle-130°$.

91-4. If $z_1 = 4-j7$ and $z_2 = 11+j5$, find the values of z_1+z_2, z_1-z_2, z_1z_2, z_1/z_2, z_1^2, $\sqrt{z_2}$, and $1/z_1$. Express the answers in both rectangular and polar notations.

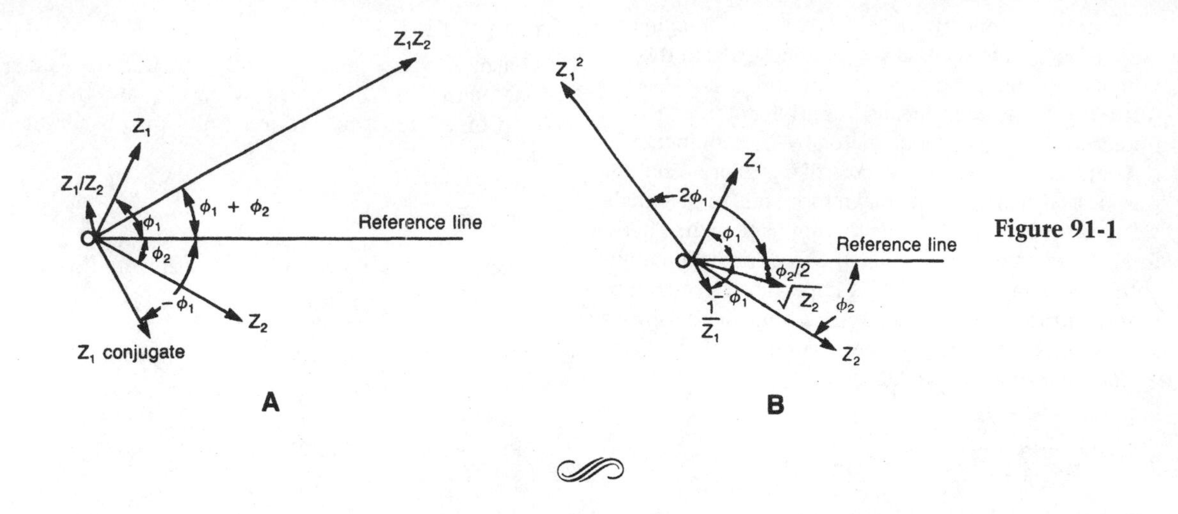

Figure 91-1

92
Analysis of a series-parallel circuit with the aid of the *j* operator

In terms of impedance phasors, the formula for the total impedance is the same as we earlier derived for comparable resistor networks. In the circuit of Fig. 92-1B, we would start by combining the parallel z_1 and z_2 phasors by using the "product-over-sum" formula. Then, to the $z_1 \| z_2$ phasor, add the series phasor z_3, and the result is the total impedance phasor, z_T. These operations are performed with the rules of complex algebra as outlined in chapters 90 and 91.

After obtaining the total impedance phasor, the source voltage (e) is divided by z_T to obtain the total current, i_T. Subsequently, you can calculate the individual voltages across the components, the branch currents, and the powers associated with each current.

MATHEMATICAL DERIVATIONS

In Figs. 92-1 A and B,

$$\text{Phasor } z_1 \| z_2 = \frac{z_1 \times z_2}{z_1 + z_2} \tag{92-1}$$

Total impedance phasor,

$$z_T = z_3 + z_1 \| z_2 \tag{92-2}$$

$$= z_3 + \frac{z_1 \times z_2}{z_1 + z_2}$$

Total current phasor,

$$i_T = \frac{e}{z_T} \tag{92-3}$$

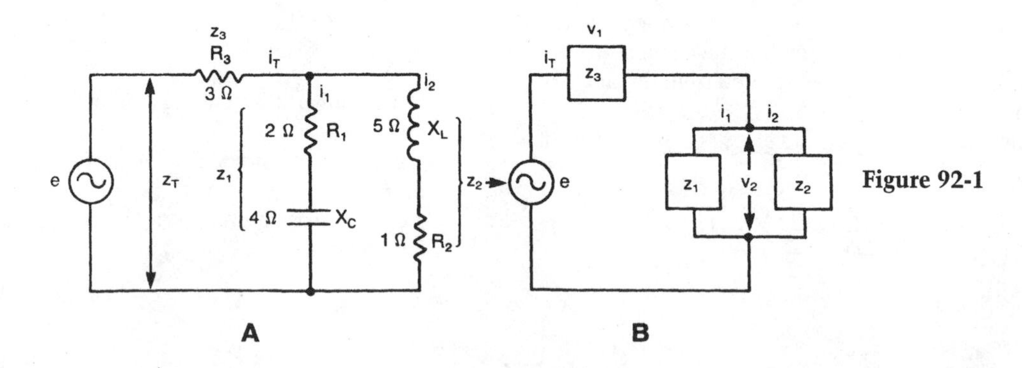

Figure 92-1

Voltage phasor,

$$v_1 = i_T \times z_3 \quad (92\text{-}4)$$

Voltage phasor,

$$v_2 = e - v_1 \quad (92\text{-}5)$$

Current phasor,

$$i_1 = \frac{v_2}{z_1} \quad (92\text{-}6)$$

Current phasor,

$$i_2 = \frac{v_2}{z_2} \quad (92\text{-}7)$$

Example 92-1

In Fig. 92-1A, $z_1 = 2 - j4\ \Omega$, $z_2 = 1 + j5\ \Omega$, $z_3 = 3 + j0\ \Omega$, $e = 7\ \angle 60°$ V. Calculate the values of z_T and i_T.

Solution

$$\text{Phasor } z_1 \| z_2 = \frac{z_1 \times z_2}{z_1 + z_2} \quad (92\text{-}1)$$

$$= \frac{(2 - j4) \times (1 + j5)}{(2 - j4) + (1 + j5)}$$

$$= \frac{(2 - j4) \times (1 + j5)}{3 + j1}$$

Converting to polar notation,

Phasor, $z_1 \| z_2$

$$= \frac{(\sqrt{2^2+4^2} \angle \text{inv tan } (-4/2))(\sqrt{1^2+5^2} \angle \text{inv tan } (5/1))}{\sqrt{3^2+1^2} \angle \text{inv tan } (1/3)}$$

$$= \frac{4.47 \angle -63.4° \times 5.10 \angle 78.7°}{3.16 \angle 18.4°}$$

$$= 7.21 \angle -3.1°$$

$$= 7.19 - j\,0.39\ \Omega$$

Phasor,

$$z_T = z_3 + z_1 \| z_2 \quad (92\text{-}2)$$

$$= 3 + 7.19 - j\,0.39$$

$$= 10.19 - j\,0.39\ \Omega \text{ (rectangular notation)}$$

$$= 10.20\ \angle -2.19°\ \Omega \text{ (polar notation)}$$

Phasor,

$$i_T = \frac{e}{z_T} \quad (92\text{-}3)$$

$$= \frac{7\ \angle 60°\ \text{V}}{10.20\ \Omega\ \angle -2.19°}$$

$$= 0.69\ \angle[(+60°) - (-2.19°)]$$

$$= 0.69\ \angle 62.19°\ \text{A}.$$

PRACTICE PROBLEMS

92-1. In Fig. 92-2, find the value of the z_T phasor.

92-2. In Practice Problem 92-1, calculate the value of the i_T phasor.

92-3. In Practice Problem 92-1, calculate the values of phasors v_1 and v_2.

92-4. In Practice Problem 92-1, calculate the values of phasors i_1 and i_2.

92-5. Determine the total power dissipated in the circuit of Fig. 92-2.

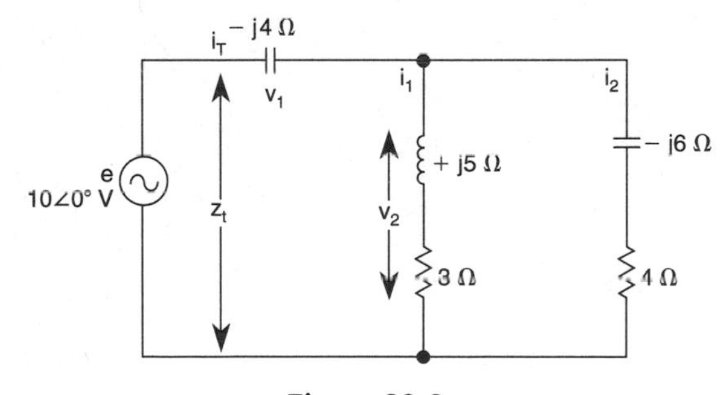

Figure 92-2

93
Analysis of a parallel branch circuit with the aid of the *j* operator

In terms of the impedance phasors, the formula for the total impedance is the same as you earlier derived for comparable resistor networks. In the circuit of Fig. 93-1B, you would use the reciprocal formula to obtain the total impedance phasor, z_T. The calculation of z_T is obtained by using the rules of complex algebra as outlined in chapters 90 and 91.

The branch currents i_1, i_2, etc., are calculated by dividing the source voltage (e) by each of the branch impedances. The products $e \times i_1$, $e \times i_2$, etc. will then provide the individual powers associated with the branches (after allowing for the phase differences between the voltage and the currents).

MATHEMATICAL DERIVATIONS

In Figs. 93-1A and B,

$$\frac{1}{z_T} = \frac{1}{z_1} + \frac{1}{z_2} + \frac{1}{z_3} + \frac{1}{z_4} + \frac{1}{z_5} + \frac{1}{z_6} \qquad (93\text{-}1)$$

Total impedance phasor,

$$z_T = \frac{1}{\dfrac{1}{z_1} + \dfrac{1}{z_2} + \dfrac{1}{z_3} + \dfrac{1}{z_4} + \dfrac{1}{z_5} + \dfrac{1}{z_6}} \qquad (93\text{-}2)$$

Total current phasor,

$$i_T = \frac{e}{z_T} \qquad (93\text{-}3)$$

Branch current phasors,

$$i_1 = \frac{e}{z_1},\ i_2 = \frac{e}{z_2},\ \text{etc.} \qquad (93\text{-}4)$$

Example 93-1

In Fig. 93-1A, $z_1 = 3 + j4$, $z_2 = 5 - j2$, $z_3 = j10 - j3$, $z_4 = 0 - j4$, $z_5 = 0 + j5$, $z_6 = 8 + j0$, $e = 8\angle 25°$ V. Find the values of the phasors z_T and i_T.

Solution

The total impedance phasor (z_T) is given by,

$$\frac{1}{z_T} = \frac{1}{z_1} + \frac{1}{z_2} + \frac{1}{z_3} + \frac{1}{z_4} + \frac{1}{z_5} + \frac{1}{z_6} \qquad (93\text{-}1)$$

$$\frac{1}{z_T} = \frac{1}{3 + j4} + \frac{1}{5 - j2} + \frac{1}{j10 - j3} + \frac{1}{0 - j4} + \frac{1}{0 + j5} + \frac{1}{8 + j0}$$

$$= \frac{3 - j4}{3^2 + 4^2} + \frac{5 + j2}{5^2 + 2^2} + \frac{1}{j7} + \frac{1}{-j4} + \frac{1}{j5} + \frac{1}{8}$$

$$= \frac{3 - j4}{25} + \frac{5 + j2}{29} - j0.142 + j0.25 - j0.2 + 0.125$$

$$= 0.12 - j0.16 + 0.172 + j0.069 - j0.142 + j0.25 - j0.2 + 0.125$$

$$= 0.417 - j0.183$$

$$= 0.455\angle -23.7° \text{ S.}$$

Total impedance phasor,

$$z_T = \frac{1}{0.455 \angle -23.7°} = 2.20\ \angle 23.7°\ \Omega$$

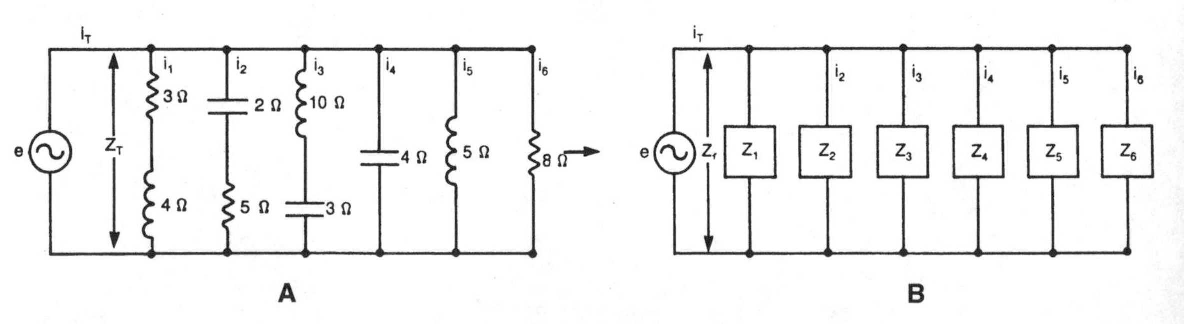

Figure 93-1

Total current,

$$i_T = \frac{e}{z_T} \qquad (93\text{-}3)$$

$$= \frac{8\ \angle 25°\ \text{V}}{2.20\ \angle +23.7°\ \Omega}$$

$$= 3.64\ \angle[(+25°) - (+23.7°)]\ \text{A}$$

$$= 3.64\ \angle 1.3°\ \text{A}.$$

PRACTICE PROBLEMS

93-1. In Fig. 93-2, find the value of the total impedance phasor, z_T.

93-2. In Practice Problem 93-1, find the value of the current, i_T.

93-3. In Practice Problem 93-1, calculate the phasor values of the currents, i_1, i_2, and i_3.

93-4. Find the amount of the total power dissipated in the circuit of Fig. 93-2.

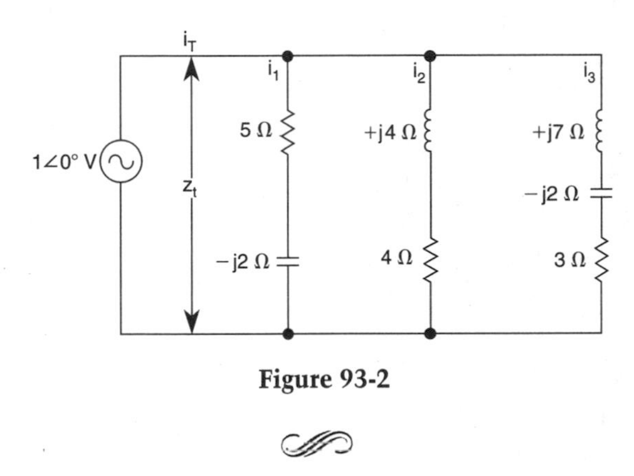

Figure 93-2

94
Kirchhoff's laws for ac circuits

Chapters 26 through 33 covered the various network theorems, and their application to dc circuit analysis. If the same theorems are to be used in sine-wave ac circuit analysis, you must recognize the added difficulties of reactance, impedance, and phase difference, all of which do not exist in dc circuits. Fortunately, these differences can be resolved by the use of the *j* operator, and the rules of complex algebra that appeared in chapters 89 through 91. You are then able to solve network problems without any reference to phasor diagrams. In such problems, all the voltages and currents are assumed to have the same frequency.

In networks that involve more than one voltage (or current) source, these sources are not necessarily in phase. Therefore, each source (for example, 8∠75°V or 3∠−40°A) has its own polar angle that is related to a reference sine wave.

MATHEMATICAL DERIVATIONS

Kirchhoff's voltage law (KVL)

The phasor sum of the source voltages and the voltage drops around any closed electrical loop is always zero.

Kirchhoff's current law (KCL)

The phasor sum of the currents existing at any electrical junction point is zero.

In dc circuits, you adopted a convention in order to distinguish between positive and negative dc voltages as they appeared in the algebraic equations. However, you cannot assign a constant direction to an alternating current or a fixed polarity to an alternating voltage. Instead, give an instantaneous polarity ("+" or "−") to a source voltage, then indicate the associated direction of the (electron) current. Figure 94-1 illustrates these points.

Example 94-1

In the circuit of Fig. 94-1, obtain the value of the current i_3.

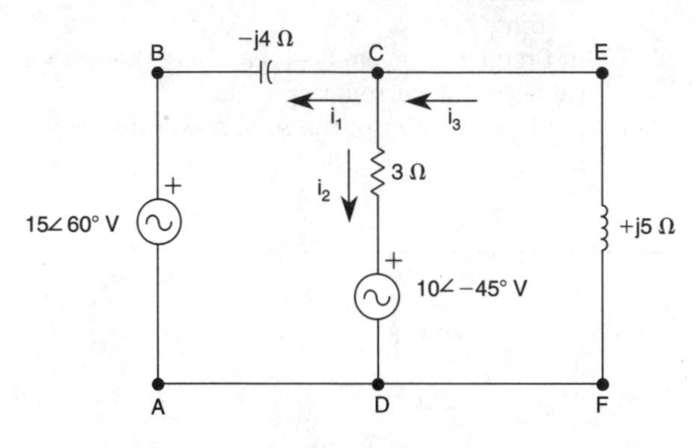

Figure 94-1

Solution

The KVL equation for the loop ABCDA is:

$$i_1 \times (-j4) - i_2 \times (3) = 15\angle 60° - 10\angle -45°$$
$$= 7.5 + j12.99 - 7.07 + j7.07$$
$$= 0.43 + j20.06$$

Note that the two voltage sources are opposing, in accordance with their polarities, as shown.
The KVL equation for the loop CEFDC is:

$$i_2 \times 3 + i_3 \times j5 = 10\angle -45° = 7.07 - j7.07$$

At the junction point (C) the KCL equation is:

$$i_3 = i_1 + i_2$$

or,

$$i_2 = i_3 - i_1$$

The two preceding equations yield,

$$3i_3 - 3i_1 + i_3 \times j5 = 7.07 - j7.07$$
$$-3i_1 + i_3\,(3 + j5) = 7.07 - j7.07 \qquad (94\text{-}1)$$

From the KVL equation for the loop ABCDA,

$$i_1 \times (-j4) - (i_3 - i_1) \times 3 = 0.43 + j20.06$$
$$i_1 \times (3 - j4) - 3i_3 = 0.43 + j20.06$$

Adding the last two equations gives,

$$i_1 \times (-j4) + i_3 \times (j5) = 7.5 + j12.99 \qquad (94\text{-}2)$$

Multiplying Equation 94-1 by $-j4$, and Equation 94-2 by -3,

$$i_1 \times j12 + i_3(-j12 + 20) = -j28.28 - 28.28 \qquad (94\text{-}3)$$
$$i_1 \times j12 - i_3 \times j15 = -22.2 - j38.97 \qquad (94\text{-}4)$$

Subtracting Equation 94-4 from Equation 94-3,

$$i_3(-j12 + 20 + j15) = -5.78 + j10.69$$
$$i_3 = \frac{-5.78 + j10.69}{20 + j3}$$

Then,

$$i_3 = \frac{12.15\angle 118.4°}{20.22\angle 8.53°} = 0.60\angle 109.9° \text{ A}$$

PRACTICE PROBLEMS

94-1. In Fig. 94-2, calculate the values of the currents i_1, i_2, and i_3.

94-2. In Practice Problem 94-1, determine the values of the voltages v_1, v_2, and v_3.

94-3. In Practice Problem 94-2, calculate the amount of the total power dissipated in the circuit.

94-4. In Practice Problem 94-1, determine the individual powers drawn from the e_1 and e_2 sources. Compare the sum of these powers with the answer to Practice Problem 94-3.

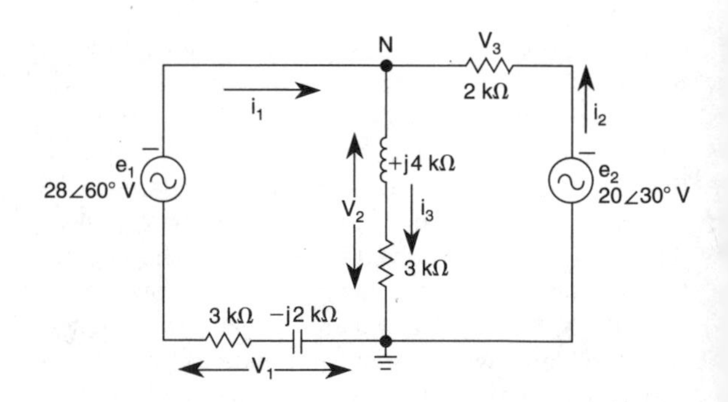

Figure 94-2

95
Mesh-current analysis for ac circuits

Chapter 27 explored the use of mesh-current analysis as applied to dc circuits. The same principles apply to ac circuits, so that you can replace the three branch currents of Fig. 94-1 with the two mesh currents i_1 and i_2 (Fig. 95-1). You will then be involved only with two simultaneous equations, rather than the three equations that appeared in the Kirchhoff analysis.

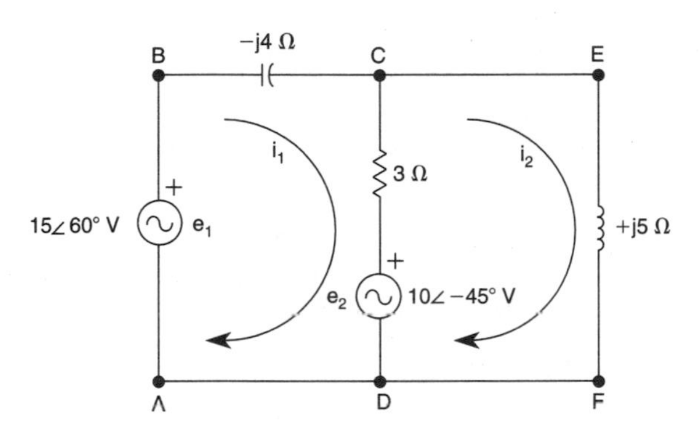

Figure 95-1

MATHEMATICAL DERIVATIONS

For the circuit of Fig. 95-2, the following mesh equations are written down simply by inspection:

$$i_1 (z_1 + z_2) - i_2 z_2 = e_1$$

$$- i_1 z_2 + i_2 (z_2 + z_3 + z_4) - i_3 z_4 = e_2$$

$$- i_2 z_4 + i_3 (z_4 + z_5 + z_6) = - e_3$$

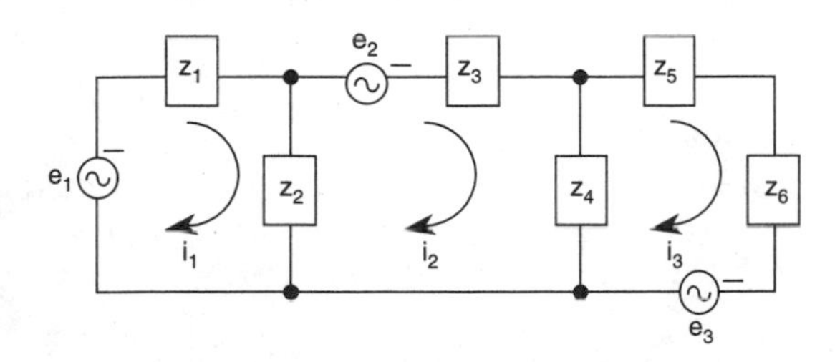

Figure 95-2

Example 95-1

In Fig. 95-1 use the method of mesh-current analysis to determine the value of i_2.

Solution

As far as the polarities of the voltage sources are concerned, you can observe the normal KVL convention. In the mesh ABCDA:

$$i_1 (3 - j4) - i_2 \times 3 = -15\angle 60° + 10\angle -45°$$

$$i_1(3 - j4) - i_2 \times 3 = - 7.5 - j12.99 + 7.07 - j7.07$$

$$= - 0.43 - j20.06$$

In the mesh CEFDC:

$$-3i_1 + i_2 \times (3 + j5) = -10\angle -45° = -7.07 + j7.07$$

Multiply the first equation by 3 and the second equation by $3 - j4$. This yields,

$$i_1(9 - j12) - i_2 \times 9 = - 1.29 - j60.18$$

$$- i_1(9 - j12) + i_2(3 + j5)(3 - j4) =$$

$$(-7.07 + j7.07)(3 - j4)$$

$$- i_1(9 - j12) + i_2(29 + j3) = 7.07 + j49.49$$

Adding the two preceding equations yields,

$$i_2(20 + j3) = 5.78 - j10.69$$

Therefore,

$$\text{current, } i_2 = \frac{5.78 - j10.69}{20 + j3}$$

$$= \frac{12.15\angle -61.6°}{20.22\angle 8.35°} = 0.60\angle -69.1° \text{ A}$$

Note that the mesh current (i_2) is 180° out of phase with the Kirchhoff branch current (i_3) in Example 94-1. This is the result of assuming opposite directions for these two currents.

PRACTICE PROBLEMS

95-1. In Fig. 95-3, calculate the values of the mesh currents i_1 and i_2.

95-2. In Practice Problem 95-1, obtain the value of the (electron flow) current which flows in the direction of N to G.

95-3. In Practice Problem 95-1, determine the total amount of power dissipated in the circuit of Fig. 95-2.

95-4. In Practice Problem 95-1, calculate the individual powers from the e_1 and e_2 sources. Compare the sum of these powers with the answer to Practice Problem 95-3.

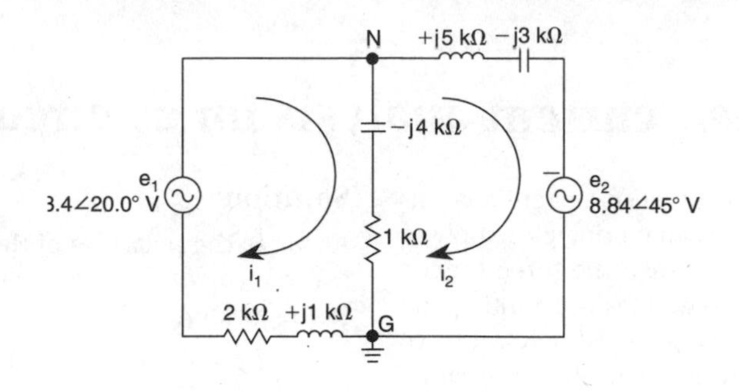

Figure 95-3

96
The superposition theorem for ac circuits

This theorem was applied to dc currents in chapter 29. For ac circuits, the theorem can be restated as follows:

If a network of linear impedances contains more than one alternating source, the current flowing at any point is the phasor sum of the currents that would flow at that point if each alternating source were considered separately, with all other alternating sources replaced by impedances equal to their internal impedances. This would involve replacing each voltage source by a short circuit and each current source by an open circuit.

MATHEMATICAL DERIVATIONS

The current flowing at any point is:

$$i = i_1 + i_2 + i_3 \ldots + i_n \qquad (96\text{-}1)$$

where,

i_1 = current flowing at the same point because of the alternating voltage (e_1) with all other sources replaced by their internal impedances.

i_2 = current flowing at the same point because of the alternating voltage (e_2) with all other sources replaced by their internal impedances, and so on.

Example 96-1

In the circuit of Fig. 96-1, use the superposition theorem to determine the value of the current i_L.

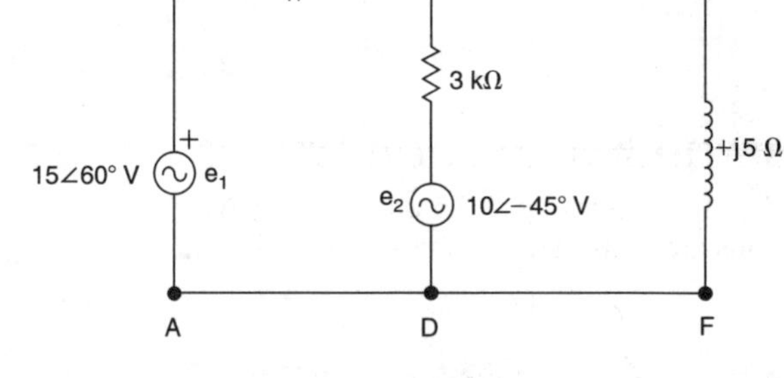

Figure 96-1

Note: This is the same circuit as appeared in Figs. 94-1 and 95-1.

Solution

Step 1. Replace the 10∠−45°-V source by a short circuit (Fig. 96-2A). The total impedance (Z_{1T}), then presented to the 15∠−60°-V source is:

$$Z_{1T} = -j4 + \frac{3 \times j5}{3 + j5} = -j4 + \frac{j15(3 - j5)}{3^2 + 5^2}$$

$$= -j4 + \frac{75 + j45}{34}$$

$$= -j4 + 2.21 + j1.32$$

$$= 2.21 - j2.68\ \Omega$$

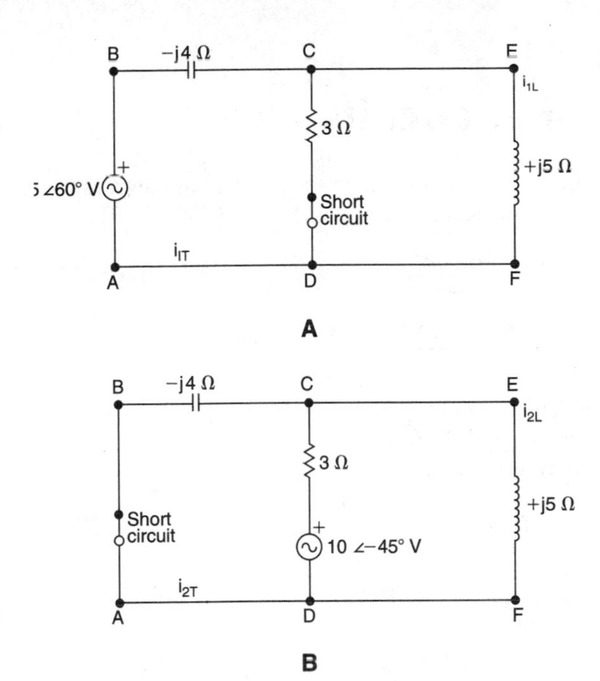

Figure 96-2

Then,

$$i_{1T} = \frac{15\angle 60^\circ \text{ V}}{2.21 - j2.28\ \Omega}$$

and,

$$i_{1L} = \frac{15\angle 60^\circ}{2.21 - j2.28} \times \frac{3}{3 + j5}$$

$$= \frac{45\angle 60^\circ}{3.47\angle -50^\circ \times 5.83\angle 59.04^\circ}$$

$$= 2.22\angle 51.46^\circ \text{ A}$$

Step 2. Replace the 15∠60°-V source by a short circuit (Fig. 96-2B). The total impedance (Z_{2T}) presented to the 10∠−45°-V source is:

$$Z_{2T} = 3 + \frac{(-j4) \times j5}{j5 - j4} = 3 + \frac{20}{j1} = 3 - j20\ \Omega$$

Then,

$$i_{2T} = \frac{10\angle -45^\circ \text{ V}}{3 - j20\ \Omega}$$

and,

$$i_{2L} = \frac{10\angle -45^\circ}{3 - j20} \times \frac{(-j4)}{-j4 + j5} = \frac{-40\angle -45^\circ}{3 - j20}$$

$$= \frac{-40\angle -45^\circ}{20.22\angle -81.47^\circ}$$

$$= -1.98\angle 36.47^\circ \text{ A}$$

Using the principle of superposition,

$$i_L = i_{1L} + i_{2L} = 2.22\angle 51.46^\circ - 1.98\angle 36.47^\circ$$

$$= 1.38 + j1.736 - 1.592 - j1.177$$

$$= -0.212 + j0.559$$

$$= 0.6\angle 110^\circ \text{ A}$$

This is the same value as was obtained for the Kirchhoff branch current (i_3) in Example 94-1.

PRACTICE PROBLEMS

96-1. In Fig. 96-3, use the superposition theorem to determine the value of the current i_2.

96-2. In Practice Problem 96-1, calculate the values of the current i_1 and the (electron flow) current flowing in the direction of G to N.

96-3. In Practice Problem 96-1, determine the amount of the total power dissipated in the circuit.

96-4. In Practice Problem 96-1, determine the amounts of power delivered to the circuit by the sources e_1 and e_2. Compare the sum of these powers with the answer to Practice Problem 96-3.

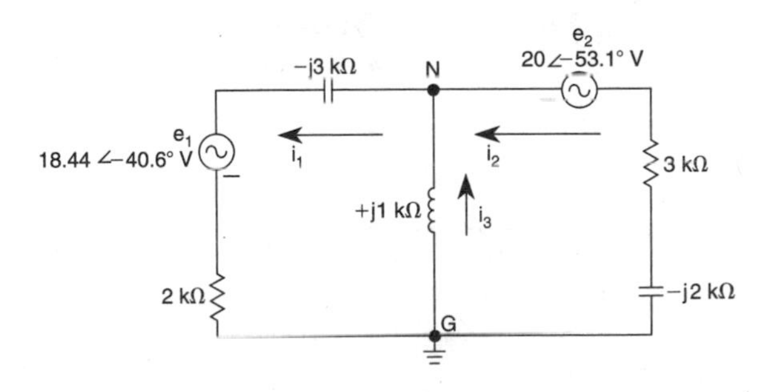

Figure 96-3

97
Nodal analysis for ac circuits

Chapter 28 demonstrated the application of nodal analysis to dc circuits. For ac circuits, you will have to modify the analysis in order to take into account the various phase relationships between the currents entering and leaving the node.

MATHEMATICAL DERIVATIONS

At a node, the algebraic phasor sum of the source currents entering equals the phasor sum of the currents leaving through the impedances.

As an example, redraw the circuit of Fig. 94-1 so that it appears as shown in Fig. 97-1A. Point N is now a node.

Example 97-1

In Fig. 97-1, determine the voltage at point N.

Solution

Regard the capacitive reactance of $-j6\ \Omega$ as the internal impedance of the $15\angle 60°$-V source. Similarly, the resistance of 3 Ω is the internal impedance of the $10\angle -45°$-V source. Convert both voltage sources into their equivalent current generators (Fig. 97-1B).

$$i_1 = \frac{15\angle 60°\ \text{V}}{-j4} = \frac{15\angle 60°\ \text{V}}{4\angle -90°} = 3.75\angle 150°\ \text{A}$$

$$i_2 = \frac{10\angle -45°\ \text{V}}{3} = 3.333\angle -45°\ \text{A}$$

The currents i_1 and i_2 are leaving the node, but the currents i_3, i_4 and i_5 are entering the point N. Therefore, the nodal equation is:

$$3.75\angle 150° + 3.333\angle -45° = \frac{v_N}{j5} + \frac{v_N}{-j4} + \frac{v_N}{3}$$

where v_N is the alternating voltage at the node.

Then,

$$-3.248 + j1.875 + 2.357 - j2.357$$
$$= v_N(-j0.2 + j0.25 + 0.333)$$
$$v_N(0.333 + j0.05) = -0.891 - j0.482$$
$$v_N = \frac{1.013\angle -151.6°}{0.337\angle 8.5°}$$
$$\approx 3\angle -160°\ \text{V}$$

Check:

Current through the inductor,

$$i_3 = \frac{v_N}{j5}$$
$$= \frac{3\angle -160°\ \text{V}}{5\angle 90°\ \Omega}$$
$$= 0.6\angle -250°$$
$$= 0.6\angle 110°\ \text{A}$$

This value agrees with the results obtained through the previous methods of analysis.

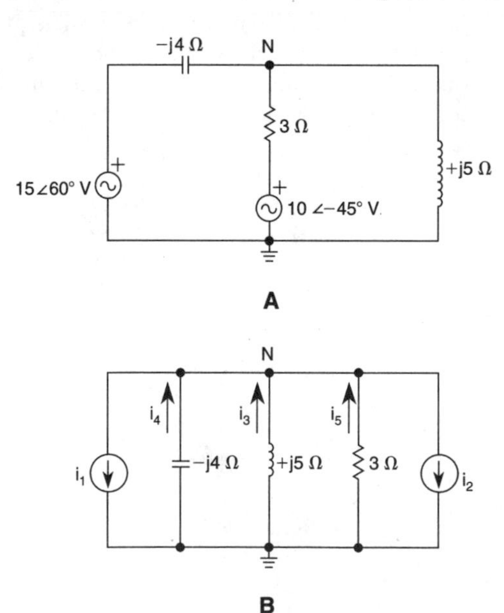

Figure 97-1

PRACTICE PROBLEMS

97-1. In Fig. 97-2 use nodal analysis to find the value of the voltage at the node N.

97-2. In Practice Problem 97-1, calculate the value of the current, i_3.

97-3. In Practice Problem 97-1, calculate the total amount of power dissipated in the circuit.

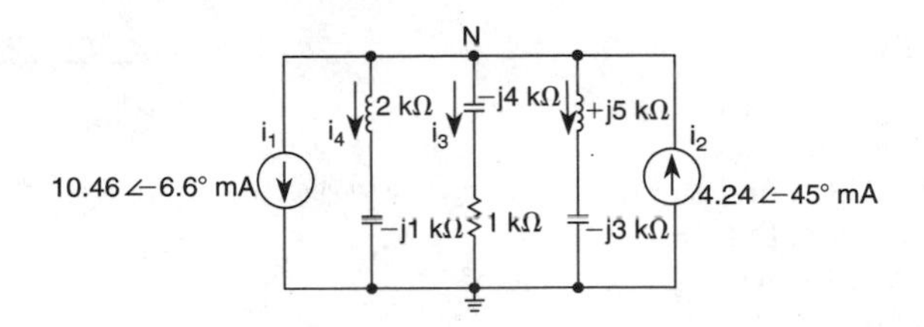

Figure 97-2

98
Millman's theorem for ac circuits

Millman's theorem for the solution of ac circuits can be stated as follows:

> *Any number of constant alternating current sources, that are connected directly in parallel, can be combined into a single alternating-current source whose total current is the phasor sum of the individual source currents, and whose total internal impedance is the result of combining the individual source impedances in parallel.*

MATHEMATICAL DERIVATIONS

$$i_T = i_1 + i_2 + i_3 \ldots + i_n \qquad (98\text{-}1)$$

where i_T = total current of the single alternating source, and $i_1, i_2, i_3 \ldots i_n$ = individual source currents.

$$z_{iT} = \cfrac{1}{\cfrac{1}{z_{i1}} + \cfrac{1}{z_{i2}} + \cfrac{1}{z_{i3}} \ldots + \cfrac{1}{z_{in}}} \qquad (98\text{-}2)$$

where z_{iT} = internal impedance of the single equivalent source, and $z_{i1}, z_{i2}, z_{i3} \ldots z_{in}$ = internal impedances of the individual sources.

Example 98-1

In the circuit of Fig. 98-1 (Fig. 97-1B), use Millman's theorem to determine the value of the current, i_3.

Solution

The currents i_1 and i_2 flow in the same direction. Consequently, the value (i_T) of the total equivalent current generator is:

$$i_T = i_1 + i_2 = 3.75\angle 150° + 3.333\angle -45°$$
$$= -0.891 - j0.482 \text{ A}$$

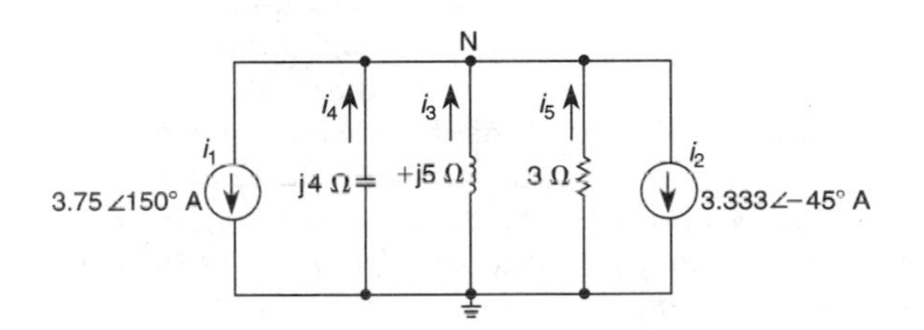

Figure 98-1

The total impedance (z_T) of $-j4\ \Omega \parallel j5\ \Omega \parallel 3\ \Omega$ is given by,

$$\frac{1}{z_T} = \frac{1}{-j4} + \frac{1}{j5} + \frac{1}{3} = +j0.25 - j0.2 + 0.333$$
$$= 0.333 + j0.05$$

Using the current division rule (CDR),

$$\text{Current, } i_3 = i_T \times \frac{z_T}{j5} = \frac{-0.891 - j0.482}{j5(0.333 + j0.05)}$$
$$= \frac{1.013\angle -151.6°}{-0.25 + j1.667}$$
$$= \frac{1.013\angle -151.6°}{1.686\angle 98.5°} \approx 0.6\angle 110° \text{ A}$$

This is the same result that was achieved with all of the other analytical methods.

PRACTICE PROBLEMS

98-1. In Fig. 98-2, use Millman's theorem to determine the value of the current, i_3.

98-2. In Practice Problem 98-1, find the total power dissipated in the circuit of Fig. 98-2.

98-3. In Practice Problem 98-1, what is the value of the potential at the point N?

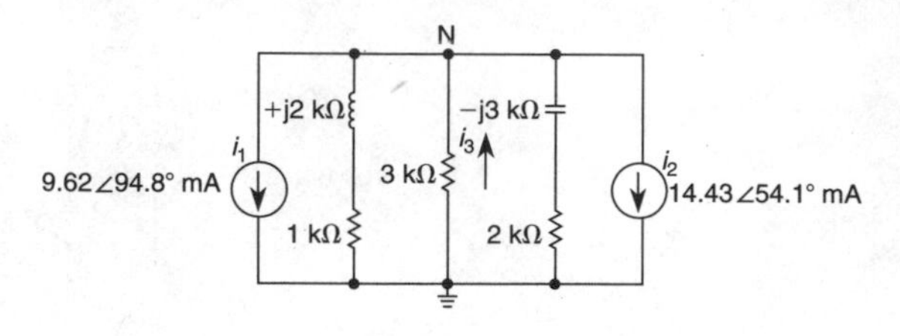

Figure 98-2

99 Thévenin's theorem for ac circuits

The procedure in "Théveninizing" an ac circuit is the same as outlined in chapter 31. In the case of ac circuits, the statement of the theorem is as follows:

> *The current in a load impedance, connected between two terminals (X and Y) of a network of impedances and alternating sources, is the same as if that load impedance were connected to a single alternating source whose output is the open-circuit voltage as measured between X and Y and whose internal impedance is the impedance of the network looking back into terminals X and Y, with all alternating sources replaced by impedances equal to their internal impedances.*

MATHEMATICAL DERIVATIONS

$$\text{Load current, } i_L = \frac{e_{TH}}{z_{TH} + z_L} \qquad (99\text{-}1)$$

$$\text{Load voltage, } v_L = \frac{e_{TH} \times z_L}{z_{TH} + z_L} \qquad (99\text{-}2)$$

where,

e_{TH} = Thévenin's alternating source voltage (V)

z_{TH} = Thévenin's source impedance (Ω)

z_L = load impedance (Ω)

Example 99-1

In Fig. 99-1A, derive the Thévenin equivalent circuit between the terminals X and Y, then obtain the load current in its polar form.

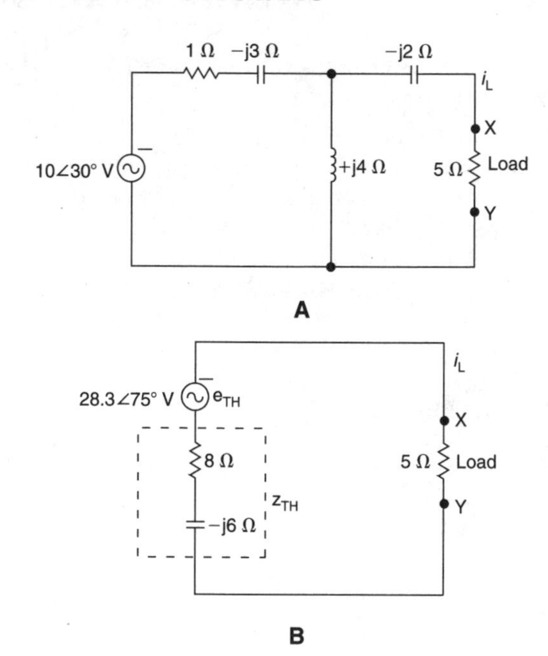

Figure 99-1

Solution

Step 1. Remove the load and calculate the open-circuit voltage between X and Y. Using the voltage division rule (VDR),

$$\text{Thévenin voltage, } e_{TH} = 10\angle 30° \times \frac{j4}{(1 - j3) + j4}$$

$$= 10\angle 30° \times \frac{4\angle 90°}{1 + j1}$$

$$= \frac{10\angle 30° \times 4\angle 90°}{\sqrt{2}\angle 45°}$$

$$= 28.3\angle 75° \text{ V}$$

Step 2. Replace the voltage source by a short-circuit, and calculate the value of the impedance phasor between X and Y.

$$\text{Thévenin impedance, } z_{TH} = -j2 + \frac{j4 \times (1 - j3)}{j4 + (1 - j3)}$$

$$= -j2 + \frac{j4 + 12}{1 + j1}$$

$$= -j2 + \frac{(j4 + 12)(1 - j1)}{2}$$

$$= -j2 + 8 - j4$$

$$= 8 - j6 \ \Omega$$

Step 3. Replace the load in the Thévenin equivalent circuit of Fig. 99-1B. The load current is then:

$$i_L = \frac{e_{TH}}{z_{TH} + R_L} = \frac{28.3\angle 75°}{8 - j6 + 5} = \frac{28.3\angle 75°}{13 - j6}$$

$$= \frac{28.3\angle 75°}{14.3\angle -24.78°}$$

$$= 1.98\angle 99.78° \text{ A}$$

PRACTICE PROBLEMS

99-1. In Fig. 99-2, Thévenize the circuit to the left of the points X and Y, and determine the value of e_{TH} and z_{TH}.

99-2. In Practice Problem 99-1, determine the value of the load current, i_1.

99-3. In Practice Problem 99-2, calculate the amount of power dissipated in the load.

99-4. In Fig. 99-3, Thévenize the circuit to the left of the points X and Y. Determine the values of e_{TH} and z_{TH}

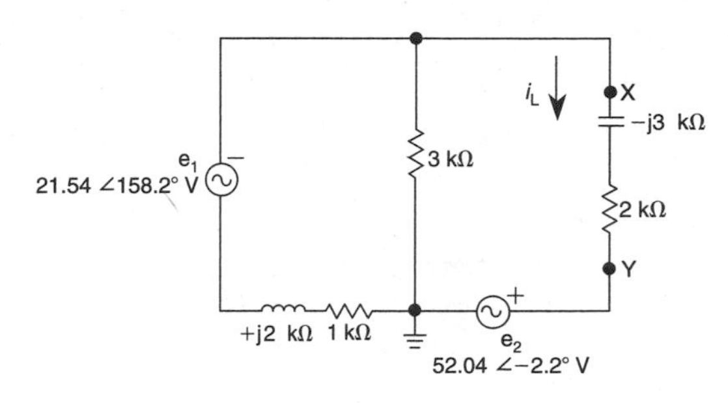

Figure 99-2

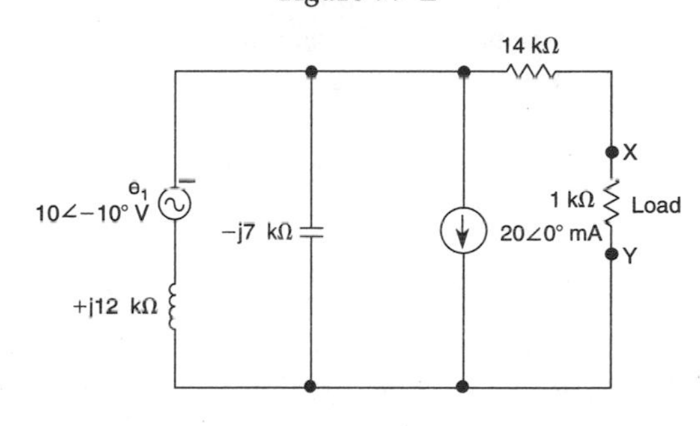

Figure 99-3

100
Norton's theorem for ac circuits

Norton's theorem was applied to dc circuits in chapter 32. For ac circuits, there is no difference in the procedure required for "Nortonizing" a complex network of alternating sources and impedances so that the equivalent Norton alternating source again will be of the constant-current type with its associated impedance in parallel.

The formal statement of Norton's theorem for ac circuits is as follows:

The alternating current, in a load connected between two output terminals (X and Y) of a complex network containing alternating sources and impedances, is the same as if the load were connected to a constant alternating source whose current (i_N) is equal to the short-circuit current measured between X and Y. This constant alternating current source (i_N) is placed in parallel with an impedance (z_N) which is equal to the impedance of the network looking back into terminals X and Y, with all sources replaced by impedances equal in value to their internal impedances.

The last part of this statement involves substituting short circuits for all constant alternating-voltage sources and open circuits for all constant alternating-current sources.

MATHEMATICAL DERIVATIONS

$$\text{Load current, } i_L = \frac{i_N \times z_N}{z_N + z_L} \quad (100\text{-}1)$$

$$\text{Load voltage, } v_L = \frac{i_N \times z_N \times z_L}{z_N + z_L} \quad (100\text{-}2)$$

where

i_N = Norton alternating source current (A)
z_N = Norton source impedance (Ω).

Example 100-1

In Fig. 100-1A, derive the Norton equivalent circuit between the terminals X and Y and then obtain the load current in its polar form.

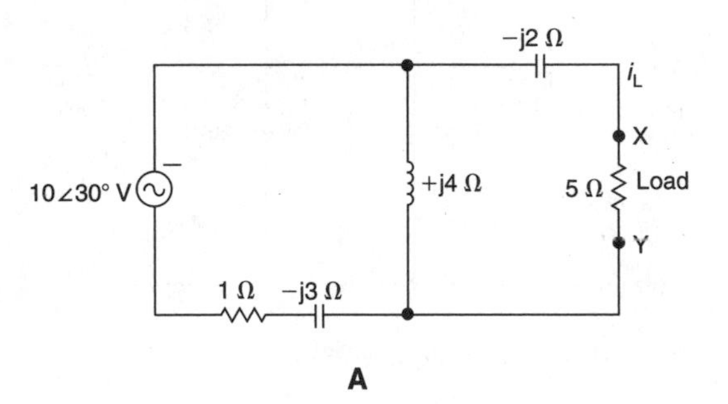

A

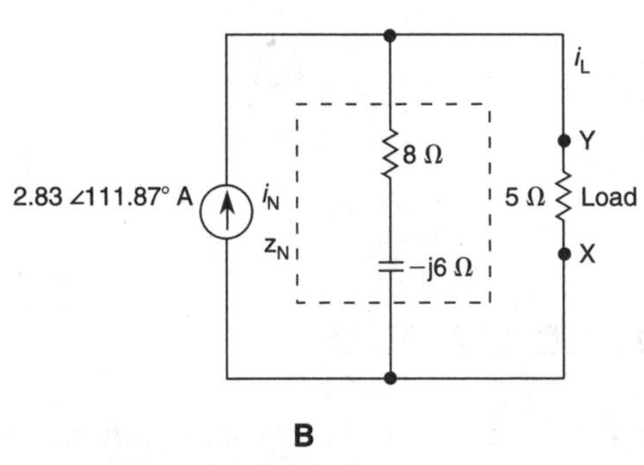

B

Figure 100-1

Solution

Step 1. Remove the load and replace it by a short circuit. The total impedance then presented to the voltage source is:

$$z_T = 1 - j3 + \frac{j4 \times (-j2)}{j4 + (-j2)}$$

$$= 1 - j3 + \frac{8}{j2} = 1 - j3 - j4$$

$$= 1 - j7\ \Omega$$

The current drawn from the source is:

$$i_T = \frac{10\angle 30°}{1 - j7}\ \text{A}$$

and the short circuit current between the terminals X and Y is:

$$\text{Norton current, } i_N = \frac{10\angle 30°}{1 - j7} \times \frac{j4}{j4 + (-j2)}$$

$$= \frac{20\angle 30°}{7.07\angle -81.87°}$$

$$= 2.83\angle 111.87°\ \text{A}$$

Step 2. The Norton impedance (z_N) is the same as the Thévenin impedance (z_{TH}) and, therefore, $z_N = 8 - j6\ \Omega$ (Fig. 100-1B).

Step 3. Remove the short circuit, and replace the load in the Norton equivalent circuit of Fig. 16-7. By the current division rule (CDR),

$$\text{Load current, } i_L = i_N \times \frac{8 - j6}{8 - j6 + 5}$$

$$= 2.83\angle 111.87° \times \frac{8 - j6}{13 - j6}$$

$$= \frac{2.83 \angle 111.87° \times 10 \angle -36.87°}{14.32 \angle -24.78°}$$

$$= 1.98 \angle 99.78°\ \text{A.}$$

This is the same result as obtained with Thévenin's theorem in Chapter 99.

PRACTICE PROBLEMS

100-1. In Fig. 100-2, Nortonize the circuit to the left of points X and Y. Determine the values of i_N and z_N.

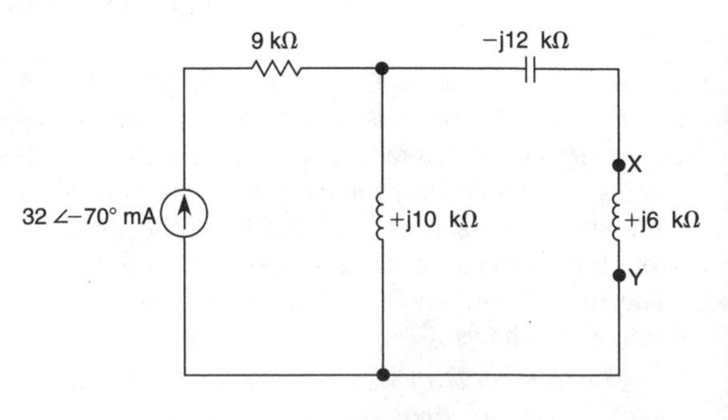

Figure 100-2

100-2. In Practice Problem 100-1, calculate the value of the current flowing through the 6-kΩ inductor.

100-3. In Fig. 100-3, Nortonize the circuit *both* to the left and to the right of points X and Y. Determine the values of i_N and z_N.

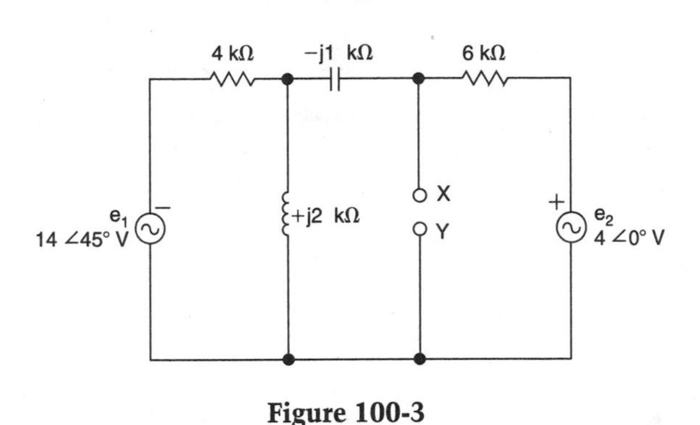

Figure 100-3

100-4. In Fig. 100-4, Nortonize the circuit between points X and Y. Determine the values of i_N and z_N.

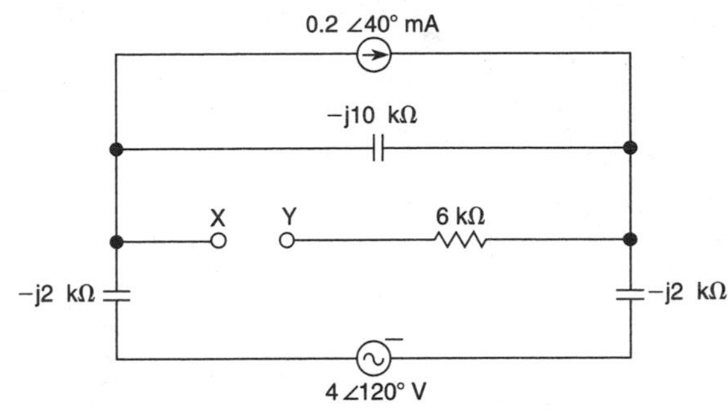

Figure 100-4

101 Delta⇄wye transformations for ac circuits

Chapter 33 delineated the equations for delta⇄wye transformations which involved resistors in dc circuits. The same equations can now be applied to ac circuits, provided that you replace the resistances with the required impedance phasors.

MATHEMATICAL DERIVATIONS

Delta-to-wye transformation (Fig. 101-1)

$$z_A = \frac{z_{AB}\, z_{CA}}{z_{AB} + z_{BC} + z_{CA}} \tag{101-1}$$

$$z_B = \frac{z_{BC}\, z_{AB}}{z_{AB} + z_{BC} + z_{CA}} \tag{101-2}$$

$$z_C = \frac{z_{CA}\, z_{BC}}{z_{AB} + z_{BC} + z_{CA}} \tag{101-3}$$

Wye-to-delta transformation (Fig. 101-1)

$$z_{AB} = \frac{z_A z_B + z_B z_C + z_C z_A}{z_C} \tag{101-4}$$

$$z_{BC} = \frac{z_A z_B + z_B z_C + z_C z_A}{z_A} \tag{101-5}$$

$$z_{CA} = \frac{z_A z_B + z_B z_C + z_C z_A}{z_B} \tag{101-6}$$

Example 101-1

In Fig. 101-2A, use a delta→wye transformation to obtain the phasor expression for the current, i_T.

Solution

Referring to the WXY delta formation,

$z_{WX} = 2\ \text{k}\Omega,\ z_{XY} = 1 + j3\ \text{k}\Omega\ ,\ z_{YW} = +j4\ \text{k}\Omega.$

Then,

$$z_W = \frac{2 \times (j4)}{2 + 1 + j3 + j4} \tag{101-1}$$

$$= \frac{j8}{3 + j7}$$

$$= \frac{j8\,(3 - j7)}{3^2 + 7^2}$$

$$= \frac{56 + j24}{58} = 0.966 + j0.414 \text{ k}\Omega$$

$$z_X = \frac{2\,(1 + j3)}{3 + j7} \qquad (101\text{-}2)$$

$$= \frac{(2 + j6)\,(3 - j7)}{3^2 + 7^2}$$

$$= \frac{48 + j4}{58} = 0.828 + j0.069 \text{ k}\Omega$$

$$z_Y = \frac{(1 + j3) \times (j4)}{3 + j7} \qquad (101\text{-}3)$$

$$= \frac{(-12 + j4)\,(3 - j7)}{3^2 + 7^2}$$

$$= \frac{-8 + j96}{58} = -0.138 + j1.655 \text{ k}\Omega$$

From the equivalent circuit of Fig. 101-2B,

Total impedance,

$$z_T = \frac{(3.966 - j1.586) \times (0.828 - j2.931)}{(3.966 - j1.586) + (0.828 - j2.931)}$$
$$-0.138 + j1.655$$

$$= \frac{4.271 \angle -21.80° \times 3.046 \angle -74.23°}{4.794 - j4.517}$$
$$-0.138 + j1.655$$

$$= \frac{4.271 \angle -21.80° \times 3.046 \angle -74.23°}{6.591 \angle -43.30°}$$
$$-0.138 + j1.655$$

$$= 1.97 \angle -52.73° - 0.138 + j1.655$$

$$= 1.193 - j1.568 - 0.138 + j1.655$$

$$= 1.055 + j0.087 = 1.059 \angle 4.7° \text{ k}\Omega.$$

Total current,

$$i_T = \frac{10 \angle 40° \text{ V}}{1.059 \angle 4.7° \text{ k}\Omega} = 9.44 \angle 35.3° \text{ mA}$$

PRACTICE PROBLEMS

101-1. Calculate the total amount of the power dissipated in the circuit of Fig. 101-2A.

101-2. In the circuit of Fig. 101-2A, convert the TWYX wye formation (W is the wye point) into its delta equivalent, and then calculate the value of the total impedance presented to the 10 ∠ 40°-V source.

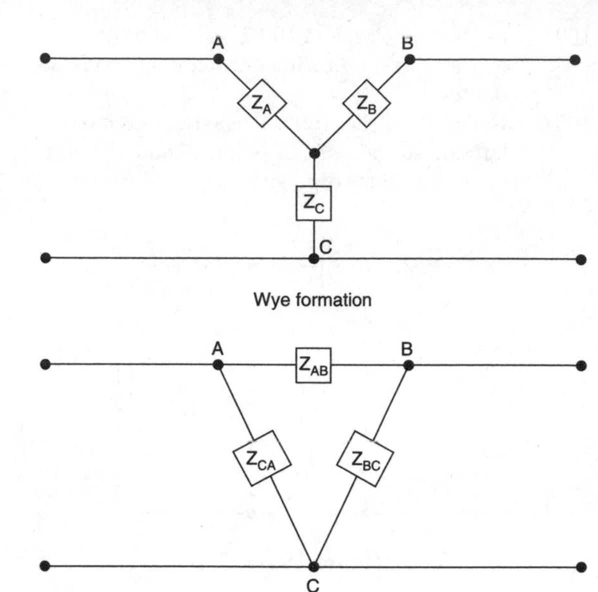

Figure 101-1

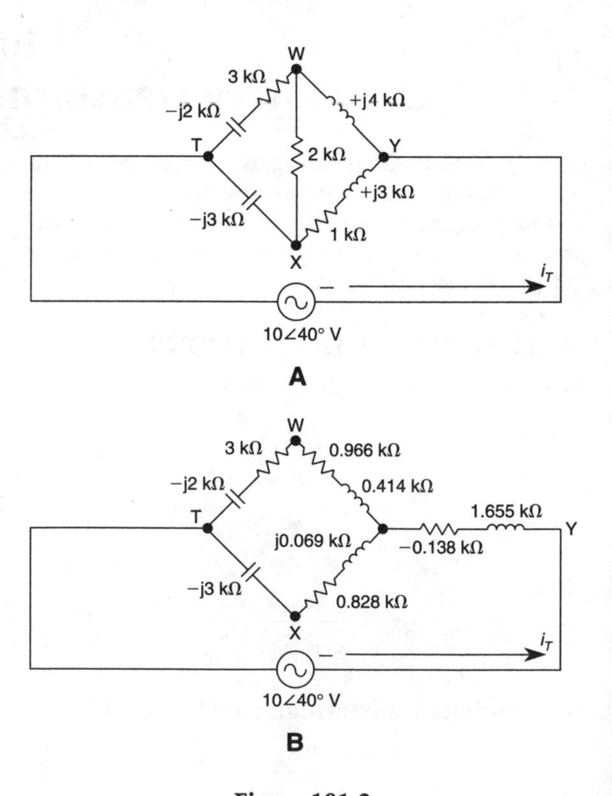

Figure 101-2

102
Reciprocity theorem

The reciprocity theorem is equally true for dc and ac circuits that contain only linear passive elements and a single source. However, this theorem is mainly applied to alternating circuits so that it was *not* included in the network theorems of chapters 26 through 33.

The *reciprocity theorem* can be stated as follows:

"An ac network of linear impedances is supplied with current from an alternating voltage source, which is connected between two points, X and Y. The current flowing through a particular impedance is measured by an ac ammeter, which is inserted between two other points, A and B. If the position of the alternating source and the ammeter are interchanged (in other words, the source is connected between the points A and B, while the ammeter is inserted between the points X and Y), the reading of the ammeter is unchanged. If the ammeter possesses an appreciable internal impedance, this impedance remains between the points A and B, and is regarded as being in series with the voltage source. In the same way, the internal impedance of the source remains between the points X and Y, and is regarded as being in series with the ammeter."

This theorem is independent of the complexity of the network. For example, it allows you to deduce that the directional properties of a receiving antenna are the same as those of the identical antenna when it is used for transmission purposes.

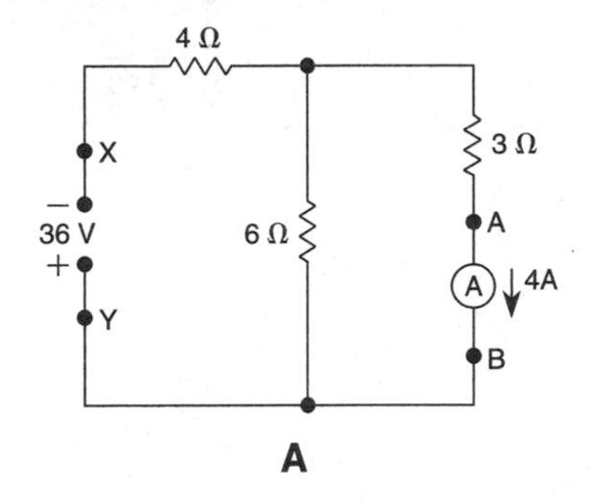

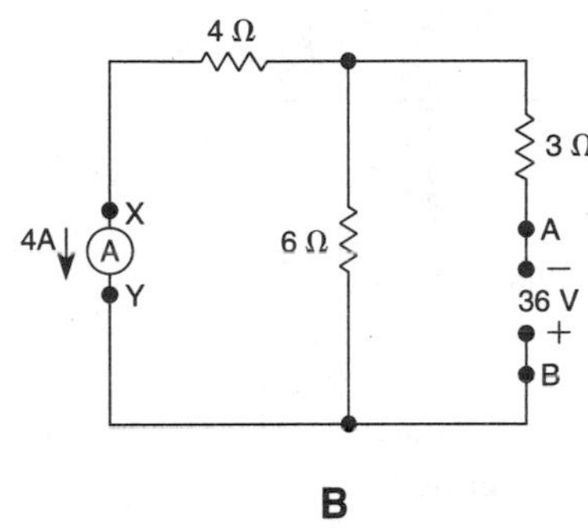

Figure 102-1

MATHEMATICAL DERIVATIONS

To verify the reciprocity theorem, analyze the dc circuit of Fig. 102-1A. When the source and the ammeter are interchanged, the result is the circuit of Fig. 102-1B.

In Fig. 102-1A

Total resistance presented to the source,

$$R_T = 4 + \frac{6 \times 3}{6 + 3}$$

$$= 4 + 2$$

$$= 6\ \Omega$$

Total current,

$$I_T = \frac{36\ \text{V}}{6\ \Omega} = 6\ \text{A}.$$

Reading of the ammeter,

$$A = 6\text{A} \times \frac{6\ \Omega}{6\ \Omega + 3\ \Omega} = 4\ \text{A}$$

In Fig. 102-1B

Total resistance presented to the source,

$$R_T = 3 + \frac{6 \times 4}{6 + 4}$$

$$= 5.4\ \Omega.$$

Total current,

$$I_T = \frac{36\ \text{V}}{5.4\ \Omega} = 6\frac{2}{3}\ \text{A}.$$

Reading of the ammeter,

$$A = 6\frac{2}{3}\ \text{A} \times \frac{6\ \Omega}{4\ \Omega + 6\ \Omega}$$

$$= 4\ \text{A}.$$

The reading of the ammeter (*A*) is the same for both circuits. Note that, however, when the two circuits are compared, the currents through the individual resistors are not the same.

Example 102-1

In Figs. 102-2A and B verify the principle of the reciprocity theorem.

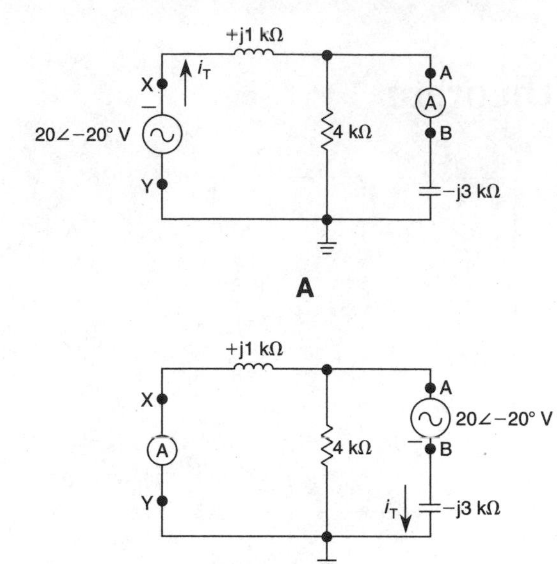

Figure 102-2

Solution

In Fig. 102-2A, the total impedance (z_T) presented to the alternating source is given by,

$$z_T = +j1 + \frac{4 \times (-j3)}{4 + (-j3)}$$

$$= +j1 + \frac{-j12}{4 - j3}$$

$$= \frac{3 - j8}{4 - j3} \text{ k}\Omega$$

Current,

$$i_T = 20 \angle -20° \times \frac{4 - j3}{3 - j8} \text{ mA}$$

By the current division rule,
Reading of the ammeter,

$$A = 20 \angle -20° \times \frac{4 - j3}{3 - j8} \times \frac{4}{4 - j3}$$

$$= \frac{80 \angle -20°}{8.544 \angle -69.4°} = 9.36 \angle 49.4° \text{ mA}$$

In Fig. 102-2B, the source and the ammeter have been interchanged. The new total impedance (z_T) presented to the alternating current source is,

$$z_T = -j3 + \frac{4 \times j1}{4 + j1}$$

$$= \frac{3 - j8}{4 + j1} \text{ k}\Omega$$

Total current,

$$i_T = 20 \angle -20° \times \frac{4 + j1}{3 - j8} \times \frac{4}{4 + j1}$$

$$= \frac{80 \angle -20°}{3 - j8} = 9.36 \angle 49.4° \text{ mA.}$$

The reading of the ammeter (A) is the same in both circuits of Figs. 102-2A, B.

PRACTICE PROBLEM

102-1. In Fig. 102-3, determine the value of the voltage drop, e. Interchange e and the current source, then verify the reciprocity theorem.

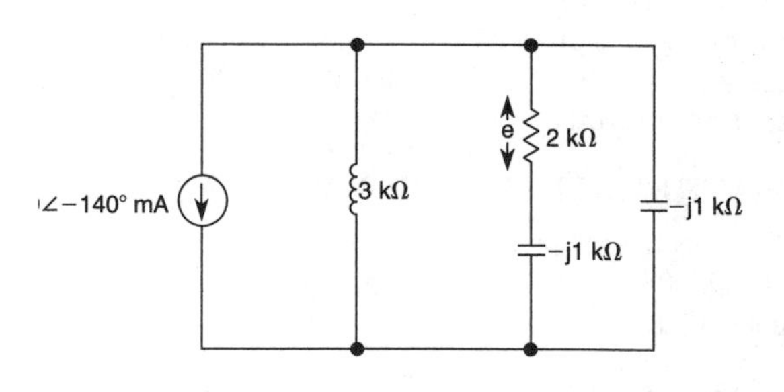

Figure 102-3

103 Radio frequency transformers

In the power and audio transformers of chapter 88, the inductances of the primary and secondary windings were on the order of henrys. These windings each contained a large number of copper turns, which were wound on a common iron core. Because of the low reluctance of the iron, the flux was largely confined to the core so that there was very little flux leakage, and the value of the coupling factor (k) was nearly unity. These low-frequency transformers suffered from the following losses:

1. The *copper loss*, which was the power dissipated in the resistances of the primary and secondary windings.
2. The *iron loss*, which was dissipated in the core. This lost energy was subdivided into:
   (a) the *eddy-current loss*; this was caused by the moving magnetic flux which cut the core and induced circulating currents within the metal. This loss was reduced by laminating the core into thin slices, with each slice insulated from its neighbor.
   (b) the *hysteresis loss*, which was caused by the rapid magnetizing, demagnetizing, and remagnetizing of the core during the cycles of the primary current.

The iron losses increase dramatically as the frequency is raised, Therefore, at radio frequencies any use of a solid iron core is prohibited. Consequently, there is a considerable flux leakage so that the coupling factor, (k) is typically less than 0.1, and the turns ratio of the primary and secondary windings has little significance.

At the high end of the RF spectrum, the primary and secondary windings have self-inductances of only a few microhenrys. Each winding has a small number of turns, which are separately wound on a nonmagnetic core (such as plastic); the separation between the windings then determines the degree of mutual coupling and the value of the coupling factor. The schematic symbol for this type of RF transformer is .

At frequencies around 1 MHz, the self-inductances of the primary and secondary windings are of the order of 100 μH. Such a transformer uses an *iron-dust core* in which granules are insulated from each other, and then compressed to form a solid slug. The value of the mutual inductance can be increased by inserting the slug deeper into the windings. An alternative type of core uses a ceramic material known as *ferrite*. The magnetic properties of ferrite are similar to those of iron, but unlike iron, ferrite is an insulator, which helps to reduce the eddy current loss. The schematic symbol for this transformer is .

MATHEMATICAL DERIVATIONS

In Fig. 103-1A, the primary sinewave current (I_p) creates an alternating magnetic flux that cuts the secondary coil and induces the secondary voltage, E_s. As a result of connecting the secondary load, there is a secondary current (I_s), which, in turn, is responsible for inducing a voltage into the primary coil.

The following analysis illustrates a further use of the operator j.

The KVL equation for the primary circuit is,

$$E_p - j\omega M I_s = I_p z_p \tag{103-1}$$

where z_p is the impedance of the primary circuit.

The voltage "$-j\omega M I_s$" is induced into the primary coil as the result of the current flowing in the secondary circuit.

The KVL equation for the secondary circuit is,

$$E_s = -j\omega M I_p = I_s z_s$$

where z_s is the total impedance of the secondary circuit.

Therefore,

$$I_s = \frac{-j\omega M I_p}{z_s} \tag{103-2}$$

Substituting for I_s in Equation 103-1,

$$E_p - j\omega M\left(\frac{-j\omega M I_p}{z_s}\right) = I_p z_p$$

or,

$$\frac{E_p}{I_p} = z_p + \frac{\omega^2 M^2}{z_s} \tag{103-3}$$

The equivalent primary circuit corresponding to Equation 103-3, is shown in Fig. 103-1B.

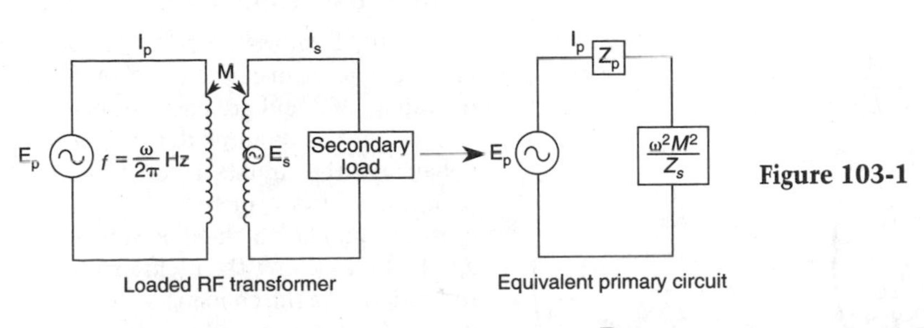

Figure 103-1

The quantity "ω^2M^2/z_s" is called the *reflected impedance*, which is reflected from the secondary circuit into the primary circuit; it is the apparent value of the impedance that is introduced into the primary circuit as a result of the presence of the secondary load.

If $z_p = (R_p + jX_p)$, and $z_s = R_s + jX_s$,

$$\frac{E_p}{I_p} = (R_p + jX_p) + \frac{\omega^2M^2}{R_s + jX_s}$$

$$= (R_p + jX_p) + \frac{\omega^2M^2\,(R_s - jX_s)}{R_s^2 + X_s^2}$$

Then,

$$\frac{E_p}{I_p} = R_p + \left(\frac{\omega^2M^2R_s}{R_s^2 + X_s^2}\right) + j\left(X_p - \frac{X_s\omega^2M^2}{R_s^2 + X_s^2}\right)$$

The effect of the secondary load is to *increase* the primary resistance, and *decrease* the primary reactance, provided that X_p and X_s are of the same nature. For example, if X_s is inductive, then the reflected reactance is capacitive, and vice-versa.

Now, examine the effects of reflected impedance in power and RF transformers.

Ideal power transformer

An ideal power transformer possesses the following features:

1. The primary and secondary coils have large inductance values, which are totally contained within the windings. The inductive reactances are very much greater than the primary and secondary resistances so that $\omega L_p \gg R_p$ and $\omega L_s \gg R_s$.
2. There are no copper and iron losses.
3. There are no capacitive reactances in the primary and secondary circuits.
4. There is no flux leakage so that $M = \sqrt{L_pL_s}$.

Under these conditions, that total primary impedance is given by,

$$z_{PT} = \left(R_p + \frac{\omega^2M^2R_s}{R_s^2 + X_s^2}\right) + j\left(X_p - \frac{\omega^2M^2X_s}{R_s^2 + X_s^2}\right)$$

$$= \left(R_p + \frac{\omega^2L_pL_sR_s}{R_s^2 + \omega^2L_s^2}\right) + j\omega\left(L_p - \frac{\omega^2L_pL_sL_s}{R_s^2 + \omega^2L_s^2}\right)$$

$$= \left(R_p + \frac{\dfrac{L_p}{L_s} \times R_s}{1 + \dfrac{R_s^2}{\omega^2L_s^2}}\right) + j\omega\left(L_p - \frac{L_p}{1 + \dfrac{R_s^2}{\omega^2L_s^2}}\right) \qquad (103\text{-}4)$$

The increase in the primary resistance is caused by the resistance reflected from the secondary circuit to the primary circuit. Therefore,

Reflected resistance,

$$R_r = \frac{\dfrac{L_p}{L_s} \times R_s}{1 + \dfrac{R_s^2}{2\omega^2L_s^2}}$$

$$\approx \frac{L_p}{L_s} \times R_s \text{ because } \omega L_s >> R_s$$

$$= R_s \times \left(\frac{N_p}{N_s}\right)^2 \qquad (103\text{-}5)$$

This is the same result obtained in the discussion on power and audio transformers (chapter 88).

Total primary reactance,

$$X_{PT} = \omega L_p - \frac{\omega L_p}{1 + \dfrac{R_s^2}{\omega^2L_s^2}}$$

By using the Binomial Expansion (chapter 209),

Total primary reactance,

$$X_{PT} = \omega L_p - \omega L_p\left(1 - \frac{R_s^2}{\omega^2L_s^2} + \ldots \text{ neglected terms}\right)$$

$$= L_p \times \frac{R_s^2}{\omega^2L_s^2} \qquad (103\text{-}6)$$

The primary reactance has been reduced to a low value so that the primary and secondary currents maintain an approximately constant flux. For example, if the value of R_s is reduced, the secondary current is increased, the primary reactance is lowered, and the primary current is also increased.

RF transformer coupling with tuned primary and secondary circuits

Figure 103-2 shows identical primary and secondary circuits that are magnetically coupled by the mutual inductance (M), and are each resonant to the same frequence, f_o. It is required to consider the changes in the behavior of the circuits as the values of M and the coupling factor k are increased. Theoretically, these increases could be obtained by starting with the coils far apart, then moving them closer together. In practice, the variation in the coupling factor is determined by the depth of penetration achieved by the iron-dust slugs.

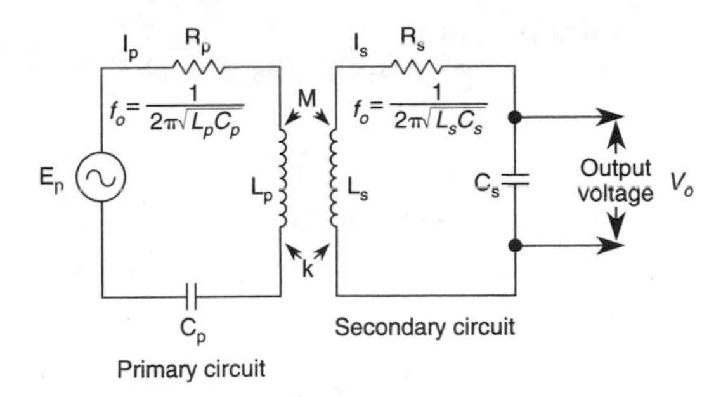

Figure 103-2

Low values of coupling factor ($k \approx 0.005$)

Under these conditions of *loose* coupling, the voltage induced in the secondary winding is low so that there is only a small secondary current, which will have little effect on the primary circuit. The shapes of the response curves will not differ appreciably from those obtained with the circuits tuned and excited separately.

Coupling factor of $0.005 < k < 0.01$

When k is increased, the voltage induced in the secondary winding rises and there is a greater secondary current. At the frequency, (f_o), both primary and secondary circuits are resistive and more resistance will be reflected from the secondary circuit into the primary circuit. This has two effects; the primary resonant current falls and the effective Q value is reduced, so that the response curve broadens. These conditions continue until the coupling factor reaches a certain critical value, at which the reflected resistance from the secondary (at the resonant frequency, f_o), is equal to the primary resistance. This is a matched condition, so that there is a maximum RF power transfer from the primary circuit to the secondary circuit. At critical coupling,

$$R = \frac{\omega_o^2 M^2}{R}$$

or,

$$R = \omega_o M = \omega_o kL$$

Therefore, critical $k = \dfrac{R}{\omega_o L} = \dfrac{1}{Q}$ (103-7)

If the primary and secondary circuits have different values of Q,

$$\text{critical } k = \frac{1}{\sqrt{Q_p Q_s}}$$

Primary current response at the resonant frequency f_o,

$$I_p = \frac{E_p}{2R}.$$

Correspondingly, secondary current response,

$$I_s = \frac{\omega_o M I_p}{R} = I_p = \frac{E_p}{2R}.$$

Coupling factor of the order of 0.01 and above

With *tight* coupling, the value of the reflected impedance is large near resonance. Consequently, the primary and secondary circuits have a considerable effect on each other. Because the primary and secondary circuits are tuned to the same frequency, (f_o), the nature of the reactance in both circuits is the same for any frequency. Consequently, at frequencies below resonance, both circuits are capacitive; the secondary reactance will reflect into the primary circuit as a certain value of inductive reactance. At a particular frequency (f_1), which depends on the value of the coupling factor, this reflected inductive reactance cancels the primary's reactance to produce a resonant "hump" in the primary response curve. A second resonant hump occurs at another frequency, (f_2) such that $f_2 - f_o = f_o - f_1$. This effect of tight coupling is known as *double-humping*, or *split-tuning*, which occurs in both primary and secondary circuits (Figs. 103-3A, B). At the resonant frequency (f_o), both

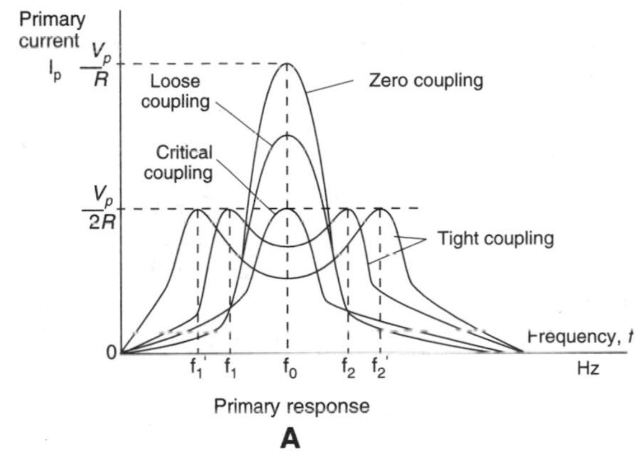

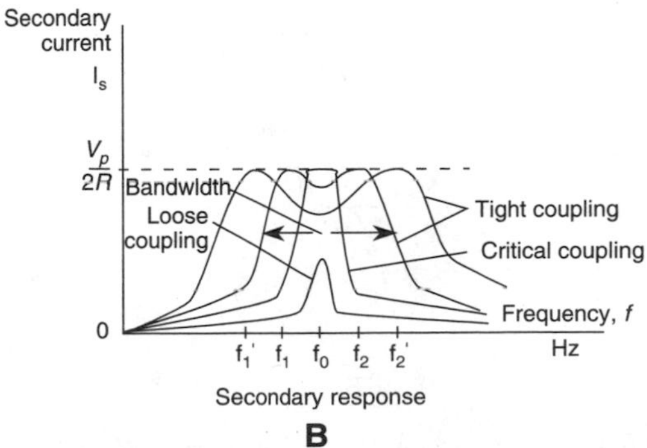

Figure 103-3

circuits are resistive. Because resistance reflects as resistance, the resonant responses at f_o in both the primary and secondary circuits decrease as k is increased.

The mathematical analysis of split-tuning is as follows:

Total impedance of the primary circuit,

$$z_{PT} = z_p + \frac{\omega^2 M^2}{z_s}$$

At any frequency $f = \frac{\omega}{2\pi}$, $z_p = z_s = R + jX$

where $X = \omega L - \frac{1}{\omega C}$,

Total primary impedance,

$$z_{PT} = R + jX + \frac{\omega^2 M^2}{R + jX}$$

$$= R\left(1 + \frac{\omega^2 M^2}{R^2 + X^2}\right) + jX\left(1 - \frac{\omega^2 M^2}{R^2 + X^2}\right)$$

Resonant "humps" will occur when,

$$1 - \frac{\omega^2 M^2}{R^2 + X^2} = 0$$

or,

$$\omega M = \sqrt{R^2 + X^2} \qquad (103\text{-}8)$$

It follows that resonant "humps" cannot occur until $\omega M > R$ so that they will not appear for low values of M and k.

At a resonant hump, the total primary impedance,

$$z_{PT} = R\left(1 + \frac{\omega^2 M^2}{R^2 + X^2}\right) = R\,(1 + 1) = 2R$$

and the primary current is $E_p/2R$, which is independent of the value of k.

The corresponding secondary current,

$$i_s = \frac{-j\omega M I_p}{z_s}$$

$$= \frac{-j\omega M I_p}{R + jX}$$

Effective value of the secondary current,

$$I_s = \frac{\omega M I_p}{\sqrt{R^2 + X^2}}$$

$$= I_p$$

$$= \frac{E_p}{2R}$$

At the resonant "humps", the primary and the secondary current responses are the same.

Determination of the resonant "hump" frequencies, f_1 and f_2

At the resonant "hump",

$$\omega M = \sqrt{R^2 + X^2}$$

Assume that $X \gg R$ when the resonant "humps" occur. Then,

$$\omega_1 M = X = \frac{1}{\omega_1 C} - \omega_1 L$$

at the frequency, f_1.

It follows that,

$$\frac{1}{C} = \omega_1^2 (L + M)$$

or,

$$\omega_1^2 = \frac{1}{(L + M)\,C} = \frac{1}{LC(1 + k)} = \frac{\omega_o^2}{1 + k}$$

Therefore,

$$\omega_1 = \frac{\omega_o}{\sqrt{1 + k}}$$

Resonant "hump" frequency, $f_1 = \frac{f_o}{\sqrt{1 + k}}$ Hz

Similarly, the resonant "hump" frequency (f_2) above f_o is given by

$$f_2 = \frac{f_o}{\sqrt{1 - k}} \text{ Hz}$$

By using the Binomial Expansion (chapter 209),

$$\text{frequency, } f_1 = \frac{f_o}{\sqrt{1 + k}}$$

$$= f_o\,(1 + k)^{-1/2}$$

$$= f_o\,\left(1 - \frac{1}{2}k + \text{neglected terms}\right)$$

$$\text{frequency, } f_2 = f_o\,(1 - k)^{-1/2}$$

$$= f_o\,\left(1 + \frac{1}{2}k + \text{neglected terms}\right)$$

Frequency separation between the resonant "humps" is given by,

$$f_2 - f_1 = k f_o \text{ Hz} \qquad (103\text{-}9)$$

Bandwidth at the half-power points,

$$\Delta f = \sqrt{2} k f_o \qquad (103\text{-}10)$$

Transformer coupling with tuned primary/tuned secondary circuits is used to provide the wide bandwidth required by the intermediate amplifiers of certain receivers. In AM communications receivers, the necessary bandwidth is only a few kilohertz so that the coupling factor is reduced to below the critical value.

By contrast, TV and radar receivers need wide bandwidths. Therefore, the coupling factor is increased above the critical level so that split-tuning occurs. A typical IF amplifier circuit is shown in Fig. 103-4; through network analysis, this circuit can be reduced to the basic arrangement of Fig. 103-2.

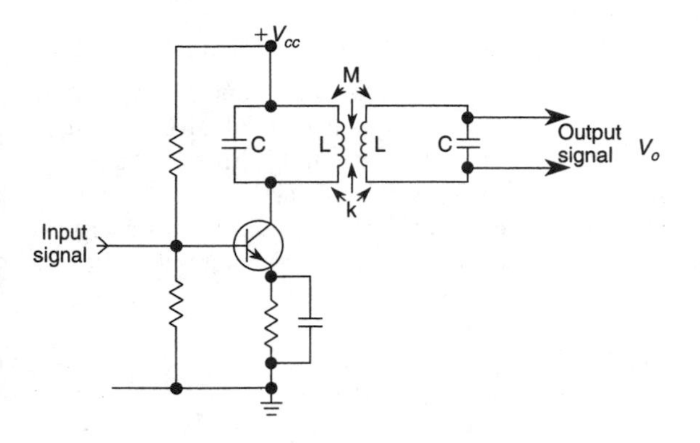

Figure 103-4

If the reactance at resonance is comparable with the resistance, a more accurate result is,

$$f_2 - f_1 = kf_o\left(1 - \frac{1}{2k^2Q^2}\right) \text{ Hz} \qquad (103\text{-}11)$$

More important still is the separation between the "humps" in the output voltage (V_o) across the secondary capacitor.

This separation is equal to

$$kf_o\left(1 - \frac{1}{k^2Q^2}\right) \text{ Hz} \qquad (103\text{-}12)$$

The analysis of this form of transformer coupling has illustrated a major application of the j operator.

RF transformer coupling with untuned primary and tuned secondary circuits

The circuitry of this stage is shown in Fig. 103-5A. Referring to the equivalent circuit of Fig. 103-5B, L_p is of the order of 50 μH so that $\omega L_p \approx 300\ \Omega$, and can be neglected in comparison with R_p. Therefore, at the resonant frequency (f_o),

$$\text{Primary current, } I_p = \frac{E_p}{R_p + \dfrac{\omega_o^2 M^2}{R_s}}$$

Voltage in the secondary winding,

$$E_s = \omega_o M I_p$$

$$= \frac{\omega_o M E_p}{R_p + \dfrac{\omega_o^2 M^2}{R_s}}$$

Output voltage across the secondary capacitor (V_o) is given by,

$$V_o = Q\,E_s$$

$$= \frac{\omega_o M\, R_s\, Q}{R_p R_s + \omega_o^2 M^2} \times E_p$$

Voltage gain,

$$G_v = \frac{V_o}{E_p} = \frac{\omega_o M\, R_s\, Q}{R_p R_s + \omega_o^2 M^2} \qquad (103\text{-}13)$$

The voltage gain (G_v) is at its maximum value (as M is varied) if $1/G_v$ is at its minimum value.

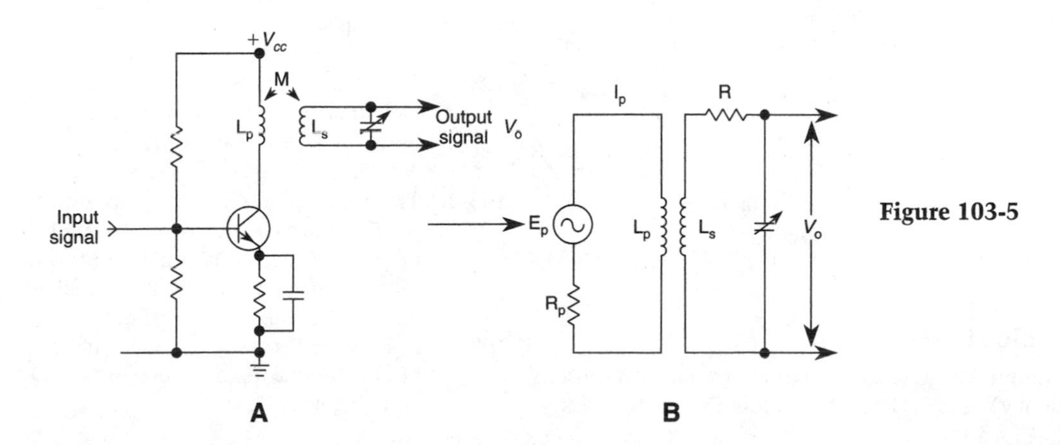

Figure 103-5

$$\frac{1}{G_v} = \frac{1}{\omega_o Q\, R_s}\left[\frac{R_pR_s + \omega_o M^2}{M}\right]$$

$$= \frac{1}{\omega_o Q\, R_s}\left[2\omega_o\sqrt{R_pR_s} + \frac{(\sqrt{R_pR_s} - \omega_o M)^2}{M}\right]$$

$1/G_v$ is at its minimum value if $\omega_o M = \sqrt{R_pR_s}\ \Omega$.

Then,

$$G_{v\max} = \frac{1}{2}\,\frac{\omega_o M\, Q}{R_s}$$

Effective Q of the secondary circuit is given by,

$$Q_{eff} = \frac{\omega_o L_s}{R_s + \dfrac{\omega_o^2 M^2}{R_p}} \qquad (103\text{-}14)$$

If G_v is at its maximum value, $\omega_o^2M^2 = R_pR_s$.

Then,

$$Q_{eff} = \frac{\omega_o L_s}{R_s + R_s} = \frac{\omega_o L_s}{2\,R_s} = \frac{1}{2} \times \text{unloaded value of } Q.$$

This will not fulfill the stage's selectivity requirement. However, a typical value of M is 25 μH, in which case

$$\frac{\omega_o^2 M^2}{R_p} \ll R_s$$

and Q_{eff} is approximately equal to the unloaded Q value. In the design of Q (in a circuit such as shown in Fig. 103-5A), a compromise must be reached between the voltage gain and the degree of selectivity, with the latter as a dominant factor. The advantage of this type of transformer coupling, over the single tuned circuit, is the control that the coil spacing exercises over the selectivity.

Example 103-1

A video IF amplifier uses transformer coupling with tuned primary and secondary coils. If the Q of both primary and secondary circuits is 20, determine the critical value of the coupling factor.

Solution

The critical value of the coupling factor is,

$$\frac{1}{Q} = \frac{1}{20} = 0.05 \qquad (103\text{-}7)$$

Example 103-2

In Example 103-1, the intermediate frequency (center frequency) is 44 MHz. If the coupling factor is increased to 1.25 times the critical value, calculate the value of the bandwidth.

Solution

Coupling factor, $k = 1.25 \times 0.05 = 0.0625$

Bandwidth $= 1.414 \times 0.0625 \times 44$ (103-10)

$= 3.9$ MHz, rounded off.

Example 103-3

A video amplifier stage in a radar receiver is operating at an intermediate frequency of 30 MHz and uses "split-tuning" to achieve a bandwidth of 4 MHz. Assuming the primary and secondary circuits are identical, and the coupling factor is 1.3 times its critical value, determine the Q value of each circuit.

Solution

The value of the coupling factor is

$$k = \frac{\text{bandwidth}}{1.414 \times f_o} = \frac{4}{1.414 \times 30} \qquad (103\text{-}10)$$

Critical coupling factor,

$$\frac{k}{1.3} = \frac{4}{1.3 \times 1.414 \times 30}$$

The value of Q for each circuit is

$$\frac{1}{\text{critical coupling factor}} = \frac{1.3 \times 1.414 \times 30}{4}$$

$= 14$, rounded off. (103-7)

PRACTICE PROBLEMS

103-1. Two identical L,C,R circuits are mutually coupled, and are each resonant to 44 MHz. If the required bandwidth is 4 MHz, which is achieved by "split-tuning," calculate the necessary value for the coupling factor.

103-2. Two L,C,R circuits are mutually coupled and have unloaded Q factors of 40 and 50. Determine the k value necessary to achieve critical coupling.

103-3. Two identical L,C,R circuits are mutually coupled, and are each resonant to 500 kHz. If the mutual inductance is 5 μH and the RF resistance of each circuit is 25 Ω, calculate the value of the reflected resistance.

103-4. In Practice Problem 103-3, calculate the value of the mutual inductance for which critical coupling will occur.

104
The practical single-phase alternator

In chapter 61, the basic alternator contained a revolving armature whose conductors cut the flux of a stationary magnetic field. The alternating voltage generated was then taken through slip-rings to the load. The contacts between these slip-rings and the carbon brushes were subject to friction wear and sparking; moreover, they were liable to arc over at high voltages. These problems are largely overcome in the revolving field alternator of Fig. 104-1A.

In the revolving field alternator, a dc source drives a direct current through the slip-rings, brushes, and the windings on a rotor, which is driven around by mechanical means. This creates a rotating magnetic field that cuts the conductors embedded in the surrounding stator. An alternating voltage then appears between the ends, S (start) and F (finish) of the stator winding. Because S and F are fixed terminals, there are no sliding contacts and the whole of the stator winding can be continuously insulated.

The alternator of Fig. 104-1A has two poles so that one complete rotation of the rotor generates one cycle of ac in the stator. Therefore, in order to generate 60 Hz, the rotor must turn at 60 revolutions per second ($60 \times 60 = 3600$ rpm). However, in the four-pole machine of Fig. 104-1B, one revolution produces two cycles of the ac voltage. Therefore, it would only require a speed of 1800 rpm to generate a 60-Hz output.

The power delivered to a resistive load by a single-phase alternator is fluctuating at twice the line frequency (chapter 67). This presents a problem because of the load on the mechanical source of energy. Therefore, the necessary torque will not be constant.

MATHEMATICAL DERIVATIONS

For the revolving field alternator,

Generated frequency,

$$f = \frac{N \times p}{60} \text{ hertz} \qquad (104\text{-}1)$$

where: N = rotor speed (rpm)
p = number of pole pairs.

Example 104-1

In a revolving field alternator, a six-pole rotor is rotating at 1000 rpm. What is the value of the generated frequency?

Solution

Generated frequency,

$$f = \frac{N \times p}{60} \qquad (104\text{-}1)$$

$$= \frac{1000 \times (6/2)}{60}$$

$$= 50 \text{ Hz.}$$

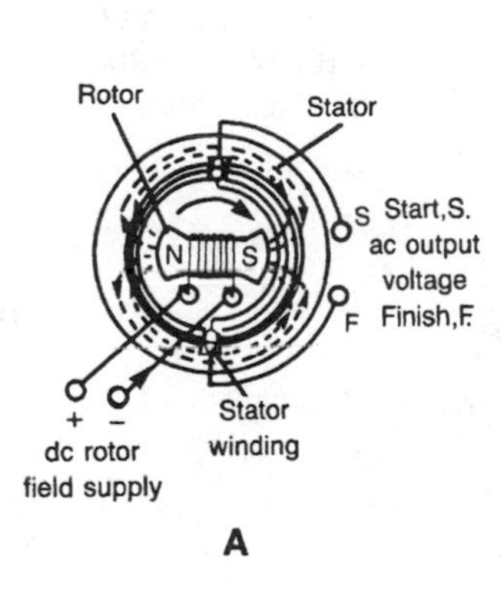

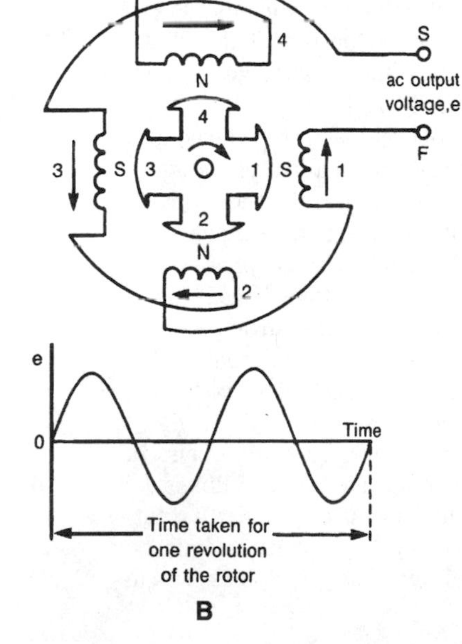

Figure 104-1

PRACTICE PROBLEMS

104-1. A revolving field alternator has an output voltage of 110 V, 60 Hz. If the rotor speed is increased by 50%, what is the alternator's new output voltage?

104-2. An alternator's output is 220 V, 50 Hz. If the rotor speed is 1000 rpm, how many poles are mounted on the rotor?

104-3. A revolving field alternator with a 16-pole rotor generates an output voltage of 120 V, 440 Hz. Calculate the speed of the rotor.

104-4. In a revolving field alternator, the speed of its 6-pole rotor is 1500 rpm. Calculate the value of the generated frequency.

105
The practical two-phase alternator

In this type of revolving field alternator, two identical coils are mounted on the stator with their axes separated by 90°. Figure 105-1A shows a simplified arrangement of the windings that are cut by the magnetic flux of a two-pole rotor. The EMFs e_1 and e_2 induced in the windings will therefore be equal in magnitude, but 90° out of phase (Fig. 105-1B).

The fact that e_2 leads e_1 by 90° is a result of the assumed clockwise movement of the rotor. This phase relationship is commonly represented by drawing the two stator windings 90° apart (Fig. 105-1C). Each of the windings can be connected to separate loads, but it is more convenient to use a neutral line so that the number of lines required is reduced from four to three, as in Fig. 105-1D. Assuming identical (balanced) resistive loads across the windings, the total instantaneous power of the two-phase alternator is constant, as opposed to the fluctuating power output of the single phase machine. Note that this advantage is only achieved when the loads are balanced.

If the effective current of each load is I_{rms}, the neutral line current is $1.414\, I_{rms}$ (see mathematical derivations). The total load current in the three lines is, therefore, $(2 + \sqrt{2})\, I_{rms} = 3.414\, I_{rms}$. If the same two balanced loads were connected in parallel across a single-phase alternator, the total current in the two supply lines is $4\, I_{rms}$. Assuming that a neutral line is used, the single-phase alternator requires more copper than the comparable two-phase generator when supplying a given voltage to a particular value of total load resistance.

If the two voltages are connected to two pairs of coils, whose axes are perpendicular, the two fluxes associated with the coils will combine to produce a rotating magnetic field whose angular velocity is equal to that of the alternating voltage. This principle is used in ac induction and synchronous motors. Compared with a single-phase alternator, a high-power two-phase generator requires a smaller stator, is more efficient, and is less subject to vibration.

MATHEMATICAL DERIVATIONS

In Fig. 105-1B,

Phase voltage,

$$e_1 = E_{max} \sin \omega t \text{ volts} \quad (105\text{-}1)$$

Phase voltage,

$$e_2 = E_{max}\left(\sin \omega t + \frac{\pi}{2}\right) \quad (105\text{-}2)$$

$$= E_{max} \cos \omega t \text{ volts}$$

where:

ω = angular frequency, ($2\pi f$ rad./s)
E_{max} = peak value of the phase voltage (V)
e_1, e_2 = instantaneous phase voltages (V)

Assuming balanced resistive loads,
Instantaneous power,

$$p_1 = \frac{e_1^2}{R_L} = \frac{E^2_{max} \sin^2 \omega t}{R_L} \text{ watts} \quad (105\text{-}3)$$

Instantaneous power,

$$p_2 = \frac{e_2^2}{R_L} = \frac{E^2_{max} \cos^2 \omega t}{R_L} \text{ watts} \quad (105\text{-}4)$$

Total instantaneous power,

$$p_1 + p_2 = \frac{E^2_{max}}{R_L} \text{ watts} \quad (105\text{-}5)$$

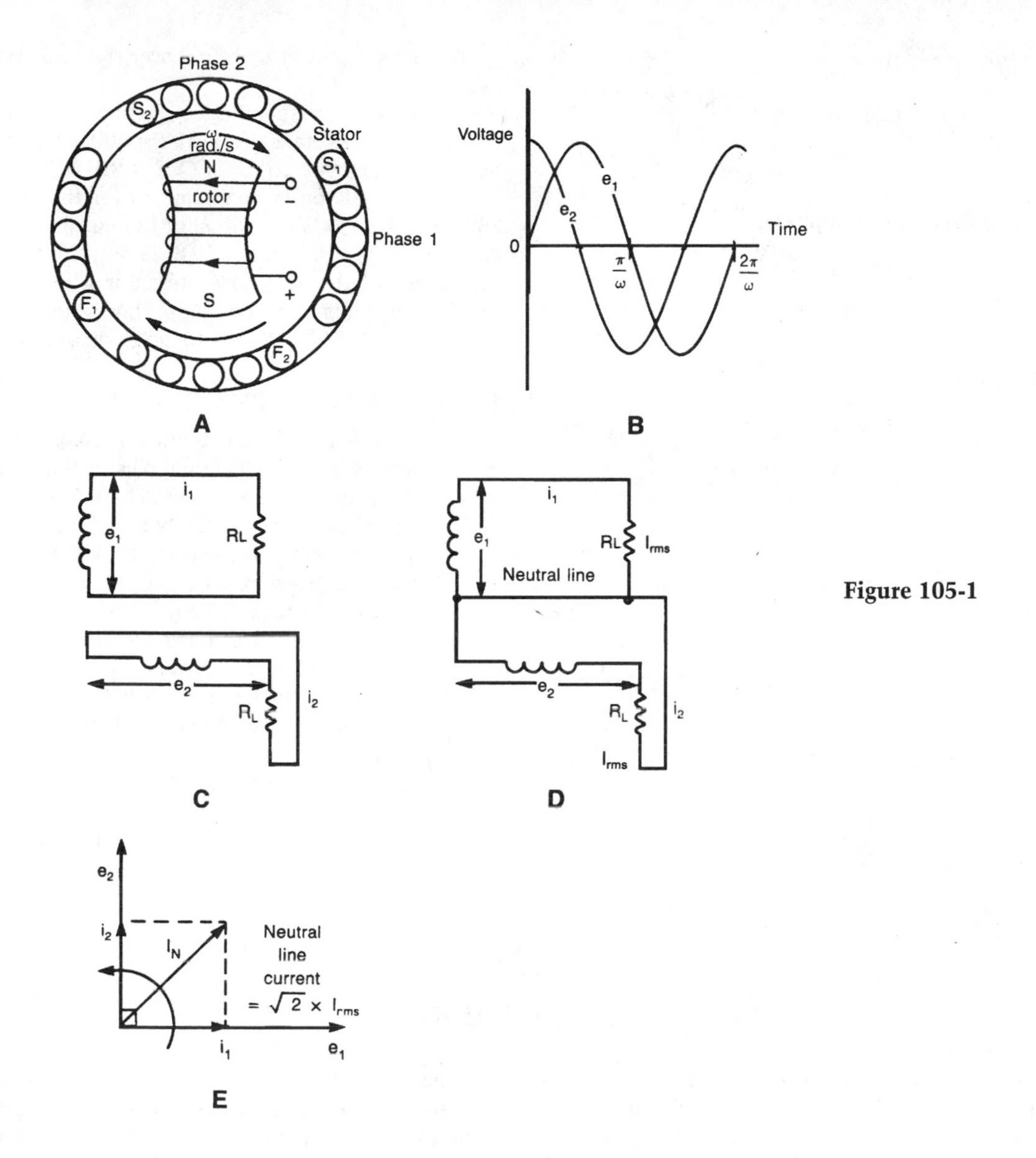

Figure 105-1

The total instantaneous power is independent of the time and is, therefore, a constant.

In Fig. 105-1E,

Assuming balanced resistive loads,

Effective value of the neutral line current,

$$I_N = \sqrt{I^2_{\text{rms}} + I^2_{\text{rms}}} \qquad (105\text{-}6)$$

$$= \sqrt{2}\, I_{\text{rms}} \text{ amperes}$$

where: I_{rms} is the effective load current for each phase.

Total current in the three lines (including the neutral line):

$$2\, I_{\text{rms}} + \sqrt{2}\, I_{\text{rms}} = \left(2 + \sqrt{2}\right) I_{\text{rms}}$$

$$= 3.414\, I_{\text{rms}} \text{ amperes}$$

Example 105-1

A two-phase alternator has a four-pole rotor whose speed is 1800 rpm. What is the value of the generated frequency? The effective voltage of each phase is 240 V, and the two resistive loads are balanced so that each has an effective resistance of 80 Ω. Calculate the current in the neutral line and the instantaneous power output of the alternator.

Solution

$$\text{Generated frequency} = \frac{1800 \times 2}{60}$$

$$= 60 \text{ Hz.}$$

Load current for each phase,

$$I_{\text{rms}} = \frac{240 \text{ V}}{80 \ \Omega}$$

$$= 3 \text{ A rms.}$$

Neutral line current,

$$I_N = \sqrt{2}\, I_{\text{rms}} \qquad (105\text{-}6)$$

$$= \sqrt{2} \times 3$$

$$= 4.242 \text{ A rms.}$$

Total instantaneous power,

$$p_1 + p_2 = \frac{E^2_{\text{max}}}{R_L} \qquad (105\text{-}5)$$

$$= \frac{(\sqrt{2} \times 240 \text{ V})^2}{80 \ \Omega}$$

$$= 1440 \text{ W.}$$

The average power delivered over the cycle is also 1440 W.

Note: For the corresponding single-phase alternator, the total load resistance is 80 Ω/2 = 40 Ω. The maximum instantaneous power is $(\sqrt{2} \times 240 \text{ V})^2/40\ \Omega$ = 2880 W so that the average power over the cycle is 2880 W/2 = 1440 W. The total of the currents in the two supply lines is 2 × 240 V/40 Ω = 12 A rms; this compares with the two-phase alternator, where the total of the currents in the two supply lines and the neutral line is (2 × 3) + 4.242 = 10.242 A rms.

PRACTICE PROBLEMS

105-1. A two-phase alternator has a six-pole rotor whose speed is 1000 rpm. What is the value of the generated frequency? The effective voltage of each phase is 220 V and the total average power over the cycle is 1.5 kW. If the two resistive loads are balanced, calculate the effective resistance value of each load.

105-2. In Practice Problem 105-1, calculate the effective value of the neutral line current.

105-3. In Practice Problem 105-1, what is the value of the effective current in each of the two supply lines?

106 The three-phase alternator

The three-phase alternator shown in Fig. 106-1A has three single-phase windings that are equally spaced on the stator so that the voltage induced in each winding is 120°(2π/3 radians) out of phase with the voltage induced in the other two windings (Fig. 106-1B). These windings are independent of each other and could be connected to three separate loads. Such connections would require a six-line system (Fig. 106-1C).

Assume that the three loads are balanced in the sense that they will all have the same ohms value and the same phase angle. It follows that they would all have identical power factors, which in this example are assumed to be lagging. The three load currents will also be equal in magnitude, but 120° out of phase so that their phasor sum is zero (Fig. 106-2A). It is then possible to join the corresponding ends of the three windings to a common point and to connect the other ends to separate terminals. A single neutral line is attached to the common point to produce a four-wire wye (Y) system (Fig. 106-2B).

Apart from the advantage of reducing the number of wires from six to four, the theoretical current carried by the neutral line is zero. The thin neutral line only requires a low current capacity because the current it carries only exists as a result of some imbalance between the loads. Alternatively, the neutral line can be replaced by a ground return. Therefore, the savings in the copper required by the supply lines is greater for a balanced three-phase system than for a balanced two-phase system.

The mathematical derivations show that the total instantaneous power for the balanced three-phase system is constant and, therefore, independent of the time.

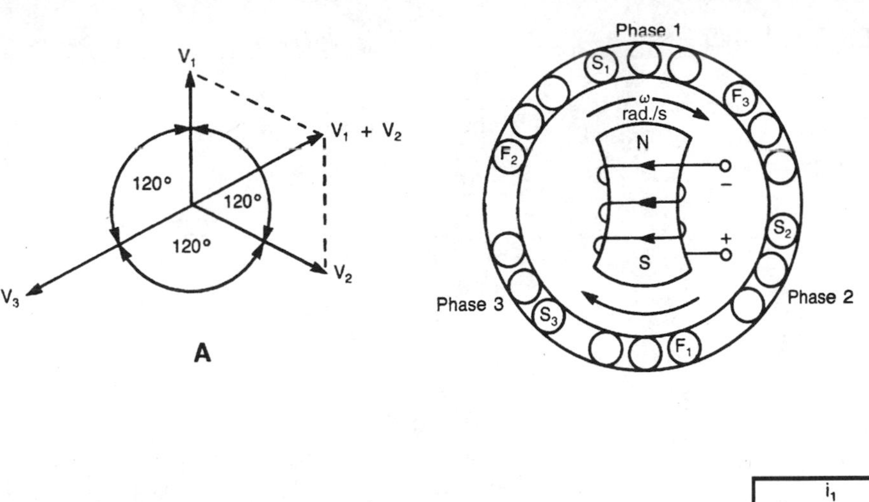

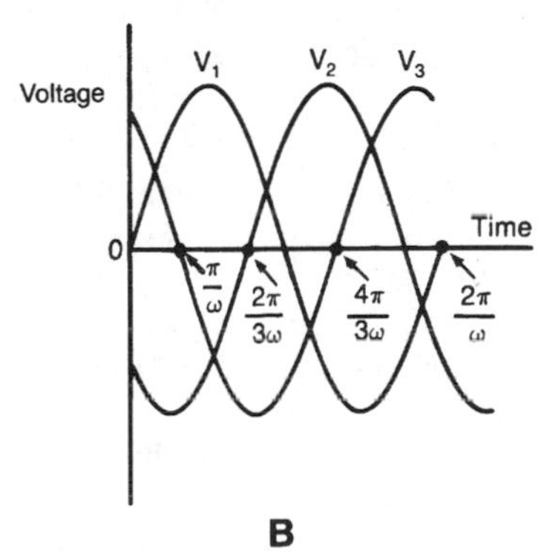

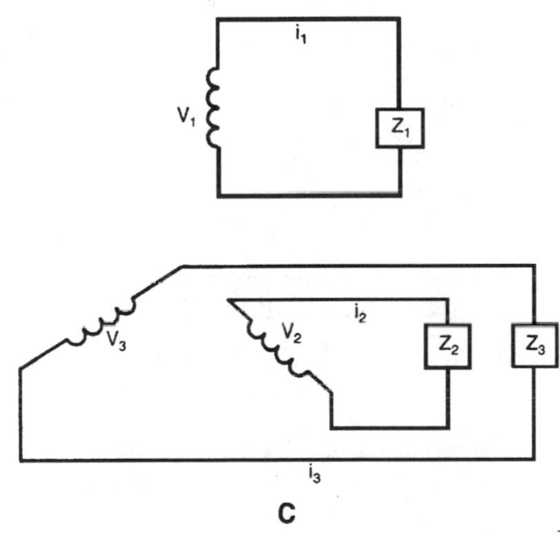

Figure 106-1

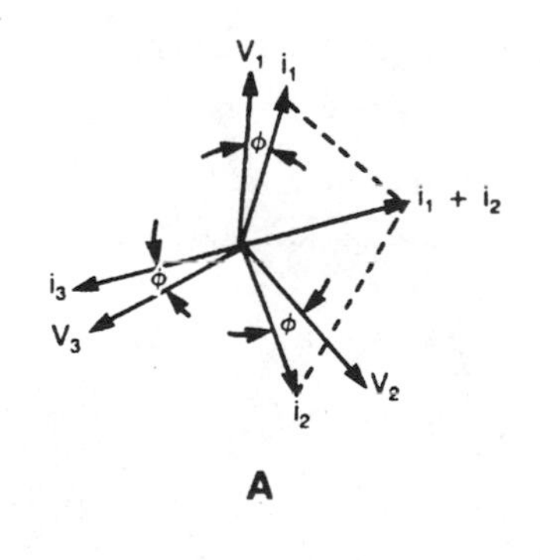

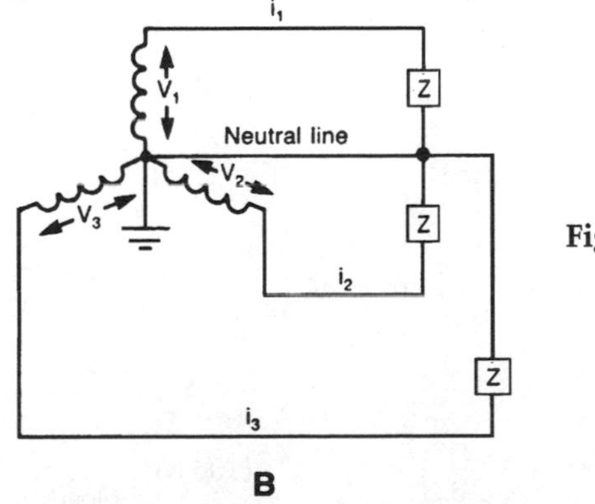

Figure 106-2

MATHEMATICAL DERIVATIONS

In Fig. 106-1B, the phase voltages v_1, v_2, and v_3 are shown as three phasors that are equally separated by $2\pi/3$ radians. If v_1 is used as the reference phasor, and the rotor is assumed to be revolving in the clockwise direction,

Phase #1 voltage,

$$v_1 = E_{max} \sin \omega t \text{ volts} \tag{106-1}$$

Phase #2 voltage,

$$v_2 = E_{max} \sin\left(\omega t - \frac{2\pi}{3}\right) \tag{106-2}$$

$$= -\frac{E_{max} \sin \omega t}{2} - \frac{\sqrt{3}}{2} E_{max} \cos \omega t \text{ volts}$$

Phase #3 voltage,

$$v_3 = E_{max} \sin\left(\omega t - \frac{4\pi}{3}\right) \tag{106-3}$$

$$= -\frac{E_{max} \sin \omega t}{2} + \frac{\sqrt{3}}{2} E_{max} \cos \omega t \text{ volts}$$

where:

E_{max} = peak value of the phase voltage (V)
ω = angular frequency (rad/s)

By adding Equations 106-1, 106-2, and 106-3,

$$v_1 + v_2 + v_3 = 0 \tag{106-4}$$

For three balanced loads with the same lagging power factor, each load current will lag its corresponding phase voltage by the same phase angle, ϕ (Fig. 106-2A).
It follows that:

$$i_1 + i_2 + i_3 = 0 \tag{106-5}$$

If each of the balanced loads is equal to a resistance (R_L), the instantaneous powers delivered by the three phases are:

Instantaneous power,

$$p_1 = \frac{E^2_{max}}{R_L} \sin^2 \omega t \text{ watts} \tag{106-6}$$

Instantaneous power,

$$p_2 = \frac{E^2_{max}}{R_L} \sin^2\left(\omega t - \frac{2\pi}{3}\right) \tag{106-7}$$

$$= \frac{E^2_{max}}{R_L}\left(\frac{-\sin \omega t + \sqrt{3} \cos \omega t}{2}\right)^2 \text{ watts}$$

Instantaneous power,

$$p_3 = \frac{E^2_{max}}{R_L} \sin^2\left(\omega t + \frac{2\pi}{3}\right) \tag{106-8}$$

$$= \frac{E^2_{max}}{R_L}\left(\frac{\sin \omega t + \sqrt{3} \cos \omega t}{2}\right)^2 \text{ watts}$$

Total instantaneous power,

$$p_T = p_1 + p_2 + p_3 \tag{106-9}$$

$$= \frac{E^2_{max}}{R_L}\left[\sin^2 \omega t + \frac{\sin^2 \omega t}{4} + \frac{\sin^2 \omega t}{4} + \frac{3}{4}\cos^2 \omega t + \frac{3}{4}\cos^2 \omega t\right]$$

$$= \frac{3E^2_{max}}{2R_L} = \frac{3E^2_{rms}}{R_L} \text{ watts}$$

The total instantaneous power, (p_T) for a balanced three-phase system is, therefore, constant and is independent of the time.

Example 106-1

A three-phase alternator has a four-pole rotor whose speed is 1500 rpm. What is the value of the generated frequency? The effective voltage of each phase is 120 V and the resistive loads are balanced so that each has an effective resistance of 60 Ω. Calculate the total of the line and neutral currents, and determine the total instantaneous power output of the alternator.

Solution

Generated frequency,

$$f = \frac{1500 \times 2}{60} = 50 \text{ Hz.}$$

Load current for each phase,

$$I_L = \frac{120 \text{ V}}{60 \ \Omega} = 2 \text{ A.}$$

Neutral line current is *zero* for a balanced load system.

Total of the load and neutral currents

$$I_T = 3 \times 2 \text{ A} = 6 \text{ A.}$$

Total instantaneous power,

$$p_T = \text{average power over the cycle} \tag{106-9}$$

$$= \frac{3E^2_{rms}}{R}$$

$$= 3 \times \frac{(120 \text{ V})^2}{60 \ \Omega}$$

$$= 720 \text{ W.}$$

Note: For the equivalent two-phase alternator, the load on each phase would be $2 \times 60 \ \Omega/3 = 40 \ \Omega$. The current for each phase is 120 V/40 Ω = 3 A and the neutral line current is $3 \times \sqrt{2} = 4.242$ A. The total of the line and neutral currents is $(2 \times 3) + 4.242 = 10.242$ A.

Total instantaneous power,

$$
\begin{aligned}
p_T &= \text{average power over the cycle} \\
&= 2 \times \frac{(120\ \text{V})^2}{40\ \Omega} \\
&= 720\ \text{W}.
\end{aligned}
$$

With the comparable single-phase alternator, the total load resistance would be 60 Ω/3 = 20 Ω. The total current in the supply lines is 2 × 120 V/20 Ω = 12 A and the average power over the cycle is $(120\ \text{V})^2/20\ \Omega = 720$ W.

PRACTICE PROBLEMS

106-1. A three-phase alternator has a six-pole rotor whose speed is 1200 rpm. What is the value of the generated frequency? The effective voltage of each phase is 110 V and the total average power is 1.8 kW. If the three resistive loads across the phases are balanced, calculate the effective resistance of each load.

106-2. In Practice Problem 106-1, what is the effective value of the neutral line current?

106-3. In Practice Problem 106-1, what is the effective current in each of the three supply lines?

107 The line voltage of the three-phase wye connection

The alternator EMF applied across one of the three balanced loads in chapter 106, was the voltage that existed between ground (or the neutral line) and the line connected to the load. This EMF is normally called the *phase voltage*, V_P. However, with three-phase systems, it is normal practice to make use of the line voltage (V_L) as well as the phase voltage. As illustrated in Fig. 107-1A, the line voltage is the alternating EMF that exists between points X and Y, and is related to the voltages generated in the first and second phases of the stator windings.

To determine the relationship between V_L and V_P, it is important to realize that the correct stator terminals must be connected to the common point if the phase voltages are to be represented as 120° ($2\pi/3$ radians) apart. For example, if the two terminals of one phase winding are reversed, its phase voltage is shifted by 180°. The phase voltages v_x, v_y, and v_z are the alternating (phasor) voltages monitored at points X, Y, Z, with respect to the common point O. Consequently, the (phasor) line voltage (v_{xy}) is the alternating voltage at point X with respect to point Y and is, therefore,

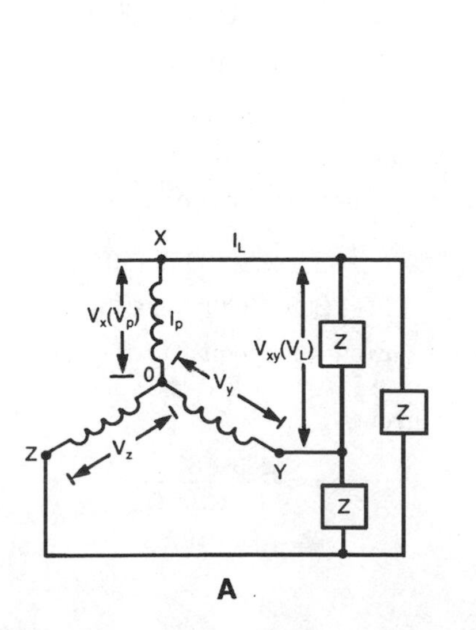

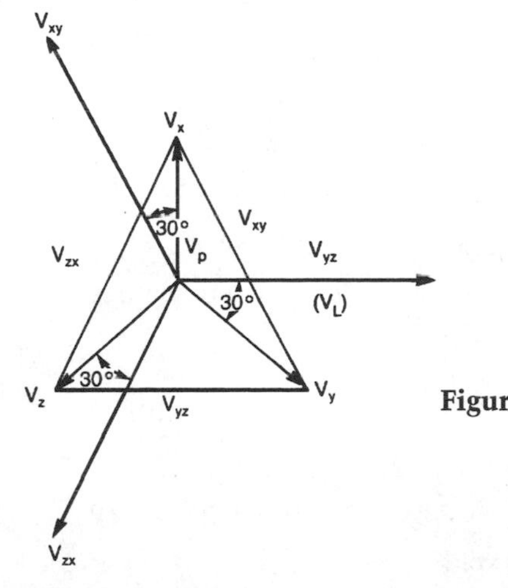

Figure 107-1

the voltage difference between the two points (Fig. 107-1B). The phasor line voltages are

$$v_{xy} = v_x - v_y$$

$$v_{yz} = v_y - v_z$$

$$v_{zx} = v_z - v_x$$

By adding together these three equations, it follows that the phasor sum of the line voltages is zero. The mathematical derivations will also prove that the phasors of the three line voltages are 120° apart and are shifted by 30° from the phasors of the phase voltages.

When three balanced loads are connected between the lines, the phasors of the three line currents are also equally spaced by 120°. Because each line is directly connected to a terminal of a phase winding, the line current (i_L) must be equal to the phase current, i_P.

MATHEMATICAL DERIVATIONS

In Fig. 107-1B, let v_x be the reference phase voltage so that:

Phase voltage,

$$v_x = E \angle 0 \text{ volts} \quad (107\text{-}1)$$

Therefore,
Phase voltage,

$$v_y = E \angle -2\pi/3 \text{ volts} \quad (107\text{-}2)$$

Phase voltage,

$$v_z = E \angle +2\pi/3 \text{ volts} \quad (107\text{-}3)$$

where E is the effective value of the phase voltage. Then,

Line voltage,

$$v_{xy} = v_x - v_y \quad (107\text{-}4)$$
$$= E \angle 0 - E \angle -2\pi/3$$
$$= E - E[(-1/2) + j(-\sqrt{3}/2)]$$
$$= E\,(3/2 + j\sqrt{3}/2)$$
$$= \sqrt{3}E \angle \pi/6 = \sqrt{3}E \angle 30° \text{ volts}$$

This means that the line voltage (v_{xy}) has a magnitude of $\sqrt{3}E$ and leads v_x by 30°.

Line voltage,

$$v_{yz} = v_y - v_z \quad (107\text{-}5)$$
$$= E \angle -2\pi/3 - E \angle +2\pi/3$$
$$= E[(-1/2) + j(-\sqrt{3}/2) - E[(-1/2) + j(+\sqrt{3}/2)]$$
$$= \sqrt{3}E \angle -\pi/2 \text{ volts}$$

The line voltage (v_{yz}) has a magnitude of $\sqrt{3}E$ volts and leads v_y by 30°.

Line voltage,

$$v_{zx} = v_z - v_x \quad (107\text{-}6)$$
$$= E \angle +2\pi/3 - E \angle 0$$
$$= E[(-1/2) + j(+\sqrt{3}/2) - E$$
$$= E[(-3/2) + j(+\sqrt{3}/2)]$$
$$= \sqrt{3}E \angle 5\pi/6 \text{ volts}$$

The line voltage (v_{zx}) has a magnitude of $\sqrt{3}E$ volts and leads v_z by 30°. Equations 107-4, 107-5, 107-6 all show that the line voltage is $\sqrt{3} \times$ the phase voltage (or 1.732 × the phase voltage). In addition, the line phasors are 120° apart and each line voltage phasor is shifted by 30° from one of the phase voltage phasors.

Example 107-1

In a three-phase system, the reference phase voltage is 120 ∠50° volts, 60 Hz. Balanced loads, each equal to 15 ∠20° Ω, are connected between the supply lines. Express the line voltages and the phase currents in their polar forms.

Solution

The phase voltages are 120 ∠50° volts, 120 ∠(50° + 120°) = 120 ∠170° volts and 120 ∠(50° − 120°) = 120 ∠ − 70° volts.

The line voltages are:

$$v_{xy} = \sqrt{3}E \angle (50° + 30°) \quad (107\text{-}4)$$
$$= 1.732 \times 120 \angle 80°$$
$$= 209 \angle 80° \text{ volts.}$$

Line voltage,

$$v_{yz} = \sqrt{3}E \angle (170° + 30°) \quad (107\text{-}5)$$
$$= 1.732 \times 120 \angle 200°$$
$$= 209 \angle -160° \text{ volts.}$$

Line voltage,

$$v_{zx} = \sqrt{3}E \angle (-70° + 30°) \quad (107\text{-}6)$$
$$= 1.732 \times 120 \angle -40°$$
$$= 209 \angle -40° \text{ volts.}$$

The phase currents are equal to the line currents, which are:

Phase current,

$$i_x = \frac{209 \angle 80° \text{ V}}{15 \angle 20° \ \Omega}$$
$$= 13.9 \angle 60° \text{ A.}$$

Phase current,

$$i_y = \frac{209 \angle -160° \text{ V}}{15 \angle 20° \ \Omega}$$
$$= 13.9 \angle 180° \text{ A.}$$

Phase current,

$$i_z = \frac{209 \angle -40° \text{ V}}{15 \angle 20° \ \Omega}$$
$$= 13.9 \angle -60° \text{ A}$$

The phasors of the three-phase currents are also equally separated by 120°.

PRACTICE PROBLEMS

107-1. In a three-phase wye system, the loads are balanced and are connected between the supply lines. If the effective line voltage is 380 V, calculate the effective value of the phase voltage.

107-2. In Practice Problem 107-1, the balanced loads are entirely resistive and each has a value of 40 Ω. Calculate the effective value of each phase current.

107-3. In a three-phase wye system, the reference phase voltage is 110 V ∠−30° V, 50 Hz. Balanced loads, each equal to 10∠ − 20° Ω, are connected between the supply lines. Express the line voltages in their polar forms.

107-4. In Practice Problem 107-3, express the phase currents in their polar forms.

107-5. In Practice Problem 107-3, determine the total power output of the three-phase system.

108 The three-phase delta connection

Because the sum of the three instantaneous phase voltages is at all times zero, it is possible to connect the phase windings in a delta (Δ) formation. This is illustrated in Fig. 108-1A and provided the correct connections are made, there will be zero circulating current in the delta loop. Three supply lines can then be connected to the corners of the delta formation and these are joined to three loads also arranged in delta. It is clear that the voltage applied across any one of the loads must be the same as the voltage generated in the corresponding phase winding across which that particular load is connected. Therefore, in the three-phase delta system, the line voltage (V_L) is equal to the phase voltage, V_P.

By contrast with the voltage relationship, a particular line current phasor (i_x) is associated with two phase currents, i_{yx} and i_{xz}. Each of the load currents will then be equal to its corresponding phase current. However, each line current will be equal to the phasor difference between two of the load currents so that the related equations are:

$$i_x = i_{xz} - i_{yx}$$
$$i_y = i_{yx} - i_{zy}$$
$$i_z = i_{zy} - i_{xz}$$

The sum of the three line currents is therefore zero; this is true, irrespective of whether the loads are bal-

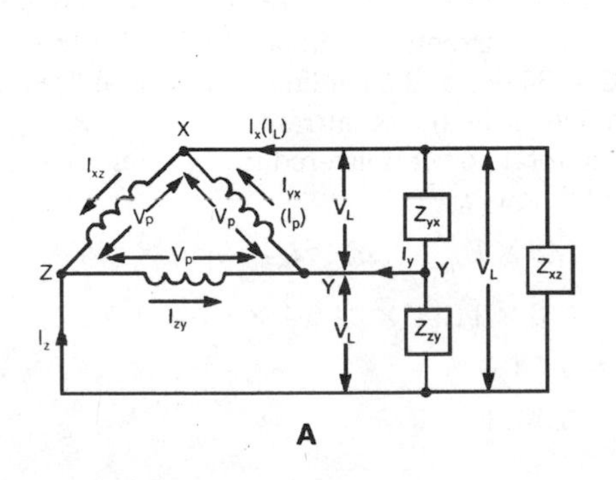

A

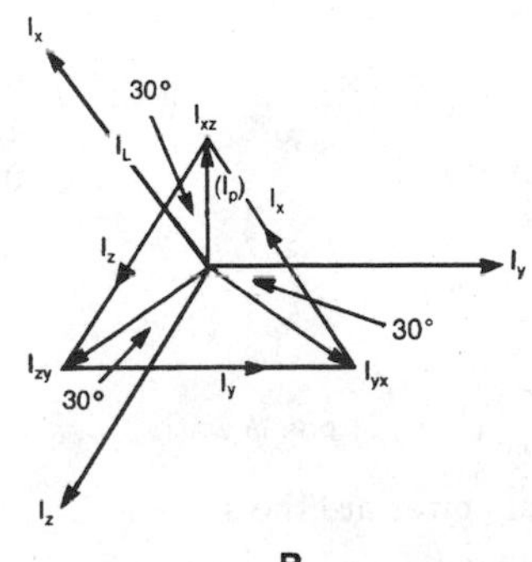

B

Figure 108-1

anced or not. However, if the loads are balanced, the currents in the three windings are separated from each other by a phase difference of 120°. The relationships between the line and phase currents of the delta system are then comparable with the equations relating the line and phase voltages in the wye arrangement (Chapter 107). Therefore, in the delta system, the line current (I_L) is equal to $\sqrt{3}$ × the phase current, I_P (or 1.732 × I_P). In addition, each line current is shifted by 30° from one of the phase currents (Fig. 108-1B).

The total power output from the alternator can be measured in terms of the line voltage and current values, together with the power factor for each of the balanced loads. As the mathematical derivations will show, the expression for the total power is the same for both wye and delta connections.

MATHEMATICAL DERIVATIONS

In a balanced wye (or delta) system the power in each load is the same. Therefore, the total power is:

Total power, (108-1)

$$P_T = 3 \times E_{rms}\ (\text{load}) \times I_{rms}\ (\text{load}) \times \cos\phi \text{ watts}$$

where cos ϕ is the power factor for each of the balanced loads.

For the wye connection:

$$E_{rms}\ (\text{line}) = \sqrt{3} \times E_{rms}\ (\text{load})$$

and

$$I_{rms}\ (\text{line}) = I_{rms}\ (\text{load}) \quad (108\text{-}2)$$

Combining Equations 108-1 and 108-2,

Total power,

$$P_T = 3 \times \frac{E_{rms}\ (\text{line})}{\sqrt{3}} \times I_{rms}\ (\text{line}) \times \cos\phi \quad (108\text{-}3)$$

$$= \sqrt{3} \times E_{rms}\ (\text{line}) \times I_{rms}\ (\text{line}) \times \cos\phi \text{ watts}$$

For the delta connection:

$$E_{rms}\ (\text{line}) = E_{rms}\ (\text{load})$$

and

$$I_{rms}\ (\text{line}) = \sqrt{3}\, I_{rms}\ (\text{load})$$

Combining Equations 108-1 and 108-3,

Total power,

$$P_T = 3 \times E_{rms}\ (\text{line}) \times \frac{I_{rms}\ (\text{line})}{\sqrt{3}} \times \cos\phi \quad (108\text{-}4)$$

$$= \sqrt{3} \times E_{rms}\ (\text{line}) \times I_{rms}\ (\text{line}) \times \cos\phi \text{ watts}$$

The expressions for the total power are the same for both the wye and delta connections.

Example 108-1

In a three-phase delta-connected alternator, the phase voltages are 110 ∠0° V, 110 ∠120° V and 110 ∠−120° V. Each of the three balanced loads has an impedance of 5.5 ∠25° Ω. Calculate the values of the line currents and the total power delivered from the three-phase alternator.

Solution

The phase currents are:

$$I_{xz} = \frac{110\ \angle 0°\ \text{V}}{5.5\ \angle 25°\ \Omega} = 20\ \angle{-25°}\ \text{A}$$

$$I_{zy} = \frac{110\ \angle 120°\ \text{V}}{5.5\ \angle 25°\ \Omega} = 20\ \angle 95°\ \text{A}$$

$$I_{yx} = \frac{110\ \angle{-120°}\ \text{V}}{5.5\ \angle 25°\ \Omega} = 20\ \angle{-145°}\ \text{A}$$

The line currents are:

$$i_x = i_{xz} - i_{yx}$$

$$= 20\ \angle{-25°} - 20\ \angle{-145°}$$

$$= 18.1 - j8.45 + 16.4 + j11.47$$

$$= 34.5 + j3.02$$

$$= 34.6\ \angle 5°\ \text{A}.$$

The same result can be derived from

$$i_x = 20 \times \sqrt{3}\ \angle[(-25°) + (+30°)] = 34.6\ \angle 5°\ \text{A}$$

Similarly,

$$i_y = 34.6\ \angle[(+95°) + (+30°)]$$

$$= 34.6\ \angle 125°\ \text{A}$$

$$= -19.84 + j28.35\ \text{A}$$

and,

$$i_z = 34.6\ \angle[(-145°) + (+30°)]$$

$$= 34.6\ \angle{-115°}\ \text{A}$$

$$= -14.66 - j31.37\ \text{A}.$$

The sum of the line currents is $(+34.5 + j3.02) + (-19.84 + j28.35) + (-14.66 - j31.37) = 0$ A. Each of the line currents has a magnitude of $20\sqrt{3} = 20 \times 1.732 = 34.6$ A and each line current is shifted by 30° from one of the phase currents.

The total power delivered from the alternator is:

Total power,

$$P_T = \sqrt{3} \times E_{rms}\ (\text{line}) \times I_{rms}\ (\text{line}) \times \cos\phi \quad (108\text{-}3)$$

$$= \sqrt{3} \times 110\ \text{V} \times 34.6\ \text{A} \times \cos 25°$$

$$= 5974\ \text{W}$$

$$= 5.974\ \text{kW}.$$

Example 108-2

In a three-phase delta connected alternator, the phase voltages are 160 $\angle 80°$ V, 160 $\angle -160°$ V, and 160 $\angle -40°$ V. The corresponding three unbalanced loads are 20 $\angle 20°$ Ω, 40 $\angle -30°$ Ω, and 80 $\angle 130°$ Ω. Calculate the values of the phase and line currents.

Solution

The phase currents are:

$$\frac{160 \angle 80° \text{ V}}{20 \angle 20° \ \Omega} = 8 \angle 60° \text{ A.}$$

$$\frac{160 \angle -160° \text{ V}}{40 \angle -30° \ \Omega} = 4 \angle -130° \text{ A}$$

$$\frac{160 \angle -40° \text{ V}}{80 \angle 130° \ \Omega} = 2 \angle -170° \text{ A}$$

The corresponding line currents are:

$$\begin{aligned} 8 \angle 60° - 2 \angle -170° &= (4 + j6.928) - (-1.97 - j0.347) \\ &= 5.97 + j7.27 \\ &= 9.41 \angle 50.6° \text{ A} \end{aligned}$$

$$\begin{aligned} 2 \angle -170° - 4 \angle -130° &= (-1.97 - j0.347) \\ &\quad - (-2.57 - j3.064) \\ &= 0.60 + j2.72 \\ &= 2.79 \angle 77.6° \text{ A} \end{aligned}$$

$$\begin{aligned} 4 \angle -130° - 8 \angle 60° &= (-2.57 - j3.064) - (4 + j6.928) \\ &= -6.57 - j9.99 \\ &= 11.96 \angle -123.3° \text{ A} \end{aligned}$$

Although the loads are unbalanced, the total of the three line currents is:

$$(5.97 + j7.27) + (0.60 + j2.72) + (-6.57 - j9.99) = 0$$

PRACTICE PROBLEMS

108-1. In a three-phase delta system, the loads are balanced and are connected between the supply lines. If the effective line voltage is 220 V, what is the effective value of an individual phase voltage?

108-2. In Practice Problem 108-1, the balanced loads are entirely resistive and each has a value of 20 Ω. What is the effective value of each line current?

108-3. In a three-phase, delta-connected alternator, the phase voltages are 220$\angle 30°$ V, 220$\angle 150°$ V, and 220$\angle -90°$ V. Each of the three balanced loads has an impedance of 10$\angle 35°$ Ω. Obtain the phasor expressions for the line currents.

108-4. In Practice Problem 108-3, calculate the power output of the three-phase system.

109 The decibel

On many occasions, electronics and communications are concerned with the transmission of alternating power from one position to another. The various lines and items of equipment (which constitute the transmission system) introduce both power gains and losses.

Consider a network that joins an alternating source (for example, an RF generator) to a load. Let the input power be P_i and the output power be P_o. The ratio of the output power to the input power is the *power ratio* P_o/P_i (Fig. 109-1A). A network, such as an attenuator, introduces a loss, in which case P_o/P_i is less than unity. By contrast, an amplifier provides a gain so that P_o/P_i is greater than unity. If a number of these networks are connected in cascade (Fig. 109-1B), and the individual power ratios are known, the overall power

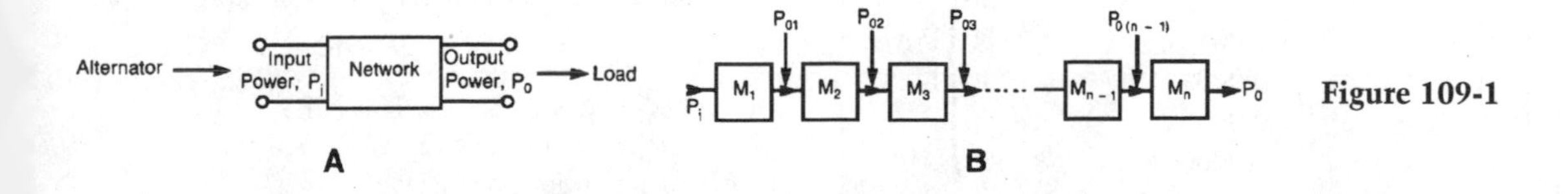

Figure 109-1

ratio is obtained by multiplying together the individual power ratios.

In a complex system that contains a variety of networks, each contributing a gain or loss, the calculation of the overall power ratio can become extremely laborious. To simplify the calculation, the individual power ratios are expressed by a logarithmic unit, enabling the algebraic sum to be employed in place of multiplication. The logarithmic unit used is the *decibel* (abbreviated to dB) and the power gain (or loss) of a network is then expressed as $10 \log (P_o/P_i)$ dB, where "log" is used as an abbreviation for *common logarithm to the base 10.*

The common logarithm of a number is the power (or exponent) to which the base (10) must be raised in order that the result is equal to the value of the number. For example, $\log 100000 = 5$ because $10^5 = 100000$ whereas $\log 0.01 = \log(1/100) = -2$ because $10^{-2} = 1/10^2 = 1/100$. Therefore, a power gain of 100000 is equivalent to $10 \log 100000 = 10 \times 5 = 50$ dB, and a power loss ratio of 0.01 corresponds to $10 \log 0.01 = 10 \times (-2) = -20$ dB. If an attenuator whose power loss is 0.01 is followed by an amplifier with a power gain ratio of 100000, the overall power ratio is $100000 \times 0.01 = 1000$ or +30 dB. The same result is obtained by taking the algebraic sum of +50 dB and −20 dB [(+50 dB) + (−20 dB) = +30 dB]. Corresponding values of power ratios and decibels are shown in Table 109-1.

Other useful decibel values and their equivalent power ratios are summarized in Table 109-2.

Table 109-1. Values of decibels and corresponding voltage, current, and power ratios.

Voltage E_o/E_i ratio, or current I_o/I_i ratio,	Power ratio, P_o/P_i	N decibels (dB)	
1000	1000000	60	Power Gain
316	100000	50	
100	10000	40	
31.6	1000	30	
10	100	20	
3.16	10	10	
1	1	0	
0.316	0.1	−10	Power Loss
0.1	0.01	−20	
0.0316	0.001	−30	
0.01	0.0001	−40	
0.00316	0.00001	−50	
0.001	0.000001	−60	

Table 109-2. Particular values of decibels and corresponding power ratios.

Power Ratio P_o/P_i	Number of Decibels (dB)
5	7
2	3
1.26	1
0.794	−1
1/2 = 0.5	−3
1/5 = 0.2	−7

A larger logarithmic unit is the *bel*, which is equivalent to 10 decibels, and named after Alexander Graham Bell. The decibel was originally related to acoustics and was regarded as the smallest change in sound intensity that could be detected by the human ear.

The decibel is essentially a unit of power ratio, and not of absolute power but if some standard reference level of power is assumed to represent 0 dB, then any value of absolute power can be expressed as so many dB above (or below) this reference standard. Various other standards are occasionally encountered but the standard most commonly adopted is 1 mW (0.001 W), which is delivered along a standard 600 Ω program transmission line. This means that one milliwatt of single tone audio power is 0 dBm (dB with respect to 1 mW). Then, one watt, for example, is $10 \log (1 \text{ W}/1 \text{ mW}) = 10 \log 1000 = 30$ dBm, and one microwatt is $10 \log (1 \mu\text{W}/1 \text{ mW}) = 10 \log (1/1000) = -30$ dBm.

Zero VU (volume unit) has the same reference level of 1 mW. It therefore appears that the dBm and the VU are identical. However, the use of the dBm is normally confined to single frequencies (tones), but the VU is reserved for complex audio signals, such as speech or music. When it is desired to compare the powers developed in equal input and output resistances, it is sufficient to use their associated voltages and currents. The power gain (or loss) in decibels is then equal to 20 times the common logarithm of the voltage or current ratio.

MATHEMATICAL DERIVATIONS

In Fig. 109-1A,

The power gain (or loss) of a network is expressed as:

$$N = 10 \log (P_o/P_i) \text{ dB} \qquad (109\text{-}1)$$

where: N = power gain or loss (dB)
P_o = output power (W)
P_i = input power (W)

Positive and negative values of N respectively represent power gains and losses.

Power ratio,

$$\frac{P_o}{P_i} = \text{inv log}\,\frac{N}{10} = 10^{N/10} \qquad (109\text{-}2)$$

In Fig. 109-1B,

Overall power ratio,

$$\frac{P_o}{P_i} = \frac{P_{o1}}{P_i} \times \frac{P_{o2}}{P_{o1}} \times \frac{P_{o3}}{P_{o2}} \ldots \times \frac{P_o}{P_{o(n-1)}} \qquad (109\text{-}3)$$

Total dB gain,

$$N_T = N_1 + N_2 + N_3 \ldots + N_N \qquad (109\text{-}4)$$

where N_T is the algebraic sum of the individual gains and losses.

Absolute power

Absolute power,

$$P = 10 \log \frac{P}{1\text{ mW}} \text{ dBm}$$

$$N \text{ dBm} = 10^{N/10} \text{ mW} \qquad (109\text{-}5)$$

Voltage and current ratios

Consider equal input and output resistances for a particular stage, such as an amplifier or an attenuator. If the input and output voltages have rms values of E_i and E_o and are associated with input and output rms currents of I_i and I_o,

Input power,

$$P_i = E_i \times I_i = I_i^2 \times R = E_i^2/R \text{ watts} \qquad (109\text{-}6)$$

Output power,

$$P_o = E_o \times I_o = I_o^2 \times R = E_o^2/R \text{ watts} \qquad (109\text{-}7)$$

Therefore, the power gain in decibels is given by,

$$N = 10 \log \left(\frac{P_o}{P_i}\right) \qquad (109\text{-}8)$$

$$= 10 \log \left(\frac{E_o^2/R}{E_i^2/R}\right) \text{ or } 10 \log \left(\frac{I_o^2 R}{I_i^2 R}\right)$$

$$= 10 \log (E_o^2/E_i^2) \text{ or } 10 \log (I_o^2/I_i^2)$$

$$= 20 \log \left(\frac{E_o}{E_i}\right) \text{ or } 20 \log \left(\frac{I_o}{I_i}\right) \text{ dB}$$

The power gain in decibels is equal to 20 log (current ratio) or 20 log (voltage ratio) provided that the input and output resistances are the same.

If the input and output powers are associated with input and output impedances of $Z_i = R_i + jX_i$ and $Z_o = R_o + jX_o$ respectively, the formulas $N = 20 \log (I_o/I_i)$ and $N = 20 \log (V_o/V_i)$ can still be used provided $R_i = R_o$. If the input and output resistances are not equal, the formulas then become:

$$N = 10 \log \left(\frac{E_o^2/R_o}{E_i^2/R_i}\right) \qquad (109\text{-}9)$$

$$= 20 \log \left(\frac{E_o}{E_i}\right) - 10 \log \left(\frac{R_o}{R_i}\right) \text{ dB}$$

and,

$$N = 10 \log \left(\frac{I_o^2 R_o}{I_i^2 R_i}\right) \qquad (109\text{-}10)$$

$$= 20 \log \left(\frac{I_o}{I_i}\right) + 10 \log \left(\frac{R_o}{R_i}\right) \text{ dB}$$

Example 109-1

1. Convert power ratios of (a) 273 and (b) 0.0469 into decibels.
2. Convert (a) +37 dB and (b) −14.6 dB into their corresponding power ratios.

Solution

1. (a) Number of decibels,

$$N = 10 \log \left(\frac{P_o}{P_i}\right) \text{ dB} \qquad (109\text{-}1)$$

$$= 10 \log 273$$

$$= 10 \times 2.436$$

$$= +24.36 \text{ dB}$$

(b) Number of decibels,

$$N = 10 \log \left(\frac{P_o}{P_i}\right) \text{ dB} \qquad (109\text{-}1)$$

$$= 10 \log 0.0469$$

$$= -13.3 \text{ dB}$$

2. (a) Power ratio,

$$\frac{P_o}{P_i} = \text{inv log} \left(\frac{N}{10}\right) \qquad (109\text{-}2)$$

$$= \text{inv log} \left(\frac{37}{10}\right)$$

$$= \text{inv log } 3.7 = 5012$$

(b) Power ratio,

$$\frac{P_o}{P_i} = \text{inv log}\,\frac{N}{10} \qquad (109\text{-}2)$$

$$= \text{inv log} \left(\frac{-14.6}{10}\right)$$

$$= 0.0347$$

Example 109-2

In Fig. 109-2, calculate the value of the overall power ratio. Express each of the individual power ratios, as well as the overall power ratio, in terms of dB.

Solution

Overall power ratio,

$$N_T = \frac{P_{o1}}{P_i} \times \frac{P_{o2}}{P_{o1}} \times \frac{P_{o3}}{P_{o2}} \times \frac{P_{o4}}{P_{o3}} \times \frac{P_o}{P_{o4}} \qquad (109\text{-}3)$$

$$= M_1 \times M_2 \times M_3 \times M_4 \times M_5$$

$$= 0.215 \times 20.3 \times 0.0246 \times 0.251 \times 25.2$$

$$= 0.679$$

The overall power ratio is less than unity and therefore, represents a loss of 10 log 0.679 = −1.69 dB.

In terms of decibels,

$$N_1 = 10 \log M_1 = 10 \log 0.215 = -6.68 \text{ dB.}$$

$$N_2 = 10 \log M_2 = 10 \log 20.3 = +13.07 \text{ dB.}$$

$$N_3 = 10 \log M_3 = 10 \log 0.0246 = -16.09 \text{ dB.}$$

$$N_4 = 10 \log M_4 = 10 \log 0.251 = -6.00 \text{ dB.}$$

$$N_5 = 10 \log M_5 = 10 \log 25.2 = +14.01 \text{ dB.}$$

$$N_T = 10 \log M_T = 10 \log 0.679 = -1.68 \text{ dB.}$$

Then,

$$N_T = N_1 + N_2 + N_3 + N_4 + N_5$$

$$= (-6.68) + (+13.07) + (-16.09) + (-6.00) + (+14.01)$$

$$= -1.69 \text{ dB.}$$

The overall gain (or loss) in dB for a number of cascaded stages is the algebraic sum of the decibel gains and losses in the individual stages.

Example 109-3

1. Express power levels of (a) 6.38 W and (b) 26.7 μW, in terms of dBm.
2. What power levels are represented by (a) +17 dBm, and (b) −8.6 VU?

Solution

1. (a) A power level of 6.38 W is equivalent to 6380 mW. Therefore,

$$6.38 \text{ W} = 10 \log(6380 \text{ mW/1 mW})$$
$$= 38.05 \text{ dBm.}$$

(b) A power level of 26.7 μW is equivalent to 0.0267 mW. Therefore,

$$26.7\ \mu\text{W} = 10 \log(0.0267 \text{ mW/1 mW})$$
$$= -15.73 \text{ dBm.}$$

2. (a) Power level = inv log(+17 dBm/10) = 50.12 mW.
   (b) Power level = inv log(−8.6 VU/10) = 0.138 mW = 138 μW.

Example 109-4

1. Assuming that the input and output resistances are equal, convert (a) +43.7 dB and (b) −27.6 dB, to their corresponding voltage (or current) ratios.
2. Assuming equal input and output resistances, convert (a) a voltage ratio of 56.3 and (b) a current ratio of 0.00353, into decibels.

Solution

1. (a) Voltage (or current) ratio,

$$\frac{V_o}{V_i} = \text{inv log}\left(\frac{43.7}{20}\right)$$
$$= 153.$$

(b) Voltage (or current) ratio,

$$\frac{V_o}{V_i} = \text{inv log}\left(\frac{-27.6}{20}\right)$$
$$= 0.0417.$$

2. (a) Number of decibels,

$$N = 20 \log\left(\frac{E_o}{E_i}\right) \qquad (109\text{-}8)$$
$$= 20 \log 56.3$$
$$= 35.01 \text{ dB.}$$

(b) Number of decibels,

$$N = 20 \log\left(\frac{I_o}{I_i}\right) \qquad (109\text{-}8)$$
$$= 20 \log 0.00353$$
$$= -49.04 \text{ dB.}$$

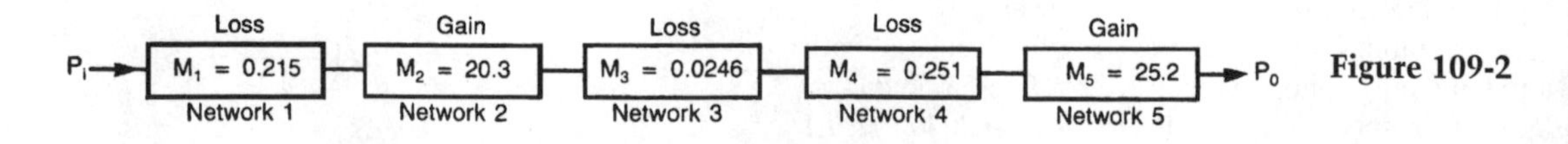

Figure 109-2

Example 109-5

The input resistance of an amplifier is 1.3 kΩ and its output resistance is 17 kΩ. If the amplifier voltage gain is 35, calculate its power gain in dB.

Solution

Power gain,

$$\frac{P_o}{P_i} = 20 \log \left(\frac{E_o}{E_i}\right) - 10 \log \left(\frac{R_o}{R_i}\right) \qquad (109\text{-}9)$$

$$= 20 \log 35 - 10 \log \left(\frac{17}{1.3}\right)$$

$$= 30.88 - 11.2$$

$$= 19.68 \text{ dB}$$

PRACTICE PROBLEMS

109-1. Convert power ratios of 4.75×10^2 and 8.26×10^{-4} into their equivalent numbers of decibels.

109-2. Convert +27.8 dB and −16.7 dB into their equivalent power ratios.

109-3. Three cascaded stages have gains of +13.7 dB, −6.9 dB, and +4.3 dB. What is the overall gain in dB, and express this result as a power ratio.

109-4. What are the absolute power values of +27 dBm and −17.6 dBm?

109-5. Convert 3.82 kW and 14.76 μW into their equivalent dBm values.

109-6. Convert +26.7 dB and −18.8 dB into their equivalent voltage ratios. Assume that the input and output resistances have equal values.

109-7. Convert voltage ratios of 12.7 and 0.00347 into their equivalent decibel values. Assume that the input and output resistances have equal values.

109-8. Convert +7.8 VU into its equivalent audio power level.

109-9. In Practice Problem 109-8, it is assumed that the audio power is being conveyed along a 600-Ω standard program transmission line. Calculate the effective value of the audio voltage on the line.

109-10. A stage has a voltage gain of 21.6, an input resistance of 8.7 kΩ and an output resistance of 1.4 kΩ. Calculate the stage's gain in decibels.

110
The neper

The decibel is, fundamentally, a unit of power ratio, but it can be used to express *current ratios*, when the resistive components of the impedances (through which the currents flow) are equal and *voltage ratios*, when the conductive components of those impedances are equal. The *neper* is a unit of current ratio, but it can be used to express power ratios when the resistive components of the impedances are equal.

The loss of power in an electrical network is known as *attenuation*. Attenuation can be measured using either the decibel or the neper notation.

If the power entering a network is P_i and the power leaving is P_o, then the attenuation in decibels is defined as $10 \log (P_o/P_i)$. If the current entering a network is I_i and the current leaving is I_o, then the attenuation in nepers is defined as $\ln (I_o/I_i)$ (where **ln** is the natural, or Napierian, logarithm, which is based on the exponential quantity, $e = 2.7183\ldots$; see Fig. 110-1.

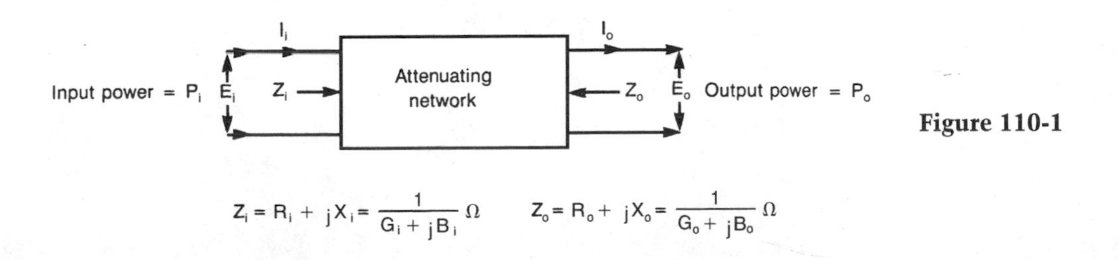

Figure 110-1

Because of its derivation from the exponential e, the neper is the most convenient unit for expressing attenuation in theoretical work. By contrast, the decibel is defined in terms of common logarithms, and is a more convenient unit in practical calculations that use the decimal system.

MATHEMATICAL DERIVATIONS

Attenuation in dB,

$$N = 10 \log \left(\frac{P_o}{P_i}\right) \qquad (110\text{-}1)$$

$$= 20 \log \left(\frac{I_o}{I_i}\right), \text{ provided that } R_i = R_o \qquad (110\text{-}2)$$

$$= 20 \log \left(\frac{E_o}{E_i}\right), \text{ provided that } G_i = G_o \qquad (110\text{-}3)$$

Attenuation in nepers,

$$N' = \ln \left(\frac{I_o}{I_i}\right) \qquad (110\text{-}4)$$

$$= \ln \left(\frac{E_o}{E_i}\right), \text{ provided that } Z_i = Z_o$$

$$= \frac{1}{2} \ln \left(\frac{P_o}{P_i}\right), \text{ provided that } R_i - R_o \qquad (110\text{-}5)$$

If the resistive components of the impedances at the input and output of the network are equal, the amount of attenuation may be readily converted from one unit to another.

Attenuation in dB,

$$N = 20 \log \left(\frac{I_o}{I_i}\right) \qquad (110\text{-}6)$$

$$= 20 \ln \left(\frac{I_o}{I_i}\right) \times \log e$$

$$= 8.686 \times \ln \left(\frac{I_o}{I_i}\right)$$

$$= 8.686 \times (\text{attenuation in nepers})$$

Therefore, the attenuation in dB = 8.686 × (attenuation in nepers) or the attenuation in nepers = 0.1151 × (attenuation in dB), provided that $R_i = R_o$.

Example 110-1

The input to an electrical network is 27 mA and the output current is 75 μA. Assuming that the input and output resistances are equal, calculate the attenuation of the network in (a) decibels and (b) nepers.

Solution

(a) Network attenuation,

$$\text{Current ratio} = \frac{I_o}{I_i} = \frac{75\ \mu\text{A}}{27\ \text{mA}} \qquad (110\text{-}2)$$

$$= 2.77 \times 10^{-3}$$

Attenuation in dB,

$$N = 20 \log (2.77 \times 10^{-3})$$

$$= -51.15\ \text{dB}$$

(b) Attenuation in nepers,

$$N' = \ln (2.77 \times 10^{-3}) \qquad (110\text{-}4)$$

$$= -5.889\ \text{nepers.}$$

Check:

$$\frac{\text{Attenuation in dB}}{\text{Attenuation in nepers}} = \frac{51.15}{5.889}$$

$$= 8.686, \text{ rounded off.}$$

PRACTICE PROBLEMS

110-1. Convert −28.6 dB and +17.4 dB into nepers.

110-2. Convert +3.49 nepers and −5.73 nepers into decibels.

110-3. The input current to an amplifier is 48 μA and the output current is 3.5 mA. Assuming that the input and output resistances are the same, calculate the amplifer's gain in nepers and decibels.

111
Waveform analysis

Any single-valued, finite, and continuous function, $f(t)$ having a period of $2\pi/\omega$ seconds, may be expressed in the following form:

FOURIER'S THEOREM

$$f(t) = a_0 + A_1 \sin(\omega t + \theta_1) + A_2 \sin(2\omega t + \theta_2) + \ldots \quad (111\text{-}1)$$

where: $\omega = 2\pi f$ radians per second
t = time in seconds

Because,

$$A \sin(\omega t + \theta) = A \sin \omega t \cos \theta + A \cos \omega t \sin \theta = a \cos \omega t + b \sin \omega t$$

where: $a = A \sin \theta$
$b = A \cos \theta$,

the Fourier expansion can be expressed as:

$$f(t) = a_0 + a_1 \cos \omega t + a_2 \cos 2\omega t + \ldots + b_1 \sin \omega t + b_2 \sin 2\omega t + \ldots \quad (111\text{-}2)$$

This means that the waveform can be regarded as being composed of a mean level (a_0), together with fundamental sine and cosine waves, as well as their harmonics.

In order to use the expansion to analyze a complex waveform, it is necessary to determine the values of the coefficients $a_0, a_1, a_2 \ldots, b_1, b_2 \ldots$. This is done by using integral calculus, which is outlined in the mathematical derivations. However, the following is the quoted result for the symmetrical square wave (Fig. 111-1A), which is frequently encountered in communications. For such a square wave whose mean level is zero, the expansion is:

$$f(t) = \frac{2D}{\pi}\left[\sin \omega t + \frac{1}{3}\sin 3\omega t + \frac{1}{5}\sin 5\omega t + \ldots + \frac{1}{n}\sin n\omega t + \ldots\right] \quad (111\text{-}3)$$

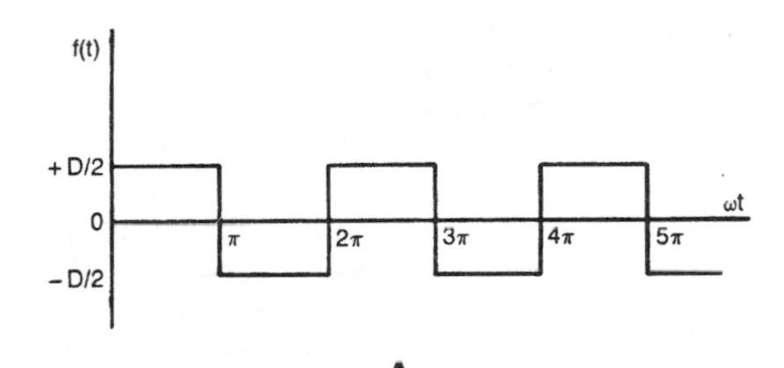

A

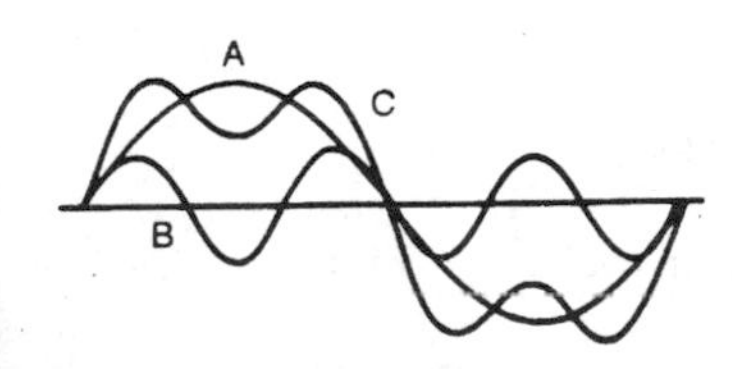

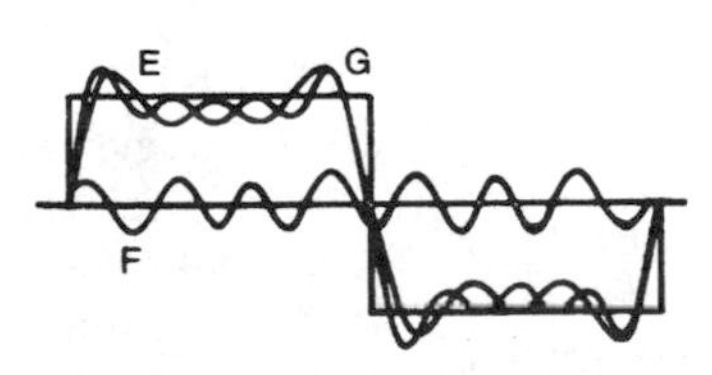

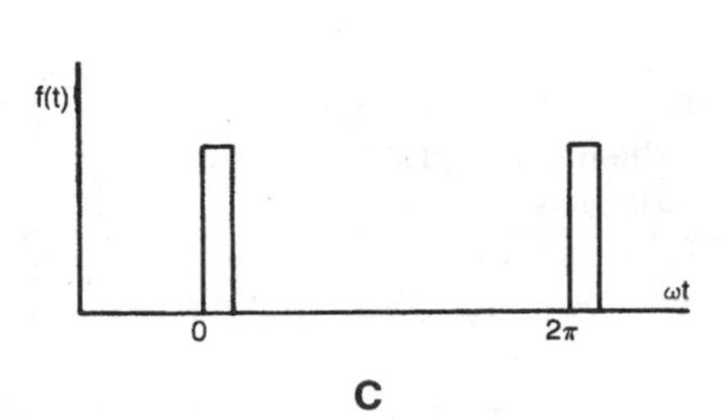

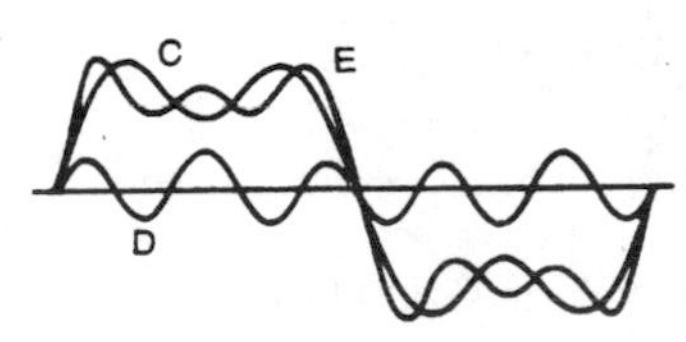

A Fundamental
B 3rd Harmonic
C Fundamental plus 3rd harmonic
D 5th harmonic
E Fundamental plus 3rd & 5th harmonics
F 7th harmonic
G Fundamental plus 3rd, 5th, & 7th harmonics

B

C

Figure 111-1

As a result of the symmetry of the square wave, this expansion contains neither cosine terms nor even harmonic sine terms.

Notice that the amplitude of the nth harmonic is $1/n$ so that the amplitudes of the higher harmonics only decrease slowly. It follows that the symmetrical square wave contains a large number of strong harmonics.

The expansion of equation 111-3 can be verified by adding the fundamental sine-wave component and its odd harmonics. This is called *waveform synthesis*, which is illustrated in Fig. 111-1B. However, waveform "G" is far from being a symmetrical square, and many more odd harmonics would be necessary before achieving a good approximation to the required waveform.

If the square waveform is made very asymmetrical, its appearance is that of a repeating pulse of short duration (Fig. 111-1C); as an example this would be the modulation waveform in a pulsed radar set. Both odd and even harmonics then appear in the expansion, but their amplitudes decrease very slowly. This waveform is, therefore, extremely rich in harmonics and to achieve a good approximation in its synthesis, you might well have to include harmonics higher than the thousandth.

MATHEMATICAL DERIVATIONS

Figure 111-2 shows a square waveform that is a single-valued repeating function of time with a period of $2\pi/\omega$ seconds; consequently, it can be analyzed by Fourier's Theorem.

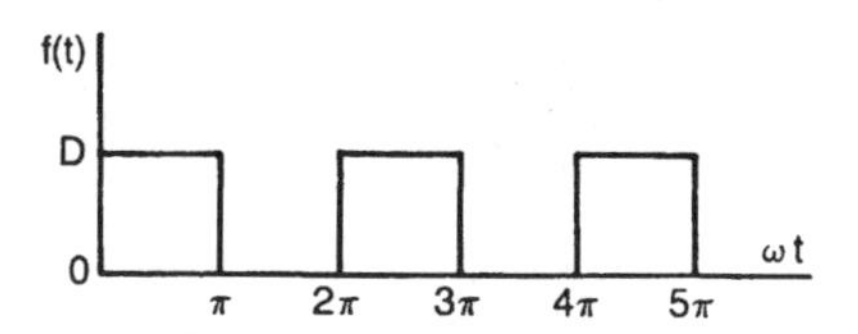

Figure 111-2

From $\omega t = 0$ to $\omega t = 2\pi$, the equation of the function is $f(t) = D$. From $\omega t = \pi$ to $\omega t = 2\pi$, the equation of the function is $f(t) = 0$.

Then,

$$a_0 = \frac{1}{2\pi}\int_0^{2\pi} f(t)\, d(\omega t) \qquad (111\text{-}4)$$

$$= \frac{1}{2\pi}\int_0^{\pi} f(t)\, d(\omega t) + \frac{1}{2\pi}\int_{\pi}^{2\pi} f(t)\, d(\omega t)$$

$$= \frac{1}{2\pi} \times D \times \pi + 0 = \frac{D}{2}$$

$$a_n = \frac{1}{\pi}\int_0^{2\pi} f(t)\cos n\omega t\, d(\omega t) \qquad (111\text{-}5)$$

$$= \frac{1}{\pi}\left[\int_0^{\pi} D\cos n\omega t\, d(\omega t) + \int_{\pi}^{2\pi} 0 \times \cos n\omega t\, d(\omega t)\right]$$

$$= \frac{1}{\pi}\left[\frac{D\sin n\omega t}{n}\right]_0^{\pi}$$

$$= 0$$

Therefore, all cosine terms are zero.

$$b_n = \frac{1}{\pi}\int_0^{2\pi} f(t)\sin n\omega t\, d(\omega t) \qquad (111\text{-}6)$$

$$= \frac{1}{\pi}\left[\int_0^{\pi} D\sin n\omega t\, d(\omega t) + \int_{\pi}^{2\pi} 0 \times \sin n\omega t\, d(\omega t)\right]$$

$$= \frac{1}{\pi}\left[\frac{-D\cos n\omega t}{n}\right]_0^{\pi}$$

$$= \frac{D}{n\pi}(1 - \cos n\pi)$$

When n is an odd number, $(1 - \cos n\pi) = 2$. When n is an even number, $(1 - \cos n\pi) = 0$.

The required equation for the square wave of Fig. 111-2 is

$$f(t) = \frac{D}{2} + \frac{2D}{\pi}\left(\sin \omega t + \frac{1}{3}\sin 3\omega t + \frac{1}{5}\sin 5\omega t + \ldots \frac{1}{n}\sin n\omega t + \ldots\right) \qquad (111\text{-}7)$$

For the square wave of Fig. 111-1A, the average value is zero and its Fourier equation is

$$f(t) = \frac{2D}{\pi}\left(\sin \omega t + \frac{1}{3}\sin 3\omega t + \frac{1}{5}\sin 5\omega t + \ldots + \frac{1}{n}\sin n\omega t + \ldots\right) \qquad (111\text{-}8)$$

which matches with equation 111-3.

Shown below are the Fourier's series for other non-sinusoidal waveforms. In each case, D is the peak value of the waveform.

Negative sawtooth waveform (Fig. 111-3)

$$f(t) = \frac{2D}{\pi}\left(\sin \omega t + \frac{1}{2}\sin 2\,\omega t + \frac{1}{3}\sin 3\omega t + \frac{1}{4}\sin 4\omega t + \ldots\right)$$

Positive sawtooth waveform (Fig. 111-4)

$$f(t) = \frac{2D}{\pi}\left(\sin \omega t - \frac{1}{2}\sin 2\,\omega t\right.$$

$$+ \frac{1}{3}\sin 3\omega t - \frac{1}{4}\sin 4\omega t + \ldots \Big)$$

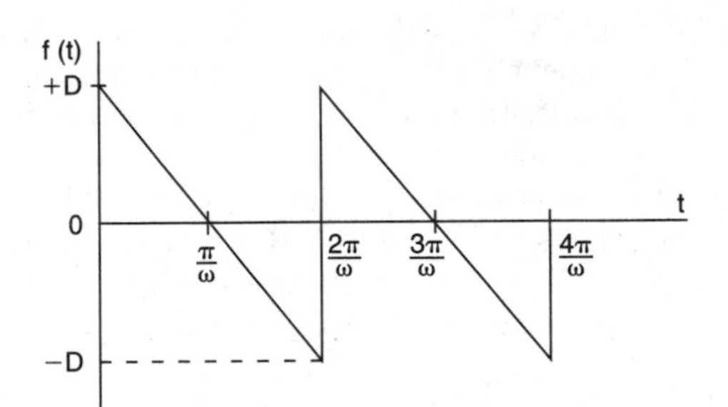

Figure 111-3

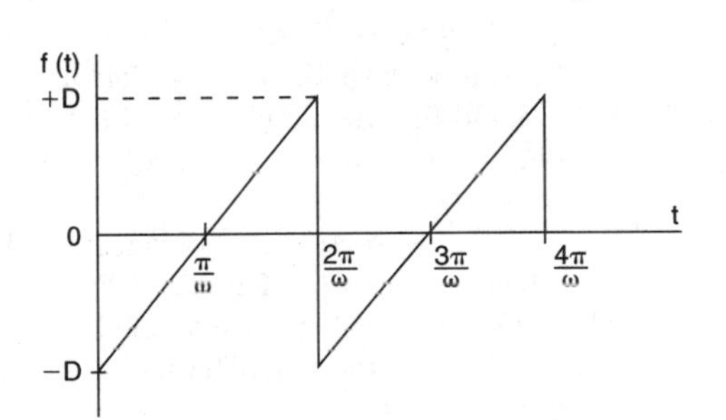

Figure 111-4

Triangular waveform (Fig. 111-5)

$$f(t) = \frac{8D}{\pi^2}\left(\cos \omega t + \frac{\cos 3\omega t}{3^2} + \frac{\cos 5\omega t}{5^2} + \frac{\cos 7\omega t}{7^2} + \ldots\right)$$

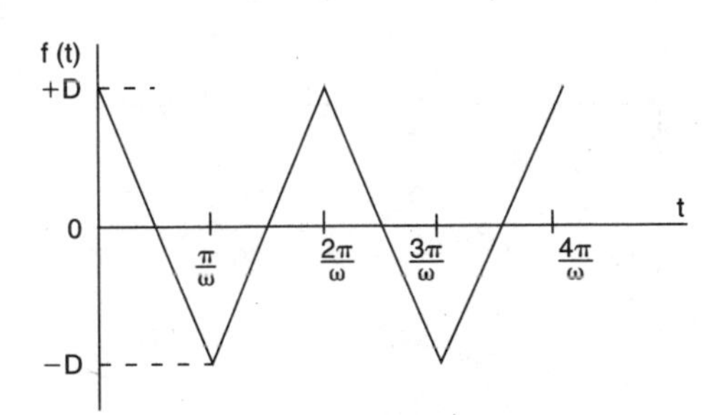

Figure 111-05

Half-wave rectification (Fig. 111-6)

$$f(t) = \frac{2D}{\pi}\left(\frac{1}{2} + \frac{\pi}{4}\sin \omega t - \frac{1}{1 \times 3}\cos 2\omega t - \frac{1}{3 \times 5}\cos 4\omega t - \frac{1}{5 \times 7}\cos 6\omega t \ldots\right)$$

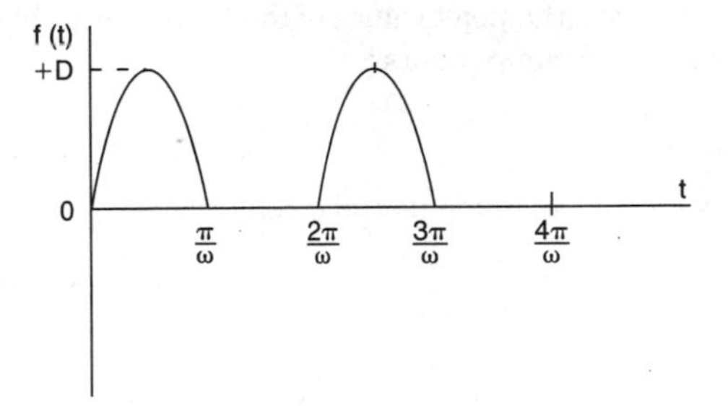

Figure 111-6

Full-wave rectification (Fig. 111-7)

$$f(t) = \frac{4D}{\pi}\left(\frac{1}{2} - \frac{1}{1 \times 3}\cos 2\omega t - \frac{1}{3 \times 5}\cos 4\omega t - \frac{1}{5 \times 7}\cos 6\omega t \ldots\right)$$

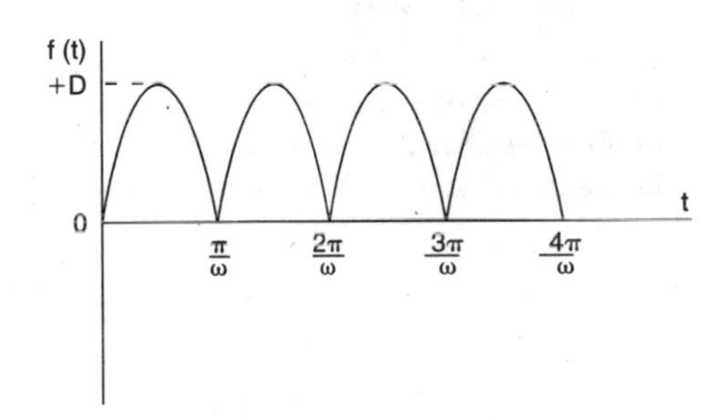

Figure 111-7

Modified sawtooth waveform (Fig. 111-8)

$$f(t) = \frac{D}{4} - \frac{2D}{\pi^2}\left(\cos \omega t + \frac{\cos 3\omega t}{3^2} + \frac{\cos 5\omega t}{5^2} + \frac{\cos 7\omega t}{7^2} + \ldots\right) + \frac{D}{\pi}\left(\sin \omega t - \frac{\sin 2\omega t}{2} + \frac{\sin 3\omega t}{3} - \ldots\right)$$

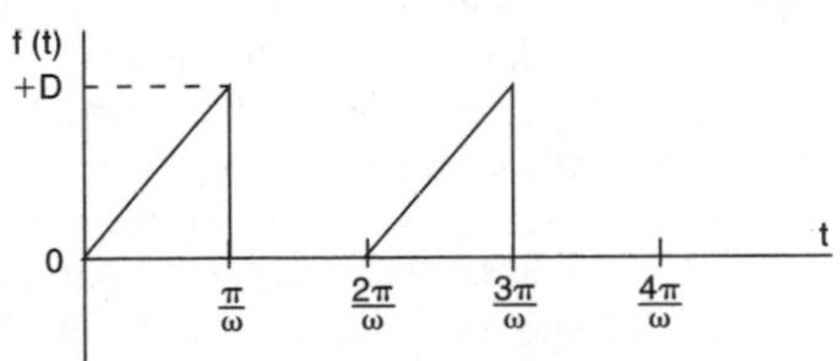

Figure 111-8

Example 111-1

A symmetrical square wave has top and bottom levels of +7 V and −7 V. In the Fourier analysis of the waveform, what are the peak values of the fundamental and fifth harmonic components?

Solution

Peak value of the fundamental component,

$$b_1 = \frac{2D}{\pi} \qquad (111\text{-}7)$$

$$= \frac{2 \times (2 \times 7)\ \text{V}}{\pi}$$

$$= 8.91\ \text{V}.$$

Peak value of the fifth harmonic component,

$$b_5 = \frac{8.91\ \text{V}}{5} \qquad (111\text{-}7)$$

$$= 1.78\ \text{V}.$$

PRACTICE PROBLEMS

111-1. A symmetrical square waveform has top and bottom levels of +8 V and 0 V (Fig. 111-2). In the Fourier analysis of the waveform, what is the value of the dc level and what are the peak values of the fundamental and harmonic components up to and including the seventh harmonic?

111-2. In the negative sawtooth waveform of Fig. 111-3, the peak value of the fundamental component is 12 V. What are the peak values of the sawtooth waveform, and its second and third harmonic components?

111-3. The positive sawtooth waveform of Fig. 111-4 has a peak value of 15 V and a frequency of 1 kHz. Determine the peak values of all components whose frequencies are less than 5000 Hz.

111-4. The triangular waveform of Fig. 111-5 has a peak value of 8 V. What are the peak values of the fundamental, third, and fifth harmonic components?

111-5. The (half-wave) rectified ac waveform of Fig. 111-6 has a peak value of 100 V. Calculate the dc level and find the amplitudes of the fundamental, second, fourth, and sixth harmonic components.

111-6. The (full-wave) rectified ac waveform of Fig. 111-7 has a peak value of 100 V. Calculate the value of the dc level and the amplitude of the second, fourth, and sixth harmonic components.

112
Nonsinusoidal voltages and currents in ac circuits

If a nonsinusoidal voltage source (e) is applied across a series circuit (Fig. 112-1), the resulting current contains fundamental and harmonic components (provided that all of these components are contained in the voltage waveform).

When the series ac circuit contains both inductors and capacitors, their reactances must be calculated separately for the fundamental and harmonic frequencies, in order to obtain the values of the corresponding currents. In other words, for each frequency the current must be recalculated. When the values of all current components are known, you can obtain the effective value of the nonsinusoidal current drawn from the source.

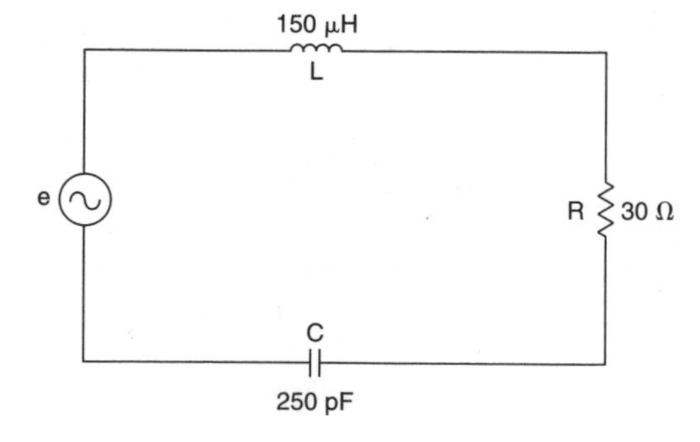

Figure 112-1

The analysis of a parallel LCR circuit is similar to that of the series circuit. The individual branch currents for the fundamental and harmonic components are calculated first. By using the rules of complex algebra, the branch currents are combined to create the supply current that is drawn from the nonsinusoidal voltage source.

When a nonsinusoidal voltage source is applied across a series-parallel circuit such as that of Fig. 112-2, you can determine the fundamental current components in each of the two branches (the reactance values correspond to the fundamental frequency). The fundamental line current is then the phasor sum of the branch currents. The procedure is repeated for each of the harmonic components. Finally, the various currents are combined to obtain the effective value of the line current drawn from the source.

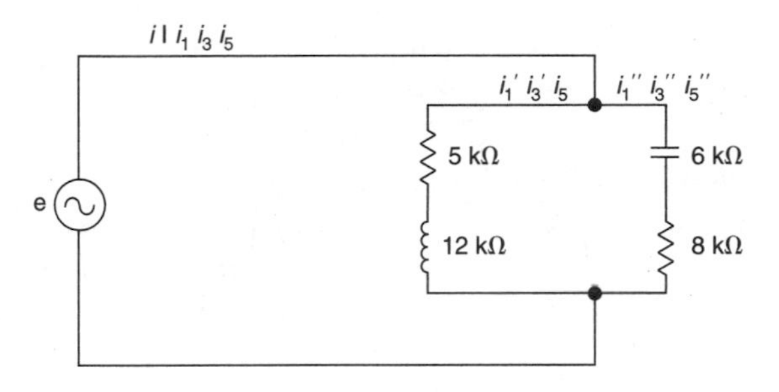

Figure 112-2

MATHEMATICAL DERIVATIONS

Effective (rms) value of a nonsinusoidal voltage or current

You have already learned that a nonsinusoidal wave (in general) can be expressed in terms of a dc level together with fundamental and harmonic components. Assume that this wave is in the form of a source voltage that is connected across a resistor, R. Each component provides its own power dissipation and, because the individual powers dissipated are additive, you can derive an equation for the effective voltage of the source. This value would be the reading of an ac voltmeter connected across the source, provided that the instrument can respond to the frequencies of the higher harmonics. Total power dissipated,

$$P_T = \frac{V_{dc}^2}{R} + \frac{V_1^2}{R} + \frac{V_2^2}{R} + \text{- - - -watts}$$

where:

V_{dc} = dc level of the voltage source (V)
V_1 = effective value of the fundamental component (V)
V_2 = effective value of the second harmonic component (V)

But,

$$P_T = V_T^2/R \text{ watts}$$

where V_T = effective value of the voltage source (V)

Therefore,

$$V_T^2 = V_{dc}^2 + V_1^2 + V_2^2 + \text{- - -}$$

or,

$$V_T = \sqrt{V_{dc}^2 + V_1^2 + V_2^2 + \text{- - -}} \text{ volts}$$

A similar analysis reveals that,

$$I_T = \sqrt{I_{dc}^2 + I_1^2 + I_2^2 + \text{- - -}} \text{ amperes}$$

where I_T is the effective value of the nonsinusoidal current.

Example 112-1

A nonsinusoidal source has a voltage that is represented by the equation $e = 100 \sin \omega t - 80 \sin (2\omega t + 40°) + 40 \sin (3\omega t - 20°)$ V, where the frequency, $f = \omega/2\pi = 400$ kHz. If this voltage is applied across the series circuit of Fig. 112-1, obtain the equation *of* the nonsinusoidal current.

Solution

Step 1. The fundamental component.

Inductive reactance,

$$X_L = 2 \times \pi \times f \times L$$
$$= 2 \times \pi \times 400 \times 10^3 \times 150 \times 10^{-6}$$
$$= 377\ \Omega$$

Capacitive reactance,

$$X_C = \frac{1}{2 \times \pi \times f \times C}$$
$$= \frac{1}{2 \times \pi \times 400 \times 10^3 \times 250 \times 10^{-12}}$$
$$= 1592\ \Omega$$

Total impedance,

$$z = 30 + j377 - j1592$$
$$= 30 - j1215$$
$$= 1215 \angle -88.6°\ \Omega$$

Peak value of the fundamental current,

$$i_1 = \frac{100 \text{ V}}{1215 \angle -88.6°\ \Omega}$$
$$= 0.082 \angle 88.6° \text{ A}$$

Effective value of the fundamental current,

$$I_1 = 0.082 \times 0.707$$
$$= 0.058 \text{ A}$$

Step 2. The second harmonic component.

Inductive reactance,

$$X_L = 2 \times 377 = 754\ \Omega$$

Capacitive reactance,

$$X_C = \frac{1592}{2} = 796\ \Omega$$

Total impedance,

$$z = 30 + j754 - j796$$
$$= 30 - j42$$
$$= 51.6 \angle -54.5°\ \Omega$$

Peak value of the second harmonic current,

$$i_2 = \frac{-80 \angle 40°\ \text{V}}{51.6 \angle -54.5°} = -1.55 \angle 94.5°$$
$$= 1.55 \angle -85.5°\ \text{A}$$

Effective value of the second harmonic current,

$$I_2 = 0.707 \times 1.55 = 1.10\ \text{A}$$

Step 3. The third harmonic component.

Inductive reactance,

$$X_L = 3 \times 377 = 1131\ \Omega$$

Capacitive reactance,

$$X_C = \frac{1592}{3} = 531\ \Omega$$

Total impedance,

$$z = 30 + j1131 - j531\ \Omega$$
$$= 30 + j600$$
$$= 600.7 \angle 87.1°\ \Omega$$

Peak value of the third harmonic current,

$$i_3 = \frac{40 \angle -20°\ \text{V}}{600.7 \angle 87.1°\ \Omega}$$
$$= 0.067 \angle -107.1° \text{A}$$

Effective value of the third harmonic current,

$$I_3 = 0.707 \times 0.067$$
$$= 0.047\ \text{A}$$

Total nonsinusoidal current,

$$i = 0.082 \sin(\omega t + 88.6°) + 1.55 \sin(2\omega t - 85.50°)$$
$$+ 0.067 \sin(3\omega t - 107.1°)\ \text{A}$$

PRACTICE PROBLEMS

112-1. In example 112-1, determine the effective value of the nonsinusoidal current. What is the total amount of power dissipated in the circuit?

112-2. In Fig. 112-2, the nonsinusoidal voltage source is expressed by the equation $e = 14.14 \sin(\omega t + 20°) + 3.535 \sin(3\omega t + 40°) - 1.44 \sin(5\omega t - 60°)$ V, where $\omega = 2\pi f = 1000$ rad/s. Obtain the equation of the nonsinusoidal source current, i.

112-3. In Practice Problem 112-2, what is the effective value of the source current?

112-4. In Practice Problem 112-2, calculate the total amount of the power dissipated in the circuit.

3
PART
Solid-state devices and their associated circuits

113
The pn junction diode

The modern power rectifier device is the silicon diode, which basically uses a pn junction to provide the "one-way" action. The p-type material is produced by doping pure silicon with an acceptor, or a trivalent impurity, such as indium, gallium, or boron. The n-type silicon is created by doping with a donor, or pentavalent impurity, of arsenic, antimony, or phosphorus. Although both n-type and p-type materials are electrically neutral, the majority charge carriers in the n-type and p-type semiconductors are respectively, negative electrons and positive holes.

When the silicon pn junction is formed, there is a movement of the majority carriers across the junction. At room temperature, this creates an internal potential barrier of about 0.7 V across a depletion region, which is in the immediate vicinity of the junction and is devoid of majority charge carriers. In order to forward-bias the junction, it is necessary to connect the positive terminal of a voltage source to the p region and the negative terminal to the n region (Fig. 113-1A). However, the applied bias must exceed the 0.7 V internal potential before appreciable forward current will flow (Fig. 113-1B).

For a general-purpose silicon diode, with the p region as the anode and the n region as the cathode, a forward voltage, (V_F), of about 1 V will correspond to a forward current, (I_F), of several hundred milliamperes. The small voltage drop across the conducting silicon diode is its real advantage when compared with other rectifier devices; even with high forward currents, the power dissipation will be low.

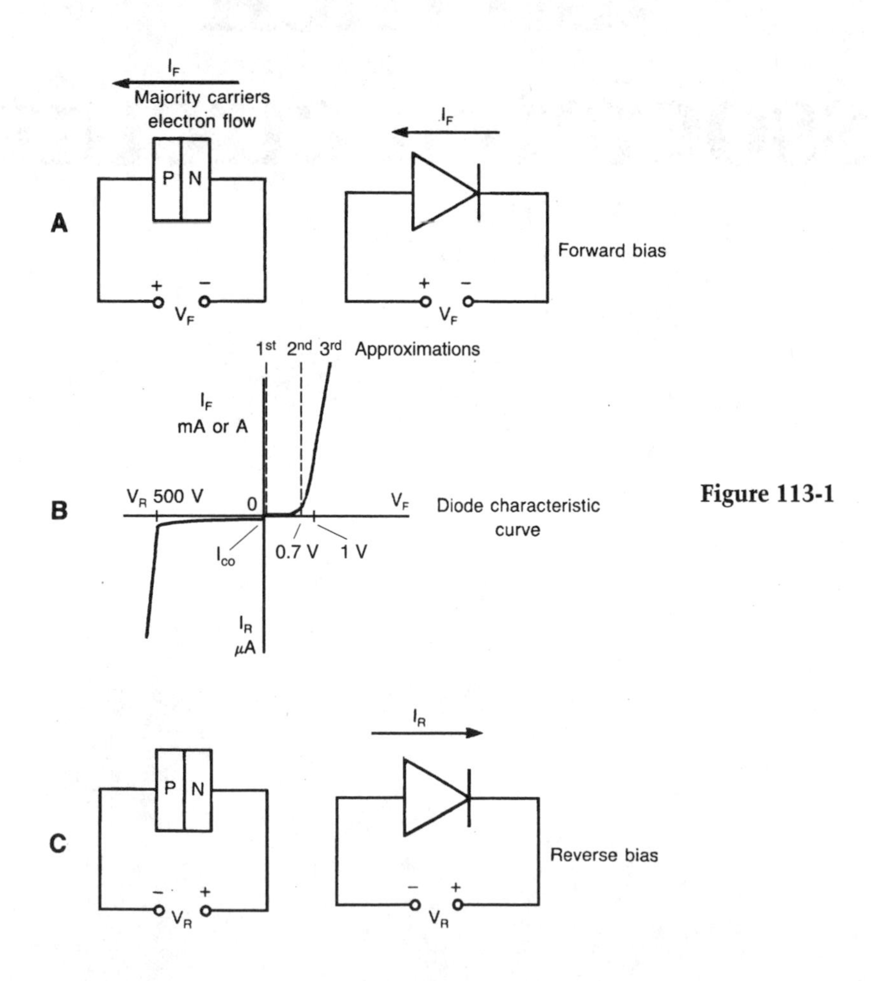

Figure 113-1

To create a reverse bias (Fig. 113-1C), the positive terminal of a voltage source is connected to the n region (cathode) and the negative terminal to the p region (anode). If only the majority carriers were involved, there would be zero reverse current; however, thermal energy creates minority carriers (electrons in the p region and holes in the n region) so that there is a small reverse current, (I_{co}), whose value rises rapidly and saturates at about 1 μA. However, the saturated value of I_{co} depends on the diode's temperature and will approximately double for every 10° Celsius rise. Typically, a reverse bias, (V_R), of about 250 V to 500 V can be applied with only a small reverse current, I_R. However, with excessive reverse voltage there is a breakdown point at which there is a rapidly increasing flow of reverse current; this action is the principle behind the zener diode, which is used for voltage regulation purposes.

MATHEMATICAL DERIVATIONS

In the first (and roughest) approximation, the forward-biased diode is regarded as a short circuit. In a second approximation, the 0.7 V drop across the diode is taken into account. For the third approximation, we include the bulk resistance of the semiconductor material from which the diode is made. The equation for the third approximation is:

$$I_F = \frac{V_F - 0.7\text{ V}}{R_B} \text{ amperes} \qquad (113\text{-}1)$$

where: I_F = forward current (A)
V_F = forward voltage (V)
R_B = bulk resistance (Ω)

Power dissipated in the diode,

$$P_D = V_F \times I_F \text{ watts} \qquad (113\text{-}2)$$

When the diode is reverse biased,

$$\text{Reverse resistance} = \frac{V_R}{I_R} \text{ megohms} \qquad (113\text{-}3)$$

where: V_R = reverse voltage (V)
I_R = reverse current (μA)

Example 113-1

For a particular silicon diode, V_F is 0.9 V and the bulk resistance is 0.5 Ω. What is the corresponding value of I_F?

Solution

Forward current,

$$I_F = \frac{V_F - 0.7\text{ V}}{R_B} \qquad (113\text{-}1)$$

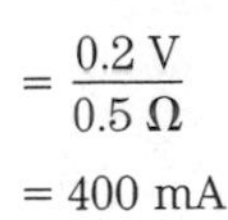

$$= \frac{0.2\text{ V}}{0.5\ \Omega}$$

$$= 400\text{ mA}$$

PRACTICE PROBLEMS

113-1. A silicon diode whose bulk resistance is 20 Ω, is connected in series with a 1-kΩ resistor across a 10-Vdc source. Calculate (a) the value of the current that flows in the circuit, and (b) the voltage drop across the 1-kΩ resistor.

113-2. In Practice Problem 113-1, what is the amount of power dissipated in the diode?

113-3. For a particular silicon diode I_R = 4 μA when V_R = 20 V. Calculate the value of the reverse resistance.

113-4. In Practice Problem 113-3, the diode is connected in series with a 1-MΩ resistor across a 24-Vdc source. What is the voltage drop across the 1-MΩ resistor?

113-5. In Fig. 113-2, it is assumed that there is a 0.7-V drop across the diode. Calculate the current flowing through the 15-kΩ resistor.

113-6. In Fig. 113-3, it is assumed that there is a 0.7-V drop across either diode when it is conducting. Calculate the value of the output voltage, V_o.

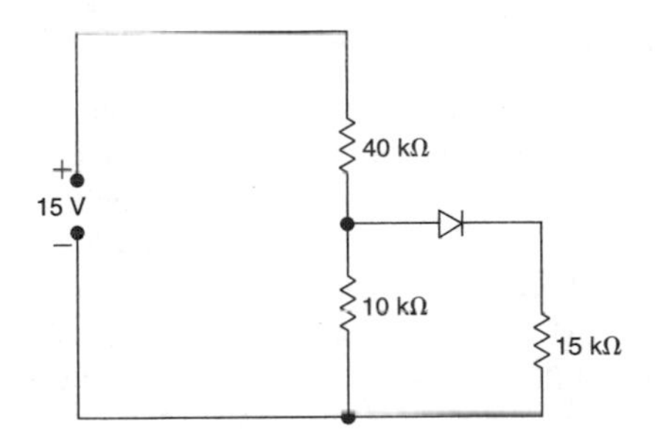

Figure 113-2

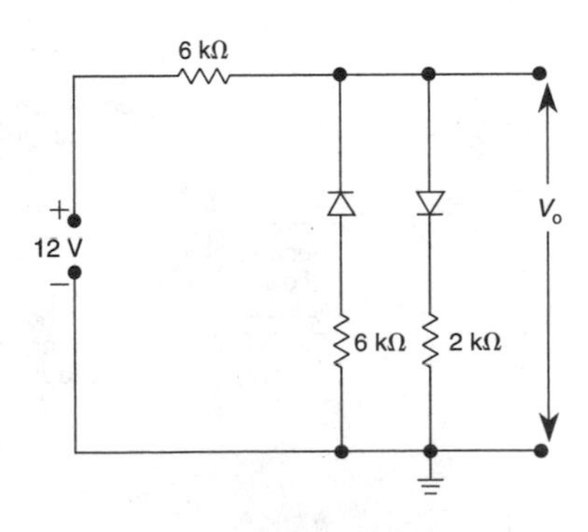

Figure 113-3

114
Half-wave rectifier circuits

The circuits of communications systems, such as transmitters and receivers, normally need dc voltages for their operations. With mobile systems, these voltages can be derived from batteries or motor generators; but for stationary communications equipment it is customary to obtain the required power from the commercial 60-Hz supply. It is therefore necessary to convert the ac 110-V rms line voltage to the required steady dc level. This function is fulfilled by a power supply unit, which might contain a power transformer to step the 60-Hz voltage up or down to the necessary level. The secondary voltage is then applied to a rectifier circuit which converts the ac sine wave into a fluctuating dc output. Before such an output is normally of any use, it must be applied to a low-pass filter so that the result is a steady dc level with an acceptable ripple superimposed. Finally, the unit might contain some form of regulation circuit so that the dc level is stabilized against variations in the load current drawn from the power supply, and against fluctuations in the ac line voltage.

SINGLE PHASE HALF-WAVE RECTIFIER CIRCUIT

The basic *half-wave rectifier* circuit is illustrated in Fig. 114-1. The secondary ac voltage of the power transformer is applied across the series combination of the diode and the load (R_L). Normally R_L is not just a resistor, but it represents the combination of a number of circuits (for example, amplifiers) to which the dc output is supplied. The value of R_L is the result of dividing the output dc voltage by the total dc load current drawn from the power supply.

During the secondary voltage's "positive" half cycle, which is shown as a solid line, it is assumed that the top of the secondary voltage is positive with respect to ground. The silicon diode is forward biased so that an electron flow (of the order of mA or A) will occur and a voltage will be developed across the load. If the small voltage drops across the diode and the resistance of the secondary winding are ignored, the half cycle of the voltage across R_L will be a replica of the secondary half cycle.

During the other "negative" half cycle, which is indicated by the broken line, the top of the secondary winding is negative with respect to ground, the diode is reverse biased, and only a small current (of the order of μA) will flow. If this current is ignored, the voltage across R_L is zero. The negative half cycle then appears across the diode, and care must be taken not to exceed its reverse voltage rating.

The load voltage across R_L can be described as a fluctuating positive dc output, which is composed of half-cycle "pulses" with an average value of 0.318 × the peak value (chapter 63). In order to obtain a negative output with respect to ground, the diode connections must be reversed.

A battery composed of secondary cells can be charged from a fluctuating dc voltage, but such an output cannot be used to operate communications stages, such as amplifiers, oscillators, and mixers. These stages require a dc level which must fluctuate as little as possible. To achieve this objective, the fluctuating output from the rectifying device is passed to a low-pass filter, whose output is subsequently applied across the load. For this purpose, the most commonly used filters are:

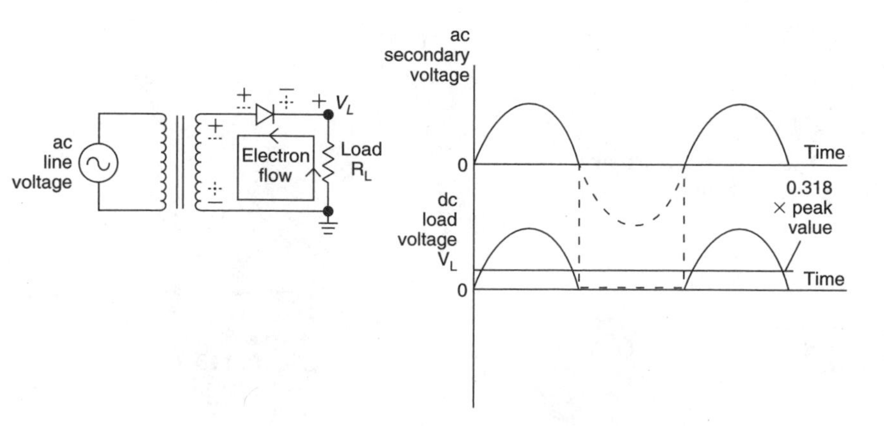

Figure 114-1

1. The *capacitor input* filter, which is normally used in situations where either the load current is small, or it is virtually constant (as in a communications receiver); and,
2. The *choke input* filter, which appears in power supplies designed to accommodate wide variations in a heavy load current; such may be the case in communications transmitters. However, the choke input filter can only be used with full-wave rectifier circuits (chapter 115).

Figure 114-2 shows a half-wave rectifier circuit that has been modified to include the input capacitor (C) of a low-pass filter. When the diode conducts during part of the "positive" half-cycle, its current charges the capacitor towards the *peak* value of the secondary voltage. The charging path of the capacitor travels through the secondary of the transformer and the conducting diode. The resistance of this path is small so that the corresponding time constant is low and the capacitor charges rapidly. When the secondary voltage falls below the potential difference (V_C) to which the capacitor has been charged (point X), the diode ceases to conduct, and the capacitor discharges through the load, R_L. However, the value of C is chosen so that the time constant of the capacitor's discharge is high compared with the period of the ac line voltage. For example, if $C = 40\ \mu F$ and $R_L = 5\ k\Omega$, their time constant is $CR_L = 5 \times 10^3 \times 40 \times 10^{-6} s = 0.2 s$, which is 12 times greater than the period of a 60-Hz supply. The operation of the circuit depends on the short time constant for the charging of the capacitor, and the high time constant for its discharge.

As a result of the time constant provided by C and R_L, the capacitor only discharges slowly and its voltage loss is small over the interval from point X to point Y. At point Y, the secondary voltage again rises above the potential difference across the capacitor; the diode once more conducts and rapidly recharges the capacitor.

As illustrated, the charging current consists of short-duration pulses, but the pulse's peak current and its average level over the complete cycle must not exceed the diode's rated values. Notice that in Fig. 114-2, the voltage drop across the silicon diode has been ignored so that the waveform of V_C follows the secondary voltage during the charging interval from point Y to point X. Therefore, the V_C waveform contains a steady level, which is close to the peak value of the secondary voltage. For example, if the secondary voltage is 30 V rms, with a peak value of $30 \times 1.414 = 42.42$ V, the output dc voltage under load conditions might be 38 V.

Both sides of the dc level contain a fluctuation, which is referred to as the *ripple*. The waveform of the ripple is complex, but the rate at which it repeats is the same as the frequency of the ac secondary voltage. Consequently, if the line frequency is 60 Hz, the ripple's fundamental frequency component will also be 60 Hz. The amount of the ripple is measured by:

$$\text{Percentage ripple} = \frac{\text{ripple's fundamental voltage (rms)}}{\text{dc output voltage}} \times 100\%$$

The lower the level of the percentage ripple, the better is the filtering action of the capacitor. It would

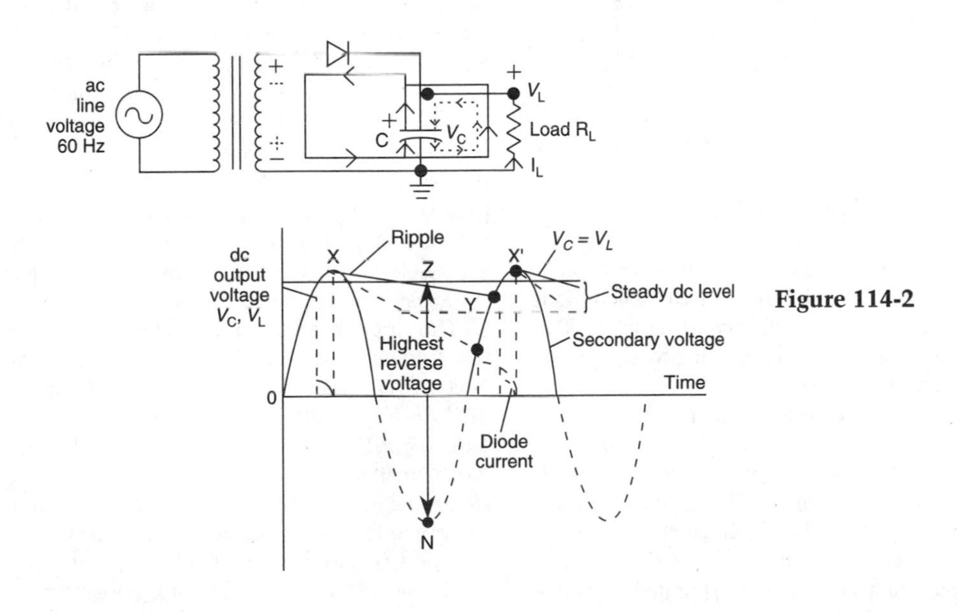

Figure 114-2

appear that we could indefinitely lower the ripple by raising the value of the capacitor. However, not only might this increase the cost and physical size of the capacitor, it would require a larger diode current during the recharging interval. This illustrates one of the "trade-offs" that must be considered in the operation of a power supply.

In the V_C waveform of Fig. 114-2, point *Z* is of particular interest. At this instant, the diode is not conducting, but the secondary's "negative" peak combines with the value of V_C to produce the highest reverse voltage across the diode. The amount of this reverse voltage (represented by the line *ZN*) is approximately equal to twice the peak value of the secondary voltage and must not exceed the diode's rating.

If the secondary voltage is 1000 V rms, the maximum reverse voltage under open-circuit conditions is $2 \times 1000 \times 1.414 = 2828$ V. Instead of using a special single diode with a reverse voltage rating of more than 3000 V, it is possible to connect diodes of the same type in series (Fig. 114-3). The voltage of 2828 V would then be divided between the reverse resistances of the three diodes. However, even for the same type of diode, the values of these reverse resistances can vary widely. The resistors (R) have the same high values, and are inserted to ensure that the maximum reverse voltage is equally distributed between the diodes. The capacitors (C) provide a bypass path for voltage surges that otherwise can cause excessive forward currents to flow in the diodes.

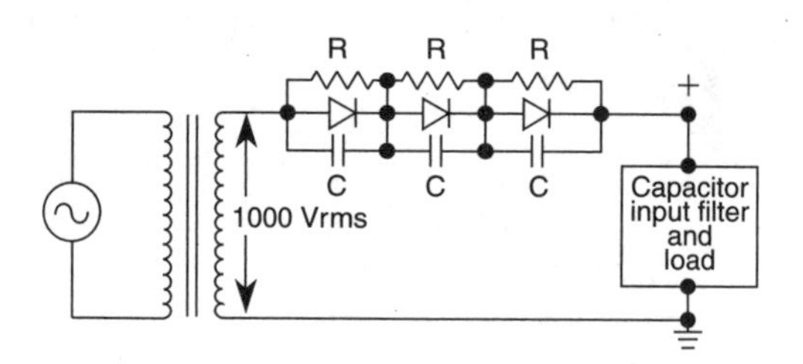

Figure 114-3

As opposed to the series arrangement, diodes can also be connected in parallel in order that each diode may carry a share of the total forward current. This will enable the total diode dissipation to be divided among the individual diodes.

In most cases, the filtering action provided by a single capacitor is inadequate and a *pi* (π) *section* (Fig. 114-4A) is then used to lower the ripple percentage. The section is completed by the filter choke (L_1) and the filter capacitor, C_1. The input capacitor voltage (V_C), which consists of the mean dc level and the ac ripple, is applied across the series-parallel combination of L_1, C_1, and R_L with the output developed across C_1 and R_L in parallel (Fig. 114-4B). Assume that the dc level of V_C is 100 V, and the peak-to-peak value of the ripple is 10 V, so that V_C fluctuates between 95 V and 105 V. For a 60-Hz line voltage, the iron-cored filter choke typically has an inductance value on the order of 10 H and its dc resistance is 50 Ω. Because the capacitors are operating in a dc circuit (with no reversal of polarity), it is customary to use the electrolytic type having values on the order of 20 μF to 40 μF and with leakage resistances of several MΩ.

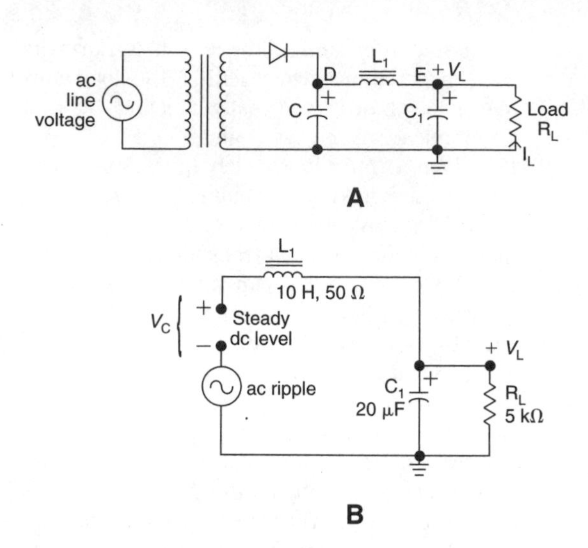

Figure 114-4

As far as the 100-Vdc level is concerned, it will be divided between the 50-Ω resistance of L_1, and the parallel combination of C_1's leakage resistance and the load resistance, R_L. Assuming that $R_L = 5$ kΩ, the leakage resistance of C_1 can be ignored, and the output dc voltage by the VDR (voltage division rule) is $100 \text{ V} \times 50\ \Omega/(5\text{ k}\Omega + 50\ \Omega) = 99$ V. Consequently, there is only a 1-Vdc voltage drop across the filter choke, and the potential at D (with respect to ground) is greater than the potential at E by that amount. The final dc output voltage can be connected across a voltage divider network in order to supply a number of different voltages to various loads.

The ripple has a fundamental 60-Hz component; at this frequency, a 10-H filter choke would have an inductive reactance of $X_{L1} = 2 \times \pi \times 60 \text{ Hz} \times 10 \text{ H} = 3770\ \Omega$. If $C_1 = 20\ \mu$F, its capacitive reactance, $X_{C1} = 1/(2 \times \pi \times 60 \text{ Hz} \times 20 \times 10^{-6} \text{ F}) = 130\ \Omega$; this is much smaller than the value of the parallel resistance (R_L) whose effect may therefore be ignored as far as the ripple is concerned. The peak-to-peak ripple voltage appearing in the output is $10 \text{ V} \times 130\ \Omega/(3770\ \Omega - 130\ \Omega) = 0.36$ V so that the ripple percentage has been reduced by a factor of 28. This result is an approximation because the ripple waveform contains harmonics that the low-pass filter attenuates even more than the fundamental frequency.

By doubling the values of L_1 and C_1, the ripple percentage would be lowered by a factor of four. However, it would be more effective and economical to repeat the filter (Fig. 114-5), which would reduce the ripple percentage by a further factor of 28 without any appreciable fall in the dc output voltage.

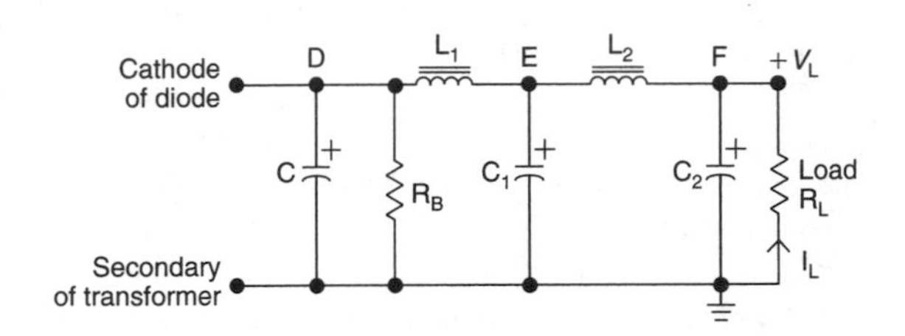

Figure 114-5

A cheaper and smaller alternative to the filter choke is a resistor whose value is comparable to the choke's reactance. However, there could be a considerable dc voltage drop across the filter resistor, so that a resistor is normally confined to those cases where the load current is small.

The next point to discuss is the degree of *regulation* provided by a half-wave power supply with a capacitor input filter. In Fig. 114-2, assume that there is an increase in load current at the instant represented by point X. This increase in load current is accompanied by a decrease in the load resistance (R_L) so that the value of the time constant (R_LC) is lowered, and the capacitor discharges more rapidly. The results are threefold:

1. There is an increase in the ripple percentage, but this can probably be reduced to an acceptable level by the low-pass filter.
2. The pulse of the diode current has a greater peak value, and a longer duration.
3. The dc output voltage falls appreciably, as the load current is increased. Consequently the percentage of voltage regulation, as defined below, is high.

Percentage voltage regulation

$$= \frac{E_{NL} - E_{FL}}{E_{FL}} \times 100\%$$

where: E_{NL} = dc output voltage under no load, or minimum load, conditions (V),

E_{FL} = dc output voltage under full-load conditions (V).

Ideal regulation corresponds to a zero percentage. The graph of the output voltage versus the load current would then be a horizontal line.

The third statement indicates that the regulation of a capacitor input filter is poor, as a result of the action of the filter itself. In addition, the regulation will be worsened by the additional dc voltage drops across the filter chokes L_1 and L_2. Therefore, in Fig. 114-5, the highest positive voltage (with respect to ground) occurs at point D, the intermediate value exists at the point E, and the lowest value would be measured across R_L (point F). The difference between the voltages (with respect to ground) at points D and E, is the voltage drop across L_1 and the voltage drop across L_2 accounts for the difference between the voltages at points E and F. A typical regulation curve for a half-wave rectifier with a capacitor input filter is shown in Fig. 114-6.

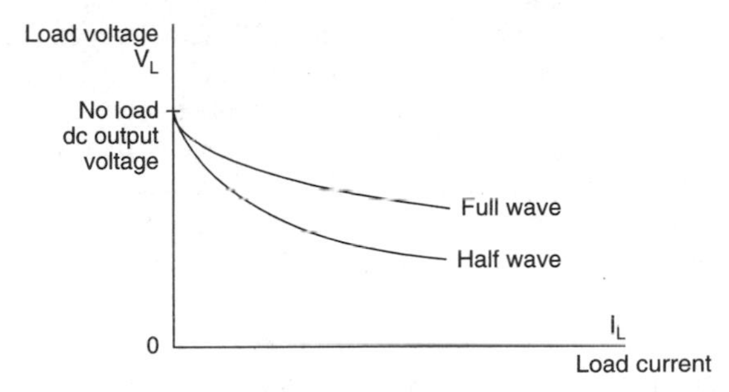

Figure 114-6

Finally, *a point of safety*! When the ac power is shut off and the load is removed from the rectifier circuit, the capacitors (C, C_1 and C_2) can remain charged for an appreciable time. To avoid possible shock to a technician carrying out repair work, it is customary to include a bleeder resistor (R_B), which will rapidly discharge the capacitors when the load is removed. Such a resistor has a high value, so it has no appreciable shunting effect on the load. In fact, if the bleeder resistor burns out, there will be no noticeable change in the dc output voltage.

THREE PHASE HALF-WAVE RECTIFIER CIRCUIT

The most common arrangement for a three-phase transformer (Fig. 114-7) is a *delta*-connected primary and *wye*-connected secondary (chapters 107 and 108). In the three-phase half-wave rectifier, it is the usual practice to ground the common wye point. Because the three ac voltages are separated in phase by 120°, the diodes (D1, D2, and D3) conduct in turn and the input capaci-

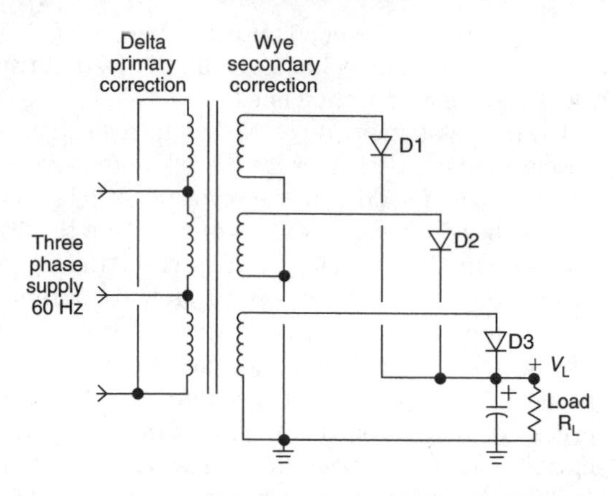

Figure 114-7

tor (C) is charged three times toward the peak value of the phase voltage during each cycle of the ac supply. As a result, the amount of ripple is small, and (for a 60-Hz supply) the frequency of the fundamental ripple component is 180 Hz. This high value of ripple frequency has the advantage of reducing the component values required for the low-pass filter. In addition, the regulation of the dc output voltage will be superior to that of a comparable full-wave single-phase rectifier.

EXTRA HIGH TENSION (EHT) SUPPLY

The purpose of an EHT supply is to provide the high voltage responsible for the horizontal acceleration of the electron beam in a cathode ray tube (CRT). Such is the case with the electrostatic CRT in an oscilloscope and the electromagnetic CRT in a TV receiver. The required voltage is several kilovolts, but the beam current is only about 1 mA so that the effective load (R_L) is of the order of MΩ. For example, if the horizontal acceleration voltage is 16 kV, and the beam current is 0.8 mA, R_L = 16 kV/0.8 mA = 20 MΩ. Adequate rectification will be provided by a half-wave circuit using a diode with a high reverse voltage rating. This will be followed by a capacitor input filter section containing a 1-MΩ, 2-W filter resistor, whose dc voltage drop is less than 1 kV. The high values of R_L and the filter resistor allow the use of relatively low values for the filter capacitors. Note that a full-wave circuit with a center-tapped secondary (chapter 115) is not suitable for an EHT supply. For the same dc output, the total full-wave secondary voltage would need to be twice that of the half-wave secondary, and this could result in severe insulation problems.

There are a number of rectifier circuits that produce dc outputs of several kilovolts. If the 60-Hz ac line voltage is used, the step-up transformer can be followed by a voltage multiplier. However, the lower frequency generally results in a power supply, which is physically large and costly, mostly because of the insulation problems that increase rapidly with high voltages.

A second method is to use a high-frequency oscillator, part of whose circuitry is the primary of a step-up transformer (or autotransformer). Because the frequency is high, the transformer will have a simplified design (with a core of air, powdered iron, or ferrite). In addition, the filtering problems will be greatly reduced.

The most common method of generating the EHT voltage for a TV tube, is to use the *flyback* portion of the horizontal sawtooth sweep voltage whose frequency is 15750 Hz (Fig. 114-8). The high voltage is generated over a short period of time by reducing to zero the large current flowing through the lower (AB) part of the autotransformer. For example, if a maximum current of 300 mA flows through a 100-mH inductor, and this current is reduced to zero in 6 μs, the magnitude of the average induced voltage is,

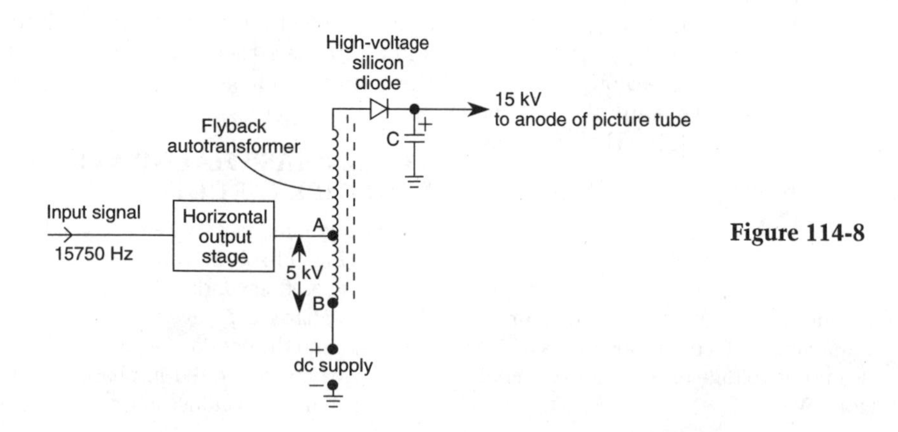

Figure 114-8

Induced voltage,

$$VL = L \times \frac{\Delta I}{\Delta T}$$

$$= \frac{100 \times 10^{-3}\,\text{H} \times 300 \times 10^{-3}\,\text{A}}{6 \times 10^{-6}\,\text{s}}$$

$$= 5000\ \text{V}$$

$$= 5\ \text{kV}.$$

If the autotransformer has a step-up ratio of 1:3, the total voltage applied to the half-wave rectifier is 3 × 5 = 15 kV. If higher voltages are required, insulation difficulties present a problem, and the best solution is to replace the half-wave circuit with a voltage multiplier. The flyback method is economical and simple because the power required by the EHT supply is provided by the horizontal output stage.

MATHEMATICAL DERIVATIONS

Single phase half-wave rectifier

Capacitor input filter

Open-circuit dc output voltage = *peak* value of the secondary voltage (114-1)

time constant, $CRL >>$ period of ac line voltage (114-2)

Ripple frequency = line frequency (114-3)

Maximum reverse voltage ≈ 2 × peak value of the secondary voltage across the diode (114-5)

Percentage ripple, (114-6)

$$= \frac{\text{rms value of ripple's fundamental output}}{\text{dc output voltage}} \times 100\%$$

Percentage regulation, (114-7)

$$= \frac{E_{NL} - E_{FL}}{E_{FL}} \times 100\%$$

where: E_{NL} = dc output voltage under no load or minimum load conditions (V)
E_{FL} = dc output voltage under full-load conditions (V)

Three phase half-wave rectifier

Capacitor input filter

Open-circuit dc output voltage = peak value of the secondary phase voltage (114-8)

Maximum reverse voltage across each diode = 2 × peak value of the secondary phase voltage (114-9)

Ripple frequency = 3 × line frequency (114-10)

EHT supply

$$\text{dc output voltage} = NL\,\frac{\Delta I}{\Delta T}\ \text{volts} \quad (114\text{-}11)$$

where: N = step-up ratio of the autotransformer
L = primary inductance of the autotransformer (H)
$\frac{\Delta I}{\Delta T}$ = rate of the primary current decay (A/s)

Example 114-1

In Fig. 114-9, what is the value of the voltage at the point (X) under open-circuit load conditions? Suggest a suitable voltage rating for the capacitor, C. What is the highest value of the reverse voltage across the diode under open-circuit load conditions? Under loaded conditions, what is the frequency of the fundamental ripple component in the output voltage? If the full load output voltage is −65 Vdc and the peak-to-peak value of the fundamental ripple component is 2 V, what are the values of the ripple percentage and the percentage of regulation?

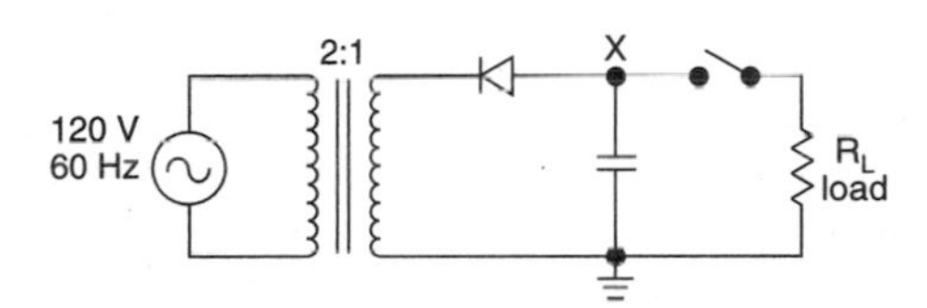

Figure 114-9

Solution

Voltage at the point, X,

$$V_X = -\frac{120 \times 1.414}{2} \quad (114\text{-}1)$$

$$\approx -85\ \text{V}$$

Suitable voltage rating for the capacitor,

C = 150 WVdc

Highest reverse voltage,

$$V_{R\max} = 2 \times 85\ \text{V} \quad (114\text{-}5)$$

$$= 170\ \text{V}$$

Frequency of fundamental ripple component = 60 Hz (114-3)

Ripple percentage, (114-6)

$$= \frac{(2/2) \times 0.707}{65} \times 100\% = 1.1\%$$

Percentage regulation, (114-7)

$$= \frac{85 - 65}{65} \times 100\% = 31\%$$

Example 114-2

In Fig. 114-7, the primary phase voltage is 220 V, 60 Hz; and the step-up ratio of the transformer is 1:3. Calculate the values of the open-circuit dc output voltage and the highest reverse voltage across each diode. Under loaded conditions, what is the frequency of the fundamental ripple component?

Solution

Open-circuit dc output voltage, (114-8)

$$E_{NL} = 220 \times 3 \times 1.414$$
$$= 933 \text{ V}$$

Highest reverse voltage across each diode, (114-9)

$$V_R = 2 \times 933$$
$$= 1866 \text{ V}$$

Ripple frequency, (114-10)

$$f = 3 \times 60$$
$$= 180 \text{ Hz}$$

Example 114-3

In Fig. 114-7, the required output voltage is 18 kV. The primary inductance of the autotransformer is 150 mH, and the maximum primary current of 250 mA is reduced to zero in 5 μs. Determine the necessary step-up ratio for the autotransformer.

Solution

Step-up ratio,

$$N = 1\text{:}\ \frac{\text{dc output voltage}}{L \times \Delta I/\Delta T} \qquad (114\text{-}11)$$

$$= 1\text{:}\ \frac{18000 \text{ V}}{150 \times 10^{-3}\text{H} \times 250 \times 10^{-3}\text{A}/5 \times 10^{-6}\text{s}}$$

$$= 1\text{:}2.4$$

PRACTICE PROBLEMS

114-1. In Fig. 114-2, the ac line voltage is 110 V, 60 Hz and the transformer's turns ratio is 1:3. Under open-circuit conditions, what are the values of the dc output voltage and the highest reverse voltage across the diode?

114-2. In Fig. 114-6, the primary phase voltage is 220 V, 50 Hz and the turns ratio for each phase is 1:10. Under open-circuit conditions, what are the values of the dc output voltage and the highest reverse voltage across each diode? Under loaded conditions, what is the frequency of the fundamental ripple component?

114-3. A fly-back autotransformer has a step-up ratio of 1:4. The maximum primary current of 400 mA is reduced to zero in 8 μs. If the dc output voltage is 16 kV, calculate the required value for the primary inductance of the autotransformer.

115
Full-wave rectifier circuits

By contrast to the half-wave rectifier, the basic full-wave circuit (Fig. 115-1) uses two diodes and the secondary winding of the power transformer carries a center tap that is usually grounded.

SINGLE-PHASE FULL-WAVE RECTIFIER CIRCUIT

During the solid line or "positive" half-cycle of the secondary voltage, the "top" of the secondary is positive (with respect to the grounded center tap), and the "bottom" carries an equal negative voltage. Consequently, the diode *D*1 conducts in the forward direction, and *D*2 is reverse-biased. The direction of the "solid" electron flow shows that the output is a positive voltage with a value equal to one half of the peak value measured across the complete secondary winding. For the "negative" half-cycle (broken line), the polarities are reversed so that *D*2 conducts and *D*1 is cut off (ignoring any reverse current flow). However, the direction of the electron flow through R_L is unchanged, so that the dc voltage is still positive. The output waveform is a fluctuating dc, consisting of positive alternations, each of which is one-half of the total secondary voltage. Ignoring any reverse flow and the voltage drops across *D*1, *D*2 when conducting in the forward direction, the average of the voltage across R_L is 63.7% of its peak value; this is equal to 31.8% of the peak

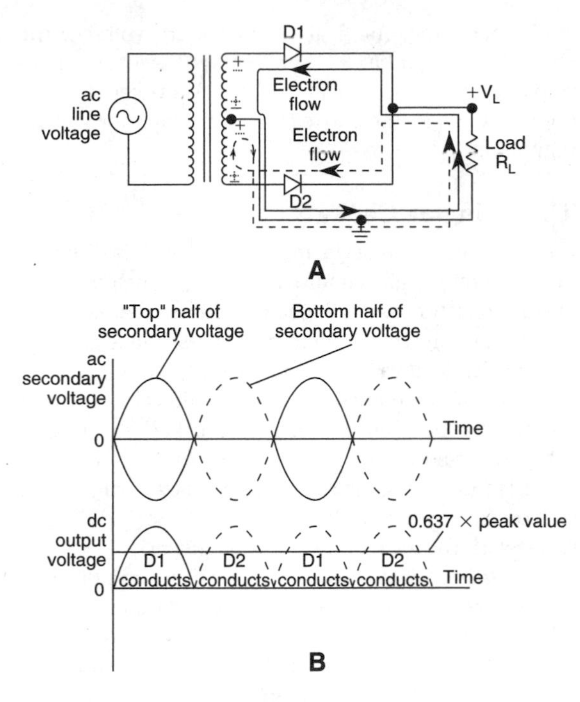

Figure 115-1

value for the complete secondary winding and is the same result as that obtained for the half-wave rectifier.

When a capacitor input filter is included (Fig. 115-2), the input capacitor charges toward the peak value of 1/2

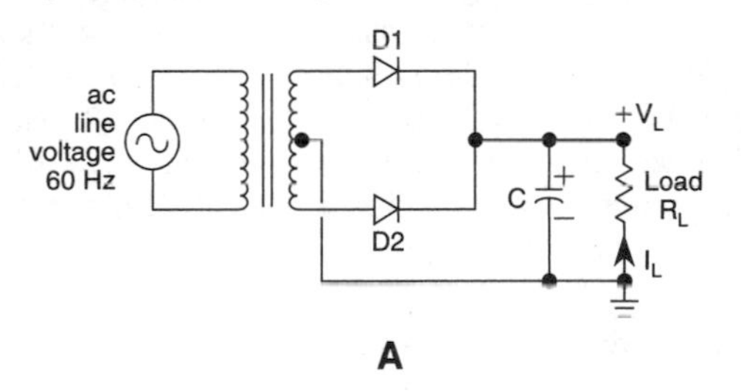

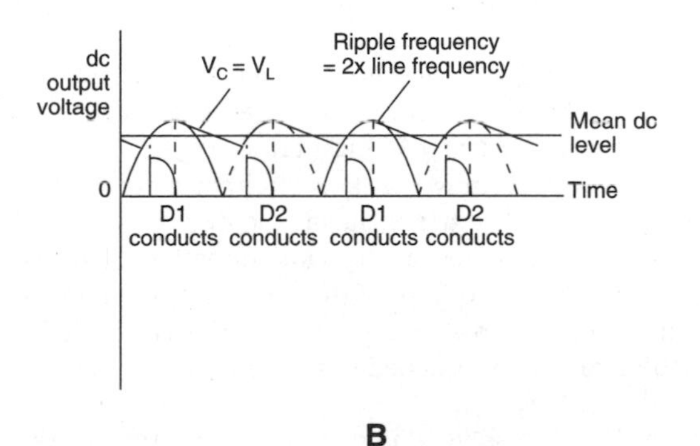

Figure 115-2

× the total secondary voltage and is recharged twice during each cycle of the line voltage. Therefore, if the line voltage is 60 Hz, the rate at which the ripple voltage repeats is 120 Hz.

When one of the diodes is reverse-biased under open-circuit load conditions and with a capacitor input filter, its reverse voltage is equal to the peak value of the full secondary voltage.

With the same secondary voltage, the discharge time of the full-wave circuit is shorter than in the half-wave circuit. Consequently, the full-wave ripple percentage is less and the regulation in the full-wave circuit is superior to that of the half-wave rectifier. However, because of the center tap required, you can only obtain the same dc output voltage by doubling the secondary voltage in the full-wave arrangement. This might cause insulation problems if you require an output of several kV. Therefore, the half-wave arrangement is preferred for high dc voltages associated with a light load current (for example, the horizontal accelerating voltage for a cathode ray tube). One other point of comparison; the half-wave rectifier can be operated directly from the ac line voltage, but the full-wave arrangement requires a power transformer.

In summary, the capacitor input filter provides a high dc output voltage and a low ripple percentage; however its degree of voltage regulation is poor.

The choke input filter

With a capacitor input filter in a full-wave rectifier circuit, the diodes conduct alternately in short duration pulses but for most of the cycle, the input capacitor discharges slowly through the load. This action is responsible for the poor regulation of the capacitor input filter. Therefore, it is used in power supplies where the load current is constant or the current varies, but always has a low value.

The choke input filter (Fig. 115-3A) is used in power supplies where the load current is high and it varies over a wide range. An amplifier, operated under Class B conditions, is an example of such a load. The action of the choke input filter requires that a forward current is always flowing through one diode. This means that, in the full-wave rectifier circuit, one diode starts to conduct as soon as the other diode is cut off. It also follows that the choke input filter action *cannot* occur with the half-wave arrangement.

Because the conducting diode can be regarded as being equivalent to a low resistance, the full-wave rectified voltage pulses are applied to the equivalent circuit of Fig. 115-4B. The mathematical derivations show that, in order for the choke input filter to operate correctly, $L > R_L/(6\pi f)$ H, where f is the frequency of the line voltage.

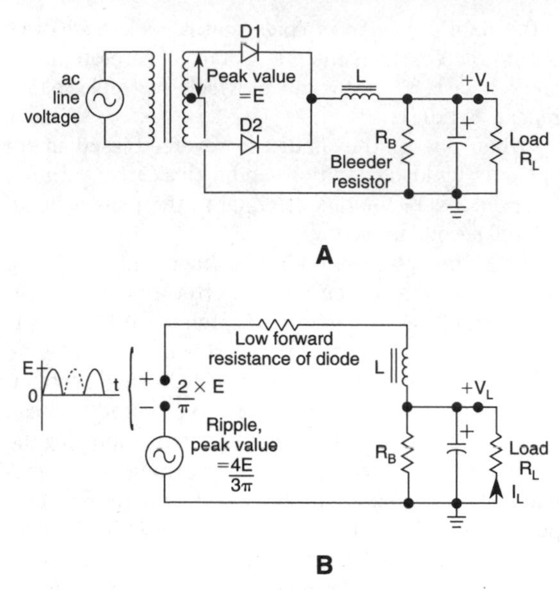

Figure 115-3

This indicates that there is a certain critical value of the inductance needed to operate the choke input filter successfully. Deduce this fact by gradually lowering the value of L; eventually, the choke would become a short circuit and the rectifier would assume the properties of the capacitor input filter. It also follows that, for a given value of L, the load current must exceed a certain minimum value in order to maintain good regulation. Therefore, a choke input filter cannot be operated under open circuit conditions, and a certain minimum load must be provided by the bleeder resistor. In order to keep the choke's value down to a reasonable level, this resistor must carry a much larger current and have a higher wattage rating than the discharge resistor of a capacitor input filter. A comparison between the regulation curves for the two types of filter is shown in Fig. 115-4, where the open-circuit dc output voltage is the same for each filter.

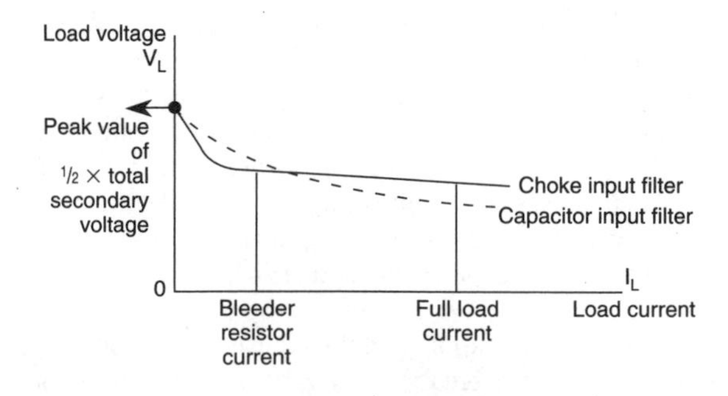

Figure 115-4

Summarizing, for a given secondary voltage, the choke input filter has a lower dc output level and more ripple than a capacitor input filter. However, over its operating range, the choke filter has a superior voltage regulation.

The Swinging Choke

The required value of the input filter choke is different for minimum-load conditions (which includes the bleeder resistor) and full-load conditions. For example, assume that the total minimum load resistance is 15 kΩ and the full-load value is 2 kΩ. Then, the required critical value of inductance, for a 60-Hz line voltage, varies between $15000/1130 \approx 15$ H and $2000/1130 \approx 2$ H. One solution would be a constant-inductance 20-H choke, but it is more economical in terms of size, weight, and cost to use a swinging choke whose value of inductance varies with the level of the dc load current.

A *constant-inductance choke* has a gap cut in its core to prevent partial saturation of the iron when the full-load current is flowing. In the *swinging choke*, the size of the gap is reduced so that with minimum-load current conditions, a high inductance (for example, 20 H) is available with fewer turns required. Under full-load conditions, the iron partially saturates and the inductance falls to a low value (for example, 4 H). The critical inductance is exceeded under both minimum- and full-load conditions so that the choke input filter will provide good regulation over its complete operating range. The physical size of the swinging choke, whose inductance varies between 4 H and 20 H, is about the same size as that of an equivalent constant 8-H choke. In addition, the fewer turns of the swinging choke will mean a lower resistance so that the voltage regulation is further improved.

Bridge Rectifier Circuit

The conventional full-wave arrangement (described in the previous section), is really a two-phase half-wave circuit in which each diode rectifies one-half of the secondary voltage, and these two ac voltages are 180° out of phase. The "true" full-wave rectifier is the bridge circuit, which is illustrated in Figs. 115-5A, B, C. The same circuit has been drawn in two different ways. Notice that, in the bridge arrangement, the opposite pairs of diodes are connected in the same direction. The two diodes joined to the "top" of the secondary winding are connected in opposite directions and the same applies to the two diodes attached to the "bottom" of the secondary winding.

During the "solid" line half-cycle, the "top" of the secondary winding is assumed to be positive, with respect to the bottom. The diodes D1 and D2 conduct

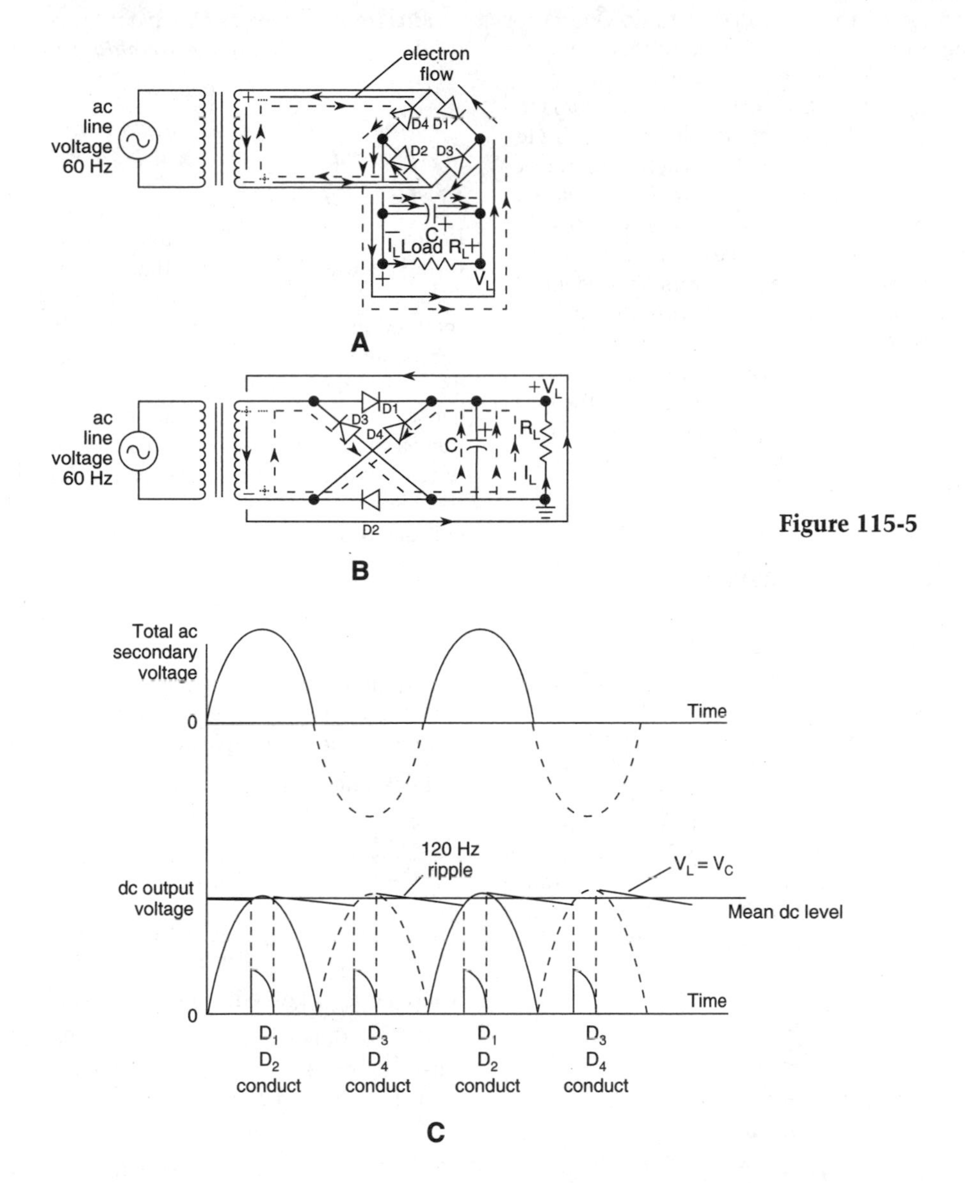

Figure 115-5

and the diodes D3 and D4 are reverse-biased. The direction of the electron flow, as indicated, results in a positive output voltage across the load, R_L. For the other ("broken" line) half cycle, the conditions are reversed; the diodes D3 and D4 conduct, but the diodes D1 and D2 are now reverse-biased. However, the electron flow associated with R_L remains unchanged so that the polarity of the dc output voltage is still positive. The result is similar to the full-wave rectifier circuit described earlier, but in the bridge arrangement, the *whole* of the secondary voltage is rectified during each half-cycle. Because the full-wave rectifier, with its center-tap secondary, only rectifies *half* of the secondary voltage during each half cycle, the dc output from the bridge circuit will be twice that of the conventional full-wave rectifier for the same total secondary voltage. Of course, this advantage is achieved at the expense of using four diodes, as opposed to two.

Because the input capacitor (C) of the bridge circuit is charged twice for each cycle of the line voltage, the fundamental component of its ripple will have a frequency that is a second harmonic of the line frequency. As previously described, the ripple can be reduced by including a low-pass filter.

The bridge rectifier circuit does not require a center-tapped secondary winding; therefore, no transformer is necessary and the input ac voltage can be directly connected to the bridge rectifier. One major use of such a rectifier is in voltmeters that contain a moving coil meter movement. Because this type of movement only responds directly to dc, an ac voltage must first be applied to a rectifier (such as the bridge circuit). The rectified dc output then activates the meter movement and the voltmeter can be calibrated so that the scale reads the effective, peak or peak-to-peak values of the measured ac voltage.

When one pair of diodes is reverse-biased under open-circuit load conditions with a capacitor input filter, the total reverse voltage is equal to twice the peak value of the secondary voltage. However, the reverse voltage is shared between two diodes so that the maximum reverse voltage across each diode is equal to the secondary voltage's peak value. The same result was obtained for the conventional rectifier with the center-tapped secondary winding.

THREE PHASE FULL-WAVE RECTIFIER CIRCUIT

In the full-wave three phase rectifier circuit of Fig. 115-6, the common wye point is not grounded. For each one of the three line voltages, four of the six diodes form a full-wave bridge rectifier so that, for each cycle of the supply voltage, the input capacitor (C) will be charged six times toward the peak value of the *line* voltage. For a 60-Hz supply, the fundamental ripple frequency is 360 Hz, so that the required filter values are smaller when compared with the three-phase half-wave circuit. Apart from the improved filtering, the full-wave circuit will also have better voltage regulation.

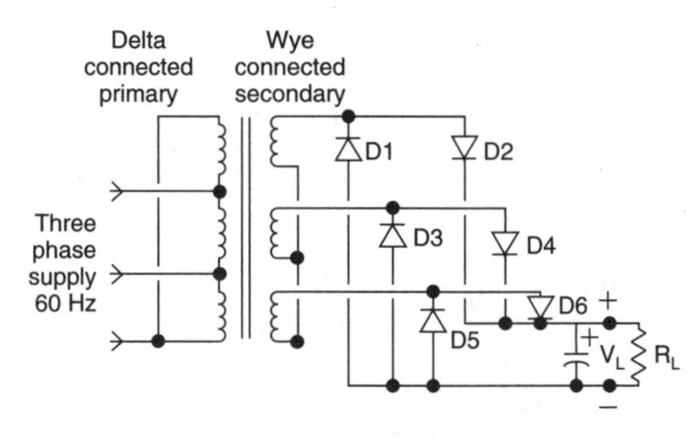

Figure 115-6

MATHEMATICAL DERIVATIONS

Single Phase Full-Wave Rectifier

Capacitor input filter

Dc output voltage under open-circuit conditions $= 1/2 \times$ *peak value of the secondary voltage* (115-1)

Maximum reverse voltage across each diode $=$ *peak value of the secondary voltage* (115-2)

Under loaded conditions,

the frequency of the fundamental ripple component $= 2 \times$ *line frequency* (115-3)

Choke input filter

Referring to Fig. 115-3B, an analysis of the input waveform (chapter 111) reveals that the mean dc level is $(2/\pi) \times E = 0.637 \times E$, where E is the peak value of the rectified pulses. The frequency of the fundamental component is twice that of the input line voltage and its peak value is equal to $(4/3\pi)E$ (chapter 111). Because the value of R_L is high compared to the resistances of the diode and the filter choke, the voltage division rule shows that large load variations cause little change in the output dc voltage; as a result, the circuit has good voltage regulation. In Fig. 115-3B,

$$\text{dc level of current} = \frac{\frac{2}{\pi} \times E}{R_L} = \frac{2E}{\pi R_L} \tag{115-4}$$

For the fundamental ac component, the resistance of the filter choke (L) is many times greater than the reactance of the filter capacitor (C); as a result,

Peak value of the ac current,

$$= \frac{\frac{4}{3\pi} \times E}{2 \times \pi \times 2f \times L} \tag{115-5}$$

$$= \frac{E}{3\pi^2 fL}$$

where f = the line frequency.

For continuous current flow, the dc current level must be greater than, or equal to, the peak value of the ac current. This yields,

$$\frac{2E}{\pi R_L} \geqslant \frac{E}{3\pi^2 fL}$$

or,

$$L \geqslant \frac{R_L}{6\pi f} \tag{115-6}$$

$$L \geqslant \frac{R_L}{1130}\ \text{H}$$

for a 60-Hz supply.

Bridge rectifier circuit

Capacitor input filter

dc output voltage under open-circuit conditions $=$ *peak value of the secondary voltage* (115-7)

$$\textit{Maximum reverse voltage across each diode} = \textit{peak value of the secondary voltage} \quad (115\text{-}8)$$

Under loaded conditions,

$$\textit{the frequency of the fundamental ripple component} = 2 \times \textit{line frequency} \quad (115\text{-}9)$$

Three phase full-wave rectifier

Capacitor input filter

$$\textit{dc output voltage under open-circuit conditions} = \textit{peak value of the line voltage} \quad (115\text{-}10)$$

$$\textit{Maximum reverse voltage across each diode} = \textit{peak value of the line voltage} \quad (115\text{-}11)$$

Under loaded conditions,

$$\textit{the frequency of the fundamental ripple component} = 6 \times \textit{line frequency} \quad (115\text{-}12)$$

Example 115-1

In Fig. 115-7 what are the values of the potential at point X, and the highest reverse voltage across each diode under no-load (open-circuit) conditions? What is the frequency of the fundamental ripple component under loaded conditions? If the full-load dc output voltage is 135 V, what is the percentage of the regulation?

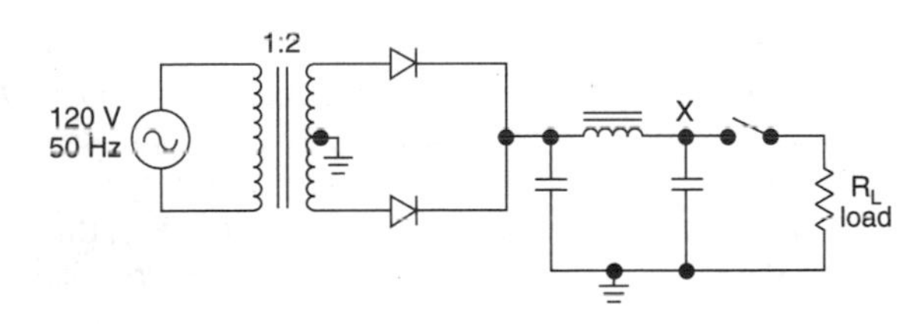

Figure 115-7

Solution

$$\text{Potential at } X = \frac{120 \times 2 \times 1.414}{2} = 170 \text{ V} \quad (115\text{-}1)$$

$$\text{Highest reverse voltage} = 2 \times 170 \text{ V} = 340 \text{ V} \quad (115\text{-}2)$$

$$\text{Frequency of the fundamental ripple component} = 2 \times 50 = 100 \text{ Hz} \quad (115\text{-}3)$$

$$\text{Percentage of regulation} = \frac{170 - 135}{135} \times 100\% = 26\%$$

Example 115-2

A full-wave rectifier circuit with a center-tapped secondary winding operates from a 110-V, 60-Hz supply and uses a choke input filter with a 20-kΩ bleeder resistor. If the load resistance is equivalent to 3 kΩ, calculate the values of the choke's critical inductance under minimum load and full-load conditions. Suggest the values of inductance for a suitable swinging choke.

Solution

Under minimum load conditions,

$$\text{Critical value of inductance} = \frac{20000}{1130} = 17.7 \text{ H} \quad (115\text{-}6)$$

Under full-load conditions,

$$\text{Total load resistance} = 3 \text{ k}\Omega \parallel 20 \text{ k}\Omega = 2.61 \text{ k}\Omega$$

$$\text{Critical value of inductance} = \frac{2610}{1130} = 2.31 \text{ H} \quad (115\text{-}6)$$

A 4- to 20-H swinging choke would be suitable.

Example 115-3

A bridge rectifier circuit includes a capacitor input filter and a 220-V, 50-Hz supply. If the turns ratio of the power transformer is 1:3, what is the dc output voltage under no-load conditions and the maximum reverse voltage across each diode? Under loaded conditions, what is the frequency of the fundamental ripple component?

Solution

$$\text{dc output voltage} = 220 \times 3 \times 1.414 = 933 \text{ V} \quad (115\text{-}7)$$

$$\text{Maximum reverse voltage} = 933 \text{ V} \quad (115\text{-}8)$$

$$\text{Frequency of the fundamental ripple component} = 2 \times 50 = 100 \text{ Hz} \quad (115\text{-}9)$$

Example 115-4

A three phase full-wave rectifier circuit uses a capacitor input filter and is operated from a 50-Hz supply. The transformer has a delta-connected primary and a wye-connected secondary with a turns ratio of 1:4. If the primary phase voltage is 110 V, calculate the dc output voltage under no load conditions and the peak reverse voltage across each diode. Under loaded conditions, what is the frequency of the fundamental ripple component?

Solution

$$\text{dc output voltage} = 110 \times 4 \times 1.732 \times 1.414 = 1078 \text{ V} \quad (115\text{-}10)$$

Maximum reverse voltage = 1078 V (115-11)

Frequency of fundamental ripple component = 6 × 50 = 300 Hz (115-12)

PRACTICE PROBLEMS

115-1. A full-wave rectifier circuit, with a center-tapped secondary, contains a capacitor input filter and is operated from a 220-V, 50-Hz supply. If the output open-circuit dc voltage is 78 V, calculate the required value for the transformer's turns ratio.

115-2. In Practice Problem 115-1, the full-wave circuit is replaced by a bridge rectifier. Assuming that the input ac voltage and output dc voltage remain unchanged, calculate the new value for the turns ratio.

115-3 A three-phase full-wave rectifier circuit contains a capacitor input filter and a power transformer with delta-connected primary windings and wye-connected secondary windings. If the open-circuit dc output voltage is 550 V and the transformer's turns ratio is 1:3, calculate the value of the 60-Hz primary phase voltage.

115-4. In Practice Problem 115-3, the filter choke has a reactance of 3800 Ω at the frequency of the fundamental ripple component. What is the value of the choke's inductance?

116 Voltage multiplier circuits

Doubler circuits enable the output dc voltage to be greater than the peak value of the secondary voltage. Such arrangements are used when: (a) it is not desirable to operate a transformer at the required voltage level for a conventional full-wave rectifier or (b) a dc voltage of about 250 V is required from the 110-V, 60-Hz line voltage without the use of a transformer.

FULL-WAVE DOUBLER CIRCUIT

In the *full-wave doubler* circuit of Fig. 116-1, the diode (D1) is forward-biased during the solid line half cycle; conversely, the diode (D2) conducts in the forward direction for a part of the other (broken line) half cycle. The capacitors (C1 and C2) are connected so that their voltages are series aiding; this combination of capacitors is also charged twice during the cycle of the line voltage. Therefore, the fundamental component of the ripple has a frequency that is twice the frequency of the line voltage. Under open-circuit load conditions, the output dc voltage is equal to twice the peak value of the secondary voltage. However, the capacitors are in series with respect to the load (R_L); this lowers the time constant of their discharge and results in relatively poor regulation.

HALF-WAVE DOUBLER CIRCUIT

By comparison, the *half-wave* (or *cascade*) *doubler* circuit (Fig. 116-2) has only a single capacitor across R_L. However, there is a common line connecting the ac input voltage and the dc output voltage so that both

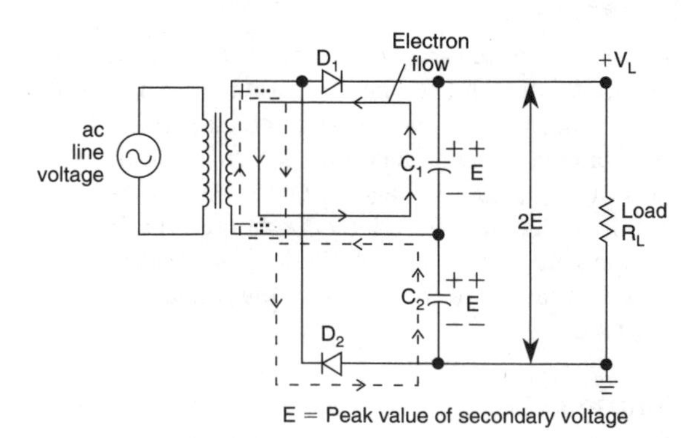

Figure 116-1

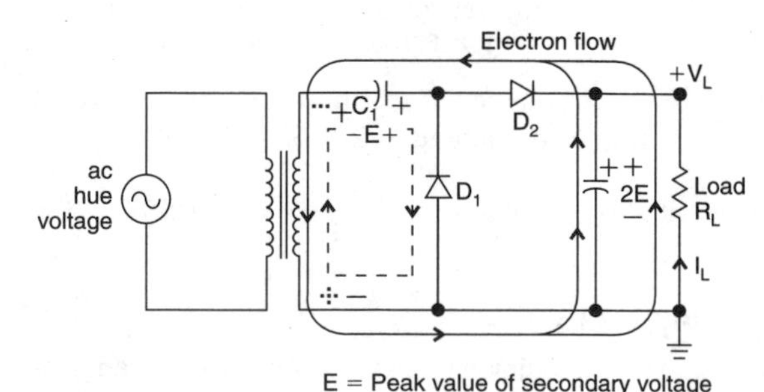

Figure 116-2

voltages can use the same ground. During the "broken-line" half cycle, the diode (D1) conducts in the forward direction and charges the capacitor (C1) toward the peak value of the secondary voltage. For the "solid line" half cycle, D1 is reverse-biased, but the secondary voltage and the voltage across C1 combine to forward-bias the diode (D2), whose current charges C2 towards twice the peak value of the secondary voltage. Because C2 is only charged once during the cycle of the line voltage, the fundamental ripple component and the line voltage will have the same frequency.

The principle of the cascade operation can be further extended to create triplers, quadruplers, and so forth. As an example, Fig. 116-3 shows a cascade quintupler. However, the higher the multiplying factor of the rectifier circuit, the worse is its regulation.

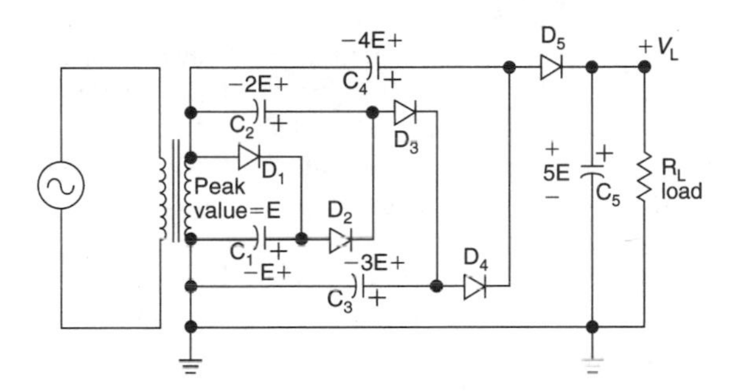

Figure 116-3

MATHEMATICAL DERIVATIONS

Full-wave doubler circuit (Fig. 116-1)

Open-circuit dc output voltage = $2E$ volts (116-1)

Maximum reverse voltage across each diode = $2E$ volts (116-2)

where: E = peak of the secondary voltage.

Under loaded conditions,

the frequency of the fundamental ripple component = 2 × *line frequency* (116-3)

Half-wave cascade doubler circuit (Fig. 116-2)

Open-circuit dc output voltage = $2E$ volts (116-4)

Maximum reverse voltage across each diode = $2E$ volts (116-5)

where: E = peak value of the secondary voltage.

Under loaded conditions,

the frequency of the fundamental ripple component = *line frequency* (116-6)

Example 116-1

In Fig. 116-1, the ac line voltage is 110 V, 60 Hz. If the transformer has a turns ratio of 1:3, calculate the values of the open-circuit dc output voltage and the maximum reverse voltage across each diode. Under loaded conditions, what is the frequency of the fundamental ripple component?

Solution

Open-circuit dc output voltage, (116-1)

$$= 2 \times 110 \times 3 \times 1.414$$

$$= 933 \text{ V}$$

Maximum reverse voltage across each diode = 933 V (116-2)

Frequency of fundamental ripple component = 2 × 60 = 120 Hz (116-3)

Example 116-2

In Fig. 116-2, the ac line voltage is 220 V, 50 Hz. If the transformer has a turns ratio of 4:1, calculate the values of the open circuit dc output voltage and the maximum reverse voltage across each diode. Under loaded conditions, what is the frequency of the fundamental ripple component?

Solution

Open-circuit dc output voltage,

$$= \frac{2 \times 220 \times 1.414}{4} \quad (116\text{-}4)$$

$$= 155 \text{ V}$$

Maximum reverse voltage across each diode = 155 V (116-5)

Frequency of the fundamental ripple component = 50 Hz (116-6)

PRACTICE PROBLEMS

116-1 The full-wave doubler circuit of Fig. 116-1, has a capacitor input filter and an open-circuit dc output voltage of 310 V. If the circuit is operated directly from a 60-Hz supply, determine the rms value of the line voltage.

116-2 A cascade tripler circuit has a capacitor input filter and a dc output voltage of 848 V. If the power transformer has a turns ratio of 1:4, calculate the value of the ac line voltage.

117
The zener diode

The basic zener diode voltage regulator circuit is shown in Fig. 117-1A. Its purpose is to stabilize the value of the load voltage, (V_L), against changes in the source voltage, (E), and/or changes in the load current, (I_L). The diode operates with reverse bias in the breakdown region (Fig. 117-1B), where as a first approximation, the diode's reverse current, (I_Z), is independent of the diode voltage, (V_Z), which is equal to the load voltage, (V_L). The value of the resistor, (R), is such that it allows the diode to operate well within the breakdown region and, at the same time, regulates the load voltage by dropping the difference between V_Z and the unregulated source voltage, E.

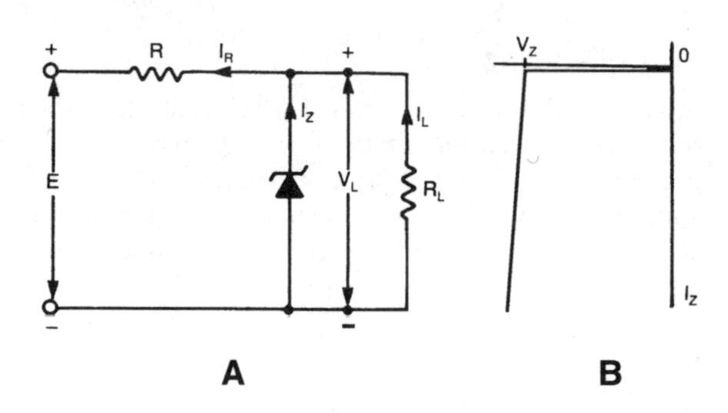

Figure 117-1

MATHEMATICAL DERIVATIONS

Using Kirchhoff's current law (KCL),

Resistor current,

$$I_R = I_L + I_Z \text{ milliamperes} \qquad (117\text{-}1)$$

In addition,

Load current,

$$I_L = \frac{V_L}{R_L} = \frac{V_Z}{R_L} \text{ milliamperes} \qquad (117\text{-}2)$$

Consequently,

Resistor current,

$$I_R = \frac{E - V_Z}{R} = \frac{E - V_L}{R} \text{ milliamperes} \qquad (117\text{-}3)$$

These equations refer to the roughest (or first) approximation. However, because the breakdown curve is not completely vertical, a second approximation will involve the zener diode's resistance, (R_Z), which is derived from the slope of its characteristic. Therefore:

Load voltage,

$$V_L = V_Z + I_Z R_Z \text{ volts} \qquad (117\text{-}4)$$

Finally, the power dissipation in the zener diode is given by:

Zener diode power dissipated,

$$P_Z = I_Z V_Z \text{ watts} \qquad (117\text{-}5)$$

The value of P_Z must not exceed the diode's power rating.

Example 117-1

In Fig. 117-1A, the zener diode's breakdown voltage is 9 V. If E = 50 V, R = 4 kΩ, R_L = 2.5 kΩ, calculate the values of I_R, I_Z, I_L, and P_Z. If E changes to 60 V, recalculate the values of I_R, I_Z, I_L, and P_Z. If R_L changes to 5 kΩ, recalculate the values of I_R, I_Z, I_L, and P_Z.

Note: Use the first (or roughest) approximation throughout and neglect any effect of R_Z.

Solution

The dropping resistor current is given by:

$$I_R = \frac{E - V_Z}{R} = \frac{50\text{ V} - 9\text{ V}}{4\text{ k}\Omega} \qquad (117\text{-}3)$$

$$= 10.25 \text{ mA}$$

The load current,

$$I_L = \frac{V_Z}{R_L} = \frac{9\text{ V}}{2.5\text{ k}\Omega} \qquad (117\text{-}2)$$

$$= 3.6 \text{ mA}$$

Zener diode current,

$$I_Z = I_R - I_L = 10.25 - 3.6 \qquad (117\text{-}1)$$

$$= 6.65 \text{ mA}$$

Power dissipation,

$$P_Z = I_Z V_Z = 6.65 \text{ mA} \times 9 \text{ V} \qquad (117\text{-}5)$$

$$= 60 \text{ mW, rounded off}$$

If E = 60 V,

$$I_R = \frac{60\text{ V} - 9\text{ V}}{4\text{ k}\Omega}$$

$$= 12.75 \text{ mA}$$

$$I_L = 3.6 \text{ mA}$$

$$I_Z = 12.75 - 3.6 = 9.15 \text{ mA}$$

$$P_Z = 9.15 \text{ mA} \times 9 \text{ V} = 82 \text{ mW}$$

The diode current has risen by 9.15 − 6.65 = 2.5 mA. The increase in the voltage drop across R is 2.5 mA × 4 kΩ = 10 V, which absorbs the change in the source voltage from 50 V to 60 V.

If $R_L = 5$ kΩ,

$$I_R = \frac{50 \text{ V} - 9 \text{ V}}{4 \text{ k}\Omega}$$

$$= 10.25 \text{ mA}$$

$$I_L = \frac{9 \text{ V}}{5 \text{ k}\Omega}$$

$$= 1.8 \text{ mA}$$

$$I_Z = 10.25 - 1.8$$

$$= 8.45 \text{ mA}$$

$$P_Z = 8.45 \text{ mA} \times 9 \text{ V}$$

$$= 76 \text{ mW, rounded off}$$

The decrease of 3.6 − 1.8 = 1.8 mA in the load current is balanced by an equal increase of 8.45 − 6.65 = 1.8 mA in the diode current. In this example, a zener diode with a power rating of ½ W would be adequate.

PRACTICE PROBLEMS

117-1. In Example 117-1, the effect of R_Z was neglected. If $R_Z = 20$ Ω, recalculate the value of V_L when $E = 50$ V and $R_L = 2.5$ kΩ.

117-2. In Practice Problem 117-1, recalculate the value of V_L if $E = 60$ V and $R_L = 2.5$ kΩ.

117-3. In Practice Problem 117-1, recalculate the value of V_L if $E = 50$ V and $R_L = 5$ kΩ.

118
The bipolar transistor

The action of the bipolar transistors involves two sets of charge carriers; these are the positive charge carriers (holes) and the negative charge carriers (electrons). Bipolar transistors are also of two types, pnp and npn, whose symbols are shown in Fig. 118-1A. Each of these types possesses three sections which are called the *emitter*, the *base*, and the *collector*; the arrow in each symbol indicates the direction of conventional flow (the direction opposite to that of the electron flow).

The bipolar transistor can be roughly compared to a triode (with the emitter, base, and collector corresponding to the cathode, control grid, and plate). For normal conditions, the voltages are applied to the transistor so as to forward-bias the emitter/base junction, but to reverse-bias the collector/base junction. For the pnp transistor (Fig. 118-1B), this requires that the emitter is positive with respect to the base which is, in turn, positive with respect to the collector. These polarities are reversed for the npn transistor. Some of the methods of producing the correct dc bias voltages for the emitter/base and collector/base junctions are discussed in chapters 119, 120, 121, and 122. Under either static or signal conditions, emitter current (I_E), base current (I_B), and collector current (I_C) flow in a bipolar transistor circuit (Fig. 118-1A). Applying Kirchhoff's current law:

$$I_E = I_C + I_B, I_C = I_E - I_B, I_B = I_E - I_C$$

Under normal operating conditions, I_B is typically less than 5% of either I_E or I_C, and I_C is only slightly less than I_E.

In the construction of an npn transistor, the emitter region is small and heavily doped, but the base is thin and lightly doped. The collector is relatively lightly doped, but physically large (because most of the power dissipated occurs in this region). It follows that the transistor is not a symmetrical device so that the emitter and the collector are not interchangeable. Under amplifier conditions, the forward bias applied to the emitter/base junction causes a large number of majority charge carriers (electrons in the conduction band) to move across the junction from the emitter into the base. There is some recombination between the electrons and the holes in the base region; as a result, there is a small external base current. However, most of the electrons pass through the base and cross into the collector region. The reverse-bias applied to the collector/base junction provides an electric field that drives

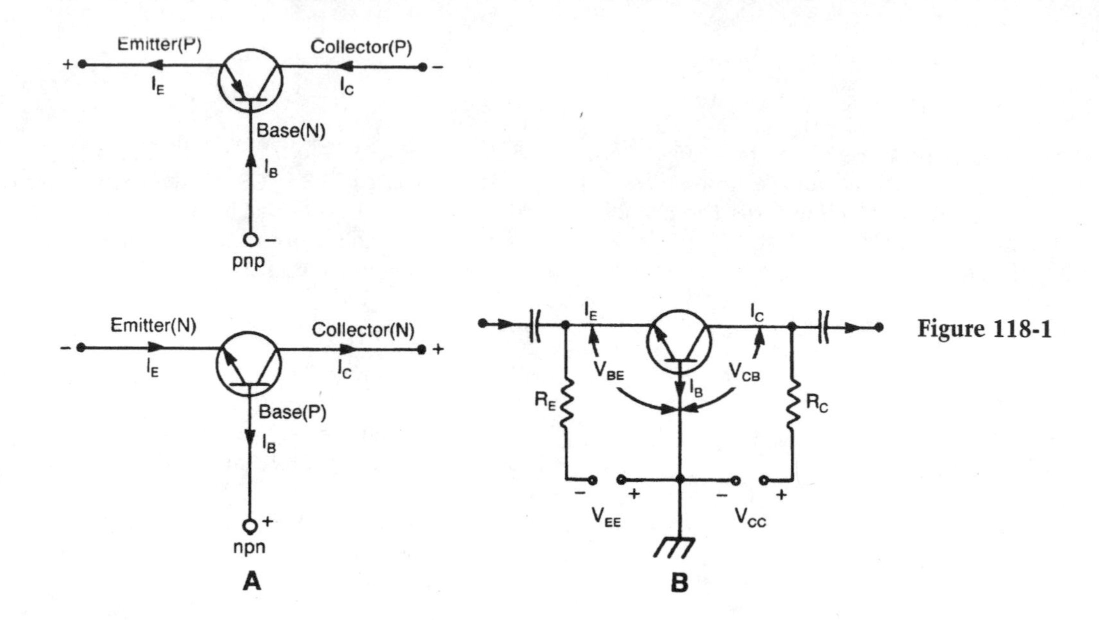

Figure 118-1

the electrons through the collector region and is responsible for producing the external flow of output current. In other words, the operation of the transistor, as an amplifying device, requires that the emitter current controls the collector current.

Apart from its operation as an amplifier, a transistor can function in two other states:

1. *Cut-off.* Both emitter/base and collector/base junctions are reverse-biased so that any collector current is extremely small.
2. *Saturation.* Both emitter/base and collector/base junctions are forward-based. The emitter current will then no longer control the collector current, but any small change in collector-base voltage will cause a very large change in collector current.

Under extreme saturation conditions, the collector current can be so large that the transistor can be destroyed. However, in some digital circuits, transistors are commonly switched between cut-off and saturation without damage to the device.

To identify the three leads at the base of the transistor, you must refer to the details of the housing. Frequently, a red or silver dot indicates the collector; in power transistors, the collector is normally connected to the casing. Four examples of the bottom views of transistors are shown in Figs. 118-2A, B, C, D.

Where the leads are arranged in a circle, lead one is found from the gap (or identifying dot), and the count from the bottom view then proceeds in the clockwise (CW) direction (Figs. 118-2A, B). The same convention is applied to a microchip (Fig. 118-3A, B), where pin 1 is located by a dot or notch. From the top view, you identify the pins by counting in the CCW direction (Fig. 118-3A); but from the bottom view, the count must be in the CW direction. The same method of counting also applies to the pins at the base of a tube.

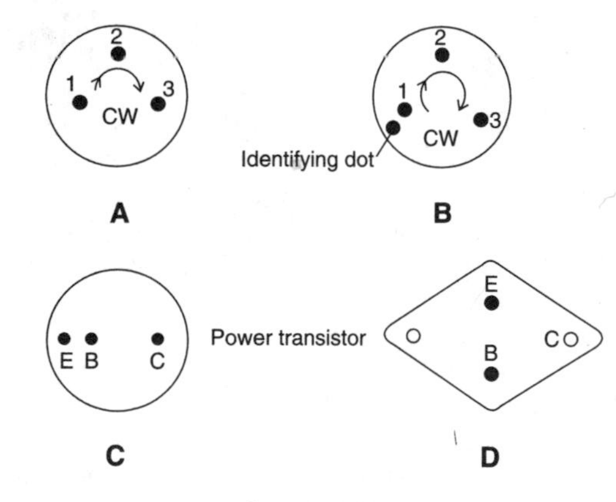

Figure 118-2

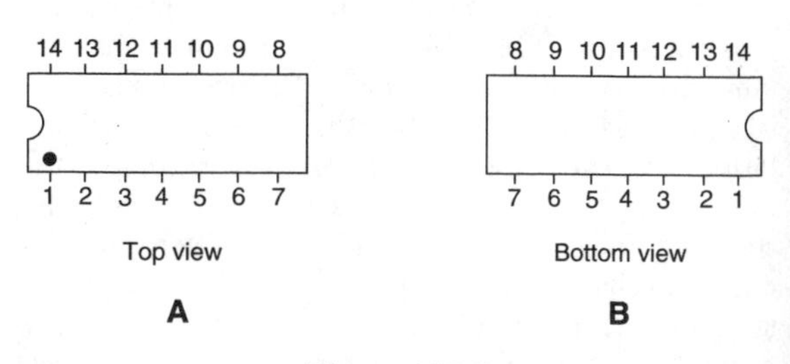

Figure 118-3

MATHEMATICAL DERIVATIONS

In the npn circuit of Fig. 118-1B, the emitter/base junction is forward-biased by V_{EE}, and the collector/base junction is reverse-biased by V_{CC}. The actual forward bias applied between the emitter and base is only a few tenths of a volt, so that:

Emitter current,

$$I_E = \frac{V_{EE} - V_{BE}}{R_E} \text{ milliamperes} \qquad (118\text{-}1)$$

or, to a first approximation,

Emitter current,

$$I_E \approx \frac{V_{EE}}{R_E} \text{ milliamperes} \qquad (118\text{-}2)$$

The voltage between the collector and base is:

$$V_{CB} = V_{CC} - (\text{the voltage drop across } R_C) \qquad (118\text{-}3)$$

$$= V_{CC} - I_C R_C \text{ volts}$$

Under quiescent conditions, the static current gain, α_{dc} is defined as the ratio of the (output) collector current, (I_C), to the (input) emitter current, I_E.

Then

Current gain,

$$\alpha_{dc} = \frac{I_C}{I_E} = \frac{I_E - I_B}{I_E} \qquad (118\text{-}4)$$

$$= 1 - \frac{I_B}{I_E}$$

and

Collector/base voltage,

$$V_{CB} = V_{CC} - I_C R_C \qquad (118\text{-}5)$$

$$= V_{CC} - \alpha_{dc} I_E R_C \text{ volts}$$

Because I_C is less than I_E, the value of α_{dc} is less than 1, and is typically 0.95 to 0.99.

In other circuit arrangements, the static current gain, (β_{dc}), is defined as the ratio of the collector current to the base current.

Therefore,

Static emitter gain,

$$\beta_{dc} = \frac{I_C}{I_B} = \frac{I_C}{I_E - I_C} = \frac{I_C/I_E}{1 - I_C/I_E} \qquad (118\text{-}6)$$

$$= \frac{\alpha_{dc}}{1 - \alpha_{dc}}$$

This yields

$$\alpha_{dc} = \frac{\beta_{dc}}{1 + \beta_{dc}} \qquad (118\text{-}7)$$

Typical values of β_{dc} vary between 50 and 200. β_{dc} is, however, an unstable quantity and changes both with temperature and the choice of the operating conditions.

Example 118-1

Assume that the circuit of Fig. 118-1B, is being operated under static conditions where: V_{EE} = 12 V, R_E = 22 kΩ, V_{BE} = 0.7 V, R_C = 15 kΩ, V_{CC} = 15 V, α_{dc} = 0.97. Calculate the values of I_E, I_C, and V_{CB}.

Solution

Emitter current,

$$I_E = \frac{V_{EE} - V_{BE}}{R_E} \qquad (118\text{-}1)$$

$$= \frac{12 \text{ V} - 0.7 \text{ V}}{22 \text{ k}\Omega}$$

$$= 0.514 \text{ mA}$$

Collector current,

$$I_C = \alpha_{dc} I_E \qquad (118\text{-}4)$$

$$= 0.97 \times 0.514$$

$$= 0.4986 \text{ mA}$$

Base current,

$$I_B = I_E - I_C$$

$$= 0.514 - 0.4986$$

$$= 0.0154 \text{ mA}$$

Collector-base voltage,

$$V_{CB} = V_{CC} - I_C R_C \qquad (118\text{-}5)$$

$$= 15 \text{ V} - 0.4986 \text{ mA} \times 15 \text{ k}\Omega$$

$$= 7.5 \text{ V}$$

PRACTICE PROBLEMS

118-1. In the circuit of Fig. 118-1B, V_{EE} = 5 V, V_{BE} = 0.7 V, V_{CC} = 20 V, R_E = 3.3 kΩ, R_C = 12 kΩ, and α_{dc} = 0.97. Obtain the values of I_E and I_C.

118-2. In Practice Problem 118-1, calculate the value of V_{CB}. Is the circuit operating under amplifier conditions?

118-3. In Practice Problem 118-1, the value of R_C is increased to 18 kΩ. Identify the conditions under which the circuit is now operating.

119
Base bias

For the normal operation of a bipolar transistor, the emitter/base junction must be forward-biased while the collector/base junction is reverse-biased. The circuit shown in Fig. 119-1A contains an npn transistor whose emitter/base junction is forward-biased by the source, V_{BB}. This voltage is connected in series with R_B between base and ground, and, for this reason, the arrangement is referred to as *base bias*. In order for the amplifier to behave in a linear manner, it is necessary that the dc or quiescent (no signal) conditions should remain as stable as possible. Unfortunately, the mathematical derivations show that base bias is sensitive to the value of β_{dc}, which is affected by temperature variations.

MATHEMATICAL DERIVATIONS

In Fig. 119-1A,

The KVL equation for the loop GBEG is,

$$-V_{BE} - I_B R_B + V_{BB} - I_E R_E = 0 \qquad (119\text{-}1)$$

However, the base current,

$$I_B = \frac{I_C}{\beta_{dc}} \approx \frac{I_E}{\beta_{dc}}$$

Therefore, the emitter current,

$$I_E = \frac{V_{BB} - V_{BE}}{R_E + R_B/\beta_{dc}} \text{ milliamperes} \qquad (119\text{-}2)$$

If Q1 is a silicon transistor,

$$V_{BE} \approx 0.7 \text{ V}.$$

This yields:

$$I_E = \frac{V_{BB} - 0.7 \text{ V}}{R_E + \dfrac{R_B}{\beta_{dc}}} \text{ milliamperes} \qquad (119\text{-}3)$$

Emitter voltage,

$$V_E = I_E R_E \text{ volts} \qquad (119\text{-}4)$$

Collector voltage,

$$V_C = V_{CC} - I_C R_C \text{ volts} \qquad (119\text{-}5)$$

Base voltage,

$$V_B = V_E + V_{BE} = V_E + 0.7 \text{ volts} \qquad (119\text{-}6)$$

Collector/base voltage,

$$V_{CB} = V_C - V_B \text{ volts} \qquad (119\text{-}7)$$

In Fig. 119-1B, source V_{BB} is removed, and the circuit is operated with a single source, V_{CC}. Equation 119-2 then becomes,

$$I_E = \frac{V_{CC} - 0.7 \text{ V}}{R_E + \dfrac{R_B}{\beta_{dc}}} \text{ milliamperes} \qquad (119\text{-}8)$$

If R_E is replaced by a short circuit (Fig. 119-1C),

$$I_E = \frac{V_{CC} - 0.7 \text{ V}}{\dfrac{R_B}{\beta_{dc}}} \qquad (119\text{-}9)$$

$$= \beta_{dc} \times \left(\frac{V_{CC} - 0.7 \text{ V}}{R_B}\right) \text{ milliamperes}$$

The emitter base current, (I_E), is proportional to β_{dc}, which is sensitive to temperature variations. If β_{dc} increases to the point where V_{CB} falls below 1 V, the transistor is saturated and linear operation is impossible.

Example 119-1

In Fig. 119-1A, $R_B = 200\ \text{k}\Omega$, $R_C = 2\ \text{k}\Omega$, $R_E = 1\ \text{k}\Omega$, $V_{BB} = 10$ V, $V_{CC} = 20$ V, $\beta_{dc} = 100$. Calculate the values of I_E, V_E, V_C, V_B, and V_{CB}.

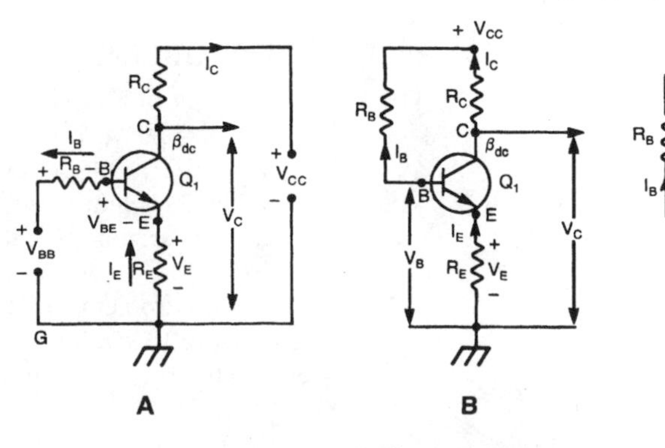

Figure 119-1

Solution

Emitter current,

$$I_E = \frac{10\text{ V} - 0.7\text{ V}}{1\text{ k}\Omega + \dfrac{200\text{ k}\Omega}{100}} = \frac{9.3\text{ V}}{3\text{ k}\Omega} \qquad (119\text{-}2)$$

$$= 3.1\text{ mA}$$

Emitter voltage,

$$V_E = 3.1\text{ mA} \times 1\text{ k}\Omega \qquad (119\text{-}4)$$

$$= +3.1\text{ V.}$$

Collector voltage,

$$V_C \approx 20\text{ V} - 3.1\text{ mA} \times 2\text{ k}\Omega \qquad (119\text{-}5)$$

$$= +\ 13.8\text{ V.}$$

Base voltage,

$$V_B = 3.1 + 0.7 \qquad (119\text{-}6)$$

$$= +\ 3.8\text{ V.}$$

Collector-to-base voltage,

$$V_{CB} = 13.8 - 3.8 \qquad (119\text{-}7)$$

$$= 10\text{ V.}$$

Example 119-2

In Fig. 119-1B, $R_B = 100\text{ k}\Omega$, $R_C = 680\ \Omega$, $R_E = 470\ \Omega$, $V_{BE} = 0.7$ V, $V_{CC} = 9$ V, $\beta_{dc} = 80$. Calculate the values of I_E, V_E, V_C, V_B, and V_{CB}.

Solution

Emitter current,

$$I_E = \frac{9\text{ V} - 0.7\text{ V}}{470\ \Omega + \dfrac{100\text{ k}\Omega}{80}} = \frac{8.3\text{ V}}{1720\ \Omega} \qquad (119\text{-}2)$$

$$= 4.83\text{ mA.}$$

Emitter voltage,

$$V_E = 4.83\text{ mA} \times 470\ \Omega \qquad (119\text{-}4)$$

$$= +\ 2.27\text{ V.}$$

Collector voltage,

$$V_C = 9\text{ V} - 4.83\text{ mA} \times 680\ \Omega \qquad (119\text{-}5)$$

$$= +\ 5.72\text{ V.}$$

Base voltage,

$$V_B = 2.27 + 0.7 \qquad (119\text{-}6)$$

$$= +\ 2.97\text{ V.}$$

Collector-to-base voltage,

$$V_{CB} = 5.72 - 2.97 \qquad (119\text{-}7)$$

$$= 2.75\text{ V.}$$

Example 119-3

In Fig. 119-1C, $R_B = 200\text{ k}\Omega$, $V_{CC} = 12$ V, $R_C = 2.2\text{ k}\Omega$, $V_{BE} = 0.7$ V, $\beta_{dc} = 50$. What is the value of V_{CB}?

Solution

Emitter current,

$$I_E = \frac{12\text{ V} - 0.7\text{ V}}{200\text{ k}\Omega/50} = \frac{11.3\text{ V}}{4\text{ k}\Omega} \qquad (119\text{-}8)$$

$$= 2.83\text{ mA.}$$

Collector voltage,

$$V_C = 12\text{ V} - 2.83\text{ mA} \times 2.2\text{ k}\Omega \qquad (119\text{-}5)$$

$$= +\ 5.77\text{ V}$$

Base voltage,

$$V_B = \ 0.7\text{ V}$$

Collector-to-base voltage,

$$V_{CB} = 5.77 - 0.7 = 5.07\text{ V}$$

Notice that if, for example, β_{dc} increased to 200, I_E would be theoretically 11.3 mA and V_C would be negative—an impossible situation. Consequently, the transistor would then have entered the saturated state.

PRACTICE PROBLEMS

119-1. In Fig. 119-2, determine the value of V_{CE}.

119-2. In Fig. 119-3, calculate the values of V_C, V_B, and V_{CE}.

119-3. In Fig. 119-4, calculate the value of V_{CB}.

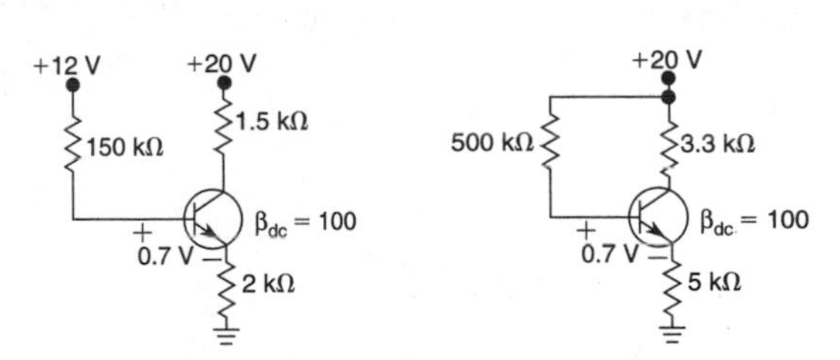

Figure 119-2 **Figure 119-3**

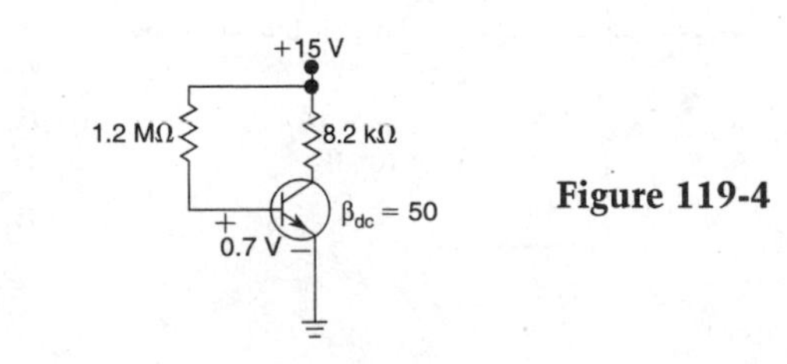

Figure 119-4

120
Voltage divider bias

The voltage divider circuit is the most important, and the most commonly used of all the base-biased arrangements. The resistors (R1 and R2) form a voltage divider (Fig. 120-1A), and are connected between the collector supply voltage (V_{CC}) and ground. The voltage drop across R2 then provides the necessary forward-bias for the emitter/base junction. The inclusion of R_E is essential, because this resistor primarily determines the value of the emitter current, I_E. It follows that the voltage divider arrangement provides a stable value for I_E that is almost independent of any fluctuations in the value of β_{dc}. In addition, this type of bias only requires a single voltage source (V_{CC}).

MATHEMATICAL DERIVATIONS

Thévenize the circuit of Fig. 120-1A to the left of the points A and ground. The equivalent circuit is then shown in Fig. 120-1B.

Thévenin generator voltage,

$$V_{TH} = V_{CC} \times \frac{R_2}{R_1 + R_2} \text{ volts} \qquad (120\text{-}1)$$

Thévenin resistance,

$$R_{TH} = \frac{R_1 \times R_2}{R_1 + R_2} \text{ ohms} \qquad (120\text{-}2)$$

Regard V_{TH} as the forward base bias voltage, and R_{TH} as the base resistor. Then from Equation 119-2,

Emitter current,

$$I_E = \frac{V_{TH} - V_{BE}}{R_E + (R_1 \times R_2)/\beta_{dc}(R_1 + R_2)} \qquad (120\text{-}3)$$

$$= \frac{V_{TH} - 0.7\text{ V}}{R_E + (R_1 \times R_2)/\beta_{dc}(R_1 + R_2)} \text{ milliamperes}$$

assuming that $V_{BE} = 0.7$ V.

Provided $R_E \gg (R_1 \times R_2)/\beta_{dc}(R_1 + R_2)$,

Emitter current,

$$I_E \approx \frac{V_{TH} - 0.7\text{ V}}{R_E} \text{ milliamperes} \qquad (120\text{-}4)$$

The emitter current is largely independent of any fluctuations in the value of β_{dc}.

Example 120-1

In Fig. 120-1A, calculate the value of I_E.

Solution

Thévenin voltage,

$$V_{TH} = V_{CC} \times \frac{R_2}{R_1 + R_2} \qquad (120\text{-}1)$$

$$= 30\text{ V} \times \frac{10\text{ k}\Omega}{20\text{ k}\Omega + 10\text{ k}\Omega}$$

$$= 10\text{ V}.$$

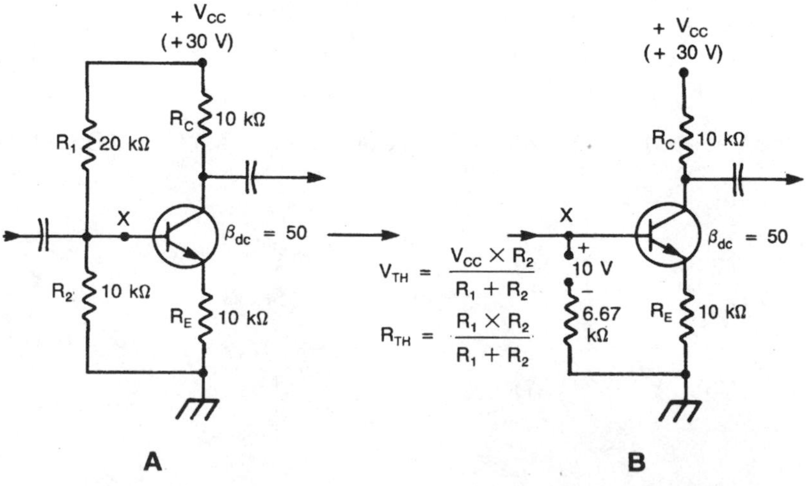

Figure 120-1

Thévenin resistance,

$$R_{TH} = \frac{R_1 \times R_2}{R_1 + R_2} \qquad (120\text{-}2)$$

$$= \frac{20 \times 10}{20 + 10}$$

$$= 6.67\ \text{k}\Omega.$$

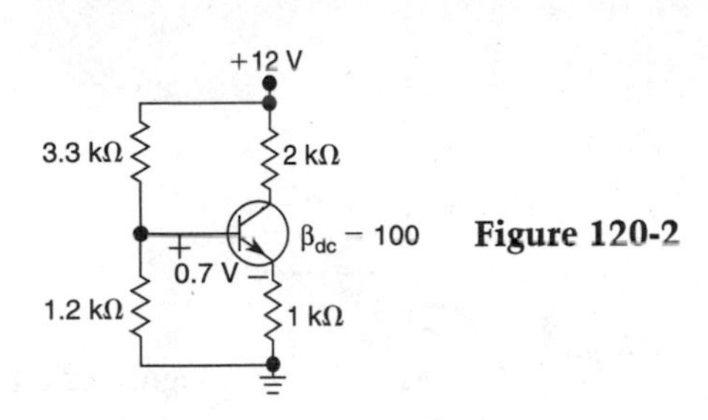

Figure 120-2

Emitter current,

$$I_E = \frac{V_{TH} - 0.7\ \text{V}}{R_E + R_{TH}/\beta_{dc}} \qquad (120\text{-}3)$$

$$= \frac{10\ \text{V} - 0.7\ \text{V}}{10\ \text{k}\Omega + 6.67\ \text{k}\Omega/50}$$

$$= \frac{9.3\ \text{V}}{10.133\ \text{k}\Omega}$$

$$= 0.918\ \text{mA}.$$

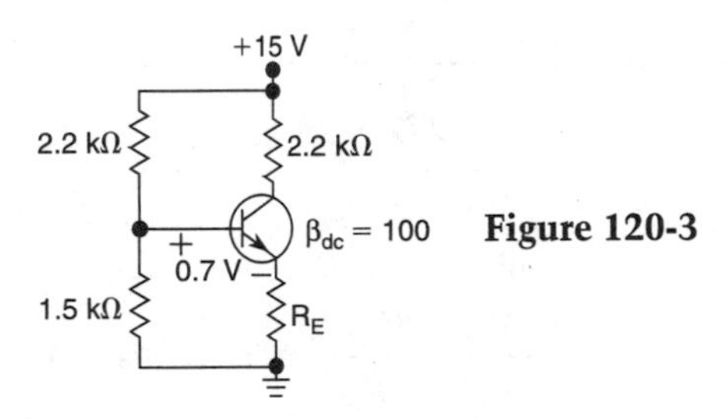

Figure 120-3

Using the approximate formula,

Emitter current,

$$I_E \approx \frac{V_{TH} - 0.7\ \text{V}}{R_E} \qquad (120\text{-}4)$$

$$= \frac{9.3\ \text{V}}{10\ \text{k}\Omega}$$

$$= 0.93\ \text{mA}.$$

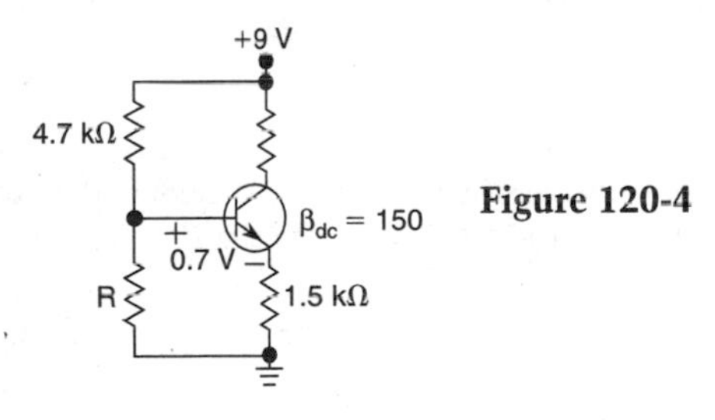

Figure 120-4

PRACTICE PROBLEMS

120-1. In Fig. 120-2, calculate the value of V_{CE}.

120-2. In Fig. 120-3, the value of the emitter current is 3 mA. Determine the required value for R_E.

120-3. In Fig. 120-4, the value of the emitter current is 2 mA. Determine the required value for R.

121 Emitter bias

Compared with the base bias method of chapter 119, the emitter bias circuit of Fig. 121-1 is a superior way to establish the dc conditions in a transistor amplifier. It does, however, require the presence of the emitter resistor (R_E) and the additional power supply, V_{EE}. The mathematical derivations will show that the ideal expression for the emitter bias does not contain a β_{dc} term so that the values of the emitter and collector currents tend to be independent of any fluctuations in β_{dc} because of temperature variations.

MATHEMATICAL DERIVATIONS

For the loop, BEGB,

$$\frac{I_E \times R_B}{\beta_{dc}} + V_{BE} + I_E R_E - V_{EE} = 0 \qquad (121\text{-}1)$$

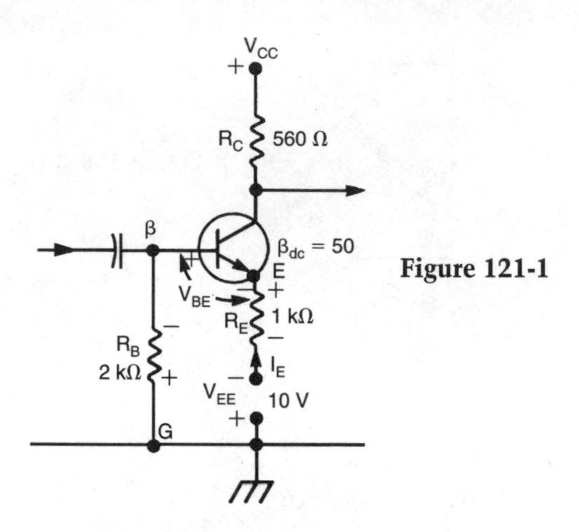

Figure 121-1

This yields:

Emitter current,

$$I_E = \frac{V_{EE} - V_{BE}}{R_E + R_B/\beta_{dc}} \text{ milliamperes} \qquad (121\text{-}2)$$

For the silicon npn transistor, $V_{BE} \approx 0.7$ V and $V_{EE} \gg V_{BE}$. In addition, you can arrange it so that $R_E \gg R_B/\beta_{dc}$. Then, the ideal expression for the emitter current is:

Emitter current,

$$I_E \approx \frac{V_{EE}}{R_E} \text{ milliamperes} \qquad (121\text{-}3)$$

This value of I_E is stable because it is independent of any fluctuations in the static current gain, β_{dc}. Base current,

$$I_B \approx \frac{I_E}{\beta_{dc}} \text{ milliamperes} \qquad (121\text{-}4)$$

Base voltage,

$$V_B = -I_B R_B \text{ volts} \qquad (121\text{-}5)$$

Example 121-1

In Fig. 121-1, calculate the values of the base and emitter potentials (use the ideal value of the emitter current).

Solution

Emitter current,

$$I_E = \frac{V_{EE}}{R_E} \qquad (121\text{-}3)$$

$$= \frac{10 \text{ V}}{1 \text{ k}\Omega}$$

$$= 10 \text{ mA.}$$

Base current,

$$I_B \approx \frac{I_E}{\beta_{dc}} \qquad (121\text{-}4)$$

$$= \frac{10 \text{ mA}}{50}$$

$$= 0.2 \text{ mA}$$

Potential at the bias,

$$V_B = -I_B R_B \qquad (121\text{-}5)$$

$$= 0.2 \text{ mA} \times 2 \text{ k}\Omega$$

$$= -0.4 \text{ V.}$$

Potential at the emitter,

$$V_E = (-0.4 \text{ V}) + (-0.7 \text{ V})$$

$$= -1.1 \text{ V.}$$

PRACTICE PROBLEMS

121-1. In Fig. 121-2, calculate the values of the collector and base potentials.

121-2. In Fig. 121-3, the value of the emitter current is 4.2 mA. What is the required value for R_E?

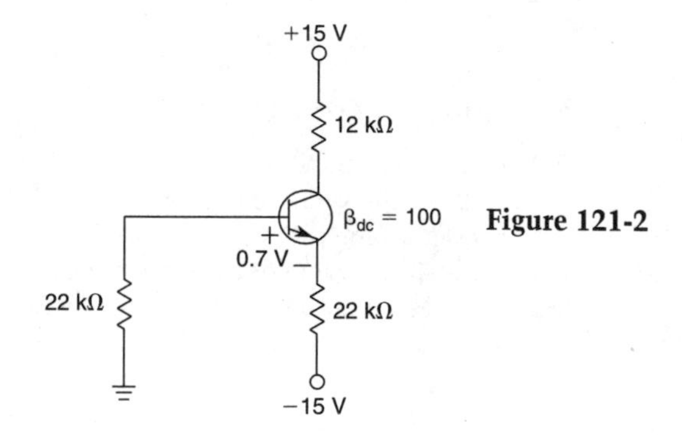

Figure 121-2

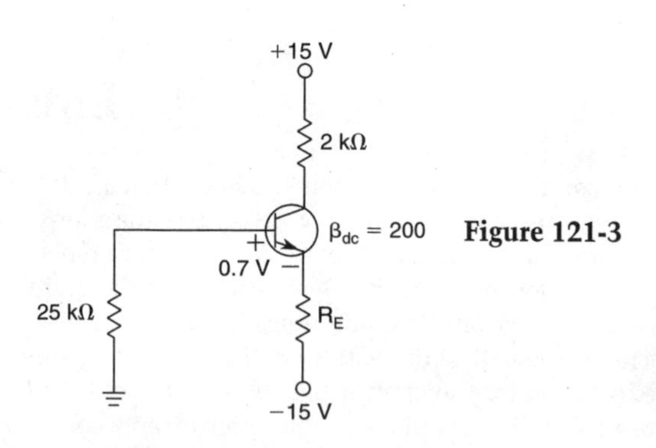

Figure 121-3

121-3. In Fig. 121-4, the value of the collector voltage is +5 V. Determine the required value for R_C.

121-4. In Fig. 121-5, the value of the emitter current is 4 mA. Find the required value for V_{EE}.

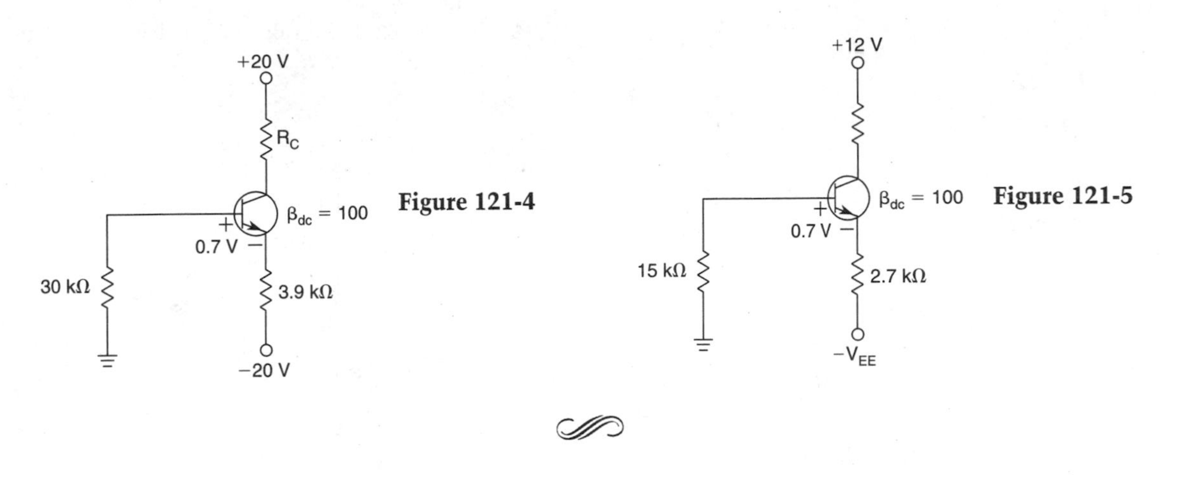

Figure 121-4

Figure 121-5

122
Collector feedback bias

Collector feedback bias (self-bias) is created by connecting a single resistor between the collector and the base (Fig. 122-1). Therefore, the forward base bias is provided by the collector voltage (V_C), and not by the collector supply voltage, $+V_{CC}$. Under amplifier conditions, the output signal is taken from the collector so that V_C is a combination of a dc voltage and an ac voltage. Consequently, R_B will not only determine the base/emitter bias, but it will also provide a path between the output collector signal and the input base signal. Because these two signals are 180° out of phase, the feedback is negative or degenerative. The result of this feedback is a reduction in the gain of the amplifier, but the circuit will also have the advantages that are outlined in chapter 150. The main disadvantage of self bias is its dependency on the value of β_{dc}.

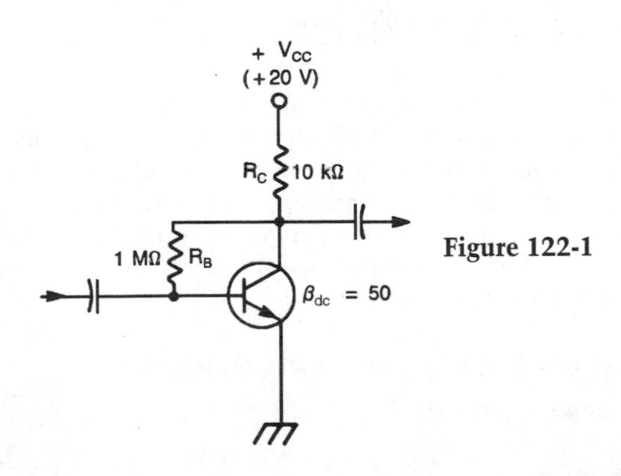

Figure 122-1

MATHEMATICAL DERIVATIONS

From the base bias equation:

Emitter current,

$$I_E = \frac{V_C - V_{BE}}{R_B/\beta_{dc}} \text{ milliamperes} \tag{122-1}$$

Collector voltage,

$$V_C = V_{CC} - (I_C + I_B)\,R_C \tag{122-2}$$
$$= V_{CC} - I_E R_C \text{ volts}$$

From Equations 122-1 and 122-2,

Emitter current,

$$I_E = \frac{V_{CC} - V_{BE}}{R_C + R_B/\beta_{dc}} \text{ milliamperes} \tag{122-3}$$

Assuming $V_{BE} = 0.7$ V,

Emitter current,

$$I_E = \frac{V_{CC} - 0.7 \text{ V}}{R_C + R_B/\beta_{dc}} \text{ milliamperes} \tag{122-4}$$

$$\approx \frac{V_{CC}}{R_C + R_B/\beta_{dc}} \text{ milliamperes} \quad (122\text{-}5)$$

if $V_{CC} \gg 0.7$ V.

Example 122-1

In Fig. 122-1, calculate the values of I_E and V_C.

Solution

Emitter current,

$$I_E = \frac{V_{CC} - 0.7 \text{ V}}{R_C + R_B/\beta_{dc}} \quad (122\text{-}4)$$

$$= \frac{20 \text{ V} - 0.7 \text{ V}}{10 \text{ k}\Omega + 1 \text{ M}\Omega/50}$$

$$= \frac{19.3 \text{ V}}{10 \text{ k}\Omega + 20 \text{ k}\Omega}$$

$$= 0.64 \text{ mA}$$

Collector voltage,

$$V_C = V_{CC} - I_E R_C \quad (122\text{-}2)$$

$$= +20 \text{ V} - 10 \text{ k}\Omega \times 0.64 \text{ mA}$$

$$= +13.6 \text{ V}$$

PRACTICE PROBLEMS

122-1. In Fig. 122-2, determine the value of V_C.

122-2. In Practice Problem 122-1, what are the values of V_{CE}, V_{CB}, and V_{BE}?

122-3. In Practice Problem 122-1, the value of β_{dc} changes from 100 to 200. Calculate the new values of I_E and V_C.

122-4. In Fig. 122-3, the value of I_E is 2.6 mA. Calculate the value of R_B.

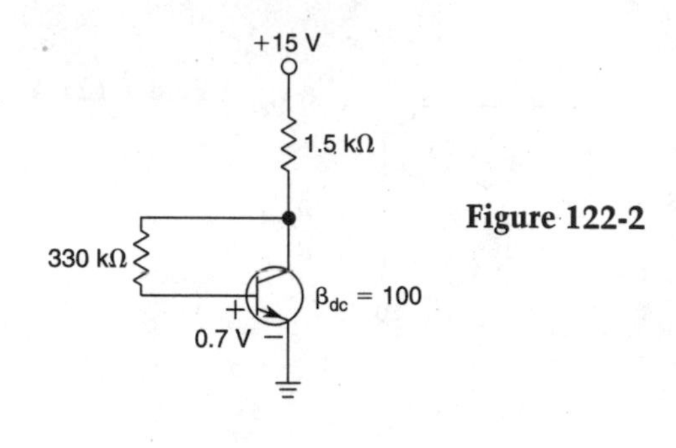

Figure 122-2

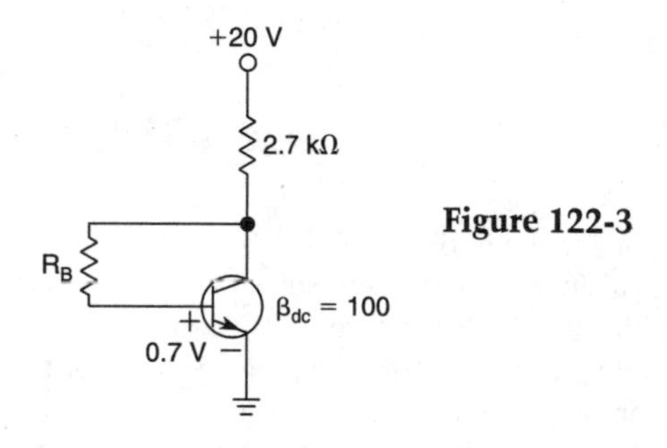

Figure 122-3

123
The common-emitter amplifier

The base-driven stage is the most frequently used type of transistor amplifier. In Fig. 123-1A, the input signal (v_b) is applied between the base of an npn transistor and ground while the amplified output signal appears between the collector and ground. As far as the signal is concerned, the equivalent circuit of Fig. 123-1B shows that the transistor behaves as a constant-current source, in association with the emitter diode's ac resistance, r_e'.

If the input signal drives the base more positive with respect to ground, the forward voltage across the emitter/base junction will be higher so that the emitter current is increased. This, in turn, will raise the level of the collector current so that there will be a greater voltage drop across the collector load, R_C. As a result, the collector voltage (v_C) will be less positive so that the input and output signals are 180° out of phase.

The expressions for such quantities as the voltage gain and the input impedance are shown in the mathematical derivations.

MATHEMATICAL DERIVATIONS

In the loop *BEGB* (Fig. 123-1B):

$$i_e(r_e' + R_E) = v_b \text{ volts} \quad (123\text{-}1)$$

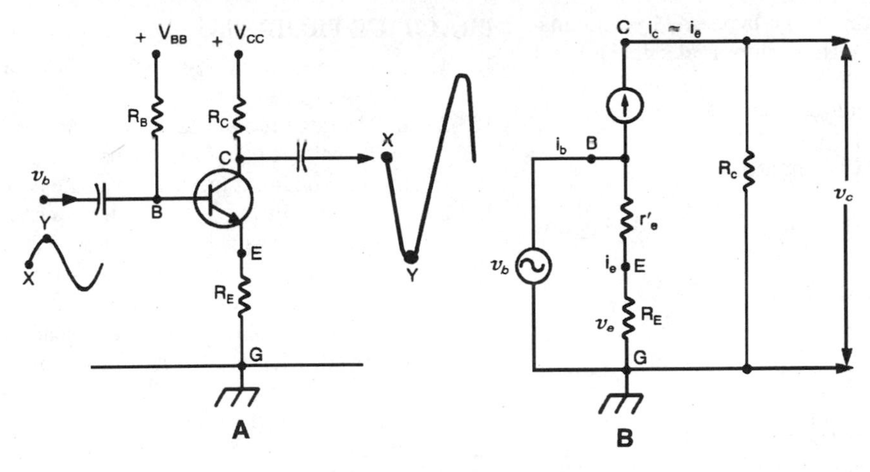

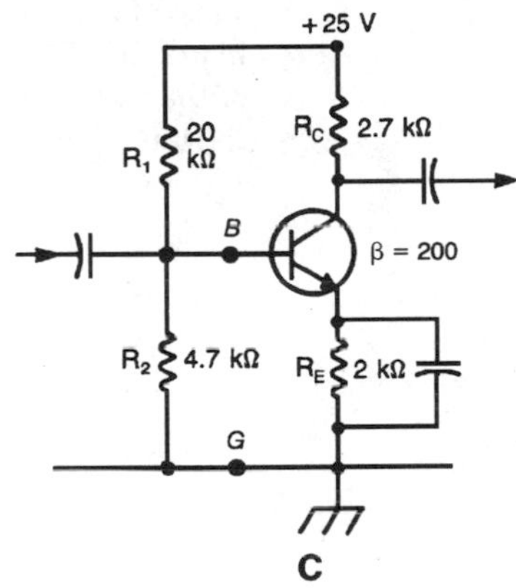

Figure 123-1

Signal emitter current,

$$i_e = \frac{v_b}{R_E + r_e'} \text{ milliamperes} \qquad (123\text{-}2)$$

Signal emitter voltage,

$$v_e = i_e R_E \text{ volts} \qquad (123\text{-}3)$$

Output signal collector voltage,

$$v_c = i_c R_C \qquad (123\text{-}4)$$

$$\approx i_e R_C \text{ volts}$$

$$\text{Voltage gain, } G_V = \frac{v_c}{v_b} \approx \frac{i_e R_C}{i_e(R_E + r_e')} \qquad (123\text{-}5)$$

$$= \frac{R_C}{R_E + r_e'}$$

Emitter-base diode's ac resistance

$$r_e' = \frac{\Delta V_{BE}}{\Delta I_E} \qquad (123\text{-}6)$$

$$= \frac{25 \text{ mV}}{I_E} \text{ ohms}$$

where: I_E = dc level of the emitter current (mA)

Equation 123-6 is obtained by differentiating Shockley's expression for a diode current.

Signal base current,

$$i_b = \frac{i_c}{\beta} \approx \frac{i_e}{\beta} \text{ milliamperes} \qquad (123\text{-}7)$$

where β is the ac current gain from base to collector and is defined by: $\beta = \Delta I_C / \Delta I_B$. For a particular transistor, the values of β and β_{dc} are comparable.

Input impedance at the base:

$$z_{in} = \frac{v_b}{i_b} \qquad (123\text{-}8)$$

$$\approx \frac{i_e(r_e' + R_E)}{i_e/\beta}$$

$$= \beta\,(r_e' + R_E) \text{ ohms}$$

If the emitter resistor (R_E) is bypassed by a suitable capacitor (Fig. 123-1C), the emitter is at signal ground and the stage can be referred to as a *common emitter* (CE) amplifier. The voltage gain of such a stage is given by:

Voltage gain of the CE amplifier,

$$G_V = \frac{R_C}{r_e'} \qquad (123\text{-}9)$$

Input impedance at the base,

$$z_{in} = \beta r_e' \text{ ohms} \qquad (123\text{-}10)$$

Example 123-1

In Fig. 123-1C, what are the values of the voltage gain from the base to the collector and the input impedance at the base?

Solution

Thévenize the circuit at the points B and G,

Thévenin voltage,

$$V_{TH} = 25\text{ V} \times \frac{4.7\text{ k}\Omega}{4.7\text{ k}\Omega + 20\text{ k}\Omega}$$

$$= 4.76\text{ V}$$

Dc level of the emitter current,

$$I_E \approx \frac{4.76\text{ V}}{2\text{ k}\Omega}$$

$$= 2.38\text{ mA}$$

Emitter-base diode's ac resistance,

$$r_e' = \frac{25\text{ mV}}{I_E} \qquad (123\text{-}6)$$

$$= \frac{25\text{ mV}}{2.38\text{ mA}}$$

$$= 10.5\ \Omega$$

Voltage gain,

$$G_V = \frac{R_C}{r_e'} \qquad (123\text{-}9)$$

$$= \frac{2.7\text{ k}\Omega}{10.5\ \Omega}$$

$$\approx 250$$

Input impedance at the base,

$$z_{in} = \beta r_e' \qquad (123\text{-}10)$$

$$= 200 \times 10.5\ \Omega$$

$$= 2.1\text{ k}\Omega$$

PRACTICE PROBLEMS

123-1. In the circuit of Fig. 123-2, determine the value of r_e'.

123-2. In Practice Problem 123-1, what are the values of the voltage gain from the base to the collector and the input impedance at the base?

123-3. In Practice Problem 123-1, the input signal at the base is a sine wave with a peak-to-peak value of 2 mV. Calculate the rms value of the signal output from the collector.

123-4. In Practice Problem 123-1, the capacitor C_E is removed. What are the new values of the voltage gain from base to collector and the input impedance at the base?

123-5. In the circuit of Fig. 123-3, determine the value of r_e'.

123-6. In Practice Problem 123-5, what are the values of the voltage gain from base to collector and the input impedance at the base?

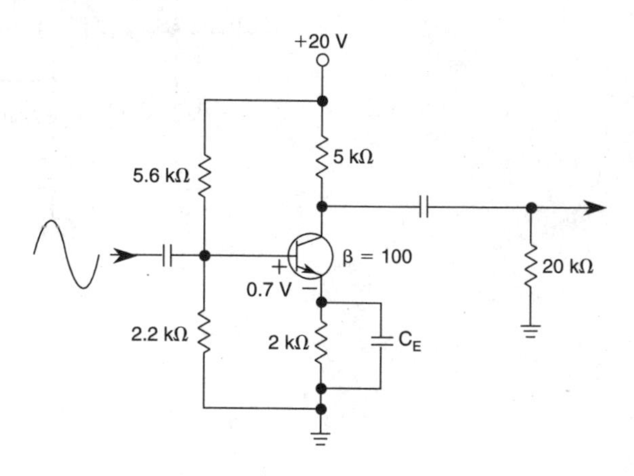

Figure 123-2

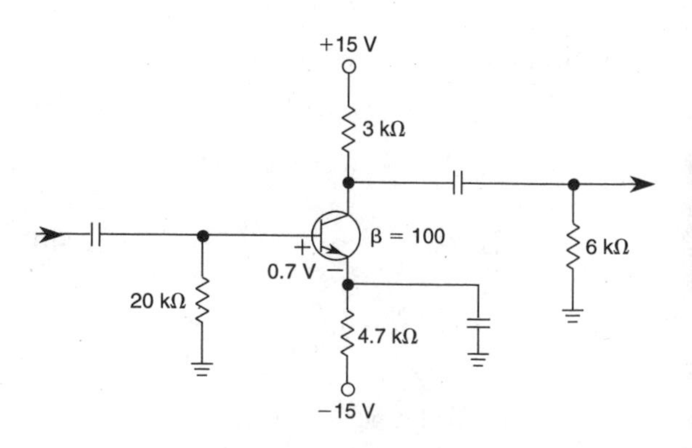

Figure 123-3

124
The common-base amplifier

In the *common base* arrangement, the base of the npn transistor is grounded and the input signal (v_e) is applied to the emitter (Fig. 124-1A). In the ac equivalent circuit of Fig. 124-1B, the output signal (v_c) appears between the collector and ground.

If the input signal drives the emitter (*n* region) more positive, the forward voltage on the emitter/base junction is lower so that the emitter current is reduced. In turn, this will decrease the collector current so that the voltage drop across the collector load will be less. The collector voltage (v_c) will therefore be more positive so that the input and the output signals are in phase. This is in contrast with the common emitter arrangement (chapter 123), where the input and output signals are 180° out of phase.

The expressions for the voltage gain and the input impedance of the CB amplifier appear in the mathematical derivations.

MATHEMATICAL DERIVATIONS

In Fig. 124-1B:

Input emitter signal,

$$v_e = i_e r_e' \text{ volts} \quad (124\text{-}1)$$

where r_e' is the ac resistance of the emitter-base diode. The value of r_e' is given by 25 mV/I_E (Equation 123-6) where I_E is the dc level of the emitter current.

Output collector signal,

$$v_c \approx i_e R_C \text{ volts} \quad (124\text{-}2)$$

Voltage gain from emitter to collector,

$$G_V = \frac{v_c}{v_e} = \frac{i_e R_C}{i_e r_e'} \quad (124\text{-}3)$$

$$= \frac{R_C}{r_e'}$$

The expression for the voltage gain is identical to that of the *CE* arrangement (Equation 123-9).

The input impedance at the emitter is given by,

$$z_{in} = \frac{v_e}{i_e} = \frac{i_e r_e'}{i_e} \quad (124\text{-}4)$$

$$= r_e' \text{ ohms}$$

Therefore, the input impedance of the CB amplifier is extremely low, when compared with the input impedance ($\beta r_e'$) at the base of the CE amplifier. This is the main disadvantage of the CB arrangement, whose main use is normally confined to high-frequency oscillator circuits.

Example 124-1

In Fig. 124-1A, calculate the values of the voltage gain from the emitter to the collector and the input impedance at the emitter.

Solution

Dc level of the emitter current,

$$I_E \approx \frac{10 \text{ V}}{10 \text{ k}\Omega}$$

$$= 1 \text{ mA.}$$

Emitter-base diode ac resistance,

$$r_e' = \frac{25 \text{ mV}}{I_E} \quad (124\text{-}1)$$

$$= \frac{25 \text{ mV}}{1 \text{ mA}}$$

$$= 25 \ \Omega.$$

Voltage gain from the emitter to the collector,

$$G_V = \frac{R_c}{r_e'} \quad (124\text{-}3)$$

$$= \frac{5 \text{ k}\Omega}{25 \ \Omega}$$

$$= 200.$$

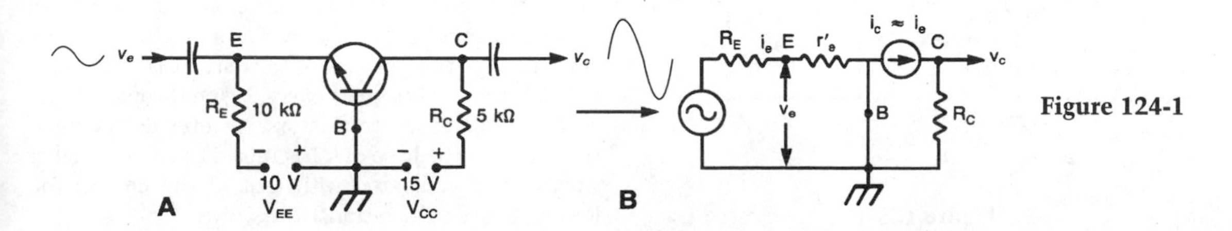

Figure 124-1

Input impedance at the emitter,

$$z_{in} = r_e' \qquad (124\text{-}4)$$

$$= 25\ \Omega.$$

PRACTICE PROBLEMS

124-1. In the circuit of Fig. 124-2, determine the values of r_e' and V_{CE}.

124-2. In Practice Problem 124-1, calculate the value of the voltage gain from the emitter to the collector.

124-3. In Practice Problem 124-1, the input signal (V_i) is a sine wave with a peak-to-peak value of 2 mV. Determine the rms value of the output signal, V_o. What is the phase relationship between V_o and V_i?

124-4. If the 20-kΩ resistor is removed in Practice Problem 124-1, recalculate the voltage gain, as defined by the ratio of V_o to V_i.

124-5. In Practice Problem 124-1, what is the value of the input impedance at the emitter?

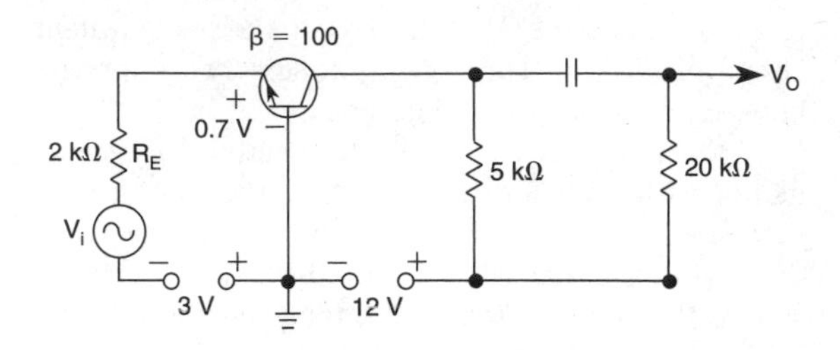

Figure 124-2

125
The emitter follower

The *emitter follower* is sometimes referred to as the *common collector* (CC) stage. In Fig. 125-1A, the input signal is applied between the base and ground, while the output signal is taken from the emitter and ground; the collector is normally joined directly to the V_{CC} supply. As is shown in the mathematical derivations, the voltage gain from the base to the emitter is less than unity, but the CC arrangement has a high input impedance and a low output impedance. Therefore, the emitter follower is primarily used for impedance matching because it allows a high-impedance source to feed a low-impedance load. In this way, it behaves as a matching transformer, but with the advantages of a better frequency response, greater simplicity, and lower cost. This also means that the emitter follower can sometimes provide some degree of isolation between two stages.

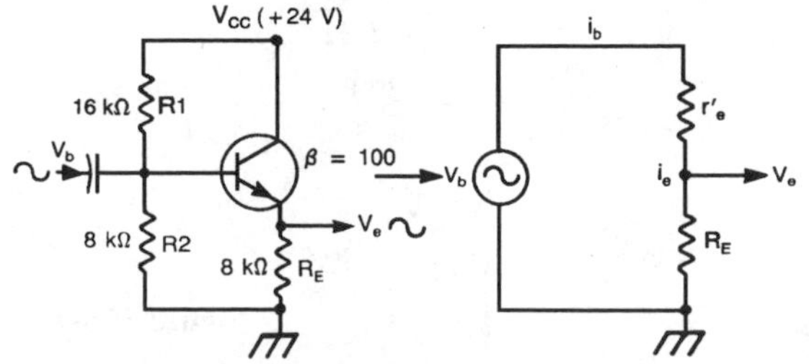

Figure 125-1

Assume that the input signal is driving the p-type base more positive with respect to ground. This increases the forward voltage applied to the emitter/base junction so that the emitter current increases. This increase of current passes through the emitter load resistor, R_E; as a result, the emitter voltage becomes more positive with respect to ground. The emitter output signal is, therefore, in phase with the input base signal. In other words, the emitter output "follows" the base input. However, the whole of the output signal is applied as negative (or degenerative) feedback to the input signal. This is the reason for the high input impedance, the low output impedance, and the voltage gain of less than unity.

The Darlington pair

The emitter-follower configuration is commonly used in an arrangement that is known as a *Darlington pair* (Fig. 125-2). The purpose of this circuit is to obtain a very large overall (signal) current gain through the direct coupling of two transistors. As a result, the input resistance at the base of Q1 is raised to a much higher value, when compared with that of the emitter follower, using only one transistor.

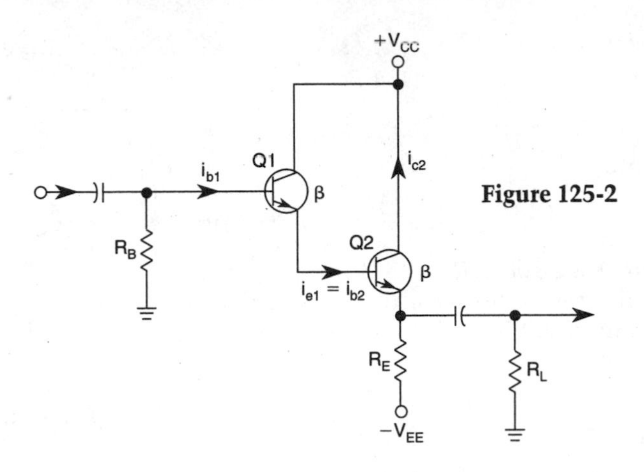

Figure 125-2

MATHEMATICAL DERIVATIONS

In the equivalent circuit of Fig. 125-1B:

Input base signal,

$$v_b = i_e(r_e' + R_E) \text{ volts} \quad (125\text{-}1)$$

where: r_e' = emitter-base diode's ac resistance ($25 \text{ mV}/I_E$ ohms)

Output emitter signal,

$$v_e = i_e \times R_E \text{ volts} \quad (125\text{-}2)$$

Voltage gain,

$$G_V = \frac{v_e}{v_b} = \frac{i_e \times R_E}{i_e(r_e' + R_E)} = \frac{R_E}{r_e' + R_E} < 1 \quad (125\text{-}3)$$

Signal base current,

$$i_b \approx \frac{i_e}{\beta} \text{ milliamperes} \quad (125\text{-}4)$$

Input impedance at the base,

$$z_{in} = \frac{v_b}{i_b} = \beta \times \frac{v_b}{i_e} \quad (125\text{-}5)$$

$$= \beta \times (r_e' + R_E) \text{ ohms}$$

Darlington pair (Fig. 125-2)

Assuming that both transistors have the same β value, $i_{c1} = i_{b1}$ under signal conditions. However, $i_{c1} \approx i_{e1} \approx i_{b2}$. Therefore,

Emitter current of Q2,

$$i_{e2} \approx i_{c2} = \beta i_{b2} \approx \beta^2 i_{b1} \text{ amperes} \quad (125\text{-}6)$$

Overall current gain,

$$\beta_T = \beta^2 \quad (125\text{-}7)$$

or,

$$\beta_T = \beta_1 \beta_2 \quad (125\text{-}8)$$

if the transistors have different β values.

Input resistance to the signal at Q1 base,

$$R_{in} \approx \beta^2 \times R_E \| R_L \quad (125\text{-}9)$$

$$= \beta^2 \times \frac{R_E R_L}{R_E + R_L} \text{ ohms}$$

Input resistance to the entire circuit of Fig. 125-2 is the parallel combination of R_{in} and R_B.

Example 125-1

In Fig. 125-1A what are the values of the voltage gain from base to emitter and the input impedance at the base?

Solution

Dc level of the emitter current,

$$I_E \approx \frac{24 \text{ V} \times 8 \text{ k}\Omega}{(8 \text{ k}\Omega + 16 \text{ k}\Omega) \times 8 \text{ k}\Omega}$$

$$= 1 \text{ mA.}$$

Emitter-base diode's ac resistance,

$$r_e' = \frac{25 \text{ mV}}{I_E}$$

$$= \frac{25 \text{ mV}}{1 \text{ mA}}$$

$$= 25\ \Omega.$$

Voltage gain from the base to the emitter,

$$G_V = \frac{R_E}{r_e' + R_E} \quad (125\text{-}3)$$

$$= \frac{8 \text{ k}\Omega}{25\ \Omega + 8 \text{ k}\Omega}$$

$$= 0.997.$$

Input impedance at the base,

$$z_{in} = \beta \times (r_e' + R_E) \quad (125\text{-}5)$$

$$= 100 \times (25\ \Omega + 8\ \text{k}\Omega)$$

$$= 802.5\ \text{k}\Omega.$$

Example 125-2

In Fig. 125-2, each transistor has a β of 40, $R_E = 15\ \text{k}\Omega$, and $R_L = 1\ \text{k}\Omega$. What is the value of the input resistance to the signal at the base of Q1?

Solution

Input resistance at the base of Q1,

$$R_{in} = \beta^2 \times R_E \| R_L \quad (125\text{-}9)$$

$$= 40^2 \times \frac{15 \times 1}{16}$$

$$= 1500\ \text{k}\Omega$$

$$= 1.5\ \text{M}\Omega$$

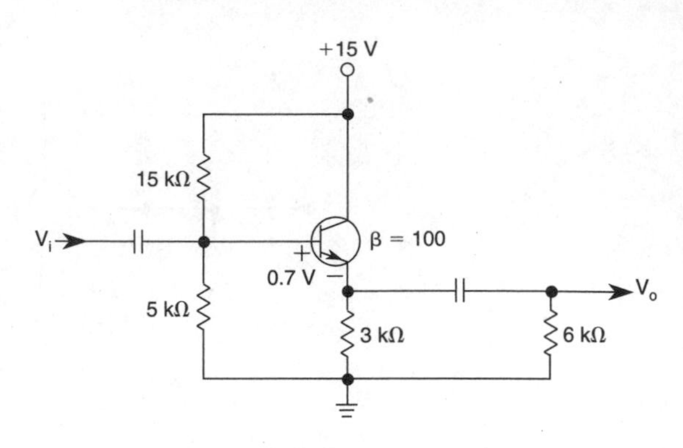

Figure 125-3

PRACTICE PROBLEMS

125-1. In the circuit of Fig. 125-3, determine the values of r_e' and V_{CE}.

125-2. In Practice Problem 125-1, calculate the value of the voltage gain from the base to the emitter.

125-3. In Practice Problem 125-1, the input signal has a peak value of 100 mV. Determine the rms value of the output signal, V_o. What is the phase relationship between V_o and V_i?

125-4. If the 6-kΩ resistor is removed from the circuit of Fig. 125-3, recalculate the voltage gain, as is defined by the ratio of V_o and V_i.

125-5. In Practice Problem 125-1, determine the value of the input impedance at the base.

126
Class-A power amplifier

Figure 126-1A illustrates a receiver's output stage, whose load is a low-impedance loudspeaker. The step-down transformer is used to match the speaker to the required value of the collector load. Under optimum conditions, the circuit will deliver audio power to the loudspeaker with the minimum amount of signal distortion.

The collector characteristics of a power transistor are shown in Fig. 126-1B. Initially, the quiescent (or no-signal) conditions are analyzed to determine the dc value (V_{CEQ}) of the collector/emitter voltage. If the resistance of the transformer's primary winding is ignored, a vertical dc load line can be drawn and a convenient Q point can be selected to determine the dc level (I_{CQ}) of the collector current. The effective collector load of the loudspeaker is given by:

$$R_L \times \left(\frac{N_P}{N_S}\right)^2$$

and this determines the slope of the ac load line.

When the input signal is applied to the base, the operating conditions move up and down the ac load line. To allow for the greatest input signal, it is important that Q is approximately in the center of this line. Such a condition is known as *class-A* operation in which the collector current flows throughout the cycle of the base signal voltage. However, the amount of the input signal is limited by saturation on the positive alternation (point Q_1), and by cut-off clipping on the negative alternation (point Q_2). You must also be careful that the transistor's maximum V_{CE} and power ratings are not exceeded at any point on the operating section of

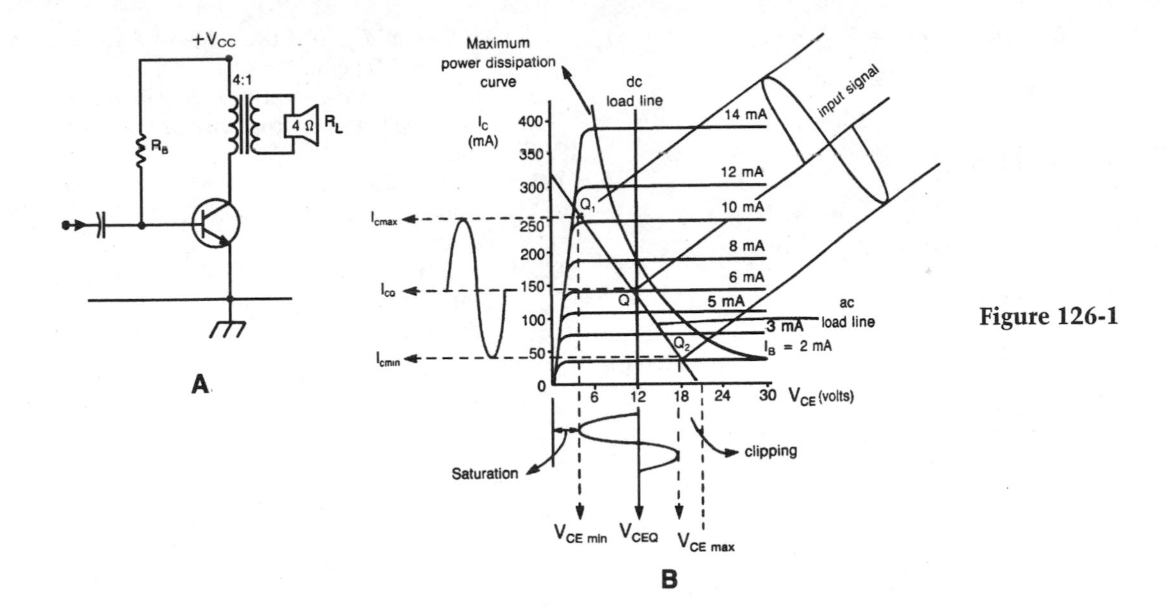

Figure 126-1

the ac load line. This is ensured by keeping the load line below the curve that represents the transistor's maximum power rating.

MATHEMATICAL DERIVATIONS

In Fig. 126-1B:

Effective ac collector load,

$$r_c = \left(\frac{N_P}{N_S}\right)^2 \times R_L \text{ ohms} \qquad (126\text{-}1)$$

The ac load line is derived from the relationship:

$$\frac{\Delta V_{CE}}{\Delta I_C} = r_c \text{ ohms} \qquad (126\text{-}2)$$

Peak voltage fluctuation at the collector,

$$v_{cpeak} = \frac{V_{CEmax} - V_{CEmin}}{2} \text{ volts} \qquad (126\text{-}3)$$

Peak fluctuation in the collector current,

$$i_{cpeak} = \frac{I_{Cmax} - I_{Cmin}}{2} \text{ milliamperes} \qquad (126\text{-}4)$$

Signal power output,

$$p_O = \frac{v_{cpeak}}{\sqrt{2}} \times \frac{i_{cpeak}}{\sqrt{2}} \qquad (126\text{-}5)$$

$$= \frac{v_{cpeak} \times i_{cpeak}}{2}$$

$$= v_{crms} \times i_{crms} \text{ milliwatts}$$

dc power input,

$$= I_{CQ} \times V_{CC} \text{ milliwatts} \qquad (126\text{-}6)$$

Percentage efficiency,

$$\eta = \frac{\textit{signal power output}}{\textit{dc power input}} \times 100\% \qquad (126\text{-}7)$$

$$= \frac{v_{crms} \times i_{crms}}{I_{CQ} \times V_{CC}} \times 100\%$$

Example 126-1

In Figs. 126-1A and B, the dc base current value is 6 mA and the input signal creates a peak base current swing of 4 mA. Determine the values of the signal power output, the dc power input, and the percentage efficiency.

Solution

Effective collector load,

$$r_c = \left(\frac{N_P}{N_S}\right)^2 \times R_L \qquad (126\text{-}1)$$

$$= 4^2 \times 4\ \Omega$$

$$= 64\ \Omega$$

Ignoring the dc voltage drop across the resistance of the primary winding, quiescent collector voltage, V_{CEQ} = 12 V.

At the Q point, the collector current, I_{CQ} = 140 mA.

$$Dc\ power\ input = V_{CEQ} \times I_{CQ} \quad (126\text{-}6)$$

$$= 12\text{ V} \times 140\text{ mA}$$

$$= 1.68\text{ W}$$

From Fig. 126-1B,

$$V_{CEmax} = 19\text{ V}, V_{CEmin} = 4\text{ V},$$

$$I_{Cmax} = 260\text{ mA}, I_{Cmin} = 40\text{ mA}$$

Signal power output,

$$p_o = \frac{v_{cpeak} \times i_{cpeak}}{2} \quad (126\text{-}5)$$

$$= \frac{(19\text{ V} - 4\text{ V}) \times (260\text{ mA} - 40\text{ mA})}{2 \times 2 \times 2}$$

$$= 0.41\text{ W}$$

Percentage efficiency,

$$\eta = \frac{signal\ power\ output}{dc\ power\ input} \times 100\% \quad (126\text{-}7)$$

$$= \frac{0.41\text{ W}}{1.68\text{ W}} \times 100\%$$

$$= 24.6\%$$

PRACTICE PROBLEMS

126-1. In the circuit of Fig. 126-2, the condition for a centered Q point is $r_c + r_e = V_{CEQ}/V_{CQ}$. If $R_e' = 250\text{ mV}/I_{CQ}$, find the value of n to obtain a centered Q point.

126-2. In Practice Problem 126-1, calculate the value of the voltage gain from the base to the collector.

126-3. In Practice Problem 126-1, determine the amount of power drawn from the 20-Vdc source.

126-4. In Practice Problem 126-1, the peak-to-peak value of the input signal is 200 mV. Calculate the percentage efficiency.

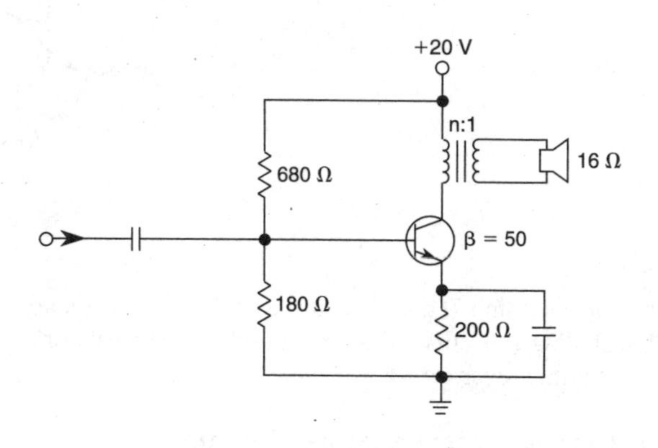

Figure 126-2

127
The class-B push-pull amplifier

Figure 127-1A illustrates a *class-B* push-pull audio amplifier. In this figure, the input signals to the two bases are *180° out of phase*; the output transformer is connected between the collectors so that the collector currents drive in opposite directions through the primary winding. As a result, the input signal to the loudspeaker is determined by the value of $i_{c1} - i_{c2}$.

In the class-A operation of chapter 126, the Q point was located at the middle of the ac load line so that the collector current flowed continuously throughout the cycle of the input signal. Although the distortion was kept to a minimum, the efficiency of the class-A circuit was only on the order of 20% to 30%. To increase the efficiency up to between 50% and 60%, you can operate the circuit under class-B conditions, whereby the Q point is theoretically lowered to the bottom of the ac load line (Fig. 127-1B) so that $V_{CEQ} = V_{CC}$ and $I_{CQ} = 0$. In other words, the forward bias on each transistor is zero. However, there is unacceptable (cross-over) distortion near the cut-off region so that, in practice, each transistor is provided with a small forward "trickle" bias by the voltage divider, R1, R2.

As a further compromise between distortion and efficiency, you can operate an audio push-pull circuit in *class-AB*, where the corresponding Q point on the ac load line lies approximately halfway between the two points that represent class A and class B (Fig. 127-1B).

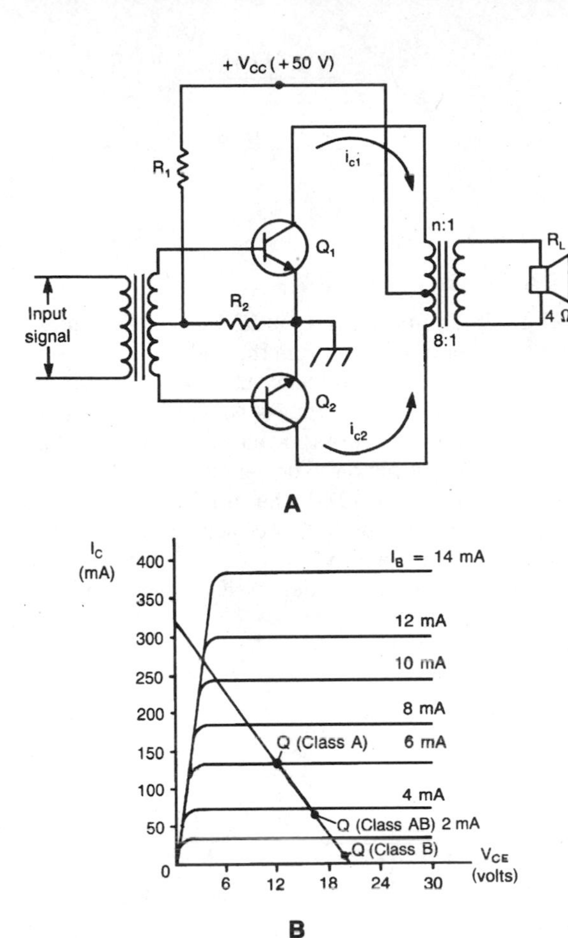

Figure 127-1

Under class-B conditions, each transistor conducts for about one half cycle of its input signal and (because the two base signals are 180° out of phase) the transistors conduct alternatively; in other words, there is a "cross-over" between the two transistors. The two half cycles then combine to produce the complete sine-wave output that appears across the loudspeaker.

Compared with a single power transistor operated under class-A conditions, the push-pull amplifier offers the following advantages:

1. There is no resultant dc current to cause partial saturation of the output transformer's iron core. If the core were to partially saturate, the output signal would be distorted.
2. Because of nonlinearity in the transistor characteristics, the collector current waveforms cannot be pure sine waves and, therefore, they are distorted to some extent. This distortion is in the form of harmonic components, both odd and even. The *even* harmonic components of i_{c1} and i_{c2} are in phase; therefore, they will cancel in the output transformer. The elimination of *even* harmonic distortion is the major advantage of the push-pull arrangement.
3. The efficiency (50% to 60%) of the class-B push-pull stage is much higher than that of the single-transistor class-A amplifier (20% to 30%). Therefore, for a given dc power input, the push-pull stage can provide a greater signal power output.
4. The maximum input signal voltage to the class-B push-pull stage is approximately four times greater than the maximum input signal to the class-A amplifier. All of these advantages are achieved at the expense of using a *matched* pair of transistors. If one transistor becomes defective, both should be replaced.

MATHEMATICAL DERIVATIONS

Under quiescent conditions,

$$V_{CEQ} = V_{CC} \text{ volts} \qquad (127\text{-}1)$$

$$I_{CQ} \approx 0 \qquad (127\text{-}2)$$

Effective load as seen by each collector,

$$r_c = \left(\frac{n}{2}\right)^2 R_L \text{ ohms} \qquad (127\text{-}3)$$

Voltage gain from base to collector,

$$G_V = \frac{v_c}{v_b} = \frac{r_c}{R_e'} \qquad (127\text{-}4)$$

where: R_e' is the emitter-base diode resistance for a large signal input.

Saturation current at the top end of the ac load line,

$$I_{C(\text{sat})} = \frac{V_{CEQ}}{r_c} = \frac{V_{CC}}{r_c} \text{ amperes} \qquad (127\text{-}5)$$

Maximum load power output is given by:
Output power,

$$p_0 = \frac{V_{CEQ} \times I_{C(\text{sat})}}{2} \text{ watts} \qquad (127\text{-}6)$$

Example 127-1

In Fig. 127-1A, the R_e' of each transistor is 1 Ω. What are the values of the maximum power output and what is the voltage gain from base to collector?

Solution

Effective load seen by each collector is:

$$r_c = \left(\frac{n}{2}\right)^2 \times R_L = 4^2 \times 4 \qquad (127\text{-}3)$$

$$= 64\ \Omega$$

$$V_{CEQ} = V_{CC} = 50\ \text{V} \qquad (127\text{-}1)$$

$$I_{C(\text{sat})} = \frac{V_{CEQ}}{r_c} = \frac{50\ \text{V}}{64\ \Omega} \qquad (127\text{-}5)$$

$$= 0.78\ \text{A}$$

Voltage gain from base to collector,

$$G_V = \frac{v_c}{v_b} = \frac{r_c}{R_e'} = \frac{64\ \Omega}{1\ \Omega} \qquad (127\text{-}4)$$

$$= 64$$

Maximum signal power output,

$$p_o = \frac{V_{CEQ} \times I_{C(\text{sat})}}{2} \qquad (127\text{-}6)$$

$$= \frac{50\ \text{V} \times 0.78\ \text{A}}{2}$$

$$= 19.5\ \text{W}$$

PRACTICE PROBLEMS

127-1. In Example 127-1, estimate the total amount of power drawn from the 50-V source.

127-2. In Practice Problem 127-1, assume that the signal power output is equal to its maximum value. What are the values of the percentage efficiency and the power dissipation?

127-3. In Example 127-1, the turns ratio is changed from 8 : 1 to 10 : 1. Recalculate the values of the voltage gain from base to collector and recalculate the maximum signal power output.

128
The class-C RF power amplifier

Figure 128-1 shows a class-C RF power amplifier that is operated at a high efficiency of 80% or more; in other words, four-fifths of the dc power drawn from the V_{CC} supply is converted into the RF signal power output.

The circuit uses signal bias in which base current flows either side of the positive alternation's peak. This base current charges the capacitor (C_B), but when the current ceases, the capacitor discharges through R_B. However, the time constant R_BC_B is high, compared to the period of the input signal. As a result, the dc voltage across R_B is approximately equal to the peak value of the input signal, and applies a *reverse*-bias to the emitter/base junction. Take care that this reverse-bias does not exceed the emitter-to-base breakdown voltage.

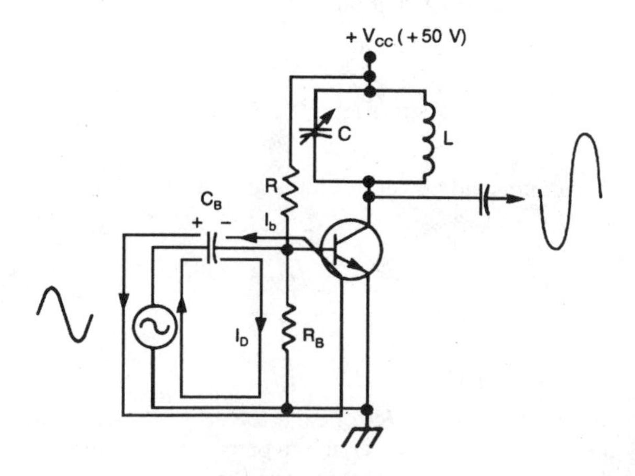

Figure 128-1

Because of the action of the signal bias, the collector current consists of narrow pulses at the same repetition rate as the frequency of the input signal. When these pulses are applied to the high-Q resonant tuned circuit, the result is a circulating current that oscillates between L and C and develops a complete sine wave for the output voltage between the collector and ground. Only a small mean level of collector current is drawn from the V_{CC} supply so that the amplifier's efficiency is high.

Frequency multipliers

For reasons of frequency stability, it is common practice to operate a transmitter RF oscillator stage at low frequency and low power, then build up the frequency and the power in the following stages. To increase the frequency, multiplier stages (sometimes called *har-*

monic generators) are used. Their circuitry is similar to that of Fig. 128-1, but their class-C operation is achieved with a bias level, which is several times the cutoff value. This reduces the angle of the output current flow and the efficiency, but increases the strength of the harmonic content. Instead of the tank circuit being tuned to the fundamental frequency of the input base signal, the tank circuit can be tuned to a particular harmonic that is contained in the output current waveform. To a first approximation, the output voltage of the multiplier stage will be an undistorted sine wave, whose frequency is a whole number multiple (twice for a doubler stage, three times for a tripler, etc.) of the frequency of the input signal on the base. Because the output frequency from the collector circuit is different from the input frequency to the base circuit, frequency-multiplier stages are not required to be neutralized.

For a doubler stage, the angle of the output current flow is typically 90°, the bias level is approximately 10 times the cutoff value, and the efficiency is 50%. A tripler stage has about the same efficiency, but the bias level is approximately 20 times the cutoff value so that the angle of the output current flow is reduced to 75°.

MATHEMATICAL DERIVATIONS

Quiescent collector-emitter voltage,

$$V_{CEQ} = V_{CC} \text{ volts} \qquad (128\text{-}1)$$

Maximum possible voltage swing at the collector extends from the saturated value of V_{CE} to $2 \times V_{CC}$. Knowing the value of $V_{CE(\text{sat})}$ (≈ 1 V), the signal power output is given by:

RF power output,

$$p_o = \frac{(0.707 \times V_{CC})^2}{r_d} \qquad (128\text{-}2)$$

$$= \frac{V^2_{CC}}{2r_d} \text{ watts}$$

where: r_d = dynamic resistance of the tuned circuit (Ω).

The power dissipation is given by:

$$P_D = \frac{V_{CE(\text{sat})} \times V_{CC}}{2r_d} \qquad (128\text{-}3)$$

$$= \frac{V_{CE(\text{sat})}}{V_{CC}} \times p_o \text{ watts}$$

Example 128-1

In Fig. 128-1, the input signal has a frequency of 2 MHz and the value of C is 100 pF. Calculate the required values for the inductor L, which has an RF resistance of 30 Ω. Calculate the values of the Q and the dynamic resistance of the LC circuit. Assuming that the complete load line is used, what are the values of the RF power output and the power dissipation ($V_{CE(\text{sat})} \approx 1$ V)?

Solution

Inductance,

$$L = \frac{1}{4 \times \pi^2 \times f_r^2 C} \text{ H}$$

$$= \frac{10^6}{4 \times \pi^2 \times (2 \times 10^6)^2 \times 100 \times 10^{-12}} \text{ H}$$

$$= 63.3\ \mu\text{H}$$

$$Q = \frac{1}{R} \times \sqrt{\frac{L}{C}}$$

$$= \frac{1}{30} \times \sqrt{\frac{63.3 \times 10^{-6}}{100 \times 10^{-12}}}$$

$$= 26.5$$

Dynamic resistance,

$$r_D = \frac{L}{CR}$$

$$= \frac{63.3 \times 10^{-6}}{100 \times 10^{-12} \times 30}\ \Omega$$

$$= 21.1\ \text{k}\Omega$$

RF power output,

$$p_o = \frac{V^2_{CC}}{2r_d} \qquad (128\text{-}2)$$

$$= \frac{(50\ \text{V})^2}{2 \times 21.1\ \text{k}\Omega}$$

$$\approx 60\ \text{mW}$$

Power dissipation,

$$P_D = \frac{V_{CE(\text{sat})}}{V_{CC}} \times p_o \qquad (128\text{-}3)$$

$$= \frac{1\ \text{V}}{50\ \text{V}} \times 60\ \text{mW}$$

$$= 1.2\ \text{mW}$$

PRACTICE PROBLEMS

128-1. Figure 128-2 illustrates a class-C RF amplifier to which a 1.5-MHz signal is applied. Calculate the required value for the capacitor, C. Determine the values of Q, and the dynamic resistance of the tank circuit.

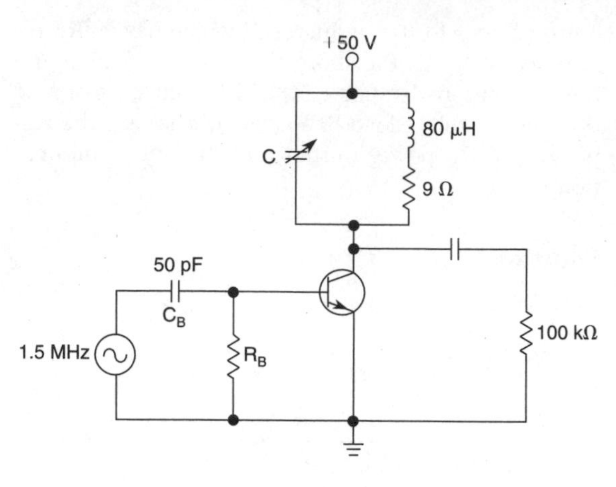

Figure 128-2

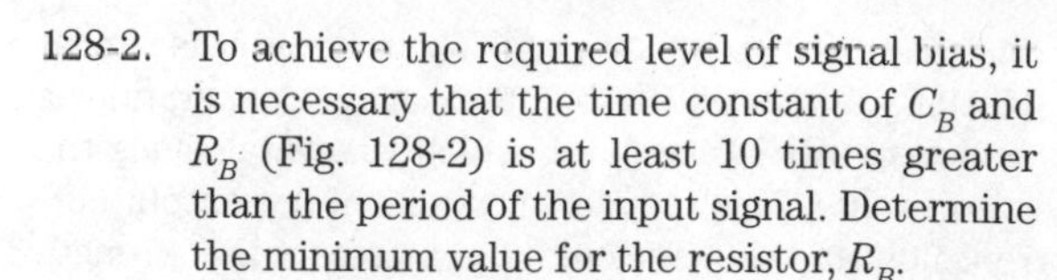

128-2. To achieve the required level of signal bias, it is necessary that the time constant of C_B and R_B (Fig. 128-2) is at least 10 times greater than the period of the input signal. Determine the minimum value for the resistor, R_B.

128-3. In Practice Problem 128-1, determine the maximum value of the signal power output and the corresponding value of the power dissipation.

128-4. From the results obtained in Practice Problem 128-3, calculate the value of the mean dc current level drawn from the 50-V supply.

129
Voltage regulator circuits

The zener diode is primarily used in power supplies where the load current is small and does not vary over a wide range. By using a transistor in conjunction with a zener diode, the current range can be increased and the regulation percentage can be reduced. Such electronic voltage regulators are referred to as either *shunt* or *series regulators*; these classifications are determined by the circuit position of the transistor, in relation to the load.

Shunt voltage regulator

The basic *shunt regulator* circuit is shown in Fig. 129-1, where the npn transistor (Q1) is in parallel across the load. The collector is directly joined to the positive side of the regulated dc output voltage and the emitter is connected to the negative (or grounded) side through the zener diode, CR1. This diode is biased to its breakdown voltage by resistor R_2, and it therefore maintains a constant reference level at Q1's emitter. The base is connected to the wiper of the potentiometer (R_1), which is adjusted so that the base is positive with respect to the fixed emitter voltage. The emitter/base junction is therefore forward-biased, and the transistor will conduct with its collector current value determined by the R_1 setting. Because the collector current flows through the series resistor R, the potentiometer's setting also determines the level of the dc output voltage.

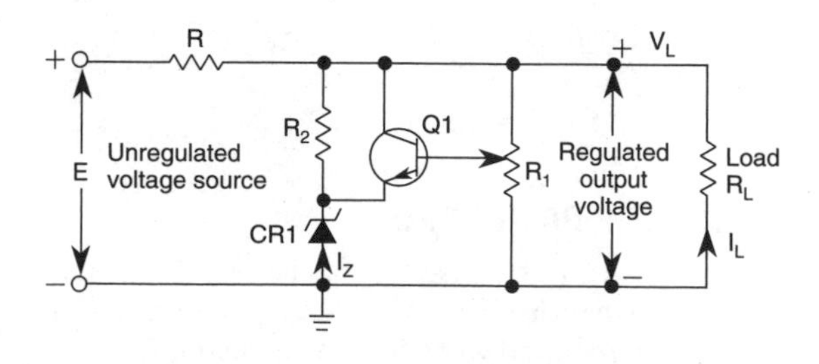

Figure 129-1

If the unregulated input voltage rises (or if the load current falls), there will be a slight increase in the dc output voltage. This will raise the forward bias that is applied to the base/emitter junction of Q1. As a result, the collector current will be greater and this increase will compensate for the original change in the values of E or I_L. Similarly, if (for any reason) the dc output voltage is tending to fall, the forward bias to Q1 is reduced, the collector current is lower, and its decrease provides the necessary compensation. Therefore, the action is similar to that of the zener diode circuit in chapter 117. However, because a small variation in forward bias can provide relatively large changes in collector current, the current range capability of the shunt-regulator circuit is equal to the zener diode's current range multiplied by the β factor of the transistor.

Under full-load conditions, the *efficiency* of the shunt regulator is high because the current through Q1 is at its minimum level. Moreover, if a short circuit occurs across the load, the shunt regulator will not be subjected to excessive current.

Series voltage regulator

The basic regulator is shown in Fig. 129-2, where the npn transistor Q1 is connected in series with the load, and also acts as a variable resistor. In conjunction with R_1, the zener diode operates at its breakdown voltage. Therefore, it maintains a constant potential at the base of Q1.

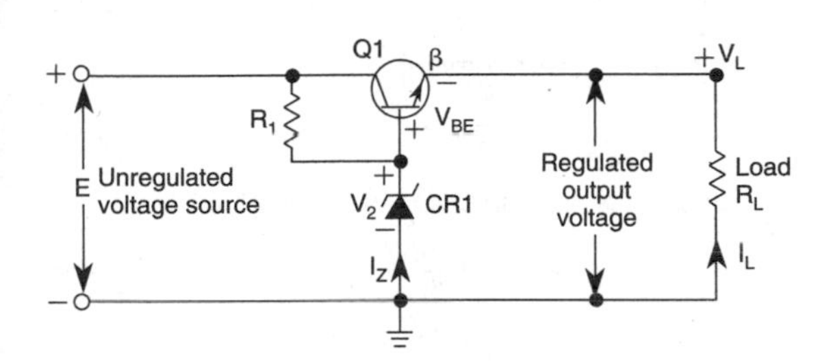

Figure 129-2

When the unregulated source voltage is applied across the series combination of Q1 and R_L, the fixed base voltage is such that the emitter/base junction is forward-biased. Therefore, the transistor conducts, and the dc output is then equal to the unregulated source voltage, less the collector/emitter drop across Q1. If the source voltage increases, the value of V_L also increases slightly, and this reduces the forward bias to Q1's emitter/base junction. As a result, a smaller emitter current is driven through R_L, and this compensates for the tendency of V_L to increase. Alternatively, you can consider that Q1's equivalent collector/emitter resistance has increased so that, by the VDR, the increased voltage drop across the transistor will compensate for the rise in the source voltage.

Now, consider the effect of an increase in the load current. This will cause a slight fall in the dc output voltage, which is the same as Q1's emitter potential. The resultant increase in the forward bias will reduce the collector/emitter resistance. It follows that the voltage drop across the transistor will be less, and this will counteract any decrease in the output voltage.

The operation of the series regulator is based on the principle of negative feedback (chapter 150). The resultant degree of *circuit stability* is the main advantage of series regulators over shunt regulators; consequently, series regulators are more widely used. However, the circuit of Fig. 129-2 has no inherent protection against excessive current; if a short circuit occurs across the output voltage, the transistor will be severely overloaded. Another disadvantage of the basic regulator circuit is its inability to respond quickly to small voltage changes. In Fig. 129-3, this problem is reduced by including another transistor (Q2), which detects and amplifies the small voltage changes so that the response time of the series regulator is reduced.

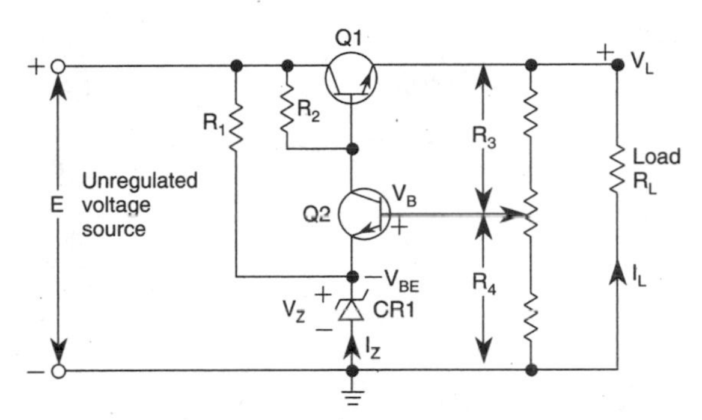

Figure 129-3

The varistor voltage regulator circuit

The *varistor* is a semiconductor device that does not obey Ohm's Law. This doped semiconductor material has both majority and minority charge carriers, but the minority current is only important at the higher voltages; these characteristics result in a nonlinear voltage-versus-current characteristic, which is illustrated in Fig. 129-4. As the voltage is increased, the varistor's equivalent resistance falls and there is a negative coefficient of resistance with respect to voltage. The varistor can then be used to regulate a voltage source, as shown in Fig. 129-5A.

If the unregulated voltage increases, the resistance of the varistor falls. The varistor current then exceeds its normal value and there will be a greater voltage drop across the series resistor, R_S. This absorbs most of the rise in the unregulated voltage and the output voltage is only slightly above its required value.

The thermistor voltage regulator circuit

The *thermistor* is another semiconductor device that is doped to provide a considerable number of minority

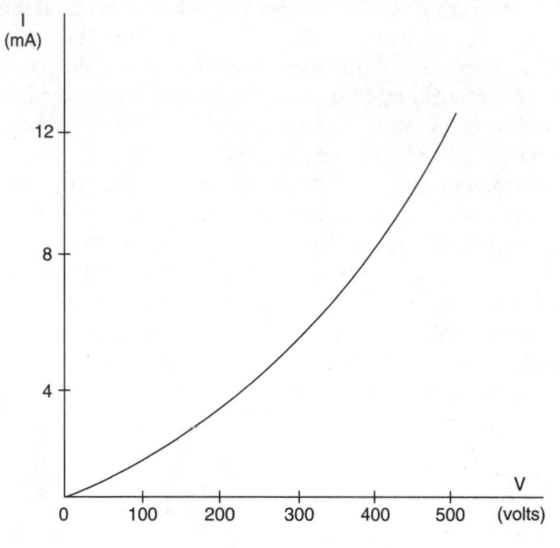

Figure 129-4

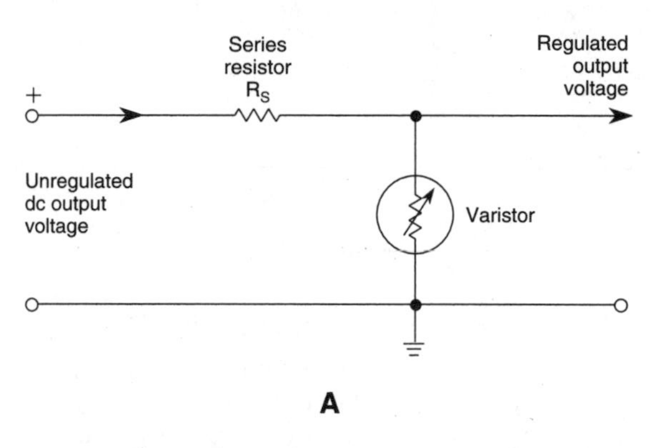

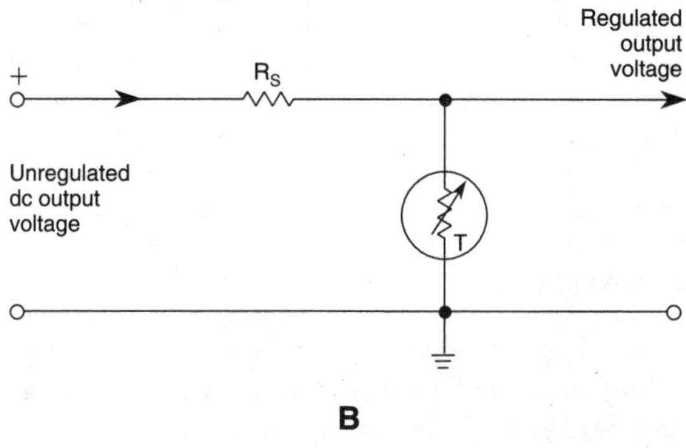

Figure 129-5

charge carriers as a result of thermal energy. It follows that the thermistor has a well-defined negative temperature coefficient of resistance over its operating range. Assume that the thermistor replaces the varistor in the voltage regulator circuit of Fig. 129-5B. When the unregulated voltage rises, there is an increase in the thermistor's power dissipation so that its temperature rises and its resistance falls. The thermistor current then exceeds its normal value and the greater voltage drop across R_S absorbs most of the rise in the unregulated voltage.

In addition to voltage regulation, thermistors are used to protect circuits against surges, and to stabilize the operation of certain transistor amplifiers.

MATHEMATICAL DERIVATIONS

In Fig. 129-2:

Voltage drop across the resistor R_1,

$$V_{R1} = E - V_Z \text{ volts} \tag{129-1}$$

Load voltage,

$$V_L = V_Z - V_{BE} \text{ volts} \tag{129-2}$$

Collector/emitter voltage,

$$V_{CE} = E - V_L \text{ volts} \tag{129-3}$$

Power dissipated in the transistor Q1,

$$P_D = I_C V_{CE} \tag{129-4}$$

$$= I_L (E - V_L) \text{ watts}$$

Change in the zener current as a result of a change in load current,

$$I_Z = -\frac{\Delta I_L}{\beta} \text{ amperes} \tag{129-5}$$

Change in the zener voltage,

$$V_Z = I_Z \times R_Z \text{ volts} \tag{129-6}$$

where: R_Z = zener diode's resistance.

In Fig. 129-3,

Voltage at the base of Q2,

$$V_B = V_Z + V_{BE} \text{ volts} \tag{129-7}$$

Load voltage,

$$V_L = V_B \times \frac{R_3 + R_4}{R_4} \text{ volts} \tag{129-8}$$

Example 129-1

In Fig. 129-2, CR1 is a zener diode with a breakdown voltage of 8 V. If $E = 15$ V, $V_{BE} = 0.7$ V, and $I_L = 500$ mA, what are the values of (a) the voltage drop across R_1, (b) V_L, (c) V_{CE} and, (d) the power dissipated in the transistor?

Solution

(a) $V_{R1} = 15 - 8 = 7$ V (129-1)

(b) $V_L = 8 - 0.7 = 7.3$ V (129-2)

(c) $V_{CE} = 15 - 7.3 = 7.7$ V (129-3)

(d) $P_D = 7.7 \text{ V} \times 500 \text{ mA}$ (129-4)

$= 3.85$ W

Example 129-2

In Example 129-1, load current I_L increases by 200 mA. If $\beta = 80$ and $R_Z = 20\ \Omega$, calculate the change in the load voltage.

Solution

Decrease in zener current,

$$I_Z = \frac{200}{80} = 2.5 \text{ mA} \quad (129\text{-}5)$$

Decrease in load voltage,

$$V_Z = 2.5 \text{ mA} \times 20\ \Omega = 50 \text{ mV} \quad (129\text{-}6)$$

PRACTICE PROBLEMS

129-1. In Fig. 129-3, $E = 60$ V, $V_Z = 10$ V, $V_{BE} = 0.7$ V, $R_3 = 820\ \Omega$, and $R_4 = 200\ \Omega$. Calculate the values of V_{R1}, V_B, and V_L.

129-2. In Practice Problem 129-1, $R_L = 100\ \Omega$. Determine the power dissipation in the transistor, Q1.

130
The junction field-effect transistor

The action of the pnp (or npn) bipolar transistor involves two types of charge carrier—the electron and the hole. The FET is a unipolar transistor because its operation requires only one charge carrier, which can be either the electron or the hole.

The two common types of FET are the *junction field-effect transistor* (*JFET*) and the *metal-oxide semiconductor field-effect transistor* (*MOSFET*); the latter is sometimes referred to as an *insulated-gate field-effect transistor* (*IGFET*).

The JFET is essentially a doped silicon bar, which is referred to as a *channel* and behaves as a resistor. The doping can either be p-type or n-type so that there are either p-channel or n-channel JFETs. At the ends of the channel are two terminals that are referred to as the source and the drain. When a particular drain/source voltage (V_{DS}) is applied between the end terminals, the amount of current flow (I_D) between the source and the drain depends on the channel's resistance. The value of this resistance is controlled by a gate that can either consist of two n-type regions diffused into a p-type channel or of two p-type regions diffused into an n-type channel. In either case, the two regions are commonly joined to provide a single gate (two separate gates can be used in some mixer circuits). Cutaway views of both types, with their schematic symbols, are shown in Fig. 130-1A and B. In these schematic symbols, the vertical line can be regarded as the channel; the arrow then points toward an n-channel, but away from a p-channel. The gate line can either be symmetrically positioned with respect to the source and the drain, or can be drawn closer to the source.

If a reverse-bias voltage is applied between the gate and the source of an n-channel JFET, depletion layers will surround the two p-regions that form the gate. If the reverse-bias is increased, the depletion layers will spread more deeply into the channel until they almost touch. The channel's resistance will then become extremely high so that the gate current is very small.

The reverse-biasing of the gate/source junction can be compared to applying a negative voltage to a triode's grid relative to its cathode. Like the tube, the FET is a voltage-controlled device in the sense that only the input voltage to the gate controls the output drain current. This is in contrast to the bipolar transistor, where the base/emitter junction is forward biased; the input voltage then controls the input current, which in turn determines the output current.

With reverse-biasing of the gate/source junction, very little gate current flows. This means that the input impedance to a JFET is on the order of several megohms; this is a definite advantage for the FET over the bipolar transistor, whose input impedance is relatively low. However, when compared with the JFET, the output current of a bipolar transistor is much more

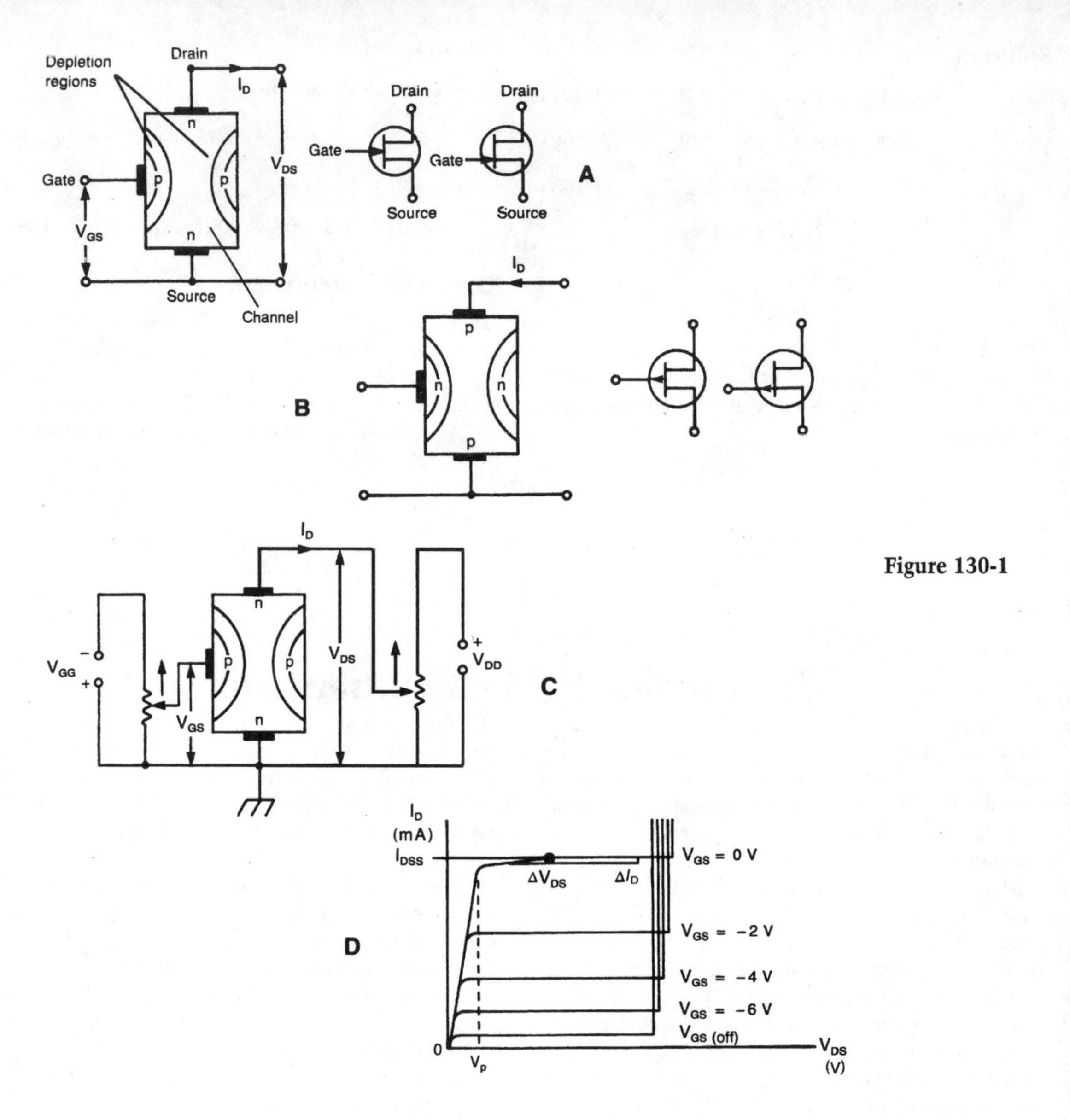

Figure 130-1

sensitive to changes in the input voltage; the result is a lower voltage gain that is available from the JFET.

In Fig. 130-1C, the voltage (V_{DS}) between the source and the drain of the n-channel JFET, is gradually increased from zero; at the same time, the voltage between gate and source (V_{GS}) is 0 V, which is referred to as the shorted gate condition. Initially, the available channel is broad so that the drain current (I_D) is directly proportional to V_{DS} and it rises rapidly as V_{DS} is increased. However, the drain voltage creates a reverse-bias on the junction between the channel and the gate. The increase in V_{DS} causes the two depletion regions to widen until, finally, they almost come into contact. This occurs when V_{DS} equals a value called the *pinch-off voltage*, V_P (Fig. 130-1D). The available channel is then very narrow so that the drain current is limited (or pinched-off). Further raising of V_{DS} above the pinch-off value will only produce a small increase in the drain current. This situation continues until the drain voltage equals $V_{DS(max)}$; at this point, an avalanche effect takes place and the JFET breaks down. Over the operating range between V_P and $V_{DS(max)}$, the approximately constant value of the drain current (with the shorted gate) is referred to as the I_{DSS} (drain-to-source current with shorted gate).

For each different negative value of V_{GS}, a different drain current curve can be obtained. This family of curves is illustrated in Fig. 130-1D. Ultimately, V_{GS} can be sufficiently negative so that the drain current is virtually cut off and equal to zero; this value of gate/source voltage is, therefore, referred to as $V_{GS(off)}$.

At the cutoff condition, the depleted layers nearly touch; this also occurred when V_{DS} was equal to V_P. Therefore, $V_{GS(off)}$ has the same value as V_P, although $V_{GS(off)}$ is a negative voltage and V_P is positive.

The *transconductance* curve is the graph of the drain current (I_D), plotted against the gate-to-source voltage (V_{GS}), while maintaining the drain-to-source voltage (V_{DS}) at a constant level. For example, in Fig. 130-2A, the drain voltage is set to 12 V and the gate is initially shorted to the source so that $V_{GS} = 0$. The recorded drain current would then equal the value of I_{DSS}. If the reverse gate voltage is now increased from zero, the drain current will fall until, ultimately, cut-off is reached when $V_{GS} = V_{GS(off)}$.

MATHEMATICAL DERIVATIONS

The shape of the transconductance curve is considered to be a parabola (Fig. 130-2B) so that there is a mathematical relationship between I_D and V_{GS}.

Drain current,

$$I_D = I_{DSS}\left(1 - \frac{V_{GS}}{V_{GS(off)}}\right)^2 \text{ milliamperes} \qquad (130\text{-}1)$$

In this equation, both V_{GS} and $V_{GS(off)}$ are considered to be negative voltages.

In a FET amplifier, the control that the gate voltage exercises over the drain current is measured by the transconductance, g_m. At a particular point (P) on the curve, the transconductance is defined by:

$$g_m = \frac{\Delta I_D}{\Delta V_{GS}} \text{ milliamperes per volt} \qquad (130\text{-}2)$$

and is normally measured in microsiemens (μS). By differentiating the expression for I_D, with respect to V_{GS},

Transconductance,

$$g_m = g_{m0}\left(1 - \frac{V_{GS}}{V_{GS(off)}}\right) \text{ microsiemens} \qquad (130\text{-}3)$$

where: g_{m0} = transconductance for the shorted gate condition.

The value of g_{m0} is $-2I_{DSS}/V_{GS(off)}$; because $V_{GS(off)}$ is a negative voltage, g_{m0} is a positive quantity.

Example 130-1

A junction field-effect transistor has a pinch-off voltage equal to 4 V. If $I_{DSS} = 12$ mA, find the values of I_D and g_m when $V_{GS} = -2$ V.

Solution

$$V_{GS(off)} = -V_P = -4 \text{ V.}$$

Drain current,

$$I_D = I_{DSS}\left(1 - \frac{V_{GS}}{V_{GS(off)}}\right)^2 \qquad (130\text{-}1)$$

$$= 12 \times \left(1 - \frac{(-2)}{(-4)}\right)^2$$

$$= 3 \text{ mA.}$$

$$g_{m0} = -\frac{2I_{DSS}}{V_{GS(off)}} = \frac{-2 \times 12}{-4} = 6 \text{ mA/V}$$

$$= 6000\ \mu\text{S.}$$

Transconductance,

$$g_m = g_{m0}\left[1 - \left(\frac{V_{GS}}{V_{GS(off)}}\right)\right] \qquad (130\text{-}3)$$

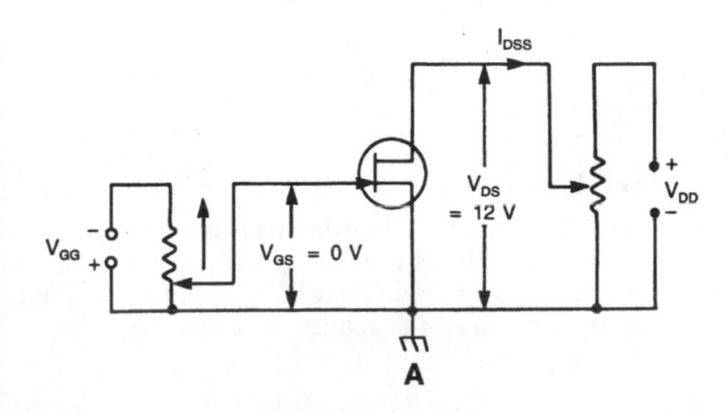

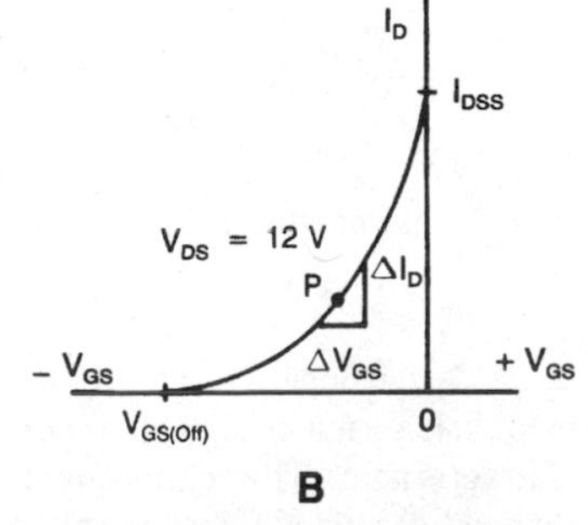

Figure 130-2

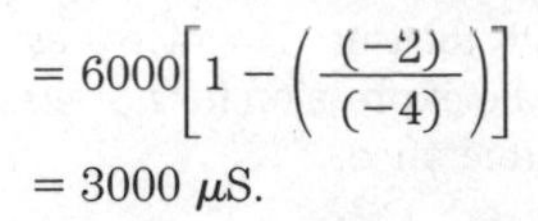

$$= 6000\left[1 - \left(\frac{(-2)}{(-4)}\right)\right]$$

$$= 3000\ \mu\text{S}.$$

PRACTICE PROBLEMS

130-1. Figure 130-3 illustrates the transconductance curve of a JFET. What are the values of I_{DSS}, V_P, and g_{m0}?

130-2. Obtain the equation for the transconductance curve of Fig. 130-3. What are the values of I_D and g_m for (a) $V_{GS} = -2$ V and, (b) $V_{GS} = -4$ V?

130-3. A JFET has a drain current of 6 mA. If $I_{DSS} = 12$ mA and $V_{GS} = -5$ V, what is the value of V_{GS}?

130-4. Figure 130-4 illustrates the drain characteristics of a JFET. What are the values of I_{DSS}, V_P, and g_{m0}? Over what voltage range should V_{DS} be confined, in order to ensure operation on the flat portion of the $V_{GS} = 0$ V curve?

130-5. Obtain the equation for the transconductance curve of the JFET in Practice Problem 130-4. What are the values of I_D and g_m for (a) $V_{GS} = -2$ V and, (b) $V_{GS} = -4$ V?

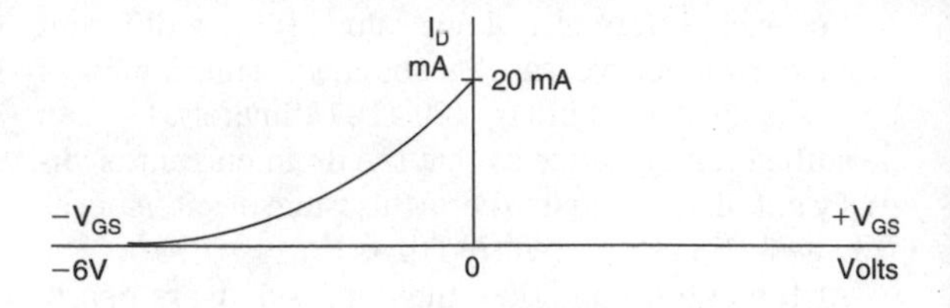

Figure 130-3

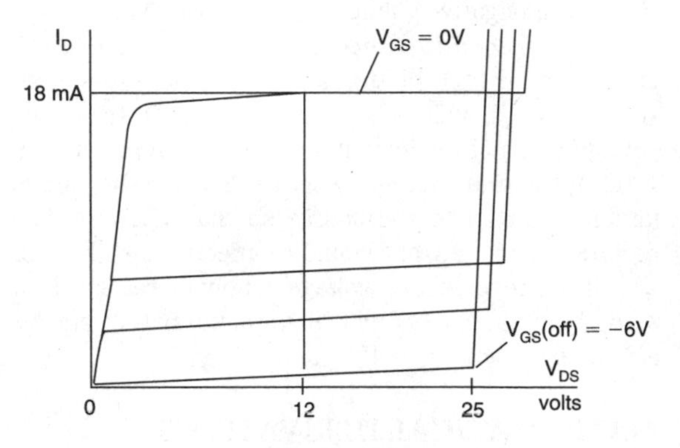

Figure 130-4

131
Methods of biasing the JFET

The biasing methods that follow are applied to n-channel FETs, although p-channel FETs can be similarly biased by reversing the polarities of the dc supply voltages.

MATHEMATICAL DERIVATIONS

The gate bias

This method is similar to the base bias of the bipolar transistors. In Fig. 131-1A, the drain current is given by:

$$I_D = \frac{V_{GG} - V_{GS}}{R_S} \qquad (131\text{-}1)$$

$$\approx \frac{V_{GG}}{R_S} \text{ milliamperes} \qquad (131\text{-}2)$$

if $V_{GG} >> V_{GS}$.

The equation $I_D \approx V_{GG}/R_S$ means that V_{GG} and R_S can be chosen to establish a value of I_D that is independent of the JFET characteristics. The requirement for a separate V_{GG} supply can be avoided by using voltage divider bias (Fig. 131-1B). By comparing Figs. 131-1A and B and using Thévenin's Theorem,

Equivalent gate supply voltage,

$$V_{GG} = \frac{V_{DD} R_2}{R_1 + R_2} \text{ volts} \qquad (131\text{-}3)$$

Equivalent gate resistance,

$$R_G = \frac{R_1 R_2}{R_1 + R_2} \text{ ohms} \qquad (131\text{-}4)$$

Self-bias

Self-bias (Fig. 131-1C) can be compared with the cathode bias used with tubes. The gate voltage is zero, but the gate is biased negative with respect to the source by the amount of the voltage drop across R_S. Therefore, gate-source voltage,

$$V_{GS} = I_D R_S \text{ volts} \qquad (131\text{-}5)$$

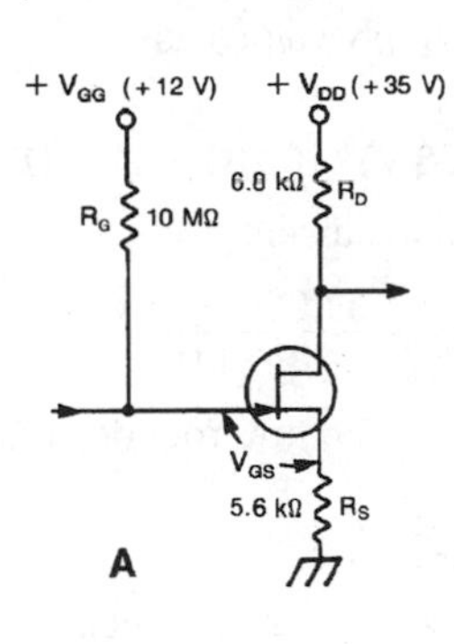

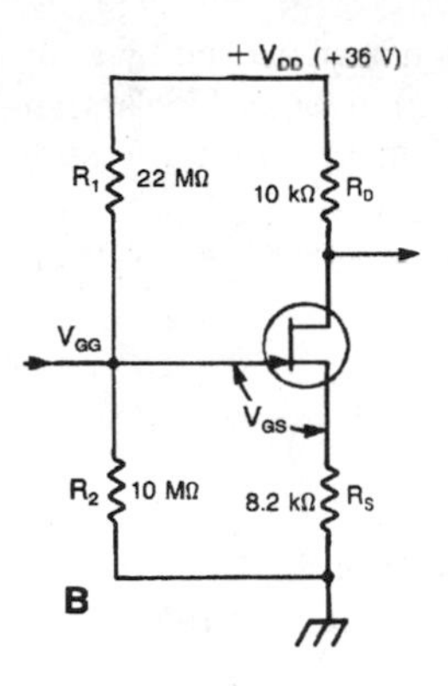

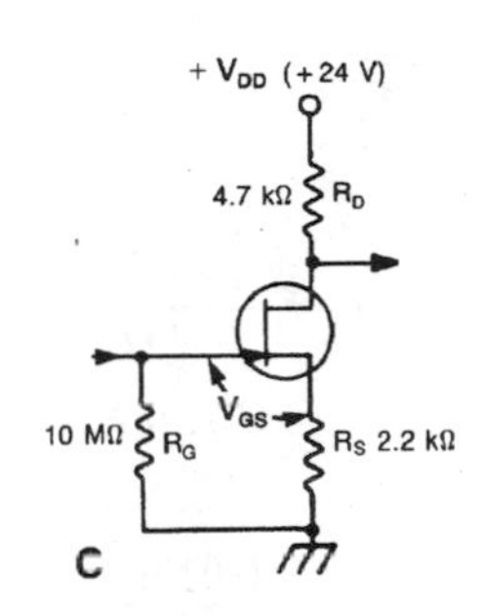

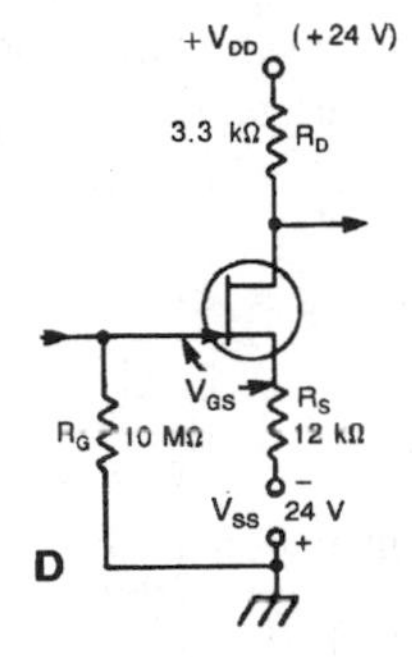

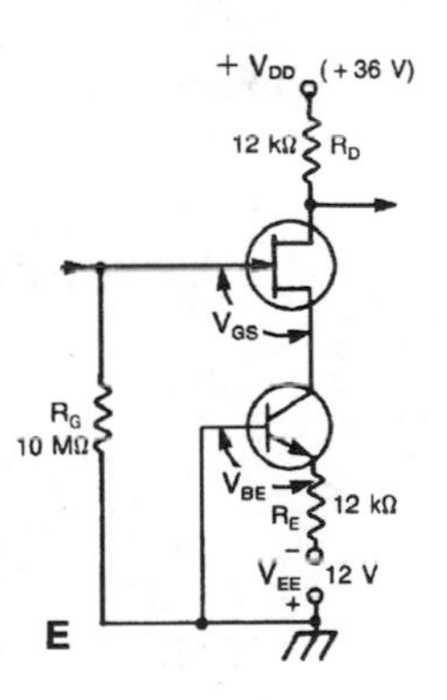

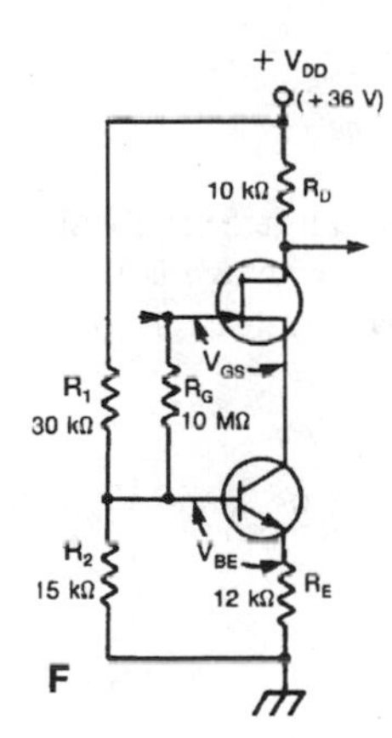

Figure 131-1

Self-bias, although a simple arrangement, does not swamp out V_{GS}. Therefore, the bias point depends on the characteristics of the JFET; this is its principal disadvantage.

Source bias

Source bias (Fig. 131-1D) is comparable with the emitter bias of a bipolar transistor. Once again the purpose is to swamp out the value of V_{GS} and to achieve a drain current that is virtually independent of the JFET characteristics. Because the gate is at ground potential, the equation for the drain current is:

$$I_D = \frac{V_{SS} - V_{GS}}{R_S} \tag{131-6}$$

$$\approx \frac{V_{SS}}{I_S} \text{ milliamperes}$$

if $V_{SS} \gg V_{GS}$.

It is easy to swamp out V_{GS} if a large negative voltage is available. The disadvantage of the source bias is obvious in that it requires two separate supplies.

If a large negative voltage is not available, the problem can be solved by replacing R_S with a bipolar transistor in the current-source bias circuit of Fig. 131-1E.

When the base/emitter junction is controlled by a forward-bias action, the emitter current of the bipolar transistor will equal the drain current of the FET so that:

$$I_D = I_E = \frac{V_{EE} - V_{BE}}{R_E} \qquad (131\text{-}7)$$

$$\approx \frac{V_{EE}}{R_E} \text{ milliamperes}$$

if $V_{EE} >> V_{BE}$.

Because $V_{BE} = 0.7$ V, and only varies by 0.1 V from one transistor to another, a high drain current can be fixed by using a low voltage source for V_{EE}. If the low negative voltage is not available, the bipolar transistor can be forward-biased by the voltage divider circuit of Fig. 131-1F. Then, by Thévenin's Theorem,

Drain current,

$$I_D = I_E = \frac{\dfrac{V_{DD} \times R_2}{R_1 + R_2} - V_{BE}}{R_E} \qquad (131\text{-}8)$$

$$\approx \frac{V_{DD} \times R_2}{(R_1 + R_2)R_E} \text{ milliamperes}$$

provided,

$$\frac{V_{DD} \times R_2}{R_1 + R_2} \gg V_{BE}$$

Current-source bias has the advantage of providing the best swamping action; its disadvantage is the requirement for the additional bipolar transistor.

Example 131-1

In Fig. 131-1A, calculate the values of I_D and V_{DS} (ignore V_{GS}).

Solution

Drain current,

$$I_D = \frac{V_{GG}}{R_S} = \frac{12 \text{ V}}{5.6 \text{ k}\Omega} \qquad (131\text{-}2)$$

$$= 2.14 \text{ mA}.$$

Drain-source voltage,

$$V_{DS} = 35 \text{ V} - (6.8 \text{ k}\Omega + 5.6 \text{ k}\Omega) \times 2.14 \text{ mA}$$

$$= 8.5 \text{ V, rounded off.}$$

Example 131-2

In Fig. 131-1B, calculate the value of I_D if $V_{GS} = -2$ V.

Solution

Effective gate supply voltage is

$$(10 \text{ M}\Omega \times 36 \text{ V})/(10 \text{ M}\Omega + 22 \text{ M}\Omega) = 11.25 \text{ V}. \qquad (131\text{-}3)$$

Then, the drain current,

$$I_D = \frac{11.25 \text{ V} + 2 \text{ V}}{8.2 \text{ k}\Omega} \qquad (131\text{-}4)$$

$$= 1.6 \text{ mA, rounded off.}$$

Example 131-3

In Fig. 131-1C, $V_{GS} = -2$ V. Calculate the values of I_D and V_D.

Solution

Drain current,

$$I_D = \frac{2 \text{ V}}{2.2 \text{ k}\Omega} \qquad (131\text{-}5)$$

$$= 0.91 \text{ mA}$$

Drain voltage,

$$V_D = +24 \text{ V} - 0.91 \text{ mA} \times 4.7 \text{ k}\Omega$$

$$= +24 \text{ V} - 4.277 \text{ V}$$

$$= +19.7 \text{ V, rounded off.}$$

Example 131-4

In Fig. 131-1D, $V_{GS} = -2$ V. Calculate the values of I_D and V_{DS}.

Solution

Drain current,

$$I_D = \frac{24 \text{ V} + 2 \text{ V}}{12 \text{ k}\Omega} \qquad (131\text{-}6)$$

$$= 2.17 \text{ mA}.$$

Drain voltage,

$$V_D = +24 \text{ V} - 2.17 \text{ mA} \times 3.3 \text{ k}\Omega$$

$$= +24 \text{ V} - 7.16 \text{ V}$$

$$= +16.8 \text{ V, rounded off.}$$

Source voltage,

$$V_S = +2 \text{ V}.$$

Drain-to-source voltage,

$$V_{DS} = +16.8 \text{ V} - (+2 \text{ V})$$

$$= 14.8 \text{ V}$$

Example 131-5

In Fig. 131-1E, calculate the values of I_D and V_{DS} (assume that $V_{GS} = -2$ V).

Solution

Assuming that the V_{BE} of the bipolar transistor is 0.7 V,

Drain current,

$$I_D = I_E = \frac{12\text{ V} - 0.7\text{ V}}{12\text{ k}\Omega} \qquad (131\text{-}7)$$

$$= 0.94\text{ mA}$$

Then, $V_E = -0.7$ V.

If $V_{GS} = 2$ V, $V_S = V_C = +2$ V, and $V_{CE} = 2.7$ V.

Drain potential, $V_D = +36\text{ V} - 0.94\text{ mA} \times 12\text{ k}\Omega = +24.7$ V.

Drain-to-source voltage, $V_{DS} = 24.7 - 2.0 = 22.7$ V.

Example 131-6

In Fig. 131-1F, calculate the value of V_{DS} (assume $V_{GS} = -2$ V).

Solution

Because of the voltage divider action, the potential at the point A is,

$$+36\text{ V} \times \frac{15\text{ k}\Omega}{15\text{ k}\Omega + 30\text{ k}\Omega}$$

$$= +12\text{ V}$$

which is equal to V_G. Because $V_{GS} = -2$ V, $V_S = +14$ V. The emitter current,

$$I_E = I_D = \frac{12\text{ V} - 0.7\text{ V}}{12\text{ k}\Omega} \qquad (131\text{-}8)$$

$$= 0.94\text{ mA}$$

The drain potential is $+36\text{ V} - 10\text{ k}\Omega \times 0.94\text{ mA} = +26.6$ V and $V_{DS} = 26.6 - 14 = 12.6$ V.

PRACTICE PROBLEMS

131-1. In Fig. 131-1A, the values are changed to $V_{GG} = +10$ V, $V_{DD} = +38$ V, $R_G = 12$ MΩ, $R_D = 5.6$ kΩ, and $R_S = 4.7$ kΩ. If $V_{GS} = -12$ V, recalculate the values of I_D and V_{DS}.

131-2. In Fig. 131-1B, the values are changed to $V_{DD} = +30$ V, $R_1 = 15$ MΩ, $R_2 = 3.3$ MΩ, $R_D = 6.8$ kΩ, and $R_S = 4.7$ kΩ. If $V_{GS} = -2$ V, recalculate the values of I_D and V_{DS}.

131-3 If Fig. 131-1C, the values are changed to $V_{DD} = +30$ V, $R_D = 5.6$ kΩ, $R_S = 1.5$ kΩ, and $R_G = 12$ MΩ. If $V_{GS} = -2$ V, recalculate the values of I_D and V_{DS}.

131-4. In Fig. 131-1D, the values are changed to $V_{DD} = +30$ V, $V_{SS} = -30$ V, $R_D = 4.7$ kΩ, $R_G = 12$ MΩ, and $R_S = 15$ kΩ. If $V_{GS} = -2$ V, recalculate the values of I_D and V_{DS}.

131-5. In Fig. 131-1E, the values are changed to $V_{DD} = +30$ V, $V_{EE} = -10$ V, $R_D = 10$ kΩ, $R_G = 12$ MΩ, and $R_E = 10$ kΩ. If $V_{GS} = -2$ V, recalculate the values of I_D and V_{DS}.

131-6. In Fig. 131-1F, the values are changed to $V_{DD} = +30$ V, $R_D = 8.2$ kΩ, $R_1 = 33$ kΩ, $R_2 = 18$ kΩ, $R_G = 12$ MΩ, and $R_E = 10$ kΩ. If $V_{GS} = -2$ V, recalculate the values of I_D and V_{DS}.

132
The JFET amplifier and the source follower

The principles of the JFET amplifier are shown in Fig. 132-1A (lowercase letters are used to indicate signal levels). The signal (v_g) to be amplified is applied between the gate and the source, and produces variations (I_D) in the drain current. The resultant voltage variations across the drain load (r_d) produce voltage variations of opposite polarity between gate and ground. This output signal (v_d) is 180° out of phase with the input signal. The same phase inversion occurs with the common emitter arrangement of a bipolar transistor amplifier and the grounded cathode tube stage.

MATHEMATICAL DERIVATIONS

Assuming that r_s is bypassed by a capacitor, to avoid negative or degenerative feedback, the JFET behaves as a current source of value $g_m v_g$ in parallel with a very high resistance (r_{ds}), which is the reciprocal of the drain characteristics' slope (Fig. 132-1B). Normally, r_{ds} exceeds 100 kΩ and can be neglected in the analysis. Then:

Output signal,

$$v_d = g_m v_g \times r_d \text{ volts} \qquad (132\text{-}1)$$

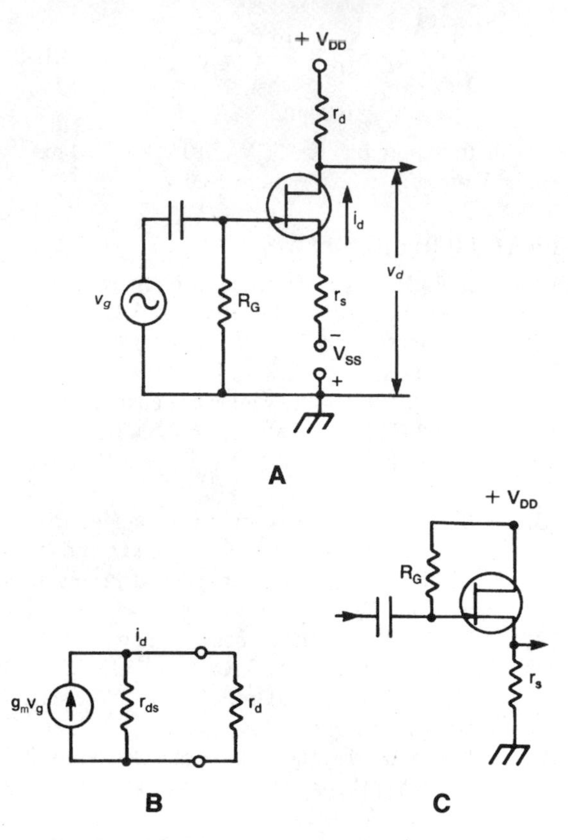

Figure 132-1

and,

$$\textit{Voltage gain} = \frac{v_d}{v_g} = g_m r_d \qquad (132\text{-}2)$$

If the source resistance (r_s) is not bypassed, negative feedback is introduced into the circuit and the voltage gain is reduced to:

$$\textit{Voltage gain} = \frac{r_d}{r_s + \dfrac{1}{g_m}} \qquad (132\text{-}3)$$

Provided $r_s >> 1/g_m$,

$$\textit{Voltage gain} = \frac{r_d}{r_s} \qquad (132\text{-}4)$$

In the source follower circuit of Fig. 132-1C, $r_d = r_s$ so that:

$$\textit{Voltage gain} = \frac{v_s}{v_g} = \frac{r_s}{r_s + \dfrac{1}{g_m}} \to 1 \qquad (132\text{-}5)$$

if $r_s >> 1/g_m$.

The low voltage gain is due to the negative feedback developed across r_s. However, this also results in the source follower having an extremely high input impedance, but a low output impedance. The same properties are possessed by the cathode follower and common collector circuits.

Example 132-1

In Fig. 132-2, the g_m of the JFET is 6000 μS. Calculate the voltage gain from gate to drain (neglect the value of r_{ds}). If the capacitor (C) is removed, what is the new value of the voltage gain?

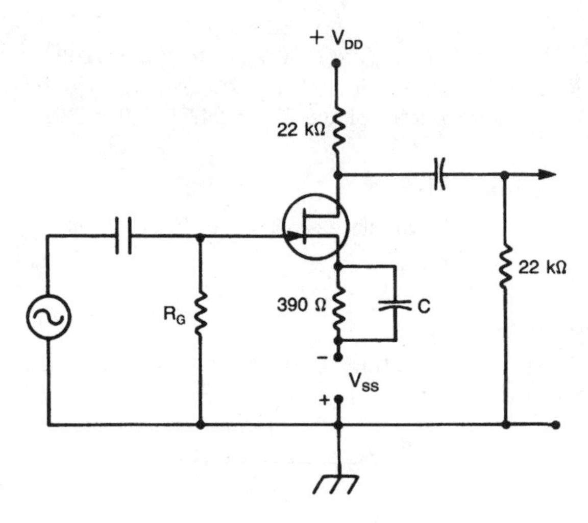

Figure 132-2

Solution

Total drain load, $r_d = 22\ \text{k}\Omega \parallel 22\ \text{k}\Omega = 11\ \text{k}\Omega$.

$$\begin{aligned}\textit{Voltage gain} &= g_m r_d \qquad (132\text{-}2)\\ &= 6000\ \mu\text{S} \times 11\ \text{k}\Omega\\ &= 66.\end{aligned}$$

When C is removed,

$$\begin{aligned}\textit{New voltage gain} &= \frac{r_d}{r_s + \dfrac{1}{g_m}} \qquad (132\text{-}3)\\ &= \frac{11000}{390 + 167}\\ &= 20, \text{ rounded off.}\end{aligned}$$

Example 132-2

In Fig. 132-3, calculate the value of the voltage gain from gate to source.

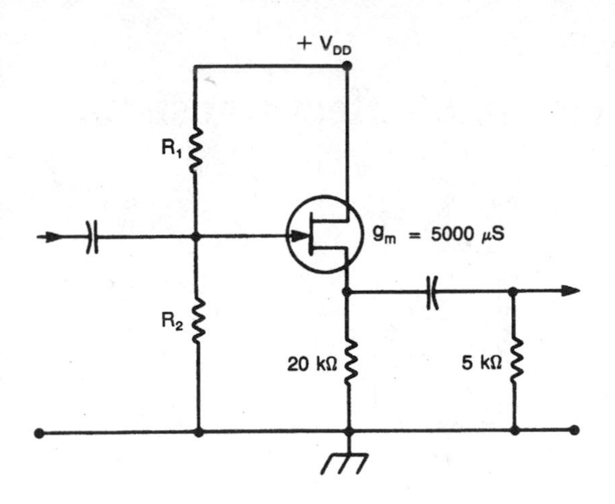

Figure 132-3

Solution

Total source load, $r_s = 20\ \text{k}\Omega \parallel 5\ \text{k}\Omega = 4\ \text{k}\Omega$.

$$\textit{Voltage gain} = \frac{r_s}{r_s + \dfrac{1}{g_m}} \qquad (132\text{-}5)$$

$$= \frac{4000}{4000 + 200}$$

$$= 0.95.$$

PRACTICE PROBLEMS

132-1. In the circuit of Fig. 132-4, the transconductance of the JFET is 2500 μS. What is the value of the voltage gain from gate to drain?

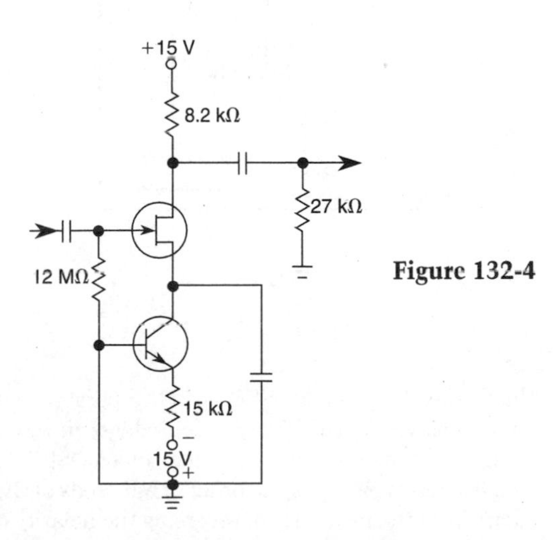

Figure 132-4

132-2. In the circuit of Fig. 132-5, the g_{m0} of the JFET is 4000 μS while the I_{DSS} is 2 mA. Calculate the value of the voltage gain from gate to drain.

132-3. The JFET in Fig. 132-6 has a g_{m0} of 4000 μS and its I_{DSS} is 8 mA. What is the value of the transconductance at the dc operating point? Neglect the value of V_{GS}.

132-4. In Practice Problem 132-3, what is the value of the voltage gain from gate to source?

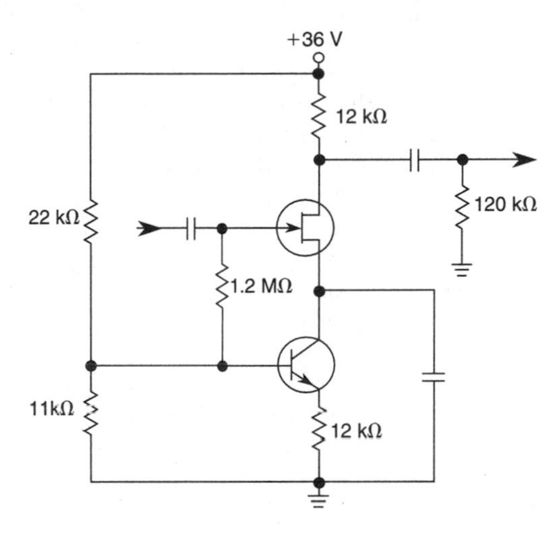

Figure 132-5

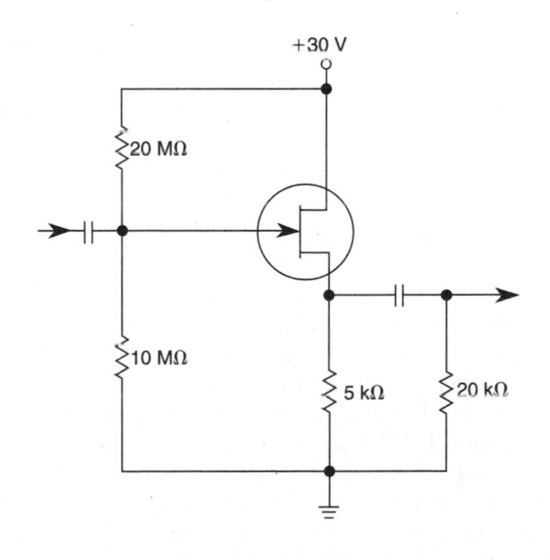

Figure 132-6

133
The metal-oxide semiconductor field-effect transistor

To avoid excessive gate current, it is necessary to apply a negative bias to the gate of a JFET. However, with a MOSFET the gate current is still virtually zero even if the gate is positive with respect to the source. This is made possible by the MOSFET's construction.

In Fig. 133-1A, the n-channel MOSFET has only one p-region, which is called the substrate. The substrate can have its own terminal (four-terminal device) or it can be internally connected to the source (three-terminal device); in this chapter it is assumed that the source and the substrate are either internally or externally connected. This is illustrated in the schematic symbol of Fig. 133-1B. As with the JFET, the arrow in the symbol points toward the n-channel. The action of the substrate is to reduce the width of the channel through which the electrons pass from source to drain. On the other side of the narrow channel, a thin layer of silicon dioxide (a metal oxide) is deposited and acts as an insulator. A metallic gate is then spread over the opposite surface of the silicon dioxide layer. Because the gate is now insulated from the channel, the device is sometimes referred to as an *insulated-gate FET* (*IGFET*).

In the operation of the n-channel MOSFET, the drain supply voltage (V_{DD}) causes electrons to flow from source to drain through the channel, and the gate voltage controls the channel's resistance. If a negative voltage is applied to the gate with respect to the source (Fig. 133-1C), electrostatic induction will cause positive charges to appear in the channel. These charges will be in the form of positive ions that have been created by the repulsion of conduction-band electrons away from the gate; in other words, the number of conduction-band electrons existing in the n-channel has been reduced or depleted. If the gate is made increasingly negative, there will be fewer and fewer conduction-band electrons available until, ultimately, the MOSFET is cut off. This action is very similar to that of the JFET; because a negative gate causes a depletion of conduction electrons, this manner of operating a MOSFET is called the *depletion mode*.

Because the channel and the gate are insulated from each other, it is possible to apply a positive voltage to the MOSFET gate (Fig. 133-1D). The result will be to induce negative charges into the n-channel; these will be in the form of additional conduction-band electrons that are drawn into the channel by the action of the positive gate. The total number of conduction-band electrons has, therefore, been increased or enhanced; consequently, this manner of operation for the MOSFET is called the *enhancement mode*.

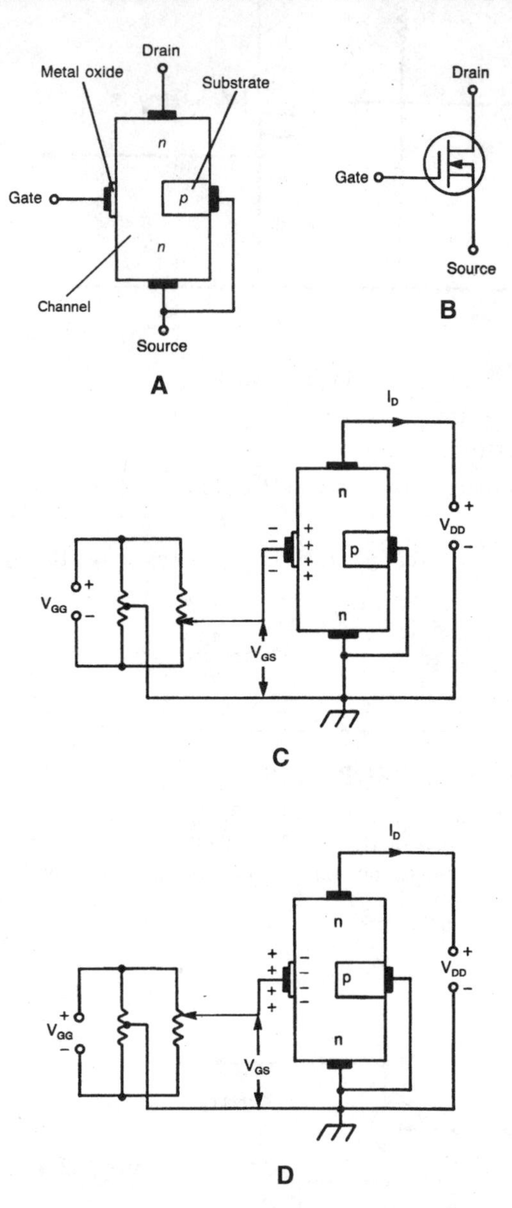

Figure 133-1

Unlike the JFET, the MOSFET can be operated with either a positive or a negative gate voltage; in either mode of operation, the input resistance of a MOSFET is very high and is typically of the order of hundreds of GΩ.

Apart from the necessity of reversing the polarity of the drain and gate supply voltages, the operation of p-channel and n-channel MOSFETs is identical.

MATHEMATICAL DERIVATIONS

The appearance of the MOSFET's characteristic curves is similar to those of the JFET, and it is illustrated in Figs. 133-2A and B. The only difference is the extension of the gate voltage into the positive region of the enhancement mode. The transconductance curve is still a parabola with its equation:

Drain current,

$$I_D = I_{DSS}\left(1 - \frac{V_{GS}}{V_{GS(off)}}\right)^2 \text{ milliamperes} \qquad (133\text{-}1)$$

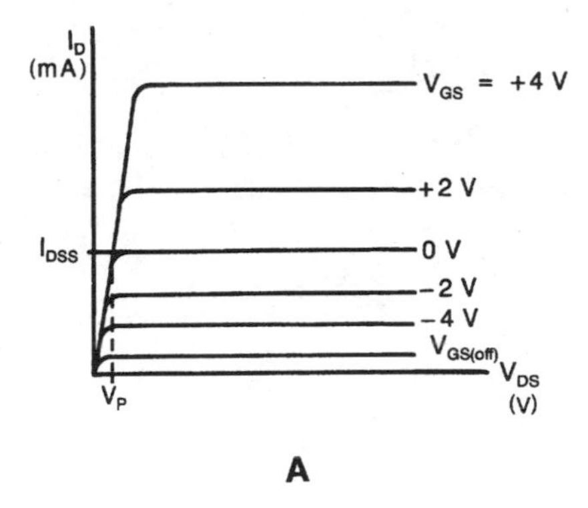

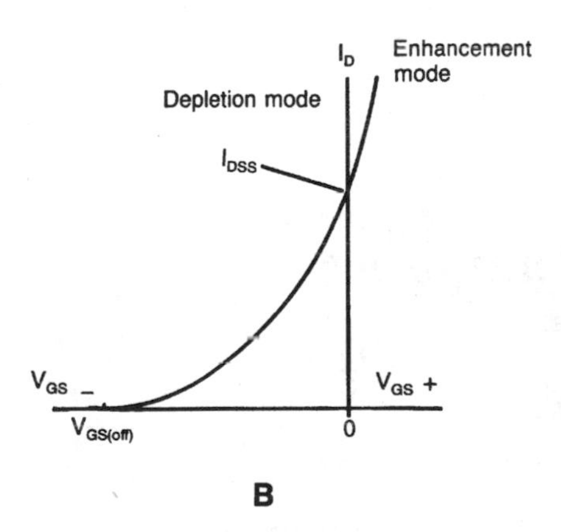

Figure 133-2

In this equation, $V_{GS(off)}$ is always a negative voltage, but V_{GS} may either be positive or negative.

The type of MOSFET, which may operate either in the depletion or the enhancement mode, conducts when $V_{GS} = 0$ with a drain current value equal to I_{DSS}. For this reason the device is called a normally on or *depletion-enhancement* (DE) MOSFET. Such a MOSFET can be operated with zero bias because the dc operating point can be chosen at $V_{GS} = 0$. When the signal is applied to the gate, the operation of the MOSFET will swing back and forth between the depletion and the enhancement modes.

Example 133-1

A DE MOSFET has values of $I_{DSS} = 10$ mA and $V_{GS(off)} = -4$ V. What are the values of I_D and g_{mO} at (a) $V_{GS} = -2$ V, and (b) $V_{GS} = +1$ V?

Solution

(a) $V_{GS} = -2$ V.

Drain current,

$$I_D = 10\left(1 - \frac{(-2)}{(-4)}\right)^2 = 2.5 \text{ mA} \qquad (133\text{-}2)$$

Transconductance when $V_{GS} = 0$ V,

$$g_{mO} = \frac{-2\,I_{DSS}}{V_{GS(off)}} = \frac{-2 \times 10}{-4} \qquad (133\text{-}3)$$

$$= 5 \text{ mS} = 5000\ \mu\text{S}.$$

Transconductance,

$$g_m = 5000\left(1 - \frac{(-2)}{(-4)}\right) \qquad (133\text{-}4)$$

$$= 2500\ \mu\text{S}.$$

(b) $V_{GS} = +1$ V.

Drain current,

$$I_D = 10\left(1 - \frac{(+1)}{(-4)}\right)^2 \qquad (133\text{-}5)$$

$$= 10 \times \left(\frac{5}{4}\right)^2$$

$$= 15.6 \text{ mA}.$$

Transconductance,

$$g_m = 5000\left(1 - \frac{(+1)}{(-4)}\right) \qquad (133\text{-}6)$$

$$= 6250\ \mu\text{S}.$$

PRACTICE PROBLEMS

133-1. Figure 133-3 illustrates the transconductance curve of a MOSFET. What are the values of (a) I_{DSS} and V_P, and (b) I_D and g_m, when $V_{GS} = -2$ V?

133-2. The circuit of Fig. 133-4 contains a MOSFET whose transconductance curve is illustrated in Fig. 133-3. Calculate the value of V_{DS}.

133-3. Figure 133-5 shows the drain characteristics of a MOSFET. Determine the values of (a) V_P and I_{DSS}, and (b) g_{m0} and g_m, when $V_{GS} = +2$ V.

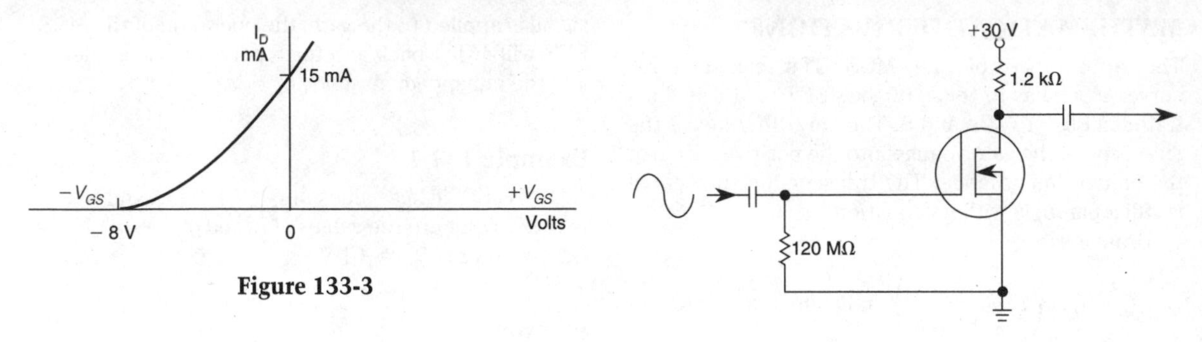

Figure 133-3

Figure 133-4

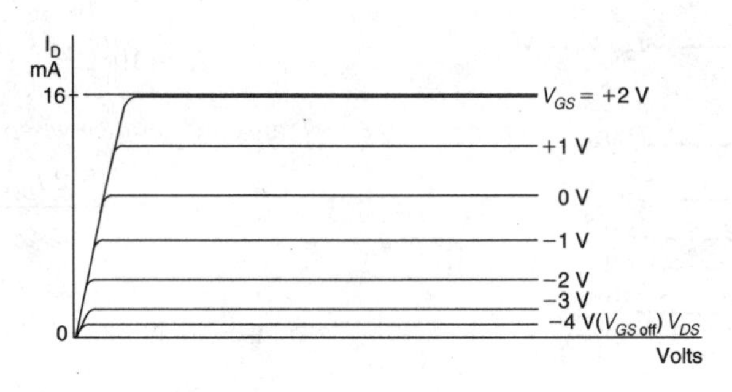

Figure 133-5

134 The DE MOSFET amplifier

The gate of the DE MOSFET can be biased either positively or negatively with respect to the source. Working in either the depletion mode or the enhancement mode can be provided by both voltage divider and source bias. However, self-bias and current-source bias can only be used to operate in the depletion mode.

Like the JFET amplifier, the DE MOSFET behaves like a constant-current source with a value of $g_m v_g$ in parallel with the very high value of r_{ds}. The equations for the signal output and the voltage gain are, therefore, the same for both the JFET and the MOSFET amplifiers (chapter 132).

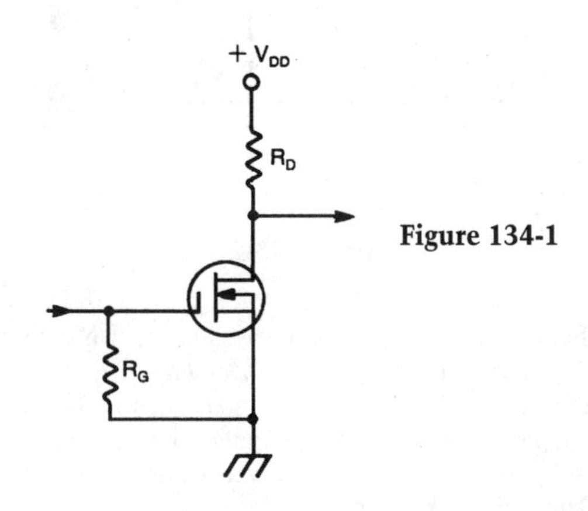

Figure 134-1

MATHEMATICAL DERIVATIONS

The DE MOSFET can be operated with zero bias, which is achieved in the circuit of Fig. 134-1. It follows that:

Drain-source voltage,

$$V_{DS} = V_{DD} - I_{DSS} R_D \text{ volts} \qquad (134\text{-}1)$$

Provided $V_{DS} > V_p$, the DE MOSFET will automatically operate on the nearly flat section of the $V_{GS} = 0$ V drain curve. This signal can also be directly coupled to the gate without the necessity for a coupling capacitor; this will allow a flat response for the amplifier at low frequencies.

Example 134-1

A DE MOSFET has values of I_{DSS} = 10 mA, $V_{GS(off)}$ = −4 V, and is used in the circuit of Fig. 134-1. If R_D = 1.5 kΩ, calculate the values of V_{DS} and the amplifier's voltage gain (neglect any effect of r_{ds}).

Solution

The circuit is operating with zero bias and, therefore, the drain current, $I_D = I_{DSS}$ = 10 mA. It follows that,

$$V_{DS} = V_{DD} - I_{DSS} R_D \tag{134-1}$$

$$= 24 \text{ V} - 10 \text{ mA} \times 1.5 \text{ k}\Omega$$

$$= 9 \text{ V}.$$

The value of V_{DS} exceeds the pinch-off voltage, V_P = 4 V; therefore, the DE MOSFET will operate on the flat section of the drain curve for V_{GS} = 0 V.

From chapter 130,

$$g_{m0} = \frac{-2\, I_{DSS}}{V_{GS(off)}}$$

$$= \frac{-2 \times 10}{-4}$$

$$= 5 \text{ mS} = 5000\ \mu\text{S}.$$

Voltage gain,

$$G_V = g_{m0} \times r_d$$

$$= 5000\ \mu\text{S} \times 1.5 \text{ k}\Omega$$

$$= 7.5.$$

PRACTICE PROBLEMS

134-1. In the circuit of Fig. 134-2, calculate the value of the voltage gain from gate to drain.

134-2. In Practice Problem 134-1, the input signal has a peak-to-peak value of 0.2 V. Calculate the rms value of the signal output.

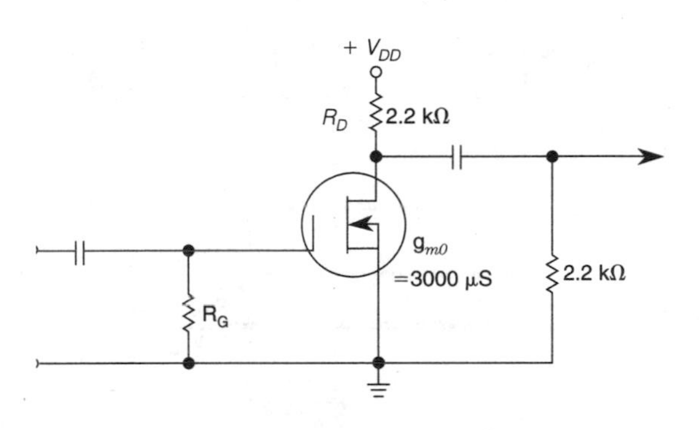

Figure 134-2

135 The enhancement-only or E MOSFET

In this type of MOSFET, there is no longer a continuous n-channel between the drain and the source. The substrate stretches all the way across to the metal oxide layer (Fig. 135-1A) so that no current can flow if the MOSFET is in the depletion mode. When V_{GS} = 0 V, there will be a small number of conduction-band electrons created in the substrate by thermal energy, but the current flow because of the drain supply is still extremely small. Consequently, the E MOSFET is also referred to as a *normally off MOSFET*. This "normally off" condition is shown in the schematic symbol (Fig. 135-1B) by the broken line which represents the channel. Because Fig. 135-1A shows an n-channel E MOSFET, the arrow in the schematic symbol points toward the channel; with a p-type MOSFET, the arrow would point away from the channel.

To produce an appreciable flow of drain current, it is necessary to apply a positive voltage to the gate. If this voltage is low, the charges induced in the substrate are negative ions, which are created by filling holes in the p-substrate with valence electrons. When the positive gate voltage is increased above a certain *minimum* threshold level, ($V_{GS} > V_{GS(th)}$), the additional induced charges are conduction band electrons that exist in a thin n-type inversion layer next to the metal oxide and allow an appreciable flow of electrons from source to drain.

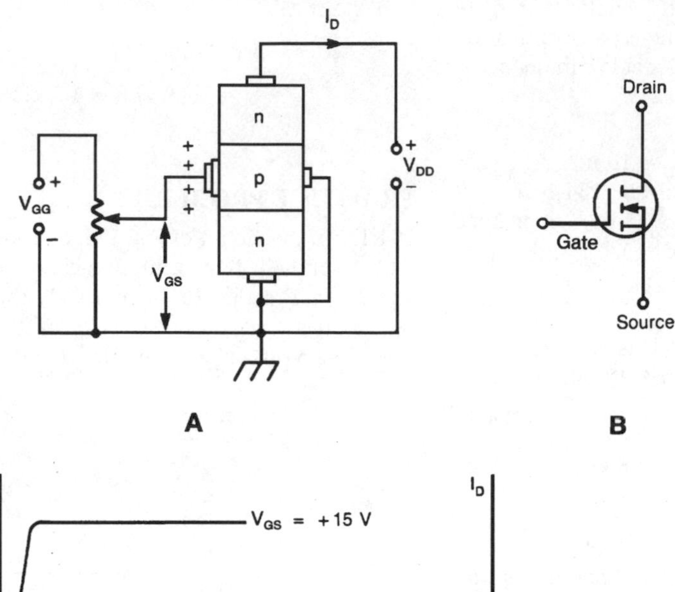

Figure 135-1

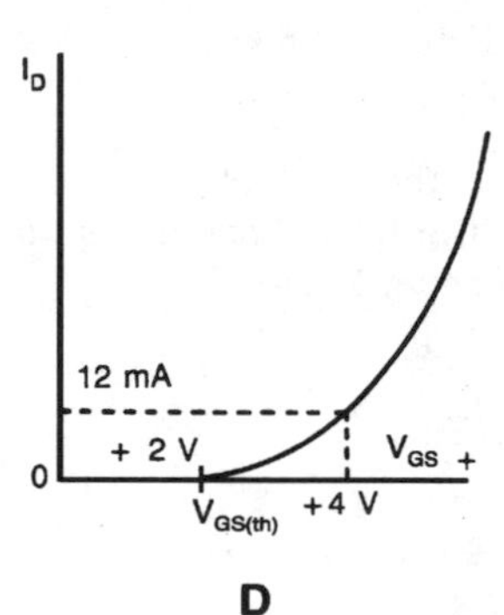

MATHEMATICAL DERIVATIONS

In Fig. 135-1C, each drain curve represents a fixed positive value of V_{GS}. For the lowest curve, $V_{GS} = +V_{GS(\mathrm{th})}$ and the E MOSFET is virtually cut off.

The transconductance curve is again parabolic in shape, with its vertex at $V_{GS(\mathrm{th})}$ (Fig. 135-1D). The curve's equation is:

Drain current,

$$I_D = k\,(V_{GS} - V_{GS(th)})^2 \text{ milliamperes} \qquad (135\text{-}1)$$

where k is a constant value for a particular MOSFET.

The range of the threshold voltage is typically +1 V to +5 V. To obtain the value of k, a point on the transconductance curve must be specified. For example, if $V_{GS(\mathrm{th})} = +2$ V and $I_D = 12$ mA when $V_{GS} = +4$ V, it follows from Equation 135-1,

$$12 \text{ mA} = k\,(4 \text{ V} - 2 \text{ V})^2$$

$$k = 0.003.$$

The equation for the transconductance (g_m) is:

$$g_m = 2\,k\,(V_{GS} - V_{GS(th)}) \text{ siemens} \qquad (135\text{-}2)$$

Example 135-1

For an E MOSFET, $k = 0.001$ and $V_{GS(th)} = +2.5$ V. What are the values of I_D and g_m at the point on the transconductance curve for which $V_{GS} = +5.5$ V?

Solution

Drain current,

$$\begin{aligned} I_D &= k\,(V_{GS} - V_{GS(th)})^2 \qquad (135\text{-}1)\\ &= 0.001\,(5.5 \text{ V} - 2.5 \text{ V})^2 \text{ A}\\ &= 9 \text{ mA.}\end{aligned}$$

Transconductance,

$$\begin{aligned} g_m &= 2\,k\,(V_{GS} - V_{GS(th)}) \qquad (135\text{-}2)\\ &= 2 \times 0.001\,(5.5 \text{ V} - 2.5 \text{ V}) \text{ S}\\ &= 6000\ \mu\text{S.}\end{aligned}$$

PRACTICE PROBLEMS

135-1. Figure 135-2 illustrates the transconductance curve for an E MOSFET whose k value is 0.002. Determine the value of I_D corresponding to $V_{GS} = +5$ V.

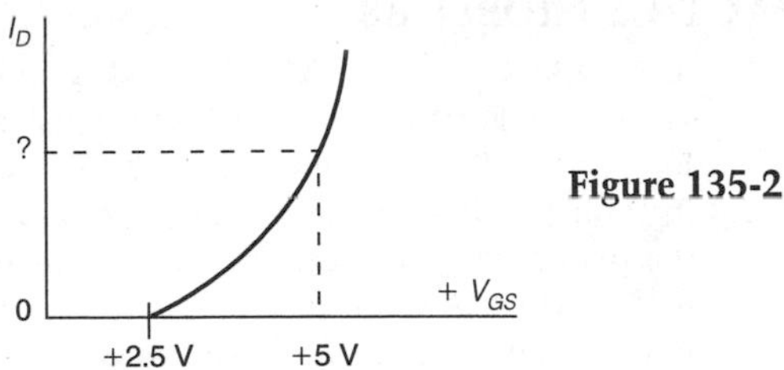

Figure 135-2

135-2. In Practice Problem 135-1, calculate the value of g_m if $V_{GS} = +4.5$ V.

135-3. Figure 135-3 illustrates the drain characteristics of an E MOSFET. Determine the k value for this MOSFET.

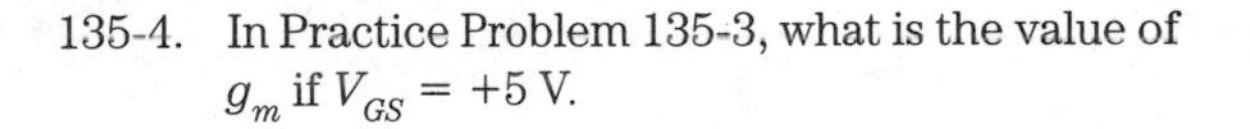

135-4. In Practice Problem 135-3, what is the value of g_m if $V_{GS} = +5$ V.

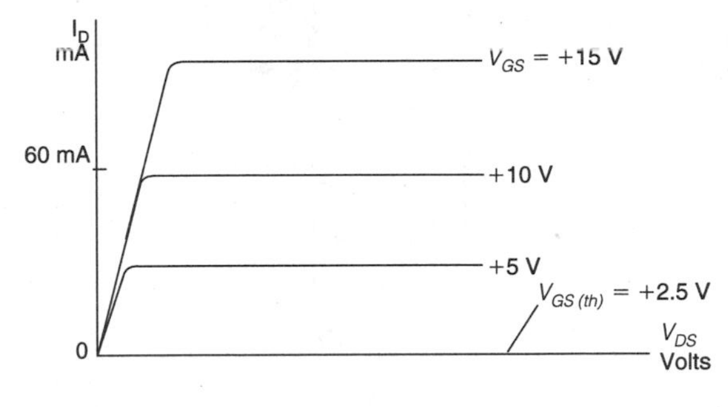

Figure 135-3

136 The E MOSFET amplifier

E MOSFETs can use voltage divider and source bias but not self and current-source bias, which can only provide operation in the depletion mode. However, there is another circuit called *drain feedback bias*, which is only suitable for the enhancement mode (Fig. 136-1A). A high-value resistor of several MΩ is connected between the drain and the gate. The voltage drop across R_G, because of the gate current, is negligible so that $V_{DS} = V_{GS} = V_{DD} - I_D R_D$. By arranging that the value of V_{DS} is well above the pinch-off-voltage, the operation of the MOSFET will occur on the nearly flat section of the drain curve. Once I_D, V_{DD}, and V_{GS} have been determined, the value of R_D can be chosen to provide the required operating conditions.

The action of drain feedback bias for the E MOSFET is similar to the bipolar's collector feedback or self-bias, which compensates for changes that occur in the FET's characteristics. For example, if I_D is tending to decrease, both V_{DS} and V_{GS} will increase, and this will level off the tendency for I_D to fall. The disadvantage of this type of bias is the degenerative feedback, which occurs from the drain output to the gate input. This has the effect of reducing the amplifier's voltage gain.

Because the gate of an E MOSFET operates with a positive voltage on the gate, it is possible to use direct coupling between the amplifier stages (Fig. 136-1B); not only is the circuitry simple, but it has the advantage of an excellent flat response at low frequencies. The positive bias required by one stage is provided by the dc drain voltage of the previous stage.

The equations of the E MOSFET amplifier are the same as those for the JFET and DE MOSFET amplifiers that were previously mentioned in chapters 132 and 134. For a common source stage, the voltage gain from gate to drain is $g_m r_d$.

Example 136-1

For a particular E MOSFET, $k = 0.001$ and $V_{GS(th)} = +2.5$ V. It is required to bias this MOSFET to the point where $V_{GS} = +5.5$ V. If the drain feedback circuit of Fig. 136-1A is used, calculate the required value of R_D. Determine the value of the voltage gain from gate to drain.

Solution

Drain current,

$$I_D = 0.001\ (5.5\text{ V} - 2.5\text{ V})^2\text{ A}$$

$$= 9\text{ mA}.$$

Transconductance,

$$g_m = 2 \times 0.001\ (5.5\text{ V} - 2.5\text{ V})\text{ S}$$

$$= 6\text{ mS}$$

$$= 6000\ \mu\text{S}.$$

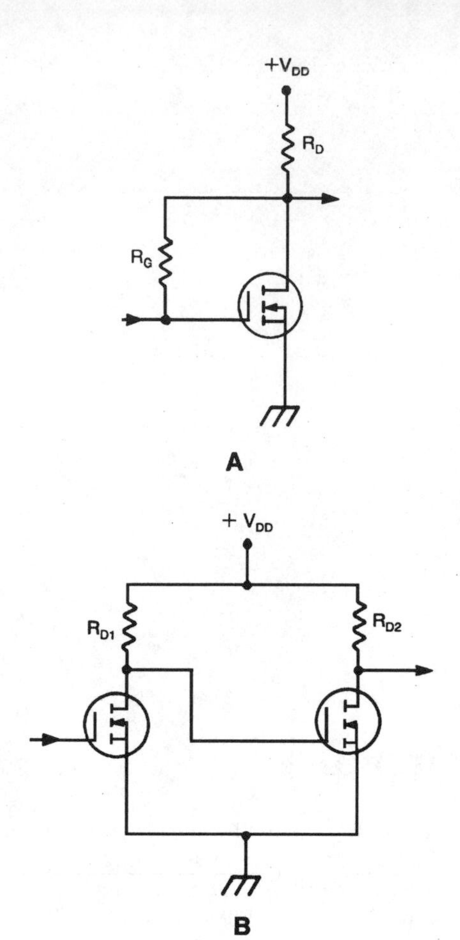

Figure 136-1

Drain load resistor,

$$R_D = \frac{24\text{ V} - 5.5\text{ V}}{9\text{ mA}}$$

$$\approx 2\text{ k}\Omega.$$

Voltage gain from gate to drain,

$$G_v = g_m r_d$$

$$= 6000\ \mu\text{S} \times 2\text{ k}\Omega$$

$$= 12.$$

PRACTICE PROBLEMS

136-1. In Fig. 136-1A, $V_{DD} = +20$ V, $R_D = 4$ kΩ, and $R_G = 100$ MΩ. For the E MOSFET, $k = 0.002$ and $V_{GS(th)} = +2.0$ V. Determine the value of V_{GS}.

136-2. In Problem 136-1, what is the value of g_m at the operating point? Determine the value of the voltage gain from gate to drain.

136-3. In the circuit of Fig. 136-2, the value of g_m at the operating point is 4000 μS. Calculate the value of the voltage gain from gate to drain.

136-4. In the source-follower circuit of Fig. 136-3, the value of g_m at the operating point is 6000 μS. Calculate the value of the voltage gain from gate to source.

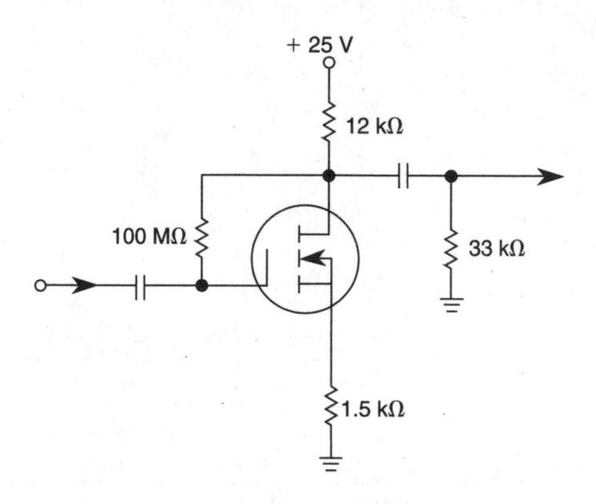

Figure 136-2

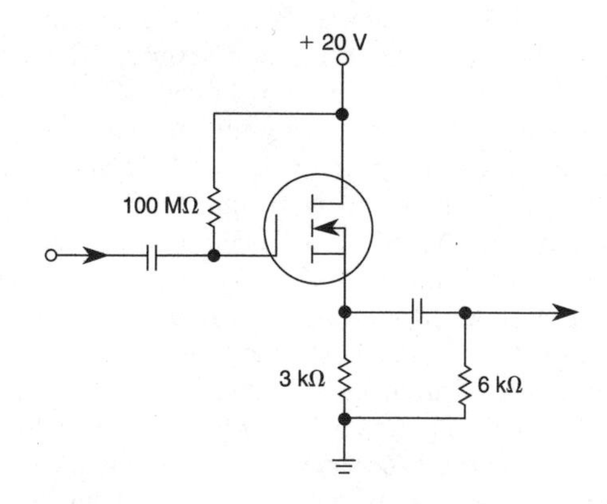

Figure 136-3

137
Operational amplifiers

Operational amplifiers are designed to simulate certain mathematical operations such as addition, subtraction, differentiation, and integration. In Fig. 137-1A, the input signal (V_i) is fed through an impedance (Z_i) to a solid-state linear amplifier whose voltage gain is $-A$ [the minus sign indicates that the output signal, V_o, is inverted with respect to the input signal, such as would occur with some form of common-emitter arrangement]. Z_f provides feedback from the output circuit to the input circuit, and the combination of Z_f and Z_i can be regarded as a voltage divider between V_o and V_i.

The circuit in Fig. 137-1B is a *noninverting* operational amplifier whose feedback action is similar to that of the inverting type. Figures 137-1C and D show operational amplifiers that are capable of performing addition and subtraction; in these cases, Z_f and Z_i are resistors.

In the operational amplifier of Fig. 137-1E, C_f is a capacitor and the amplifier is then capable of *integration*. Owing to the *Miller effect*, the input capacitance will be equal to $(A + 1)C_f$. If the input signal (V_i) is a step voltage as shown, the voltage at the amplifier input terminal will slowly rise; the output signal (V_o) will be an amplified linear fall, which will represent the result of integrating V_i. If C_f is replaced by a suitable inductor (L_f) the operational amplifier will be capable of *differentiation*.

The circuit of Fig. 137-1F shows a *noninverting comparator*, which is used to determine whether an input voltage (V_i) is greater or less than the reference voltage, V_{ref}. When V_i is greater than V_{ref}, the comparator circuit is driven into positive saturation, and the output (V_o) is equal to $+V_{sat}$, which is about 1 V less than +V. By contrast, when V_i drops below V_{ref}, the

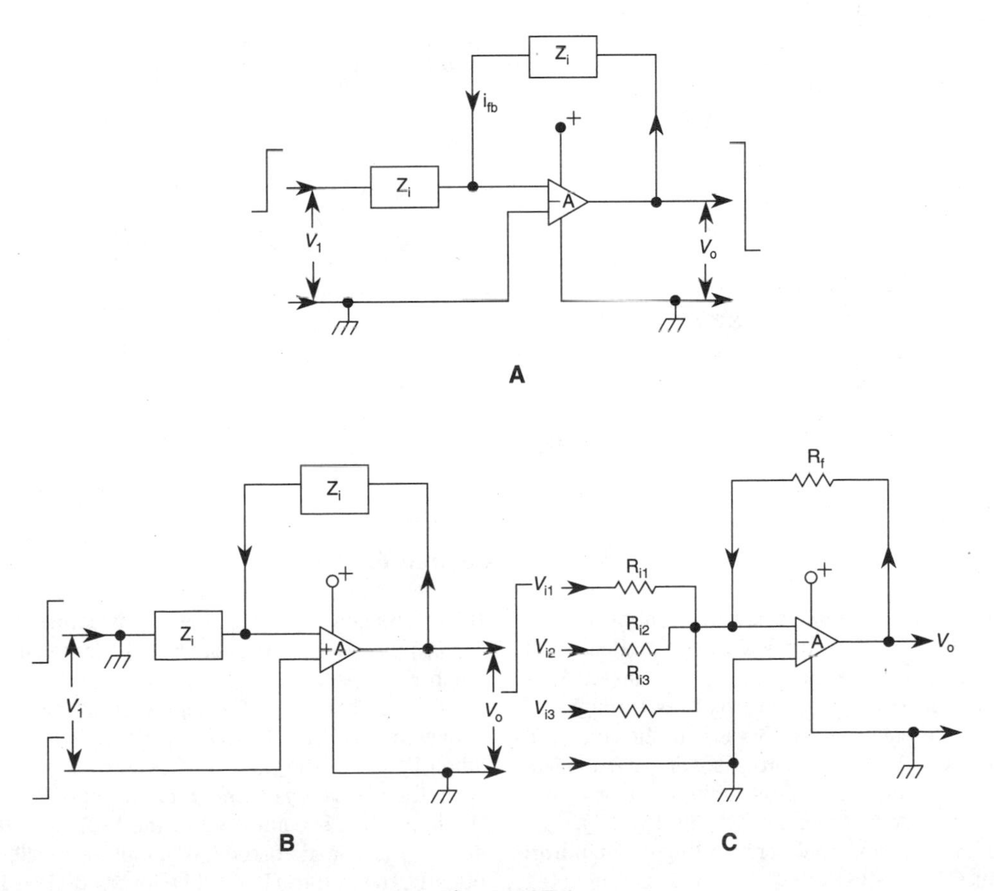

Figure 137-1

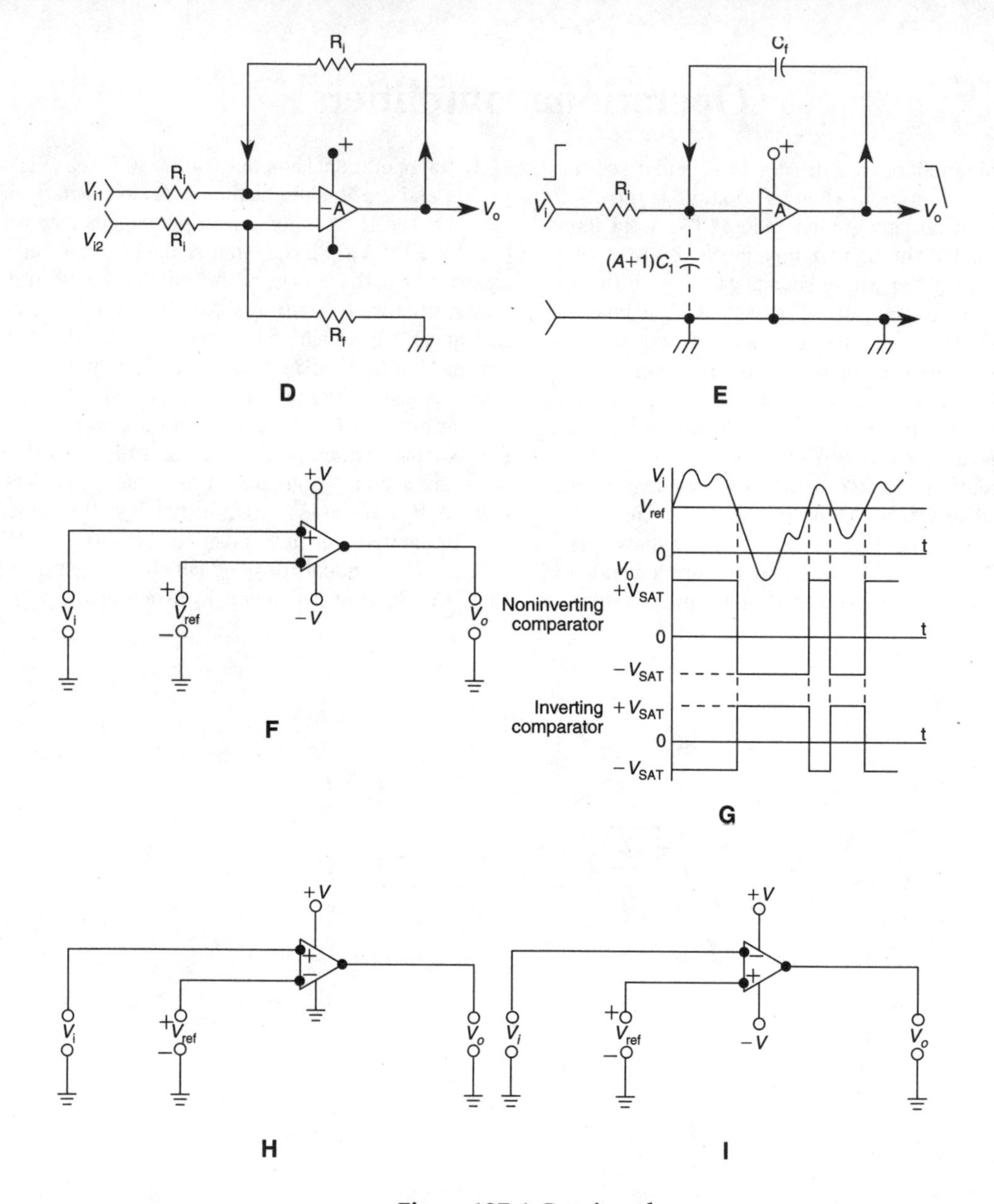

Figure 137-1 Continued.

comparator is driven into negative saturation and its output (V_o) is $-V_{sat}$ (about 1 V above $-$ V). Because it is virtually impossible to hold the input steady at $V_i = V_{ref}$, V_i will not rest at V_{ref}, but will pass through the level of the reference voltage. Because of the large gain of the op amp (which is not providing linear amplification), a variation of 1 mV or less, in the value of V_i, will cause V_o to change by several volts (Fig. 137-1G).

To summarize, for the noninverting comparator: the output (V_o) is equal to $+V_{sat}$ when V_i is connected to the "+" input terminal of the op amp, and it exceeds the reference voltage (V_{ref}), which is connected to the "−" input terminal. By contrast, V_o is equal to $-V_{sat}$ when V_i drops below V_{ref}.

A simplified form of a noninverting comparator is shown in Fig. 137-1H. When $V_i > V_{ref}$, $V_o = +V_{sat}$; but when $V_i < V_{ref}$, the value of V_o is zero.

In the *inverting comparator* circuit of Fig. 137-1H, the input (V_i) is connected to the "−" input terminal and V_{ref} is joined to the "+" input terminal. Consequently, the output (V_o) will be inverted (Fig. 137-1G) when compared with that of the circuit in Fig. 137-1F.

The circuit of Fig. 137-2 is called a voltage-follower because the output signal (V_o) equals the input signal V_i. The purpose of such a circuit is to couple a signal associated with a high output impedance, to a low impedance load, with a voltage gain of near unity. Consequently, the voltage-follower is basically an op amp that is used for impedance matching.

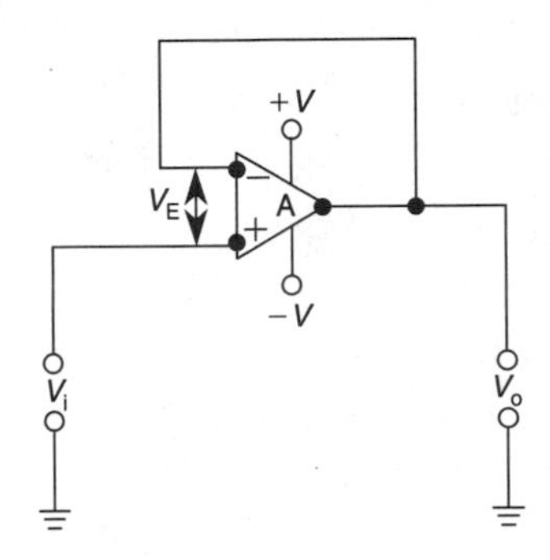

Figure 137-2

Note that there is no feedback resistor; therefore, there is 100% negative feedback to the inverting terminal. When a signal is applied to the noninverting input terminal, the op amp at once causes the inverting input (V_E) to fall to nearly zero so that V_i and V_o are virtually equal. For example, if V_i is +10 V, V_o might be 9.999 V; the inverting input is −0.001 V, which allows for an open loop gain of 10000. The same principles apply to both dc and ac voltage followers.

MATHEMATICAL DERIVATIONS

In Fig. 137-1A, the open loop gain (A) of the op amp is very large so that there is only a small voltage at the "−" input terminal. It follows that the voltage across Z_f is virtually equal to the output voltage (V_o), while the negative feedback voltage across Z_i is equal to V_i. Because the same feedback current (i_{fb}) flows through Z_f and Z_i,

$$V_o = i_{fb} \times Z_f$$

$$V_i = -i_{fb} \times Z_i$$

Voltage gain with feedback,

$$A' = \frac{V_o}{V_i} = -\frac{Z_f}{Z_i} \qquad (137\text{-}1)$$

In Fig. 137-1B, Z_f and Z_i form a voltage divider circuit for the output voltage. The negative feedback voltage across Z_i is virtually equal to the input voltage, V_i. Then,

$$V_i = V_o \times \frac{Z_i}{Z_i + Z_f}$$

Voltage gain with feedback,

$$A' = \frac{V_o}{V_i} = \frac{Z_i + Z_f}{Z_i} = 1 + \frac{Z_f}{Z_i} \qquad (137\text{-}2)$$

In the summing amplifier (Fig. 137-1C), the gain for each input signal is $-R_f/R_i$,

Output signal,

$$V_o = -R_f\left(\frac{V_{i1}}{R_{i1}} + \frac{V_{i2}}{R_{i2}} + \frac{V_{i3}}{R_{i3}}\right) \text{ volts} \qquad (137\text{-}3)$$

For the differential amplifier (Fig. 137-1D),

Output signal,

$$V_o = -\frac{R_f}{R_i}(V_{i1} - V_{i2}) \text{ volts} \qquad (137\text{-}4)$$

For the noninverting comparator (Fig. 137-1F),

$$\text{When } V_i = V_{ref}, \qquad (137\text{-}5)$$

$$V_o = 0$$

$$\text{When } V_i > V_{ref}, \qquad (137\text{-}6)$$

$$V_o = +V_{sat} \text{ volts}$$

$$\text{When } V_i < V_{ref}, \qquad (137\text{-}7)$$

$$V_o = -V_{sat} \text{ volts}$$

For the inverting comparator (Fig. 137-1H),

$$\text{When } V_i = V_{ref}, \qquad (137\text{-}8)$$

$$V_o = 0$$

$$\text{When } V_i > V_{ref}, \qquad (137\text{-}9)$$

$$V_o = -V_{sat} \text{ volts}$$

$$\text{When } V_i < V_{ref}, \qquad (137\text{-}10)$$

$$V_o = +V_{sat} \text{ volts}$$

For the voltage follower (Fig. 137-2),

$$V_o = A \times V_E \qquad (137\text{-}11)$$

$$V_i + V_E = V_o$$

$$\text{Because } V_E \to 0 \text{ V}, \qquad (137\text{-}12)$$

$$V_i = V_o$$

Example 137-1

In Fig. 137-1A, Z_f is a 150-kΩ resistor and Z_i is a 10-kΩ resistor. Assuming that A is large, what is the gain of this operational amplifier?

Solution

Operational amplifier gain,

$$A' = -\frac{R_f}{R_i} \qquad (137\text{-}1)$$

$$= -\frac{150}{10}$$

$$= -15$$

Example 137-2

In Fig. 137-1B, Z_f is a 150-kΩ resistor and Z_i is a 10-kΩ resistor. Assuming that A is large, calculate the gain of the operational amplifier.

Solution

Operational amplifier gain,

$$A' = 1 + \frac{R_f}{R_i} \qquad (137\text{-}2)$$

$$= 1 + \frac{150}{10}$$

$$= +16$$

Example 137-3

In Fig. 137-1C, V_{i1}, V_{i2}, and V_{i3} are step voltages of +2 mV, +5 mV, and +7 mV, respectively; $R_f = 1$ MΩ and each of the resistors R_{i1}, R_{i2}, and R_{i3} has a value of 0.5 MΩ. Determine the value of the output signal.

Solution

Voltage gain for each signal,

$$A' = -\frac{R_f}{R_i}$$

$$= -\frac{1\ \text{M}\Omega}{0.5\ \text{M}\Omega}$$

$$= -2$$

Output signal,

$$V_o = -2\,(2 + 5 + 7)$$

$$= -28\ \text{mV}$$

Example 137-4

In Fig. 137-1D, $R_f = 120$ kΩ and $R_i = 10$ kΩ. V_{i1} and V_{i2} are step voltages of +15 mV and +10 mV respectively. What is the value of the output signal?

Solution

Voltage gain for each signal,

$$A' = -\frac{R_f}{R_i}$$

$$= -\frac{120\ \text{k}\Omega}{10\ \text{k}\Omega}$$

$$= -12$$

Output signal,

$$V_o = -12 \times (15 - 10)$$

$$= -60\ \text{mV}.$$

Example 137-5

In Fig. 137-1F, $V_i = +27$ V, $V_{ref} = +13$ V, $V = 15$ V. What is the value of V_o?

Solution

The circuit is that of a noninverting comparator, which is driven into positive saturation. Therefore, V_o is approximately +14 V.

Example 137-6

In Fig. 137-2, $V_i = -7$ V, $V = 15$ V. What is the value of V_o?

Solution

The circuit is that of a voltage-follower so that $V_o = V_i = -7$ V.

PRACTICE PROBLEMS

137-1. In Fig. 137-1A, the operational amplifier has a closed loop gain of −18. If $R_f = 180$ kΩ, calculate the value of R_i.

137-2. In Fig. 137-1B, the operational amplifier has a closed loop gain of +23. If $R_i = 12$ kΩ, calculate the value of R_f.

137-3. In Fig. 137-1C, the value of V_o is −21 mV. If $V_{i1} = +5$ mV, $V_{i2} = -2$ mV, $V_{i3} = +4$ mV, and $R_{i1} = R_{i2} = R_{i3} = 25$ kΩ, calculate the value of R_f.

137-4. In Fig. 137-1D, the value of V_o is −25 mV. If $V_{i1} = +18$ mV, $V_{i2} = -7$ mV, and $R_{i1} = R_{i2} = 15$ kΩ, calculate the value of R_f.

137-5. In the comparator circuit of Fig. 137-1H, $V_i = 3$ V, $V_{ref} = 7$ V, and $V = 15$ V. What is the value of V_o?

138
Positive or regenerative feedback

Positive feedback can be used to increase the gain of an amplifier circuit. This is illustrated in Fig. 138-1A, where V_i is the input signal from the preceding stage. Such a signal would be applied between the base of a transistor (or the control grid of a tube) and ground. The output signal (V_o) appears between collector (or plate) and ground and a fraction, β (beta), of this output signal is then fed back to the input circuit so that this feedback voltage ($+\beta V_o$) is in phase with V_i. β is called the *feedback factor*, which can either be expressed as a decimal fraction or as a percentage.

In order for the feedback to be positive, there must be a total of 360° phase shift (equivalent to zero phase shift) around the feedback loop [base → collector → base (or grid → plate → grid)]. The total signal voltage when applied between base and emitter (or control grid and cathode) is the sum of the input signal (V_i) and the positive feedback voltage $+\beta V_o$. The voltage gain (open loop gain) of the active device is A and the mathematical derivations will show that $A' = A/(1 - A\beta)$, where A' is the overall voltage gain with the positive feedback present (closed loop gain). There are then three possible conditions in the circuit:

1. If A and β are chosen so that the value of $A\beta$ is less than 1, then A' is greater than A and the amplifier's gain has been increased as the result of the positive feedback. Such is the case with the so-called *regenerative amplifier*.
2. If $A\beta = 1$ (for example $A = 10$ and $\beta = 0.1$ or 10%), A' is theoretically infinite. This means that the circuit can provide a continuous output, without any input signal from the previous stage. This is the condition in a stable oscillator.
3. If $A\beta > 1$, the oscillator is unstable. The output (V_o) then increases, which tends to reduce the value of A until the equilibrium condition of $A\beta = 1$ is reached.

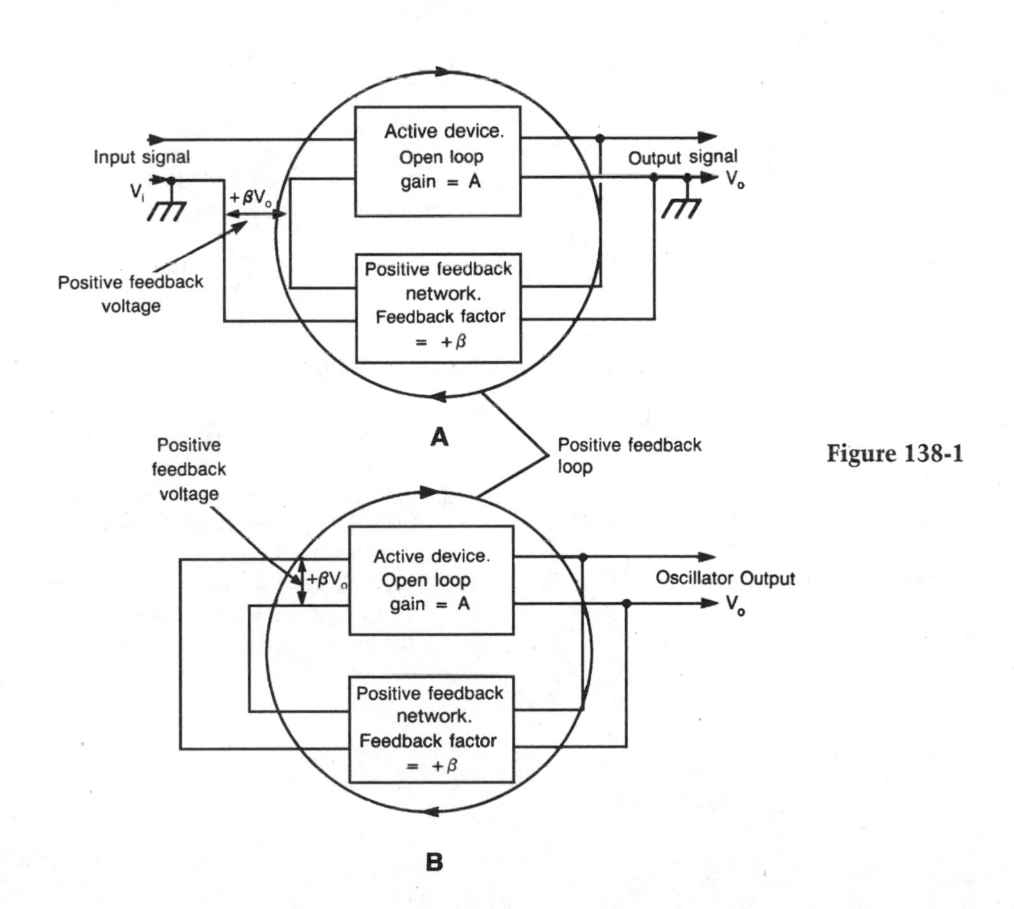

Figure 138-1

The condition for oscillation is therefore $A\beta = 1\angle 0°$; this is sometimes referred to as the *Barkhausen* or *Nyquist criterion.* The inclusion of "$\angle 0°$" in the polar value of $A\beta$ means that the resultant phase shift around the loop is zero degrees; consequently, the feedback is positive. By contrast, an angle of 180° would indicate that the feedback is negative.

Figure 138-1B shows the principle of positive feedback in an oscillator circuit. If we assume that the input signal is 1 V rms, and the voltage gain of the active device is 10, the output signal is 10 V rms; if the feedback factor is 1/10 or 10%, the 10 V output signal will be responsible for creating the 1-V input signal (this does *not* mean that there is only 9 V left of the output signal). This argument sounds rather like the chicken and the egg; the question arises, "how does the circuit get started in the first place?" The answer is that all active devices are inherently noisy. Because the noise is spread throughout the frequency spectrum, it will contain a component at the frequency of oscillation. This component will trigger the positive feedback network so that the oscillation will increase until the equilibrium condition of $A\beta = 1\angle 0°$ is reached.

MATHEMATICAL DERIVATIONS

For a regenerative amplifier:

Total input signal between base and emitter,

$$V_{BE} = V_i + \beta V_o \text{ volts} \quad (138\text{-}1)$$

Output signal,

$$V_o = A \times (V_i + \beta V_o) \text{ volts} \quad (138\text{-}2)$$

This yields,

$$A' = \frac{V_o}{V_i} = \frac{A}{1 - A\beta} \quad (138\text{-}3)$$

where A' is the overall voltage gain with the positive feedback present (closed loop gain).

For a stable oscillator,

$$A\beta = 1\angle 0° \quad (138\text{-}4)$$

$$\text{Positive feedback factor, } \beta = \frac{1}{A} \quad (138\text{-}5)$$

Positive feedback percentage $= \beta \times 100\%$

Example 138-1

An amplifier provides an open loop gain of 15 and has a positive feedback factor of 3%. What is the value of the closed loop gain?

Solution

Closed loop gain,

$$A' = \frac{A}{1 - A\beta} \quad (138\text{-}3)$$

$$= \frac{15}{1 - \dfrac{15 \times 3}{100}}$$

$$= \frac{15}{0.55}$$

$$= 27.3$$

Example 138-2

An amplifier has an open loop gain of 50. What percentage of positive feedback is required to sustain a stable oscillation?

Solution

Positive feedback factor,

$$\beta = \frac{1}{A} \quad (138\text{-}5)$$

$$= \frac{1}{50}$$

Positive feedback percentage,

$$= \beta \times 100 \quad (138\text{-}6)$$

$$= \frac{1}{50} \times 100$$

$$= 2\%.$$

PRACTICE PROBLEMS

138-1. An amplifier has a closed loop gain of 50 and a positive feedback percentage of 4%. What is the value of the open loop gain?

138-2. An amplifier has an open loop gain of 20 and a closed loop of 60. What is the value of the positive feedback factor?

138-3. An oscillator has a positive feedback percentage of 4.7%. What is the required open loop gain of the amplifier that forms part of the oscillator?

139
The Wien bridge oscillator

This oscillator (Fig. 139-1) uses two common-emitter stages so that there is theoretically zero phase shift between a signal voltage on the base of Q1 and the output voltage at the collector of Q2. The feedback loop is completed by the Wien filter consisting of R1, R2, C1, and C2; therefore, in order for the feedback to be positive, the input voltage to the filter (V_i) and the output voltage from the filter (V_o) must be in phase.

MATHEMATICAL DERIVATIONS

Let the phase angle of the entire network (Fig. 139-2) be ϕ_1, while the phase angle of the parallel R2/C2 combination is ϕ_2. Then, i leads V_i by ϕ_1 and V_o lags i by ϕ_2. If V_o is in phase with V_i, the phase angle ϕ_1 must be equal to the phase angle ϕ_2.

Impedance phasor,

$$z_2 = \frac{R_2 \times \dfrac{1}{j\omega_o C_2}}{R_2 + \dfrac{1}{j\omega_o C_2}}$$

$$= \frac{R_2}{\omega_o^2 C_2^2 R_2^2 + 1} - j\,\frac{\omega_o C_2 R_2^2}{\omega_o^2 C_2^2 R_2^2 + 1}$$

Impedance phasor,

$$z_t = \left(R_1 + \frac{R_2}{\omega_o^2 C_2^2 R_2^2 + 1}\right) - j\left(\frac{1}{\omega_o C_1} + \frac{\omega_o C_2 R_2^2}{\omega_o^2 C_2^2 R_2^2 + 1}\right) \qquad (139\text{-}2)$$

Phase angle,

$$\phi_2 = \text{inv tan}\left(\frac{-\omega_o C_2 R_2^2}{R_2}\right) = \text{inv tan}\,(-\omega_o C_2 R_2) \qquad (139\text{-}3)$$

Phase angle,

$$\phi_1 = \text{inv tan}\left[\frac{-\dfrac{1}{\omega_o C_1} - \dfrac{\omega_o C_2 R_2^2}{\omega_o^2 C_2^2 R_2^2 + 1}}{R_1 + \dfrac{R_2}{\omega_o^2 C_2^2 R_2^2 + 1}}\right] \qquad (139\text{-}4)$$

If V_o is in phase with V_i,

$$\omega_o C_2 R_2 = \frac{\dfrac{1}{\omega_o C_1} + \dfrac{\omega_o C_2 R_2^2}{\omega_o^2 C_2^2 R_2^2 + 1}}{R_1 + \dfrac{R_2}{\omega_o^2 C_2^2 R_2^2 + 1}}$$

$$\omega_o C_2 R_2 R_1 = \frac{1}{\omega_o C_1}$$

$$\omega_o^2 = \frac{1}{C_1 R_1 C_2 R_2}$$

Frequency of oscillation,

$$f_o = \frac{\omega_o}{2\pi} = \frac{1}{2\pi\sqrt{R_1 R_2 C_1 C_2}}\ \text{Hz} \qquad (139\text{-}5)$$

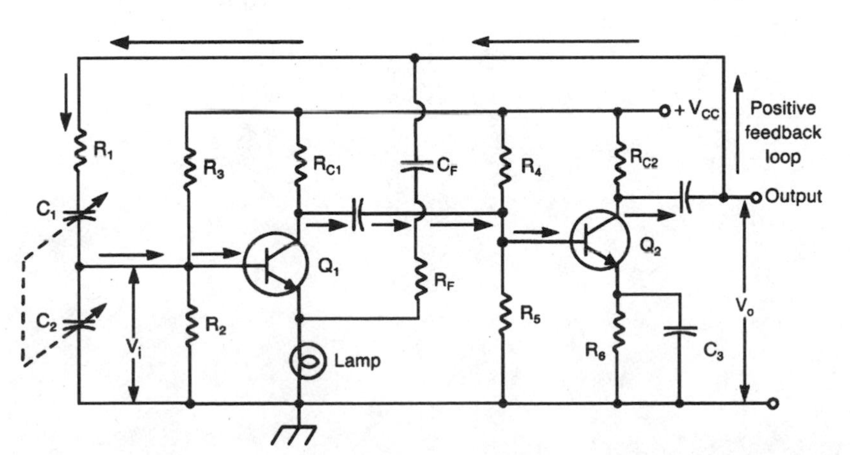

Figure 139-1

If $R_1 = R_2 = R$ and $C_1 = C_2 = C$,

$$f_o = \frac{1}{2\pi RC} \text{ Hz} \tag{139-6}$$

At this frequency of oscillation,

$$\beta = \frac{V_o}{V_i} = \frac{z_2}{z_1 + z_2}$$

$$= \frac{\dfrac{R}{\omega_o{}^2C^2R^2+1} - j \times \dfrac{\omega_o CR^2}{\omega_o{}^2C^2R^2+1}}{\left(R + \dfrac{R}{\omega_o{}^2C^2R^2+1}\right) - j\left(\dfrac{1}{\omega_o C} + \dfrac{\omega_o CR^2}{\omega_o{}^2C^2R^2+1}\right)}$$

Because $\omega_o CR = 1$,

$$\beta = \frac{\dfrac{R}{2} - j \times \dfrac{R}{2}}{\left(R + \dfrac{R}{2}\right) - j\left(R + \dfrac{R}{2}\right)} \tag{139-7}$$

$$= \frac{\dfrac{R}{2}(1 - j1)}{\dfrac{3R}{2}(1 - j1)} = \frac{1}{3}$$

The mathematical derivations show that the combined voltage gain of Q1 and Q2 must be equal to three. This is not practical and, therefore, the oscillator circuit contains negative feedback, provided by R_f and the lamp, which form the bridge circuit in conjunction with R1, C1, R2, and C2. The combined gain of Q1 and Q2 without the negative feedback can then be high, but their gain with feedback will equal three under stable conditions, which can be determined by the operating resistance of the lamp.

Like the *RC* phase-shift oscillator (chapter 140), the Wien bridge circuit is especially suitable for the generation of low-frequency sine waves with good stability and a lack of harmonic distortion.

Example 139-1

In Fig. 139-1, $C_1 = C_2 = 8200$ pF, $R_1 = R_2 = 12$ kΩ. Determine the frequency of the oscillation.

Solution

Frequency of oscillation,

$$f_o = \frac{0.159}{RC} \text{ Hz} \tag{139-1}$$

$$= \frac{0.159}{12 \times 10^3 \times 8200 \times 10^{-12}}$$

$$= \frac{1.59 \times 10^{-1}}{1.2 \times 8.2 \times 10^{-5}}$$

$$= \frac{15900}{1.2 \times 8.2} \text{ Hz}$$

$= 1.62$ kHz, rounded off.

PRACTICE PROBLEMS

139-1. In Fig. 139-1, the oscillator generates a frequency of 10 Hz. If $R_1 = R_2 = 150$ kΩ, calculate the value of C_1, if $C_1 = C_2$.

139-2. In Fig. 139-2, the oscillator generates a frequency of 120 kHz. If $C_1 = C_2 = 200$ pF, calculate the value of R_1, if $R_1 = R_2$.

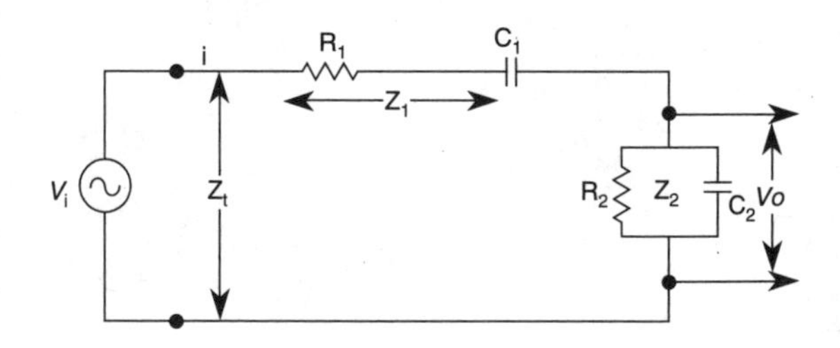

Figure 139-2

140
The RC phase-shift oscillator

This type of oscillator (Fig. 140-1) is capable of generating a sine wave that is relatively free of harmonic distortion; its frequency output can range from less than 1 Hz to a few hundred kHz.

The circuit uses the common-emitter configuration so that there is 180° phase change from base to collector; the feedback loop is completed by the RC phase shift network that contains a minimum number of three sections (as shown). For the feedback to be positive, the network must therefore provide a further 180° shift (ignoring any effect of the transistor circuitry); it would appear, at first glance, that each RC section should contribute a 60° shift. However, this simple approach ignores the shunting effect of one section on another. The results of a precise analysis are given in the mathematical derivations.

MATHEMATICAL DERIVATIONS

Ignoring any loading effect of the transistor on the RC network of Fig. 140-2, the mesh equations are:

Solving for i_3, by the method of determinants (chapter 196). See Eq. 140-1.

Output voltage,

$$v_o = i_3 R$$

Refer to Eq. 140-2.

If v_o and v_i are 180° out of phase, the coefficient of "j" in Equation 140-2 must be equal to zero.

Therefore,

$$\frac{-6R^2}{\omega C} + \frac{1}{\omega^3 C^3} = 0$$

$$\frac{1}{\omega^2 C^2} = 6R^2$$

$$6\omega^2 C^2 R^2 = 1$$

$$\omega^2 = \frac{1}{6R^2C^2}$$

$$\omega = 2\pi f = \frac{1}{\sqrt{6}RC}$$

Frequency of oscillation,

$$f = \frac{1}{2\pi\sqrt{6}RC} \text{ Hz} \qquad (140\text{-}3)$$

Substituting $1/\omega^2C^2 = 6R^2$ in the "real" part of Equation 140-2,

$$v_i R^3 = v_o\,(R^3 - 5R \times 6R^2)$$

$$= -29R^3 v_o$$

$$\begin{aligned}
i_1\left(R - \frac{j}{\omega C}\right) - i_2 R \qquad\qquad\qquad & - v_i = 0 \\
- i_1 R + i_2\left(2R - \frac{j}{\omega C}\right) - i_3 R \qquad & = 0 \\
- i_2 R + i_3\left(2R - \frac{j}{\omega C}\right) & = 0
\end{aligned} \qquad (140\text{-}1)$$

$$\frac{i_3}{\begin{vmatrix} R - \frac{j}{\omega C} & -R & -v_i \\ -R & 2R - \frac{j}{\omega C} & 0 \\ 0 & -R & 0 \end{vmatrix}} = \frac{-1}{\begin{vmatrix} R - \frac{j}{\omega C} & -R & 0 \\ -R & 2R - \frac{j}{\omega C} & -R \\ 0 & -R & 2R - \frac{j}{\omega C} \end{vmatrix}}$$

$$\frac{i_3}{-v_i R^2} = \frac{-1}{\left(R - \frac{j}{\omega C}\right)\left[\left(2R - \frac{j}{\omega C}\right)^2 - R^2\right] - R^2\left(2R - \frac{j}{\omega C}\right)^2}$$

$$v_i R^3 = v_o\left[\left(R - \frac{j}{\omega C}\right)\left(3R^2 - j\frac{4R}{\omega C} - \frac{1}{\omega^2 C^2}\right) - 2R^3 + j\frac{R^2}{\omega C}\right]$$

$$= v_o\left(R^3 - j\frac{4R^2}{\omega C} - \frac{R}{\omega^2 C^2} - j\frac{3R^2}{\omega C} - \frac{4R}{\omega^2 C^2} + \frac{j1}{\omega^2 C^3} + \frac{jR^2}{\omega C}\right) \tag{140-2}$$

$$= v_o\left(R^3 - j\frac{6R^2}{\omega C} + \frac{j1}{\omega^3 C^3} - \frac{5R}{\omega^2 C^2}\right)$$

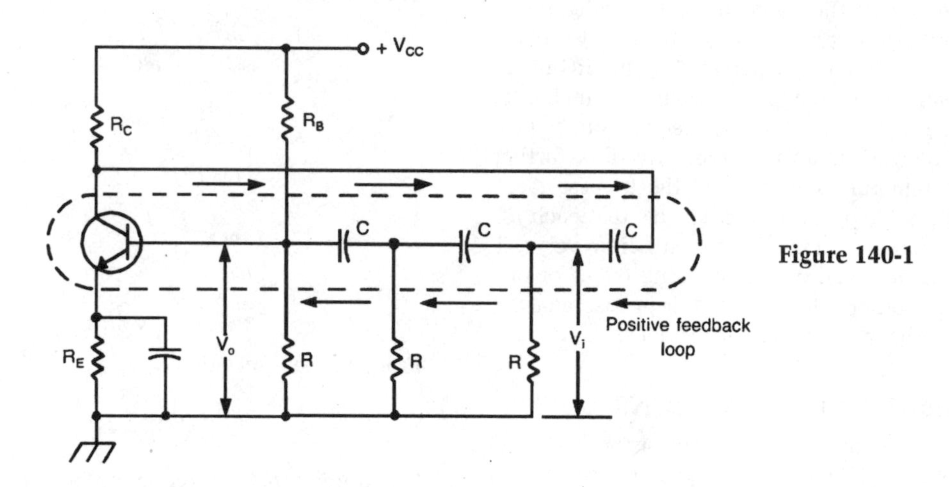

Figure 140-1

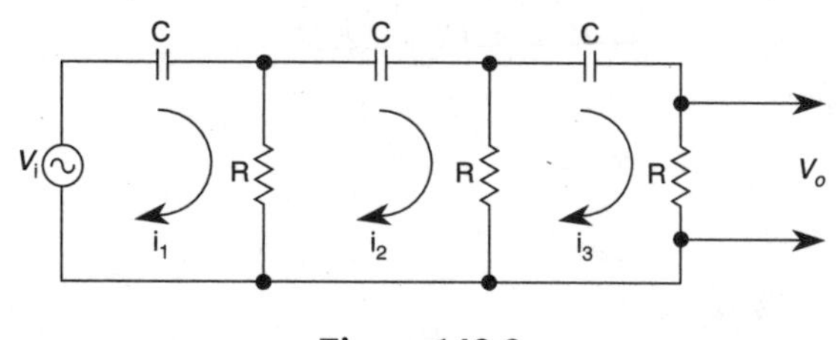

Figure 140-2

Therefore,

$$\frac{v_o}{v_i} = -\frac{1}{29} \tag{140-4}$$

The negative sign indicates that v_i and v_o are 180° out of phase.

It follows that if oscillation is to be sustained, the open-loop voltage gain of the common-emitter amplifier must be 29. Notice that in Equation 140-3, f_o is inversely proportional to C (and not $\sqrt{C}$ as in an LC oscillator).

If a four-section RC feedback network is used:
Frequency of oscillation,

$$f_o = \frac{1}{2\pi RC}\ \text{Hz} \tag{140-5}$$

Attenuation factor,

$$\beta = \frac{1}{18.4} \tag{140-6}$$

Example 140-1

In Fig. 140-1, $C = 0.01\ \mu\text{F}$, $R = 10\ \text{k}\Omega$. Calculate the frequency of the oscillation.

Solution

Frequency of oscillation,

$$f_o = \frac{0.159}{\sqrt{6}RC} \tag{140-3}$$

$$= \frac{1.59 \times 10^{-1}}{\sqrt{6} \times 0.01 \times 10^{-6} \times 10^4}\ \text{Hz}$$

$$= \frac{1590}{\sqrt{6}}$$

$= 650$ Hz, rounded off.

PRACTICE PROBLEMS

140-1. An RC phase-shift oscillator uses a four-section feedback network. If $R = 220\ \text{k}\Omega$ and $C = 0.1\ \mu\text{F}$, calculate the frequency of the oscillation. What is the gain provided by the CE configuration of the transistor?

140-2. The oscillator of Fig. 140-1 generates an output frequency of 1 kHz. If $R = 6.8\ \text{k}\Omega$, what is the required value for C?

140-3. The oscillator of Fig. 140-1 generates an output frequency of 1 Hz. If $C = 0.1\ \mu\text{F}$, what is the required value for R?

141
The multivibrator

The multivibrator circuit of Fig. 141-1, consists of two common-emitter stages that are cross-connected for positive feedback. The resultant instability causes the transistors to cut on and off alternately so that approximate square-wave voltage outputs appear at the collectors; the base waveforms have a sawtooth appearance.

When the transistor Q1 has been driven to the cutoff condition as a result of the positive feedback action, the base potential (e_{b1}) is approximately equal to $-V_{CC}$. The capacitor C2 will then discharge through R_{B2} and Q2 so that e_{b1} will rise toward V_{CC} with a time constant approximately equal to $R_{B2}C_2$ seconds.

When e_{b1} becomes slightly positive, Q1 will switch on and the positive feedback action will drive Q2 to the cutoff position. Because e_{b1} approximately reaches its halfway mark in rising from $-V_{CC}$ to a slightly positive potential (on its way toward $+V_{CC}$), Q1 is cut off for a time interval of approximately $0.7R_{B2}C_2$ seconds.

MATHEMATICAL DERIVATIONS

$$\textit{Multivibrator frequency} = \frac{1}{\textit{total period}}$$

$$= \frac{1}{0.7(R_{B1}C_1 + R_{B2}C_2)}$$

$$= \frac{1}{1.4R_BC} \text{ hertz} \qquad (141\text{-}1)$$

if the multivibrator is symmetrical with $C_1 = C_2 = C$ farads, and $R_{B1} = R_{B2} = R_B$ ohms.

Example 141-1

In Fig. 141-1, $R_{B1} = R_{B2} = 100\ \text{k}\Omega$, and $C_1 = C_2 = 0.01\ \mu\text{F}$. What is the approximate frequency of the multivibrator?

Solution

Multivibrator frequency,

$$f_o = \frac{1}{1.4R_BC} \text{ Hz} \qquad (141\text{-}1)$$

$$= \frac{1}{1.4 \times 0.01 \times 10^{-6} \times 10^5} \text{ Hz}$$

$= 714$ Hz, rounded off.

PRACTICE PROBLEMS

141-1. In Fig. 141-1, $R_{B1} = 47\ \text{k}\Omega$, $R_{B2} = 100\ \text{k}\Omega$, $C_1 = 0.004\ \mu\text{F}$, and $C_2 = 0.01\ \mu\text{F}$. What is the approximate frequency of the multivibrator?

141-2. The approximate frequency of the multivibrator in Fig. 141-1 is 12.5 kHz. If $R_{B1} = R_{B2} = 10\ \text{k}\Omega$, what is the value of C_1, if $C_1 = C_2$?

141-3. The approximate frequency of the multivibrator in Fig. 141-1 is 110 Hz. If $C_1 = C_2 = 0.1\ \mu\text{F}$, what is the value of R_{B1}, if $R_{B1} = R_{B2}$?

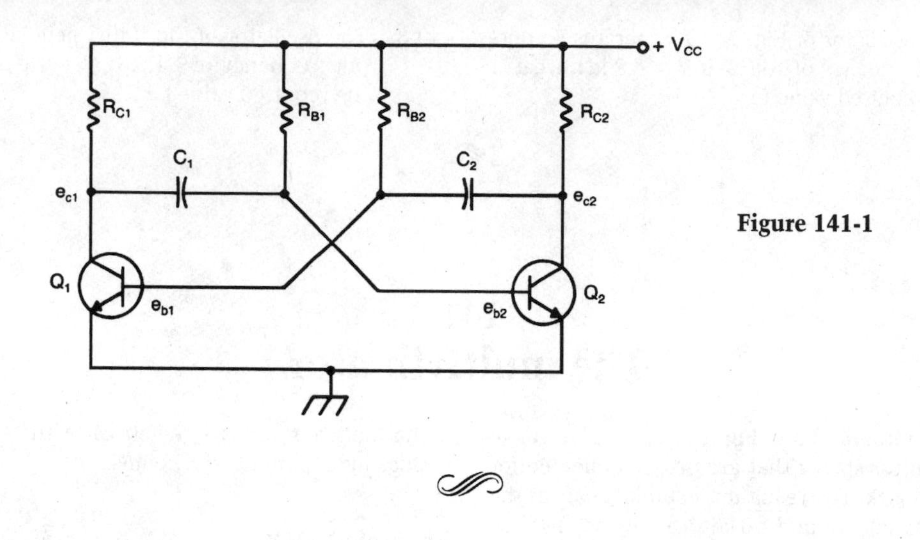

Figure 141-1

4
PART
Tubes and their associated circuits

142
The triode tube—static characteristics

The diode, which is the simplest form of thermionic tube, is limited to a single function, namely rectification (the conversion of an ac voltage to a dc voltage). In the triode (or three-electrode tube), the flow of the electrons from the heated cathode to the plate is controlled by means of an additional electrode, which is interposed between the cathode and the plate (Fig. 142-1A). This electrode is called the *control grid* (G1) because of its form taken in early examples of such tubes. Its modern form is in the shape of a thin spiral of wire or an open mesh. This grid is commonly operated at a negative potential, relative to the cathode, so that it attracts no electrons to itself and there is no flow of grid current; however, it tends to repel those electrons that are being attracted toward the positive plate.

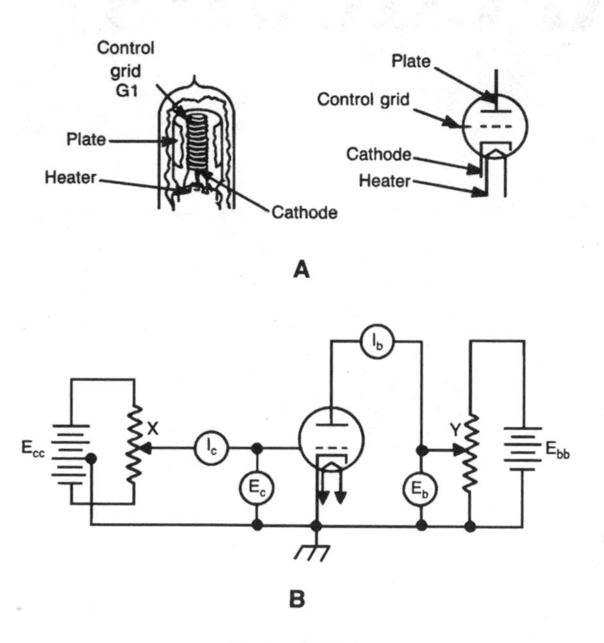

Figure 142-1

The number of electrons reaching the plate (per second) is determined mainly by the electric field near the cathode, and hardly at all by the rest of the field in the remaining space between the cathode and the plate. Near the cathode, the electrons are travelling slowly compared to those that have already moved some distance toward the plate. Therefore, the electron density in the interelectrode space will be high near the cathode, but will decrease toward the plate. The total space charge will be concentrated near the cathode because, once an electron has left this region, it contributes to the space charge for only a very brief interval of time. Therefore, the space current in the triode is determined by the electric field near the cathode, produced by the combined effect of the plate and grid potentials.

When the grid is made sufficiently negative, with respect to the cathode, all of the emitted electrons will be repelled back to the cathode and no plate current will flow. Therefore, the plate current, $I_b = 0$ and the tube is said to be *cut off*.

It is clear that the value of the plate current is governed by both the grid voltage and the plate voltage. You can say that the triode is a *voltage controlled device*, such as the field-effect transistor (FET). For a particular tube, you need to know:

1. The degree of control that the plate voltage exercises over the plate current, while maintaining the grid voltage constant.
2. The degree of control that the grid voltage exercises over the plate current, while maintaining the plate voltage constant.
3. The relative controls of the grid and the plate in maintaining a constant level of plate current.

This information is covered in chapters 143, 144, and 145. It is derived from static characteristic curves that show the interaction between the variables I_b (plate current), E_c (grid voltage), and E_b (plate voltage). The word *static* means that the curves are obtained under controlled laboratory conditions, in which two of the three quantities are varied during the experiment, but the third quantity is kept constant (Fig. 142-1B). By contrast, under dynamic conditions, such as those that occur in an amplifier circuit, all three variables are changing simultaneously.

MATHEMATICAL DERIVATIONS

For a symmetrical grid structure, it can be shown that the electric field near the cathode is proportional to $E_c + E_b/\mu$, where E_c and E_b are the grid and plate potentials as measured relative to the cathode, which is assumed to be grounded. The amplification factor (μ) (chapter 145) is a "constant" that is determined by the geometry of the tube. The total, or plate current (I_b) varies with $E_c + E_b/\mu$ in exactly the same way as the plate current varies with the plate voltage for a space-charge limited diode. Therefore,

Plate current,

$$I_b = K\left(E_c + \frac{E_b}{\mu}\right)^{3/2} \text{ milliamperes} \quad (142\text{-}1)$$

where K is a constant that is determined by the dimensions of the tube.

This yields,

$$I_b = 0$$

when,

$$E_c = -\frac{E_b}{\mu} \text{ volts} \quad (142\text{-}2)$$

Consequently, the value of the cutoff bias on the grid is $-E_b/\mu$ volts.

Example 142-1

A triode has an amplification factor of 40, and its plate voltage is +240 V. What is the value of its cutoff bias?

Solution

$$\text{Cutoff bias} = -\frac{E_b}{\mu} \quad (142\text{-}2)$$

$$= -\frac{240 \text{ V}}{40}$$

$$= -6 \text{ V}.$$

PRACTICE PROBLEMS

142-1. For a particular triode, $K = 0.0035$, $E_c = -2$ V, $E_b = +180$ V, and $\mu = 40$. Determine the value of the plate current in milliamperes.

142-2. In Practice Problem 142-1, the value of E_c is changed to −5 V. Assuming that all other values remain unchanged, obtain the new value of the plate current.

142-3. In Practice Problem 142-1, the value of E_b is changed to +70 V. Assuming that all other values remain unchanged, calculate the new value of the plate current.

142-4. In Practice Problem 142-1, what is the value of the cut-off bias?

143 Ac and dc plate resistances

The ac plate resistance (r_p) is a static parameter that measures the control that the plate voltage (E_b) exercises over the plate current (I_b). Figure 143-1B shows a typical family of plate characteristic curves (chapter 142); these are the graphs of the plate current versus the plate voltage, and are obtained from the experimental set-up of Fig. 143-1A. Assume that the grid voltage (E_c) is 2 V negative with respect to the grounded cathode; this is achieved by adjusting the setting of the potentiometer X. At the same time, adjust the potentiometer (Y) to apply +200 V to the plate. From the plate characteristics, observe that the corresponding plate current is 10 mA (point A). The dc plate resistance (R_p) of the triode is then 200 V/10 mA = 20 kΩ; of course, this resistance value is not a constant because the triode is a nonlinear device and, therefore, its dc resistance depends on the particular operating point that you have selected.

Under dynamic (amplifier) conditions, the plate voltage and the plate current are continuously changing, so that we are more concerned with the (small) change in the plate current that is produced by a corresponding (small) change in the plate voltage. For example, if you increase the plate voltage by 20 V (ΔE_b) to +220 V, the plate current could increase from 10 mA to 12 mA; therefore, the corresponding change in plate current is 12 − 10 = 2 mA (ΔI_b). The ac plate resistance (r_p) is then defined as the ratio of $\Delta E_b : \Delta I_b$ = 20 V : 2 mA = 10 kΩ (the ac plate resistance value is less than the dc resistance value). This means that a 10 V change in the plate voltage produces a 1 mA change in the plate current.

Because a high r_p value means that a large change in the plate voltage only produces a small change in the plate current, the ac plate resistance is an *inverse* measurement of the control that the plate voltage exercises over the plate current. In strict mathematical terms, the ac plate resistance is the reciprocal of the characteristic's slope at the operating point A.

The value of the ac plate resistance is not a constant, but depends on the chosen operating point.

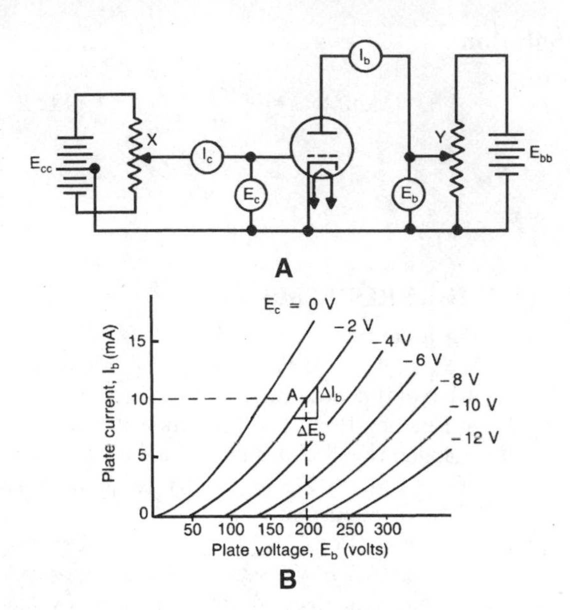

Figure 143-1

MATHEMATICAL DERIVATIONS

Plate resistance, dc,

$$R_p = \frac{E_b}{I_b} \text{ ohms} \qquad (143\text{-}1)$$

Plate resistance, ac,

$$r_p = \frac{\Delta E_b}{\Delta I_b} \text{ ohms} \qquad (143\text{-}2)$$

while maintaining the value of E_c at a constant level.

At the operating point A,

ac plate resistance,

$$r_p = \left(\frac{\delta E_b}{\delta I_b}\right)(E_c \text{ constant}) \text{ ohms} \qquad (143\text{-}3)$$

From Equation 142-1, $I_b = K(E_c + E_b/\mu)^{3/2}$

Therefore,

$$\left(\frac{\delta I_b}{\delta E_b}\right)(E_c \text{ constant}) = \frac{3K}{2\mu}\left(E_c + \frac{E_b}{\mu}\right)^{1/2}$$

or,

ac plate resistance,

$$r_p = \left(\frac{\delta E_b}{\delta I_b}\right)(E_c \text{ constant}) \qquad (143\text{-}4)$$

$$= \frac{2\mu}{3K}\left(E_c + \frac{E_b}{\mu}\right)^{-1/2} \text{ ohms}$$

Equation 143-4 shows that the value of r_p depends on the values of E_c and E_b at the chosen operating point.

Example 143-1

On a family of plate characteristics, the operating point is specified by $E_c = -3$ V, $I_b = 6$ mA, $E_b = 180$ V. Without altering the grid voltage, the plate voltage is reduced to 160 V, and the plate current then falls to 4.5 mA. What are the values of the dc and ac plate resistances?

Solution

Plate resistance, dc,

$$R_p = \frac{E_b}{I_b} \qquad (143\text{-}1)$$

$$= \frac{180 \text{ V}}{6 \text{ mA}}$$

$$= 30 \text{ k}\Omega.$$

Plate resistance, ac,

$$r_p = \frac{\Delta E_b}{\Delta I_b} \qquad (143\text{-}2)$$

$$= \frac{180 \text{ V} - 160 \text{ V}}{6 \text{ mA} - 4.5 \text{ mA}}$$

$$= \frac{20 \text{ V}}{1.5 \text{ mA}}$$

$$= 13.33 \text{ k}\Omega.$$

PRACTICE PROBLEMS

143-1. A triode's operating point is defined by $E_b = +250$ V, $I_b = 10$ mA, and $E_c = -4$ V. When the plate voltage is varied from +240 V to + 260 V, while maintaining the grid voltage at −4 V, the plate current changes from 9.4 mA to 10.6 mA. Calculate the value of the triode's ac plate resistance.

143-2. In Practice Problem 143-1, determine the value of the triode's dc plate resistance.

143-3. At a particular operating point of a triode, the ac plate resistance is 8 kΩ. If the plate voltage is changed by 2 V, determine the corresponding change in the plate current.

143-4. In Fig. 143-1B, estimate the value of the ac plate resistance at the operating point where $I_b = 5$ mA, and $E_c = -8$ V.

143-5. In Practice Problem 143-4, estimate the value of the dc plate resistance at the same operating point.

143-6. For a particular triode, $K = 0.0025$, $\mu = 30$, $E_c = -1.5$ V, and $E_b = +120$ V. Calculate the value of the ac plate resistance.

143-7. In Practice Problem 143-6, the value of E_c is changed to -5 V. Assuming that all other values remain unchanged, recalculate the value of the ac plate resistance.

144 Transconductance, g_m

The plate current of a triode tube is simultaneously determined by the voltages that exist on the plate and the grid, relative to the cathode. In chapter 143 you observed the control that the plate voltage exercised over the plate current; this was governed by the value of the ac plate resistance r_p. Now, explore the degree of control that the grid electrode exercises over the plate current.

In the experimental set-up of Fig. 144-1A, the setting of the potentiometer (X) is adjusted so that the grid voltage (E_c) is 2 V negative with respect to the grounded cathode. At the same time, the potentiometer (Y) is adjusted to apply +200 V to the plate. On the plate characteristics of Fig. 144-1B, the corresponding plate current is 10 mA, and you have arrived at the operating point A. The same operating point is shown on the transfer characteristic of Fig. 144-1C; this characteristic is the graph of the plate current (I_b) versus the grid voltage (E_c) while maintaining a constant level of the plate voltage E_b.

Now, alter the setting of the potentiometer (X) so that the grid is 3 V negative with respect to the cathode; at the same time, the potentiometer (Y) is unchanged. The plate current drops from 10 mA to 8 mA (for example) so that a 1-volt *change* on the grid (ΔE_c) is responsible for a 2 mA *change* (ΔI_b) in the

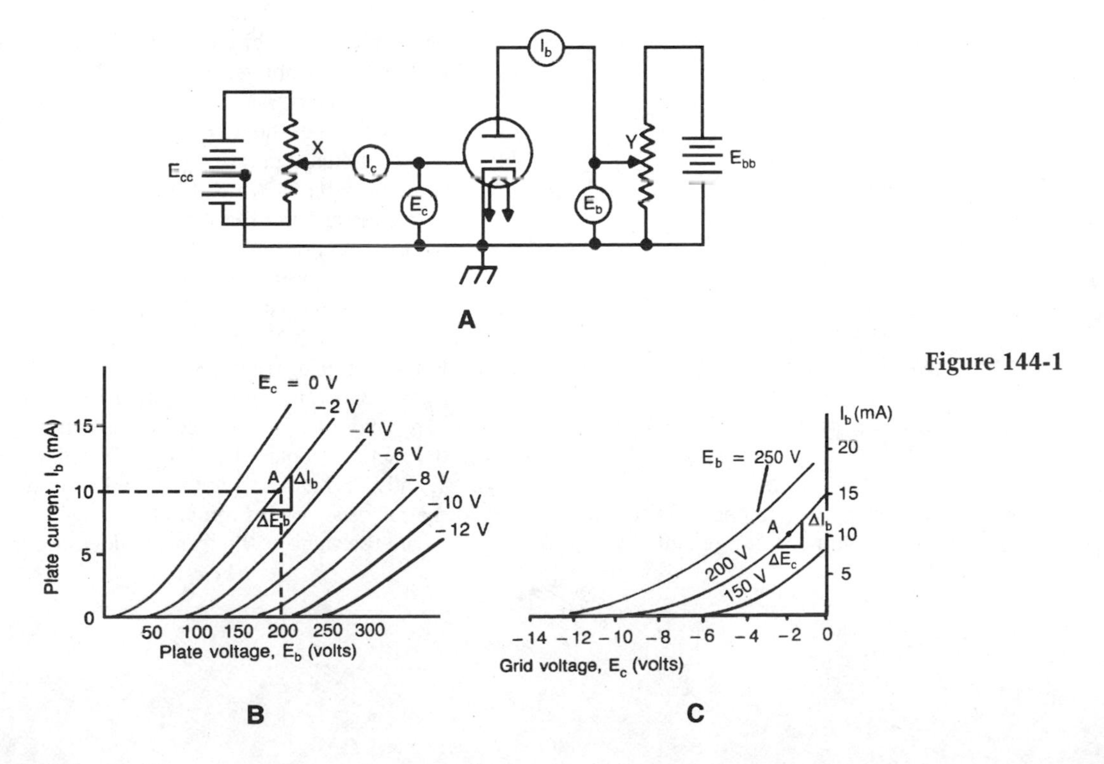

Figure 144-1

value of the plate current. The *transconductance* (g_m) is then defined as the ratio of $\Delta I_b : \Delta I_c$, while maintaining the value of E_b at a constant level. This ratio of current to voltage is a *conductance* (letter symbol, G) which is measured in siemens (S).

In this example, the value of g_m is 2 mA/V, or 2000 μS (however, the tube manuals still refer to micromhos, in which one mho has the same value as one siemens). The transconductance is a direct measure of the control that the grid exercises over the plate current because a high value of g_m means that a small change in the grid voltage produces a shift in the plate current that would, otherwise, only be achieved by a relatively large change in the plate voltage. In strict mathematical terms, the value of g_m is equal to the slope of the transfer characteristic at the operating point A.

Like the value of the ac plate resistance, the magnitude of the transconductance is not a constant, but is dependent on the chosen operating point.

MATHEMATICAL DERIVATIONS

Transconductance,

$$g_m = \frac{\Delta I_b}{\Delta E_c} \text{ siemens} \qquad (144\text{-}1)$$

while maintaining the value of E_b at a constant level.

Transconductance,

$$g_m = \left(\frac{\delta I_b}{\delta E_c}\right) (E_b \text{ constant}) \qquad (144\text{-}2)$$

From Equation 142-1,

$$I_b = K\left(E_c + \frac{E_b}{\mu}\right)^{3/2}$$

Therefore,

Transconductance,

$$g_m = \left(\frac{\delta I_b}{\delta E_c}\right) (E_b \text{ constant}) \qquad (144\text{-}3)$$

$$= \frac{3K}{2}\left(E_c + \frac{E_b}{\mu}\right)^{1/2} \text{ siemens}$$

This shows that the value of g_m depends on the values of E_c and E_b at the chosen operating point.

Example 144-1

On the plate characteristics of a triode, the operating point is specified by $E_c = -3$ V, $I_b = 6$ mA, and $E_b = 180$ V. Without changing the plate voltage, the grid voltage is raised to -2 V, and the plate current increases to 9 mA. What is the value of the transconductance?

Solution

Change in the plate current,

$$\Delta I_b = 9 - 6$$
$$= 3 \text{ mA.}$$

Change in the grid voltage,

$$\Delta E_c = -2 - (-3)$$
$$= 1 \text{ V.}$$

Transconductance,

$$g_m = \frac{\Delta I_b}{\Delta E_c} \qquad (144\text{-}1)$$

$$= \frac{3 \text{ mA}}{1 \text{ V}}$$

$$= 3 \text{ mA/V}$$

$$= 3000\ \mu\text{S}$$

PRACTICE PROBLEMS

144-1. A triode's operating point is defined by $E_b = +250$ V, $I_b = 10$ mA, and $E_c = -4$ V. If the plate voltage is held constant, and the control grid voltage is shifted from -3.8 V to -4.2 V, the change in the plate current is 0.8 mA. Calculate the value of the triode's transconductance.

144-2. At a particular operating point of a triode, the value of g_m is 2500 μS. If the plate voltage is held constant and the grid voltage is changed by 0.5 V, determine the corresponding change in the plate current.

144-3. In Fig. 144-1C, estimate the value of g_m at the point for which $I_b = 5$ mA, $E_b = 250$ V.

144-4. For a particular triode, $K = 0.002$, $E_c = -2$ V, $E_b = +200$ V, and $\mu = 40$. Calculate the value of g_m under these operating conditions.

144-5. In Practice Problem 144-4, E_c is changed to -4 V, and E_b is held constant at $+200$ V. Assuming that all other values remain unchanged, recalculate the value of the transconductance.

145
Amplification factor

The amplification factor (μ) is a static parameter that compares the relative controls that the plate voltage (E_b) and the grid voltage (E_c) exercise over the plate current (I_b). Its value depends on the geometry of the triode and in particular, the size and spacing of the electrodes. Because the control grid is wound close to the cathode, its effect on the plate current is greater than that of the plate, which is positioned further from the cathode.

Referring to Fig. 145-1A, assume that the potentiometer (X) is set so that the grid is 2 V negative with respect to the cathode. At the same time, adjust the potentiometer (Y) to apply +200 V to the plate. From the plate characteristics illustrated in Fig. 145-1B, observe that the plate current is 10 mA (point A). If the potentiometer (X) is now reset to apply −3 V to the grid, the plate current will be reduced to 8 mA (point B). However, you can restore the current to 10 mA if you increase the plate voltage from +200 V to +220 V (point C). Therefore, we can deduce that a 1 V change on the grid can be compensated by a 20 V change on the plate. This means that the grid is 20 times more effective than the plate in controlling the plate current. Therefore, at the operating point (A) the value of the amplification factor (μ) is 20. Notice that μ has no units because it merely compares two voltage changes.

Triodes can be classified in terms of their μ values. Low-μ triodes have an amplification factor of less than 10 and they are primarily power amplifier tubes. By comparison, medium-μ ($\mu \approx 20$) and high-μ ($\mu \approx 100$) triodes are used for the voltage amplification of small signals.

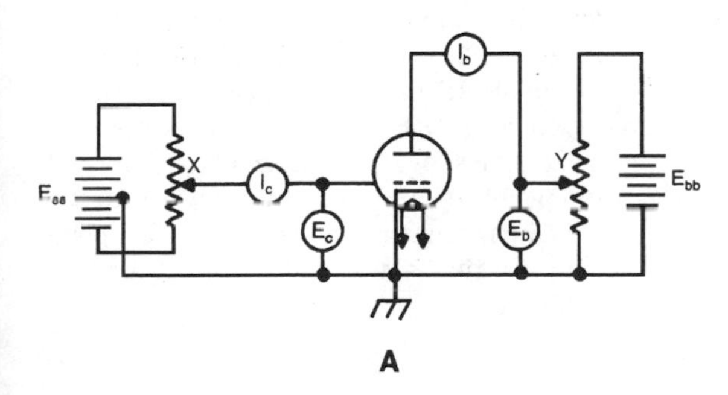

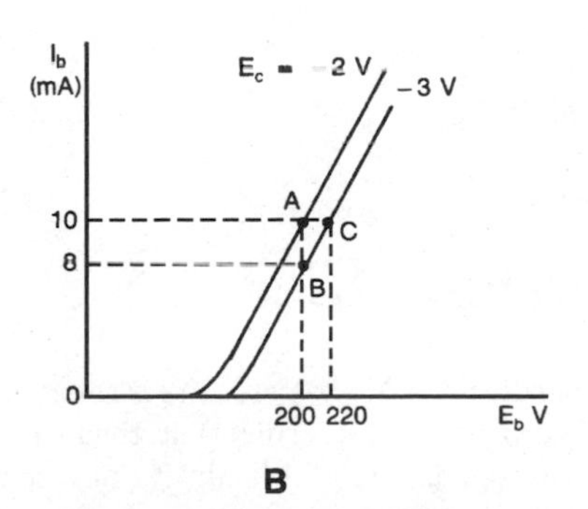

Figure 145-1

MATHEMATICAL DERIVATIONS

Amplification factor,

$$\mu = \frac{\Delta E_b}{\Delta E_c}, \text{ keeping } I_b \text{ constant} \qquad (145\text{-}1)$$

Transconductance,

$$g_m = \frac{\Delta I_b}{\Delta E_c}, \text{ keeping } E_b \text{ constant} \qquad (144\text{-}1)$$

Plate resistance,

$$r_p = \frac{\Delta E_b}{\Delta I_b}, \text{ keeping } E_c \text{ constant} \qquad (143\text{-}2)$$

Then,

$$\mu = \frac{\Delta E_b}{\Delta E_c} \qquad (145\text{-}2)$$

$$= \frac{\Delta E_b}{\Delta I_b} \times \frac{\Delta I_b}{\Delta E_c}$$

$$= r_p \times g_m$$

This relationship is only true provided the values of μ, r_p, and g_m are all measured at the same operating point. Alternatively, from Equations 143-4 and 144-3,

$$r_p = \frac{2\mu}{3K}\left(E_c + \frac{E_b}{\mu}\right)^{-1/2}$$

and,

$$g_m = \frac{3K}{2}\left(E_c + \frac{E_b}{\mu}\right)^{1/2}$$

then,

$$r_p \times g_m = \mu \qquad (145\text{-}2)$$

Example 145-1

In Fig. 145-1A the potentiometers (X and Y) are set to provide the initial operating point: $E_b = +250$ V, $E_c = -6$ V, and $I_b = 8$ mA. When E_c is changed to -4 V, I_b increases to 12 mA. The plate current is restored to 8 mA by reducing the plate voltage to +200 V. Calculate the value of the amplification factor.

Solution

Change in the plate voltage, $\Delta E_b = 250 - 200 = 50$ V. Corresponding change in the grid voltage, $\Delta E_c = 6 - 4 = 2$ V.

Amplification factor,

$$\mu = \frac{\Delta E_b}{\Delta E_c} = \frac{50 \text{ V}}{2 \text{ V}} \qquad (145\text{-}2)$$

$$= 25$$

Notice that the value of μ is not a constant, but it depends to some degree on the chosen operating point.

Example 145-2

In Example 145-1, what is the value of the grid cutoff bias if $E_b = +250$ V?

Solution

Grid cutoff bias,

$$E_c = -\frac{E_b}{\mu} \qquad (142\text{-}2)$$

$$= -\frac{250 \text{ V}}{25}$$

$$= -10 \text{ V}$$

Example 145-3

At a triode's operating point, $r_p = 15$ kΩ and $g_m = 4000$ μS. What is the value of the amplification factor?

Solution

Amplification factor,

$$\mu = r_p \times g_m \qquad (145\text{-}2)$$

$$= 15 \times 10^3 \times 4000 \times 10^{-6}$$

$$= 60.$$

Notice that μ is actually a negative number because, if the grid is made more *negative*, the plate voltage must be made more *positive* in order to maintain the same level of plate current.

PRACTICE PROBLEMS

145-1. A triode's operating point is defined by $E_b = +250$ V, $I_b = 10$ mA, and $E_c = -4$ V. At this point, it is determined that $r_p = 16.7$ kΩ, and $g_m = 2000$ μS. For the same point, calculate the value of the amplification factor.

145-2. In Practice Problem 145-1, the control grid voltage is changed from -4 V to -3.6 V. To what value must the plate voltage be shifted in order to restore the plate current to the 10-mA level?

145-3. In Practice Problem 145-1, the plate voltage is changed from +250 V to +275 V. To what value must the grid voltage be shifted in order to restore the plate current to the 10-mA level?

145-4. In Fig. 144-1B, estimate the value of the amplification factor (μ) in the vicinity of the operating point defined by $I_b = 5$ mA and $E_c = -10$ V.

145-5. In Fig. 144-1C, estimate the value of the amplification factor in the vicinity of the operating point that is defined by $I_b = 5$ mA and $E_b = 200$ V.

146
The triode tube as an amplifier

Figure 146-1 shows the triode as an amplifier tube in the simplest possible way. The signal to be amplified is some type of alternating voltage (e_{in}) (for example, sine wave, pulse, sawtooth, square wave) which is applied to the control grid. E_{cc} is a steady dc voltage that is supplied from a "C" battery, and is referred to as the bias. The value of bias is such that, throughout the cycle of the signal, the grid is always negative with respect to the cathode. The plate is maintained at a high positive potential by the "B" battery, which provides a

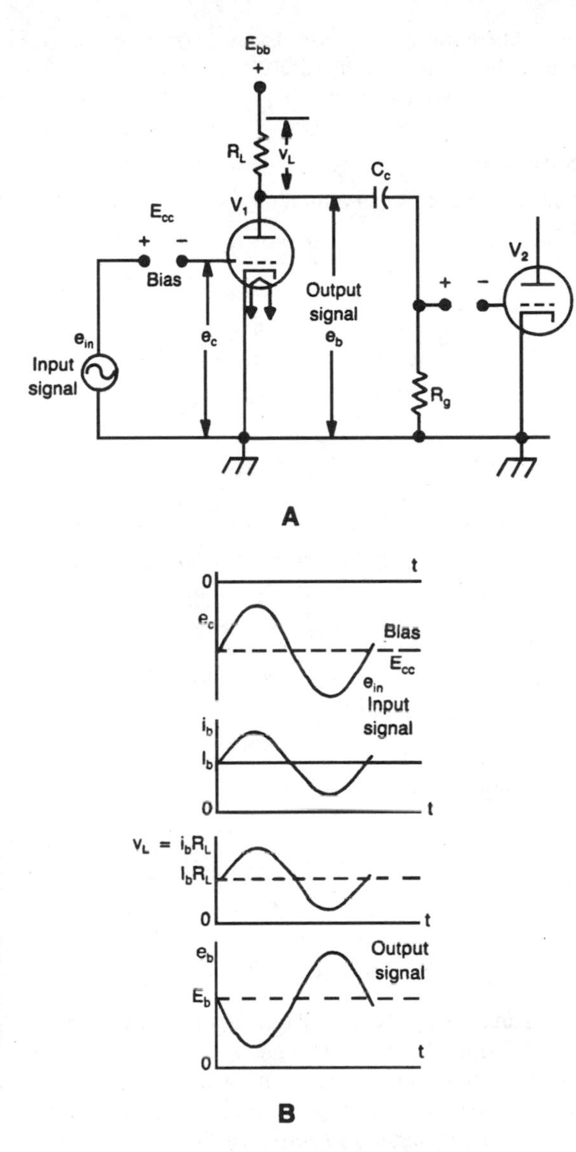

Figure 146-1

voltage (E_{bb}) in series with the tube and the resistor R_L. Plate current will flow, and there will be a voltage drop across the amplifier's load, R_L. Consequently, the voltage at the plate (E_b), with respect to the grounded cathode, will always be less than the value of E_{bb}.

Under quiescent (dc) conditions [when no signal is being applied to the grid (e_{in} = 0 V)], the bias voltage (E_{cc}) will determine the value of the steady plate current (I_b) and the steady plate voltage, E_b. With signal conditions, an alternating voltage is applied to the control grid. This creates a fluctuating plate current, which will now contain an alternating current component, i_b. This component will develop an alternating voltage drop across R_L. At all times, because the sum of the voltages across the load and the tube must equal the fixed value of E_{bb}, an alternating component (e_b) will appear in the waveform of the plate voltage. This fluctuation in the voltage at the plate is the amplified output signal. To determine the actual voltages and currents in the circuit, combine the dc levels with the signal values by using the principle of superposition (chapter 29).

Notice that, under the dynamic conditions in an amplifier circuit, the three quantities e_c, e_b, and i_b are varying simultaneously. This is in contrast with the laboratory conditions that were used when developing the tube's static parameters. In the procedures for calculating these parameters, two of the quantities were varied, and the third quantity was kept constant.

When the grid voltage (e_c) is becoming less negative, the plate current (i_b) is increasing, and there is a greater voltage drop across the load, R_L. Therefore, the plate voltage (e_b) across the tube decreases so that e_c and e_b are 180° out of phase. This is often described as the phase inversion that occurs as the signal is transferred through the tube from the input control grid to the output plate.

MATHEMATICAL DERIVATIONS

Quiescent (dc) conditions

Plate voltage,

$$E_b = E_{bb} - I_b R_L \text{ volts} \tag{146-1}$$

Signal conditions

The plate current (I_b) is a function of the plate voltage (E_b) and the control grid voltage E_c. Therefore,

$$I_b = f(E_b, E_c) \tag{146-2}$$

$$\Delta I_b = \left(\frac{\delta I_b}{\delta E_b}\right)(E_b \text{ constant}) \times \Delta E_b$$

$$+\left(\frac{\delta I_b}{\delta E_c}\right)(E_b \text{ constant}) \times \Delta E_c$$

$$i_b = \frac{1}{r_p} \times e_b + g_m \times e_c \text{ amperes}$$

But,

$$e_b = -i_b R_L \text{ volts} \tag{146-3}$$

where the negative sign indicates the 180° phase change between the input and output signals. Combining Equations 146-2 and 146-3,

$$i_b = \frac{-i_b R_L}{r_p} + g_m \times e_c$$

This yields,

$$i_b = \frac{r_p g_m e_c}{r_p + R_L} = \frac{-\mu e_c}{r_p + R_L} \text{ amperes} \qquad (146\text{-}4)$$

Equation 146-4 indicates that the amplifier behaves as a constant voltage source whose open circuit output is $-\mu e_c$, and whose internal resistance is r_p (Fig. 146-2).

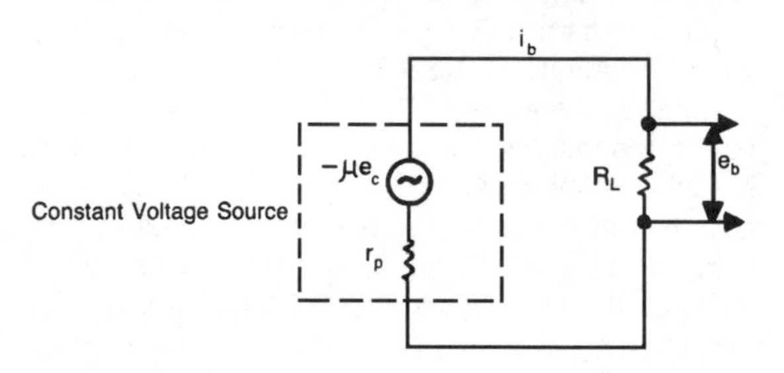

Figure 146-2

The output signal is,

$$e_b = i_b R_L = -\frac{\mu R_L e_c}{r_p + R_L} \text{ volts} \qquad (146\text{-}5)$$

Voltage gain,

$$G_V = \frac{\textit{output signal}}{\textit{input signal}} = \frac{e_b}{e_c} = \frac{\mu R_L}{r_p + R_L} \qquad (146\text{-}6)$$

The negative sign in the expression G_V is dropped because μ itself is actually a negative number. Normally, the value of R_L is about five times greater than the value of r_p, so that the voltage gain approaches the level of μ.

It is common practice to connect the output signal to the next stage for further amplification. This is achieved by a coupling capacitor (C_c), which blocks the dc level of the plate voltage from being applied to the grid of the following stage. If this capacitor shorts, the tube V2 is saturated, and the output signal of the next stage is reduced to a very low level.

The use of the plate load resistor and the coupling capacitor is referred to as RC coupling. However, the following grid resistor (R_g) is effectively in parallel with the plate load (R_L); this reduces the effective load, so that the voltage gain is lowered to a value given by

Stage gain,

$$G_V = \frac{\mu \times R_L \| R_g}{r_p + R_L \| R_g} \qquad (146\text{-}7)$$

Example 146-1

A triode amplifier has an ac plate resistance of 5 kΩ, an amplification factor of 25, and a plate load resistor of 33 kΩ. What is the voltage gain of the amplifier? If the amplifier is now connected to the next stage, whose grid resistor has a value of 100 kΩ, what is the new value of the voltage gain?

Solution

Voltage gain of the amplifier,

$$G_V = \frac{\mu R_L}{r_p + R_L} \qquad (146\text{-}6)$$

$$= \frac{25 \times 33\text{ k}\Omega}{5\text{ k}\Omega + 33\text{ k}\Omega}$$

$$= \frac{25 \times 33}{38}$$

$$= 21.7$$

When the amplifier is connected to the next stage,

$$\text{Effective load} = 33\text{ k}\Omega \| 100\text{ k}\Omega$$

$$= \frac{33 \times 100}{100 + 33}$$

$$= \frac{33 \times 100}{133}$$

$$= 24.8\text{ k}\Omega.$$

New voltage gain,

$$G_V = \frac{25 \times 24.8\text{ k}\Omega}{5\text{ k}\Omega + 24.8\text{ k}\Omega}$$

$$= \frac{25 \times 24.8}{29.8}$$

$$= 20.8$$

As an example (Fig. 146-1B), assume that $E_{cc} = -5$ V, $I_b = 8$ mA, $R_L = 10$ kΩ, and $E_{bb} = +250$ V. Notice that capital letters represent the steady quiescent values, but lowercase letters are used for the signal values. The dc voltage drop across the load is 8 mA × 10 kΩ = 80 V; therefore, E_b = +250 − 80 = +170 V (with respect to the grounded cathode).

When a sine-wave voltage (e_{in}) of 1.5 V peak is applied to the control grid, the triode's assumed parameters are such that the fluctuation in the plate current waveform has a peak value of 3 mA. The plate current, therefore, varies between 8 + 3 = 11 mA and 8 − 3 = 5 mA; these extremes correspond to plate voltages of + 250 V − (11 mA × 10 kΩ) = + 140 V, and + 250 V − (5 mA × 10 kΩ) = +200 V. The plate voltage waveform then consists of a + (200 + 140)/2 V = +170-Vdc level, together with an alternating component (output signal) of (200 − 140)/2 = 30 V peak.

The input signal has a peak value of 1.5 V, and the voltage gain (G_V) of the amplifier is 30 V/1.5 V = 20.

Because G_V is a voltage ratio, it has no units. The waveforms of e_c, i_b and e_b are shown in Fig. 146-1B.

PRACTICE PROBLEMS

146-1. In Fig. 146-1A, $E_{cc} = -4$ V, $E_{bb} = +200$ V, $R_L = 10$ kΩ, and $I_b = 18$ mA. Calculate the value of the plate voltage with respect to ground. What is the value of the plate dissipation for the tube?

146-2. In Practice Problem 146-1, the values of μ and r_p (at the tube's operating point) are 30 and 5 kΩ respectively. Calculate the value of the amplifier's voltage gain.

146-3. In Practice Problem 146-2, the amplifier is connected to a following stage, which contains a 470-kΩ grid resistor. Calculate the value of the stage gain from the grid of the first stage to the grid of the second stage.

146-4. In Practice Problem 146-3, the input sinewave signal is a single tone with a peak value of 3 V and a frequency of 1 kHz. Calculate the peak value of the voltage signal swing at the plate.

146-5. In Practice Problem 146-4, determine the signal power output from the amplifier.

146-6. In Practice Problem 146-5, determine the value of the dc power input to the amplifier. Calculate the value of the plate efficiency as defined by:

$$\frac{\textit{signal power output}}{\textit{dc power input}} \times 100\%$$

147 Dynamic characteristics

The mutual characteristics so far considered have shown how I_b varies with E_c, provided that E_b is kept constant. Similarly, in the case of the plate characteristics, there was the proviso that E_c be kept constant. These characteristics are known as the "static characteristics," and give certain information about the tube itself, making possible the choice of suitable tubes and suitable working conditions for any particular purpose.

If a tube is connected in a particular circuit (as in Fig. 147-1A) with a plate load resistance (R_L), then if the potential on the grid is varied, the potential on the plate is also varied. This did not, however, occur under the conditions when the static mutual characteristics were plotted with a constant plate voltage. When the plate current changes, so does the voltage drop across R_L; because the plate voltage is applied from a source of

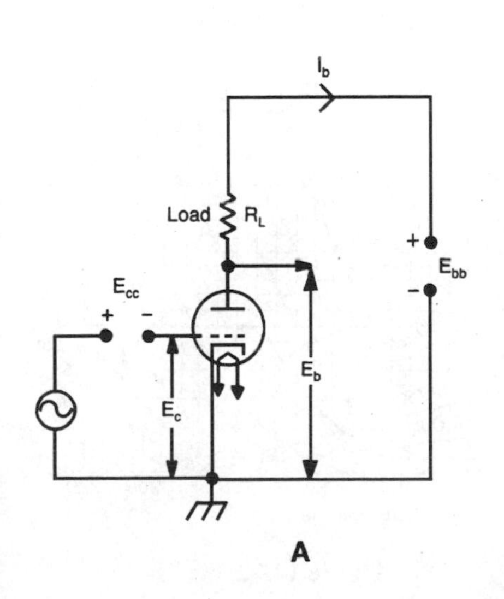

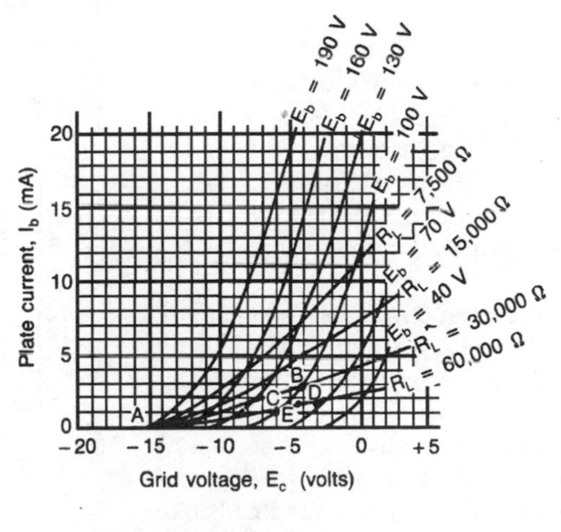

Figure 147-1

constant voltage E_{bb}, plate voltage E_b will change. To get a true picture of what is happening, a set of characteristics is required giving the variations of I_b with E_c, subject to the simultaneous and consequent variations of E_b (the extent of which will vary with the load resistance). Such a family of characteristics is called a set of *dynamic* mutual characteristics, and would be indicative not of the tube itself, but of the tube when connected to a particular value of the plate load. This would appear to necessitate a set of dynamic characteristics for every value of load resistance; fortunately, the dynamic characteristics that correspond to any particular value of load resistance can be deduced from the static characteristics. For this reason, only the static characteristics are found on tube data sheets.

Consider the case where the available plate supply voltage (E_{bb}) is 190 volts and the tube's static mutual characteristics are as illustrated in Fig. 147-1B. For simplicity, assume a purely resistive load of 30,000 Ω.

Now, when $I_b = 0$, there will be no potential drop across the plate load, and $E_b = 190$ volts. From the static characteristic corresponding to 190 volts, it is seen that $I_b = 0$ corresponds to $E_c = -15$ volts. The point *A* is therefore on the dynamic characteristic.

When $I_b = 3$ mA, the potential drop across the resistive load will be 90 volts so that $E_b = 100$ volts; the point *B*, corresponding to $E_b = 100$ volts and $I_b = 3$ mA, will therefore lie on the dynamic characteristic. In the same way, by assuming other values for I_b, the dynamic characteristic corresponding to a resistive plate load of 30,000 ohms (or for any other value of resistance) can be plotted completely. A number of these dynamic characteristics are shown in Fig. 147-1B.

These facts are at once apparent:

(a) The characteristics are (for the most part) very straight, except for slight curvature near the cutoff region.

(b) The higher the load resistance, the less is the slope of the dynamic characteristic, and vice-versa.

(c) The smaller the load resistance, the more nearly does the dynamic characteristic coincide with the static characteristic for $E_b = 190$ volts and the greater is the curvature at the lower end.

After the dynamic characteristic corresponding to an available plate supply voltage of 190 V and a load resistance of 30,000 ohms has been deduced, the operating point on that characteristic (for example, $E_c = -4.5$ volts, corresponding to a plate current of 2.8 mA) can be chosen. Now, suppose that an alternating voltage is applied to the grid, in addition to the steady bias of −4.5 volts. Let the peak value of this signal be 1.5 volts; then the grid potential will vary between −6 volts and −3 volts. It is now apparent that, for the given load of 30,000 Ω, the variations in the plate current will all lie on the dynamic mutual characteristic corresponding to that value of load resistance. Therefore, I_b will vary between 2.3 mA and 3.3 mA; that is, a total variation of 1 mA, or 0.5 mA on either side of the "no signal" (or quiescent) current of 2.8 mA. The change in the plate current will be proportional to the change in the grid voltage, and an undistorted signal will result, provided that the dynamic characteristic is straight throughout the range of the variation of the grid voltage. For this reason, the operating point is chosen in the center of the straight portion of the characteristic, which lies in the range of negative values of grid potential. This allows the maximum voltage signal to be applied to the grid without causing distortion. Generally speaking, the voltage on the grid, relative to the cathode, must always be sufficiently negative to prevent the flow of grid current and yet, on the other hand, not so negative as to cause operation over the lower curved portion of the dynamic characteristic.

With a signal of 1.5 volts peak voltage on the grid, an alternating anode current of 0.5 mA peak value flows in the load resistance of 30,000 Ω, thereby developing a peak voltage of 15 volts across the load. The voltage gain is therefore 15 V/1.5 V = 10.

THE LOAD LINE

Corresponding to the dynamic mutual characteristics, are the *load lines* of the family of the static plate characteristics. Figure 147-2 shows the static plate characteristics for the same triode tube, together with the load lines for the various values of the plate current and the plate voltage. This implies a minimum level of distortion. In choosing a load, a value must be selected such that the corresponding load line makes equal intercepts on

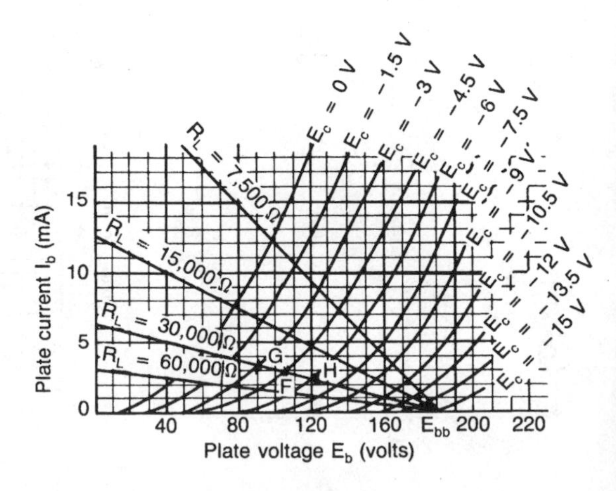

Figure 147-2

the plate characteristics. The selection of a load line making equal intercepts is equivalent to choosing a dynamic mutual characteristic with a straight portion.

From Fig. 147-1A, it is clear that $E_b = E_{bb} - I_b R_L$. If R_L and E_b are constants, the two variables (I_b and E_b) can be plotted in the form of a graph. This is a straight line that will clearly pass through the points given by: (1) $E_b = E_{bb}$, $I_b = 0$, and (2) $E_b = 0$, when $I_b = E_{bb}/R_L$. This line is called the *load line* for the particular load considered and the E_{bb} supply available. Consider the load line for $R_L = 30{,}000\ \Omega$ and $E_b = 190$ volts. Assume that $E_c = -4.5$ volts. The -4.5 volts plate characteristic (E_c) meets the 30,000-Ω load line at point F, corresponding to $E_b = 106$ volts and $I_b = 2.8$ mA. This is taken as the operating point for $E_c = -4.5$ volts.

Now, let a signal of peak value 1.5 volts be applied to the grid so that E_c will vary between -3 volts and -6 volts (points G and H). From the points of intersection of the corresponding characteristics with the 30,000-Ω load line, it can be seen that I_b varies between 3.3 mA and 2.3 mA, and E_b between 91 and 121 volts. Consequently, for equal swings of grid voltage about the standing bias, there are approximately equal swings in the values of the plate voltage, which has an amplification of 10. This is known as the *voltage gain* (G_V), whose value depends on the value of the load resistance.

Assume that the load is now changed to 60,000 Ω. Again, choosing $E_c = -4.5$ volts, you can see that a 1.5-volt peak signal will cause an alternating plate current of peak value 0.275 mA (points C and D, Fig. 147-1B) about the steady value of 1.60 mA (point E). This gives a peak alternating voltage of 16.5 volts across the 60,000-Ω load so that the voltage gain increases to 11.

The higher the load resistance, the higher is the voltage gain but the voltage gain is always less than the amplification factor (μ) of the tube, as derived from its static characteristics. In this example, the amplification factor of the tube is 11.5.

Just as the static transfer and plate characteristics are exactly equivalent (as far as imparting information about the tube itself), the dynamic mutual characteristic and the load line are equivalent ways of expressing the behavior of the tube with a given resistive plate load. It is, however, somewhat easier to detect inequality of the intercepts on the load line than it is to detect a slight curvature of the dynamic characteristic; therefore, the load line method of choosing operating conditions is the one that is usually used.

Example 147-1

Referring to Fig. 147-1B, $E_{bb} = 190$ V and $R_L = 15$ kΩ. If $E_{cc} = -6$ V, what is the dc level of the plate current? A signal, whose rms value is 1.414 V, is applied to the control grid. What are the values of the amplifier's voltage gain and what is the amount of dc power drawn from the E_{bb} supply?

Solution

From the dynamic mutual characteristic corresponding to $R_L = 15$ kΩ, $I_b = 4$ mA when $E_{cc} = -6$ V. A sine-wave signal, whose rms value is 1.414 V, has a peak value of 1.414 V × 1.414 = 2 V. The grid voltage (E_c), therefore, swings between $-6 + (-2) = -8$ V and $-6 - (-2) = -4$ V.

The plate current corresponding to $E_c = -8$ V is 2.8 mA. The plate current corresponding to $E_c = -4$ V is 5.2 mA. The peak value of the plate current fluctuation is 4 mA − 2.8 mA = 1.2 mA (or 5.2 mA − 4 mA = 1.2 mA). The peak value of the output signal = 1.2 mA × 15 kΩ = 18 V. The amplifier's voltage gain = 18 V/2 V = 9. The dc power drawn from the E_{bb} supply = 4 mA × 190 V = 760 mW.

Example 147-2

Referring to Fig. 147-2, $E_{bb} = 190$ V and $R_L = 15$ kΩ. If $E_c = -6$ V, what is the dc level of the plate current? A sine-wave signal, whose rms value is 1.414 V, is applied to the control grid. What is the (rms) value of the output signal voltage? Calculate the amplifier's voltage gain.

Solution

From Fig. 147-2, the dc level of the plate current is 6 mA and the corresponding plate voltage is 134 V. The peak value of the signal input is 1.414 V × 1.414 = 2 V. The grid voltage (E_c), therefore, swings between $-6 + (-2) = -8$ V and $-6 - (-2) = -4$ V. The plate voltage corresponding to $E_c = -8$ V is 152 V. The plate voltage corresponding to $E_c = -4$ V is 116 V. The peak value of the plate voltage fluctuation is 152 V − 134 V = 18 V (or 134 V − 116 V = 18 V). Therefore, the rms value of the output signal is 18 V × 0.707 = 12.73 V. The amplifier's voltage gain = 18 V/2 V = 9.

These results are the same as those obtained in Example 147-1.

PRACTICE PROBLEMS

147-1. In Fig. 147-1B, $E_{bb} = 190$ V, $R_L = 30$ kΩ, and $E_{cc} = -7$ V. What are the values of the plate current's dc level and what is the amount of the dc power drawn from the plate supply voltage?

147-2. In Practice Problem 147-1, a sine-wave signal with a peak-to-peak value of 4 V is applied to the control grid. Estimate the peak values of the swings in the plate voltage and the plate current.

147-3. In Practice Problem 147-2, what is the value of the voltage gain and what is the percentage of plate efficiency?

147-4. In Fig. 147-2, E_{bb} = 190 V and R_L = 30 kΩ. If E_c = −4.5 V, what is the value of the dc level of plate current and what is the amount of dc power drawn from the plate supply voltage?

147-5. In Practice Problem 147-4, a sine-wave voltage signal whose peak-to-peak value is 6 V is applied to the control grid. Estimate the peak values of the swings in the plate voltage, and in the plate current.

147-6. In Practice Problem 147-5, what are the values of the voltage gain and what is the percentage of plate efficiency?

148 Types of bias

The *bias* is a dc voltage that is applied in series with the signal between the control grid and the cathode. The polarity of this voltage is such as to make the grid negative, with respect to the cathode, and thereby establish the operating point on the dynamic transfer characteristic (Fig. 148-1A). The position of this point determines the amplifier's class of operation. The various classes of operation are as follows:

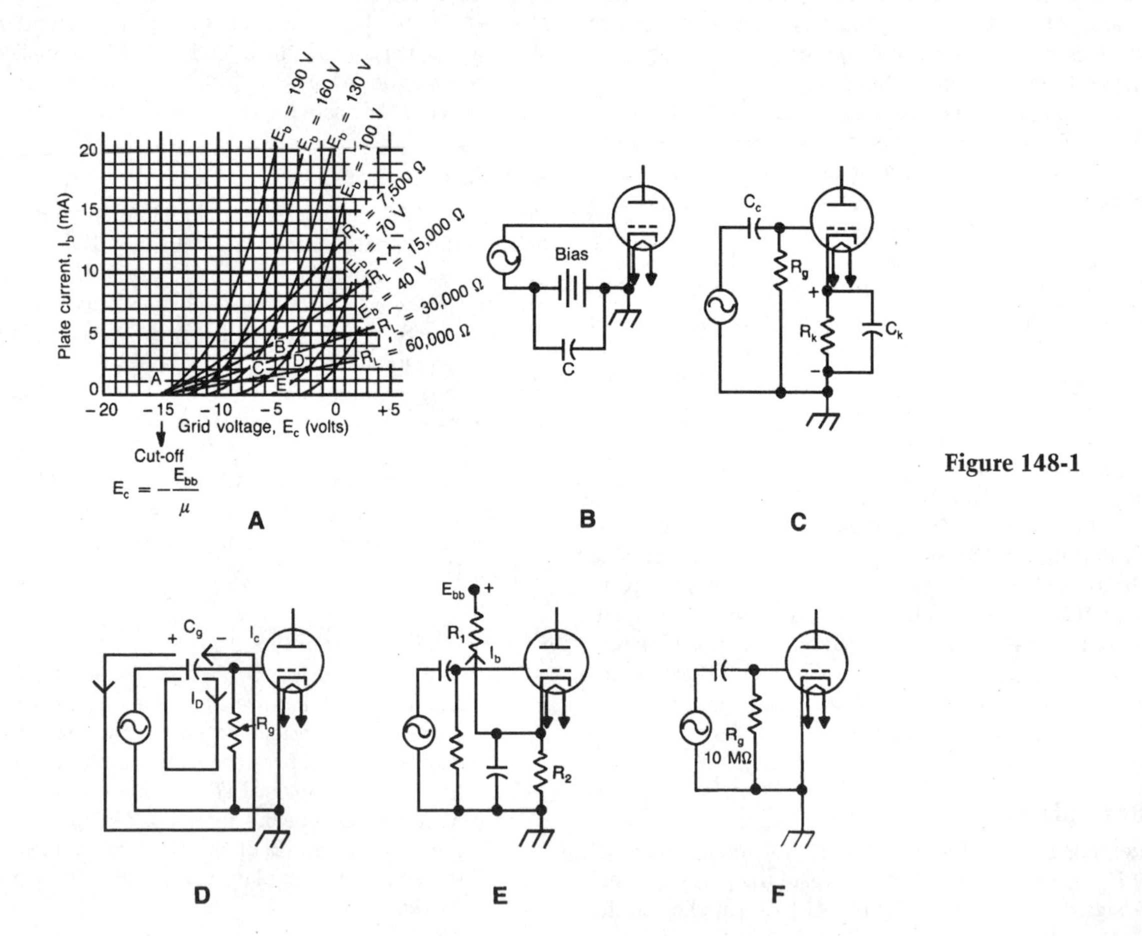

Figure 148-1

Class A

This requires that the bias point is approximately halfway between $E_c = 0$ V and the cutoff point ($E_c = -E_{bb}/\mu$ V). For this reason, class-A operation is sometimes referred to as *midpoint bias*, which requires a value of approximately $-E_{bb}/(2\mu)$ volts. This type of operation ensures minimum distortion of the output signal waveform. Therefore, it is commonly used in audio amplifiers for speech and music. However, the corresponding dc level of plate current is high so that the plate efficiency:

$$\frac{\textit{signal power output from the plate}}{\textit{dc power input from the } E_{bb} \textit{ supply}} \times 100\%$$

is low, and is typically less than 25%. This is of little importance when the signal levels are small (such as in the audio stages immediately following a microphone or in the RF voltage amplifiers of a receiver).

Class B

The triode is biased to the projected cutoff point so that the plate current only flows during the positive half cycle in the input signal; this means that there is considerable distortion in the output signal waveform. However, the dc level of the plate current is low; therefore, it is possible to achieve practical plate efficiencies of 50% to 60%. This type of operation is used in audio push-pull stages that use two matched tubes; the circuit arrangement is such that some of the distortion associated with one tube is cancelled by an opposite distortion created by the other tube. Class B is also used in RF linear stages that are capable of amplifying amplitude-modulated (AM) signals. Because distortion is not a major factor, the amplitude of the signal is as large as possible and, at its positive peak, the grid is driven positive with respect to the cathode so that grid current flows.

Classes AB-1 and AB-2

The bias point lies approximately halfway between class-A and class B operation. The distortion is more severe when compared with class-A operation, but it is possible to achieve plate efficiencies on the order of 35%. One of the main uses of class-AB is in audio push-pull stages; if there were a cascaded series of such amplifiers, the early ones would be biased in class A, subsequent stages would operate in class AB, and the final stage would be a class-B amplifier. When comparing class AB-2 with class AB-1, class AB-2 allows a larger input signal so that at its positive peak, grid current flows. This allows a greater signal power output, but at the expense of increased distortion.

Class C

The bias point is typically between two and four times the cutoff value so that the plate current only flows for about one quarter to one third of the input cycle. The distortion is very severe, but the plate efficiency is as high as 85%. However, this distortion is not a problem because class C is used in RF power amplifiers that have tank circuits as their loads. The tank circuit can then be tuned to resonance at the input signal frequency and unwanted frequencies will then be eliminated. However, in frequency multiplier stages (such as doublers and triplers), the tank circuit can be tuned to the required harmonic. The doubler stage requires a bias of 5 or 10 times the cutoff value and the bias for a tripler stage is 10 to 20 times the cutoff value. These frequency multipliers therefore have a lower plate efficiency (about 50%) than that of the standard class-C power amplifier (up to 85%).

In previous chapters, the bias has been provided by a "C" battery; such a solution is cumbersome and expensive because the battery must periodically be replaced. The battery is bypassed by a capacitor so that none of the signal is lost across the battery's internal resistance (Fig. 148-1B).

An alternative solution is the use of *cathode bias*, which consists of the resistor R_k (Fig. 148-1C). The dc current passes through R_k and develops the bias voltage between the cathode and ground. The direction of the electron flow is from ground to the cathode, which therefore carries a positive potential. Because there is no dc voltage drop across the grid resistor (R_g), the grid is negative with respect to the cathode by the amount of the cathode bias. For example, if the amount of cathode bias required is -5 V (grid relative to cathode), and the corresponding plate current is 2.5 mA, the required value of R_k is 5 V/2.5 mA = 2 kΩ. Because the amount of the cathode bias depends on the dc plate current level, this type of bias is only suitable for class-A or class-AB operation.

Because the signal component of the plate current also passes through R_k, there will be a signal voltage across the bias resistor. This signal voltage is 180° out of phase with the input signal; therefore, it represents negative or degenerative feedback. The result is to reduce the amplifier's voltage gain. To prevent this degeneration, you can include a bypass or decoupling capacitor (C_k), which offers a low reactance (compared with the value of R_k) to the frequencies that are contained in the input signal. Taking music as an example, the frequency range will extend from below 100 Hz to several kHz. A common criterion is to arrange that the reactance of C_k does not exceed ⅒ of the value of R_k at the *lowest* frequency to be bypassed. For audio amplifiers, the values of C_k are normally several microfarads; for receiver RF amplifiers, C_k is 0.01 μF or less.

Grid-leak bias (Fig. 148-1D) is commonly used in class-C RF power amplifier stages and in oscillators. The amount of the bias depends on the amplitude of the input signal that causes grid current to flow at the peak of its positive half cycle. This grid current then charges C_g toward the peak value of the input signal. When the grid current ceases, C_g discharges slowly through R_g, but their time constant is high compared with the period of the input signal. As a result of this action, there is developed across R_g a dc bias level that is determined by the magnitude of the input signal; consequently, this type of bias is sometimes referred to as *signal bias*. If the drive from the preceding stage fails, no signal bias will be generated so that some RF amplifiers include a fixed safety battery bias (or cathode bias) in order to limit the plate current to a "safe" value. Provided that the input signal is sufficiently large, this form of bias can be used to provide class-C operation.

Bleeder bias (Fig. 148-1E) uses a voltage divider (R1and R2), which is connected between E_{bb} and ground. Ignoring the flow of the cathode current through R2, the positive potential on the cathode due to the bleeder current (I_b) is $+ E_{bb} \times R_2/(R_1 + R_2)$ volts. Theoretically, this form of bias could provide any class of operation but in practice, the amount of bias is limited by the maximum dc voltage that can be safely applied between the cathode and the heater.

Contact bias is possible with certain amplifier tubes. The control grid requires a narrow pitch for its spiral of wire and is also wound very close to the cathode. As a result, there is an accumulation of the space charge around the control grid, which carries a bias of approximately -0.5 V to -1.0 V. To maintain this bias (Fig. 148-1F), most of the space charge must not be allowed to leak away to ground through the grid resistor R_g. This is achieved by increasing the value of R_g from a typical value of 470 kΩ to about 10 MΩ. Clearly, contact bias is only suitable for small signals and it is rarely used nowadays.

MATHEMATICAL DERIVATIONS

Cathode bias

Value of cathode bias resistor,

$$R_k = \frac{E_c}{I_k} \text{ ohms} \qquad (148\text{-}1)$$

where: E_c = the value of the required bias (V)
I_k = dc level of the cathode current (A)

Reactance of the cathode bypass capacitor (C_k) is one-tenth the value of the cathode bias resistor. Therefore,

$$\frac{R_k}{10} = \frac{1}{2\pi f C_k}$$

This yields,

$$C_k = \frac{10}{2\pi f R_k} \text{ F} = \frac{10^7}{2\pi f R_k} \ \mu\text{F} \qquad (148\text{-}2)$$

where: f = lowest frequency to be bypassed.

Grid-leak (signal) bias

Value of the grid leak or signal bias,

$$E_c = -I_{c1} \times R_g \text{ volts} \qquad (148\text{-}3)$$

where: R_g = total of all the resistances associated with the grid current (Ω)
I_{c1} = dc level of the control grid current (A).

Bleeder bias

Value of the bleeder bias,

$$E_c = E_{bb} \times \frac{R_2}{R_1 + R_2} \text{ volts} \qquad (148\text{-}4)$$

This formula neglects the flow of the cathode current through the resistor, R2.

Example 148-1

A triode is used in an audio amplifier stage that covers the frequency range of 50 Hz to 15 kHz. The dc level of the plate current is 7 mA and the cathode bias resistor has a value of 1.2 kΩ. What is the amount of the cathode bias and what is the required value for the cathode bypass capacitor?

Solution

Amount of the cathode bias,

$$E_c = I_b \times R_k \qquad (148\text{-}1)$$

$$= 7 \text{ mA} \times 1.2 \text{ k}\Omega$$

$$= 8.4 \text{ V.}$$

Value of the cathode bypass capacitor,

$$C_k = \frac{10^7}{2\pi f R_k} \ \mu\text{F} \qquad (148\text{-}2)$$

$$= \frac{10^7}{2 \times \pi \times 50 \times 1.2 \times 10^3} \mu\text{F}$$

$$= 21 \ \mu\text{F.}$$

A 20-μF, 10-WVdc capacitor would be adequate.

Example 148-3

An RF amplifier uses grid-leak bias with a grid resistor of 47 kΩ. If the level of the control grid current is 1.2 mA, calculate the value of the bias. What is the approximate rms value of the input sine-wave signal?

Solution

Value of the bias,

$$E_c = -I_{c1} \times R_g \quad (148\text{-}3)$$
$$= -1.2 \text{ mA} \times 47 \text{ k}\Omega$$
$$= -56.4 \text{ V}.$$

The level of the bias is approximately equal to the peak value of the signal. Therefore, the rms value of the input signal,

$$e_{rms} \approx 0.707 \times 56.4 \text{ V}$$
$$= 40 \text{ V}.$$

PRACTICE PROBLEMS

148-1. In Fig. 148-1C, the dc level of plate current is 6 mA. If the grid cutoff voltage is −10 V, suggest a suitable value of R_k for class-A operation.

148-2. In Practice Problem 148-1, the lowest frequency contained in the input signal is 100 Hz. Suggest a suitable value for the cathode bypass capacitor.

148-3. In Practice Problem 148-2, the grid resistor has a value of 470 kΩ. Suggest a suitable value for the coupling capacitor, C_C.

148-4. In Fig. 148-1D, the input signal is a sine wave with a rms value of 17 V. Estimate the value of the grid-leak bias.

148-5. In Fig. 148-1D, the grid-leak resistor (R_g) has a value of 56 kΩ. If the dc level of grid current is 1.3 mA, what is the value of the grid-leak bias? Estimate the rms value of the input signal voltage.

148-6. In Practice Problem 148-5, the value of C_g is 200 pF. Estimate the frequency of the input signal.

148-7. In Fig. 148-1E, E_{bb} = +200 V, R_1 = 33 kΩ, and R_2 = 1.2 kΩ. Ignoring the flow of the cathode current through the resistor (R_k) calculate the value of the bleeder bias.

149 The development of the amplifier tube

Within the triode are the cathode, the control grid, and the plate. These electrodes represent conducting surfaces that are separated by the vacuum dielectric. Consequently, there are three interelectrode capacitances whose values are normally on the order of a few picofarads. Referring to Fig. 149-1A, these capacitances are:

1. The grid-cathode capacitance, (C_{gk}), which is effectively in parallel with the input signal circuit. This is called the tube's *input capacitance*.
2. The plate-cathode capacitance, (C_{pk}), which is effectively in parallel with R_L as far as the signal is concerned. This is called the tube's *output capacitance*.
3. The plate-grid capacitance, (C_{pg}), which allows the output signal to drive a current *back* into the input circuit.

INSTABILITY OF THE TRIODE

Of these three capacitances, the C_{pg} has the most important effect, which is covered in greater detail. Clearly, the C_{pg} , particularly at high frequencies, provides a path between the plate and the grid circuits so that the output signal (at the plate) can drive a current back into the grid circuit and develop a *feedback* voltage across the impedance of this circuit; this is commonly referred to as the *Miller effect*. If this feedback voltage is of sufficient magnitude and if it has the correct phase (positive feedback), the circuit might cease to function as an amplifier and become an oscillator. This will occur if the plate load (the tank circuit L1, C1) behaves inductively. However, if the plate tank circuit is capacitive, the phase of the feedback voltage is reversed (negative feedback) and the amplifier's gain is reduced. Neither of these effects will occur when the plate tank circuit is at resonance and behaves resis-

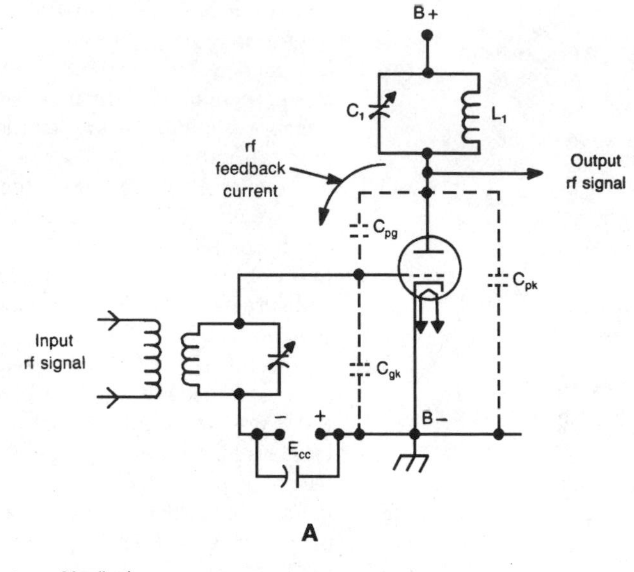

Figure 149-1

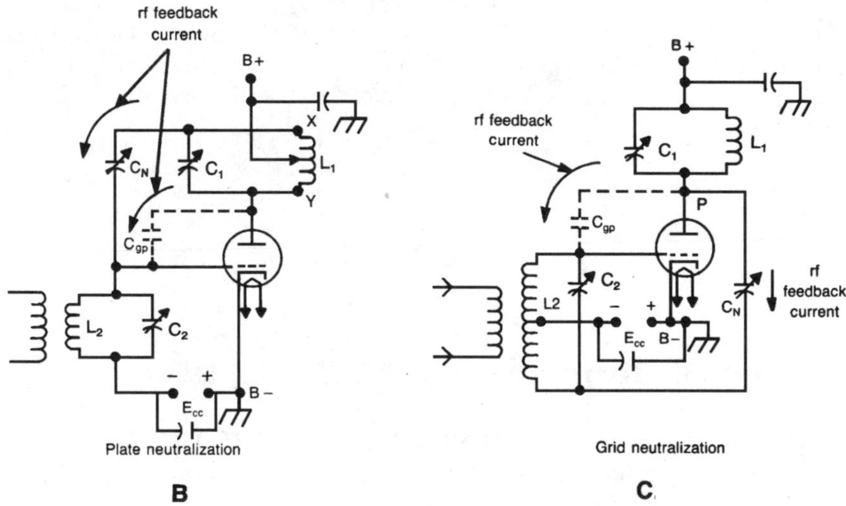

tively but this circuit might drift to become either inductive or capacitive. The triode is therefore unstable as an RF amplifier unless the circuit is neutralized.

Typical neutralization circuits are shown in Figs. 149-1B and C. Here, the purpose is not to eliminate the feedback through the C_{gp}, but to cancel its effect by an opposite feedback through the neutralizing capacitor C_N. In the *plate* neutralization circuit of Fig. 149-1B, a center tap on the *plate* tank coil, L1, is connected to $B+$ and is, therefore, effectively at RF ground. The points (X and Y) are at opposite ends of the coil L1 and, therefore, the RF potentials at these points are equal in magnitude, but 180° out of phase. The feedback from the voltage at Y through the interelectrode capacitance (C_{gp}) is then cancelled by the opposite feedback from the voltage at X through the neutralizing capacitor, C_N. During the neutralizing procedure the value of C_N is varied until complete cancellation is achieved.

In the grid neutralizing circuit of Fig. 149-1C, the grid coil (L2) is center-tapped and is at RF ground. The feedback from the point P through the C_{gp} is taken to the top of the grid coil, and the feedback from the same point through C_N is applied to the bottom of L2. The neutralizing capacitor can then be varied until the overall feedback to the grid circuit is zero.

Historically, the evolution of the various screen grid tubes are a direct consequence of attempts to reduce the plate-control grid capacitance to such an extent that the tube would be stable when used in an RF amplifier circuit. However, some screen-grid tubes have other advantages over triodes, and are commonly used where a voltage amplifier at audio or radio frequencies is required.

THE SCREEN GRID TUBE: THE TETRODE

The first screen-grid tube, or tetrode, was originally introduced to overcome the ill-effects of the grid-plate capacitance; these become apparent when an unneutralized triode is used as an RF amplifier. The screen-grid tube has two grids between the cathode and the plate; the grid nearer the cathode performs exactly the same function as the grid in the triode and is referred to as the "control grid" (G1). The additional grid acts as an electrostatic screen between the control grid and the plate. It is therefore called the *screen-grid* or *screen* (G2). The screen is maintained at a high positive potential (approaching that of the plate) and has a considerable effect on the electron stream between the control grid and the plate.

Consider the electric field between the electrodes in terms of its flux lines. If the screen was a solid metal plate that is maintained at a potential equal to that of the plate, the flux lines that leave the cathode and grid would terminate on the screen; there would be no electric field in the space between the screen grid and the plate. Consequently, there would be no capacitance between the plate and the screen grid, nor between the control grid and the plate. Now, consider that the screen grid is made in the form of a close mesh and that it is maintained at a potential not necessarily equal to, but approaching, that of the plate. The screening effect will be considerable, but not perfect although with a fine mesh structure, it will be practically so. The result is that there will be capacitance between the pairs of electrodes: control grid and screen grid, screen grid and plate, plate and control grid (although the grid-plate capacitance will be very much reduced from that in the triode). In commercial types of screen grid tubes, the residual control grid-plate capacitance varies from 0.001 pF to 0.02 pF, as compared with 2 pF to 8 pF for a triode.

Figure 149-2 shows a typical plate characteristic for a tetrode, drawn under conditions of constant control grid voltage (E_{c1}) and constant screen voltage (E_{c2}). When the plate potential is zero, all the emitted electrons are attracted to the screen, which gives it a fairly high screen current (I_{c2}); the plate current (I_b) will be zero. Now, if the plate potential is increased, some of the electrons that pass through the mesh of the screen are carried on by their momentum and come under the influence of the plate, to which they are attracted; thus, the plate current will increase with a greater plate potential. Because of the shielding effect of the screen, however, the potential of the plate will have very little effect on the electric field in the vicinity of the cathode; an increase in plate potential will not appreciably increase the total space or cathode current. Any increase in plate current will, therefore, be at the expense of the screen current.

As the plate potential increases, so also will the velocity of the electrons on their arrival at the plate. One effect of bombarding the plate with fast-moving electrons is that other electrons may be ejected by the force of impact. The quantity of these ejected electrons (or *secondary electrons,* as they are usually called) will vary with the material of the plate and the velocity of the electrons reaching the plate from the cathode *(primary electrons).* In certain circumstances, as many as 10 secondary electrons can be liberated by one fast-moving primary electron. This phenomenon of *secondary emission* also occurred in the diode and the triode, but in those cases the secondary electrons were attracted back into the plate surface and they had no effect on the tube. With the tetrode, however, the velocity of the primary electrons is sufficiently high to cause secondary emission while the plate is at a lower potential than the screen grid. It follows that there is a tendency for the screen to collect these secondary electrons that are emitted from the plate. The result is an increase in the screen current at the expense of the plate current. A further in-

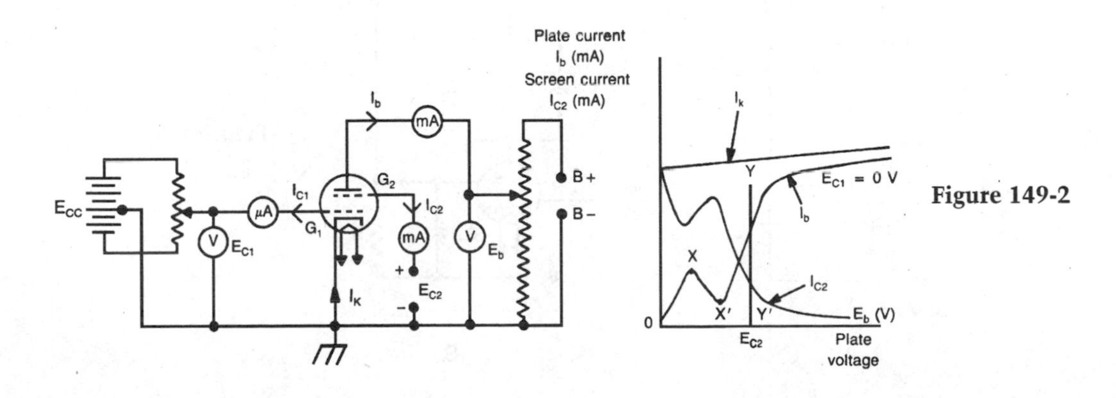

Figure 149-2

crease in the plate potential will increase the velocity of the primary electrons and, therefore, will increase the emission of the secondary electrons. If the screen is still at a higher potential than the plate, it will collect practically all of these slow-moving secondary electrons, with the result that the plate current will actually decrease with a greater plate potential.

This state of affairs is represented by the portion of the plate characteristic *XX′* (Fig. 149-2). Under the operating conditions that control this portion of the characteristic, the tube behaves as a *negative resistance device* because a *decrease* in the plate voltage causes an *increase* in the plate current.

If the plate potential is still further increased, the majority of the secondary electrons will no longer be attracted to the screen. More and more will be drawn back into the plate and the plate current will once more increase with a greater plate potential (at the expense of a decreasing screen current). The portion of the tetrode characteristic that is useful for most purposes is that portion well to the right of the vertical line *YY′* in Fig. 149-2. In this region, the curve becomes practically straight, and the plate current is nearly independent of the plate voltage (which indicates a very high value for the ac plate resistance, r_p). The effect of the control grid, however, is practically the same as if the screen and the plate together form the collecting electrode—that is to say, the transconductance is of the same order as for a triode.

The required value of the screen voltage is commonly obtained by connecting the screen to *B*+ by means of a dropping resistor of suitable value (or a potential divider) as shown in Figs. 149-3A and B; the value of R_{sg} can be calculated from the fact that the screen current is normally about one fifth to one tenth of the value of the plate current.

With an alternating signal applied between the cathode and the control grid, there will be fluctuations in the screen current, just as there are fluctuations in the plate current. The effect of the fluctuating screen voltage is overcome by connecting the screen grid to the cathode through a capacitor, C_{sg}. This capacitor represents a negligible reactance at high frequencies, and the screen grid and the cathode will be virtually at the same alternating potential. There will be no coupling between the plate and the control grid circuits, apart from the very small residual control grid-plate capacitance.

Because of the restriction on the working part of the characteristic imposed by secondary emission, the screen grid tetrode is of limited use as a power amplifier; its use as a voltage amplifier is also limited because it can handle only a very small input signal. These tubes are virtually obsolete, although high power tetrodes have been developed for the output stages of UHF TV transmitters.

THE PENTODE TUBE

One method of reducing the secondary emission effect is the introduction of an additional electrode, in the form of a third grid, which is positioned between the screen and the plate. This third grid (G3) is called the *suppressor,* and the resulting five-electrode tube is referred to as a *pentode* (Fig. 149-4A). The suppressor is given a negative potential, relative to the plate and the screen grid, and this prevents the low-velocity secondary electrons from reaching the screen. At the same time, the suppressor is usually built of open-mesh wire so that it does not interfere appreciably with the passage of the high-velocity primary electrons toward the plate. The suppressor grid is usually connected to the cathode but, because other connections might be needed, the lead to the suppressor grid may be brought out of the tube to a

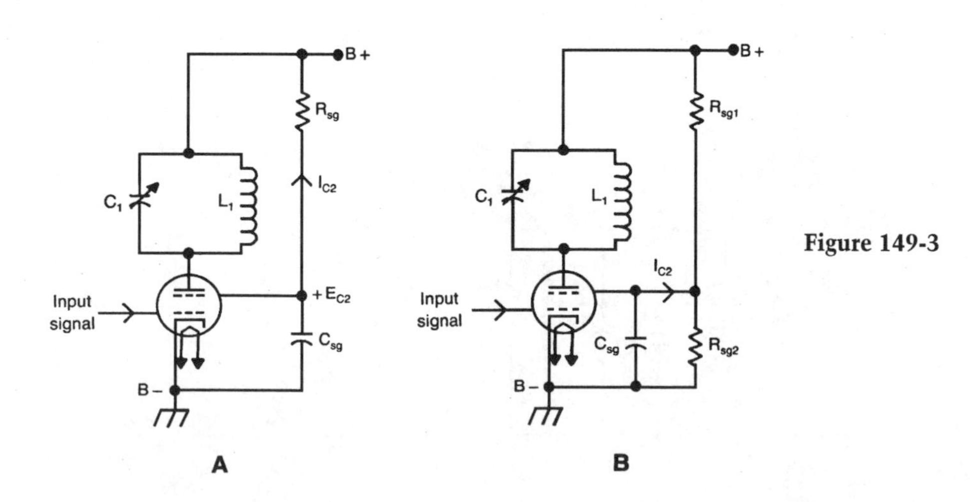

Figure 149-3

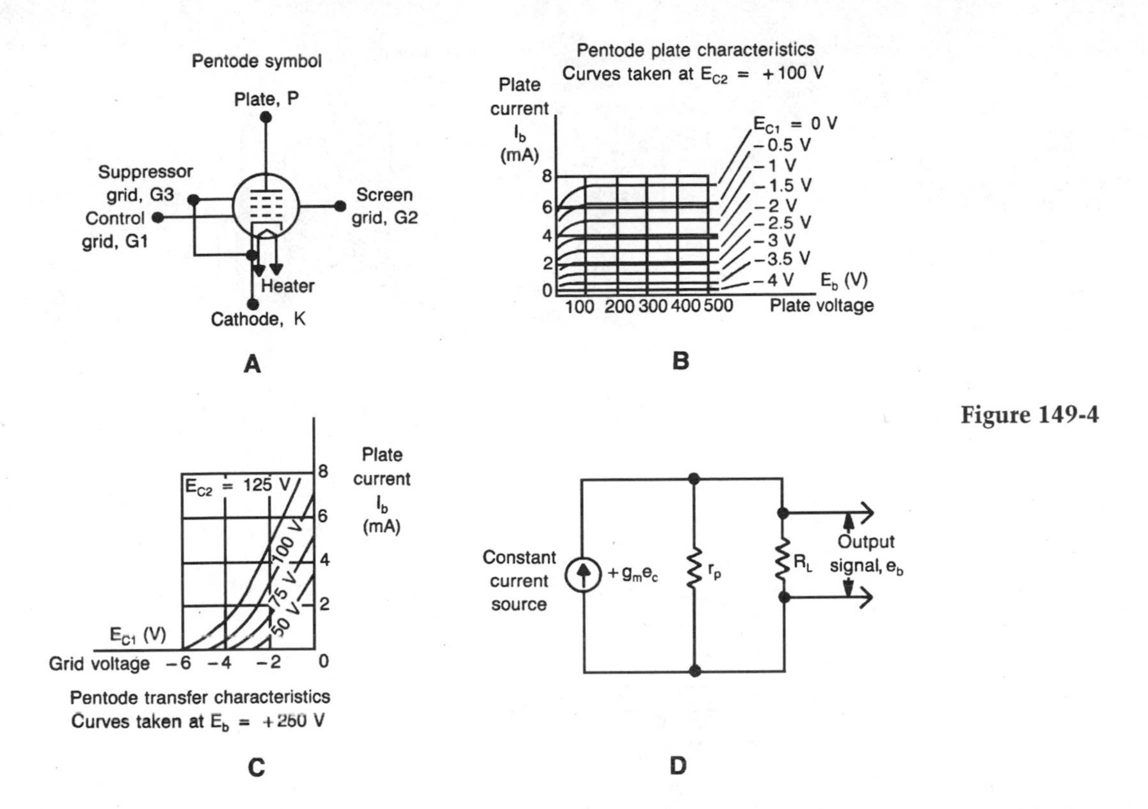

Figure 149-4

separate pin and the connection is made externally. In certain cases, where a pentode is suitable only as a power amplifier, the connection is made internally.

Figure 149-4B illustrates the pentode's plate characteristics, which do not contain the negative resistance section associated with the tetrode. The transfer characteristics are displayed in Fig. 149-4C, and the approximate values of the tube's parameters are r_p = 1.5 MΩ, g_m = 2400 μS, and μ = 3600. With such high values of r_p and μ, it is preferable to regard the tube in the equivalent amplifier circuit as a constant-current generator (Fig. 149-4D).

THE BEAM POWER TUBE

A second method of reducing the effects of secondary emission, is to include additional electrodes between the screen grid and the plate. These electrodes are connected to the cathode inside the tube and they will repel the electron stream. The electrodes are arranged so that they concentrate the electron stream into a comparatively narrow beam. For this reason, they are usually referred to as *beam-forming plates* (Fig. 149-5A). The concentration of the electrons into this beam, combined with a large distance between the screen grid and the plate, gives an intensified space-charge effect in the screen grid-plate space that will repel the secondary electrons back into the plate's surface. The screen current is made small by having an open-meshed screen, together with optical alignment of the control grid and screen grid. Such a tube is referred to as a *beam power tube* and its plate characteristics are shown in Fig. 149-5B.

THE VARIABLE MU-PENTODE

The *variable mu-pentode* is a pentode in which the control grid is made to have an asymmetrical structure. This is normally done by making the pitch of the control grid vary along its length, the mesh of the grid being closer at the ends, rather than the center. The result is that various parts of the tube cut off with different grid bias voltages so that the overall cutoff comes gradually, rather than abruptly.

Figure 149-6 shows a set of transfer characteristics for a typical variable mu-pentode, plotted for a constant plate voltage, but a variable screen voltage. With curved characteristics, the value of the g_m will depend on the chosen bias point. Because the voltage gain, $G_V = g_m R_L$, this will enable the voltage gain of an amplifier to be varied over a wide range by changing the bias voltage on the control grid.

MATHEMATICAL DERIVATIONS

Screen grid voltage,

$$E_{c2} = E_{bb} - I_{c2} \times R_{SG} \text{ volts} \qquad (149\text{-}1)$$

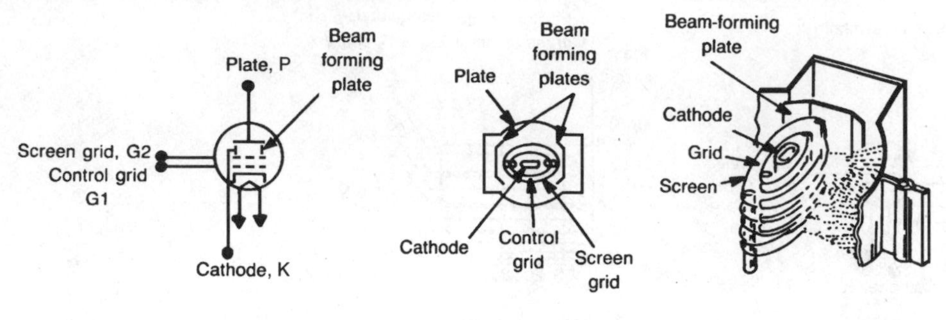

A

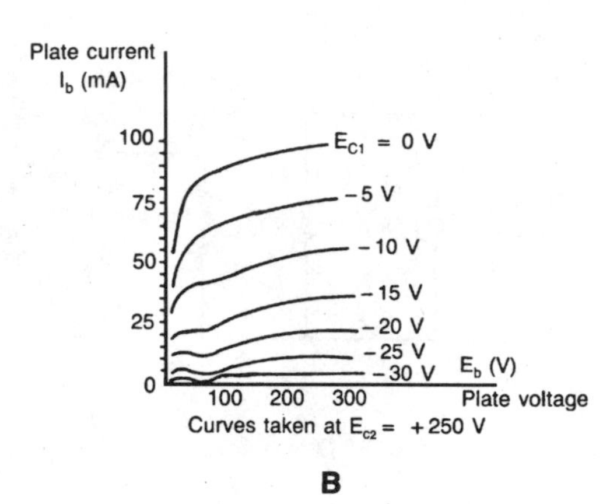

B

Figure 149-5

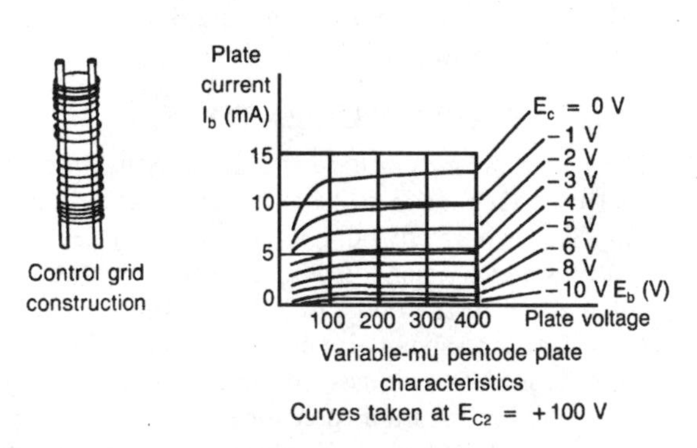

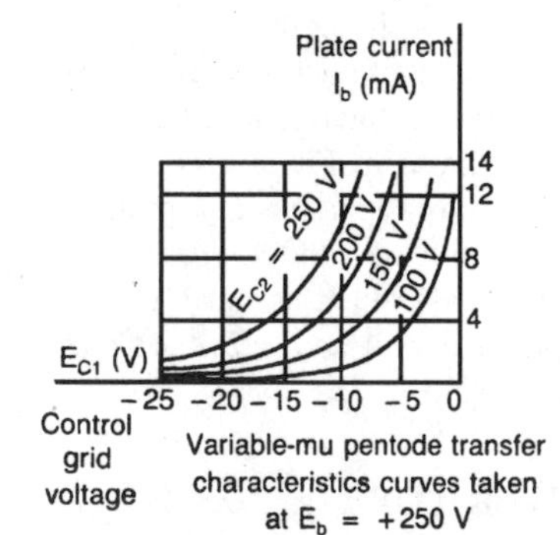

Figure 149-6

Cathode current,

$$I_K = I_b + I_{c2}(+I_{c1}) \text{ milliamperes} \qquad (149\text{-}2)$$

The control grid current (I_{c1}) would be present in a power amplifier, but not in a voltage amplifier.

In Fig. 149-4D

Output signal:

$$e_b = g_m e_c \times (r_p \| R_L) \qquad (149\text{-}3)$$

$$= \frac{g_m e_c}{\dfrac{1}{r_p} + \dfrac{1}{R_L}}$$

$$\approx g_m R_L e_c \text{ volts if } r_p >> R_L$$

Voltage gain of the pentode amplifier is,

$$G_V = \frac{e_b}{e_c} = g_m R_L \qquad (149\text{-}4)$$

Example 149-1

A pentode tube is operated in a voltage amplifier circuit under the following conditions: $I_b = 3$ mA, $R_L = 33$ kΩ, $R_{SG} = 120$ kΩ, $E_{bb} = +250$ V, and transconductance $g_m = 2700$ μS. If the dc level of the screen current is one-fifth of the plate current value, calculate the values of the screen potential and the amplifier's voltage gain.

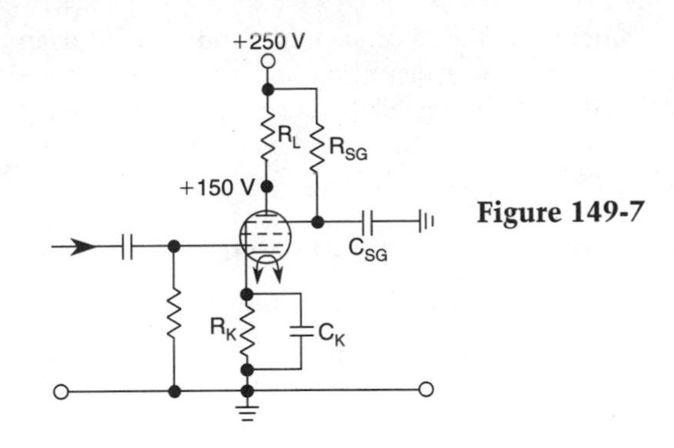

Figure 149-7

Solution

Level (dc) of screen current, $I_{c2} = 3$ mA/5 $= 0.6$ mA.
Screen grid voltage,

$$E_{c2} = E_{bb} - I_{c2} \times R_{SG} \qquad (149\text{-}1)$$

$$= 250 \text{ V} - 0.6 \text{ mA} \times 120 \text{ k}\Omega$$

$$= +178 \text{ V}.$$

Voltage gain,

$$G_V = g_m R_L \qquad (149\text{-}4)$$

$$= 2700 \times 10^{-6} \times 33 \times 10^3$$

$$= 89.$$

PRACTICE PROBLEMS

149-1. Figure 149-7 illustrates the circuit of an audio voltage amplifier. Under quiescent (or no-signal) conditions, $E_{bb} = +250$ V, $E_b = +150$ V, and $I_b = 3$ mA. It is required to operate the screen at two thirds of the value of the plate voltage. If the screen grid current is one fifth of the plate current, suggest a suitable value for the screen resistor, R_{SG}.

149-2. In Practice Problem 149-1, the control grid needs a bias voltage of −3 V. Calculate the required value for the cathode resistor, R_K. If the lowest frequency contained in the audio signal is 80 Hz, estimate the values of C_K and C_{SG}.

149-3. In Practice Problem 149-1, what is the value of the load resistor, R_L? If the transconductance is 2500 μS, calculate the value of the voltage gain.

149-4. In Fig. 149-5B, estimate the value of the transconductance at the operating point defined by $E_b = +250$ V, $E_{c2} = +250$ V, and $E_{c1} = -15$ V.

149-5. In Fig. 149-6, estimate the values of the transconductance at the two operating points defined by $E_b = +250$ V, $E_{c2} = +150$ V, and $E_{c1} = -5$ V; $E_b = +250$ V, $E_{c2} = +150$ V, and $E_{c1} = -15$ V.

150
Negative feedback

The introduction of negative feedback in an audio amplifier requires that a fraction, β of the output signal is fed back in opposition to the input signal. In contrast with the disadvantage of reducing the amplifier gain, negative (or degenerative) feedback provides the following features:

1. Stabilization of the amplifier voltage gain against changes in the parameters of the active device, such as a transistor or tube.
2. Reduction in the amplitude distortion caused by nonlinearity in the characteristics of the active device.
3. Reduction in frequency and phase distortion produced by the junction and stray capacitances.
4. Reduction in noise.
5. Changes in the amplifier's input and output impedances.

Notice that the second, third, and fourth advantages refer to the distortion and the noise created within the amplifier itself. Negative feedback has no effect on the noise and the distortion that are fed in from the previous stage.

MATHEMATICAL DERIVATIONS

Voltage gain with negative feedback

In the block diagram of Fig. 150-1, the signal applied between the grid and the cathode is the sum of the input audio signal (V_i) between the control grid and ground, and the negative feedback voltage ($-\beta V_o$). Therefore,

$$\text{Signal between the grid and the cathode} = V_i - \beta V_o \text{ volts} \qquad (150\text{-}1)$$

where: V_o = output audio signal.

Output audio signal,

$$V_o = A \times (\text{signal between } G \text{ and } K) \qquad (150\text{-}2)$$
$$= A(V_i - \beta V_o) \text{ volts}$$

where: A = open loop gain.

Amplifier gain with negative feedback,

$$A' = \frac{V_o}{V_i} = \frac{A}{1 + A\beta} \qquad (150\text{-}3)$$

If $A\beta$ is appreciably greater than 1,

$$A' \to \frac{A}{A\beta} = \frac{1}{\beta} \qquad (150\text{-}4)$$

which is independent of A; consequently, the value of A' will be little affected by any changes in the parameters of the active device.

Because,

$$A' = \frac{A}{1 + A\beta'} \qquad (150\text{-}5)$$

$$A = \frac{A'}{1 - A'\beta}$$

and,

$$\beta = \frac{1}{A'} - \frac{1}{A} = \frac{A - A'}{AA'} \qquad (150\text{-}6)$$

The feedback factor, β, can either be expressed as a fraction or as a percentage.

The two basic negative feedback circuits are shown in Figs. 150-2A and B. Figure 150-2A represents negative voltage feedback in which the audio output signal voltage (V_o) is divided between R1 and R2 (C is only a dc blocking capacitor). The voltage ($-\beta V_o$) which is developed across R1, is applied as degenerative feedback to the input signal, V_i.

The feedback factor is given by,

Feedback factor,

$$\beta = \frac{R_1}{R_1 + R_2} \qquad (150\text{-}7)$$

The voltage gain of the amplifier with feedback is,

$$A' = \frac{A}{1 + A\beta} \qquad (150\text{-}8)$$

Open loop gain,

$$A = \frac{\mu R_L}{r_p + R_L}$$

This expression for the voltage gain without feedback (A) applies when the value of $R_1 + R_2$ is sufficiently large so that the equivalent value of the plate load resistance is not appreciably affected by the shunting action of the two resistors.

Current feedback, as shown in Fig. 150-2B, uses the signal component of the plate current to develop the degenerative feedback voltage across the cathode resistor (R_K); β is then equal to R_K/R_L and the voltage gain of the amplifier with feedback,

$$A' = \frac{A}{1 + A\beta}$$

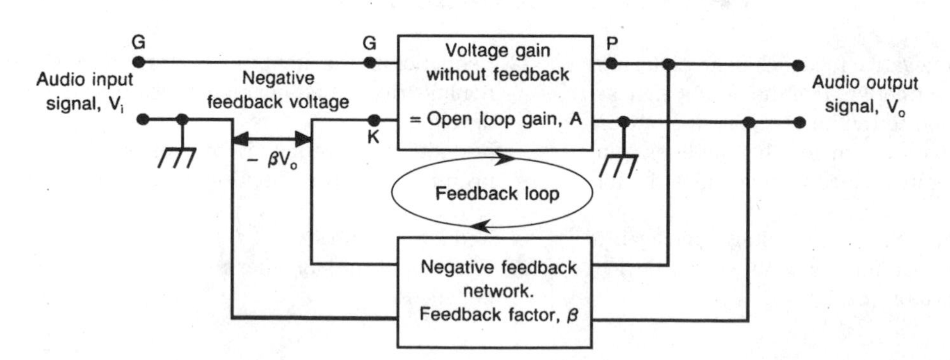

Figure 150-1

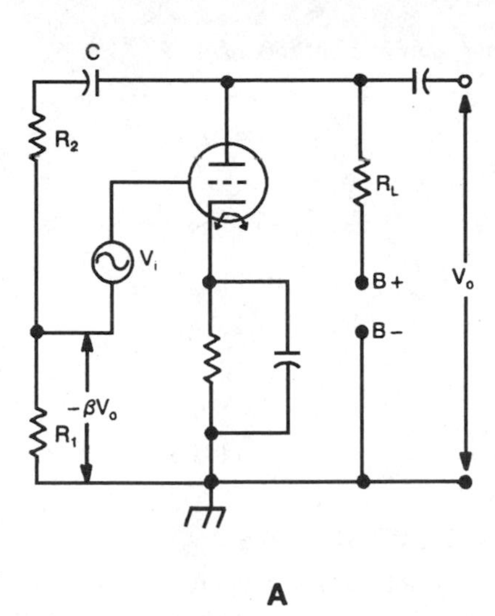

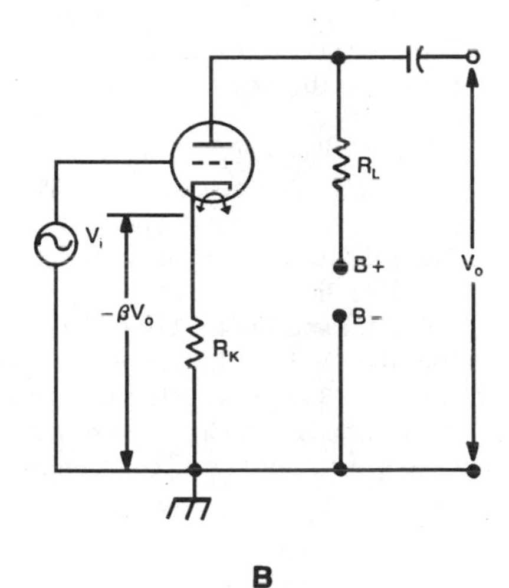

Figure 150-2

where,

$$A = \frac{\mu R_L}{r_p + R_K + R_L} \qquad (150\text{-}9)$$

The last two equations yield,

$$A' = \frac{\mu R_L}{r_p + R_L + R_K(1 + \mu)} \qquad (150\text{-}10)$$

Example 150-1

The voltage gain of an audio amplifier without negative feedback is 35. If 20% degenerative feedback is introduced into the circuit, calculate the value of the amplifier gain with feedback.

Solution

Because $A = 35$ and $\beta = {}^{20}\!/_{100} = 0.2$, the gain with feedback is:

$$A' = \frac{A}{1 + A\beta}$$

$$= \frac{35}{1 + (35 \times 0.2)}$$

$$= \frac{35}{1 + 7} = \frac{35}{8}$$

$$= 4.4, \text{ rounded off.}$$

Example 150-2

The voltage gain of an audio amplifier with degenerative feedback is 8. If the negative feedback factor is $1/10$, what is the amplifier gain without feedback?

Solution

Because $A' = 8$ and $\beta = 1/10$, the gain without feedback is:

$$A = \frac{A'}{1 - A'\beta} \qquad (150\text{-}5)$$

$$= \frac{8}{1 - (8 \times 0.1)} = \frac{8}{0.2}$$

$$= 40.$$

Example 150-3

The voltage gain of an audio amplifier without feedback is 25. When degenerative feedback is introduced into the circuit, the voltage gain falls to 10. What is the feedback percentage?

Solution

Because $A = 25$ and $A' = 10$, the feedback factor is:

$$\beta = \frac{A - A'}{AA'} \qquad (150\text{-}6)$$

$$= \frac{25 - 10}{25 \times 10}$$

$$= \frac{15}{250}$$

$$\textit{Feedback percentage} = \frac{15 \times 100}{250} = 6\%.$$

Example 150-4

In Fig. 150-3, what is the negative feedback percentage? If the triode's amplification factor is 24 and its ac plate resistance is 11 kΩ, what is the voltage gain with feedback?

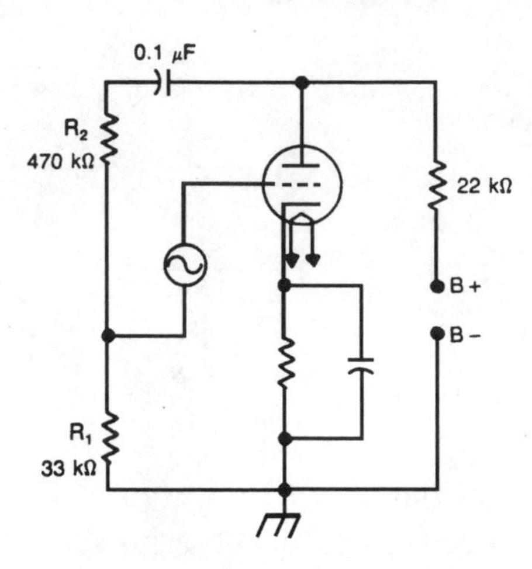

Figure 150-3

Solution

Feedback factor,

$$\beta = \frac{R_1}{R_1 + R_2} \qquad (150\text{-}7)$$

$$= \frac{33\ \text{k}\Omega}{470\ \text{k}\Omega + 33\ \text{k}\Omega}$$

$$= \frac{33}{503}$$

Feedback Percentage

$$= \frac{33 \times 100}{503} = 6.6\%, \text{ rounded off.}$$

Voltage gain without feedback,

$$A = \frac{\mu R_L}{r_p + R_L}$$

$$= \frac{24 \times 22}{11 + 22}$$

$$= 16$$

Voltage gain with feedback,

$$A' = \frac{16}{1 + (16 \times 0.066)}$$

$$= 7.9, \text{ rounded off.}$$

PRACTICE PROBLEMS

150-1. An amplifier has an open-loop gain of 65 and a negative feedback factor of 0.05. Calculate the value of the closed-loop gain.

150-2. The open- and closed-loop gains of an amplifier are, respectively, 72 and 15. Determine the value of the negative feedback percentage.

150-3. An amplifier has a closed-loop gain of 15 and its negative feedback percentage is 4%. Obtain the value of the amplifier's open-loop gain.

150-4. In Fig. 150-2A, the open-loop gain of the amplifier is 22. If $R_1 = 27$ kΩ and $R_2 = 330$ kΩ, calculate the value of the closed-loop gain.

150-5. In Fig. 150-2B, $R_K = 1.2$ kΩ, $R_L = 20$ kΩ, $\mu = 25$, and $r_p = 8$ kΩ. Calculate the values of the current negative feedback percentage, the open-loop gain, and the closed-loop gain.

151 The grounded-grid triode circuit and the cathode-follower

The previous chapters have covered the operation of grounded-cathode amplifiers in which the input signal is applied between the control grid and ground, and the output signal appears between the plate and ground. However, there are two other possible arrangements:

1. The *grounded-grid* amplifier. In this circuit, the input signal is applied between cathode and ground. The grid is grounded (as far as the signal is concerned) and the output signal appears between the plate and ground.
2. The *cathode-follower* or *grounded-plate* circuit. In this arrangement, the input signal is applied between the grid and ground. The plate is grounded (as far as the signal is concerned), but is still connected to the $B+$ supply. The output signal appears between the cathode and ground.

THE GROUNDED-GRID TRIODE CIRCUIT

This circuit (Figs. 151-1A and B) is commonly used for RF amplification in the VHF and UHF bands; one example is the first stage of a TV receiver. Its advantages are twofold:

1. It contributes less noise than an amplifier using a conventional screen grid which suffers from partition noise because of the random effect in the division of the electron stream between the screen grid and the plate circuits.
2. The circuit does not require neutralization because the grounded grid provides an electrostatic screen between the output plate circuit and the input cathode circuit. This is particularly advantageous at high frequencies, where the adjustment of the neutralizing circuit is more critical. In addition, the elimination of the neutralizing capacitor reduces the input and output capacitances of the stage. Unlike the conventional grounded-cathode amplifier, there is no phase change between the input signal at the cathode and the output signal on the plate. Both of these signal voltages are 180° out of phase with the signal component of the plate current.

The value of the stage's input impedance is much lower than that of the conventional grounded-cathode amplifier that was covered earlier. This low input impedance is the main disadvantage of the grounded-grid circuit because it limits the amplification that can be obtained.

THE CATHODE-FOLLOWER CIRCUIT

The cathode-follower circuit is shown in Fig. 151-1C, and is an example of negative voltage feedback, with a feedback factor (β) of unity. The features of the circuit are:

1. There is no phase change between the cathode-ground output signal and the grid-ground input signal (the cathode output "follows" the grid input).
2. The voltage gain of the circuit is less than unity. This voltage gain is given by:
   Voltage gain with feedback,
   $$A' = \frac{A}{A+1}$$
   where voltage gain without feedback,
   $$A = \frac{\mu R_k}{r_p + R_k}$$
   assuming that no load is connected across R_k.
3. A higher input impedance, and a lower output impedance, as compared with a conventional grounded-cathode amplifier.

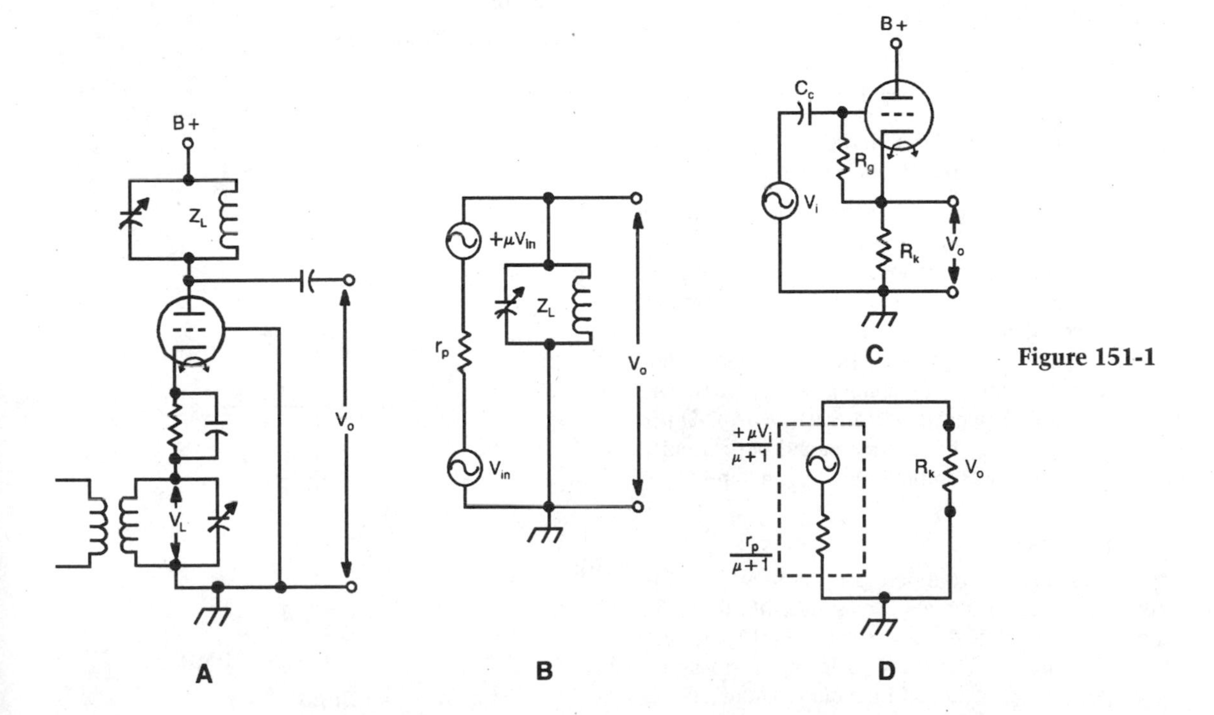

Figure 151-1

This circuit is an impedance matching device, and may be used to match a high impedance source to a low impedance load. The cathode-follower can therefore provide some degree of isolation between a stage with a low input impedance and a preceding stage with a high output impedance.

MATHEMATICAL DERIVATIONS

In the equivalent circuit of Fig. 151-1B,

Plate signal current,

$$i_b = \frac{(\mu + 1)V_i}{r_p + Z_L} \text{ amperes} \quad (151\text{-}1)$$

Output signal voltage,

$$V_o = i_b Z_L = \frac{(\mu + 1)Z_L}{r_p + Z_L} \times V_i \text{ volts} \quad (151\text{-}2)$$

Voltage gain,

$$G_V = \frac{V_o}{V_i} = \frac{(\mu + 1)Z_L}{r_p + Z_L} \quad (151\text{-}3)$$

Input impedance,

$$Z_{in} = \frac{V_i}{i_b} = \frac{r_p + Z_L}{\mu + 1} \text{ ohms} \quad (151\text{-}4)$$

In the circuits of Figs. 151-1C and D,

The input impedance of the cathode-follower circuit (Z_{in}) consists of the parallel combination of:

$$(A + 1)R_g, \frac{C_{gk}}{1 + A}, \text{ and } C_{gp}$$

where: $A = \mu R_k/(r_p + R_k)$, and C_{gk}, C_{gp}, and C_{pk} are the interelectrode capacitances.

Output impedance (Z_o) consists of the parallel combination of R_k, $r_p/(\mu + 1)$ and C_{pk}. These values of the input and output impedances can be compared with the equivalent expressions for the grounded-cathode arrangement.

Input impedance (Z_{in}) of the grounded-cathode circuit is the result of combining R_g, C_{gk}, and $(A + 1)C_{pg}$ in parallel, where $A = \mu R_L/(r_p + R_L)$. Output impedance (Z_o) of the grounded-cathode circuit is equal to the parallel combination of R_L, r_p, and C_{pk}.

Example 151-1

The plate load of a grounded-grid triode amplifier is a tank circuit with a dynamic resistance of 20 kΩ. The tube has an amplification factor (μ) of 15 and an ac plate resistance (r_p) of 8 kΩ. Calculate the values of the stage's voltage gain and its input resistance.

Solution

Voltage gain,

$$G_V = \frac{(\mu + 1)Z_L}{r_p + Z_L} \quad (151\text{-}3)$$

$$= \frac{16 \times 20}{8 + 20}$$

$$= \frac{16 \times 20}{28}$$

$$= 11.$$

Input resistance,

$$R_{in} = \frac{r_p + Z_L}{\mu + 1} \quad (151\text{-}4)$$

$$= \frac{8 + 20}{16}$$

$$= \frac{28}{16}$$

$$= 1.8 \text{ k}\Omega.$$

Example 151-2

In Fig. 151-1C, $R_k = 10$ kΩ, $R_g = 1$ MΩ, $r_p = 8$ kΩ, and $\mu = 20$. What is the voltage gain of the cathode-follower? Calculate the values of the stage's input and output resistances.

Solution

Voltage gain without feedback,

$$A = \frac{\mu R_k}{r_p + R_k}$$

$$= \frac{20 \times 10 \text{ k}\Omega}{8 \text{ k}\Omega + 10 \text{ k}\Omega}$$

$$= 11.1$$

Voltage gain with feedback,

$$A' = \frac{A}{A + 1}$$

$$= \frac{11.1}{11.1 + 1}$$

$$= \frac{11.1}{12.1}$$

$$= 0.92$$

Input resistance,

$$R_{in} = (A + 1)R_g$$

$$= (11.1 + 1) \times 1 \text{ M}\Omega$$

$$= 12.1 \text{ M}\Omega.$$

The output resistance consists of R_k(10 kΩ) in parallel with $r_p/(\mu + 1) = 8/(20 + 1)$ kΩ = 381 Ω.
Output resistance,

$$R_o = 381\ \Omega \| 10000\ \Omega$$

$$= \frac{381 \times 10000}{381 + 10000}$$

$$= 367\ \Omega.$$

PRACTICE PROBLEMS

151-1. A grounded-grid triode has an amplification factor of 20 and its ac plate resistance is 7 kΩ. The plate load is a tank circuit whose dynamic resistance is 25 kΩ. Determine the values of the amplifier's voltage gain and its input resistance.

151-2. A triode has an amplification factor of 18, an ac plate resistance of 8 kΩ, and interelectrode capacitances of C_{pg} = 5 pF, C_{gk} = 3 pF, and C_{pk} = 2 pF. This tube is used in a cathode-follower circuit for which R_k = 12 kΩ and R_g = 1 MΩ (Fig. 151-1C). Obtain the values of the cathode-follower's closed-loop voltage gain, its input resistance and its output resistance.

151-3. In Practice Problem 151-2, calculate the values of the circuit's input and output capacitances.

5
PART
Principles of radio communications

152
Amplitude modulation—percentage of modulation

Modulation means controlling some feature of the RF carrier by the audio signal in such a way that the receiver is able to reproduce the information (speech, music, etc.). Because the carrier can be regarded as a high-frequency sine wave, it only possesses three features that can be modulated—amplitude, frequency, and phase (such quantities as period and wavelength are directly related to the frequency). In *amplitude modulation*, the instantaneous amplitude (voltage) of RF wave is linearly related to the instantaneous magnitude (voltage) of the AF signal and the rate of amplitude variation is equal to the modulating frequency.

In Fig. 152-1A, the modulating signal is a single sine wave (or test tone), which is used to amplitude modulate an RF carrier. The extent of modulation is measured by the percentage of modulation as defined by: $(E_{max} - E_o) \times 100/E_o\%$. This expression is for positive peaks (upward modulation), where E_o is the amplitude of the unmodulated carrier. The percentage modulation on negative peaks (downward modulation) is $(E_o - E_{min}) \times 100/E_o\%$ (Fig. 152-1B).

For a variety of reasons, the modulation envelope for a symmetrical tone contains asymmetrical distortion that is referred to as *carrier shift*. In negative carrier shift, the troughs of the modulation envelope are greater than the peaks and this will cause a decrease (compared with the unmodulated value) in the dc plate current reading of a plate modulated class-C stage. Similarly, with positive carrier shift, the envelope's peaks exceed the troughs and the plate current will increase.

Increasing the amplitude of the test tone raises the percentage modulation, improves the signal-to-noise ratio at the receiver, and results in a higher audio power output. However, if the modulation percentage is increased to 100% and beyond (overmodulation), the modulation envelope is severely distorted; this distortion will ultimately appear in the output from the loudspeaker. Overmodulation also results in the generation of spurious sidebands that will cause interference to adjacent channels by heterodyning with the sidebands of those channels. Figure 152-2A shows the AM waveforms for various modulation percentages.

Other than measuring the percentage modulation from the AM waveform, it is common practice to use a trapezoidal display on an oscilloscope. This is achieved by feeding the AM signal to the vertical deflecting (Y) plates and feeding the audio signal to the horizontal deflecting (X) plates. In addition to measuring percentage modulation, this method will clearly indicate whether there exists the required linear relationship between the instantaneous amplitude of the RF carrier and the instantaneous magnitude of the tone signal. Nonlinearity will appear either as a "barrel" or a "pincushion effect" on the trapezoidal display (Fig. 152-2B). The percentage of modulation is given by $[(P - Q)/(P + Q)] \times 100\%$ so that for 50% modulation, $P = 3Q$ (*not* $2Q$). When AM is being used to transmit speech or music, the percentage of modulation varies from instant to instant. Permitted values can then be quoted for:

1. The minimum percentage modulation on average peaks of the audio signal.
2. The maximum modulation percentage on negative modulation peaks.
3. The maximum modulation on positive modulation peaks.
4. The maximum carrier shift.

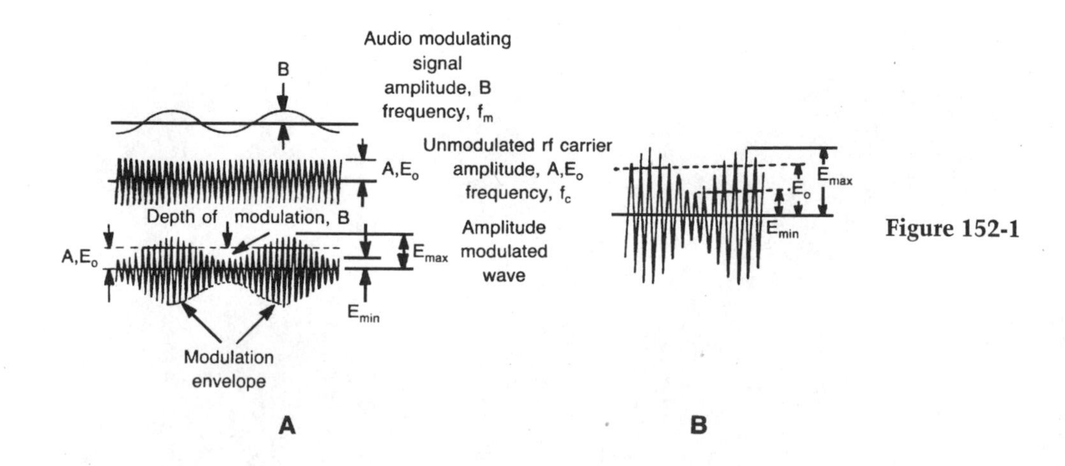

Figure 152-1

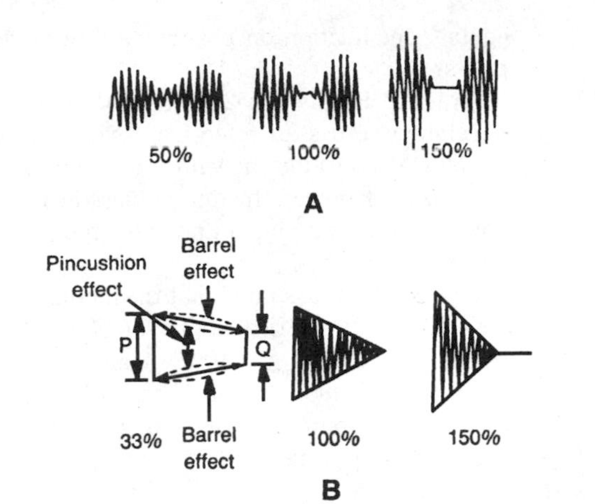

Figure 152-2

MATHEMATICAL DERIVATIONS

In Fig. 152-1

Percentage modulation on positive peaks,

$$= \frac{E_{max} - E_o}{E_o} \times 100\% \qquad (152\text{-}1)$$

Percentage modulation on negative peaks,

$$= \frac{E_o - E_{min}}{E_o} \times 100\% \qquad (152\text{-}2)$$

If the modulation is symmetrical about the E_o level,

$$E_o = \frac{E_{max} + E_{min}}{2} \text{ volts} \qquad (152\text{-}3)$$

Equations 152-1 and 152-3 yield:

Percentage modulation,

$$m = \frac{E_{max} - \dfrac{(E_{max} + E_{min})}{2}}{E_o} \times 100\% \qquad (152\text{-}4)$$

$$= \frac{E_{max} - E_{min}}{2E_o} \times 100\%$$

$$= \frac{E_{max} - E_{min}}{E_{max} + E_{min}} \times 100\% \qquad (152\text{-}5)$$

Degree of modulation,

$$m = \frac{\textit{percentage modulation}}{100} \qquad (152\text{-}6)$$

$$= \frac{(A + B) - (A - B)}{(A + B) + (A - B)}$$

$$= \frac{B}{A}$$

Percentage of carrier shift = (152-7)

$$\frac{\textit{dc level (mod.)} - \textit{dc level (unmod.)}}{\textit{dc level (unmod.)}} \times 100\%$$

Trapezoidal display:

Percentage of modulation = (152-8)

$$\frac{P - Q}{P + Q} \times 100\%$$

Example 152-1

In Fig. 152-1, it is assumed that the modulation envelope is symmetrical about the E_o level. If $E_{max} = 800$ V and $E_o = 500$ V, what is the value of the percentage of modulation? Calculate the value of the instantaneous peak power if the AM signal is developed across an effective load of 100 Ω.

Solution

Percentage modulation (152-1)

$$= \frac{(E_{max} - E_o)}{E_o} \times 100$$

$$= \frac{800 - 500}{500} \times 100$$

$$= 60\%.$$

This instantaneous peak voltage, E_{max}, is 800 V; therefore, the peak power is:

$$\frac{E^2_{max}}{R} = \frac{(800 \text{ V})^2}{100\ \Omega}$$

$$= 6400 \text{ W}.$$

Notice that the unmodulated carrier power is $(500\ \text{V})^2/100\ \Omega = 2500$ W and the instantaneous minimum power is $(200\ \text{V})^2/100\ \Omega = 400$ W.

The next chapter will show that the average power in this AM signal is 2950 W, which is different from the arithmetic mean of (6400 + 400)/2 = 3400 W. This difference is caused by the fact that the instantaneous power curve is not symmetrical about the unmodulated carrier power level.

Example 152-2

A trapezoidal display (Fig. 152-2) is used to measure the percentage modulation of the AM signal in Example 152-1. If $P = 3$ cm, calculate the value of Q.

Solution

Because the percentage of modulation is 60%,

$$\frac{P - Q}{P + Q} \times 100 = 60 \qquad (152\text{-}8)$$

Therefore:

$$\frac{3 - Q}{3 + Q} = 0.6$$

or,

$$Q = \frac{1.2}{1.6} = 0.75 \text{ cm.}$$

Example 152-3

When unmodulated, the dc plate current in the final stage of an AM broadcast is 5.4 A. When modulated by a symmetrical audio tone, the plate current falls to 5.2 A. What is the percentage of the negative carrier shift?

Solution

Percentage of negative carrier shift

$$= \frac{5.2 - 5.4}{5.4} \times 100\% \qquad (152\text{-}7)$$

$$= -3.7\%.$$

PRACTICE PROBLEMS

152-1. In Fig. 152-1, $E_{max} = 700$ V, $E_o = 400$ V, and $E_{min} = 150$ V. What are the values of the percentage modulation on positive and negative peaks?

152-2. In Practice Problem 152-1, the AM signal is developed across an effective resistive load of 250 Ω. Determine the values of the instantaneous peak power, the unmodulated carrier power, and the instantaneous minimum power.

152-3. In Fig. 152-1 it is assumed that the modulation envelope is symmetrical about the E_o level. If $E_o = 600$ V and $E_{min} = 100$ V, calculate the percentage of modulation. If this AM signal produces a trapezoidal display in which $P = 3$ cm, what is the length of Q?

152-4. In Fig. 152-2A, $E_{max} = 400$ V, $E_o = 150$ V, and $E_{min} = 0$ V. Assuming that the modulating signal is a single tone, what is the value of the percentage modulation?

152-5. When unmodulated, the dc level of the plate current in a transmitter's final stage is 9.3 A. The plate current rises to 9.7 A when the transmitter is amplitude modulated by a symmetrical audio tone. Calculate the percentage of positive carrier shift.

153 AM sidebands and bandwidth

Figure 153-1 illustrates the result of amplitude modulating an RF carrier (amplitude A, frequency f_c) by an audio test tone (amplitude B, frequency f_m). As previously covered, the degree of modulation, $m = B/A$ (Equation 152-6). In the mathematical analysis shown, it is revealed that the AM (voltage) signal is entirely RF in nature (no audio voltage is present) and contains the following three components:

1. An upper sideband component whose amplitude is $B/2 = mA/2$ and whose frequency is $f_c + f_m$.
2. A carrier component whose amplitude is A and whose frequency is f_c.
3. A lower sideband component, whose amplitude is $B/2 = mA/2$ and whose frequency is $f_c - f_m$.

When these three RF components are literally added together, the resultant waveform is that of the AM signal.

As an example, let a 100-V 1-MHz carrier be 50% amplitude modulated by a 1-kHz test tone. The act of modulation creates the sidebands that contain the signal's intelligence so that the AM signal is composed of: (1) the upper sideband with an amplitude of $0.5 \times 100/2 = 25$ V at a frequency of 1 MHz + 1 kHz = 1001

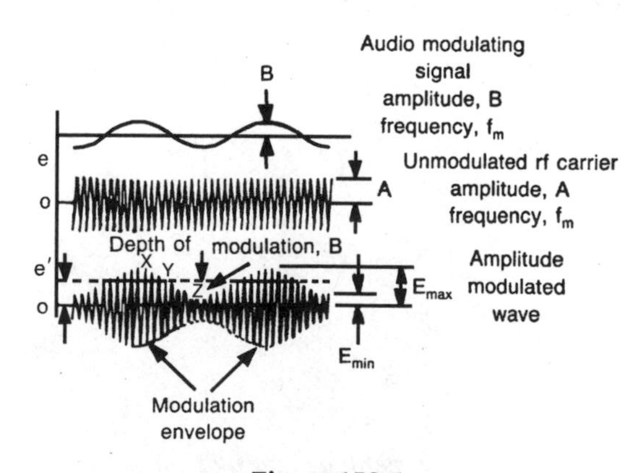

Figure 153-1

kHz, (2) the carrier component with an amplitude of 100 V at a frequency of 1 MHz = 1000 kHz, and (3) the lower sideband whose amplitude is 25 V at a frequency of 1 MHz − 1 kHz = 999 kHz. The instantaneous maximum voltage is 100 + 25 + 25 = 150 V while the minimum voltage is 100 − 25 − 25 = 50 V. The spectrum analyzer display of this AM signal is shown in Fig. 153-2.

The AM signal is accommodated within a certain *bandwidth* that is defined as the frequency difference between the upper and lower sidebands. In this example, the bandwidth required is 1001 − 999 = 2 kHz, which is equal to twice the frequency of the modulating tone. When the signal is speech or music that contains many instantaneous frequencies of varying amplitudes, each audio component will produce a pair of sidebands, and the bandwidth occupied by the AM signal will be the frequency difference between the highest upper sideband and the lowest lower sideband transmitted. Consequently, the bandwidth is equal to twice the *highest* audio modulating frequency.

In addition to the waveform (Fig. 153-1) and spectrum (Fig. 153-2) representations, the AM signal can be shown in terms of phasor diagrams (Fig. 153-3). The corresponding times in the waveform and the phasor diagrams are indicated by the points X, Y, and Z.

Each sideband component contains a certain RF power that is determined by the degree of modulation (m) and the unmodulated carrier power, P_c. The expressions for the sideband and carrier powers are shown in the mathematical derivations.

Experimentally, the sideband structure of an AM signal (modulated by a constant amplitude tone) can be displayed on a spectrum analyzer that has a CRT (Fig. 153-4). The output of a swept frequency generator is applied to the X plates so that the horizontal axis of the CRT measures frequency and not time. The sidebands and the carrier are displayed as vertical positive "spikes," which are measured in decibels; their frequencies are determined from the calibrated horizontal scale.

MATHEMATICAL DERIVATIONS

The voltage equation of the unmodulated carrier wave is:

$$e = A \sin 2\pi f_c t \text{ volts} \qquad (153\text{-}1)$$

The expression for the modulation envelope of the AM signal is $A + B \sin 2\pi f_m t$. Consequently, the voltage equation of the AM signal is:

$$\begin{aligned} e' &= (A + B \sin 2\pi f_m t) \sin 2\pi f_c t. \\ &= A(1 + m \sin 2\pi f_m t) \sin 2\pi f_c t. \\ &= A \sin 2\pi f_c t + mA \sin 2\pi f_m t \sin 2\pi f_c t. \\ &= A \sin 2\pi f_c t + \quad \rightarrow \text{Carrier} \\ &\frac{mA}{2} \cos 2\pi (f_c - f_m)t - \quad \rightarrow \text{LSB} \\ &\frac{mA}{2} \cos 2\pi (f_c + f_m) \text{ volts} \rightarrow \text{USB} \end{aligned}$$

$$\textit{Peak value of the maximum instantaneous voltage} = A(1 + m) \text{ volts} \qquad (153\text{-}2)$$

$$\textit{Peak value of the minimum instantaneous voltage} = A(1 - m) \text{ volts} \qquad (153\text{-}3)$$

$$\textit{Bandwidth} = (f_c + f_m) - (f_c - f_m) = 2f_m \text{ hertz} \qquad (153\text{-}4)$$

Let the carrier and the sidebands be associated with the same resistive load, R. Converting the peak values to their rms values,

Carrier power,

$$P_c = \frac{\left(\frac{A}{\sqrt{2}}\right)^2}{R} = \frac{A^2}{2R} \text{ watts} \qquad (153\text{-}5)$$

Upper sideband power,

$$P_{USB} = m^2 \frac{\left(\frac{A}{\sqrt{2}}\right)^2}{4R} = \frac{m^2A^2}{8R} \text{ watts} \qquad (153\text{-}6)$$

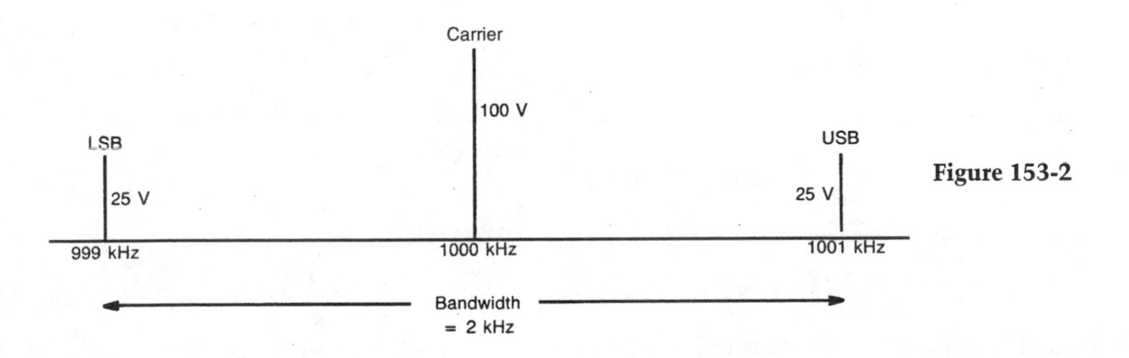

Figure 153-2

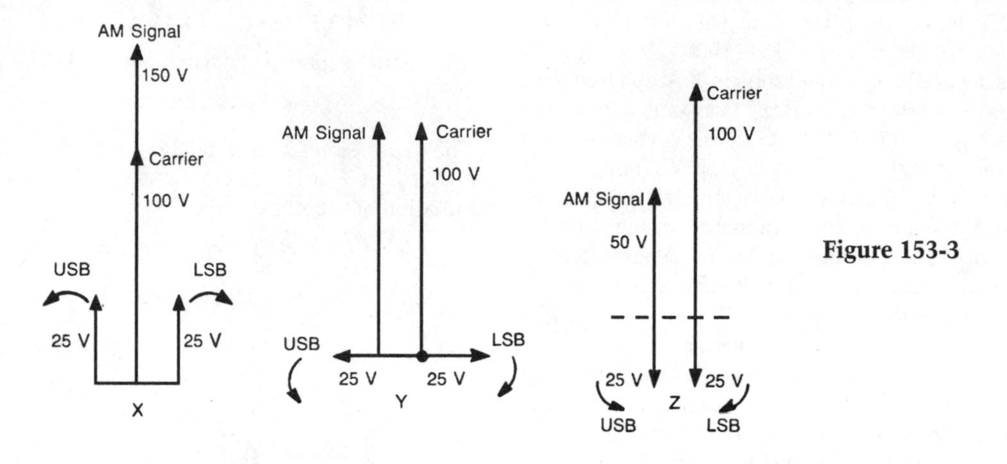

Figure 153-3

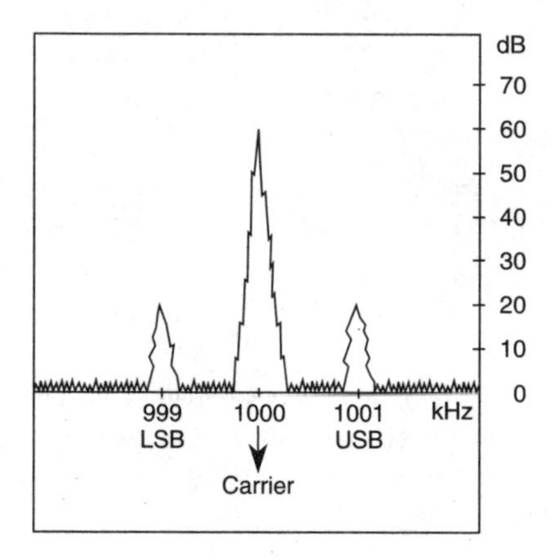

Figure 153-4

Lower sideband power,

$$P_{LSB} = \frac{m^2A^2}{8R} \text{ watts} \tag{153-7}$$

Total sideband power,

$$P_{SB} = 2 \times \frac{m^2A^2}{8R} = \frac{m^2A^2}{4R} \tag{153-8}$$

$$= \frac{1}{2} m^2P_c \text{ watts}$$

Peak instantaneous power,

$$P_{\max} = \frac{A^2(1 + m)^2}{2R} \text{ watts} \tag{153-9}$$

Minimum instantaneous power,

$$P_{\min} = \frac{A^2(1 - m)^2}{2R} \text{ watts} \tag{153-10}$$

Total RF power (carrier and sidebands),

$$P_T = P_c + \frac{1}{2} m^2P_c \tag{153-11}$$

$$= P_c\left(1 + \frac{1}{2} m^2\right) \text{ watts}$$

In the particular case of 100% modulation, $m = 1$ so that the total sideband power $= 0.5\,P_c$ and the total RF power is $1.5\,P_c$. Consequently, one third of the total RF power is concentrated in the sidebands and two thirds in the carrier.

Example 153-1

A 2-kW 1.3-MHz carrier is 80% amplitude modulated by a 2-kHz test tone. Calculate the power in each of the sidebands and their frequencies. What is the bandwidth of the AM signal?

Solution

Frequency of the upper sideband,

$$f_{USB} = 1.3 \text{ MHz} + 2 \text{ kHz}$$

$$= 1302 \text{ kHz.}$$

Frequency of the lower sideband,

$$f_{LSB} = 1.3 \text{ MHz} - 2 \text{ kHz}$$

$$= 1298 \text{ kHz.}$$

Bandwidth,

$$f_{USB} - f_{LSB} = 2 \times 2 \text{ kHz} = 4 \text{ kHz.} \tag{153-4}$$

Total sideband power,

$$P_{SB} = \frac{1}{2} \times (0.8)^2 \times 2000 \tag{153-8}$$

$$= 640 \text{ W.}$$

Power in each sideband,

$$\frac{P_{SB}}{2} = \frac{640}{2} \qquad (153\text{-}6,\ 153\text{-}7)$$

$$= 320 \text{ W}.$$

Example 153-2

A 1-kW carrier is 60% modulated by a 1-kHz test tone. If the modulation is increased to 80%, what is the percentage increase in the total sideband power?

Solution

Total sideband power, 60% modulation,

$$P_{SB} = \frac{1}{2} \times (0.6)^2 \times 1000 \qquad (153\text{-}8)$$

$$= 180 \text{ W}$$

Total sideband power, 80% modulation,

$$P_{SB} = \frac{1}{2} \times (0.8)^2 \times 1000$$

$$= 320 \text{ W}$$

Percentage increase in the total sideband power,

$$\frac{320 - 180}{180} \times 100 = \frac{1400}{18}$$

$$= 78\%, \text{ rounded off.}$$

PRACTICE PROBLEMS

153-1. A 3-kW 1.8-MHz carrier is 70% amplitude modulated by a 1.5-kHz audio signal. Calculate the values of the RF power in the carrier and in each of the sidebands. Determine the values of the peak and the minimum instantaneous powers.

153-2. In Practice Problem 153-1, a strong second-harmonic component is introduced into the audio signal. Determine the frequencies of the sidebands and the required bandwidth.

153-3. If a transmitter is 85% modulated by an audio tone, what percentage of the total RF power is concentrated in the sidebands?

153-4. A 2-kW carrier is 75% amplitude modulated by a 2-kHz test tone. If the percentage of modulation is decreased from 75% to 50%, calculate the percentage decrease in the total RF power.

153-5. If a transmitter is 40% modulated by an audio tone, what percentage of the unmodulated power is the total sideband power?

153-6. A 10-MHz RF carrier is amplitude modulated by an audio signal that contains frequencies that range from 100 Hz to 5 kHz. What is the value of the bandwidth occupied by the emission?

154
Percentage changes in the antenna current and the sideband power because of amplitude modulation

The act of amplitude modulation creates the additional sideband power; therefore, the total RF power that is associated with the antenna is greater under modulated conditions. The higher the degree of modulation, the greater is the amount of the sideband power.

The antenna current is directly proportional to the square root of the total RF power and will increase with the degree of modulation. Problems involving the percentage changes in the antenna current and the sideband power, can be solved by assuming convenient and compatible values for the unmodulated carrier power, the antenna current, and the effective antenna resistance. A suitable set of values is:

Unmodulated carrier power = 100 W

Unmodulated antenna current = 1 A

Effective antenna resistance = 100 Ω

As an example, if the modulation percentage produced by a single tone is 50%, the total sideband power is $1/2 \times 0.5^2 \times 100$ W = 12.5 W and the total RF power is 100 + 12.5 W = 112.5 W. The new antenna current is $\sqrt{112.5 \text{ W}/100\ \Omega} = \sqrt{1.125} = 1.0607$ A. If the percentage modulation is now increased to 100%, the new sideband power = $1/2 \times 1^2 \times 100$ W = 50 W, the new total RF power is 100 + 50 = 150 W, and the new antenna current is $\sqrt{150 \text{ W}/100\ \Omega} = \sqrt{1.5} = 1.2247$ A.

Consequently, when compared with the unmodulated condition, the antenna current increases by 6.07% for 50% modulation, and 22.47% for 100% modulation. If the percentage modulation is increased from 50% to 100%, the sideband power increases by:

$$\frac{50\text{ W} - 12.5\text{ W}}{12.5\text{ W}} \times 100\% = 300\%$$

and the percentage increase in the antenna current is:

$$\frac{1.2247 - 1.0607}{1.0607} \times 100\% = 15.46\%$$

MATHEMATICAL DERIVATIONS

With double-sideband amplitude modulation produced by a single audio tone (Fig. 154-1),

Total sideband power, $P_{SB} = 0.5\, m^2 P_c$ watts

Total RF power $= P_c(1 + 0.5\, m^2)$ watts (154-1)

If I_A is the unmodulated antenna current and I_A' is the value of the modulated antenna current,

$$I_A' = I_A \times \sqrt{\frac{P_c(1 + 0.5\, m^2)}{P_c}} \qquad (154\text{-}2)$$

$$= I_A \times \sqrt{1 + 0.5\, m^2} \text{ amperes}$$

Increase in antenna current,

$$I_{\text{inc}} = I_A' - I_A \qquad (154\text{-}3)$$

$$= I_A \times \left(\sqrt{1 + 0.5\, m^2} - 1\right) \text{ amperes}$$

% increase in antenna current,

$$\%_{\text{inc}} = \left(\frac{I_A' - I_A}{I_A}\right) \times 100\% \qquad (154\text{-}4)$$

$$= \left(\sqrt{1 + 0.5\, m^2} - 1\right) \times 100\%$$

If the degree of the amplitude modulation produced by a single tone is increased from m_1 to m_2,

Increase in antenna current,

$$I_{\text{inc}} = I_A\left[\sqrt{1 + 0.5\, m_2^2} - \sqrt{1 + 0.5\, m_1^2}\right]$$

% increase in antenna current,

$$\%_{inc} = \frac{I_A\left(\sqrt{1 + 0.5\, m_2^2} - \sqrt{1 + 0.5\, m_1^2}\right)}{I_A\sqrt{1 + 0.5\, m_1^2}} \times 100\% \qquad (154\text{-}5)$$

$$= \left(\sqrt{\frac{2 + m_2^2}{2 + m_1^2}} - 1\right) \times 100\%$$

Similarly, if the degree of modulation is decreased from m_2 to m_1, the percentage decrease in the antenna current is given by:

$$\%_{\text{dec}} = \left(1 - \sqrt{\frac{2 + m_1^2}{2 + m_2^2}}\right) \times 100\% \qquad (154\text{-}6)$$

If the degree of modulation is increased from m_1 to m_2,

% increase in P_{SB},

$$\%_{\text{inc}} = \left(\frac{0.5\, m_2^2 P_c - 0.5\, m_1^2 P_c}{0.5\, m_1^2 P_c}\right) \times 100\% \qquad (154\text{-}7)$$

$$= \left(\frac{m_2^2 - m_1^2}{m_1^2}\right) \times 100\%$$

$$= \left(\frac{m_2^2}{m_1^2} - 1\right) \times 100\%$$

If the degree of modulation is decreased from m_2 to m_1, the percentage decrease in the total sideband power is:

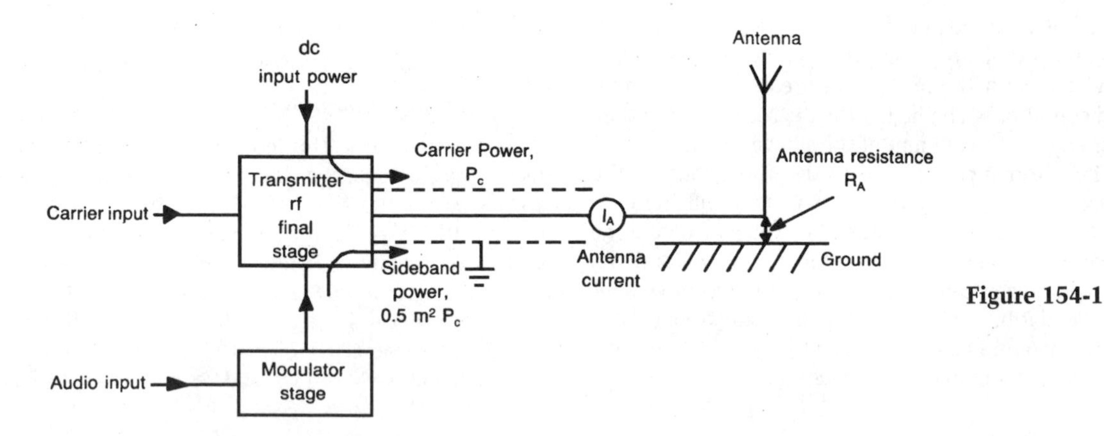

Figure 154-1

$$\% \text{ decrease in } P_{SB} = \left(\frac{m_2^2 - m_1^2}{m_2^2}\right) \times 100\% \quad (154\text{-}8)$$

$$= \left(1 - \frac{m_1^2}{m_2^2}\right) \times 100\%$$

Example 154-1

An RF carrier is 70% amplitude modulated (double-sideband) by a single tone. What is the percentage increase in the antenna current, as compared with its unmodulated value?

Solution

% increase in antenna current,

$$\%_{inc} = (\sqrt{1 + 0.5m^2} - 1) \times 100\% \quad (154\text{-}4)$$

$$= (\sqrt{1 + 0.5 \times 0.7^2} - 1) \times 100\%$$

$$= 11.6\%.$$

Example 154-2

An RF carrier is double-sideband amplitude modulated by a single tone. If the degree of modulation is increased from 0.50 to 0.80, what are the percentage increases in the antenna current and the total sideband power?

Solution

% increase in antenna current,

$$\%_{inc} = \left(\sqrt{\frac{2 + m_2^2}{2 + m_1^2}} - 1\right) \times 100\% \quad (154\text{-}5)$$

$$= \left(\sqrt{\frac{2 + 0.64}{2 + 0.25}} - 1\right) \times 100\%$$

$$= \left(\sqrt{\frac{2.64}{2.25}} - 1\right) \times 100\%$$

$$= 8.32\%$$

$$\% \text{ increase in } P_{SB} = \frac{m_2^2 - m_1^2}{m_1^2} \times 100\% \quad (154\text{-}7)$$

$$= \frac{0.8^2 - 0.5^2}{0.5^2} \times 100\%$$

$$= \frac{0.64 - 0.25}{0.25} \times 100\%$$

$$= \frac{0.39}{0.25} \times 100\%$$

$$= 156\%.$$

PRACTICE PROBLEMS

154-1. An RF carrier is (double sideband) amplitude modulated by a single tone. If the percentage modulation is decreased from 80% to 50%, what are the percentage decreases in the antenna current and the total sideband power?

154-2. An RF carrier is 70% (double sideband) amplitude modulated by a single tone. If the tone is removed, calculate the percentage decrease in the total RF power.

154-3. An RF carrier is (double sideband) amplitude modulated by a single tone. If the total sideband power is 40% of the modulated carrier power, what is the percentage increase in the antenna current when compared with its unmodulated value?

154-4. An RF carrier is (double sideband) amplitude modulation by a single tone. If there is a 15% increase in the antenna current when compared with its unmodulated value, what percentage of the total RF power is concentrated in the sidebands?

155 Plate modulation

It is appropriate to consider a tube version of this circuit because modulation primarily occurs in the final stage of a commercial AM transmitter. The RF power level of this stage is frequently on the order of 10 kW or more and no solid-state devices are capable of operating at this level.

The plate modulated RF power amplifier of Fig. 155-1 is a class-C amplifier in which the audio signal

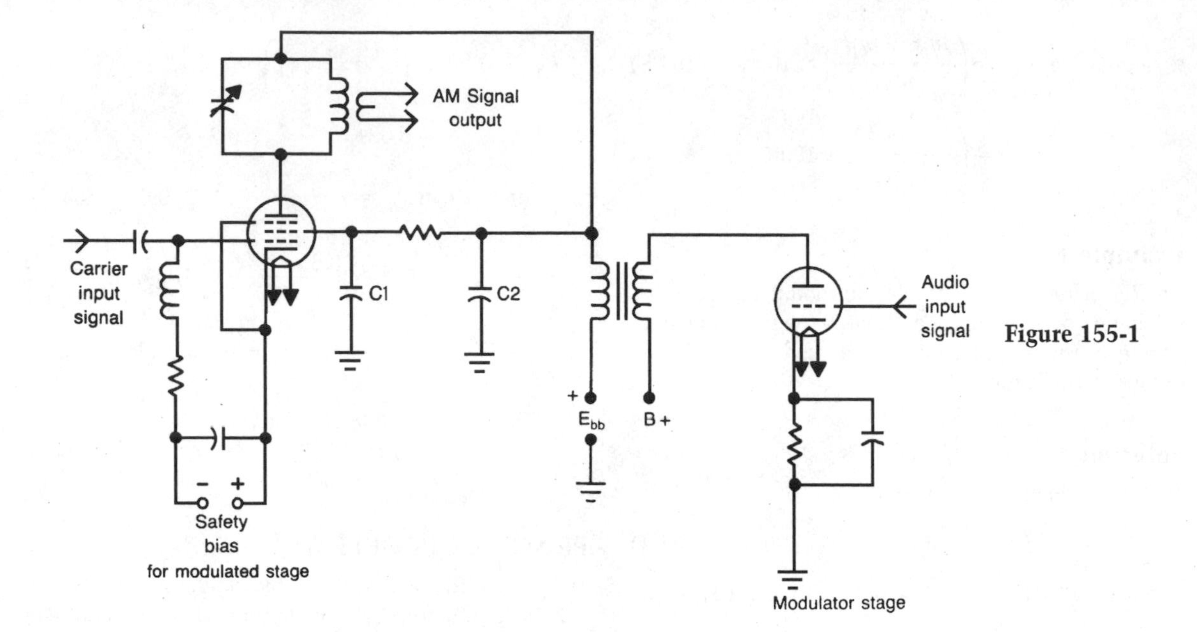

Figure 155-1

output is superimposed on the dc plate supply voltage (E_{bb}); this total voltage is then applied to the plate circuit and corresponds to the required modulation envelope of the AM signal.

The level of the carrier signal input, the value of the control grid bias, and the impedance of the plate load tank circuit can be adjusted so that the instantaneous amplitude of the AM signal output is approximately equal to the instantaneous voltage that is applied to the plate circuit. This will provide the necessary linear relationship between the instantaneous amplitude of the AM signal and the instantaneous magnitude of the audio signal.

Modulation can occur in the plate circuit of the transmitter final RF stage (high-level modulation) or in an earlier stage, where the modulator will be required to provide less audio power (low-level). However, if low-level modulation is used, the following stages cannot be operated under class-C conditions for high plate efficiency or to provide frequency multiplication, but they must be linear class-AB or class-B amplifiers.

If a beam power tube is used for the rf stage, both the plate and the screen grid are modulated by the audio signal (C1 and C2 are RF, but not AF, bypass capacitors). By modulating the screen grid, excessive screen current is prevented at the trough of the modulation cycle.

In plate modulation, the power in the RF sidebands is obtained from the modulator stage and the carrier power is derived from the E_{bb} supply. The modulator is therefore required to deliver to the RF stage that amount of power that is sufficient to generate the sidebands, while taking into account the percentage value of the class-C plate efficiency.

MATHEMATICAL DERIVATIONS

If the audio signal is a single tone and the required degree of modulation is m, the total double sideband power is $0.5\,m^2P_c$, where P_c is the amount of carrier power developed in the plate circuit. Then, carrier power,

$$P_c = \frac{E_{bb} \times I_b \times \eta_R}{100} \text{ watts} \qquad (155\text{-}1)$$

where: η_R = plate efficiency percentage of the RF final stage.
E_{bb} = plate supply voltage of the RF final stage (V).
I_b = dc level of plate current for the RF final stage (A).

The audio power output required from the modulator stage is:

$$\textit{Modulator power output} = \qquad (155\text{-}2)$$

$$\frac{0.5\,m^2P_c}{\eta_R} \times 100 = \frac{50\,m^2P_c}{\eta_R} \text{ watts}$$

The modulator stage of Fig. 155-1 is a single tube audio power amplifier that will operate under class-A conditions with a relatively low plate efficiency, η_M.

Plate input power dc to the modulator stage,

$$= \frac{\text{audio power input}}{\eta_M} \times 100 \qquad (155\text{-}3)$$

$$= \frac{50\ m^2 P_c}{\eta_R \times \eta_M} \times 100$$

$$= \frac{5000\ m^2 P_c}{\eta_R \times \eta_M}$$

$$= \frac{50\ m^2 \times E_{bb} \times I_b}{\eta_M} \text{ watts}$$

Plate dissipation of the modulator stage,

$$= (\textit{modulator's dc input power}) \times \left(1 - \frac{\eta_M}{100}\right)$$

$$= \left(\frac{100}{\eta_M} - 1\right) \times \frac{50\ m^2 P_c}{\eta_R} \text{ watts} \qquad (155\text{-}4)$$

The effective load connected across the secondary of the output modulator's transformer is E_{bb}/I_b ohms. The turns ratio of the transformer must be used to match this effective load to the optimum plate load required by the modulator tube. For example, if the modulator tube is a triode, the optimum plate load is $2r_p$, where r_p is the triode's ac plate resistance.

Example 155-1

The plate-modulated final RF stage of an AM transmitter has a dc supply voltage of 800 V and a dc plate current level of 1.5 A. The stage is operated in class-C, with a plate efficiency of 75%, and is 80% modulated by a single tone. The modulator stage uses two triodes in parallel and has a plate efficiency of 30%. What are: (a) the total power in the sidebands, (b) the modulator output power, (c) the required dc power input to the modulator, (d) the required turns ratio for the modulator output transformer (if each triode modulator tube has an ac plate resistance of 800 Ω), and (e) the plate dissipation in each of the triodes?

Solution

(a) The *dc plate input power* $= E_{bb} \times I_b = 800 \times 1.5 = 1200$ W. The RF carrier power developed in the plate circuit of the final RF stage is given by:

$$\text{Carrier power, } P_c = 1200 \times \frac{75}{100}$$

$$= 900 \text{ W}$$

Total double sideband power,

$$P_{SB} = 0.5\ m^2 P_c$$

$$= 0.5 \times (0.8)^2 \times 900$$

$$= 288 \text{ W.}$$

(b) The modulator output must provide the sideband power, taking into account the class-C plate efficiency of the final RF stage.

$$\textit{Modulator output power} = \frac{P_{SB}}{\eta_R} \times 100 \qquad (155\text{-}2)$$

$$= \frac{288}{75} \times 100$$

$$= 384 \text{ W}$$

(c) The plate efficiency of the modulator stage is 30%. The dc plate power input to the modulator is:

$$\frac{\textit{Modulator output power}}{\eta_M} \times 100 = \frac{384}{30} \times 100 \qquad (155\text{-}3)$$

$$= 1280 \text{ W}$$

Alternatively, the dc plate power input to the modulator is:

$$\frac{50\ m^2 E_{bb} I_b}{\eta_M} = \frac{50 \times (0.8)^2 \times 800 \times 1.5}{30}$$

$$= 1280 \text{ W}$$

$$= 1.28 \text{ kW}$$

(d) The effective load on the transformer secondary is:

$$\frac{E_{bb}}{I_b} = \frac{800 \text{ V}}{1.5 \text{ A}}$$

$$= 533\ \Omega$$

Effective r_p of the two triodes in parallel = 800/2 = 400 Ω.

Optimum load required by the triodes = 2 × 400 = 800 Ω. *Required turns ratio for the modulator transformer* = $\sqrt{800/533}$ = 1.23 : 1.

(e) The total plate dissipation of the modulator stage is:

$$1.28 \times \frac{70}{100} = 0.896 \text{ kW} \qquad (155\text{-}4)$$

$$= 896 \text{ W}$$

The plate dissipation in each triode is:

$$\frac{896}{2} = 450 \text{ W, rounded off.}$$

Note that, if the triode had been connected in push-pull and not in parallel, the plate dissipation in each triode would still have been half of the total plate dissipation in the modulator stage.

PRACTICE PROBLEMS

155-1. In Example 155-1, calculate the value of the total plate dissipation in the final stage.

155-2. In Example 155-1, the plate supply voltage to the modulator stage is 700 V. Calculate the dc level of current flowing through each of the triodes.

155-3. In Example 155-1, the percentage of double-sideband modulation falls from 80% to 40%. Recalculate the values of: (a) the total power in the sidebands, (b) the modulator output power, (c) the required dc power input to the modulator, and (d) the plate dissipation in each of the triodes. Assume that the plate efficiency of the modulator stage falls from 30% to 7.5%.

156
The AM transmitter

Figure 156-1 shows the block diagram of an AM transmitter that is capable of double-sideband voice communications (A3E: Appendix G). The RF (radio frequency) section is responsible for generating the required carrier power and the AF (audio frequency) section provides the audio power, which is necessary to achieve amplitude modulation of the carrier (chapter 154).

The RF section is divided into the following blocks.

THE OSCILLATOR STAGE

The FCC assigns a certain carrier frequency to a particular station; at the same time, the FCC requires that the output frequency is maintained within narrow tolerance limits. For example, in the AM broadcast band, a station is assigned a carrier frequency of 980 kHz with an allowed tolerance of ±20 Hz. Consequently, the output frequency must not rise above 980 kHz + 20 Hz = 980.02 kHz, nor fall below 980 kHz − 20 Hz = 979.98 kHz. The narrow tolerance is obeyed by using a crystal oscillator and enclosing the crystal in a thermostatically controlled oven (chapter 163).

In addition to quoting the permitted variation in hertz, the frequency tolerance can also be measured as a percentage (%), or in parts per million (ppm) of the carrier frequency. These methods of measurements are related by:

$$\textit{Percentage of tolerance} = \frac{\textit{parts per million}}{10^4}$$

$$= \text{ppm} \times 10^{-4}\,\%$$

$$\textit{Parts per million} = \textit{percentage} \times 10^4 \text{ ppm.}$$

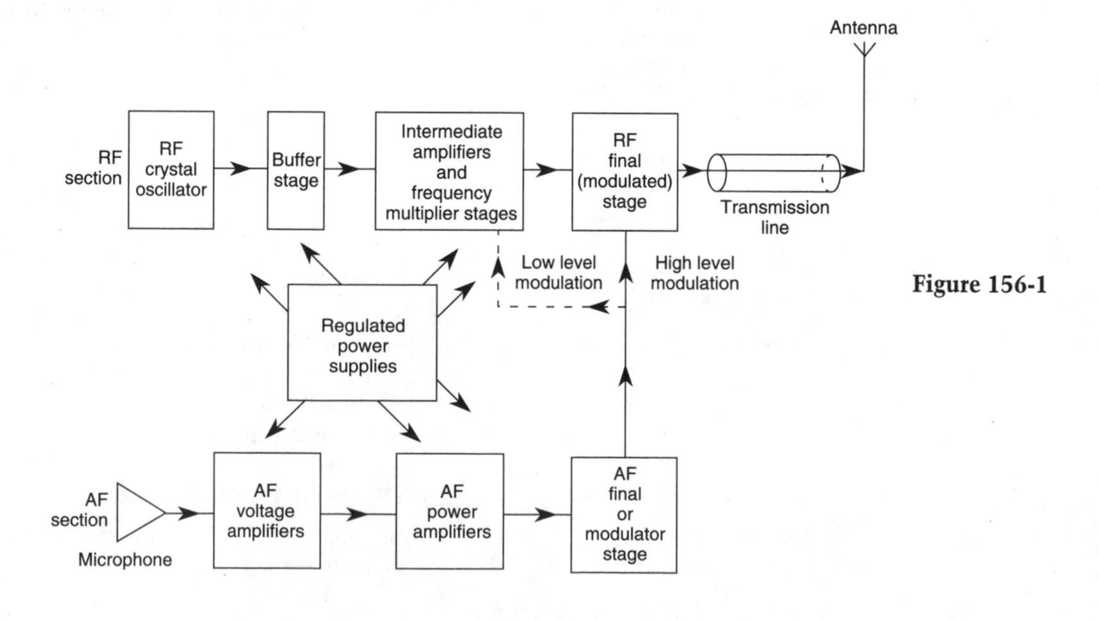

Figure 156-1

For example, if a 2-MHz carrier has a percentage tolerance of 0.002%, its tolerance is $0.002 \times 10^4 = 20$ ppm, or $2 \times 20 = 40$ Hz.

The oscillator stage is operated under conditions that may lie between class-AB and class-C, and provides a power output that can extend up to a few watts. In addition, the oscillator is normally operated from a highly regulated and well filtered dc power supply.

THE BUFFER STAGE

To improve the frequency stability still further, the oscillator is followed by a buffer stage, which has a very high input impedance and is typically operated under class-AB conditions. The power gain of the buffer stage is low, but it presents a constant (rather than a varying load) to the oscillator. This is an important factor in stabilizing the frequency generated by the oscillator stage.

THE INTERMEDIATE RF POWER AMPLIFIERS

The main function of these stages is to increase the power level of the carrier. Because efficiency is of vital importance, each stage is invariably operated under class-C conditions (chapter 128) and it uses a tank circuit as the collector (or plate) load (Fig. 156-2). Because of feedback from the output circuit to the input circuit, such stages might need to be neutralized.

If the transmitter is operating in the VHF band, it is impossible to find a crystal that will oscillate at the required frequency. For example, let the required carrier frequency be 100 MHz. Because the upper frequency limit for a crystal's fundamental vibration is only about 25 MHz, you cannot use class-C amplifiers alone for the intermediate stages, but you must instead include some frequency multipliers (chapter 128). In this example, one solution is to use a 12.5-MHz crystal and then insert three doubler stages in the intermediate amplifiers. The total product factor is $2 \times 2 \times 2 = 8$ and, therefore, the output frequency is 8×12.5 MHz = 100 MHz. Because the output frequency from a doubler or a tripler is a multiple of the input frequency, there is no effective feedback from the collector circuit; consequently, frequency multipliers do not need to be neutralized. Notice that if the tolerance of the output frequency is +2 kHz, the tolerance at the crystal oscillator is +2 kHz/8 = +250 Hz. In other words, the amount of tolerance is changed, but the percentage of the tolerance, +2 kHz/100 MHz × 100 = +250 Hz/12.5 MHz × 100 = +0.002% remains unchanged. The actual deviation of the output frequency from its assigned value is measured by the percentage of variance (see Mathematical Derivations).

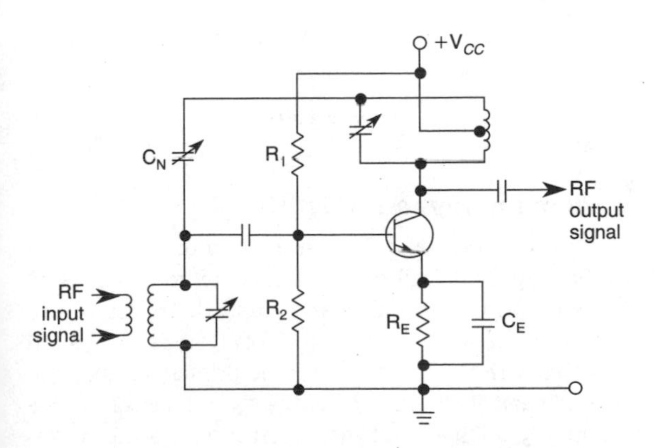

Figure 156-2

THE RF FINAL STAGE

This is the most important stage in the entire RF section because it is responsible for delivering the rated power output of the transmitter to the antenna system. The FCC assigns an authorized power output to a particular station but, at the same time, there is a certain power tolerance. For example, the power tolerance for AM standard broadcast stations extends from 5% above to 10% below; so that for a station with an assigned power of 10 kW, the actual operating power can extend up to 10.5 kW or down to 9 kW. This operating power is determined by the direct method as described in chapter 160. In addition, readings of the supply voltage for the final stage and its associated dc level of current are entered in the station log. The dc current meter must be shunted by an RF bypass capacitor so that the reading is not affected by the RF components that are contained in the current waveform.

Because the final stage produces the highest level of RF power, this amplifier must be operated under class-C conditions with an efficiency of up to 90%. This will inevitably create harmonic components (chapter 111) of the carrier frequency. However, harmonic radiation from the antenna must be kept to a minimum in order to prevent interference to other stations. The following methods can be used to reduce the harmonics to a level that is normally at least 40 dB below the carrier power:

1. The use of a push-pull RF circuit for the final stage. Such an arrangement will eliminate all even harmonics of the carrier frequency.
2. The insertion of a Faraday shield between the tank coil and the coupling coil to the transmission line (Fig. 156-3). This screen is grounded so that the capacitive coupling between the two coils is considerably reduced and will no longer permit the transfer of high-frequency harmonic components to the antenna system.

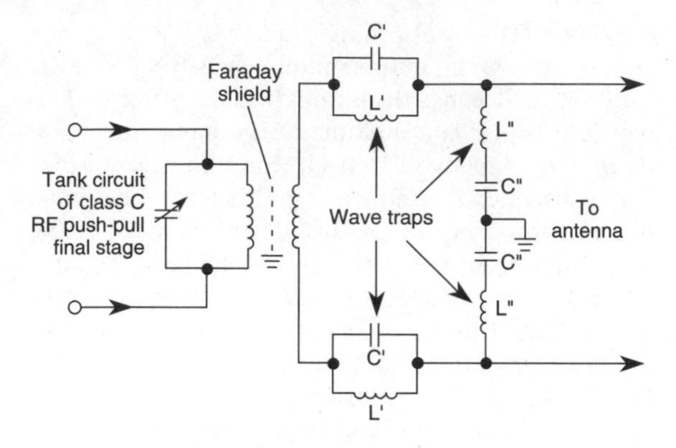

Figure 156-3

3. If a certain strong harmonic component appears across the coupling coil, it can be prevented from reaching the transmission line by the use of series and parallel wave traps (Fig. 156-3). At the particular harmonic frequency, the parallel wave trap (L'/C') behaves as a theoretical open circuit, but the series wave trap (L"/C") acts as a short circuit.
4. The bias on the final stage can be reduced so that the angle of the output current flow is increased.

The antenna system is a load on the tank circuit of the final stage and lowers its Q; as a result, the responses to the harmonic components are increased. The degree of coupling between L_1 and L_2 is therefore adjusted to compromise between the power transfer to the antenna system and the harmonic response; a loaded Q value on the order of 10 to 15 is typical.

The RF section (as described) merely generates the carrier which, by itself, contains no information. However, the carrier can be keyed to produce "mark" and "space" intervals for the transmission of code (emission designation for telegraphy, A1A). For voice communication, the transmitter needs an AF section, which is divided into the following stages.

THE MICROPHONE

Every microphone basically consists of a diaphragm (which vibrates in the presence of a sound wave) and some form of transducer that is capable of converting the mechanical vibration into an electrical signal. Some microphones may have their diaphragms stretched in order to improve the frequency response; this is achieved by making the diaphragm's natural resonant frequency higher than the upper limit of the audio range.

Dynamic (or moving coil) microphone (Fig. 156-4)

A circular coil, with turns passing between the poles of a permanent magnet, is joined with the diaphragm. When the diaphragm vibrates, the coil cuts the magnetic flux so that there is an induced voltage that is the electrical equivalent of the sound wave. This type of microphone is commonly used in broadcast studios.

The features of this type of microphone include:

- Low output impedance.
- Wide frequency response.
- No dc power supply necessary.
- Relatively low level of output.
- Relatively unaffected by humidity and temperature.
- Low hum.
- Light weight, but rugged construction.

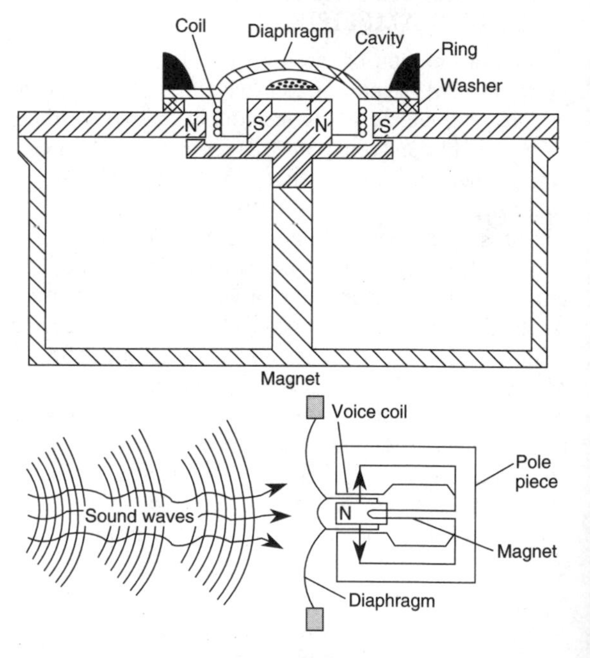

Figure 156-4

Carbon microphone (Fig. 156-5)

A "button" containing fine carbon granules is attached to the diaphragm. Vibration of the diaphragm varies its pressure on the granules, and alters the resistance of the "button." A low-voltage (6 V) battery is connected in series with the "button" and the primary of an audio transformer. The primary current, and therefore the secondary voltage, will then correspond to the movement of the diaphragm created by the sound wave.

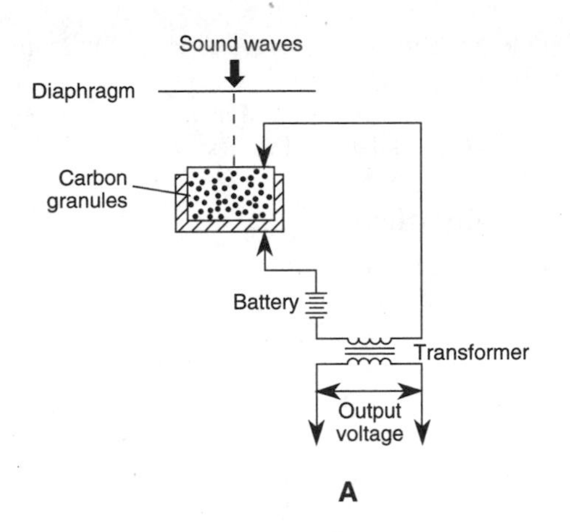

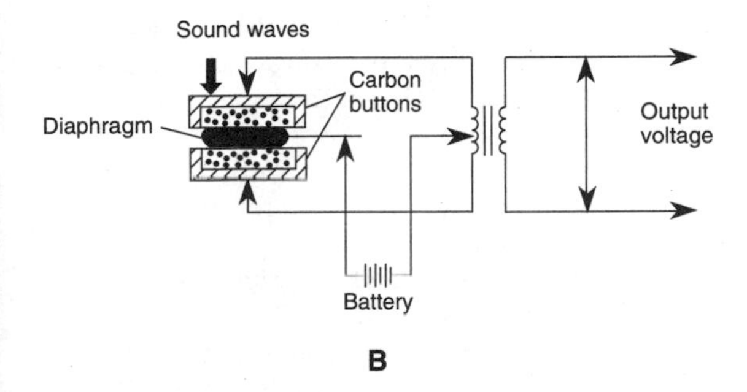

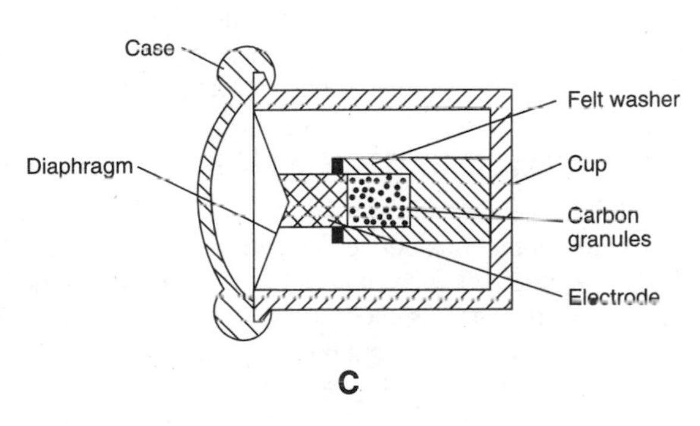

Figure 156-5

The carbon microphone features include:

- Very high sensitivity that is associated with a high level of output.
- Poor frequency response, which is the main reason why the carbon microphone is no longer widely used.
- Background hissing present in the microphone's output.
- Carbon granules subject to "packing" with excessive moisture (the microphone must not be allowed to get wet).
- Low output impedance (11 Ω or less).
- Inexpensive.

Crystal microphone (Fig. 156-6)

In the crystal microphone, the transducer is commonly a series of Rochelle salt crystals, which display the piezoelectric effect, and are wax-impregnated in an airtight container. Crystal microphone features include:

- Adversely affected by excessive heat (the microphone must be protected from hot sunlight), shock, and humidity (Rochelle salt dissolves in water).
- Good frequency response.
- Absence of background noise.
- High impedance output, which can allow hum pickup.
- No power supply required.

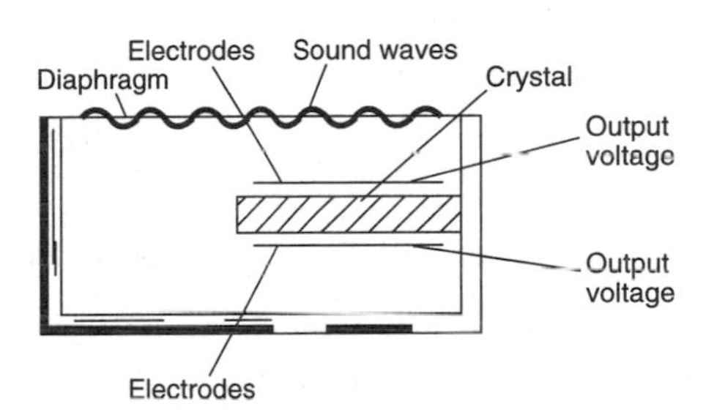

Figure 156-6

Ceramic microphone

The principle and construction of the ceramic microphone elements are similar to the crystal microphone, except for the use of certain ceramic materials, as opposed to Rochelle salt. The ceramic materials act as transducers in the same way as the crystals, but they are less affected by extremes of temperature, humidity, and shock.

Velocity (ribbon) microphone (Fig. 156-7)

The velocity (ribbon) microphone is commonly used in a broadcast studio. The diaphragm is a very light aluminum ribbon, which is allowed to move freely in a magnetic field. When the sound wave causes a pressure difference on opposite sides of the ribbon, a velocity gradient causes the ribbon to vibrate. The corresponding audio electrical signal is then induced into the ribbon.

The output impedance of this type of microphone, is extremely low so that the signal is fed to a step-up transformer. This not only raises the output level to a practical value, but also matches the low microphone

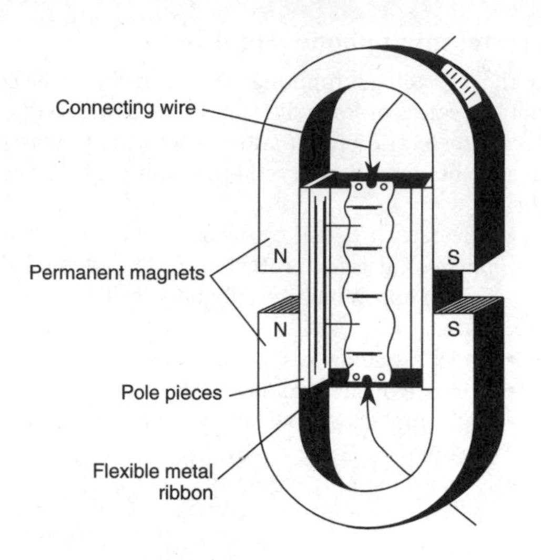

Figure 156-7

impedance to the surge impedance of a standard audio line. Velocity microphone features include:

- Blasting or wind can affect the ribbon, which is delicate; its use is confined to an indoor studio.
- No power supply required.
- Low level of background noise.
- Good frequency response.
- High level of output.
- Not affected by temperature and humidity.

THE AUDIO AMPLIFIERS

The output from a microphone is typically on the order of millivolts and, therefore, it is necessary to build up this signal by utilizing a number of audio voltage amplifiers. The early stages are operated in class-A with single active devices, but the later stages are more likely to be class-A or class-AB push-pull amplifiers (chapter 127).

In the block diagram of Fig. 156-1, the sideband power is supplied from the final audio amplifier, which is termed the "modulator stage" (chapter 154) and is commonly operated in a class-B push-pull arrangement. Such a stage is preceded by one (or more) audio power amplifiers.

High- and low-level modulation

When the output of the modulator is fed to the RF final stage, the result is "high-level" modulation, which has the disadvantage of requiring a considerable amount of audio power. It is possible to generate the AM signal in one of the intermediate amplifiers ("low-level" modulation); less audio power is needed, but none of the following stages can be operated in the class-C mode or can be used for frequency multiplication. After low-level modulation has occurred, all following amplifiers must be linear stages, which are normally operated in class-B mode, with its lower efficiency when compared with class-C modulation (Fig. 156-8).

Intermodulation

Assume that you have a transmitter (Tx_1), which is operating on a carrier frequency (f_1); in close proximity is a second transmitter (Tx_2) whose carrier frequency is f_2. It is possible for the Tx_2 transmission to be intercepted by the Tx_1 antenna so that both RF signals will exist in the first transmitter's RF final stage, which is operated under class-C conditions. A complex form of mixing occurs; and there will be a large number of intermodulation products, whose carrier frequencies are of the form $\pm mf_1 \pm nf_2$ (where m and n equal 0, 1, 2, 3, 4, . . .). The same type of mixing will, of course, occur in the final stage of the second transmitter.

All of these harmonic sum and difference components will carry one, the other, or both, of the modulating signals and may cause severe interference to other transmissions.

The effects of intermodulation can be reduced by inserting appropriate wave traps in the antenna circuits of the two transmitters. For example, in the antenna circuit of the transmitter (Tx_1), there can be a wave trap that is tuned to the carrier frequency (f_2); consequently, very little of the f_2 transmission is passed through to the final stage of the first transmitter.

TWO-WAY COMMUNICATION SYSTEMS

The following methods can be used to establish two-way communications between a pair of stations:

1. *Simplex* At each station the transmitter and the receiver use the same antenna. Both stations operate on the *same* frequency so that simultaneous transmission in both directions is not possible.
2. *Duplex* The stations operate on two *different* frequencies so that simultaneous communication in both directions is allowed.

MATHEMATICAL DERIVATIONS

Percentage of variance, (156-1)

$$= \frac{\textit{unmodulated carrier frequency} \sim \textit{Assigned frequency}}{\textit{assigned frequency}} \times 100\%$$

Percentage of tolerance, (156-2)

$$= \frac{\textit{maximum permitted frequency deviation}}{\textit{assigned frequency}} \times 100\%$$

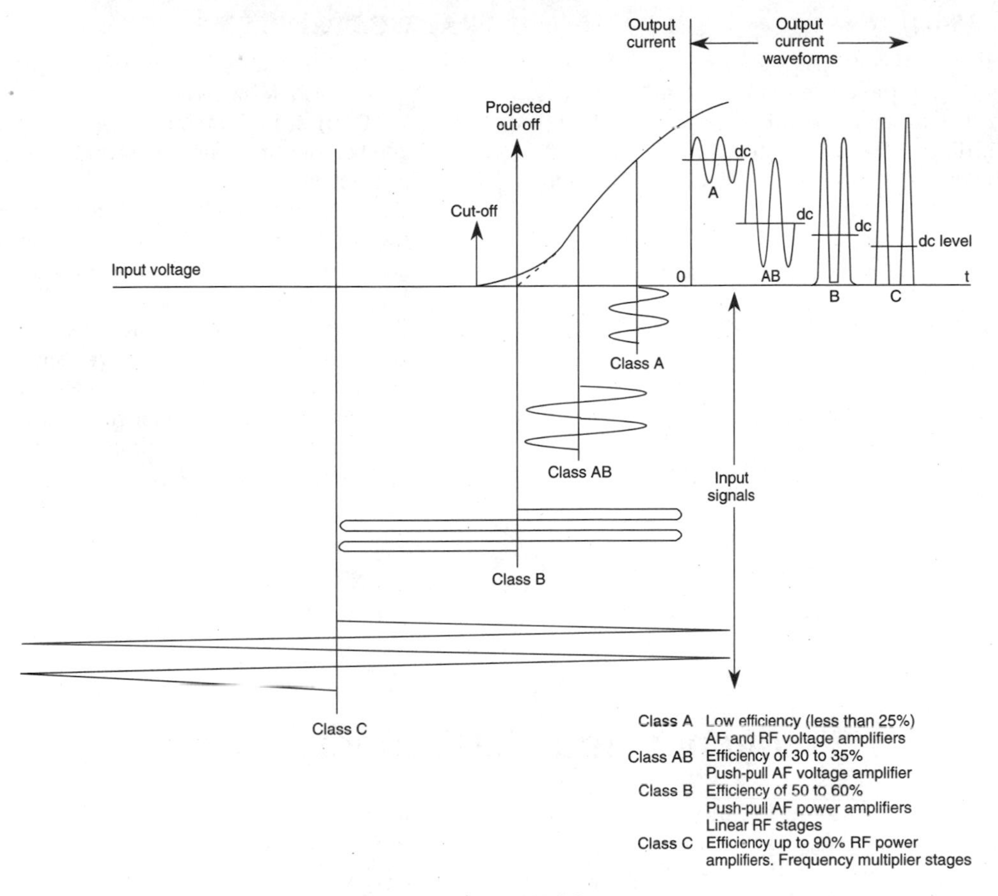

Figure 156-8

Frequency deviation at the oscillator, (156-3)

$$= \frac{\text{output deviation in the RF final stage}}{\text{product factor of the multiplier stages}}$$

Example 156-1

An AM transmitter has an unmodulated carrier frequency of 50.0005 MHz and an assigned frequency of 50 MHz. If the percentage of tolerance is 0.002%, is the transmitter's output frequency within tolerance?

Solution

Percentage of variance,

$$= \frac{50.0005 - 50}{50} \times 100\% \qquad (156\text{-}1)$$

$$= 0.001\%$$

Because the percentage of variance is less than the percentage of tolerance, the transmitter's output frequency is within tolerance.

Example 156-2

An 8-MHz oscillator is followed by two doubler stages and a tripler stage. If the permitted tolerance is ±0.002%, calculate the maximum allowed value of the output frequency.

Solution

Transmitter's assigned output frequency,

$$= 2 \times 2 \times 3 \times 8 \text{ MHz}$$

$$= 96 \text{ MHz.}$$

Maximum deviation of the output frequency,

$$= \pm\frac{0.002}{100} \times 96 \text{ MHz} \qquad (156\text{-}2)$$

$$= \pm 1920 \text{ Hz}$$

Maximum allowed output frequency,

$$= 96 \text{ MHz} + 1.92 \text{ kHz}$$

$$= 96.00192 \text{ MHz.}$$

Example 156-3

The dc supply voltage for an AM transmitter's RF final stage is 400 V and the current drawn from this supply is 800 mA. The antenna has a load resistance of 70 Ω and the reading of the antenna current meter is 2 A. What is the overall percentage efficiency for the combination of the final stage and the transmission line?

Solution:

$$\textit{Dc input power} = 400 \text{ V} \times 800 \text{ mA} = 320 \text{ W}.$$

$$\textit{Antenna RF power} = (2 \text{ A})^2 \times 70\ \Omega = 280 \text{ W}.$$

$$\textit{Overall percentage efficiency} = \frac{280 \text{ W}}{320 \text{ W}} \times 100\% = 87.5\%.$$

PRACTICE PROBLEMS

156-1. An AM transmitter has an assigned carrier frequency of 25 MHz. If the percentage of variance is ±0.001%, by how much does the unmodulated carrier frequency differ from the assigned frequency?

156-2. The frequency tolerance of an AM broadcast transmitter is ±20 Hz. If the assigned frequency is 1.5 MHz, calculate the value of the percentage tolerance.

156-3. The carrier frequency of an AM transmitter is 120 MHz, with a percentage tolerance of ±0.002%. The intermediate amplifier contains one doubler and two tripler stages. What is the value of the maximum permitted frequency deviation at the oscillator stage?

157
Frequency and phase modulation

With frequency modulation, the instantaneous frequency of the RF wave is varied in accordance with the modulating signal, while the amplitude of the RF wave is kept constant. The RF power output and the antenna current of the FM transmitter are therefore independent of the modulation.

FREQUENCY MODULATION

In the FM wave (Fig. 157-1), the instantaneous amount of the *frequency shift* or *deviation* (away from its average unmodulated value) is linearly related to the instantaneous *magnitude* (voltage) of the modulating signal; while the rate at which the frequency deviation (f_d) occurs, is equal to the modulating frequency. Note carefully, that the amount of the frequency shift in the RF wave is independent of the modulating frequency. Therefore, if they are of the same amplitude, modulating tones of 200 Hz and 400 Hz will provide the same amount of frequency shift in the FM wave. However, in FM and TV broadcast transmitters, the higher audio frequencies above 800 Hz are progressively accentuated (preemphasized) in order to improve their signal-to-noise ratio at the receiver. The degree of preemphasis is measured by the time constant of an RC circuit whose audio output increases with frequency. This time constant is specified by the FCC as 75 microseconds. In order to restore the tonal balance, the receiver discriminator output is fed to an RC deemphasis circuit with the same time constant.

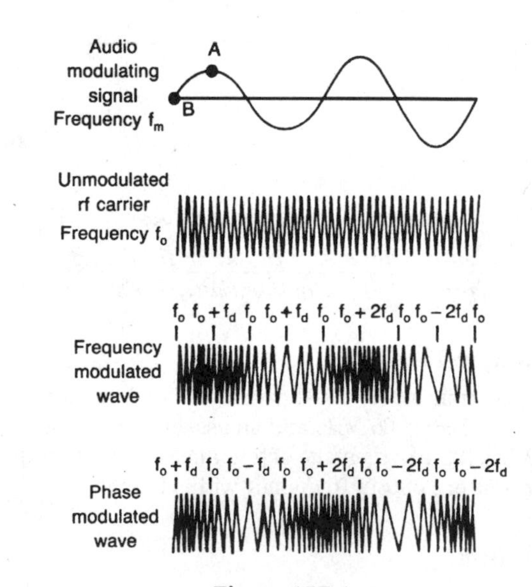

Figure 157-1

When 100% modulation of an FM wave occurs, the amount of frequency shift reaches the maximum value allowed for the particular communications system. This maximum value is called the highest *fre-*

quency deviation which, in the FM broadcast service (emission designation F3E), is a shift of 75 kHz on either side of the unmodulated carrier frequency. The output frequency swing for 100% modulation is therefore ±75 kHz. However, in the aural transmitter of a TV broadcast station, 100% modulation corresponds to a frequency deviation of 25 kHz and in the Public Safety Radio Services, the frequency deviation is only 5 kHz. The percentage modulation, and the amount of the frequency shift, are directly proportional so that 40% modulation in the FM broadcast service would correspond to an output frequency swing of $\pm 40 \times 75/100 = \pm 30$ kHz.

For FM the degree of modulation is defined by the *modulation index*:

$$\text{Modulation index, } m = \frac{f_d}{f_m}$$

where: f_m = modulating frequency
f_d = frequency deviation

During the transmission of speech and music, the value of the instantaneous modulation index can vary from less than 1 to over 100. However, for a particular system, there is a certain modulation index value, which is called the *deviation ratio* and is determined by:

$$\textit{Deviation ratio} = \frac{f_d \text{ corresponding to 100\% modulation}}{\text{highest transmitted value of } f_m}$$

For commercial FM broadcasting, the transmitted audio range is 50 Hz to 15 kHz. Because the frequency deviation value for 100% modulation is 75 kHz, the deviation ratio is 75 kHz/15 kHz = 5, which is the typical value for a *wide band* FM system. By contrast, marine FM transmitters use a narrow-band system, in which the frequency deviation for 100% modulation is 5 kHz and the highest audio frequency is 3 kHz (adequate for voice communication). Consequently, the value of the deviation ratio is only 5 kHz/3 kHz = 1.67.

PHASE MODULATION

In *phase modulation* (Fig. 157-1), the instantaneous phase of the RF carrier wave is varied in accordance with the modulating signal, while the amplitude of the RF wave is kept constant. The instantaneous amount of the phase shift (away from its unmodulated value) is linearly related to the instantaneous magnitude (voltage) of the modulating signal. Because a rate of change in phase is equivalent to a shift in frequency, the PM wave is similar in appearance to the FM wave; in fact, they cannot be distinguished, except by reference to the modulating signal. The important differences are:

1. With PM, the output frequency swing is proportional to the product of the amplitude and the frequency of the modulating signal, but in FM, the swing is proportional to the amplitude alone. Therefore, if the carrier is being modulated by a single tone and the tone amplitude and the frequency are both doubled, the output frequency swing in the PM signal would be quadrupled (neglecting any effect of preemphasis).
2. Relative to the cycle of the modulating signal, the instantaneous maximum (and/or minimum) frequency positions are 90° apart in the FM and PM waves. In Fig. 157-1, point *A* in the modulating signal corresponds to the instantaneous maximum frequency in the FM wave. For the PM wave the maximum frequency occurs at point *B*, which is one-fourth of a cycle (or 90°) from point *A*.

The main importance of PM is its use in the *indirect method* of creating FM. For reasons of frequency stability (and to avoid the requirement for an AFC system) it is common practice in many FM transmitters to use a master crystal oscillator, whose frequency is fixed and which therefore cannot be directly modulated. The modulated stage comes after the crystal oscillator and consists of a circuit that actually creates phase modulation. However, before reaching the modulated stage, the audio signal is passed through a correction network (not to be confused with the preemphasis circuit), whose voltage output is inversely proportional to frequency and which introduces an approximate 90° phase shift. Then, as far as the undistorted audio signal *input to* the correction network is concerned, the output from the modulated stage is the required FM signal, but with regard to the *output from* the correction network, the modulated stage produces PM. Clearly FM and PM are closely related and each is an example of *angle* modulation.

MATHEMATICAL DERIVATIONS

The equation of an FM wave is

$$f(t) = A \cos [\omega_o t + m \sin \omega_m t] \qquad (157\text{-}1)$$

where: A = peak value of the unmodulated carrier voltage (V)
ω_o = angular frequency of the carrier (rad/s)
ω_m = angular frequency of the audio modulating tone (rad/s)
m = modulation index

Then,

$$f(t) = A \cos(\omega_o t + m \sin \omega_m t) \quad (157\text{-}2)$$
$$= A \cos \omega_o t \cos(m \sin \omega_m t)$$
$$- A \sin \omega_o t \sin(m \sin \omega_m t)$$

Now,

$$\cos(m \sin \omega_m t) = \quad (157\text{-}3)$$
$$J_o(m) + 2[J_2(m) \cos 2\omega_m t + J_4(m) \cos 4\omega_m t + \ldots]$$

and,

$$\sin(m \sin \omega_m t) = \quad (157\text{-}4)$$
$$2[J_1(m) \sin \omega_m t + J_3(m) \sin 3\omega_m t + \ldots]$$

where: $J_n(m)$ = Bessel function coefficients
for n = 0, 1, 2, 3, . . . etc.

Substituting Equations 157-3 and 157-4 in Equation 157-2,

$$f(t) = A[J_o(m) \cos \omega_o t \quad (157\text{-}5)$$
$$+ J_1(m)[\cos(\omega_o + \omega_m)t - \cos(\omega_o - \omega_m)t]$$
$$+ J_2(m)[\cos(\omega_o + 2\omega_m)t + \cos(\omega_o - 2\omega_m)t]$$
$$+ J_3(m)[\cos(\omega_o + 3\omega_m)t - \cos(\omega_o - 3\omega_m)t]$$
$$+ J_4(m)[\cos(\omega_o + 4\omega_m)t + \cos(\omega_o - 4\omega_m)t],$$

. . . etc.

Remembering that the modulating signal is only a single sine wave, the most outstanding conclusion drawn from Equation 157-5 is that the modulated carrier consists of an infinite number of sidebands above and below the carrier frequency. In practice, the sidebands are limited to those greater than 1% of the *unmodulated carrier voltage.*

As an example, assume that an unmodulated carrier has a frequency of 100 MHz and a peak value of 1 V. The modulating signal is a tone of 15 kHz and it produces a modulation index of 5, so the frequency deviation is 5 × 15 = 75 kHz.

From a list of Bessel Functions:

$J_0(5) = -0.178$
$J_1(5) = -0.328$
$J_2(5) = +0.047$
$J_3(5) = +0.365$
$J_4(5) = 0.391$ } significant sidebands
$J_5(5) = 0.261$
$J_6(5) = 0.131$
$J_7(5) = 0.053$
$J_8(5) = 0.018$ significant sideband limit

$J_9(5) = 0.005$ insignificant sideband

It can be seen that there are eight pairs of significant sidebands, so that the effective bandwidth is 15 × 8 × 2 kHz = 240 kHz. The bandwidth for commercial FM broadcast is 200 kHz; by contrast, the narrow-band marine transmitter only has a bandwidth of 20 kHz.

Table 157-1 has been compiled from Bessel Function values. It shows the number of pairs of significant sidebands for values of the modulation index, m.

Table 157-1

Modulation index m	Number of pairs of sidebands
1	3
2	4
3	6
4	7
5	8
6	9
7	10
8	$m + 4$

Note: once $J_0(x)$ and $J_1(x)$ have been computed, the remainder can be calculated from the relationship:

$$J_{n+1}(x) + J_{n-1}(x) = \frac{2n}{x} J_n(x) \quad (157\text{-}6)$$

For example if $n = 3$ and $x = 5$,

$$J_4(5) + J_2(5) = 0.047 + 0.391 = 0.438$$
$$\frac{2 \times 3}{5} J_3(5) = 1.2 \times 0.365 = 0.438.$$

Example 157-1

For test purposes, a 10-kW FM broadcast transmitter is 80% modulated by a 2-kHz tone. Including the carrier and sidebands, what is the total power in the FM signal? Calculate the value of the output frequency swing.

Solution

Because the RF power in an FM signal is independent of the modulation, the total power remains at 10 kW. In the FM broadcast service, 100% modulation corresponds to a frequency deviation of 75 kHz. The frequency swing for 80% modulation is ±(80/100) × 75 = ±60 kHz.

Example 157-2

The aural transmitter of a TV broadcast station is 60% modulated by a 500-Hz test tone. The amplitude and the frequency of the tone are both halved. What is the output frequency swing in the FM signal?

Solution

In the aural transmitter of a TV broadcast station, 100% modulation corresponds to a frequency deviation of 25 kHz. Because the percentage modulation is directly proportional to the tone amplitude (the effect of the preemphasis can be neglected at 500 Hz and 250 Hz), but is independent of its frequency, the new modulation frequency will be 30%, which will correspond to an output frequency swing of:

$$\pm \frac{30}{100} \times 25 = \pm 7.5 \text{ kHz}$$

Example 157-3

A 50-kW FM broadcast station is instantaneously 40% modulated by an audio frequency of 5 kHz. What is (a) the value of the modulation index and (b) the bandwidth of the FM signal?

Solution

(a) For an FM broadcast station, 40% modulation corresponds to the frequency shift of:

$$\frac{40}{100} \times 75 = 30 \text{ kHz}$$

The modulation index is:

$$\frac{30 \text{ kHz}}{5 \text{ kHz}} = 6$$

(b) From Table 157-1, the number of significant sidebands on either side of the carrier is 9. Because adjacent sidebands are separated by the audio frequency of 5 kHz, the bandwidth is 2 × 9 × 5 = 90 kHz.

Example 157-4

An FM transmitter is operating on 102 MHz and the oscillator stage is being directly modulated by a 3-kHz test tone so that its frequency swing is ±2 kHz. The oscillator stage generates 4250 kHz when it is unmodulated. If 100% modulation corresponds to a frequency swing of ±75 kHz, what is the percentage modulation produced by the test tone?

Solution

The total product factor of the frequency multiplier stages is 102 MHz/4250 kHz = 24. The output frequency swing = ±2 × 24 = 48 kHz and the percentage modulation is (48/75) × 100% = 64%.

Example 157-5

When an FM broadcast transmitter is modulated by a 450-Hz tone, the output frequency swing is ±30 kHz. If the tone frequency is lowered to 225 Hz, but the amplitude is unchanged, what is the new percentage of modulation?

Solution

For tones of 450 Hz and 225 Hz, the effect of preemphasis can be neglected. Therefore, the change in the tone frequency does not alter the modulation percentage, which remains at (30/75) × 100% = 40%.

PRACTICE PROBLEMS

157-1. In the TV broadcast of a sound signal, the range of transmitted audio frequencies lies between 50 Hz and 15 kHz. Calculate the value of the deviation ratio.

157-2. In the Public Safety Radio services, the highest audio frequency transmitted is 3 kHz. Calculate the value of the deviation ratio.

157-3. A 400-Hz audio tone provides 40% modulation of a TV sound signal. Calculate the values of the modulation index and the associated bandwidth.

157-4. In Practice Problem 157-3, the frequency of the tone is changed from 400 to 200 Hz. Calculate the new value of the bandwidth.

157-5. An FM transmitter is operating on 96 MHz, while its crystal oscillator frequency is 8 MHz. If an audio tone provides 80% modulation, and 100% modulation corresponds to a frequency swing of ±75 kHz, what is the amount of the frequency swing at the oscillator stage?

158
The FM transmitter

As covered in chapter 157, there are two methods of frequency modulating an RF carrier in a communications system.

THE DIRECT METHOD

The *direct method* is illustrated in Fig. 158-1. You will observe that the modulation process is carried out at the oscillator stage, which must therefore be of the variable-frequency type (such as a Colpitts oscillator), and not crystal-controlled. Modulating at this early stage has two advantages:

1. The power level is low so that the microphone is only followed by one or two audio voltage amplifiers.
2. Unlike AM signals, an FM signal can be fed to a class-C stage, which can either be an amplifier or a frequency multiplier. The instantaneous RF shift controlled by the audio signal is then multiplied by the same product factor as the carrier frequency. As an example, an FM transmitter has an output carrier frequency of 94.5 MHz, with an instantaneous frequency deviation of 60 kHz. If the intermediate stages contain two doublers and a tripler, the product factor is $2 \times 2 \times 3 = 12$. The variable frequency oscillator (VFO) generates $94.5/12 = 7.875$ MHz and the frequency deviation of the oscillator is only $60/12 = 5$ kHz.

Now, look at the stages and circuits which appear in a direct FM transmitter, but not in an AM transmitter (chapter 156). These include the following:

The JFET reactance modulator

The circuit of Fig. 158-2A is a solid-state version of the earlier reactance tube modulator. The RF voltage (v_o) is developed across the VFO's tank circuit, and is applied to the phase shift arrangement of C and R in series. If $X_C >> R$, the current (i) will lead v_o by approximately 90°; however, the gate voltage (v_g) across the capacitor (R) is in phase with i and, therefore, leads v_o by 90°. Because the drain current (i_d) is in phase with v_g (Fig. 158-2B), i_d leads v_o by nearly 90°; thus, the JFET presents a certain value of capacitive reactance and a corresponding capacitance to the voltage, v_o. This equivalent capacitance is in parallel with the oscillator's tank circuit and therefore, will determine the instantaneous value of the generated frequency.

When the audio signal is fed to the gate, it amplitude modulates the drain current and then produces variations in the JFET's equivalent capacitance. For example, at the peak (X) of the audio tone, the drain current is instantaneously at its maximum value so that the JFET's equivalent capacitance rises to a maximum. As a result, the total capacitance associated with the tank circuit increases to its highest level and the instantaneous generated frequency is at its minimum value, $f_c - f_d$. A full analysis of the reactance modulator is included in the Mathematical Derivations.

The JFET circuit, as described, behaves as a capacitive reactance modulator. However, if R and C are interchanged and $R >> X_C$, v_g lags v_o by nearly 90° and the circuit will become an inductive reactance modulator.

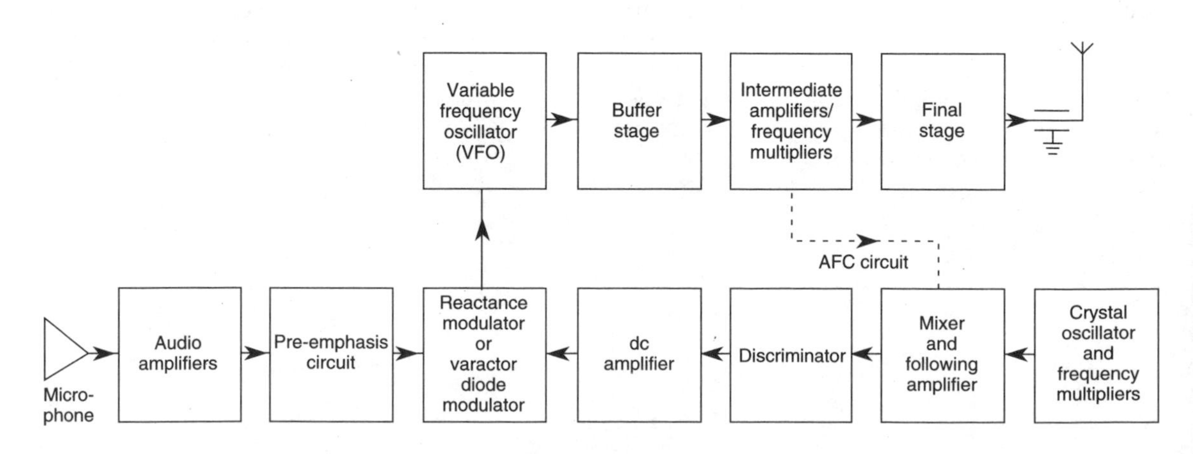

Figure 158-1

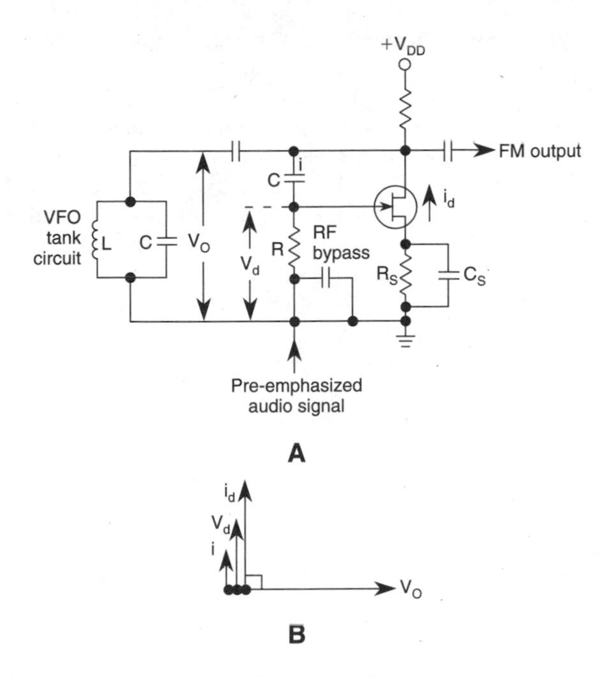

Figure 158-2

The varactor diode modulator

This circuit also provides frequency modulation by the direct method. The varactor diode is reverse-biased and as the amount of bias is increased, the depletion layer widens and the junction capacitance decreases. These semiconductor devices (which are used to behave as voltage-controlled variable capacitances) are referred to as either *varactor* or *varicap diodes*. The characteristic of a typical varactor diode, with its accepted symbols, is illustrated in Fig. 158-3. The range of the reverse bias (v_r) is limited between zero and the point at which avalanche breakdown occurs. Notice that there is a sharp initial drop in the value of the junction (or transition) capacitance (C_T), and that the characteristic of C_T versus v_r has a marked curvature.

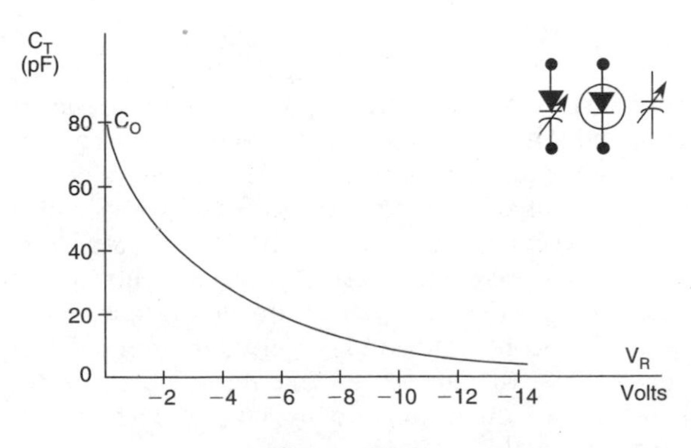

Figure 158-3

It is this degree of curvature that allows the varactor diode to be used for frequency multiplication in the microwave region.

Figure 158-4 shows the varactor diode circuit for generating an FM carrier by the direct method. The voltage divider (R_1 and R_2) provides the necessary reverse bias for the diode (whose capacitance is in parallel with the oscillator tank circuit and therefore determines the instantaneous value of the output frequency). The audio signal varies the diode's reverse bias and the corresponding fluctuations in C_T are responsible for generating the required FM signal. At the point (X) on the audio waveform, the reverse bias is increased, the value of C_T is reduced, and the instantaneous output frequency rises to its highest value. Provided that $C_T << C$, the RF shift is directly proportional to the value of C_T (see Mathematical Derivations).

The radio-frequency choke (L1) and the capacitor (C2) form a low pass filter whose purpose is to prevent interference between the RF and AF circuits. The capacitor (C2) is an AF and RF bypass capacitor so that there is no signal feedback to the dc bias supply.

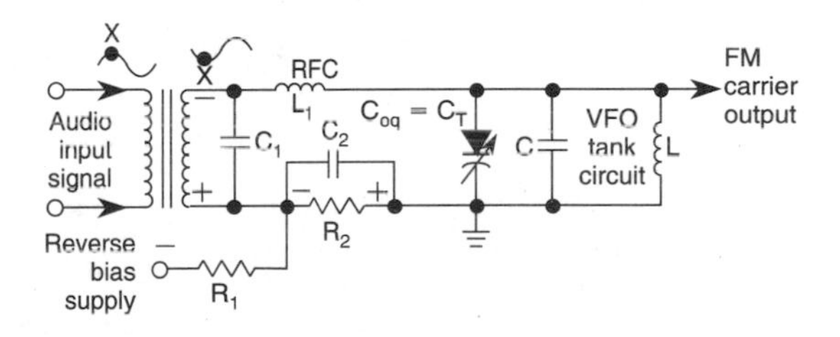

Figure 158-4

Pre-emphasis circuit

The high-frequency components of speech contribute most toward intelligibility, but unfortunately, these components generally have the lower amplitudes. In addition, noise in an FM system has more effect on the higher modulating frequencies. It follows that, if the audio frequencies are gradually accentuated at the transmitter, their signal-to-noise ratios are improved at the receiver. After the FM signal is detected by the discriminator (chapter 159), the same audio frequencies must receive a corresponding degree of gradual attenuation in order to restore the correct tonal balance to the voice signal.

The gradual accentuation of the audio frequencies at the transmitter is referred to as *pre-emphasis* and the corresponding attenuation at the receiver is called *de-emphasis*. A common circuit that is designed to achieve pre-emphasis is shown in Fig. 158-5A. If the modulating frequency is increased, the reactance of the capacitor (C) falls so that the impedance of the parallel R,C combination is reduced. Consequently, the audio voltage drop across this impedance decreases and the output voltage will increase. The amount of pre-emphasis is measured by the decibel gain between a high-frequency output voltage and the output voltage at 30 Hz. When the decibel gain-versus-frequency graph (Fig. 158-5B) is plotted, the result is the pre-emphasis curve, whose shape is determined by the RC time constant of 75 kΩ × 0.001 μF = 75 μs; this value of the time constant is the FCC standard for FM commercial broadcast and the TV broadcast of the aural (sound) signal. At the audio frequency of 2120 Hz, the pre-emphasis gain is 3 dB; at 15 kHz, the gain rises to 17 dB.

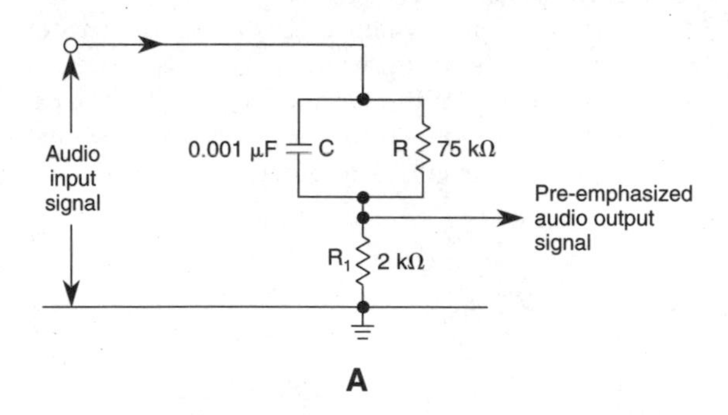

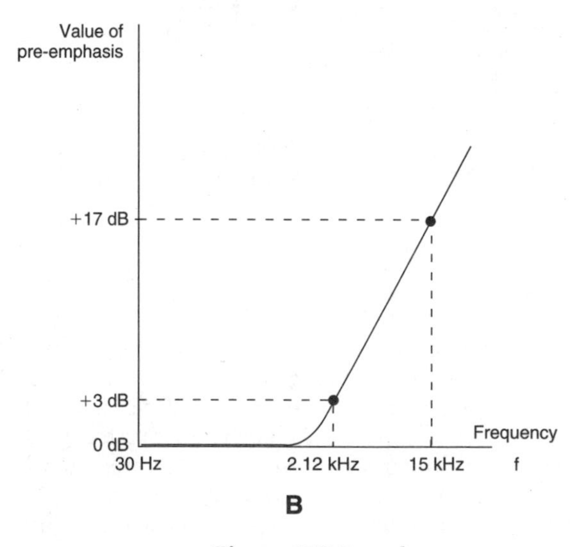

Figure 158-5

The AFC circuit

In the direct method, the audio information produces frequency variations in the oscillator's output. For commercial purposes, it is essential that the oscillator has a high degree of frequency stability that can be achieved through an AFC (automatic frequency control) circuit. In Fig. 158-2, a small amount of the FM signal from the intermediate stages is passed to a mixer, which is also fed with a signal from a crystal oscillator and its harmonic generator. The mixer's output is a signal at the difference frequency, which is typically between 10% and 20% of the VFO's frequency. This difference frequency is amplified before being fed to a Foster-Seeley discriminator (chapter 159). If the difference frequency is at its correct value, the output from the discriminator is zero. When the difference frequency rises above the correct value, the discriminator output is a dc voltage that is amplified before being passed to the varactor diode; this diode is connected across the VFO's tank circuit and is only used for AFC purposes. The result is to drive the carrier's frequency back to its assigned value. In order that the AFC circuit shall not be affected by the modulating signal, the discriminator's output circuit has a high time constant and therefore follows the slow fluctuations in the carrier frequency, but not the rapid frequency changes that are caused by modulation.

THE INDIRECT METHOD

The main problem with the direct method is the complexity of the AFC circuit that is required to achieve the necessary degree of frequency stability. Such stability is mainly determined by the drifts in the crystal oscillator and the input tuned circuit of the discriminator. The *indirect method* overcomes these difficulties by using a highly stable crystal oscillator to generate the unmodulated carrier; the modulation process is subsequently carried out at a later stage, which produces phase modulation (Fig. 158-6). As explained in chapter 157, the phase modulation can then be converted to the required frequency modulation by the audio correction network.

The phase modulator

A simplified version of a phase modulator circuit is shown in Fig. 158-7. The RF input signal (e_i) from the buffer stage drives a current (i) through the capacitor (C) and the collector tank circuit across which a voltage (e_c) is developed. Because C has a low value (on the order of a few picofarads), its reactance is so high that i and e_c both lead v_i by 90°. At the same time, e_i is fed to the base of Q1 so that the collector current (i_c) is in phase with v_i and passes through the tank circuit L_1C_1. A second voltage (e_m) will therefore be developed across the tank circuit, and the modulator's output signal is the resultant of e_c and e_m, which are 90° out of phase.

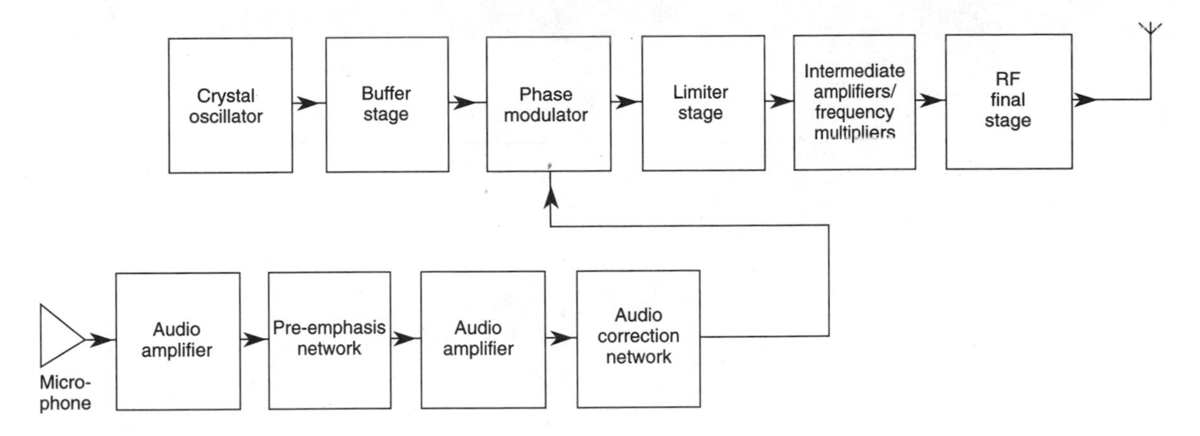

Figure 158-6

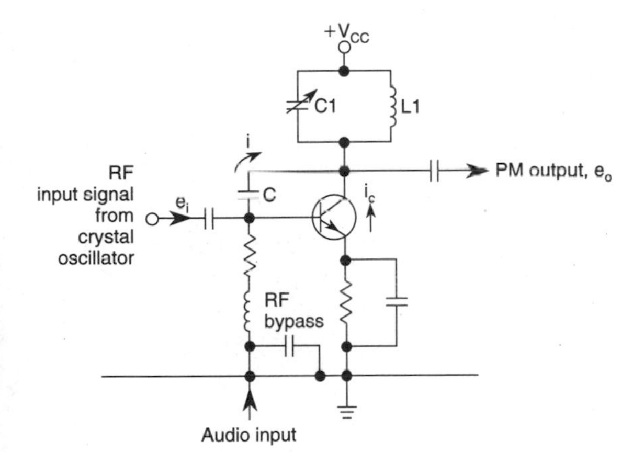

Figure 158-7

When the audio signal is fed to the Q1 base, it amplitude modulates the collector current and the value of e_m; however, the value of e_c remains unchanged. The effect of modulation by a single tone is shown in Fig. 158-8A. Provided that the phase angle (ϕ) is small (Fig. 158-8B), its value is directly proportional to the amplitude of the tone. At the peak (X) of the tone cycle, the phase angle has its maximum value. Therefore, the rate of change of phase is zero. It follows that the instantaneous RF shift is also zero, and the instantaneous frequency is equal to the unmodulated value, f_c. Consequently, the output signal (e_o) is phase modulated and not frequency modulated (chapter 157). Conversion from PM to FM is achieved by an RC correction network (Fig. 158-9); this circuit provides a 90° shift for all the components in the audio signal and it has an output voltage whose value is inversely proportional to the modulating frequency.

Notice that the output from the phase modulator fluctuates in amplitude. These amplitude variations are removed by the following limiter stage.

MATHEMATICAL DERIVATIONS

Varactor diode

Junction capacitance,

$$C_T = \frac{\epsilon A}{W_d} \text{ picofarads} \qquad (158\text{-}1)$$

where: ϵ = permittivity of the depletion layer
A = effective area of the pn junction (m^2)
W_d = width of the depletion layer (m)
C_T = junction, depletion region, or transition capacitance (pF)

Junction capacitance,

$$C_T = \frac{C_o}{(1 + V_r/V_o)^n} \text{ picofarads} \qquad (158\text{-}2)$$

where: C_o = value of capacitance corresponding to zero bias (pF)
V_r = reverse bias (V)
V_o = value of the knee voltage (0.7 V for silicon)
n = an exponent whose values are, respectively, 0.50 and 0.33 for alloy and diffused junctions.

JFET reactance modulator

In Fig. 158-2, the gate RF signal voltage is,

$$v_g = iR = v_o \times \frac{R}{R - j\dfrac{1}{\omega C}}$$

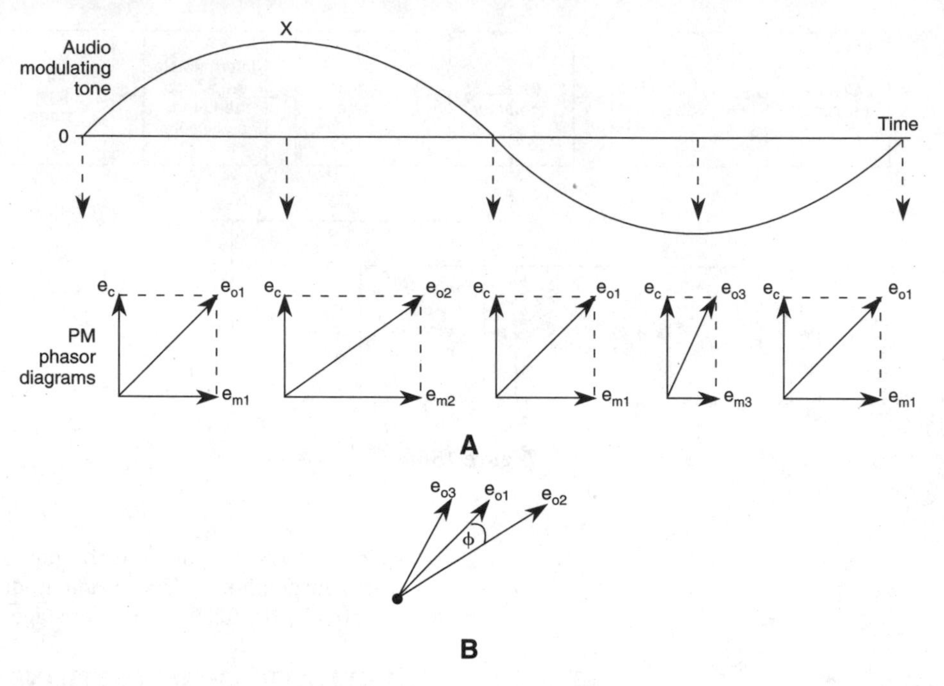

Figure 158-8

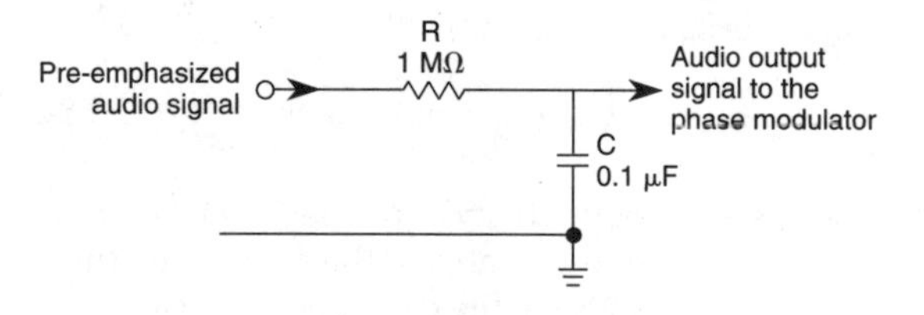

Figure 158-9

Drain current,

$$i_d = g_m v_g = v_o \times \frac{g_m R}{R - j\frac{1}{\omega C}}$$

If $X_C >> R$,

$$i_d \approx j v_o g_m \omega RC. \qquad (158\text{-}3)$$

Equivalent reactance presented to the tank circuit is,

$$X_{eq} = \frac{v_o}{i_d} = \frac{1}{j\omega g_m RC} \text{ ohms } (\Omega)$$

Equivalent capacitance in parallel with the tank circuit is,

$$C_{eq} = g_m RC \text{ picofarads (pF)} \qquad (158\text{-}4)$$

Notice that the C_{eq} depends on the JFET's transconductance, whose value is determined by V_{GS}. In addition, C_{eq} can be adjusted to the required capacitance by changing the values of R and C.

If $X_C = aR$ at the carrier frequency (f_c),

$$\frac{1}{2\pi f_c C} = aR$$

or,

$$RC = \frac{1}{2\pi a f_c}$$

Substituting for RC in Equation 158-4,

$$C_{eq} = g_m RC = \frac{g_m}{2\pi a f_c}$$

where a is a number whose value is typically between 5 and 10.

Varactor diode modulator

Instantaneous frequency generated by the variable oscillator is,

$$f = \frac{1}{2\pi\sqrt{L(C + C_T)}}$$

$$= \frac{1}{2\pi\sqrt{LC}}\left(1 + \frac{C_T}{C}\right)^{-1/2}$$

$$= f_c\left(1 + \frac{C_T}{C}\right)^{-1/2}$$

where: L = inductance of the tank circuit (H)
C = capacitance of the tank circuit (F)
C_T = junction capacitance of the varactor diode (F)
f_c = unmodulated carrier frequency (Hz)

Using the Binomial Expansion (chapter 209), instantaneous frequency,

$$f = f_c\left(1 - \frac{C_T}{2C} + ---- \text{ neglected terms}\right)$$

provided $C >> C_T$.

Then,

$$f = f_c\left(1 - \frac{C_T}{2C}\right) = f_c - f_d.$$

Frequency deviation,

$$f_d = \frac{f_c C_T}{2C} \text{ hertz.} \qquad (158\text{-}5)$$

Therefore, the frequency deviation (f_d) is directly proportional to the junction capacitance C_T.

Example 158-1

In the direct FM method, a 5-MHz oscillator is modulated by a 300-Hz tone, which produces a swing of ±1 kHz. If the intermediate amplifier's contain two triplers and a doubler stage, what is the value of the modulation index at the transmitter's output stage?

Solution

Frequency deviation at the output stage,

$$f_d = 3 \times 3 \times 2 \times 1 \text{ kHz} = 18 \text{ kHz.}$$

Modulating frequency,

$$f_m = 300 \text{ Hz.}$$

Modulation index at the output stage,

$$m = \frac{f_d}{f_m} = \frac{18 \text{ kHz}}{300 \text{ Hz}} = 60.$$

Example 158-2

In an FM transmitter that uses the indirect method, a 4-V 3-kHz audio tone at the input to the correction network, produces a frequency deviation of 500 Hz at the output of the phase modulator. What is the value of the frequency deviation produced by a 4-V 6-kHz tone?

Solution

As far as the input signal to the correction network is concerned, the output from the phase modulator is an FM signal whose deviation is independent of the modulating frequency. Therefore, the frequency deviation will be unchanged at 500 Hz.

PRACTICE PROBLEMS

158-1. In Fig. 158-2A, the value of R is one seventh of C's reactance. If the transconductance of the JFET is 9000 μS, obtain the value of the equivalent capacitance (C_{eq}) if the oscillator's carrier frequency is 6 MHz.

158-2. A 600-Hz audio tone is applied to a phase modulator and produces a frequency deviation of 2.4 kHz. Determine the value of the modulation index. If the amplitude of the tone is unchanged, but the frequency is raised to 5.4 kHz, obtain the new value of the frequency deviation.

158-3. A 400-Hz tone with an amplitude of 2.4 V is applied to an FM modulator and produces a modulation index of 60. Calculate the value of the maximum frequency deviation. If the amplitude of the tone is increased to 3.2 V, but its frequency is reduced to 250 Hz, obtain the new value of the modulation index.

159
AM and FM receivers

THE AM RECEIVER

Figure 159-1 represents the block diagram of an AM superhet receiver in the MF band. The purpose of the mixer and the local oscillator is to achieve frequency conversion. In this process, the receiver is tuned to a particular signal and the carrier frequency is then converted to a fixed intermediate frequency (IF), which is 455 kHz in the AM broadcast band. Therefore, for all incoming carrier frequencies, the IF stages always operate on the same fixed intermediate (low) frequency and can be designed to provide high selectivity and sensitivity. Moreover, because the RF amplifier, local oscillator, and IF stages all operate at different frequencies, there is little danger of self-oscillation as a result of unwanted feedback.

As indicated in Fig. 159-1, the RF tuned circuits of the initial amplifier, the mixer and the local oscillator, are mechanically ganged; thus, as the RF amplifier and the input circuit of the mixer are tuned to the desired signal frequency, the local oscillator stage generates its own frequency and produces a continuous RF output that differs in frequency from the incoming carrier by the value of the intermediate frequency. The term *intermediate* means that the value of the IF lies between the RF and AF frequencies.

The output of the oscillator and the incoming signal are both fed to the mixer stage, which contains a nonlinear device. In general, when two signals of different frequencies are mixed together (this mixing process is sometimes called *heterodyning*, or *beating*), the result contains a component at their difference frequency, which is equal to the intermediate value. The mixer output circuit can therefore be tuned to the intermediate frequency and this will automatically eliminate the unwanted components that are created by the mixing process. During the frequency conversion, the modulation originally carried by the wanted signal is transferred to the intermediate-frequency component in the mixer's output. However, if the local oscillator frequency is greater than the incoming signal frequency (known as *tracking above*), the upper sidebands that are associated with the incoming carrier frequency will become the lower sidebands in the intermediate frequency amplifiers, and vice-versa.

In an AM receiver, the local oscillator always tracks above (in order to reduce the physical size required for the local oscillator tuning capacitor). However, in TV and FM receivers, which operate in the VHF and UHF bands, the local oscillator can track either above or below.

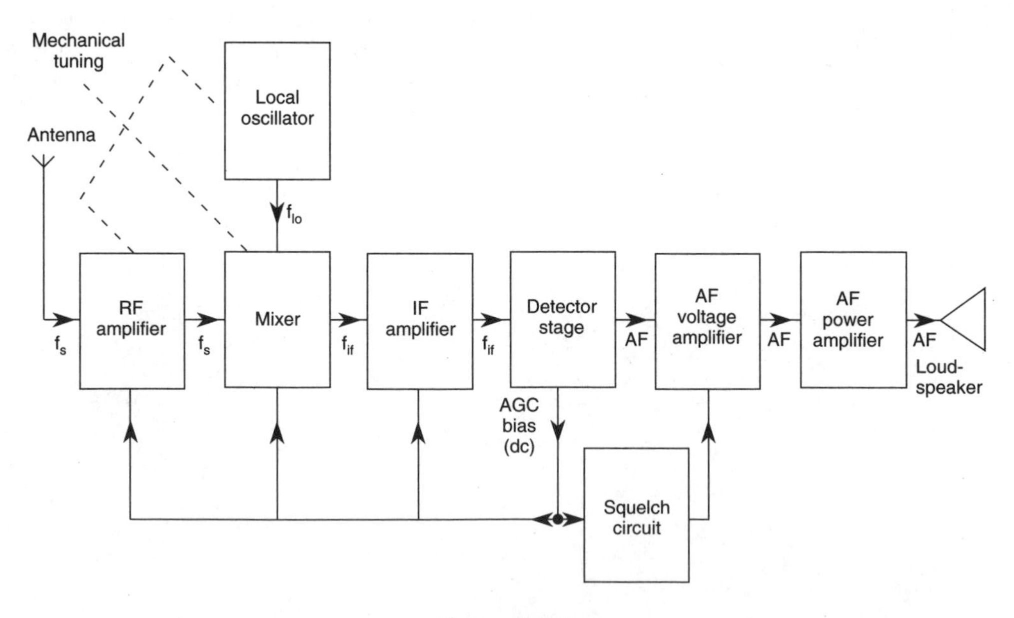

Figure 159-1

Image channel

Consider that the superheterodyne receiver is tuned to a carrier frequency of 640 kHz in the MF band. Because the IF is 455 kHz, the local oscillator will generate 640 + 455 = 1095 kHz. Assume that there exists an unwanted signal, whose carrier frequency is 1095 + 455 = 1550 kHz. If this signal is sufficiently strong to reach the mixer stage, it will beat with the local oscillator output, and its carrier frequency will also be converted to the IF. Once this has occurred, the IF stages will be unable to separate the wanted and the unwanted signals; both will be detected and both will be heard on the loudspeaker. The unwanted signal of 1550 kHz is an example of an *image channel* because its frequency is the image of the wanted signal in the "mirror" of the local oscillator frequency (i.e., the frequency of the wanted signal is the same amount below the local oscillator frequency as the image channel is above). The frequency of an image channel, therefore, differs from the wanted carrier frequency by an amount equal to twice the intermediate frequency.

Detector stage

The detector stage is the essential stage of any AM receiver. Into the detector (or demodulator) stage (Fig. 159-2) is fed the amplitude modulated IF signal; emerging from the detector is the audio information, such as voice, music, or a simple tone for test purposes. In other words, the detector stage reverses the modulation process that is carried out at the transmitter. In the crystal set of the 1920's, it was the combination of the galena crystal and the tungsten "cat's whisker" that acted as a rectifying contact and provided the necessary nonlinearity for detection to occur.

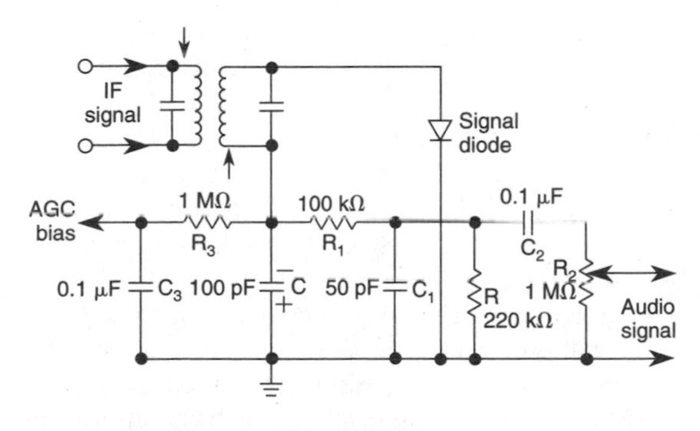

Figure 159-2

Referring to Fig. 159-2, the signal diode is the nonlinear device that will pass pulses of current when the AM signal is applied. The diode current charges the capacitor (C) which subsequently discharges through R and R1 with a time constant of 100 pF × (220 kΩ + 100 kΩ) = 32 μs. This time constant is more than 10 times greater than the period of the IF (period = 1/(455 kHz) = 2.2 μs), but is about 30 times less than the period of the 1-kHz modulating tone (period = 1/(1 kHz) = 1000 μs), which amplitude modulates the IF carrier. Consequently, the voltage (V_C) across the capacitor (C) follows the envelope of the AM signal. As shown in the waveforms of Fig. 159-3, V_C contains the audio tone, a dc level, and an RF ripple. The ripple is attenuated by the low-pass filter (R1, C1), while the dc level is blocked by the C2 capacitor. As a result, the audio tone appears across the 1-MΩ potentiometer, which acts as a volume control. Subsequently, the tone signal is built up to loudspeaker strength by a combination of audio voltage and power amplifiers.

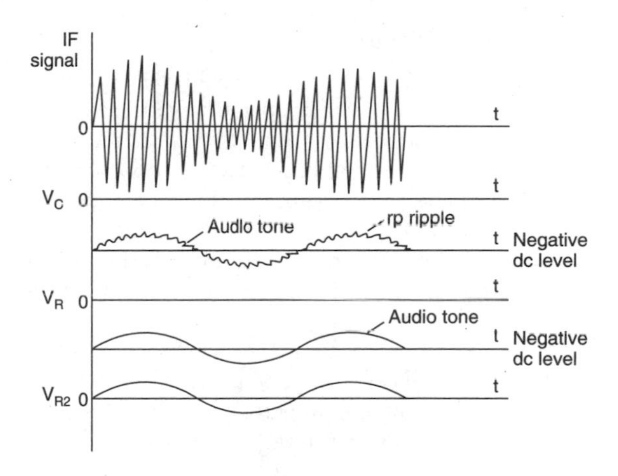

Figure 159-3

Automatic gain control (AGC)

Because of changes in propagation conditions, the carrier strength at the receiver fluctuates at a rate that is typically about once every two seconds. This variation can cause objectionable fading in the sound output from the loudspeaker. To overcome this problem, the *automatic gain control (AGC)* system, which is sometimes called *automatic volume control (AVC)*, maintains an approximately constant signal voltage at the detector. This is achieved by biasing the circuits of the RF, IF, and mixer stages with a dc voltage that is derived from the detector circuit. As a result, any increase in the carrier signal slightly increases the bias. This tends to counteract the increased signal by reducing the receiver's amplification (and vice-versa). In other words, the signal at the detector does increase slightly, but not nearly as much as it would have done if the AGC circuit had not been present. In addition to smoothing out fluctuations in the signal strength, the AGC system allows the receiver to be tuned from a weak signal to a strong signal without the danger of loudspeaker "blasting" (or from a strong signal to a weak signal without adjusting the manual volume control).

The advantages of using an AGC system are obtained at the expense of receiver sensitivity; this loss in sensitivity can present a problem when the receiver is trying to pick up a weak signal.

In the detector circuit of Fig. 159-2, the AGC bias is developed across capacitor C. However, the voltage (V_C) contains the audio signal and an RF ripple; these unwanted components are removed by the R3/C3 filter, which has a time constant of the order of 0.1 s. This time constant enables the AGC bias to follow the fluctuations in the carrier signal strength.

The squelch circuit

In the absence of a carrier, the AGC level is virtually zero so that the receiver is operating with maximum sensitivity. Consequently, the noise within the receiver's bandwidth is amplified to an objectionable degree. This can be an annoying situation for an operator who is required to use the receiver on a continuous basis. To eliminate this problem, a *squelch* (muting or quieting) circuit cuts off the receiver's audio output when no carrier is present.

A typical squelch circuit is shown in Fig. 159-4. A dc amplifier (Q1) controls the operation of the first amplifier, which follows the detector stage. Once the AGC bias falls below a certain level, the transistor (Q2) is forward-biased so that its collector current flows through the R1 resistor. The voltage drop across this resistor provides sufficient reverse bias to cut off the transistor Q2 so that the excessive noise is prevented from reaching the loudspeaker. The low level of AGC bias at which Q1 conducts is determined by the setting of the potentiometer R2.

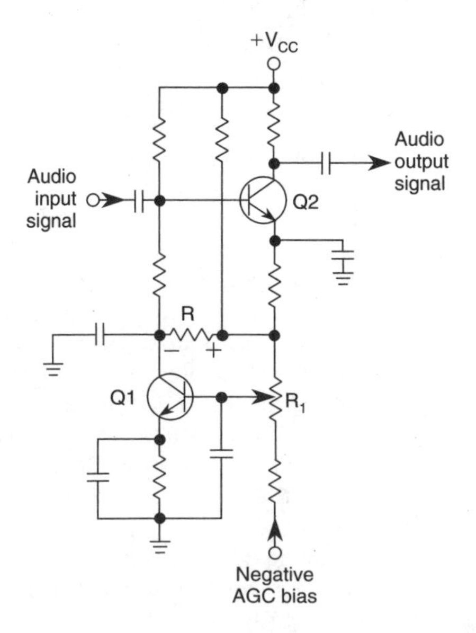

Figure 159-4

Crossmodulation

Apart from adjacent channel and image-channel interference, an AM receiver might experience the phenomenon of *cross-modulation*, in which the modulation of an unwanted transmission is transferred to the carrier of the wanted signal. This effect can occur either externally to the receiver or in the receiver itself. In the external case, the wanted signal and a strong unwanted signal induce RF voltages in metal structures, which are in the vicinity of the receiving antenna. If there is a discontinuity in one of these structures (for example, a rusty joint), there exists a rectifying contact that will provide some degree of nonlinearity. When the nonlinear current-versus-voltage equation contains a third-order term, the result is crossmodulation. This is shown in the Mathematical Derivations, where the assumed current-versus-voltage equation is $i = a + bV + cV^2 + dV^3 + ----$ and the third order, or cubic term (dV^3), contains the crossmodulation components. When the crossmodulation component at the wanted signal is reradiated from the metal structure, it is intercepted by the receiving antenna, after which it is impossible to eliminate the unwanted modulation and both signals appear on the loudspeaker.

Internal crossmodulation occurs in the first RF stage of the receiver. If this amplifier has an output current versus input voltage relationship, which contains a third-order term, a crossmodulation component will be passed through the following IF stages. To avoid this effect, the shape of the current versus voltage curve should be as close as possible to a parabola; this is a second-order curve with an equation of the form $i = a + bV + cV^2$ so that no third-order term is present.

It is important to realize that the unwanted modulation will only appear in the receiver's output if the wanted carrier signal is present. A striking example is provided by "musical morse," in which the interfering music modulation has been transferred to the morse carrier. Consequently, the music will go "on" and "off" with the "mark" and "space" intervals of the morse signal. This is in contrast with adjacent- or image-channel interference; in these instances, the music is heard continuously in the background.

THE FM RECEIVER

Figure 159-5 shows the diagram of an FM superheterodyne receiver. Although the principle of frequency conversion is the same as that in the AM receiver, the two types of receivers have a number of significant differences:

1. The FM signal occupies a wide bandwidth that must be accommodated by the IF stages. The value of the intermediate frequency itself must be much higher; it is 10.7 MHz in the commercial FM broadcast band, which extends from 88 to 108 MHz. Because the image channel is separated from the wanted channel by $2 \times 10.7 = 21.4$ MHz, no FM station is the image of another FM station. However, a commonly used aeronautical frequency is 121.5 MHz, which can cause image interference to FM reception.
2. A high IF value reduces the amount of freedom from adjacent-channel interference, and a low IF value increases the possibility of image-channel problems. A *double-conversion system* attempts to have "the best of both worlds." The first frequency conversion typically uses a high IF of about 25 MHz to further reduce image-channel interference. This is then followed by a second conversion with a low IF of approximately 150 kHz to achieve a high degree of freedom from adjacent-channel interference.
3. The final IF amplifier is a limiter stage, which removes all amplitude variations from the FM signal before the signal is passed to the discriminator.
4. The AM detector is replaced by the combination of the discriminator and de-emphasis circuitry.

The FM limiter

External noise, such as static, primarily amplitude modulates the FM signal. By contrast, internal receiver noise from the active devices and the circuitry creates both amplitude and phase modulation. Virtually all of the AM noise is removed by the limiter stage and the PM noise can be reduced by using a wide band (as opposed to a narrow band) system.

The schematic of a typical limiter stage is shown in Fig. 159-6A. The dropper resistor (R) ensures that the JFET is operated with a low drain supply voltage. Therefore, it is easily saturated by weak FM signals of the order of 0.5 V. In addition, the flow of the gate current charges up the capacitor (C_g), which subsequently discharges through R_g. The time constant (R_gC_g) is sufficiently short to allow the signal bias to follow rapid fluctuations in the FM signal amplitude; as a result, there is a limiting effect on the positive peaks of the input signal (Fig. 159-6B). Amplitude variations on the negative peaks are limited by the cut-off action of the JFET. To summarize, the limiter stage has a low gain, but constant output, and is operated from a reduced dc supply voltage. Notice that the JFET in Fig. 159-6A is neutralized by capacitance C_N.

The presence of the limiter appears to eliminate the necessity for an AGC system. However, such a system might be needed to avoid overdriving the limiter stage. Any strong input signal to the limiter might reduce the

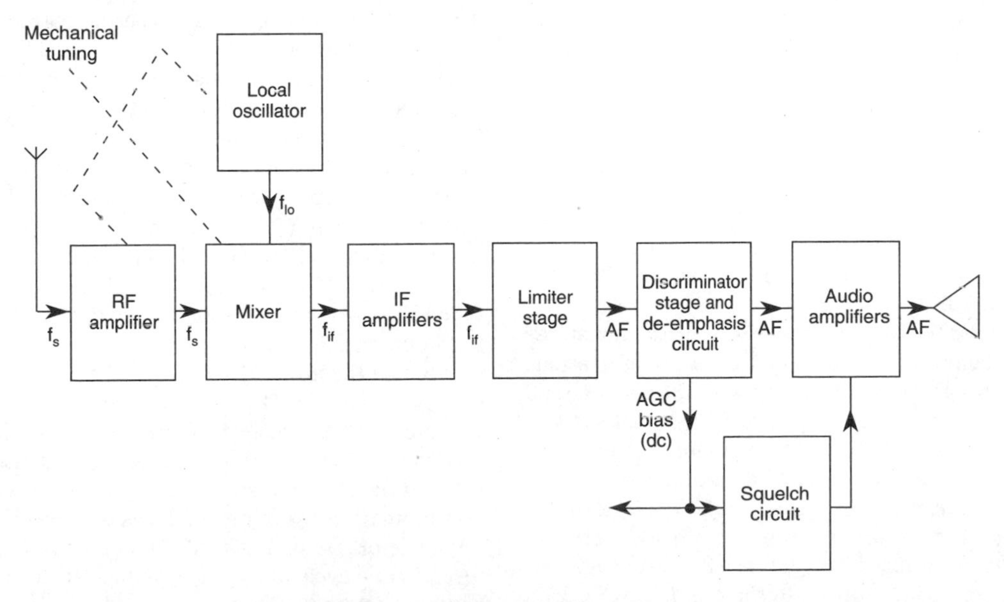

Figure 159-5

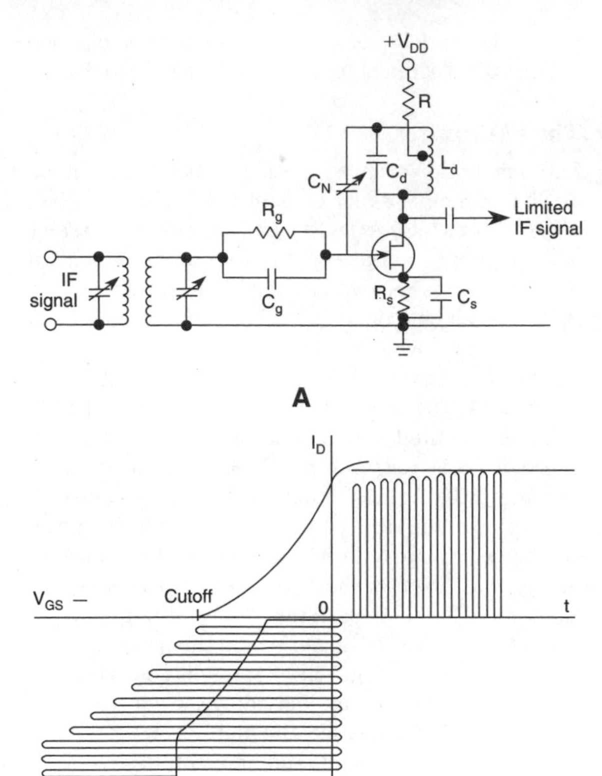

Figure 159-6

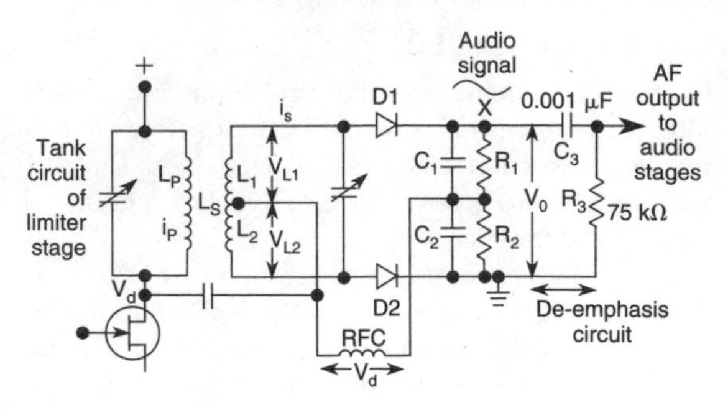

Figure 159-7

angle of drain current flow to the point where the output to the discriminator starts to fall off. In addition, an AGC bias voltage can be used to operate a squelch circuit.

The discriminator stage

One of the most common types of FM detector is the Foster-Seeley discriminator (Fig. 159-7). The function of this circuit is to provide an instantaneous audio output voltage, which is directly proportional to the amount of instantaneous frequency deviation from the value of the IF (10.7 MHz).

The operation of the Foster-Seeley discriminator is best explained in terms of phasor diagrams. Assume that the IF signal is modulated by an audio test tone and that the instantaneous frequency to the input signal is equal to the value of the IF; it is therefore required that the output voltage (v_o) must be instantaneously zero. The horizontal reference phasor (Fig. 159-8A) is the RF voltage (v_d) at the drain of the limiter's JFET. The primary current (i_p) lags v_d by 90° and the secondary voltage (v_s) (mutually induced in the coil, L_s) lags i_p by a further 90° so that v_p and v_s are 180° out of phase (assuming that the primary and secondary coils are wound in the same sense). Because the secondary circuit is resonant at the incoming frequency, the secondary current (i_s) will be in phase with v_s, and the voltage (v_L) across L_s will lead i_s by 90°. With respect to the secondary's center tap, v_{L1} and v_{L2} are 180° out of phase and the voltage across the radio frequency choke is virtually equal to v_d. The RF voltage (v_{D1}) is applied to the diodc (D1) and is the phasor sum of v_{L1} and v_d. Similarly, the voltage (v_{D2}) is applied to the diode (D2) and is the phasor sum of v_{L2} and v_d. Because v_{D1} and v_{D2} have the same magnitude, the two diodes conduct equally and charge the capacitors (C1 and C2) to the same voltage. However, the voltages across these capacitors are series-opposing so that the output voltage (v_o) is instantaneously zero.

When the instantaneous input frequency rises to the value of $f_{if} + f_d$, the directions of the v_d, i_p, and v_s phasors remain unchanged (Fig. 159-8B). However, the secondary circuit of L_s and C_s now behaves inductively with respect to the input frequency; consequently, i_s lags v_s by the angle ϕ. The voltage (v_{D1}) is now larger than v_{D2} so that D1 conducts more than D2. The capacitor (C1) is charged to a higher voltage than C2 and the instantaneous output voltage (v_o) is at its maximum positive value.

Now, let the instantaneous input frequency fall to $f_{if} - f_d$ (Fig. 159-8C). The secondary circuit becomes capacitive so that i_s leads v_s by the angle ϕ. Diode D2 carries the larger current so that v_{C2} is greater than v_{C1}, and the instantaneous output voltage (v_o) is at its maximum negative value. It follows that one complete sequence of the input signal's frequency variation will result in one cycle of the audio voltage at the discriminator's output point, X.

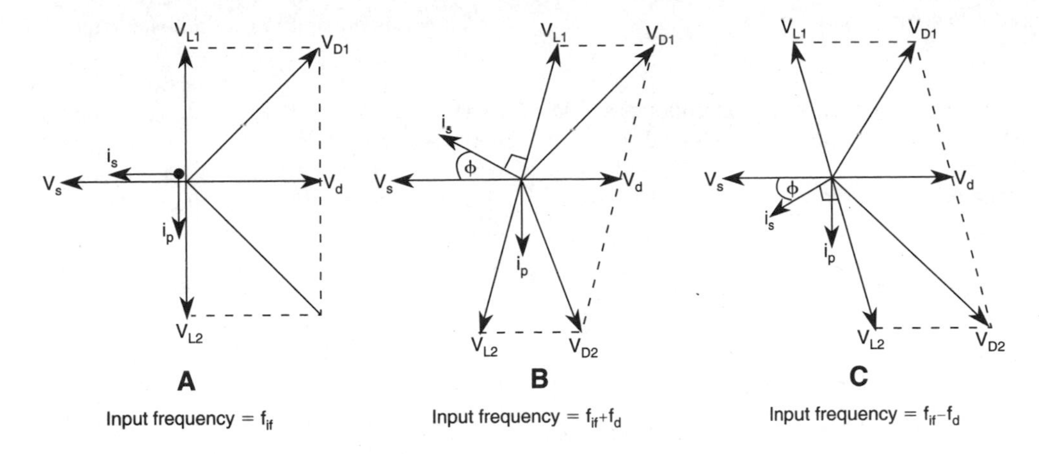

Figure 159-8

As described in chapter 158, the modulating signal at the transmitter is pre-emphasized in order to accentuate the higher audio frequencies, and thereby improve their signal-to-noise ratios. Therefore, it is necessary to restore the tonal balance in the receiver by applying the discriminator output voltage (v_o) to the de-emphasis circuit (C3/R3) with its time constant of 75 kΩ × 0.001 μF = 75 μs. If the audio frequency increases, the reactance of the capacitor (C3) decreases, and the output to the audio amplifiers is reduced. The graph of the attenuation versus frequency for the de-emphasis circuit is illustrated in Fig. 159-9; as required, the pre-emphasis and de-emphasis curves are "mirror" images.

MATHEMATICAL DERIVATIONS

If the local oscillator "tracks above" the incoming frequency,

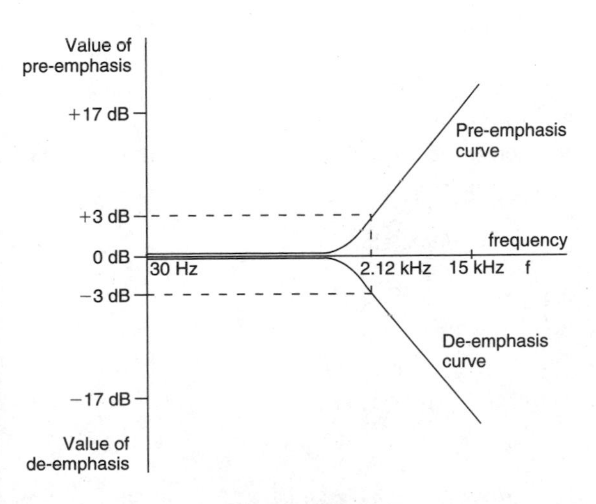

Figure 159-9

$$f_{Lo} = f_s + f_{if} \qquad (159\text{-}1)$$

$$f_s = f_{Lo} - f_{if}$$

$$f_{if} = f_{Lo} - f_s$$

If the local oscillator "tracks below" the incoming frequency:

$$f_{Lo} = f_s - f_{if} \qquad (159\text{-}2)$$

$$f_s = f_{Lo} + f_{if}$$

$$f_{if} = f_s - f_{Lo}$$

where: f_s = wanted signal frequency (kHz)
f_{Lo} = local oscillator frequency (kHz)
f_{if} = intermediate frequency (kHz)

The image channel frequency is given by:

Tracking above,

$$\textit{Image frequency} = f_s + (2 \times f_{if}) \qquad (159\text{-}3)$$

$$= f_{Lo} + f_{if}$$

Tracking below,

$$\textit{Image frequency} = f_s - (2 \times f_{if}) \qquad (159\text{-}4)$$

$$= f_{Lo} - f_{if}$$

Mixing (heterodyning)

Additive mixing

Let the radio frequency signal and the local oscillator signal be applied in series to the base of a junction transistor. Because of the nonlinearity of this active device, the collector current can be represented by the following series:

$$i_c = a + bv_b + cv_b^2 + dv_b^3 + --- \qquad (159\text{-}5)$$

where,

$$v_b = A \sin\omega_s t + B \sin\omega_{Lo} t$$

Because v_b is the sum of the radio frequency and local oscillator signals, the process is referred to as *additive mixing*.

Then, the collector current,

$$i_c = a + \quad (159\text{-}6)$$

$$b(A \sin\omega_s t + B \sin\omega_{Lo} t) +$$

$$c(A \sin\omega_s t + B \sin\omega_{Lo} t)^2 +$$

$$d(A \sin\omega_s t + B \sin\omega_{Lo} t)^3 + ---$$

The expansion of "$c(A \sin\omega_s t + B \sin\omega_{Lo} t)^2$" reveals the following terms:

1. $cA^2\sin^2\omega_s t = ca^2(1 - \cos\ 2\omega_s t)/2$, which contains a component at the second harmonic $(2f_s)$ of the signal frequency.
2. $2cAB\sin\omega_s t\sin\omega_{Lo} t = cAB(\cos(\omega_{Lo} - \omega_s)t - \cos(\omega_{Lo} + \omega_s)t)$. This shows the existence of components at the sum frequency, $f_{Lo} + f_s$, and the difference frequency, $f_{Lo} - f_s$.
3. $cB^2\sin^2\omega_{Lo} t = cB^2(1 - \cos 2\omega_{Lo} t)/2$ which contains a component at the second harmonic $(2f_{Lo})$ of the local oscillator frequency.

The collector current, therefore, contains the frequencies f_s, $f_{Lo} - f_s$ $(= f_{if})$, $f_{Lo} + f_s$, $2f_{Lo}$, $2f_s$. The other terms of the expansion will involve a large number of additional harmonics and unwanted frequencies.

Multiplicative mixing

Consider a dual-gate DE MOSFET in which the RF signal is applied to one gate and the local oscillator signal is passed to the other gate. The local oscillator signal can then vary the values of the transconductance and the gain for the RF signal.

Drain current,

$$i_d = g_{m2} \times v_{g1} \quad (159\text{-}7)$$

where,

$$g_{m2} = g_m(1 + \sin\omega_{Lo} t)$$

$$v_{g1} = v_g(1 + \sin\omega_s t)$$

Then,

$$i_d = g_m v_g\ (1 + \sin\omega_{Lo} t)(1 + \sin\omega_s t)$$

$$= g_m v_g[1 + \sin\omega_s t + \sin\omega_{Lo} t + \sin\omega_{Lo} t\ \sin\omega_s t]$$

$$= g_m v_g\ [1 + \sin\omega_s t + \sin\omega_{Lo} t +$$

$$\frac{1}{2}\cos(\omega_{Lo} - \omega_s)t - \frac{1}{2}\cos(\omega_{Lo} + \omega_s)t]$$

The drain current only contains components at the frequencies f_s, f_{Lo}, $f_{Lo} - f_s$ $(= f_{if})$, $f_{Lo} + f_s$. This is the principal advantage of multiplicative mixing over additive mixing. However, it is fair to point out that further nonlinearity in the electronic mixing process usually produces a number of additional unwanted frequencies. The general expression for a frequency (f) contained in a mixer's output is,

$$f = \pm mf_s \pm nf_{Lo} \quad (159\text{-}8)$$

where: $m = 0, 1, 2, 3$ -------
$n = 0, 1, 2, 3$ -------

Crossmodulation

Let the nonlinear current versus voltage relationship be,

$$i = a + bv + cv^2 + dv^3 + -------- \quad (159\text{-}9)$$

The voltage,

$$v = E_1\sin\omega_1 t(1 + m_1\sin\omega_{m1} t) \quad (159\text{-}10)$$

$$+ E_2\sin\omega_2 t(1 + m_2\sin\omega_{m2} t)$$

where: E_1 = peak value of wanted carrier (V)
ω_1 = angular frequency of wanted carrier (rad/s)
m_1 = degree of modulation for the wanted signal.
ω_{m1} = angular frequency of the wanted modulation (rad/s)
E_2 = peak value of interfering carrier (V)
ω_2 = angular frequency of unwanted carrier (rad/s)
m_2 = degree of modulation for the interfering signal
ω_{m2} = angular frequency of the interfering modulation (rad/s)

The third-order term,

$$dv^3 = d\ [E_1\sin\omega_1 t(1 + m_1\sin\omega_{m1} t) +$$

$$E_2\sin\omega_2 t(1 + m_2\sin\omega_{m2} t)]^3$$

Because $(x + y)^3 = x^3 + 3x^2y + 3xy^2 + y^3$, the expression of the third-order term contains the following:

$$3dE_1E_2^2\sin\omega_1 t\ \sin^2\omega_2 t \times \quad (159\text{-}11)$$

$$(1 + m_1\sin\omega_{m1} t)(1 + m_2\sin\omega_{m2} t)^2$$

$$= \frac{3}{2}\ dE_1E_2^2\sin\omega_1 t(1 - \cos 2\ \omega_2 t) \times$$

$$(1 + m_1\sin\omega_{m1} t)(1 + m_2\sin\omega_{m2} t)^2$$

Part of this expression is,

$$(159\text{-}12)$$

$$\frac{3}{2}\ dE_1E_2^2\sin\omega_1 t(1 + m_1\sin\omega_{m1} t + 2m_2\sin\omega_{m2} t)$$

These terms represent the crossmodulation component, which is modulated by both ω_{m1} and ω_{m2}, and whose angular carrier frequency is ω_1. Consequently, modulation of the unwanted transmission has been transferred to the wanted carrier. Similarly, other terms in the expansion of dv^3 show that the wanted modulation has been transferred to the unwanted carrier; this is the reason why the phenomenon is called crossmodulation. Notice that the relevant terms are directly proportional to the square of E_2 so that a strong interfering signal is most likely to produce an appreciable amount of crossmodulation. In addition, the term $2m_2 \text{Sin}\omega_{m2}t$ indicates that the percentage for the unwanted modulation has been doubled.

Example 159-1

A superheterodyne AM broadcast receiver with an IF of 455 kHz is tuned to receive a station on 670 kHz. What are the frequencies of the local oscillator and a possible image channel?

Solution

In an AM broadcast receiver, the frequency of the local oscillator is tracking above the wanted signal frequency.

Local oscillator frequency,

$$f_{Lo} = f_s + f_{if} \qquad (159\text{-}1)$$

$$= 670 + 455$$

$$= 1125 \text{ kHz.}$$

$$\textit{Image-channel frequency} = f_s + (2 \times f_{if}) \qquad (159\text{-}3)$$

$$= 670 + 2 \times 455$$

$$= 1580 \text{ kHz.}$$

Example 159-2

An 800-kHz radio-frequency signal and a 1300-kHz local-oscillator signal are applied in series to an additive mixer circuit, which has a square-law voltage/current relationship. List the frequencies that are contained in the mixer's output current.

Solution

Frequencies contained in the mixer's output current,

$$f_s = 800 \text{ kHz.}$$

$$f_{Lo} = 1300 \text{ kHz.}$$

$$f_{Lo} - f_s = 1300 - 800 = 500 \text{ kHz.}$$

$$f_{Lo} + f_s = 1300 + 800 = 2100 \text{ kHz.}$$

$$2f_s = 2 \times 800 = 1600 \text{ kHz.}$$

$$2f_{Lo} = 2 \times 1300 = 2600 \text{ kHz.}$$

PRACTICE PROBLEMS

159-1. A commercial FM broadcast receiver is experiencing interference from a strong aeronautical transmission whose frequency is 121.5 MHz. To which frequency is the receiver tuned?

159-2. An AM broadcast receiver is experiencing image-channel interference from a station whose carrier frequency is 1720 kHz. What is the frequency output of the local oscillator?

160
Calculation of operating power by the direct and indirect methods

The operating power of a transmitter is the actual power supplied to the antenna system and is not the same as the authorized power. The difference between the authorized and operating power is associated with the station's *power tolerance*. For example, if an AM broadcast station has an authorized power of 10 kW, its power tolerance extends from +5% to −10%. Therefore, the operating power can legitimately lie between 9 kW and 10.5 kW. The operating power can be calculated by either the direct method or by the indirect method.

THE DIRECT METHOD

Using the direct method (Fig. 160-1A), a transmitter's operating power is determined from the RF current (I_A) that is delivered by the transmitter to the antenna

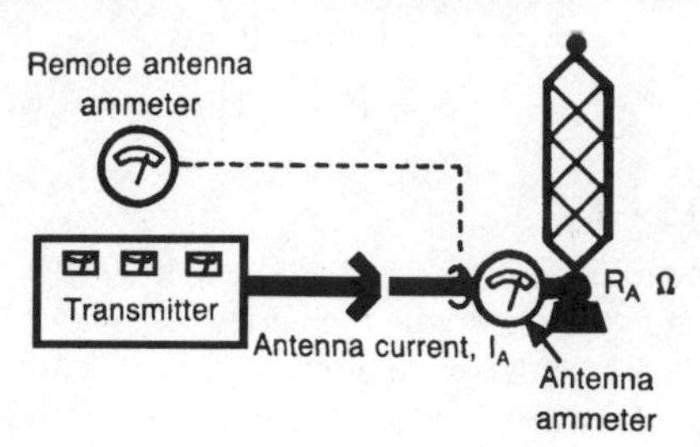

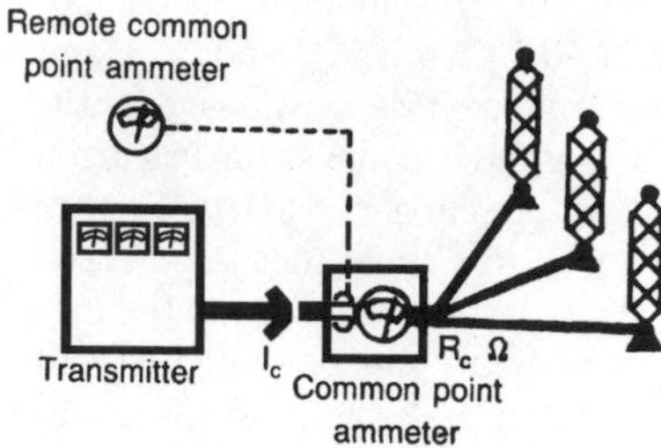

A

Plate voltage
Plate current
Final power amplifier

E_{bb} × I_b × F = P

Input power × Efficiency = Operating power

B

Figure 160-1

and the antenna's effective load at resonance (R_A). The operating power in watts is then given by the expression (*antenna current*)^2 × *antenna resistance*.

Directional AM broadcast stations use multiple radiating elements. The operating power of these stations, as determined by the direct method, is equal to the product of the resistance common to all of the antenna towers (the common point resistance, R_c) and the square of the current common to all of the antenna towers (common point current, I_c).

THE INDIRECT METHOD

The indirect method (Fig. 160-1B) involves the plate input power to the final RF stage and an efficiency factor (F), which is normally expressed as a decimal fraction, but can be given as a percentage. The operating power is then calculated from the value of the *dc plate supply voltage* × *dc level of the plate current* × *efficiency factor*.

For a standard AM broadcast station, the efficiency factor depends on the method of modulation (plate, grid, or low level) and the maximum rated carrier power output. By contrast, the value of the efficiency factor for FM stations is normally supplied by the manufacturer.

MATHEMATICAL DERIVATIONS

Direct method

Operating power,

$$P = I_A^2 R_A \text{ watts} \tag{160-1}$$

This yields,

$$I_A = \sqrt{\frac{P}{R_A}} \text{ amperes} \tag{160-2}$$

and,

$$R_A = \frac{P}{I_A^2} \text{ ohms} \tag{160-3}$$

where: I_A = antenna current in amperes (A)
R_A = effective antenna resistance (Ω)

For a directional AM station,

$$\textit{Operating power} = I_c^2 R_c \text{ watts} \tag{160-4}$$

where: I_c = common point current (A)
R_c = common point antenna resistance (Ω).

Indirect method

Operating power,

$$P = E_{bb} \times I_b \times F \text{ watts} \tag{160-5}$$

where: E_{bb} = plate supply voltage (V)
I_b = dc level of the plate current (A)

Example 160-1

The RF current delivered to an antenna is 8 A and the antenna's effective resistance is 125 Ω. What is the value of the operating power, as calculated by the direct method?

Solution

$$\textit{Operating power} = I_A^2 \times R_A \text{ watts} \quad (160\text{-}1)$$
$$= (8\text{ A})^2 \times 125\ \Omega$$
$$= 8000\text{ W}$$
$$= 8\text{ kW}.$$

Example 160-2

The final RF stage of a standard broadcast station has a plate supply voltage of 2500 V and a dc level of plate current equal to 1.5 A. The stage uses plate modulation with an efficiency factor of 0.7. What is the value of the operating power, as determined by the indirect method?

Solution

$$\textit{Operating power} = E_{bb} \times I_b \times F \quad (160\text{-}5)$$
$$= 2500\text{ V} \times 1.5\text{ A} \times 0.7$$
$$= 2625\text{ W}.$$

PRACTICE PROBLEMS

160-1. If a transmitter's operating power is 6.5 kW and the antenna's effective resistance is 300 Ω, what is the value of the RF current delivered to the antenna's feedpoint?

160-2. A transmitter's RF final stage has a plate supply voltage of 3.5 kV and a dc level of plate current equal to 2.5 A. If the stage uses plate modulation and the operating power is 7 kW, calculate the value of the efficiency factor.

160-3. A system of five antenna towers has a common point RF current of 15.3 A. If the transmitter's operating power is 5 kW, determine the value of the antenna's common point resistance.

161
The field strength of an antenna—effective radiated power

The electric field strength (field intensity) (ε) at a particular position from a transmitting antenna is measured in millivolts (or microvolts) per meter and determines the signal voltage induced in a receiving antenna. The field strength is directly proportional to the antenna current and to the square root of the radiated power, but it is inversely proportional to the distance from the transmitting antenna. For example, a thin center-fed dipole has a radiation resistance of 73.2 ohms so that if the RF power delivered to the antenna is 1 kilowatt, the antenna current is $\sqrt{1000\text{ W}/73.2\ \Omega} = 3.696$ A. The field strength at a position of one mile (or 1609.3 meters) from the antenna is 60×3.696 A/1609.3 m = 0.1378 volts per meter or 137.8 millivolts per meter. This value is used as the FCC standard to determine the gain of more sophisticated antennas.

In displaying the radiation pattern of a transmitting antenna, positions of equal field strength are joined together to form a particular *field strength contour* (Fig. 161-1).

If parasitic elements, such as reflectors and directors, are attached to a simple dipole (or more antennas are added to produce an array), the radiated RF power is concentrated in particular directions and the antenna gain is increased. This gain can either be considered in terms of the field strength or the RF power.

From measurements taken at a particular position, the field gain is the ratio of the field strength produced by the complex antenna system to the field strength created by a simple ideal dipole (assuming that this dipole is fed with the same RF power as the complex antenna). Because the square of the field intensity is directly proportional to the radiated power, the power gain of the antenna is equal to the square of the field gain and is either expressed as a power ratio or in decibels.

The *effective radiated power* (ERP) in a particular direction is calculated from the product of the RF power delivered to the antenna and its power gain ratio.

MATHEMATICAL DERIVATIONS

In Fig. 161-1,

Electric field strength,

$$\varepsilon = \frac{60 \times I_A}{d} \text{ volts per meter} \quad (161\text{-}1)$$

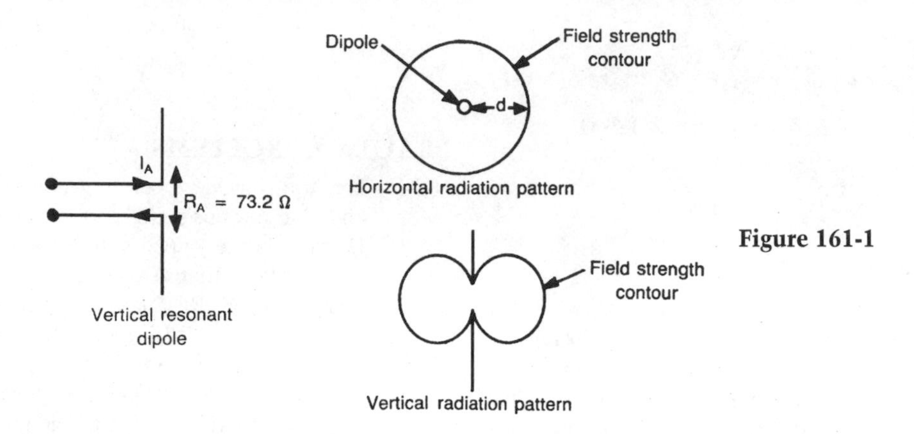

Figure 161-1

where: I_A = current at the antenna's feedpoint (A)
d = distance from the transmitting antenna (m)

Antenna current,

$$I_A = \sqrt{\frac{P_A}{R_A}} \text{ amperes} \qquad (161\text{-}2)$$

where: P_A = antenna power (W)
R_A = antenna's radiation resistance (Ω)

Effective radiated power,

(161-3)

ERP = (*RF power delivered to the antenna*) × (*antenna power gain ratio*) watts

Antenna power gain (161-4)

= 10 log (*antenna power gain ratio*) decibels

= 20 log (*antenna field gain ratio*) decibels

Example 161-1

At a position that is three miles from a transmitting antenna, the field strength is 650 μV/m. If the antenna current is doubled, what is the new field strength (a) at the same position and (b) at a position that is five miles from the antenna?

Solution

(a) The field strength is directly proportional to the antenna current. The new field strength is therefore 2 × 650 = 1300 μV per meter = 1.3 mV per meter.
(b) The field strength is inversely proportional to the distance from the transmitting antenna. Therefore, the new field strength at a position 5 miles from the transmitting antenna is 1300 × 3/5 = 780 μV per meter.

Example 161-2

The RF power input to a transmitting antenna is 8.7 kW. If the antenna power gain is 8 dB, what is the value of the effective radiated power?

Solution

Antenna power gain ratio = inv log (8/10) = 6.309.

Effective radiated power = 8.7 × 6.309 = 54.9 kW.

PRACTICE PROBLEMS

161-1. A measurement is made of the field strength at a certain distance from a transmitter's antenna. If the transmitter power is now doubled, by what factor will the field strength (at the same position) be multiplied?

161-2. At a position that is five miles from the antenna of a 5-kW transmitter, the field strength is 400 μV per meter. If the power is now reduced to 3.5 kW, what is the new distance from the antenna of the 600-μV per-meter contour?

161-3. At a position that is three miles from a transmitter's antenna, the field intensity is 15 mV per meter. The antenna current is now increased to 25 A and the new field intensity at the same position is 20 mV per meter. Calculate the value of the original current.

161-4. A 25-kW transmitter produces a field strength of 25 mV per meter at a position that is five miles from the antenna. When the transmitter's power input is increased, the new field strength at the same position is 35 mV per meter. Calculate the percentage increase in the transmitter's output power.

161-5. A transmitter delivers 25 kW of RF carrier power to the input of a coaxial cable. If the cable's loss is 0.35 dB and the antenna's field gain is 2.5, calculate the value of the effective radiated power.

161-6. A transmitter's RF carrier power output to a transmission line is 10 kW. The transmission line's loss is 0.4 dB and the effective radiated power is 50 kW. Calculate the dB value of the antenna's power gain.

162 The single-sideband system

The sidebands in an AM wave represent the audio information; in order to convey this information, it is not essential that the carrier and both sets of sidebands are transmitted.

THE SINGLE-SIDEBAND SYSTEM

There are various types of AM emission and each type must be considered separately as regards the power content of the sidebands and the bandwidth. For example in the single-sideband suppressed carrier (SSSC) system of Fig. 162-1, designated J3E (formerly A3J), all of the RF power is concentrated in one set of sidebands (either the upper set or the lower set). This improves the signal-to-noise ratio at the receiver as well as reducing the bandwidth required. However, when compared with double-sideband operation (A3E), the SSSC system demands a higher degree of frequency stability for the carrier generated in the transmitter's oscillator and also requires that a carrier component be reinserted at the receiver.

The advantage of power saving with the SSSC system lies in the fact that the whole of the RF carrier power can be concentrated in one set of sidebands, rather than being distributed over the carrier and two sets of sidebands; this is the reason for the improvement in the signal-to-noise ratio at the receiver.

For example, if a 100-W carrier is 100% (A3E) modulated by a single tone, the power in each of the sidebands is 25 W. If the whole of the 100-W unmodulated carrier power is concentrated in one sideband, the power advantage is equivalent to:

$$10 \log_{10}\left(\frac{100\text{ W}}{25\text{ W}}\right) = 10 \log_{10} 4$$

$$= 6\text{ dB}$$

This power advantage will result in the improved signal-to-noise ratio at the receiver.

Other single-sideband emission designations are:

- R3E Single-sideband with reduced (or pilot) carrier.
- H3E Single-sideband with full carrier power.

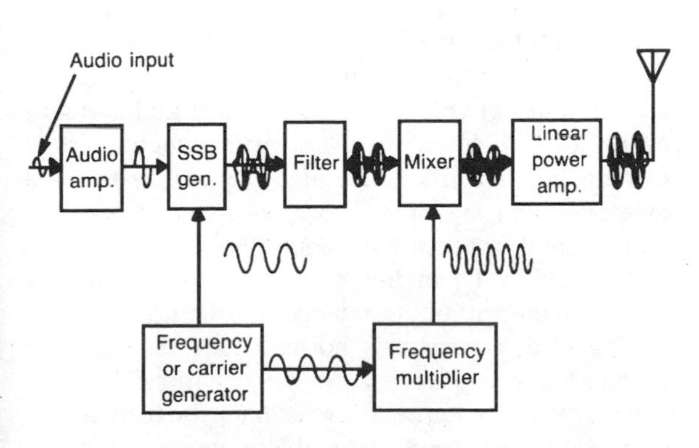

Figure 162-1

MATHEMATICAL DERIVATIONS

Let the unmodulated carrier power be P_c watts and let the carrier be 100% double sideband modulated by a single audio tone.

Total double sideband power,

$$P_{SB} = \frac{1}{2} \times m^2 \times P_c \text{ watts}$$

$$= \frac{1}{2} \times 1^2 \times P_c$$

$$= P_c/2 \text{ watts.}$$

Total RF power with double sideband operation = $P_c + P_c/2 = 3P_c/2$ watts.

Power in one sideband = $P_c/4$ watts.

Percentage power saving with the SSSC system

$$= \frac{3P_c/2 - P_c/4}{3P_c/2} \times 100\%$$

$$= \frac{6 - 1}{6} \times 100\%$$

$$= 83\frac{1}{3}\%$$

Example 162-1

The unmodulated carrier of a 1-MHz transmitter is 5 kW. The carrier is then 80% double-sideband amplitude modulated by a 600-Hz tone. Calculate the value of the total RF power. If the carrier and one sideband are now suppressed, what is the percentage saving in the transmitter's RF power?

Solution

Double sideband

Total sideband power,

$$P_{SB} = \frac{1}{2} \times m^2 \times P_c$$

$$= \frac{1}{2} \times 0.8^2 \times 5$$

$$= \frac{1}{2} \times 0.64 \times 5$$

$$= 1.6 \text{ kW}$$

Total RF power = 5 + 1.6 = 6.6 kW.

Single sideband

Total power in the single sideband,

$$\frac{P_{SB}}{2} = 1.6 \text{ kW}/2$$

$$= 800 \text{ W}.$$

Percentage saving in total RF power

$$= \frac{6.6 - 0.8}{6.6} \times 100\%$$

$$= 88\%.$$

PRACTICE PROBLEMS

162-1. In H3E emission, a 2-kW carrier is 100% modulated by a 3-kHz tone. What are the values of the bandwidth and the power in the transmitted sideband?

162-2. In Practice Problem 162-1, the emission is changed to R3E. What is the new value of the bandwidth?

162-3. In Practice Problem 162-1, the carrier is totally suppressed and only one sideband is transmitted (J3E). If all of the carrier power is concentrated into the one sideband, calculate the power advantage as a ratio and in decibels.

163 The piezoelectric crystal

When a crystal (such as quartz) vibrates, an alternating voltage appears between a pair of opposite faces; this is called the *piezoelectric effect.* The fundamental frequency of vibration for a particular crystal is inversely proportional to the crystal's thickness and also depends on the type of cut. The crystal can be compared with an LCR circuit, with the inductance, capacitance, and resistance related to equivalent physical properties of the crystal. When used in an oscillator circuit, the crystal is placed between metal surfaces (the holder) to which the positive feedback voltage is applied. Electrically, the holder can be regarded as a capacitance (typically a few pF) in parallel with the LCR circuit of the crystal. This equivalent circuit of the crystal and its holder is shown in Fig. 163-1A.

Although a crystal's fundamental frequency is primarily determined by the thickness of the cut, the frequency can be varied by connecting a small trimmer capacitor across the crystal and its holder. In addition, the vibration of the crystal is complex and contains overtone frequencies; these are virtually harmonics of the fundamental frequency and it is mainly the odd overtones that are produced. The upper limit of the fundamental frequency is about 20 to 25 MHz and is determined by the minimum thickness to which a crystal can be cut without the danger of fracturing.

The Q of a crystal is normally several thousand (compared with 300 or less for a conventional LCR circuit). This is one important reason for the high degree of frequency accuracy and stability obtainable from a

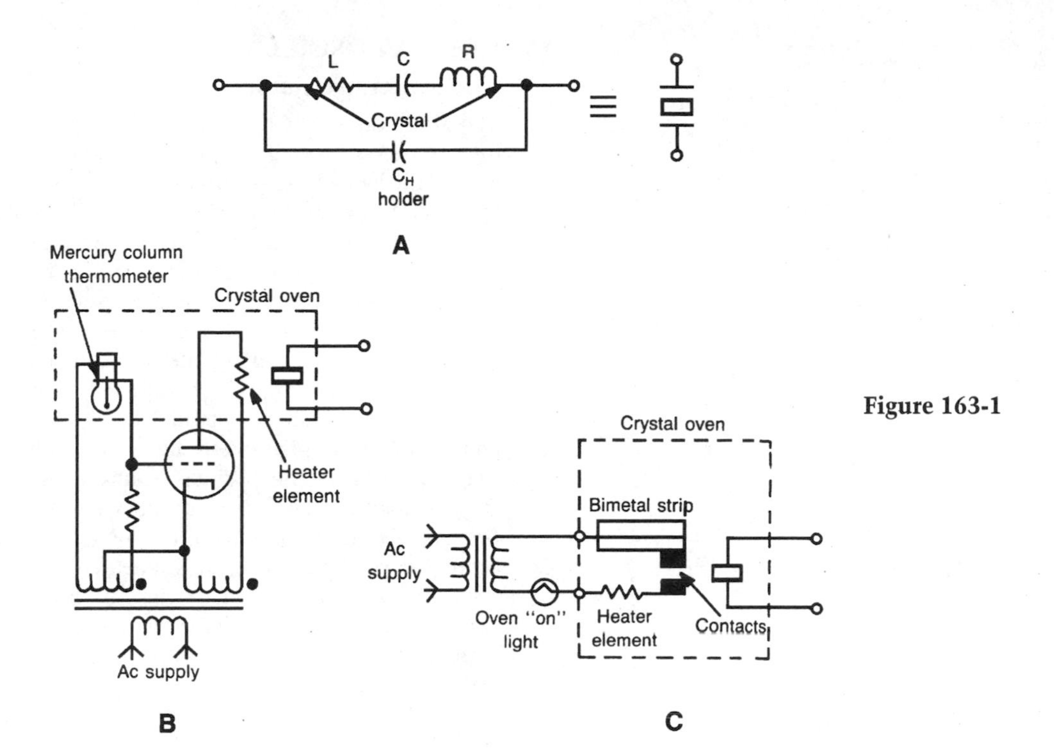

Figure 163-1

crystal oscillator. Another reason is the crystal's low temperature coefficient of frequency.

A change in crystal temperature can cause a change in resonant frequency. The crystal's temperature coefficient can be used to calculate the amount of frequency shift that can be expected from a given change in temperature. This coefficient is measured in Hz per MHz per degree celsius (parts per million per degree celsius) and it can be either positive or negative. For example, if a crystal is marked as $-30/10^6/°C$, it indicates that if the temperature increases by one degree celsius, the decrease (because of the negative coefficient) will be 30 Hz for every megahertz of the crystal's operating frequency. Accordingly, the change in the crystal's operating frequency is Δf = *value of the temperature coefficient* (either positive or negative) × *crystal frequency* (in MHz) × *temperature change*(°C), which is positive if the temperature is increasing, and negative if the temperature is decreasing. Note that if the temperature coefficient is negative and the temperature is decreasing, the result will be an increase in the frequency.

Because the frequency change (Δf) occurs at the crystal oscillator stage, the change in the output frequency of the final RF stage will be Δf × *the total product factor of the frequency multiplier stages* (if any).

To improve the frequency stability, the crystal can be enclosed in a thermostatically controlled oven. For broadcast stations, the FCC has specified certain temperature tolerances at the crystal position within the oven. If an X-cut or a Y-cut crystal is used, the permitted tolerance is only ±0.1°C, but with a low temperature-coefficient crystal, the tolerance is ±1.0°C.

The form of thermostat can either be a mercury thermometer type, or the less sensitive, (but simpler and cheaper), thermocouple variety (Figs. 163-1B and C). In Fig. 163-1B, any increase of temperature above the tolerance limit will cause the triode to be permanently cut off, so that the heater element will carry no current. If the temperature in the oven (Fig. 163-1C) falls below the tolerance limit, the bimetal strip will cause the contacts to close and complete the circuit for the heater element.

When using crystals, precautions must be taken to ensure that excessive voltage is not developed across the crystal because this might cause the crystal to fracture. If the crystal becomes dirty, it must be removed from its holder, held by the edges (not the face) and then washed in either alcohol or soap and water.

RESPONSE CURVE OF THE CRYSTAL AND ITS HOLDER

To obtain the response curve of I_S versus f (Fig. 163-2), combine the holder current (I_H) with the crystal current (I_X), and take their phase relationships into ac-

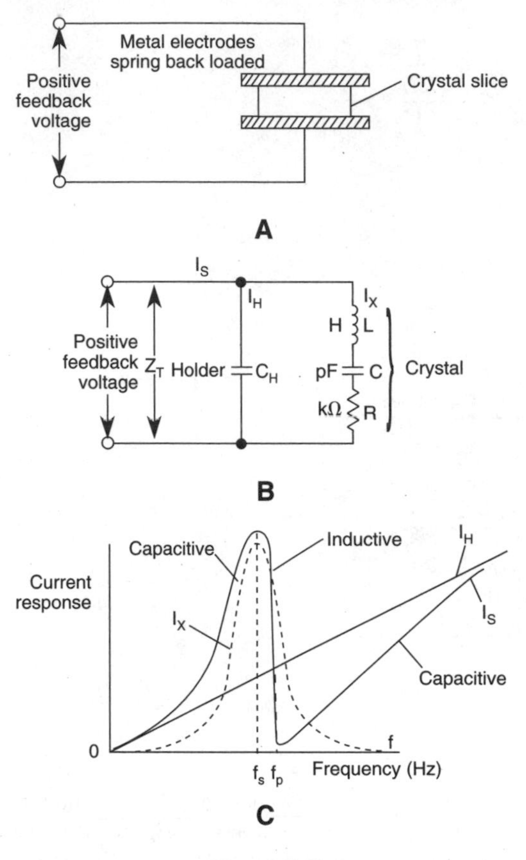

Figure 163-2

count. For a capacitor, there is a linear relationship between its current and the frequency; for the crystal, however, the current response has the familiar "bell" shape that is associated with a series LCR circuit. Below resonance, the crystal behaves capacitively but above resonance, it is inductive. After taking the phasor sum of I_X and I_H to produce I_S, it is revealed that the combination of the crystal and the holder has two resonant frequencies. There is a series resonant frequency (f_s) and a parallel resonant frequency (f_p), which (because of the crystal's high Q) are close together. As the frequency is increased, the combination is capacitive up to the point of series resonance (f_s), but it behaves inductively over the narrow region between f_s and the point of parallel resonance, f_p. Above parallel resonance, the combination is again capacitive. In some oscillator circuits, the crystal and its holder can function in either the series mode or in the parallel mode but in other circuits, the combination of the crystal and its holder operates at some position on the response curve between f_s and f_p, and therefore behaves inductively. The expressions for f_s and f_p are shown in the Mathematical Derivations.

CRYSTAL OSCILLATORS

The Miller oscillator

The JFET version of the Miller oscillator is shown in Fig. 163-3A. In order to provide positive feedback around the loop of gate → drain → gate, it is necessary that both the drain tank circuit and the crystal (with its holder) should behave inductively with respect to the generated frequency. This is illustrated in the phasor diagram of Fig. 163-3B. The gate oscillation (v_g) is the horizontal reference phasor and because the drain tank circuit is tuned slightly above the generated frequency, the drain circuit (i_d) will lag v_g. The voltage (v_L) across the drain tank circuit leads i_d by nearly 90° and is 180° out of phase with the output drain oscillation, v_d. Because the positive feedback capacitance (C_{gd}) is small, the feedback current (i_f) leads v_d by approximately 90°. To complete the loop, the feedback current develops the original voltage (v_g) across the crystal-holder combination, which is oscillating between f_s and f_p and is, therefore, acting as a high value of inductive reactance.

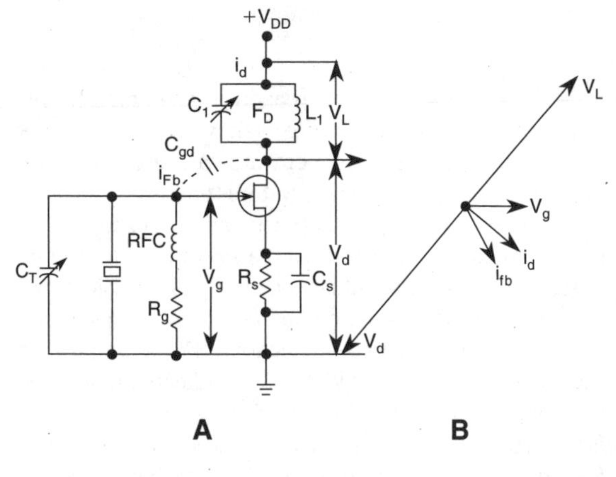

Figure 163-3

Signal bias for the JFET is provided by the combination of R_g and the capacitance of the crystal/holder; the circuit also contains source bias that is provided by R_s and C_s. The radio frequency choke (RFC) is included to reduce the shunting effect of R_g across the crystal; such a shunting effect would tend to lower the crystal's Q and to increase the amount of frequency drift.

The trimmer capacitance (C_T) is capable of "pulling" the crystal frequency slightly; this is a common method of calibrating a crystal oscillator against a superior frequency standard. Other factors that affect the output frequency are the type of crystal material, the size and nature of the cut, as well as the value of the dc supply voltage (V_{DD}).

The Miller oscillator is a popular crystal oscillator because the vibration is confined to the input circuit. Consequently, for a given excitation of the crystal, this type of oscillator will create the greatest power output because the feedback occurs from the drain to the gate, and not through the crystal itself. The Miller oscillator is also reliable because the crystal is located in the gate circuit and is less subject to stresses and strains that might cause it to crack and fail.

The Pierce oscillator

The JFET version of this circuit is shown in Fig. 163-4. The crystal is now connected between drain and gate so that it is more subject to mechanical stress than in the Miller circuit. Consequently, the main application of the Pierce circuit is in low-power RF oscillators. The crystal's holder in conjunction with the gate resistor (R_g) generates signal bias and C is a blocking capacitor to reduce the amount of dc voltage developed across the crystal.

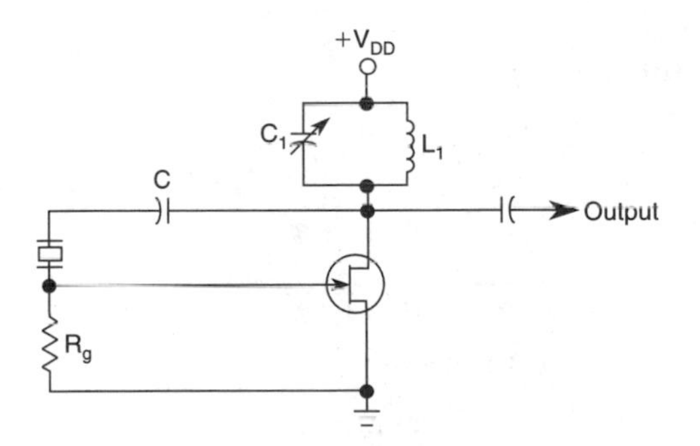

Figure 163-4

To provide the positive feedback, the crystal/holder combination must behave inductively and, therefore, oscillates at some position on the response curve between f_s and f_p. The indirect feedback path is through the gate-source capacitance and phasor analysis shows that the gate tank circuit must be *capacitive* for the circuit to oscillate. Because of the presence of the active device's output capacitance, the gate circuit is naturally capacitive. Consequently, the tank circuit is not essential and it can be replaced by either a resistor or an RF choke. If this replacement is made, the Pierce oscillator is less subject to dynamic instability than the Miller circuit.

Summarizing, the advantages of the Pierce oscillator are its lack of need for an oscillator tank circuit, and its ability to oscillate easily over a broad range of frequencies by using a number of different crystals. The disadvantage is its comparatively low power operation so that it is mainly used in crystal calibrators, test equipment, receivers, and transmitters that have an output power of only a few watts.

MATHEMATICAL DERIVATIONS

If the crystal is operating at a frequency of f MHz, and the temperature coefficient is Δf Hz/MHz/°C (Δf is either positive or negative),

$$\textit{Frequency shift at the oscillator stage} = \Delta f \times f \times \Delta T \text{ Hz} \qquad (163\text{-}1)$$

where ΔT = temperature change (°C).

If the frequency-multiplier stages have a product factor (m),

$$\textit{Output frequency shift} = m \times (\Delta f \times f \times \Delta T) \text{ Hz} \qquad (163\text{-}2)$$

Response curve of the crystal and holder

Neglecting the equivalent resistance (R), which owing to the crystal's high Q, is relatively small, the total impedance (z_T) of the circuit in Fig. 163-2 is given by:

$$z_T = \frac{\dfrac{-j}{\omega C_H}\left(j\omega L - \dfrac{j}{\omega C}\right)}{j\omega L - \dfrac{j}{\omega C} - \dfrac{j}{\omega C_H}} = \frac{-j(\omega^2 LC - 1)}{\omega C C_H\left(\omega^2 L - \dfrac{C + C_H}{C \times C_H}\right)}$$

At the series resonant frequency (f_s) z_T is theoretically zero. Therefore,

$$\omega_s^{\,2} = \frac{1}{LC}$$

This indicates that L resonates with the crystal's equivalent capacitance, C, and

$$f_s = \frac{1}{2\pi\sqrt{LC}} \text{ hertz} \qquad (163\text{-}3)$$

At the resonant frequency (f_p), z_T is theoretically infinite. Then,

$$\omega_p^{\,2} = \frac{1}{L \times \dfrac{C \times C_H}{C + C_H}}$$

This equation indicates that L resonates with the series combination of C and C_H, and

$$f_p = \frac{1}{2\pi\sqrt{L \times \dfrac{C \times C_H}{C + C_H}}}$$

$$= \frac{\sqrt{1 + \dfrac{C}{C_H}}}{2\pi\sqrt{LC}}$$

$$= f_s\sqrt{1 + \frac{C}{C_H}} \text{ hertz}$$

Therefore, $f_p > f_s$. Expanding by the Binomial Theorem (chapter 209),

$$f_p = f_s\left(1 + \frac{C}{2C_H} + - - - \text{neglected terms}\right)$$

because $C_H >> C$,

$$f_p = f_s\left(1 + \frac{C}{2C_H}\right) \approx 1.004\ f_s$$

because the ratio $C{:}C_H$ is of the order of 1:125.

The separation between f_s and f_p is

$$f_p - f_s = f_s\left(1 + \frac{C}{2C_H}\right) - f_s$$

$$= f_s \times \frac{C}{2C_H} \text{ hertz}$$

$$\approx 0.4\% \text{ of } f_s$$

If a trimmer capacitor (C_T) is added in parallel with the crystal,

$$f_p = f_s\left(1 + \frac{C}{2(C_H + C_T)}\right)$$

Therefore, the value of f_p is lowered and the separation between f_s and f_p is reduced.

Example 163-1

A Y-cut crystal has an operating frequency of 3750 kHz at a temperature of 20° C. If the crystal temperature coefficient is $-15/10^6$/°C, what is the operating frequency at 25° C?

Solution

Because 3750 kHz = 3.75 MHz, the change in frequency is given by:

$$\Delta f = -15 \times 3.75 \times (25 - 20)$$

$$= -281.25 \text{ Hz}$$

The operating frequency at 25° C = 3750 kHz − 281.25 Hz = 3749.719 kHz, rounded off.

Example 163-2

A transmitter has an operating frequency of 22 MHz and uses a 2725-kHz crystal enclosed in an oven with the temperature at 65° C. If the crystal temperature coefficient is $+5/10^6$/°C, and the oven temperature decreases to 63° C, what is the transmitter's output frequency?

Solution

The total product factor of the frequency multiplier stages is 22 MHz/2725 kHz = 8. The change in the crystal operating frequency is $\Delta f = +5 \times 2.725 \times (63 - 65) = -27.25$ Hz. The change in the transmitter output frequency is $8 \times \Delta f = 8 \times (-27.25) = -218$ Hz. The transmitter output frequency is 22 MHz − 218 Hz = 21999.782 kHz.

Example 163-3

For a particular crystal, the equivalent values of inductance, capacitance, and Q are (respectively) 115 H, 0.02 pF and 15000. If the holder capacitance is 2.5 pF, calculate the values of f_s, f_p, and R.

Solution

Series resonant frequency,

$$f_s = \frac{1}{2\pi\sqrt{LC}} \qquad (163\text{-}3)$$

$$= \frac{1}{2\pi\sqrt{115 \times 0.02 \times 10^{-12}}} \text{ Hz}$$

$$= 104.8 \text{ kHz.}$$

Parallel resonant frequency,

$$f_p = \frac{1}{2\pi\sqrt{L \times \dfrac{C \times C_H}{C + C_H}}}$$

$$= \frac{1}{2\pi\sqrt{\dfrac{115 \times 0.02 \times 2.5 \times 10^{-12}}{252}}} \text{ Hz}$$

$$= 105.2 \text{ kHz.}$$

Equivalent resistance,

$$R = \frac{2\pi f_s L}{Q} \qquad (82\text{-}2)$$

$$= \frac{2\pi \times 104.8 \times 10^3 \times 115}{15000}$$

$$= 5000\ \Omega.$$

PRACTICE PROBLEMS

163-1. The holder and stray capacitance for a crystal's mounting is 200 times the equivalent capacitance of the crystal. If f_s = 10 MHz, calculate the frequency separation between f_s and f_p.

163-2. For a particular crystal in its mounting, L = 0.1 H, C = 0.1 pF, and R = 100 Ω. The holder and stray capacitance in parallel with the crystal amounts to 15 pF. Calculate the values of f_s, the crystal's Q, and the frequency separation between f_s and f_p.

163-3. A transmitter's RF oscillator contains an 8-MHz quartz crystal with a temperature coefficient of +10 Hz/MHz/°C. Between the oscillator and the final stage there are frequency multipliers that involve two doublers and a tripler. If the temperature increases by 2.5°C, what is the change of frequency in the final stage?

163-4. A crystal has an operating frequency of 2500 kHz at a temperature of 40°C. If the crystal's temperature coefficient is $-12/10^6$/°C, what is its operating frequency at 39.9°C?

164
The frequency monitor

The frequency tolerance of an AM broadcast station is ±20 Hz. The FCC requires that every broadcast station has a means of determining the number of Hz that the actual radiated carrier frequency differs from the assigned value. Such a device has to be approved by the FCC and is known as a *frequency monitor*. This auxiliary unit contains a crystal-controlled oscillator that is entirely separate from the AM transmitter. The amount by which the carrier frequency is too high (or too low) is shown on a frequency-deviation indicator (Fig. 164-1A); this must be checked at intervals during the broadcast day and the readings must be entered in the transmitter log.

The basic principle of the frequency monitor is shown in the block diagram of Fig. 164-1B. A crystal oscillator, with a thermostatically controlled oven, produces a stable frequency that differs from the authorized carrier frequency by 1 kHz. The output of the crystal oscillator is fed to a mixer stage that is also supplied with a sample of the carrier voltage, which is derived from an unmodulated stage of the transmitter. The frequency of the mixer's output to the discrimina-

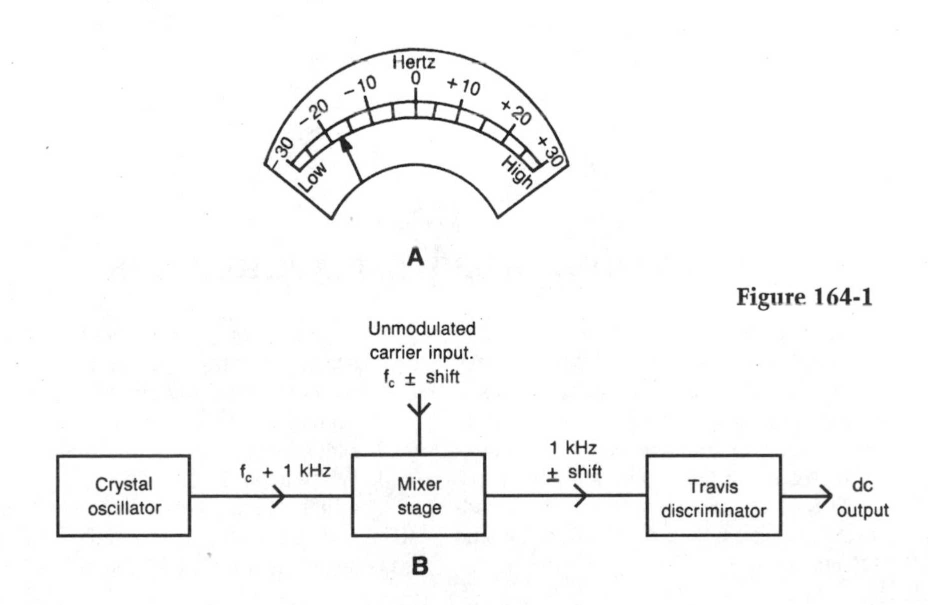

Figure 164-1

tor circuit is the difference of 1 kHz ± any shift of the carrier frequency from its assigned value. The discriminator is of the tuned high (1050 Hz)/tuned low (950 Hz) [or Travis] variety so that the magnitude and polarity of its dc output is determined by the instantaneous shift in the carrier frequency. This dc output is then passed to the monitor's indicating device with its center zero scale.

Periodically, the accuracy of the frequency monitor is measured. An external check is made of the radiated carrier frequency, and the result is compared with the simultaneous reading of the frequency monitor. The error in the frequency monitor can then be corrected by adjusting the monitor oscillator frequency (although the oscillator is crystal-controlled, the frequency can be varied a few Hz by adjusting the value of a trimmer capacitor across the crystal).

MATHEMATICAL DERIVATIONS

The error in the frequency monitor is equal to the deviation (positive for "high," negative for "low") from the assigned frequency as recorded by the frequency monitor, minus the deviation (positive for "high," negative for "low"), as is indicated by the external frequency check (Example 164-1).

Example 164-1

An external frequency check of an AM broadcast station carrier reveals that, at a particular time, the carrier frequency was 8 Hz low. The transmitter log showed, for the same time, the carrier frequency was 10 Hz high. What was the error in the station frequency monitor?

Solution

Error in the frequency monitor

$$= +10 \text{ Hz} - (-8 \text{ Hz})$$
$$= +18 \text{ Hz or 18 Hz high.}$$

Example 164-2

At a particular time, the log of an AM broadcast station showed that the carrier frequency was 10 Hz low. An external frequency check made at the same time, indicated that the carrier frequency was 7 Hz low. What was the error in the frequency monitor reading?

Solution

The error in the frequency monitor was -10 Hz $-$ (-7 Hz) = -3 Hz, or the frequency monitor was reading 3 Hz low.

PRACTICE PROBLEMS

164-1. An external frequency check of an AM broadcast station's carrier reveals that, at a particular time, the carrier frequency was 10 Hz high. The transmitter log showed that, for the same time, the carrier frequency was 14 Hz high. What was the error in the station's frequency monitor?

164-2. At a particular time, the log of an AM broadcast station showed that the carrier frequency was 6 Hz high. An external frequency check made at the same time, indicated that the carrier frequency was 5 Hz low. What was the error in the frequency monitor's reading?

165
Television broadcast frequencies

The composite video signal of a television broadcast contains the picture information together with the blanking, synchronizing, and equalizing pulses. When the composite signal amplitude-modulates the RF carrier, a white picture element produces minimum modulation and a black element creates maximum modulation. This system is called *negative transmission*, which has the advantages of providing greater power efficiency and reducing the effects of noise.

The modulation percentages produced by the various parts of the composite signal are:

1. Synchronizing and equalizing pulses, 100%.
2. Blanking pulses, 75 ± 2.5%.
3. Black level, 70%, approximately.
4. White level, 12.5 ± 2.5%.

The high video frequencies, which extend up to 4 MHz, represent the fine detail in the picture to be transmitted. If a double-sideband system were used,

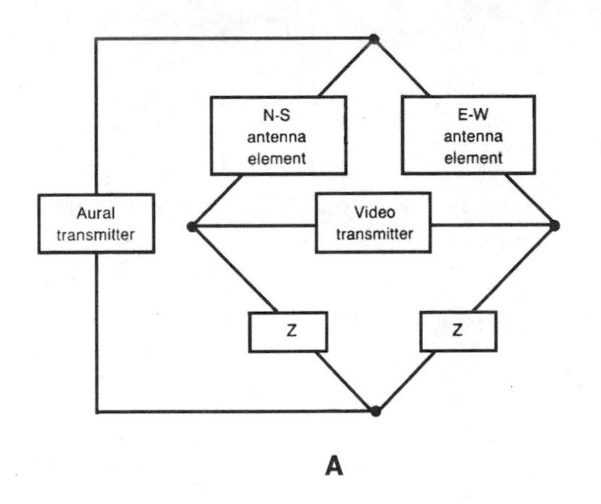

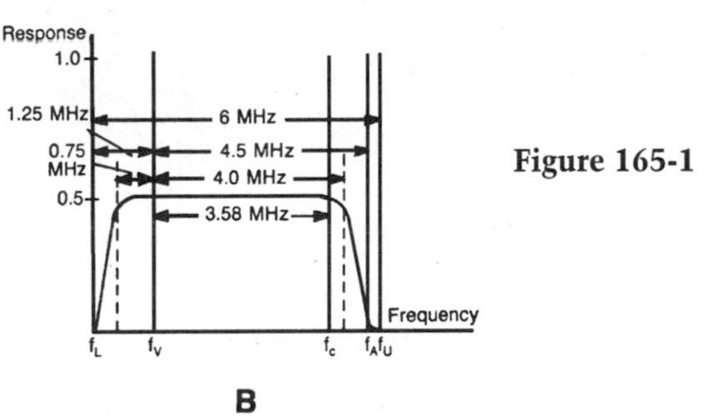

Figure 165-1

the total bandwidth required would be 8 MHz, which would exceed the allocated channel width of 6 MHz. To reduce the bandwidth, a double-sideband system is used for video frequencies up to 750 kHz, but only the upper sidebands are transmitted for the remaining frequencies. The result is vestigial sideband transmission (C3F), which is produced by the filter system following the RF power amplifier.

The modulated signal is finally radiated as a horizontally polarized wave from a VHF/UHF antenna, generally of the turnstile variety. The use of a horizontal antenna system eliminates some of the effects of noise, which is primarily vertically polarized.

The sound, or aural, transmitter uses frequency modulation, with 100% modulation corresponding to a frequency deviation of 25 kHz. The transmitted audio range is 50 Hz to 15 kHz so that the deviation ratio is 20 kHz/15 kHz = 1.667. Pre-emphasis and corresponding de-emphasis are used, with a time constant of 75 microseconds. A common antenna system is used for both the AM video signal and the FM aural signal, so it is necessary to avoid interaction between the two transmitters. This is achieved by means of a *diplexer* (Fig. 165-1A), that is essentially a balanced bridge. The north-south and east-west antenna elements are two arms of the bridge which is balanced by equal impedances in the other arms. The output of the video transmitter is fed to one pair of opposite corners of the bridge while the aural transmitter is connected to the other pair. With the bridge balanced, both transmitters will activate the antenna elements, but neither will interact with the other.

MATHEMATICAL DERIVATIONS

The video carrier frequency (f_V) has a tolerance of ±1 kHz and is located 1.25 MHz above the channel's lower frequency limit (f_L) so that

$$f_V = (f_L + 1.25) \text{ MHz} \qquad (165\text{-}1)$$
$$= (f_U - 4.75) \text{ MHz}$$

where f_U is the channel's upper frequency limit. These frequencies are shown in Fig. 165-1B, which represent a channel's response curve. The aural carrier frequency (f_A) also has a frequency tolerance of ±1 kHz and is positioned 0.25 MHz below the channel's upper frequency limit. Therefore,

$$f_A = (f_U - 0.25) \text{ MHz} \qquad (165\text{-}2)$$
$$= (f_L + 5.75) \text{ MHz}$$
$$= (f_V + 4.5) \text{ MHz}$$

As a result of the frequency conversion process in the superheterodyne TV receiver, the video and aural intermediate frequencies are respectively 45.75 MHz and 41.25 MHz; their frequency separation is still 4.5 MHz. These values are independent of the channel selected.

In a color transmission, the positioning of the video and aural carriers is the same as in the monochrome transmission, but the color information (chrominance signal) is used to amplitude modulate a chrominance subcarrier with its frequency (f_C) located 3.579545 MHz ± 10 Hz (commonly rounded off to 3.58 MHz) above the video carrier frequency (f_V). Then,

$$f_C = (f_V + 3.579545) \text{ MHz} \qquad (165\text{-}3)$$
$$= (f_L + 4.829545) \text{ MHz}$$
$$= (f_U - 1.170455) \text{ MHz}$$
$$= (f_A - 0.920455) \text{ MHz}$$

Example 165-1

The width of TV channel 4 extends from 66 to 72 MHz. What are the frequencies of the video carrier, the aural carrier, and the chrominance subcarrier?

Solution

Video carrier frequency,

$$f_V = f_L + 1.25 \qquad (165\text{-}1)$$

$$= 66 + 1.25$$

$$= 67.25 \text{ MHz}.$$

Aural carrier frequency,

$$f_A = f_U - 0.25 \qquad (165\text{-}2)$$

$$= 72 - 0.25$$

$$= 71.75 \text{ MHz}.$$

Chrominance subcarrier frequency,

$$f_C = f_L + 4.83 \qquad (165\text{-}3)$$

$$= 66 + 4.83$$

$$= 70.83 \text{ MHz}.$$

Note: The TV receiver's local oscillator frequency for channel 4 is 67.25 + 45.75 = 113 MHz.

PRACTICE PROBLEM

165-1. The frequency band of channel 7 is 174 to 180 MHz. What are the frequencies of the video carrier, the aural carrier, the chrominance subcarrier, and the receiver's local oscillator?

166
Distributed constants of transmission lines

In previous chapters, you have been concerned with so-called "lumped" circuitry, where the three electrical properties (resistance, inductance, capacitance) were related to specific components. For example, a resistor was regarded as being a "lump" of resistance and all connecting wires were considered to be perfect conductors. By contrast, the resistance (R) associated with a two-wire transmission line is not concentrated into a "lump," but is "distributed" along the entire length of the line. The distributed constant of resistance is therefore measured in the basic unit of ohms per meter, rather than in ohms.

Because the straight wires of the parallel line are conducting surfaces separated by an insulator or dielectric, the line will possess the distributed constants of self-inductance L (henrys per meter) and capacitance C (farads per meter). In addition, no insulator is perfect; consequently, there is another distributed constant G (siemens per meter), which is the leakage conductance between the wires. These four distributed constants are illustrated in Fig. 166-1A; their order of values in a practical line are R mΩ/m, L μH/m, C pF/m, and G nS/m.

In physical terms, the properties of resistance and leakage conductance will relate to the line's power loss in the form of heat dissipated and will, therefore, govern the degree of attenuation, measured in decibels per meter (dB/m). The self-inductance results in a magnetic field surrounding the wires, and the capacitance means that an electric field exists between the wires (Fig. 166-1B); these L and C properties determine the line's behavior in relation to the frequency.

The two-wire ribbon line (that connects an antenna to its TV receiver) is often referred to as a 300-Ω line. But what is the meaning of the 300 Ω? Certainly, you cannot find this value with an ohmmeter, so you are led to the conclusion that it only appears under working conditions when the line is being used to convey RF power. In fact, the 300-Ω value refers to the line's *surge or characteristic impedance*, whose letter symbol is Z_o. The surge impedance is theoretically defined as the input at the RF generator to an infinite length of the line (Figs. 166-1C and D). In the equivalent circuit of Fig. 166-1D, the C and G line constants will complete a path for current to flow so that an effective current (I) will be drawn from the RF generator, whose effective output voltage is E. Then, the input impedance is,

$$Z_{in} = Z_o = \frac{E_{\text{rms}}}{I_{\text{rms}}} \ \Omega.$$

The mathematical derivations show that an RF line behaves resistively and it has a surge impedance of $\sqrt{L/C}$ Ω; this value must also depend on the line's physical construction.

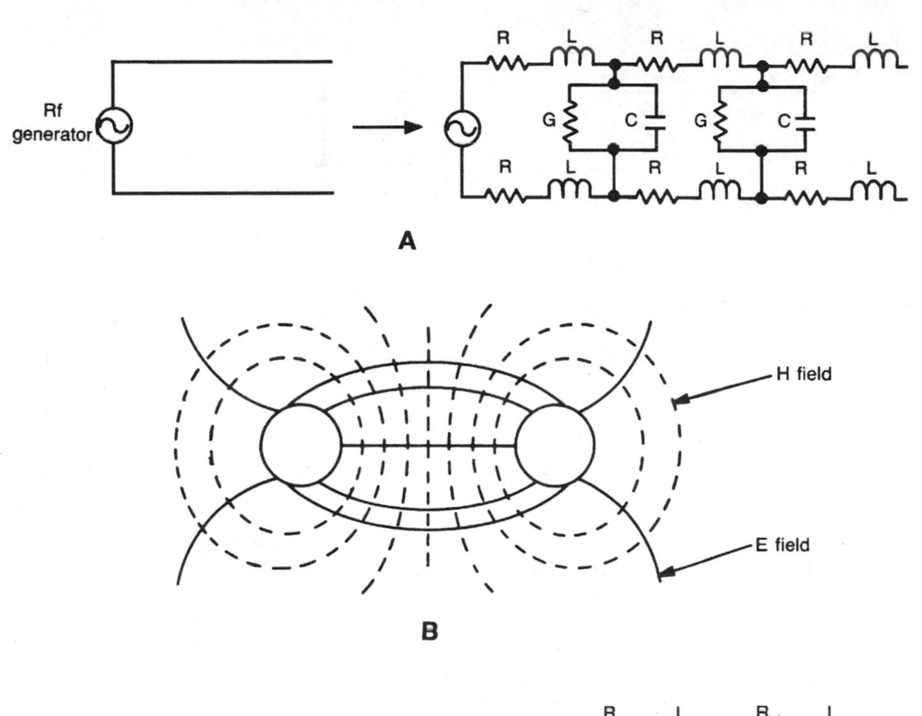

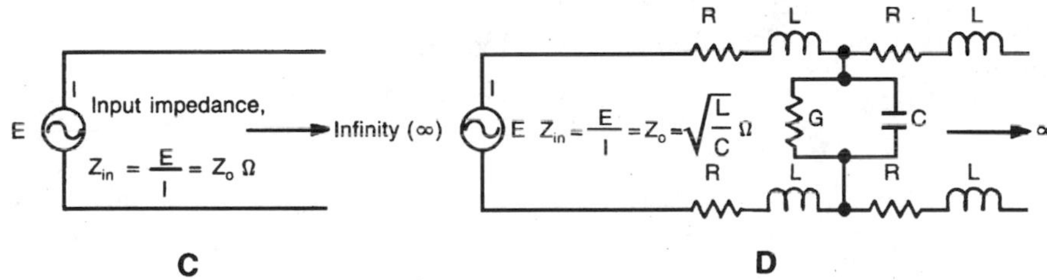

Figure 166-1

MATHEMATICAL DERIVATIONS

It can be shown that the surge impedance (Z_o) is a complex quantity that is given by:

Surge impedance,

$$Z_o = \sqrt{\frac{R + j\omega L}{G + j\omega C}}\ \text{ohms} \qquad (166\text{-}1)$$

$$= \sqrt[4]{\frac{R^2 + \omega^2 L^2}{G^2 + \omega^2 C^2}}\ \angle\left[\frac{1}{2}\left(\text{inv tan}\ \frac{\omega L}{R} - \text{inv tan}\ \frac{\omega C}{G}\right)\right]\ \text{ohms}$$

where: $\omega = 2\pi f$, angular frequency (rad/s)
f = frequency of the RF generator (Hz).

For a low-loss line operating at radio frequencies (RF), $\omega L >> R$ and $\omega C >> G$. Then,

Surge impedance,

$$Z_o \rightarrow \sqrt{\frac{L}{C}}\ \angle 0°\ \Omega \qquad (166\text{-}2)$$

Two-wire line or twin lead (Fig. 166-2A)

Surge impedance,

$$Z_o = \frac{276}{\sqrt{\epsilon_r}} \log\left(\frac{2S}{d}\right)\ \text{ohms} \qquad (166\text{-}3)$$

where: S = spacing between the conductors (m)
d = diameter of each conductor (m)
ϵ_r = relative permittivity of the dielectric.

Coaxial cable (Fig. 166-2B)

Surge impedance,

$$Z_o = \frac{138}{\sqrt{\epsilon_r}} \log\left(\frac{D}{d}\right)\ \text{ohms} \qquad (166\text{-}4)$$

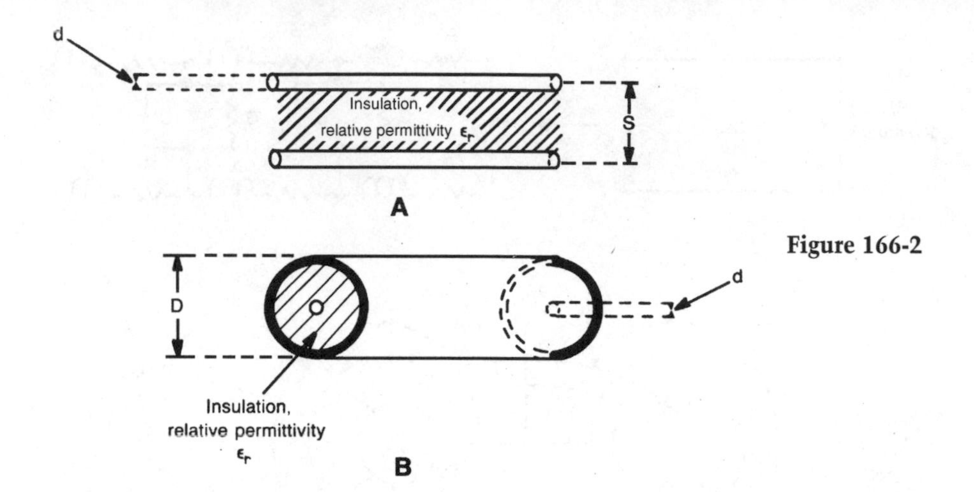

Figure 166-2

where: D = inner diameter of the outer conductor (m)
d = outer diameter of the inner conductor (m)

Example 166-1

Each conductor of a twin-lead transmission line has a radius of 1.5 mm and the spacing between the centers of the conductors is 0.8 cm. If the dielectric is air, what is the value of the line's surge impedance?

Solution

Surge impedance,

$$Z_o = \frac{276}{\sqrt{\epsilon_r}} \log\left(\frac{2S}{d}\right) \tag{166-3}$$

$$= \frac{276}{\sqrt{1}} \log\left(\frac{2 \times 0.8 \times 10^{-2}\ \text{m}}{2 \times 1.5 \times 10^{-3}\ \text{m}}\right)$$

$$= 276 \log 5.333$$

$$\approx 200\ \Omega.$$

Example 166-2

The outer conductor of a coaxial cable has an inner diameter of 1.75 cm and the inner conductor has an outer diameter of 2.5 mm. If teflon (relative permittivity = 2.1) is used as the insulator between the conductors, what is the value of the cable's surge inpedance?

Solution

Surge impedance,

$$Z_o = \frac{138}{\sqrt{\epsilon_r}} \log\left(\frac{D}{d}\right) \tag{166-4}$$

$$= \frac{138}{\sqrt{2.1}} \log\left(\frac{1.75 \times 10^{-2}\ \text{m}}{2.5 \times 10^{-3}\ \text{m}}\right)$$

$$= \frac{138}{\sqrt{2.1}} \log 7$$

$$\approx 81\ \Omega.$$

Example 166-3

The primary constants of an RF transmission line are R = 50 mΩ/m, L = 1.6 μH/m, C = 7.5 pF/m, and G = 10 nS/m. If the frequency of the generator feeding the line is 300 MHz, what is the value of the line's surge impedance?

Solution

Surge impedance,

$$Z_o = \sqrt{\frac{L}{C}} \tag{166-2}$$

$$= \sqrt{\frac{1.6 \times 10^{-6}}{7.5 \times 10^{-12}}}$$

$$= 462\ \Omega.$$

PRACTICE PROBLEMS

166-1. In Example 166-1, the radius of each conductor and the spacing between the conductors are both doubled. What is the new value of the surge impedance?

166-2. In Example 166-2, the two diameters are both halved. What is the new value of the surge impedance?

166-3. In Example 166-3, the frequency of the generator is reduced from 300 MHz to 10 kHz. Calculate the new value of the surge impedance.

167
The matched line

Referring to Fig. 167-1A, consider the conditions that exist at the position (X, Y) on the infinite line. Because there is still an infinite length to the right of (X, Y), the input impedance at this position looking down the line will be equal to the value of Z_o. Consequently, if the section of the line to the right of X, Y is removed and replaced by a resistive load, whose value in ohms is the same as that of the surge impedance, it will appear to the generator as if it is still connected to an infinite line and the input impedance at the generator will remain equal to Z_o.

When a line is terminated by a resistive load of value Z_o, the line is said to be "matched" to the load. Under matched conditions, the line is most effective in conveying RF power from the generator (for example, a transmitter) to the load (such as an antenna).

Examine in detail what happens on a matched line. Travelling sine waves of voltage and current start out from the RF generator and move down the line in phase. Because of the small amount of attenuation present on the line, the effective (rms) values of the voltage and current decay slightly, but at all positions $V_{rms}/I_{rms} = Z_o$ (Fig. 167-1B).

On arrival at the termination, the power contained in the voltage and current waves is completely absorbed by the load. It is important to realize that power is being conveyed down the line and is *not* being dissipated and lost as heat in the surge impedance.

In previous chapters, you have assumed that in a series circuit consisting of a source, a two-wire line, and a load, the current was instantaneously the same throughout the circuit. This can only be regarded as true provided that the distances involved in the circuit are small compared with the wavelength of the output from the source. The wavelength whose letter symbol is the Greek lambda (λ) is defined as the distance between two consecutive identical states in the path of the wave. For example, the wavelength of a wave in

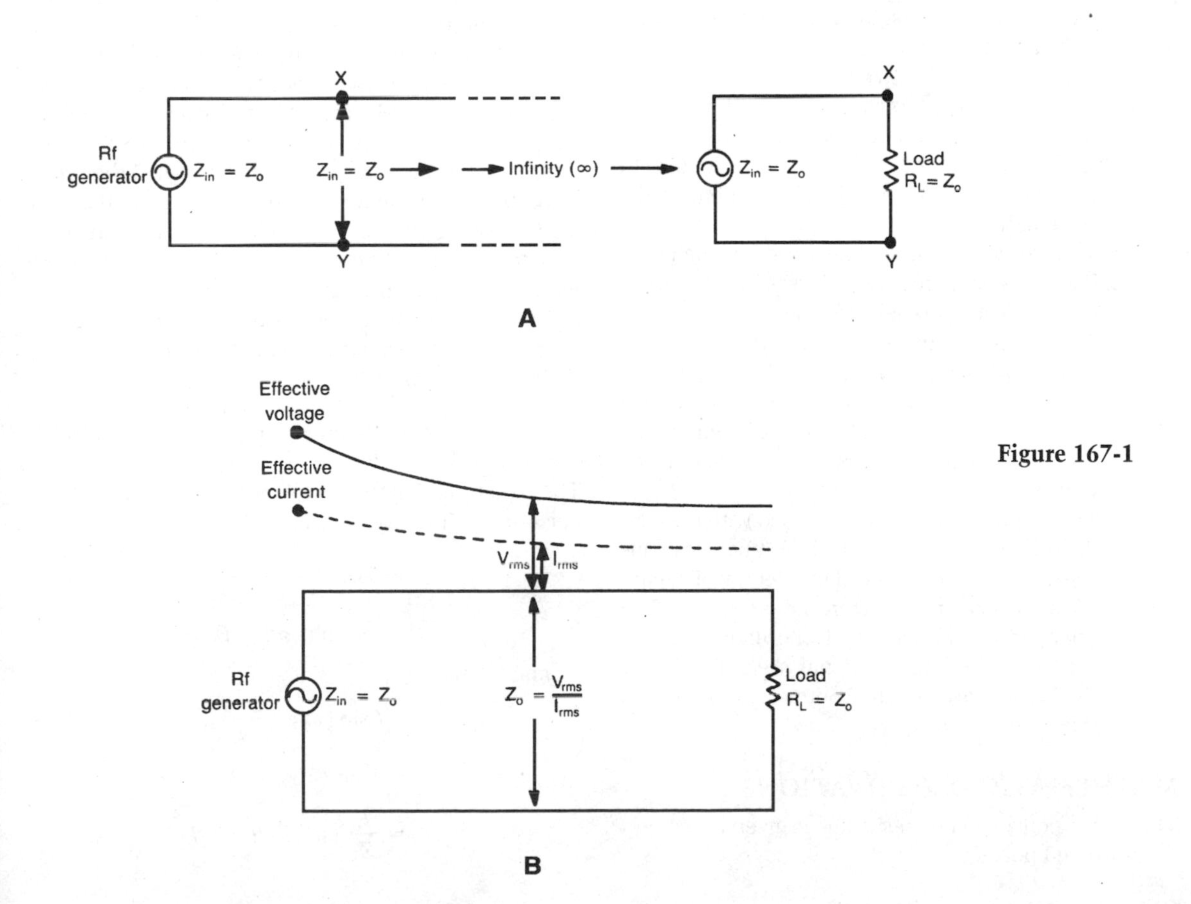

Figure 167-1

water is the distance between two neighboring crests (or troughs). On a matched transmission line, the wavelength is the distance between two adjacent positions, where identical voltage (and current) conditions occur instantaneously. Travelling (or progressive) waves exist on the line because time is involved in propagating RF energy from the source to the load.

In free space, the velocity of an electromagnetic wave (radio wave) is a constant that is approximately equal to 3×10^8 m/s (the speed of light). However, on a transmission line, the velocity is the speed at which the voltage and current waves, as well as the electric and magnetic fields, travel toward the matched load. This velocity is always less than the speed of light (c). The ratio of the velocity on the line to the velocity in free space is called the *velocity factor* (δ). The value of δ varies from 0.66 for certain types of coaxial cable to 0.975 for an air insulated two-wire line. It follows that, because the velocity is the product of the frequency and the wavelength, the wavelength on the line is shorter than the wavelength in free space.

The features of the matched line can be summarized as follows:

1. Travelling waves of the voltage and current move down the line in phase and their power is completely absorbed by the load.
2. The ratio of the effective voltage to the effective current is constant over the entire line and is equal to the surge impedance, Z_o.
3. The input impedance at the generator is equal to the surge impedance and is independent of the line's length.
4. The power losses on the line are subdivided into:
   a. Radiation and induction losses, which are a problem with parallel wire lines.
   b. The dielectric hysteresis loss that increases with frequency and depends on the type of insulator used. At microwave frequencies of a few GHz, the dielectric loss is the ultimate reason for abandoning coaxial lines and using waveguides instead.
   c. The copper loss that is associated with the conductor's resistance. At high frequencies, this loss is increased by the skin effect, which confines most of the electron flow to the surface (skin) of a conductor and, therefore, reduces the available cross-sectional area. The larger the surface area of the conductors, the less is this type of loss.

MATHEMATICAL DERIVATIONS

The power conveyed down the line is given by:

Conveyed power,

$$P = V_{rms} \times I_{rms} \quad (167\text{-}1)$$
$$= I^2_{rms} \times Z_o$$
$$= V^2_{rms}/Z_o \text{ watts}$$

Attenuation constant,

$$\alpha = 9.8\left(\frac{R}{2Z_o} + \frac{GZ_o}{2}\right) \text{ decibels per meter} \quad (167\text{-}2)$$

Phase velocity,

$$v = \frac{\lambda}{T} = f \times \lambda \text{ meters per second} \quad (167\text{-}3)$$

Wavelength,

$$\lambda = \frac{v}{f} \text{ meters} \quad (167\text{-}4)$$

Frequency,

$$f = \frac{v}{\lambda} \text{ hertz} \quad (167\text{-}5)$$

where: T = period (s)

On a transmission line, consider the generator's output as the reference voltage. At position A, which is a distance of $\lambda/4$ from the generator, the instantaneous voltage will lag by 90° on this reference level while at distances of $\lambda/2$ and $3\lambda/4$, (positions B and C), the voltages will be respectively 180° out of phase and 270° lagging (90° leading). A particular phase condition will then travel down at a speed, which is called the *phase velocity* (v); this movement of a phase condition is illustrated in Fig. 167-2. Consequently, over a distance of 1 meter, there will be an angular difference, which is measured by the phase shift constant (β). The unit of β is the radian per meter. Because there must be a difference of 360° (2π radians) over a distance of λ meters,

Phase shift constant,

$$\beta = \frac{2\pi}{\lambda} \text{ radians per meter} \quad (167\text{-}6)$$

The equations for the travelling waves of voltage and current on a loss-free line are

$$\text{v} = E \sin\left(\omega t - \frac{2\pi d}{\lambda}\right) \quad (167\text{-}7)$$
$$= E \sin(2\pi ft - \beta d)$$

and

$$\text{i} = I \sin\left(\omega t - \frac{2\pi d}{\lambda}\right) \quad (167\text{-}8)$$
$$= I \sin(2\pi ft - \beta d)$$
$$= \frac{E}{Z_o} \sin(2\pi ft - \beta d)$$

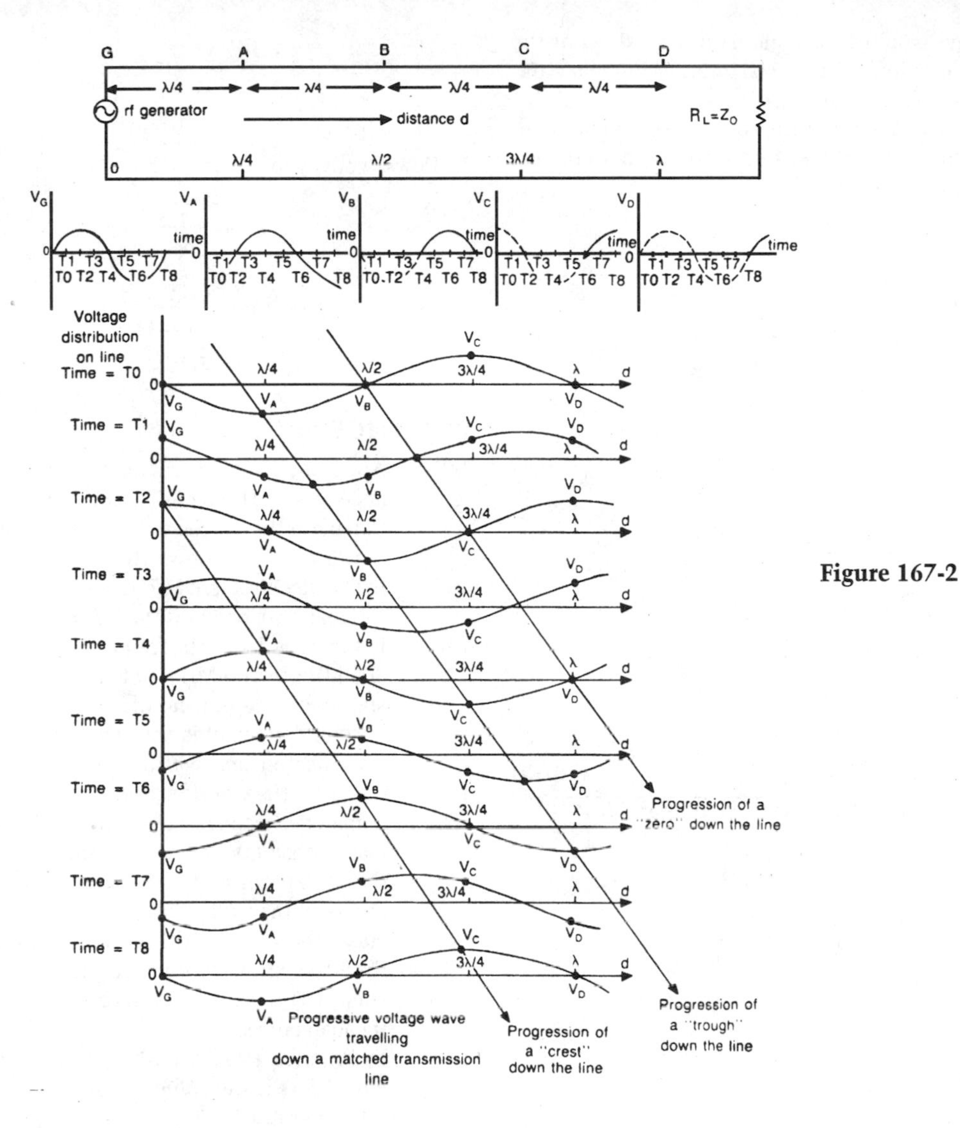

Figure 167-2

where: v = instantaneous value of the voltage wave (V)
i = instantaneous value of the current wave (A)
E = peak value of the voltage wave (V)
I = peak value of the current wave (A)
ω = angular frequency (rad/s)
f = frequency (Hz)
β = phase shift constant (rad/m)
d = distance from the generator (m)
t = time (s)
Z_o = surge impedance (Ω).

Mathematically, it can be shown that in a low-loss line,

$$\beta = \omega\sqrt{LC} = 2\pi f\sqrt{LC} \text{ radians per meter} \quad (167\text{-}9)$$

Therefore,
Phase velocity,

$$v = f \times \lambda \quad (167\text{-}10)$$

$$= \frac{2\pi f}{\beta}$$

$$= \frac{\omega}{\beta}$$

$$= \frac{1}{\sqrt{LC}} \text{ meters per second}$$

Example 167-1

The primary constants of an RF transmission line are R = 50 mΩ/m, L = 1.6 μH/m, C = 7.5 pF/m, and G = 10 nS/m. If the frequency of the generator feeding into the line is 300 MHz, calculate the values of the line's (a) surge impedance, (b) attenuation constant, (c) phase-shift constant, (d) wavelength, and (e) phase velocity. What is the amount of the power conveyed down this matched line if the load voltage is 160 V rms?

Solution

(a) Surge impedance,

$$Z_o = \sqrt{\frac{L}{C}} = \sqrt{\frac{1.6 \times 10^{-6}\ \text{H}}{7.5 \times 10^{-12}\ \text{F}}} = 462\ \Omega$$

(b) Attenuation constant,

$$\alpha = \frac{R}{2Z_o} + \frac{GZ_o}{2} \qquad (167\text{-}2)$$

$$= \frac{50 \times 10^{-2}}{2 \times 462} + \frac{10 \times 10^{-9} \times 462}{2}$$

$$= 54 \times 10^{-6} + 2.3 \times 10^{-6}$$

$$= 56.3 \times 10^{-6} \text{ nepers per meter}$$

$$= 490 \times 10^{-6}\ \text{dB/m}$$

(c) Phase shift constant,

$$\beta = \omega\sqrt{LC} \qquad (167\text{-}9)$$

$$= 2 \times \pi \times 300 \times 10^6 \times \sqrt{1.6 \times 10^{-6} \times 7.5 \times 10^{-12}}$$

$$= 6.53\ \text{rad/m}$$

(d) Line's wavelength,

$$\lambda = \frac{2\pi}{\beta} \qquad (167\text{-}6)$$

$$= \frac{2\pi}{6.53}$$

$$= 0.96\ \text{m}$$

(e) Phase velocity,

$$v = f \times \lambda \qquad (167\text{-}3)$$

$$= 300 \times 10^6 \times 0.96$$

$$= 288 \times 10^6\ \text{m/s.}$$

Power conveyed to the load,

$$P = \frac{V^2_{rms}}{Z_o} \qquad (167\text{-}1)$$

$$= \frac{(160\ \text{V})^2}{462\ \Omega}$$

$$= 55\ \text{W}$$

PRACTICE PROBLEMS

167-1. A matched coaxial cable, whose surge impedance is 450 Ω, carries a line current of 3.5 A. Calculate the values of the line voltage and the power conveyed down the line.

167-2. In Practice Problem 167-1, what are the values of the resistive load and the L:C ratio for the line?

167-3. In Example 167-1 the frequency is changed from 300 MHz to 100 MHz. Calculate the new phase shift constant and the wavelength on the line.

167-4. A 400-MHz signal is travelling along a matched transmission line with a velocity factor of 0.8. Calculate the wavelength on the line in meters.

167-5. A matched coaxial cable has the following primary constants: $R = 50$ mΩ/m, $L = 1.55$ μH/m, $C = 8.10$ pF/m, and $G = 8$ nS/m. A 50-MHz signal arrives at the load with a power level of 80 W. Calculate the values of (a) the surge impedance, (b) the attenuation constant, (c) the phase shift constant, (d) the wavelength on the cable, and (e) the load current.

167-6. In Practice Problem 167-5, the cable is 10 m long. What is the value of the input power from the generator?

168
The unmatched line—reflection coefficient

If an RF line is terminated by a resistive load that is not equal to the surge impedance, the generator will still send voltage and current waves down the line in phase, but their power will only be partially absorbed by the load. A certain fraction of the voltage and current waves that arrive (are incident) at the load will be reflected back toward the generator. At any position on the line, the instantaneous voltage (or current) will be the resultant of the incident and reflected voltage (or current) waves; these combine to produce so-called *standing waves*. In the extreme cases of open and short circuited lines, no power can be absorbed by the "load" and total reflection will occur.

As the incident voltage wave moves down the line toward the open circuit, you can derive the changes in the instantaneous standing wave for each subsequent interval of one eighth of the period. The instantaneous standing wave is illustrated in Fig. 168-1 for the times: $T, T + 1/8f, T + 1/4f, \ldots T + 7/8f$. Finally, all instantaneous voltage standing waves are collected together in one representation (Fig. 168-2A). At the distances of 0, $\lambda/2$, and λ from the open circuit termination, there are *fixed* positions of maximum voltage variation with time; such variations are called voltage *antinodes* or *loops*. However, at the distances of $\lambda/4$ and $3\lambda/4$, there are *stationary* positions where the voltage is at all times zero; these are called the voltage *nodes* or *nulls*. The stationary points that occur with the standing wave, are in contrast with the traveling wave, where, for example, a zero-voltage condition would move down the line at a speed equal to the phase velocity.

It is customary to represent the standing wave in terms of its effective or rms value (V_s), which is expressed by.

$$V_s = 2\,E \cos \frac{2\pi x}{\lambda}$$

and is illustrated in Fig. 168-2B. Notice that only the magnitude of this expression is considered and that no positive or negative sign is involved.

Now, examine the current (broken line) standing wave distribution on an open circuited line. At the open circuit termination, the circuit must reverse its direction. Therefore, you have an "out of phase" reflection in which the wave is extended beyond the open circuit, then phase shifted by 180°, and afterwards folded back to provide the reflected wave. Incident and reflected waves are then combined to produce the current standing wave at times: T, $T + 1/8f$, $T + 1/4f$, . . . $T + 7/8f$ (Fig. 168-3). Finally, all eight instantaneous current standing waves are collected together in one presentation (Fig. 168-4).

The variation of the impedance along the open-circuited line is shown in Fig. 168-5. One is struck with the similarity to the behavior of series and parallel LC circuits. Remember that the impedance response of the series circuit changes from capacitive through resistive (low value) to inductive. By contrast, the paral-

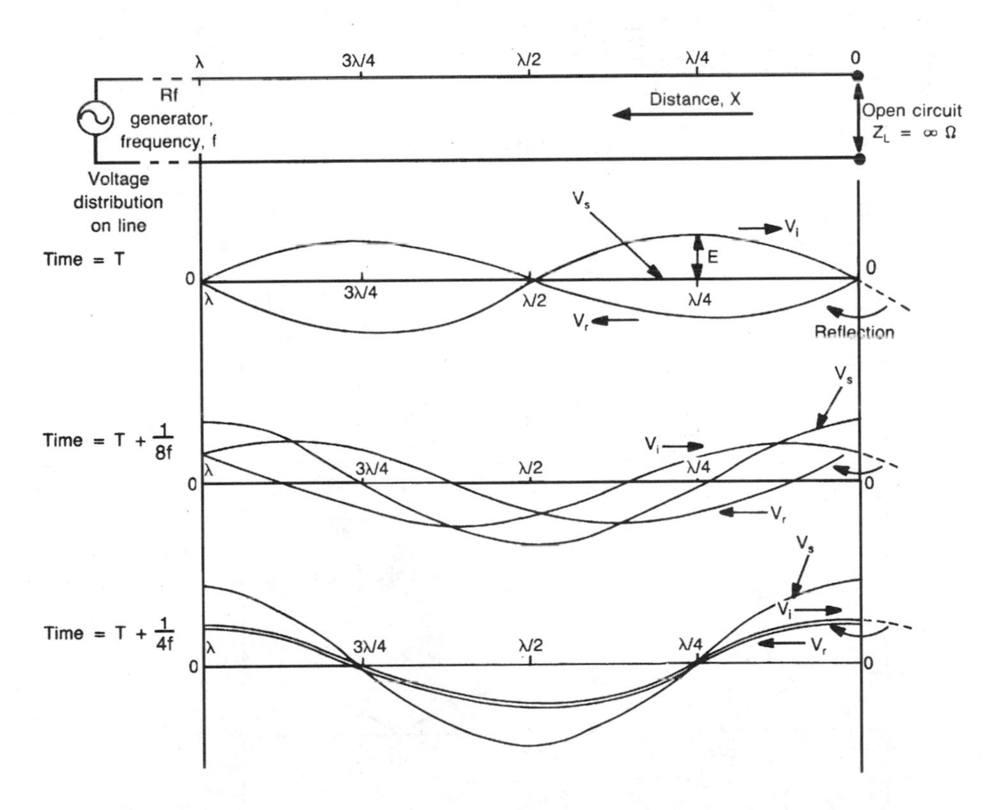

Figure 168-1

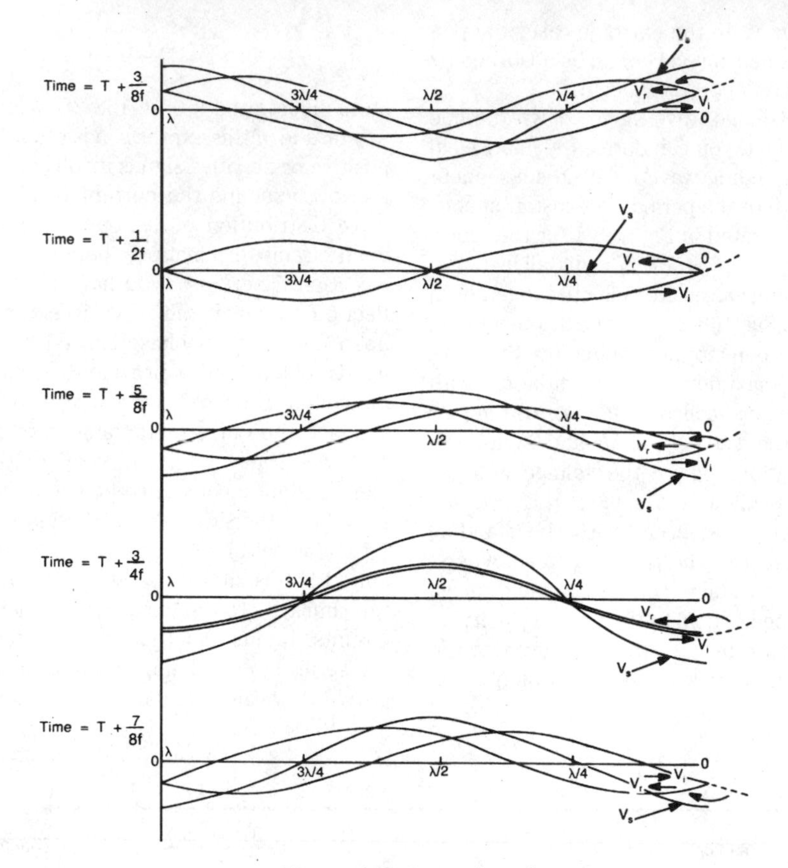

Figure 168-1 continued.

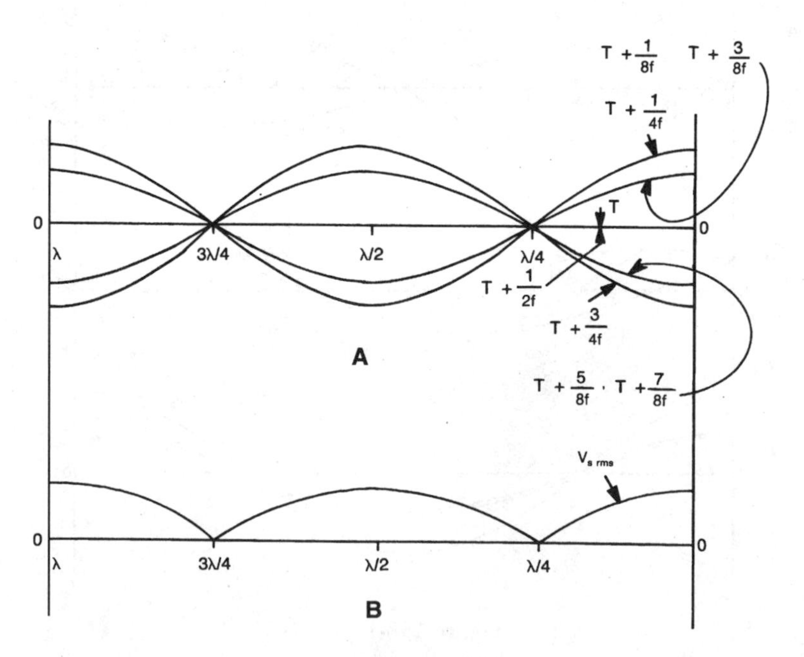

Figure 168-2

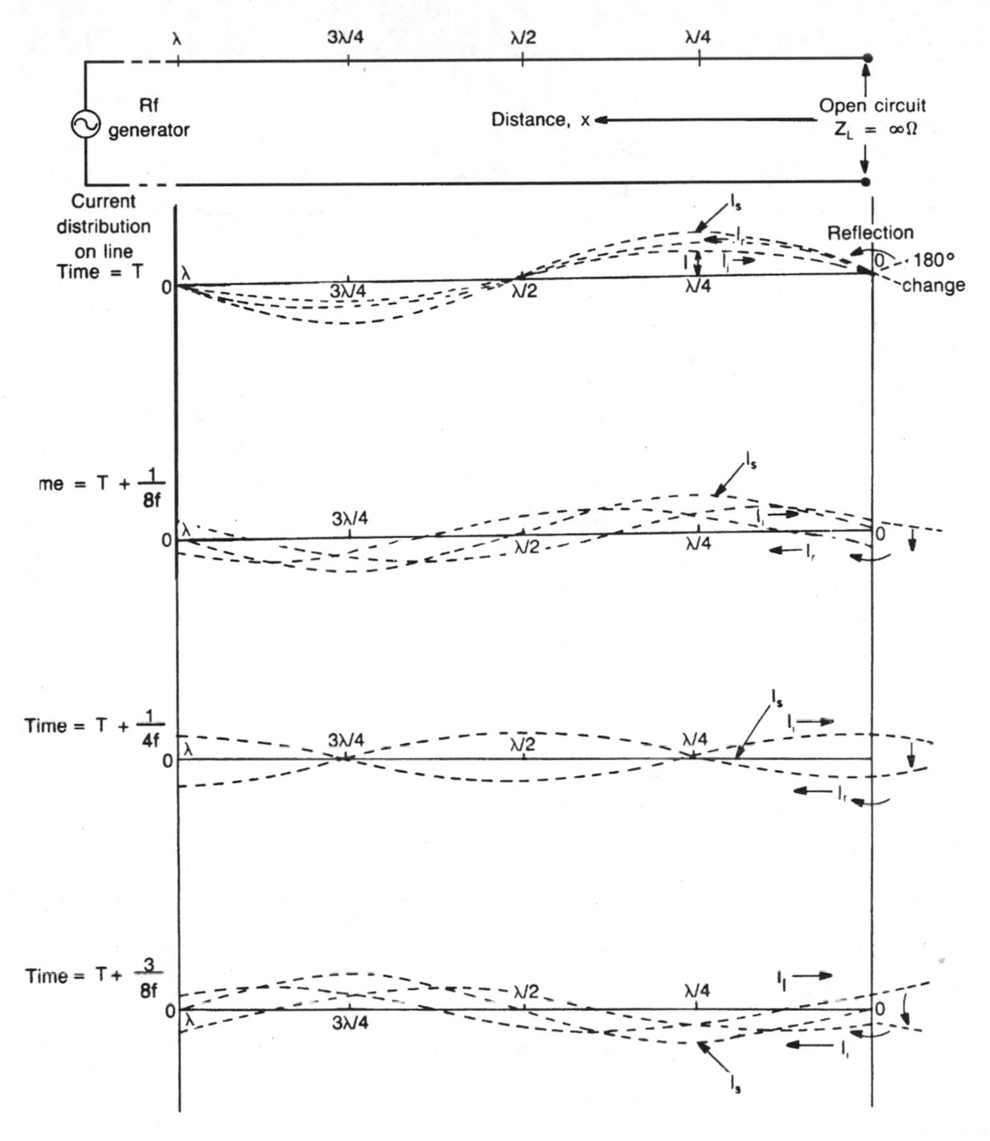

Figure 168-3

lel circuit has an impedance response that varies from inductive through resistive (high value) to capacitive. These equivalent LC circuits have been included in Fig. 168-5. Because the line exhibits resonant properties, it can be referred to as *tuned;* by contrast, the matched line is called "flat, untuned, or nonresonant."

Now, examine the effect of terminating a lossless line with a resistive load that is not equal to the surge impedance. The amount of reflection will be reduced so that the nodes and antinodes will be less pronounced. Neglecting any losses on the line, the effective voltage and current distributions for the three possible cases of a resistive load, namely $R_L > Z_o$, $R_L = Z_o$, and $R_L < Z_o$, are shown in Fig. 168-6.

MATHEMATICAL DERIVATIONS

It is important to realize that a standing wave represents a sinusoidal distribution both with time and distance. The incident traveling wave can be represented by,

$$v_i = E \sin\left(2\pi ft + \frac{2\pi x}{\lambda}\right) \text{ volts} \qquad (168\text{-}1)$$

where: v_i = instantaneous value of the incident voltage wave (V)

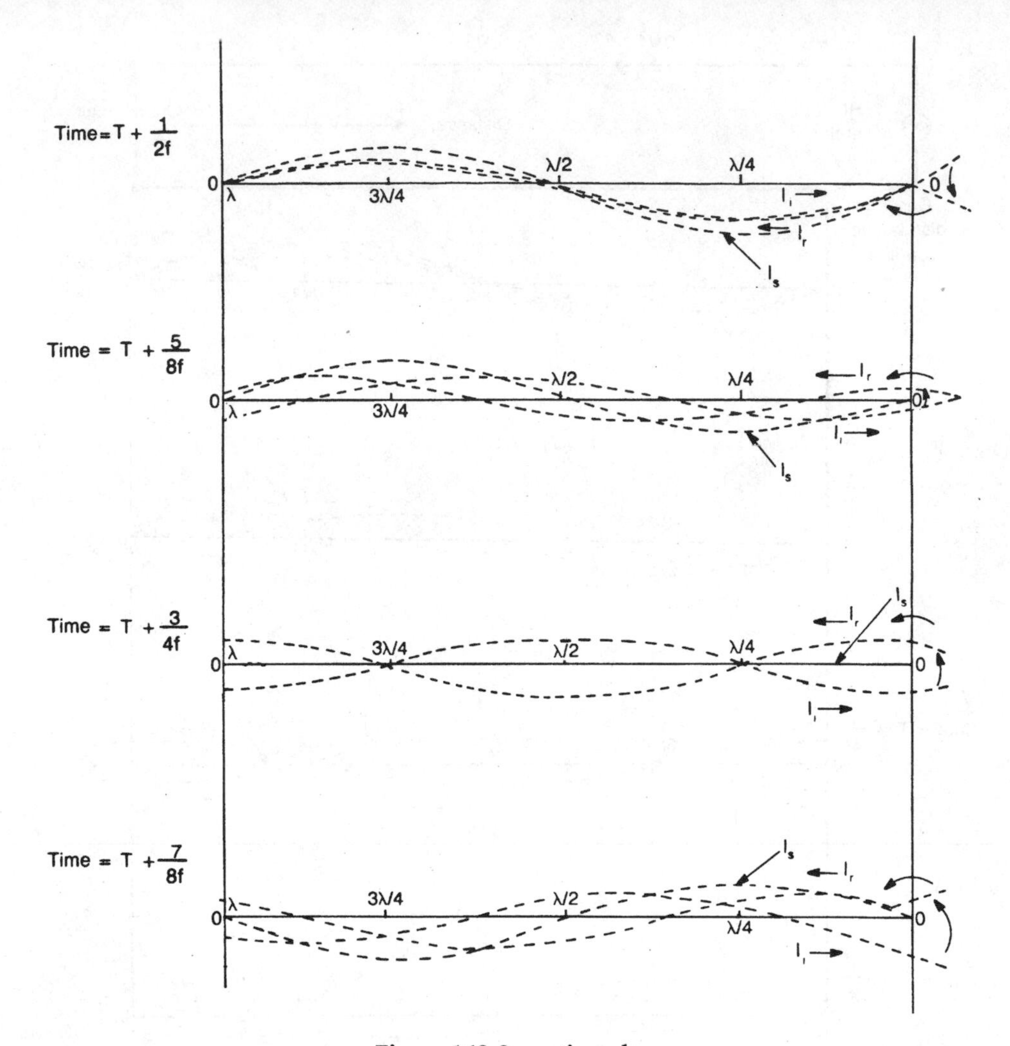

Figure 168-3 continued.

E = peak value of the incident voltage wave (V)
f = frequency (Hz)
x = distance measured from the open circuit termination (m)
λ = wavelength existing on line (m)

The equation of the reflected wave is,

$$v_r = E \sin\left(2\pi ft - \frac{2\pi x}{\lambda}\right) \text{ volts} \qquad (168\text{-}2)$$

where: v_r = instantaneous value of the reflected voltage wave (V)

The instantaneous standing wave voltage (v_s) is:

$$v_s = v_i + v_r = E\left[\sin\left(2\pi ft + \frac{2\pi x}{\lambda}\right) + \sin\left(2\pi ft - \frac{2\pi x}{\lambda}\right)\right] \qquad (168\text{-}3)$$

$$= 2\,E \sin 2\pi ft \cos \frac{2\pi x}{\lambda} \text{ volts}$$

(Sinusoidal distribution with time) (Sinusoidal distribution with distance)

The incident current wave that is in phase with the incident voltage, is presented by:

$$i_i = I \sin\left(2\pi ft + \frac{2\pi x}{\lambda}\right) \qquad (168\text{-}4)$$

$$= \frac{E}{Z_o} \sin\left(2\pi ft + \frac{2\pi x}{\lambda}\right) \text{ amperes}$$

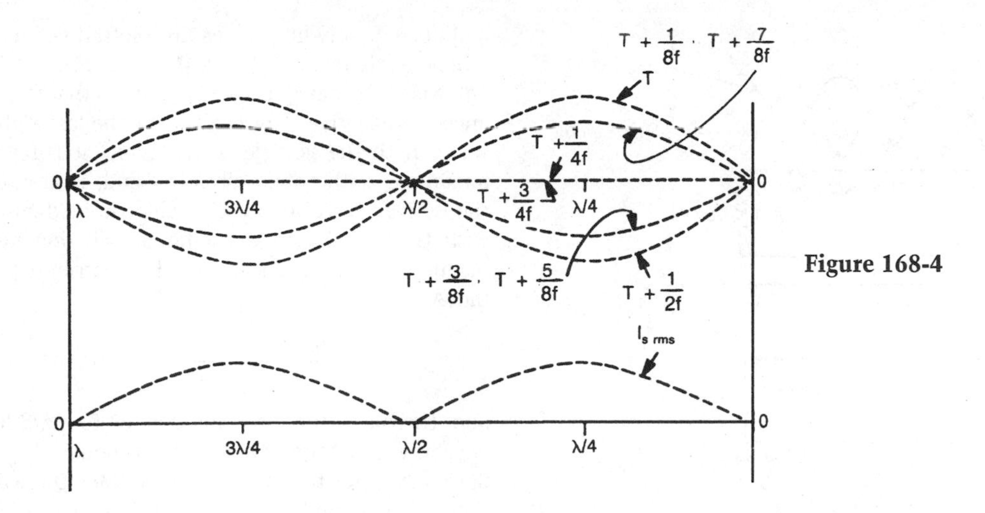

Figure 168-4

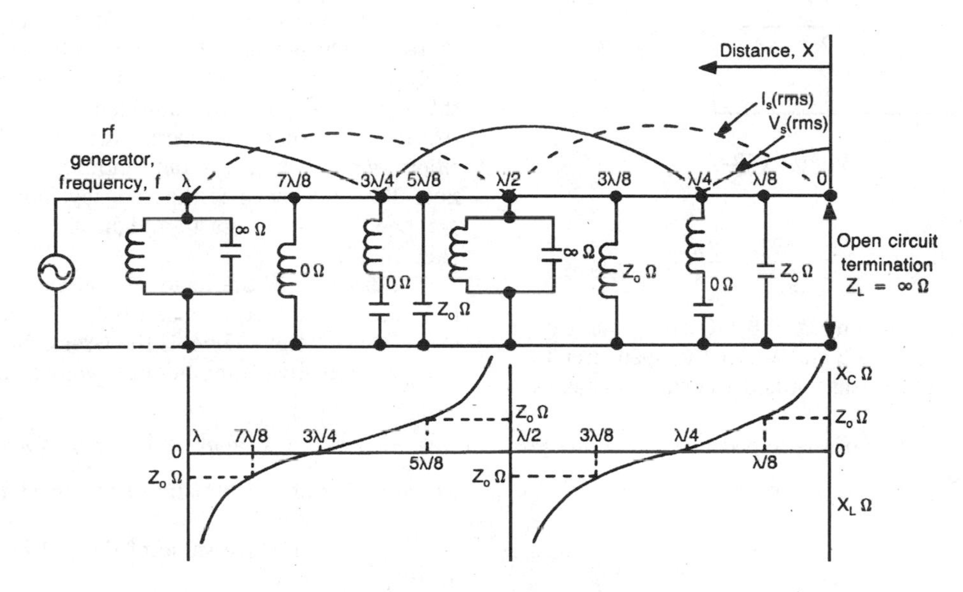

Figure 168-5

where: i_i = instantaneous value of the incident current wave (A)

I = peak value of the incident current wave (A)

The equation of the reflected current wave is:

$$i_r = -I \sin\left(2\pi ft - \frac{2\pi x}{\lambda}\right) \qquad (168\text{-}5)$$

$$= -\frac{E}{Z_o} \sin\left(2\pi ft - \frac{2\pi x}{\lambda}\right) \text{ amperes}$$

where the minus sign preceding I indicates the "out of phase" reflection, and i_r = instantaneous value of the reflected current wave (A).

The instantaneous standing wave current (i_s) is:

$$i_s = i_i + i_r \qquad (168\text{-}6)$$

$$= I\left[\sin\left(2\pi ft + \frac{2\pi x}{\lambda}\right) - \sin\left(2\pi ft - \frac{2\pi x}{\lambda}\right)\right]$$

$$= \frac{2E}{Z_o} \cos 2\pi ft \sin \frac{2\pi x}{\lambda} \text{ amperes}$$

Notice that, because of the "sin $2\pi ft$" and "cos $2\pi ft$" terms, v_s and i_s are 90° out of phase in terms of time. The rms value of the current standing wave (I_s) has a magnitude of,

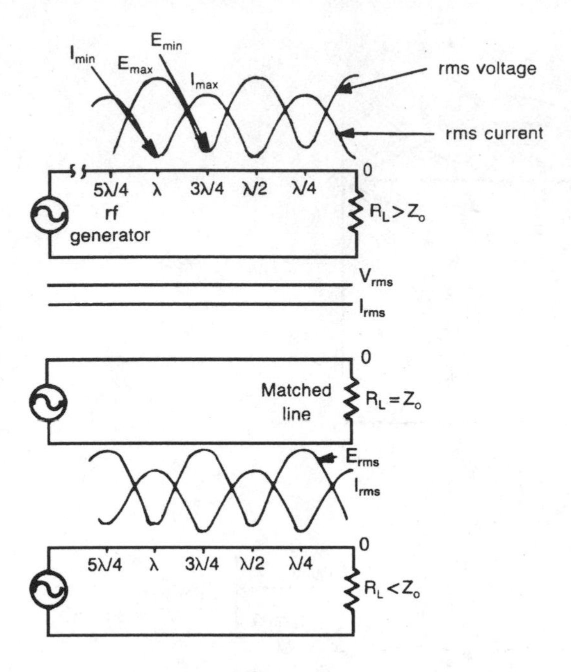

Figure 168-6

$$I_s = \frac{2E}{Z_o} \sin \frac{2\pi x}{\lambda},$$

which is illustrated in Fig. 168-4. Current nulls exist at distances of zero, λ/2 and λ from the open circuit termination, and current antinodes occur at distances of λ/4 and 3λ/4.

The magnitude of the impedance (Z) at any position is given by:

$$Z = \frac{v_s}{i_s} \qquad (168\text{-}7)$$

$$= \frac{2E \cos \frac{2\pi x}{\lambda}}{\frac{2E}{Z_o} \sin \frac{2\pi x}{\lambda}}$$

$$= Z_o \cot \frac{2\pi x}{\lambda} \text{ ohms}$$

Because v_s and i_s are 90° out of phase, the nature of this impedance must be reactive. But at what positions on the line is the impedance capacitive and at what positions is it inductive? Start by considering the impedance at the position, which is at the distance of λ/8 from the open circuit termination (Fig. 168-7). From this position, the values of the instantaneous voltage and current standing waves are plotted versus a time scale that contains: T, $T + 1/4f$, $T + 1/2f$, $T + 3/4f$, $T + 1/f$. It is clear that i_s leads v_s by 90° so that the impedance is capacitive. Physically, over the last eighth of a wavelength the voltage distribution, starting from a maximum at the open circuited end, is greater than the current distribution. The electric field associated with the distributed capacitance will dominate the magnetic field produced by the distributed inductance. If $x = \lambda/8$,

$$\cot \frac{2\pi x}{\lambda} = \cot \frac{\pi}{4} = \cot 45° = 1$$

and, therefore, the input impedance to a λ/8 line is a capacitive reactance whose value is equal to Z_o. A similar analysis for the open circuited 3λ/8 line will show that i_s lags v_s and that the input impedance is an inductive reactance of value Z_o. At the λ/4 position, there is a voltage null and a current antinode. Theoretically, the impedance is zero, but on a practical line, the impedance would be equivalent to a low value of resistance. The conditions are reversed for the λ/2 position, where there is a voltage antinode and a current null. The impedance is theoretically infinite, but in practice, it is equivalent to a high value of resistance (Fig. 168-5).

The fraction of the incident voltage and current reflected at the load is called the *reflection coefficient* whose letter symbol is the Greek rho (ρ). If $R_L > Z_o$, the effective standing-wave voltage at the load is,

$$V_L = V_i + V_r = V_i + \rho V_i = V_i(1 + \rho) \text{ volts} \qquad (168\text{-}8)$$

where: V_i = effective value of the incident voltage wave.
V_r = effective value of the reflected voltage wave.

The effective standing wave current through the load is,

$$I_L = I_i + I_r = I_i - \rho I_i = I_i(1 - \rho) \text{ amperes} \qquad (168\text{-}9)$$

The negative sign indicates the "out of phase" reflection of the current wave, compared with the "in phase" reflection of the voltage wave.

Then,

$$R_L = \frac{V_L}{I_L} = \frac{V_i(1 + \rho)}{I_i(1 - \rho)} \qquad (168\text{-}10)$$

$$= \frac{Z_o(1 + \rho)}{1 - \rho} \text{ ohms}$$

This yields,

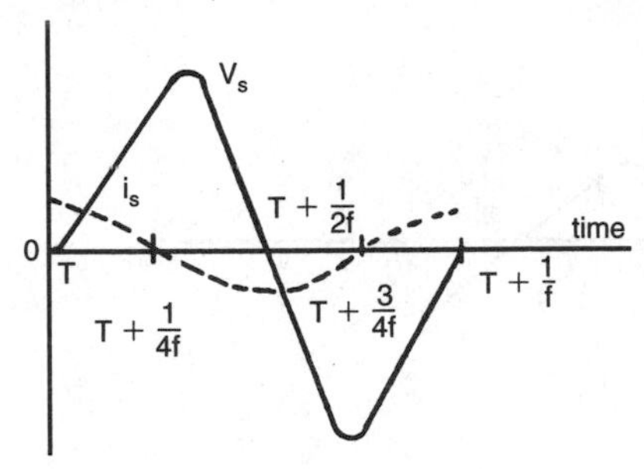

Distance of λ/16 from open circuited end. Impedance is a high value of capacitive reactance.

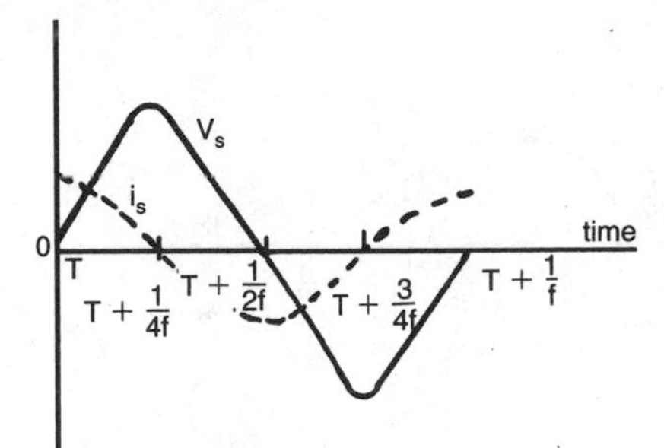

Distance of λ/8 from open circuited end. Impedance is a capacitive reactance whose value is equal to the lines surge impedance, Z_o.

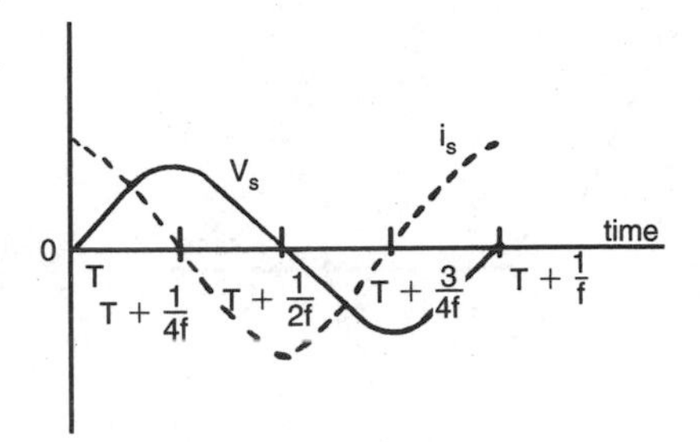

Distance of 3λ/16 from open circuited end. Impedance is a low value of capacitive reactance.

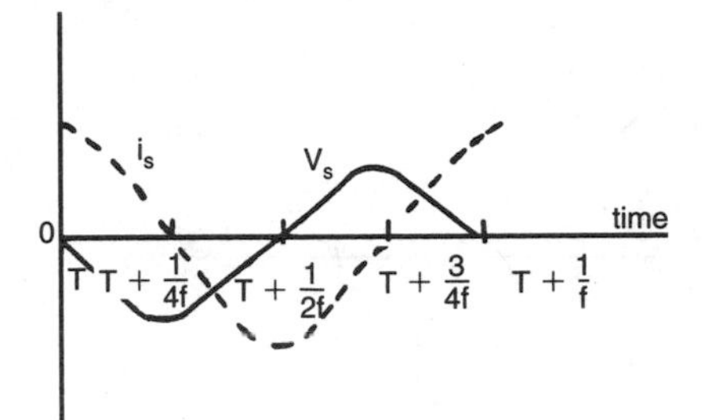

Distance of 5λ/16 from open circuited end. Impedance is a low value of inductive reactance.

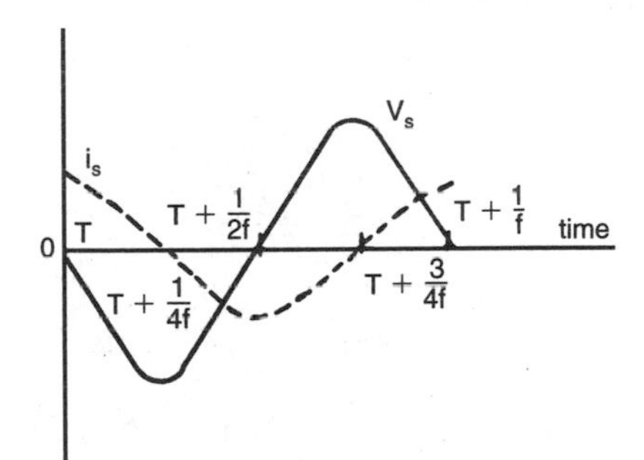

Distance of 3λ/8 from open circuited end. Impedance is an inductive reactance whose value is equal to the line's surge impedance, Z_o.

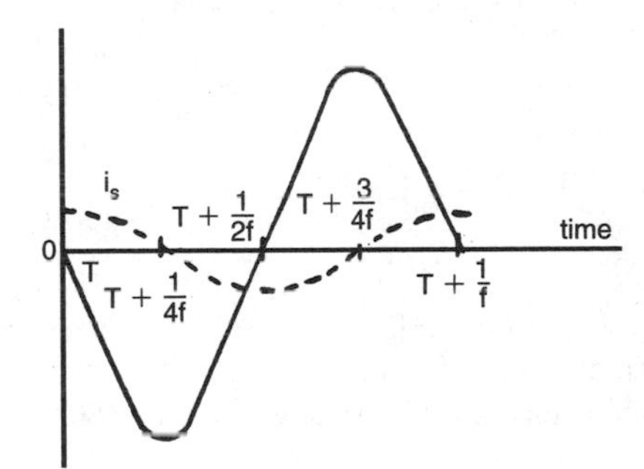

Distance of 7λ/16 from open circuited end. Impedance is a high value of inductive reactance.

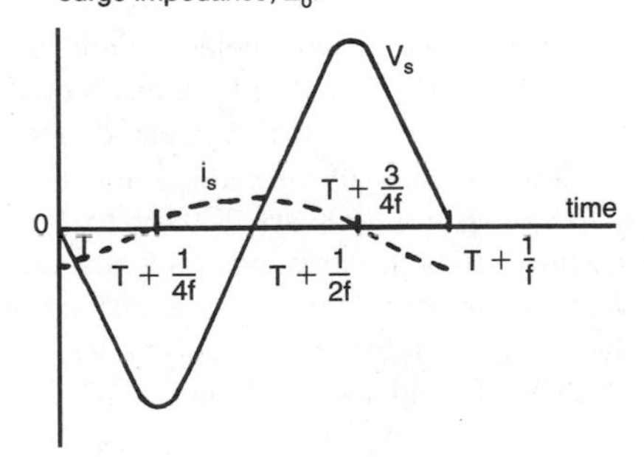

Distance of 9λ/16 from open circuited end. Impedance is a high value of capacitive reactance.

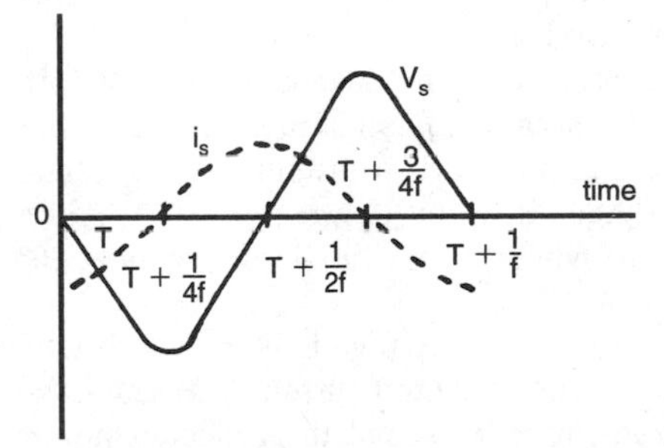

Distance of 5λ/8 from open circuited end. Impedance is a capacitive reactance whose value is equal to the line's surge impedance, Z_o.

Figure 168-7

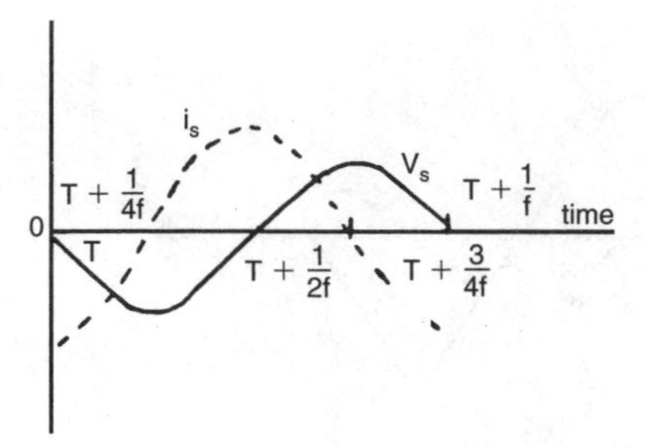

Distance of 11λ/16 from open circuited end. Impedance is a low value of capacitive reactance.

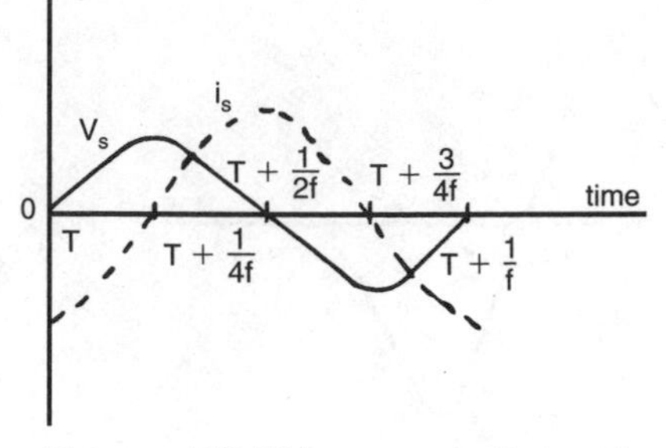

Distance of 13λ/16 from open circuited end. Impedance is a low value of inductive reactance.

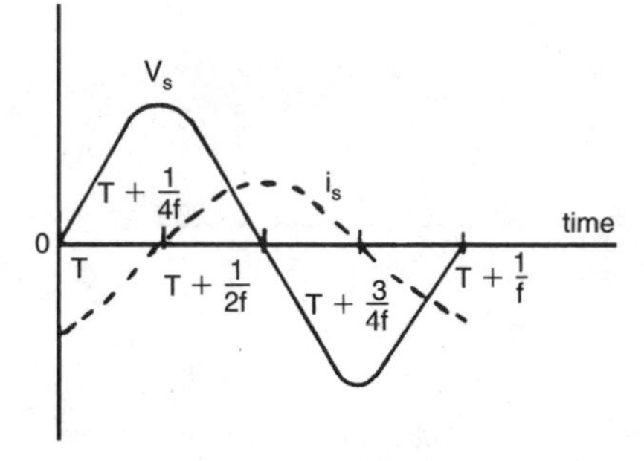

Distance of 7λ/8 from open circuited end. Impedance is an inductive reactance equal in value to the line's surge impedance, Z_o.

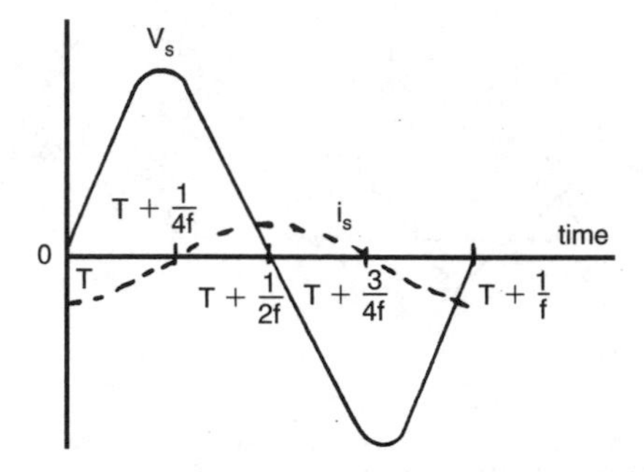

Distance of 15λ/16 from open circuited end. Impedance is a high value of inductive reactance.

Figure 168-7 continued.

$$\rho = \frac{R_L - Z_o}{R_L + Z_o} \qquad (168\text{-}11)$$

If: $R_L = \infty$ (open circuited line), $\rho = +1$
$R_L = Z_o$ (matched line), $\rho = 0$
$R_L = 0$ (short circuited line), $\rho = -1 = 1\angle 180°$

The negative sign indicates that when $R_L < Z_o$, the current wave is reflected "in phase," but the voltage reflection is "out of phase." These results are only true for purely resistive loads.

If the load is a general impedance possessing both resistance and reactance, ρ is not just a number, but is a complex quantity with a magnitude P (capital rho), which ranges in value between 0 and 1, and a phase angle (θ) whose value can lie between $+180°$ and $-180°$.

Because the reflected voltage is P times the incident voltage and the reflected current is P times the incident current, it follows that the reflected power (P_r) is P^2 times the incident power, P_i.

Then, the power absorbed by the load, P_L is:

$$P_L = P_i - P_r = (1 - P^2)\,P_i \qquad (168\text{-}12)$$

$$= \frac{1 - P^2}{P^2}\,P_r$$

$$= \frac{4S}{(1 + S)^2}\,P_i \text{ watts}$$

where: S = the voltage standing wave ratio (see chapter 169).

Example 168-1

A transmission line with an air dielectric and a surge impedance of 50 Ω is terminated by a load consisting of 40-Ω resistance in series with 65-Ω capacitive reactance. Calculate the value of the reflection coefficient. When an RF generator delivers 120 W to the line, which is assumed to be lossless, what are the amounts of the reflected power and the power absorbed by the load? If the wavelength on the line is 3 m, what is the distance between the adjacent voltage nulls?

Solution

Load impedance,

$$Z_L = 40 - j65\ \Omega.$$

Reflection coefficient,

$$\rho = \frac{Z_L - Z_o}{Z_L + Z_o} \qquad (168\text{-}11)$$

$$= \frac{(40 - j65) - 50}{50 + (40 - j65)}$$

$$= \frac{-10 - j65}{90 - j65}$$

$$= \frac{65.76\angle -98.75°}{111.01\angle -35.84°}$$

$$= 0.6\angle -63°$$

Magnitude of the reflection coefficient, P = 0.6

Reflected power,

$$P_r = P^2 \times P_i$$

$$= (0.6)^2 \times 120$$

$$= 43.2 \text{ W.}$$

Power absorbed by the load,

$$P_L = P_i - P_r$$

$$= 120 - 43.2$$

$$= 76.8 \text{ W.}$$

Distance between two adjacent voltage nulls

$$= \lambda/2$$

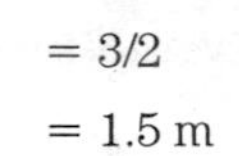

$$= 3/2$$

$$= 1.5 \text{ m}$$

PRACTICE PROBLEMS

168-1. An RF generator delivers 200 W to a line, which is assumed to be lossless. If the amount of reflected power is 15 W, calculate the values of the reflection coefficient's magnitude and the power absorbed by the load.

168-2. If the power absorbed by a load is 250 W and the reflected power is 32 W, calculate the value of the voltage reflection coefficient's magnitude.

168-3. A 150-Ω line is terminated by a load consisting of a 170-Ω resistance in series with a 135-Ω inductive reactance. What is the value of the reflection coefficient?

168-4. In Practice Problem 168-3, an RF generator delivers 270 W to the line, which is assumed to be lossless. What are the values of the reflected power and the power absorbed by the load?

168-5. In Practice Problem 168-3, a voltage antinode and an adjacent voltage null are separated by 0.3 m. Determine the value of the wavelength on the line.

169 Voltage standing-wave ratio

Referring to Fig. 169-1, the effective value of a voltage antinode is E_{max}, and the effective value of an adjacent voltage node is E_{min}. The degree of standing waves is measured by the voltage standing wave ratio (VSWR), whose letter symbol is S. The VSWR is defined as the ratio $E_{max} : E_{min}$, which is also equal in magnitude to $I_{max} : I_{min}$. E_{max} is the result of an in-phase condition between the incident and reflected voltages, and E_{min} is the result of a 180° out-of-phase situation.

The presence of standing waves on a practical transmission line has the following disadvantages:

1. The incident power reaching the termination is not fully absorbed by the load. The difference between the incident power and the load power is the power reflected back towards the generator.
2. The voltage antinodes can break down the insulation (dielectric) between the conductors and cause arc-over.
3. Because power losses are proportional to the square of the voltage and the square of the current, the attenuation of the practical line increases due to the presence of standing waves.
4. The input impedance at the generator is a totally unknown quantity that varies with the length of the line and the frequency of the generator.

Note that VSWR can be abbreviated to SWR (standing wave ratio).

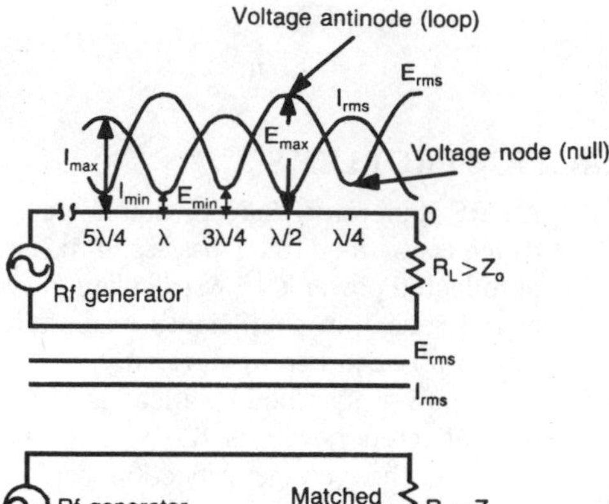

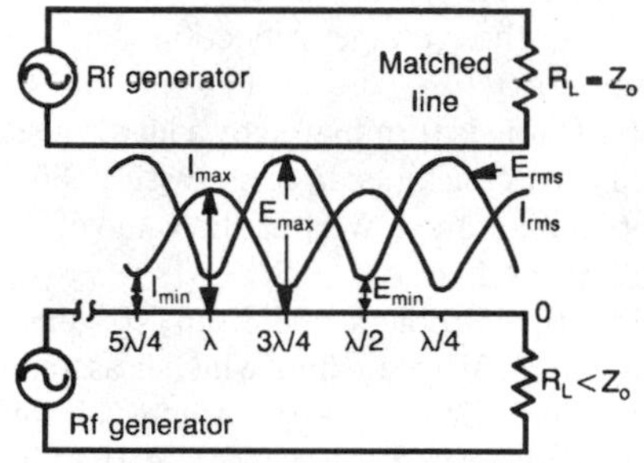

Figure 169-1

MATHEMATICAL DERIVATIONS

Maximum effective voltage,

$$E_{max} = V_i + V_r = V_i(1 + \text{P}) \text{ volts} \quad (169\text{-}1)$$

Minimum effective voltage,

$$E_{min} = V_i - V_r = V_i(1 - \text{P}) \text{ volts} \quad (169\text{-}2)$$

Then,

VSWR,

$$S = \frac{E_{max}}{E_{min}} = \frac{1 + \text{P}}{1 - \text{P}} \quad (169\text{-}3)$$

If $\text{P} = 0$ (matched line), $S = 1$

If $\text{P} = 1$ (open or short-circuited line), $S = \infty$

The VSWR is just a number whose value ranges from 1 to ∞. On a practical matched system, a value of S, which is less than 1.2, is normally regarded as acceptable. It is sometimes preferable to measure the VSWR in decibels, in which case VSWR = 20 log S dB.

Equation 169-3 yields,

$$\text{P} = \frac{S - 1}{S + 1} \quad (169\text{-}4)$$

If a resistive load (R_L) is greater than Z_o, the magnitude of the reflection coefficient is given by:

$$\text{P} = \frac{R_L - Z_o}{R_L + Z_o} \quad (169\text{-}5)$$

Combining Equations 169-4 and 169-5,

$$S = \frac{1 + \text{P}}{1 - \text{P}} = \frac{R_L}{Z_o} \quad (169\text{-}6)$$

If R_L is less than Z_o,

$$\text{P} = \frac{Z_o - R_L}{Z_0 + R_L}$$

and,

$$S = \frac{Z_o}{R_L} \quad (169\text{-}7)$$

At the position where the voltage antinodes and the current nodes coincide, the impedance of the line is a maximum and is resistive.

Maximum impedance,

$$Z_{max} = \frac{E_{max}}{I_{min}} = \frac{E_i(1 + \text{P})}{I_i(1 - \text{P})} \quad (169\text{-}8)$$

$$= Z_o \frac{(1 + \text{P})}{(1 - \text{P})} = SZ_0 \text{ ohms}$$

At the voltage node position, there is a minimum resistive impedance which is given by:

$$Z_{min} = \frac{E_{min}}{I_{max}} = \frac{E_i(1 - \text{P})}{I_i(1 + \text{P})} \quad (169\text{-}9)$$

$$= Z_o \times \frac{(1 - \text{P})}{(1 + \text{P})} = \frac{Z_o}{S} \text{ ohms.}$$

Example 169-1

An RF generator delivers 100 W to a 50-Ω loss-free line. If the line is terminated by a 40-Ω resistive load, what are the values of the standing wave ratio, the reflection coefficient, the reflected power, and the load power?

Solution

Voltage standing wave ratio,

$$S = \frac{Z_o}{R_L} \quad (169\text{-}7)$$

$$= \frac{50}{40}$$

$$= 1.25$$

Reflection coefficient,

$$\text{P} = \frac{S - 1}{S + 1} \quad (169\text{-}4)$$

$$= \frac{1.25 - 1}{1.25 + 1}$$

$$= 0.11$$

Reflected power,

$$P_r = P^2 \times P_i$$

$$= (0.11)^2 \times 100$$

$$= 1.2 \text{ W}$$

Load power,

$$P_L = P_i - P_r$$

$$= 100 - 1.2$$

$$= 98.8 \text{ W}$$

PRACTICE PROBLEMS

169-1. On a 150-Ω lossless transmission line, the SWR value is found to be 2.5. Calculate the value of the reflection coefficient's magnitude. What are the values of the maximum and minimum impedances on the line?

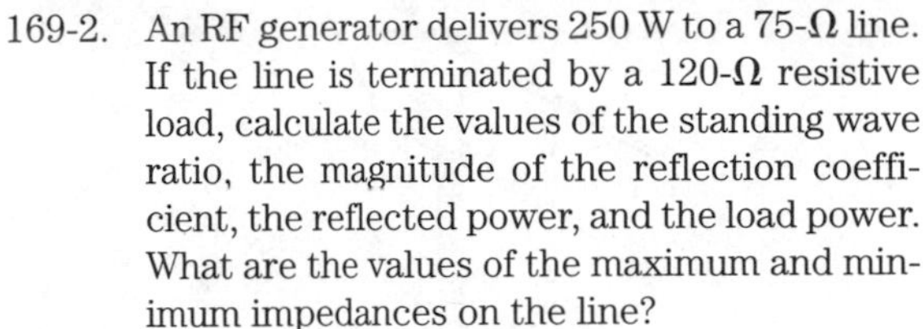

169-2. An RF generator delivers 250 W to a 75-Ω line. If the line is terminated by a 120-Ω resistive load, calculate the values of the standing wave ratio, the magnitude of the reflection coefficient, the reflected power, and the load power. What are the values of the maximum and minimum impedances on the line?

169-3. On a lossless mismatched line, a voltage antinode has a value of 260 V and the adjacent null is 180 V. The positions of the antinode and the null are separated by 0.2 m. What are the values of the SWR and the wavelength?

169-4. On a lossless mismatched line, the maximum impedance is 350 Ω, and the minimum impedance is 200 Ω. Calculate the values of the SWR, the reflection coefficient's magnitude, and the surge impedance.

169-5. A lossless 100-Ω transmission line is terminated by a load that consists of a 150-Ω resistance and an 80-Ω inductive reactance. Determine the value of the SWR.

170
The quarter-wave line

When a quarter-wave line is shortened at one end and is excited to resonance by the source frequency applied at the other end, standing waves of voltage and current appear on the line. At the short circuit, the voltage is zero and the current is a maximum. At the input end, the current is nearly zero and the voltage is at its peak. The input impedance to the resonant quarter-wave line is extremely high; therefore, this section behaves as an insulator. Figure 170-1A shows a quarter-wave section of line acting as a standoff insulator (support stub) for a two-wire transmission line. At the particular frequency that makes the section a quarter-wavelength line, the stub acts as a highly efficient insulator. However, at higher frequencies the stub is no longer resonant and will behave as a capacitor; such behavior causes a mismatch and standing waves will appear on the two-wire line. At the second harmonic frequency, the stub is λ/2 long and, because there is no impedance transformation over the distance of one-half wavelength, the stub will place a short circuit across the main line. No power at the second harmonic frequency can then be transferred down the line. In this way, a section of line that is λ/4 long at a fundamental frequency and is shorted at one end, will have the effect of filtering out all even harmonics.

The quarter-wave section can also be used to match two nonresonant lines with different surge impedances (see mathematical derivations).

MATHEMATICAL DERIVATIONS

When a λ/4 line of surge impedance (Z_o) is terminated by a load (Z_L) the input impedance (Z_{in}) is given by,

$$Z_{in} = \frac{Z_o^2}{Z_L} \text{ ohms} \qquad (170\text{-}1)$$

This yields,

$$Z_o = \sqrt{Z_{in} \times Z_L} \text{ ohms} \qquad (170\text{-}2)$$

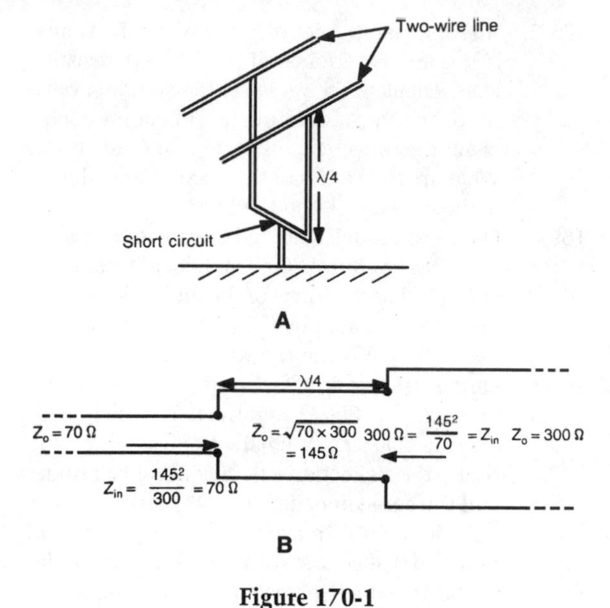

Figure 170-1

This formula can be used to derive some results that have already been obtained in previous chapters. For example,

1. If $Z_L = \infty\ \Omega$ (an open circuit), $Z_{in} = 0\ \Omega$ (a short circuit).
2. If $Z_L = Z_o\ \Omega$ (matched conditions), $Z_{in} = Z_o\ \Omega$.
3. If $Z_L = 0\ \Omega$ (a short circuit), $Z_{in} = \infty\ \Omega$ (an open circuit).

Example 170-1

In Fig. 170-1B, what is the required value for the surge impedance of the λ/4 matching section?

Solution

It is required to match 70-Ω and 300-Ω nonresonant lines. The λ/4 matching section has a surge impedance given by:

Surge impedance,

$$Z_o = \sqrt{Z_{in} \times Z_L} \tag{170-2}$$

$$= \sqrt{70 \times 300}$$

$$= 145\ \Omega.$$

The effective load on the 300-Ω line (as seen from the generator) is $145^2/70 = 300\ \Omega$, while the 70 Ω line is effectively connected to a $145^2/300 = 70$-Ω termination. Both nonresonant lines are matched and the standing waves only exist on the λ/4 section.

PRACTICE PROBLEMS

170-1. A λ/4 matching section with a surge impedance of 86 Ω is connected at one end to an effective resistive load of 210 Ω. Determine the input impedance at the other end of the section.

170-2. A support stub is λ/4 long at a frequency of 10 MHz. What impedance will the stub present to the line at frequencies of (a) 20 MHz, (b) 5 MHz, and (c) 30 MHz?

170-3. The 3λ/4 section of a transmission line is terminated by a short circuit. When a generator is added to the other end of the line, what is the nature of the input impedance at the generator?

170-4. In Practice Problem 170-3, describe the "lumped" circuit to which the 3λ/4 section is equivalent.

171
The Hertz antenna

An antenna is defined as an efficient radiator of electromagnetic energy (radio waves) into free space. The same principles apply to both transmitting and receiving antennas, although the RF power levels for the two antennas are completely different. The purpose of a transmitting antenna is to radiate as much RF power as possible either in all directions (omnidirectional antenna) or in a specified direction (directional antenna). By contrast, the receiving antenna is used to intercept an RF signal voltage that is sufficiently large compared to the noise existing within the receiver's bandwidth.

Referring to transmission lines, the main disadvantage of the twin-lead is its radiation loss. However, provided that the separation between the leads is short compared with the wavelength, the radiation loss is small because the fields associated with the two conductors will tend to cancel out. Referring to Fig. 171-1A, the equal currents (i), which exist over the last resonant quarter wavelength of an open-circuited transmission line, will instantaneously flow in opposite directions so that their resultant magnetic field is weak. However, if each $\lambda/4$ conductor is twisted back through 90° (Fig. 171-1B), the currents (i) are now instantaneously in the same direction so that the surrounding magnetic field is strong.

In addition to the currents, the conductors carry standing-wave voltage distributions with their associated electric fields. The standing-wave voltage and current distributions over the complete half wavelength are shown in Fig. 171-1C. At the center feedpoint, the effective voltage is at its minimum level and the effective current value has its maximum value. The effective voltage distribution is drawn on opposite sides of the two sections to indicate that these sections instantaneously carry opposite polarities. In other words, when a particular point in the top section carries a positive voltage (with respect to ground), the corresponding point in the bottom section has an equal negative voltage. The distributed inductance and capacitance (Fig. 171-1D) together form the equivalent of a series resonant LC circuit. Notice, however, that the distributed capacitance exists between the two quarter-wave sections and that the ground is not involved in this distribution.

By bending the two $\lambda/4$ sections outward by 90°, you have formed the half-wave ($\lambda/2$) dipole, also known as the *Hertz* (Heinrich Hertz, 1857–1894) *antenna*. This antenna is resonant at the frequency for which it is cut. For example, at a frequency of 100 MHz, which lies within the FM commercial broadcast band of 88 to 108 MHz, the length required for the Hertz antenna is 1.43 m. Such an antenna would be made from two thin conducting rods, each 0.716 m long, and positioned remote from ground. At the frequency of 100 MHz, the antenna behaves as a series resonant circuit, with a Q of approximately 10. Therefore, the thin dipole is capable of operating effectively within a narrow range that is centered on the resonant frequency. However, you must remember that if the operating frequency is below the resonant frequency, the antenna will appear to be too short and will behave capacitively. Likewise at frequencies above resonance, the antenna will be too long, and will be inductive.

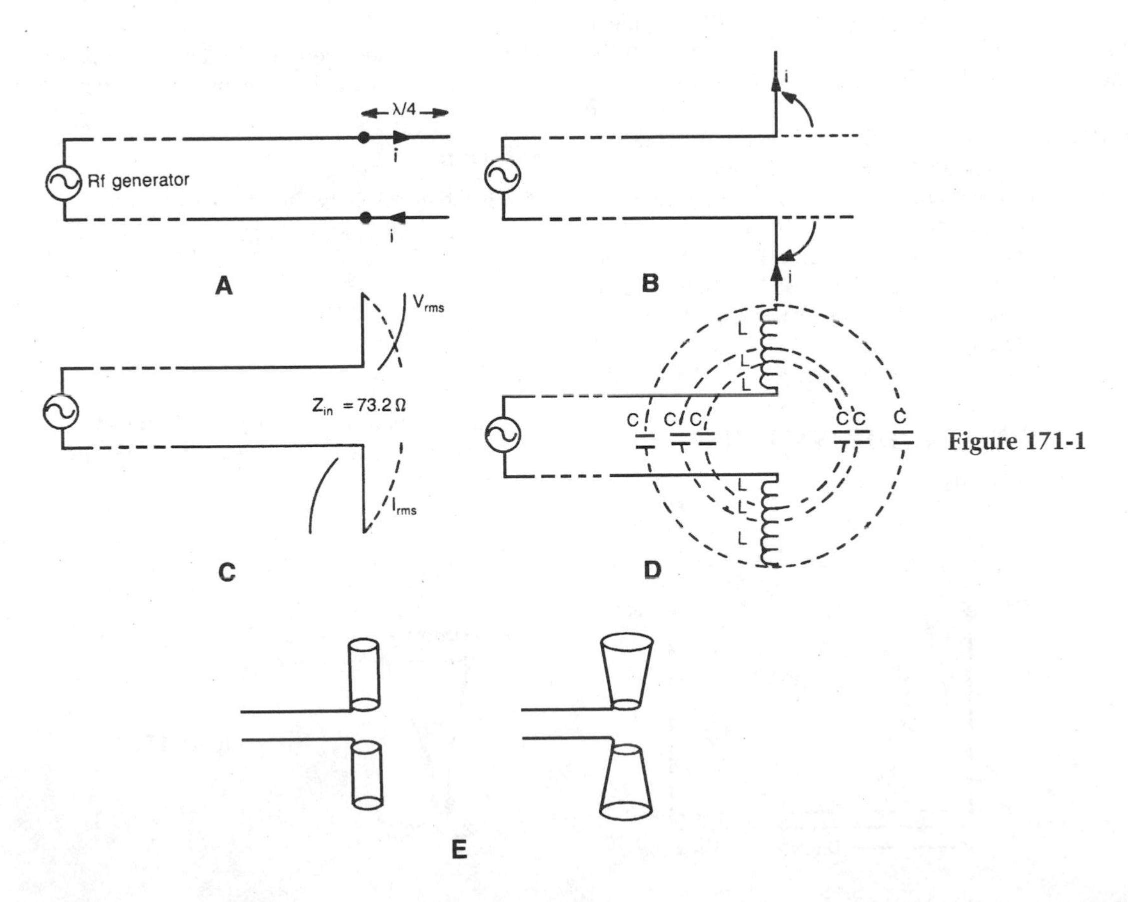

Figure 171-1

If it is required to operate a dipole over a wide range of frequencies, it is necessary to "broad band" the antenna by lowering its Q without changing its resonant frequency. As an example, if you want to operate a dipole satisfactorily over the range of 125 to 175 MHz, the antenna should be cut to the mid-frequency of 150 MHz and it should possess a Q of 150/(175 − 125) = 3. Because the resonant frequency of a series LC circuit is given by,

$$f_r = \frac{1}{2\pi\sqrt{LC}}$$

while:

$$Q = \frac{1}{R}\sqrt{\frac{L}{C}}$$

you must lower the distributed inductance (L) and increase the distributed capacitance (C). The solution is to shorten the antenna, while at the same time increase its surface area. This gives rise to such "broad band" shapes as the cylindrical and biconical dipoles (Fig. 171-1E).

When the Hertz dipole is resonant and the RF power is applied to the center of the antenna, the input impedance at the feedpoint is a low resistance that, mathematically, can be shown to have a value of 73.2 Ω (for this reason, you will often hear of a "70-Ω dipole"). This is referred to as the "radiation resistance" of the dipole and it is the ohmic load that the half-wave antenna represents at resonance. To achieve a matched condition, the dipole should be fed with a 70-Ω line.

As you move from the center of the antenna toward the ends, the impedance increases from approximately 35 Ω (balanced either side with respect to ground) to about 2500 Ω (not infinity because of the "end" capacitance effects). It is therefore possible to select points on the antenna, where the impedance can be matched by a gradual taper to a line whose Z_o is not 70 Ω. Such an arrangement is known as a *delta feed* (Fig. 171-2).

MATHEMATICAL DERIVATIONS

Electrical wavelength,

$$\lambda = \frac{300}{f} \text{ meters} = \frac{984}{f} \text{ feet} \qquad (171\text{-}1)$$

Electrical half-wavelength,

$$\frac{\lambda}{2} = \frac{150}{f} \text{ meters} = \frac{492}{f} \text{ feet} \qquad (171\text{-}2)$$

where: f = frequency (MHz).

The voltage and current waves on an antenna travel at a speed that is typically 5% slower than the velocity of light. Therefore, the *physical* half wavelength to which the Hertz antenna should be cut, is shorter than the *electrical* half wavelength.

Physical half-wavelength,

$$l = \frac{143}{f(\text{MHz})} \text{ meters} = \frac{468}{f(\text{MHz})} \text{ feet} \qquad (171\text{-}3)$$

The RF power (P) at the feedpoint is given by:

$$P = I_A^2 \times R_A \text{ watts} \qquad (171\text{-}4)$$

where: I_A = effective RF current at the center of the antenna (A).
R_A = radiation resistance (Ω).

Example 171-1

What is the physical length of a dipole that is resonant at a frequency of 3 GHz in the microwave region?

Solution

Physical half-wavelength,

$$l = \frac{143}{3000 \text{ MHz}} \text{ m} \qquad (171\text{-}3)$$

$$= 4.67 \text{ cm.}$$

Example 171-2

The current at the frequency of a resonant dipole is 3.2 A. What is the amount of RF power delivered to the antenna?

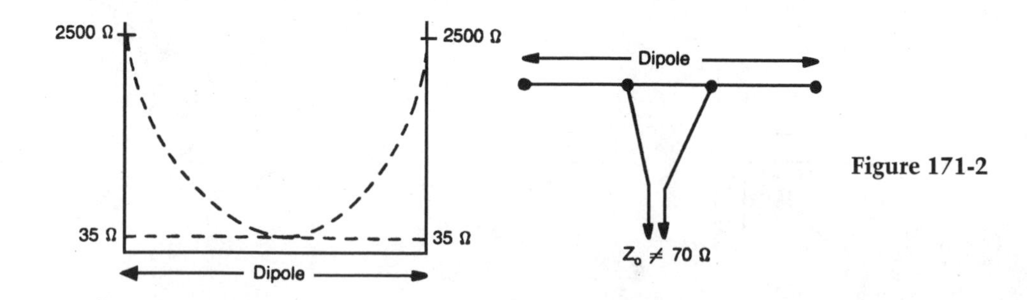

Figure 171-2

Solution

Antenna power,

$$P = I_A^{\,2} \times R_A \qquad (171\text{-}4)$$

$$= (3.2 \text{ A})^2 \times 73.2\ \Omega$$

$$= 750 \text{ W}.$$

PRACTICE PROBLEMS

171-1. What is the physical length in feet of a Hertz dipole that is designed to resonate at 145 MHz?

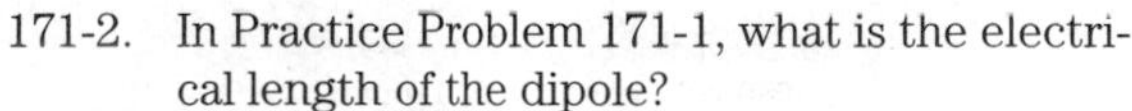

171-2. In Practice Problem 171-1, what is the electrical length of the dipole?

171-3. A resonant Hertz dipole is center fed by a 100-Ω twin line. What is the value of the SWR on the line, which is assumed to be lossless? If the line is to be matched to the load of the dipole, determine the value of the surge impedance for the $\lambda/4$ matching section (chapter 170).

171-4. A thin-rod Hertz dipole is cut to be resonant at 156 MHz. If the transmitter that feeds the dipole, is operating at 130 MHz, describe the nature of the antenna load.

172
Parasitic elements

A vertical Hertz dipole has a circular (omnidirectional) radiation pattern in the horizontal plane. The directional properties can be modified by adding parasitic elements to the antenna system. These parasitic elements are metallic structures (such as rods) that are not electrically connected (nondriven), and are placed in the vicinity of the driven dipole. The radiated field from the driven dipole induces a voltage and current distribution in the parasitic element, which then reradiates the signal. At any position in space, the total field strength will be the phasor resultant of the field radiated from the dipole and the reradiated field from the parasitic element.

As a simple example, consider a vertical $\lambda/2$ (physical half-wavelength) dipole that is separated by a quarter-wavelength from a parasitic element (rod) that is also $\lambda/2$ long (Fig. 172-1A). Because of this separation, the field (ε_{DP}) reaching the parasitic element from the dipole will lag the dipole's field (ε_D) by 90°. Because the parasitic element is resonant, it behaves resistively, and the current (I_P) will be in phase with ε_{DP} (Fig. 172-1B). The reradiated field (ε_P) lags I_P by 90° and is, therefore, 180° out of phase with ε_D. Again, because of the $\lambda/4$ separation, the reradiated field reaching the dipole (ε_{PD}) lags ε_P by 90°. In direction X, the resultant field (ε_X) is the phasor combination of ε_D and ε_{PD}; the magnitude of ε_X is the same as that of ε_Y, which is the resultant of ε_{DP} and ε_P in the direction of Y. However, in the two directions, which are perpendicular to the plane of the paper and passing through the point (O), ε_D and ε_P are 180° out of phase and the two fields will tend to cancel. The result is to change the shape of the horizontal radiation pattern from a circle (omnidirectional) into two "ovals" or lobes (bidirectional) such that the area of the two ovals is equal to the area of the original circle (Fig. 172-1C). In fact, this treatment is oversimplified because it ignores the reduction in the field strength as a result of the separation between the dipole and parasitic element. This is taken into account in the field pattern.

Let the parasitic element now be increased in length so that it is more than $\lambda/2$ long. Therefore, it behaves inductively (Fig. 172-2A). The parasitic current (I_P) lags ε_{DP} so that the various phase relationships are modified, as in Fig. 172-2B. The magnitude of ε_X is now greater than ε_Y so that the major radiation lobe is pointed toward the X direction, while the minor lobe is in the Y direction (Fig. 172-2C). The parasitic element is, therefore, behaving as a *reflector*, because its effect is to increase the radiation in the forward direction (from the parasitic rod toward the dipole), at the expense of the radiation in the backward direction. By comparing the field strengths along the axes of the two lobes, you can calculate the "front-to-back" ratio as defined by:

$$\textit{Front-to-back ratio} = \frac{\textit{forward field strength}}{\textit{backward field strength}}$$

For a single reflector rod, the front-to-back ratio can be adjusted up to a value of 3:1. This value takes into account the reduction in field strength between the dipole and the parasitic elements, as well as the separa-

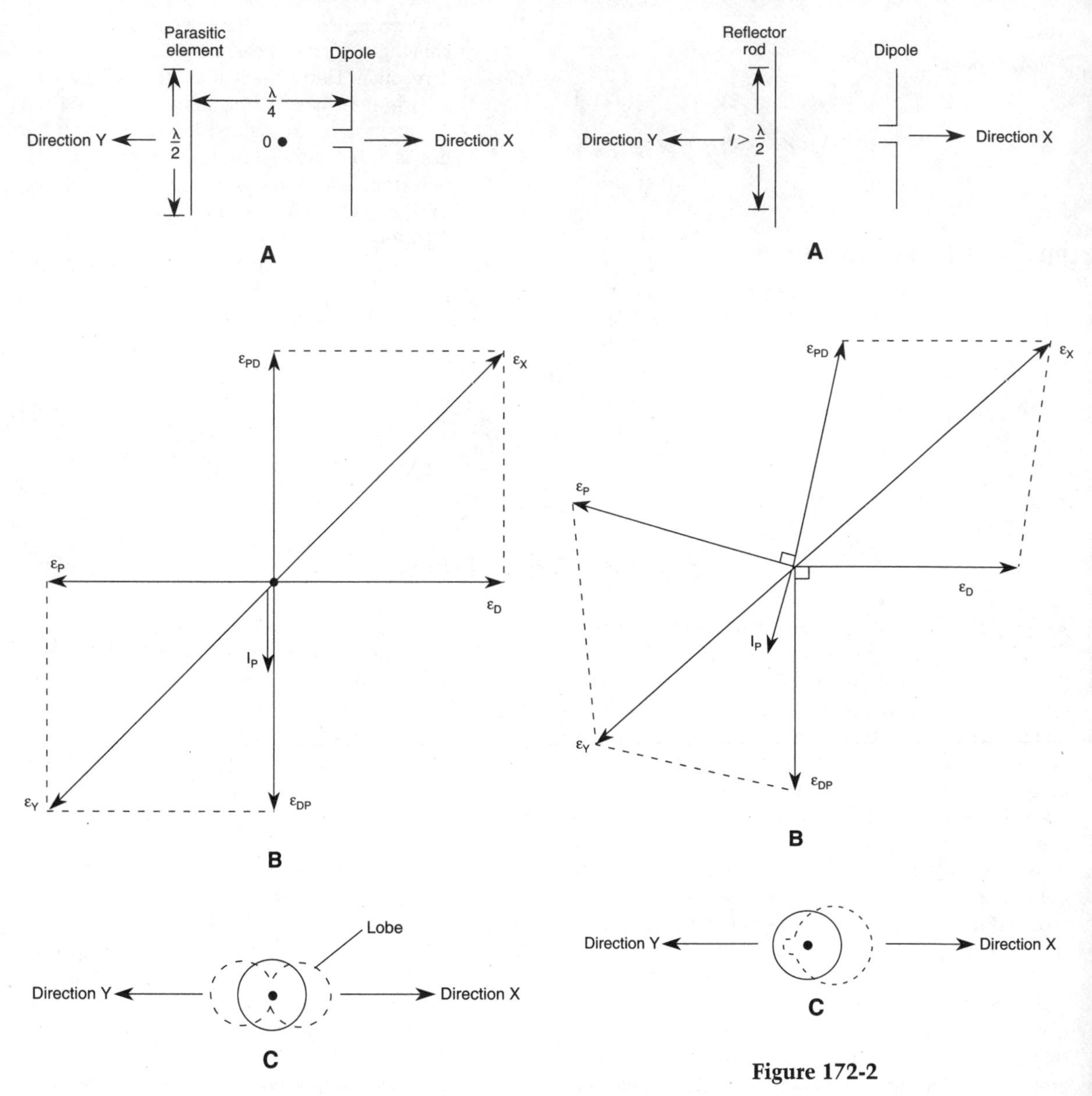

Figure 172-1

Figure 172-2

tion distance and the length of the reflector rod, which are typically optimized to 0.15λ and 0.55λ, respectively.

If the length of the parasitic rod is reduced to less than half of the physical wavelength (Fig. 172-3A), the rod behaves capacitively (I_P leads $\boldsymbol{\varepsilon}_{DP}$) and $\boldsymbol{\varepsilon}_X$ is greater than $\boldsymbol{\varepsilon}_Y$ (Fig. 172-3B). The parasitic element now behaves as a *director* because it "directs" the radiation toward the major lobe (Fig. 172-3C). Typically, the length of the director rod is 0.45 of the physical wavelength and its separation from the driven dipole is 0.1λ.

A combination of a driven dipole, a reflector rod, and a director rod forms a *Yagi array* (Fig. 172-4A), which has a power gain of 5 to 7 dB. The power gain can be increased further by adding more director rods. With the major lobe increased at the expense of the minor lobe, the Yagi system is a unidirectional antenna.

It is common practice to use a folded dipole (Fig. 172-4B), which has a lower Q than a thin rod and can, therefore, be operated effectively over a wider range of frequencies. The input resistance of a resonant folded dipole is approximately 300 Ω and, therefore, it can be conveniently attached to a 300-Ω twin line.

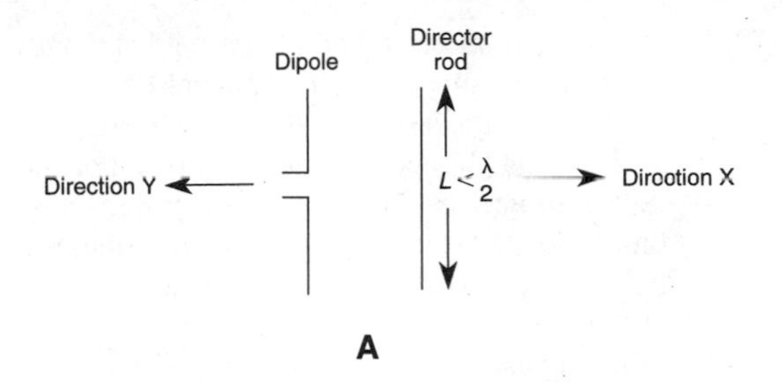

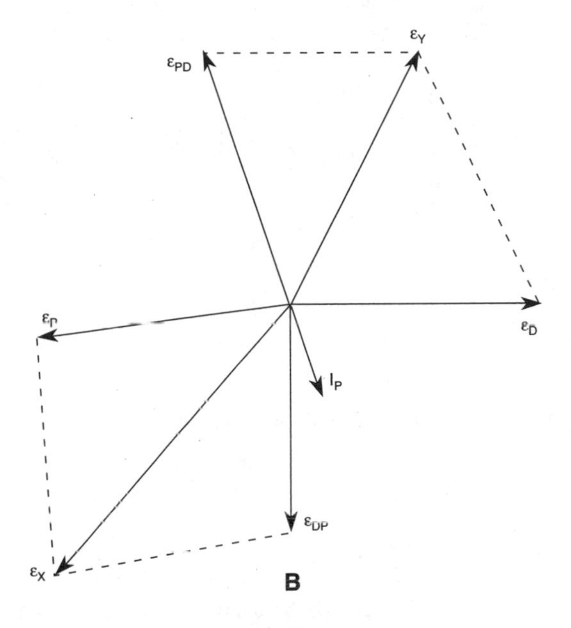

Figure 172-3

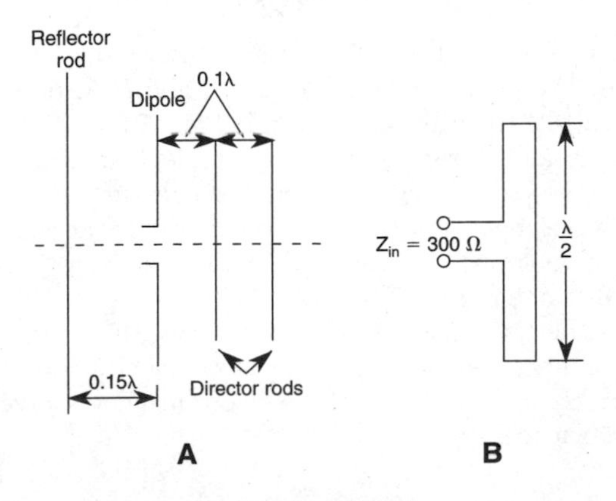

Figure 172-4

MATHEMATICAL DERIVATIONS

The Yagi antenna

Length of dipole (172-1)

$$= 0.5 \times \textit{physical wavelength}$$

Length of reflector rod (172-2)

$$= 0.55 \times \textit{physical wavelength}$$

$$= 1.1 \times \textit{length of dipole}$$

Separation between the dipole and reflector rod (172-3)

$$= 0.15 \times \textit{physical wavelength}$$

$$= 0.3 \times \textit{length of dipole}$$

Length of director rod (172-4)

$$= 0.45 \times \textit{physical wavelength}$$

$$= 0.9 \times \textit{length of dipole}$$

Separation between dipole and director rod (172-5)

$$= 0.1 \times \textit{physical wavelength}$$

$$= 0.2 \times \textit{length of dipole}$$

"Front-to-back" ratio (172-6)

$$= \frac{\textit{forward field strength}}{\textit{backward field strength}}$$

$$\textit{Antenna power gain} = 5 \text{ to } 7 \text{ dB} \quad (172\text{-}7)$$

Example 172-1

Design a Yagi antenna that is tuned to operate at a frequency of 200 MHz. Assuming that the input power to the antenna is 50 W and the antenna power gain is 6 dB, determine the value of the effective radiated power.

Solution

Length of dipole,

$$l = \frac{468}{200} = 2.34 \text{ feet} \quad (172\text{-}1)$$

Length of reflector rod (172-2)

$$= 1.1 \times 2.34 = 2.57 \text{ feet}$$

Separation between dipole and reflector rod (172-3)

$$= 0.3 \times 2.34 = 0.70 \text{ feet}$$

Length of director rod (172-4)

$$= 0.9 \times 2.34 = 2.11 \text{ feet}$$

Separation between dipole and director rod (172-5)

$= 0.2 \times 2.34 = 0.47$ feet

Effected radiated power $= 50$ inv $\log(6/10)$

≈ 200 W.

PRACTICE PROBLEMS

172-1. A thin-rod dipole has a length of 0.84 m. Determine the required lengths of the reflector rod and a single director rod. At which frequency is this antenna array designed to operate?

172-2. Measurements are taken of the field strength pattern that surrounds the transmitting array of Practice Problem 172-1. At a certain distance along the axis of the forward lobe, the field strength is 350 μV/m. At the same distance along the axis of the backward lobe, the field strength is 70 μV/m. Determine the value of the "front-to-back" ratio.

172-3. When a basic transmitting dipole is used, the field strength at a certain position is 325 μV/m. After the addition of reflector and director rods, the field strength at the same position (which lies on the axis of the forward lobe) increases to 975 μV/m. Calculate the power gain in dB of the antenna array.

173 Electromagnetic (EM) waves in free space

Because of the voltage and current distribution, there are electric and magnetic fields in the vicinity of an antenna (Fig. 173-1A). Because the instantaneous voltage and current on the antenna are 90° out of phase, the same phase relationship applies to the E and H fields (Fig. 173-1B). These fields are continuously expanding out from the antenna, and collapsing back with the speed of light. However, because the action is not instantaneous, the collapse will only be partial so that the closed electric and magnetic loops will be left in space (Fig. 173-1C). These loops represent the *radiated* electromagnetic energy that is propagated into free space, and they travel with the velocity of light. Those flux lines that collapse back into the antenna represent the *induction field*, which is strong only in the immediate vicinity of the antenna.

The fields in space around a half-wave antenna are shown in Fig. 173-2. The radiated E and H fields of the electromagnetic (EM) wave are in time phase, but the two sets of flux lines are 90° apart in space. A vertical Hertz antenna radiates an electric field with vertical flux lines and a magnetic field with horizontal flux lines. In addition, the two sets of flux lines are each at right angles to the direction in which the electromagnetic energy is being propagated. Such a wave is of the *TEM* type (transverse electric, transverse magnetic), which occurs only in free space and on a conventional transmission line (but not in a waveguide).

The plane that contains the E field and the direction of propagation is referred to as the *plane of polarization.* A vertical Hertz antenna will radiate a vertically polarized wave with a vertical E field and a horizontal H field. Similarly, a horizontally polarized wave has a horizontal E field, a vertical H field, and is associated with a horizontal antenna. As practical examples, AM broadcast stations have antenna systems that radiate vertically polarized waves while TV broadcast stations (and many microwave systems) use horizontal polarization.

MATHEMATICAL DERIVATIONS

Figure 173-3 illustrates the various features of a TEM wave. The peak conditions of the E and H fields in the path of the wave are shown in Fig. 173-3A. Positions X and Z represent adjacent identical conditions; therefore, they are separated by a distance of one wavelength. For the instantaneous fields at position X to travel to position Z would require a time equal to one period (Fig. 173-3C). It follows that the wave travels a distance of one wavelength (λ) in a time equal to one period, T. Because *velocity = distance/time*, the velocity of the electromagnetic wave in free space,

$$c = \frac{\lambda}{T} = \lambda \times \frac{1}{T} = f\lambda$$

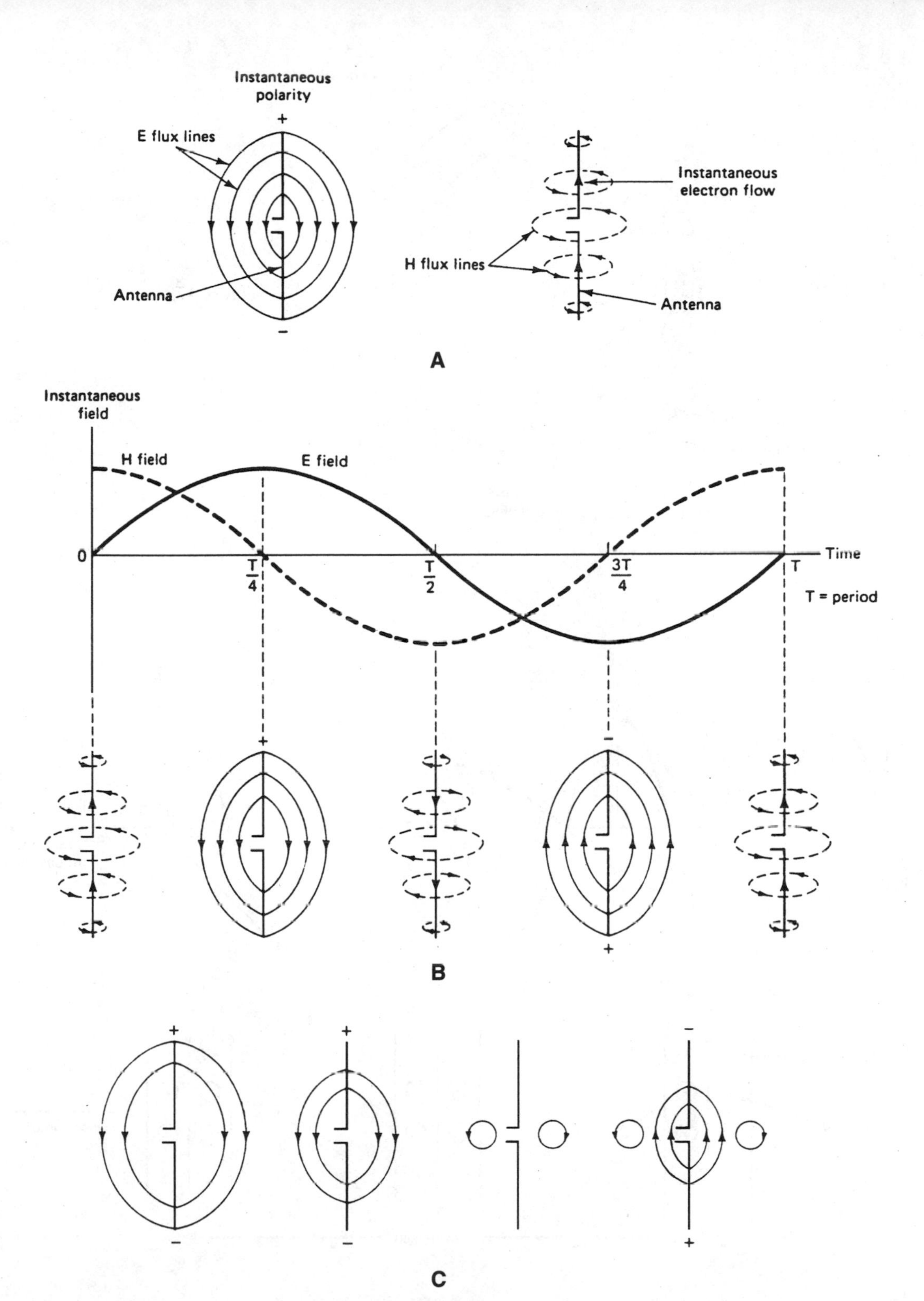
Instantaneous
polarity
+
E flux lines
Antenna
−
Instantaneous
electron flow
H flux lines
Antenna
A
Instantaneous
field
H field
E field
0
T/4
T/2
3T/4
T
Time
T = period
+
−
+
B
+
−
+
−
−
+
C

Figure 173-1

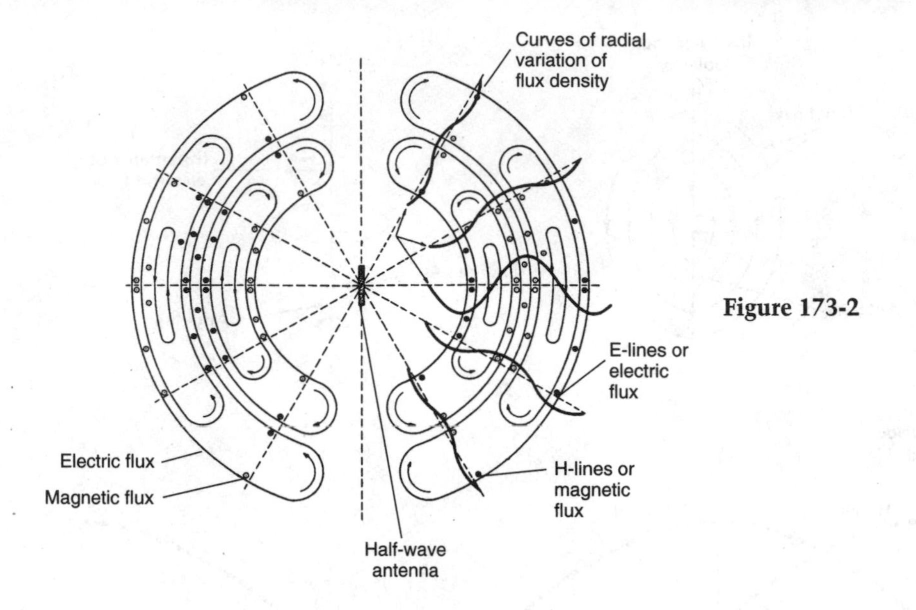

Curves of radial variation of flux density
E-lines or electric flux
Electric flux
Magnetic flux
H-lines or magnetic flux
Half-wave antenna

Figure 173-2

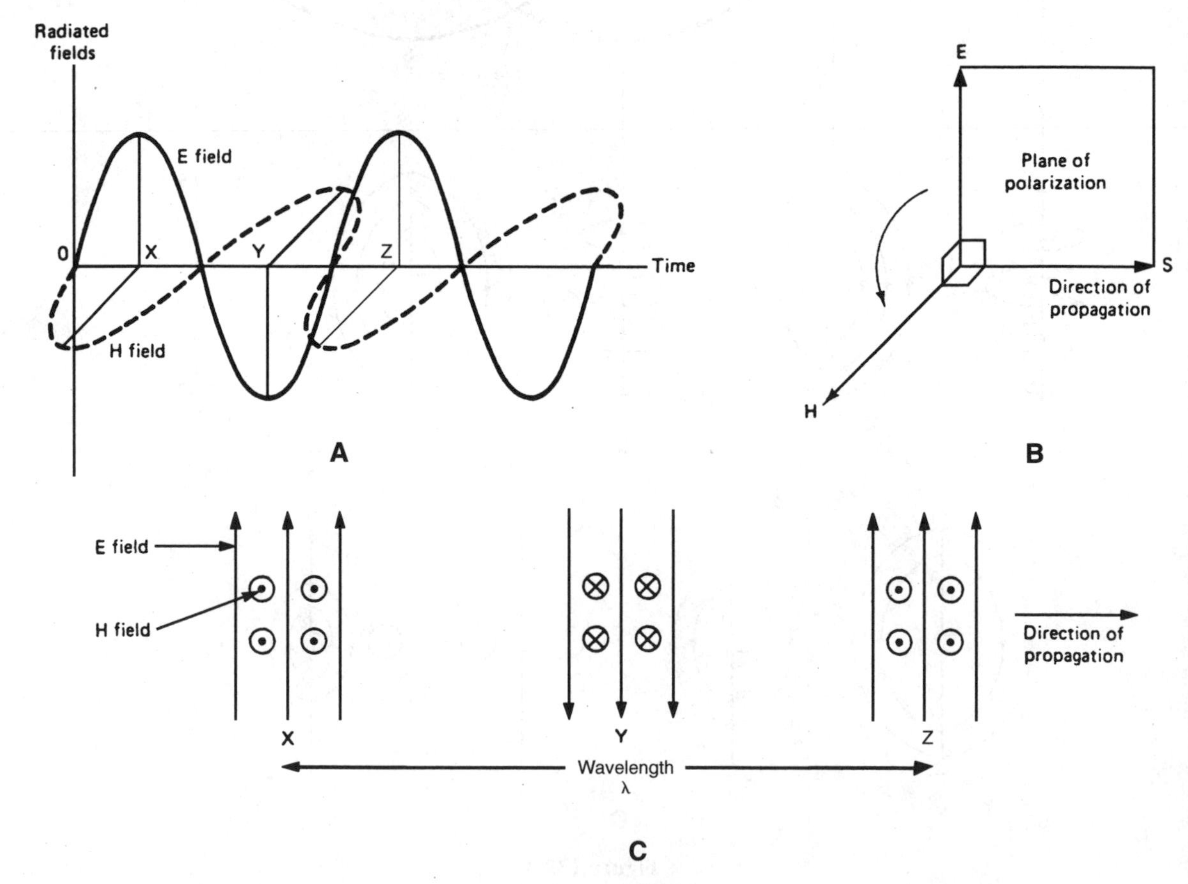

Radiated fields
E field
0
X
Y
Z
Time
H field
A
E
Plane of polarization
S
Direction of propagation
H
B
E field
H field
Direction of propagation
X
Y
Z
Wavelength
λ
C

Figure 173-3

Therefore,

$$f = \frac{c}{\lambda}, \text{ and } \lambda = \frac{c}{f} \qquad (173\text{-}1)$$

With the SI system of units, the *permeability* of free space is $\mu_o = 4 \times 10^{-7}$ H/m, and the *permittivity* of free space is $\epsilon_o = 8.85 \times 10^{-12}$ F/m. In 1865, Clerk-Maxwell predicted that the velocity of *all* electromagnetic waves in free space would be given by,

$$c = \frac{1}{\sqrt{\mu_o \times \epsilon_o}} \approx 3 \times 10^8 \text{ m/s} \qquad (173\text{-}2)$$

The value of c is equal to the velocity of light, which is another example of an electromagnetic wave. If the wave is traveling through a medium whose relative permittivity is ϵ_r, the velocity is reduced and is equal to $c/\sqrt{\epsilon_r}$.

For free space, the velocity of all electromagnetic waves is a constant. Because this velocity is equal to the product of the frequency and the wavelength, either a wavelength or a frequency scale can be used to distinguish one signal from another.

Mathematically, electric field E (measured in volts per meter), magnetic field H (measured in amperes per meter), and the transfer of EM energy in a particular direction represent a "right-handed" system of vectors. The transfer of EM energy is measured by the *Poynting vector (S)*, which represents the amount of radiated energy passing through a unit of area (m^2) in a unit of time (1 s). E, H, and S (in that order) form a right-handed system of vectors. If the instantaneous E direction is rotated through 90° to lie along the instantaneous H direction, the direction of S is that of a right-handed screw, which is subjected to the same rotation (Fig. 173-3B). S is the vector product of E and H. In terms of units: volts/meter (E) × amperes/meter (H) = volts × amperes/(meter)^2 = watts/(meter)^2 = joules per square meter per second (S).

The ratio of E (volts per meter) to H (amperes per meter) must have the units of ohms. Clerk-Maxwell showed that for the EM wave in free space,

$$\frac{E}{H} = \sqrt{\frac{\mu_o}{\epsilon_o}} = \eta_o \qquad (173\text{-}3)$$

where, η_o = the intrinsic impedance of free space (Ω).

From Equation (173-2),

$$c = \frac{1}{\sqrt{\mu_o \epsilon_o}} = 3 \times 10^8 \text{ m/s}$$

Therefore,

$$\frac{1}{\sqrt{\epsilon_o}} = \sqrt{\mu_o} \times 3 \times 10^8$$

Combining Equations 173-2 and 173-3 gives us:

Intrinsic impedance,

$$\eta_o = \sqrt{\frac{\mu_o}{\epsilon_o}} = \sqrt{\mu_o} \times \sqrt{\mu_o} \times 3 \times 10^8$$

$$= 4\pi \times 10^{-7} \times 3 \times 10^8$$

$$= 120\pi \approx 377 \ \Omega.$$

Example 173-1

What is a wavelength in free space of a microwave signal whose frequency is 5.2 GHz?

Solution

The wavelength is free space,

$$\lambda = \frac{30}{5.2} = 5.77 \text{ cm} \qquad (173\text{-}1)$$

Example 173-2

A radio wave traveling in free space has an electric field intensity (E) of 500 mV/m. The corresponding value of the wave's magnetic field intensity (H), is 1.326 mA/m. Find (a) the E:H ratio, and (b) the Poynting vector, S.

Solution

The value of the ratio,

$$E{:}H = \eta_o = \frac{500 \text{ mV/m}}{1.326 \text{ mA/m}} = 377 \ \Omega$$

The value of the Poynting vector,

$$S = 500 \text{ mV/m} \times 1.326 \text{ mA/m}$$

$$= 663 \ \mu\text{W/m}^2$$

PRACTICE PROBLEMS

173-1. What is the frequency of an infrared signal whose wavelength in free space is 23 μm?

173-2. A radio wave traveling in free space has a magnetic field intensity of 0.274 mA/m. Find the values of the electric field intensity and the Poynting vector.

174
The Marconi antenna

At an operating frequency of 2 MHz, the required physical length for a resonant Hertz dipole is 143.1/2 = 71.6 m. It is difficult to position a vertical antenna of this size so that it is remote from the ground. However, the dipole could be in the form of a long wire antenna, slung horizontally between two towers from which the antenna is insulated. Clearly, the vertical Hertz antenna is not a practical proposition at low frequencies, and it is replaced by the $\lambda/4$ *unipole* or *Marconi antenna*. The Marconi type can be regarded as a Hertz dipole, in which the lower half is replaced by a nonradiating ground *image antenna* (Fig. 174-1A). This means that, unlike the Hertz antenna, ground is an integral part of the Marconi antenna system. This also means that the Hertz dipole is balanced with both $\lambda/4$ sections mounted remote from ground, but the Marconi unipole is unbalanced because part of the antenna is the ground itself. In the same way, the twin lead (with neither conductor grounded) is a balanced transmission line, but a coaxial cable is unbalanced (with its outer conductor grounded at intervals along its length).

The distributed inductance of the Marconi antenna is associated with the vertical $\lambda/4$ rod, and the distributed capacitance exists between the rod and ground (Fig. 174-1B). Therefore, it is impossible to operate with a horizontal Marconi antenna; this compares with the Hertz dipole, which can be mounted either vertically or horizontally. Consequently, all Marconi antennas radiate only vertically polarized waves.

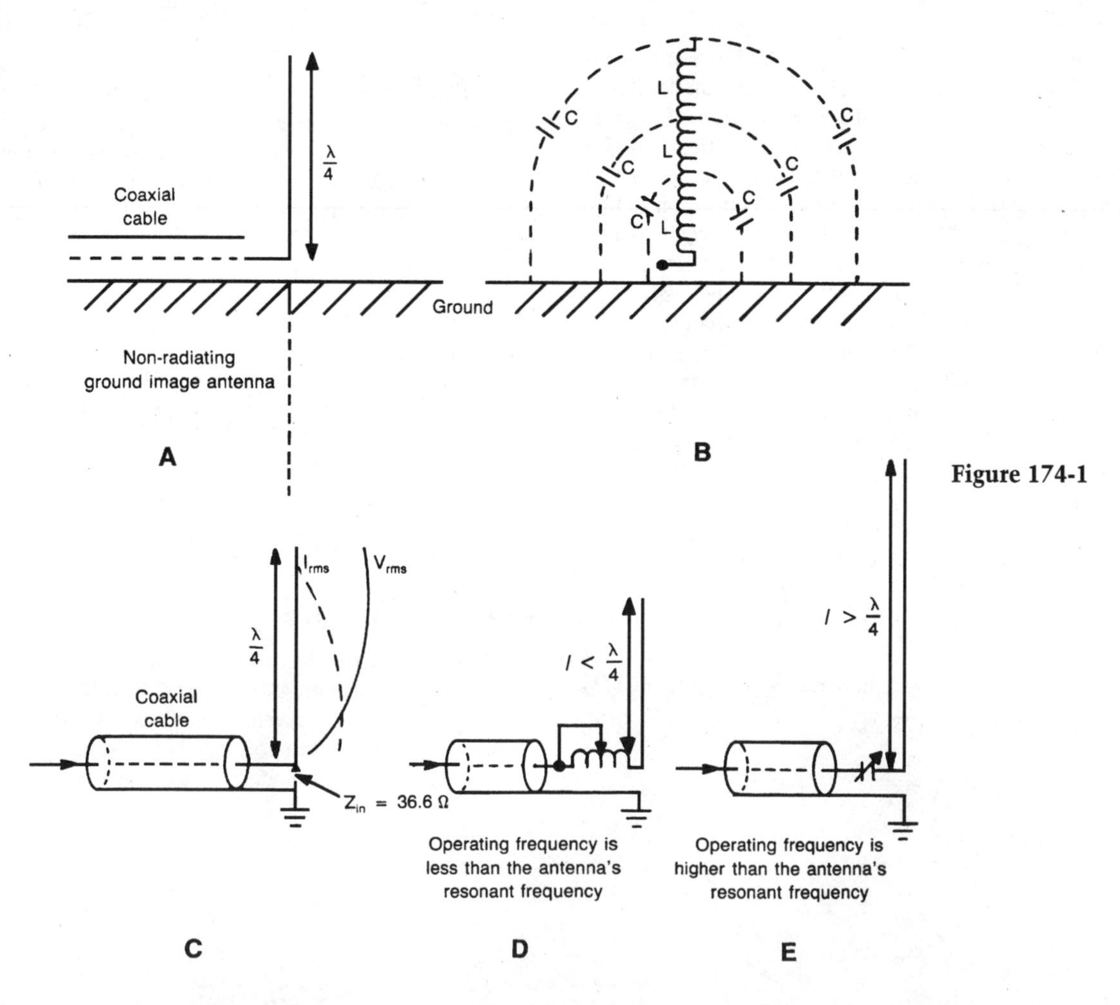

Figure 174-1

The effective voltage and current distribution on the resonant Marconi antenna is comparable with the distribution on the upper half of the vertical Hertz dipole (Fig. 174-1C). If the $\lambda/4$ antenna is end fed by an unbalanced coaxial cable, the radiation resistance is $73.2/2 = 36.6\ \Omega$, which is the effective resistive load of a resonant Marconi antenna. If the operating frequency is below the resonant value, the unipole is too short and it behaves capacitively. However, the antenna can be tuned to resonance by adding an inductor in series (Fig. 174-1D). When operating above the resonant frequency, the antenna is too long, behaves inductively, and can be tuned to resonance with the aid of a series capacitor (Fig. 174-1E).

One type of practical Marconi antenna is the vertical *whip* (a rod that is flexible to a limited extent). As an example, an end-fed whip antenna is commonly mounted on the top of an automobile so that the roof can act as the required ground plane. If the antenna is positioned in the vicinity of the rear bumper, most of the ground plane will be provided by the road surface. If it is required to operate a whip antenna at the top of a building, it is necessary to provide the antenna with an apparent ground or counterpoise. This normally consists of a wire structure mounted just beneath the feedpoint of the antenna and connected to the outer conductor of the coaxial cable. The main distributed capacitance then exists between the whip and the counterpoise, which itself acts as a large capacitance in relation to ground. For this reason, the counterpoise should be normally larger in size than the antenna and it must be well insulated from ground.

Another type of vertical radiator is a steel tower, which can be tapered to optimize the current distribution. Such towers can either be end-fed or shunt-fed as shown in Fig. 174-2. With *shunt-feeding* the bottom of the tower is connected directly to ground, and the center conductor of the coaxial cable is joined to a point, (P), on the tower (where the resistive component of the antenna's impedance can be matched to the surge impedance of the coaxial cable). With the grounded end behaving as a short circuit, the impedance at the point, (P), is inductive and the reactance is cancelled by the capacitor, C. The dc resistance between P and ground is virtually zero.

To improve the quality of the ground and, therefore, reduce ground losses, it is common practice to add a ground system that consists of a number of bare copper conductors, arranged radially, and connected to a center point beneath the antenna; the center point is then joined to the outer conductor of the coaxial cable. These ground radials are from $\lambda/10$ to $\lambda/2$ in length and are buried a short distance down; in addition, the ground can be further improved by laying copper mats beneath the surface. The ultimate purpose of the radial ground system is to improve the *antenna efficiency*, defined as the ratio of the RF power radiated from the antenna (as useful electromagnetic energy) to the RF power applied to the antenna feedpoint.

A tower can be either self-supporting, or supported by guy wires. Such wires can pick up some of the live transmitted radiation; reradiation from the guy wires will then occur and this will interfere with the EM wave transmitted from the antenna. To reduce the reradiation effect to a minimum, an insulator is connected to each end of the guy wire, then intermediate insulators are included with a spacing of about a tenth of the wavelength apart. These insulators are normally of the porcelain "egg" type, so if one of the insulators breaks, the guy wire still provides mechanical support.

If, at low frequencies, the vertical antenna is too short for mechanical reasons, the amount of useful vertically polarized radiation can be increased by using "top loading," examples of which are shown in Fig. 174-3. Because the current is zero at the open end of the antenna, the inverted "L" arrangement will have

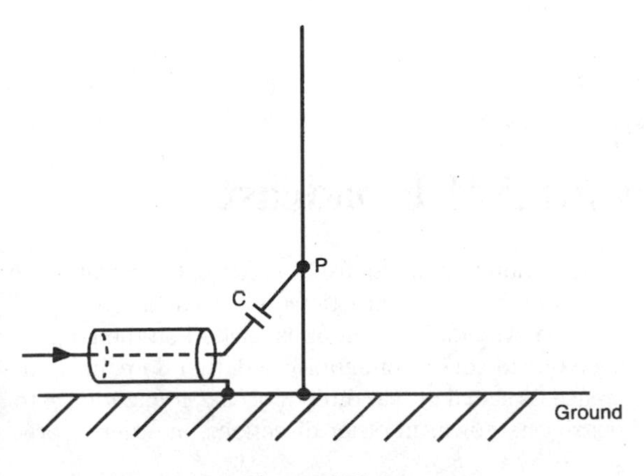

Figure 174-2

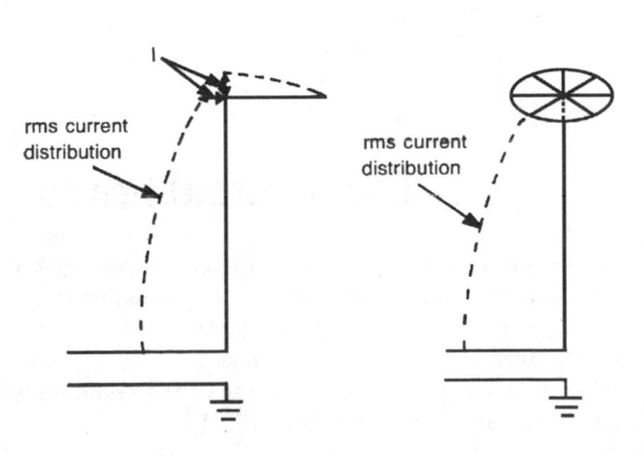

Figure 174-3

the result of increasing the effective current distributed over the antenna's vertical section. A more practical solution is the metallic spoked wheel that will increase the amount of the distributed capacitance to ground, and therefore will lower the antenna's resonant frequency to the required value.

MATHEMATICAL DERIVATIONS

Electrical quarter wavelength,

$$\lambda/4 = \frac{75.2}{f} \text{ meters} = \frac{246}{f} \text{ feet} \qquad (174\text{-}1)$$

Physical length of the Marconi antenna

$$l = \frac{71.6}{f} \text{ meters} = \frac{234}{f} \text{ feet} \qquad (174\text{-}2)$$

where: f = frequency (MHz).

The radiation resistance of the Marconi antenna = 36.6 ohms. (174-3)

Power delivered to an end-fed resonant Marconi antenna,

$$P_A = I_A^2 R_A \text{ watts} \qquad (174\text{-}4)$$

where: I_A = effective current at the antenna's feedpoint (A)
R_A = radiation resistance (Ω).

Example 174-1

What is the physical length of a Marconi antenna that is resonant at 7.5 MHz?

Solution

$$\textit{Physical length} = \frac{71.6}{f} \qquad (174\text{-}2)$$

$$= \frac{71.6}{7.5}$$

$$= 9.54 \text{ m.}$$

Example 174-2

The effective current at the feedpoint of a resonant Marconi antenna is 8.7 A. Assuming that the antenna is end-fed, what is the amount of RF power that is delivered to the antenna?

Solution

Antenna power,

$$P_A = I_A^2 \times R_A \qquad (174\text{-}4)$$

$$= (8.7\,A)^2 \times 36.6\ \Omega$$

$$= 2770 \text{ W.}$$

PRACTICE PROBLEMS

174-1. What is the physical length (in feet) of a Marconi unipole that is designed to resonate at 7.5 MHz?

174-2. In Practice Problem 174-1, what is the electrical length of the antenna?

174-3. A Marconi antenna is designed to resonate at 12 MHz. If the transmitter that feeds the antenna is operating at 15 MHz, describe the nature of the antenna load.

174-4. An antenna tower is 175-ft high and is operated at a frequency of 835 kHz. What percentage of the electrical wavelength is the height of the tower?

174-5. Two antenna towers that are operated at 950 kHz are separated by 150 electrical degrees. What is the distance in feet between the towers?

175
Directional antennas for AM broadcast

An AM broadcast station that radiates a signal equally well in all horizontal directions, uses a nondirectional antenna. It follows that such a station only requires one antenna tower. This antenna is most effective when it is located near the center of the community that it is licensed to serve (Fig. 175-1).

A station that is required to control the radiation of its signal in certain directions, must use a directional antenna system. The station is located on the edge or even outside of its community and, therefore, must increase its signal in that direction; also, it might have to restrict its power in other directions, in order to pre-

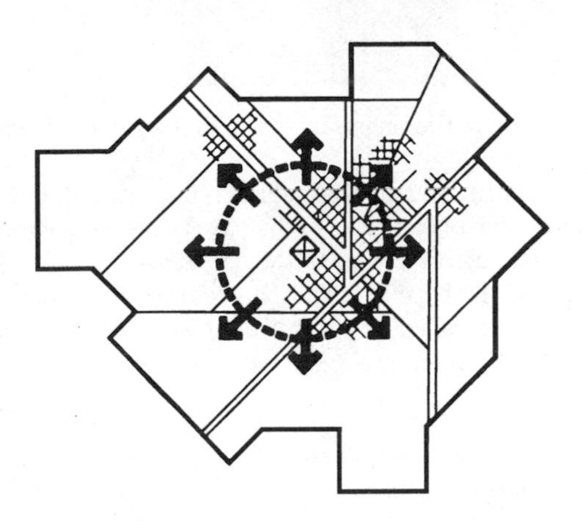

Figure 175-1

vent interference with a second station operating on the same frequency. This type of directional antenna must have two (or more) towers.

Figure 175-2 shows two stations (A and B) serving different communities but operating on the same frequency. Station A is located outside its community and, therefore, its antenna must direct most of its signal toward the specified area. However, in order to prevent interference with Station B, the antenna system must be designed to radiate only a small amount of RF power toward the B community.

Many factors influence the design of a directional antenna system that will produce the desired radiation pattern. Ideally, the pattern is determined by:

1. Size and shape of the towers.
2. Number of towers.
3. Placement of the towers.
4. Phase angle of the current in each tower.
5. Relative magnitudes of the currents in the towers.

In practice, the radiation pattern is also influenced by:

1. Surrounding terrain.
2. Ground moisture.
3. Nearby structures such as water towers and power lines.
4. Weather.

Notice that after a directional antenna system has been constructed, the only two factors (which are readily controllable) are the relative phase angle and the magnitude of the current in each tower. As a result of these controls, the signals radiated from the towers will differ in strength and phase. In some directions, these signals can practically cancel one another, but in other directions, they will reinforce. The main design problem is to produce the cancellations and reinforcements in the desired directions.

MATHEMATICAL DERIVATIONS

For a nondirectional (omnidirectional) station, the transmitter supplies the RF power directly to the base of the single tower antenna. The operating power is then given by:

$$\textit{Operating power} = I_A^2 R_A \text{ watts} \qquad (175\text{-}1)$$

where: I_A = antenna base current (A)
R_A = antenna resistance as measured at the base (Ω)

With a directional station, it is necessary that the RF power is correctly split between the towers. Figure

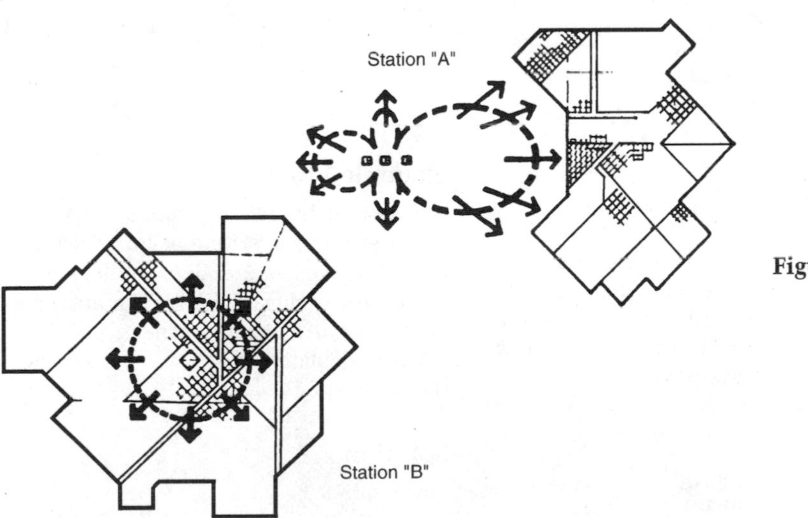

Figure 175-2

175-3 shows the phasor device which divides and distributes the RF power. The current flowing from the transmitter to the phasor, is the common point current (I_C). This common point current is used to calculate the operating power of a directional AM station.

$$\textit{Operating power} = I_C^2 R_C \text{ watts} \quad (175\text{-}2)$$

where: I_C = common point current (A)
R_C = common point antenna resistance as measured at the phasor (Ω).

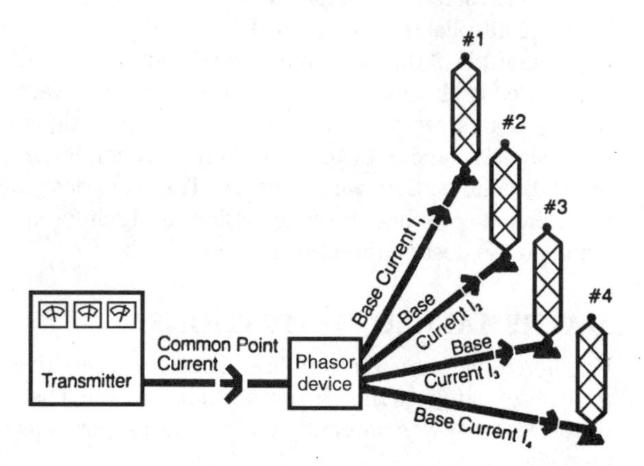

Figure 175-3

The current that flows from the phasor to each tower is called the *antenna base current* and it can be measured with an RF ammeter installed at the base of the tower.

Usually, the RF power is divided unequally; thus, the base currents at the various towers are different. For example, a set of readings taken from the base currents in Fig. 175-3, might be:

Tower	Base current
#1	2.45 A
#2	5.9 A
#3	6.6 A
#4	4.7 A

In addition to controlling the amount of current flowing to each tower, the phasor introduces a time shift, which is revealed by a phase angle and is measured in degrees. It is normal practice to identify one of the towers as the *reference tower*, whose phase angle is 0°. Figure 175-3 shows the reference as Tower #3; then, the phase angle values for the four-tower array might be:

Tower	Phase angle
#1	+18°
#2	−144°
#3	0°
#4	+105°

The value of the phase angle lies in the range of +180° to −180°.

Because the antenna radiation pattern depends on the relative magnitudes of the base currents, a variable is necessary to compare the current of a particular tower with the current of the reference tower. This variable is called the *antenna base current ratio*, which is defined by:

$$\textit{Antenna base current ratio for a particular tower} = \frac{\textit{antenna base current reading of that tower}}{\textit{base current reading of the reference tower}} \quad (175\text{-}3)$$

Assuming that Tower #3 is again the reference, the values of the antenna base current ratios in this example are: Tower #1, 2.45/6.6 = 0.37; Tower #2, 5.9/6.6 = 0.89; Tower #3, 6.6/6.6 = 1; Tower #4, 4.7/6.6 = 0.71. These ratios, either calculated or read on an antenna monitor, must not deviate by more than 5% from the values shown on the station license.

Example 175-1

The directional antenna system of a standard broadcast station consists of five towers, whose common point resistance is 85 Ω. If the associated common point current is 12.5 A, calculate the value of the operating power.

Solution

$$\textit{Operating power} = I_C^2 R_C \quad (175\text{-}2)$$
$$= (12.5\text{ A})^2 \times 85\ \Omega$$
$$= 13281.25\text{ W}$$
$$= 13.3\text{ kW, rounded off.}$$

Example 175-2

A standard broadcast station has a directional antenna system consisting of four towers. Tower #1 is the reference tower and its antenna base current has the correct value of 2.5 A. The antenna currents of Towers #2, 3, and 4 are respectively, 3.5 A, 4.0 A, and 2.1 A. Calculate the antenna base current ratios for these towers.

Solution

Tower #2:

$$\textit{Antenna base current ratio} = \frac{3.5}{2.5} = 1.4 \quad (175\text{-}3)$$

Tower #3:

$$\textit{Antenna base current ratio} = \frac{4.0}{2.5} = 1.6$$

Tower #4:

$$\textit{Antenna base current ratio} = \frac{2.1}{2.5} = 0.84$$

Example 175-3

In Example 175-2, the phase angles of the towers, are: Tower #1, 0°; Tower #2, +32°; Tower #3, −112°; and Tower #4, −47°. If Tower #3 is now regarded as the reference tower, what are the new values for the phase angles?

Solution

Tower #1, +112°; Tower #2, +32° − (−112°) = +144°; Tower #3, 0°; and Tower #4, −47° − (−112°) = +65°.

PRACTICE PROBLEMS

175-1. In Example 175-2, between what limits must the antenna base current ratio of Tower #3 lie?

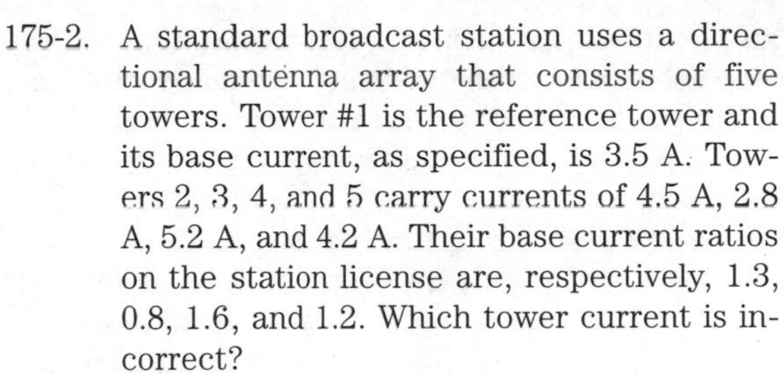

175-2. A standard broadcast station uses a directional antenna array that consists of five towers. Tower #1 is the reference tower and its base current, as specified, is 3.5 A. Towers 2, 3, 4, and 5 carry currents of 4.5 A, 2.8 A, 5.2 A, and 4.2 A. Their base current ratios on the station license are, respectively, 1.3, 0.8, 1.6, and 1.2. Which tower current is incorrect?

175-3. A standard broadcast station has a directional antenna system consisting of five towers. As measured, the antenna currents for Towers #1, 2, 3, 4, and 5 are respectively, 3.8 A, 2.8 A, 1.7 A, 4.2 A, and 5.3 A. It is specified that the tower currents must be in the proportion of 1.36 : 1.00 : 0.61 : 1.50 : 1.79. Which tower current is furthest away from its correct value?

175-4. In Practice Problem 175-3, Tower #2 is regarded as the reference tower. By what percentage does the base current ratio of Tower #5 deviate from its specified value?

176 The loop antenna

The loop antenna (Fig. 176-1A) is primarily used for direction-finding purposes in the medium frequency (MF) band. In other words, the loop (which can be a single turn but, more probably, consists of several turns) enables you to determine the bearing of a transmitter. At these frequencies, the method of propagation uses the ground (or surface) wave, which requires that the radio signal is vertically polarized. Consequently, only the vertical sections (AB, CD) of the loop antenna will pick up a signal. Ideally, there should be

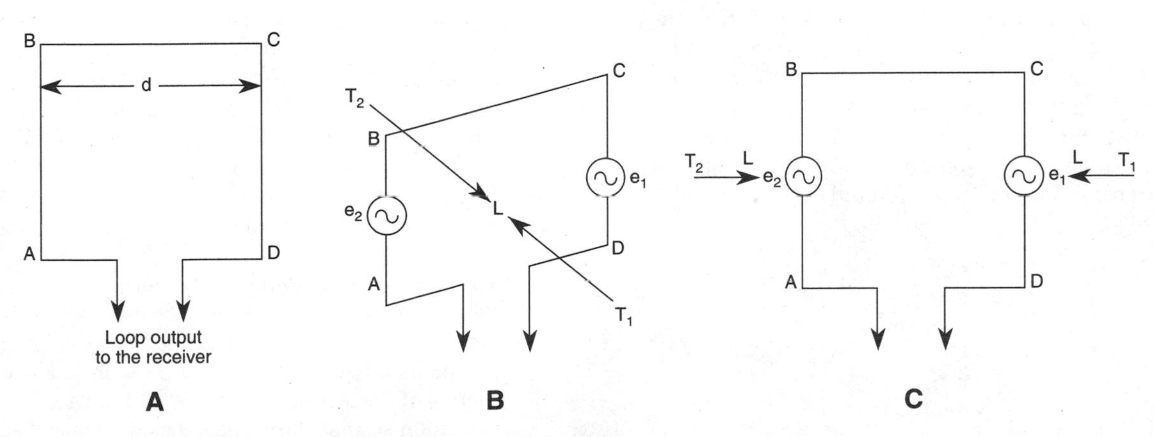

Figure 176-1

no voltages induced in the horizontal sections. The loop antenna is capable of being manually rotated and its output is passed to a receiver that is tuned to the operating frequency of the transmitter, whose bearing is to be determined.

Assume that the vertically polarized radio wave is arriving broadside to the loop antenna; or in other words, the direction of propagation (from T_1 to L) is at right angles to the plane of the vertical loop (Fig. 176-1B). The radio wave will then reach both vertical sections at the same time; the RF voltages (e_1, e_2) induced in these sections will then be equal in magnitude and phase so that the loop output, $e_1 - e_2$, is zero. However, the loop's output will also be zero if the direction of propagation is from T_2 to L; the radiation pattern of the loop is therefore bidirectional in the horizontal plane.

If the radio wave travels in direction T_1L (Fig. 176-1C) and arrives end-on to the loop, the wave will arrive at the vertical section CD before it reaches the vertical section AB. Consequently, although e_1 and e_2 will be equal in magnitude, e_1 will lead by e_2 by the maximum phase difference (ϕ). However, ϕ is a small angle because, for any practical size of the loop, the width $d << \lambda$ (the wavelength of the RF wave). For example, if d = 1 m and the frequency of f = 1 MHz, the wavelength (λ) is 300 m, and ϕ = 360°/300 = 1.2°. As the mathematical derivations show, the loop's output is a maximum when the wave arrives end-on to the loop (either from the direction T_1L or T_2L). The only distinction between these two maximum outputs lies in the fact that they are 180° out of phase.

The vertical loop's complete radiation pattern in the horizontal plane consists of two circles. This is illustrated in Fig. 176-2, which gives a "bird's eye" view of the loop. Notice that, for the null positions of zero loop output, there is a maximum rate of change of the received signal with the bearing; however, for the positions of maximum loop output, the corresponding rate of change is a minimum. It follows that the nulls are well-defined and will be used in the direction finding procedure.

Assume that you have picked up a signal from a transmitter. The loop is rotated for a null and this rotation will cause a simultaneous movement of a needle over a calibrated bearing scale. When the null has been found, the needle will indicate the transmitter's bearing. Or will it? This is a problem. If the loop is now rotated through 180°, a second null will be found, and there will be a new reading for the bearing. These two readings are 180° apart; one is the transmitter's *true bearing* and the other is the so-called *reciprocal bearing*. Which is which?

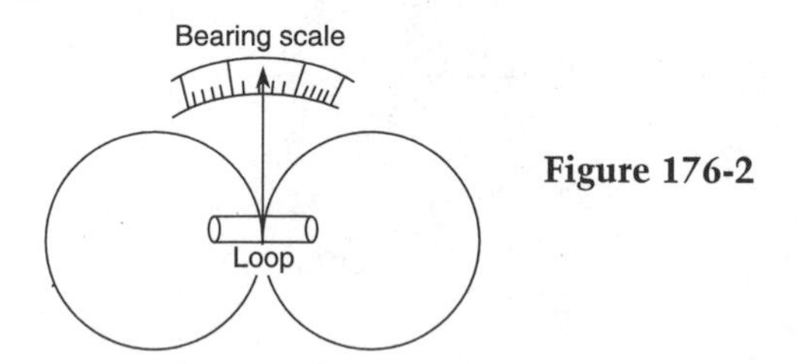

Figure 176-2

The process of distinguishing the true and reciprocal bearing is called *sensing*. A vertical antenna is placed at (or near to) the center of the loop (Fig. 176-3). The signal output from this antenna is phase shifted by 90°, and then is combined with the output signal from the loop. A full explanation of the sensing procedure is given in the mathematical derivations.

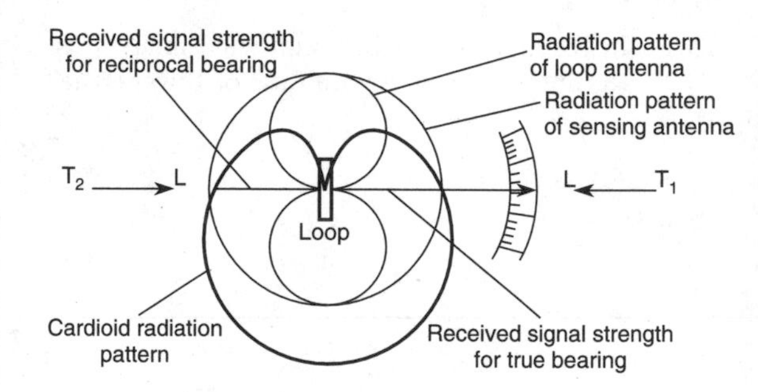

Figure 176-3

A single loop might be too unwieldy and cumbersome for a rapid manual rotation. An alternative is to use two fixed, crossed loops (with one loop in the N/S plane, and the other in the E/W plane) so that the two loops are at right angles; this arrangement is known as the *Bellini-Tosi system*. The signal outputs from the loops are fed to two coils, whose axes are at right angles. Mutually coupled to the fixed coils is another coil that can be rotated to a null position; a needle that is controlled by the third coil's movement, indicates the true (or reciprocal) bearing of the transmitter. The unit that contains the three coils is referred to as a *goniometer*. Errors in the bearing are caused by:

1. *Unbalanced distributed capacitance between the loop and ground*. This can be avoided by using a loop of many turns which are enclosed in a hollow metal screen. At some position, a small section of the screen is removed and replaced by insulation so that the screen does not behave as a short-circuited turn.

2. *Horizontally polarized down-coming sky waves that induce signal voltages in the top and bottom sections of the loop.* It can then be impossible to locate a null or if a null is found, the bearing indicated will be in error. The error can be eliminated by using an *Adcock antenna* (Fig. 176-4), which has no separation between its horizontal sections. However, unlike the loop, the Adcock antenna cannot consist of a number of turns; therefore, its sensitivity is lower.

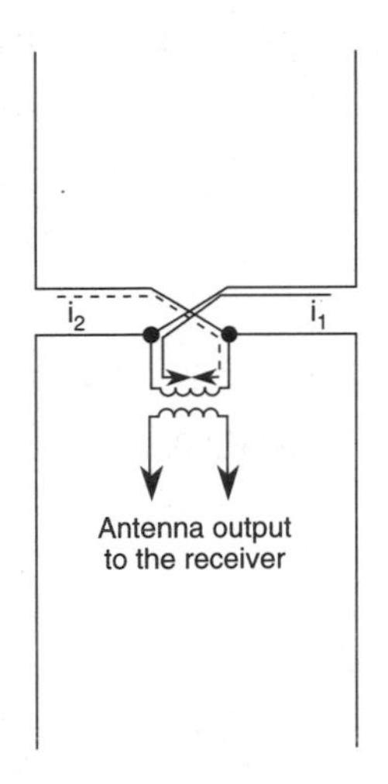

Figure 176-4

MATHEMATICAL DERIVATIONS

In Fig. 176-1C, let the radio wave arrive end-on to the loop in the direction of T_1 to L. Then, the maximum loop output signal is:

$$= e_1 - e_2 = e \sin(\omega t + \phi) - e \sin \omega t \qquad (176\text{-}1)$$

$$= 2e \cos\left(\omega t + \frac{\phi}{2}\right) \sin \frac{\phi}{2}$$

$$\approx e\phi \cos(\omega t + \frac{\phi}{2}) \text{ volts.}$$

where $\phi = 2\pi d/\lambda$ radians and is a small angle
d = width of loop (m)
ω = angular frequency (rad./s)
λ = wavelength of the radio wave (m)

If the radio wave arrives end-on to the loop in the direction T_2 to L,

Maximum loop output signal

$$= e_2 - e_1$$

$$= -2e \cos\left(\omega t + \frac{\phi}{2}\right) \sin \frac{\phi}{2}$$

$$\approx -e\phi \cos\left(\omega t + \frac{\phi}{2}\right)$$

$$= e\phi \cos\left(\omega t + \frac{\phi}{2} + \pi\right) \text{ volts}$$

This result shows that the two maximum loop output signals are 180° out of phase.

If the wave arrives at an angle (θ) to the plane of the loop (Fig. 176-5), the distance between the two vertical sections in the path of the wave is reduced from d to $d \cos\theta$ so that the *loop output signal* $\approx e\phi \cos\theta \cos(\omega t + \phi/2)$.

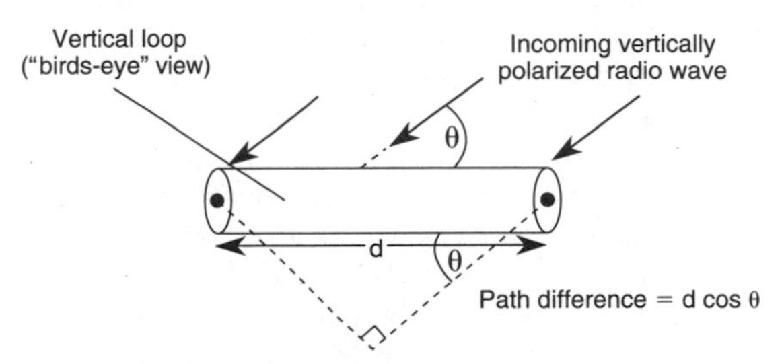

Figure 176-5

In terms of the effective value,

$$\textit{rms value of loop output signal} = E \cos\theta \qquad (176\text{-}2)$$

where E is the effective value of $e\phi \cos(\omega t + \phi/2)$.

If $r_1 = E \cos\theta$ is plotted using polar coordinates (chapter 199), the result is a pair of circles (Fig. 176-2). The voltages represented by the two circles are 180° out of phase; this is taken care of because the value of $\cos\theta$ is negative in the second and third quadrants.

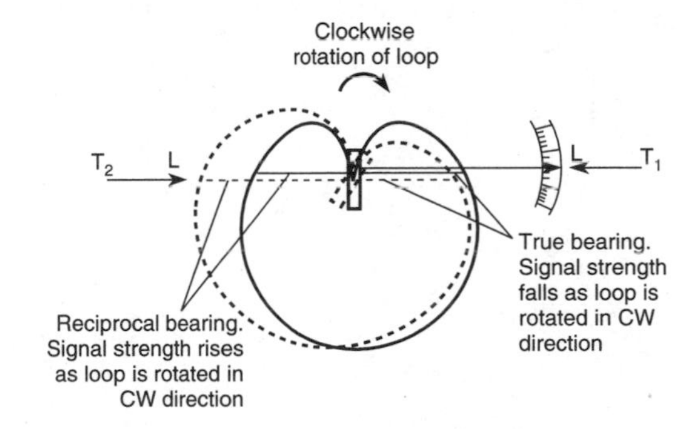

Figure 176-6

Let the vertical "sense" antenna be placed at the center of the loop; the output signal from this antenna is $e \sin(\omega t + \phi/2)$, which is 90° out of phase with the loop's output. Because the wavelength is so long, the sense antenna can actually be placed at any convenient position. If the sense antenna output signal is advanced in phase by 90°, and is reduced in magnitude so that its peak value is only $e\phi$, the new sensing signal is $e\phi \sin(\omega t + \phi/2 + \pi/2) = e\phi \cos(\omega t + \phi/2)$. This is in phase with the voltage of one circle in the loop's radiation pattern, but 180° out of phase with the voltage of the other circle.

In polar coordinates, the equation of the sensing signal ($r_2 = E$) means that the horizontal radiation pattern of the vertical antenna is a single circle. Combining the loop signal with the sensing voltage, the polar equation becomes,

$$r = r_1 + r_2 = E + E\cos\theta = E(1 + \cos\theta) \quad (176\text{-}3)$$

The graph that is associated with this equation is a cardioid, or heart-shaped, curve (chapter 199), which is used to distinguish between the transmitter's true and reciprocal bearings (Figs. 176-3, 6). The procedure is as follows:

1. After the direction finding receiver has picked up the transmitter's signal, rotate the loop for a null and note the indicated bearing.
2. Switch in the sensing antenna. The signal strength will immediately jump from zero to the sensing value.
3. Rotate the loop slightly clockwise. If the signal strength falls, the indicated bearing is true but if the signal strength rises, the bearing is a reciprocal and must be corrected.

Notice that this rule is entirely arbitrary and depends on the location of the bearing needle, on whether the output of the sensing antenna is advanced (or retarded) by 90°, etc.

The direction-finding equipment (as described) only indicates the bearing of the transmitter, not its location. However, if two separate direction-finding stations take simultaneous bearings, and these bearings are plotted on a map, the point of interception of the bearing lines indicates the approximate position of the transmitter. Still better is to use three stations; the bearing lines then form a "cocked hat" within which the transmitter lies.

Example 176-1

A vertical loop of 100 turns has a height of 0.8 m, and a width of 1 m. An 800-kHz radio wave, whose electric field intensity is 500 μV/m, arrives from a horizontal direction that is end-on to the plane of the loop. Calculate the signal output from the loop.

Solution

Wavelength: $\lambda = \dfrac{300}{0.8} = 375$ m

$$\textit{Loop output} = E\cos\theta \quad (176\text{-}2)$$

$$= \frac{100 \times 500 \times 0.8 \times 2\pi \times 1}{375}$$

$$= 818\ \mu\text{V}$$

PRACTICE PROBLEMS

176-1. In Example 176-1, determine the phase difference in degrees between the two voltages induced in the vertical sections.

176-2. The loop of Example 176-1 is rotated through 35° in the clockwise direction. Calculate the new signal output from the loop.

177 Propagation

The word *propagation* is derived from the Latin verb "propagare—to travel," and describes the various ways by which a radio wave "travels" from the transmitting antenna to the receiving antenna. Start by considering a long half-wave antenna, which is fed at one end by a coaxial cable with its outer conductor grounded. The three-dimensional radiation pattern of the wave resembles a large "doughnut" that is lying on the ground with the antenna at its center (Fig. 177-1A). Part of the radiated wave moves downward and outward in contact with the ground, and is affected by the conditions at (and below) the surface of the earth. This component of the radiation is referred to as the *ground* (or *surface*) *wave*.

Higher up the antenna, the radiation is little affected by the ground conditions. This portion is called the *space* (or *direct*) *wave*, which travels in (practically) a straight line from the transmitting antenna to the receiving antenna (direct line propagation). Operating in conjunction with the space wave is the ground-reflected wave (Fig. 177-1B) so that the signal arrives at the receiving antenna through two different paths.

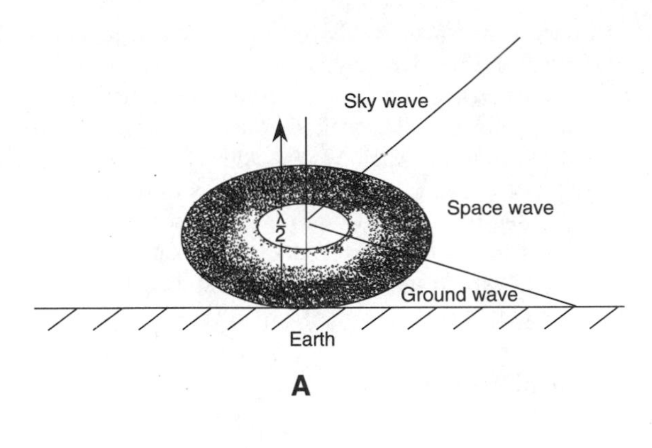

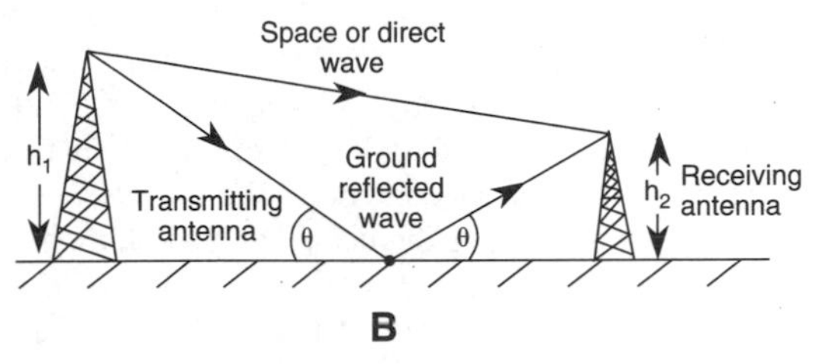

Figure 177-1

Toward the top of the antenna, the radiation moves outward and upward to the *ionosphere*, which is an ionized layer of gas (primarily hydrogen) that extends from about 35 miles to 250 miles above the earth's surface. This portion of the radiation is referred to as the *sky* (or *indirect*) *wave*, which is refracted by the ionosphere back to earth.

The ground, space, and sky waves are illustrated in Fig. 177-2. Primarily, the ground wave is used for long-distance communication by high-power transmitters at relatively low frequencies (VLF, LF, and MF); as an example, the signals from commercial AM stations are carried by the ground wave. The space wave mainly operates at very high frequencies (VHF, UHF, and SHF) for both short-distance and long-distance communications; practical examples are TV and FM broadcasts, as well as communication systems that use satellites. The sky wave, which travels through the ionosphere, is used for long-distance fixed service and mobile communications, and operates in the short wave (HF) bands.

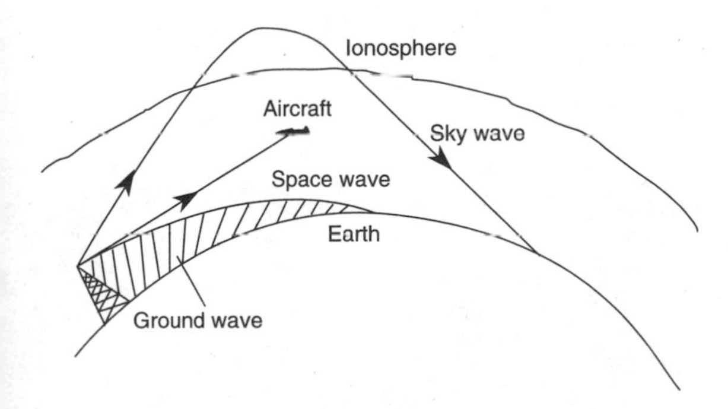

Figure 177-2

GROUND WAVE

The ground wave can only be used with vertically polarized signals that are radiated from a vertical antenna system. As the wave passes over and through the ground, there is a velocity difference between the portions that travel through the air, and through the earth. As a result, the wave front leans forward and there is an electric field component that is parallel to the earth's surface. This creates a voltage stress such that ground currents flow, and energy is absorbed from the wave. At the same time, the magnetic field component of the radio wave cuts the earth and induces eddy currents so that further energy is absorbed. With a horizontally polarized wave, the losses would be enormous and the wave could be completely wiped out at a short distance from the transmitting antenna.

For a vertically polarized wave, the attenuation caused by ground losses depends on the conductivity of the ground and on the frequency used. The attenuation rises rapidly as the frequency is increased so that the ground-wave transmission is limited to frequencies between 12 kHz and 2 MHz. The lower end of this spectrum is used for long distance submarine communications with extremely high-power transmitters. The wavelengths of these broadcasts are on the order of miles so that the ground wave can be *diffracted* (or bent) around the earth's surface and the signal can be received well beyond the horizon by a submarine operating at periscope depth. In fact, providing that the transmitter power is sufficient (up to 1 MW), the ground-wave coverage can be worldwide. The disadvantage of using ground wave under these conditions is the need for enormous antenna systems, which are relatively inefficient; as an example, one such system is slung between two mountains that are 20 miles apart and has an efficiency of about 15%.

Because the electrical properties of the earth (along which the ground wave travels) are relatively constant, the signal strength from a particular station does not vary greatly at any given point. This is essentially true in most localities and is the main advantage in using ground wave, as opposed to sky wave, for long-distance communications. Exceptions would be those areas with distinct rainy and dry seasons; in these cases, the difference in the amount of moisture causes the soil's conductivity to change. The conductivity of salt water is approximately 5000 times greater than that of dry soil. Submarine transmitters can therefore, be built as close to the ocean's edge as possible to take advantage of the superior ground-wave propagation over salt water.

Summarizing, the range of ground-wave propagation depends on:

1. The plane of polarization. Only vertically polarized waves can be propagated over any appreciable distance.
2. The RF power output from the transmitting antenna.
3. The signal frequency. The degree of attenuation, caused by ground losses, increases rapidly as the frequency is raised.
4. The type of ground over which the surface wave travels. The attenuation is much less if the wave travels over salt water, as opposed to desert.

SPACE WAVE

The space wave is the radiation component that travels in a relatively straight line from the transmitting antenna to the receiving antenna. This method of propagation is the one most commonly used in modern communications; everyday examples are TV and FM broadcasts, as well as radar and microwave links. The frequencies used in these systems extend from 30 MHz upwards; at these high frequencies, the ground-wave range is negligible. Radio waves in this part of the frequency spectrum cannot *normally* be refracted back to earth by the ionosphere so that sky-wave propagation is usually impossible.

If the transmitting and the receiving antennas are both located near the ground, the range of the space wave is limited by the curvature of the earth (Fig. 177-3A). For example, the "hump" of the Atlantic ocean between England and America is approximately 200 miles high; therefore, communication by space wave between these two countries can only occur via satellite. Figure 177-3B shows the limiting "line-of-sight" condition in which the line joining the tops of the antennas just grazes the earth's surface. The formula for the optical range (D) is given in the Mathematical Derivations.

Space waves are readily reflected from large metal obstacles; the received signal will be the result of the space wave travelling directly from the transmitting antenna, and from all those radio waves that have been reflected from the various obstacles. Because all of these waves have taken different paths to the receiving antenna, their signals are not in phase. The result is various forms of interference. One example is TV "ghosting," which appears as a double-image pattern on the receiver screen.

Tropospheric ducting

Unusual ranges, well beyond the predictions of line-of-sight transmission, can be caused by abnormal atmospheric conditions a few miles above the earth. These conditions occur in the *troposphere*, which extends from the earth's surface to a height of approximately 35,000 feet. Under normal conditions, the warmest air is found near the surface of the ocean, and the temperature subsequently decreases with height. However, in the tropics, a situation can occur where pockets of warm air are trapped between layers of cooler air. The boundaries between the warm air and the cooler air are called *temperature inversions*; these are capable of refracting space waves that would otherwise continue their line-of-sight propagation. The result is a high tropospheric duct, in which the space wave can be trapped and transmitted for hundreds of miles. These high-level ducts typically start at heights of 500 to 1000 feet, and extend an additional 500 to 1000 feet into the atmosphere. The signal can then be received by an antenna located in the duct (such as an aircraft), or an antenna on the earth's surface, provided that the lower temperature inversion has disappeared (Fig. 177-4A). Under other tropical conditions, the temperature of the air above the surface of the sea will initially increase with the elevation but at a certain height, there will be an inversion and the temperature will subsequently decrease. The result is a *surface duct*, which can extend to an elevation of a few hundred feet. A space wave can be trapped in the surface duct and can be received well beyond the horizon (Fig. 177-4B).

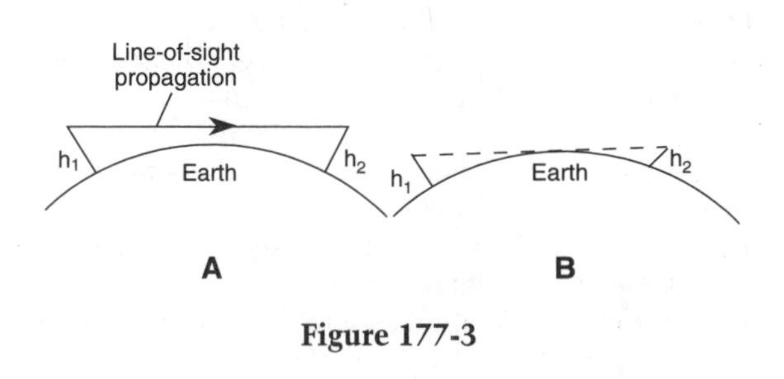

Figure 177-3

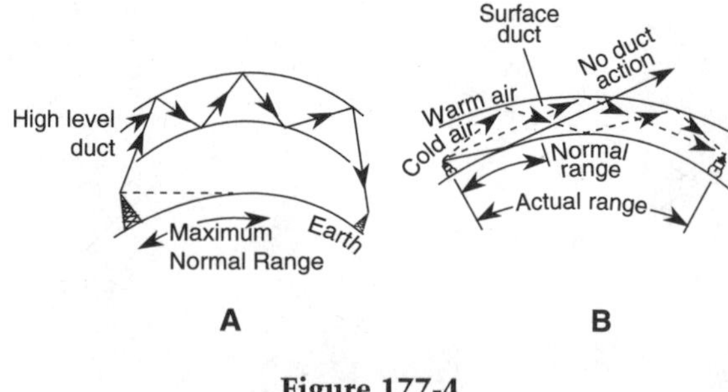

Figure 177-4

Little use can be made of ducts for reliable communications, because their occurrence and duration cannot be predicted. Both surface and high-level ducts represent *ano*malous *prop*agation (sometimes referred to as "anaprop") in which ranges far in excess of normal are experienced.

Tropospheric scatter

Tropospheric scatter is a method of propagation that finds an increasing use in modern communications and relies on conditions in the troposphere at an elevation of a few thousand feet. Because of discontinuities of temperature and humidity in the troposphere, a *scattering region* is formed as shown in Fig. 177-5. Unlike the creation of ducts, the scattering region always exists and is capable of returning a small fraction of the transmitted power back to earth. An analogy of this process is to consider a stream of water that is directed toward a screened skylight. Most of the transmitted energy travels straight out into space (through the screen), and only a small amount is scattered in the required forward direction; there is also some scattering in other undesired directions so that this energy will also not be intercepted by the receiving system. Typically, the received power is only 10^{-6} to 10^{-9} times the transmitted power. Consequently, high-power transmitters and extremely sensitive receivers are essential to operate a reliable tropospheric scatter communication system.

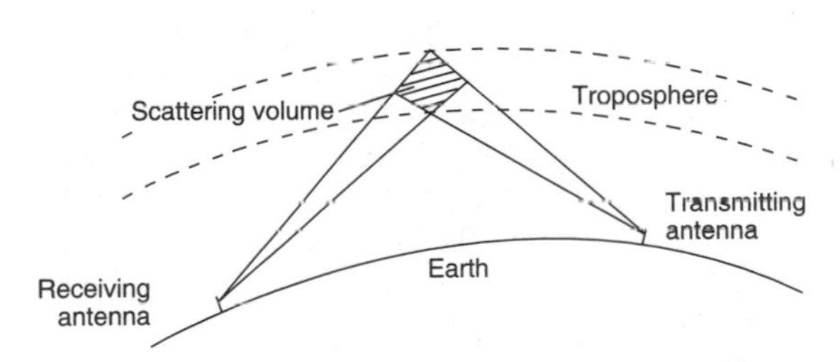

Figure 177-5

The possible frequency range for this propagation method extends from 300 MHz to 10 GHz, although the most commonly used frequencies are 900 MHz, 2 GHz, and 5 GHz. This wide range is possible because the attenuation only rises slowly as the frequency is increased. The minimum range is about 50 miles and is determined by the size of the scattering volume. The maximum range is 400 to 500 miles and is the geometrical result of the average height of the region from which the energy is scattered.

Apart from the high level of transmitted power, the main disadvantage of tropospheric scatter is the severe fading that is caused by atmospheric changes, and the many possible paths by which the energy travels through the scattering volume. The problem of fading can be overcome by some form of *diversity reception* in which a number of intercepted signals are combined at the receiving system, or in which the strongest of these signals at any instant is automatically selected. The following types of diversity reception are commonly used:

Frequency diversity

The same information is simultaneously transmitted on a number of separate frequencies that suffer different, and varying, degrees of fading.

Space diversity

A number of receiving antennas are positioned several wavelengths apart. The strongest signal is automatically selected and fed to the receiver.

SKY WAVE

The upper parts of the earth's atmosphere absorb large quantities of radiant energy from the sun, which not only heats the atmosphere, but it also produces some ionization in the form of free electrons and positive ions.

When an electromagnetic wave strikes an atom, it is capable of moving an electron from an inner to an outer orbit. When this occurs, the electron has absorbed energy from the wave. If the frequency of this incident wave is sufficiently high, such as in the ultraviolet region, an electron might be knocked completely out of the atom. When this occurs, a positively charged atom (or ion) remains in space together with a free electron. The rate of ion and free electron formation depends upon the density of the atmosphere and the intensity of the ultraviolet wave. As the ultraviolet wave produces positive ions and free electrons, its intensity diminishes. Therefore, the ionized region will tend to form into a layer because of the following three physical relationships of the upper atmosphere: (a) forming few positive ions and free electrons because of the less dense atmosphere when the ultraviolet radiation is most intense; (b) forming more positive ions and free electrons because of the more dense atmosphere when the ultraviolet wave is of moderate strength; and (c) again forming few positive ions and free electrons because of the low intensity of the ultraviolet wave in the most dense atmosphere. This relationship between ultraviolet intensity, rate of ionization, and atmospheric density is shown in Fig. 177-6.

The formation of the ions and free electrons is not, in itself, sufficient reason to account for the existence of an ionized layer; the positive ions and free electrons tend to recombine because of the inherent attraction

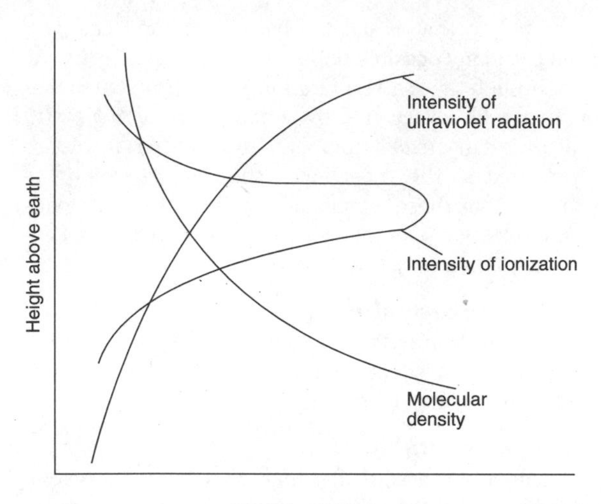

Figure 177-6

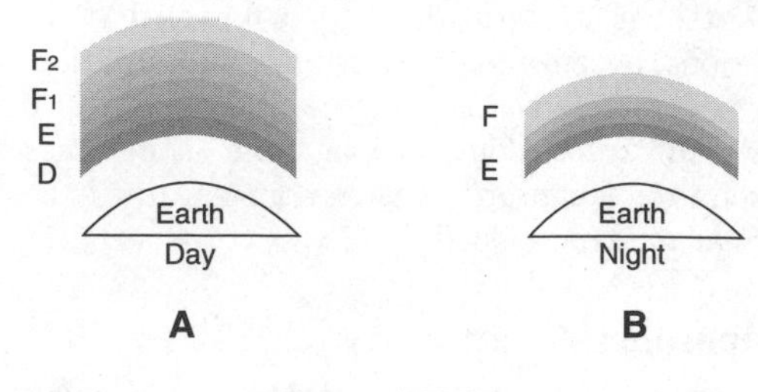

Figure 177-7

of unlike charges. The recombination rate is directly related to the molecular density of the atmosphere because the more dense the atmosphere, the smaller is the mean free path of the electrons. The recombination rate is also directly proportional to the density of the positive ions and the free electrons. Therefore, as the ultraviolet waves continue to produce ions and free electrons, a free-electron density will be reached where the recombination rate just equals the rate of formation. In this state of equilibrium, a free-electron density exists for every set of given conditions.

The formation of more than one layer is explained by the existence of different ultraviolet wave frequencies. The lower frequency ultraviolet waves tend to produce a higher altitude layer, expending all of their energy at this height. The higher frequency ultraviolet waves tend to penetrate deeper into the atmosphere before producing appreciable ionization. In addition to the ultraviolet waves from the sun, particle radiation, cosmic radiation, and meteors all produce ionization in the earth's atmosphere, particularly at the higher altitude layers. Moreover, there is no mixing process in the ionosphere and, consequently, gases are arranged in layers according to their different densities; each gas requires a particular ultraviolet intensity to produce a strong ionization density.

The three principal layers formed in the daytime are called the *E*, *F_1* and *F_2 layers* (Fig. 177-7A). In addition to these regular layers, there is a region below the E layer, which is responsible for much of the daytime attenuation of high-frequency radio waves. Called the *D region* or *layer*, it lies between heights of 50 and 90 km. The heights of the maximum density of the regular E and F_1 layers are relatively constant, with only small diurnal and seasonal changes. The F_2 layer is more variable with typical heights lying within the range 200 to 400 km. The F layers are composed of the lightest gas (hydrogen) in an ionized state.

At night (Fig. 177-7B), the F_1 and F_2 layers join to form a single *F* layer. The regular E layer is governed closely by the amount of ultraviolet radiation from the sun and, at night, tends to decay uniformly with time. The D layer almost entirely disappears at night as a result of recombination.

An anomalous ionization, termed *sporadic E*, is often present in the E region in addition to the regular ionization. "Sporadic E" ionization usually exhibits the characteristics of patches of intense ionization, which can appear anywhere in the height range of 90 to 130 km. These patches can be from 1 km to several hundred km across. The occurrence of sporadic E is quite unpredictable and, although very high frequencies are regularly returned, there is no possibility of predicting the conditions so that they cannot be used for reliable communications.

In contrast with the unpredictable sporadic conditions, the E layer consistently represents a scattering region that is comparable with the troposphere conditions already described in this chapter. However, with ionospheric scatter, the attenuation increases rapidly with frequency so that the usable range only extends from approximately 30 to 100 MHz. The minimum range is determined by the size of the scattering volume and is typically on the order of 250 miles. The maximum range is about 1500 miles and is the geometrical result of the E layer's average height.

The changes that can occur in the ionosphere can be loosely tabulated:

1. *Diurnal changes*. Day and night changes in the height, and the density, of the layers.
2. *Seasonal changes*. These are obviously very much tied in with the geographical position. For example, the ionosphere is very weak and irregular near the earth's poles.
3. *Sporadic E*. Patches of intense ionization in the E layer.
4. *Sudden ionospheric disturbances (SIDs) or Dellinger fade-outs*. These are caused by in-

tense ultraviolet radiation given out by the sun during a solar flare. The result of a SID is a sudden large increase in the ionic density of the highly absorptive D region and an increase in the ionic density of the moderately absorptive E layer.

5. *Magnetic storms* should not be confused with SIDs, although they both have the same effect of reducing the probability of any communications by means of sky waves. Magnetic storms are apparently caused by the emission of particles by the sun. These particles are emitted at the same time as the flare but being of finite mass, they take up to about 36 hours to arrive after a SID. A magnetic storm could last for several days, with its appearance being very sudden, and the recovery of the layers to normal being very slow. As the emitted particles are mostly magnetic, the effects of a magnetic storm are most severe in the geomagnetic polar regions. Magnetic storms are likely to occur during periods of maximum sunspot activity, which occurs in a regular 11-year pattern. At the height of the 11-year sun spot cycle, the emission of ultraviolet waves from the sun is considerably greater than during years of the "quiet sun." During periods of intense solar activity, the ionization is considerably greater and much higher frequencies are regularly returned in the HF spectrum. However, communications are more likely to be interrupted by SIDs and magnetic storms.

When sky-wave propagation is used for communication, the electromagnetic wave from the antenna is transmitted toward the ionized layer at an oblique angle. The incident wave is then apparently reflected back at a certain angle toward the receiving antenna. Actually, the wave is not reflected, although this term is commonly used for convenience; in fact, the wave is bent back toward the earth by refraction, just as a prism refracts light.

The *skip distance* (Fig. 177-8A) is defined as the separation between the transmitting antenna and the point where the first sky wave is returned to the earth's surface. If the frequency is increased, the refraction in the ionosphere decreases and the skip distance increases. Between the positions where the intensity of the ground wave has fallen below the noise level and the places where the sky waves are first returned, there is a *skip* (or silent) *zone* in which no reception is possible (Fig. 177-8B). The extent of the silent zone can be reduced by lowering the frequency. However, if the ground and sky-wave coverages overlap, the two received signals will have taken different paths and the result will be a form of interference fading.

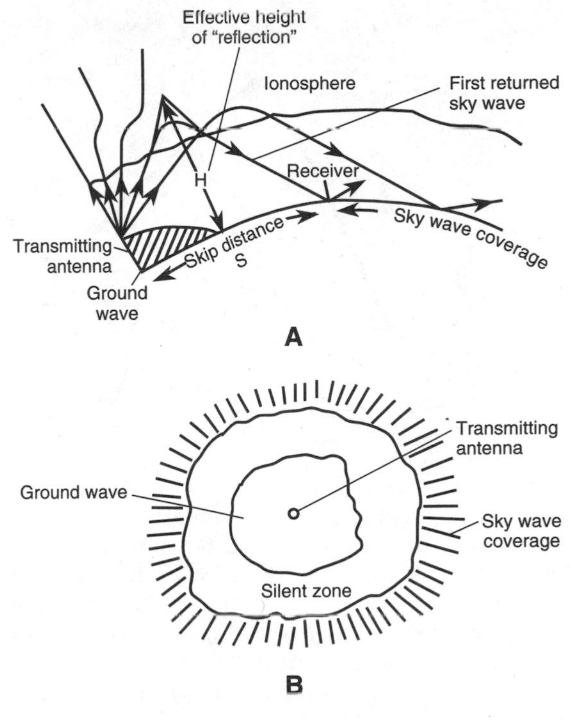

Figure 177-8

If the receiver is located at the position where the first sky wave is returned, the condition of the *maximum usable frequency (MUF)* occurs. If the transmitter frequency is increased, the skip distance will increase, and the receiver will be placed in the silent zone. Because the intensity of the ionosphere decreases at night, it is common practice to lower the frequency used for night-time communications.

Multihop transmission

The sky waves described so far have been single-hop transmissions, where the wave has been refracted only once in the ionosphere before being picked up by the receiver. However, it is possible for the wave to be reflected by the earth back into the ionosphere so that the refraction occurs for a second time (Fig. 177-9). This process can be repeated for a number of times and is referred to as *multihop transmission*, which can be used to transmit over long distances within the frequency range of 9 to 30 MHz. However, in general, single-hop transmissions result in greater field intensities than the multihop type.

As illustrated in Fig. 177-9, it is possible for a receiver to pick up the signal by both single-hop and multihop transmission. Because the two paths are different there will be interference fading, which can be

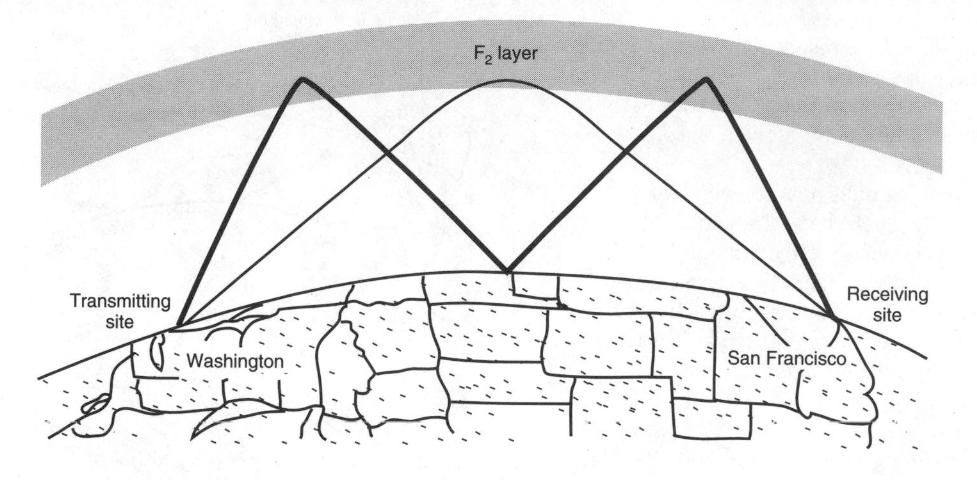

Figure 177-9

overcome by using diversity reception. Over extremely long distances there can be interference between two multihop paths, one of which has travelled the "short way" around the earth's surface while the other has taken the "long way". This form of interference can be avoided by using a *rhombic* antenna, whose one major lobe is directed at a specific angle toward the ionosphere.

Multihop communications can involve a number of different time zones, which use *Universal Time Coordinated (UTC)* as a reference. This reference uses a 24-hour clock (military time), and is the time at 0° Longitude, which passes through the Greenwich Observatory in England. In other words, Universal Time Coordinated is the same as Greenwich Mean Time (GMT). For example, the time on the Pacific coast of North America is eight hours behind GMT; consequently, 3 P.M. PST (Pacific Standard Time) is equivalent to 11 P.M. at Greenwich, or 2300 hours UTC.

Other USA time zones are Mountain Standard Time (MST), Central Standard Time (CST), and Eastern Standard Time (EST). The relationships between all of these zones and UTC are:

PST + 8 hours = UTC
MST + 7 hours = UTC
CST + 6 hours = UTC
EST + 5 hours = UTC

Absorption

A radio wave entering an ionized layer interchanges energy with the free electrons and the ions. If the ions do not collide with gas molecules or other ions, all of the energy transferred to the ionosphere is reconverted back to electromagnetic energy and the wave continues to be propagated with undiminished intensity. On the other hand, where ions and electrons engage in collisions, they dissipate some of the energy which they have acquired from the wave so that the result is a certain degree of attenuation. This attenuation, or *absorption*, is proportional to the product of the number of ions and the collision frequency. Therefore, the attenuation is ordinarily greatest in the region where the product of ion intensity and the collision frequency is greatest. In fact, most absorption occurs in the D layer, which lies closest to the earth. Considerably less absorption occurs in the E layer, and very little in the layers above. It follows that absorption is restricted mainly to the hours of daylight and falls to a low value at night. In general, absorption (during daylight hours) increases to a maximum value at around 1.5 MHz and then decreases rapidly with an increase in frequency.

Maximum usable frequency (MUF)

The basis for all frequency planning is the layer's critical frequency (f_o), and measurements are collected systematically by ionospheric sounding centers throughout the world. The practical parameter, however, is the *maximum usable frequency (MUF)*, of which there are several definitions; in each case, MUF designates the highest frequency, which, in a particular set of conditions, can be used to propagate radio waves over a given route.

Predictions for f_o and *MUF* for the lower layers can be obtained from charts for any given sunspot condition and a knowledge of the sun's elevation. The *MUF*, which applies to the F_2 layer, is not so easily determined because it obeys complex laws and its correlation with solar activity and geographical factors is not so close.

However, figures are published based on a maximum usable frequency that will give a reasonable prospect of good communication. The published figures are the *Optimum Working Frequencies (OWF)*, and in general, these figures are about 85 to 90% of the maximum usable frequency; this allows for slight changes in the ionosphere.

Lowest usable frequency (LUF)

The *LUF* is the lower limiting frequency, which will provide satisfactory communication for a given link. The *LUF* is the frequency at which the received field intensity just equals the required field intensity for reception. The received field intensity depends upon the receiver antenna system, the path length, and the absorption, which generally increases as the frequency is decreased. The required field intensity for reception depends on noise limitations at the site; these generally decrease as the frequency is raised.

From the previous descriptions, it is clear that the MUF is a natural limitation, but the LUF is a man-made limitation. It is possible that conditions can arise where the *LUF* is higher than the *MUF*; in this situation, propagation is impossible.

PROPAGATION UNDER THE VARIOUS FREQUENCY BANDS

The following frequency bands were named as the result of the Atlantic City Conference (1947).

Band	Frequency
Very Low Frequency (VLF)	3 to 30 kHz
Low Frequency (LF)	30 to 300 kHz
Medium Frequency (MF)	300 kHz to 3 MHz
High Frequency (HF)	3 to 30 MHz
Very High Frequency (VHF)	30 to 300 MHz
Ultra High Frequency (UHF)	300 MHz to 3 GHz
Super High Frequency (SHF)	3 to 30 GHz
Extremely High Frequency (EHF)	30 to 300 GHz

VLF and LF bands

Propagation in the VLF and LF bands mainly depends on the use of the ground wave. Vertically polarized radiation is invariably used and the coverage can be worldwide, provided that the transmitter's power is sufficient. Sky waves are refracted back by the ionosphere's D layer and on their return to earth, will establish further ground waves. Notice that the *LOng RAnge Navigation* system (LORAN C) operates on a frequency of 100 kHz in the LF band. At ever greater distances, the Omega radio navigation system is assigned a frequency of 10.2 kHz in the VLF band.

MF band

The range of an MF ground wave can extend to several hundred miles. For frequencies up to 2 MHz, the sky waves are returned by the E layer. During day conditions, the absorption is high but at night, the sky wave can be received well beyond the ground-wave range. However, if the sky waves are returned within the ground-wave coverage, severe interference fading might result.

HF band

The ground-wave range in the HF band is on the order of 100 miles or less. Sky waves normally pass through the D and E layers to be refracted back by the F layer(s). Single-hop transmissions can cover distances of 2000 to 2500 miles with frequencies up to 20 MHz. Multihop transmission can cover any distance in the frequency range of 10 to 30 MHz.

Sky-wave communication can be blacked out by sudden ionospheric disturbances, which are caused by solar flares, and the arrival of charged particles emitted by the sun.

VHF and higher frequency bands

The ground-wave coverage can virtually be ignored. Communication is by space wave, with its range primarily limited by "line-of-sight" conditions. Above 30 MHz (and particularly above 300 MHz), atmospheric noise, such as static, is less than in the lower frequency bands.

Within this frequency range, the radio waves normally pass straight through the ionosphere. However, at the peak of the sunspot cycle, VHF waves can be refracted back to earth so that ranges far in excess of normal are experienced; this can also occur as the result of "sporadic E" conditions.

Anomalous and unpredictable propagation conditions frequently occur as the result of high-level or surface ducting in the troposphere. However, reliable communications can be established through the scattering process in either the troposphere or the ionosphere.

MATHEMATICAL DERIVATIONS

Space wave

Figure 177-10 shows the limiting line-of-sight condition in which the line joining the tops of the antennas just grazes the earth's surface. It is assumed that the earth is a sphere with a radius (R) of approximately 4000 statute miles.

By the theorem of Pythagoras (chapter 201),

$$R^2 + d_1^2 = (R + h_1)^2 = R^2 + 2Rh_1 + h_1^2$$

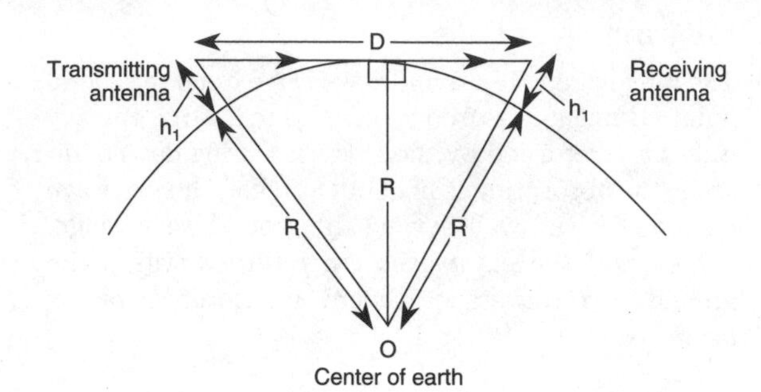

Figure 177-10

or,

$$d_1^2 = 2Rh_1 + h_1^2$$

Because,

$$2Rh_1 >> h_1^2,$$

$$d_1^2 \approx 2Rh_1$$

$$d_1 \approx \sqrt{2R} \times \sqrt{h_1}$$

Similarly,

$$d_2 \approx \sqrt{2R} \times \sqrt{h_2}$$

Maximum optical range,

$$D = d_1 + d_2$$

$$= \sqrt{2R}(\sqrt{h_1} + \sqrt{h_2})$$

If the antenna heights (h_1 and h_2) are measured in feet,

$$D = \sqrt{\frac{2 \times 4000}{5280}}(\sqrt{h_1} + \sqrt{h_2}) \qquad (177\text{-}1)$$

$$= 1.23(\sqrt{h_1} + \sqrt{h_2}) \text{ statute miles}$$

$$= 1.06(\sqrt{h_1} + \sqrt{h_2}) \text{ nautical miles}$$

In practice, the space wave is refracted by the earth's atmosphere so that 10% to 15% must be added to the optical range. As an example, calculate the maximum range when the transmitting and the receiving dipoles are mounted on board two ships, and each dipole is 64 feet above the surface of the sea. The optical range is 1.23 ($\sqrt{64} + \sqrt{64}$) = 1.23 × 16 = 19.7 land miles and, after allowing an additional 10% for atmospheric refraction, the maximum radio range is approximately 22 land miles (or 20 sea miles). However, if one of these ships is communicating with an aircraft flying at 10000 feet, the optical range increases to 1.23 $(\sqrt{64} + \sqrt{10000})$ = 1.23 × 108 = 133 land miles, and the maximum radio range is about 145 miles. For the same reason, it is customary to increase the possible range and service area by locating TV and FM antenna systems on mountain tops. In addition to the line-of-sight transmission and the atmospheric refraction, there is a further small increase in the range because of the diffraction around the earth's surface. In particular, if a space wave in the UHF band passes over a mountain whose peak has a sharp edge, part of the wave is diffracted around the edge and down the far side of the mountain. This effect is known as *knife-edge diffraction* or (wrongly) as *knife-edge refraction.*

Sky wave

As far as propagation is concerned, the major effect of ionization is to reduce the refractive index (μ) of the ionosphere. This causes the wave (travelling from a medium of high to a medium of low) to be refracted in accordance with Snell's optical law. Referring to Fig. 177-11,

$$\mu = \frac{\sin i}{\sin r} \qquad (177\text{-}2)$$

where: i = angle of incidence measured relative to the normal.
r = angle of refraction measured relative to the normal.

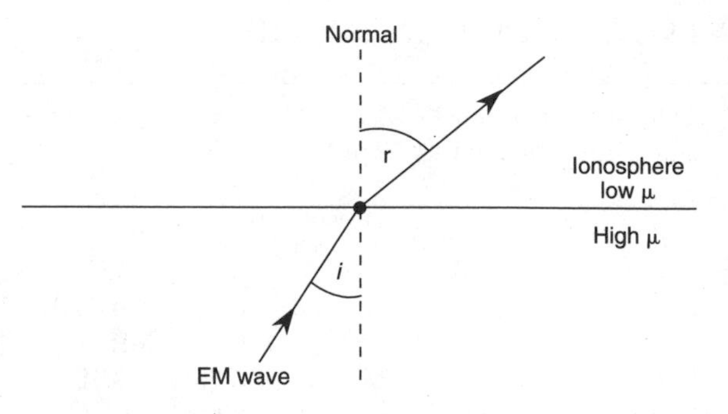

Figure 177-11

When the radio wave approaches the layer at an oblique angle, the upper part of the wave front will pass through a region of stronger ionization, compared with the lower part. Because the wave's phase velocity increases with a greater ionization density, the top of the wave front will move faster than the bottom and refraction will occur.

The refractive index of the ionosphere is a function of the frequency (f) of the wave and is given by,

$$\mu = \sqrt{1 - \frac{81N}{f^2}} \qquad (177\text{-}3)$$

where N = the number of free electrons per cubic centimeter of the ionized region.

For a given frequency, μ decreases as the density of ionization increases toward the center of the ionized layer.

At a certain value of μ, depending on the angle of incidence, $\sin r = 1$ and $r = 90°$ (Fig. 177-12). This makes the path of the wave perpendicular to the earth's radius and further refraction will cause the wave eventually to leave the ionized layer at the same angle as it entered. The further refraction occurs provided that the top of the wave is still travelling in a region of greater ionospheric density.

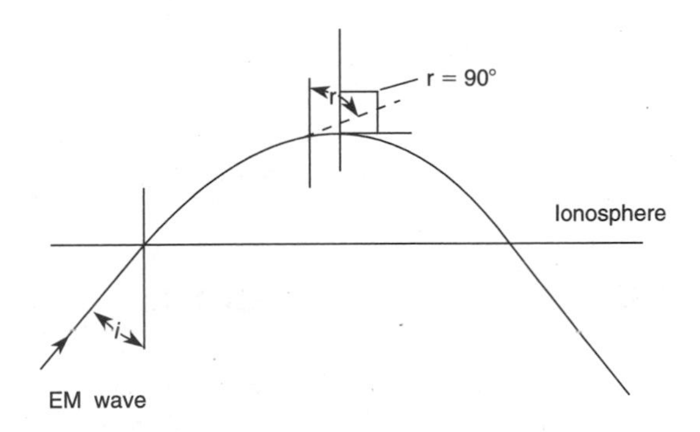

Figure 177-12

In the special case of a vertically incident wave, μ must reach a value of zero for the wave to be returned to earth. The highest value at which this occurs in a given layer is called the *critical frequency* (f_o) of the layer.

From Equation 177-3,

$$0 = \sqrt{1 - \frac{81N_{max}}{f_o^2}}$$

This yields,

$$f_o = \sqrt{81N_{max}} \text{ Hertz} \qquad (177\text{-}4)$$

where N_{max} = maximum free electron density in the layer.

From a knowledge of the critical frequency of a layer, the highest frequency, at an oblique angle of incidence (i), which will be returned to earth, can be calculated. Because $r = 90°$, $\sin r = 1$, and:

$$\sin i = \mu = \sqrt{1 - \frac{f_o^2}{f^2}}$$

$$\sin^2 i = \mu^2 = 1 - \frac{f_o^2}{f^2}$$

This yields,

$$\cos^2 i = \frac{f_o^2}{f^2}$$

and,

$$f = f_o \sec i \qquad (177\text{-}5)$$

This is known as the *Secant Law*, which is the basic equation used in choosing operating frequencies for a given transmission path. It is only true for a flat earth layer and must be modified for accurate calculations. Assuming that such conditions exist, the length (S) of the skip distance (Fig. 177-8A) is given by:

$$S = 2H \tan i \qquad (177\text{-}6)$$

$$= 2H\sqrt{\left(\frac{f}{f_o}\right)^2 - 1} \text{ miles}$$

where H = effective height of reflection (mi).

The Secant Law shows that, for a particular frequency, there is a certain minimum angle of incidence for which the first sky wave is returned to the earth. As the angle is increased, the sky wave will penetrate the ionospheric layer to a lesser depth. However, the length of the path through the ionosphere will increase and the absorption will be greater. Consequently, there is a limit to the reception zone that is covered by the sky wave.

Example 177-1

The antenna of a TV broadcast station is located at the top of a mountain whose height is 10000 feet. Neglecting diffraction, calculate the maximum reception range for a receiving antenna with a height of 25 feet.

Solution

Maximum optical range,

$$D = 1.23\left(\sqrt{h_1} + \sqrt{h_2}\right) \qquad (177\text{-}1)$$

$$= 1.23\left(\sqrt{10000} + \sqrt{25}\right)$$

$$\approx 130 \text{ land miles.}$$

Adding 15% because of the atmospheric refraction,

Maximum reception range = 1.15 D

$\approx$ 150 land miles.

Example 177-2

The maximum free electron density in an ionized layer is 7.9×10^{11} electrons per cubic centimeter. Calculate the value of the critical frequency. For a frequency of 17.5 MHz, calculate the angle of incidence for which the first returned sky wave is refracted back to earth.

Solution

Critical frequency,

$$f_o = \sqrt{81N_{max}} \qquad (177\text{-}4)$$

$$= \sqrt{81 \times 7.9 \times 10^{11}} \text{ Hz}$$

$$\approx 8.0 \text{ MHz.}$$

Minimum angle of incidence,

$$i = \text{inv.cos.} \frac{8.0}{17.5} \qquad (177\text{-}5)$$

$$\approx 63°$$

PRACTICE PROBLEMS

177-1. In Example 177-2, the effective height of reflection is 110 miles. Calculate the approximate value of the skip distance.

177-2. A space wave is required to have a maximum optical range of 90 land miles. If the receiving antenna has a height of 50 feet, calculate the required height for the transmitting antenna.

177-3. If the value of a critical frequency is 7.5 MHz, what is the maximum value of the required electron density. If the required angle of incidence is 60°, what is the maximum frequency for which there will be a first returned sky wave (assuming that the electron density remains the same).

178 Microwave antenna with the parabolic reflector

At microwave frequencies, it is possible to design highly directive antenna systems of reasonable size. One of the most common systems uses a paraboloid as a "dish" reflector with a flared waveguide termination placed at the focus (Fig. 178-1A). If the paraboloid were infinitely large, the result of the reflection would be to produce a unidirectional beam with no spreading. However, with a paraboloid of practical dimensions, there is some degree of spreading, and the radiation pattern then contains a major lobe (and a number of minor lobes) as illustrated in Fig. 178-1B. The full lobe is, of course, three dimensional so that it is only showing the pattern in one plane.

Referring to the narrow major lobe produced by the antenna with the paraboloid reflector, the beamwidth in the horizontal plane is measured by the angle P_1OP_2 (where P_1 and P_2 are the half-power points). In radar systems, typical beamwidths are of the order of 1° to 5°. Because of the construction of the paraboloid, the horizontal and vertical beamwidths are not necessarily the same. For a microwave link the normal beamwidth is 2°; with such narrow beamwidths the power gains of the antenna systems are extremely high.

For a paraboloid reflector to be effective, the value of d must be greater than 10 wavelengths. Consequently, such reflectors would not be practical in the VHF band because of their excessive size.

Satellite systems are operated in the microwave region. When a shipborne parabolic antenna is aimed at a particular satellite in order to receive a signal, there are two angles to consider—the *angle of elevation* in the vertical plane and the *angle of azimuth* in the horizontal plane. The angle of azimuth is the bearing as measured from true North.

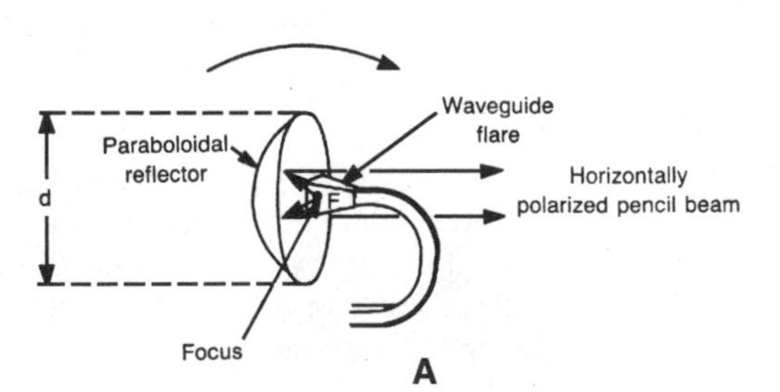

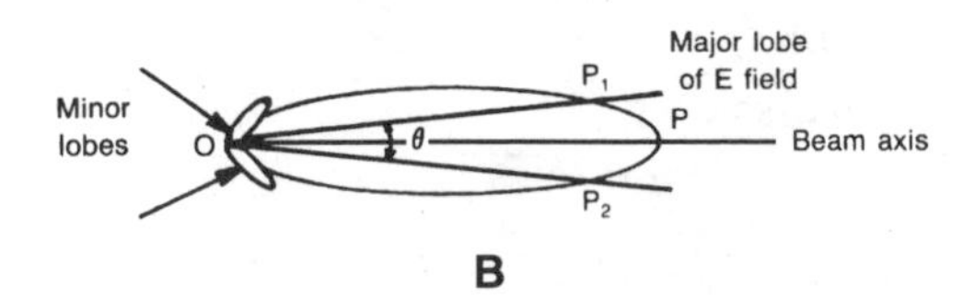

Figure 178-1

MATHEMATICAL DERIVATIONS

The beamwidth is given by:

Beamwidth angle,

$$\theta = \frac{70\lambda}{d} \text{ degrees} \qquad (178\text{-}1)$$

With respect to the half-wave dipole standard, the power gain (G_p) is given by,

$$G_p = 6\left(\frac{d}{\lambda}\right)^2 \qquad (178\text{-}2)$$

where: λ = wavelength (m)
d = diameter of paraboloid (m).

Example 178-1

An X-band microwave antenna system is operating at 10 GHz. If the parabolic reflector has a diameter of 1 m, what are the values of the beamwidth and the antenna power gain?

Solution

Wavelength,

$$\lambda = \frac{30}{10 \text{ GHz}}$$

$$= 3 \text{ cm}$$

Beamwidth angle,

$$\theta = \frac{70\lambda}{d} \tag{178-1}$$

$$= \frac{70 \times 0.03 \text{ m}}{1 \text{ m}}$$

$$= 2.1°$$

Power gain,

$$G_p = 6\left(\frac{d}{\lambda}\right)^2 \tag{178-2}$$

$$= 6 \times \left(\frac{1 \text{ m}}{0.03 \text{ m}}\right)^2$$

$$= 6667.$$

PRACTICE PROBLEMS

178-1. An S-band microwave antenna system is operating at 3 GHz. If the parabolic reflector has a diameter of 10 ft, what is the value of the beamwidth of the radiated major lobe?

178-2. In Practice Problem 178-1, what is the value of the power gain? If the RF power fed to the antenna is 120 W, calculate the value of the effective radiated power.

179 Propagation in a rectangular waveguide

The first question is: "Does the energy move straight down a rectangular waveguide as a TEM wave?" The answer is "no." If you enclosed a rectangular waveguide around a TEM wave, you could not obey the necessary boundary conditions. These conditions are:

1. There can be no E-field component that is parallel to and existing at an inner surface of the waveguide.
2. There can be no H-field component that is perpendicular to an inner surface of a waveguide.

In fact, the electromagnetic energy progresses down the guide by a series of reflections off the internal surface of the narrow dimension (Fig. 179-1A). At each reflection, the angles (θ) of incidence and reflection are equal.

Consider a TEM wave that approaches a plane (flat metal surface) at an angle θ (Fig. 179-1B). The dark lines with their arrows represent the directions of the incident and reflected wavefronts, which are moving with the free space velocity (virtually at the speed of light). The full lines are $\lambda/2$ apart in free space and they represent the incident H-flux lines. The broken lines are also $\lambda/2$ apart and they are used to show the reflected H-field. The symbols for the incident and reflected E-fields are respectively (⊗, ⊙) and (×, ·). These incident and reflected waves are of the same amplitude, but they have a phase reversal of 180°.

The fields of the incident and reflected waves are superimposed and must be combined to produce the resultant E- and H-field patterns. At the metal surface itself, the phase reversal causes cancellation between the incident and the reflected E-fields so that there are no resultant lines that are parallel to the surface.

When the incident and the reflected waves are combined, the resulting pattern of the E- and H-fields (between the lines XX' and YY') is exactly the same as can be derived from a support stub analogy. Therefore, you can infer that the EM wave must progress down the guide by a series of reflections off the narrow dimension.

So far, so good. Now, you must find out the factors that determine the angle of incidence, and the velocity with which the energy progresses down the guide. To do this, extract the triangle LMN from Fig. 179-1B, and display its magnified form in Fig. 179-1C. As the wavefront moves from P to N (a distance of $\lambda/2$), the field pattern progresses a greater distance from M to N. The distance MN is a half wavelength of the field pattern as

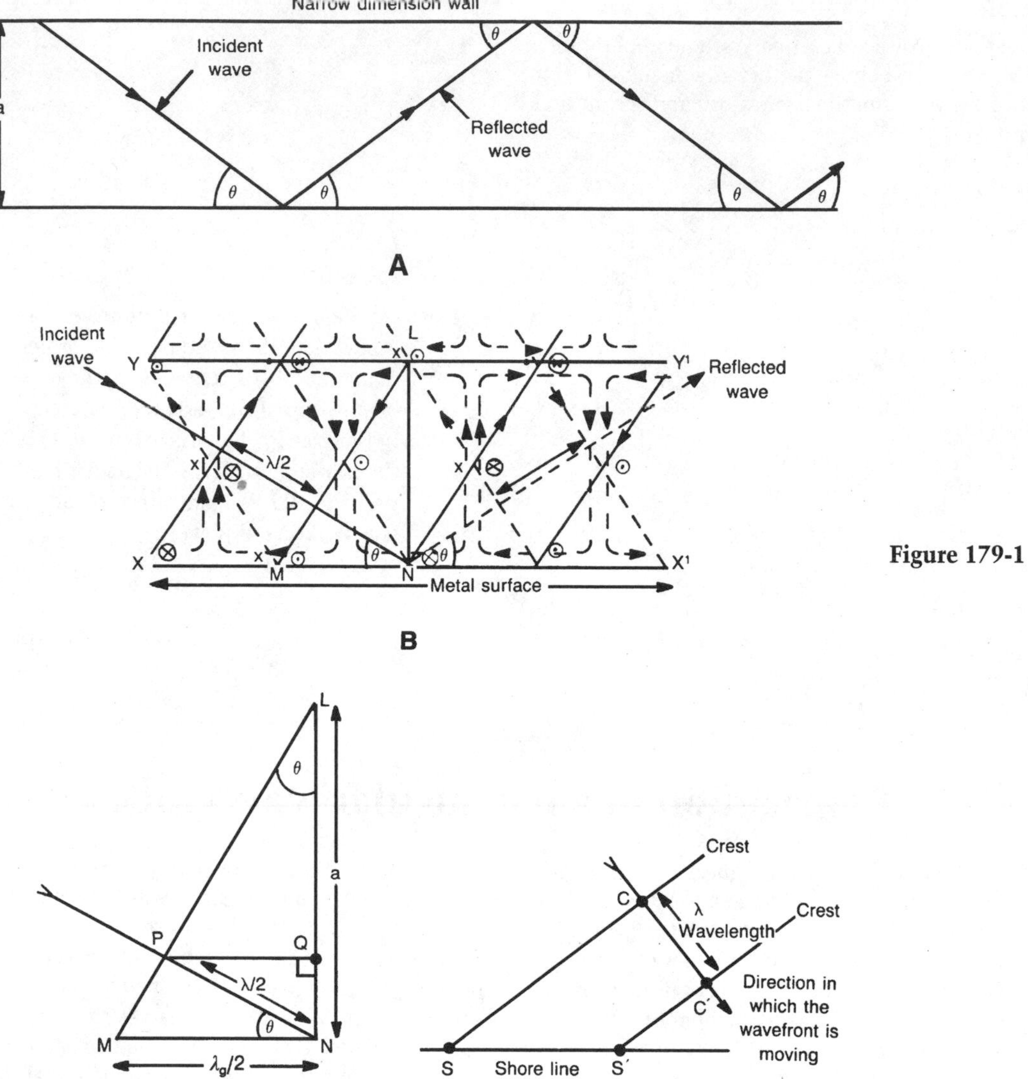

Figure 179-1

it exists inside the waveguide, and is termed $\lambda_g/2$ (where λ_g is the guide wavelength). For an analogy, think of sea waves approaching the shore line at an angle (Fig. 179-1D). As the wavefront moves from one crest to the next (the distance CC′, one wavelength), the pattern of the crests at the shore line covers the greater distance, *SS′*.

For a given rectangular waveguide, the wide (a) and the narrow (b) dimensions are fixed. If the frequency is lowered, the wavelength is longer and the value of λ is increased. In the limiting cut-off condition, the cutoff wavelength equals $2a$ and $\theta = 90°$; the wave will then bounce back and forth between the narrow dimensions, and will not progress down the guide. When the frequency is raised, the wavelength decreases and the value of θ is lowered. This could be continued indefinitely; however, at a frequency of approximately $1.9\, f_c$, the narrow (b) dimension comes into play and limits the top frequency at which the waveguide can successfully operate.

MATHEMATICAL DERIVATIONS

In the right-angled triangle LPN,

$$\sin\theta = \frac{PN}{LN} = \frac{\lambda/2}{a} = \frac{\lambda}{2a} \qquad (179\text{-}1)$$

In the right-angled triangle MPN,

$$\cos\theta = \frac{PN}{MN} = \frac{\lambda/2}{\lambda_g/2} = \frac{\lambda}{\lambda_g} \qquad (179\text{-}2)$$

Therefore,

$$\tan\theta = \frac{\sin\theta}{\cos\theta} = \frac{\lambda/2a}{\lambda/\lambda_g} = \frac{\lambda_g}{2a} \qquad (179\text{-}3)$$

Equation 179-3 shows that, if the frequency is decreased, θ rises, and the guide wavelength (λ_g) increases. In the cutoff condition, $\theta = 90°$ and the guide wavelength (λ_g) is infinitely long.

Using the trigonometric relationship $\sin^2\theta + \cos^2\theta = 1$, you can write,

$$\left(\frac{\lambda}{2a}\right)^2 + \left(\frac{\lambda}{\lambda_g}\right)^2 = 1$$

This yields,

$$\lambda = \frac{\lambda_g}{\sqrt{1 + \left(\frac{\lambda_g}{2a}\right)^2}} \qquad (179\text{-}4)$$

and,

$$\lambda_g = \frac{\lambda}{\sqrt{1 - \left(\frac{\lambda}{2a}\right)^2}} \qquad (179\text{-}5)$$

Remembering that,

$$\lambda = c/f \text{ and } \lambda_c = 2a = c/f_c, \qquad (179\text{-}6)$$

$$\lambda_g = \frac{\lambda}{\sqrt{1 - \left(\frac{f_c}{f}\right)^2}}$$

where λ_c and f_c are, respectively, the cutoff wavelength and the cutoff frequency.

In the time that the wavefront moves from P to N at the speed of light (c), the field pattern progresses from M to N (a distance of $\lambda_g/2$) with the *phase* velocity (v_ϕ). During the same time interval, the electromagnetic energy has moved down the guide a distance PQ. This physical movement of energy occurs at the *group* velocity (v_g).

Because

$$MN = PN \sec\theta \text{ and } PQ = PN\cos\theta,$$

Phase velocity,

$$v_\phi = \frac{c}{\cos\theta} = c\sec\theta \qquad (179\text{-}7)$$

and group velocity,

$$v_g = c\cos\theta \qquad (179\text{-}8)$$

Therefore,

$$v_\phi \times v_g = \frac{c}{\cos\theta} \times c\cos\theta = c^2$$

where v_ϕ, v_g, and c are all measured in meters per second.

At the cutoff condition, the group velocity is zero and the phase velocity is infinite. This result does not contravene any physical laws because the phase velocity involves the *apparent* movement of a field pattern and not the movement of any physical quantity. As the frequency is raised, v_ϕ decreases and v_g increases; both of these velocities approach the velocity of light (c). For an X-band waveguide, typical graphs of v_g and v_ϕ versus frequency are shown in Fig. 179-2.

You are left to consider the waveguide's phase-shift constant, (β). A phase shift of 2π radians occurs over a distance equal to the guide wavelength (λ_g). Therefore:

Phase shift constant,

$$\beta = \frac{2\pi}{\lambda_g} \qquad (179\text{-}9)$$

where: β = phase shift constant (rad/m)
λ_g = guide wavelength (m).

These equations refer to propagation in the dominant TE_{10} mode. On the numbering system, the first subscript ("1") indicates the number of half-wave E patterns, which occur across the wide dimension; the second subscript ("0") refers to the number of half-wave E patterns across the narrow dimension. The dominant mode is normally used because it allows for the shortest wide dimension over a given frequency range.

In chapter 173, it was determined that the intrinsic impedance (η_o) of free space was 120 π = 377 Ω. But what impedance does a waveguide possess? For waveguides, we refer to the wave impedance, which is comparable with the surge impedance of transmission lines. However, this wave impedance (η) depends on the particular mode of operation. For example:

All TE modes: The wave impedance,

$$\eta = \frac{\eta_o \times \lambda_g}{\lambda} \text{ ohms} \qquad (179\text{-}10)$$

All TM modes: The wave impedance,

$$\eta = \frac{\eta_o \times \lambda}{\lambda_g} \text{ ohms} \qquad (179\text{-}11)$$

Notice that the value of the wave impedance depends on the values of η_o, the frequency, and the wide a dimension. This means that the concept of wave impedance cannot be used to match two waveguides sections with different narrow b dimensions.

Consider two waveguide sections that have the same wide dimensions, frequency, and the same mode

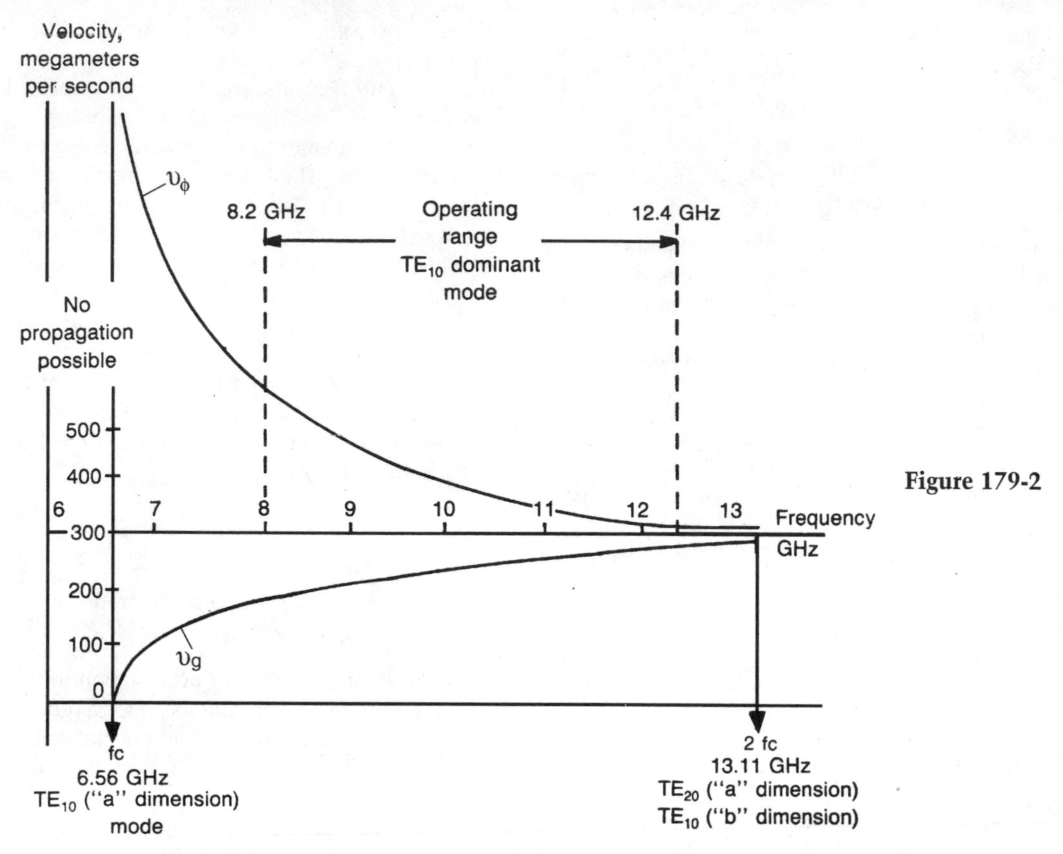

Figure 179-2

of operation. If the narrow dimensions are different, the two sections can be matched only if their characteristic impedances are made equal. The characteristic impedance of a waveguide is defined by,

$$\textit{Characteristic impedance} = \frac{\eta_o \times \pi b \lambda_g}{2a\lambda} \text{ ohms} \tag{179-12}$$

The match can be achieved by using a taper to adjust the values of b and a until the characteristic impedances of the sections are the same. In another application, it is possible to flare out a guide until the factor $\pi b\lambda_g/(2a\lambda)$ is equal to unity. The characteristic impedance is then matched to the intrinsic impedance of free space (Fig. 178-1).

As far as possible, waveguides should be kept free from moisture, which alters the value of the wave impedance so that there is some degree of mismatch and the SWR value increases. To reduce the effect of moisture, some waveguides are filled with pressurized nitrogen gas. To prevent any accumulation of moisture, long horizontal runs of waveguide are normally not used. If such runs are unavoidable, the waveguide is slightly tilted and there is a drain hole at the lower end.

Example 179-1

A practical rectangular waveguide is designed to operate over the X-band, 8.2 to 12.4 GHz. The inner guide dimensions are 2.286 cm and 1.143 cm. If the transmitted frequency is 10 GHz, find the values of the free space wavelength, the cut-off wavelength, the cut-off frequency, the angle of incidence, the guide wavelength, the phase velocity, the group velocity, the phase-shift constant, the wave impedance, and the characteristic impedance.

Solution

Free space wavelength,

$$\lambda = \frac{c}{f}$$

$$= \frac{3 \times 10^8 \text{ m/s}}{10 \times 10^9 \text{ Hz}}$$

$$= 0.03 \text{ m}$$

$$= 3 \text{ cm.}$$

Cut-off wavelength,

$$\lambda_c = 2a = 2 \times 2.286$$

$$= 4.572 \text{ cm}$$

Cut-off frequency,

$$f_c = \frac{c}{\lambda_c}$$

$$= \frac{3 \times 10^8 \text{ m/s}}{4.572 \times 10^{-2} \text{ m}} \text{ Hz}$$

$$= 6.562 \text{ GHz}$$

Angle of incidence,

$$\theta = \text{inv sin } \frac{\lambda}{2a} \quad (179\text{-}1)$$

$$= \text{inv sin } \frac{3}{4.572}$$

$$= 41.01°$$

Guide wavelength,

$$\lambda_g = \frac{\lambda}{\cos \theta} \quad (179\text{-}2)$$

$$= \frac{3}{\cos 41.01°}$$

$$= 3.98 \text{ cm}$$

Phase velocity,

$$v_\phi = \frac{c}{\cos \theta} \quad (179\text{-}7)$$

$$= \frac{300}{\cos 41.01°}$$

$$= 397.6 \text{ megameters per second.}$$

Group velocity,

$$v_g = c \cos \theta \quad (179\text{-}8)$$

$$= 300 \cos 41.01°$$

$$= 226.4 \text{ megameters per second.}$$

Phase shift constant,

$$\beta = \frac{2\pi}{\lambda_g} \quad (179\text{-}9)$$

$$= \frac{2\pi}{3.98 \times 10^{-2} \text{ m}}$$

$$= 159 \text{ radians per meter.}$$

Wave impedance,

$$\eta = \frac{\eta_o \times \lambda_g}{\lambda} \quad (179\text{-}10)$$

$$= \frac{377 \times 3.98}{3.0}$$

$$= 500 \ \Omega.$$

Characteristic impedance (179-12)

$$= \frac{\eta_o \pi b \lambda_g}{2a\lambda}$$

$$= 377 \times \frac{\pi \times 1.143 \times 3.98}{2 \times 2.286 \times 3.0}$$

$$= 393 \ \Omega.$$

PRACTICE PROBLEMS

179-1. An S-band waveguide has internal dimensions of 7.214 cm and 3.404 cm. What are the cut-off wavelengths and frequencies for (a) the TE_{10} mode in the wide dimension, (b) the TE_{20} mode in the wide dimension, and (c) TE_{10} mode in the narrow dimension.

179-2. In Practice Problem 179-1, the frequency of the signal feeding the waveguide is 3.3 GHz. For the dominant mode, what are the values of (a) the wavelength in free space, (b) the angle of incidence, (c) the guide wavelength, (d) the group velocity, (e) the phase velocity, and (f) the phase shift constant.

179-3. In Practice Problem 179-2, what are the values of the wave and characteristic impedances?

179-4. The internal dimensions of a rectangular waveguide are 7.2 cm and 3.6 cm. Determine the frequency range over which only the dominant TE_{10} mode will be propagated down the waveguide.

179-5. A rectangular waveguide is excited by a 5.8-GHz signal. If the group velocity is $0.88 \times c$ (the velocity of light), what is the length of the wide dimension ?

179-6. The angle of incidence is 25° for an EM wave propagating down a rectangular guide in the TE_{10} mode. If the wide dimension is 2.8 cm, calculate the frequency of the EM wave.

180
Circular waveguides

Although rectangular guides are used almost exclusively in radar systems, there are specific cases in which circular waveguides find their application. A good example occurs in a radar antenna system, which is required to revolve, relative to the stationary transmitter and receiver. When one waveguide section is rotated relative to another, it is impossible to maintain continuity with rectangular guides; the only solution is to use a circular guide.

The dominant mode of a circular waveguide is illustrated in Fig. 180-1A. For this mode, the cutoff wavelength is 1.71 times the inner diameter (d) of the waveguide. Remember that the cutoff wavelength of a rectangular guide is twice the wide dimension when operating in the TE_{10} dominant mode. For the same cutoff frequency, $1.71d = 2a$ or $d = 2a/1.71 = 1.17a$. Thus, the circular guide is larger than the corresponding rectangular waveguide—an obvious disadvantage.

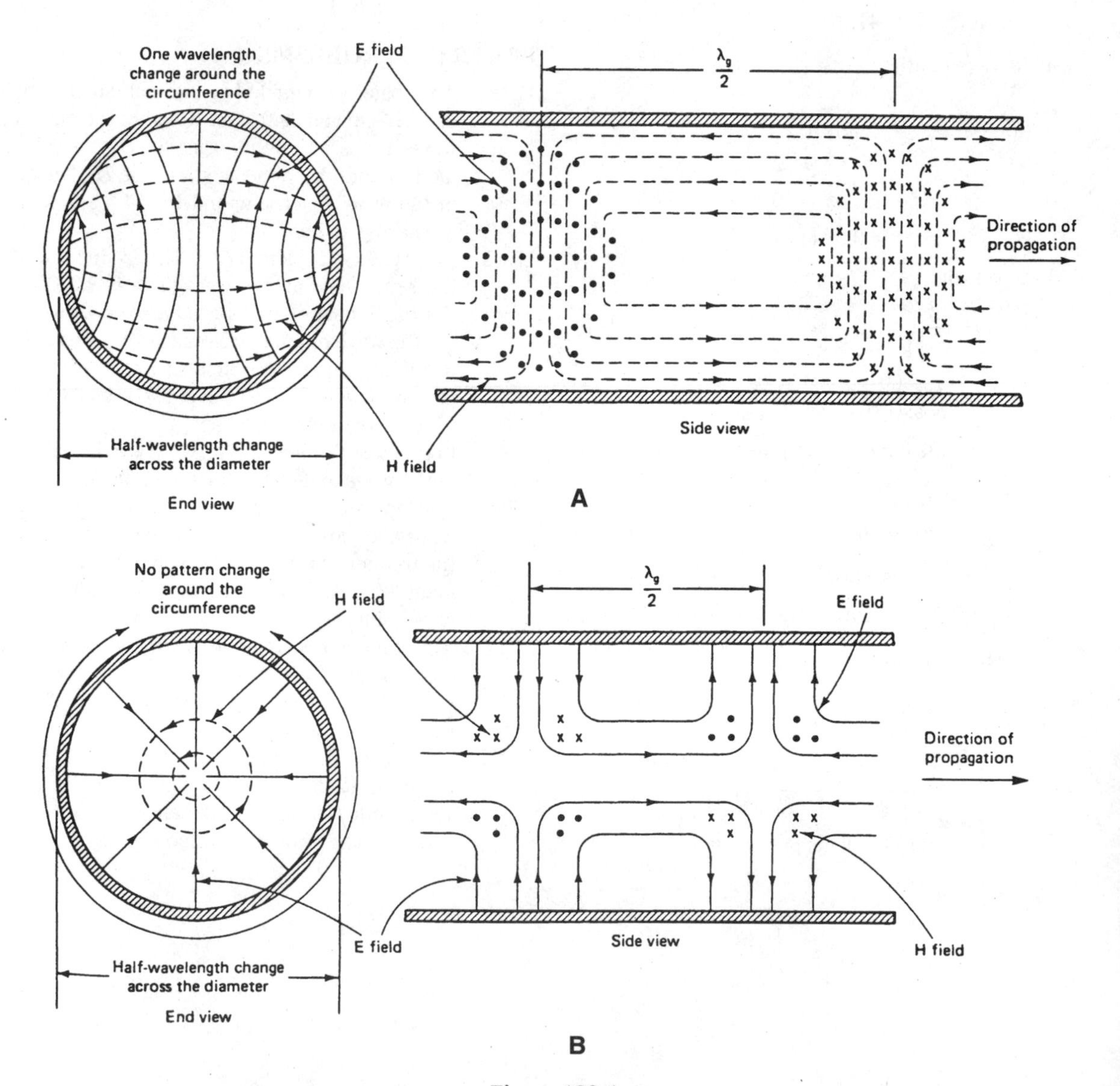

Figure 180-1

However, the circular waveguide suffers from a far more serious problem. In a rectangular guide, the directions of the E and H lines can be referred to the directions of the narrow and wide dimensions. With a circular guide, no such references exist. Consequently, the wave's plane of polarization tends to rotate as the wave moves down the guide. It can then be difficult to remove the energy from the guide because, for example, an E probe requires that the direction of the probe is parallel to the direction of the E lines.

To classify the various modes that can exist in a circular guide, again divide the modes into TE and TM. In the numbering system, the first subscript indicates the number of full-wave E patterns that are encountered as you move around the circumference; the second subscript refers to the number of half-wave E patterns across the diameter. For the *dominant mode*, the electric flux lines are entirely transverse. There is one full-wave pattern around the circumference and one half-wave pattern across the diameter; the full designation of the dominant mode is, therefore, TE_{11} (Fig. 180-1A).

When the center conductor is removed from a coaxial cable to create a circular waveguide, the mode of operation is designated as TM_{01} (Fig. 180-1B), whose cutoff wavelength is only 1.31 times the guide's diameter. This mode is of particular interest because its E and H patterns are compatible with the fields that exist for the TE_{10} dominant mode of the rectangular guide. This allows you to convert from a rectangular guide to a circular guide, carry out a rotation, then convert back to a rectangular guide (Fig. 180-2).

Summarizing, the use of the circular (rather than a rectangular) guide is limited by its increased size and the problem associating with the twisting of the wave's polarized plane.

THE FERRITE ISOLATOR

Another use of a circular waveguide section occurs in the *isolator* device, which has a low forward loss, but a high reverse loss. This effect is the result of the nonreciprocal phase shift created by a ferrite material that is the heart of the isolator circuit. In 1845, Michael Faraday demonstrated that the plane of a linearly polarization light wave is rotated when the light is passed through certain materials in a direction parallel to the flux lines of an external magnetic field. The same effect is produced in the microwave region when operating with ferrite materials. Such materials are transparent to electromagnetic waves, possess excellent magnetic properties, and have very high values of specific resistance.

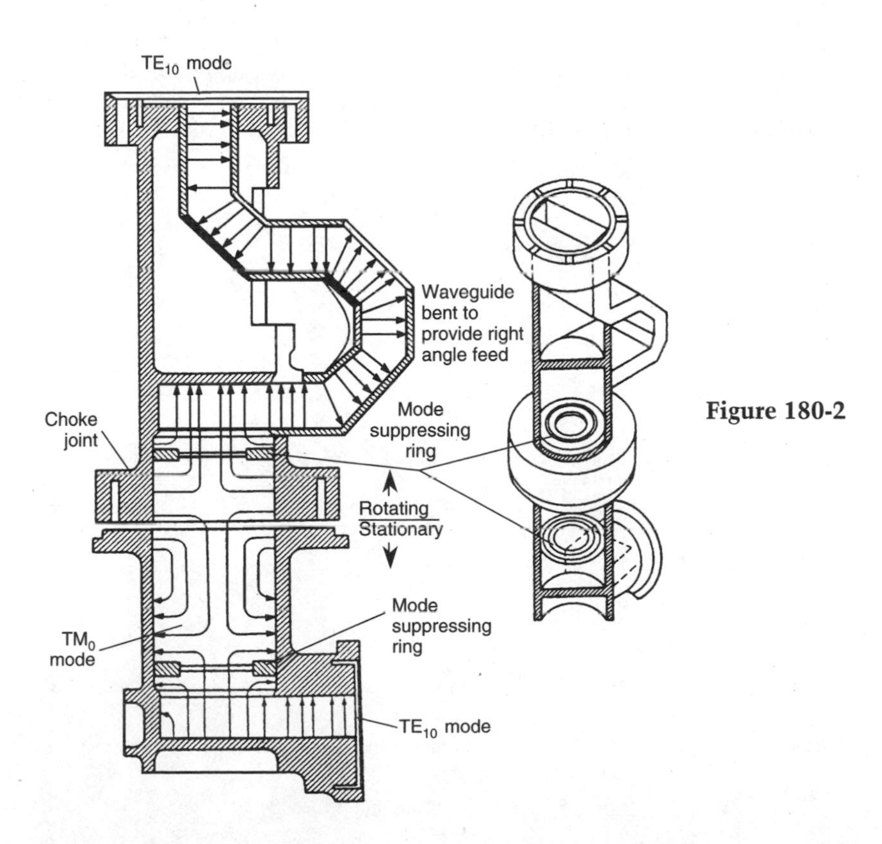

Figure 180-2

The basic principle of a microwave isolator is illustrated in Fig. 180-3A. At the input end, there is a rectangular waveguide, which (relative to the plane of the paper) is fed by an EM-wave with vertical electric flux lines. A transition is now made to a circular waveguide operating in the TE_{11} mode, and a resistive attenuator vane is inserted parallel to the original rectangular guide's wide dimension. Because the E-lines are perpendicular to the plane of the vane, no attenuation occurs.

The next step is to pass the wave through the ferrite specimen that is situated in the presence of the external magnetic field. The result is to twist the plane of polarization through 45°, in the clockwise direction. The emerging electric field is also at right-angles to the second resistive vane so that the attenuation of the isolator in the *forward* direction has been kept to a minimum. Finally, there is another conversion from a circular to a rectangular waveguide whose orientation corresponds to the emerging wave's plane of polarization.

If you now attempt to send the same signal as a reverse wave back through the isolator, the plane of polarization will be twisted another 45° in the *same* direction as the *shift* of the forward wave (Fig. 180-3B). Consequently, when the reverse wave emerges from the ferrite specimen the E-lines will be parallel to the first attenuator vane, and severe attenuation is the result. Typically, the low forward loss is about 1 dB or less, whereas the high reverse loss is 20 dB or more. If the external magnetic field is supplied by an electromagnet, whose current can be varied, the characteristics of the isolator can be changed. One purpose of an isolator is to buffer a microwave source from the effects of a varying load.

THE CIRCULATOR

The properties of the ferrite specimen used in the isolator can be adapted to the *circulator*. There are many applications of circulators, but the principle is to establish various entry/exit points, or *ports*, where the RF power can either be fed or extracted.

Figure 180-4 shows a circulator that uses the principle of the 45° Faraday rotation. The results of the circulator are:

1. When a vertically polarized wave enters port 1, its plane of polarization is rotated through 45° by the ferrite specimen and the wave then leaves through port 2. Port 3 and port 4 represent incompatible junctions.
2. A 45° polarized wave entering port 2 is horizontally polarized after passing through the ferrite specimen and will, therefore, exit from port 3. Likewise, a wave entering port 3 has its plane of polarization rotated through 45° and only leaves through port 4.
3. A wave entering port 4 will be vertically polarized after leaving the ferrite specimen and will then emerge from port 1.

These results are summarized in the circular network diagram of Fig. 180-5. As a simple rule, a 90° rotation from the entrance port, in the clockwise direction of the circular arrow, will automatically locate the single

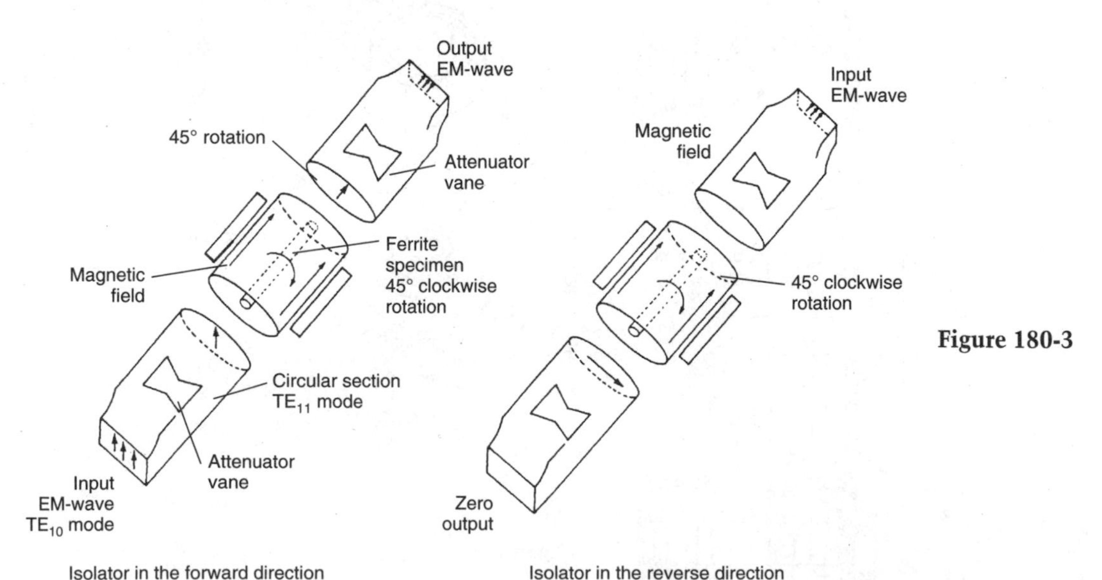

Figure 180-3

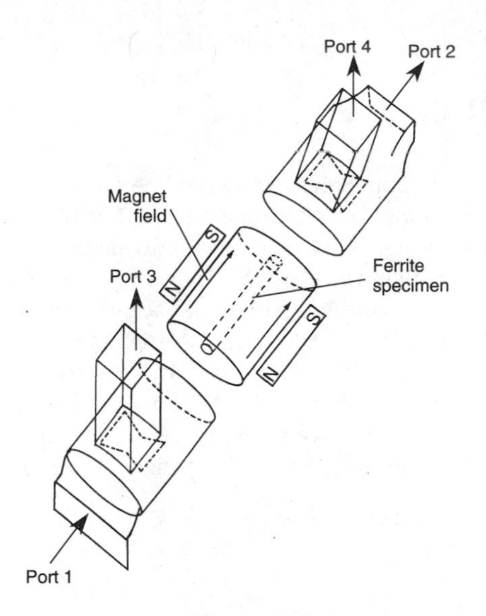

Figure 180-4

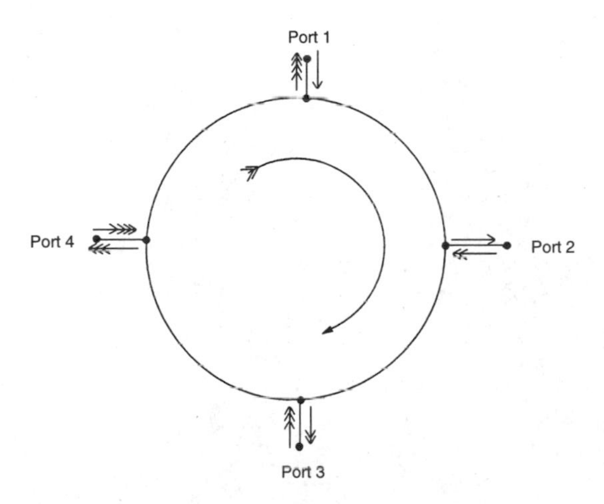

Figure 180-5

exit port. Such an arrangement could be used as the *duplexer* of a radar system, which uses the same antenna for both the transmitter and the receiver.

MATHEMATICAL DERIVATIONS

Circular Waveguides:

TE_{11} dominant mode:

Cut-off wavelength,

$$\lambda_c = 1.71\,d$$

TM_{01} mode:

Cut-off wavelength,

$$\lambda_c = 1.31\,d$$

where d = inner diameter of the waveguide (cm).

Example 180-1

A circular waveguide has an inner diameter of 4.5 cm. Calculate its cut-off frequency when it is operated in (a) the TE_{11} mode, and (b) the TM_{01} mode.

Solution

(a) For the TE_{11} mode, the cut-off wavelength,

$$\lambda_c = 1.71 \times 4.5 = 7.695 \text{ cm}$$

The cut-off frequency,

$$f_c = \frac{30}{7.695} = 3.9 \text{ GHz}$$

(b) For the TM_{01} mode, the cut-off wavelength,

$$\lambda_c = 1.31 \times 4.5 = 5.895 \text{ cm}$$

The cut-off frequency,

$$f_c = \frac{30}{5.895} = 5.1 \text{ GHz}$$

PRACTICE PROBLEMS

180-1. A circular waveguide has an internal diameter of 5.276 cm. What is the value of the cut-off frequency in the TE_{11} dominant mode?

180-2. A circular waveguide has an internal diameter of 3.815 cm. What is the value of the cut-off frequency in the TM_{10} mode?

180-3. A circular waveguide and a rectangular waveguide have the same cut-off frequency, and are both operated in their dominant modes. Assuming that the rectangular waveguide's wide dimension is twice the narrow dimension ($a = 2b$), what is the ratio of the circular waveguide's cross-sectional area to that of the rectangular waveguide?

181
The 50-Ω Smith chart

In chapter 168, you were able to determine the conditions that existed at the end of the mismatched line. However, you were in no way able to calculate the value of the impedance existing at any position on the line. Such calculations involve a level of mathematics that is beyond the scope of this book. However, you can obtain reasonably accurate results by the use of the *Smith Chart*, which was originally developed in 1939. In the example of Fig. 181-1, the chart refers to a particular surge impedance of 50 Ω.

First, the impedance along the line must vary in a periodic manner because on a lossless line, two identical conditions of voltage (and current) are separated by a half wavelength. Therefore, notice that each of the two outermost circular scales represent a total distance of 0.5 λ; in other words, one complete revolution around the chart is equivalent to moving a distance of one half wavelength along the line. If actual distances are given, you must convert them to equivalent wavelengths by knowing the frequency.

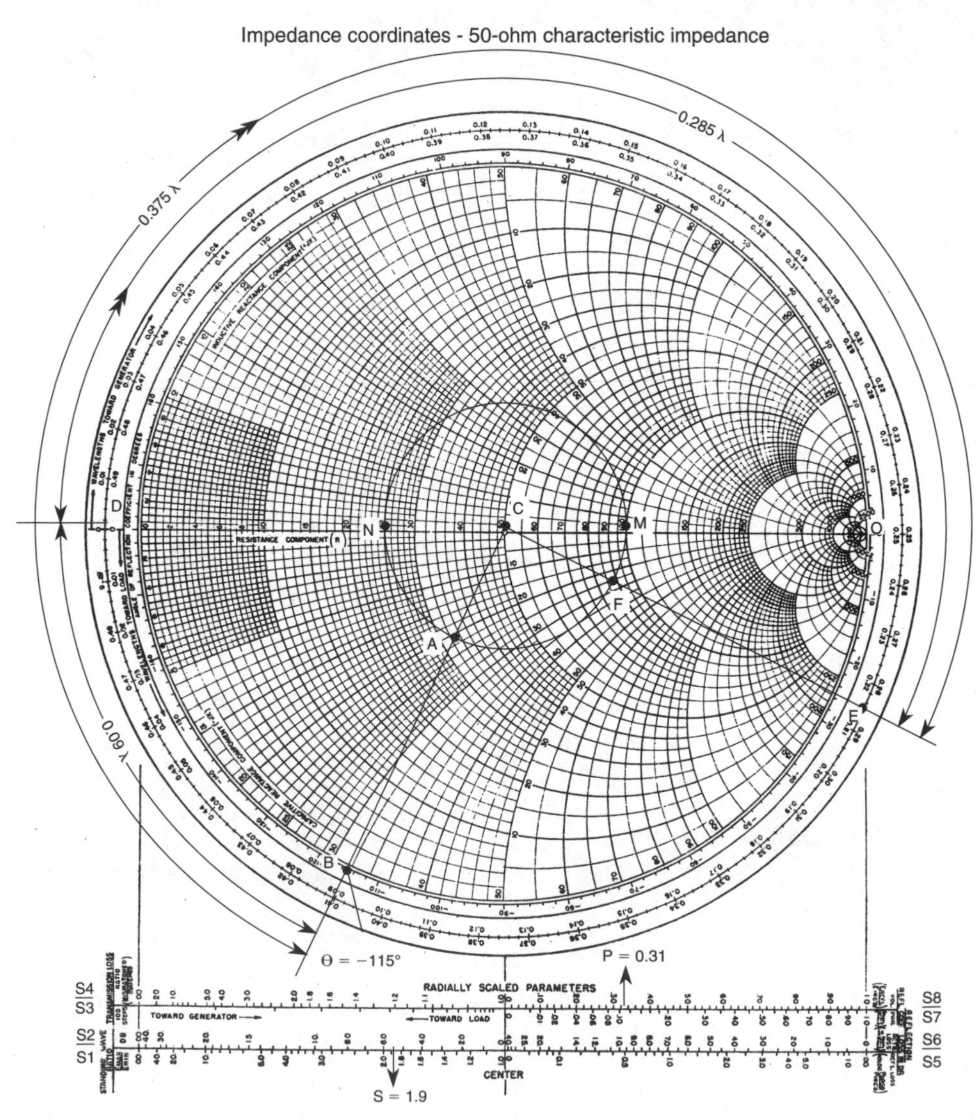

Figure 181-1

For example, if the frequency is 7.5 GHz, the wavelength is 30/7.5 = 4 cm and the distance of 2.8 cm would correspond to 2.8/4.0 = 0.7λ.

The chart itself is composed of a coordinate grid that consists of two sets of circles; examples of these circles appear in Fig. 181-1. The first set are complete circles that represent the resistive components of impedance. All of these circles touch one another at the common point O. The other sets of circles are, in fact, partial arcs which also mutually touch at point O. The arcs to the right and left of the center line, respectively, represent the inductive and capacitive reactance components of the impedance.

The *resistance circles* and the *reactance arcs* always intersect at right angles (90°). The point that relates to a particular impedance is found by first locating the two circles, which represent the particular resistance and reactance values. The point of intersection of these circles corresponds to the impedance. Of course, if the resistance and reactance values do not exactly correspond to the values of the circles, the impedance point must be found by interpolating between the circles.

MATHEMATICAL DERIVATIONS

To illustrate the Smith chart, consider a problem in which a transmission line with an air dielectric has a surge impedance of 50 Ω, and is terminated by a load that consists of a 33-Ω resistance in series with a 21-Ω capacitive reactance. Calculate the values of the reflection coefficient and the VSWR.

Load impedance,

$$z_L = 33 - j21\ \Omega$$

Reflection coefficient,

$$\begin{aligned}\rho &= \frac{z_L - z_o}{z_L + z_o}\\ &= \frac{(33 - j21) - 50}{(33 - j21) + 50}\\ &= \frac{-17 - j21}{83 - j21}\\ &= \frac{27\ \angle -129^\circ}{86\ \angle -14^\circ}\\ &= 0.31\ \angle -115^\circ\end{aligned}$$

The magnitude (P) of the reflection coefficient is 0.31 and the angle (θ) of the reflection coefficient is −115°.

Voltage standing wave ratio,

$$\begin{aligned}S &= \frac{1 + P}{1 - P}\\ &= \frac{1 + 0.31}{1 - 0.31}\\ &= \frac{1.31}{0.69}\\ &= 1.90.\end{aligned}$$

Now, use the 50-Ω Smith chart. By interpolation, the 33-Ω resistance circle and the 21-Ω capacitive reactance arc intersect at the impedance point A. Now, draw a line through A and the center of the chart C; this line is then extended to the scale that measures the angle of the reflection coefficient. At point B, the angle is found to be −115°, which agrees with the mathematical result (Fig. 181-1).

The next step is to draw a circle whose center is C, and which passes through point A. This is known as the *VSWR circle* because its radius is equal to the magnitude of the VSWR. The circle will also enable you to determine the value of the impedance at any position on the line.

To find the magnitude of the reflection coefficient, measure the distance CA along the radial scale S8, which is marked as "Refl. Coef., VOL(E REFL/E INCD)." The value of P is shown to be 0.31. Using the same procedure, on the radial scale, S1 (standing-wave ratio E_{max}/E_{min}), the corresponding value of S is 1.9. It is obvious that these results have been obtained much more quickly and simply with the Smith chart than with the mathematical analysis.

Next, expand the example by assuming that the load is 21.5 cm from a 7.5-GHz generator. The distance of 21.5 cm is equivalent to 21.5/4 = 5.375λ. To find the input impedance at the generator, remember that 5λ represents 10 complete revolutions around the Smith chart; consequently, only consider the remaining 0.375λ. Starting at point B, you must move around the chart in the clockwise direction "toward the generator." When point D is reached, you have covered a distance of 0.090λ. Therefore, you still have to travel a further 0.375λ − 0.090λ = 0.265λ, which brings you to point E. The line CE intersects the VSWR circle at point F, which represents the input impedance at the generator. Reading off the coordinates at point F, the value of this input impedance is 85 − j25 Ω. Points M and N represent, respectively, the maximum and minimum (resistive) impedances of 96 Ω and 26 Ω.

Example 181-1

A transmission line with an air dielectric, and a surge impedance of 50 Ω, is terminated by a load consisting of a 40-Ω resistance in series with a 65-Ω capacitive reactance. What are (a) the phasor value of the reflection coefficient, and (b) the magnitude of the SWR?

Solution

Plot the value of the load impedance on the 50-Ω chart. Determine the distance on the chart between its center and the load impedance point. Mark off this distance on the radial scale S8; the magnitude of the voltage reflection coefficient is 0.59 (Fig. 181-2). Mark off the same distance on the S1 scale; the value of the SWR is 3.88.

Draw a line from the center of the chart to the load impedance point, and extend the line to the scale marked "Angle of Reflection Coefficient in Degrees." The angle is shown as −63° so that the reflection coefficient is 0.59 ∠−63°.

Note that, if the load impedance is changed to a 40-Ω resistance in series with a 65-Ω *inductive* reactance, the values of the voltage reflection coefficient and the SWR remain equal to 0.59 and 3.88, respectively, but the angle of the reflection coefficient is now +63° (Fig. 181-2).

PRACTICE PROBLEMS

181-1. A 50-Ω loss-free line is terminated by a load, which consists of a 35-Ω resistance in series with a 75-Ω inductive reactance. Use the 50-Ω Smith chart to find the values of the voltage reflection coefficient and the standing-wave ratio, together with the maximum and minimum impedances that exist on the line.

181-2. In Practice Problem 181-1, the power at the load is 80 W. Determine the values of the reflected power and the power absorbed by the load.

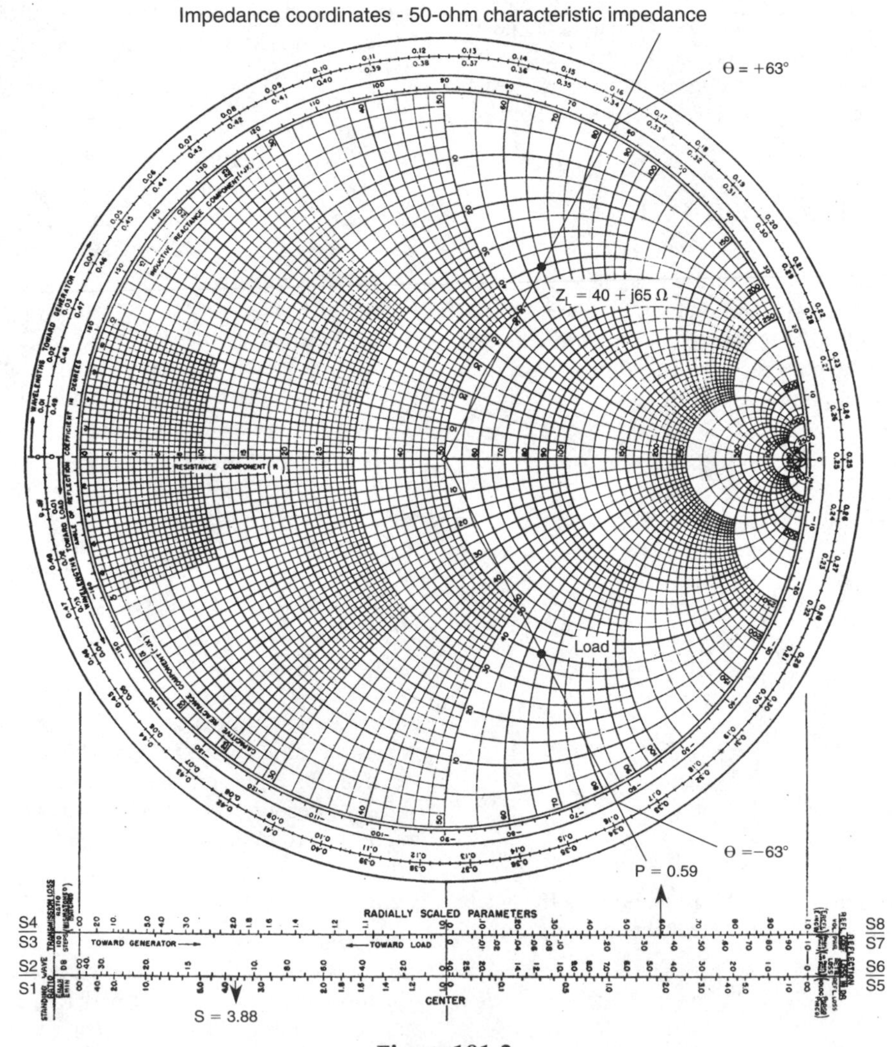

Figure 181-2

181-3. In Practice Problem 181-1, the generator is located 77 cm from the load. If the line is operating at a frequency of 2 GHz, obtain the value of the input impedance at the generator.

181-4. In Example 181-1, a 100-MHz generator delivers 90 W to the line, which is assumed to be lossless. What are the amounts of the reflected power and the power absorbed by the load? What is the distance between two adjacent voltage nulls on the line?

182 The universal Smith chart

In chapter 181, it would appear that you need a new chart for each value of the surge impedance. However, if (in the example of chapter 180) you doubled the load impedance to $2 \times (33 - j21) = 66 - j42\ \Omega$, the position of this impedance point on a 100-Ω chart would be identical to point A (Fig. 181-1) on the 50-Ω chart. In other words, the position of an impedance point is determined by the ratios of the resistive and reactive component values to the value of the surge impedance. It follows that you can develop a universal Smith chart (Fig. 182-1), in which all impedance values are divided by the magnitude of the characteristic impedance before being entered on the chart.

The process of dividing a load impedance by the line's characteristic impedance is called *normalization*; in the examples, the normalized value of $66 - j42\ \Omega$ for $Z_o = 100\ \Omega$ is $(66 - j42)/100 = 0.66 - j0.42$ (0.66 is the normalized resistance, R/Z_o and $-j0.42$ is the normalized reactance, $-jX_c/Z_o$). This is the same result that you obtain for $(33 - j21)/50 = 0.66 - j0.42$, with $Z_o = 50\ \Omega$. When $0.66 - j0.42$ is entered on the universal chart, its position is the same as that of point A on the 50-Ω chart. However, you must remember that when a solution is finally obtained on the universal chart, the coordinates are read off and then *denormalized* (multiplied by Z_o) to obtain the actual value of the impedance.

Apart from the fact that the universal chart can be used with any value of surge impedance, it can also be used for *admittance coordinates*, where the full circles represent normalized conductances (G/Y_o), with the surge admittance $Y_o = 1/Z_o$ and the arcs signify the normalized susceptances $-j(B_L/Y_o)$ to the left of the center line and $+j(B_C/Y_o)$ on the right. The admittance coordinates will be used for analyzing the use of stubs (chapter 184).

MATHEMATICAL DERIVATIONS

The following is the complete list of the radial scales, together with the quantities that these scales measure.

S1 Voltage standing-wave ratio,

$$S = \frac{1 + P}{1 - P}$$

S2 Voltage standing-wave ratio in dB = $20 \log S$

S3 Transmission-loss in dB = $10 \log P$

S4 Loss coefficient (on an unmatched line as compared with the same line when matched)

$$= \frac{1 + S^2}{2S}$$

S5 Reflection loss in dB in $10 \log(1 - P^2)$

S6 Return loss in dB = $20 \log P$

S7 Power reflection coefficient

$$= \frac{\textit{reflected power}}{\textit{incident power}} = P^2$$

S8 Voltage reflection coefficient

$$= \frac{\textit{reflected voltage}}{\textit{incident voltage}} = \frac{S - 1}{S + 1}$$

Example 182-1

A 100-Ω transmission line, with an air dielectric, is terminated by a load of $50 - j80\ \Omega$. Determine the values of VSWR, the reflection coefficient, and the percentage of the reflected power. If the power incident at the load is 100 mW, calculate the power in the load. If the generator frequency is 3 GHz and the line is 73-cm long, find the input impedance at the generator. What are the values of the maximum and minimum impedances existing on the line?

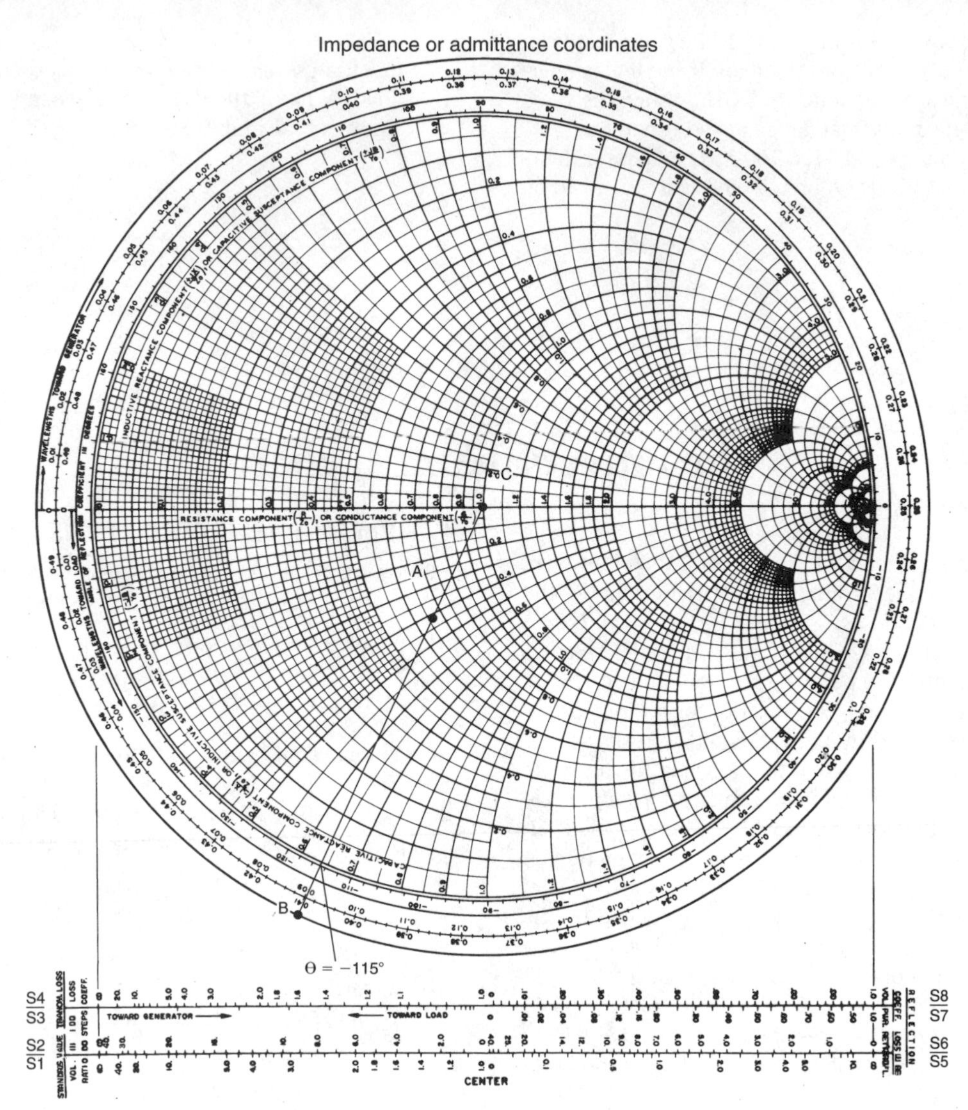

Figure 182-1

Solution

Normalize the value of the load so that $Z_{LN} = 50 - j80)/100 = 0.5 - j0.8$. Plot Z_{LN} on the universal Smith chart (Fig. 182-2) and draw the VSWR circle through Z_{LN}.

Mark off a length that is equal to the circle's radius on the S8 scale. Then, the reflection coefficient $P = 0.56$. Following the same procedure on the S7 scale, the percentage of the reflected power is $P^2 \times 100 = 0.31 \times 100 = 31\%$. Then, the reflected power is $(31/100) \times 100 = 31$ mW and the power in the load is $100 - 31 = 69$ mW.

Notice that the reflected power is $10 \log(31/100) = -5.08$ dB, with respect to the incident power; the load power, when compared with the incident power, is $10 \log(69/100) = -1.61$ dB. By marking off a length equal to the VSWR circle's radius on the S5 scale (REFL Loss), you can obtain the value of 1.61 dB. Carrying out the same procedure for the S6 scale (RET'N Loss) gives you the answer of 5.08 dB. By measuring off the same length on the S1 scale, the value of the VSWR, $S = 3.6$.

From the center of the chart draw a line through Z_{LN}, and extend the line to the peripheral scales. Because all distances on the chart are measured in terms of wavelength, it is necessary to determine the line's wavelength, which is given by,

$$\lambda = \frac{c}{f} = \frac{3 \times 10^{10}}{3 \times 10^{9}} \text{ cm} = 10 \text{ cm}$$

Therefore, 73 cm is equivalent to $73/10 = 7.3\lambda$.

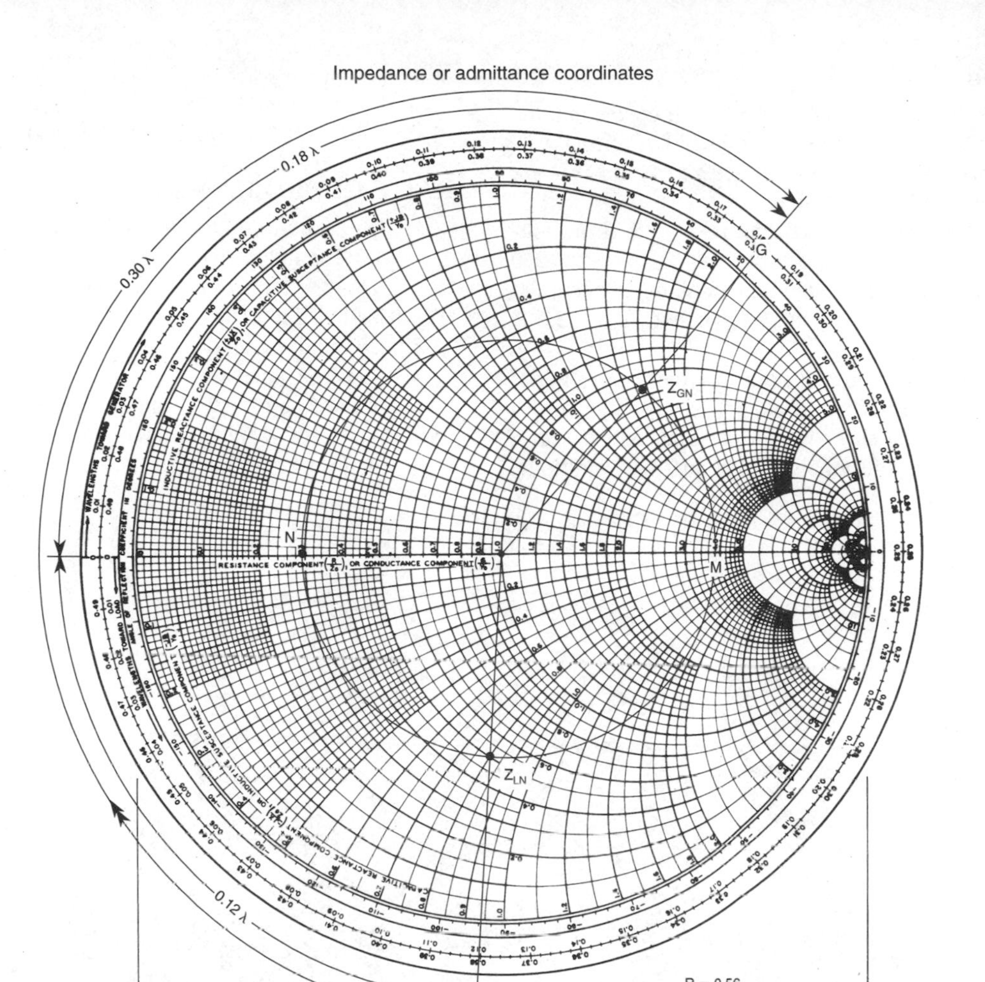

Figure 182-2

Identical impedance values on a mismatched line repeat every half wavelength, which is the distance covered by one complete revolution on the Smith chart. A distance of 7.3 will require 14 complete revolutions (7.0λ), together with an additional rotation of 0.3λ in the clockwise direction (toward the generator).

On the inner peripheral wavelength scale, the distance from the Z_{LN} position to the null position (N) is 0.12λ. Then, travel toward the generator a further 0.3λ − 0.12 λ = 0.18λ on the outermost peripheral scale and arrive at point G.

Draw a line from G to the center of the chart. Point Z_{GN} of intersection between this line and the VSWR circle represents the normalized input impedance at the generator. Therefore, $Z_{GN} = 1.17 + j1.45$ and the denormalized value of this input impedance is $(1.17 + j1.45) \times 100 = 117 + j145\ \Omega$. The minimum and maximum impedance values on the line, respectively, occur at points N and M. At point N, the normalized value is $0.28 + j0$ so that the minimum impedance is $0.28 \times 100 = 28\ \Omega$ (or Z_o/S). The normalized value at point M is $3.6 + j0$, so that the maximum impedance is $3.6 \times 100 = 360\ \Omega$ (or $S \times Z_o$).

Example 182-2

The VSWR on a 200-Ω transmission line is 2.5. If a null point is located at a position which is 3.38λ from the load, find Z_L, Z_{max}, Z_{min} and give the value of the VSWR in decibels.

Solution

On the $S1$ scale, determine the length corresponding to the VSWR of 2.5, then draw the VSWR circle with this length as its radius (Fig. 182-3); using the $S2$ scale, the VSWR is read off as 8 dB.

Point N represents the normalized value of the voltage null impedance, Z_{min}; therefore, $Z_{min} = 0.4 \times 200 = 80\ \Omega$. Similarly, the maximum impedance occurs at the voltage antinode (point M) so that $Z_{max} = 2.5 \times 200 = 500\ \Omega$.

Z_{LN} is found by moving a distance of 3.38λ from the null position in the counterclockwise direction toward the load. The first three wavelengths are equivalent to six complete revolutions on the chart; the additional 0.38λ on the inner wavelength scale brings us to point Z_{LN} on the VSWR circle. The "denormalized" load impedance is then $(0.65 + j0.7) \times 200 = 130 + j140\ \Omega$.

PRACTICE PROBLEMS

182-1. On a loss-free line, the VSWR is found to be 4.3. If a null is located at a distance of 2.632λ from the load, obtain the value of the normalized load impedance.

182-2. An unknown load is connected across a 200-Ω loss-free line, and the value of the SWR is found to be 1.5. The unknown load is then replaced by a known impedance of $40 + j60\ \Omega$; as a result of this change, there is no shift in

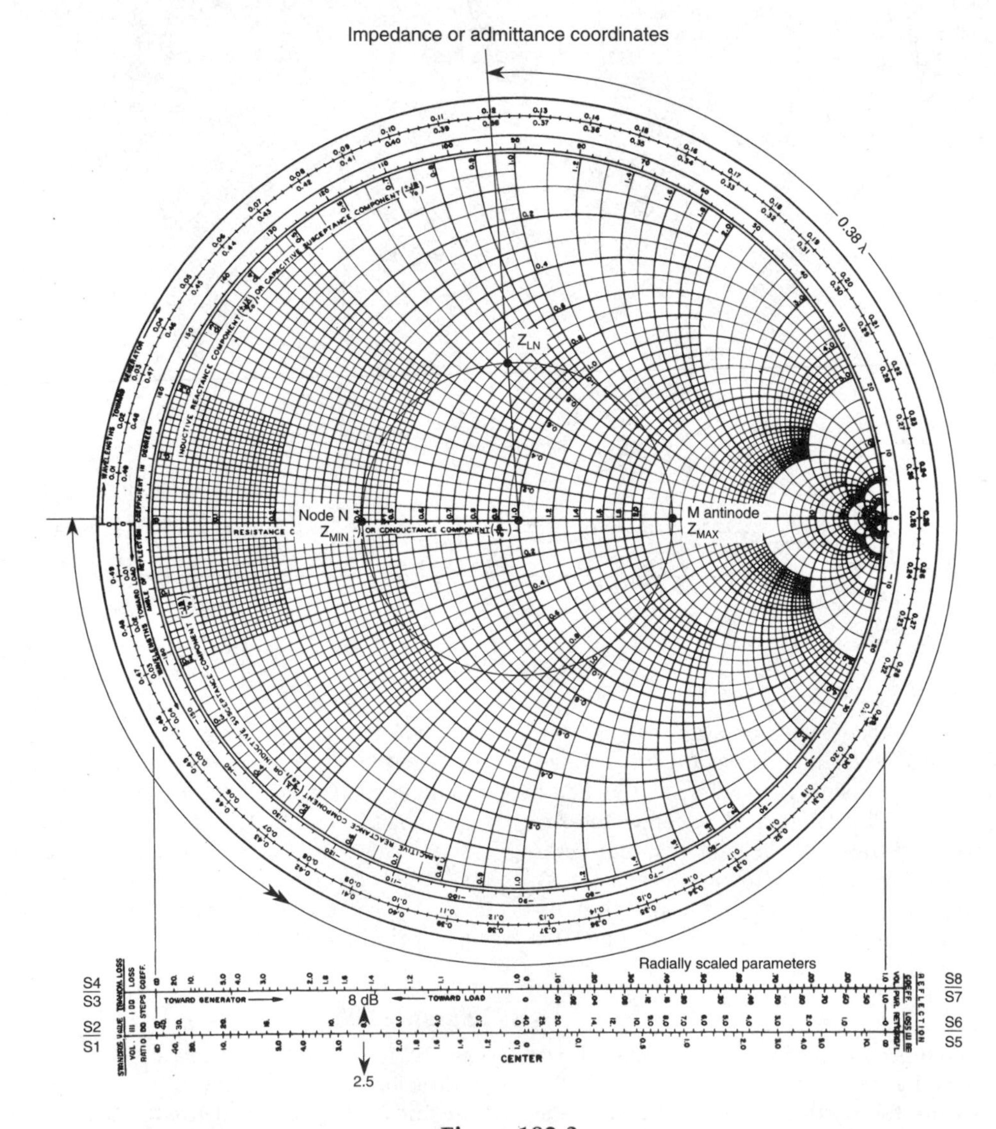

Figure 182-3

the positions of the voltage nulls on the line. Determine the impedance value of the unknown load.

182-3. A 500-Ω loss-free line is 25.75λ in length. If the line is terminated by a 200-Ω resistive load, determine the value of the input impedance at the generator.

182-4. A 300-Ω loss-free line is terminated by a load that consists of a 210-Ω resistance in series with a 450-Ω capacitive reactance. Use the universal Smith chart to find the values of the reflection coefficient and the standing-wave ratio, together with the maximum and minimum impedances that exist on the line.

183
Determination of the load impedance

In Example 182-2, the load impedance was obtained from measurements of the VSWR value and by the distance between the positions of a null and the load. However, at microwave frequencies, it is difficult to determine the position of the load with sufficient accuracy; at 10 GHz, an error of only 3 mm is equivalent to 0.1λ (one fifth of the Smith chart's circumference). A more practical method of finding the load impedance involves using accurate indirect measurements.

MATHEMATICAL DERIVATIONS

Figure 183-1 illustrates a common method of determining the value of an unknown load impedance. The position of a voltage null is located with the line terminated by the load. The load is then removed, and is replaced by a short circuit. The result is a shift in the position of the null (Fig. 183-1A). According to the nature of the load (inductive or capacitive), the direction of the shift can be either toward the generator or the load, and the amount of shift must be 0.25λ or less. If the shift is either zero or exactly 0.25λ, the load is a resistance whose value is, respectively, less than or greater than Z_o (Fig. 183-1B). Knowing the amount and the direction of the null shift, Z_{LN} can be determined by carrying out the procedure outlined in the following example.

Example 183-1

The VSWR on a 50-Ω line is 1.8 and a null point is located at a certain position on the line. When the load is replaced by a short circuit, the position on the null point shifts a distance of 0.15λ *toward the load*. Find Z_L, as well as the magnitude and the phase angle of the reflection coefficient. How far from the load is the first voltage maximum?

Solution

On scale S1, determine the length corresponding to a VSWR of 1.8. Then, draw the VSWR circle with this length as its radius (Fig. 183-2). From the null position (point N), travel a distance of 0.15λ on the inner wavelength scale in the counterclockwise direction toward the load. Arriving at point L, draw a line from L to the center of the chart. The point of intersection between the line and the VSWR circle is the normalized value of the load impedance. From the impedance coordinates, $Z_{LN} = 1 - j0.6$.

The "denormalized" load impedance is $Z_L = Z_{LN} \times Z_o = (1 - j0.6) \times 50 = 50 - j30\ \Omega$.

The line passing through L intersects the angle, $\theta°$, of the reflection coefficient scale at $-72°$. Marking off the radius of the VSWR circle on the radial scale (S8) shows that point $P = 0.28$ and, therefore, the reflection coefficient, $\rho = 0.29 \angle -72°$.

The voltage maximum nearest the load exists at point M. The required distance is found by moving from L to M on the wavelength scales in the clockwise direction toward the generator. Moving from L to N on the inner wavelength scale is equivalent to a distance of 0.15λ, and moving from N to M on the outer wavelength scale is a distance of 0.25λ. The total distance is, therefore, 0.15λ + 0.25λ = 0.4λ.

PRACTICE PROBLEMS

183-1. The value of the VSWR on a loss-free 300-Ω line is 2.5, and the position of a null point is first located. When the load is replaced by a short circuit, the position of the null point shifts a distance of 0.12λ toward the generator. Determine the value of the load impedance.

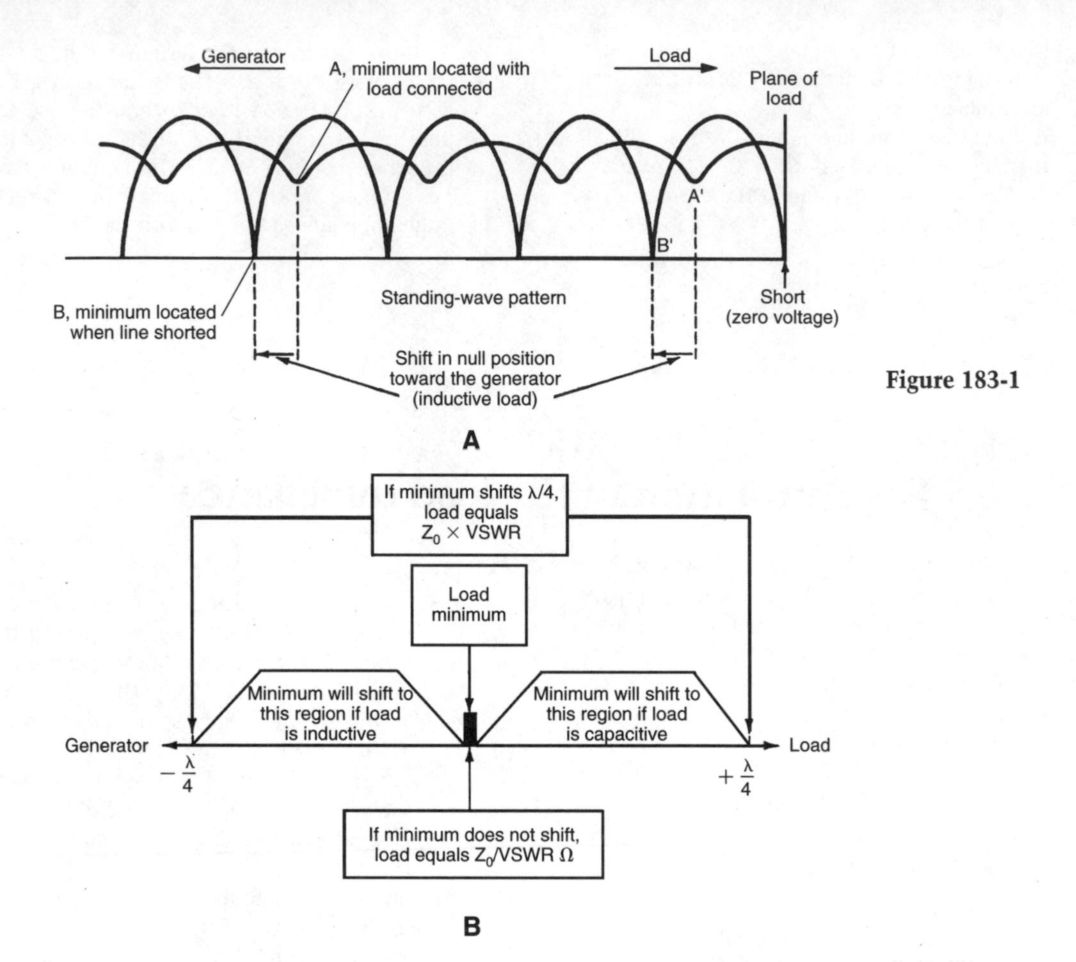

Figure 183-1

183-2. In Practice Problem 183-1, what is the value of the reflection coefficient's phasor? In terms of the wavelength, how far from the load is the first voltage antinode?

183-3. The value of the VSWR on a loss-free 100-Ω line is 2.0. When the unknown load is replaced by a short circuit, the position of a voltage null shifts a distance of $\lambda/8$ toward the load. What is the value of the load impedance?

183-4. If, in Practice Problem 183-3, the replacement of the unknown load by the short circuit causes the position of a voltage null to shift by a distance of exactly $\lambda/4$, what is the new value of the load impedance?

183-5. The value of the VSWR on a loss-free 100-Ω line is 5.0. When the unknown load is replaced by a short circuit, there is no shift in the position of a voltage null. Determine the value of the load impedance?

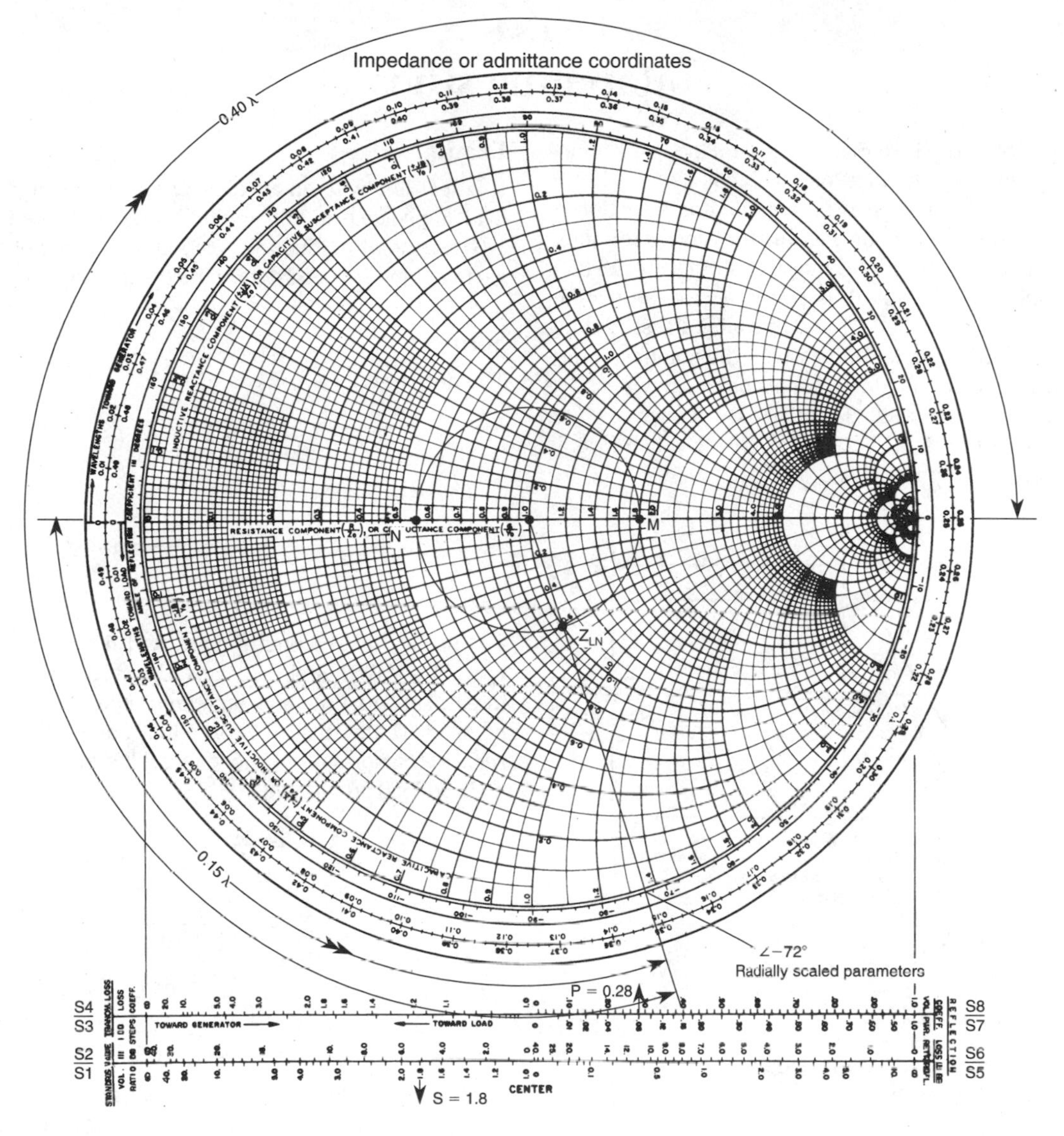

Impedance or admittance coordinates
0.40 λ
0.15 λ
N
M
Z_{LN}
∠−72°
Radially scaled parameters
P = 0.28
S = 1.8
S1
S2
S3
S4
S5
S6
S7
S8
CENTER
TOWARD GENERATOR
TOWARD LOAD

Figure 183-2

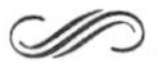

184
Matching stubs

One use of a stub is to match a general load impedance to the surge impedance of the cable that feeds the load. The stub itself can be regarded as a section of transmission line, about one-quarter wavelength long, which is terminated by a movable short (Fig. 184-1). The single movable stub is placed across (in parallel with) the twin-line feeder and by sliding the stub along the line, a position is found where the combination of the stub and the load represents an entirely resistive impedance whose value is equal to the feeder's Z_o. The purpose of the Smith chart (Fig. 184-2), therefore, is to determine the lengths, L_1 and L_2.

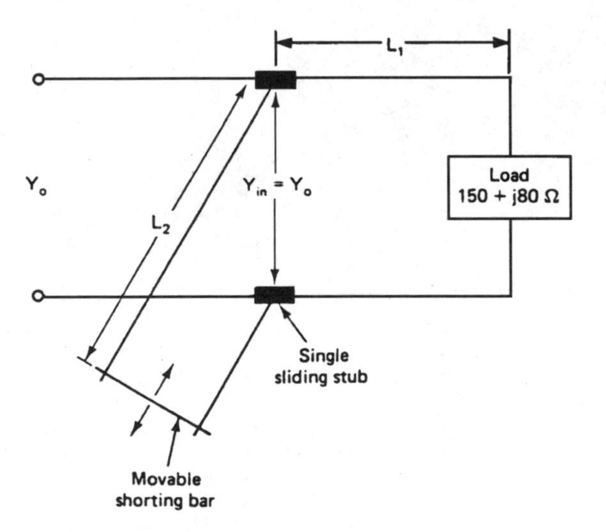

Figure 184-1

The application of a single movable stub is limited to the two-wire line. For a coaxial cable, you are frequently required to use two fixed stubs, which are typically $3\lambda/8$ apart. It is then necessary to use the Smith chart (Fig. 184-3) to determine the lengths of the stubs to achieve the required impedance match.

MATHEMATICAL DERIVATIONS

The stubs are in parallel across the transmission line and, consequently, you must use conductance, susceptance, and admittance coordinates on the Smith chart.

Normalized load admittance,

$$Y_{LN} = \frac{1}{Z_{LN}} \qquad (184\text{-}1)$$

Example 184-1

A 200-Ω line with an air dielectric is terminated by a load impedance of $150 + j80\ \Omega$, and is excited by a 1-GHz generator. Find the position on the line of a single matching stub and determine the length of the stub.

Solution

The normalized load, Z_{LN} is,

$$\frac{150 + j80}{200} = 0.75 + j0.4$$

The wavelength,

$$\lambda = \frac{c}{f} = \frac{3 \times 10^{10}}{1 \times 10^{9}} \text{ cm}$$

$$= 30 \text{ cm.}$$

Plot Z_{LN} on the Smith chart, and draw the corresponding VSWR circle. Because the stub is across (or in parallel with) the line, an admittance analysis must be used. The normalized load admittance is the reciprocal of the normalized impedance. Therefore, the normalized load admittance,

$$Y_{LN} = \frac{1}{Z_{LN}} \qquad (184\text{-}1)$$

$$= \frac{1}{0.75 + j0.4}$$

$$= \frac{0.75 - j0.4}{0.75^2 + 0.4^2}$$

$$= 1.04 - j0.55.$$

When you plot Y_{LN}, you find that its position is at the *opposite* end of the diameter passing through Z_{LN}.

From the position of Y_{LN}, move around the VSWR circle in the clockwise direction (toward the generator) until you reach the point of intersection (P) on the VSWR circle for which the normalized conductance $G/Y_o = 1$. The distance covered in this move is equivalent to the length $L_1 = 0.152\lambda + 0.146\lambda = 0.298\lambda$, or $L_1 = 0.298 \times 30 = 8.94$ cm. At position P, the normalized admittance load is $Y_{PN} = 1 + j0.54$, which represents a normalized conductance in parallel with a normalized capacitive susceptance.

To achieve a match with the line's surge impedance, the stub must eliminate the normalized capacitive susceptance by contributing an equal amount of normalized inductive susceptance; specifically, $-j0.54$. This value is then entered at point S on the chart. The short at the

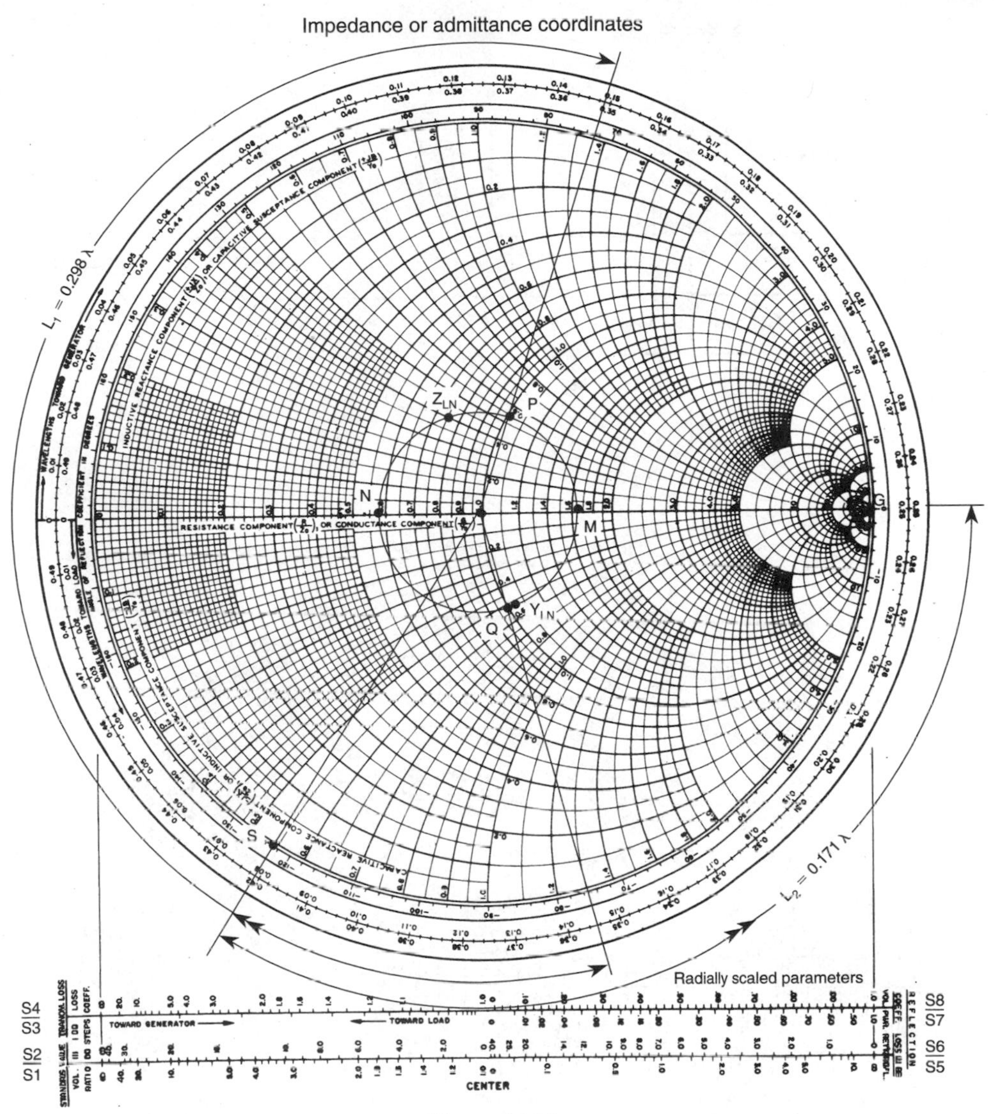

Figure 184-2

end of the stub has infinite conductance, and is represented by point G. Because this short is the stub's load termination, you must travel from S to G on the inner wavelength scale in a counterclockwise direction (toward the load). The length of the stub, L_2 is, therefore, $0.25\lambda - 0.079\lambda = 0.171\lambda = 0.171 \times 30$ cm $= 5.13$ cm.

Another solution to the problem is represented by point Q. However, the distance L_1 would then be very short (although any multiple of $\lambda/2$ could be added to L_1) and distance L_2 would be much longer.

Example 184-2

A load is to be matched to a 100-Ω coaxial cable line by means of two fixed stubs that are $3\lambda/8$ apart (Fig. 184-3). At the reference plane that corresponds to the position of the first stub, the admittance is $0.016 - j0.008$ siemens. Find the length of each stub.

Solution

The surge admittance of the line is $1/100\ \Omega = 0.01$ S. Therefore, the normalized admittance Y_{N1} at the reference plane of the first stub is $(0.016 - j0.008)/0.01 = 1.6 - j0.8$.

Enter Y_{N1} on the Smith chart (Fig. 184-4) and observe that it lies on the 1.6 normalized conductance circle. Rotate this circle through three-quarters of a revolution in the clockwise direction; this rotation corresponds to the $3\lambda/8$ spacing between the stubs. The

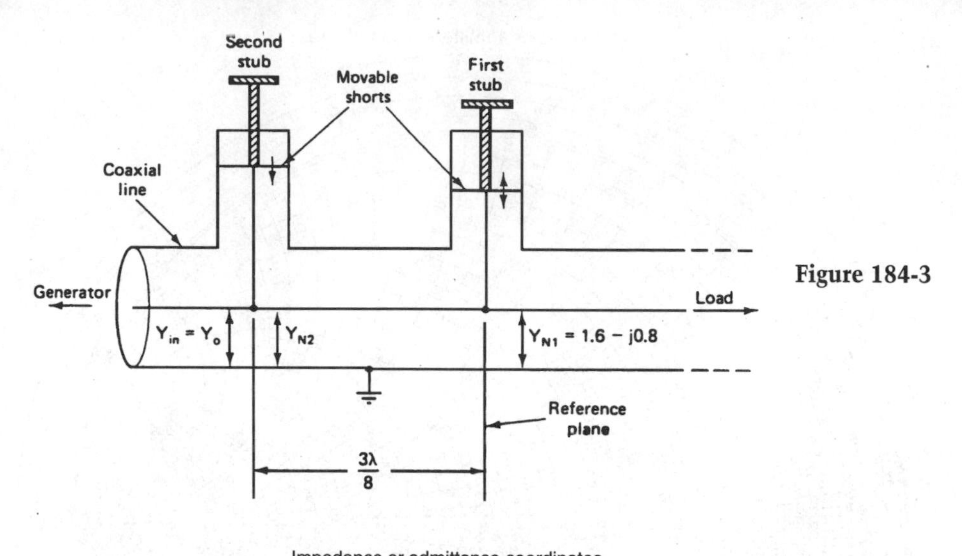

Figure 184-3

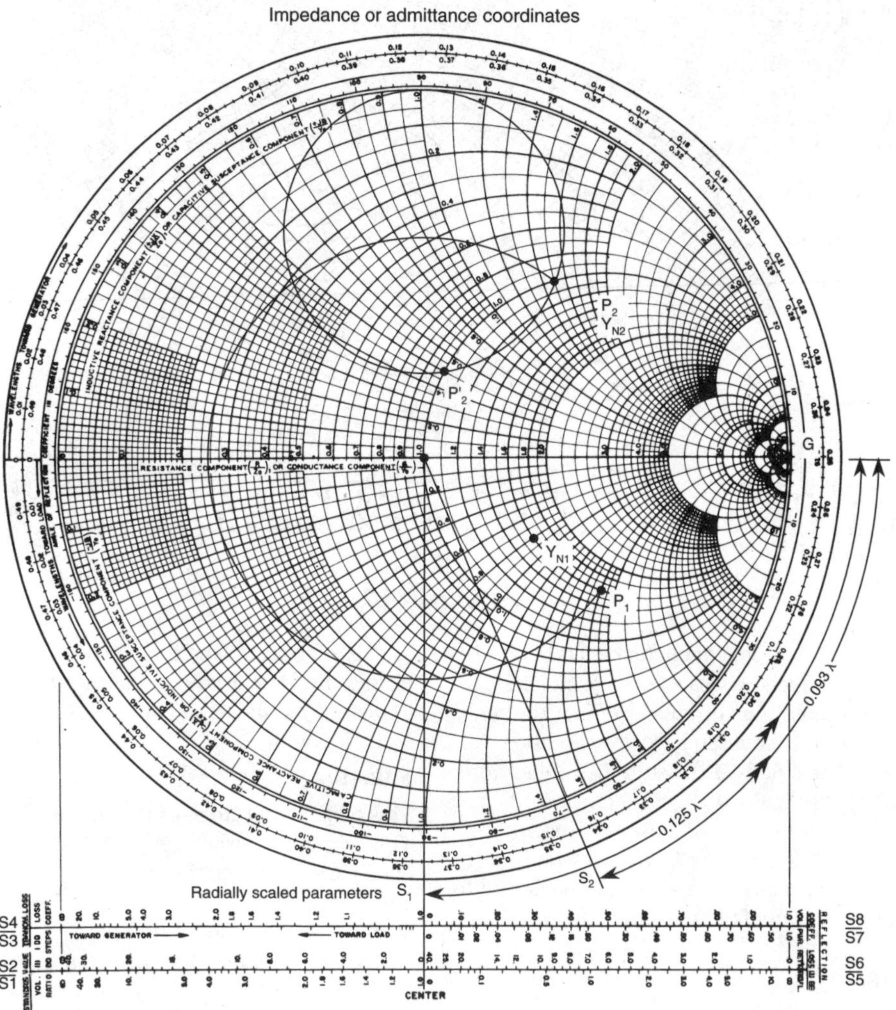

Figure 184-4

center of the circle after the rotation will lie on the line that passes through the points $+j1.0$ and $-j1.0$.

Determine the point (P_2) of intersection between the newly constructed circle and the circle for which $G/Y_o = 1$. Draw the VSWR circle through P_1 and P_2, then P_1 and P_2 are $3\lambda/8$ apart. Normalized admittance at P_2 is $Y_{N2} = 1 + j1.5$. Normalized admittance at P_1 is $1.6 - j1.8$.

The effect of the first stub is to change the susceptance ($-j0.8$) of Y_{N1} at the reference plane to the susceptance ($-j1.8$) at point P_1. Therefore, this stub must contribute a susceptance of $-j1.8 - (-j0.8) = -j0.10$ which corresponds to point S_1. The short circuit that terminates the stub is point G. The length of the first stub is, therefore, $0.25\lambda - 0.125\lambda = 0.125\lambda$.

The second stub must change the normalized susceptance from its value ($+j1.5$) at point P_2 to the center of the chart, where the match to the line's surge impedance occurs, and the normalized susceptance is zero. This stub must, therefore, contribute a normalized susceptance of $-j1.5$, which is entered at point S_2. The length of the second stub is measured from S_2 to G; this distance is equal to $0.25\lambda - 0.156\lambda = 0.094\lambda$. Notice that there is a second possible solution provided by the point of intersection, P_2'.

Clearly, there is a limitation to the use of the two stubs, which are $3\lambda/8$ apart. If the normalized conductance G/Z_o is greater than 2, there will be no point of intersection (such as point P_2), and no solution can be found, except through the use of three fixed stubs.

PRACTICE PROBLEMS

184-1. A load is to be matched to a 100-Ω line by means of two fixed stubs, which are separated by a quarter wavelength. At the reference plane corresponding to the position of the first stub, the admittance is $0.008 - j0.014$ S. Find the length of each stub in terms of the wavelength. Discuss any limitation on the use of the two stubs that are $\lambda/4$ apart.

184-2. A 50-Ω loss-free line is terminated by a load impedance of $70 - j35$ Ω, and is excited by a 600-MHz generator. How far from the load is the position of a single matching stub? Determine the length (in centimeters) of the stub.

184-3. In Practice Problem 184-2, the frequency is reduced to 450 MHz. Find the new position of the stub, and its new length.

185 Transmission line losses

When there are appreciable losses associated with the line, the incident voltage and current waves are attenuated as they travel from the generator toward the load. Assuming that the load is mismatched to the line, the reflected voltage and current waves will be further attenuated as they move back from the load toward the generator. Therefore, the VSWR value will be highest at the load position, and will gradually decrease in the direction of the generator. Consequently, as you move from the load toward the generator, you must visualize that the VSWR value will represented by a decreasing spiral, rather than a circle. On the Smith chart (Fig. 185-1), the change in the value of the VSWR (scale S1) can be used to determine the line's one-way attenuation in decibels (scale S3, attenuation in decibel steps).

Example 185-1

On a slotted line the VSWR is measured and found to be 3.6. When the load is removed and replaced by a short, the position of the null point on the line moves 0.11λ toward the generator and the measured VSWR is 8.6. Find the one-way attenuation in decibels between this null position and the load, the value of the VSWR at the load (with the load connected), and the normalized value of the load.

Solution

When the "short" is placed at the end of the line, the VSWR at the position of the short is a maximum, and is theoretically infinite; this corresponds to the maximum radial distance on the S1 scale. However, the

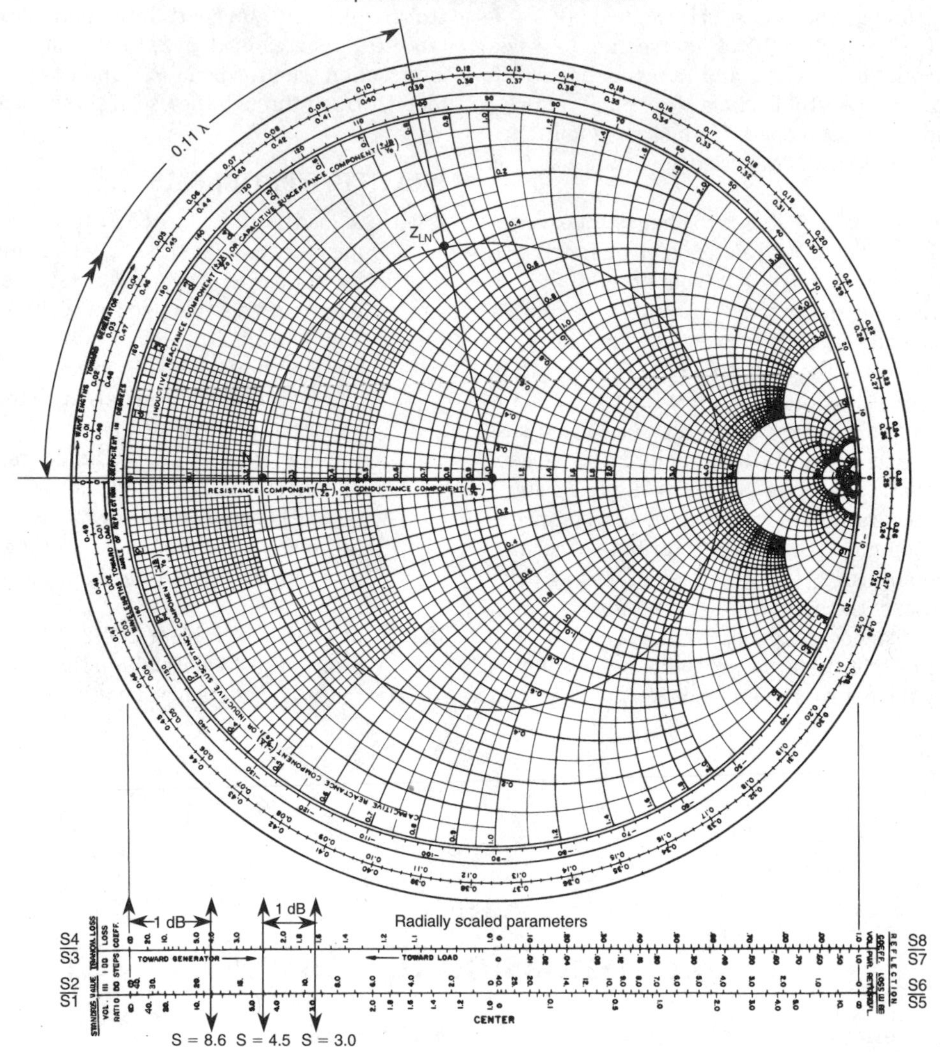

Figure 185-1

VSWR as measured at the null position is only 8.6. Therefore, the change from $S = \infty$ to $S = 8.6$ is caused by the line's attenuation. After locating the $S = \infty$ and $S = 8.6$ positions on the S1 scale, mark off the same radial distances on the S3 scale. The difference between these distances is approximately equivalent to a one-way attenuation of 1.0 dB.

On the S1 and S3 scales mark off the radial distances for the VSWR of 3.0. Increase the distance on the S3 scale by an amount that corresponds to the one-way attenuation of 1.0 dB, and then apply the same increase in distance to the S1 scale. The result is a VSWR of approximately 4.5 at the load.

Draw the VSWR circle for $S = 4.5$. On the outer wave-length scale, move clockwise through a distance of 0.11λ from the null position N toward the generator. When point P is reached, draw a straight line from P to the center of the chart. The intersection of this line and the VSWR circle reveals the normalized load impedance Z_{LN} to be $0.35 + j0.77$.

Example 185-2

It is desired to achieve a VSWR of 2.5 on an unmatched line by the introduction of attenuation. If the VSWR at the load is 5.4, find the required one-way attenuation in decibels. Determine the value of the loss coefficient for the VSWR of 5.4.

Solution

Locate the $S = 2.5$ and $S = 5.4$ positions on the S1 scale, and mark off the same radial distances on the S3 scale. The difference between these distances is equivalent to a one-way attenuation of approximately 2.0 dB (Fig. 185-2).

A more accurate estimate of the attenuation can be obtained by using the S6 scale (return loss in dB), which represents the two-way loss. On this scale, $S = 2.5$ and $S = 5.4$ correspond to losses of 7.2 and 3.2 dB, respectively. The one-way attenuation is, therefore, $(7.2 - 3.2)/2 = 2.0$ dB. On scale S4, mark off the radial distance for $S = 5.4$. The corresponding loss coefficient is approximately 2.8.

When using the universal Smith chart, it is important to remember that any change in the frequency will alter the values of the normalized reactance and/or susceptance. This, in turn, will change the position of the normalized impedance and/or admittance points so that the answers in the solution will be different.

PRACTICE PROBLEM

185-1. At a certain null position (N) on an unmatched line, the VSWR is measured and found to be equal to 1.9. When the load is removed and replaced by a short circuit, the position of the null shifts 0.13λ toward the load and the measured VSWR increases to 4.4. Find the one-way attenuation between N and the load, the value of the VSWR at the load (with the load connected), and find the normalized value of the load.

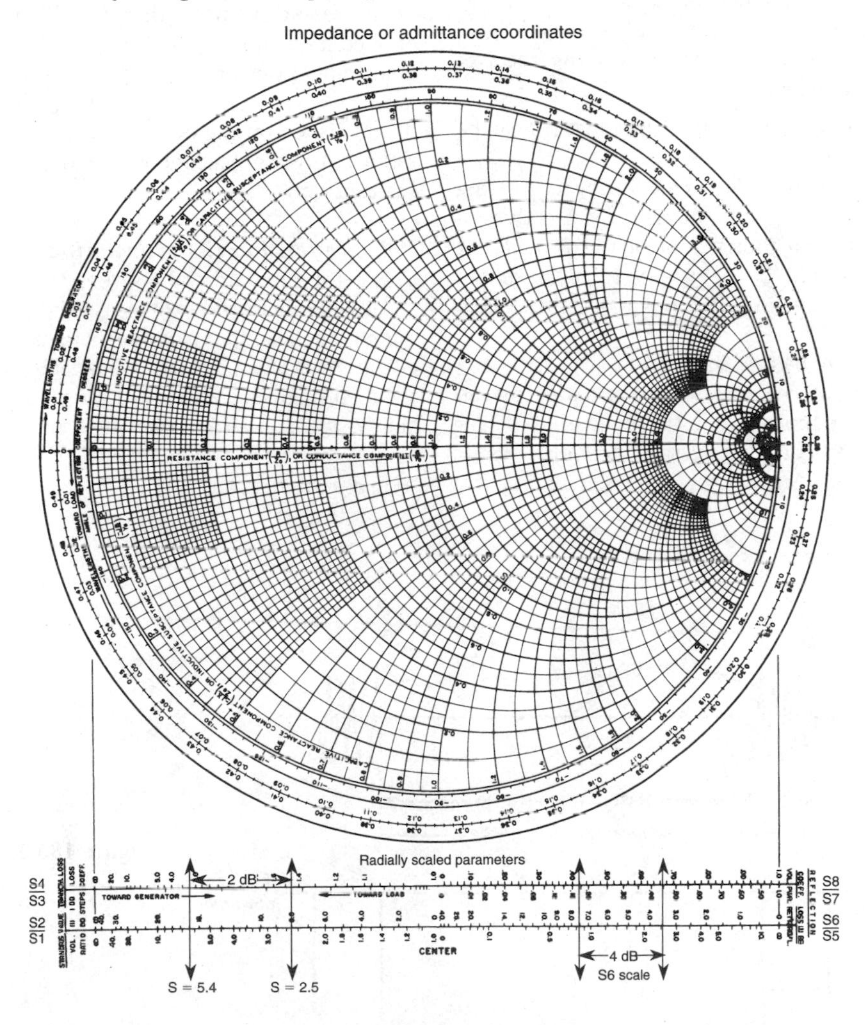

Figure 185-2

186
Parameters of a pulsed radar set

The sensitivity, or the maximum detection range, of a pulsed radar set is proportional to the following expression:

$$\sqrt[4]{\frac{P_t \times G_t \times A_r \times \delta}{P_r}}$$

where: P_t = transmitter peak power output (W)
G_t = gain or directivity of the transmitting antenna.
A_r = effective area of the receiving antenna.
δ = a factor governed by the reflecting properties of the target, namely size, conductivity, and inclination.
P_r = the minimum detectable received signal power that must exceed the receiver noise level.

Therefore, the maximum range of the radar set can be increased by:

1. *Raising the transmitter power output.* Bearing in mind that the maximum range is directly proportional to the fourth root of the transmitter power, it would be necessary to increase the transmitted power by 16 times in order to double the range.
2. *Reducing the beamwidth.*
3. *Reducing the receiver noise*, primarily by the design of the first stage in the receiver.

The main parameters that are associated with a pulsed radar set (Fig. 186-1) are:

Pulse shape
Pulses of various shapes can be used in radar systems, but for practical reasons, the rectangular shape is usually chosen.

Pulse duration (width or length)
This is the duration of the rectangular pulse and is normally measured in microseconds. For most radar sets, the pulse duration lies between 0.1 and 10 microseconds. The pulse width is a factor in determining the minimum range obtainable because the echo pulse and the transmitted pulse must not overlap.

Increasing the pulse duration will increase the energy content of the pulse and will increase the maximum range. However, those targets whose difference in range corresponds to the pulse duration, cannot be distinguished on the display; for example, a pulse duration of 0.1 microsecond will cover approximately 160 yards of range.

Pulse repetition time (or period)
This is the time interval between the leading edges of successive rectangular pulses. It is a factor in determining the maximum range obtainable because the echoes from a particular pulse must be received before the next pulse is transmitted. However, for range reliability, the pulse repetition time must be kept sufficiently short to allow a number of pulses to be received from a particular target.

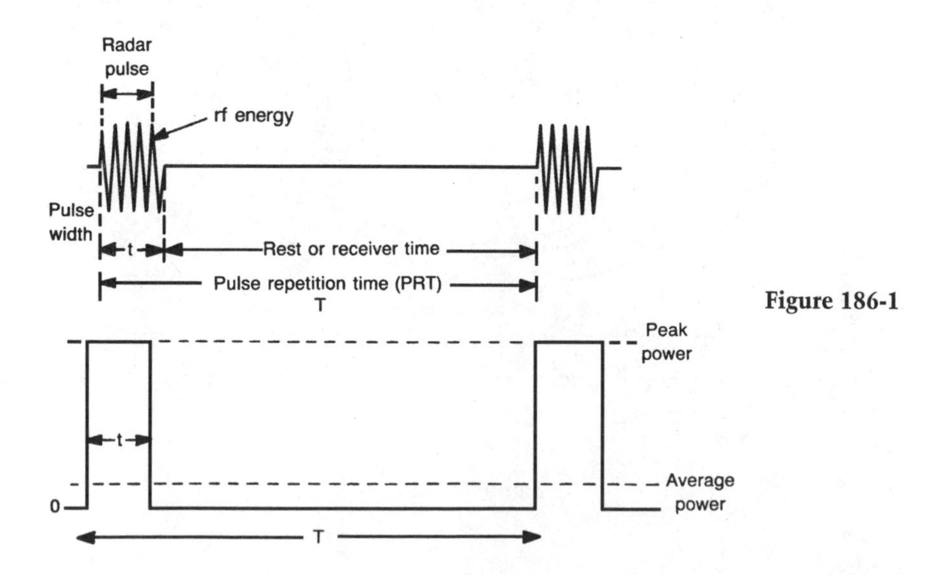

Figure 186-1

Pulse repetition frequency (or rate)
This parameter is also known as the pulse frequency, and is the reciprocal of the pulse repetition time. The values of the pulse frequency generally lie between 250 Hz and 5000 Hz. The higher the pulse frequency, the greater is the intensity of the echoes on the display. The result will be an improvement in the target definition.

Peak power
The peak power is the average power transmitted during the duration of the pulse. This can vary from several kilowatts to a few megawatts. It has been quoted that the maximum range available is directly proportional to the fourth root of the peak power so that to triple the range, it would require the transmitter power to be increased by 81 times.

Average power
This is the mean power over the pulse repetition period. It follows that the ratio of the average power to the peak power is the same as the ratio of the pulse duration to the pulse repetition time. Both of these ratios are equal to the duty cycle, which has no units. The value of the duty cycle typically ranges from 0.01 to 0.0001.

Power gain of the antenna
The maximum range is proportional to the square root of the antenna's power gain. Increasing the range by raising the power gain reduces the beamwidth and improves the bearing resolution.

Antenna rotation rate
This rate is limited by the number of received pulses required for each target. The higher the antenna rotation rate, the less will be the intensity of the echoes on the display.

Radio frequency
The efficiency of surface detection increases with the value of the radio frequency. Moreover, as the radio frequency is increased, the sizes of the antenna and the waveguide elements decrease. The parabolic reflectors used with the centimeter wavelengths can provide very narrow pencil beams that are especially suitable for accurate bearing measurements. The beamwidth is proportional to the ratio of the wavelength to the diameter of the parabolic reflector. It is defined as the angle between those two directions, in which the received signal power falls to half of the maximum value, which occurs along the axis of the main beam.

The radio frequencies that are allocated to marine radar sets are in the gigahertz range, which corresponds to wavelengths that are on the order of centimeters.

MATHEMATICAL DERIVATIONS

$$\frac{\textit{Average power}}{\textit{Peak power}} = \frac{\textit{pulse duration}}{\textit{pulse repetition time}} = \textit{duty cycle} \quad (186\text{-}1)$$

$$\textit{Beamwidth} = \frac{70\lambda}{d} \text{ degrees} \quad (186\text{-}2)$$

where: λ = wavelength (m)
d = diameter of paraboloid reflector (m)

Frequency,

$$f = \frac{30}{\lambda(\text{cm})} \text{ gigahertz} \quad (186\text{-}3)$$

Wavelength,

$$\lambda = \frac{30}{f(\text{GHz})} \text{ centimeters} \quad (186\text{-}4)$$

$$\textit{Pulse duration} \quad (186\text{-}5)$$

$$= \frac{\textit{average power} \times \textit{pulse repetition time}}{\textit{peak power}}$$

$$= \frac{\textit{average power}}{\textit{peak power} \times \textit{pulse frequency}}$$

$$= \textit{duty cycle} \times \textit{pulse repetition time}$$

$$= \frac{\textit{duty cycle}}{\textit{pulse frequency}} \text{ seconds}$$

$$\textit{Pulse repetition time} \quad (186\text{-}6)$$

$$= \frac{\textit{pulse duration} \times \textit{peak power}}{\textit{average power}}$$

$$= \frac{\textit{pulse duration}}{\textit{duty cycle}} \text{ seconds}$$

$$\textit{Pulse repetition frequency} \quad (186\text{-}7)$$

$$= \frac{\textit{average power}}{\textit{pulse duration} \times \textit{peak power}}$$

$$= \frac{\textit{duty cycle}}{\textit{pulse duration}} \text{ hertz}$$

$$\textit{Average power} \quad (186\text{-}8)$$

$$= \frac{\textit{peak power} \times \textit{pulse duration}}{\textit{pulse repetition time}}$$

$$= \textit{peak power} \times \textit{pulse duration} \times \textit{pulse frequency}$$

$$= \textit{peak power} \times \textit{duty cycle} \text{ watts}$$

$$\textit{Peak power} \quad (186\text{-}9)$$

$$= \frac{\textit{average power} \times \textit{pulse repetition time}}{\textit{pulse duration}}$$

$$= \frac{\textit{average power}}{\textit{pulse duration} \times \textit{pulse frequency}}$$

$$= \frac{\textit{average power}}{\textit{duty cycle}} \text{ watts}$$

(186-10)

$$\textit{Duty cycle} = \textit{pulse duration} \times \textit{pulse frequency}$$

$$= \frac{\textit{average power}}{\textit{peak power}}$$

Example 186-1

A radar set has a peak power of 0.5 MW and a pulse duration of 2 μs. If the pulse frequency is 500 Hz, what are the values of the average power and the duty cycle?

Solution

Average power (186-8)

$$= \textit{peak power} \times \textit{pulse duration} \times \textit{pulse frequency}$$

$$= 0.5 \times 10^6 \times 2 \times 10^{-6} \times 500$$

$$= 500 \text{ W.}$$

(186-10)

$$\textit{Duty cycle} = \textit{pulse duration} \times \textit{pulse frequency}$$

$$= 2 \times 10^{-6} \times 500 = 1 \times 10^{-3} = 0.001$$

Note that the duty cycle is also given by:

$$\textit{Duty cycle} = \frac{\textit{average power}}{\textit{peak power}} \quad (186\text{-}1)$$

$$= \frac{500}{0.5 \times 10^6}$$

$$= 0.001$$

Example 186-2

A radar set has a peak power output of 750 kW and a duty cycle of 0.0005. What is the average power?

Solution

$$\textit{Average power} = \textit{peak power} \times \textit{duty cycle} \quad (186\text{-}8)$$

$$= 750 \times 10^3 \times 0.0005$$

$$= 375 \text{ W.}$$

Example 186-3

The wavelength of a radar transmission in free space is 9.6 centimeters. What is the corresponding radio frequency?

Solution

Radio frequency (in GHz) (186-3)

$$= \frac{30}{\textit{wavelength} \text{ (in centimeters)}}$$

$$= \frac{30}{9.6}$$

$$= 3.125 \text{ GHz}$$

Example 186-4

The peak power of a radar pulse is 1 MW and its average power is 500 W. If the pulse duration is 2 μs, what are the values of the pulse repetition time and the pulse frequency?

Solution

Pulse repetition time (186-6)

$$= \frac{\textit{pulse duration} \times \textit{peak power}}{\textit{average power}}$$

$$= \frac{2 \times 10^{-6} \times 10^6}{500}$$

$$= \frac{1}{250} \text{ s}$$

$$= 4000 \ \mu\text{s.}$$

$$\textit{Pulse frequency} = \frac{1}{\textit{pulse repetition time}}$$

$$= \frac{1}{4000 \times 10^{-6}}$$

$$= 250 \text{ Hz.}$$

Example 186-5

The length (duration) of a radar pulse is 5 μs, the peak power is 800 kW, and the average power is 2000 W. What is the value of the pulse frequency?

Solution

(186-7)

$$\textit{Pulse frequency} = \frac{\textit{average power}}{\textit{pulse duration} \times \textit{peak power}}$$

$$= \frac{2000}{5 \times 10^{-6} \times 800 \times 10^3}$$

$$= 500 \text{ Hz.}$$

Example 186-6

The pulse repetition time of a radar set is 2 ms and the duty cycle is 5×10^{-4}. What is the pulse duration?

Solution

(186-5)

$$\textit{Pulse duration} = \textit{duty cycle} \times \textit{pulse repetition time}$$

$$= 5 \times 10^{-4} \times 2 \times 10^{-3} \text{ s}$$
$$= 1 \ \mu\text{s}.$$

Example 186-7

The average power of a radar pulse is 1 kW, the pulse width is 2 μs, and the pulse frequency is 400 Hz. What is the pulse's peak power?

Solution

$$\textit{Peak power} = \frac{\textit{average power}}{\textit{pulse duration} \times \textit{pulse frequency}} \quad (186\text{-}9)$$
$$= \frac{1000}{2 \times 10^{-6} \times 400} \text{ W}$$
$$= 1.25 \text{ MW}.$$

Example 186-8

The peak power of a pulse radar set is increased by 50%. Assuming that all other parameters remain the same, what will be the increase in the maximum range?

Solution

The maximum range is proportional to the fourth root of the transmitter peak power output. The maximum range is, therefore, multiplied by a factor of $\sqrt[4]{1.5} = 1.1068$. Because $\sqrt[4]{1.5}$ is approximately 1.11, the range is increased by 11%.

PRACTICE PROBLEMS

186-1. The frequency of a radar transmission in free space is 4.57 GHz. Calculate the value of the wavelength.

186-2. A radar set has a pulse duration of 1.5 μs and a duty cycle of 0.0005. What is the value of the pulse repetition time?

186-3. In Practice Problem 186-2, what is the value of the pulse repetition frequency?

186-4. A radar set has a peak power of 1 MW, a pulse duration of 1.5 μs, and a pulse repetition time of 2000 μs. What is the value of the average power?

186-5. In a pulsed radar transmission, the average power is 600 W, the pulse repetition time is 1500 μs, and the pulse duration is 1.5 μs. Calculate the value of the peak power.

186-6. In Practice Problem 186-5, what is the value of the duty cycle?

186-7. A pulsed radar transmission has a pulse duration of 2 μs, and a pulse frequency of 750 Hz. What is the value of the duty cycle?

186-8. How many yards of range are approximately covered by a pulse duration of 0.5 μs?

187
Radar pulse duration and the discharge line

The pulse duration (length, width) is determined in the modulator unit (Fig. 187-1), which commonly uses an artificial discharge line consisting of a number of L,C sections. The total discharge time is then equal to the duration of the negative pulse applied to the cathode of the magnetron.

MATHEMATICAL DERIVATIONS

Pulse duration, $t = 2N \times \sqrt{LC}$ seconds (187-1)

where: N = number of L,C sections.
L = inductance of each section (H).
C = capacitance of each section (F).

Example 187-1

A modulator unit has an artificial discharge line of six sections, each of which contains an inductance of 6.5 μH and a capacitance of 4000 pF. What is the value of the pulse duration?

Solution

Pulse duration,

$$t = 2N \times \sqrt{LC} \quad (187\text{-}1)$$
$$= 2 \times 6 \times \sqrt{6.5 \times 10^{-6} \times 4000 \times 10^{-12}} \text{ s}$$
$$= 1.9 \ \mu\text{s}.$$

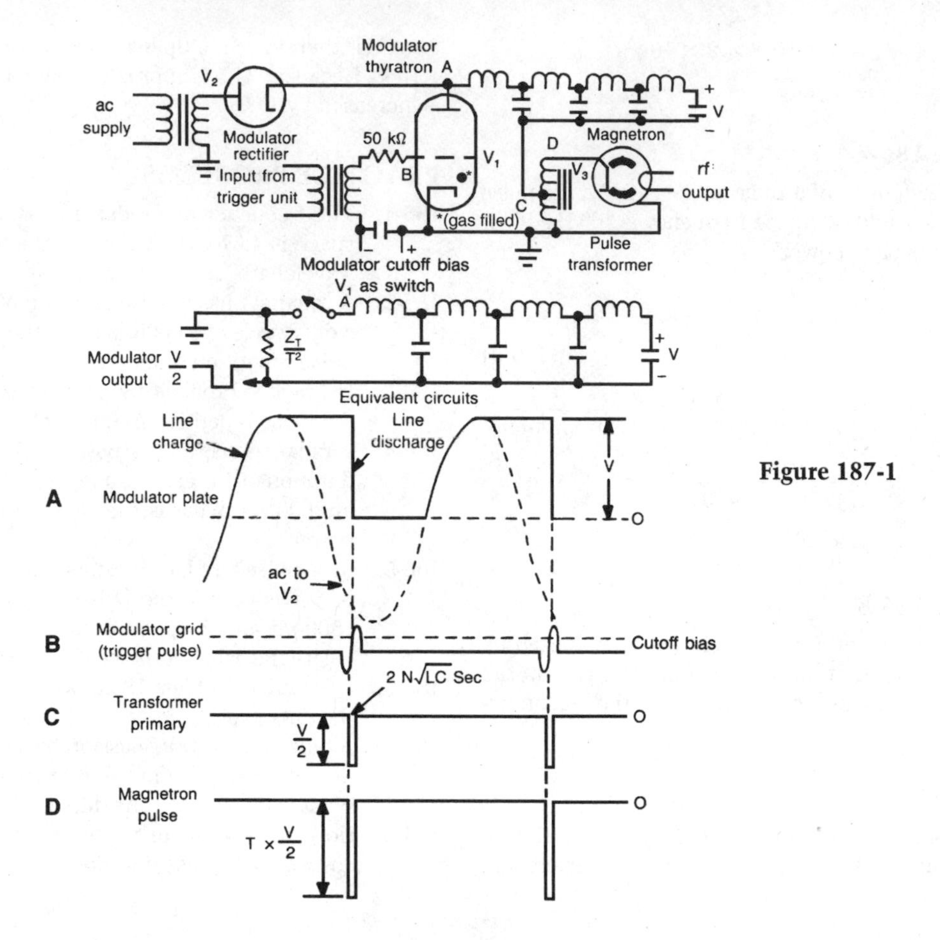

Figure 187-1

PRACTICE PROBLEMS

187-1. A modulator unit has an artificial discharge line of 10 sections, each of which has an inductance of 10 μH. If the pulse duration is 2 μs, calculate the value of the capacitance for each section.

187-2. A modulator unit has an artificial discharge line of five sections, each of which has a capacitance of 2000 pF. If the pulse duration is 1 μs, calculate the value of the inductance for each section.

188
Bandwidth of the radar receiver—intermediate frequency and video amplifier stages

To achieve accurate ranging, it is necessary to preserve the shape of the rectangular pulse and, in particular, its leading edge. This requires a wide bandwidth for the radar receiver's intermediate amplifier stages, in order to pass the numerous harmonics associated with a short-duration pulse (chapter 111).

Because a wide bandwidth is required, the gain of each IF amplifier stage is low so that the number of

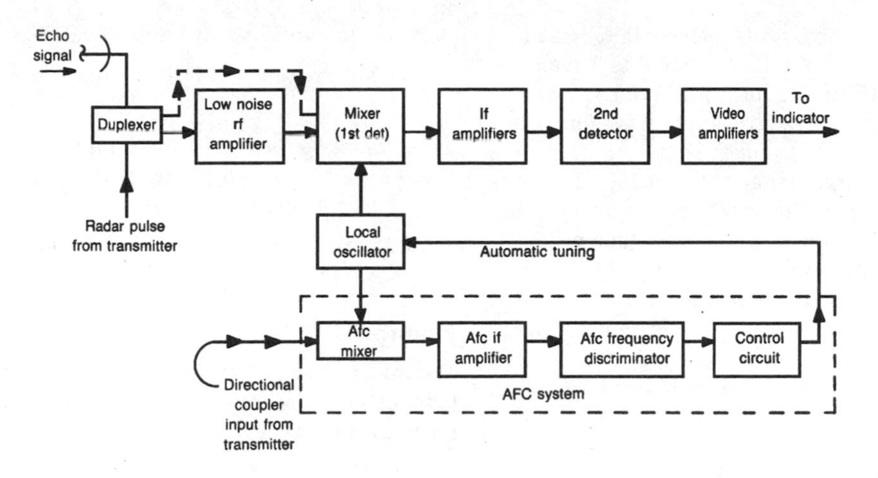

Figure 188-1

such stages in a radar receiver can be five or more. (Fig. 188-1.)

MATHEMATICAL DERIVATIONS

$$\textit{Bandwidth} = \frac{2}{\textit{pulse duration}\ \text{(microseconds)}}\ \text{MHz} \quad (188\text{-}1)$$

$$\textit{Pulse duration} = \frac{2}{\textit{bandwidth}\ \text{(MHz)}}\ \text{microseconds} \quad (188\text{-}2)$$

Example 188-1

A marine radar set has a pulse duration of 0.5 μs. What is the required bandwidth for the receiver's IF stages?

Solution

Required bandwidth (188-1)

$$= \frac{2}{\textit{pulse duration}\ (\mu s)}$$

$$= \frac{2}{0.5}$$

$$= 4\ \text{MHz}.$$

PRACTICE PROBLEMS

188-1. A marine radar set has a pulse duration of 1.0 μs. If the value of the intermediate frequency is 30 MHz, calculate the Q factor for the IF stages.

188-2. The IF stages of a radar set have a Q factor of 30, and the intermediate frequency is 60 MHz. What is the value of the pulse duration?

188-3. The bandwidth of the IF stages in a radar receiver is 1 MHz. What is the value of the pulse duration?

189
Radar ranges and their corresponding time intervals

A radar pulse is an electromagnetic wave that travels with the velocity of light in air. At standard temperature and pressure, this velocity is 2.997×10^8 meters per second (or 161700 nautical miles per second, where 1 nautical mile ≈ 6080 feet ≈ 2027 yards).

It follows that: 1 nautical (sea) mile = 1.15 × 1

statute (land) mile (or 1 statute mile = 0.87 × 1 nautical mile). A velocity of 1 knot is equal to 1 nautical mile per hour (1.15 statute miles per hour).

The time taken for a radar pulse to travel one nautical mile is 6.184 microseconds. Therefore, the total time interval for a target range of 1 nautical mile is the total time for the pulse to reach the target and the echo to return to the receiver; this total time will be 2 × 6.184 = 12.368 microseconds (Fig. 189-1).

The distance traveled by a radar pulse in 1 microsecond = 0.161711 nautical miles, which is approximately 328 yards. Therefore, a pulse duration of 1 microsecond will cover a range interval of 328/2 = 164 yards.

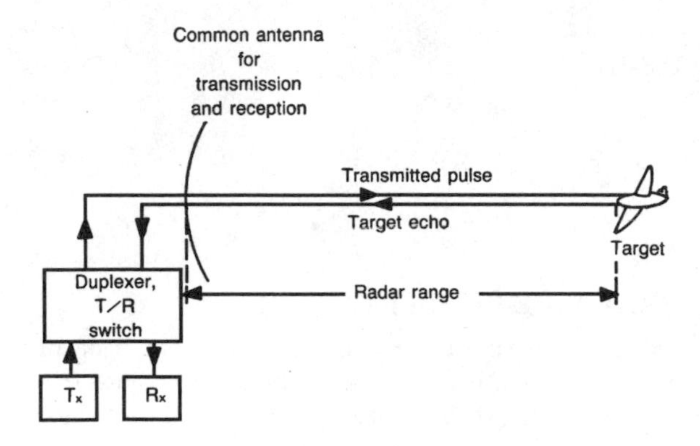

Figure 189-1

MATHEMATICAL DERIVATIONS

Radar range (in nautical miles) (189-1)

$$= \frac{\textit{total round-trip time interval } (\mu s)}{12.37}$$

Total time interval for the pulse (μs) (189-2)
= 12.37 × *radar range* (in nautical miles)
to travel to the target and back.

Distance to the target (in nautical miles) = (189-3)

$$\frac{\textit{time for transmitted pulse to reach target } (\mu s)}{6.18}$$

Time taken by the transmitted pulse to reach target (μs) = 6.18 × *distance to target* (in nautical miles) (189-4)

Total distance travelled by the pulse to the target and back to the receiver = 2 × *radar range* (189-5)

Example 189-1

The range of a target is 8.7 nautical miles. What is the time in microseconds for the transmitted pulse to reach the target?

Solution

The time taken for the transmitted pulse to reach the target is 6.18 × 8.7 = 53.8 microseconds.

Example 189-2

A radar pulse travels to the target and back to the receiver in a total time interval of 83 microseconds. What is the range of the target in nautical miles?

Solution

Range of the target = 83/12.37 = 6.7 nautical miles.

PRACTICE PROBLEMS

189-1. For a radar range of 7.6 nautical miles, what is the total time interval for the pulse to travel to the target and back?

189-2. If the time for a transmitter pulse to reach the target is 53 μs, what is the distance to the target in nautical miles?

189-3. If a target's radar range is 5.8 nautical miles, what is the total distance travelled by the pulse to the target and back to the receiver?

190
Frequency-modulated and continuous-wave radar systems

Apart from the use of short-duration RF pulses, there are two other possible radar systems. In FM radar, the RF wave is modulated by a sawtooth. The difference between the instantaneous frequencies being transmitted and received will then be a direct measure of the reflecting object's range. The most common use of this system is in the radio altimeter, where the reflecting object is the earth as seen from the aircraft. The instantaneous frequency will be an accurate measurement of the vertical distance between the ground and the aircraft.

In a CW radar system, the velocity of a target (but not its range) can be determined. The principle is based on the *Doppler effect*, in which the frequency of the signal radiated from the reflecting object is shifted when the object is moving relative to the receiver. An example of such a system is a police radar speed-trap (Fig. 190-1).

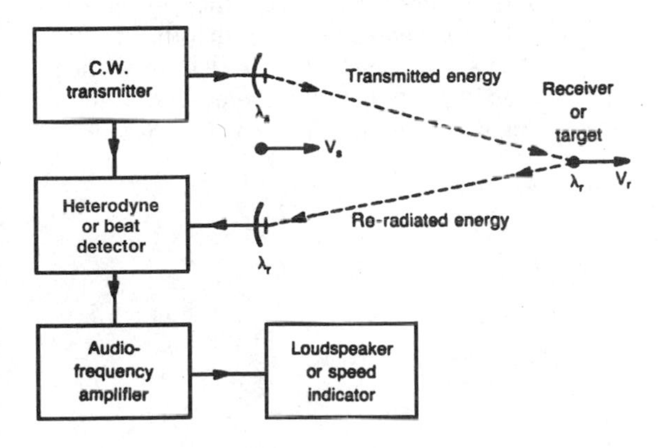

Figure 190-1

MATHEMATICAL DERIVATIONS

Let the source be moving with a component of velocity ($+v_s$) in the direction of propagation; the result will be to produce a change in the wavelength, and hence a shift in the frequency at the receiver. If λ_s is the wavelength and t is the period ($= 1/f_s$) of the radiation from the source when stationary:

New wavelength,

$$\lambda_r = \lambda_s - v_s \times t \text{ meters} \qquad (190\text{-}1)$$

Velocity of propagation $= c = f_s\lambda_s \qquad (190\text{-}2)$

$$= f_r(\lambda_s - v_s \times t)$$

meters per second

where f_r is the received frequency, λ_r is the received wavelength, and c is the velocity of an electromagnetic wave (300,000,000 meters per second or 984,000,000 feet per second). Then:

Received frequency,

$$f_r = \frac{c}{\lambda_s - v_s \times t} = \frac{c}{\lambda_s - \dfrac{v_s}{f_s}} \qquad (190\text{-}3)$$

$$= f_s \times \left(\frac{c}{c - v_s}\right) \text{ hertz}$$

If the source is stationary, but the receiver is moving with a component of velocity ($+v_r$) in the direction of propagation, there is (as regards the receiver) an apparent change in the velocity of propagation.

Apparent change in the velocity of propagation

$$= c - v_r$$

$$= \lambda_s \times f_r$$

This yields:

Received frequency,

$$f_r = \frac{c - v_r}{\lambda_s} = f_s \times \frac{(c - v_r)}{c} \text{ hertz} \qquad (190\text{-}4)$$

If the source and the receiver are both moving,

Received frequency,

$$f_r = f_s \times \left(\frac{c - v_r}{c - v_s}\right) \text{ hertz} \qquad (190\text{-}5)$$

The Doppler frequency shift $= f_r - f_s \qquad (190\text{-}6)$

$$= f_s \times \left(\frac{c - v_r}{c - v_s} - 1\right)$$

$$= f_s \times \left(\frac{v_s - v_r}{c - v_s}\right)$$

$$\approx f_s \frac{(v_s - v_r)}{c} \text{ hertz}$$

because c is very much greater than v_s.

The frequency shift depends on the source frequency and the relative velocity between the source and the receiver in the direction of propagation. If the relative velocity is 1 mile per hour and f_s = 1 GHz.

Doppler frequency shift (190-7)

$$= \frac{10^9 \times 5280}{9.84 \times 10^8 \times 3600} \text{ Hz}$$

$$\approx 1.5 \text{ Hz per mph per GHz}$$

For a CW radar set, the two-way propagation shift is $2 \times 1.5 = 3.0$ Hz per mph per GHz.

The same principles apply to a sonar set whose radiated acoustic pulse travels with a velocity of approximately 5000 ft/sec in seawater. If the relative velocity between the source and the receiver is 1 knot (6080 ft/hr), and $f_s = 1$ kHz:

(190-8)

$$\textit{Doppler shift} = \frac{1000 \text{ Hz} \times 6080 \text{ ft/hr}}{3600 \text{ s/hr} \times 5000 \text{ ft/s}} = 0.34 \text{ Hz}$$

For two-way propagation,

(190-9)

$$\textit{Doppler shift} = 2 \times 0.34 \approx 0.7 \text{ Hz per knot per kHz}$$

Example 190-1

A CW radar set is operating on a frequency of 10 GHz. If the target is moving in the direction of propagation with a velocity of 60 mph, calculate the amount of the Doppler shift at the receiver.

Solution

Two-way Doppler shift $= 3.0 \times 10 \times 60$ (190-7)

$= 1800$ Hz

Example 190-2

A sonar set is operating on a frequency of 9 kHz. If the relative velocity between the frigate and a submarine (in the direction of propagation) is 18 knots, what is the amount of the two-way Doppler shift?

Solution

Two-way Doppler shift $= 0.7 \times 18 \times 9$ (190-9)

$= 110$ IIz

PRACTICE PROBLEMS

190-1. A stationary CW radar set is operating on a frequency of 8 GHz. If a target produces a Doppler shift of 1680 Hz at the receiver, calculate the speed of the target in the direction of propagation.

190-2. A sonar set is operating on a frequency of 12 kHz. A destroyer detects the presence of a submarine and determines that the amount of the two-way shift is 140 Hz. Determine the relative velocity between the destroyer and the submarine, in the direction of propagation.

6

PART

Introductory mathematics for electronics

191
Common fractions

If a pie (Fig. 191-1) is cut into eight equal slices, each slice is one eighth of the whole pie. Mathematically, "one eighth" is a common fraction that can be written as 1/8. The number (1) above the line is called the fraction's *numerator*, and the number (8) below the line is the *denominator*. The single slice represents the division of the whole pie into eight equal parts; consequently, 1/8 can also be regarded as the result of dividing 1 by 8 so that the line is used to indicate the division of the numerator by the denominator.

If the value of the numerator is less than the value of the denominator, you have a *proper fraction*. Examples such as 2/7, 3/5, and 6/11 are all proper fractions; in every case, the value of a proper fraction is less than 1.

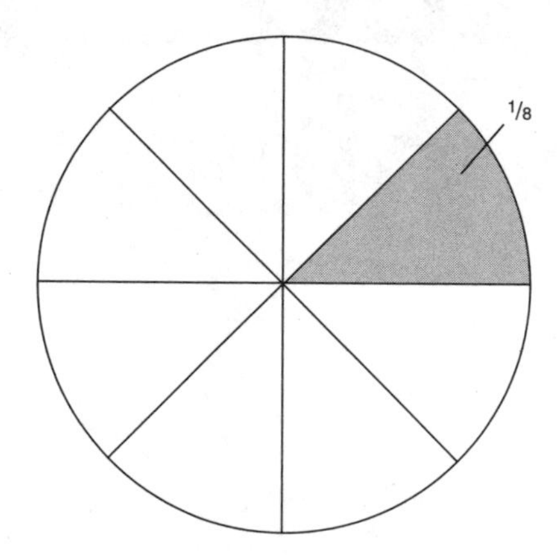

Figure 191-1

In Fig. 191-2, the total of the shaded areas amounts to two complete pies and five eighths of a third pie; mathematically, this would be written as 2 + 5/8, or 2 5/8. This combination of the whole number (2), and the proper fraction (5/8), is called a *mixed number*. However, in terms of eighths, the two complete pies are equivalent to 16/8, and the total of the shaded areas is 21/8. In this form, the numerator (21) is greater than the denominator (8), and you have an *improper fraction*, whose value is always greater than 1. It follows that

$$\underset{\text{Mixed number}}{\underset{\downarrow}{2\frac{5}{8}}} = \underset{\text{Improper fraction}}{\underset{\downarrow}{\frac{21}{8}}}$$

Conversion from a mixed number to an improper fraction

Multiply the whole number by the denominator, but then add in the value of the proper fraction's numerator. The result is the numerator of the improper fraction, and the denominator is unchanged. For example,

$$4\frac{3}{7} = \frac{4 \times 7 + 3}{7} = \frac{28 + 3}{7} = \frac{31}{7}$$

Conversion from an improper fraction to a mixed number

Divide the numerator of the improper fraction by the denominator. The result is a whole number and a remainder, which becomes the numerator of the accompanying proper fraction. For example,

$$35 \div 11 = 3 \text{ with the remainder of } 2$$

Therefore,

$$\frac{35}{11} = 3\frac{2}{11}$$

RULES OF COMMON FRACTIONS

In Fig. 191-3, the shaded area covers six slices of the pie, which is divided into eight equal parts; it follows that the area represents the proper fraction, 6/8. However, it is quite clear that the shaded area is also three-quarters of the pie, so:

$$\frac{6}{8} = \frac{3}{4}$$

and,

$$\frac{6}{8} = \frac{2 \times 3}{2 \times 4}$$

Therefore, the numerator (3) and the denominator (4) are multiplied by the same number (2) without altering the value of the fraction. As a formal statement:

Rule A

"For either a proper or an improper fraction, the numerator and the denominator can be multiplied by the same whole number without altering the value of the fraction."

As further examples,

$$\frac{2}{5} = \frac{2 \times 2}{5 \times 2} = \frac{4}{10}$$
$$= \frac{2 \times 3}{5 \times 3} = \frac{6}{15}$$
$$= \frac{2 \times 5}{5 \times 5} = \frac{10}{25}$$

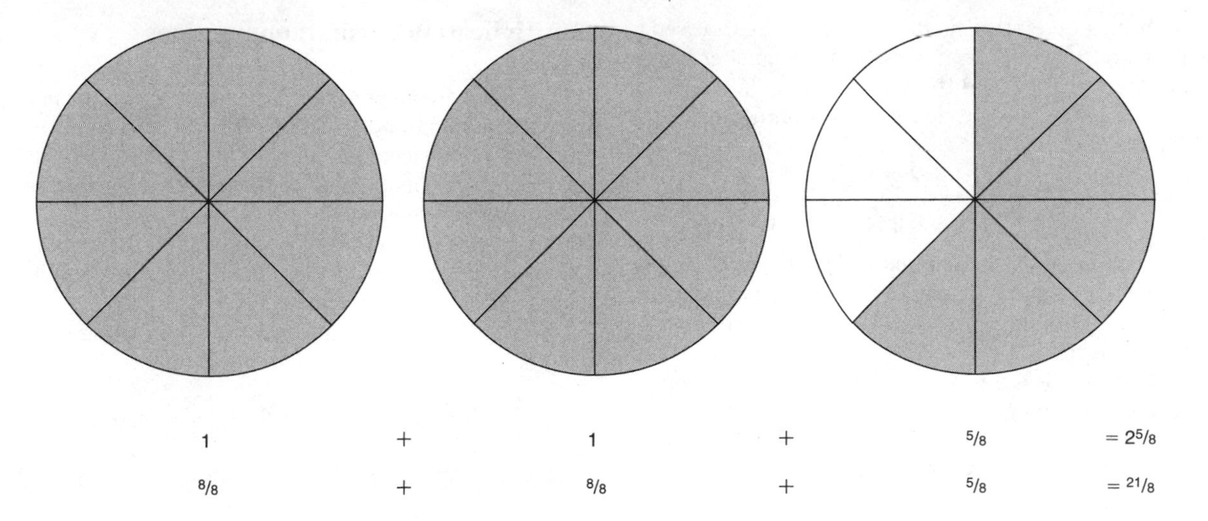

Figure 191-2

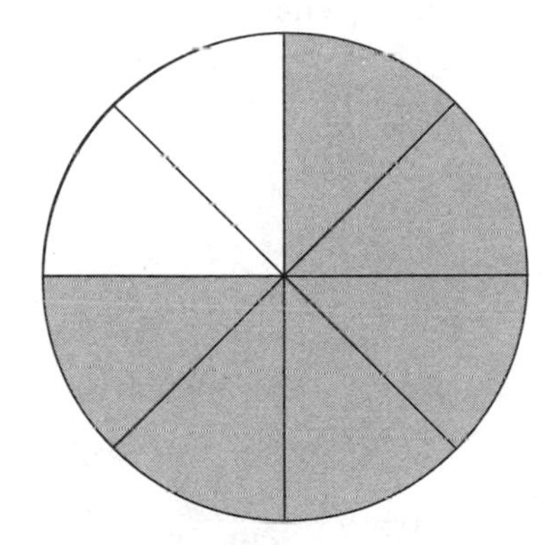

Figure 191-3

and,

$$\frac{11}{7} = \frac{11 \times 2}{7 \times 2} = \frac{22}{14}$$

$$= \frac{11 \times 3}{7 \times 3} = \frac{33}{21}$$

$$= \frac{11 \times 5}{7 \times 5} = \frac{55}{35}$$

It is not surprising that the value of the fractions remains unchanged because if, for example, both numerator and denominator are multiplied by 3, it is equivalent to multiplying the original fraction by 3/3 or 1.

Returning to the equivalence between six-eighths and three-quarters,

$$\frac{6}{8} = \frac{6 \div 2}{8 \div 2} = \frac{3}{4}$$

Therefore, the numerator (6) and the denominator (8) are divided by the same number (2), without altering the value of the fraction. As a formal statement:

Rule B

"For either a proper or an improper fraction, the numerator and the denominator can be divided by the same number without altering the value of the fraction."

As further examples,

$$\frac{12}{27} = \frac{12 \div 3}{27 \div 3} = \frac{4}{9}$$

$$\frac{35}{15} = \frac{35 \div 5}{15 \div 5} = \frac{7}{3}$$

$$\frac{120}{252} = \frac{120 \div 2}{252 \div 2}$$

$$= \frac{60}{126}$$

$$= \frac{60 \div 2}{126 \div 2}$$

$$= \frac{30}{63}$$

$$= \frac{30 \div 3}{63 \div 3}$$

$$= \frac{10}{21}$$

The last example involved repeated division until the final numerator (10) and denominator (21) had no common factor, except 1. The original fraction was then reduced to its *lowest terms*. An alternative method of reduction to the lowest terms requires that both the numerator and denominator are expressed in terms of their *prime factors* (a prime number has no factors, except 1 and the number itself; 2, 3, 5, 7, 11, 13, 17, 19, 23, 29, 31, and so on are prime numbers). Then,

$$\frac{120}{252} = \frac{\cancel{2} \times \cancel{2} \times 2 \times \cancel{3} \times 5}{\cancel{2} \times \cancel{2} \times \cancel{3} \times 3 \times 7} = \frac{10}{21}$$

Corresponding factors are cancelled out as shown; the remaining numerators and denominators are then separately multiplied out to produce the fraction in its lowest terms.

Notice that Rules A and B are not true if the multiplying or dividing number is zero.

MATHEMATICAL DERIVATIONS

Addition and subtraction of common fractions

Each of the fractions to be added or subtracted must be converted to the same *lowest common denominator, LCD* (the lowest common multiple of the individual denominators).

Rule C

"To add or subtract fractions, convert all the fractions to the same LCD, then combine all the numerators and, if necessary, reduce the answer to its lowest terms. If the answer is an improper fraction, it can be converted to a mixed number."

Rule C is illustrated in the following examples:

1. $1\frac{6}{7} + 4\frac{5}{21}$

The LCD of 7 and 21 is 21. As a first step, combine the whole numbers. Therefore,

$$1\frac{6}{7} + 4\frac{5}{21} = 5 + \frac{6 \times 3}{7 \times 3} + \frac{5}{21}$$

$$= 5 + \frac{18 + 5}{21}$$

$$= 5 + \frac{23}{21} = 5 + 1\frac{2}{21} = 6\frac{2}{21}$$

2. $4\frac{9}{10} - 3\frac{3}{4} = 1 + \frac{18 - 15}{20} = 1\frac{3}{20}$

3. $3\frac{7}{9} - 1\frac{4}{5} + \frac{16}{45} + \frac{8}{15}$. The LCD is 45.

$$= 2 + \frac{35 - 36 + 16 + 24}{45}$$

$$= 2 + \frac{39}{45} = 2\frac{13}{15}$$

Multiplication of common fractions

Rule D

"To multiply common fractions, express all mixed numbers as improper fractions. Then, cancel out all common factors contained in the numerators and the denominators. Finally, multiply together the remaining numerators to obtain the answer's numerator and carry out the same process for the remaining denominators. If the answer is an improper fraction, it can be converted to a mixed number."

It is important to emphasize that mixed numbers cannot be directly multiplied. For example,

$$2\frac{2}{3} \times 3\frac{1}{5} \text{ does } \textit{not} \text{ equal } 6 + \frac{2 \times 1}{3 \times 5} = 6\frac{2}{15}.$$

The correct answer is

$$\frac{8}{3} \times \frac{16}{5} = \frac{128}{15} = 8\frac{8}{15}$$

Rule D is illustrated in the following examples:

1. $\frac{3}{5} \times \frac{10}{21} = \frac{\overset{1}{\cancel{3}} \times \overset{2}{\cancel{10}}}{\underset{1}{\cancel{5}} \times \underset{7}{\cancel{21}}} = \frac{2}{7}$

2. $1\frac{4}{7} \times 2\frac{5}{22} = \frac{\overset{1}{\cancel{11}} \times \overset{7}{\cancel{49}}}{\underset{1}{\cancel{7}} \times \underset{2}{\cancel{22}}} = \frac{7}{2} = 3\frac{1}{2}$

3. $\frac{5}{9} \times \frac{3}{10} \times 1\frac{1}{2} = \frac{\overset{1}{\cancel{5}} \times \overset{1}{\cancel{3}} \times \overset{1}{\cancel{3}}}{\underset{1}{\cancel{9}} \times \underset{2}{\cancel{10}} \times 2} = \frac{1}{4}$

Division of common fractions

Because $3 \times 2 = 6$, it follows that $3 = 6 \div 2 = 6/2$.

Therefore, because $7/5 \times 5/7 = 1$, $7/5 = 1 \div 5/7 = \frac{1}{\frac{5}{7}}$.

When 1 is divided by a number, the result is the *reciprocal* of the number; for example, the reciprocal of 3 is 1/3. With the common fraction 5/7, its reciprocal is $1 \div 5/7 = 7/5$ so that the numerator and the denominator are inverted. The product of a fraction and its reciprocal is always equal to 1.

Rule E

"When dividing one common fraction by another, express any mixed numbers as improper fractions. Obtain the reciprocal of the denominator fraction and multiply this reciprocal by the numerator fraction with the aid of Rule D."

Rule E is illustrated in the following examples:

$$1.\quad \frac{\frac{2}{3}}{\frac{5}{6}} = \frac{2}{3} \times \frac{1}{\frac{5}{6}} = \frac{2}{\underset{1}{\not 3}} \times \frac{\overset{2}{\not 6}}{5} = \frac{4}{5}$$

$$2.\quad \frac{2\frac{7}{8}}{4\frac{5}{16}} = \frac{\frac{23}{8}}{\frac{69}{16}} = \frac{\overset{1}{\not 23}}{\underset{1}{\not 8}} \times \frac{\overset{2}{\not 16}}{\underset{3}{\not 69}} = \frac{2}{3}$$

$$3.\quad \frac{3\frac{3}{7}}{4} = \frac{\overset{6}{\not 24}}{7} \times \frac{1}{\underset{1}{\not 4}} = \frac{6}{7}$$

Example 191-1

For a series string of resistors $R_T = R_1 + R_2 + R_3$ (chapter 9). If $R_1 = 3\ 2/3\ \Omega$, $R_2 = 4\ 3/10\ \Omega$, and $R_3 = 6\ 7/24\ \Omega$, obtain the value of R_T.

Solution

$$R_T = R_1 + R_2 + R_3$$

$$= 3\frac{2}{3} + 4\frac{3}{10} + 6\frac{7}{24}$$

$$= 13 + \frac{2}{3} + \frac{3}{10} + \frac{7}{24}$$

The LCD is $3 \times 2 \times 5 \times 2 \times 2 = 120$.

Therefore,

$$R_T = 13 + \frac{80 + 36 + 35}{120} \qquad \text{Rule C}$$

$$= 13 + \frac{151}{120}$$

$$= 13 + 1\frac{31}{120} = 14\frac{31}{120}\ \Omega$$

Example 191-2

For a series string of resistors $E = V_1 + V_2 + V_3$ (chapter 9). If $E = 10$ V, $V_1 = 3\ 2/15$ V, and $V_2 = 4\ 5/12$ V, obtain the value of V_3.

Solution

$$V_3 = E - V_1 - V_2 \qquad \text{Rule C}$$

$$= 10 - 3\frac{2}{15} - 4\frac{5}{12}$$

$$= 3 - \frac{8}{60} - \frac{25}{60}$$

$$= 3 - \frac{33}{60}$$

$$= 3 - \frac{11}{20}$$

$$= 2 + 1 - \frac{11}{20} = 2\frac{9}{20}\ \text{V}$$

Notice that it was necessary to borrow "1" from the whole number "3."

Example 191-3

For two resistors in parallel, the total equivalent resistance,

$$R_T = \frac{R_1 \times R_2}{R_1 + R_2}$$

(chapter 15). If $R_1 = 3\ 1/2\ \Omega$ $R_2 = 4\ 2/3\ \Omega$, determine the value of R_T.

Solution

$$R_T = \frac{3\frac{1}{2} \times 4\frac{2}{3}}{3\frac{1}{2} + 4\frac{2}{3}}$$

$$= \frac{\frac{7}{2} \times \frac{14}{3}}{\frac{7}{2} + \frac{14}{3}} \qquad \text{Rule D}$$

$$= \frac{\frac{49}{3}}{\frac{21 + 28}{6}} \qquad \text{Rule C}$$

$$= \frac{49}{3} \times \frac{6}{49} = 2\ \Omega \qquad \text{Rule E}$$

PRACTICE PROBLEMS

191-1. Reduce the following common fractions to their lowest terms:

$$\frac{21}{49}, \frac{36}{48}, \frac{26}{91}, \frac{70}{105}$$

191-2. Convert the following improper fractions into mixed numbers:

$$\frac{191}{18}, \frac{1531}{42}, \frac{835}{18}, \frac{6758}{29}$$

191-3. Convert the following mixed numbers into improper fractions:

$$5\frac{7}{16}, 55\frac{6}{7}, 13\frac{13}{79}, 15\frac{9}{35}$$

191-4. Simplify each of the following to a common fraction in its lowest terms:

(a) $\frac{3}{4} + \frac{7}{6} + 2\frac{7}{12}$

(b) $1\frac{5}{6} + \frac{2}{13} - 1\frac{1}{8}$

(c) $3\frac{7}{9} - 1\frac{4}{5} + \frac{16}{45} + \frac{8}{15}$

191-5. Simplify each of the following to a common fraction in its lowest terms:

(a) $\frac{5}{8} \times 4\frac{2}{3}$

(b) $6 \times 5\frac{1}{2} \times \frac{1}{3}$

(c) $3\frac{2}{3} \times 1\frac{1}{2} \times 5\frac{3}{8} \times 4\frac{1}{6} \times 3\frac{1}{11}$

191-6. Simplify each of the following to a common fraction in its lowest terms:

(a) $1\frac{3}{7} \div \frac{5}{14}$

(b) $3\frac{4}{9} \div \frac{11}{18}$

(c) $\frac{1}{13} \div 1\frac{7}{39}$

191-7. Determine the reciprocals of the following common fractions:

$$\frac{3}{11}, 2\frac{2}{3}, 5\frac{2}{5}$$

192
Decimal numbers

A number system is any set of symbols or characters used for the purpose of enumerating objects and performing mathematical computations (such as addition, subtraction, multiplication, and division). The most commonly used system is the *Hindu-Arabic*, which has the digits 0, 1, 2, 3, 4, 5, 6, 7, 8, and 9. There are a total of 10 digits and this is a decimal (or base-10) system.

The acceptance of two basic concepts has greatly simplified mathematical computations and has led to the development of modern number systems. These two concepts are:

1. The use of zero to signify the absence of a value.
2. The principle of positional value.

The principle of *positional value* consists of assigning to a digit a value that depends on its position within a given number. For example, the digit 6 has a different value in each of the decimal numbers 876, 867, and 687. In the first number (876), the digit 6 has its base value of 6. For the second number (867), the digit 6 has a value of 60. In the third number (687), the value of the digit 6 is 600. Sometimes a position within a given number does not have a value. However, if the position is totally disregarded, there is no way to distinguish between two different numbers, such as 706 and 76. Therefore, the zero is used to signify that a particular position, within a certain number, has no value assigned.

The uses of zero and positional value have greatly simplified counting and mathematical computations; consequently, these concepts are used in all modern number systems.

You have seen that the value of a particular digit depends on its position within a number. Figure 192-1 illustrates the number 3007406.089002, where the (decimal) point separates the whole number part from the decimal fraction. This number is written as three-million, seven-thousand, four-hundred, six *and* eight-hundredths, nine-thousandths, two-millionths. The connecting word "and" is only used to join the whole number to the fraction. Alternatively, a number such as 34.782 can be read as thirty-four "point" seven eight two. In this number, there are three digits to the right of the point, and these are referred to as three *decimal places.*

MATHEMATICAL DERIVATIONS

Rule A

"When a decimal number is *multiplied* by 10, the point is moved one place to the *right*. It follows that if

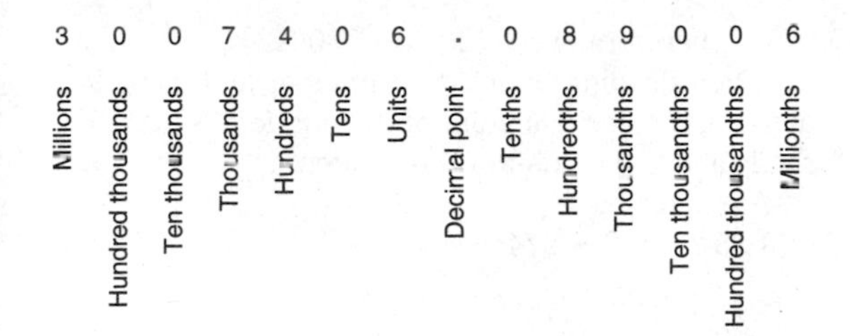

Figure 192-1

a decimal number is multiplied by, for example, 10000 ($10 \times 10 \times 10 \times 10$), the point is moved four places (corresponding to the four zeros) to the right. If there are insufficient digits to allow this operation to be performed, zeros can be added, as necessary, to the right of the number."

Rule A is illustrated in the following examples:

1. $375.86 \times 10 = 375.86 = 3758.6.$
   1 zero → 1 place
2. $0.0459 \times 1000 = 0.0459 = 45.9.$
   3 zeros → 3 places
3. $16.385 \times 100000 = 16.38500 = 1638500$
   5 zeros → 5 places (2 zeros added to the right)

Rule B

"When a decimal number is *divided* by 10, the point is moved one place to the *left*. It follows that if a decimal number is divided by, for example, 10,000 ($10 \times 10 \times 10 \times 10$), the point is moved four places (corresponding to the four zeros) to the left. If there are insufficient digits to allow this operation to be performed, zeros can be added, as necessary, to the left of the number."

Rule B is illustrated in the following examples:

1. $375.86 \div 10 = \frac{375.86}{10} = 375.86 = 37.586$
   1 zero → 1 place
2. $0.0459 \div 1000 = \frac{0.0459}{1000} = 0000.0459$
   3 zeros → 3 places
   $= 0.0000459$
3. $16.385 \div 100000 = \frac{16.385}{100000} = 000016.35$
   5 zeros → 5 places
   $= 0.0001635$

Addition of decimal numbers

Rule C

"Write down the decimal numbers so that all points are vertically aligned. Carry out the addition process and place the answer's decimal point directly beneath all the other points."

Rule C is illustrated in the following examples:

1. 67.35 + 0.00062 + 637.4531

```
          67.35
   +       0.00062
   +     637.4531
Answer   704.80372
```

2. 0.384 + 1.00459 + 0.000378

```
         0.384
   +     1.00459
   +     0.000378
Answer   1.388968
```

Subtraction of decimal numbers

Rule D

"Write down the decimal numbers so that one point is directly beneath the other. Carry out the subtraction process and place the answer's decimal point directly beneath the other two. If necessary, zeros can be added to the right to either number."

Rule D is illustrated in the following examples:

1. 16.783 − 7.9745

```
        16.7830   ←—(one zero added to the right)
   −     7.9745
Answer   8.8085
```

2. 3.7153 − 0.008742

```
         3.715300 ←—(two zeros added to the right)
   −     0.008742
Answer   3.706558
```

3. 0.069347 − 0.00195

```
         0.069347
   −     0.001950 ←—(one zero added to the right)
Answer   0.067397
```

Multiplication of decimal numbers

Rule E

"Count up the total number of decimal places in the two numbers that are to be multiplied. Carry out the multiplication process in the same manner as with whole numbers. In the answer, locate the position of the decimal point by counting off the same total number of places from the *left*. If there are insufficient digits, as many zeros as necessary can be included to the left of the answer."

Rule E is illustrated in the following examples:

1. 37.05 × 7.8.

   2 places 1 place

   There is a total of three decimal places in these two numbers.

```
   3705
  × 78
  29640
 25935
 288.990
```

counting off 3 places

Therefore, 37.05 × 7.8 = 288.99

2. 0.00372 × 0.451. Total number of decimal places is 5 + 3 = 8.

```
       372
     × 451
       372
     1860
    1488
0.00167772
```

counting off 8 places

Therefore, 0.00372 × 0.451 = 0.00167772

Division of decimal numbers

When dividing by a whole number and using short division, the decimal point of the answer must be placed directly beneath the point of the number being divided.

As examples,

1. 17.453 ÷ 7. In mathematical language, 17.453 is the *dividend*, 7 is the *divisor*, and the answer is the *quotient*.

```
        7)17.45300 . . .
Answer     2.49328 . . .
```

Because the decimal number is not exactly divisible by 7, zeros are added to the right of the dividend, until you obtain the desired number of decimal places.

Therefore, 17.453 ÷ 7 = 2.49328 . . .

2. 0.00372 ÷ 3

```
        3)0.00372
Answer    0.00124
```

Therefore, 0.00372 ÷ 3 = 0.00124

When dividing by a whole number and using long division, the decimal point of the quotient is placed directly above the point of the dividend.

As examples,

1. 8.752108 ÷ 274

```
     0.031942  Answer
274)8.752108
     8 22
       532
       274
       2581
       2466
        1150
        1096
          548
          548
          . . .
```

Therefore, 8.752108 ÷ 274 = 0.031942.

2. 0.0069814 ÷ 118

```
     0.0000591644  Answer
118)0.0069814000
         590
         1081
         1062
           194
           118
            760
            708
             520
             472
              480
              472
                8
```

Therefore, 0.0069814 ÷ 118 = 0.0000591644 . . .

Again, the dividend is not exactly divisible; the zeros are added to the right of the dividend in order to obtain the desired number of decimal places in the quotient.

Rule F

"When the divisor is involved with a decimal fraction (for example, 0.86493 ÷ 3.657), the divisor is made a whole number by shifting the decimal point the required number of places to the right. In this example, the point would be shifted three places to the right so that 3.657 is converted to 3.657. = 3657. To maintain a correct answer, the decimal point must also be shifted three places to the right in the dividend, which becomes 0.86493 = 864.93. This process has resulted in multiplying both dividend and divisor by 1000. Finally, the quotient is obtained by the long-division method previously outlined."

Rule F is illustrated in the following examples:

1. $0.86493 \div 3.657 = 864.93 \div 3657$

```
          0.2365  Answer
3657)864.9300
       731 4
       133 53
       109 71
        23 820
        21 942
         1 9780
         1 8285
           1495
```

Therefore, $0.86493 \div 3.657 = 0.2365 \ldots$

2. $0.007641 \div 0.03$

In both numbers, the decimal point is shifted two places to the right. Then, $0.007641 \div 0.03 = 0.7641 \div 3$

```
3)0.7641
  0.2547
```

Therefore, $0.007641 \div 0.03 = 0.2547$.

Conversion from a common fraction to a decimal fraction

Divide the numerator by the denominator using either short or long division, as is appropriate.

As examples,

1. $\frac{3}{5} = \frac{3.0}{5} = 0.6$

It follows that, the mixed number 2 3/5 is equivalent to 2.6

2. $\frac{7}{37}$

```
    0.189
37)7.000
    3 7
    3 30
    2 96
      340
      333
        7
```

Therefore, $7/37 = 0.189 \ldots$

Conversion from a decimal fraction to a common fraction

First, determine the denominator of the decimal fraction. For example, 0.375 means three tenths, seven hundredths, five thousandths (or three hundred, seventy five thousandths). Therefore, 0.375 = 375/1000 and 1000 is the required denominator. Then, reduce the common fraction to its lowest terms so that 0.375 = 375/1000 = 3/8.

As further examples,

1. $0.64 = \frac{64}{100} = \frac{16}{25}$

2. $0.004 = \frac{4}{1000} = \frac{1}{250}$

3. $0.8 = \frac{8}{10} = \frac{4}{5}$

Reciprocal of a decimal number

The reciprocal of a decimal number is the result of dividing 1 by the number. For example, the reciprocal of 0.5 is 1/0.5 = 10/5 = 2. The reciprocal of 2 is 1/2 = 0.5. Consequently, if the reciprocal of a first number is a second number, then the reciprocal of the second number is the first number.

As further examples,

1. The reciprocal of 1.25 is $\frac{1}{1.25} = \frac{100}{125} = \frac{4}{5} = 0.8$.

2. The reciprocal of 1473 is $\frac{1}{1473}$.

```
        0.0006789  Answer
1473)1.0000000
        8838
        11620
        10311
         13090
         11784
          13060
          11784
           1276
```

Therefore, the reciprocal of 1473 is 0.0006789, rounded off.

Percentages

The word "percentage" (symbol %) is derived from the Latin "per centum," which means that a whole is divided into 100 equal parts. For example 83¢ is 83 equal parts of the complete whole which is 100¢ or $1.00. Consequently, 83¢ is 83% of $1.00.

Rule G

"To convert either a common fraction or a decimal fraction to a percentage, the fraction is multiplied by 100."

Rule G is illustrated by the following examples:

1. $\frac{1}{8} = \frac{1}{8} \times 100\% = 12\frac{1}{2}\%$
2. $0.721 = 0.721 \times 100\% = 72.1\%$
3. $2\frac{3}{4} = \frac{11}{4} \times 100\% = 275\%$
4. $0.00495 = 0.00495 \times 100\% = 0.495\%$

Ratios

A ratio is used to compare the values of two quantities that are measured in the same units. One quantity is divided by the other, and their quotient (which has no units) is then compared with unity. For example, if the primary winding of the transformer has 12000 turns, and the secondary winding has 4000 turns, the turns ratio is

$$\frac{12000}{4000} = \frac{3}{1}$$

This ratio is written as 3:1.

As further examples,

1. The power input to an amplifier is 2.5 watts and the power output is 12.5 watts.

$$\textit{Power ratio} = \frac{12.5}{2.5} = \frac{5}{1} = 5{:}1$$

2. The power input to an attenuator is 8 W and the power output is 2 W.

$$\textit{Power ratio} = \frac{2}{8} = 0.25{:}1 \text{ or } 1{:}4$$

Example 192-1

In a network of parallel resistors, $I_T = I_1 + I_2 + I_3$ (chapter 15). If $I_1 = 3.752$ amperes, $I_2 = 0.456$ amperes, and $I_3 = 9.71$ amperes, calculate the value of I_T.

Solution

$$I_T = 3.752 + 0.456 + 9.71$$

$$\begin{array}{r} 3.752 \\ +\ 0.456 \\ +\ \underline{0.71\ \ } \\ 13.918 \end{array}$$

Therefore, $I_T = 13.918$ amperes.

Example 192-2

In a network of two parallel conductors, $G_T = G_1 + G_2$ (chapter 15). If $G_T = 0.00357$ siemens and $G_1 = 0.000863$ siemens, calculate the value of G_2.

Solution

$$G_2 = G_T - G_1 = 0.00357 - 0.000863$$

$$\begin{array}{r} 0.003570 \\ -\ \underline{0.000863} \\ 0.002707 \end{array}$$

Therefore, $G_2 = 0.002707$ siemens.

Example 192-3

One of the Ohm's Law relationships is $E = I \times R$ (chapter 5). If $I = 0.00678$ amperes and $R = 3900$ ohms, calculate the value of E in volts.

Solution

$$\begin{aligned} E &= 0.00678 \times 3900 \\ &= 0.00678 \times 100 \times 39 \\ &= 0.678 \times 39 \text{ V} \end{aligned}$$

There are three decimal places. Thus,

$$\begin{array}{r} 678 \\ \underline{39} \\ 6102 \\ \underline{2034\ } \\ 26.442 \end{array}$$

Therefore, $E = 26.442$ volts.

Example 192-4

One of the power relationships is $I = P/E$ (chapter 4). If $P = 0.0384$ watt and $E = 2.74$ volts, calculate the value of I.

Solution

$$I = \frac{0.0386}{2.74} = \frac{3.86}{274} \text{ ampere.}$$

```
      0.01408 Answer
 274)3.86000
     2 74
     1 120
     1 096
        2400
        2192
         208
```

Therefore, $I = 0.0141$ ampere, rounded off.

Example 192-5

Conversions:

(a) Convert 0.575 into a common fraction.
(b) Convert 7/25 into a decimal fraction.
(c) What is the reciprocal of 0.0125?
(d) Convert 0.00384 into a percentage.
(e) Convert 75% into a common fraction.
(f) What percentages are 6 of 30 and 30 of 6?
(g) If the input voltage to a stage is 0.025 V and the output voltage is 10 V, calculate the ratio of the output voltage to the input voltage.

Solution

(a) $0.575 = \frac{575}{1000} = \frac{115}{200} = \frac{23}{40}$

(b) $\frac{7}{25} = \frac{7.00}{25}$

$$\begin{array}{r} 0.28 \text{ Answer} \\ 25\overline{)7.00} \\ \underline{5\,0} \\ 2\,00 \\ \underline{2\,00} \\ \ldots \end{array}$$

Therefore, $\frac{7}{25} = 0.28$.

(c) Reciprocal of 0.0125 is:

$$\frac{1}{0.0125} = \frac{1.0000}{0.0125} = \frac{10000}{25} = 80.$$

(d) $0.00384 = 0.00384 \times 100\% = 0.384\%$

(e) $75\% = \frac{75}{100} = \frac{3}{4}$

(f) $Percentage = \frac{6}{30} \times 100\% = 20\%$

$Percentage = \frac{30}{6} \times 100\% = 500\%$

(g) $Ratio = \frac{output\ voltage}{input\ voltage} = \frac{10}{0.025} = 400{:}1$

PRACTICE PROBLEMS

192-1. Convert the following common fractions into decimal fractions:

$$\frac{2}{5}, \frac{3}{16}, \frac{3}{40}, 3\frac{3}{8}.$$

192-2. Convert the following decimal fractions into common fractions in their lowest terms: 0.15625, 0.46875, 5.3125, 3.0005.

192-3. Calculate the values of the following:
(a) $3.005 + 1.9 + 2.35$
(b) $9 + 0.001 + 0.0002$
(c) $3.0005 - 0.375$
(d) $0.15 - 0.67 + 0.5$
(e) $-0.00025 - 0.0013 + 6.21$

192-4. Calculate the values of the following:
(a) 2.2×3.5
(b) 5000000×0.000001
(c) 0.0007×0.0175

192-5. Calculate the values of the following:
(a) $23.9 \div 0.187$
(b) $\frac{4 \times 0.5}{0.2}$
(c) $\frac{0.0087 \times 195}{25.47 \times 4.38}$
(d) $\frac{1}{2.3 \times 1.7}$

192-6. Express the following numbers as a percentage: (a) 0.83, (b) 0.004, (c) 1.63, (d) 3/5, (e) 5/8.

192-7. Express the following percentages as decimal fractions: (a) 19%, (b) 97.3%, (c) 18 9/10%, (d) 7%, (e) 1/4%.

192-8. Express the values of the following as decimal numbers: (a) 20% of 150, (b) 35% of 200, (c) 0.25% of 1000, (d) 11/2% of 50, (e) 0.003% of 20000.

192-9. The value of the resistances (R_1 and R_2) are, respectively, 390 Ω and 910 Ω. Calculate the value of the ratio, $R_2{:}R_1$.

193
Scientific notation

The molecule of water is the smallest quantity that still retains all the properties of water. However, water is a compound of two gases, oxygen, and hydrogen, whose atoms are combined to produce the molecule. Oxygen and hydrogen are examples of *elements* that are capable of a separate existence. More than 100 different elements have been identified. Of these, the first 92 occur naturally in the earth and the others have been discovered as the results of scientific experiments. The *atom* of an element is the smallest particle that still retains all the properties of the element.

The structure of the hydrogen atom contains a core, called the *nucleus*, where most of the atom's mass is concentrated. This nucleus of the hydrogen atom is the subatomic *proton* particle, which is positively charged. Revolving around the nucleus is a much lighter particle, called the *electron*, which carries an equal negative charge. The practical unit of charge is the coulomb (chapter 2); a certain number of electrons must therefore carry a total amount of charge equal to 1 coulomb.

These figures are:

1. 6,240,000,000,000,000,000 electrons (approximately) carry a charge of one coulomb.
2. The mass of the electron is (approximately) 0.000,000,000,000,000,000,000,000,000,000,910,96 kilogram.
3. The mass of the proton is (approximately) 0.000,000,000,000,000,000,000,000,001,672,6 kilogram.
4. The radius of the electron's circular path is approximately 0.000,000,000,053 meter.

Clearly, these numbers are far too cumbersome to be written out in full, so you must express them in some form of mathematical shorthand. The one most commonly used is called *scientific notation*, which has been developed through the use of exponents, sometimes referred to as *powers*.

The base 10, when raised to the exponent (or power) N, is written is 10^N; then, if N is a positive whole number, 10^N means 10 multiplied by itself N times. As an example, $10^6 = 10 \times 10 \times 10 \times 10 \times 10 \times 10 = 1{,}000{,}000$, or 1 million. Therefore, 56380000. can be written as $5.6383 \times 10{,}000{,}000 = 5.6383 \times 10^7$; the number 5.638 is called the coefficient and the exponent "7" is found by counting the number of places the decimal point is shifted to the left. When numbers are written in scientific notation, there is a single figure before the decimal point, followed by such decimal places as are required by the number's accuracy. This decimal coefficient (which is greater than or equal to 1, but is less than 10) must then be multiplied by 10, raised to the required exponent.

As further examples,

1. $1378546200. = 1.379 \times 10^9$
   nine places
   This answer has been rounded off to four significant figures. Similarly,
2. $236.74 = 2.37 \times 10^2$, rounded off to three significant figures.
   two places
3. $37.586 = 3.759 \times 10^1$, rounded off to four significant figures.
   one place
   Note that 10^1 has the same value as 10.

When calculating a result from a formula, all numbers entered into the formula should be expressed in scientific notation. This makes the calculation easier and also allows the answer to appear in scientific notation for possible substitution into another formula. As a result, most electronic calculators display their results in scientific notation.

MATHEMATICAL DERIVATIONS

Addition and subtraction of numbers

When you add 5370 and 284, the result is 5370 + 284 = 5654. If these numbers are expressed in scientific notation,

$$5370 + 284 = 5654$$

$$(5.37 \times 10^3) + (2.84 \times 10^2) = 5.654 \times 10^3$$

or,

$$5.37 \times 10^3 + 0.284 \times 10^3 = (5.37 + 0.284) \times 10^3 = 5.654 \times 10^3.$$

Rule A

"To carry out addition or subtraction using scientific notation, the numbers must be expressed in terms of the same exponent; the coefficients are then either added or subtracted, but the exponent is unchanged."

As examples,

$$1.\ 874000 + 23860 = 8.74 \times 10^5 + 0.2386 \times 10^5 = (8.74 + 0.2386) \times 10^5 = 8.9786 \times 10^5$$

2. $346000 - 65700 = 3.346 \times 10^7 - 0.00657 \times 10^7$

$$= (3.346 - 0.00657) \times 10^7$$

$$= 3.33943 \times 10^7$$

Multiplication of numbers

You know that $300 \times 20{,}000 = 6{,}000{,}000$; if these numbers are expressed in scientific notation:

$$300 \times 20{,}000 = 6{,}000{,}000$$

$$(3 \times 10^2) \times (2 \times 10^4) = 6 \times 10^6$$

Notice that the individual exponents "2" and "4" are *added* (*not* multiplied) to produce the exponent "6" in the answer.

Rule B

"When multiplying numbers that are expressed in scientific notations, the coefficients are multiplied, but you take the algebraic sum of the exponents."

The *algebraic sum* means that you must take into account the signs (+ or −) of the exponents. The meaning of the negative exponent is explained later.

As a further example,

$34.2 \times 7{,}531{,}000 \times 8740$

$$= 3.42 \times 10^1 \times 7.531 \times 10^6 \times 8.74 \times 10^3$$

$$= 3.42 \times 7.531 \times 8.74 \times 10^1 \times 10^6 \times 10^3$$

$$= 225.1 \times 10^{1+6+3}$$

$$= 2.251 \times 10^2 \times 10^{10}$$

$$= 2.251 \times 10^{12}$$

The answer is rounded off to four significant figures because of the four significant figures in the number 7,531,000; the accuracy of the answer is, therefore, comparable with the accuracy of the problem.

When any number is raised to the exponent 2, you obtain the *square* of the number. The square of the number is the result of multiplying a number by itself so that $10^2 = 10 \times 10 = 100$ and $8^2 = 8 \times 8 = 64$. Remember that the square of a negative number is always positive. For example, $(-8)^2 = (-8) \times (-8) = +64$. Consequently, you can say that the minimum value of a squared number is zero.

If a number is expressed in scientific notation, the square of the number is obtained by squaring the coefficient, but *doubling* the exponent. For example, $(7543)^2 = (7.543 \times 10^3)^2 = 7.543^2 \times 10^{3 \times 2} = 56.9 \times 10^6 = (5.69 \times 10^1) \times 10^6 = 5.69 \times 10^7$.

When calculating the area of a rectangle, it is necessary to multiply the length by the width. If the length is 5 m and the width is 3 m, the area is 5 m × 3 m = 15 square meters; square meters are abbreviated as m^2. Because 100 centimeters = 1 meter, 1 square meter = 100 centimeters × 100 centimeters = 10,000 square centimeters; using abbreviations, $1\ m^2 = 10^4\ cm^2$.

Division of numbers

If you divide 6,000,000 by 20,000, the answer is 300. Using scientific notation,

$$\frac{6{,}000{,}000}{20{,}000} = 300$$

$$\frac{6 \times 10^6}{2 \times 10^4} = 3 \times 10^2$$

In the operation of division, the exponent "4" is subtracted from the exponent "6" so that the exponent in the answer is 6 − 4 = "2."

Rule C

"When dividing numbers that are expressed in scientific notation, the coefficient of the dividend is divided by the coefficient of the divisor, but the exponent of the divisor is algebraically subtracted from the exponent of the dividend."

The following example illustrates the rules of both multiplication and division:

$$\frac{1740 \times 43{,}520{,}000}{13.52 \times 278000}$$

$$= \frac{1.74 \times 10^3 \times 4.352 \times 10^7}{1.352 \times 10^1 \times 2.78 \times 10^5}$$

$$= \frac{1.74 \times 4.352 \times 10^{10}}{1.352 \times 2.78 \times 10^6}$$

$$= 2.015 \times 10^{10-6}$$

$$= 2.015 \times 10^4$$

One special case. If you divide 3000 by 3000, the answer must, of course, be 1. However, using scientific notation,

$$\frac{3000}{3000} = \frac{3 \times 10^3}{3 \times 10^3} = 1 \times 10^{3-3} = 1 \times 10^0$$

Consequently, $10^0 = 1$. This means that a number, such as 4.39, can be written in scientific notation as 4.39×10^0.

Meaning of the negative exponent

Divide 500 by 2,500,000; the result is 0.0002. However, if you use scientific notation and subtract the exponents (Rule C),

$$\frac{500}{2{,}500{,}000} = \frac{5 \times 10^2}{2.5 \times 10^6} = 2 \times 10^{2-6} = 2 \times 10^{-4}$$

Because $0.0002 = 2 \times 10^{-4}$, you can use the negative exponent to express, in scientific notation, all numbers that are less than 1. For example,

$$0.00000174 = 1.74 \times 10^{-6},$$

six places

with the value of the negative exponent equal to the number of places that the decimal point is shifted to the *right*. If you again divide 500 by 2,500,000,

$$\frac{500}{2{,}500{,}000} = \frac{2}{10{,}000} = \frac{2}{10^4} = 2 \times \frac{1}{10^4}$$

Comparing the answers "2×10^{-4}" and "$2 \times 1/10^4$," it follows that $1/10^4 = 10^{-4}$; therefore, $1/10^{-4} = 10^4$. You can then say that 10^4 and 10^{-4} are reciprocals, with their product $10^4 \times 10^{-4}$ equal to 10^0 or 1.

When a number is expressed in scientific notation, its reciprocal is obtained by taking the reciprocal of the coefficient, and changing the sign of the exponent. For example, the reciprocal of 4×10^4 is $(1/4) \times 10^{-4} = 0.25 \times 10^{-4} = 2.5 \times 10^{-5}$.

Table 193-1 shows the application of positive and negative exponents to scientific notation. The combined use of positive and negative exponents is illustrated in the following example:

$$\frac{149 \times 0.000763}{0.000000125 \times 52.7}$$

$$= \frac{1.49 \times 10^2 \times 7.63 \times 10^{-4}}{1.25 \times 10^{-7} \times 5.27 \times 10^1}$$

$$= \frac{1.49 \times 7.63 \times 10^{(+2)+(-4)}}{1.25 \times 5.27 \times 10^{(-7)+(+1)}}$$

$$= \frac{1.73 \times 10^{-2}}{1 \times 10^{-6}}$$

$$= 1.73 \times 10^{(-2)-(-6)}$$

$$= 1.73 \times 10^4$$

Raising a number to a power

Many of our electronics formulas involve squares and higher powers so that you need to know how to use scientific notation in raising a number to a particular power. If you square 3000, the result is $3000 \times 3000 = 9{,}000{,}000$. By using scientific notation:

$$(3000)^2 = (3 \times 10^3)^2 = 9 \times 10^6$$

Therefore, the coefficient is squared, but the exponent "3" is *multiplied* by the power "2" to equal the answer's exponent of "6." This rule of multiplication applies to both positive and negative exponents.

Table 193-1. Positive and negative exponents.

Exponents of 10	Meaning
10^6	$\times$ 1,000,000
10^5	$\times$ 100,000
10^4	$\times$ 10,000
10^3	$\times$ 1,000
10^2	$\times$ 100
10^1	$\times$ 10
10^0	$\times$ 1
$10^{-1} = \frac{1}{10^1}$	$\div$ 10 or $\times$ 0.1
$10^{-2} = \frac{1}{10^2}$	$\div$ 100 or $\times$ 0.01
$10^{-3} = \frac{1}{10^3}$	$\div$ 1000 or $\times$ 0.001
$10^{-4} = \frac{1}{10^4}$	$\div$ 10,000 or $\times$ 0.0001
$10^{-5} = \frac{1}{10^5}$	$\div$ 100,000 or $\times$ 0.00001
$10^{-6} = \frac{1}{10^6}$	$\div$ 1,000,000 or $\times$ 0.000001

Rule D

"When a number is expressed in scientific notation, the number is raised to a certain power by raising the coefficient to that power, but multiplying the exponent by the value of the power."

For example, $(2.75 \times 10^{-3})^4 = 2.75^4 \times 10^{-3\times4} = 57.2 \times 10^{-12} = 5.72 \times 10^1 \times 10^{-12} = 5.72 \times 10^{-11}$.

As further examples,

1. $(324)^4 = (3.24 \times 10^2)^4 = (3.24)^4 \times 10^{2\times4}$
   $= 110 \times 10^8 = 1.10 \times 10^2 \times 10^8$
   $= 1.10 \times 10^{10}$, rounded off.
2. $(0.000743)^3 = (7.43 \times 10^{-4})^3 = (7.43)^3 \times 10^{-12}$
   $= 410 \times 10^{-12} = 4.10 \times 10^2 \times 10^{-12}$
   $= 4.10 \times 10^{(+2)+(-12)}$
   $= 4.10 \times 10^{-10}$

Obtaining the square root of a number

As far as electronics formulas are concerned, square roots are every bit as important as squares. For example, the square root of 49, written as $\sqrt{49}$, is a number such that the square of that number is 49. Therefore, $\sqrt{49} = +7$ or -7 because $7^2 = 7 \times 7 = 49$, and $(-7)^2 = (-7) \times (-7) = 49$; for most problems in electronics, you only need to consider the positive square root.

As a further example, $\sqrt{6{,}250{,}000} = 2500$. To obtain the same result by scientific notation,

$$\sqrt{6{,}250{,}000} = \sqrt{6.25 \times 10^6} = \sqrt{6.25} \times 10^3 = 2.5 \times 10^3.$$

Rule E

"When taking the square root of a number expressed in scientific notation, square root the coefficient, but halve the value of the exponent. For example,

$$\sqrt{3.87 \times 10^{-10}} = \sqrt{3.87} \times 10^{-10/2} = \sqrt{3.87} \times 10^{-5} = 1.97 \times 10^{-5}.$$

There is one problem with taking square roots:

$$\sqrt{16{,}000{,}000} = \sqrt{1.6 \times 10^7} = \sqrt{1.6} \times 10^{7/2} = 1.26 \times 10^{7/2} = 1.26 \times 10^{3.5}$$

Although this is, in fact, a perfectly correct answer, it is, as yet, difficult to interpret the value of "$10^{3.5}$." The problem is resolved by rewriting 1.6×10^7 as 16.0×10^6; then,

$$\sqrt{16.0 \times 10^6} = \sqrt{16.0} \times 10^{6/2} = 4.0 \times 10^3.$$

Consequently, if when expressing the original number in scientific notation, you come up with an odd-numbered exponent, the number must be rewritten with two digits before the decimal point. The exponent will then be an even number so that after division by 2, the exponent in the answer is a whole number and not a fraction. This method also applies to negative exponents, as illustrated in the following examples:

1. $\sqrt{0.0327} = \sqrt{3.27 \times 10^{-2}} = \sqrt{3.27} \times 10^{-2/2} = 1.81 \times 10^{-1}$
2. $\sqrt{0.0000327} = \sqrt{3.27 \times 10^{-5}} = \sqrt{32.7 \times 10^{-6}} = \sqrt{32.7} \times 10^{-6/2} = 5.72 \times 10^{-3}$.

The meaning of the fractional exponent

You know that $5 \times 5 = \sqrt{25} \times \sqrt{25} = 25 = 25^1$. Assume that fractional exponents obey the same rules as whole number exponents; it follows that $\sqrt{25} = 25^{1/2}$ because $25^{1/2} \times 25^{1/2} = 25^{1/2 + 1/2} = 25^1$. Therefore, $25^{1/2}$ means the square root of 25, $25^{1/3}$ is the cube root of 25, $25^{1/4}$ is the fourth root and so on.

Rule F

"When taking the nth root of a number expressed in scientific notation, you take the nth root of the coefficient and divide the exponent (either positive or negative) by n. It might be necessary to adjust the value of the coefficient until it is exactly divisible."

Rule F is illustrated in the following examples:

1. $\sqrt[3]{270000000} = \sqrt[3]{2.7 \times 10^7} = \sqrt[3]{27 \times 10^6} = 27^{1/3} \times 10^{6/3} = 3.0 \times 10^2$
2. $\sqrt[5]{0.00032} = \sqrt[5]{3.2 \times 10^{-4}} = \sqrt[5]{32 \times 10^{-5}} = 32^{1/5} \times 10^{-5/5} = 2.0 \times 10^{-1}$.

Table 193-2. Prefixes used in electronics.

Prefix	Symbol	Scientific notation	Meaning
Tera-	T	$\times 10^{12}$	$\times$ 1,000,000,000,000
Giga-	G	$\times 10^9$	$\times$ 1,000,000,000
Mega-	M	$\times 10^6$	$\times$ 1,000,000
kilo-	k	$\times 10^3$	$\times$ 1,000
centi-	c	$\div 10^2$ or 10^{-2}	$\div$ 100 or $\times$ 0.01
milli-	m	$\div 10^3$ or $\times 10^{-3}$	$\div$ 1000 or $\times$ 0.001
micro-	μ	$\div 10^6$ or $\times 10^{-6}$	$\div$ 1,000,000 or $\times$ 0.000001
nano-	n or ν	$\div 10^9$ or $\times 10^{-9}$	$\div$ 1,000,000,000 or $\times$ 0.000000001
pico- or micromicro-	p $\mu\mu$	$\div 10^{12}$ or $\times 10^{-12}$	$\div$ 1,000,000,000,000 or $\times$ 0.000000000001
*femto-	f	$\div 10^{15}$ or $\times 10^{-15}$	$\div$ 1,000,000,000,000,000 or $\times$ 0.000000000000001
*atto-	a	$\div 10^{18}$ or $\times 10^{-18}$	$\div$ 1,000,000,000,000,000,000 or $\times$ 0.000000000000000001

*Rarely used in electronics.

Previously, you have expressed $\sqrt{16{,}000{,}000}$ as $1.26 \times 10^{3.5}$, and also as 4.0×10^3. Now, you know the meaning of a fractional exponent,

$$1.26 \times 10^{3.5} = 1.26 \times 10^3 \times 10^{1/2}$$
$$= 1.26 \times 10^3 \times 3.16$$
$$= 4.0 \times 10^3$$

Therefore, the two answers are the same.

Prefixes

As well as scientific notation, you commonly use word prefixes in electronics to represent both large and small quantities. Such prefixes are based on the metric system, and the ones most often used are shown in Table 193-2.

When using the formulas that occur in electronics, all quantities must be entered in terms of their basic units. As an example, if the value of a voltage is 8 millivolts, it is entered into the formula as 8×10^{-3} volt.

Example 193-1

In a network of three parallel resistors, $I_T = I_1 + I_2 + I_3$. If $I_1 = 67.5$ mA, $I_2 = 0.374$ A, and $I_3 = 1.28$ mA, calculate the value of I_T.

Solution

$$I_1 = 67.5 \text{ mA} = 67.5 \times 10^{-3} \text{ A.}$$
$$I_2 = 0.374 \text{ A} = 374.0 \times 10^{-3} \text{ A.}$$
$$I_3 = 1.28 \text{ mA} = 1.28 \times 10^{-3} \text{ A.}$$
$$\text{Then, } I_T = (67.5 + 374.0 + 1.28) \times 10^{-3}$$
$$= 442.78 \times 10^{-3} \text{ A}$$
$$= 442.78 \text{ mA.}$$

Example 193-2

In a network of three series resistors, $R_T = R_1 + R_2 + R_3$. If $R_T = 1.36$ MΩ, $R_1 = 680$ kΩ, and $R_2 = 560$ kΩ, calculate the value of R_3.

Solution

$$R_3 = R_T - R_1 - R_2$$
$$= 1.36 \times 10^6 - 680 \times 10^3 - 560 \times 10^3$$
$$= 1360 \times 10^3 - 680 \times 10^3 - 560 \times 10^3$$
$$= (1360 - 680 - 560) \times 10^3$$
$$= 120 \times 10^3\ \Omega = 120 \text{ k}\Omega.$$

Example 193-3

One of the Ohm's Law relationships is $E = I \times R$. If $I = 13.72$ mA, and $R = 470$ Ω, calculate the value of E in volts.

Solution

$$E = 13.72 \times 10^{-3} \text{ A} \times 470\ \Omega$$
$$= 1.372 \times 10^1 \times 10^{-3} \times 4.70 \times 10^2$$
$$= 6.4484 \times 10^{(+1)+(-3)+(+2)}$$
$$= 6.4484 \times 10^0$$
$$= 6.4484 \text{ V.}$$

Example 193-4

One of the Ohm's Law relationships is $I = P/E$. If $P = 375$ mW and $E = 14.3$ V, calculate the value of I in milliamps.

Solution

$$I = \frac{375 \times 10^{-3} \text{ W}}{14.3 \text{ V}}$$
$$= \frac{3.75 \times 10^2 \times 10^{-3}}{1.43 \times 10^1}$$
$$= \frac{3.75}{1.43} \times 10^{(+2)+(-3)-(+1)}$$
$$= 2.62 \times 10^{-2} \text{ A}$$
$$= 26.2 \times 10^{-3} \text{ A}$$
$$= 26.2 \text{ mA.}$$

Example 193-5

For three resistors in parallel, $1/R_T = 1/R_1 + 1/R_2 + 1/R_3$. If $R_1 = 680$ Ω, $R_2 = 820$ Ω, and $R_3 = 1.2$ kΩ, calculate the value of R_T.

Solution

$$\frac{1}{R_T} = \frac{1}{680} + \frac{1}{820} + \frac{1}{1200}$$
$$= \frac{1}{6.8 \times 10^2} + \frac{1}{8.2 \times 10^2} + \frac{1}{1.2 \times 10^3}$$
$$= 0.147 \times 10^{-2} + 0.122 \times 10^{-2} + 0.833 \times 10^{-3}$$
$$= 1.47 \times 10^{-3} + 1.22 \times 10^{-3} + 0.833 \times 10^{-3}$$
$$= 3.523 \times 10^{-3}.$$

Taking the reciprocal of both sides of the equation,

$$R_T = \frac{1}{3.523 \times 10^{-3}} = 0.284 \times 10^3 = 284\ \Omega.$$

Example 193-6

One of the Ohm's Law relationships is $P = E^2/R$. If $E =$ 63.2 V and $R = 470\ \text{k}\Omega$, calculate the value of P.

Solution

$$P = \frac{(63.2\ \text{V})^2}{470 \times 10^3\ \Omega}$$

$$= \frac{(6.32 \times 10^1)^2}{4.7 \times 10^5}$$

$$= \frac{39.9}{4.7} \times 10^2 \times 10^{-5}$$

$$= 8.5 \times 10^{-3}\ \text{W}$$

$$= 8.5\ \text{mW}.$$

Example 193-7

One of Ohm's Law relationships is $E = \sqrt{P \times R}$. If $P =$ 125 mW and $R = 330\ \Omega$, calculate the value of E.

Solution

$$E = \sqrt{125 \times 10^{-3}\ \text{W} \times 330\ \Omega}$$

$$= \sqrt{1.25 \times 10^2 \times 10^{-3} \times 3.3 \times 10^2}$$

$$= \sqrt{1.25 \times 3.3 \times 10^1} = \sqrt{4.125 \times 10^1}$$

$$= \sqrt{41.25} = 6.42\ \text{V}.$$

Example 193-8

A diode's voltage/current relationship is $I = kE^{3/2}$ amperes. If $k = 0.0035$ and $E = 12$ V, calculate the value of I.

Solution

$$I = 0.0015 \times 12^{3/2}$$

$$= 0.0015 \times \sqrt{12^3}$$

$$= 0.0015 \times 41.6$$

$$= 1.5 \times 10^{-3} \times 41.6\ \text{A}$$

$$= 62.4\ \text{mA}.$$

PRACTICE PROBLEMS

193-1. Convert the following decimal numbers to scientific notation: 5280, 0.1377, 186280, 0.000029809, 0.01745, 32.1578, 0.00367, 365.237, 2997960000, 96517.

193-2. Convert the following into decimal numbers: 3.7×10^0, 4.73×10^{-3}, 6.943×10^4, 3.857×10^{-1}, 5.34×10^{-6}.

193-3. Express the answers in scientific notation for the following:

(a) $(4.27 \times 10^5) + (7.13 \times 10^3) + (3.15 \times 10^4)$

(b) $(2.42 \times 10^{-9}) + (7.13 \times 10^{-8}) + (8.21 \times 10^{-7})$

(c) $(6.54 \times 10^1) + (3.43 \times 10^0) + (8.33 \times 10^{-1})$

193-4. Express the answers in scientific notation for the following:

(a) $8.3 \times 10^4 - 4.7 \times 10^3$

(b) $4.3 \times 10^4 - 1.8 \times 10^5$

(c) $3.63 \times 10^{-3} - 2.02 \times 10^{-4}$

(d) $2.99 \times 10^{-7} - 8.74 \times 10^{-6}$

193-5. Express the answers in scientific notation for the following:

(a) $(3.55 \times 10^2) \times (4.77 \times 10^4)$

(b) $(5.90 \times 10^3) \times (6.61 \times 10^7)$

(c) $(2.55 \times 10^{12}) \times (3.13 \times 10^{-5})$

(d) $(9 \times 10^{-3}) \times (7 \times 10^{-5}) \times (8 \times 10^4)$

(e) $(5 \times 10^3) \times (8 \times 10^{-12}) \times (4 \times 10^8) \times (3 \times 10^3)$

193-6. Express the answers in scientific notation for the following:

(a) $\dfrac{3.5 \times 10^6}{2.3 \times 10^2}$

(b) $\dfrac{6 \times 10^7}{8 \times 10^{10}}$

(c) $\dfrac{(4 \times 10^2)(7 \times 10^7)}{7 \times 10^6}$

(d) $\dfrac{(3.5 \times 10^6)(2.3 \times 10^3)}{(4.7 \times 10^{-4})}$

(e) $\dfrac{(8 \times 10^{-3})(6 \times 10^{-7})(7 \times 10^5)(3 \times 10^7)}{(4 \times 10^9)(4 \times 10^3)(5 \times 10^{-10})}$

193-7. Express the answers in scientific notation for the following:

(a) $(4 \times 10^3)^3$

(b) $(6 \times 10^6)^4$

(c) $(2.3 \times 10^{12})^2$

(d) $\sqrt{5.1 \times 10^3}$

(e) $\sqrt{2.7 \times 10^8}$

(f) $\sqrt{7.2 \times 10^{11}}$

193-8. Express the answers in scientific notation for the following:

(a) $\dfrac{1}{4.7 \times 10^3}$

(b) $\dfrac{1}{6.7 \times 10^{-5}}$

(c) $\dfrac{1}{2.45 \times 10^0}$

(d) $\dfrac{1}{7.42 \times 10^{-1}}$

(e) $\dfrac{1}{8.31 \times 10^1}$

194
Linear equations of the first degree

In electronics, the letter "I" is used as the symbol for current, which is measured in amperes (whose unit symbol is A). The value of I can be unknown or any general value can be assumed by I; for example, I can stand for 8 A or 3 mA. Therefore, I is called a *general* (or *literal*) *number*.

An equation involves one or more literal numbers and states that the two sides (members) of the equation are equal. The equation's left-hand side and right-hand side are separated by the equality sign (=). As an example,

$$\underbrace{4I - 3}_{\text{left-hand side}} \underset{\text{equality sign}}{\overset{\downarrow}{=}} \underbrace{2I + 5}_{\text{right-hand side}}$$

The sides of the equation can be interchanged without affecting its validity. If $I = 6$; the left-hand side of the equation is equal to $4 \times 6 - 3 = 21$ and the right-hand side equals $2 \times 6 + 5 = 17$. Because the two sides are *not* equal in value, the equation is not *satisfied* by substituting $I = 6$. However, if you substitute $I = 4$; $4I - 3 = 4 \times 4 - 3 = 13$, and $2I + 5 = 2 \times 4 + 5 = 13$. Therefore, the equation *is* satisfied. In this situation, you have a conditional equation that is satisfied by only *one* value of the literal number. Solving conditional equations means finding the value (or values) for the unknown literal number so that the equation is satisfied.

In this chapter, you will only be concerned with linear equations of the first degree. In such equations, there are no squares or higher-order terms of the single literal number.

MATHEMATICAL DERIVATIONS

The following operations can be performed on both sides of an equation without affecting the equality:

A. Equal numbers can be added to both sides. For example, if:

$$x = 4,$$

then,

$$x + 3 = 4 + 3 = 7.$$

B. Equal numbers can be subtracted from both sides. For example, if:

$$x = 11,$$

then,

$$x - 5 = 11 - 5 = 6.$$

C. Both sides can be multiplied by the same number. For example, if:

$$x = 3,$$

then,

$$4x = 4 \times 3 = 12.$$

During this process, remember the rules for multiplying the signs of the numbers; namely:

$$\text{“+”} \times \text{“+”} = \text{“+”}$$

$$\text{“+”} \times \text{“−”} = \text{“−”}$$

$$\text{“−”} \times \text{“+”} = \text{“−”}$$

$$\text{“−”} \times \text{“−”} = \text{“+”}$$

Notice that, if both sides are multiplied by "-1," the signs of all the terms in the equation will be changed without affecting the equality.

If decimal fractions appear in the equation, both sides are multiplied by the appropriate multiple of 10 so that the fractions are converted into whole numbers.

D. Both sides can be divided by the same number. For example, if:

$$10x = 20$$

$$\frac{10x}{5} = \frac{20}{5}$$

$$2x = 4$$

$$\frac{2x}{2} = \frac{4}{2}$$

$$x = 2.$$

When carrying out this process, remember the rules for dividing the signs of the numbers; namely:

"+" ÷ "+" = "+"

"+" ÷ "−" = "−"

"−" ÷ "+" = "−"

"−" ÷ "−" = "+"

E. Both sides can be raised to the same exponent (power). For example, if:

$$x = 3,$$

then,

$$x^3 = 3^3 = 27.$$

F. The same root of both sides can be taken. For example, if:

$$x = 27,$$

then,

$$\sqrt[3]{x} = \sqrt[3]{27} = 3.$$

G. A term can be transferred from one side to the other provided the sign of the term is changed. For example, if:

$$2x - 5 = 3$$

Adding +5 to both sides (operation A),

$$2x - 5 + 5 = 3 + 5$$

$$2x = 3 + 5 = 8$$

The term "−5" has been transferred from the left-hand side to the right-hand side, where it appears as "+5."

As a general principle, those terms that contain the unknown literal number, are transferred to the left-hand side of the equation, and other terms are collected on the right-hand side. If parentheses are involved, these must be multiplied out before collecting the terms.

As examples:

1. $5I - 10 - 3I + 3 = -1$

$2I - 7 = -1$

Transfer "−7" from the left-hand side to the right-hand side, where it appears as "+7"

$$2I = -1 + 7 = 6$$

Divide both sides by 2,

$$\frac{2I}{2} = I = \frac{6}{2} = 3$$

Solution: $I = 3$.
Check: Substitute $I = 3$ in the original equation.

Left-hand side $= 5 \times 3 - 10 - 3 \times 3 + 3$
$= 15 - 10 - 9 + 3$
$= -1$

2. $E + 1 = 2(E - 3) - 3(E - 1)$

$E + 1 = 2E - 6 - 3E + 3$

$E + 1 = -E - 3$

Transfer " +1" from the left-hand side to the right-hand side, where it appears as " −1."
Transfer " $-E$" from the right-hand side to the left-hand side, where it appears as " $+E$." Then,

$$E + E = -3 - 1$$

$$2E = -4$$

Divide both sides by 2,

$$\frac{2E}{2} = E = -\frac{4}{2} = -2$$

Solution: $E = -2$.
Check: Substitute $E = -2$ in the original equation.

Left-hand side $= -2 + 1 = -1$

Right-hand side $= 2(-2 - 3) - 3(-2 - 1)$
$= -2 \times 5 + 3 \times 3 = -1.$

3. $1.8R + 0.63 = 0.$

Multiply both sides by 100

$1.8R \times 100 + 0.63 \times 100 = 0 \times 100$

$180R + 63 = 0$

Transfer "+63" from the left-hand side to the right-hand side, where it appears as "−63." Then,

$$180R = -63,$$

Divide both sides by 180,

$$\frac{180R}{180} = R = -\frac{63}{180} = -\frac{7}{20} = -0.35$$

Solution: $R = -0.35$.

Fractional equations

A fractional equation contains fractional coefficients of the literal unknown number. To solve such equations, the fractions must first be removed. Therefore, both sides are multiplied by the lowest common denominator (LCD). The resulting equation is then solved by using the operations previously described.

As examples:

1. $\frac{I}{12} + \frac{I}{15} = 18$

The lowest common denominator is 60.

$$\frac{60I}{12} + \frac{60I}{15} = 60 \times 18 = 1080$$

$$5I + 4I = 1080$$

$$9I = 1080$$

$$\frac{9I}{9} = I = \frac{1080}{9} = 120$$

Solution: $I = 120$.

2. $\frac{E-1}{2} - \frac{3-E}{4} = 2$

The lowest common denominator is 4. The terms above each of the division lines are grouped so that these groups must be placed within parentheses, when carrying out the multiplication process to clear the fractions.

$$\frac{4(E-1)}{2} - \frac{4(3-E)}{4} = 4 \times 2 = 8$$

$$2(E-1) - 1(3-E) = 8$$

$$2E - 2 - 3 + E = 8$$

$$3E = 8 + 2 + 3 = 13$$

$$\frac{3E}{3} = E = \frac{13}{3} = 4\frac{1}{3}$$

Solution: $E = 4\frac{1}{3}$

3. $\frac{1}{2R} - \frac{1}{3R} = \frac{1}{4R} - 1$

The lowest common denominator is $12R$.

$$\frac{12R \times 1}{2R} - \frac{12R \times 1}{3R} = \frac{12R \times 1}{4R} - 12R \times 1$$

$$6 - 4 = 3 - 12R$$

$$6 - 4 - 3 = -12R$$

$$-1 = -12R$$

$$12R = 1.$$

$$\frac{12R}{12} = R = \frac{1}{12}$$

Solution: $R = \frac{1}{12}$

4. $\frac{3R+4}{5} - \frac{7R-3}{2} = \frac{R-16}{4}$

The lowest common denominator is 20.

$$\frac{20(3R+4)}{5} - \frac{20(7R-3)}{2} = \frac{20(R-16)}{4}$$

$$4\,(3R+4) - 10\,(7R-3) = 5\,(R-16)$$

$$12R + 16 - 70R + 30 = 5R - 80$$

$$-58R + 46 = 5R - 80$$

$$-58R - 5R = -80 - 46$$

$$-63R = -126$$

$$63R = 126$$

$$\frac{63R}{63} = R = \frac{126}{63} = 2$$

Solution: $R = 2$.

Check: Substitute $R = 2$ in the original equation.

$$\text{Left-hand side} = \frac{3 \times 2 + 4}{5} - \frac{7 \times 2 - 3}{2}$$

$$= 2 - 5.5 = -3.5$$

$$\text{Right-hand side} = \frac{2 - 16}{4} = -\frac{14}{4} = -3.5$$

Formulas

Formulas are equations that illustrate some basic law, or principle, encountered in electronics. A formula contains mostly literal numbers; in its standard form, there is one literal number on the left-hand side of the equation. This literal number is called the *subject* of the equation. For example, the formula for the inductive reactance (chapter 69) is:

$$X_L = 2\pi fL$$

where: X_L is the subject of the formula, and is the inductive reactance measured in ohms (Ω).

π = circular constant (3.1415926 . . .)

f = frequency measured in hertz (Hz)

L = inductance measured in henrys (H)

If the values of f and L are known, you can calculate the value of X_L. For example, if $f = 800$ kHz, and $L = 150\ \mu$H,

$$X_L = 2 \times \pi \times 800 \times 10^3 \times 150 \times 10^{-6}$$

$$= 2 \times \pi \times 8 \times 10^2 \times 10^3 \times 1.5 \times 10^2 \times 10^{-6}$$

$= 2 \times \pi \times 8 \times 1.5 \times 10^1$

$= 754\ \Omega$

However, if you know the values of X_L and f, and wish to find the value of L, the formula is transposed so that L is the new subject. This is achieved by using the same operations that are already outlined in this chapter.

$$X_L = 2\pi f L$$

Divide both sides by $2\pi f$.

$$\frac{X_L}{2\pi f} = \frac{2\pi f L}{2\pi f} = L$$

Interchanging the sides of the equation,

$$L = \frac{X_L}{2\pi f}\ \text{H}$$

Examples of other transpositions are illustrated in the following examples:

1. Make I the subject of the formula $P = I^2R$ (chapter 5).

$$P = I^2R$$

Divide both sides by R,

$$\frac{P}{R} = \frac{I^2R}{R} = I^2$$

Interchange both sides of the equation,

$$I^2 = \frac{P}{R}$$

Take the square root of both sides,

$$\sqrt{I^2} = I = \sqrt{\frac{P}{R}}$$

Solution: $I = \sqrt{\frac{P}{R}}$ A

2. Make f the subject of the formula,

$$X_C = \frac{1}{2\pi f C}\ \text{(chapter 70)},$$

$$X_C = \frac{1}{2\pi f C}\ \Omega$$

Multiply both sides by f,

$$f \times X_C = \frac{f \times 1}{2\pi f C} = \frac{1}{2\pi C}$$

Divide both sides by X_C,

$$\frac{f \times X_C}{X_C} = f = \frac{1}{2\pi C \times X_C}$$

Solution: $f = \frac{1}{2\pi C X_C}$ Hz

3. Make X_C the subject of the formula (chapter 73),

$$Z = \sqrt{R^2 + X_C^2}\ \Omega$$

Square both sides of the equation,

$$Z^2 = \left(\sqrt{R^2 + X_C^2}\right)^2 = R^2 + X_C^2$$

Transfer the term "R^2" from the right-hand side to the left-hand side,

$$Z^2 - R^2 = X_C^2$$

Interchange both sides of the equation,

$$X_C^2 = Z^2 - R^2$$

Take the square root of both sides,

$$\sqrt{X_C^2} = X_C = \sqrt{Z^2 - R^2}$$

Solution: $X_C = \sqrt{Z^2 - R^2}\ \Omega$

4. Make C the subject of the formula (chapter 75),

$$f = \frac{1}{2\pi\sqrt{LC}}\ \text{Hz}$$

Square both sides of the equation,

$$f^2 = \frac{1^2}{(2\pi\sqrt{LC})^2} = \frac{1}{4\pi^2 LC}$$

Multiply both sides by C,

$$Cf^2 = \frac{C}{4\pi^2 LC} = \frac{1}{4\pi^2 L}$$

Divide both sides by f^2,

$$\frac{Cf^2}{f^2} = C = \frac{1}{f^2 \times 4\pi^2 L} = \frac{1}{4\pi^2 f^2 L}$$

Solution: $C = \frac{1}{4\pi^2 f^2 L}$ F

5. Make C the subject of the equation (chapter 82),

$$Q = \frac{1}{R}\sqrt{\frac{L}{C}}$$

Multiply both sides by R,

$$R \times Q = R \times \frac{1}{R}\sqrt{\frac{L}{C}} = \sqrt{\frac{L}{C}}$$

Square both sides of the equation,

$$R^2Q^2 = \left(\sqrt{\frac{L}{C}}\right)^2 = \frac{L}{C}$$

Multiply both sides by C,

$$C \times R^2Q^2 = C \times \frac{L}{C} = L$$

$$C R^2Q^2 = L$$

Divide both sides by R^2Q^2,

$$\frac{C R^2Q^2}{R^2Q^2} = C = \frac{L}{R^2Q^2}$$

Solution: $C = \frac{L}{R^2Q^2}$ F

Example 194-1

Solve the following equations for x:

(a) $4 - 4(x - 5) = 2(2 - x) - 6$,

(b) $1.4\,(x - 3) - 0.4\,(2x - 1) = -0.2$,

(c) $\frac{3}{2x} - 4 = 3 - \frac{9}{x}$

Solution

(a) $4 - 4(x - 5) = 2(2 - x) - 6$

$$4 - 4x + 20 = 4 - 2x - 6$$

$$-4x + 2x = 4 - 6 - 4 - 20$$

$$-2x = -26$$

$$2x = 26$$

$$\frac{2x}{2} = x = \frac{26}{2} = 13$$

$$x = 13$$

(b) $1.4(x - 3) - 0.4(2x - 1) = -0.2$

Multiply both sides by 10,

$$14(x - 3) - 4(2x - 1) = -2$$

$$14x - 42 - 8x + 4 = -2$$

$$6x = -2 + 42 - 4 = 36$$

$$\frac{6x}{6} = x = \frac{36}{6} = 6$$

$$x = 6$$

(c) $\frac{3}{2x} - 4 = 3 - \frac{9}{x}$

Multiply both sides by the LCD which is $2x$,

$$\frac{2x \times 3}{2x} - 2x \times 4 = 2x \times 3 - \frac{2x \times 9}{x}$$

$$3 - 8x = 6x - 18$$

$$-6x - 8x = -3 - 18 = -21$$

$$-14x = -21$$

$$14x = 21$$

$$\frac{14x}{14} = x = \frac{21}{14} = 1.5$$

$$x = 1.5$$

Example 194-2

Solve the following equations for x:

(a) $\frac{x + 1}{2} + \frac{x + 2}{3} = 14 - \frac{x - 5}{4}$

(b) $\frac{1 + 0.2x}{3} - \frac{2 - 0.3x}{7} = 0.19$

(c) $\frac{x + 5}{2\frac{1}{2}} - \frac{x - 1}{3\frac{1}{3}} = 1$

Solution

(a) $\frac{x + 1}{2} + \frac{x + 2}{3} = 14 - \frac{x - 5}{4}$

Multiply both sides by the LCD of 12,

$$\frac{12(x + 1)}{2} + \frac{12(x + 2)}{3} = 12 \times 14 - \frac{12(x - 5)}{4}$$

$$6(x + 1) + 4(x + 2) = 12 \times 14 - 3(x - 5)$$

$$6x + 6 + 4x + 8 = 168 - 3x + 15$$

$$10x + 14 = 168 - 3x + 15$$

$$10x + 3x = -14 + 168 + 15 = 169$$

$$13x = 169$$

$$\frac{13x}{13} = x = \frac{169}{13} = 13$$

$$x = 13.$$

(b) $\frac{1 + 0.2x}{3} - \frac{2 - 0.3x}{7} = 0.19$

Multiply both sides by the LCD of 21,

$$\frac{21(1 + 0.2x)}{3} - \frac{21(2 - 0.3x)}{7} = 21 \times 0.19$$

$$= 3.99$$

$$7(1 + 0.2x) - 3(2 - 0.3x) = 3.99$$

$$2.3x = 3.99 - 1 = 2.99$$

$$\frac{2.3x}{2.3} = x = \frac{2.99}{2.3} = 1.3$$

$$x = 1.3$$

(c) $\frac{x + 5}{2\frac{1}{2}} - \frac{x - 1}{2\frac{1}{3}} = 1$

$$\frac{x + 5}{\frac{5}{2}} - \frac{x - 1}{\frac{10}{3}} = 1$$

$$\frac{2(x + 5)}{5} - \frac{3(x - 1)}{10} = 1$$

Multiply both sides by the LCD of 10.

$$\frac{10 \times 2\,(x+5)}{5} - \frac{10 \times 3(x-1)}{10} = 10 \times 1 = 10$$

$$4(x+5) - 3(x-1) = 10$$

$$4x + 20 - 3x + 3 = 10$$

$$x = 10 - 20 - 3 = -13$$

$$x = -13$$

Example 194-3

(a) For two resistors in parallel,

$$R_T = \frac{R_1 R_2}{R_1 + R_2} \text{ (chapter 15).}$$

Transpose the formula to make R_1 the subject.

(b) Make X_C the subject of the formula,

$$Z = \sqrt{R^2 + (X_L - X_C)^2} \text{ (chapter 75).}$$

(c) Make R_i the subject of the formula,

$$P_L = \frac{E^2 R_L}{(R_i + R_L)^2} \text{ (chapter 23).}$$

Solution

(a) $R_T = \frac{R_1 R_2}{R_1 + R_2}$

Multiply both sides by $R_1 + R_2$.

$$(R_1 + R_2)R_T = \frac{(R_1 + R_2)\,(R_1 R_2)}{R_1 + R_2} = R_1 R_2$$

$$R_1 R_T + R_2 R_T = R_1 R_2$$

$$R_2 R_T = R_1 R_2 - R_1 R_T$$

$$R_1(R_2 - R_T) = R_2 R_T$$

$$R_1 = \frac{R_2 R_T}{R_2 - R_T}$$

(b) $Z = \sqrt{R^2 + (X_L - X_C)^2}$

Square both sides of the equation,

$$Z^2 = \left(\sqrt{(R^2 + (X_L - X_C)^2)}\right)^2 = R^2 + (X_L - X_C)^2$$

$$Z^2 - R^2 = (X_L - X_C)^2$$

Take the square root of both sides,

$$\sqrt{Z^2 - R^2} = \sqrt{(X_L - X_C)^2} = X_L - X_C$$

$$X_C + \sqrt{Z^2 - R^2} = X_L$$

$$X_C = X_L - \sqrt{Z^2 - R^2}$$

(c) $P_L = \frac{E^2 R_L}{(R_i + R_L)^2}$

Multiply both sides by $(R_i + R_L)^2$,

$$P_L \times (R_i + R_L)^2 = E^2 R_L$$

Divide both sides by P_L,

$$(R_i + R_L)^2 = \frac{E^2 R_L}{P_L}$$

Take the square root of both sides,

$$R_i + R_L = \sqrt{\frac{E^2 R_L}{P_L}}$$

$$R_i = \sqrt{\frac{E^2 R_L}{P_L}} - R_L$$

PRACTICE PROBLEMS

194-1. Solve for x in the following equations:

(a) $9(x + 1) + 4(x + 1) = 104$

(b) $3(x + 1) + 2(x + 2) = 22$

(c) $x - 1/2 = 2x + 1$

(d) $2x - 4 = 5x + 2$

194-2. Solve for x in the following equations:

(a) $\frac{1}{x} + \frac{2}{x} + \frac{3}{x} = 2$

(b) $\frac{1}{x} + 1 = \frac{9}{x} - 3$

(c) $\frac{x+1}{3x} = \frac{2}{3} - \frac{2}{x}$

(d) $\frac{2}{x} + \frac{2}{x+1} = \frac{5}{x}$

194-3. Solve for C_3 in the formula,

$$\frac{1}{C_T} = \frac{1}{C_1} + \frac{1}{C_2} + \frac{1}{C_3}$$

194-4. Solve for N in the formula,

$$L = \frac{N^2 A\mu}{l}$$

194-5. Solve for α in the formula,

$$R_{T°C} = R_{20°C}\left[1 + \alpha\,(T°C - 20°C)\right]$$

194-6. Solve for d in the equation,

$$y = \frac{LlV_y}{2dV_x}$$

194-7. Solve for C in the equation,

$$Q = \frac{1}{R}\sqrt{\frac{L}{C}}$$

194-8. Solve for $V_{GS(\text{off})}$ in the equation,

$$I_D = I_{DSS}\left(1 - \frac{V_{GS}}{V_{GS(\text{off})}}\right)^2$$

195
Simultaneous equations

CONSIDER THE FOLLOWING PROBLEM:

Two tanks (A and B) respectively contain 120 gallons and 500 gallons of water. If water is allowed to run into tank A at the rate of 10 gallons per minute and to run out of tank B at the rate of 20 gallons per minute, find a formula for the number (N) of gallons in each tank after t minutes have elapsed.

Tank A

After t minutes, $10t$ gallons of water have run into tank A. Therefore, $N = 120 + 10t$ gallons.

Tank B

After t minutes, $20t$ gallons of water have run out of tank B. Therefore, $N = 300 - 20t$ gallons.

Because the amount of water in tank A is increasing while that in tank B is decreasing, there must come a *single* moment when the two tanks contain equal amounts of water. This means that there is one value of t which makes the value of N for tank A the same as the value of N for tank B. At this moment, and *only* at this moment, the equations $N = 120 + 10t$ (tank A) and $N = 300 - 20t$ (tank B) are true for each tank; therefore, they are called *simultaneous equations*. Strictly speaking, they are simultaneous *linear* equations because the two variables N and t are not involved with squares and/or higher-order terms.

If $N = 120 + 10t$ and $N = 300 - 20t$ are simultaneous equations,

$$120 + 10t = 300 - 20t$$
$$30t = 180$$
$$t = 6 \text{ minutes}$$

Then,

$$N = 120 + 10t = 120 + 60 = 180 \text{ gallons}$$

or,

$$N = 300 - 20t = 300 - 120 = 180 \text{ gallons}$$

This shows that after 6 minutes have elapsed, there are equal amounts of water, namely 180 gallons, in each of the two tanks. At no other time will this be true.

MATHEMATICAL DERIVATIONS

There are two basic methods for solving linear simultaneous equations with two (or more) variables:

Method 1. Solution by substitution

Considering the following simultaneous equations,

$$7y - 2x = 4 \qquad \text{(A)}$$
$$3y - x = 1 \qquad \text{(B)}$$

Step 1 The substitution method involves choosing one of the equations, then using it to solve for one unknown in terms of the second unknown. Looking at the two equations, it is simpler to choose Equation B and then solve for x in terms of y.

Because,

$$3y - x = 1,$$

Then,

$$x = 3y - 1.$$

Step 2 The result, $x = 3y - 1$, of Step 1 is substituted in Equation A so that,

$$7y - 2(3y - 1) = 4$$
$$y = 2$$

It follows that,

$$x = 3y - 1 = 6 - 1 = 5.$$

As a check, the results of $x = 5$, $y = 2$ should be substituted in Equation A, not Equation B.

Check:

Equation A: $7y - 2x = 7 \times 2 - 2 \times 5 = 14 - 10 = 4.$

Example 195-1

Solve the simultaneous equations $3y - x = 11$, $2y - 3x = 5$ by the substitution method.

Solution

The equations are:

$$3y - x = 11 \quad (A)$$

$$2y - 3x = 5 \quad (B)$$

Step 1 From Equation A,

$$3y - x = 11$$

$$x = 3y - 11$$

Step 2 Substituting $x = 3y - 11$ in Equation B,

$$2y - 3(3y - 11) = 5$$

$$-7y = -28$$

$$y = 4$$

From Equation A,

$$x = 3y - 11 = 3 \times 4 - 11 = 1$$

The solution is $x = 1$, $y = 4$.

Check:

Equation B: $2y - 3x = 2 \times 4 - 3 \times 1 = 8 - 3 = 5.$

The substitution method can be extended to three variables, with three simultaneous equations, by using one equation to solve for one "unknown" in terms of the other two "unknowns." This result is then substituted in each of the other equations so that you are left with only two equations and two unknowns. With the substitution method, it is possible to reduce any number of simultaneous linear equations down to two equations with two unknowns. However, this process can be extremely tedious, especially if the coefficients of the variables involve awkward numbers. In such circumstances, it might be preferable to use the second method, which requires the addition or subtraction of simultaneous equations.

Method 2. Solution by addition or subtraction of simultaneous equations

Consider the following simultaneous equations,

$$3x - 2y = 11 \quad (A)$$

$$5x + 2y = 29 \quad (B)$$

The result of adding the left side of Equation A, to the left side of Equation B, must equal the result of adding the right sides. When you carry out this operation, the term "$-2y$" cancels the term "$+2y$," and you are left with a single equation in x.

Therefore,

$$(3x - 2y) + (5x + 2y) = 11 + 29$$

$$8x = 40$$

$$x = 5.$$

Substituting $x = 5$ in Equation A,

$$3 \times 5 - 2y = 11$$

$$2y = 4$$

$$y = 2.$$

Solution is $x = 5$, $y = 2$.

Check:

Equation B: $5x + 2y = 5 \times 5 + 2 \times 2 = 25 + 4 = 29.$

As another example, solve the simultaneous equations,

$$2x - 5y = 27 \quad (A)$$

$$2x + 3y = 3 \quad (B)$$

If we *subtract* Equation B from Equation A, the terms in x will vanish. Therefore,

$$(2x - 5y) - (2x + 3y) = 27 - 3$$

$$-8y = 24$$

$$y = -3$$

Substituting $y = -3$ in Equation A,

$$2x - 5 \times (-3) = 27$$

$$2x = 27 - 15 = 12$$

$$x = 6$$

Solution is $x = 6$, $y = -3$

Check:

Substituting $x = 6$, $y = -3$ in Equation B,

$$2x + 3y = 2 \times 6 + 3 \times (-3) = 3$$

The process of removing one of the unknowns is called *elimination*. In the first example, you eliminated y, and in the second example you eliminated x. It does not matter which of the unknowns is eliminated; always choose the one that involves the simpler procedure.

In these examples, the coefficients of one of the variables in the two equations were numerically equal. But what do you do when this is not the case? To show how this difficulty is resolved, consider a third example, in which the simultaneous equations are:

$$7x - 6y = 20 \quad (A)$$

$$3x + 4y = 2 \quad (B)$$

The lowest common multiple of 6 and 4 is 12. Therefore, if you multiply each side of Equation A by 2, and each side of Equation B by 3, you will obtain equations in which the coefficients of y are numerically equal to 12. This is an easier process than to make the coefficients of x equal because you would then have to multiply Equation A by 3 and Equation B by 7.

Multiplying Equation A by 2 yields,

$$14x - 12y = 40 \quad \text{(C)}$$

Multiplying Equation B by 3 yields,

$$9x + 12y = 6 \quad \text{(D)}$$

Adding Equations C and D,

$$(14x - 12y) + (9x + 12y) = 40 + 6$$

$$23x = 46$$

$$x = 2$$

Substituting $x = 2$ in Equation B,

$$3 \times 2 + 4y = 2$$

$$4y = -4$$

$$y = -1$$

The solution is $x = 2$, $y = -1$.

Check:

Equation A: $7x - 6y = 7 \times 2 - 6 \times (-1)$

$$= 14 + 6 = 20$$

General rules for the elimination method

1. First decide which of the unknowns is easier to eliminate.
2. When one unknown has been found, obtain the other by substituting in the simplest equation that contains this unknown.
3. When checking, use the original equation that was not involved with rule 2.

Linear simultaneous equations can be solved more quickly and directly by the use of determinants; the procedure involved is covered in chapter 186. A graphical method for solving two linear simultaneous equations appears in chapter 189.

Example 195-2

Solve the simultaneous equations $0.5x - 0.7y = 2$, $1.8x - 2.2y = 8.8$, using the method of elimination.

Solution

$$0.5x - 0.7y = 2 \quad \text{(A)}$$

$$1.8x - 2.2y = 8.8 \quad \text{(B)}$$

Multiply Equation A by 1.8,

$$0.9x - 1.26y = 3.6 \quad \text{(C)}$$

Multiply Equation B by 0.5,

$$0.9x - 1.1y = 4.4 \quad \text{(D)}$$

Subtract Equation D from Equation C,

$$-0.16y = -0.8$$

$$y = \frac{0.8}{0.16} = 5.$$

Substituting the value of y in Equation A,

$$0.5x - 0.7 \times 5 = 2$$

$$0.5x = 5.5$$

$$x = \frac{5.5}{0.5} = 11$$

Check:

Equation B: $1.8x - 2.2y = 1.8 \times 11 - 2.2 \times 5$

$$= 19.8 - 11$$

$$= 8.8$$

Example 195-3

A circuit is analyzed by the use of Kirchhoff's Laws. The following simultaneous equations are derived:

$$4.7I_1 + 10I_3 = 12 \quad \text{(A)}$$

$$3.3I_2 + 10I_3 = 19.8 \quad \text{(B)}$$

$$I_2 = I_3 - I_1$$

Determine the values of I_1, I_2, and I_3 that represent currents measured in milliamperes.

Solution:

Substituting for I_2 in Equation B,

$$3.3(I_3 - I_1) + 10I_3 = 19.8$$

$$-3.3I_1 + 13.3I_3 = 19.8 \quad \text{(D)}$$

Multiplying Equation A by 3.3 and Equation D by 4.7,

$$15.51I_1 + 33I_3 = 39.6 \quad \text{(E)}$$

$$-15.51I_1 + 62.5I_3 = 93.06 \quad \text{(F)}$$

Adding Equations E and F,

$$95.51I_3 = 132.66$$

$$I_3 = \frac{132.66}{95.51} = 1.39 \text{ mA}$$

Substituting for I_3 in Equation B,

$$3.3I_2 = 19.8 - 10I_3 = 19.8 - 13.9 = 5.9$$

$$I_2 = \frac{5.9}{3.3} = 1.79 \text{ mA}$$

Then,

$$I_1 = I_3 - I_2 = 1.39 - 1.79 = -0.4 \text{ mA.}$$

The negative sign indicates that the assumed direction for the current (I_1) was incorrect.

Example 195-4

In the mesh current analysis of a dc circuit, the following simultaneous equations were obtained,

$$10i_1 - 3i_2 = 4 \quad \text{(A)}$$

$$-3i_1 + 7i_2 = 5 \quad \text{(B)}$$

Determine the values of the mesh currents, i_1 and i_2, which are measured in milliamperes.

Solution

Multiplying Equation A by 7, and Equation B by 3,

$$70i_1 - 21i_2 = 168 \quad \text{(C)}$$

$$-9I_1 + 21i_2 = 15 \quad \text{(D)}$$

Adding Equations C and D,

$$61i_1 = 183$$

Mesh current,

$$i_1 = \frac{183}{61} = 3 \text{ mA}$$

Substituting for i_1 in Equation B,

$$-3 \times 3 + 7i_2 = 5$$

$$7i_2 = 14$$

Mesh current,

$$i_2 = \frac{14}{7} = 2 \text{ mA}$$

PRACTICE PROBLEMS

195-1. Solve the following simultaneous equations by the substitution method:

(a) $x + y = 13$

$x - y = 1$

(b) $x = 3y$

$x = 15 - 2y$

(c) $3x + y = 11$

$x + y = 7$

(d) $2x - y = 3$

$x + y = 9$

195-2. Solve the following simultaneous equations by the elimination method:

(a) $x + 2y = 8$

$2x + y = 7$

(b) $7x - y = 2$

$6x = y$

(c) $x = 2y + 1$

$3x = 5(y + 1)$

(d) $4y = 3x + 2$

$2y + x + 1 = 0$

195-3. Solve the following simultaneous equations:

$2x + y - 4z = -11$

$x - 2y - 2z = -18$

$-2x + 3y + 4z = 31$

195-4. A voltage divider is composed of two resistors in series. The sum of their resistances is 100,000 Ω. The ratio of the resistances must be 4:1. What is the value of each resistor?

196
Determinants

The general form of two simultaneous linear equations is:

$$ax + by + c = 0 \quad \text{(A)}$$

$$dx + ey + f = 0 \quad \text{(B)}$$

where,

x, y are the unknowns
a, d are the coefficients of x
b, e are the coefficients of y
c, f are numbers

Using the elimination method, we can solve for x. Multiply Equation A by e, and Equation B by b. Then,

$$aex + bey + ec = 0 \quad (C)$$

$$dbx + bey + bf = 0 \quad (D)$$

Subtracting Equation D from Equation C,

$$x(ae - db) + (ec - bf) = 0$$

or,

$$\frac{x}{(bf - ec)} = \frac{1}{(ae - db)} \quad (E)$$

Again using the elimination method, you can solve for y. Multiply Equation A by d, and Equation B by a. Then,

$$adx + bdy + dc = 0 \quad (F)$$

$$adx + aey + af = 0 \quad (G)$$

Subtracting Equation G from Equation F,

$$y(db - ae) + (dc - af) = 0$$

$$\frac{-y}{(af - dc)} = \frac{1}{(ae - db)} \quad (H)$$

Combine Equations E and H.

$$\frac{x}{(bf - ec)} = \frac{-y}{(af - dc)} = \frac{1}{(ae - db)} \quad (J)$$

In its determinant form, Equation J is rewritten as:

$$\frac{x}{\begin{vmatrix} b & c \\ e & f \end{vmatrix}} = \frac{-y}{\begin{vmatrix} a & c \\ d & f \end{vmatrix}} = \frac{1}{\begin{vmatrix} a & b \\ d & e \end{vmatrix}}$$

$\begin{vmatrix} b & c \\ e & f \end{vmatrix}$ is a *second-order determinant*

(two rows and two columns) whose value is $b \times f - e \times c$. Notice that the line "\" links the positive product (bf), while the negative product (ec) is associated with the "/" line. *Determinants* (unlike the matrices of chapter 197) always have the same number of rows and columns. A first-order determinant contains only a single number, which is the determinant's value; as an example, $| 7 | = 7$.

The advantage of the determinant method is that the determinants themselves can be directly written down by inspecting the equations. For example, solve the equations,

$$3x - 7y + 15 = 0$$

$$-5x + 2y + 4 = 0$$

The divisor of x is the second-order determinant, containing only those four numbers that are *not* the coefficients of x. The divisor of y is the determinant that contains only those four numbers that are not the coefficients of y. Finally, the divisor of "1" is the determinant that only contains the coefficients of x and y. Notice that the signs of the dividends (x, $-y$, and 1) are alternately positive and negative. Then,

$$\frac{x}{\begin{vmatrix} -7 & 15 \\ 2 & 4 \end{vmatrix}} = \frac{-y}{\begin{vmatrix} 3 & 15 \\ -5 & 4 \end{vmatrix}} = \frac{1}{\begin{vmatrix} 3 & -7 \\ -5 & 2 \end{vmatrix}}$$

$$\frac{x}{(-7) \times 4 - 15 \times 2} = \frac{-y}{3 \times 4 - 15 \times (-5)}$$

$$= \frac{1}{3 \times 2 - (-7) \times (-5)}$$

$$\frac{x}{-58} = \frac{-y}{87} = \frac{1}{-29}$$

Solution:

$$x = \frac{-58}{-29} = 2$$

$$y = \frac{-87}{-29} = 3$$

MATHEMATICAL DERIVATIONS

Three simultaneous linear equations with three unknowns (x, y, and z) can be written in their general form as:

$$ax + by + cz + d = 0$$

$$ex + fy + gz + h = 0$$

$$ix + jy + kz + l = 0$$

The determinant solution is:

$$\frac{x}{\begin{vmatrix} b & c & d \\ f & g & h \\ j & k & l \end{vmatrix}} = \frac{-y}{\begin{vmatrix} a & c & d \\ e & g & h \\ i & k & l \end{vmatrix}} = \frac{z}{\begin{vmatrix} a & b & d \\ e & f & h \\ i & j & l \end{vmatrix}} = \frac{-1}{\begin{vmatrix} a & b & c \\ e & f & g \\ i & j & k \end{vmatrix}}$$

Each of the divisors is a *third-order determinant* (three rows, three columns), whose value is determined as follows:

$$\begin{vmatrix} b & c & d \\ f & g & h \\ j & k & l \end{vmatrix} = b \times \begin{vmatrix} g & h \\ k & l \end{vmatrix} - c \times \begin{vmatrix} f & h \\ j & l \end{vmatrix} + d \times \begin{vmatrix} f & g \\ j & k \end{vmatrix}$$

$$= b(gl - hk) - c(fl - hj) + d(fk - gj)$$

$$= (bgl + chj + dfk) - (dgj + bhk + cfl)$$

The last form of expressing the determinant's value suggests another method of evaluating the determinant. Follow the third-order determinant by repeating the first two columns:

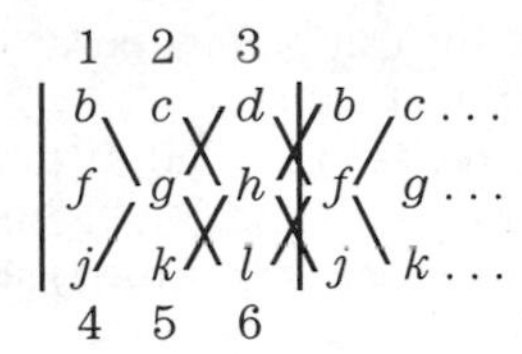

Now, draw lines that pass through three letters. There are six such lines with three "\" lines and three "/" lines. As you have noticed previously, "\" lines represent positive products, but negative products are associated with "/" lines. Then,

$$\text{Determinant's value} = (\underbrace{bgl}_{\text{line 1}} + \underbrace{chj}_{\text{line 2}} + \underbrace{dfk}_{\text{line 3}}) - (\underbrace{dgj}_{\text{line 4}} + \underbrace{bhk}_{\text{line 5}} + \underbrace{cfl}_{\text{line 6}})$$

The expression is the same as that previously obtained. Notice, however, that this method cannot be applied to fourth- and higher-order determinants.

At first glance, it appears that the process of obtaining the determinant's value, is both lengthy and tedious. However, determinants have a number of properties that allow a determinant to be simplified before calculating its value. The following properties of determinants are stated without formal proofs:

A. If any two rows (or any two columns) are interchanged, the numerical value of the determinant is unchanged, but its sign is reversed.

For example,

$$\begin{vmatrix} 2 & 1 & 6 \\ 3 & 2 & 4 \\ 5 & 7 & 8 \end{vmatrix} = 2 \times (2 \times 8 - 4 \times 7) - 1(3 \times 8 - 5 \times 4) + 6(3 \times 8 - 4 \times 5)$$
$$= 2 \times (-12) - 1 \times 4 + 6 \times 11 = +38$$

but,

$$\begin{vmatrix} 2 & 1 & 6 \\ 5 & 7 & 8 \\ 3 & 2 & 4 \end{vmatrix} = 2 \times (7 \times 4 - 8 \times 2) - 1(5 \times 4 - 3 \times 8) + 6(5 \times 2 - 7 \times 3)$$
$$= 2 \times 12 - 1(-4) + 6(-11) = -38$$

and,

$$\begin{vmatrix} 2 & 6 & 1 \\ 3 & 4 & 2 \\ 5 & 8 & 7 \end{vmatrix} = 2 \times (4 \times 7 - 2 \times 8) - 6(3 \times 7 - 2 \times 5) + 1(3 \times 8 - 4 \times 5)$$
$$= 2 \times 12 - 6 \times 11 + 1 \times 4 = -38$$

B. If all of the determinant's rows and columns are interchanged (the first column becomes the first row and so on), the value of the determinant is unchanged.

For example,

You already know that,

$$\begin{vmatrix} 2 & 1 & 6 \\ 3 & 2 & 4 \\ 5 & 7 & 8 \end{vmatrix} = +38$$

Interchanging all rows and columns,

$$\begin{vmatrix} 2 & 3 & 5 \\ 1 & 2 & 7 \\ 6 & 4 & 8 \end{vmatrix} = 2(2 \times 8 - 7 \times 4) - 3(1 \times 8 - 7 \times 6) + 5(1 \times 4 - 2 \times 6)$$
$$= 2(-12) - 3(-34) + 5(-8)$$
$$= -24 + 102 - 40 = 38$$

C. If all of the numbers in any column (or row) are zero, the value of the determinant is zero.

For example,

$$\begin{vmatrix} 2 & 1 & 6 \\ 0 & 0 & 0 \\ 5 & 7 & 8 \end{vmatrix} = 2(0 \times 8 - 0 \times 7) - 1(0 \times 8 - 0 \times 5) + 6(0 \times 7 - 0 \times 5)$$
$$= 0$$

and,

$$\begin{vmatrix} 2 & 0 & 6 \\ 3 & 0 & 4 \\ 5 & 0 & 8 \end{vmatrix} = 2(0 \times 8 - 4 \times 0) - 0(3 \times 8 - 4 \times 5) + 6(3 \times 0 - 0 \times 5)$$
$$= 0$$

D. One column can be added to or subtracted from another column without affecting the value of the determinant. This same property applies to any two rows.

In the determinant,

$$\begin{vmatrix} 2 & 1 & 6 \\ 3 & 2 & 4 \\ 5 & 7 & 8 \end{vmatrix} = +38,$$

add column 2 to column 1. Then,

$$\begin{vmatrix} 2+1 & 1 & 6 \\ 3+2 & 2 & 4 \\ 5+7 & 7 & 8 \end{vmatrix} = \begin{vmatrix} 3 & 1 & 6 \\ 5 & 2 & 4 \\ 12 & 7 & 8 \end{vmatrix}$$
$$= 3(2 \times 8 - 4 \times 7) - 1(5 \times 8 - 4 \times 12) + 6(5 \times 7 - 2 \times 12)$$
$$= 3(-12) - 1(-8) + 6 \times 11$$
$$= 36 + 8 + 66 = +38$$

In the same determinant, subtract row 2 from row 3. Then,

$$\begin{vmatrix} 2 & 1 & 6 \\ 3 & 2 & 4 \\ 5-3 & 7-2 & 8-4 \end{vmatrix} = \begin{vmatrix} 2 & 1 & 6 \\ 3 & 2 & 4 \\ 2 & 5 & 4 \end{vmatrix}$$

$$= 2(2 \times 4 - 4 \times 5) - 1(3 \times 4 - 4 \times 2) + 6(3 \times 5 - 2 \times 2)$$

$$= 2(-12) - 1 \times 4 + 6 \times 11$$

$$= -24 - 4 + 66 = +38$$

E. If any row or column is multiplied by a common factor, the value of the determinant is multiplied by the same factor.

In the determinant,

$$\begin{vmatrix} 2 & 1 & 6 \\ 3 & 2 & 4 \\ 5 & 7 & 8 \end{vmatrix} = +38,$$

multiply column 2 by -3; then,

$$\begin{vmatrix} 2 & 1 \times (-3) & 6 \\ 3 & 2 \times (-3) & 4 \\ 5 & 7 \times (-3) & 8 \end{vmatrix} = \begin{vmatrix} 2 & -3 & 6 \\ 3 & -6 & 4 \\ 5 & -21 & 8 \end{vmatrix}$$

$$= 2((-6) \times 8 - 4 \times (-21)) - (-3)(3 \times 8 - 4 \times 5) + 6(3 \times (-21) - (-6) \times 5)$$

$$= 2 \times 36 + 3 \times 4 + 6 \times (-33)$$

$$= 72 + 12 - 198$$

$$= -114 = (-3) \times (+38)$$

In the same determinant, multiply row 3 by -2. Then,

$$\begin{vmatrix} 2 & 1 & 6 \\ 3 & 2 & 4 \\ 5 \times (-2) & 7 \times (-2) & 8 \times (-2) \end{vmatrix}$$

$$= \begin{vmatrix} 2 & 1 & 6 \\ 3 & 2 & 4 \\ -10 & -14 & -16 \end{vmatrix}$$

$$= 2(2 \times (-16) - 4 \times (-14)) - 1(3 \times (-16) - 4 \times (-10)) + 6(3 \times (-14) - 2 \times (-10))$$

$$= 2 \times 24 - 1 \times (-8) + 6 \times (-22)$$

$$= 48 + 8 - 132 = -76 = (-2) \times (-38)$$

The common factor can be any positive or negative whole number or fraction.

From the properties D and E, it follows that the value of a determinant is zero if the corresponding numbers of any two columns (or rows) are identical or in proportion.

For example,

$$\begin{vmatrix} 3 & -2 & 4 \\ 1 & 5 & -1 \\ -6 & 4 & -8 \end{vmatrix} = -2 \times \begin{vmatrix} 3 & -2 & 4 \\ 1 & 5 & 1 \\ 3 & -2 & 4 \end{vmatrix}$$

$$= -2 \times \begin{vmatrix} 3-3 & -2-(-2) & 4-4 \\ 1 & 5 & -1 \\ 3 & -2 & 4 \end{vmatrix}$$

$$= 2 \times \begin{vmatrix} 0 & 0 & 0 \\ 1 & 5 & -1 \\ 3 & -2 & 4 \end{vmatrix}$$

$$= 2 \times 0 = 0$$

Check:

$$\begin{vmatrix} 3 & -2 & 4 \\ 1 & 5 & -1 \\ -6 & 4 & -8 \end{vmatrix}$$

$$= 3(5 \times (-8) - (-1) \times 4) - (-2)(1 \times (-8) - (-1)(-6)) + 4(1 \times 4 - 5 \times (-6))$$

$$= 3 \times (-36) + 2 \times (-14) + 4 \times 34$$

$$= -108 - 28 + 136 = 0$$

F. If the numbers in any row or column are each split into the sum of two numbers (which may be either positive or negative whole numbers or fractions), the determinant can be expressed as the sum of two determinants.

For example,

$$\begin{vmatrix} 2 & 1 & 6 \\ 3 & 2 & 4 \\ 5 & 7 & 8 \end{vmatrix} = \begin{vmatrix} 2 & 1 & 6 \\ 3 & 2 & 4 \\ 6+(-1) & 5+2 & 1+7 \end{vmatrix}$$

$$= \begin{vmatrix} 2 & 1 & 6 \\ 3 & 2 & 4 \\ 6 & 5 & 1 \end{vmatrix} + \begin{vmatrix} 2 & 1 & 6 \\ 3 & 2 & 4 \\ -1 & 2 & 7 \end{vmatrix}$$

Check:

$$\begin{vmatrix} 2 & 1 & 6 \\ 3 & 2 & 4 \\ 5 & 7 & 8 \end{vmatrix} = +38$$

$$\begin{vmatrix} 2 & 1 & 6 \\ 3 & 2 & 4 \\ 6 & 5 & 1 \end{vmatrix} + \begin{vmatrix} 2 & 1 & 6 \\ 3 & 2 & 4 \\ -1 & 2 & 7 \end{vmatrix}$$

$$\begin{aligned} &= 2(2 \times 1 - 4 \times 5) \\ &- 1(3 \times 1 - 6 \times 4) \\ &+ 6(3 \times 5 - 6 \times 2) \\ &+ 2(2 \times 7 - 4 \times 2) \\ &- 1(3 \times 7 - 4(-1) \\ &+ 6(3 \times 2 - 2(-1)) \\ &= -36 + 21 + 18 + 12 - 25 + 48 = +38 \end{aligned}$$

G. If any row (or column) is multiplied by a common factor, the result can be added to (or subtracted from) another row (column) without affecting the value of the determinant.

For example,

$$\begin{vmatrix} 2 & 1 & 6 \\ 3 & 2 & 4 \\ 5 & 7 & 8 \end{vmatrix} = \begin{vmatrix} 2 & 1 & 6 \\ 3 & 2 & 4 \\ 5-(2 \times 2) & 7-(2 \times 1) & 8-(2 \times 6) \end{vmatrix}$$

$$= \begin{vmatrix} 2 & 1 & 6 \\ 3 & 2 & 4 \\ 1 & 5 & -4 \end{vmatrix}$$

In this process, the top row was multiplied by 2; the result was then subtracted from the bottom row.

Check:

$$\begin{vmatrix} 2 & 1 & 6 \\ 3 & 2 & 4 \\ 5 & 7 & 8 \end{vmatrix} = +38$$

$$\begin{vmatrix} 2 & 1 & 6 \\ 3 & 2 & 4 \\ 1 & 5 & -4 \end{vmatrix} \begin{aligned} &= 2(2(-4) - 4 \times 5) \\ &\quad - 1(3(-4) - 4 \times 1) \\ &\quad + 6(3 \times 5 - 2 \times 1) \\ &= -56 + 16 + 78 = +38 \end{aligned}$$

When simplifying a determinant, the strategy is to create as many zeros as possible without producing any unwieldy numbers. This will reduce the number of terms required to obtain the determinant's value. The example that follows illustrates the principles of simplification.

Example 196-1

Simplify the following fourth-order determinant:

$$\begin{vmatrix} -1 & 6 & 4 & -12 \\ 3 & 4 & -7 & -6 \\ -1 & -3 & 4 & 6 \\ 1 & 2 & -2 & 8 \end{vmatrix}$$

Solution

The fourth column has a common factor of 2. Therefore, the value of the determinant is:

$$2 \times \begin{vmatrix} -1 & 6 & 4 & -6 \\ 3 & 4 & -7 & -3 \\ -1 & -3 & 4 & 3 \\ 1 & 2 & -2 & 4 \end{vmatrix}$$

Subtract Row #3 from Row #1.

$$= 2 \times \begin{vmatrix} 0 & 9 & 0 & -9 \\ 3 & 4 & -7 & -3 \\ -1 & -3 & 4 & 3 \\ 1 & 2 & -2 & 4 \end{vmatrix}$$

Add Column #4 to Column #2.

$$= 2 \times \begin{vmatrix} 0 & 0 & 0 & -9 \\ 3 & 1 & -7 & -3 \\ -1 & 0 & 4 & 3 \\ 1 & 6 & -2 & 7 \end{vmatrix}$$

$$= 2 \times (+9) \times \begin{vmatrix} 3 & 1 & -7 \\ -1 & 0 & 4 \\ 1 & 6 & -2 \end{vmatrix}$$

$$= +18 \times [(3 \times (-24) - 1(-2) - 7(-6)]$$

$$= +18 \times (-28) = -504.$$

Example 196-2

Using the determinant method, solve the following equations for I_1, I_2, and I_3.

$$\begin{aligned} 3I_1 - 2I_2 - 4I_3 + 1 &= 0 \\ 4I_1 + 3I_2 + I_3 &= 0 \\ 6I_1 + 2I_2 - 3I_3 + 4 &= 0 \end{aligned}$$

Solution

The determinant solution is:

$$\frac{I_1}{\begin{vmatrix} -2 & -4 & 1 \\ 3 & 1 & 0 \\ 2 & -3 & 4 \end{vmatrix}} = \frac{-I_2}{\begin{vmatrix} 3 & -4 & 1 \\ 4 & 1 & 0 \\ 6 & -3 & 4 \end{vmatrix}}$$

$$= \frac{I_3}{\begin{vmatrix} 3 & -2 & 1 \\ 4 & 3 & 0 \\ 6 & 2 & 4 \end{vmatrix}} = \frac{-1}{\begin{vmatrix} 3 & -2 & -4 \\ 4 & 3 & 1 \\ 6 & 2 & -3 \end{vmatrix}}$$

$$\frac{I_1}{-2 \times 4 + 4 \times 12 - 11}$$

$$= \frac{-I_2}{3 \times 4 + 4 \times 16 - 1 \times 18}$$

$$= \frac{I_3}{3 \times 12 - 2 \times 16 - 10}$$

$$= \frac{-1}{3 \times (-11) + 2(-18) - 4(-10)}$$

$$\frac{I_1}{29} = \frac{-I_2}{58} = \frac{I_3}{58} = \frac{-1}{-29}$$

$$I_1 = \frac{29}{29} = 1$$

$$I_2 = \frac{-58}{29} = -2$$

$$I_3 = \frac{58}{29} = 2$$

It is rarely worthwhile simplifying a third-order determinant because, in general, only six terms are involved. However, the general fourth-order determinant contains 24 terms and a fifth-order determinant contains 120 terms. Such determinants should be simplified as much as possible in order to reduce the number of terms required in obtaining the determinant's value.

A good example of the use of determinants appears in the analysis of the RC phase shift oscillator (chapter 140).

PRACTICE PROBLEMS

196-1. Use the method of determinants to solve the following simultaneous equations for i_1, i_2, and i_3:

$$4i_1 - 3i_2 - 5i_3 = 13$$

$$-2i_1 + 6i_2 - 3i_3 = -21$$

$$7i_1 + i_2 + 4i_3 = 23$$

196-2. Evaluate the following determinant:

$$\begin{vmatrix} 1 & 0 & 0 & 0 & 0 \\ -4 & 7 & 0 & 1 & 1 \\ -2 & 1 & -1 & 1 & -1 \\ -1 & 2 & 4 & 0 & 3 \\ 1 & 4 & 0 & 1 & 1 \end{vmatrix}$$

196-3. Use the method of determinants to solve the following simultaneous equations for i_1, i_2, i_3, and i_4:

$$3i_1 - 2i_2 + i_3 - 5i_4 = -6$$

$$4i_1 + i_2 - 6i_3 + 2i_4 = -13$$

$$-7i_1 + 5i_2 - 2i_3 + 3i_4 = 24$$

$$5i_1 + 4i_2 - 8i_3 + 4i_4 = -10$$

197 Matrices

In chapters 195 and 196, you solved linear simultaneous equations by the substitution, elimination, and determinant methods. Now, look at another method, which allows the equations to be written down in a form of shorthand that is referred to as *matrix representation*. The value of any one of the unknowns can then be directly expressed as a quotient of two determinants. A solution that uses matrices is particularly applicable to circuits which require mesh analysis (chapters 27 and 95).

When compared with a determinant (chapter 196), a matrix has the following properties:

(a) A matrix has no particular value, but is merely an array of elements or numbers which obey a given set of rules.

(b) A matrix is not required to have equal numbers of rows and columns.

For example, a car dealer decides to list the vehicles that are available in ascending order of price:

47 cars in the price range of $5,000 to $10,000

58 cars in the price range of $10,001 to $15,000
38 cars in the price range of $15,001 to $20,000
42 cars in the price range of $20,001 (upwards)

When expressed as a row matrix, the inventory would be shown as:

$$\text{Inventory}, I = (\,47\ 58\ 38\ 42)$$

The one rule of this matrix is that the cost of the car increases as you move across from left to right. Because there is only a single rule, you have a linear array (a one-dimensional matrix) with four columns, but only one row; this is called a *1 × 4 row matrix*. By contrast, you could rearrange the same inventory as a 4 × 1 column matrix with 4 rows and only one column.

$$I = \begin{pmatrix} 47 \\ 58 \\ 38 \\ 42 \end{pmatrix}$$

Here, the rule states that the price increases from the top to the bottom of the matrix. Notice the use of parentheses, ( ), or brackets [ ], to enclose a matrix, as compared to vertical lines, | |, for a determinant.

Going back to the car dealer, he/she might wish to refine the inventory in terms of the cars' engine capacities (such as 1.5 liter, 2.0 liter, 2.5 liter, 3.0 liter, above 3.0 liter). This involves the introduction of a second rule so that you have a two-dimensional 4 × 5 column matrix with 4 rows and 5 columns.

Increasing engine capacity →

$$\text{Increasing cost} \downarrow \quad I = \begin{pmatrix} 35 & 10 & 1 & 1 & 0 \\ 19 & 29 & 7 & 2 & 1 \\ 8 & 12 & \textcircled{12} & 3 & 3 \\ 2 & 5 & 10 & 15 & 10 \end{pmatrix}$$

In this matrix, the circled (12) indicates that there are 12 cars in stock, each in the price range of $15,001 to $20,000 and with an engine capacity of 2.5 liters.

By adding more rules, you can increase the number of dimensions but it is difficult to represent such matrices in a two-dimensional plane, such as a flat sheet of paper. Fortunately, for mesh and nodal analysis, only one- and two-dimensional matrices are required.

The following is a representation of a general two-dimensional matrix, which is of the order of $j \times k$ (j rows and k columns):

$$I = \begin{pmatrix} i_{11} & i_{12} & i_{13} & \cdots & i_{1e} & \cdots & i_{1k} \\ i_{21} & i_{22} & i_{23} & \cdots & i_{2e} & \cdots & i_{2k} \\ i_{31} & i_{32} & i_{33} & \cdots & i_{3e} & \cdots & i_{3k} \\ | & | & | & \cdots & | & \cdots & | \\ | & | & | & \cdots & | & \cdots & | \\ | & | & | & \cdots & | & \cdots & | \\ i_{f1} & i_{f2} & i_{f3} & \cdots & i_{fe} & \cdots & i_{fk} \\ | & | & | & \cdots & | & \cdots & | \\ | & | & | & \cdots & | & \cdots & | \\ | & | & | & \cdots & | & \cdots & | \\ i_{j1} & i_{j2} & i_{j3} & \cdots & i_{je} & \cdots & i_{jk} \end{pmatrix} \qquad (197\text{-}1)$$

The general element of this matrix is the number, i_{fe}. The first suffix (f) indicates the row in which the element is located and the second suffix (e) shows the column that contains this element. If $j = 1$, you have a one-dimensional single-row matrix, whereas if $k = 1$, the matrix is a one-column array. In the case of $j = k$, there are as many rows as there are columns, and the result is a square matrix.

As an example,

$$I = \begin{pmatrix} 3 & -7 & 4 & 2 & 0 \\ 6 & 5 & -2 & 1 & 8 \\ 2 & -3 & -4 & 9 & 3 \\ 4 & 5 & 1 & 2 & 6 \end{pmatrix}$$

This is a two-dimensional matrix of the order 4 × 5. Examples of the various elements are:

Element $i_{23} = -2$
Element $i_{32} = -3$
Element $i_{44} = 2$
Element $i_{11} = 3$
Element $i_{45} = 6$

MATHEMATICAL DERIVATIONS

Multiplication of matrices

Assume that, in network analysis, you obtain one matrix that consists of resistance values, and a second matrix that consists of current values. It is reasonable to suppose that if you are able to multiply the resistance matrix by the current matrix, the result will be a matrix whose elements are values of voltage. However, two matrices must obey a certain condition in order for multiplication to be possible. Cramer's rule is as follows:

"Multiplication of matrix R by matrix I is only possible if matrix R has the same number of *columns* as matrix I has *rows*."

The product matrix is expressed as,

$$E = RI$$

In the general case, R is a matrix of the order $j \times k$, I is a matrix of the order $k \times 1$, and E is a matrix of the order $j \times 1$. The product matrix (E) has the same number of rows as matrix R, and the same number of columns as matrix I.

The question arises "how can the element of E be determined from the elements of R and I?" For a general element, such as e_{cd},

$$e_{cd} = r_{c1}i_{1d} + r_{c2}i_{2d} + r_{c3}i_{3d} + \ldots + r_{ck}i_{kd} \qquad (197\text{-}2)$$

Each element of row c in matrix R is multiplied by the corresponding element of column d in matrix I. The sum of these products is then equal to element e_{cd} in matrix E. Equation 197-2 clearly shows why the number of columns in matrix R must be the same as the number of rows in matrix I. If this were not so, some of the elements would be left out of the multiplication process. This also means that the sequence of the product matrix is important; as specified, $E = RI$, but the expression $E = IR$, cannot be evaluated.

As an example, determine product matrix E if:

$$E = RI = \begin{pmatrix} 3 & 4 \\ 1 & 5 \\ 2 & -3 \end{pmatrix} \begin{pmatrix} -1 & -2 \\ -4 & 3 \end{pmatrix}$$

Multiplication is possible because the first matrix has two columns, and the second matrix has two rows. The product matrix has two columns and three rows.

Then,

$$\begin{pmatrix} e_{11} & e_{12} \\ e_{21} & e_{22} \\ e_{31} & e_{32} \end{pmatrix} = \begin{pmatrix} 3 & 4 \\ 1 & 5 \\ 2 & -3 \end{pmatrix} \begin{pmatrix} -1 & -2 \\ -4 & 3 \end{pmatrix}$$

Using Equation 197-2,

$$e_{11} = r_{11}i_{11} + r_{12}i_{21}$$
$$= 3 \times (-1) + 4 \times (-4) = -19$$
$$e_{12} = r_{11}i_{12} + r_{12}i_{22}$$
$$= 3 \times (-2) + 4 \times 3 = 6$$
$$e_{21} = r_{21}i_{11} + r_{22}i_{21}$$
$$= 1 \times (-1) + 5 \times (-4) = -21$$
$$e_{22} = r_{21}i_{12} + r_{22}i_{22}$$
$$= 1 \times (-2) + 5 \times 3 = 13$$
$$e_{31} = r_{31}i_{11} + r_{32}i_{21}$$
$$= 2 \times (-i) + (-3) \times (-4) = 10$$
$$e_{32} = r_{31}i_{12} + r_{32}i_{22}$$
$$= 2 \times (-2) + (-3) \times 3 = -13$$

Therefore,

$$\text{Matrix } E = \begin{pmatrix} -19 & 6 \\ -21 & 13 \\ 10 & -13 \end{pmatrix}$$

Matrix representation for mesh analysis

The simultaneous equations for mesh analysis (chapters 27 and 95) can be written in the form:

$$\begin{array}{ccccccccccc} r_{11}i_1 & + & r_{12}i_2 & + & r_{13}i_3 & + & \ldots & + & r_{1n}i_n & = & e_1 \\ r_{21}i_1 & + & r_{22}i_2 & + & r_{23}i_3 & + & \ldots & + & r_{2n}i_n & = & e_2 \\ r_{31}i_1 & + & r_{32}i_2 & + & r_{33}i_3 & + & \ldots & + & r_{3n}i_n & = & e_3 \\ | & & | & & | & & & & | & & | \\ | & & | & & | & & & & | & & | \\ | & & | & & | & & & & | & & | \\ | & & | & & | & & & & | & & | \\ r_{n1}i_1 & + & r_{n2}i_2 & + & r_{n3}i_3 & + & \ldots & + & r_{nn}i_n & = & e_n \end{array}$$

where: n = number of meshes
$i_1, i_2 \ldots i_n$ = n unknown mesh currents
$e_1, e_2, e_3, \ldots e_n$ = n known total source mesh voltages
$r_{11}, r_{12} \ldots r_{nn}$ = the resistance elements of which there are n^2 in number.

If the resistance elements are arranged in a matrix, there are n rows and n columns so that this matrix is square and of the order $n \times n$. The current and voltage matrices would each have n rows and one column so that their order is $n \times 1$. Consequently, the resistance matrix times the current matrix obeys Cramer's rule for the multiplication process; the product voltage matrix will contain n rows and 1 column. Therefore,

$$\begin{pmatrix} r_{11}r_{12}r_{13}\ldots r_{1n} \\ r_{21}r_{22}r_{23}\ldots r_{2n} \\ r_{31}r_{32}r_{33}\ldots r_{3n} \\ | \;\; | \;\; | \qquad | \\ | \;\; | \;\; | \qquad | \\ | \;\; | \;\; | \qquad | \\ | \;\; | \;\; | \qquad | \\ r_{n1}r_{n2}r_{n3}\ldots r_{nn} \end{pmatrix} \begin{pmatrix} i_1 \\ i_2 \\ i_3 \\ | \\ | \\ | \\ | \\ i_n \end{pmatrix} = \begin{pmatrix} e_1 \\ e_2 \\ e_3 \\ | \\ | \\ | \\ | \\ e_n \end{pmatrix} \qquad (197\text{-}3)$$

Once the mesh equations have been written down in matrix form, it is required to find the value(s) of one or more mesh currents. If you elect to find the value of i_j, this value is expressed as the quotient of two determinants. The denominator determinant contains the same elements as the resistance matrix. The numerator determinant is also the same as the resistance matrix, *except* that the jth column is replaced by the column in voltage matrix E.

Referring back to Equation 197-3,

$$i_j = \frac{\begin{vmatrix} r_{11} r_{12} r_{13} \ldots e_1 \ldots r_{1n} \\ r_{21} r_{22} r_{23} \ldots e_2 \ldots r_{2n} \\ r_{31} r_{32} r_{33} \ldots e_3 \ldots r_{3n} \\ | \quad | \quad | \qquad | \qquad | \\ | \quad | \quad | \qquad | \qquad | \\ | \quad | \quad | \qquad | \qquad | \\ | \quad | \quad | \qquad | \qquad | \\ r_{n1} r_{n2} r_{n3} \ldots e_n \ldots r_{nn} \end{vmatrix}}{\begin{vmatrix} r_{11} r_{12} r_{13} \ldots r_{1j} \ldots r_{1n} \\ r_{21} r_{22} r_{23} \ldots r_{2j} \ldots r_{2n} \\ r_{31} r_{32} r_{33} \ldots r_{3j} \ldots r_{3n} \\ | \quad | \quad | \qquad | \qquad | \\ | \quad | \quad | \qquad | \qquad | \\ | \quad | \quad | \qquad | \qquad | \\ | \quad | \quad | \qquad | \qquad | \\ r_{n1} r_{n2} r_{n3} \ldots r_{nj} \ldots r_{nn} \end{vmatrix}} \qquad (197\text{-}4)$$

Notice the use of vertical lines that indicate the presence of determinants, rather than matrices. The denominator determinant is often given the symbol "Δ"; whichever current is found, the value of Δ is always the same.

Rules of mesh analysis using a matrix representation

1. Assign to each mesh a clockwise current that has the direction of the electron flow. Assign a suffix to each current of 1 through n in any order.
2. Referring to Equation 197-3, $i_1, i_2 \ldots i_n$ are the unknown mesh currents.

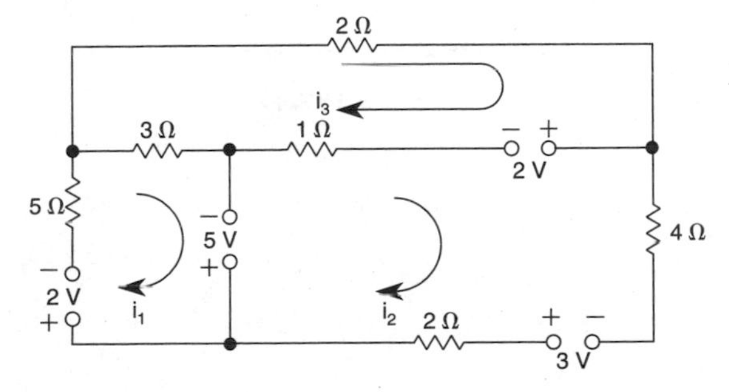

Figure 197-1

3. r_{jj} is the positive sum of the resistances in the jth mesh.
   r_{jk} is the negative sum of the resistances shared by the jth and kth loops. Notice that $r_{jk} = r_{kj}$.
4. e_j is the sum of the voltage sources in the jth loop. The sign of a voltage is positive if the mesh current leaves the negative terminal of the source. Likewise, the voltage is negative if the mesh current leaves the positive terminal.

Consider the network of Fig. 197-1. With three loops, the matrix representation is:

$$\begin{pmatrix} r_{11}\, r_{12}\, r_{13} \\ r_{21}\, r_{22}\, r_{23} \\ r_{31}\, r_{32}\, r_{33} \end{pmatrix} \begin{pmatrix} i_1 \\ i_2 \\ i_3 \end{pmatrix} = \begin{pmatrix} e_1 \\ e_2 \\ e_3 \end{pmatrix}$$

$r_{11} = 3 + 5 = 8\ \Omega$, $r_{22} = 1 + 4 + 2 = 7\ \Omega$,
$r_{33} = 2 + 1 + 3 = 6\ \Omega$.
$r_{12} = r_{21} = 0\ \Omega$, $r_{13} = r_{31} = -3\ \Omega$, $r_{23} = r_{32} = -1\ \Omega$.
$e_1 = +2 - 5 = -3$ V, $e_2 = +5 - 2 - 3 = 0$ V, $e_3 = 2$ V

Then,

$$\begin{pmatrix} 8 & 0 & -3 \\ 0 & 7 & -1 \\ -3 & -1 & 6 \end{pmatrix} \begin{pmatrix} i_1 \\ i_2 \\ i_3 \end{pmatrix} = \begin{pmatrix} -3 \\ 0 \\ 2 \end{pmatrix}$$

$$\Delta = \begin{vmatrix} 8 & 0 & -3 \\ 0 & 7 & -1 \\ -3 & -1 & 6 \end{vmatrix} = 8 \times 41 - 3 \times 31 = 265$$

$$i_1 = \frac{\begin{vmatrix} -3 & 0 & -3 \\ 0 & 7 & -1 \\ 2 & -1 & 6 \end{vmatrix}}{\Delta} = \frac{-3 \times 41 - 3 \times (-14)}{265}$$

$$= \frac{-81}{265} = -0.306 \text{ A}$$

$$i_2 = \frac{\begin{vmatrix} 8 & -3 & -3 \\ 0 & 0 & -1 \\ -3 & 2 & 6 \end{vmatrix}}{\Delta} = \frac{8 \times (+2) + 3 \times (-3)}{265}$$

$$= \frac{7}{265} = 0.0264 \text{ A}$$

$$i_3 = \frac{\begin{vmatrix} 8 & 0 & -3 \\ 0 & 7 & 0 \\ -3 & -1 & 2 \end{vmatrix}}{\Delta} = \frac{8 \times 14 - 3 \times 21}{265} = \frac{49}{265}$$

$$= 0.185 \text{ A}$$

The negative sign for the current i_1 means that the direction for this mesh current (electron flow) as indicated in Fig. 197-1, is incorrect and should be reversed.

Example 197-1

Using a matrix representation, obtain the values of the mesh currents in Fig. 197-2.

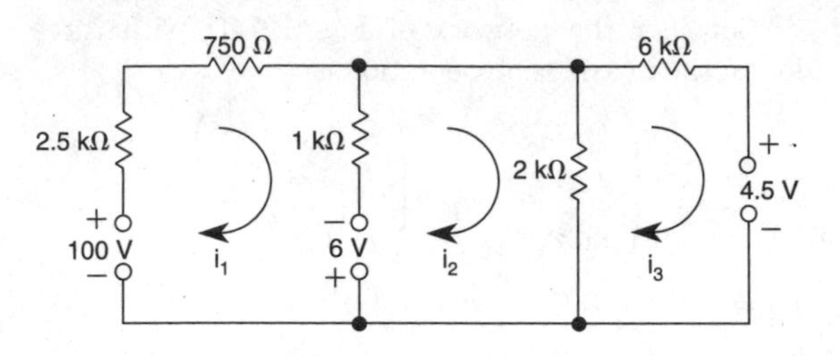

Figure 197-2

Solution

The matrix representation is:

$$\begin{pmatrix} 4.25 & -1 & 0 \\ -1 & 3 & -2 \\ 0 & -2 & 8 \end{pmatrix} \begin{pmatrix} i_1 \\ i_2 \\ i_3 \end{pmatrix} = \begin{pmatrix} -106 \\ 6 \\ 4.5 \end{pmatrix}$$

By Cramer's rule:

$$i_1 = \frac{\begin{vmatrix} -106 & -1 & 0 \\ 6 & 3 & -2 \\ 4.5 & -2 & 8 \end{vmatrix}}{\begin{vmatrix} 4.25 & -1 & 0 \\ -1 & 3 & -2 \\ 0 & -2 & 8 \end{vmatrix}} = \frac{-2060}{77} = -26.8 \text{ mA}$$

$$i_2 = \frac{\begin{vmatrix} 4.25 & -106 & 0 \\ -1 & 6 & -2 \\ 0 & 4.5 & 8 \end{vmatrix}}{77} = \frac{-606}{77} = -7.87 \text{ mA}$$

$$i_3 = \frac{\begin{vmatrix} 4.25 & -1 & -106 \\ -1 & 3 & 6 \\ 0 & -2 & 4.5 \end{vmatrix}}{77} = \frac{-108}{77} = -1.40 \text{ mA}$$

The electron flows of all three mesh currents are in the opposite directions to those indicated in Fig. 197-2.

PRACTICE PROBLEMS

197-1. Write down the matrix equation for the circuit of Fig. 197-3, and then solve for the mesh currents i_1, i_2, i_3.

197-2. Write down the matrix equation for the circuit of Fig. 197-4, and then solve for the mesh currents i_1, i_2, and i_3

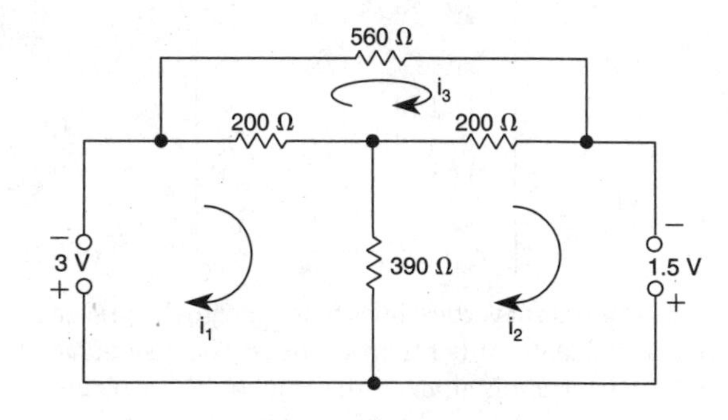

Figure 197-3

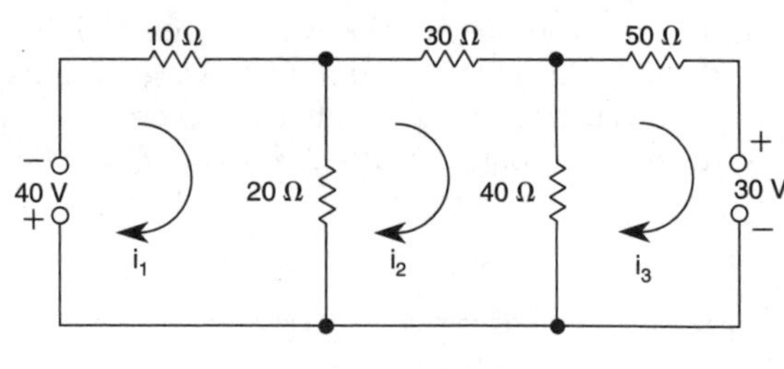

Figure 197-4

198
Quadratic equations

First-degree equations, such as $5x + 3 = 0$ and $2x - 7 = 0$ involve one unknown, and only have one solution. If $5x + 3 = 0, x = -3/5 = -0.6$ and if $2x - 7 = 0, x = 7/2 = 3.5$. Now, let $A = 5x + 3$ and $B = 2x - 7$; you can then consider the possible solutions of the equation $A \times B = 0$.

$$A \times B = 0$$

$$(5x + 3)(2x - 7) = 0$$

$$10x^2 - 35x + 6x - 21 = 0$$

$$10x^2 - 29x - 21 = 0$$

If $A \times B = 0$, either $A = 0$ or $B = 0$. If $A = 0$, $x = -3/5$ while if $B = 0$, $x = 3.5$. Consequently, there are *two* possible solutions to $10x^2 - 29x - 21 = 0$, which is referred to as a *quadratic* or *second-degree equation*. Such an equation always contains a term that involves the square of the single unknown, but there are no higher order terms present (such as x^3, x^4, and so on). The general form of the quadratic equation is:

$$ax^2 + bx + c = 0 \text{ or } x^2 + \frac{bx}{a} + \frac{c}{a} = 0,$$

where a is a positive coefficient. This is a valid assumption because an equation (such as $-3x^2 + 7x - 4 = 0$) can be transformed into $3x^2 - 7x + 4 = 0$.

In this example $a = +10$, $b = -29$, and $c = 21$.

When you test the two solutions,

1. If $x = -3/5$, then $10x^2 - 29x - 21 = 10 \times 9/25 + 29 \times 3/5 - 21 = 0$.
2. If $x = 3\ 1/2$, then $10x^2 - 29x - 21 = 10 \times 49/4 - 29 \times 7/2 - 21 = 0$.

The solutions of a quadratic equation are known as its *roots*. If the roots are designated as α and β, either $x = \alpha$ or $x = \beta$, in which case $x - \alpha = 0$, or $x - \beta = 0$.

In either case,

$$(x - \alpha)(x - \beta) = 0$$

or,

$$x^2 - (\alpha + \beta)\, x + \alpha\beta = 0$$

If you compare this equation with the general form of:

$$x^2 + \frac{bx}{a} + \frac{c}{a} = 0,$$

it follows that,

1. The sum of the roots, $\alpha + \beta = -\, b/a$, which is the coefficient of x with the sign changed.
2. The product of the roots, $\alpha\beta = c/a$, namely the constant term (which is not involved with the unknown).

From this information, the following can be deduced:

1. c positive, b positive. Both roots are negative.
2. c positive, b negative. Both roots are positive.
3. c negative, b positive. Roots are of opposite sign, and the negative root is numerically larger.
4. c negative, b negative. Roots are of opposite sign, and the positive root is numerically larger.
5. If c is zero, one root is zero.
6. If b is zero, the two roots are numerically equal, but opposite in sign.

MATHEMATICAL DERIVATIONS

Methods of solving quadratic equations

Factorization

The products of two first degree functions of x is a quadratic function of x. For example,

$$(2x - 5)(3x + 4) = 2x(3x + 4) - 5(3x + 4)$$
$$= 6x^2 + 8x - 15x - 20$$
$$= 6x^2 - 7x - 20$$

In order to factorize a quadratic equation, it must be expressed so that the above procedure can be worked backwards.

To solve the quadratic equation $8x^2 + 10x + 3 = 0$ by the factorization process, replace $+ 10x$ by two equivalent terms whose product is $8x^2 \times 3 = 24x^2$. Because the product is positive, the terms have the same sign; because their sum is positive, each term is positive. Therefore,

$$8x^2 + 10x + 3 = 0$$

$$8x^2 + 4x + 6x + 3 = 0$$

$$\downarrow \quad \downarrow$$

terms

$$4x(2x + 1) + 3(2x + 1) = 0$$

$$(2x + 1)(4x + 3) = 0$$

Either,

$$2x + 1 = 0, x = -\frac{1}{2}$$

or,

$$4x + 3 = 0, x = -\frac{3}{4}$$

To solve the quadratic equation $x^2 - 22x + 96 = 0$, replace $-22x$ by two equivalent terms whose product is $x^2 \times 96 = 96x^2$. Because the terms have the same sign and each term is negative, the terms are $-6x$ and $-16x$. Then,

$$x^2 - 22x + 96 = 0$$

$$x^2 - 6x - 16x + 96 = 0$$

$$x(x - 6) - 16(x - 6) = 0$$

$$(x - 16)(x - 6) = 0$$

Either,

$$x - 16 = 0, x = 16$$

or,

$$x - 6 = 0, x = 6.$$

The quadratic equation $6x^2 - 11x - 10 = 0$ can be factorized by replacing $-11x$ with two equivalent terms whose product is $6x^2 \times (-10) = -60x^2$. Because the product is negative, the terms have opposite signs and because their sum is negative, the numerically larger term is also negative.

From inspection, the terms are $4x$ and $-15x$,

$$6x^2 - 11x - 10 = 0$$

$$6x^2 + 4x - 15x - 10 = 0$$

$$2x(3x + 2) - 5(3x + 2) = 0$$

$$(2x - 5)(3x + 2) = 0$$

Either,

$$2x - 5 = 0, x = 2.5$$

or,

$$3x + 2 = 0, x = -\frac{2}{3} = -0.67, \text{ rounded off.}$$

The quadratic equation $12x^2 - x - 35 = 0$ can be factorized by replacing $-x$ with two equivalent terms whose product is $12x^2 \times (-35) = -420x^2$. Because the product is negative, the terms have opposite signs and because their sum is positive, the numerically larger term is also positive.

From inspection, the terms are $+21x$ and $-20x$,

$$12x^2 - x - 35 = 0$$

$$12x^2 - 20x + 21x - 35 = 0$$

$$4x(3x - 5) + 7(3x - 5) = 0$$

$$(4x + 7)(3x - 5) = 0$$

Either,

$$4x + 7 = 0, x = -7/4$$

or,

$$3x - 5 = 0, x = +5/3$$

Completing the square

As a first step, attempt to solve a quadratic equation by factorization. However, if no simple factors can be found, you may turn to the method known as "completing the square."

An expression, such as $(x + a)^2 = x^2 + 2ax + a^2$, is referred to as a "perfect square." In order to make "$x^2 + 2ax$" into a perfect square, you must add a^2. The procedure involves dividing the coefficient of x (namely $2a$) by 2 and then squaring the result. As an example, you would complete the square for $x^2 - 5x$ by adding $(-5/2)^2 = 25/4$ so that $x^2 - 5x + 25/4 = (x - 5/2)^2$.

The quadratic equation $2x^2 - 9x + 5 = 0$ has no simple factors. The solution is achieved by "completing the square," and is as follows:

$$2x^2 - 9x + 5 = 0$$

$$x^2 - \frac{9x}{2} + \frac{5}{2} = 0$$

$$x^2 - \frac{9x}{2} = -\frac{5}{2}$$

To "complete the square," add

$$\left(\frac{-9}{2 \times 2}\right)^2 = \frac{81}{16}$$

to both sides of the equation. Then,

$$x^2 - \frac{9x}{2} + \frac{81}{16} = \frac{81}{16} - \frac{5}{2} = \frac{41}{16}$$

$$\left(x - \frac{9}{4}\right)^2 = \frac{41}{16}$$

Taking the square root of both sides of the equation,

$$x - \frac{9}{4} = \pm\frac{\sqrt{41}}{4}$$

$$x = \frac{\pm\sqrt{41} + 9}{4}$$

Either,

$$x = \frac{+\sqrt{41} + 9}{4} = 3.85$$

or,

$$x = \frac{-\sqrt{41} + 9}{4} = 0.65$$

Formula

The method of "completing the square" is used to find the formula for solving the general equation, $ax^2 + bx + c = 0$. The solution is as follows:

$$ax^2 + bx + c = 0$$

$$x^2 + \frac{bx}{a} = -\frac{c}{a}$$

In order to "complete the square" on the left side of the equation, the term

$$\left(\frac{b}{2a}\right)^2 = \frac{b^2}{4a^2}$$

must be added to both sides. Then,

$$x^2 + \frac{bx}{a} + \frac{b^2}{4a^2} = \frac{b^2}{4a^2} - \frac{c}{a} = \frac{b^2 - 4ac}{4a^2}$$

$$\left(x + \frac{b}{2a}\right)^2 = \frac{b^2 - 4ac}{4a^2}$$

Taking the square root of both sides of the equation,

$$x + \frac{b}{2a} = \frac{\pm\sqrt{b^2 - 4ac}}{2a}$$

Therefore,

$$x = \frac{\pm\sqrt{b^2 - 4ac} - b}{2a} \qquad (198\text{-}1)$$

This is the formula for solving the general quadratic equation.

The quantity $b^2 - 4ac$ is called the *discriminant* of the equation. The value of the discriminant falls into one of the three following categories:

1. If $b^2 > 4ac$, the discriminant is positive and the two roots are numerically unequal.

   As an example, if $3x^2 + 8x + 2 = 0$,

$$a = 3, b = 8, c = 2$$

$$x = \frac{\pm\sqrt{64 - 4 \times 3 \times 2} - 8}{2 \times 3}$$

$$= \frac{\pm\sqrt{40} - 8}{6}$$

$$= \frac{\pm\sqrt{10} - 4}{3}$$

Either,

$$x = \frac{+\sqrt{10} - 4}{3} = -0.28$$

or,

$$x = \frac{-\sqrt{10} - 4}{3} = -2.39$$

2. If $b^2 = 4ac$, the value of the discriminant is zero, so that the two roots are identical, and you therefore have a repeated root. For the quadratic equation: $4x^2 - 12x + 9 = 0$, $a = 4$, $b = -12$, and $c = 9$ so that b^2 (144) $= 4ac$ ($4 \times 4 \times 9 = 144$).

   Then,

$$x = \frac{\pm\sqrt{0} - (-12)}{2 \times 4} = 1.5$$

   This result is verified by factoring the equation into $(2x - 3)^2 = 0$.

3. If $b^2 < 4ac$, the discriminant is negative; therefore, the square root of the discriminant cannot be calculated as an ordinary ("real") number. However, if you invent a quantity (j), which is defined by $j^2 = -1$, the roots of the equation can be expressed in terms of j. For example, if $x^2 + 4 = 0$, $x = \pm\sqrt{-4} = \pm\sqrt{j^2 4} = \pm j2$. The full meaning of j is explained in chapter 89.

Consider the quadratic equation:

$$x^2 + x + 1 = 0$$

in which $a = 1, b = 1, c = 1$.

Then,

$$x = \frac{\pm\sqrt{1^2 - 4 \times 1 \times 1} - 1}{2}$$

$$= \frac{\pm\sqrt{-3} - 1}{2}$$

$$= \frac{\pm j\sqrt{3} - 1}{2}$$

Either,

$$x = \frac{+j\sqrt{3} - 1}{2} = -0.5 + j\,0.866$$

or,

$$x = \frac{-j\sqrt{3} - 1}{2} = -0.5 - j\,0.866$$

It is interesting to note that each of these solutions is a cube root of 1; because if:

$$x^3 = 1$$

$$(x - 1)(x^2 + x + 1) = 0$$

Either,

$x - 1 = 0, x = 1$ (the obvious cube root of 1)

or,

$$x^2 + x + 1 = 0$$

Verification:

$$\text{If } x = +j\frac{\sqrt{3}}{2} - \frac{1}{2}, x^2 = \left(j\frac{\sqrt{3}}{2} - \frac{1}{2}\right)^2$$

$$= -\frac{3}{4} - j\frac{\sqrt{3}}{2} + \frac{1}{4} = -\frac{1}{2} - j\frac{\sqrt{3}}{2}$$

$$\text{and } x^3 = \left(-\frac{1}{2} - j\frac{\sqrt{3}}{2}\right)\left(+j\frac{\sqrt{3}}{2} - \frac{1}{2}\right)$$

$$= \frac{1}{4} + \frac{3}{4} = 1$$

Example 198-1

Use the factorization method to solve the quadratic equation $12x^2 + 19x - 21 = 0$.

Solution

Replace $+\ 19x$ with two equivalent terms, whose product is $12x^2 \times (-21) = -252x^2$. Inspection reveals that the two equivalent terms are $-\ 9x$ and $28x$. Therefore,

$$12x^2 + 19x - 21 = 0$$

$$12x^2 - 9x + 28x - 21 = 0$$

$$3x(4x - 3) + 7(4x - 3) = 0$$

$$(3x + 7)(4x - 3) = 0$$

Either,

$$3x + 7 = 0, x = -7/3$$

or,

$$4x - 3 = 0, x = 3/4$$

Example 198-2

Use the "completing the square" method to solve the quadratic equation $4x^2 - 9x - 7 = 0$.

Solution

$$4x^2 - 9x - 7 = 0$$

$$x^2 - \frac{9x}{4} = \frac{7}{4}$$

$$x^2 - \frac{9x}{4} + \left(\frac{9}{8}\right)^2 = \frac{7}{4} + \left(\frac{9}{8}\right)^2$$

$$\left(x - \frac{9}{8}\right)^2 = \frac{193}{64}$$

$$x - \frac{9}{8} = \frac{\pm\sqrt{193}}{8}$$

Either,

$$x = \frac{+\sqrt{193} + 9}{8} = \frac{13.9 + 9}{8} = 2.85$$

or,

$$x = \frac{-\sqrt{193} + 9}{8} = -0.61$$

Example 198-3

Use the formula method to solve the quadratic equation $-10x^2 + 16x - 4 = 0$.

Solution

Factoring the equation leads to,

$$-2(5x^2 - 8x + 2) = 0$$

Therefore,

$$5x^2 - 8x + 2 = 0 \qquad (198\text{-}1)$$

$$a = +5, b = -8, c = +2$$

$$x = \frac{\pm\sqrt{(-8)^2 - 4 \times 5 \times 2} - (-8)}{2 \times 5}$$

$$= \frac{\pm\sqrt{64 - 40} + 8}{10}$$

Either,

$$x = \frac{+\sqrt{24} + 8}{10} = +1.29$$

or,

$$x = \frac{-\sqrt{24} + 8}{10} = +0.31$$

Notice that the graphical method for solving a quadratic equation appears in chapter 199.

PRACTICE PROBLEMS

198-1. Use the factorization method to obtain the roots of the quadratic equation, $6x^2 - 19x - 77 = 0$.

198-2. Solve the quadratic equation, $4\pi^2 f^2 LC + 2\pi fCR - 1 = 0$, for the variable, f.

198-3. Solve the quadratic equation, $5x^2 - 17x + 7 = 0$, by the method of "completing the square."

198-4. Use the formula method to solve the quadratic equation, $2x^2 - 5x + 6 = 0$.

199
Graphs

On many occasions in electronics, the relationship between two variables (such as voltage, current, power, and so on) is illustrated by means of a graph. From the shape of the graph, you can often predict the type of equation that relates the two variables. Conversely, if you know the equation, you have the ability to visualize the appearance of the graph.

By drawing two axes, the position of any point in a plane may be fixed by its two *coordinates*. These two axes are known as the x-axis (denoted by $0x$) and the y-axis (denoted by $0y$). The two axes intersect at the point (0), which is referred to as the origin.

In Fig. 199-1, *PM* is perpendicular (at right angles) to $0x$; then the "x-coordinate" of the point (*P*) is $OM = X$. Similarly, *PN* is perpendicular to $0y$ and the "y-coordinate" of the point (*P*) is $ON = Y$. The point *P* is then identified as point (*X*,*Y*). Because *ONPM* is a rectangle, point *P* is determined by a system of *rectangular coordinates*. Notice that the origin (*O*) is point (0,0).

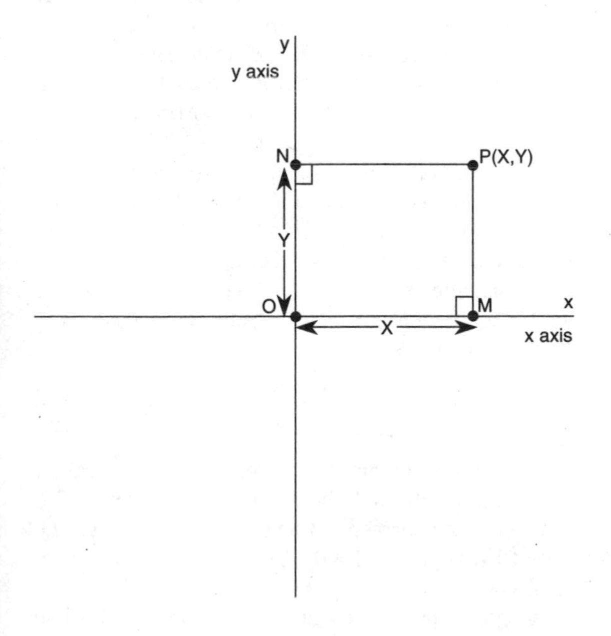

Figure 199-1

If point (*P*) lies to the left of $0y$, the *X* coordinate is considered to be negative. Likewise, *Y* is negative if *P* is below $0x$. The scales used for the two axes can be the same, but in general, they are different. Figure 199-2 uses different scales and illustrates the coordinates of four points, with one point in each quadrant.

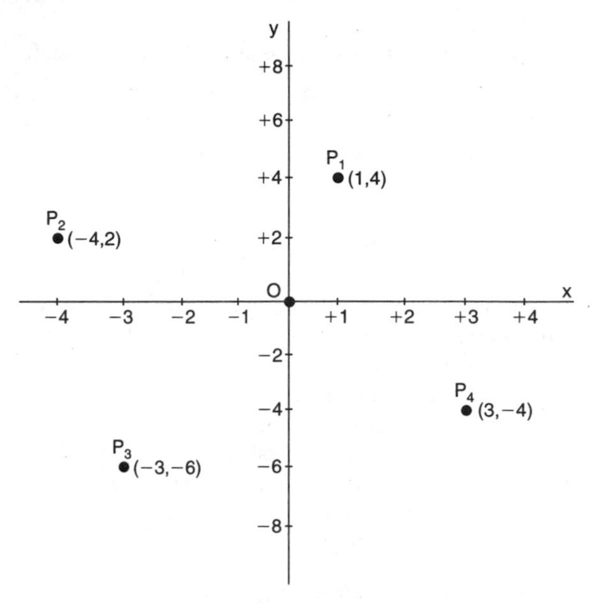

Figure 199-2

If the values of x and y for point *P* are unrestricted, *P* can be anywhere in the plane. When, however, the values are restricted by stating some definite relationship between x and y, *P* can only lie in certain positions. If these positions are joined together, the result is the graph of y versus x.

For a normal relationship (in the form of an equation) between x and y, all of the possible positions of *P* will lie on a curve such that a curve (graph) of some sort exists for every equation. There are two types of equations:

1. An *explicit* equation, such as

$$y = \frac{3x^3 - 7}{1 + x},$$

where y is directly given as a function of x.

2. An *implicit* equation, such as

$$3x^2y - 4x + 7y^2 + 5 = 0,$$

in which y is not expressed directly in terms of x.

It is common to regard x as the controlled variable, and y as the dependent variable; in other words, the value of y depends on the value of x.

MATHEMATICAL DERIVATIONS

The following are examples of graphs and their related equations.

The straight line

A *first-degree* equation (of the form $ax + by + c = 0$) always corresponds to a straight line graph. For this reason, $ax + by + c = 0$ is also referred to as a *linear equation*.

If $ax + by + c = 0$, which is an implicit relationship,

$$y = \frac{-ax}{b} + \frac{-c}{a}$$

or,

$$y = mx + k$$

where $m = -a/b$ and $k = -c/a$

The equation is now in its explicit form. If x changes in value by 1 unit, there is a change of m units in the value of y. Therefore, m is equal to the slope of the straight-line graph.

If m is positive, the line has a positive (or "/") slope; however, if m is negative, the slope is "\".

If $x = 0$, $y = k$, and the straight line must cut the y axis at the point $(0,k)$; if $y = 0$, $x = -k/m$, and the line cuts the x axis at the point $(-k/m,0)$. The straight line graph is then drawn to pass through these two points.

Examples of straight line graphs

1. $y = 3x + 6$. The line has a positive slope of 3 and it passes through points (0,6) and −2,0).
2. $2y = -4x + 5$. The line has a negative slope of $-4/2 = -2$ and passes through points (0,5/2) and (5/4,0).
3. $y = 3$ is a horizontal line, parallel to the x axis, and it passes through point (0,3).
4. $x = -3$ is a vertical line, parallel to the y axis, and it passes through point (−3,0).
5. The lines $x = 0$ and $y = 0$ are, respectively, the y and x axes.

   All these straight line graphs are illustrated in Fig. 199-3.

In the equations of electronics, various letter symbols replace y and x. Consider the following examples:

1. Ohm's Law equation ($I = E/R$) where R is a constant (chapter 5); in this equation, I (current) and E (voltage), respectively, replace y and x. When $E = 0$ and $I = 0$, the graph of I versus E is a straight line that passes through the origin (0,0) and has a positive slope of $1/R$. This straight line graph indicates that I and E are directly proportional (Fig. 199-4A).
2. Cell's terminal voltage ($V_L = E - I_L R_i$), where E and R_i are constant (chapter 22). The graph of V_L versus I_L is a straight line that passes through the points $(0,E)$, $(E/R_i,0)$ and has a negative slope of $-R_i$ (Fig. 199-4B).

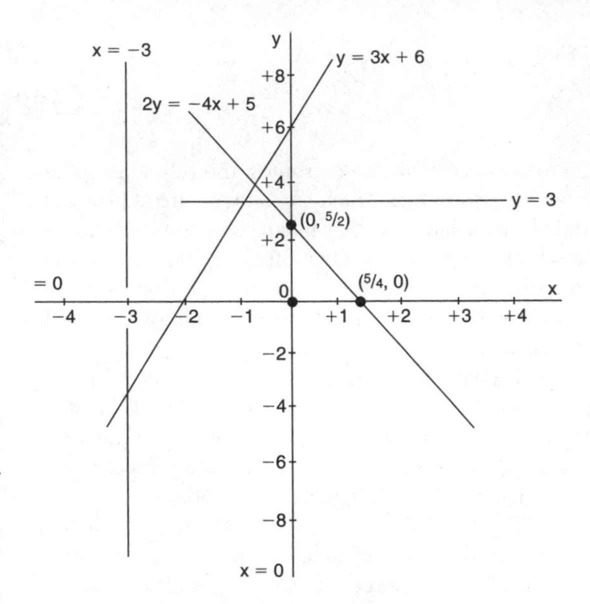

Figure 199-3

3. Inductive reactance ($X_L = 2\pi fL$) (chapter 69), where π and L are constants. The graph of X_L versus f is a straight line, passing through the origin and with a slope of $2\pi L$ (Fig. 199-4C). This straight-line relationship indicates that the values of X_L and f are directly proportional.
4. Resistance, $R_{T°C} = R_{20°C}\ [1 + \alpha\ (T°C - 20°C)]$ (chapter 7) where $R_{20°C}$ and α are constants. The graph of $R_{T°C}$ versus T°C (Fig. 199-4D) passes through the points $(20°C, R_{20°C})$, $(-1/\alpha + 20°C, 0)$, and has a positive slope of $\alpha R_{20°C}$.

Straight-line graphs can also be used to solve linear simultaneous equations. As an example, consider the following equations:

$$x + 2y - 8 = 0 \quad \text{(A)}$$

$$4x - 3y + 1 = 0 \quad \text{(B)}$$

For Equation A, the graph of y versus x is the straight line that passes through the points (0,4) and (8,0). For Equation B, the corresponding line passes through the points (0,1/3) and (−1/4,0). These graphs are shown in Fig. 199-5.

The point that represents the solution of the two equations must satisfy each equation. This point must lie on each line and is, therefore, the point of intersection, in this case (2,3). Hence, the solution is $x = 2$, $y = 3$.

The circle

Consider the second-degree equation $x^2 + y^2 = a^2$, which can be rewritten as $\sqrt{x^2 + y^2} = a$. Referring to Fig. 199-6A, $ON = X$ and $PN = Y$; by Pythagoras' Theo-

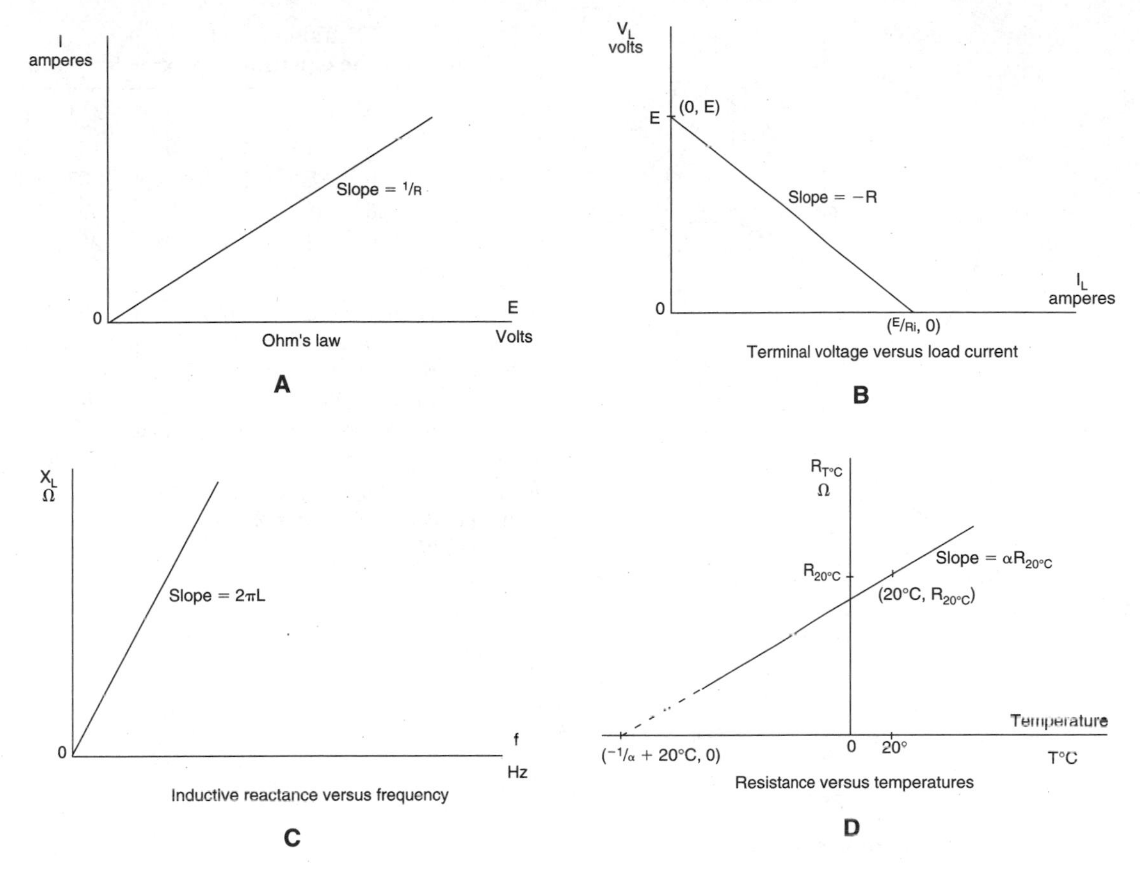

Figure 199-4

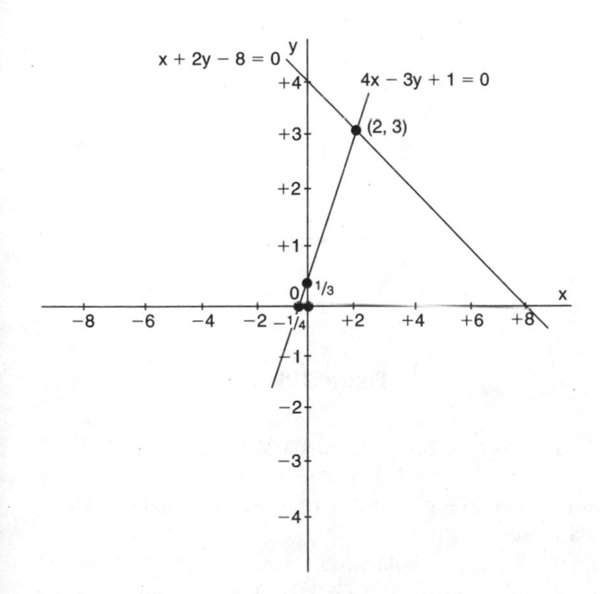

Figure 199-5

rem (chapter 201), $OP = \sqrt{X^2 + Y^2} = a$. Therefore, P is any point that is at a distance (a) from the origin, O. Consequently, the graph of equation $x^2 + y^2 = a^2$ is a circle, whose radius is a and whose center is the origin.

For the second-degree equation $(x - 2)^2 + (y - 1)^2 = 4 = 2^2$, the graph is a circle whose center is the point (2,1) and whose radius is 2 (Fig. 199-6B).

In communications, the circle is the horizontal radiation pattern emanating from a vertical antenna (chapter 161).

The parabola

The *parabola* is another type of curve whose graph is associated with a second-degree equation. Geometrically, the parabola is defined in terms of a point (the *focus*) and a line (the *directrix*). In Fig. 199-7, point P is given by $PF = PD$. If all possible positions of P are joined, the result is a parabola whose axis of symmetry is horizontal.

The explicit equation $y = ax^2 + bx + c$ has a graph which is a parabola with a vertical axis. As an example,

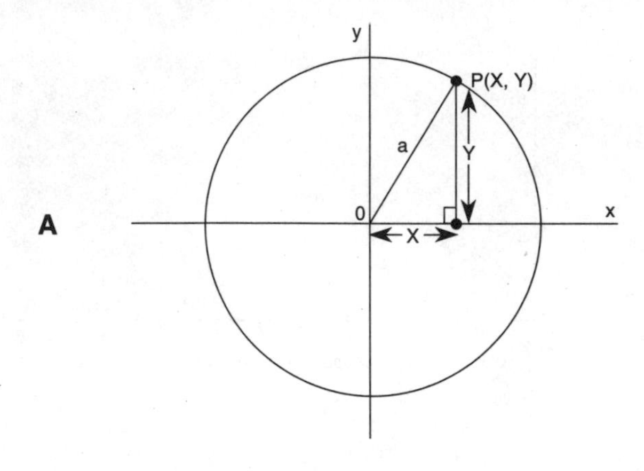

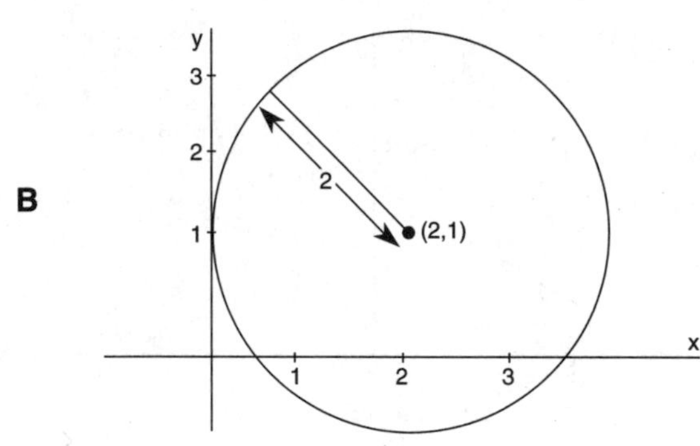

Figure 199-6

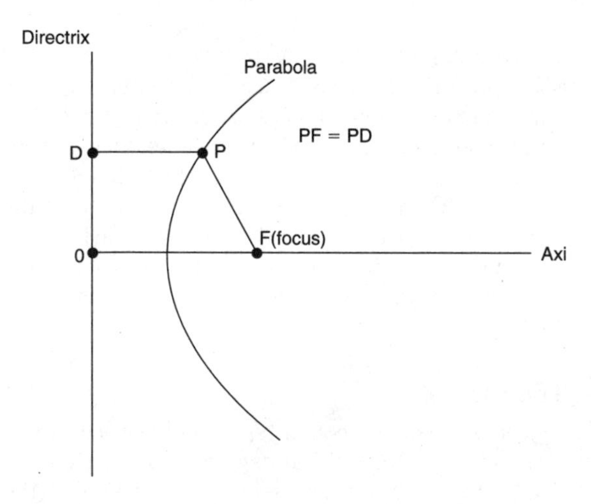

Figure 199-7

consider the equation $y = x^2 - 5x + 6$ and plot the graph over the range $x = -1$ to $x = 5$. Start by preparing the following (Table 199-1):

Table 199-1.
Values for the equation, $y = x^2 - 5x + 6$.

x	-1	0	1	2	3	4	5
x^2	1	0	1	4	9	16	25
$-5x$	5	0	-5	-10	-15	-20	-25
6	6	6	6	6	6	6	6
y	12	6	2	0	0	2	6

The graph of y versus x is shown in Fig. 199-8. This parabola cuts the x axis ($y = 0$) at the points (2,0) and (3,0), where $y = 0$ and $y = x^2 - 5x + 6$ are simultaneously satisfied. Therefore, these points must provide the solutions to the quadratic equation $x^2 - 5x + 6 = 0$, namely $x = 2$ or $x = 3$ (chapter 198). Because of the curve's symmetry, the minimum point (M) of the parabola occurs when $x = 2.5$ and $y = 2.5^2 - 5 \times 2.5 + 6 = -0.25$.

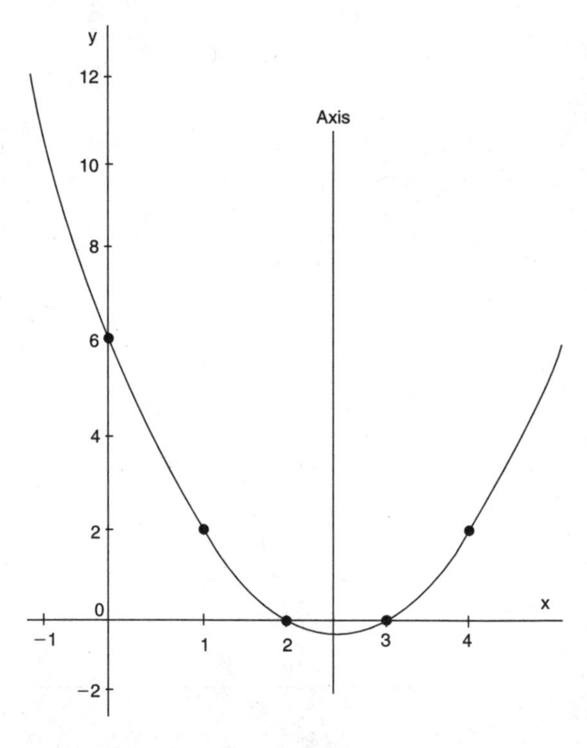

Figure 199-8

For comparison purposes, the graphs of $y = x^2 + x - 6$, $y = x^2 - 2x + 1$, $y = x^2$, $y^2 = x$, $y = x^2 + x + 1$ are shown in Fig. 199-9. To comment on each of these parabolas:

1. The 'A' parabola $y = x^2 + x - 6$ cuts the x axis at points (2,0) and (−3,0). The solutions to quadratic equation $x^2 + x - 6 = 0$ are either $x = 2$

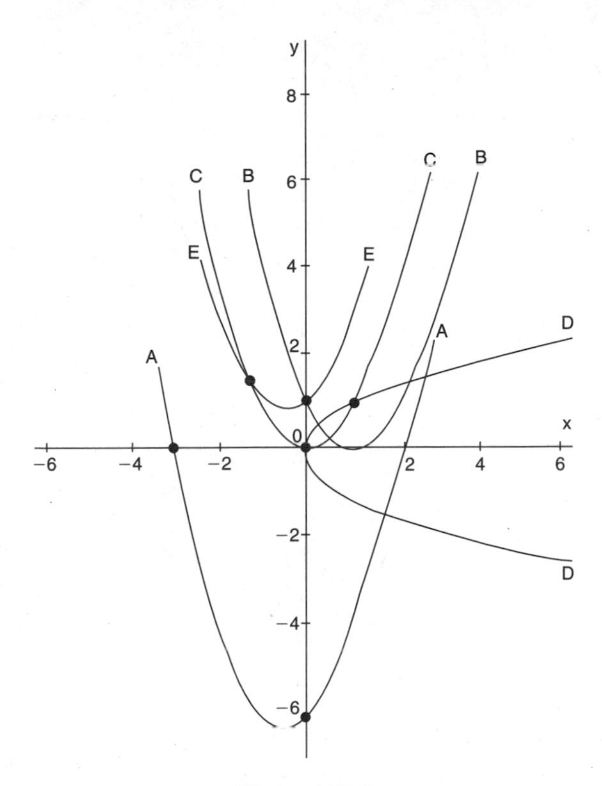

Figure 199-9

or $x = -3$; therefore, the roots have opposite signs. The minimum point occurs when $x = -1/2$ and $y = (-1/2)^2 + (-1/2) - 6 = -6\ 1/4$.

Notice that the curve for equation $y = -x^2 - x + 6$ would mean interchanging $-y$ for $+y$, so that the parabola would be "upside down;" to be more precise, the two parabolas would be the mirror images of each other in the line of the x axis.

2. The 'B' parabola $y = x^2 - 2x + 1$ touches the x axis at point (1,1); in other words, the line $y = 0$ is a tangent to the parabola at this point. The solution to the quadratic equation $x^2 - 2x + 1 = (x - 1)^2 = 0$, is a repeated root of $x = 1$.
3. The 'C' parabola $y = x^2$ touches the x axis at the origin. Line $x = 0$ is the vertical axis of the parabola. The solution to the quadratic equation $x^2 = 0$ is a repeated root of $x = 0$.
4. By interchanging x and y in the equation $y = x^2$, you obtain another parabola, $x = y^2$. This new 'D' parabola has a horizontal axis, which is the line $y = 0$. At the origin, the y axis is a tangent to the parabola.
5. The 'E' parabola $y = x^2 + x + 1$ does not cut or touch the x axis. Consequently, there are no real roots for quadratic equation $x^2 + x + 1 = 0$ (chapter 198).

When a parabola is revolved around its axis, the practical application of this surface of revolution is the paraboloidal dish reflector, which is used for some microwave antennas (chapter 178).

The rectangular hyperbola

Many curves approach infinity along a straight line, which is known as an *asymptote*. Notice that a curve that goes to infinity along an asymptote must also come from infinity along the same line.

Of particular interest in electronics is the rectangular *hyperbola*, which has two asymptotes. One equation whose graph is a rectangular hyperbola, is $xy = c^2$ which may be rewritten as $y = c^2/x$, or $x = c^2/y$; these equations indicate that x and y are inversely proportional. If $x = 0$ and y is infinite, the line $x = 0$ (the y axis) is an asymptote. Similarly, if $y = 0$ and x is infinite, the x axis is a second asymptote. Because the two asymptotes are at right angles, the curve is called a *rectangular* hyperbola (as opposed to other hyperbolas whose asymptotes are inclined, but are not 90° apart). The graph for $xy = c^2$ is shown in Fig. 199-10A.

You will recall that the graph for equation $x^2 + y^2 = a^2$ is a circle. However, the graph for equation $x^2 - y^2 = a^2$ is another rectangular hyperbola whose asymptotes are $y = x$ and $y = -x$ (Fig. 199-10B).

In electronics, the equation for the capacitive reactance is

$$X_C = \frac{1}{2\pi fC},$$

where π and C are constants. Therefore, X_C is inversely proportional to f and the graph of X_C versus f is a rectangular hyperbola (Fig. 199-10C).

To summarize an important point: a graph that is a straight line through the origin indicates that there is a relationship of direct proportion between the two variables. By contrast, a rectangular hyperbola of the form $xy = c^2$ means that the two variables are inversely proportional.

The ellipse

A curve that is occasionally encountered in electronics and communications is the *ellipse*. For example, certain specialized waveguides have an elliptical cross-section and some amplifiers use load "lines" that are, in fact, ellipses. When compared with the circle ($x^2 + y^2 = a^2$) and the rectangular hyperbola ($x^2 - y^2 = a^2$), the corresponding equation for the ellipse is $x^2/a^2 + y^2/b^2 = 1$ (Fig. 199-11). The ellipse has two foci (F_1 and F_2) and for any point (P) on the ellipse: $PF_1 + PF_2 = 2a$, where a is the length of the semi-major axis and the length of the semi-minor axis is b. The ellipse

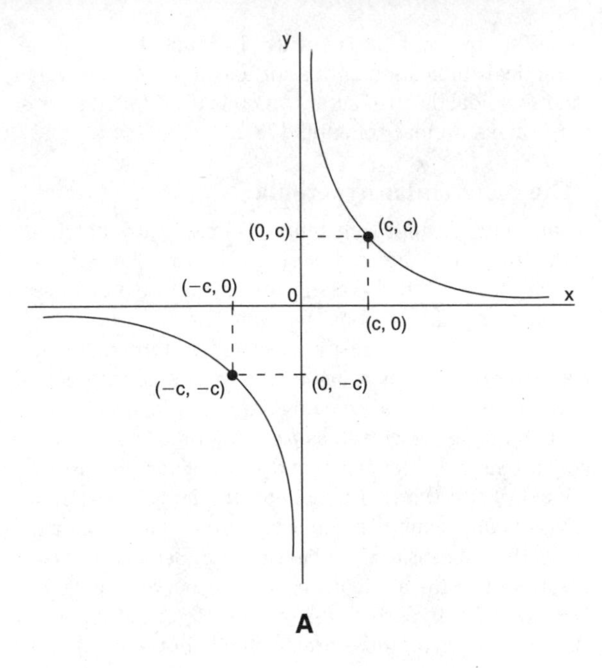

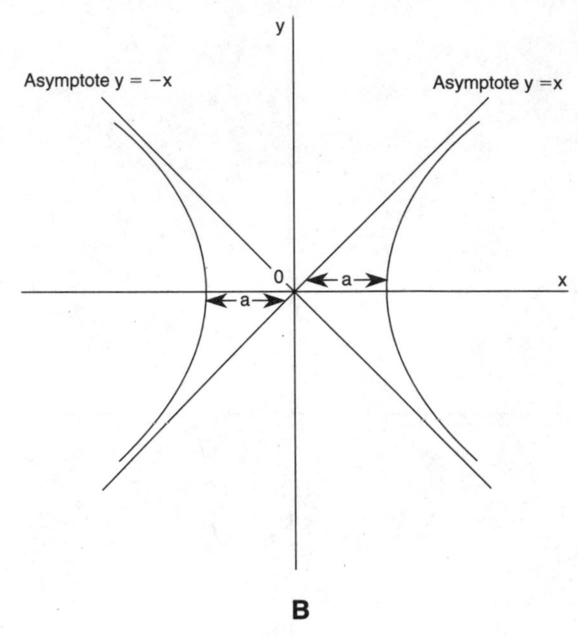

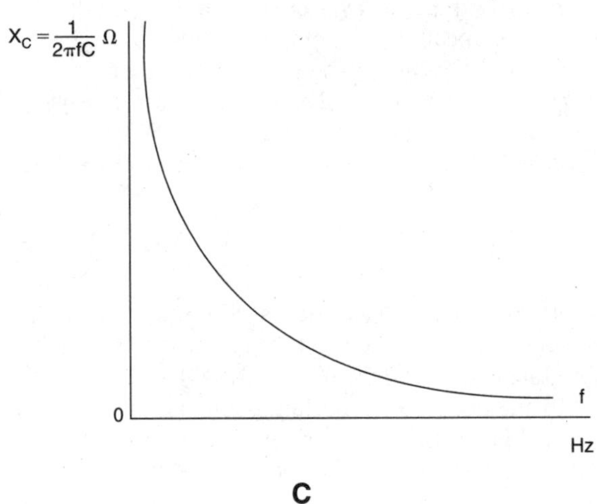

Figure 199-10

passes through points $(a,0)$, $(-a,0)$, $(0,b)$ and $(0,-b)$ and the coordinates of the foci are $F_1\ (+\sqrt{a^2 - b^2},0)$ and $F_2\ (-\sqrt{a^2 - b^2},0)$.

Polar coordinates

So far, you have only covered the use of rectangular (sometimes called *Cartesian*) coordinates. However, there is another system of coordinates that can determine the position of a point in a plane. In Fig. 199-12, there is an origin (O) and a horizontal reference line. The position of the point (P) can then be fixed by the length (r) of the line OP and the angle (θ) between OP and the horizontal reference line. This is the *polar system* of coordinates in which the point (P) is identified as (r,θ). As an example, the polar coordinates of P in Fig. 199-13, are (2,30°), which corresponds to rectangular coordinates of (1.732,1). Therefore, you can convert from polar to rectangular coordinates and vice-versa.

Polar to rectangular conversion

The equations are:

$$x = r\cos\theta$$

$$y = r\sin\theta$$

$$\frac{y}{x} = \tan\theta$$

The meanings of the trigonometric functions can be found in chapters 202 through 208.

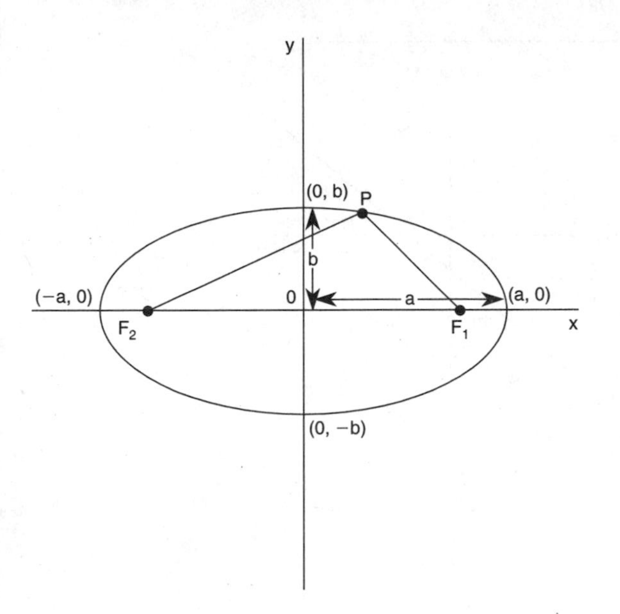

Figure 199-11

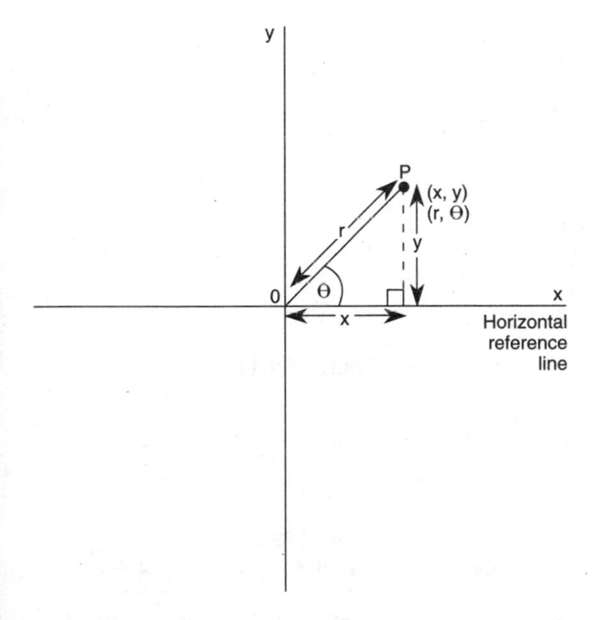

Figure 199-12

Rectangular to polar conversion

$$r = +\sqrt{x^2 + y^2}$$

$$\theta = \text{inv. tan } \frac{y}{x}$$

Examples of graphs with their corresponding polar equations are shown in Figs. 199-14A, B, C.

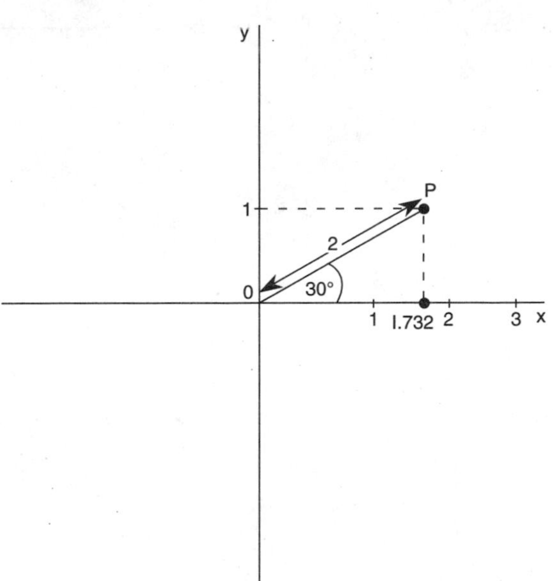

Figure 199-13

In Fig. 199-14A, the polar equation is $r(2 \cos\theta - 3 \sin\theta) = 5$ corresponds to:

$$2r \cos \theta - 3r \sin\theta = 5$$

$$2x - 3y = 5$$

The polar equation $r = 2 \sec\theta$ can be rewritten as $r \cos\theta = 2$, or $x = 2$. The polar equation $r = 3 \csc\theta$ corresponds to $r \sin\theta = 3$, or $y = 3$.

In Fig. 199-14B, polar equation $r = a$ can be converted to $\sqrt{x^2 + y^2} = a$ or $x^2 + y^2 = a^2$, which is the rectangular equation of a circle.

In Fig. 199-14C, the polar equation $r = 4 \cos\theta$ can be converted to:

$$\sqrt{x^2 + y^2} = \frac{4x}{r} = \frac{4x}{\sqrt{x^2 + y^2}}$$

$$x^2 + y^2 = 4x$$

$$(x^2 - 2)^2 + y^2 = 4$$

This is the equation of a circle that has point (2,0) as its center and it has a radius of 2.

In Fig. 199-4D, polar equation $r = 4 \cot\theta \csc\theta$ can be converted to:

$$r^2\sin^2\theta = 4r \cos\theta$$

$$y^2 = 4x$$

This is the equation of a parabola that passes through the origin, and with the line $y = 0$ as its axis.

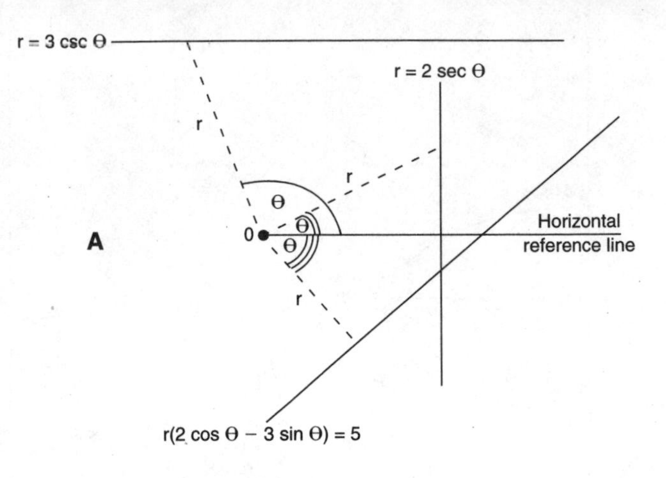

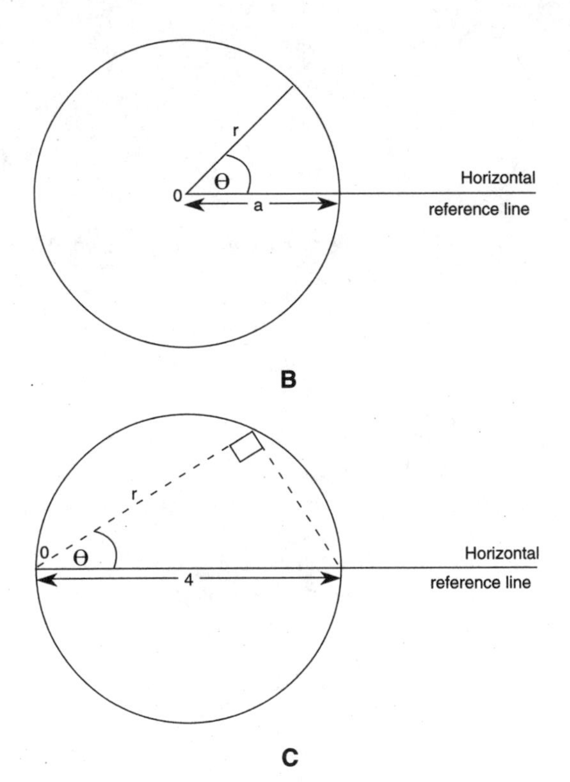

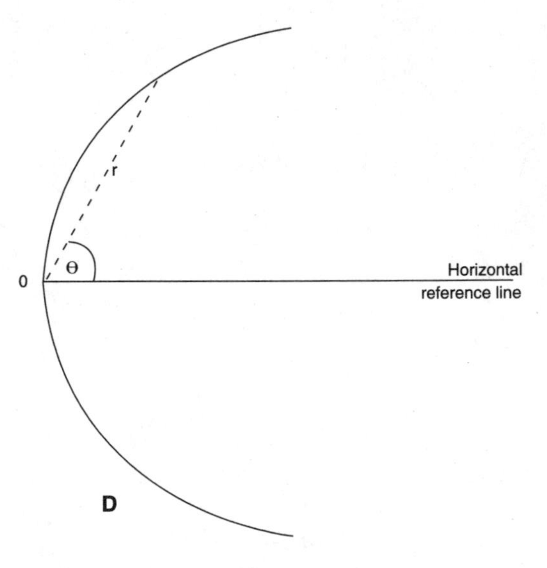

Figure 199-14

Example 199-1

Plot graph y versus x for $y = x^2 + 7x + 7$ over the range $x = 0$ to $x = -5$. Use the curve to solve quadratic equation $x^2 + 5x + 4 = 0$.

Solution

Corresponding values of x and y are shown in Table 199-2. In Fig. 199-15, draw the line for $y = 2x + 3$; this line passes through points (−1 1/2,0) and (0,3). The points where the line $y = 2x + 3$ cuts the parabola $y = x^2 + 7x + 7$, are the solutions to the quadratic equation $x^2 + 5x + 4 = 0$

Table 199-2.
Values for the equation, $y = x^2 - 7x + 7$.

x	0	−1	−2	−3	−4	−5
x^2	0	1	4	9	16	25
$-7x$	0	−7	−14	−21	−28	−35
7	7	7	7	7	7	7
y	7	1	−3	−5	−5	−3

Example 199-2

Plot the graph of r versus θ for the equation $r = 2(1 + \cos\theta)$.

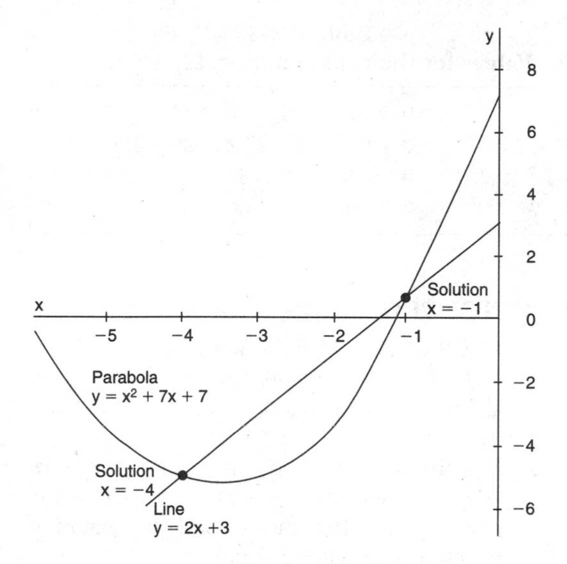

Figure 199-15

Solution

Corresponding values of r and θ are shown in Table 199-3. The graph shown in Fig. 199-16 is the *cardioid* or "heart-shaped" curve. In communications, the cardioid is important when distinguishing between the true and reciprocal bearings of a transmitter. This occurs when using a loop antenna for direction-finding purposes (chapter 176).

Example 199-3

In the equation,

$$I = \frac{E}{Z} = \frac{E}{\sqrt{R^2 + \left(2\pi f L - \dfrac{1}{2\pi f C}\right)^2}},$$

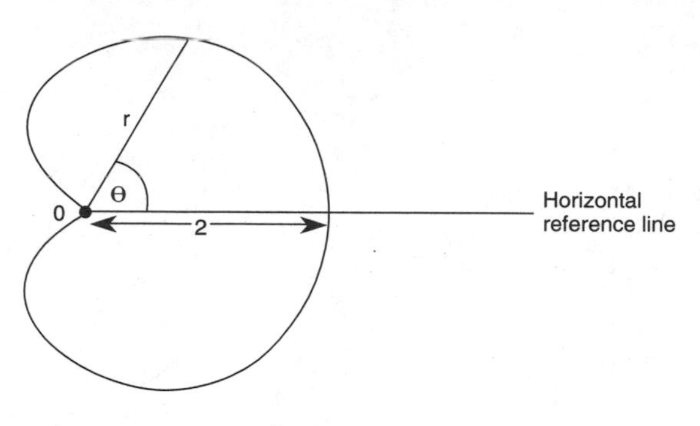

Figure 199-16

where: $E = 1$ V, $R = 100\ \Omega$, $L = 100\ \mu$H, $C = 250$ pF. Plot the graph of I versus f for the values of $f = 800$, 850, 900, 1000, 1050, 1100, 1150, and 1200 kHz.

Solution

Corresponding values of I and f are shown in Table 199-4. The graph is shown in Fig. 199-17 and has a bell-shaped appearance. It is the response curve of a series L, C, R circuit (chapter 75).

Example 199-4

In $P_L = EI_L - I_L^2R_i$ (chapter 22), $E = 24$ V and $R_i = 4\ \Omega$. Plot the graph of P_L versus I_L for the values of $I_L = 0, 1, 2, 3, 4, 5,$ and 6 A.

Solution

Corresponding values of P_L and I_L are shown in Table 199-5. $P_L = EI_L - I_L^2R_i$ can be compared with $y = ax^2 + bx + c$, where $y = P_L$, $x = I_L$, $a = -R_i$, $b = E$, and $c = 0$. Consequently, the graph is a parabola, as shown in Fig. 199-18. Notice that the parabola has its maximum value for P_L (36 W) when $I_L = 3$ A.

Table 199-3. Values for the equation, $r = 2(1 + \cos\theta)$.

θ	0°	60°	90°	120°	180°	240°	270°	300°	360°
$\cos\theta$	1	0.5	0	−0.5	−1	−0.5	0	0.5	1
$1 + \cos\theta$	2	1.5	1	0.5	0	0.5	1	1.5	2
r	4	3	2	1	0	1	2	3	4

Table 199-4. Values for the equation, $I = E\Big/\sqrt{R^2 + \left(2\pi f L - \dfrac{1}{2\pi f C}\right)^2}$.

f kHz	800	850	900	950	1000	1050	1100	1150	1200
$2\pi fL\ \Omega$	503	534	565	597	628	660	691	723	754
$1/(2\pi fC)\ \Omega$	796	749	707	670	637	606	579	554	531
$Z\ \Omega$	310	237	174	124	100	114	150	196	244
I mA	3.22	4.22	5.75	8.06	10.0	8.77	6.67	5.10	4.10

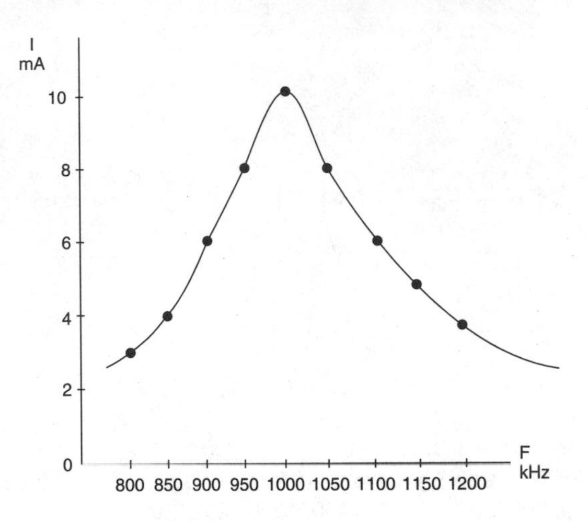

Figure 199-17

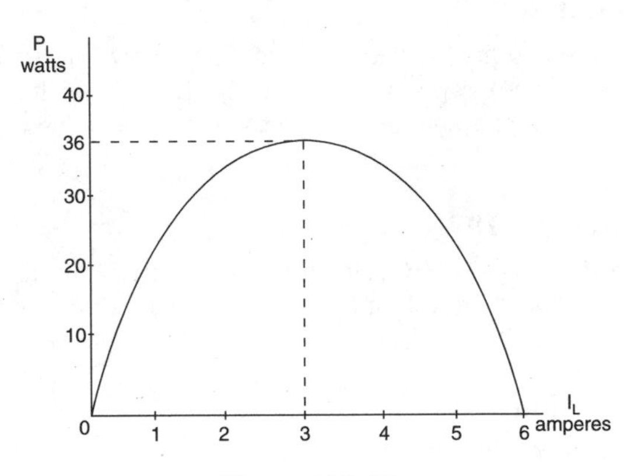

Figure 199-18

Table 199-5.
Values for the equation, $P_L = EI_L - I_L^2R_i$.

I_L A	0	1	2	3	4	5	6
$E \times I_L$ W	0	24	48	72	96	120	144
$-I_L^2 \times R_i$ W	0	–4	–16	–36	–64	–100	–144
P_L W	0	20	32	36	32	20	0

PRACTICE PROBLEMS

199-1. Plot the graphs for the lines $3x - y - 7 = 0$ and $-2x + 5y + 9 = 0$. Graphically solve the simultaneous equations, $3x - y = 7$, and $2x - 5y = 9$.

199-2. Plot the graph for the parabola $y = 2x^2 - 5x - 7$ over the range $x = -2$ to $x = +4$. Use this parabola to obtain the roots of the quadratic equation $x^2 - 3x - 2 = 0$.

199-3. Plot the graph of the circle $x^2 + y^2 = 25$. What is the equation of the tangent at point (3,4)?

199-4. Plot the graphs of the rectangular hyperbola $xy = 36$ and the line $y = 4x$. What are the points of intersection between the line and the curve?

199-5. Plot the graph of the ellipse, $16x^2 + 25y^2 = 400$. Determine the foci of this ellipse.

199-6. In Practice Problem 199-5, obtain the polar equation of the same ellipse.

200
Angles and their measurement

It is impossible to solve many problems in electricity and electronics without a knowledge of angles. For example, in ac analysis, two alternating quantities can be separated by a certain phase difference; this difference is normally measured in terms of an angle.

DEFINITION OF AN ANGLE

Consider two straight, but very thin rods or lines *OA*, *OB* (Fig. 200-1A), which are pivoted at *O*. Holding *OB* fixed, turn *OA* in a counterclockwise direction; *OA* then describes a positive angle, which is measured by the amount of turning through the arc, *AB*, of the circle. At the center of the circle, point *O* is the *vertex* of the angle, which measures the separation between rods *OA* and *OB*, and is denoted by $\angle AOB$, or $\angle O$. The angle described by a complete revolution of *OA* is arbitrarily divided into 360 equal parts, which are called *degrees* (symbol °).

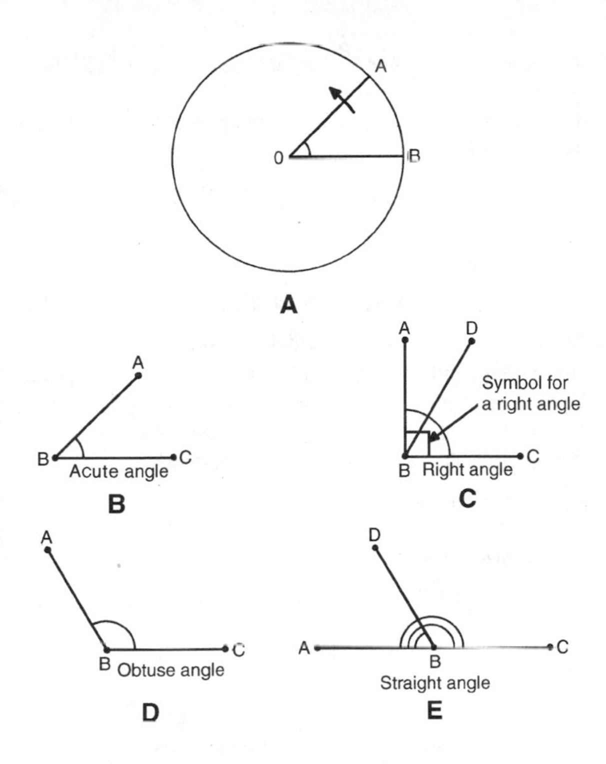

Figure 200-1

TYPES OF ANGLES

Acute angle

An *acute angle* (Fig. 200-1B) involves a rotation that is less than one-quarter of a revolution. The value of the angle then lies between 0° and 90°. Therefore, $0° < \angle ABC < 90°$.

Right angle

The *right angle* (Fig. 200-1C) is related to one-quarter of a revolution: therefore, its value is 90°. When $\angle ABC = 90°$, the lines *AB* and *BC* are said to be *perpendicular.*

If the line *BD* is drawn,

$$\angle ABD + \angle DBC = \angle ABC = 90°$$

When the sum of two angles is 90°, the angles are said to be *complementary;* it follows that $\angle ABD$ and $\angle DBC$ are complementary angles. As an example, 20° and 70° are complementary angles because 20° + 70° = 90°.

Obtuse angle

An *obtuse angle* (Fig. 200-1D) requires more than one-quarter revolution, but less than one-half revolution. The value of an obtuse angle is then greater than 90°, but less than 180°. Therefore, $90° < \angle ABC < 180°$.

Straight angle

The *straight angle* (Fig. 200-1E) is related to one-half of a revolution; consequently, its value is 180° ($\angle ABC = 180°$).

If the line *BD* is drawn,

$$\angle ABD + \angle DBC = 180°.$$

When the sum of two angles is 180°, the angles are said to be *supplementary;* it follows that $\angle ABD$ and $\angle DBC$ are supplementary angles. For example, 143° and 37° are supplementary angles because 143° + 37° = 180°.

MEASUREMENT OF ANGLES

Angles of less than one degree can be measured in terms of a decimal fraction of a degree. For example,

$$0.375° = \frac{375°}{1000} = \frac{3°}{8}.$$

Alternatively, the degree can be subdivided into 60 minutes and each minute into 60 seconds. Degrees, minutes, and seconds are respectively denoted by the symbols °, ′ and ″. Then:

$$\frac{3°}{8} = \frac{3}{8} \times 60' = 22.5' = 22'30''.$$

Scientific calculators normally have mode positions that are shown as DEG. (degrees), GRAD. (gradients), and RAD. (radians). The *gradient* (g) is based on the metric system; it is defined by $100\,g = 1$ right angle $= 90°$ or $1\,g = 0.9°$. So far, it does not appear that the gradient is in common use.

The *radian* is the scientific method of measuring angles; for example, an angle derived from solving differential calculus equations is automatically measured in radians. But what is the radian, and how is it related to the degree? In Fig. 200-2A, measure the length of each arc between P_1 and Q_1, P_2, and Q_2, and so on. Also, measure the length of each corresponding radius, then divide each length of arc by that of its radius. In the particular case when the lengths of the arc and the radius are equal, the angle *POQ* is equal to 1 radian. If you measure this angle with a protractor, you will find that the value of the angle is slightly greater than 57°.

MATHEMATICAL DERIVATIONS

If Fig. 200-2A is repeated for any angle, the ratio of arc to radius is constant. This ratio can therefore be taken as a measure of the angle. Such a system of measurement is known as *circular measurement* (^{c}). In Fig. 200-2B, the lengths of the arc and the radius are, re-

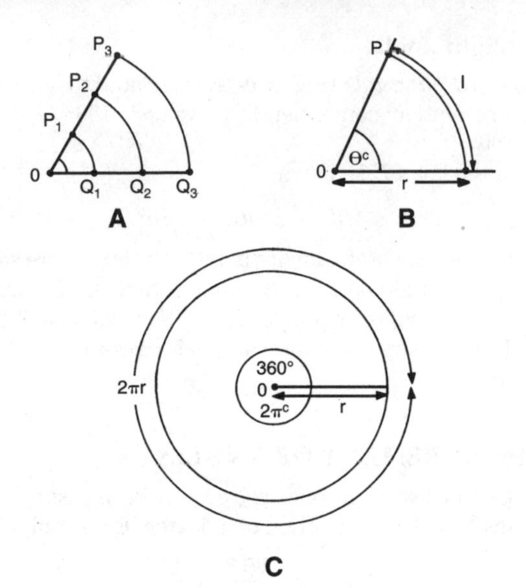

Figure 200-2

spectively, l and r; the value of the angle (θ) in radians is then given by:

$$\text{Angle } \theta^c = \frac{l}{r} \text{ radians} \quad (200\text{-}1)$$

For the complete circumference (Fig. 200-2C), the angle at the center of the circle is 360° and the length of the perimeter is $2\pi r$, where the circular constant, $\pi = 3.1415926\ldots$.

Therefore:

$$360° = \frac{2\pi r^c}{r}$$

$$= 2\pi \text{ radians}$$

This yields:

$$1 \text{ radian} = \frac{360°}{2\pi} = \frac{180°}{\pi} \approx 57.2958°$$

$$\approx 57°17'45''$$

Conversion from decimal degrees to radians

$$\theta^c = \frac{\pi^c \times \theta°}{180°} \text{ radians} \quad (200\text{-}2)$$

Conversion from radians to decimal degrees

$$\theta° = \frac{180° \times \theta^c}{\pi^c} \text{ degrees} \quad (200\text{-}3)$$

ANGULAR FREQUENCY (VELOCITY)

In Fig. 200-1A, let line OA rotate through an angle of θ radians in a time of t seconds. Then, the angular velocity (ω) is given by:

$$\text{Angular velocity, } \omega = \frac{\theta}{t} \text{ radians per second} \quad (200\text{-}4)$$

This equation should be compared with the following relationship:

$$\text{Linear velocity, } v = \frac{\text{Distance, } d}{\text{Time, } t}$$

Equation 200-4 yields:

$$\text{Angle, } \theta = \omega t \text{ radians} \quad (200\text{-}5)$$

Let line OA be a phasor, which is used to represent a sine-wave voltage (or current) whose frequency is f Hz. Then, the line will rotate at f revolutions per second, and each revolution is equivalent to an angular sweep of 2π radians. The angular frequency is given by:

$$\text{Angular frequency, } \omega = 2\pi f \text{ rad/s} \quad (200\text{-}6)$$

Example 200-1

Convert (a) 137.526° to degrees, minutes, seconds, and radians.
(b) 73°51'37" into decimal degrees and radians.
(c) 4.579^c into decimal degrees and degrees, minutes, and seconds.

Solution

(a) Using a scientific calculator,

Entry, Key	Display
0.526° [×]	0.526°
60 [=]	31.56′
[−] 31 [=]	0.56′
[×] 60 [=]	33.6″

Therefore:

$$137.526° = 137°31'33.6''$$

Entry, Key	Display
137.526° [×]	137.526°
π [=] [÷]	432.051
180 [=]	2.4003^c

$$137.526° = \frac{\pi \times 137.526^c}{180} \quad (200\text{-}2)$$

$$= 2.4003^c$$

(b)

Entry, Key		Display
37	÷	37″
60	= +	0.616′
51	= ÷	51.616′
60	=	0.8603°

Therefore:

$$73°51'37'' = 73.8603°$$

and,

$$73.8603° = \frac{\pi \times 73.8603^c}{180} \qquad (200\text{-}2)$$

$$= 1.289^c$$

(c)

Entry, Key		Display
4.579^c	×	4.579^c
180	= ÷	824.22
π	=	262.357°

$$4.579^c = \frac{180 \times 4.579°}{\pi} \qquad (200\text{-}3)$$

$$= 262.357°$$

Entry, Key		Display
0.357°	×	0.357°
60	= −	21.442′
21	= ×	0.442′
60	=	26.5″

Therefore:

$$4.579^c = 262.357° = 262°21'26.5''$$

Example 200-2

An AM broadcast station is operating on a frequency of 980 kHz. What is the value of its sine wave's angular frequency?

Solution

Angular frequency,

$$\omega = 2\pi f \qquad (200\text{-}6)$$

$$= 2 \times \pi \times 980 \times 10^3$$

$$= 6.16 \times 10^6 \text{ rad/s}$$

PRACTICE PROBLEMS

200-1. Convert 129.765° into degrees, minutes, seconds, and radians.

200-2. Convert 97° 43' 52" into decimal degrees and radians.

200-3. Convert 3.748^c into decimal degrees; and degrees, minutes, and seconds.

200-4. What are the values of the complement and the supplement of 37°?

7
PART
Intermediate mathematics for electronics

201
The theorem of Pythagoras

A plane figure that is bounded by three straight lines is called a *triangle*. Examples of various triangles are shown in Figs. 201-1A,B,C. Each of these triangles has three sides, which are denoted by the lowercase letters a, b, c. By contrast, the capital letters A, B, C represent the three interior angles. Notice that in each triangle, side a is opposite or facing angle A; side b is facing angle B, and so on.

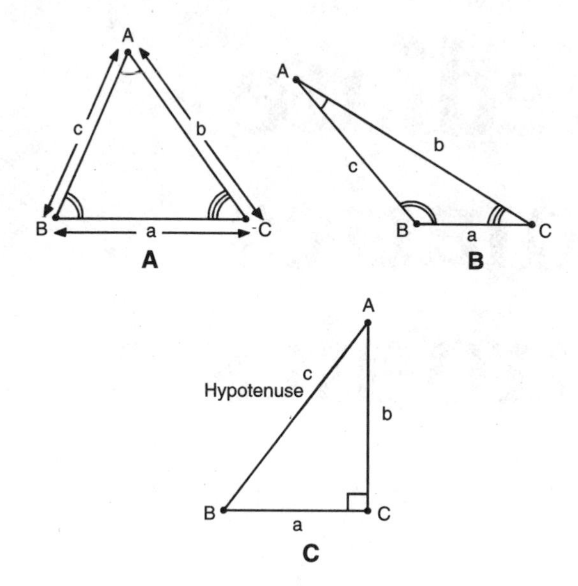

Figure 201-1

Figure 201-1A illustrates an acute-angled triangle in which all of the three interior angles are acute angles. Figure 201-1B shows an obtuse-angled triangle in which one of the interior angles (B) is obtuse. Figure 201-1C represents a right-angled triangle in which one of the interior angles (C) is a right angle (90°).

All triangles possess the following two properties:

1. The sum of the lengths of any two of the sides must exceed the length of the third side. It follows that, it is impossible to construct a triangle whose sides are 3, 4, and 8 inches; because (3 + 4) is not greater than 8.
2. The sum of the three interior angles of a triangle is 180° (π radians). Therefore:

$$A + B + C = 180°$$

For the right-angled triangle of Fig. 201-1C, $C = 90°$. Therefore, $A + B = 90°$ so that angles A and B are complementary. The longest side (c), which is facing the right angle, is called the *hypotenuse*.

A famous theorem that applies only to a right-angled triangle was discovered by Pythagoras, who lived in Sicily between the years 570 B.C. and 500 B.C. Formally stated, the theorem is as follows: "In a right-angled triangle, the square on the hypotenuse is equal to the sum of the squares on the sides that contain the right angle."

In Fig. 201-2, the theorem means that if the area (a^2) of the square $BCHG$ is added to the area (b^2) of the square $ACJI$, the result is the area (c^2) of the square $ABEF$; in equation form, $a^2 + b^2 = c^2$, or $c = \sqrt{a^2 + b^2}$. Proofs of Pythagoras' theorem are shown in the mathematical derivations.

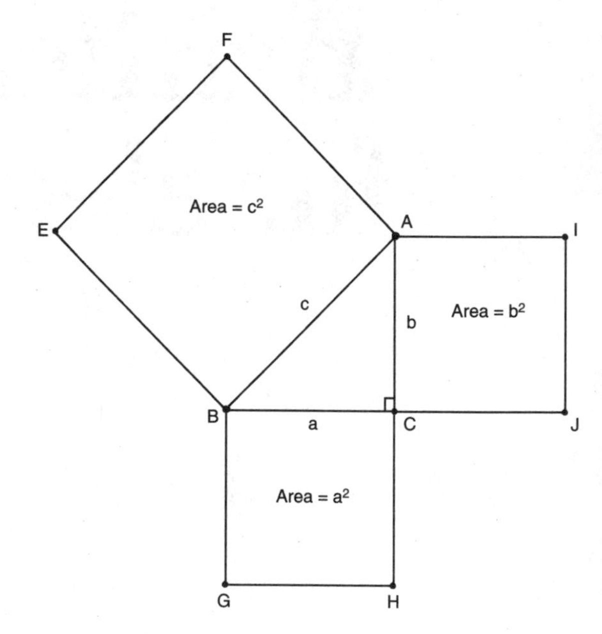

Figure 201-2

The theorem of Pythagoras is a vital tool in ac circuit analysis. For example, if the effective voltages across a resistor and an inductor in series are, respectively, 6 V and 8 V, the total voltage across this combination is not 14 V, but it is, in fact:

$$\sqrt{6^2 + 8^2} = \sqrt{36 + 64} = \sqrt{100} = 10 \text{ V}$$

(See chapter 72.) Similarly, if the values of an ac circuit's true power and reactive power are known, it is necessary to combine these values, by the Pythagorean equation, in order to determine the apparent power (see chapters 72 through 79).

MATHEMATICAL DERIVATIONS

In Fig. 201-3, triangles ABC and $A'B'C'$ are equiangular in the sense that $\angle A = \angle A'$, $\angle B = \angle B'$, and $\angle C = \angle C'$. Such triangles are said to be *similar.*

In similar triangles, the corresponding sides are in proportion. Therefore:

$$\frac{a}{a'} = \frac{b}{b'} = \frac{c}{c'} \qquad (201\text{-}1)$$

These equations can be rearranged as:

$$\frac{b}{c} = \frac{b'}{c'} \qquad (201\text{-}2)$$

$$\frac{a}{c} = \frac{a'}{c'} \qquad (201\text{-}3)$$

$$\frac{b}{a} = \frac{b'}{a'} \qquad (201\text{-}4)$$

The last three equations are important in learning about trigonometric functions (see chapters 202, 203, and 204).

Proof of Pythagoras' theorem

In Fig. 201-4, ΔABC is a right-angled triangle because $\angle C = 90°$. CD is perpendicular to AB so that Δ's ABC, ACD, and BCD are similar. In Δ's ABC, BCD:

$$\frac{x}{a} = \frac{a}{c}$$

or,

$$a^2 = cx \qquad (201\text{-}5)$$

In Δ's ABC, ACD:

$$\frac{c - x}{b} = \frac{b}{c}$$

or,

$$c^2 - cx = b^2 \qquad (201\text{-}6)$$

Substituting $cx = a^2$ in Equation 201-6,

$$c^2 - a^2 = b^2$$

or,

$$a^2 + b^2 = c^2 \qquad (201\text{-}7)$$

Equation 201-7 is the required Pythagorean relationship. Other forms of the equation are:

$$c = \sqrt{a^2 + b^2} \qquad (201\text{-}8)$$

$$a^2 = c^2 - b^2 \qquad (201\text{-}9)$$

$$a = \sqrt{c^2 - b^2} \qquad (201\text{-}10)$$

$$b^2 = c^2 - a^2 \qquad (201\text{-}11)$$

$$b = \sqrt{c^2 - a^2} \qquad (201\text{-}12)$$

Notice that there are certain right-angled triangles in which the lengths of the sides are whole numbers. Three examples are:

(1) $a = 3, b = 4, c = 5$, because $3^2 + 4^2 = 5^2$
(2) $a = 5, b = 12, c = 13$, because $5^2 + 12^2 = 13^2$ and
(3) $a = 8, b = 15, c = 17$, because $8^2 + 15^2 = 17^2$.

Example 201-1

A 1.8-kΩ resistor is connected in series with an inductor, whose reactance is 2.3 kΩ. Determine the value of the total impedance, Z (chapter 72).

Solution

Total impedance,

$$Z = \sqrt{R^2 + X_L^2} \qquad (201\text{-}8)$$

$$= \sqrt{1.8^2 + 2.3^2}$$

The impedance is 2.92 kΩ.

Entry, Key	Display
1.8 [x²] [+]	3.24
2.3 [x²]	5.29
[=]	8.53
[√x]	2.92

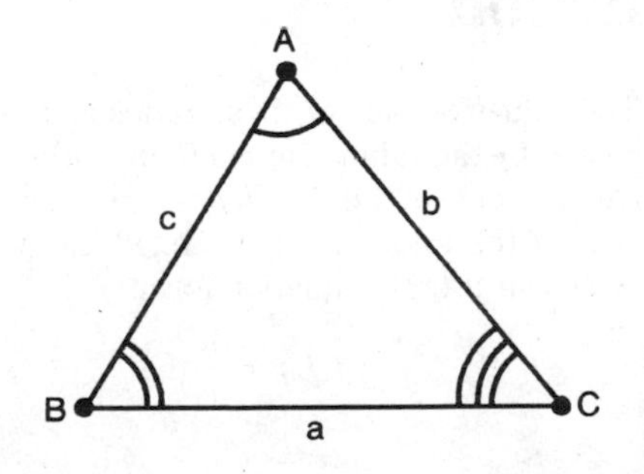

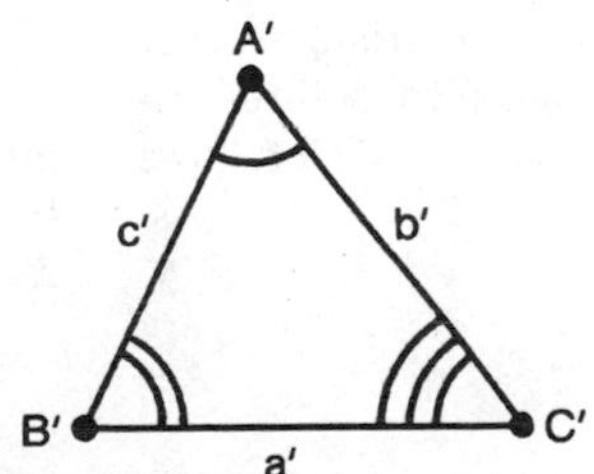

Figure 201-3

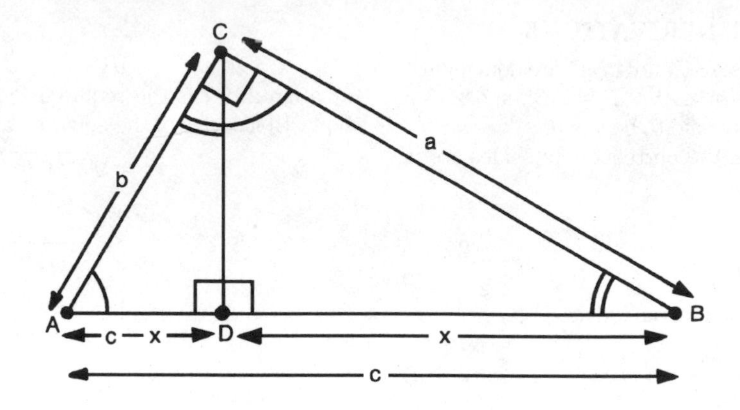

Figure 201-4

Example 201-2

In an ac circuit, the apparent power is 25.6 mVA and the reactive power is 12.8 mVAr. Determine the value of the true power.

Solution

True Power (201-10)

$= \sqrt{(Apparent\ Power)^2 - (Reactive\ power)^2}$

$= \sqrt{25.6^2 - 17.8^2}$

Entry, Key	Display
25.6 x^2 −	655.36
17.8 x^2	316.84
=	338.52
$\sqrt{x}$	18.4

The true power is 18.4 mW.

PRACTICE PROBLEMS

201-1. In $\Delta ABC, A = 43°$ and $B = 67°$. Calculate the value of angle C.

201-2. In Fig. 201-3, $a = 4.72$, $b = 5.43$, and $c = 3.74$. If $c' = 2.75$, find the values of a' and b'.

201-3. In $\Delta\ ABC$, $C = 90°$. If $c = 7.87$ and $b = 5.36$, calculate the value of a.

201-4. In ΔABC, $C = 90°$. If $a = 11.42$ and $b = 13.64$, calculate the value of c.

201-5. In ΔABC, $C = 90°$. If $a = b = 15.63$, calculate the value of c.

202
The sine function

Figure 202-1 shows three right-angled triangles that are similar because angle B is common to all three triangles, and consequently, $\angle A = \angle A' = \angle A''$. It follows from Equation 201-2 that:

$$\frac{b}{c} = \frac{b'}{c'} = \frac{b''}{c''}$$

The values of each of these ratios is determined in some way by the value of angle B. In mathematical language, the value of the ratio $b{:}c$ is a function of angle B; because if the value of B is changed, the value of the ratio $b{:}c$ will alter; in equation form:

$$f(B) = \frac{b}{c}$$

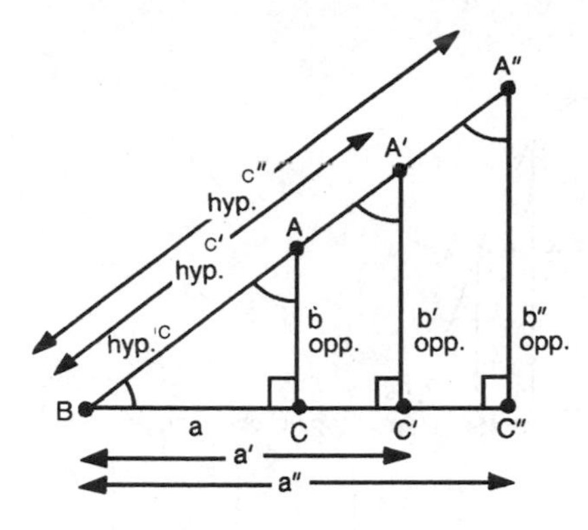

Figure 202-1

The ratio $b{:}c$ involves the circular trigonometric function *sine*, which is commonly abbreviated sin.

Therefore:

$$\sin B = \frac{b}{c}.$$

A knowledge of the sine function is important in the ac analysis of series and parallel LCR circuits, and also in the use of the j operator.

MATHEMATICAL DERIVATIONS

In Fig. 202-2 side b is facing, or *opposite*, (abbreviated opp.) to angle B, and side a is beside, or *adjacent*, (abbreviated adj.) to angle B. It can be argued that side c is also adjacent to the angle B, but side c has already been identified as the *hypotenuse* (abbreviated hyp.).

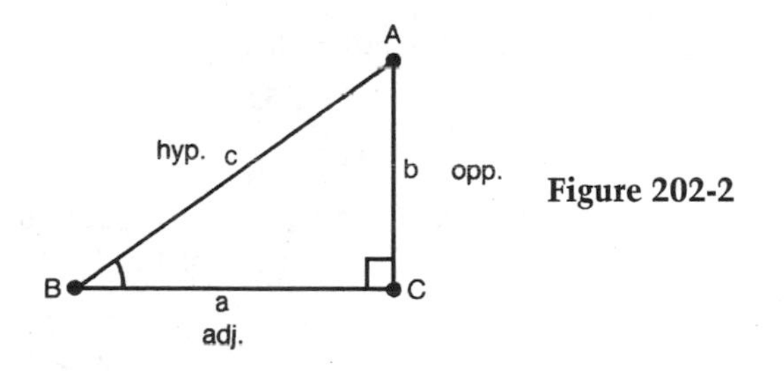

Figure 202-2

The sine of angle B is defined by:

$$\sin B = \frac{\text{opp}}{\text{hyp}} \quad (202\text{-}1)$$

$$= \frac{b}{c}$$

As an example, sin 30° = 0.5 so that the length of b is one-half the length of c. But if you know that the value of sin B is 0.5, how do you obtain the value of 30° on the calculator? If $\sin B = 0.5$,

$$\text{Angle } B = \left\{ \begin{matrix} \text{inv sin } 0.5 \\ \text{or arc sin } 0.5 \\ \text{or } \sin^{-1} 0.5 \end{matrix} \right\} = 30°$$

The terms *inverse (inv) sin, arc sin,* and $\sin^{-1}$ have the same meaning: namely "the value of the angle whose sine value is . . .." These three key combinations are used on a variety of scientific calculators. Using the inverse and sin keys,

$$\text{Angle, } B = \text{inv sin } \frac{b}{c} \quad (202\text{-}2)$$

Because $\sin B = \dfrac{b}{c}$,

$$\text{Side } b = c \sin B \quad (202\text{-}3)$$

and,

$$\text{Side } c = \frac{b}{\sin B} \quad (202\text{-}4)$$

As always, the choice of which equation to use depends on the information given and the quantity to be found.

The sine is a circular function whose value can be obtained with the aid of measurements taken from a circle. Figure 202-3 shows a circle of unit radius. As line OP rotates in the counterclockwise (or positive) direction,

$$\sin \theta = \frac{PN}{OP} = \frac{PN}{1}$$

$$= PN$$

Therefore, the length of PN is a direct measure of the value of sin θ. In this way, you can obtain the sin θ vs θ graph, which is the waveform of a sine-wave voltage (or current). For such a voltage, the instantaneous value is:

$$e = E_{max} \sin \theta$$

$$= E_{max} \sin \omega t \quad (200\text{-}5, 202\text{-}5)$$

$$= E_{max} \sin 2\pi ft \quad (200\text{-}6, 202\text{-}6)$$

Notice that the value of sin θ lies between +1 and −1.

Example 202-1

In Fig. 202-2, $b = 16.7$, and $c = 23.5$. Find the values of B, A, and a.

Solution

$$\text{Angle } B = \text{inv sin } \frac{b}{c} \quad (202\text{-}2)$$

$$= \text{inv sin } \frac{16.7}{23.5}$$

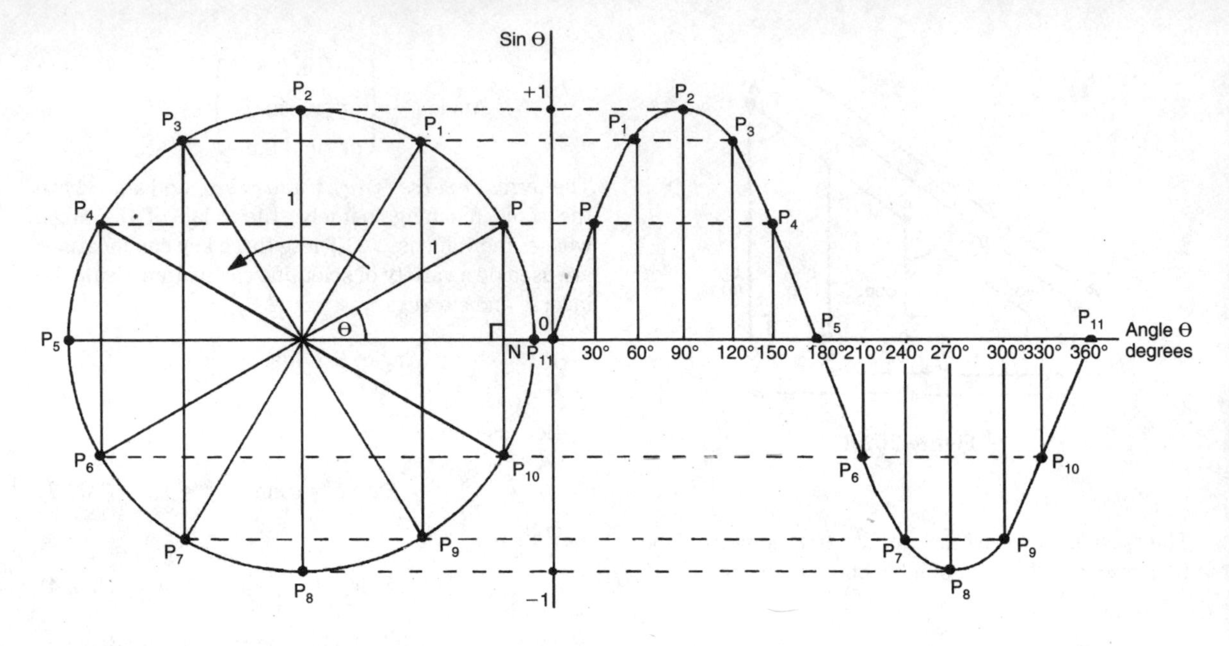

Figure 202-3

Entry, Key		Display
16.7	÷	16.7
23.5	=	0.7106
inv	sin	45.287°

The value of angle B is 45.3°, rounded off. Angles A, B are complementary.

Therefore:

$$\text{Angle } A = 90^\circ - B$$
$$= 90^\circ - 45.3^\circ$$
$$= 44.7^\circ$$

$$\text{Side } a = \sqrt{c^2 - b^2} \qquad (201\text{-}10)$$
$$= \sqrt{23.5^2 - 16.7^2}$$
$$= 16.5$$

Example 202-2

In Fig. 202-2, $B = 75.8^\circ$, and $c = 112.8$. Find the values of b, A, and a.

Solution

$$\text{Side } b = c \sin B \qquad (202\text{-}3)$$
$$= 112.8 \sin 75.8^\circ$$

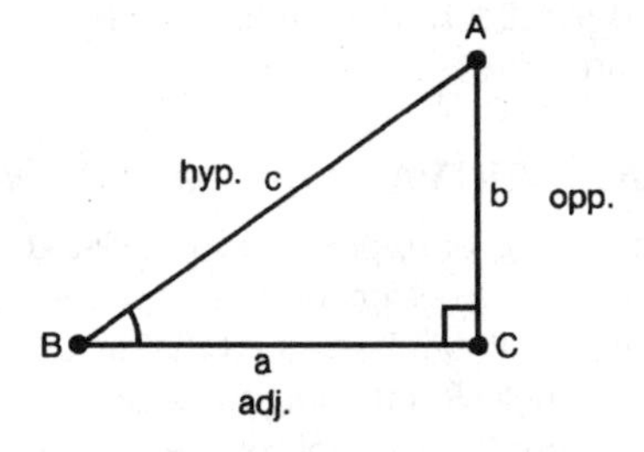

The length of the side b is 109.35, rounded off.

$$\text{Angle } A = 90^\circ - B$$
$$= 90^\circ - 75.8^\circ$$
$$= 14.2^\circ$$

$$\text{Side } a = \sqrt{c^2 - b^2} \qquad (201\text{-}10)$$
$$= \sqrt{112.8^2 - 109.35^2}$$
$$= 27.68$$

Example 202-3

In Fig. 202-2, $B = 18.6^\circ$ and $b = 69.3$. Find the values of c, A, and a.

Solution

$$\text{Hypotenuse } c = \frac{b}{\sin B} \qquad (202\text{-}4)$$
$$= \frac{69.3}{\sin 18.6^\circ}$$

Entry, Key	Display
69.3 ÷	69.3
18.6° sin	0.31896
=	217.3

The length of the hypotenuse, c, is 217.3, rounded off.

$$\text{Angle } A = 90° - B$$
$$= 90° - 18.6°$$
$$= 71.4°$$

$$\text{Side } a = \sqrt{c^2 - b^2} \qquad (201\text{-}10)$$
$$= \sqrt{217.3^2 - 69.3^2}$$
$$= 206.0, \text{ rounded off.}$$

PRACTICE PROBLEMS

202-1. In Fig. 202-2, $B = 72.4°$ and $b = 3.89$. Determine the values of c, A, and a.

202-2. In Fig. 202-2, $b = 17.4$ and $c = 23.7$. Determine the values of B, A, and a.

202-3. In Fig. 202-2, $B = 16.4°$ and $c = 34.5$. Determine the values of b, A, and a.

202-4. In Fig. 202-3, $E_{max} = 17.5$ V. Find the instantaneous voltages for $\theta = 53°$, 116°, 229°, and 317°.

203 The cosine function

Figure 203-1 shows three right-angled triangles that are similar because angle B is common to all three triangles; consequently, $\angle A = \angle A' = \angle A''$. It follows from Equation 201-3 that:

$$\frac{a}{c} = \frac{a'}{c'} = \frac{a''}{c''}$$

The value of each of these ratios is determined in some way by the value of angle B. In mathematical language, the value of the ratio $a{:}c$ is a function of angle B, because if the value of B is changed, the value of the ratio $a{:}c$ will alter; in equation form: $f(B) = a/c$. The ratio $a{:}c$ involves the circular trigonometric function, *cosine*, which is commonly abbreviated to cos.

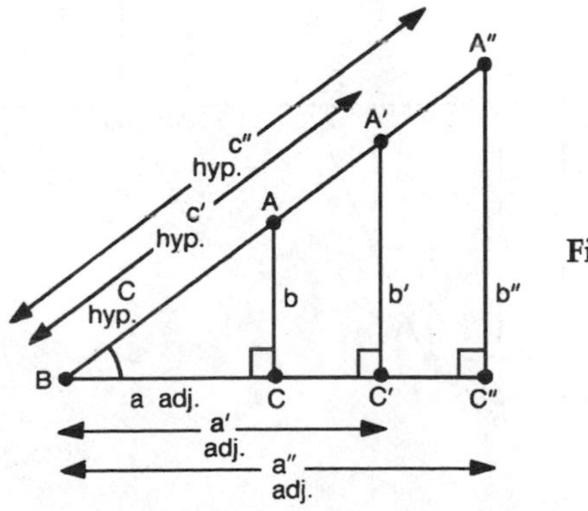

Figure 203-1

Therefore:

$$\cos B = \frac{a}{c}$$

A knowledge of the cosine function is important in the ac analysis of series and parallel LCR circuits and also in the use of the j operator. Moreover, if the source voltage and the source current differ in phase, the power factor of the circuit is equal to the cosine of the phase angle (see chapters 72 thru 79).

MATHEMATICAL DERIVATIONS

In Fig. 203-2 the cosine of the angle (B) is defined by:

$$\cos B = \frac{\text{adj}}{\text{hyp}} \qquad (203\text{-}1)$$
$$= \frac{a}{c}$$

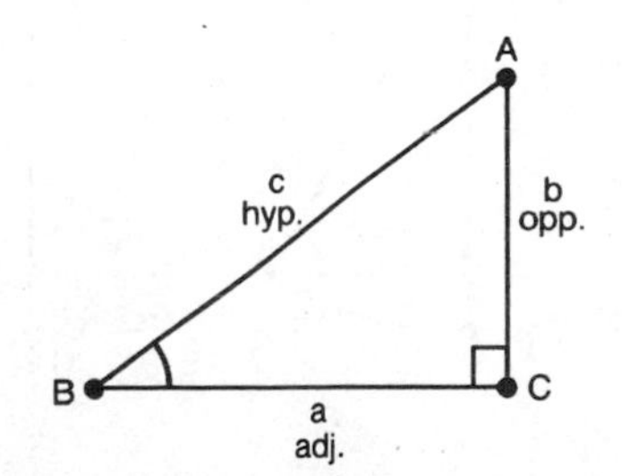

Figure 203-2

Then,

$$\text{Angle } B = \text{inv cos } \frac{a}{c} \tag{203-2}$$

Because:

$$\cos B = \frac{a}{c}$$

$$\text{Side } a = c \cos B \tag{203-3}$$

and,

$$\text{Side } c = \frac{a}{\cos B} \tag{203-4}$$

Like the sine function, the value of the cosine function can be obtained with the aid of measurements taken from a circle. Figure 203-3 shows a circle of unit radius. As the line (OP) rotates in a counterclockwise (or positive) direction,

$$\cos \theta = \frac{ON}{OP} = \frac{ON}{1} = ON$$

Therefore, the length of the line, ON, is a direct measure of the value of $\cos \theta$. In this way, you can obtain the graph of $\cos \theta$ vs θ. For comparison purposes, the graph of $\sin \theta$ vs θ has also been included. Observe that the cosine curve is 90° ahead of the sine curve so that:

$$\cos \theta = \sin (\theta + 90°) \tag{203-5}$$

and,

$$\sin \theta = \cos (\theta - 90°)$$

Alternatively, you can say that the cosine curve leads the sine curve by 90° or that the sine curve lags the cosine curve by 90°. Like the sine function, the value of the cosine function lies between +1 and −1.

Relationships between the sine and cosine functions

In Fig. 203-2,

$$\cos B = \frac{a}{c}$$

With reference to the angle A, the side opposite the angle is a.

Therefore:

$$\sin A = \frac{a}{c}$$

and,

$$\cos B = \sin A \tag{203-6}$$

$$= \sin (90° - B)$$

$$\sin A = \cos B \tag{203-7}$$

$$= \cos (90° - A)$$

The cosine of an angle is equal to the sine of its complement; the sine of an angle is equal to the cosine of its complement. As examples, cos 40° = sin 50° = 0.7660 and sin 20° = cos 70° = 0.3420.

In Fig. 203-2,

$$\sin B = \frac{b}{c}$$

and,

$$\cos B = \frac{a}{c}$$

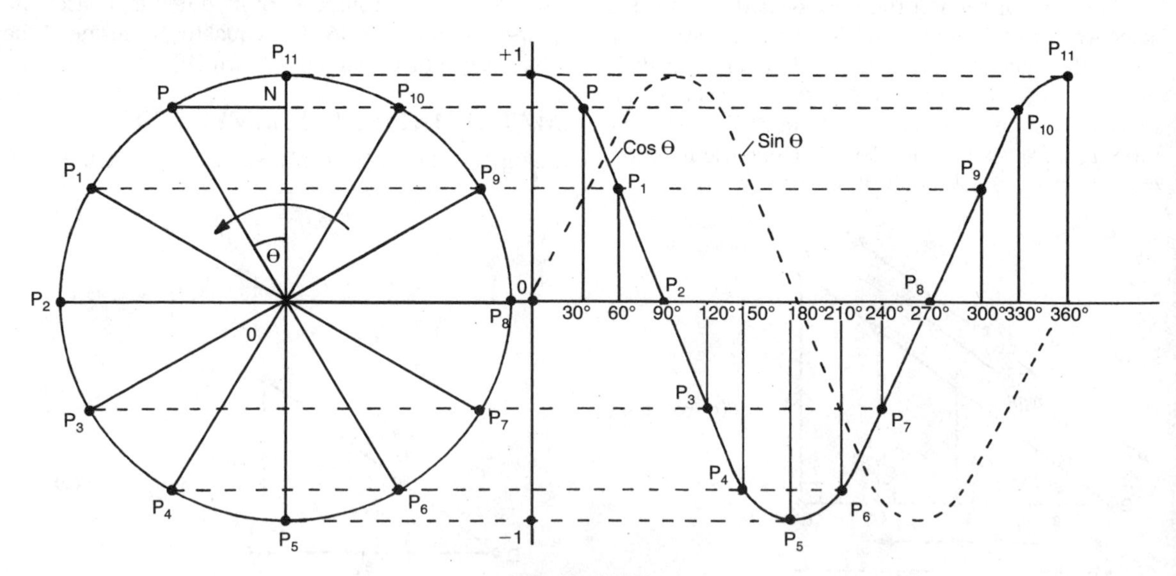

Figure 203-3

Then:

$$(\sin B)^2 + (\cos B)^2 = \frac{b^2}{c^2} + \frac{a^2}{c^2} \qquad (203\text{-}8)$$

$$= \frac{b^2 + a^2}{c^2}$$

$$= \frac{c^2}{c^2}$$

$$= 1$$

This relationship can be written as:

$$\sin^2 B + \cos^2 B = 1 \qquad (203\text{-}9)$$

where $\sin^2 B$ and $\cos^2 B$ are, respectively, the mathematical abbreviations for $(\sin B)^2$ and $(\cos B)^2$. It follows that the sine and cosine functions are not independent, but are closely related.

Example 203-1

In Fig. 203-1, $a = 73.9$ and $c = 112.5$. Find the values of B, A, and b.

Solution

$$\text{Angle } B = \text{inv cos } \frac{a}{c} \qquad (203\text{-}2)$$

$$= \text{inv cos } \frac{73.9}{112.5}$$

Entry, Key	Display
73.9 ÷	73.9
112.5 =	0.6568
inv cos	48.9°

The value of the angle B is 48.9°.

$$\text{Angle } A = 90° - 48.9°$$

$$= 41.1°.$$

$$\text{Side } b = \sqrt{112.5^2 - 73.9^2}$$

$$= 84.8$$

Example 203-2

In Fig. 203-2, $B = 12.7°$ and $c = 0.763$. Find the values of a, A, and b.

Solution

$$\text{Side } a = c \cos B \qquad (203\text{-}3)$$

$$= 0.763 \cos 12.7°.$$

Entry, Key	Display
0.763 ×	0.763
12.7° cos	0.9755
=	0.744

The length of side a is 0.744.

$$\text{Angle } A = 90° - 12.7°$$

$$= 77.3°$$

$$\text{Side } b = \sqrt{0.763^2 - 0.744^2}$$

$$= 0.169$$

Example 203-3

In Fig. 203-2, $B = 74.8°$ and $a = 27.3$. Find the values of c, A, and b.

Solution

$$\text{Hypotenuse } c = \frac{a}{\cos A} \qquad (203\text{-}4)$$

$$= \frac{27.3}{\cos 74.8°}$$

Entry, Key	Display
27.3 ÷	27.3
74.8° cos	0.2622
=	104.1

The length of the hypotenuse (c) is 104.1.

$$\text{Angle } A = 90° - 74.8°$$

$$= 15.2°$$

$$\text{Side } b = \sqrt{104.1^2 - 27.3^2}$$

$$= 100.5.$$

Example 203-4

In an ac circuit, the source voltage and the source current differ in phase by 40°. What is the value of the circuit's power factor?

Solution

$$\textit{Power factor} = \cos \phi = \cos 40°$$

$$= 0.766.$$

PRACTICE PROBLEMS

203-1. In Fig. 203-2, $B = 61.3°$ and $a = 4.91$. Determine the values of c, A, and b.

203-2. In Fig. 203-2, $a = 18.5$ and $c = 34.8$. Determine the values of B, A, and b.

203-3. In Fig. 203-2, $B = 27.5°$ and $c = 45.6$. Determine the values of a, A, and b.

203-4. If x is an acute angle and $\cos x = 0.6342$, what is the value of $\sin x$?

203-5. If $\sin x = 0.3472$, what is the value of $\cos (90° - x)$?

204 The tangent function

Figure 204-1 shows three right-angled triangles that are similar because angle B is common to all three; consequently $\angle A = \angle A' = \angle A''$. It follows that:

$$\frac{b}{a} = \frac{b'}{a'} = \frac{b''}{a''}$$

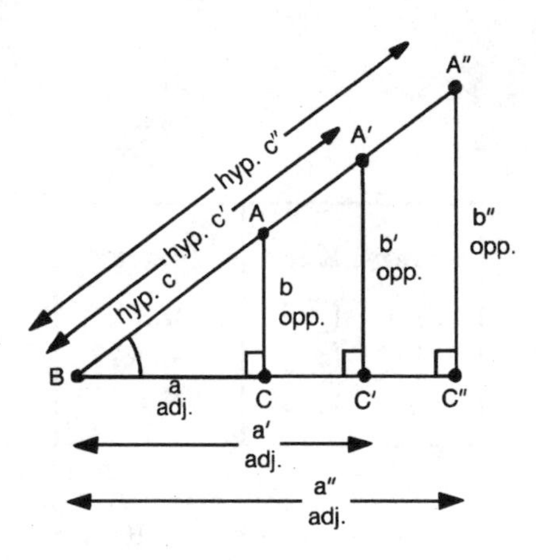

Figure 204-1

The value of each of these ratios is determined in some way by the value of angle B. In mathematical language, the value of ratio $b{:}a$ is a function of angle B. If the value of B is changed, the value of the ratio $b{:}a$ will alter; in equation form $f(B) = b/a$. Ratio $b{:}a$ involves the circular trigonometric function, *tangent*, commonly abbreviated *tan*. Therefore, $\tan B = b/a$. A knowledge of the tangent function is especially important in the use of operator j and it is also required in the ac analysis of various LCR circuits.

MATHEMATICAL DERIVATIONS

In Fig. 204-2 the tangent of angle B is defined by:

$$\tan B = \frac{\text{opp}}{\text{adj}} = \frac{b}{a} \qquad (204\text{-}1)$$

Then,

$$\text{Angle } B = \text{inv tan } \frac{b}{a} \qquad (204\text{-}2)$$

Because,

$$\tan B = \frac{b}{a}, \qquad (204\text{-}3)$$

$$\text{Side } b = a \tan B$$

and,

$$\text{Side } a = \frac{b}{\tan B} \qquad (204\text{-}4)$$

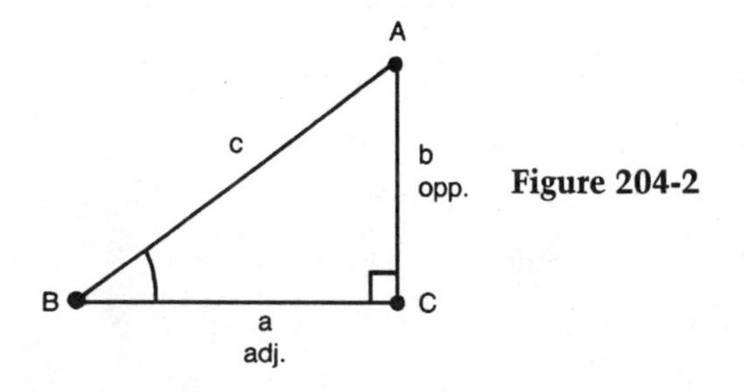

Figure 204-2

Like the sine and cosine functions, the value of the tangent function can be obtained with the aid of measurements taken from a circle. Figure 204-3 shows a circle of unit radius; at point Q, a tangent QT is drawn so that

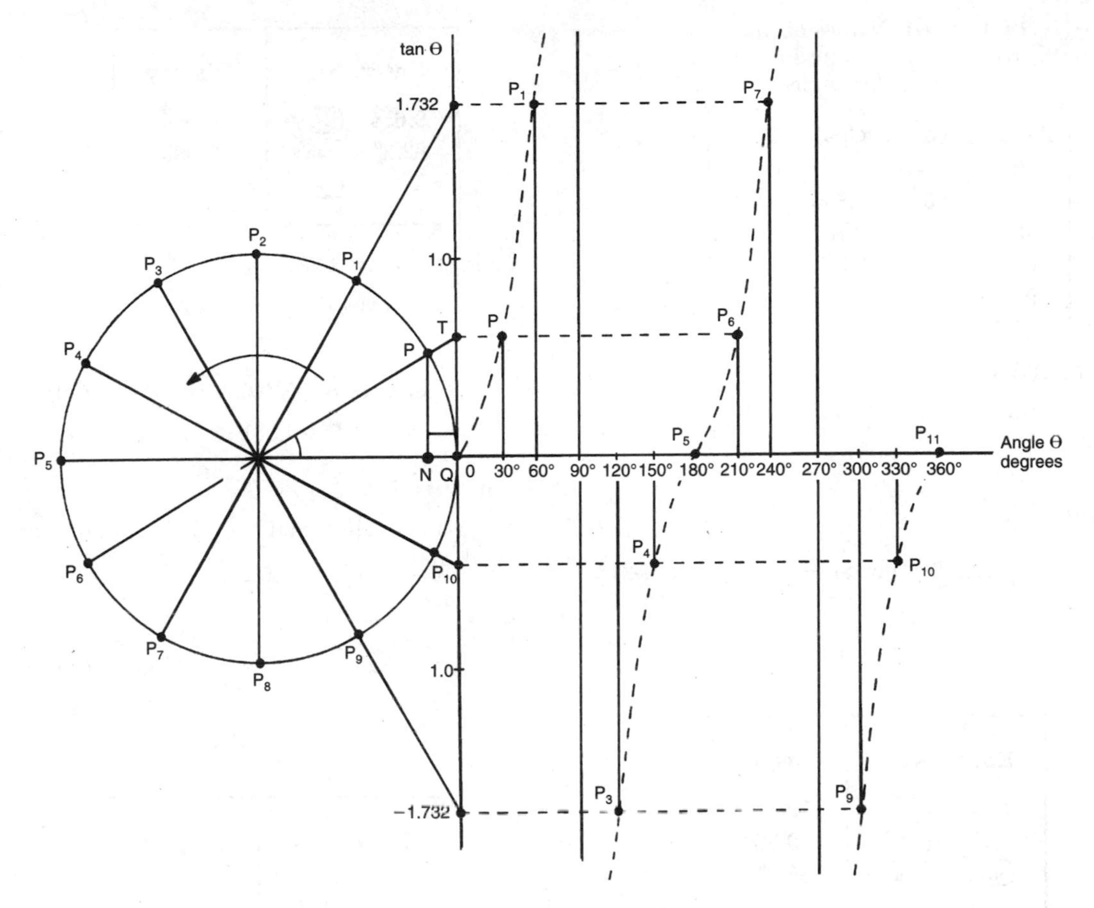

Figure 204-3

lines OQ and TQ are perpendicular to each other (a tangent to a circle touches or grazes the circle at a particular point). As line OP rotates in the counterclockwise (or positive) direction,

$$\tan\theta = \frac{TQ}{OQ} = \frac{TQ}{1} = TQ \qquad (204\text{-}5)$$

Therefore, length of line TQ is a direct measure of the value of $\tan\theta$. In this way, you can obtain the graph of $\tan\theta$ versus θ. Notice that the value of the tangent function lies between $+\infty$ and $-\infty$.

In Fig. 204-3,

$$\tan\theta = \frac{TQ}{OQ} = \frac{TQ}{OP}$$

$$\text{Angle, } \theta = \frac{\text{Arc } PQ}{OP} \text{ radians}$$

and,

$$\sin\theta = \frac{PN}{OP}$$

Because $TQ > \text{Arc } PQ > PN$,

$$\tan\theta > \theta > \sin\theta.$$

For small angles of less than 10°, TQ, arc PQ, and PN are approximately equal. Therefore:

$$\tan\theta \approx \theta^c \approx \sin\theta \qquad (204\text{-}6)$$

As an example, $\sin 5° = 0.0872$, $5° = 0.0873^c$, and $\tan 5° = 0.0875$, rounded off.

Relationship among sin, cos, and tan

In Fig. 204-2,

$$\tan B = \frac{b}{a} = \frac{b/c}{a/c} \qquad (204\text{-}7)$$

$$= \frac{\sin B}{\cos B}$$

This relationship is illustrated in Table 204-1.

Table 204-1. Values of the trigonometric functions for particular angles.

Angle	Sin	Cos	Tan
0°	0	1	0
30°	0.5	0.866	0.577
45°	0.707	0.707	1
60°	0.866	0.5	1.732
90°	1	0	∞

Example 204-1

In Fig. 204-2, $b = 72.5$, $a = 98.7$. Find the values of B, A and c.

Solution

$$\text{Angle } B = \text{inv tan } \frac{b}{a} \qquad (204\text{-}2)$$

$$= \text{inv tan } \frac{72.5}{98.7}$$

Entry, Key		Display
72.5	÷	72.5
98.7	=	0.7345
inv	tan	36.3°

The value of angle B is 36.3°.

$$\text{Angle } A = 90° - 36.3° \qquad (201\text{-}8)$$

$$= 53.7°$$

$$\text{Hypotenuse } c = \sqrt{72.5^2 + 98.7^2}$$

$$= 122.5$$

Example 204-2

In Fig. 204-2, $a = 0.654$ and $B = 82.3°$. Find the values of b, A, and c.

Solution

$$\text{Side } b = a \tan B \qquad (204\text{-}3)$$

$$= 0.654 \tan 82.3°$$

Entry, Key		Display
0.654	×	0.654
82.3°	tan	7.396
	=	4.837

The value of side b is 4.837.

$$\text{Angle } A = 90° - 82.3°$$

$$= 7.7°$$

$$\text{Hypotenuse } c = \sqrt{0.654^2 + 4.837^2} \qquad (201\text{-}8)$$

$$= 4.881$$

Example 204-3

In Fig. 204-2, $b = 116.4$, and $B = 11.7°$. Find the values of a, A, and c.

Solution

$$\text{Side } a = \frac{b}{\tan B} \qquad (204\text{-}4)$$

$$= \frac{116.4}{\tan 11.7°}$$

Entry, Key		Display
116.4	÷	116.4
11.7°	tan	0.2071
	=	562.1

The value of side a is 562.1.

$$\text{Angle } A = 90° - 11.7°$$

$$= 78.3°$$

$$\text{Hypotenuse } c = \sqrt{116.4^2 + 562.1^2} \qquad (201\text{-}8)$$

$$= 574.0$$

PRACTICE PROBLEMS

204-1. In Fig. 204-2, $B = 19.8°$ and $a = 5.02$. Determine the values of b, A, and c.

204-2. In Fig. 204-2, $b = 19.6$ and $a = 45.9$. Determine the values of B, A, and c.

204-3. In Fig. 204-3, $B = 38.6°$ and $b = 56.7$. Determine the values of a, A, and c.

204-4. If the ratio of sin B:cos B = 0.3157, find the value of B.

205
Cosecant, secant, and cotangent

The sin, cos, and tan functions are convenient multipliers for converting from one side of a right-angled triangle into another side. However, none of these three functions can convert either of the sides forming the right-angle into the hypotenuse. For example, in Fig. 205-1:

$$c = \frac{b}{\sin B} = b \times \frac{1}{\sin B},$$

$$c = \frac{a}{\sin A} = a \times \frac{1}{\sin A}$$

and,

$$c = \frac{a}{\cos B} = a \times \frac{1}{\cos B}$$

$$c = \frac{b}{\cos A} = b \times \frac{1}{\cos A}$$

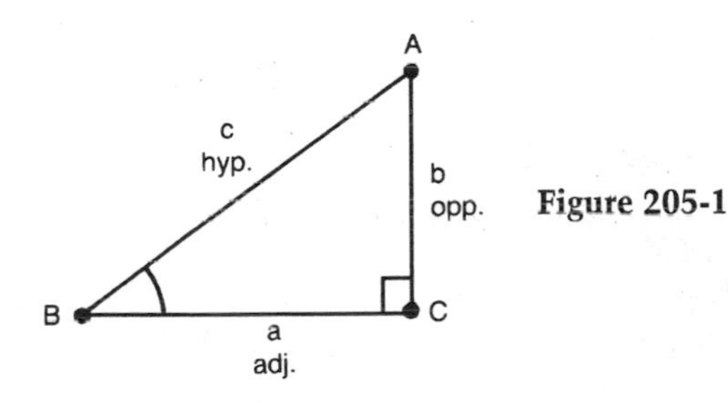

Figure 205-1

In some problems, it is more convenient to have simple multipliers to convert either side a or b into the hypotenuse. In order to do this, you need trigonometric functions, which are the reciprocals of the sine and cosine functions. The reciprocal of sine θ (1/sin θ) is termed the *cosecant* of θ, where θ is any angle. The reciprocal of cosine θ (1/cos θ) is called the *secant* of θ; similarly, the reciprocal of tangent θ (1/tan θ) is the *cotangent* of θ.

The cosecant (csc), secant (sec), and cotangent (cot) functions are less frequently encountered than the other three functions, and they do not normally appear as keys on a scientific calculator. If, for example, the value of cosecant 50° is required, the value of sine 50° is found first and the reciprocal key is used to obtain the cosecant of 50°.

The equations of electronics and communications theory rarely involve the reciprocal trigonometric functions; however, chapter 168 contains the equation:

$$Z = Z_0 \cot \frac{2\pi x}{\lambda},$$

which determined the variation of the impedance along an open-circuited transmission line.

MATHEMATICAL DERIVATIONS

In Fig. 205-1, the cosecant of B, is defined by:

$$\csc B = \frac{\textit{hypotenuse}}{\textit{opposite}} \tag{205-1}$$

$$= \frac{c}{b}$$

$$= \frac{1}{\sin B}$$

$$\text{Angle } B = \text{inv csc } \frac{c}{b} \tag{205-2}$$

$$\text{Hypotenuse } c = b \csc B \tag{205-3}$$

$$\text{Side } b = \frac{c}{\csc B} \tag{205-4}$$

The secant of B is defined by:

$$\sec B = \frac{\textit{hypotenuse}}{\textit{adjacent}} \tag{205-5}$$

$$= \frac{c}{a}$$

$$= \frac{1}{\cos B}$$

$$\text{Angle } B = \text{inv sec } \frac{c}{a} \tag{205-6}$$

$$\text{Hypotenuse } c = a \sec B \tag{205-7}$$

$$\text{Side } a = \frac{c}{\sec B} \tag{205-8}$$

The cotangent of B is defined by:

$$\cot B = \frac{\textit{adjacent}}{\textit{opposite}} \tag{205-9}$$

$$= \frac{a}{b}$$

$$= \frac{1}{\tan B}$$

$$= \frac{\cos B}{\sin B}$$

$$\text{Angle } B = \text{inv cot } \frac{a}{b} \tag{205-10}$$

$$\text{Side } a = b \cot B \tag{205-11}$$

$$\text{Side } b = \frac{a}{\cot B} \qquad (205\text{-}12)$$

Figure 205-2 shows a circle of unit radius with a tangent constructed at point P.

In ΔOPN,

$$\sin\theta = \frac{PN}{OP} = \frac{PN}{1} = PN$$

$$\cos\theta = \frac{ON}{OP} = \frac{ON}{1} = ON$$

In ΔOPQ,

$$\tan\theta = \frac{PQ}{OP} = \frac{PQ}{1} = PQ$$

$$\sec\theta = \frac{OQ}{OP} = \frac{OQ}{1} = OQ$$

In ΔOPR,

$$\cot\theta = \frac{PR}{OP} = \frac{PR}{1} = PR$$

$$\csc\theta = \frac{OR}{OP} = \frac{OR}{1} = OR$$

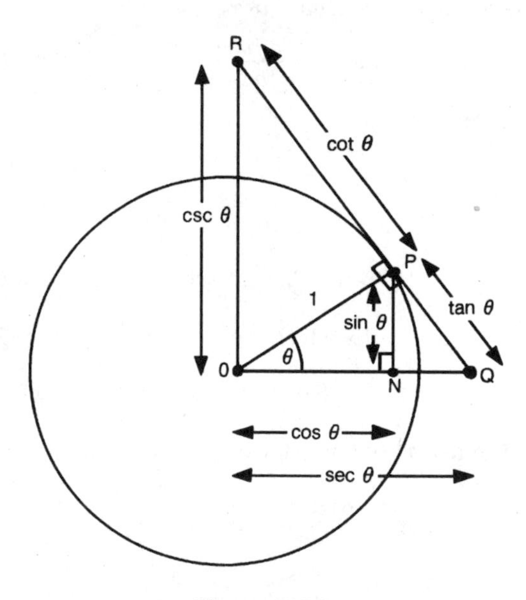

Figure 205-2

Therefore, the values of all six trigonometric functions can be obtained with the aid of measurements taken from the circle of unit radius. The graphs of csc θ, sec θ, and cot θ versus θ, can then be plotted as shown in Fig. 205-3.

Relationships involving csc, sec, and cot

From Equation 203-9:

$$\sin^2\theta + \cos^2\theta = 1$$

Therefore:

$$\frac{\sin^2\theta}{\sin^2\theta} + \frac{\cos^2\theta}{\sin^2\theta} = \frac{1}{\sin^2\theta}$$

$$1 + \cot^2\theta = \csc^2\theta \qquad (205\text{-}13)$$

This relationship is also apparent from ΔORP, where:

$$OP^2 + PR^2 = OR^2 \text{ (see Fig. 205-2).}$$

From Equation 203-9,

$$\cos^2\theta + \sin^2\theta = 1$$

Therefore:

$$\frac{\cos^2\theta}{\cos^2\theta} + \frac{\sin^2\theta}{\cos^2\theta} = \frac{1}{\cos^2\theta}$$

$$1 + \tan^2\theta = \sec^2\theta \qquad (205\text{-}14)$$

This relationship is also apparent from ΔOPQ, where:

$$OP^2 + PQ^2 = OQ^2$$

(see Fig. 205-2).

Example 205-1

Find the values of (a) csc 42.3°, (b) sec 37.4°, and (c) cot 18.6°.

Solution

$$\text{(a) } \csc 42.3° = \frac{1}{\sin 42.3°}$$

Entry, Key	Display
42.3° (sin)	0.6730
(1/x)	1.486

The value of csc 42.3° = 1.486, rounded off.

$$\text{(b) } \sec 37.4° = \frac{1}{\cos 37.4°}$$

Entry, Key	Display
37.4° (cos)	0.7944
(1/x)	1.259

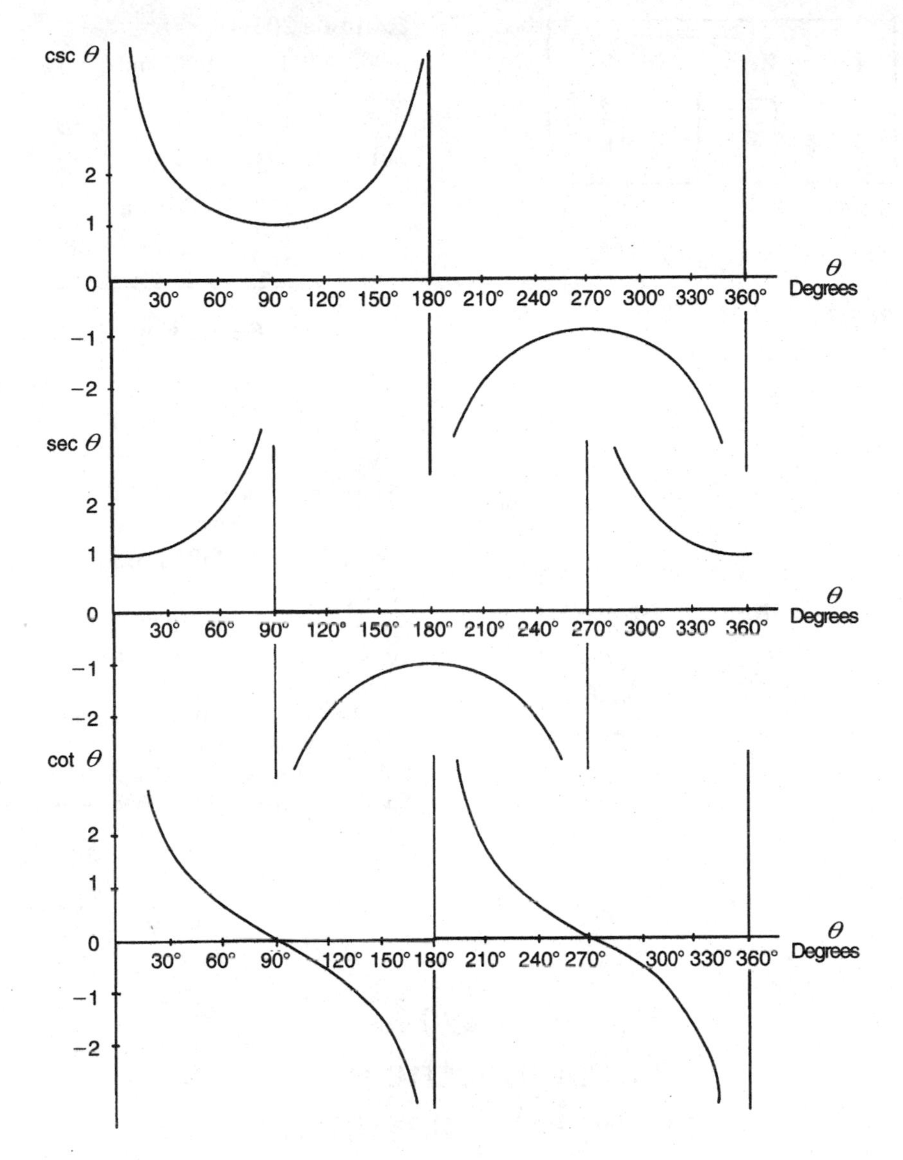

Figure 205-3

The value of sec 37.4° is 1.259.

$$\text{(c) } \cot 18.6° = \frac{1}{\tan 18.6°}$$

Entry, Key	Display
18.6° (tan)	0.3365
(1/x)	2.971

The value of cot 18.6° is 2.971.

Example 205-2

In Fig. 205-1, $B = 42.3°$ and $b = 13.2$. Find the value of c.

Solution

$$\text{Hypotenuse } c = b \csc B \qquad (205\text{-}3)$$

$$= 13.2 \csc 42.3°$$

Entry, Key	Display
13.2 ⊠	13.2
1.486 ⊟	19.6

The value of hypotenuse c is 19.6.

Example 205-3

In Fig. 205-1, $B = 37.4°$ and $a = 2.78$. Find the value of c.

Solution

$$\text{Hypotenuse } c = a \sec B \qquad (205\text{-}7)$$

Entry, Key	Display
2.78 ⊠	2.78
1.259 ⊟	3.50

The value of c is 3.50.

Example 205-4

In Fig. 205-1, $B = 18.6°$ and $b = 117.6$. Find the value of a.

Solution

$$a = b \cot B \qquad (205\text{-}11)$$

$$= 117.6 \cot 18.6°.$$

Entry, Key	Display
117.6° ⊠	117.6
2.971 ⊟	349.4

The value of side a is 349.4.

PRACTICE PROBLEMS

205-1. Find the cosecant values of 71.4°, 123.7°, 213.8°, and 317.3°.
205-2. Find the secant values of 51.4°, 111.8°, 243.6°, and 294.8°.
205-3. Find the cotangent values of 17.2°, 171.8°, 201.4°, and 283.2°.
205-4. If $1 + \cot^2\theta = 3.7436$, determine the value of θ.
205-5. If $1 + \tan^2\theta = 1.4638$, determine the value of θ.

206 Trigonometric functions for angles of any magnitude

Use Fig. 206-1 to obtain the values of the trigonometric functions for an angle of any magnitude. For positive angles, start at the horizontal reference line (OX) and rotate the line OP counterclockwise through the desired angle. This might involve a number of complete rotations, but eventually OP will lie in one of the four quadrants. As examples, OP_1 is in the first quadrant and OP_2, OP_3, and OP_4 are, respectively, in the second, third, and fourth quadrants. Perpendiculars are then dropped from points P_1, P_2, P_3, and P_4 onto the horizontal line $X'OX$ so that the right-angled triangles OP_1N_1, OP_2N_2, OP_3N_3, and OP_4N_4 are created. The same procedure is used for negative angles, except that the rotation from the reference line (OX)is in the clockwise direction.

In defining the trigonometric functions for angles of any size, it is necessary to consider the directions in which the perpendiculars (P_1N_1, P_2N_2, P_3N_3, P_4N_4) and the lines (ON_1, ON_2, ON_3, ON_4) are measured. All distances measured horizontally from left to right starting at point O are considered to be positive, and those measured in the opposite direction are taken as negative. Therefore, ON_1 and ON_4 are positive, but ON_2 and ON_3 are negative. Similarly, all distances measured vertically upward from $X'OX$ are positive; whereas all distances measured vertically downward

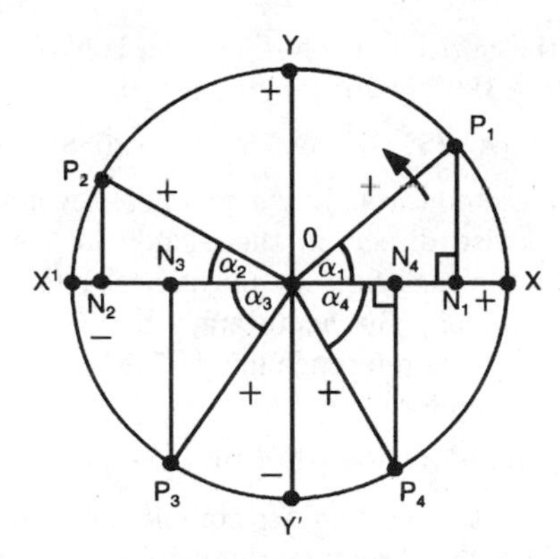

Figure 206-1

are negative. It follows that P_1N_1 and P_2N_2 are positive, but P_3N_3 and P_4N_4 are negative. By contrast, lines OP_1, OP_2, OP_3, and OP_4 are always considered to be positive.

The magnitude of the trigonometric function for any angle is equal to the value of the corresponding function for the acute angle formed by the line OP and the horizontal line $X'OX$. Suppose that, for any angle, the line OP (after its rotation) reaches the position OP_1 in the first quadrant; then:

$$\sin\theta = \frac{(+)P_1N_1}{(+)OP_1} = +\frac{P_1N_1}{OP_1}$$
$$= \sin\alpha_1$$

$$\cos\theta = \frac{(+)ON_1}{(+)OP_1} = +\frac{ON_1}{OP_1}$$
$$= \cos\alpha_1$$

$$\tan\theta = \frac{(+)P_1N_1}{(+)ON_1} = +\frac{P_1N_1}{ON_1}$$
$$= \tan\alpha_1$$

The sine, cosine, and tangent functions all have positive values in the first quadrant.

When OP reaches the position of OP_2 in the second quadrant, then:

$$\sin\theta = \frac{(+)P_2N_2}{(+)OP_2} = +\frac{P_2N_2}{OP_2}$$
$$= \sin\alpha_2$$

$$\cos\theta = \frac{(-)ON_2}{(+)OP_2} = -\frac{ON_2}{OP_2}$$
$$= -\cos\alpha_2$$

$$\tan\theta = \frac{(+)P_2N_2}{(-)ON_2} = -\frac{P_2N_2}{ON_2}$$
$$= -\tan\alpha_2$$

The sine function has a positive value in the second quadrant, but the values of the cosine and tangent functions are negative.

When OP reaches the position of OP_3 in the third quadrant, then:

$$\sin\theta = \frac{(-)P_3N_3}{(+)OP_3} = -\frac{P_3N_3}{OP_3}$$
$$= -\sin\alpha_3$$

$$\cos\theta = \frac{(-)ON_3}{(+)OP_3} = -\frac{ON_3}{OP_3}$$
$$= -\cos\alpha_3$$

$$\tan\theta = \frac{(-)P_3N_3}{(-)ON_3} = +\frac{P_3N_3}{ON_3}$$
$$= \tan\alpha_3$$

The tangent function has a positive value in the third quadrant, but the values of the sine and cosine functions are negative.

When OP reaches the position of OP_4 in the fourth quadrant, then:

$$\sin\theta = \frac{(-)P_4N_4}{(+)OP_4} = -\frac{P_4N_4}{OP_4}$$
$$= -\sin\alpha_4$$

$$\cos\theta = \frac{(+)ON_4}{(+)OP_4} = \frac{ON_4}{OP_4}$$
$$= \cos\alpha_4$$

$$\tan\theta = \frac{(-)P_4N_4}{(+)ON_4} = -\frac{P_4N_4}{ON_4}$$
$$= -\tan\alpha_4$$

The cosine function has a positive value in the fourth quadrant, but the values of the sine and tangent functions are negative. The positive and negative quadrant signs for the six trigonometric functions are summarized in Fig. 206-2.

MATHEMATICAL DERIVATIONS

If $\sin\theta = \sin\alpha$, where θ is an angle of any size and α is an acute angle, then:

$$\text{Angle } \theta = 2n\pi + \alpha \text{ radians} \qquad (206\text{-}1)$$
$$= 360n + \alpha \text{ degrees}$$

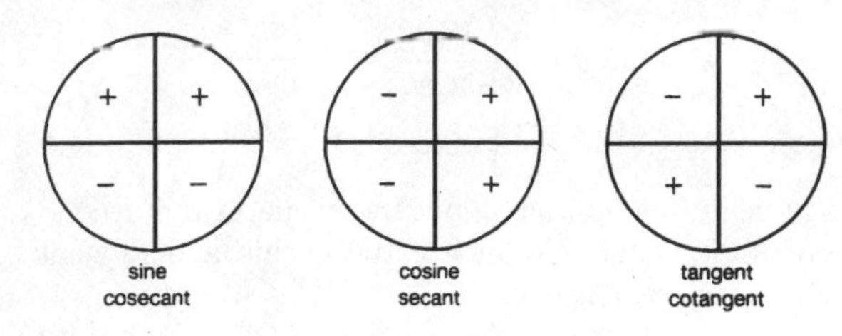

Figure 206-2

or,

$$\theta = (2n + 1)\pi - \alpha \text{ radians} \qquad (206\text{-}2)$$
$$= (2n + 1)180 - \alpha \text{ degrees}$$

where n is any positive or negative integer (whole number), or zero.

Therefore, if $\alpha = 65°$, sin 115°, sin 425°, sin 475°, sin 785°, sin 835°, and so on, all have the same value as sin 65° = 0.9063.

If $\cos \theta = \cos \alpha$, where θ is an angle of any size and α is an acute angle, then:

$$\text{Angle } \theta = 2n\pi \pm \alpha \text{ radians} \qquad (206\text{-}3)$$
$$= 360n \pm \alpha \text{ degrees}$$

Therefore, if $\alpha = 65°$, cos (−65°), cos 295°, cos 425°, cos 785°, and so on, all have the same value as cos 65° = 0.4226.

If $\tan \theta = \tan \alpha$, where θ is an angle of any size and α is an acute angle, then:

$$\text{Angle } \theta = n\pi + \alpha \text{ radians} \qquad (206\text{-}4)$$
$$= 180n + \alpha \text{ degrees}$$

Therefore, if $\alpha = 65°$, tan 245°, tan 425°, tan 605°, tan 785°, tan 965°, and so on, all have the same value as tan 65° = 2.145.

Example 206-1

Find the values of (a) sin 1137°, (b) cos 975°, (c) tan (−739°), and (d) sin (−956°).

Solution

(a) *OP* is rotated through three complete revolutions in the counterclockwise direction, then lands in the first quadrant. The acute angle between *OP* and the horizontal reference line is 1137° − (3 × 360°) = 57°. Therefore,

$$\sin 1137° = +\sin 57° = 0.8387$$

(b) *OP* is rotated through two complete revolutions in the counterclockwise direction, then lands in the third quadrant, which carries a negative sign for the cosine function. The acute angle between *OP* and the horizontal reference line is 975° − [(2 × 360°) + 180°] = 75°. Therefore,

$$\cos 975° = -\cos 75° = -0.2588$$

(c) *OP* is rotated through two complete revolutions in a clockwise direction, then lands in the fourth quadrant, which carries a negative sign for the tangent function. The acute angle between *OP* and the horizontal reference line is 739° − (2 × 360°) = 19°. Therefore,

$$\tan (-739°) = -\tan 19° = -0.3443$$

(d) *OP* is rotated through two complete revolutions in the clockwise direction, then lands in the second quadrant, which carries a positive sign for the sine function. The acute angle between *OP* and the horizontal reference line is 956° − [(2 × 360°) + 180°] = 56°. Therefore,

$$\sin (-956°) = +\sin 56° = 0.8290$$

Most, but not all, scientific calculators enable the values of the functions for any angle to be obtained directly. In the example of sin (−956°),

Entry, Key	Display
956° ±	−956°
sin	0.8290

Therefore:

$$\sin (-956°) = 0.8290$$

Example 206-2

The equation of a sine-wave voltage is $e = 2.7 \sin 2\pi ft$, where $f = 200$ kHz. What is the instantaneous value of the voltage if $t = 12\ \mu s$?

Solution

Instantaneous voltage,

$$e = 2.7 \sin (2\pi \times 200 \times 10^3 \times 12 \times 10^{-6})^c$$
$$= 2.7 \sin (15.08^c)$$
$$= 2.7 \sin (864°)$$
$$= 1.59 \text{ V}$$

PRACTICE PROBLEMS

206-1. Find the sine values of 138°, −217°, 573°, −732°, and 4.72°.

206-2. Find the cosine values of 232°, −478°, 694°, −138.2°, and −2.17°.

206-3. Find the tangent values of 98.4°, −338°, 538°, −634°, and 3.64°.

206-4. Find the value of csc 739°, sec (−417°), and cot 436°.

206-5. If the equation of an ac voltage is $e = 7.3 \cos 2\pi ft$ V, where $f = 375$ kHz, what is the instantaneous value of the voltage if $t = 1.8$ μs?

207 The sine and cosine rules

Chapters 202 through 206 have only been concerned with right-angled triangles. However, in the analysis of ac circuits, the phasor diagrams can involve general triangles. Moreover, you will need more than Pythagoras' theorem to find the resultant of two phasors that are not mutually perpendicular.

A general triangle can be precisely defined in any one of three ways. In such cases, the given information is:

- One side and two angles
- Two sides and the included angle
- Three sides

In the first case, the triangle is solved completely with the aid of the sine rule. For the second and third cases, the cosine rule is used.

MATHEMATICAL DERIVATIONS

The sine rule

In Fig. 207-1, lines AD and BE are respectively perpendicular to BC and AC.

In ΔABD:

$$\sin B = \frac{AD}{c} \tag{207-1}$$

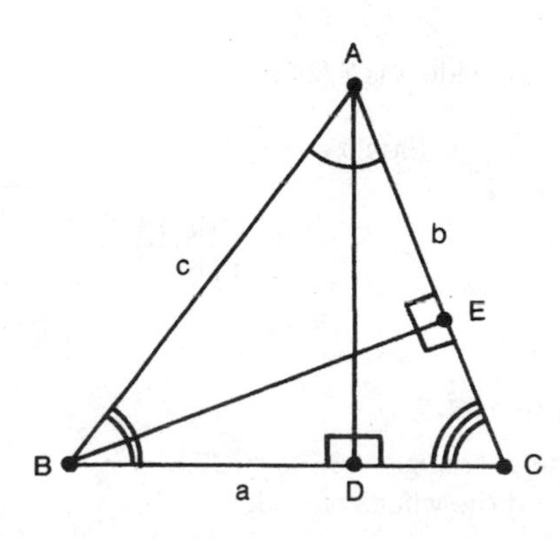

Figure 207-1

or,

$$AD = c \sin B$$

In ΔADC:

$$\sin C = \frac{AD}{b} \tag{207-2}$$

or,

$$AD = b \sin C$$

Therefore:

$$AD = c \sin B$$
$$= b \sin C$$

and,

$$\frac{c}{\sin C} = \frac{b}{\sin B} \tag{207-3}$$

In ΔABE:

$$\sin A = \frac{BE}{c} \tag{207-4}$$

or,

$$BE = c \sin A$$

In ΔBCE:

$$\sin C = \frac{BE}{a} \tag{207-5}$$

or,

$$BE = a \sin C$$

Therefore:

$$BE = c \sin A$$
$$= a \sin C$$

and,

$$\frac{c}{\sin C} = \frac{a}{\sin A} \tag{207-6}$$

Combining equations 207-3 and 207-6 yields:

$$\frac{a}{\sin A} = \frac{b}{\sin B} = \frac{c}{\sin C} \tag{207-7}$$

This important relationship is called the *sine rule* of a general triangle.

The cosine rule

In Fig. 207-2, AD is perpendicular to BC. If $BD = x$, $CD = a - x$. Then, in ΔABD:

$$AD^2 = c^2 - x^2$$

and,

$$x = c \cos B$$

In ΔACD,

$$AD^2 = b^2 - (a - x)^2$$

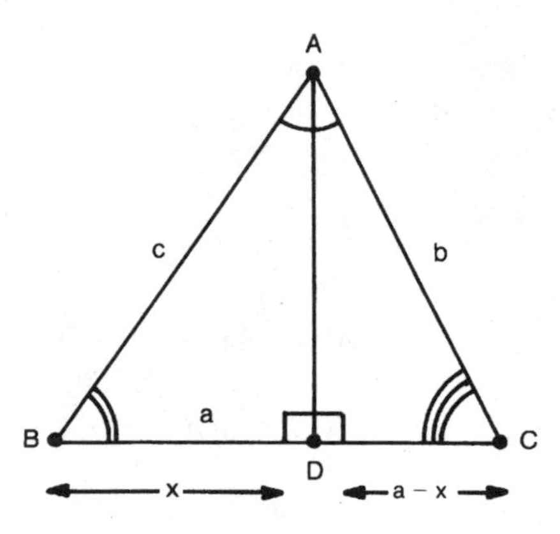

Figure 207-2

Therefore:

$$c^2 - x^2 = b^2 - (a - x)^2$$
$$= b^2 - a^2 + 2ax - x^2$$

or,

$$b^2 = a^2 + c^2 - 2ax \quad (207\text{-}8)$$
$$= a^2 + c^2 - 2ac \cos B$$

Therefore:

$$b = \sqrt{a^2 + c^2 - 2ac \cos B}$$
$$a = \sqrt{b^2 + c^2 - 2bc \cos A} \quad (207\text{-}9)$$
$$c = \sqrt{a^2 + b^2 - 2ab \cos C}$$

These relationships constitute the *cosine rule* of a general triangle. This rule will be used to find the third side of a triangle, when two sides and the included angle are given. The sine rule then allows a second angle to be found.

The relationships for the cosine rule can be rearranged as:

$$\cos A = \frac{b^2 + c^2 - a^2}{2bc} \quad (207\text{-}10)$$

$$\cos B = \frac{a^2 + c^2 - b^2}{2ac} \quad (207\text{-}11)$$

$$\cos C = \frac{a^2 + b^2 - c^2}{2ab} \quad (207\text{-}12)$$

This form of the cosine rule is used to find one of the angles when the three sides are given. The sine rule then allows a second angle to be obtained. Notice that in Equation 207-10, it is possible for $a^2 > b^2 + c^2$ so that the value of $\cos A$ is negative; angle A is then obtuse.

Example 207-1

In a general triangle ABC, $A = 37.6°$, $B = 78.7°$, and $c = 11.6$. Find the values of C, a, and b.

Solution

$$\text{Angle } C = 180° - 37.6° - 78.7°$$
$$= 63.7°$$

Then,

$$\frac{a}{\sin A} = \frac{c}{\sin C} \quad (207\text{-}7)$$

$$\text{Side } a = \frac{c \sin A}{\sin C}$$
$$= \frac{11.6 \times \sin 37.6°}{\sin 63.7°}$$

Entry, Key	Display
11.6 [×]	11.6
37.6° [sin] [÷]	7.078
63.7° [sin]	0.8956
[=]	7.895

The value of side a is 7.895.

$$\text{Side } b = \frac{c \sin B}{\sin C}$$
$$= \frac{11.6 \sin 78.7°}{\sin 63.7°}$$
$$= 12.69$$

Example 207-2

In a general triangle ABC, $b = 27.5$, $c = 47.8$, and $A = 37.2°$. Find the values of a, B, and C.

Solution

$$\text{Side } a = \sqrt{b^2 + c^2 - 2bc \cos A} \qquad (207\text{-}9)$$
$$= \sqrt{27.5^2 + 47.8^2 - 2 \times 27.5 \times 47.8 \times \cos 37.2^\circ}$$
$$= 30.7$$

Entry, Key	Display
27.5 [x²] [+]	756.25
47.8 [x²] [−]	2284.84
[(]	0
2 [×]	2
27.5 [×]	55
47.8 [×]	2629
37.2° [cos] [)]	2094
[=] [√x]	30.7

Then,

$$\frac{\sin B}{b} = \frac{\sin A}{a} \qquad (207\text{-}7)$$

Therefore:

$$\text{Angle } B = \text{inv sin}\left(\frac{b \sin A}{a}\right)$$
$$= \text{inv sin}\left(\frac{27.5 \sin 37.2^\circ}{30.7}\right)$$
$$= 32.7^\circ$$

and,

$$\text{Angle } C = 180^\circ - 37.2^\circ - 32.7^\circ$$
$$= 110.1^\circ$$

Example 207-3

In a general triangle ABC, $c = 23.5$, $b = 17.3$, and $a = 12.7$. Find the values of the angles C, B, and A.

Solution

$$\cos C = \frac{a^2 + b^2 - c^2}{2ab} \qquad (207\text{-}12)$$

Then

$$\text{Angle } C = \text{inv cos}\left(\frac{12.7^2 + 17.3^2 - 23.5^2}{2 \times 12.7 \times 17.3}\right)$$

Entry, Key	Display
12.7 [x²] [+]	161.29
17.3 [x²] [−]	299.29
23.5 [x²] [=] [÷]	−91.67
2 [÷]	−45.835
12.7 [÷]	−3.609
17.3 [=]	−0.2086
[inv] [cos]	102.04°

Therefore the angle A is obtuse and has a value of 102.04°.

Then,

$$\frac{\sin B}{b} = \frac{\sin C}{c} \qquad (207\text{-}7)$$

$$\text{Angle } B = \text{inv sin}\left(\frac{b \sin C}{c}\right)$$
$$= \text{inv sin}\left(\frac{17.3 \sin 102.04^\circ}{23.5}\right)$$
$$= 46.05^\circ$$
$$\text{Angle } A = 180^\circ - 102.04^\circ - 46.05^\circ$$
$$= 31.91^\circ$$

Example 207-4

In the parallelogram of Fig. 207-3, find the values of c and the angle $\angle ABC$.

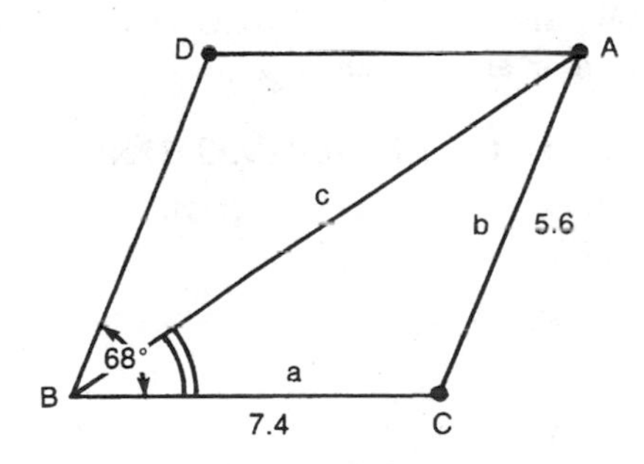

Figure 207-3

Solution

$$\angle ACB = 180^\circ - 68^\circ$$
$$= 112^\circ \text{ (an obtuse angle)}$$
$$\text{Side } c = \sqrt{a^2 + b^2 - 2ab \cos C} \qquad (207\text{-}9)$$
$$= \sqrt{7.4^2 + 5.6^2 - 2 \times 7.4 \times 5.6 \cos 112^\circ}$$
$$= 10.8$$

$$\angle ABC = \text{inv}\sin\left(\frac{5.6 \times \sin 112°}{10.8}\right)$$
$$= 28.7°$$

This example illustrates one method of obtaining the sum of two phasors (see chapter 66).

PRACTICE PROBLEMS

207-1. In Δ ABC, $A = 44°$, $B = 73°$, and $c = 4.72$. Find the values of C, a, and b.

207-2. In ΔABC, $a = 11.94$, $b = 13.42$, and $C = 117°$. Find the values of c, A, and B.

207-3. In Δ ABC, $a = 0.396$, $b = 0.543$, and $C = 37°$. Find the values of c, A, and B.

207-4. In Δ ABC, $a = 4.37$, $b = 5.49$, and $c = 7.78$. Find the values of A, B, and C.

207-5. In Δ ABC, $B = 37°$, $b = 4.36$, and $c = 5.94$. Find the values of a, A, and C.

207-6. The trigonometric expressions of two voltage phasors are $e_1 = 11.4 \sin(\omega t + 43°)$ and $e_2 = 13.7 \sin(\omega t - 14°)$. Obtain the trigonometric expressions for $e_1 + e_2$ and $e_1 - e_2$.

208
Multiple angles

Chapters 202 through 206 have contained only equations that involve single angles, such as A and B. Now, turn your attention to multiple angles that involve more than one angle; for example, $A + B$ and $A - B$ are regarded as multiple angles.

The expansions of $\sin(A + B)$ and $\cos(A + B)$ lead to other equations that are important in electronics and communications theory. The analysis of an amplitude-modulated wave depends on multiple angle equations. Such equations also enable us to obtain the true power in an ac circuit.

MATHEMATICAL DERIVATIONS

In Fig. 208-1, let OP be of unit length.

In ΔOPR,

$$\sin(A + B) = \frac{PR}{OP} = \frac{PR}{1} = PR$$

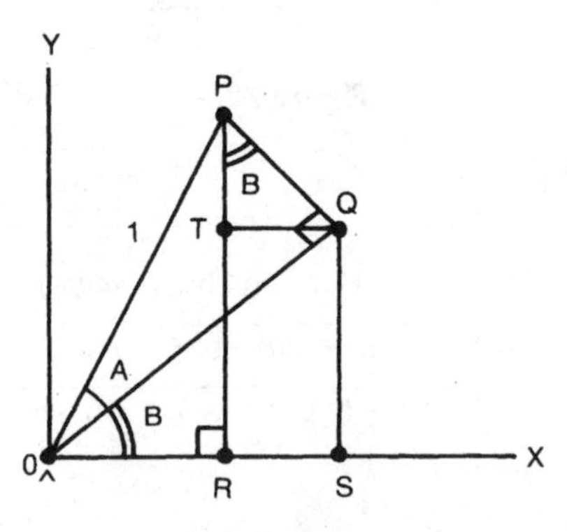

Figure 208-1

In ΔOPQ,

$$\sin A = \frac{PQ}{OP} = \frac{PQ}{1} = PQ$$

In ΔPTQ,

$$\cos B = \frac{PT}{PQ} = \frac{PT}{\sin A}$$

Therefore:

$$PT = \sin A \cos B$$

In ΔOPQ,

$$\cos A = \frac{OQ}{OP} = \frac{OQ}{1} = OQ$$

In ΔOQS,

$$\sin B = \frac{QS}{OQ} = \frac{QS}{\cos A}$$

Therefore:

$$QS = TR = \cos A \sin B$$

Because,

$$PR = PT + TR$$

$$\sin(A + B) = \sin A \cos B + \cos A \sin B \qquad (208\text{-}1)$$

In ΔOPR,

$$\cos(A + B) = \frac{OR}{OP} = \frac{OR}{1} = OR$$

In ΔOPQ,

$$\cos A = \frac{OQ}{OP} = \frac{OQ}{1} = OQ$$

In ΔOQS,

$$\cos B = \frac{OS}{OQ} = \frac{OS}{\cos A}$$

Therefore:

$$OS = \cos A \cos B$$

In ΔPTQ,

$$\sin B = \frac{QT}{PQ} = \frac{QT}{\sin A}$$

Therefore:

$$QT = \sin A \sin B$$

Because,

$$OR = OS - SR = OS - QT$$

$$\cos (A + B) = \cos A \cos B - \sin A \sin B \qquad (208\text{-}2)$$

Alternative proof

In Fig. 208-2, let OP represent a unit vector. Resolving OP along the perpendicular directions (OX, OY), the component along the direction $OX = \cos (A + B)$. The component along direction $OY = \sin (A + B)$.

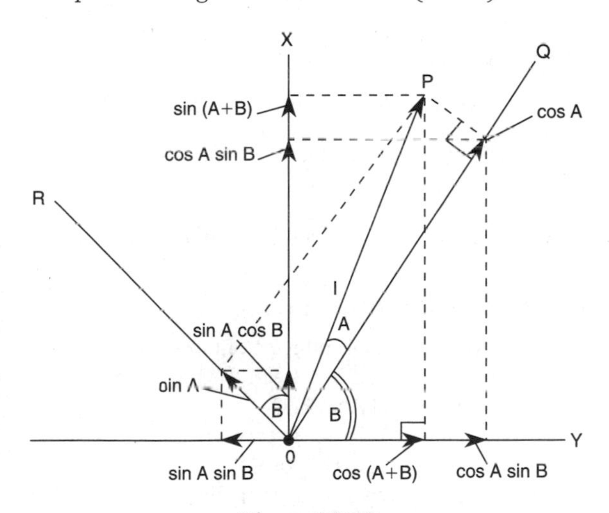

Figure 208-2

Resolving OP along the perpendicular directions (OQ, OR), the component along direction $OQ = \cos A$. The component along the direction $OR = \sin A$.

Resolving the components for the directions (OQ, OR) in direction OY, the resultant along direction $OY = \sin A \cos B + \cos A \sin B$. Therefore:

$$\sin (A + B) = \sin A \cos B + \cos A \sin B \qquad (208\text{-}1)$$

Resolving the components for the directions (OQ, OR) in the direction OX, the resultant along the direction $OX = \cos A \cos B - \sin A \sin B$. Therefore:

$$\cos (A + B) = \cos A \cos B - \sin A \sin B \qquad (208\text{-}2)$$

This vector proof is true for angles of any magnitude.

In Equation 208-1, replace $+B$ by $-B$. Then,

$$\sin (A - B) = \sin A \cos (-B) + \cos A \sin (-B)$$

or:

$$\sin (A - B) = \sin A \cos B - \cos A \sin B \qquad (208\text{-}3)$$

Adding Equations 208-1 and 208-3:

$$\sin (A + B) + \sin (A - B) = 2 \sin A \cos B \qquad (208\text{-}4)$$

Subtracting Equation 208-3 from 208-1:

$$\sin (A + B) - \sin (A - B) = 2 \cos A \sin B \qquad (208\text{-}5)$$

In Equation 208-2, replace $+B$ by $-B$. Then,

$$\cos (A - B) = \cos A \cos (-B) - \sin A \sin (-B)$$

or:

$$\cos (A - B) = \cos A \cos B + \sin A \sin B \qquad (208\text{-}6)$$

Adding Equations 208-2 and 208-6:

$$\cos (A + B) + \cos (A - B) = 2 \cos A \cos B \qquad (208\text{-}7)$$

Subtracting Equation 208-2 from Equation 208-6:

$$\cos (A - B) - \cos (A + B) = 2 \sin A \sin B \qquad (208\text{-}8)$$

In Equation 208-1, replace B by A. Then,

$$\sin (A + A) = \sin A \cos A + \cos A \sin A$$

or,

$$\sin 2A = 2 \sin A \cos A \qquad (208\text{-}9)$$

In Equation 208-2, replace B by A. Then,

$$\cos 2A = \cos A \cos A - \sin A \sin A \qquad (208\text{-}10)$$

$$= \cos^2 A - \sin^2 A$$

$$= \cos^2 A - (1 - \cos^2 A)$$

$$= 2 \cos^2 A - 1 \qquad (208\text{-}11)$$

or,

$$\cos 2A = \cos^2 A - \sin^2 A$$

$$= (1 - \sin^2 A) - \sin^2 A$$

$$= 1 - 2 \sin^2 A \qquad (208\text{-}12)$$

Similarly:

$$\sin 3A = \sin (A + 2A) = \sin A \cos 2A + \sin 2A \cos A$$

$$= \sin A (1 - 2 \sin^2 A) + 2 \sin A \cos A \times \cos A$$

$$= \sin A - 2 \sin^3 A + 2 \sin A (1 - \sin^2 A)$$

$$= 3 \sin A - 4 \sin^3 A \qquad (208\text{-}13)$$

$$\cos 3A = \cos (A + 2A) = \cos A \cos 2A - \sin 2A \sin A$$

$$= \cos A (2 \cos^2 A - 1) - 2 \sin A \cos A \times \sin A$$

$$= 2 \cos^3 A - \cos A - 2 \cos A (1 - \cos^2 A)$$

$$= 4 \cos^3 A - 3 \cos A \qquad (208\text{-}14)$$

Divide Equation 208-1 by Equation 208-2. Then:

$$\frac{\sin(A+B)}{\cos(A+B)} = \frac{\sin A \cos B + \cos A \sin B}{\cos A \cos B - \sin A \sin B}$$

$$\tan(A+B) = \frac{(\sin A \cos B + \cos A \sin B)/\cos A \cos B}{(\cos A \cos B - \sin A \sin B)/\cos A \cos B}$$

$$\tan(A+B) = \frac{\tan A + \tan B}{1 - \tan A \tan B} \qquad (208\text{-}15)$$

In Equation 208-15, replace $+B$ by $-B$. Then:

$$\tan(A-B) = \frac{\tan A + \tan(-B)}{1 - \tan A \tan(-B)}$$

$$\tan(A-B) = \frac{\tan A - \tan B}{1 + \tan A \tan B} \qquad (208\text{-}16)$$

Example 208-1

In an ac circuit, the source voltage $e = 155 \sin 337t$ V, and the corresponding source current $i = 3.7 \cos 377t$ A. Calculate the value of the average power over the cycle of the source voltage.

Solution

$$\begin{aligned} \textit{Instantaneous power} &= e \times i \\ &= 155 \sin 377t \times 3.7 \cos 377t \\ &= \frac{155 \times 3.7}{2} \sin(2 \times 377t) \\ &= 286.75 \sin 754t \text{ VAr} \qquad (208\text{-}9) \end{aligned}$$

The average value of sin $754t$ over a cycle is 0. Therefore, the average power over the cycle is zero because the load on the source is a capacitive reactance whose value is 155 V/3.7 A = 42 Ω, rounded off.

Example 208-2

In an ac circuit, the source voltage $e = 80 \sin 1000t$ V, and the corresponding source current $i = 4 \sin(1000t - 40°)$ A. Calculate the value of the average power over the complete cycle of the source voltage.

Solution

Instantaneous power,

$$p = e \times i = 80 \sin 1000t \times 4 \sin(1000t - 40°)$$

$$= \frac{80 \times 4}{2}[\cos 40° - \cos(2000t - 40°)]$$

$$= 160 \cos 40° - 160 \cos(2000t - 40°)$$

The average value of the term "$-160 \cos(2000t - 40°)$" is zero over the complete cycle of the source voltage. Therefore, the average or true power over the cycle is 160 cos 40° = 122.6 W.

Example 208-3

The equation of an amplitude-modulated wave is $e = A \sin 2\pi f_c t + B \sin 2\pi f_c t \sin 2\pi f_m t$. Analyze the wave into its separate sine and cosine components (see chapter 136).

Solution

The equation of the AM wave is:

$$\begin{aligned} e &= A \sin 2\pi f_c t + B \sin 2\pi f_c t \sin 2\pi f_m t \\ &= A \sin 2\pi f_c t + \frac{B}{2}[\cos 2\pi(f_c - f_m)t - \cos 2\pi(f_c + f_m)t] \qquad (208\text{-}8) \end{aligned}$$

The separate components are:

$A \sin 2\pi f_c t$, which is the carrier component

$\frac{B}{2} \cos 2\pi(f_c - f_m)t$, which is the lower sideband

$\frac{B}{2} \cos 2\pi(f_c + f_m)t$, which is the upper sideband.

PRACTICE PROBLEMS

208-1. Express tan $2A$ in terms of tan A.

208-2. If $\cos A - \cos B = 2 \sin 37° \sin 26°$, what are the values of A and B?

208-3. Express (a) sin 32° in terms of cos 64° and (b) cos 32° in terms of cos 64°.

208-4. If $\sin A + \sin B = 2 \sin 43° \cos 21°$, what are the values of A and B?

208-5. Express 2 sin 19° cos 19° in terms of cos 38°.

208-6. If $e_1 = 11 \sin 4000t$ and $e_2 = 11 \sin 2000t$, express $e_1 + e_2$ as the product of sine and cosine terms.

208-7. If $\tan x = \frac{\tan 57° - \tan 24°}{1 + \tan 57° \tan 24°}$, what is the value of x?

209
The binomial series

Consider the following expansions:

$$(1 + x)^1 = 1 + x$$

$$(1 + x)^2 = 1 + 2x + x^2$$

$$(1 + x)^3 = 1 + 3x + 3x^2 + x^3$$

$$(1 + x)^4 = 1 + 4x + 6x^2 + 4x^3 + x^4$$

Clearly, there is a pattern to the values of the coefficients of x, x^2, x^3, and so on. You can assume that in the expansion of $(1 + x)^5$, the coefficients of x, x^2, x^3, and x^4 are respectively $1 + 4 = 5$, $4 + 6 = 10$, $6 + 4 = 10$, $4 + 1 = 5$, so that:

$$(1 + x)^5 = 1 + 5x + 10x^2 + 10x^3 + 5x^4 + x^5$$

A general form of these expansions is given by the binomial series:

$$(1 + x)^n =$$

$$1 + \frac{nx}{1} + \frac{n(n-1)x^2}{1 \times 2} + \frac{n(n-1)(n-2)x^3}{1 \times 2 \times 3} + \ldots$$

$$+ \frac{n(n-1)(n-2)\ldots(n-r+1)x^r}{1 \times 2 \times 3 \times \ldots \times r} + \ldots + x^n$$

If $n = 5$,

$$(1 + x)^5 = 1 + \frac{5x}{1} + \frac{5 \times 4x^2}{1 \times 2} + \frac{5 \times 4 \times 3x^3}{1 \times 2 \times 3}$$

$$+ \frac{5 \times 4 \times 3 \times 2x^4}{1 \times 2 \times 3 \times 4}$$

$$+ \frac{5 \times 4 \times 3 \times 2 \times 1x^5}{1 \times 2 \times 3 \times 4 \times 5}$$

$$= 1 + 5x + 10x^2 + 10x^3 + 5x^4 + x^5$$

This is the same result as was previously deduced.

The terms of the binomial expansion can be simplified by using the *factorial* notation. "Factorial n" is written as $n!$ or $\angle n$ and is defined by:

$$n! = n \times (n-1) \times (n-2) \ldots \times 4 \times 3 \times 2 \times 1$$

As an example,

$$6! = 6 \times 5 \times 4 \times 3 \times 2 \times 1 = 720$$

The binomial expansion can then be rewritten as:

$$(1 + x)^n = 1 + \frac{nx}{1!} + \frac{n(n-1)x^2}{2!} + \qquad (209\text{-}1)$$

$$\frac{n(n-1)(n-2)x^3}{3!} + \ldots +$$

$$\frac{n(n-1)(n-2)\ldots(n-r+1)x^r}{r!} + \ldots + x^n$$

When the value of x is much less than 1 ($x << 1$), the binomial expansion can be used to simplify certain expressions. For example, $(1 + x)^6 \approx 1 + 6x$ if $x << 1$. The missing terms $15x^2$, $20x^3$, $15x^4$, $6x^5$, x^6 all have small values when compared to $6x$. If, for example, the value of x is 0.001, $(1 + x)^6 = (1 + 0.001)^6 \approx 1 + 0.006 = 1.006$; as opposed to the exact value of 1.006015020015006001. Incidentally, the binomial series has provided us with this exact value, which would not be available from some scientific calculators.

MATHEMATICAL DERIVATIONS

The term,

$$\frac{n(n-1)(n-2)\ldots(n-r+1)x^r}{r!}$$

is the general term of the expansion, and is used to find the coefficient of x^r, where r is a positive integer.

Consider the following expansion of $(1 + x)^{n+1}$:

$$(1 + x)^{n+1} = (1 + x)(1 + x)^n$$

$$= (1 + x)\left[1 + \frac{nx}{1!} + \frac{n(n-1)x^2}{2!} + \right.$$

$$\ldots + \frac{n(n-1)\ldots(n-r+2)x^{r-1}}{r-1!}$$

$$\left. + \frac{n(n-1)\ldots(n-r+1)x^r}{r!} + \ldots + x^n\right]$$

The x^r term in the expansion of $(1 + x)^{n+1}$ is:

$$\left[\frac{n(n-1)\ldots(n-r+2)}{r-1!}\right.$$

$$\left. + \frac{n(n-1)\ldots(n-r+1)}{r!}\right]x^r$$

$$= \left[\frac{n(n-1)\ldots(n-r+2)}{r-1!}\right.$$

$$\left. \times \left(1 + \frac{n-r+1}{r}\right)\right]x^r$$

$$= \frac{(n+1)(n)\ldots(n-r+2)x^r}{r!}$$

This expression is the same as would be obtained by replacing n by $n + 1$ in the coefficient of the x^r term in the expansion of $(1 + x)^n$. Consequently, if the series is true for $(1 + x)^n$, the series must be true for $(1 + x)^{n+1}$. But the series is obviously true for $n = 1$ because $(1 + x)^1 = 1 + x$. It follows that the series is true for $n = 2, n = 3$, and so on. The binomial theorem has, therefore, been proven if n is a positive integer. This form of proof is known as the *induction method.*

If n is not a positive integer, there is an infinite number of terms in the expansion. Under these conditions, the series is not valid unless the value of x lies between + 1 and − 1 $(1 > x > -1)$.

Example 209-1

Use the binomial series to expand the following:

(a) $(1 + x)^7$,

(b) $(1 - x)^5$,

(c) $\sqrt{1 + x}$,

(d) $1/(1 - x)$, and

(e) $(1 + x)^{5/2}$

Solution

(a) $$(1 + x)^7 = 1 + 7x + \frac{7 \times 6}{1 \times 2}x^2 + \frac{7 \times 6 \times 5}{1 \times 2 \times 3}x^3 + \frac{7 \times 6 \times 5 \times 4}{1 \times 2 \times 3 \times 4}x^4 + \frac{7 \times 6 \times 5 \times 4 \times 3}{1 \times 2 \times 3 \times 4 \times 5}x^5 + \frac{7 \times 6 \times 5 \times 4 \times 3 \times 2}{1 \times 2 \times 3 \times 4 \times 5 \times 6}x^6 + \frac{7 \times 6 \times 5 \times 4 \times 3 \times 2 \times 1}{1 \times 2 \times 3 \times 4 \times 5 \times 6 \times 7}x^7$$

$$= 1 + 7x + 21x^2 + 35x^3 + 35x^4 + 21x^5 + 7x^6 + x^7$$

Verification:

If $x = 1$,

$$(1 + 1)^7 = 2^7$$
$$= 1 + 7 + 21 + 35 + 35 + 21 + 7 + 1$$
$$= 128$$

(b) $$(1 - x)^5 = 1 + 5(-x) + \frac{5 \times 4}{1 \times 2}(-x)^2 + \frac{5 \times 4 \times 3}{1 \times 2 \times 3}(-x)^3 + \frac{5 \times 4 \times 3 \times 2}{1 \times 2 \times 3 \times 4}(-x)^4 + \frac{5 \times 4 \times 3 \times 2 \times 1}{1 \times 2 \times 3 \times 4 \times 5}(-x)^5$$

$$= 1 - 5x + 10x^2 - 10x^3 + 5x^4 - x^5$$

(c) $\sqrt{1 + x} = (1 + x)^{1/2}$

$$= 1 + \frac{1}{2}x + \frac{(1/2) \times (-1/2)}{1 \times 2}x^2 + \frac{(1/2) \times (-1/2) \times (-3/2)}{1 \times 2 \times 3}x^3 + \frac{(1/2) \times (-1/2) \times (-3/2) \times (-5/2)}{1 \times 2 \times 3 \times 4}x^4 \ldots$$

$$= 1 + \frac{1}{2}x - \frac{1}{8}x^2 + \frac{1}{16}x^3 - \frac{5}{128}x^4 \ldots$$

Notice that there is an infinite number of terms in the series.

If $x << 1$:

$$\sqrt{1 + x} \approx 1 + \frac{1}{2}x$$

For example, if $x = 0.004$,

$$\sqrt{1 + 0.004} = (1.004)^{1/2} \approx 1 + \frac{0.004}{2}$$
$$= 1.002$$

(d) $$\frac{1}{1 - x} = (1 - x)^{-1}$$

$$= 1 + (-1)(-x) + \frac{(-1)(-2)}{1 \times 2}(-x)^2 + \frac{(-1)(-2)(-3)}{1 \times 2 \times 3}(-x)^3 + \ldots$$

$$= 1 + x + x^2 + x^3 + \ldots$$

If $x = 2$,

$$\frac{1}{1 - x} = -1,$$

which is clearly not equal to $1 + 2 + 2^2 + 2^3 + \ldots$. The expansion is not valid when $x > 1$.

If $x << 1$:

$$\frac{1}{1 - x} = (1 - x)^{-1} \approx 1 - x$$

For example, if $x = 0.003$,

$$\frac{1}{1 - x} = \frac{1}{1 - 0.003}$$
$$= \frac{1}{0.997} \approx 1 + x$$
$$= 1 + 0.003$$
$$= 1.003$$

(e) $$(1 + x)^{5/2} = 1 + \frac{5x}{2} + \frac{(5/2)(3/2)x^2}{1 \times 2} + \frac{(5/2)(3/2)(1/2)x^3}{1 \times 2 \times 3} + \frac{(5/2)(3/2)(1/2)(-1/2)x^4}{1 \times 2 \times 3 \times 4} + \ldots$$

$$= 1 + \frac{5x}{2} + \frac{15x^2}{8} + \frac{5x^3}{16} - \frac{5x^4}{128} \cdots$$

If $x << 1$:

$$(1 + x)^{5/2} \approx 1 + \frac{5}{2}x$$

For example, if $x = 0.005$:

$$(1 + x)^{5/2} = (1.005)^{5/2} \approx 1 + \frac{5}{2} \times 0.005$$

$$= 1.0125$$

Example 209-2

On a transmission line, the propagation constant,

$$\gamma = \alpha + j\beta$$

$$= \sqrt{(R + j\omega L)(G + j\omega C)}$$

(chapter 148). If $\omega L >> R$ and $\omega C >> G$, find the values of α and β.

Solution

Propagation constant,

$$\gamma = j\omega\sqrt{LC}\sqrt{\left(1 + \frac{R}{j\omega L}\right)\left(1 + \frac{G}{j\omega C}\right)} \qquad (209\text{-}1)$$

$$= j\omega\sqrt{LC} \times \left(1 + \frac{R}{j\omega L}\right)^{1/2} \times \left(1 + \frac{G}{j\omega C}\right)^{1/2}$$

$$= j\omega\sqrt{LC}\left(1 + \frac{R}{2j\omega L} + \text{neglected terms}\right)$$

$$\times \left(1 + \frac{G}{2j\omega C} + \text{neglected terms}\right)$$

using the binomial expansion.

Therefore:

$\alpha + j\beta =$

$$j\omega\sqrt{LC}\left(1 + \frac{R}{2j\omega L} + \frac{G}{2j\omega C} + \text{neglected terms}\right)$$

$$= \frac{R}{2}\sqrt{\frac{C}{L}} + \frac{G}{2}\sqrt{\frac{L}{C}} + j\omega\sqrt{LC}$$

Equating real and imaginary parts, (see chapter 90),

Attenuation constant,

$$\alpha = \frac{R}{2}\sqrt{\frac{C}{L}} + \frac{G}{2}\sqrt{\frac{L}{C}} \text{ nepers per meter}$$

Phase shift constant, $\beta = \omega\sqrt{LC}$ radians per meter.

PRACTICE PROBLEMS

209-1. If $f_1 = \dfrac{f_o}{\sqrt{1-k}}$ and $f_2 = \dfrac{f_o}{\sqrt{1+k}}$, use the binomial series to find the value of $f_1 - f_2$. Assume that $k << 1$.

209-2. What is the binomial expansion of $(a + b)^n$?

209-3. Assuming that $\omega L_s >> R_s$, use the binomial series to find the approximate value of

$$\omega L_p - \frac{\omega L_p}{1 + \dfrac{R_S}{\omega^2 L_S{}^2}}$$

209-4. Expand the following in terms of a binomial series (state the first four terms of the series):

(a) $\sqrt{1 - \dfrac{x}{2}}$ (b) $\sqrt[4]{1 + 4x}$

209-5. Expand the following in terms of a binomial series (state the first four terms of the series):

(a) $\dfrac{1}{(1 - 2x)^3}$ (b) $\dfrac{1}{\sqrt{1 + \dfrac{x}{2}}}$ (c) $\sqrt[3]{1 + x^2}$

210
The exponential and logarithmic functions

Exponential growth and decay occur frequently in various scientific subjects. For example, trees grow exponentially because their heights increase rapidly when they are initially planted. As the tree becomes taller, its height increases less rapidly; theoretically, it would take an infinite time to achieve its maximum height. It is doubtful whether a tree could live that long! However, in a limited time period, the tree attains over 99 percent of its theoretical maximum height.

Newton's Law of Cooling involves exponential decay; this law states that the rate of heat loss from a hot body is directly proportional to the difference between the temperature of the body and the temperature of its surroundings. Consequently, as the temperature of the body approaches that of its surroundings, the rate of the heat loss lessens. Theoretically, it would take an infinite time for the temperatures of the body and its surroundings to become equal, but the body loses virtually all of its excess heat in a limited period of time.

Radioactive substances decay exponentially, so as their intensities decline, the rate of decline is reduced. To lose all of the radioactivity would theoretically require infinite time, but you can refer to the *half-life period,* which is the time taken for the radioactivity to fall to half of its initial value. For different substances, the half-life period varies from less than one microsecond to several thousand years.

In electronics, the charging of a capacitor through a resistor is an example of exponential growth, while its discharge involves exponential decay (see chapter 51). On a transmission line used in radio communications, the voltage and current waves decay exponentially as they travel along the line. The equations involving both growth and decay are explored in the mathematical derivations.

Napierian logarithms are closely related to the exponential number e (approximately 2.7183). The logarithm of a number (N) to base (a) is the exponent or power to which a must be raised to produce N. For example, the logarithm of 8 to the base 2 (written as $\log_2 8$) equals 3 because $2^3 = 8$. Napierian, or natural logarithms use the base e so that $\log_e 8 = 2.08$ approximately, because $2.7183^{2.08} \approx 8$. Notice that the natural logarithm is normally abbreviated (ln) and consequently, if $\ln N = x, N = e^x$.

MATHEMATICAL DERIVATIONS

The exponential series is derived from the Binomial Theorem.

Consider the expansion of $\left(1 + \dfrac{1}{n}\right)^{nx}$.

The rth term of this expansion is:

$$\frac{nx(nx-1)(nx-2)\ldots(nx-r+1)}{r!} \times \frac{1}{n^r}$$

As $n \to \infty$, the value of each term in parentheses in the numerator approaches nx.

Therefore, as $n \to \infty$, the rth term $\to \dfrac{x^r}{r!}$

Consequently, when $n \to \infty$,

$$\left(1 + \frac{1}{n}\right)^{nx} = 1 + \frac{x}{1!} + \frac{x^2}{2!} + \ldots + \frac{x^r}{r!} + \ldots \text{ to infinity} \qquad (210\text{-}1)$$

It can be shown that this infinite series is valid for all values of x.

Equation 210-1 can be written as

$$e^x = 1 + \frac{x}{1!} + \frac{x^2}{2!} + \ldots + \frac{x^r}{r!} + \ldots \text{ to infinity}$$

where:

$$e = \lim_{n \to \infty} \left(1 + \frac{1}{n}\right)^n \qquad (210\text{-}2)$$

If $x = 1$,

$$e^x = e^1 = e = 1 + \frac{1}{1!} + \frac{1}{2!} + \frac{1}{3!} + \ldots \text{ to infinity}$$

or

$$e = 2.71828\ldots$$

The graphs of e^x (exponential growth) and e^{-x} (exponential decay) versus x are shown in Fig. 210-1. However, the more common equation for exponential growth is $y = 1 - e^{-x}$ and the graphs of $1 - e^{-x}$ and e^{-x} versus x are plotted in Fig. 210-2 for values of x between 0 and 5 (also see Table 210-1).

Notice that the graph of $1 - e^{-x}$ versus x climbs toward a final value of 1 and is within 1 percent of this value when $x = 5$.

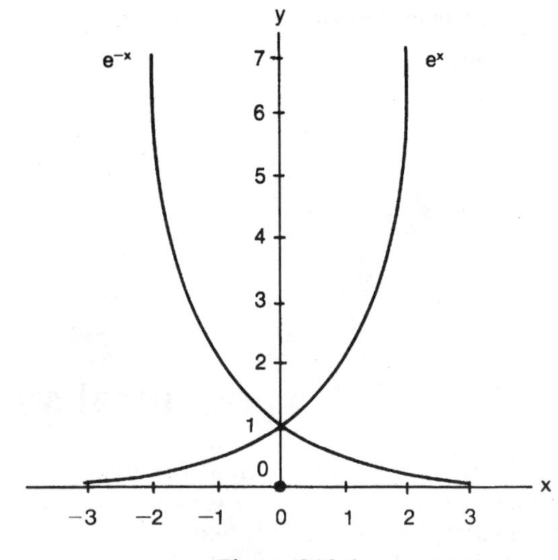

Figure 210-1

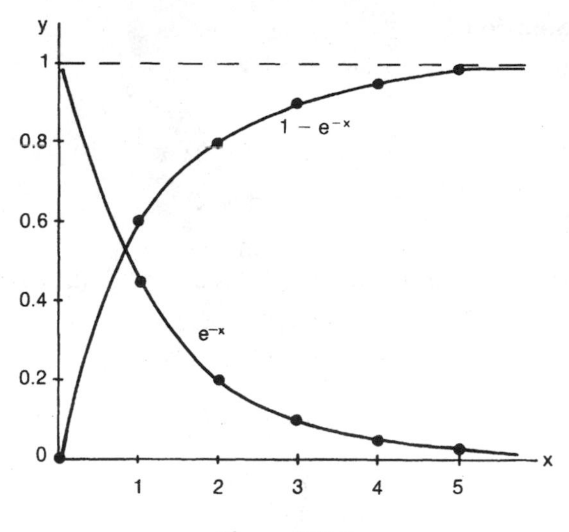

Figure 210-2

Table 210-1

x	0	1	2	3	4	5
$1 - e^{-x}$	0	0.632	0.865	0.950	0.982	0.993
e^{-x}	1	0.368	0.135	0.050	0.018	0.007

LOGARITHMS

Logarithms obtained from the solution of calculus equations are always the so-called natural logarithms, whose base is the value of e (2.71828. . .). For example, ln 17 = 2.833 . . . because $2.71828\ldots^{2.833\ldots} = 17$. By contrast common logarithms (abbreviated log) use a base of 10 so that log 17 = 1.230 . . . because $10^{1.230\ldots} = 17$. The graphs of ln x and log x versus x are shown in Fig. 210-3. Notice that logarithms cannot be used to add or subtract numbers.

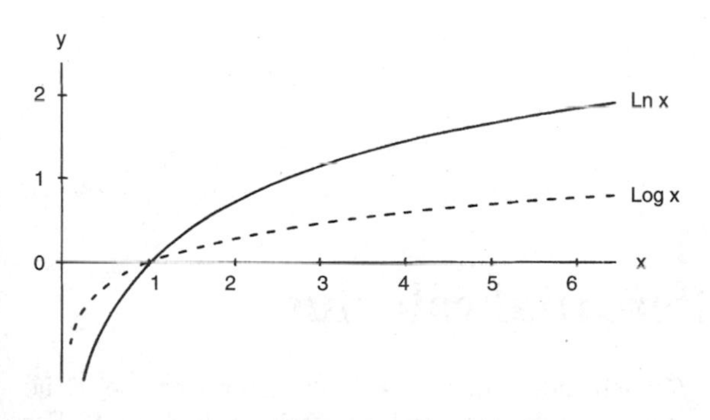

Figure 210-3

Multiplication and division by logarithms

If $N = a^x$ and $M = a^y$,

$$NM = a^x \times a^y$$
$$= a^{x+y}$$

Therefore:

$$\log_a (NM) = x + y \qquad (210\text{-}3)$$
$$= \log_a N + \log_a M$$

When *multiplying* two numbers, add their logarithms.

Similarly,

$$\log_a \frac{N}{M} = \log_a N - \log_a M \qquad (210\text{-}4)$$

When *dividing* one number by another, subtract the logarithm of the denominator from the logarithm of the numerator.

Exponents or powers

If the number M is raised to the power n.

$$M^n = (a^y)^n = a^{ny}$$

and,

$$\log_a M^n = ny = n \log_a M \qquad (210\text{-}5)$$

When raising a number to a certain power (exponent), *multiply* the logarithm of the number by the value of the exponent.

Also:

$$\log_a \frac{1}{M} = \log_a M^{-1} = -y = -\log_a M \qquad (210\text{-}6)$$

The logarithm of a number's reciprocal is equal to the negative of the number's logarithm. Notice that the logarithm of 1, to any base, is equal to 0; and the logarithm of 0, to a positive base greater than 1, is negative infinity.

Change of base

Let $\log_a N = x$ so that $N = a^x$. If you now take the logarithm of N to another base b,

$$\log_b N = \log_b a^x$$
$$= x \log_b a \qquad (210\text{-}5)$$
$$= \log_a N \times \log_b a \qquad (210\text{-}7)$$

For example,

$$\log_{10} N = \ln N \times \log_{10} e \qquad (210\text{-}8)$$
$$= \log_{10} 2.71828\ldots \times \ln N$$
$$\approx 0.4343 \times \ln N$$

This also yields,

$$\ln N \approx \frac{\log N}{0.4343}$$
$$\approx 2.302 \times \log N$$

Example 210-1

In the equation $y = 1 - e^{-x}$, find the value of x when $y = 0.9$. In the same equation, find the value of y if $x = 6$.

Solution

When $y = 0.9$,

$$e^{-x} = 1 - 0.9 = 0.1$$

so that $e^x = 10$.

Then,

$$x = \ln 10 = 2.3$$

When $x = 6$,

$$y = 1 - e^{-x} = 1 - e^{-6}$$
$$= 1 - 0.002479$$
$$= 0.997521$$

Example 210-2

In the equation $y = e^{-x}$, find the value of x when the value of $y = 7.6$.

Solution

$$\text{If } y = 7.6, e^x = \frac{1}{7.6} = 0.1316.$$

Then,

$$x = \ln 0.1316 = -2.03$$

Example 210-3

Obtain the values of: (a) $\log_2 16$, (b) $\log_2 1/32$, (c) $\log_8 128$, and (d) $\log_{25} 1/125$.

Solution

(a)
$$\log_2 16 = \log_2 2^4$$
$$= 4$$

(b)
$$\log_2 1/32 = -\log_2 32$$
$$= -\log_2 2^5$$
$$= -5$$

(c)
$$\log_8 128 = \log_8 (64 \times 2) \qquad (210\text{-}3)$$
$$= \log_8 64 + \log_8 2$$
$$= \log_8 8^2 + \log_8 8^{1/3}$$
$$= 2 + 1/3 = 2\frac{1}{3}$$
$$= \frac{7}{3}$$

(d)
$$\log_{25} 1/125 = -\log_{25} 125$$
$$= -\log_{25} (5 \times 25)$$
$$= -\log_{25} 25^{1/2} - \log_{25} 25$$
$$= -\frac{1}{2} - 1$$
$$= -\frac{3}{2}$$

PRACTICE PROBLEMS

210-1. If $67 = 113(1 - e^{-x})$, obtain the value of x.
210-2. If $37 = 83^{-x}$, obtain the value of x.
210-3. If $432 = 17^x$, obtain the value of x.
210-4. What are the values of (a) log 237 and (b) ln 237?
210-5. What are the values of:
(a) $\log_8 128$, (b) $\log_{25} 125$, (c) $\log_7 1$,
(d) $\frac{\log_{10} 216}{\log_{10} 36}$, and (e) $\frac{\log_{10} \frac{1}{9}}{\log_{10} \sqrt{3}}$?

211 Introduction to differential calculus

Differential calculus is involved with the rate of change between one variable and another. For example, if one variable (y) is a function of another variable (x), ($y = f(x)$), by how much will y increase (or decrease) if the value of x is increased by some small amount? For functions in general, the answer will depend on the initial value of x. This fact is illustrated graphically in Fig. 211-1. When the curve of $y = f(x)$ is drawn, the rate of change of y, with respect to x, is proportional to the slope of the curve at a particular point (P) corresponding to the initial value of x. If the curve is very steep at that point, a small increase in x will produce a proportionately large increase in y. By contrast, if the curve is fairly flat, the increase in y will be relatively small.

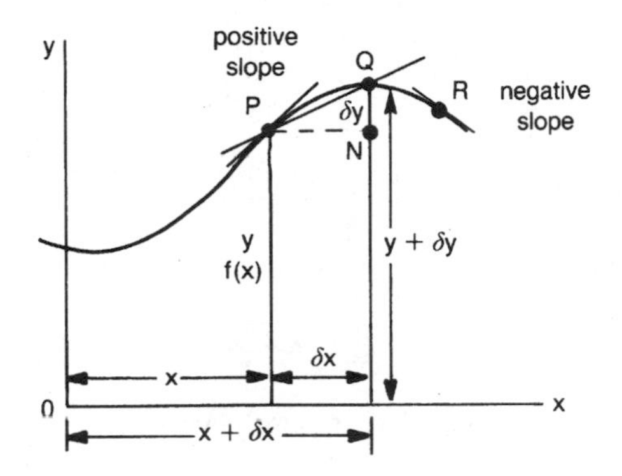

Figure 211-1

For any graph other than a straight line, the value of the slope, or the *rate of change,* varies from one point to another. If the slope is to be measured by considering the ratio of a change in y to a corresponding change in x, it is vital that these changes be as small as possible; otherwise, the ratio will not provide a true value of the curve's slope at the chosen point.

The small changes in the values of x and y are usually denoted by δx and δy (alternatively Δx, Δy) with the Greek lowercase letter δ or capital letter Δ (delta) signifying "a small change in . . . ." Notice that if δx is small, $(\delta x)^2$ will be even smaller and its value can be neglected in comparison with the value of δx. Higher powers of δx, such as $(\delta x)^3$ and $(\delta x)^4$, are smaller still.

In Fig. 211-1, the rate of change is determined by the ratio of the changes (δy, δx) in y and x or by the value of $\delta y/\delta x$. To find the slope at the point $P(x, y)$, select an adjacent point (Q), whose coordinates are ($x + \delta x$, $y + \delta y$). The value of the ratio $\delta y/\delta x$ is equal to the tangent function of the angle QPN and therefore, to the slope of the line PQ. If Q is moved closer and closer to P, the value of $\delta y/\delta x$ approximates more and more to the slope of the curve. If δy and δx both tend to zero, you have the condition for $\delta y/\delta x$ to equal the slope of the tangent line at point P.

Although both δy and δx become zero as Q reaches P, the ratio $\delta y/\delta x$ approaches some finite value, which is denoted by dy/dx. In other words, dy/dx is the limit of the ratio $\delta y/\delta x$ as δy and $\delta x \to 0$. In equation form:

$$\frac{dy}{dx} = \underset{\delta x, \delta y \to 0}{\text{Limit}} \frac{\delta y}{\delta x}$$

The quantity dy/dx is equal to the rate of change of y with respect to x; this is, in turn, equal to the value of the slope at point P. Provided the graph of y versus x is not a straight line, the value of the slope depends on the position of the point (P); therefore, dy/dx is a function of x. To find the numerical value of the slope at a particular point, the value of x for that point must be substituted in the expression for dy/dx.

It was stated that y is a function of x, as denoted by $f(x)$. The quantity dy/dx is also a function of x, and is denoted by $f'(x)$, which is the result of differentiating y with respect to x. The function $f'(x)$ is called the *first derivative,* or *first differential coefficient,* of $f(x)$. As an equation,

$$\frac{dy}{dx} = \frac{df(x)}{dx} = f'(x)$$

MATHEMATICAL DERIVATIONS

Calculation of $dy/dx = f'(x)$.

The quantity $f'(x)$ is defined as the limit of $\delta y/\delta x$ when $\delta x, \delta y \to 0$.

At the point P,

$$y = f(x) \qquad (211\text{-}1)$$

At the point Q, x increases to $x + \delta x$, and as a result, y increases to $y + \delta y$. Therefore,

$$y + \delta y = f(x + \delta x) \qquad (211\text{-}2)$$

This condition must be fulfilled in order for point Q to be on the curve.

Subtracting Equation 211-2 from Equation 211-1:

$$\delta y = f(x + \delta x) - f(x)$$

and,

$$\frac{\delta y}{\delta x} = \frac{f(x + \delta x) - f(x)}{\delta x} \qquad (211\text{-}3)$$

Then, the first derivative,

$$f'(x) = \frac{dy}{dx} \qquad (211\text{-}4)$$

$$= \underset{\delta x, \delta y \to 0}{\text{Limit}} \frac{\delta y}{\delta x}$$

$$= \underset{\delta x, \delta y \to 0}{\text{Limit}} \frac{f(x + \delta x) - f(x)}{\delta x}$$

If this procedure is repeated for point R in Fig. 211-1, the increase from x to $x + \delta x$ will be accompanied by a decrease from y to $y - \delta y$. It follows that the slope at the point R is negative.

In the particular case of $y = x^2$, the graph of y versus x is a parabola, which is illustrated in Fig. 211-2. At points $P(1,1)$ and $Q(3,9)$, tangent lines have been estimated and their slopes have been calculated. At point P, for which $x = 1$, the slope is 2; at point Q, where $x = 3$, the slope is 6. These results are satisfied by the equation $dy/dx = 2x$.

Substituting $f(x) = y = x^2$ in Equation 211-4:

$$f'(x) = \frac{dy}{dx} = \lim_{\delta x, \delta y \to 0} \frac{(x + \delta x)^2 - x^2}{\delta x}$$

$$= \lim_{\delta x, \delta y \to 0} \frac{x^2 + 2x\delta x + (\delta x)^2 - x^2}{\delta x}$$

$$= \lim_{\delta x, \delta y \to 0} 2x + \delta x$$

$$= 2x$$

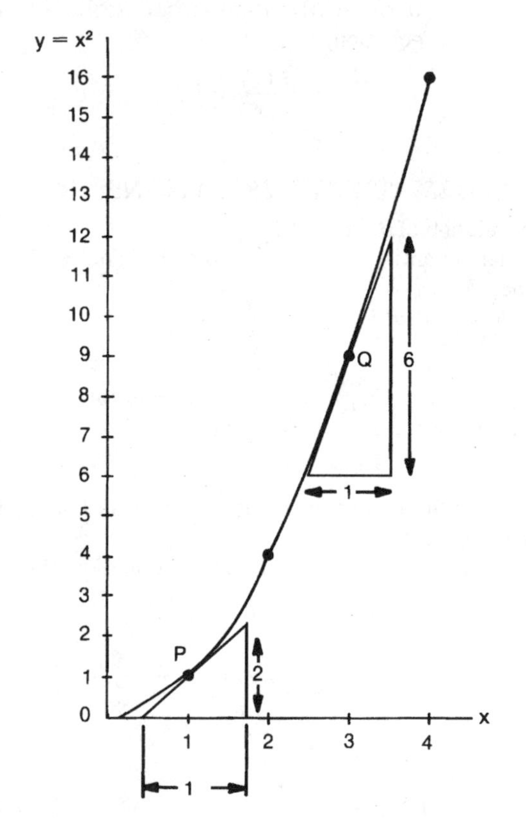

Figure 211-2

This proves the deduction that the slope of $y = x^2$ at any point is equal to $2x$. For example, at the point Q (3,9), the slope is $2 \times 3 = 6$.

Derivative of x^n

Let $f(x) = x^n$. Then,

$$\frac{\delta y}{\delta x} = \frac{(x + \delta x)^n - x^n}{\delta x}$$

$$= \frac{x^n\left(1 + \frac{\delta x}{x}\right)^n - x^n}{\delta x}$$

Expand $[1 + (\delta x/x)]^n$ by the binomial series, where n and x can have any value because $-1 < \delta x/x < 1$ (see chapter 209). Then,

$$\frac{\delta y}{\delta x} = \frac{x^n\left[1 + n\frac{\delta x}{x} + \frac{n(n-1)}{2!}\left(\frac{\delta x}{x}\right)^2 + \ldots\right] - x^n}{\delta x}$$

$$= nx^{n-1} + \frac{n(n-1)x^{n-2}}{2!} \cdot \delta x + \ldots$$

In the limit when $\delta x, \delta y \to 0$,

$$\frac{dy}{dx} = nx^{n-1} \qquad (211\text{-}5)$$

If $f(x)$ is multiplied by some constant, the constant remains after differentiation. For example, the slope of the curve $y = 5x^3$ is five times as great as that of $y = x^3$. Therefore, if $y = 5x^3$:

$$\frac{dy}{dx} = 5 \times 3 \times x^{3-1} = 15x^2$$

The equation $y = K$, where K is a constant, represents a straight line parallel to the x axis. The slope of such a line is zero so that if $y = K$, $dy/dx = 0$.

If the expression for y contains terms that are added or subtracted, each term can be differentiated separately to obtain the first derivative.

Example 211-1

Differentiate the following with respect to x:

(a) $y = -17$,

(b) $y = 7x^3 + 3x^2 - 5x + 6$,

(c) $y = 3\sqrt{x}$,

(d) $y = 11/x^2$, and

(e) $y = 5/x^{3/2}$.

In each case, find the value of the slope when $x = 1$.

Solution

(a) If $y = -17$, $\frac{dy}{dx} = 0$.

The value of the slope when $x = 1$ is zero.

(b) If $y = 7x^3 + 3x^2 - 5x + 6$,

$$\frac{dy}{dx} = 7 \times 3 \times x^{3-1} + 3 \times 2 \times x^{2-1} - 5 \times 1 \times x^{1-1} + 0$$

$$= 21x^2 + 6x - 5$$

The value of the slope when $x = 1$ is $21 + 6 - 5 = 22$

(c) If $y = 3\sqrt{x} = 3x^{1/2}$,

$$\frac{dy}{dx} = 3 \times \frac{1}{2} \times x^{1/2-1} = \frac{3}{2}x^{-1/2}$$

$$= \frac{3}{2\sqrt{x}}.$$

The value of the slope when $x = 1$ is $\frac{3}{2\sqrt{1}} = 1.5$

(d) If $y = \frac{11}{x^2} = 11x^{-2}$,

$$\frac{dy}{dx} = 11 \times (-2) \times x^{-2-1} = -22x^{-3}$$

$$= \frac{-22}{x^3}.$$

The value of the slope when $x = 1$ is $- 22/1^3 = -22$

(e) If $y = \frac{5}{x^{3/2}} = 5x^{-3/2}$

$$\frac{dy}{dx} = 5 \times \left(\frac{-3}{2}\right) \times x^{-3/2-1}$$

$$= \frac{-15}{2}x^{-5/2}$$

The value of the slope when $x = 1$ is $-15/2 \times (1)^{-5/2} = -7.5$

PRACTICE PROBLEMS

211-1. Differentiate the following with respect to x:
(a) $7x + 13$,
(b) $5x^4$,
(c) $\frac{6}{x^3}$,
(d) $\sqrt[3]{8x^2}$, and
(e) $\frac{11}{\sqrt[4]{16x^3}}$

211-2. Differentiate $f(x) = 7x^3 - 5x^2 + 11x - 3$ with respect to x.

211-3. What is the slope of $f(x) = 5x^{1/2} + 6x^{-1/3}$ at the point (1,11)?

211-4. If $f(x) = 6x^2 - 8x + 3$, determine the value of x for which

$$\frac{dy}{dx} = 0.$$

211-5. Differentiate $f(x) = \frac{x^3 - 1}{x - 1}$ with respect to x.

212
Differentiation of a function of a function

If, for example, $y = f(x) = (3x^2 - 7x)^3$, you could, with some difficulty, expand the expression, then differentiate each term separately. However, the process would be very tedious, and it is preferable to regard y as equal to u^3, where $u = 3x^2 - 7x$. In other words, y is a function of u, where u is a function of x. You can, therefore, say that y is a function of a function of x; mathematically, this could be shown as $y = F(f(x))$.

Because $\frac{\delta y}{\delta x} = \frac{\delta y}{\delta u} \times \frac{\delta u}{\delta x}$,

$$\frac{dy}{dx} = \frac{dy}{du} \times \frac{du}{dx} \qquad (212\text{-}1)$$

in the limit as $\delta y, \delta x, \delta u \to 0$.
Therefore:

$$y = u^3, \frac{dy}{du} = 3u^2 = 3(3x^2 - 7x)^2$$

$$u = 3x^2 - 7x, \frac{du}{dx} = 6x - 7$$

Consequently,

$$\frac{dy}{dx} = \frac{dy}{du} \times \frac{du}{dx}$$

$$= 3(3x^2 - 7x)^2(6x - 7)$$

If the expression for y had been expanded,

$$y = (3x^2 - 7x)^3$$

$$= (9x^4 - 42x^3 + 49x^2)(3x^2 - 7x)$$

$$= 27x^6 - 126x^5 + 147x^4 - 63x^5 + 294x^4 - 343x^3$$

$$= 27x^6 - 189x^5 + 441x^4 - 343x^3$$

$$\frac{dy}{dx} = 162x^5 - 945x^4 + 1764x^3 - 1029x^2$$

Comparing this result with:

$$\frac{dy}{dx} = 3(3x^2 - 7x)^2 \times (6x - 7)$$

$$= 3(9x^4 - 42x^3 + 49x^2) \times (6x - 7)$$

$$= 3(54x^5 - 252x^4 + 294x^3$$

$$- 63x^4 + 294x^3 - 343x^2)$$

$$= 3(54x^3 - 315x^4 + 588x^3 - 343x^2)$$

$$= 162x^5 - 945x^4 + 1764x^3 - 1029x^2$$

The two methods produce the same result, but it is much easier and quicker to obtain $3(3x^2 - 7x)^2 \times (6x - 7)$, rather than $162x^5 - 945x^4 + 1764x^3 - 1029x^2$. Moreover, if $y = (3x^2 - 7x)^{20}$, expansion is virtually impossible; but by using the function of the function method, $dy/dx = 20(3x^2 - 7x)^{19} \times (6x - 7)$.

Notice that a function, such as $y = \sqrt{3x^2 - 7x} = (3x^2 - 7x)^{1/2}$, cannot be expanded, but by using the function of a function method,

$$\frac{dy}{dx} = \frac{1}{2}(3x^2 - 7x)^{1/2 - 1} \times (6x - 7)$$

$$= \frac{6x - 7}{2\sqrt{3x^2 - 7x}}$$

MATHEMATICAL DERIVATIONS

The principle of differentiating the function of a function can be extended to any number of functions. For example, if y is a function of u, u is a function of v, and v is a function of x,

$$\frac{\delta y}{\delta x} = \frac{\delta y}{\delta u} \times \frac{\delta u}{\delta v} \times \frac{\delta v}{\delta x} \qquad (212\text{-}2)$$

then,

$$\frac{dy}{dx} = \frac{dy}{du} \times \frac{du}{dv} \times \frac{dv}{dx}$$

in the limit as $\delta y, \delta x, \delta u, \delta v \to 0$.

As an example, let $y = 7[1 + 4\sqrt{5x^3 - 8x}]^4$. Then,

$$y = 7u^4$$

$$u = 1 + 4\sqrt{v}$$

$$v = 5x^3 - 8x$$

$$\frac{dy}{du} = 28u^3$$

$$\frac{du}{dv} = 2v^{-1/2}$$

$$\frac{dv}{dx} = 15x^2 - 8$$

Therefore:

$$\frac{dy}{dx} = 28u^3 \times 2v^{-1/2} \times (15x^2 - 8)$$

$$= \frac{56[1 + 4\sqrt{5x^3 - 8x}]^3 \times (15x^2 - 8)}{\sqrt{5x^3 - 8x}}$$

Example 212-1

Differentiate the following with respect to x.

(a) $5(3x^2 + 2)^8$, (b) $8\sqrt{x^3 - 3x}$,

(c) $\dfrac{6}{(5x^3 + 6x - 2)^3}$, and (d) $4(5x^4 - 4x^2)^{1/3}$.

Solution

(a) If $y = 5(3x^2 + 2)^8$,

$$\frac{dy}{dx} = 5 \times 8 \times (3x^2 + 2)^{8 - 1} \times 6x$$

$$= 240x \times (3x^2 + 2)^7.$$

(b) If $y = 8\sqrt{x^3 - 3x} = 8(x^3 - 3x)^{1/2}$,

$$\frac{dy}{dx} = 8 \times \frac{1}{2} \times (x^3 - 3x)^{1/2 - 1} \times (3x^2 - 3)$$

$$= \frac{12(x^2 - 1)}{\sqrt{x^3 - 3x}}$$

(c) If $y = \dfrac{6}{(5x^3 + 6x - 2)^3} = 6 \times (5x^3 + 6x - 2)^{-3}$,

$$\frac{dy}{dx} = 6 \times (-3) \times (5x^3 + 6x - 2)^{-4} \times (15x^2 + 6)$$

$$= \frac{-18(15x^2 + 6)}{(5x^3 + 6x - 2)^4}.$$

(d) If $y = 4(5x^4 - 4x^2)^{1/3}$,

$$\frac{dy}{dx} = 4 \times \frac{1}{3} \times (5x^4 - 4x^2)^{1/3 - 1} \times (20x^3 - 8x)$$

$$= \frac{16(5x^3 - 2x)}{3(5x^4 - 4x^2)^{2/3}}$$

PRACTICE PROBLEMS

212-1. Obtain the first derivative, with respect to x, of $f(x) = 7(4x^3 - 5x)^{20}$.

212-2. Obtain the first derivative, with respect to x, of $f(x) = 12\sqrt[3]{3x^2 - 11}$.

212-3. What is the slope of $f(x) = 7/(3x^4 - 7x^2 + 5)^3$ at the point for which $x = -1$?

212-4. Differentiate $f(x) = 3(4x^3 - x)^{-1/2}$ with respect to x.

212-5. Obtain the first derivative, with respect to x, of $f(x) = 5[1 - 3\sqrt[3]{3x^4 - 5x^3}]^{-3}$.

212-6. If $Z = \sqrt{R^2 + \left(\omega L - \dfrac{1}{\omega C}\right)^2}$, obtain the first derivative of Z with respect to ω.

213 Differentiation of a product

If you were asked to obtain the first derivative of $(2x + 7)(3x^2 + 4)$, you could multiply out the expression, and then differentiate each term separately.

The procedure would be as follows:

$$y = (2x + 7)(3x^2 + 4)$$
$$= 6x^3 + 21x^2 + 8x + 28$$

Then,

$$\frac{dy}{dx} = 18x^2 + 42x + 8.$$

The procedure in this case is relatively simple, but if you need to obtain the derivative of $(2x + 7)^6(3x^2 + 4)^9$, multiplication would be extremely tedious and the first derivative would contain more than 20 terms. However, you already know how to differentiate $(2x + 7)^6$ and $(3x^2 + 4)^9$ by using the function of a function process, as outlined in chapter 212. Therefore, you need a formula to differentiate the product of these two functions of x.

MATHEMATICAL DERIVATIONS

Consider $y = f(x) = uv$, where u and v are both functions of x. Then,

$$\frac{\delta y}{\delta x} = \frac{(u + \delta u)(v + \delta v) - uv}{\delta x}$$
$$= \frac{uv + u\delta v + v\delta u + \delta u\delta v - uv}{\delta x}$$
$$= u\frac{\delta v}{\delta x} + v\frac{\delta u}{\delta x} + \frac{\delta u\delta v}{\delta x}$$

As δx approaches zero, so also will δu and δv. The term $\delta u\delta v/\delta x$ is then infinitesimally small and can be neglected.

In the limits as $\delta y, \delta x, \delta u, \delta v \to 0$,

$$\frac{dy}{dx} = u\frac{dv}{dx} + v\frac{du}{dx} \qquad (213\text{-}1)$$

If $y = (2x + 7)(3x^2 + 4)$, let $u = 2x + 7$ and $v = 3x^2 + 4$. Then,

$$\frac{dy}{dx} = \underbrace{(2x + 7)}_{u} \times \underbrace{6x}_{\frac{dv}{dx}} + \underbrace{(3x^2 + 4)}_{v} \times \underbrace{2}_{\frac{du}{dx}}$$
$$= 12x^2 + 42x + 6x^2 + 8$$
$$= 18x^2 + 42x + 8.$$

The result is the same as the one that was previously obtained.

If $y = (2x + 7)^6(3x^2 + 4)^9$, let $u = (2x + 7)^6$ and $v = (3x^2 + 4)^9$. Then,

$$\frac{dy}{dx} = \underbrace{(2x + 7)^6}_{u} \times \underbrace{9 \times (3x^2 + 4)^8 \times 6x}_{\frac{dv}{dx}}$$
$$+ \underbrace{(3x^2 + 4)^9}_{v} \times \underbrace{6 \times (2x + 7)^5 \times 2}_{\frac{du}{dx}}$$
$$= 54x(2x + 7)^6(3x^2 + 4)^8$$
$$+ 12(3x^2 + 4)^9(2x + 7)^5$$

Example 213-1

Differentiate the following with respect to x:

(a) $\sqrt{3x^2 + 7x} \times (8x^3 - 6)^{-2}$, and (b) $\dfrac{7x^{3/2}}{2x^3 - 5x^2}$.

Solution

(a) $y = \sqrt{3x^2 + 7x} \times (8x^3 - 6)^{-2}$

Let $u = \sqrt{3x^2 + 7x} = (3x^2 + 7x)^{1/2}$, and $v = (8x^3 - 6)^{-2}$. Then,

$$\frac{dy}{dx} = \underbrace{(3x^2 + 7x)^{1/2}}_{u} \times \underbrace{(-2) \times (8x^3 - 6)^{-3} \times 24x^2}_{\times \frac{dv}{dx}} + \underbrace{(8x^3 - 6)^{-2}}_{v}$$

$$\underbrace{\times \left(\frac{1}{2}\right) \times (3x^2 + 7x)^{-1/2} \times (6x + 7)}_{\times \frac{du}{dx}}$$

$$= \frac{-48x^2\sqrt{3x^2 + 7x}}{(8x^3 - 6)^3} + \frac{6x + 7}{2(8x^3 - 6)^2\sqrt{3x^2 + 7x}}$$

(b) $y = \dfrac{7x^{3/2}}{(2x^3 - 5x^2)^2} = 7x^{3/2} \times (2x^3 - 5x^2)^{-2}$

Let $u = 7x^{3/2}$ and $v = (2x^3 - 5x^2)^{-2}$. Then,

$$\frac{dy}{dx} = \underbrace{7x^{3/2}}_{u} \times \underbrace{(-2) \times (2x^3 - 5x^2)^{-3} \times (6x^2 - 10x)}_{\frac{dv}{dx}}$$

$$+ \underbrace{(2x^3 - 5x^2)^{-2}}_{v} \times \underbrace{7 \times 3/2 \times x^{1/2}}_{\frac{du}{dx}}$$

$$= \frac{-14x^{3/2}(6x^2 - 10x)}{(2x^3 - 5x^2)^3} + \frac{21x^{1/2}}{2(2x^3 - 5x^2)^2}$$

Example 213-2

The expression for the power dissipated in a load is:

$$P_L = \frac{E^2R_L}{(R_i + R_L)^2}$$

where E and R_i are constants. Differentiate P_L with respect to R_L.

Solution

$$\text{Power, } P_L = \frac{E^2R_L}{(R_i + R_L)^2}$$

$$= E^2R_L \times (R_i + R_L)^{-2}$$

Let $u = E^2R_L$ and $v = (R_i + R_L)^{-2}$. Then,

$$\frac{dP_L}{dR_L} = E^2R_L \times (-2) \times (R_i + R_L)^{-3}$$

$$+ (R_i + R_L)^{-2} \times E^2$$

$$= -\frac{2E^2R_L}{(R_i + R_L)^3} + \frac{E^2}{(R_i + R_L)^2}$$

$$= \frac{E^2(R_i - R_L)}{(R_i + R_L)^3}$$

PRACTICE PROBLEMS

213-1. Obtain the first derivative, with respect to x, of $f(x) = 7\sqrt[3]{4x^3 + 5x} \times (3x^4 - 8x^2)^{-3}$.

213-2. Obtain the first derivative, with respect to x, of
$$f(x) = \frac{8x^{3/4}}{\sqrt{3x^4 - 5x^2}}$$

213-3. If $\gamma = \sqrt{(R + j\omega L)(G + j\omega C)}$, obtain the first derivative of γ with respect to ω.

214
Differentiation of a quotient

In the solution to example 213-2, you differentiated the function:

$$f(R_L) = P_L = \frac{E^2R_L}{(R_i + R_L)^2}$$

by rearranging $f(R_L)$ as the product $E^2R_L \times (R_i + R_L)^{-2}$. However, as originally expressed, $f(R_L)$ is a quotient; therefore, you need to know the formula for differentiating all such quotients.

MATHEMATICAL DERIVATIONS

Let $y = u/v$, where u and v are both functions of x. Then,

$$\frac{\delta y}{\delta x} = \frac{\dfrac{u + \delta u}{v + \delta v} - \dfrac{u}{v}}{\delta x}$$

$$= \frac{\dfrac{u}{v}\left[\dfrac{1 + \dfrac{\delta u}{u}}{1 + \dfrac{\delta v}{v}}\right] - \dfrac{u}{v}}{\delta x}$$

$$= \frac{\dfrac{u}{v}\left(1 + \dfrac{\delta u}{u}\right)\left(1 + \dfrac{\delta v}{v}\right)^{-1} - \dfrac{u}{v}}{\delta x}$$

Expanding $(1 + \delta v/v)^{-1}$ by the binomial series and neglecting $(\delta v/v)^2$ as well as all other higher-order terms, then:

$$\frac{\delta y}{\delta x} = \frac{\dfrac{u}{v}\left(1 + \dfrac{\delta u}{u}\right)\left(1 - \dfrac{\delta v}{v} + \text{neglected terms}\right) - \dfrac{u}{v}}{\delta x}$$

$$\frac{\delta y}{\delta x} = \frac{\dfrac{u}{v}\left(1 + \dfrac{\delta u}{u} - \dfrac{\delta v}{v} - \dfrac{\delta u \delta v}{uv}\right) - \dfrac{u}{v}}{\delta x}$$

$$= \frac{\dfrac{\delta u}{v} - \dfrac{u\delta v}{v^2} - \dfrac{\delta u \delta v}{v^2}}{\delta x}$$

Because δy, δx, δu, $\delta v \to 0$, $\delta u \delta v/v^2 \to 0$, therefore:

$$\frac{dy}{dx} = \frac{v\dfrac{du}{dx} - u\dfrac{dv}{dx}}{v^2} \qquad (214\text{-}1)$$

This is the formula for obtaining the first derivative of a quotient.

Example 214-1

Differentiate the following with respect to x.

(a) $3x^2/(x^3 + 7)$,

(b) $(5x - 6)/(x^4 - 2x^2)$,

(c) $(2x^3 - 3x)/\sqrt{x^2 + 7}$.

Solution

(a) $y = \dfrac{3x^2}{x^3 + 7}$

Let $u = 3x^2$ and $v = x^3 + 7$. Then,

$$\frac{dy}{dx} = \frac{(x^3 + 7)6x - 3x^2 \times 3x^2}{(x^3 + 7)^2}$$

$$= \frac{-3x^4 + 42x}{(x^3 + 7)^2} \qquad (214\text{-}1)$$

(b) $y = \dfrac{5x - 6}{x^4 - 2x^2}$

Let $u = 5x - 6$ and $v = x^4 - 2x^2$. Then,

$$\frac{dy}{dx} = \frac{(x^4 - 2x^2) \times 5 - (5x - 6) \times (4x^3 - 4x)}{(x^4 - 2x^2)^2}$$

$$= \frac{-15x^4 + 24x^3 + 10x^2 - 24x}{(x^4 - 2x^2)^2}$$

(c) $y = \dfrac{2x^3 - 3x}{\sqrt{x^2 + 7}}$

Let $u = 2x^3 - 3x$ and $v = \sqrt{x^2 + 7} = (x^2 + 7)^{1/2}$. Then,

$$\frac{dy}{dx} = \frac{(x^2 + 7)^{1/2} \times (6x^2 - 3) - (2x^3 - 3x)\times 1/2 \times (x^2 + 7)^{-1/2} \times 2x}{x^2 + 7} \qquad (214\text{-}1)$$

$$= \frac{4x^4 + 42x^2 - 21}{(x^2 + 7)^{3/2}}$$

Example 214-2

If,

$$P_L = \frac{E^2R_L}{(R_i + R_L)^2},$$

differentiate P_L with respect to R_L. (Regard E and R_i as constants.)

Solution

$$\text{Power, } P_L = \frac{E^2R_L}{(R_i + R_L)^2}$$

Let $u = E^2R_L$ and $v = (R_i + R_L)^2$. Then

$$\frac{dP_L}{dR_L} = \frac{(R_i + R_L)^2 \times E^2 - E^2R_L \times 2 \times (R_i + R_L)}{(R_i + R_L)^4} \tag{214-1}$$

$$= \frac{E^2 \times (R_i^2 - R_L^2)}{(R_i + R_L)^4}$$

$$= \frac{E^2(R_i - R_L)}{(R_i + R_L)^3}$$

In this case differentiation by the quotient formula is simpler than using the product method (Example 213-2).

PRACTICE PROBLEMS

214-1. Differentiate

$$f(x) = \frac{4x + 1}{8(2x + 1)^2}$$

with respect to x, and simplify the answer as far as possible.

214-2. Differentiate

$$f(x) = \frac{3x^2 + 6x + 4}{3(x + 2)^3}$$

with respect to x, and simplify the answer as far as possible.

214-3. Differentiate

$$f(x) = \frac{2x^2 + 4x + 3}{4(x + 3)^4}$$

with respect to x, and simplify the answer as far as possible.

214-4. Differentiate

$$f(x) = \frac{\sqrt{(1 + x^2)^3}}{3x^3}$$

with respect to x, and simplify the answer as far as possible.

214-5. Differentiate

$$f(\omega) = \frac{R + j\omega L}{G + j\omega C}$$

with respect to ω.

215
Differentiation of the trigonometric functions

The voltage induced in a coil is directly proportional to the rate of change of the current flowing through the coil. If the waveform of the current in the coil is an alternating sine wave, you need to be capable of differentiating the sine function in order to determine the expression for the voltage across the coil. Only in this way can you prove the "*eLi*" relationship, namely that for an inductor (L) the voltage (e) leads the current (i) by 90°. As the result of differentiation, you can further prove the inductive reactance equation:

$$X_L = 2\pi fL\ \Omega$$

(See chapter 69.)

In an ac circuit, the current associated with a capacitor is directly proportional to the rate of change of the voltage across the capacitor. Again, you need to be capable of differentiating the sine function to determine the expression for the capacitor's current. You can then prove the "*iCe*" relationship, namely that for a capacitor (C), the current (i) leads the voltage (e) by 90°. As the result of differentiation, you can further prove the capacitive reactance equation:

$$X_C = \frac{1}{2\pi fC}\ \Omega$$

(See chapter 70.)

Once the sine and cosine functions have been differentiated, the first derivatives of other trigonometric functions can be obtained through the use of the function-of-a-function, product, and quotient formulas.

MATHEMATICAL DERIVATIONS

Let $y = \sin x$. Then,

$$\frac{\delta y}{\delta x} = \frac{\sin(x + \delta x) - \sin x}{\delta x} \tag{211-4}$$

$$= \frac{\sin x \cos \delta x + \cos x \sin \delta x - \sin x}{\delta x}$$

As δy and $\delta x \to 0$, $\cos \delta x \to 1$ and $\sin \delta x \to x$. (204-6)

Therefore:

$$\frac{dy}{dx} = \frac{\sin x \times 1 + \cos x \times \delta x - \sin x}{\delta x} \qquad (215\text{-}1)$$

$$= \cos x$$

This result can be verified by referring to Fig. 215-1.

When $x = 0°$, $y = \sin x = 0$, $dy/dx = \cos x = 1$, and the slope of $y = \sin x$ is a maximum in a positive direction. When $x = 90° = \pi^c/2$, $y = \sin x = 1$, $dy/dx = \cos x = 0$, and the slope of $y = \sin x$ is zero. When $x = 180° = \pi^c$, $y = \sin x = 0$, $dy/dx = \cos x = -1$, and the slope of $y = \sin x$ is a maximum in the negative direction.

Let $y = \cos x = \sin (\pi/2 - x)$. Then,

$$\frac{dy}{dx} = \cos\left(\frac{\pi}{2} - x\right) \times (-1) \qquad (212\text{-}1)$$

$$= -\sin x \qquad (215\text{-}2)$$

Let $y = \tan x = \sin x/\cos x$. This can be considered as a quotient. Therefore:

$$\frac{dy}{dx} = \frac{\cos x \cos x - \sin x\,(-\sin x)}{\cos^2 x} \qquad (214\text{-}1)$$

$$= \frac{\cos^2 x + \sin^2 x}{\cos^2 x}$$

$$= \frac{1}{\cos^2 x} = \sec^2 x \qquad (203\text{-}9),\ (205\text{-}5),\ (215\text{-}3)$$

The first derivatives of the six trigonometric functions are listed in Table 215-1.

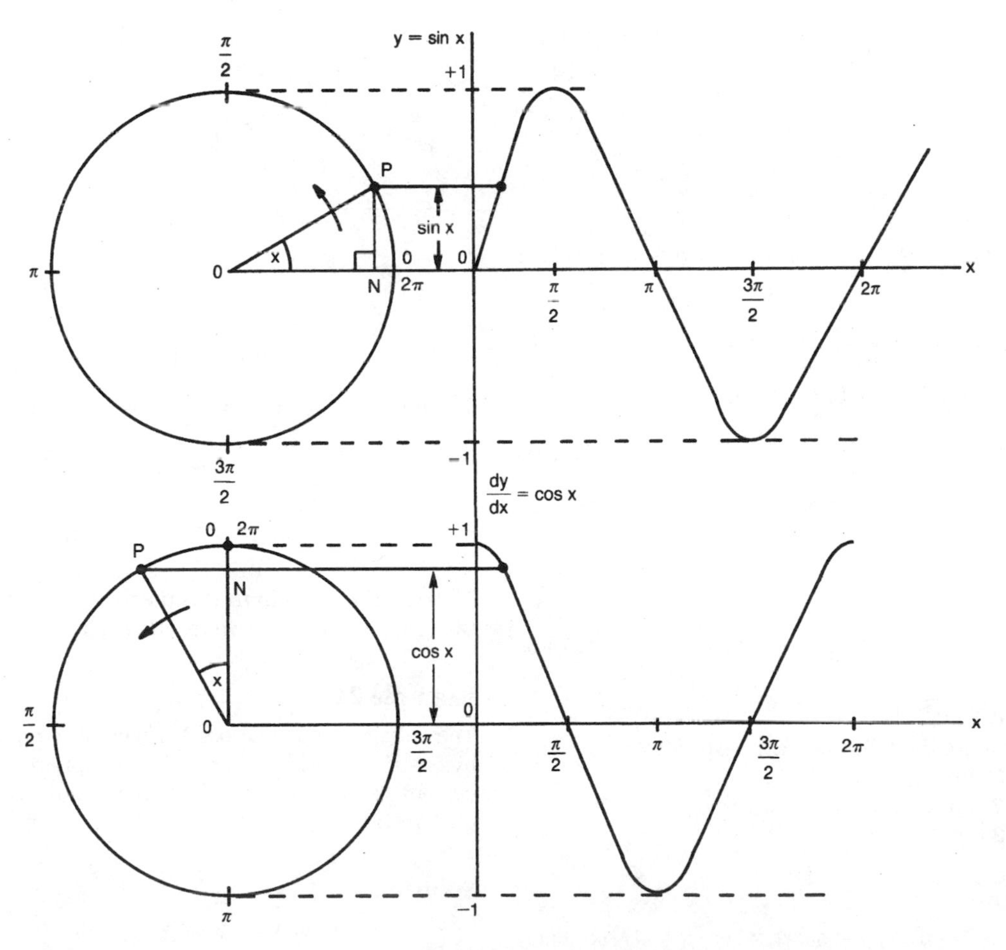

Figure 215-1

Table 215-1.
First derivatives of the trigonometric functions.

y	dy/dx
$\sin x$	$\cos x$
$\cos x$	$-\sin x$
$\tan x$	$\sec^2 x$
$\cot x$	$-\csc^2 x$
$\sec x$	$\sec x \times \tan x$
$\csc x$	$-\csc x \times \cot x$

Differentiation of the inverse trigonometric functions

If $y = \sin^{-1} x$, $x = \sin y$. Then,

$$\frac{dx}{dy} = \cos y$$

Therefore:

$$\frac{dy}{dx} = \frac{1}{\cos y} \qquad (215\text{-}4)$$

$$= \frac{1}{\sqrt{1 - \sin^2 y}}$$

$$= \frac{1}{\sqrt{1 - x^2}}$$

Likewise, if $y = \cos^{-1} x$, $x = \cos y = \sin[(\pi/2) - y]$, and

$$\frac{dy}{dx} = \frac{-1}{\sqrt{1 - x^2}} \qquad (215\text{-}5)$$

If $y = \tan^{-1} x$, $x = \tan y$. Then,

$$\frac{dx}{dy} = \sec^2 y \qquad (215\text{-}3)$$

$$= 1 + \tan^2 y \qquad (205\text{-}14)$$

$$= 1 + x^2$$

Therefore:

$$\frac{dy}{dx} = \frac{1}{1 + x^2} \qquad (215\text{-}6)$$

Example 215-1

Differentiate the following with respect to x:

(a) $5 \sin 3x$,
(b) $7 \cos 6x$, and
(c) $3 \tan 2x$.

Solution

(a) If $y = 5 \sin 3x$, $y = \sin u$, where $u = 3x$, so that y is a function-of-a-function. Therefore:

$$\frac{dy}{dx} = (5 \cos 3x) \times 3 \qquad (215\text{-}1)$$

$$= 15 \cos 3x$$

(b) If $y = 7 \cos 6x$,

$$\frac{dy}{dx} = (-7 \sin 6x) \times 6 \qquad (215\text{-}2)$$

$$= -42 \sin 6x$$

(c) If $y = 3 \tan 2x$,

$$\frac{dy}{dx} = (3 \sec^2 2x) \times 2 \qquad (215\text{-}3)$$

$$= 6 \sec^2 2x$$

Example 215-2

The alternating current flowing through an inductor is represented by $i = I_{peak} \sin \omega t$. If the inductor is connected directly across an ac source, determine the expression for the source voltage, e.

Solution

Source voltage, $e = L\dfrac{di}{dt}$

$$= L\frac{d(I_{peak} \sin \omega t)}{dt}$$

$$= \omega L\, I_{peak} \cos \omega t \qquad (215\text{-}1)$$

$$= \omega L\, I_{peak} \sin\left(\omega t + \frac{\pi}{2}\right)$$

$$= E_{peak} \sin\left(\omega t + \frac{\pi}{2}\right)$$

Therefore, e leads i by $\pi/2$ radians or 90°. Moreover,

$$E_{peak} = \omega L\, I_{peak}$$

$$\frac{E_{peak}}{I_{peak}} = \frac{E_{rms}}{I_{rms}} = X_L = \omega L = 2\pi f L\ \Omega$$

The equation $X_L = 2\pi f L\ \Omega$ is the formula for obtaining the value of the inductive reactance, X_L.

Example 215-3

The voltage applied across a capacitor is represented by $e = E_{peak} \sin \omega t$. Determine the expression for the current that is alternately charging and discharging the capacitor.

Solution

Source voltage, $e = E_{peak} \sin \omega t$

Because the charge, $q = Ce$,

$$\frac{dq}{dt} = C\frac{de}{dt}$$

The current is equivalent to the rate of change of charge. Therefore:

$$\text{Current, } i = \frac{dq}{dt} = C\frac{d(E_{\text{peak}} \sin \omega t)}{dt} \qquad (215\text{-}1)$$

$$= \omega C E_{\text{peak}} \cos \omega t$$

$$= I_{\text{peak}} \sin\left(\omega t + \frac{\pi}{2}\right)$$

Therefore, i leads e by $\pi/2$ radians or 90°. Moreover:

$$I_{\text{peak}} = \omega C E_{\text{peak}}$$

$$\frac{E_{\text{peak}}}{I_{\text{peak}}} = \frac{E_{\text{rms}}}{I_{\text{rms}}} = X_C = \frac{1}{\omega C} = \frac{1}{2\pi f C}\ \Omega$$

Equation $X_C = 1/(2\pi f C)\ \Omega$ is the formula for obtaining the value of the capacitive reactance X_C.

PRACTICE PROBLEMS

215-1. Differentiate $f(x) = -x^2 \cos x + 2\,(x \sin x + \cos x)$ with respect to x.

215-2. Differentiate $f(x) = 5 \sin^2 x$ with respect to x.

215-3. Differentiate $f(x) = 1/9\ (\sin 3x - 3x \cos 3x)$ with respect to x.

215-4. Differentiate $f(x) = (x^2 - 2)\sin x + 2x \cos x$ with respect to x.

215-5. Differentiate $f(x) = x\cos^{-1} x - \sqrt{1 - x^2}$ with respect to x.

215-6. Differentiate $f(x) = 1/2\ [(x^2 + 1)\tan^{-1} x - x]$ with respect to x.

215-7. Differentiate $f(x) =$

$$\frac{\sin 4x - 4 \sin 2x}{32} - \frac{x(\cos 4x - 2 \cos 2x)}{8}$$

with respect to x.

216 Differentiation of the exponential and logarithmic functions

The applications of the exponential and logarithmic functions to electronics and communications were explored in chapter 210. You are now going to obtain the first derivatives of these quantities. This will lead to series expansions of the sine and cosine functions.

MATHEMATICAL DERIVATIONS

The first derivative of the exponential function can be found by expressing the function in terms of its series. Each term of the series is differentiated in turn, and the resulting derivatives are then added.

If $y = e^x$, then

$$y = 1 + \frac{x}{1!} + \frac{x^2}{2!} + \frac{x^3}{3!} + \frac{x^4}{4!} + \ldots \qquad (210\text{-}2)$$

$$\frac{dy}{dx} = 0 + 1 + \frac{2x}{2!} + \frac{3x^2}{3!} + \frac{4x^3}{4!} + \ldots$$

$$= 1 + \frac{x}{1!} + \frac{x^2}{2!} + \frac{x^3}{3!} + \ldots$$

or,

$$\frac{dy}{dx} = e^x$$

Therefore:

$$\frac{d(e^x)}{dx} = e^x$$

This result can be used to find the first derivative of $\ln x$.

If $y = \ln x$, then,

$$x = e^y \qquad (216\text{-}1)$$

$$\frac{dx}{dy} = \frac{d(e^y)}{dy}$$

$$= e^y$$

Therefore:

$$\frac{dy}{dx} = \frac{1}{e^y}$$

$$= \frac{1}{x}$$

so that,

$$\frac{d(\ln x)}{dx} = \frac{1}{x} \qquad (216\text{-}2)$$

Differentiation of a phasor

If phasor $z = R\cos\theta + jR\sin\theta$, (89-6, 89-7)

$$\frac{dz}{d\theta} = -R\sin\theta + jR\cos\theta \qquad (215\text{-}1, 215\text{-}2)$$

then,

$$\frac{dz}{d\theta} = jz \qquad (216\text{-}3)$$

Differentiation of a phasor, with respect to its angle, results in the phasor being multiplied by the operator j.

Comparing Equation 216-3 with $d(e^x)/dx = e^x$,

$$\text{Phasor, } z = Re^{j\theta} \qquad (216\text{-}4)$$

This leads to,

$$e^{j\theta} = \cos\theta + j\sin\theta \qquad (216\text{-}5)$$

and,

$$e^{-j\theta} = \cos(-\theta) + j\sin(-\theta) \qquad (216\text{-}6)$$

$$= \cos\theta - j\sin\theta$$

Therefore:

$$\cos\theta + j\sin\theta = 1 + \frac{j\theta}{1!} + \frac{(j\theta)^2}{2!} + \frac{(j\theta)^3}{3!} + \frac{(j\theta)^4}{4!} + \ldots$$

$$= 1 + \frac{j\theta}{1!} - \frac{\theta^2}{2!} - \frac{j\theta^3}{3!} + \frac{\theta^4}{4!} + \ldots$$

Equating real and imaginary parts (see chapter 88),

$$\sin\theta = \theta - \frac{\theta^3}{3!} + \frac{\theta^5}{5!} - \frac{\theta^7}{7!} + \ldots \qquad (216\text{-}7)$$

and,

$$\cos\theta = 1 - \frac{\theta^2}{2!} + \frac{\theta^4}{4!} - \frac{\theta^6}{6!} + \ldots \qquad (216\text{-}8)$$

In these series expansions of $\sin\theta$ and $\cos\theta$, the angle is measured in radians. For example, if $\theta = 30° = 0.5236^c$,

$$\sin 30° = \sin 0.5236^c$$

$$= 0.5236 - \frac{0.5236^3}{6} + \frac{0.5236^5}{120} - \ldots$$

$$= 0.5236 - 0.0239 + 0.0003 - \ldots$$

$$= 0.5, \text{ rounded off}$$

$$\cos 30° = \cos 0.5236^c = 1 - \frac{0.5236^2}{2} + \frac{0.5236^4}{24} - \ldots$$

$$= 1 - 0.1371 + 0.0031 - \ldots$$

$$= 0.8660, \text{ rounded off}$$

Adding Equations 216-5 and 216-6,

$$2\cos\theta = e^{j\theta} + e^{-j\theta}$$

or,

$$\cos\theta = \frac{e^{j\theta} + e^{-j\theta}}{2} \qquad (216\text{-}9)$$

Subtracting Equation 216-6 from Equation 216-5.

$$j2\sin\theta = e^{j\theta} - e^{-j\theta}$$

or,

$$\sin\theta = \frac{e^{j\theta} - e^{-j\theta}}{j2} \qquad (216\text{-}10)$$

It follows that:

$$\tan\theta = -j \times \frac{e^{j\theta} - e^{-j\theta}}{e^{j\theta} + e^{-j\theta}} \qquad (216\text{-}11)$$

$$\sec\theta = \frac{2}{e^{j\theta} + e^{-j\theta}} \qquad (216\text{-}12)$$

$$\csc\theta = \frac{j2}{e^{j\theta} - e^{-j\theta}} \qquad (216\text{-}13)$$

$$\cot\theta = j\,\frac{e^{j\theta} + e^{-j\theta}}{e^{j\theta} - e^{-j\theta}} \qquad (216\text{-}14)$$

Example 216-1

Differentiate the following with respect to x:

(a) $y = 5e^{3x}$, (b) $y = 4\ln(5x)$,
(c) $y = \ln(\sin x)$, and (d) $y = \ln(\tan\sqrt{x^2+1})$.

Solution

(a) If $y = 5e^{3x}$, $y = e^u$, where $u = 3x$. Therefore, y is a function of a function of x. Then:

$$\frac{dy}{dx} = 5e^{3x} \times 3 \qquad (216\text{-}1)$$

$$= 15e^{3x}$$

(b) If $y = 4\ln(5x)$,

$$\frac{dy}{dx} = \frac{4}{5x} \times 5 \qquad (216\text{-}2)$$

$$= \frac{4}{x}$$

(c) If $y = \ln(\sin x)$,

$$\frac{dy}{dx} = \frac{1}{\sin x} \times \cos x \qquad (215\text{-}1, 216\text{-}2)$$

$$= \cot x$$

(d) This is a function (ln) of a function (tan) of a function ($\sqrt{\ }$) of a function ($x^2 + 1$), which is, in turn, a function of x.

Therefore, if $y = \ln(\tan\sqrt{x^2+1})$,

$$\frac{dy}{dx} = \frac{1}{\tan\sqrt{x^2+1}} \times \sec^2\sqrt{x^2+1} \quad (215\text{-}3, 216\text{-}2)$$

$$\times \frac{1}{2} \times (x^2+1)^{-1/2} \times 2x$$

$$= \frac{x}{\sqrt{x^2+1} \times \sin\sqrt{x^2+1} \times \cos\sqrt{x^2+1}}.$$

PRACTICE PROBLEMS

216-1. Differentiate, with respect to x, the following functions:
(a) $\ln x^3$, (b) $\ln(1/x)$, and (c) $\ln\sqrt{x}$.

216-2. Differentiate, with respect to x, the following functions:
(a) $\ln(\sin x)$, (b) $\ln(1-x)$, and (c) $\ln(\cos^2 x)$.

216-3. Differentiate, with respect to x, the following functions:
(a) $\ln(4x+3)$, (b) $\ln(\tan x)$, and (c) $x \ln x$.

216-4. Differentiate, with respect to x, the following functions:

(a) $\dfrac{\ln x}{x}$ and (b) $\ln\left[\sqrt{(1+x)(1-x)}\right]$

216-5. Differentiate, with respect to x, the following functions:

(a) $e^{\gamma x}$, (b) e^{-2x}, and (c) xe^x.

216-6. Differentiate, with respect to x, the following functions:

(a)e^{x^2}, (b) $x^{2x}\sin 3x$, and (c) $(x^2 - x)/e^x$.

217
The hyperbolic functions and their derivatives

The functions $(e^\theta - e^{-\theta})/2$ and $(e^\theta + e^{-\theta})/2$ have properties that are analogous to those of $\sin\theta$ and $\cos\theta$. These functions are called, respectively, the *hyperbolic sine* and the *hyperbolic cosine* of θ. Therefore,

$$\sinh\theta = \frac{1}{2}(e^\theta - e^{-\theta})$$

and,

$$\cosh\theta = \frac{1}{2}(e^\theta + e^{-\theta})$$

It follows that,

$$\cosh^2\theta - \sinh^2\theta = \frac{1}{2}(e^\theta + e^{-\theta})^2 - \frac{1}{2}(e^\theta - e^{-\theta})^2$$

$$= 1$$

(The *hyperbolic sine* (sinh) is pronounced "shine," and the hyperbolic cosine is pronounced "kosh.")

The trigonometric functions sine and cosine are called *circular functions* because they are connected with the geometry of the circle. The coordinates of any point (P) on the circle, $x^2 + y^2 = a^2$ (Fig. 217-1), can be expressed in the form ($a\cos\theta$, $a\sin\theta$) because $a^2\cos^2\theta + a^2\sin^2\theta = a^2$.

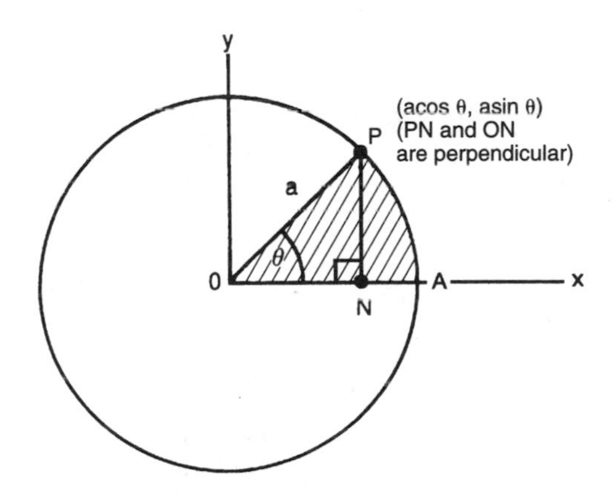

Figure 217-1

The functions $\cosh\theta$, $\sinh\theta$ are connected in a similar way with the geometry of a rectangular hyperbola, and this is the reason for the term *hyperbolic functions.* Figure 217-2 shows the rectangular hyperbola, $x^2 - y^2 = a^2$. The coordinates of any point (P) on this hyperbola can be expressed in the form ($a\cosh\theta$, $a\sinh\theta$) because $a^2\cosh^2\theta - a^2\sinh^2\theta = a^2$. For com-

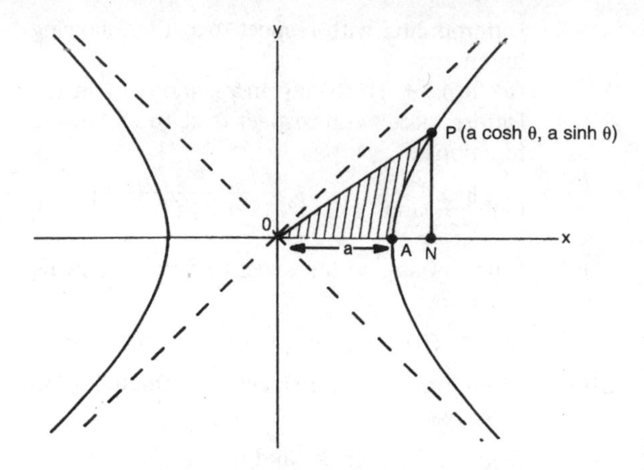

Figure 217-2

parison purposes, the shaded areas in Figs. 217-1 and 217-2 can both be expressed by $a^2\theta/2$.

The hyperbolic functions appear in the equations of transmission line theory. (See Example 217-2.)

MATHEMATICAL DERIVATIONS

Relationships between circular and hyperbolic functions

$$\sinh(-\theta) = \frac{e^{-\theta} - e^{\theta}}{2} \quad (217\text{-}1)$$

$$= -\sinh\theta$$

$$\cosh(-\theta) = \frac{e^{-\theta} + e^{\theta}}{2} \quad (217\text{-}2)$$

$$= \cosh\theta$$

Moreover,

$$\cosh\theta + \sinh\theta = e^{\theta} \quad (217\text{-}3)$$

and,

$$\cosh\theta - \sinh\theta = e^{-\theta} \quad (217\text{-}4)$$

Because $e^{\theta} = 1 + \frac{\theta}{1!} + \frac{\theta^2}{2!} + \frac{\theta^3}{3!} + \frac{\theta^4}{4!} + \ldots$ (210-2)

and $e^{-\theta} = 1 - \frac{\theta}{1!} + \frac{\theta^2}{2!} - \frac{\theta^3}{3!} + \frac{\theta^4}{4!} - \ldots$,

$$\cosh\theta = 1 + \frac{\theta^2}{2!} + \frac{\theta^4}{4!} + \ldots \quad (217\text{-}5)$$

$$\sinh\theta = \theta + \frac{\theta^3}{3!} + \frac{\theta^5}{5!} + \ldots \quad (217\text{-}6)$$

From Equations 216-5 and 217-3:

$$e^{j\theta} = \cos\theta + j\sin\theta$$

$$= \cosh j\theta + \sinh j\theta$$

Equating the "real" and "imaginary" parts, by using Equations 217-5 and 217-6,

$$\cosh j\theta = \cos\theta \quad (217\text{-}7)$$

$$\sinh j\theta = j\sin\theta$$

The following conversion rules apply:

Circular → Hyperbolic	Hyperbolic → Circular
$\sin\theta = -j\sinh j\theta$	$\sinh\theta = -j\sin j\theta$
$\cos\theta = \cosh j\theta$	$\cosh\theta = \cos j\theta$
$\sin j\theta = j\sinh\theta$	$\sinh j\theta = j\sin\theta$
$\cos j\theta = \cosh\theta$	$\cosh j\theta = \cos\theta$

The hyperbolic tangent is defined by:

$$\tanh\theta = \frac{\sinh\theta}{\cosh\theta} \quad (217\text{-}8)$$

$$= \frac{e^{\theta} - e^{-\theta}}{e^{\theta} + e^{-\theta}}$$

$$= \frac{e^{2\theta} - 1}{e^{2\theta} + 1}$$

(*Tanh* is pronounced "than").

Moreover,

$$\tanh\theta = \frac{\sinh\theta}{\cosh\theta} = \frac{-j\sin j\theta}{\cos j\theta} \quad (217\text{-}9)$$

$$= -j\tan j\theta$$

and,

$$\tan\theta = -j\tanh j\theta \quad (217\text{-}10)$$

The graphs of the hyperbolic functions are shown in Fig. 217-3.

Notice that: 1) Sinh θ lies between $-\infty$ and $+\infty$. 2) Cosh θ is always greater or equal to 1. 3) Tanh θ lies between $+1$ and -1. 4) The hyperbolic functions, unlike the circular functions, are not periodic.

The remaining hyperbolic functions are defined as follows:

$$\text{csch}\,\theta = \frac{1}{\sinh\theta} \quad (217\text{-}11)$$

$$= \frac{2}{e^{\theta} - e^{-\theta}} = j\csc j\theta$$

(*csch* is pronounced as "co-shek").

$$\text{sech}\,\theta = \frac{1}{\cosh\theta} \quad (217\text{-}12)$$

$$= \frac{2}{e^{\theta} + e^{-\theta}}$$

$$= \sec j\theta$$

("sech" is pronounced as "shek").

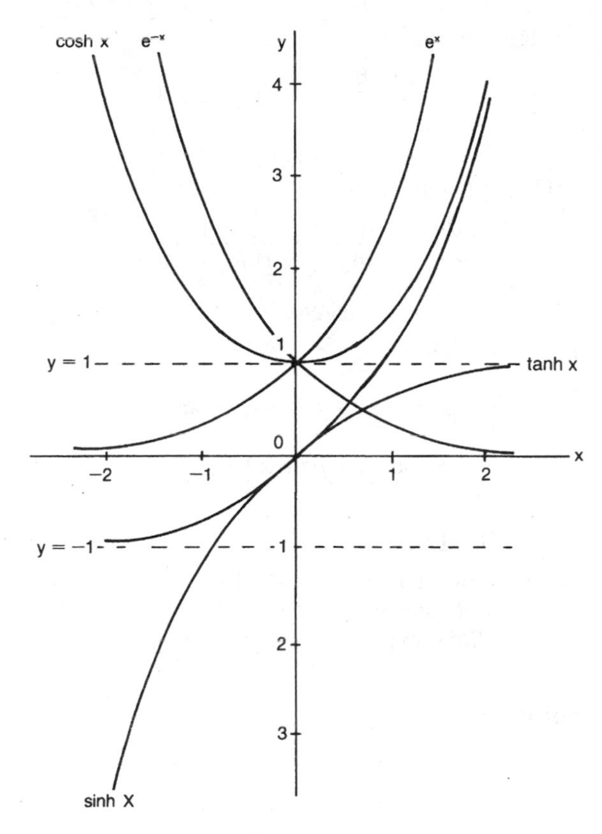

Figure 217-3

$$\coth\theta = \frac{1}{\tanh\theta} \quad (217\text{-}13)$$

$$= \frac{e^{2\theta}+1}{e^{2\theta}-1}$$

$$= j\cot j\theta$$

(*coth* is pronounced as "koth").

Hyperbolic identities

$$\sinh(\theta+\phi) = \frac{1}{2}[e^{\theta+\phi} - e^{-\theta-\phi}] \quad (217\text{-}14)$$

$$= \frac{1}{4}\left[2e^{\theta}e^{\phi} - \frac{2}{e^{\theta}e^{\phi}}\right]$$

$$= \frac{1}{4}\left[\left(e^{\theta} - \frac{1}{e^{\theta}}\right)\left(e^{\phi} + \frac{1}{e^{\phi}}\right) + \left(e^{\theta} + \frac{1}{e^{\theta}}\right)\left(e^{\phi} - \frac{1}{e^{\phi}}\right)\right]$$

$$= \sinh\theta\cosh\phi + \cosh\theta\sinh\phi$$

This result suggests that the relationships connecting hyperbolic functions are closely related to those connecting circular functions. In fact, hyperbolic identities can be written down using Osborn's Rule:

In any formula, connecting the circular functions of general angles, replace each circular function by the corresponding hyperbolic function and change the sign of every product (or implied product) of two sines.

Therefore,

$$\sinh(\theta-\phi) = \sinh\theta\cosh\phi - \cosh\theta\sinh\phi \quad (217\text{-}15)$$

$$\cosh(\theta+\phi) = \cosh\theta\cosh\phi + \sinh\theta\sinh\phi \quad (217\text{-}16)$$

$$\cosh(\theta-\phi) = \cosh\theta\cosh\phi - \sinh\theta\sinh\phi \quad (217\text{-}17)$$

$$\sinh(\theta+\phi) + \sinh(\theta-\phi) = 2\sinh\theta\cosh\phi \quad (217\text{-}18)$$

$$\sinh(\theta+\phi) - \sinh(\theta-\phi) = 2\cosh\theta\sinh\phi \quad (217\text{-}19)$$

$$\cosh(\theta+\phi) + \cosh(\theta-\phi) = 2\cosh\theta\cosh\phi \quad (217\text{-}20)$$

$$\cosh(\theta+\phi) - \cosh(\theta-\phi) = 2\sinh\theta\sinh\phi \quad (217\text{-}21)$$

$$\sinh 2\theta = 2\sinh\theta\cosh\theta \quad (217\text{-}22)$$

$$\cosh 2\phi = \cosh^2\theta + \sinh^2\theta \quad (217\text{-}23)$$

$$= 2\cosh^2\theta - 1$$

$$= 2\sinh^2\theta + 1$$

$$\tanh(\theta+\phi) = \frac{\tanh\theta + \tanh\phi}{1 + \tanh\theta \times \tanh\phi} \quad (217\text{-}24)$$

$$\tanh 2\theta = \frac{2\tanh\theta}{1+\tanh^2\theta} \quad (217\text{-}25)$$

Derivatives of hyperbolic functions

The six hyperbolic function derivatives are presented here:

If $y = \sinh\theta = (e^{\theta} - e^{-\theta})/2$,

$$\frac{dy}{dx} = \frac{d(\sinh\theta)}{d\theta} \quad (217\text{-}26)$$

$$= \frac{1}{2}(e^{\theta} + e^{-\theta})$$

$$= \cosh\theta$$

If $y = \cosh\theta = (e^{\theta} + e^{-\theta})/2$,

$$\frac{dy}{dx} = \frac{d(\cosh\theta)}{d\theta} \quad (217\text{-}27)$$

$$= \frac{1}{2}(e^{\theta} - e^{-\theta})$$

$$= \sinh \theta$$

If $y = \tanh \theta = \sinh \theta/\cosh \theta$,

$$\frac{dy}{d\theta} = \frac{d(\tanh \theta)}{d\theta} \quad (217\text{-}28)$$

$$= \frac{\cosh^2 \theta - \sinh^2 \theta}{\cosh^2 \theta}$$

$$= \frac{1}{\cosh^2 \theta}$$

$$= \text{sech}^2\, \theta$$

If $y = \text{csch}\, \theta = 1/\sinh \theta = (\sinh \theta)^{-1}$,

$$\frac{dy}{d\theta} = \frac{d(\text{csch}\, \theta)}{d\theta} \quad (217\text{-}29)$$

$$= \frac{-\cosh \theta}{\sinh^2 \theta}$$

$$= -\text{csch}\, \theta \coth \theta$$

If $y = \text{sech}\, \theta = 1/\cosh \theta = (\cosh \theta)^{-1}$,

$$\frac{dy}{d\theta} = \frac{d(\text{sech}\, \theta)}{d\theta} \quad (217\text{-}30)$$

$$= \frac{-\sinh \theta}{\cosh^2 \theta}$$

$$= -\text{sech}\, \theta \tanh \theta$$

If $y = \coth \theta = \cosh \theta/\sinh \theta$,

$$\frac{dy}{d\theta} = \frac{d(\coth \theta)}{d\theta} \quad (217\text{-}31)$$

$$= \frac{\sinh^2 \theta - \cosh^2 \theta}{\sinh^2 \theta}$$

$$= -\text{csch}^2\, \theta$$

Example 217-1

Differentiate the following with respect to θ:

(a) $y = \sinh^2 3\theta$,

(b) $y = \ln \sinh \theta$,

(c) $y = \ln \tanh \theta$, and

(d) $y = \text{sech}^2\, 2\theta$.

Solution

(a) If $y = \sinh^2 3\theta$,

$$\frac{dy}{d\theta} = 2 \sinh 3\theta \cosh 3\theta \times 3$$

$$= 3 \sinh 6\theta$$

(b) If $y = \ln \sinh \theta$,

$$\frac{dy}{d\theta} = \frac{\cosh \theta}{\sinh \theta}$$

$$= \coth \theta$$

(c) If $y = \ln \tanh \theta$,

$$\frac{dy}{d\theta} = \frac{\text{sech}^2\, \theta}{\tanh \theta} = \frac{1}{\sinh \theta \cosh \theta}$$

$$= 2\, \text{csch}\, 2\theta$$

(d) If $y = \text{sech}^2\, 2\theta$.

$$\frac{dy}{d\theta} = 2\, \text{sech}\, 2\theta \times (-\text{sech}\, 2\theta \tanh 2\theta) \times 2$$

$$= -4 \sinh 2\theta\, \text{sech}^3\, 2\theta$$

Example 217-2

The current on a transmission line is given by $I = A \cosh\gamma x + B \sinh\gamma x$. If the voltage $E = [1/(G + j\omega C)] \times dI/dx$, express E as a function of γx.

Solution

Voltage,

$$E = \frac{1}{G + j\omega C} \times \frac{dI}{dx}$$

$$= \frac{1}{G + j\omega C} \times \frac{d}{dx}(A \cosh\gamma x + B \sinh\gamma x)$$

$$= \frac{\gamma}{G + j\omega C}(A \sinh \gamma x + B \cosh \gamma x)$$

PRACTICE PROBLEMS

217-1. Differentiate, with respect to x, the following functions:

(a) $\sinh 2x$,

(b) $\cosh 3x$, and

(c) $\text{csch}^2\, 2x$.

217-2. Differentiate, with respect to x, the following functions:

(a) $\text{csch}\, x/2$,

(b) $\tanh^2 2x$, and

(c) $\coth \dfrac{1}{2x}$.

217-3. Differentiate, with respect to x, the following functions:

(a) $x - \coth x$,

(b) $\sinh x \cosh x$, and

(c) $\log(\cosh 2x)/2$.

217-4 Differentiate, with respect to x, the following functions:

(a) $\ln(\sinh 3x)/3$,

(b) $(x + \sinh x \cosh x)/2$, and

(c) $(\sinh x - x)/2$.

217-5. Differentiate, with respect to x, the following functions:

(a) $x - \coth x$,

(b) $\tan^{-1}(e^{2x})$, and

(c) $\ln(\tanh x/2)$.

218 Partial derivatives

So far in this study of calculus, you have differentiated one variable (y) with respect to another variable (x). $y = f(x)$ means that y is a function of one variable (x) and that the graph of y versus x is some curve that lies in the plane that is defined by the axes Ox, Oy. Now, consider that a new variable (z) is a function of both x and y so that $z = f(x,y)$. Instead of a curve, you must visualize a surface, such as is shown in Fig. 218-1. At point (P), you drop a perpendicular PN on to the plane xOy; point (N) is then defined by the x,y coordinates, and length (PN) is the z coordinate. Because P lies on a sloping surface, any change in the position of N will alter the length of PN. It follows that z is a function of x, y and $z = f(x,y)$.

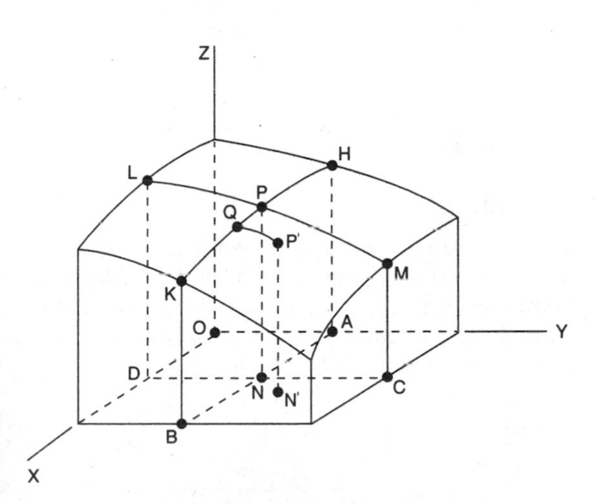

Figure 218-1

Plane $KPHANB$ is parallel to the plane that is defined by Ox and Oz. Consequently, at every point on the curve KPH, the value of the y coordinate is constant. You can obtain the slope at point (P) along the curve KPH, by differentiating z with respect to x, while regarding the value of y as constant. The result of differentiating z in this way is called the "partial differential coefficient" (or "partial derivative") and is denoted by $\partial z/\partial x$.

The plane $LPMCND$ is parallel to the plane defined by the axes Oy, Oz. At every point on curve LPM, the value of the x coordinate is constant. Therefore, you can obtain the slope at point P along curve LPM by differentiating z with respect to y, while regarding the value of x as constant. The resulting partial derivative is denoted by $\partial z/\partial y$.

MATHEMATICAL DERIVATIONS

From the introductory discussion,

$$\frac{\partial z}{\partial x} = \lim_{\delta x \to 0} \frac{f(x + \delta x, y) - f(x,y)}{\delta x} \qquad (218\text{-}1)$$

and,

$$\frac{\partial z}{\partial y} = \lim_{\delta y \to 0} \frac{f(x, y + \delta y) - f(x,y)}{\delta y} \qquad (218\text{-}2)$$

If z is a function of more than two independent variables,

$\dfrac{\partial z}{\partial x}$ = the rate at which z changes with respect to x, when all other relevant independent variables are kept constant.

$\dfrac{\partial z}{\partial y}$ = the rate at which z changes with respect to y, when all other relevant independent variables are kept constant.

Partial derivatives are evaluated in the normal manner, all variables (other than the two involved in the differentiation) being regarded as constants. For example, if,

$$z = \pi x^2 y$$

which is the formula for the volume (z) of a cylinder, in terms of its radius (x) and its length (y),

$$\frac{\partial z}{\partial x} = 2\pi xy$$

and,

$$\frac{\partial z}{\partial y} = \pi x^2$$

If a small change (δx) is made in the value of x while y is kept constant,

$$\delta z \approx \frac{\partial z}{\partial x}\,\delta x$$

Similarly, for a change δy in the value of y while x is kept constant,

$$\delta z \approx \frac{\partial z}{\partial y}\,\delta y$$

In general, if x and y are changed simultaneously,

$$\delta z \approx \frac{\partial z}{\partial x}\,\delta x + \frac{\partial z}{\partial y}\,\delta y \qquad (218\text{-}3)$$

Referring back to Fig. 218-1, the Equation 218-3 is equivalent to moving first from point P to point Q, whose x,y coordinates are $(x + \delta x, y)$. Next, travel from point Q to point P', whose x,y coordinates are $(x + \delta x, y + \delta y)$. The length of $(P'N' - PN)$ is then equal to δz.

If $z = f(x,y)$ there are four second-order partial derivates. These are:

$\dfrac{\partial^2 z}{\partial x^2}$ = the rate at which $\partial z/\partial x$ changes with respect to x when y is kept constant.

$\dfrac{\partial^2 z}{\partial y^2}$ = the rate at which $\partial z/\partial y$ changes with respect to y when x is kept constant.

$\dfrac{\partial^2 z}{\partial x \partial y}$ = the rate at which $\partial z/\partial y$ changes with respect to x when y is kept constant.

$\dfrac{\partial^2 z}{\partial y \partial x}$ = the rate at which $\partial z/\partial x$ changes with respect to y when x is kept constant.

In the example of $z = \pi x^2 y$, $\partial z/\partial x = 2\pi xy$ and $\partial z/\partial y = \pi x^2$.

Then,

$$\frac{\partial^2 z}{\partial x^2} = \frac{\partial}{\partial x}\left(\frac{\partial z}{\partial x}\right) = \frac{\partial}{\partial x}(2\pi xy) = 2\pi y.$$

$$\frac{\partial^2 z}{\partial y^2} = \frac{\partial}{\partial y}\left(\frac{\partial z}{\partial y}\right) = \frac{\partial}{\partial y}(\pi x^2) = 0.$$

$$\frac{\partial^2 z}{\partial x \partial y} = \frac{\partial}{\partial x}\left(\frac{\partial z}{\partial y}\right) = \frac{\partial}{\partial x}(\pi x^2) = 2\pi x.$$

$$\frac{\partial^2 z}{\partial y \partial x} = \frac{\partial}{\partial y}\left(\frac{\partial z}{\partial x}\right) = \frac{\partial}{\partial y}(2\pi xy) = 2\pi x.$$

Notice that $\dfrac{\partial^2 z}{\partial x \partial y}$ and $\dfrac{\partial^2 z}{\partial y \partial x}$ are both equal to $2\pi x$.

In fact, $\partial^2 z/\partial x \partial y = \partial^2 z/\partial y \partial x$ is a general result, which is true for virtually all functions that involve two independent variables.

Example 218-1

If $z = (x/y) \ln x$, determine the expressions for the first and second-order partial derivatives.

Solution

$$\frac{\partial z}{\partial y} = \frac{-x \ln x}{y^2}$$

$$\frac{\partial z}{\partial x} = \frac{1}{y}\left(x \times \frac{1}{x} + \ln x\right) = \frac{1 + \ln x}{y} = \frac{1}{y} + \frac{\ln x}{y}$$

$$\frac{\partial^2 z}{\partial y^2} = \frac{2x \ln x}{y^3}$$

$$\frac{\partial^2 z}{\partial x^2} = \frac{1}{xy}$$

$$\frac{\partial^2 z}{\partial y \partial x} = \frac{-1}{y^2} - \frac{\ln x}{y^2} = -\frac{1 + \ln x}{y^2}$$

$$\frac{\partial^2 z}{\partial x \partial y} = \frac{1}{y^2}(-\ln x - 1) = -\frac{1 + \ln x}{y^2}$$

Example 218-2

The plate current (i_b) of a triode tube is a function of the plate voltage (e_b) and the control grid voltage (e_c). Determine the approximate total change in the plate current that is caused by the instantaneous small changes in the plate and grid voltages.

Solution

$$\delta i_b \approx \frac{\partial i_b}{\partial e_b}\,\delta e_b + \frac{\partial i_b}{\partial e_c}\,\delta e_c \qquad (218\text{-}3)$$

From chapters 143 and 144,

$$\frac{\partial i_b}{\partial e_b} = \frac{1}{r_p} \text{ and } \frac{\partial i_b}{\partial e_b} = g_m.$$

Therefore,

$$\delta i_b \approx \frac{\delta e_b}{r_p} + g_m \delta e_c.$$

From chapter 146,

$$\delta e_b = -\delta i_b R_L.$$

Then,

$$\delta i_b \approx \frac{-\delta i_b R_L}{r_p} + g_m \delta e_c$$

$$\delta i_b \,(r_p + R_L) \approx r_p g_m \delta e_c.$$

From chapter 145,

$$r_p g_m = -\mu$$

Therefore,

$$i_b = \frac{-\mu\, \delta e_c}{r_p + R_L}$$

PRACTICE PROBLEMS

218-1. Obtain the values of $\partial z/\partial x$, $\partial z/\partial y$ for the following $z = f(x, y)$ functions:

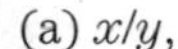

(a) x/y,
(b) $ax^2 + 2hxy + by^2$,
(c) $\tan^{-1}(y/x)$,
(d) $(x - y)/(x + y)$,
(e) $\sin^{-1}(x/y)$, and
(f) $1/\sqrt{x^2 + y^2}$.

218-2. Obtain the values of $\partial^2 z/\partial x^2$, $\partial^2 z/\partial x\partial y$, $\partial^2 z/\partial y\partial x$, $\partial^2 z/\partial y^2$ for the following $z = f(x,y)$ functions:

(a) xy,
(b) $ax^3 + 3bx^2y + cy^3$,
(c) $\ln(xy)$,
(d) e^{x+y}
(e) $x \cos y + y \cos x$, and
(f) $\sinh x \cosh 2y$.

218-3. If $z = \ln(x^2 + y^2)$, what is the value of $\partial^2 z/\partial x^2 + \partial^2 z/\partial y^2$?

218-4. If $\theta = \tan^{-1} y/x$, what is the value of $\partial^2\theta/\partial x^2 + \partial^2\theta/\partial y^2$?

218-5. If $V = x^2 + y^2 + z^2$, what is the value of $x\,\partial V/\partial x + y\,\partial V/\partial y + z\,\partial V/\partial z$?

219 Applications of derivatives: maxima, minima, and inflexion conditions

Because dy/dx is the slope of $y = f(x)$ at any point, dy/dx can be used to find the points where y is a maximum or a minimum. At the maximum and minimum points, the curve of $y = f(x)$ is instantaneously horizontal so that the slope is zero. The values of x for which y is a maximum or a minimum are the solutions to the equation $dy/dx = 0$.

For the position of a *maximum*, the slope is *positive* before the position and *negative* afterwards (Fig. 219-1); therefore, the rate of change of the slope is *negative*. This means a *negative* value for the second derivative, which is written as d^2y/dx^2, rather than $d(dy/dx)/dx$. The same notation—namely d^3y/dx^3, d^4y/dx^4 and so on—is used to express further derivatives.

For the position of a *minimum*, the slope is *negative* before the position and *positive* after (Fig. 219-2). Therefore, the rate of change of the slope is *positive*, and the value of d^2y/dx^2 is *positive*.

At a point where $d^2y/dx^2 = 0$, the rate of change of the slope is momentarily zero, and you have a stationary tangent! The simplest case of this condition is at a point of *inflexion*, where the curve crosses its tangent (Fig. 219-3). It is further necessary that d^2y/dx^2 changes its sign as x increases through its value at the point of inflexion.

Example 219-1

Examine the curve $y = x^4 - 5x^2 + 4$ for maxima, minima, and inflexion points.

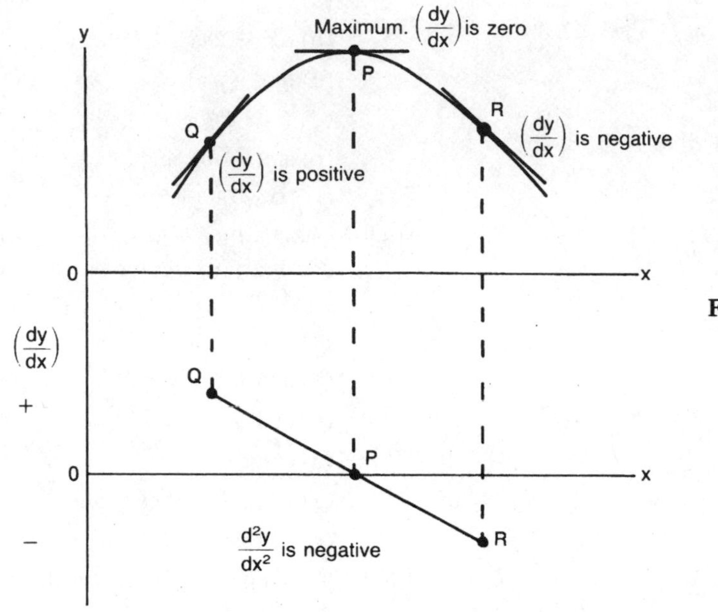

Figure 219-1

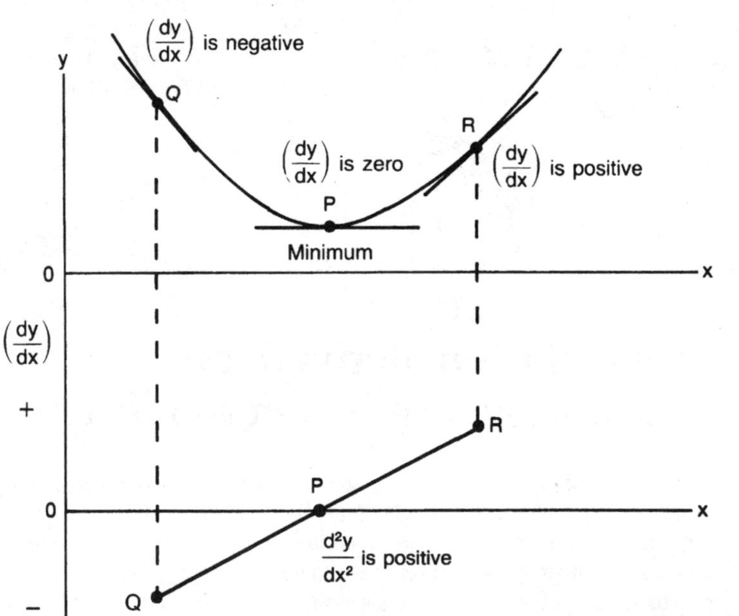

Figure 219-2

Solution

If $y = x^4 - 5x^2 + 4$,

$$\frac{dy}{dx} = 4x^3 - 10x$$

$$\frac{d^2y}{dx^2} = 12x^2 - 10.$$

Maxima or minima occur when,

$$\frac{dy}{dx} = 4x^3 - 10x = 0.$$

The solutions are $x = 0$ and $x = \pm\sqrt{5/2}$.

If $x = 0$,

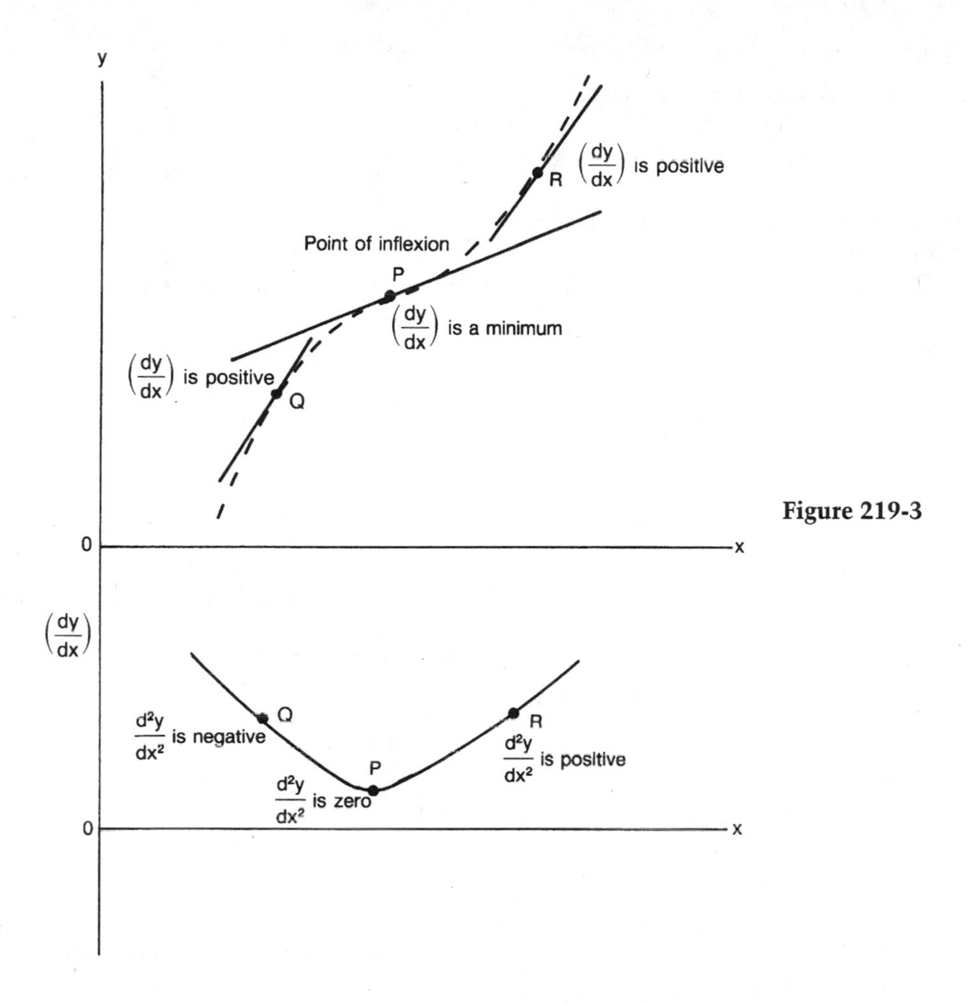

Figure 219-3

$$\frac{d^2y}{dx^2} = -10 \text{ and is negative.}$$

Therefore, a maximum occurs at the point (0,4).

If $x = \pm\sqrt{\frac{5}{2}}$,

$$\frac{d^2y}{dx^2} = 12 \times \frac{5}{2} - 10$$

$$= 20$$

and is positive.

Therefore, minima occur at the points $(+\sqrt{5/2}, -9/4)$ and $(-\sqrt{5/2}, -9/4)$.

When $d^2y/dx^2 = 0$,

$$12x^2 - 10 = 0$$

$$x = \pm\sqrt{\frac{5}{6}}$$

Because d^2y/dx^2 changes sign as x increases through each of the values $\pm\sqrt{5/6}$, there are points of inflexion at $(+\sqrt{5/6}, 19/36)$ and $(-\sqrt{5/6}, 19/36)$. The curve for $y = x^4 - 5x^2 + 4$ is shown in Fig. 219-4.

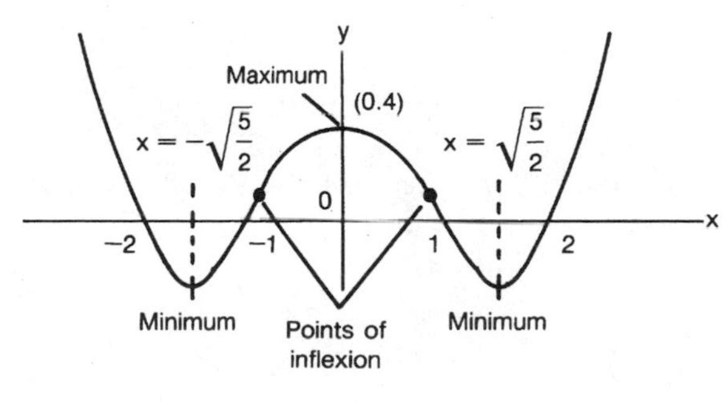

Figure 219-4

Example 219-2

The equation for the maximum power in a load is:

$$P_L = \frac{E^2R_L}{(R_i - R_L)^2}$$

where E and R_i are constants. Find the condition for maximum power in the load.

Solution

If $P_L = \frac{E^2R_L}{(R_i + R_L)^2}$,

$$\frac{dP_L}{dR_L} = E^2 \times \left[\frac{(R_i + R_L)^2 - R_L \times 2(R_i + R_L)}{(R_i + R_L)^4}\right]$$

$$= E^2 \times \frac{R_i - R_L}{(R_i + R_L)^3}$$

$$\frac{d^2P}{dR_L{}^2} =$$

$$E^2 \times \left[\frac{-(R_i + R_L)^3 - (R_i - R_L) \times 3(R_i + R_L)^2}{(R_i + R_L)^6}\right]$$

For a maximum or minimum value of P_L,

$$\frac{dP_L}{dR_L} = E^2 \times \frac{(R_i - R_L)}{(R_i + R_L)^3} = 0$$

This yields,

$$R_L = R_i$$

When $R_L = R_i$, $d^2P/dR_L{}^2$ is negative, therefore, P_L is at its maximum value.

Alternatively, the load power (from chapter 23) is

$$P_L = EI_L - I_L{}^2R_i$$

Differentiating P_L with respect to I_L,

$$\frac{dP_L}{dI_L} = E - 2I_LR_i$$

and,

$$\frac{d^2P_L}{dI_L{}^2} = -2R_i$$

Because $d^2P_L/dI_L{}^2$ is negative, P_L will reach its *maximum* value when,

$$\frac{dP_L}{dI_L} = E - 2I_LR_i = 0.$$

This yields,

$$I_L = \frac{E}{2R_i}$$

This value of I_L will only occur when $R_L = R_i$.

Example 219-3

Examine the curve $y = \sin x$ for points of inflexion.

Solution

If $y = \sin x$,

$$\frac{dy}{dx} = \cos x \qquad (215\text{-}1)$$

$$\frac{d^2y}{dx^2} = -\sin x = -y \qquad (215\text{-}2)$$

Therefore, d^2y/dx^2 changes sign whenever the curve crosses the x axis. These are points of inflexion because the values of d^2y/dx^2 at these points are zero (Fig. 219-5). Maxima occur at $x = 2n\pi + \pi/2$ radians, where n is any whole number. For these points, $dy/dx = 0$ and d^2y/dx^2 is negative. Minima exist at $x = 2n\pi - \pi/2$ radians. For these points, $dy/dx = 0$, but d^2y/dx^2 is positive.

PRACTICE PROBLEMS

219-1. For which values of x does a maximum and a minimum occur in the curve whose equation is $y = 2x^3 - 3x^2 - 36x + 10$?

219-2. For which values of x does a maximum (or a minimum) occur in the curve whose equation is $y = 4x^3 - 18x^2 + 27x - 7$?

219-3. For which value of x does a maximum occur in the equation for which $y = 10x^6 - 12x^5 + 15x^4 - 20x^3 + 20$?

219-4. If $y = x\sqrt{ax - x^2}$, determine the value of x, for which y has a maximum value.

219-5. If $y = (x - 1)^2/(x + 1)^3$, what are the maximum and minimum values of y?

219-6. If $y = (1 + x + x^2)/(1 - x + x^2)$, what are the maximum and minimum values of y?

219-7. If $y = x^2(3 - x)$, determine the point of inflexion on the curve.

219-8. Determine the points of inflexion on the curve for which $y = x^3/(a^2 + x^2)$.

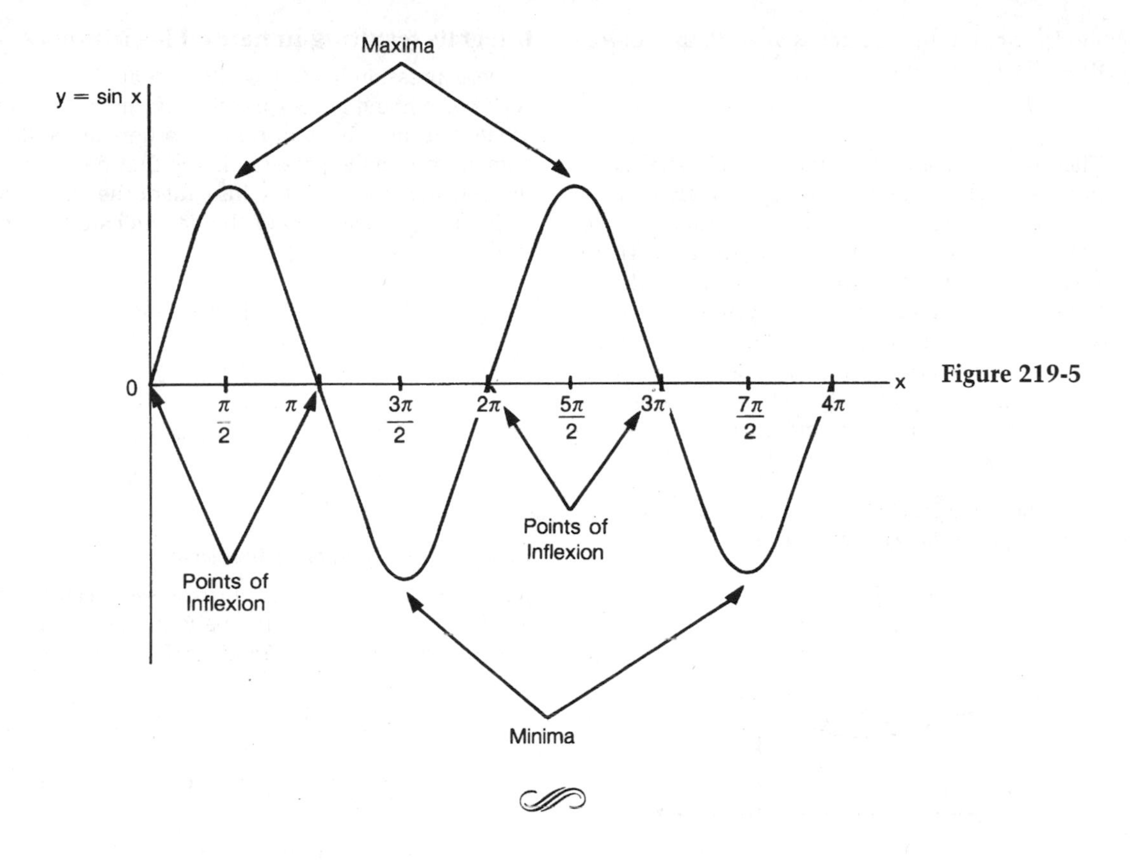

Figure 219-5

220
Introduction to integral calculus

Integration is the reverse of differentiation. Therefore, if $dy/dx = f'(x)$, y is the "integral" of $f'(x)$ with respect to x. Mathematically, this is written as:

$$y = f(x) = \int f'(x)\, dx$$

There is not a complete set of rules for integration. In fact, for certain integrals, the results are unknown. The process of integration depends on recalling the various results of differentiation.

Before proceeding further, examine in detail one interpretation of the integration symbol ($\int$). Referring to Fig. 220-1, the area $OMPN$ under the curve $y = f(x)$, and bounded by the axes and the ordinate NP, is a function of x. The area cannot be found at once, but its first derivative can. If x and y are respectively increased to $x + \delta x$ and $y + \delta y$, the area A increases by a small amount δA, which is equal to

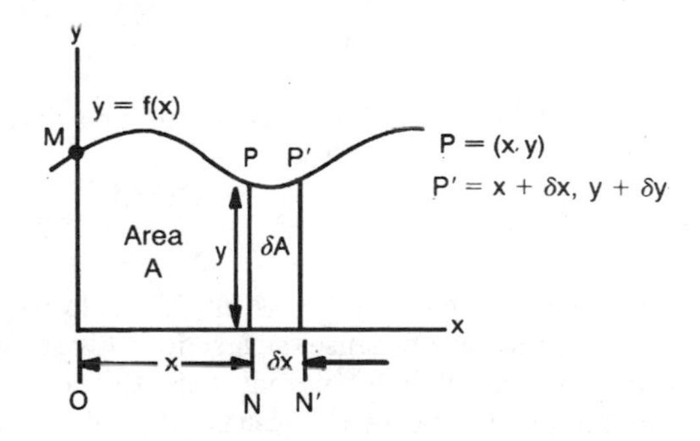

Figure 220-1

the area $PP'N'N$. Regarding this incremental area to be an approximate rectangle, $\delta A = y\delta x$, you can take the sum (the symbol $\int$ is a form of the letter S standing for "summation") of all the incremental areas to obtain the area A. In the limit as $\delta x \to 0$, the approximation disappears and,

$$\text{Area, } A = \int y\, dx = \int f(x)\, dx$$

Therefore, the area underneath the curve up to point $P = (x,y)$ is found by integration. However, A does not have an exact numerical value because no precise limits have been set for the x coordinates. Therefore, $\int f(x)\, dx$ is known as an *indefinite integral.*

Sometimes the area is required between two coordinates, such as x_1 and x_2 in Fig. 220-2. This is found by calculating the area up to x_2 and subtracting from the result the area up to x_1. If $A = \int f(x)\, dx$, the notation used for the area under consideration is:

$$[A]_{x_1}^{x_2}$$

which is the value of A when $x = x_2$ minus the value of A when $x = x_1$. This *definite* integral is written as,

$$A = \int_{x_1}^{x_2} f(x)\, dx$$

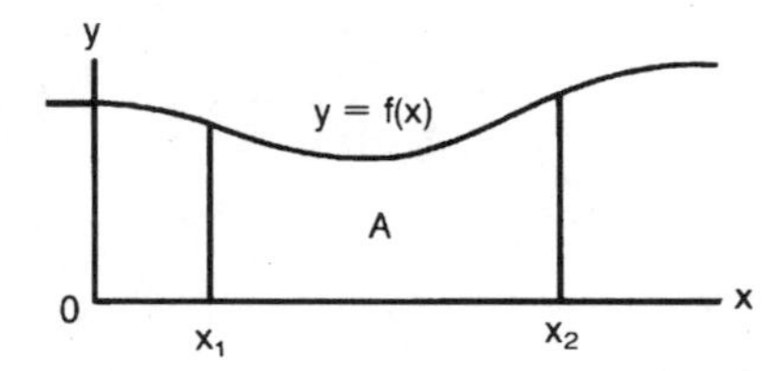

Figure 220-2

MATHEMATICAL DERIVATIONS

Integral of x^n

Consider the indefinite integral of x^n,

$$y = \int x^n\, dx$$

From a knowledge of differentiation, the first derivative of $x^{n+1} = (n+1)x^n$. Therefore:

$$y = \int x^n\, dx = \left(\frac{1}{n+1}\right) x^{n+1} + K \qquad (220\text{-}1)$$

For indefinite integrals a constant (K) is always included in the answer because the first derivative of any constant is zero. With definite integrals, the constant of integration disappears as a result of the subtraction process.

Equation 220-1 is valid for all values of n except $n = -1$. In this case, you will recall that if $y = \ln x$, $dy/dx = 1/x$. Therefore:

$$y = \int x^{-1}\, dx = \int \frac{1}{x}\, dx = \ln x + K \qquad (220\text{-}2)$$

Integrals resulting in natural logarithms

If the expression $\ln f(x)$ is differentiated, the result is $f'(x)/f(x)$, and any expression of this form can be integrated directly as $\ln f(x)$. In other words, if the numerator of an expression is the first derivative of the denominator, then the integral of the expression is the natural logarithm of the denominator. For example,

$$y = \int \tan x\, dx = \int \frac{\sin x}{\cos x}\, dx \qquad (220\text{-}3)$$

$$= -\int \frac{(-\sin x)}{\cos x}\, dx$$

$$= -\ln \cos x + K$$

$$= \ln \sec x + K$$

Integration by partial fractions

When the denominator of an expression can be factored, the integral can often be found by rearranging the expression as partial fractions. For example,

$$y = \int \frac{x+1}{x^2 + 5x + 6}\, dx$$

$$= \int \frac{x+1}{(x+3)(x+2)}\, dx$$

$$= \int \frac{2}{x+3}\, dx - \int \frac{1}{x+2}\, dx$$

$$= 2 \ln (x+3) - \ln (x+2) + K$$

Integration by parts

This method of integration is obtained from the rule for differentiating a product. From Equation 213-1, if $y = uv$:

$$\frac{dy}{dx} = \frac{d}{dx}(uv) = u\frac{dv}{dx} + v\frac{du}{dx}$$

then,

$$uv = \int \left(u\frac{dv}{dx}\right) dx + \int \left(v\frac{du}{dx}\right) dx$$

or,

$$\int \left(u\frac{dv}{dx}\right) dx = uv - \int \left(v\frac{du}{dx}\right) dx \qquad (220\text{-}4)$$

This equation represents the method of integration by parts. For example, in the integral $\int x \ln x\, dx$, let $u = \ln x$ and $dv/dx = x$ so that $v = x^2/2$. Then:

$$\int x \ln x\, dx = \frac{1}{2}x^2 \ln x - \int \frac{1}{2}x^2\left(\frac{1}{x}\right) dx$$

$$= \frac{1}{2}x^2 \ln x - \int \frac{1}{2}x\, dx$$

$$= \frac{1}{2}x^2 \ln x - \frac{1}{4}x^2 + K$$

As another example, let $u = \ln x$ and $dv/dx = 1$ in the integral $\int \ln x\, dx$. Then:

$$\int \ln x\, dx = x \ln x - \int x\left(\frac{1}{x}\right) dx \qquad (220\text{-}5)$$

$$= x \ln x - \int 1\, dx$$

$$= x \ln x - x + K$$

Trigonometric transformations

Certain products and powers of the sine and cosine functions can sometimes be integrated by converting to the forms $\sin nx$ and $\cos nx$. For example, because $\cos 2x = 1 - 2 \sin^2 x$ (Equation 208-12),

$$\int \sin^2 x\, dx = \int \frac{1}{2}(1 - \cos 2x)\, dx$$

$$= \int \frac{1}{2}\, dx - \int \frac{1}{2} \cos 2x\, dx$$

$$= \frac{x}{2} - \frac{\sin 2x}{4} + K$$

Integration by substitution

The process of integration can frequently be simplified by making a substitution, either algebraic or trigonometric. It is important to realize that the dx term must be converted into terms of the new variable. As an example, consider the integral of $x(4 + 2x^2)^6$ and notice that the derivative of x^2 is $2x$. Then let,

$$u = 4 + 2x^2$$

so that,

$$du = 4x\, dx$$

or,

$$x\, dx = \frac{du}{4}$$

then,

$$\int x(4 + 2x^2)^6\, dx = \int \frac{1}{4}u^6\, du$$

$$= \frac{u^7}{28} + K$$

$$= \frac{(4 + 2x^2)^7}{28} + K$$

If you integrate from $x = 1$ to $x = 2$, the limits must be changed when you make the substitution in the definite integral.

When $x = 1, u = 4 + 2 \times 1^2 = 6$.

When $x = 2, u = 4 + 2 \times 2^2 = 12$.

Then,

$$\int_1^2 x(4 + 2x^2)^6\, dx = \int_6^{12} \frac{1}{4}u^6\, du$$

$$= \left[\frac{u^7}{28}\right]_6^{12}$$

$$= \left[\frac{12^7}{28} - \frac{6^7}{28}\right]$$

$$= 1.27 \times 10^6, \text{ rounded off.}$$

Recommendations for trigonometric and hyperbolic substitutions

If the expression to be integrated contains:

1. $\sqrt{a^2 - x^2}$, substitute $x = a \sin \theta$ or $x = a \cos \theta$
2. $a^2 + x^2$, substitute $x = a \tan \theta$
3. $\sqrt{x^2 - a^2}$, substitute $x = a \cosh \theta$
4. $\sqrt{a^2 + x^2}$, substitute $x = a \sinh \theta$
5. $a^2 - x^2$, substitute $x = a \tanh \theta$

As an example, consider the integral:

$$\int \frac{1}{\sqrt{a^2 - x^2}}\, dx$$

Substituting $x = a \sin \theta$, $dx = a \cos \theta\, d\theta$; therefore:

$$\int \frac{1}{\sqrt{a^2 - x^2}}\, dx = \int \frac{a \cos \theta\, d\theta}{\sqrt{a^2 - a^2 \sin^2 \theta}} \qquad (220\text{-}6)$$

$$= \int d\theta$$

$$= \theta + K$$

$$= \sin^{-1}\left(\frac{x}{a}\right) + K$$

This is known as a *standard integral,* which allows similar expressions to be integrated. For example,

$$\int \frac{5}{\sqrt{4 - x^2}}\, dx = 5 \sin^{-1} \frac{x}{2} + K$$

A list of selected standard integrals is shown in Table 220-1.

Table 220-1

y	$\frac{dy}{dx}$	$\int y\,dx$
C	0	Cx
$a\,f(x)$	$a\frac{d}{dx}f(x)$	$a\int f(x)\,dx$
x^n	nx^{n-1}	$\frac{1}{n+1}x^{n+1}$ when $n \neq -1$
x^{-1}	$-x^{-2}$	$\log x$
e^{ax}	ae^{ax}	$\frac{1}{a}e^{ax}$
xe^x	$e^x(x+1)$	$e^x(x-1)$
a^x	$a^x \log_e a$	$a^x(\log_e a)$
$\log_e x$	x^{-1}	$x(\log x - 1)$
$\log_a x$	$\frac{1}{x}\log_a e$	$x \log_a\left(\frac{x}{e}\right)$
uv	$vdu + udv$	$u\int vdx - \int\left(\int vdx\right)du$
$\frac{u}{v}$	$\frac{vdu - udv}{v^2}$	$u\int\frac{1}{v}dx - \int\left(\int\frac{1}{v}dx\right)du$
$\sin ax$	$a\cos ax$	$-\frac{1}{a}\cos ax$
$\cos ax$	$-a\sin ax$	$+\frac{1}{a}\sin ax$
$\tan ax$	$a\sec^2 ax$	$-\frac{1}{a}\log_e \cos ax$
$\sin^{-1}\frac{x}{a}$	$\frac{1}{\sqrt{a^2-x^2}}$	$x\sin^{-1}\frac{x}{a} + \sqrt{a^2-x^2}$
$\cos^{-1}\frac{x}{a}$	$\frac{-1}{\sqrt{a^2-x^2}}$	$x\cos^{-1}\frac{x}{a} - \sqrt{a^2-x^2}$
$\tan^{-1}\frac{x}{a}$	$\frac{a}{a^2+x^2}$	$x\tan^{-1}\frac{x}{a} - \frac{1}{a}\log_e\sqrt{a^2+x^2}$
$\sinh ax$	$a\cosh ax$	$\frac{1}{a}\cosh ax$
$\cosh ax$	$a\sinh ax$	$\frac{1}{a}\sinh ax$
$\tanh ax$	$a\,\mathrm{sech}^2 ax$	$\frac{1}{a}\log_e \cosh ax$
$\sinh^{-1}\frac{x}{a}$	$\frac{1}{\sqrt{a^2+x^2}}$	$x\sinh^{-1}\frac{x}{a} - \sqrt{x^2+a^2}$
$\cosh^{-1}\frac{x}{a}$	$\frac{1}{\sqrt{x^2-a^2}}$	$x\cosh^{-1}\frac{x}{a} - \sqrt{x^2-a^2}$
$\tanh^{-1}\frac{x}{a}$	$\frac{a}{a^2-x^2}$	$x\tanh^{-1}\frac{x}{a} + a\log_e\sqrt{a^2-x^2}$
$\frac{1}{\sin x}$	$-\frac{\cos x}{\sin^2 x}$	$\log_e \tan\frac{x}{2}$
$\frac{1}{\cos x}$	$+\frac{\sin x}{\cos^2 x}$	$\log_e \tan\left(\frac{x}{2}+\frac{n}{4}\right) = \log_e(\tan x + \sec x)$
$\sin^2 ax$	$a\sin 2ax$	$\left(\frac{x}{2} - \frac{\sin ax \cos ax}{2a}\right) = \left(\frac{x}{2} - \frac{\sin 2ax}{4a}\right)$
$\cos^2 ax$	$-a\sin 2ax$	$\left(\frac{x}{2} + \frac{\sin ax \cos ax}{2a}\right) = \left(\frac{x}{2} + \frac{\sin 2ax}{4a}\right)$
$e^{ax}\sin\beta x$		$\frac{\alpha e^{ax}\sin\beta x - \beta e^{ax}\cos\beta x}{\alpha^2+\beta^2}$
$e^{ax}\cos\beta x$		$\frac{\alpha e^{ax}\cos\beta x + \beta e^{ax}\sin\beta x}{\alpha^2+\beta^2}$
$e^{ax}\sinh\beta x$		$\frac{\alpha e^{ax}\sinh\beta x - \beta e^{ax}\cosh\beta x}{\alpha^2-\beta^2}$
$e^{ax}\cosh\beta x$		$\frac{\alpha e^{ax}\cosh\beta x - \beta e^{ax}\sinh\beta x}{\alpha^2-\beta^2}$
$\sinh ax\sin\beta x$		$\frac{\alpha\cosh\alpha x\sin\beta x - \beta\sinh\alpha x\cos\beta x}{\alpha^2+\beta^2}$
$\cosh ax\sin\beta x$		$\frac{\alpha\sinh\alpha x\sin\beta x - \beta\cosh\alpha x\cos\beta x}{\alpha^2+\beta^2}$
$\sinh ax\cos\beta x$		$\frac{\alpha\cosh\alpha x\cos\beta x + \beta\sinh\alpha x\sin\beta x}{\alpha^2+\beta^2}$
$\cosh ax\cos\beta x$		$\frac{\alpha\sinh\alpha x\cos\beta x + \beta\cosh\alpha x\sin\beta x}{\alpha^2+\beta^2}$

Note: In this table "ln" is shown as "$\log_e$". The constant of integration is omitted from the expressions for $\int y\,dx$. The integrals of dy/dx are expressions for y.

Example 220-1

Find the area of the circle, $x^2 + y^2 = a^2$.

Solution

Consider the area (A) of one quadrant. Then,

$$A = \int_0^a \sqrt{a^2 - x^2}\,dx$$

Let $x = a\sin\theta$ so that $dx = a\cos\theta\,d\theta$. When $x = 0$, $\theta = 0$ and when $x = a$, $\theta = \pi^c/2$. Therefore:

$$A = \int_0^{\pi/2} a^2\cos^2\theta\,d\theta$$

$$= \int_0^{\pi/2} \frac{a^2}{2} \times (1 + \cos 2\theta)\,d\theta \quad (208\text{-}11)$$

$$= \frac{a^2}{2}\left[\theta + \frac{\sin 2\theta}{2}\right]_0^{\pi/2} \quad (215\text{-}1)$$

$$= \frac{a^2}{2}\left(\frac{\pi}{2} - 0 + 0 - 0\right)$$

$$= \frac{\pi a^2}{4}$$

This is the area of one quadrant of the circle. The total area of the circle is:

$$4 \times \frac{\pi a^2}{4} = \pi a^2$$

Example 220-2

Find the mean height of the curve $y = A \sin \theta$ between the limits $\theta = 0$ and $\theta = \pi$.

Solution

The mean height is determined by finding the area beneath the curve between the given limits and then dividing this area by the length (π) of the base. Therefore:

$$\textit{Mean height} = \frac{\int_0^{\pi} A \sin \theta \, d\theta}{\pi}$$

$$= \left[\frac{-A \cos \theta}{\pi}\right]_0^{\pi}$$

$$= \frac{A}{\pi}[-(-1) - (-1)]$$

$$= \frac{2A}{\pi}$$

Therefore, $2A/\pi$ is the average value of the curve $y = A \sin \theta$ between the limits 0 and π.

Example 220-3

Integrate the following with respect to x:
(a) $7x^{5/2}$,
(b) $x\sqrt{x + 1}$,
(c) $\sqrt{x/(1 - x)}$,
(d) $x \sin x$,
(e) $x^2/(1 + x^3)$,
(f) $e^{\sin x} \cos x$,
(g) xe^x, and
(h) $\sqrt{a^2 + x^2}$.

Solution

(a)
$$7x^{5/2}\, dx = \frac{7}{\frac{5}{2} + 1} x^{5/2+1} + K \qquad (220\text{-}1)$$
$$= 2x^{7/2} + K$$

(b) Let $x + 1 = z^2$ so that $dx = 2z\, dz$.
Then,

$$\int x\sqrt{x + 1}\, dx = \int (z^2 - 1) \times z \times 2z\, dz$$
$$= \int 2z^4 - 2z^2\, dz$$
$$= \frac{2}{5}z^5 - \frac{2}{3}z^3 + K$$
$$= \frac{2}{5}(x + 1)^{5/2} - \frac{2}{3}(x + 1)^{3/2} + K$$

(c) Let $x = \sin^2 \theta$ so that $dx = 2 \sin \theta \cos \theta\, d\theta$

$$\int \sqrt{\frac{x}{1 - x}}\, dx = \int \frac{\sin \theta \times 2 \sin \theta \cos \theta}{\cos \theta}\, d\theta$$
$$= \int 2 \sin^2 \theta\, d\theta = \int 1 - \cos 2\theta\, d\theta$$
$$= \theta - \frac{\sin 2\theta}{2} + K$$
$$= \theta - \sin \theta \cos \theta + K$$
$$= \sin^{-1} \sqrt{x} - \sqrt{x - x^2} + K$$

(d) (220-4)
$$\int x \sin x\, dx = -x \cos x + \int 1 \times \cos x\, dx$$
$$= -\cos x + \sin x + K$$

(e)
$$\frac{x^2}{1 + x^3} = \frac{1}{3}\int \frac{3x^2}{1 + x^3}\, dx = \frac{1}{3} \ln (1 + x^3) + K$$

(f) Let $y = \sin x$ so that $dy = \cos x\, dx$. Then,
$$\int e^{\sin x} \cos x\, dx = \int e^y\, dy = e^y + K = e^{\sin x} + K$$

(g)
$$\int x\, e^x\, dx = x\, e^x - \int 1(e^x)\, dx \qquad (220\text{-}4)$$
$$= xe^x - e^x + K$$

(h) Let $x = a \sinh \theta$ so that $dx = a \cosh \theta\, d\theta$

$$\int \sqrt{a^2 + x^2}\, dx$$
$$= \int \sqrt{a^2 + a^2 \sinh^2 \theta} \times a \cosh \theta\, d\theta$$
$$= \int a^2 \cosh^2 \theta\, d\theta$$
$$= \int \frac{a^2 (\cosh 2\theta + 1)}{2}\, d\theta \qquad (217\text{-}23)$$
$$= \frac{a^2}{4} \sinh 2\theta + \frac{a^2\theta}{2} + K$$
$$= \frac{a^2}{2} \sinh \theta \cosh \theta + \frac{a^2\theta}{2} + K \qquad (217\text{-}22)$$
$$= \frac{1}{2}x\sqrt{a^2 + x^2} + \frac{1}{2}a^2 \sinh^{-1} \frac{x}{a} + K$$

PRACTICE PROBLEMS

220-1. Evaluate the indefinite integral,

$$\int x\sqrt{x+1}\,dx.$$

220-2. Evaluate the indefinite integral,

$$\int \sqrt{1-x^2}\,dx.$$

220-3. Evaluate the definite integral,

$$\int_0^1 \frac{dx}{\sqrt{(1+x^2)^3}}.$$

220-4. Evaluate the indefinite integral,

$$\int x^2 \sin x\,dx.$$

220-5. Evaluate the definite integral,

$$\int_0^{\pi/2} x \sin x\,dx.$$

220-6. Evaluate the indefinite integral,

$$\int \frac{x}{1+x^2}\,dx.$$

220-7. Evaluate the definite integral,

$$\int_2^3 \frac{3-2x}{1-x}\,dx.$$

220-8. Evaluate the definite integral,

$$\int_0^{\pi/2} e^{\sin x} \cos x\,dx.$$

220-9. Evaluate the indefinite integral,

$$\int \frac{\sin 2x}{e^x}\,dx.$$

220-10. Evaluate the definite integral,

$$\int_0^1 \sinh^2 x.dx.$$

220-11. Evaluate the indefinite integral,

$$\int x\,\mathrm{sech}^2 x.dx.$$

220-12. Evaluate the indefinite integral,

$$\int \frac{dx}{\sqrt{9x^2-16}}.$$

220-13. Evaluate the indefinite integral,

$$\int \frac{x-8}{x^2-x-2}\,dx.$$

220-14. Evaluate the indefinite integral,

$$\int \frac{4x}{(x-1)(x+1)^2}\,dx.$$

8
PART
Digital principles

221
Number systems

A *number system* is any set of symbols or characters that is used to enumerate objects and perform mathematical computations such as addition, subtraction, multiplication, and division. All number systems are related to each other by symbols (or characters) that are commonly referred to as *digits*. Modern number systems have certain digits in common; however, these systems do not all use the same number of digits, as shown in Table 221-1.

Table 221-1. Comparison of number systems.

Decimal	Binary	Octal	Hexa-decimal	Duo-decimal
0	00000	0	0	0
1	00001	1	1	1
2	00010	2	2	2
3	00011	3	3	3
4	00100	4	4	4
5	00101	5	5	5
6	00110	6	6	6
7	00111	7	7	7
8	01000	10	8	8
9	01001	11	9	9
10	01010	12	A	t
11	01011	13	B	e
12	01100	14	C	10
13	01101	15	D	11
14	01110	16	E	12
15	01111	17	F	13
16	10000	20	10	14
17	10001	21	11	15
18	10010	22	12	16
19	10011	23	13	17
20	10100	24	14	18

The most commonly used system is the Hindu-Arabic system, which uses the digits 0, 1, 2, 3, 4, 5, 6, 7, 8, and 9. There are a total of 10 digits, so you have a *decimal* (or base-10) *system*. Because most measurements are made with this system, it is used as the basis for comparing other number systems.

Number systems in ancient times were used primarily to take measurements and keep records because mathematical computations using the Greek, Roman, and Egyptian number systems were extremely difficult. The lack of an adequate number system was probably a major factor in hampering scientific development in these early civilizations. Obviously, mathematical computations were difficult with Roman numerals where, for example, MCMLXXXIX is equivalent to 1989 in the decimal system.

The acceptance of two basic concepts has greatly simplified mathematical computations and has led to the development of modern number systems. These two concepts are the use of zero to signify the absence of a unit and the principle of *positional value*.

The principle of positional value consists of assigning to a digit a value that depends on its position within a given number. For example, the digit 6 has a different value in each of the decimal numbers 876, 867, and 687. In the first number, 876, the digit 6 has its base value of 6. In the second number, 867, the digit 6 has a value of 60 (6×10 or 6×10^1). In the third number, 687, the value of the digit 6 is 600 (6×100, or 6×10^2).

Sometimes, a position within a given number does not have a value. However, if this position is totally disregarded, there is no way to distinguish between two different numbers, such as 706 and 76. Therefore, the 0 is used to signify that a particular position within a certain number has no value assigned.

The use of 0 and positional value has greatly simplified counting and mathematical computations. Consequently, these concepts are used in all modern number systems.

MATHEMATICAL DERIVATIONS

Positional notation

The standard shorthand form of writing numbers is known as *positional notation.* The value of a particular digit depends not only on its basic value, but also on its position within a number. For example, the decimal number 2365.74 is the standard shorthand form of the quantity two thousand three hundred sixty-five, seven-tenths, four-hundredths. Expressing this number in its general form:

$$2365.74 = (2 \times 10^3) + (3 \times 10^2) + (6 \times 10^1) + (5 \times 10^0) + (7 \times 10^{-1}) + (4 \times 10^{-2})$$

In this number, the 2 carries the most weight of all the digits and is called the *most significant digit* (MSD). By contrast, the 4 carries the least weight and is referred to as the *least significant digit* (LSD).

A number can be expressed with positional notation in any system. The general form for expressing a number is:

$$N = (d_n \times r^n) + \ldots + (d_2 \times r^2) + (d_1 \times r^1) + (d_0 \times r^0) + (d_{-1} \times r^{-1}) + (d_{-2} \times r^{-2}) + \ldots + (d_{-n} \times r^{-n}) \quad (221\text{-}1)$$

where: N = the number expressed in a positional notation form

r = the base that is raised in turn to a series of exponents

d = the digits of the number system.

A base point, such as a decimal point, is not required in the general form because at the position of the point, the exponent changes from positive to negative. In the shorthand form, the base point is between the $d_0 \times r^0$ and $d_{-1} \times r^{-1}$ values.

The base

Every number system has a base with a certain value. The hexadecimal, duodecimal, decimal, octal, and binary systems have bases whose values are (respectively) 16, 12, 10, 8, and 2. The distinction between integers (whole numbers) and fractions is recognized by the position of the base point. In addition:

1. The base of a number system is equal to the number of the different characters used to indicate all the various magnitudes that a digit might represent. For example, the decimal system, with its base of 10, has 10 digits whose magnitudes are 0 through 9.
2. The value of the base is always one unit greater than the largest-value character in the system. This follows from the fact that the base is equal to the number of the characters, and the characters themselves start from zero. As an example, the highest value digit in the decimal system is 9; therefore, the base is 9 + 1 = 10.
3. The positional notation does not, by itself, indicate the value of the base. The symbol "123.41" could represent a number written in a system that has a base value of five (4 + 1) or more. To avoid confusion, numbers written in systems other than the decimal system have the base denoted by a subscript, such as 123.41_8. The base subscript is always written as a decimal number.
4. Any number can easily be multiplied or divided by the base of its number system. When multiplying a number by its base, move the base point one digit to the right of its former position. For example, $123.41_8 \times 8 = 1234.1_8$. To divide a number by its base, move the base point one digit to the left of its former position so that $123.41_8 \div 8 = 12.341_8$.
5. In any number system, the symbol "10" always equals the value of the base. This follows from the fact that the value of the base is one unit greater than the highest-value character.

Counting

In any system using positional notation, the rules for counting are the same, and are independent of the base. With the octal system as an example, the rules are:

1. Start from zero and then add 1 to the least significant digit until the series of all the basic characters in sequence is complete. Such a series is known as a *cycle*, which for the octal system, would be 0, 1, 2, 3, 4, 5, 6, 7.
2. Because 7 is the highest-value character in the octal system, the next number requires two digits. Begin the start of the two-digit numbers with 0 as the least significant digit and place a 1 to the left of the 0. Therefore, the series becomes 0, 1, 2, 3, 4, 5, 6, 7, 10, 11, 12, 13, 14, 15, 16, 17, 20, 21.
3. When a digit reaches its maximum value, replace it with a 0, then add 1 to the next more significant digit. Consequently, the series is:
   ... 16,17,20,21 ... 26,27,30,31 ... 36,37,40,41
   ... 46,47,50,51 ... 56,57,60,61 ... 66,67,70,71
   ... 76,77
4. When two or more consecutive digits reach the maximum value, replace them with 0's and add 1 to the next more significant digit. The series continues as:
   ... 76,77,100,101 ... 176,177,200,201
   ... 276,277,300,301 ... 376,377,400,401
   ... 476,477,500,501 ... 576,577,600,601
   ... 676,677,700,701 ... 767,777,1000,1001
5. Notice that, in any number system, the maximum whole number to be expressed by N digits is given by,

$$\textit{Maximum number} = r^N - 1 \quad (221\text{-}2)$$

where r = base value. For example, $777_8 = 8^3 - 1 = 511_{10}$

Example 221-1

Express (a) 375_{16}, (b) 483_{12}, (c) 726_8, and (d) 11010_2 as numbers to the base 10.

Solution

(a) *Decimal number* (221-1)

$= (3 \times 16^2) + (7 \times 16^1) + (5 \times 16^0)$

$= 768 + 112 + 5$

$= 885$

(b) *Decimal number* (221-1)

$= (4 \times 12^2) + (8 \times 12^1) + (3 \times 12^0)$

$= 576 + 96 + 3$

$= 675$

(c) *Decimal number* (221-1)

$= (7 \times 8^2) + (2 \times 8^1) + (6 \times 8^0)$

$= 448 + 16 + 6$

$= 470$

(d) *Decimal number* (221-1)

$= (1 \times 2^4) + (1 \times 2^3) + (0 \times 2^2) + (1 \times 2^1) + (0 \times 2^0)$

$= 16 + 8 + 0 + 2 + 0$

$= 26$

PRACTICE PROBLEMS

221-1. In the hexadecimal system, what is the maximum whole decimal number to be expressed by four digits?

221-2. Express (a) $EC5_{16}$, (b) $t4e_{12}$, (c) 905_8, and (d) 111000111_2 as decimal numbers.

221-3. Express 723.174_8 as a decimal number.

221-4. Express $AB6.C8_{16}$ as a decimal number.

221-5. Express $e5t.46e_{12}$ as a decimal number.

221-6. Express 1110010.1011_2 as a decimal number.

222
The binary number system

The binary number system has a base of 2 and is used in virtually all digital circuits, as well as in computers. Because the base is 2, there will only be two digits: 0 and 1. This is an enormous advantage because it is relatively easy to design electronic circuits having only two possible states. For example, a bipolar transistor can either be in the saturation mode or in the cut-off mode, and these two states can then correspond to two different output voltages from the circuit containing the transistor. However, because of temperature variations, such voltages are bound to fluctuate to a certain extent, so the two states will each correspond to a limited voltage range. This is illustrated in Fig. 222-1, where the 0 state is from 0 V to 1 V, and the 1 state exists between 9 V and 12 V; this means that the range between 1 V and 9 V is not used. The situation as described is sometimes referred to as *positive logic.* By contrast, *negative logic* means that the voltage levels are interchanged so that the range of 9 V to 12 V represents the 0 state, while the 0 V to 1 V range indicates the 1 state. These chapters are only concerned with positive logic.

The simplicity of the binary system is its advantage over other systems. For example, with a decimal system it would be necessary to design circuits capable of working with ten different voltage ranges, each of which would correspond to a digit between 0 and 9.

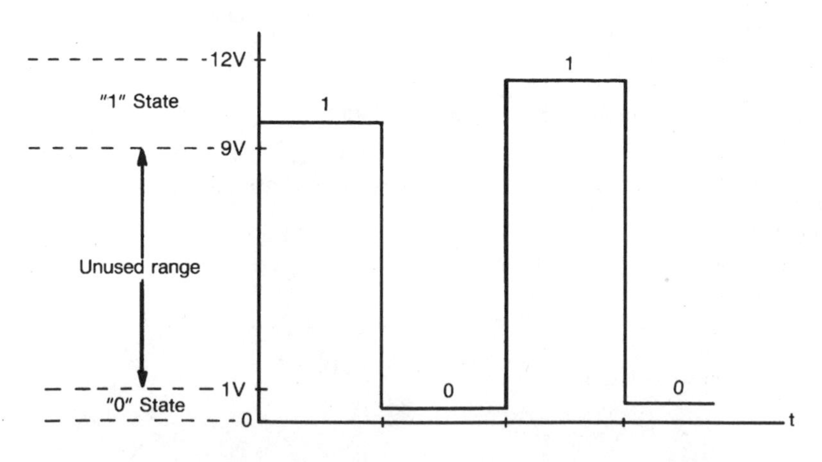

Figure 222-1

Binary numbers use positional notation, so that the value of a particular digit depends not only on the digit's value but also on its position within the number.

Using Equation 221-1,

Binary Number $\underbrace{11010}_{\text{whole number}} \cdot \underbrace{011}_{\text{fraction}}{}_2$ (base point)

$$= (1 \times 2^4) + (1 \times 2^3) + (0 \times 2^2) + (1 \times 2^1) + (0 \times 2^0) + (0 \times 2^{-1}) + (1 \times 2^{-2}) + (1 \times 2^{-3})$$

$$= 16 + 8 + 2 + 0.25 + 0.125$$

$$= 26.375_{10}$$

As an abbreviation, binary digits are referred to as *bits*. In this example whereas the most significant bit (MSB) is the leftmost 1, which has a value of $(1 \times 2^4) = 16$, the least significant bit (LSB) is the rightmost 1, which has a value of only $(1 \times 2^{-3}) = 0.125$.

In a binary number, the leftmost 1 (Fig. 222-2) is referred to as the MSB because it is multiplied by the highest coefficient. Once the MSB has been determined, all positions to the left of the MSB have no significance even though they might be occupied by zeros; this must be true because in a given number, any position unoccupied, or occupied by a zero, does not have a value assigned.

Whether or not a value is assigned, all bit positions to the right of the MSB must be occupied by a 1 or a 0 (so that one number can be distinguished from another). The bit position at the extreme right of a given number is always considered to be occupied by the LSB—even though it might contain a zero to indicate that no value has been assigned to this position.

The value of the number shown in Fig. 222-2 is $32768 + 8192 + 2048 + 64 + 16 + 8 = 43096$. This illustrates the disadvantage of the binary system, namely the large number (16) of bits required to be equivalent to a decimal number having only 5 significant digits. The maximum decimal number to be represented by N bits is $2^N - 1$ (Equation 221-2). For example, $1111_2 = 2^4 - 1 = 15_{10}$.

REPRESENTATION OF BINARY QUANTITIES

If a device has only two possible operating states, a number of such devices can be used to represent a binary quantity. A single-pole, single-throw switch has only two states: open and closed.

If an open switch is in the 0 state, and a closed switch is in the 1 state, the switches shown in Fig. 222-3A indicate the binary number $11001_2 = 25_{10}$. Alternately, you can represent the same number by punching holes in a paper tape (Fig. 222-3B). A punched hole then indicates a 1, but the absence of a hole is a 0.

Binary Counting

Beginning with a zero count, all bits in Fig. 222-4 are in the 0 state. First count: the 2^0 position changes from 0 to 1. Second count: the 2^0 position reverts back to 0, while the 2^1 position changes from 0 to 1. Third count: the 2^0 position changes to 1, and the 2^1 position remains at 1. Fourth count: the 2^0 and 2^1 position are 0, and the 2^2 position changes from 0 to 1. Observing the pattern, the 2^0 position changes from 0 to 1 or 1 to 0 with every count. In the general 2^N position, the bit will stay at 0 for 2^N counts and then change to 1 for the next 2^N counts.

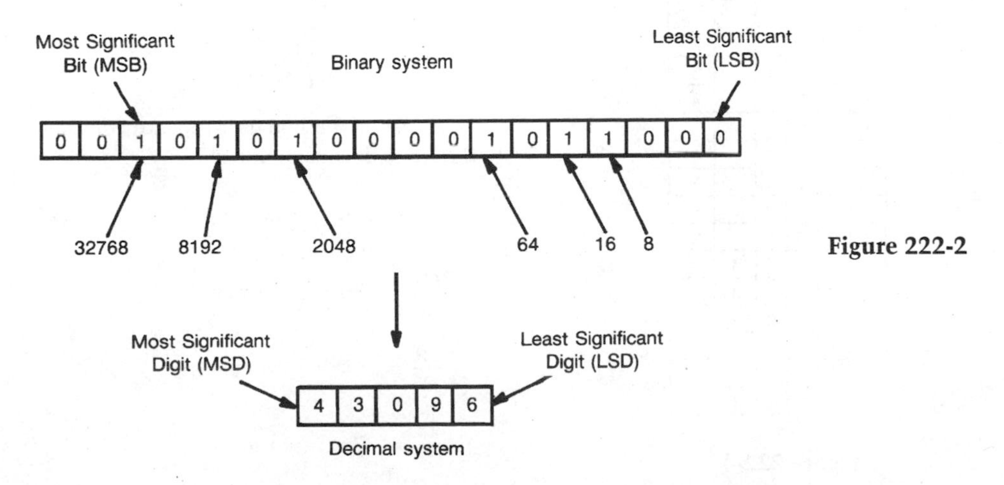

Figure 222-2

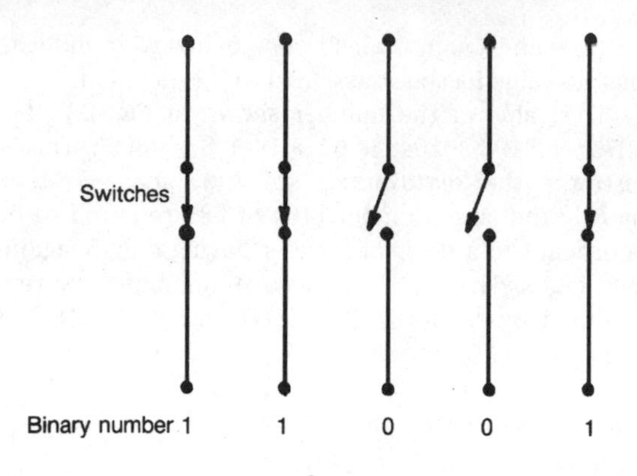

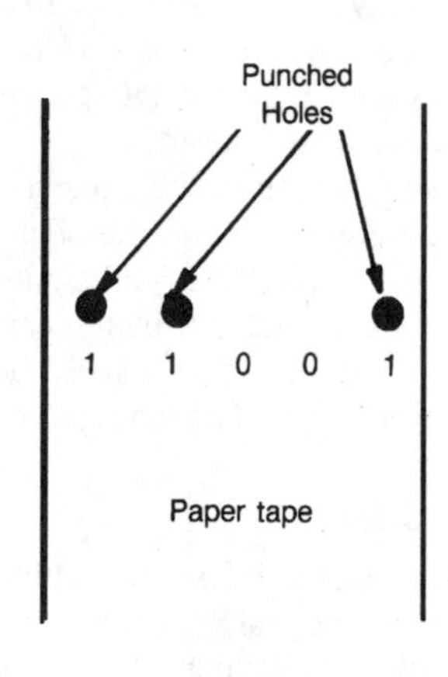

Figure 222-3

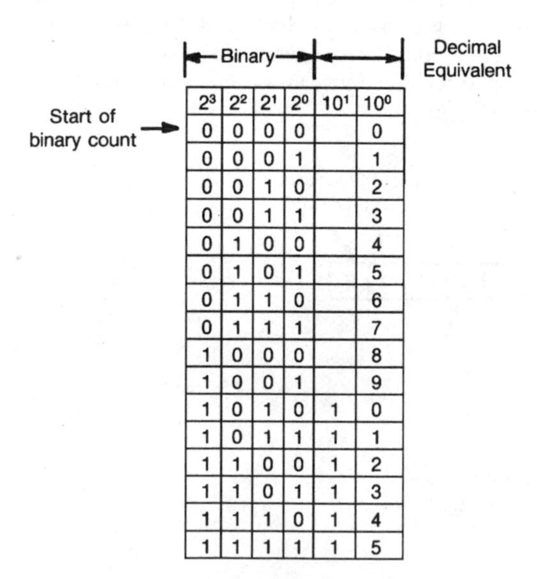

2^3	2^2	2^1	2^0	10^1	10^0
0	0	0	0		0
0	0	0	1		1
0	0	1	0		2
0	0	1	1		3
0	1	0	0		4
0	1	0	1		5
0	1	1	0		6
0	1	1	1		7
1	0	0	0		8
1	0	0	1		9
1	0	1	0	1	0
1	0	1	1	1	1
1	1	0	0	1	2
1	1	0	1	1	3
1	1	1	0	1	4
1	1	1	1	1	5

Figure 222-4

MATHEMATICAL DERIVATIONS

Binary-to-decimal conversion

The binary-to-decimal conversion can be achieved by using Equation 221-1. For example,

$$\begin{aligned} 10011_2 &= (1 \times 2^4) + (0 \times 2^3) + \\ &\quad (0 \times 2^2) + (1 \times 2^1) + (1 \times 2^0) \\ &= 16 + 2 + 1 = 19_{10} \end{aligned}$$

Alternatively, the conversion can be carried out using some calculators. First, locate the MSB and then proceed as follows:

Entry, Key	Display
1 [×] 2 [+]	2
0 [×] 2 [+]	4
0 [×] 2 [+]	8
1 [×] 2 [+]	18
1 [=]	19

This method is sometimes referred to as the *double-dabble.*

Decimal-to-binary conversions

One method of decimal-to-binary conversion is to reverse the procedure described for binary-to-decimal conversion. For example, the nearest power of two to 119_{10} is 64 (2^6). The remainder is $119 - 64 = 55$, for which the nearest power of two is 32. This leaves a new remainder of $55 - 32 = 23$. Repeating the process, $23 - 16 = 7$, $7 - 4 = 3$, $3 - 2 = 1$. Then,

$$\begin{aligned} 119_{10} &= 2^6 + 2^5 + 2^4 + 0 + 2^2 + 2^1 + 2^0 \\ &= 1 \quad 1 \quad 1 \quad 0 \quad 1 \quad 1 \quad 1 \\ &= 1110111_2 \end{aligned}$$

A second method uses repeated division by 2. Using the same decimal number 119_{10},

$$\frac{119}{2} = 59 \qquad \text{remainder} \qquad 1 \text{ (LSB)}$$

$$\frac{59}{2} = 29 \qquad \text{remainder} \qquad 1$$

$$\frac{29}{2} = 14 \qquad \text{remainder} \qquad 1$$

$$\frac{14}{2} = 7 \qquad \text{remainder} \qquad 0$$

$\frac{7}{2} = 3$ remainder 1

$\frac{3}{2} = 1$ remainder 1

$\frac{1}{2} = 0$ remainder 1 (MSB)

Then, 119_{10} is equal to 1110111_2

Binary Addition

The rules of binary addition are shown in Fig. 222-5, where $A + B = C$. Therefore:

$$0 + 0 = 0$$
$$0 + 1 = 1$$
$$1 + 0 = 1$$
$$1 + 1 = 10$$

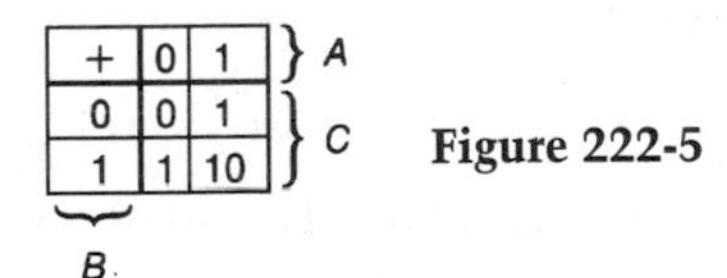

Figure 222-5

$1 + 1 = 10$ means that when 1 is added to 1, the result is 0 with a 1 carried over to the next column. As an example, add 1011_2 (11_{10}) and 11011_2 (27_{10}):

$$1011_2 = 11_{10}$$
$$+11011_2 = 27_{10}$$
$$\text{Answer} = 100110_2 = 38_{10}$$

In the second column from the right, you were faced with adding three 1's, in which case $1 + 1 + 1 = 10 + 1 = 11$.

Binary Subtraction

In Fig. 222-5, $A = C - B$ so that,

$$0 - 0 = 0$$
$$1 - 0 = 1$$
$$1 - 1 = 0$$
$$10 - 1 = 1$$

$10 - 1 = 1$ means that when 1 is subtracted from 10, 1 is borrowed from the next column to the left. As an example, subtract 10010_2 (18_{10}) from 11101_2 (29_{10}):

$$11101_2 = 29_{10}$$
$$10010_2 = 18_{10}$$
$$01011_2 = 11_{10}$$

Binary Multiplication

The rules of binary multiplication are illustrated in Fig. 222-6, where $A \times B = C$. Therefore:

$$0 \times 0 = 0$$
$$0 \times 1 = 0$$
$$1 \times 0 = 0$$
$$1 \times 1 = 1$$

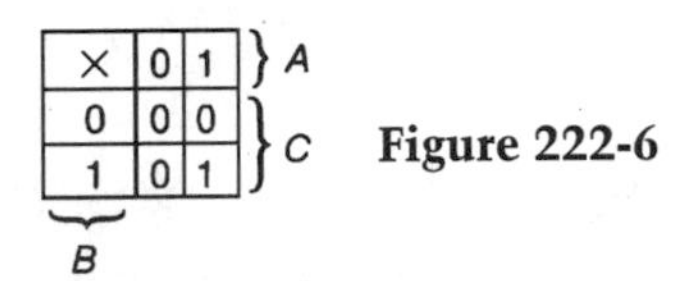

Figure 222-6

As an example, multiply 10111 (23_{10}) by 1111_2 (15_{10}):

$$10111_2 = 23_{10}$$
$$1111_2 = 15_{10}$$

```
        10111
       10111
      10111
     10111
```

$$101011001_2 = 23_{10} \times 15_{10} = 345_{10}$$

Notice that, in the third column from the right, you had to add four 1s, which is equal to 100_2 so that 1 is carried forward to the fifth column. In that column, you added five 1s; this equals 101_2 so that 1 appears in the fifth column and 1 is carried forward to the seventh column.

Binary division

Using Fig. 222-6, binary division will be defined by $A = C/B$. Therefore:

$$0 \div 0 = 0$$
$$0 \div 1 = 0$$
$$1 \div 1 = 1$$

Using the "long-division" method, divide 101011001_2 (345_{10}) by 10111_2 (23_{10}):

```
             1111
10111)101011001
       10111
       101000
        10111
        100010
         10111
         010111
          10111
              0
```

Therefore, $\dfrac{101011001_2}{10111_2} = 1111_2$

or, $$\frac{345_{10}}{23_{10}} = 15_{10}$$

Example 222-1

Convert (a) 111001101_2 to a decimal number, and (b) 377_{10} to a binary number.

Solution

(a) $$\begin{aligned} 111001101_2 &= (1 \times 2^8) + (1 \times 2^7) + (1 \times 2^6) + \\ &\quad (0 \times 2^5) + (0 \times 2^4) + (1 \times 2^3) + \\ &\quad (1 \times 2^2) + (0 \times 2^1) + (1 \times 2^0) \\ &= 256 + 128 + 64 + 8 + 4 + 1 \\ &= 461_{10} \end{aligned} \qquad (221\text{-}1)$$

Find the answer to (a) using the double-dabble method.

(b) $\dfrac{377}{2} = 188$, remainder 1 (LSB)

$\dfrac{188}{2} = 94$, remainder 0

$\dfrac{94}{2} = 47$, remainder 0

$\dfrac{47}{2} = 23$, remainder 1

$\dfrac{23}{2} = 11$, remainder 1

$\dfrac{11}{2} = 5$, remainder 1

$\dfrac{5}{2} = 2$, remainder 1

$\dfrac{2}{2} = 1$, remainder 0

$\dfrac{1}{2} = 0$, remainder 1 (MSB)

Therefore, $377_{10} = 101111001_2$

Example 222-2

What are the binary values of:
(a) $111001_2 + 1011011_2$, (b) $1011101_2 - 100111_2$, (c) $1011101_2 \times 110011_2$, and (d) $1100111_2 \div 1011_2$?

Solution

(a)
$$\begin{array}{rl} 111001_2 &= 57_{10} \\ +\ 1011011_2 &= 91_{10} \\ \hline 10010100_2 &= 148_{10} \end{array}$$

Therefore, $111001_2 + 1011011_2 = 10010100_2$

(b)
$$\begin{array}{rl} 1011101_2 &= 93_{10} \\ -\ 100111_2 &= 39_{10} \\ \hline 110110_2 &= 54_{10} \end{array}$$

Therefore, $1011101_2 - 100111_2 = 110110_2$

(c)
$$\begin{array}{rl} 1011101_2 &= 93_{10} \\ \times\ 110011_2 &= 51_{10} \\ \hline 1011101 & \\ 1011101 & \\ 1011101 & \\ 1011101 & \\ \hline 1001010000111_2 &= 93_{10} \times 51_{10} = 4743_{10} \end{array}$$

Therefore, $1011101_2 \times 110011_2 = 1001010000111_2$

(d)
$$\begin{array}{r} 1001_2 \\ 1011_2 \overline{)1100111_2} \\ \underline{1011} \\ 0001111 \\ \underline{1011} \\ 0100_2\text{, remainder} \end{array}$$

In decimal form,

$$\frac{103_{10}}{11_{10}} = 9_{10} + \text{remainder } 4_{10}$$

Therefore, $1100111_2 \div 1011_2 = 1001_2$, remainder 0100_2

PRACTICE PROBLEMS

222-1. Convert (a) 1732_{10} into a binary number and (b) 11001111101_2 into a decimal number.
222-2. Express the result of the following addition as a binary number: $1100111.101_2 + 1111001.011_2 + 11110011.111_2$.
222-3. Express the result of the following subtraction in binary form: $1111100.11_2 - 1100111.101_2$.
222-4. Express the result of the following multiplication as a binary number: $1101111_2 \times 100111_2$.
222-5. Express the result of the following division as a binary number: $10000000101011_2 \div 101101_2$.

223
Binary-coded decimal system

The binary-coded decimal (BCD) is not a true number system, but is used in digital readout meters and in other situations where speed is not important. Compared with the binary number system, BCD has the advantage of simpler conversions to and from decimal numbers, but has the disadvantage of requiring a greater number of bits. Moreover, certain complications arise when BCD numbers are added. For these reasons, BCD is not applicable to high-speed computers.

BCD makes use of groups of binary bits to represent a decimal number. Because there are 10 digits in the decimal system, only 10 groups are required, each of which contains four binary bits:

Decimal Digit	BCD Groups
0	0000
1	0001
2	0010
3	0011
4	0100
5	0101
6	0110
7	0111
8	1000
9	1001

Because a group of four binary bits has a maximum decimal value of 15, there will be six forbidden groups in BCD: 1010, 1011, 1100, 1101, 1110, and 1111.

To convert a decimal number into a BCD number, the appropriate binary group is substituted for each decimal digit. As an example,

$$\begin{array}{lllll} & 3 & 8 & 1 & \\ 381_{10} = & 0011 & 1000 & 0001 & = 001110000001 \text{ (BCD)} \end{array}$$

In true binary,

$$381_{10} = 101111101_2$$

BCD requires 12 bits, but the true binary system only needs 9 bits.

To convert a BCD number to a decimal number, separate the binary bits into groups of four. Then, write down the decimal digit that corresponds to each group. For example,

$$\begin{array}{lllll} \text{BCD } 0100100110010011 = & 0100 & 1001 & 1001 & 0011 \\ & 4 & 9 & 9 & 3 \end{array}$$

$$= 4993_{10}$$

MATHEMATICAL DERIVATIONS

BCD addition

Consider the following addition:

Decimal	BCD
132	0001 0011 0010
+ 216	+ 0010 0100 1000
348	3 4 8

By applying the normal rules of binary addition, the correct answer of 348 was obtained. However, in this example there was no "carry-over" in the decimal addition.

When the sum in a particular BCD group exceeds decimal 9, problems arise, as illustrated in the following example:

Decimal	BCD
362	0011 0110 0010
+ 456	+ 0100 0101 0110
818	0111 1011 1000
← "carry-over"	

The middle group "1011" is forbidden when operating in BCD. The difficulty is solved by adding 0110 (decimal 6) to the middle group. The result is:

Decimal	BCD
362	0111 1011 1000
+ 456	+ 0110
818	1000 0001 1000

Whenever a forbidden group is observed, 0110 must be added to that group in order to obtain the correct BCD sum. The extra requirement increases the complexity of the electronic circuitry and slows down the addition process.

Consider another example,

396	0011	1001	0110
+ 487	0100	1000	0111
883	1000	0001	1101
←← "carry-overs"		0110	0110
	1000	1000	0011
	8	8	3

There are two "carry-overs" in the decimal addition so that "0110" must be added to both the middle and right-hand groups [even though the initial addition (0001) of the middle group is not forbidden].

Example 223-1

Convert (a) 5728_{10} into its BCD equivalent, (b) convert BCD 1001011001000011 into its equivalent decimal number, and (c) add 786_{10} and 559_{10} using BCD.

Solution

(a) 5 7 2 8

5728_{10} = 0101 0111 0010 1000

= 0101011100101000 BCD

(b) BCD 1001011001000011

= 1001 0110 0100 0011

9 6 4 3

= 9643_{10}

(c) Decimal / BCD

Decimal		BCD
786_{10}		0111 1000 0110
$+ 559_{10}$		+ 0101 0101 1001
1345_{10}		1100 1101 1111
Three forbidden groups		+ 0110 0110 0110
Add "0110" to each group.	0001	0011 0100 0101
	1	3 4 5_{10}

PRACTICE PROBLEMS

223-1. Convert (a) 1573_{10} to a BCD number and (b) the BCD number 01110101010100111001 into a decimal number.

223-2. Convert (a) 111001011_2 to a BCD number and (b) the BCD number 0001010000100111 to a binary number.

223-3. Convert 352_{10} and 867_{10} to BCD numbers, then obtain their sum.

224 The octal number system

It is very simple to convert between octal (base 8) and binary (base 2) numbers. Consequently, when a system such as a computer is involved with large quantities of binary numbers, each consisting of many bits, it is more convenient for the operators to program the information using octal numbers. However, remember that the computer's digital circuits operate only with the binary system.

Although you can convert between octal and decimal numbers directly, it is preferable to use the binary system as an intermediate step (Fig. 224-1). The same principle applies to hexadecimal-to-decimal conversions.

MATHEMATICAL DERIVATIONS

Direct octal-to-decimal conversion

To achieve this conversion, use Equation 221-1. For example,

$$235.7_8 = (2 \times 8^2) + (3 \times 8^1) + (5 \times 8^0) + (7 \times 8^{-1})$$
$$= 157.875_{10}$$

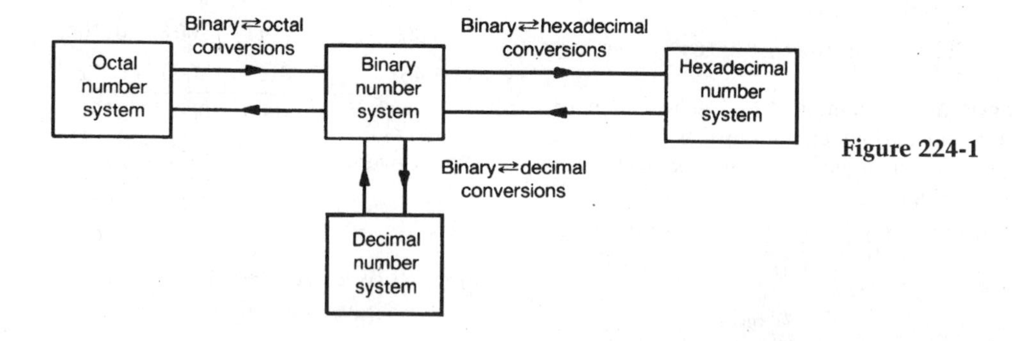

Figure 224-1

Direct decimal-to-octal conversion

To convert a decimal whole number into its equivalent octal number, use the repeated division method with 8 as the division factor. For example, 437_{10} is equivalent to:

$$\frac{437}{8} = 54, \text{ remainder } 5 \text{ (LSD)}$$

$$\frac{54}{8} = 6, \text{ remainder } 6$$

$$\frac{6}{8} = 0, \text{ remainder } 6 \text{ (MSD)}$$

Therefore, $437_{10} = 665_8$

Octal-to-binary conversion

Because the base of the binary system is 2 and the base of the octal system is $8 = 2^3$, you can convert from octal whole numbers to binary whole numbers by changing each octal digit into its three-bit binary equivalent. As an example,

$$1426_8 = \underbrace{1}_{001}\quad \underbrace{4}_{100}\quad \underbrace{2}_{010}\quad \underbrace{6}_{110}$$

$$= 1100010110_2$$

Notice that the two 0s to the left of the MSB are discarded.

Binary-to-octal conversion

To convert from binary whole numbers to octal whole numbers, reverse the octal-to-binary procedure. Starting at the binary point and moving to the left, mark off the bits in groups of 3, then express each group in its octal equivalent. If the number of bits is not exactly divisible by 3, complete the last group by adding one or two 0s to the left of the MSB. For example,

$$11001011_2 = \underbrace{\overset{\overset{\text{added "0"}}{\downarrow}}{011}}_{3}\quad \underbrace{001}_{1}\quad \underbrace{011}_{3_8}\quad \overset{\overset{\text{binary point}}{\downarrow}}{\cdot}$$

Therefore, $11001011_2 = 313_8$

Octal addition

The results of octal addition are shown in Fig. 224-2. When an addition carry is made, the amount of the carry is 8. For example, when you add 7_8 and 6_8, the sum is 13 in base 10, but 13_{10} in base 8 is $(1 \times 8^1) + (5 \times 8^0) = 15_8$. This result is shown in Fig. 224-2, where $A + B = C$. As another example,

+	0	1	2	3	4	5	6	7	} A
0	0	1	2	3	4	5	6	7	
1	1	2	3	4	5	6	7	10	
2	2	3	4	5	6	7	10	11	
3	3	4	5	6	7	10	11	12	
4	4	5	6	7	10	11	12	13	} C
5	5	6	7	10	11	12	13	14	
6	6	7	10	11	12	13	14	15	
7	7	10	11	12	13	14	15	16	
B									

Figure 224-2

$$\begin{array}{rcr} 1743_8 & = & 995_{10} \\ +\ 652_8 & = & 426_{10} \\ \hline 2615_8 & = & 1421_{10} \end{array}$$

Therefore, $1743_8 + 652_8 = 2615_8$

Octal subtraction

Because $A = C - B$, Fig. 224-2 can be used to determine the results of subtraction so that if $C = 13_8$ and $B = 6_8$, $A = 5_8$. As another example,

$$\begin{array}{rcr} 3752_8 & = & 2026_{10} \\ -\ 674_8 & = & -444_{10} \\ \hline 3056_8 & = & 1582_{10} \end{array}$$

Therefore, $3752_8 - 674_8 = 3056_8$

When a subtraction borrow is made, the amount of the borrow is 8.

Octal multiplication

The results of octal multiplication are shown in Fig. 224-3, where $A \times B = C$. When, for example, 6_8 is multiplied by 7_8, the product would be 42_{10}, which is equal

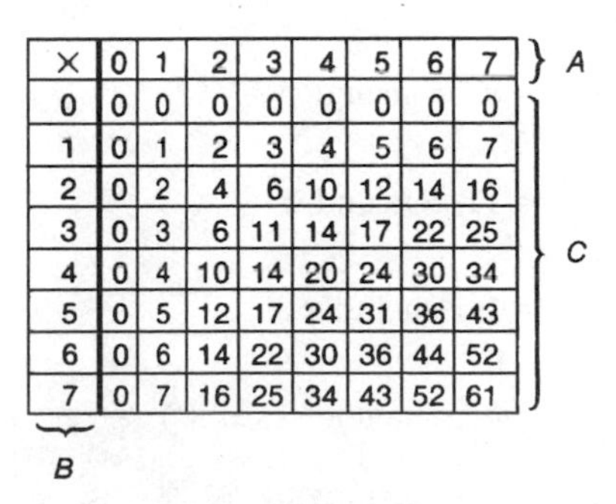

×	0	1	2	3	4	5	6	7	} A
0	0	0	0	0	0	0	0	0	
1	0	1	2	3	4	5	6	7	
2	0	2	4	6	10	12	14	16	
3	0	3	6	11	14	17	22	25	
4	0	4	10	14	20	24	30	34	} C
5	0	5	12	17	24	31	36	43	
6	0	6	14	22	30	36	44	52	
7	0	7	16	25	34	43	52	61	
B									

Figure 224-3

to $(5 \times 8^1) + (2 \times 8^0) = 52_8$. As another example, multiply 473_8 by 67_8:

$$
\begin{array}{r}
473_8 = 315_{10} \\
\times\ 67_8 = \ \ 55_{10} \\
\hline
4235\ \ \ \ \ \ \ \ \ \\
3542\ \ \ \ \ \ \ \ \ \ \ \\
\hline
41655_8 = 17325_{10}
\end{array}
$$

Therefore, $473_8 \times 67_8 = 41655_8 = 17325_{10}$

Octal division

Using Fig. 224-3, octal division is defined by $A = C/B$. When, for example, 234_8 is divided by 6_8,

$$
\begin{array}{r}
32_8 \\
6_8\overline{)234_8} \\
22\ \ \\
\hline
14 \\
14 \\
\hline
0
\end{array}
$$

Therefore, $234_8 \div 6_8 = 32_8$

Example 224-1

Convert (a) 673_8 into its equivalent binary and decimal numbers, (b) 11011001_2 into its equivalent octal number, and (c) 5693_{10} into its equivalent octal number.

Solution

(a) $673_8 = \underbrace{110}_{6}\ \underbrace{111}_{7}\ \underbrace{011}_{3} = 110111011_2$

$673_8 = (6 \times 8^2) + (7 \times 8^1) + (3 \times 8^0) = 443_{10}$

(b) $11011001_2 = \underbrace{011}_{3}\ \underbrace{011}_{3}\ \underbrace{001}_{1} = 331_8$

(c) $\dfrac{5693_{10}}{8} = 711$, remainder 5 (LSD)

$\dfrac{711}{8} = 88$, remainder 7

$\dfrac{88}{8} = 11$, remainder 0

$\dfrac{11}{8} = 1$, remainder 3

$\dfrac{1}{8} = 0$, remainder 1 (MSD)

Answer: $5693_{10} = 13075_8$

Check:

$13075_8 = 001011000111101$

$= 1011000111101_2$

$= 5693_{10}$, using the double-dabble method.

Example 224-2

What are the octal values of (a) $1672_8 + 354_8$, (b) $13875_8 - 6437_8$, (c) $132_8 \times 65_8$, (d) $765_8 \div 4_8$?

Solution

(a)
$$
\begin{array}{r}
1672_8 = 954_{10} \\
+\ 354_8 = 236_{10} \\
\hline
2246_8 = 1190_{10}
\end{array}
$$

Therefore, $1672_8 + 354_8 = 2246_8 = 1190_{10}$

(b)
$$
\begin{array}{r}
13675_8 = 6077_{10} \\
-\ 6437_8 = 3359_{10} \\
\hline
5236_8 = 2718_{10}
\end{array}
$$

Therefore, $13675_8 - 6437_8 = 5236_8 = 2718_{10}$

(c)
$$
\begin{array}{r}
132_8 = 90_{10} \\
\times\ 65_8 = 53_{10} \\
\hline
702\ \ \ \ \ \ \ \ \ \ \ \\
1034\ \ \ \ \ \ \ \ \ \ \ \ \ \\
\hline
11242_8 = 4770_{10}
\end{array}
$$

Therefore, $132_8 \times 65_8 = 11242_8$

(d)
$$
\begin{array}{r}
175_8 \\
4_8\overline{)765_8} \\
4\ \ \ \ \\
\hline
36\ \ \\
34\ \ \\
\hline
25 \\
24 \\
\hline
1, \text{remainder}
\end{array}
$$

Therefore,

$$\frac{765_8}{4_8} = 175_8 \text{ with a remainder of } 1_8.$$

Check:

$$\frac{765_8}{4_8} = \frac{501_{10}}{4_{10}} = 125_{10} \text{ (or } 175_8\text{)}$$

with a remainder of $1_{10}(= 1_8)$.

PRACTICE PROBLEMS

224-1. (a) Convert 1574.3_8 to a decimal number and (b) convert 1382_{10} to an octal number.

224-2. (a) Convert 701127_8 to a binary number and (b) convert 1110111001_2 to an octal number.

224-3. What are the octal values of (a) $2375_8 + 764_8$ and (b) $27436_8 - 5657_8$?

224-4. What are the octal values of (a) $1227_8 \times 73_8$ and (b) $151000_8 \div 214_8$?

225 The hexadecimal number system

The hexadecimal number system (base 16) requires 16 characters: the digits 0,1,2,3,4,5,6,7,8,9, followed by the letters A,B,C,D,E,F. As shown in Fig. 221-1, letters A through F are equivalent to the decimal values 10 through 15.

Although you can convert between hexadecimal and decimal numbers directly, it is possible to use the binary system as an intermediate step. (See Fig. 224-1.)

Hexadecimal numbers are frequently used in conjunction with computer memories and with various forms of programming. During such operations, it is necessary to add or subtract hexadecimal numbers.

MATHEMATICAL DERIVATIONS

Direct hexadecimal-to-decimal conversion

To obtain a hexadecimal-to-decimal conversion, use Equation 221-1. For example,

$$\begin{aligned} 1B6_{16} &= (1 \times 16^2) + (11 \times 16^1) + (6 \times 16^0) \\ &= 256 + 176 + 6 \\ &= 438_{10} \end{aligned}$$

Direct decimal-to-hexadecimal conversion

A decimal whole number is converted into its equivalent hexadecimal number using the repeated division method with 16 as the division factor. For example, 5831_{10} is equivalent to:

$$\frac{5831}{16} = 364, \text{ remainder 7 (LSD)}$$

$$\frac{364}{16} = 22, \text{ remainder 12(C)}$$

$$\frac{22}{16} = 1, \text{ remainder 6}$$

$$\frac{1}{16} = 0, \text{ remainder 1 (MSD)}$$

Therefore, $5831_{10} = 16C7_{16}$.

Hexadecimal-to-binary conversion

Because the base of the binary system is 2 and the base of the hexadecimal system is $16 = 2^4$, you can convert from hexadecimal whole numbers to binary whole numbers by changing each hexadecimal digit into its four-bit binary equivalent. As an example,

$$\begin{aligned} 3B7D_{16} &= \underbrace{0011}_{3}\ \underbrace{1011}_{B}\ \underbrace{0111}_{7}\ \underbrace{1101}_{D} \\ &= 11101101111101_2 \\ &= 15229_{10} \text{ by the "double-dabble" method.} \end{aligned}$$

It is clear that the hexadecimal system can represent large numbers more efficiently than the binary system can.

Binary-to-hexadecimal conversion

To convert from binary whole numbers to hexadecimal whole numbers, reverse the hexadecimal-to-binary procedure. Starting at the binary point and moving to the left, mark off the bits in groups of four, then express each group in its hexadecimal equivalent. If the number of bits is not exactly divisible by 4, the last group can be completed by adding one, two, or three 0s to the left of the MSB. For example,

$$11010111010011_2 = \underbrace{0011}_{3}\ \underbrace{0101}_{5}\ \underbrace{1101}_{D}\ \underbrace{0011}_{3}$$

Therefore,

$$\begin{aligned} 11010111010011_2 &= 35D3_{16} \\ &= 13779_{10}. \end{aligned}$$

Hexadecimal addition

The results of hexadecimal addition are shown in Fig. 225-1. When an addition carry is made, the amount of the carry is 16. For example, when you add D_{16} and E_{16}, the sum is $13 + 14 = 27_{10}$, which in base 16 is $(1 \times 16^1) + (11 \times 16^0) = 1B_{16}$. This result is shown in Fig. 225-1, where $X + Y = Z$. As an example,

$$\begin{array}{rcr} 3DB_{16} & = & 987_{10} \\ +\ 7A_{16} & = & 122_{10} \\ \hline \text{Answer} = 455_{16} & = & 1109_{10} \end{array}$$

Therefore, $3DB_{16} + 7A_{16} = 455_{16}$.

Hexadecimal subtraction

Because $X = Z - Y$, Fig. 225-1 can be used to determine the results of subtraction so that if $Z = ID_{16}$ and $Y = E_{16}$, $X = F_{16}$. When a subtraction borrow is made, the amount of the borrow is 16. As a further example,

$$\begin{array}{rcr} 4C8_{16} & = & 1224_{10} \\ -\ D9_{16} & = & 217_{10} \\ \hline 3EF_{16} & = & 1007_{10} \end{array}$$

Therefore, $4C8_{16} - D9_{16} = 3EF_{16}$.

+	0	1	2	3	4	5	6	7	8	9	A	B	C	D	E	F	} X
0	0	1	2	3	4	5	5	7	8	9	A	B	C	D	E	F	
1	1	2	3	4	5	6	7	8	9	A	B	C	D	E	F	10	
2	2	3	4	5	6	7	8	9	A	B	C	D	E	F	10	11	
3	3	4	5	6	7	8	9	A	B	C	D	E	F	10	11	12	
4	4	5	6	7	8	9	A	B	C	D	E	F	10	11	12	13	
5	5	6	7	8	9	A	B	C	D	E	F	10	11	12	13	14	
6	6	7	8	9	A	B	C	D	E	F	10	11	12	13	14	15	
7	7	8	9	A	B	C	D	E	F	10	11	12	13	14	15	16	
8	8	9	A	B	C	D	E	F	10	11	12	13	14	15	16	17	} Z
9	9	A	B	C	D	E	F	10	11	12	13	14	15	16	17	18	
A	A	B	C	D	E	F	10	11	12	13	14	15	16	17	18	19	
B	B	C	D	E	F	10	11	12	13	14	15	16	17	18	19	1A	
C	C	D	E	F	10	11	12	13	14	15	16	17	18	19	1A	1B	
D	D	E	F	10	11	12	13	14	15	16	17	18	19	1A	1B	1C	
E	E	F	10	11	12	13	14	15	16	17	18	19	1A	1B	1C	1D	
F	F	10	11	12	13	14	15	16	17	18	19	1A	1B	1C	1D	1E	

Y

Figure 225-1

Hexadecimal multiplication

The results of hexadecimal multiplication are shown in Fig. 225-2, in which $X \times Y = Z$. When, for example, you multiply A_{16} by D_{16}, the product is $10_{10} \times 13_{10} = 130_{10} = (8 \times 16^1) + (2 \times 10^0) = 82_{16}$. As another example, multiply $7C3_{16}$ by $A6_{16}$:

$$\begin{array}{rcr} 7C3_{16} & = & 1987_{10} \\ \times\quad A6_{16} & = & 166_{10} \\ \hline 2E92 & & \\ 4D9E\ & & \\ \hline 50872_{16} & = & 329842_{10} \end{array}$$

Therefore, $7C3_{16} \times A6_{16} = 50872_{16}$.

×	0	1	2	3	4	5	6	7	8	9	A	B	C	D	E	F	} X
0	0	0	0	0	0	0	0	0	0	0	0	0	0	0	0	0	
1	0	1	2	3	4	5	6	7	8	9	A	B	C	D	E	F	
2	0	2	4	6	8	A	C	E	10	12	14	16	18	1A	1C	1E	
3	0	3	6	9	C	F	12	15	18	1B	1E	21	24	27	2A	2D	
4	0	4	8	C	10	14	18	1C	20	24	28	2C	30	34	38	3C	
5	0	5	A	F	14	19	1E	23	28	2D	32	37	3C	41	46	4B	
6	0	6	C	12	18	1E	24	2A	30	36	3C	42	48	4E	54	5A	
7	0	7	E	15	1C	23	2A	31	38	3F	46	4D	54	5B	62	69	
8	0	8	10	18	20	28	30	38	40	48	50	58	60	68	70	78	} Z
9	0	9	12	1B	24	2D	36	3F	48	51	5A	63	6C	75	7E	87	
A	0	A	14	1E	28	32	3C	46	50	5A	64	6E	78	82	8C	96	
B	0	B	16	21	2C	37	42	4D	58	63	6E	79	84	8F	9A	A5	
C	0	C	18	24	30	3C	48	54	60	6C	78	84	90	9C	A8	B4	
D	0	D	1A	27	34	41	4E	5B	68	75	82	8F	9C	A9	B6	C3	
E	0	E	1C	2A	38	46	54	62	70	7E	8C	9A	A8	B6	C4	D2	
F	0	F	1E	2D	3C	4B	5A	69	78	87	96	A5	B4	C3	D2	E1	

Y

Figure 225-2

Hexadecimal division

Using Fig. 225-2, hexadecimal division is defined by $X = Z/Y$. When, for example, $A8_{16}$ (168_{10}) is divided by E_{16} (14_{10}), the result is C_{16} (12_{10}). As another example, divide $3F1D8_{16}$ by $B8_{16}$:

$$\begin{array}{r} 57D_{16} \\ B8_{16}\overline{)3F1D8_{16}} \\ \underline{398}\phantom{D8_{16}} \\ 59D\phantom{8_{16}} \\ \underline{508}\phantom{8_{16}} \\ 958\phantom{_{16}} \\ \underline{958}\phantom{_{16}} \\ 0\phantom{_{16}} \end{array}$$

Therefore, $3F1D8_{16} \div B8_{16} = 57D_{16}$.

Example 225-1

Convert (a) $A6B_{16}$ into its equivalent binary and decimal numbers, (b) 11100101111_2 into its equivalent hexadecimal number, and (c) 4376_{10} into its equivalent hexadecimal value.

Solution

(a) $A6B_{16} = \underbrace{A(10)}_{1010}\ \underbrace{6}_{0110}\ \underbrace{B(11)}_{1011}$

$= 101001101011_2$

$= 2667_{10}$

Alternatively,

$A6B_{16} = (10 \times 16^2) + (6 \times 16^1) + (11 \times 16^0)$

$= 2667_{10}$

(b) $11100101111_2 = \underbrace{0111}_{7}\ \underbrace{0010}_{2}\ \underbrace{1111}_{F}$

$= 72F_{16}$

(c) $\frac{4376}{16} = 273$, remainder 8

$\frac{273}{16} = 17$, remainder 1

$\frac{17}{16} = 1$, remainder 1

$\frac{1}{16} = 0$, remainder 1

Therefore, $4376_{10} = 1118_{16}$.

Example 225-2

What are the hexadecimal values of (a) $9CD_{16} + 47E_{16}$, (b) $17AB_{16} - C6F_{16}$, (c) $7A5_{16} \times 83_{16}$, and (d) $109C8_{16} \div B4_{16}$?

Solution

(a)

$$\begin{array}{rl} 9CD_{16} = & 2509_{10} \\ +\ 47E_{16} = & 1151_{10} \\ \hline E4C_{16} = & 3660_{10} \end{array}$$

Therefore, $9CD_{16} + 47E_{16} = E4C_{16}$

(b)

$$\begin{array}{rl} 17AB_{16} = & 6059_{10} \\ -\ C6F_{16} = & 2671_{10} \\ \hline D3C_{16} = & 3388_{10} \end{array}$$

Therefore, $17AB_{16} - C6F_{16} = D3C_{16}$.

(c)

$$\begin{array}{rl} 7A5_{16} = & 1957_{10} \\ \times\ 83_{16} = & 131_{10} \\ \hline 16EF & \\ 3D28\ \ & \\ \hline 3E96F_{16} = & 256367_{10} \end{array}$$

Therefore, $7A5_{16} \times 83_{16} = 3E96F_{16}$.

(d)

$$\begin{array}{r} 17A_{16} \\ B4_{16}\overline{)109C8_{16}} \\ B4\ \ \ \ \\ \hline 55C\ \ \\ 4EC\ \ \\ \hline 708\ \ \\ 708\ \ \\ \hline 0\ \ \end{array}$$

Therefore, $109C8_{16} \div B4_{16} = 17A_{16}$.

PRACTICE PROBLEMS

225-1. (a) Convert $BCD7_{16}$ to a decimal number and (b) convert 73892_{10} to a hexadecimal number.

225-2. (a) Convert $B643_{16}$ to a binary number and (b) convert 1101110011_2 to a hexadecimal number.

225-3. (a)Convert $AC38_{16}$ to an octal number and (b) convert 4775_8 to a hexadecimal number.

225-4. What are the hexadecimal values of (a) $B7A9_{16} + 374C_{16}$ and (b) $A98C6_{16} - 9346D_{16}$?

225-5. What are the hexadecimal values of (a) $CDF_{16} \times 734_{16}$ and (b) $14B824_{16} \div 6AC_{16}$?

226
The duodecimal number system

The main use of the duodecimal number system is for error detection and correction in certain digital equipment. Because the base of the system is 12, it requires 12 characters: the digits 0, 1, 2, 3, 4, 5, 6, 7, 8, 9, followed by the letters *t* and *e*, which (respectively) represent "*t*en" and "*e*leven."

MATHEMATICAL DERIVATIONS

Duodecimal-to-decimal conversion

To achieve a duodecimal-to-decimal conversion, use Equation 221-1. For example,

$$
\begin{aligned}
2t8_{12} &= (2 \times 12^2) + (10 \times 12^1) + (8 \times 12^0) \\
&= 288 + 120 + 8 \\
&= 416_{10}
\end{aligned}
$$

Decimal-to-duodecimal conversion

A decimal whole number is converted into its equivalent duodecimal number by using the repeated division method, with 12 as the division factor. For example, 5831_{10} is equivalent to:

$$\frac{5831}{12} = 485, \text{ remainder } 11(\text{e})$$

$$\frac{485}{12} = 40, \text{ remainder } 5$$

$$\frac{40}{12} = 3, \text{ remainder } 4$$

$$\frac{3}{12} = 0, \text{ remainder } 3$$

Therefore, $5831_{10} = 345\text{e}_{12}$.

Binary $\rightleftarrows$ duodecimal, octal $\rightleftarrows$ duodecimal, and hexadecimal $\rightleftarrows$ duodecimal conversions can be conveniently carried out by using the decimal system as the intermediate step. However, a second method involves the use of repeated division and is as follows:

1. Consider a number N in a nondecimal system whose base is r_1. It is required to convert N to its equivalent number in another nondecimal system, whose base is r_2.
2. Convert the base r_2 to its equivalent value in the base r_1.
3. Carry out repeated division of the number N using the base r_1.
4. Convert all remainders to the base r_2. As an example, convert 2767_8 to its equivalent number in base 12. Because $12 = 14_8$, the repeated division process is

$$\frac{2767_8}{14_8} = 177_8, \text{ remainder } 3_8 = 3_{12}$$

$$\frac{177_8}{14_8} = 12_8, \text{ remainder } 7_8 = 7_{12}$$

$$\frac{12_8}{14_8} = 0, \text{ remainder } 12_8 = t_{12}$$

Therefore, $2767_8 = t73_{12}$.

A third method consists of the following steps:

1. Perform all arithmetic operations in the desired base.
2. Express the base of the original number in terms of the desired base.
3. Multiply the number obtained in step 2 by the MSD of the original number and add the product to the digit immediately on the right of the MSD.
4. Repeat step 3 as many times as there are digits in the original number. The final sum is the required answer.

For example, convert 173_{12} to base 8.

Step 1. All arithmetic operations will be carried out in base 8.

Step 2. $12 = 14_8$

Step 3.

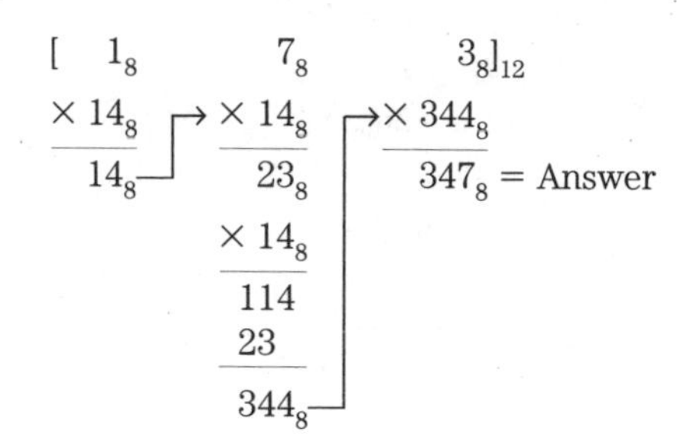

Therefore, $173_{12} = 347_8$.

Duodecimal addition

The results of duodecimal addition are shown in Fig. 226-1. When an addition carry is made, the amount of the carry is 12. For example, when you add t_{12} and e_{12}, the result is $10_{10} + 11_{10} = 21_{10}$, which in base 12 is (1

+	0	1	2	3	4	5	6	7	8	9	t	e
0	0	1	2	3	4	5	6	7	8	9	t	e
1	1	2	3	4	5	6	7	8	9	t	e	10
2	2	3	4	5	6	7	8	9	t	e	10	11
3	3	4	5	6	7	8	9	t	e	10	11	12
4	4	5	6	7	8	9	t	e	10	11	12	13
5	5	6	7	8	9	t	e	10	11	12	13	14
6	6	7	8	9	t	e	10	11	12	13	14	15
7	7	8	9	t	e	10	11	12	13	14	15	16
8	8	9	t	e	10	11	12	13	14	15	16	17
9	9	t	e	10	11	12	13	14	15	16	17	18
t	t	e	10	11	12	13	14	15	16	17	18	19
e	e	10	11	12	13	14	15	16	17	18	19	1t

A (top row), B (left column), C (body)

Figure 226-1

$\times\ 12^1) + (9 \times 12^0) = 19_{12}$. This is shown in Fig. 226-1, where $A + B = C$.

As another example,

$$\begin{array}{rcr} \text{8te2}_{12} & = & 15398_{10} \\ +\ \text{9e4}_{12} & = & 1432_{10} \\ \hline \text{98t6}_{12} & = & 16830_{10} \end{array}$$

Therefore, $\text{8te2}_{12} + \text{9e4}_{12} = \text{98t6}_{12}$.

Duodecimal subtraction

Because $A = C - B$, Fig. 226-1 can be used to determine the results of the subtraction; if $C = 19_{12}$ and $B = e_{12}$, $A = t_{12}$. When a subtraction borrow is made, the amount of the borrow is 12. As a further example,

$$\begin{array}{rcr} \text{t64}_{12} & = & 1516_{10} \\ -\ \text{2e9}_{12} & = & 429_{10} \\ \hline 767_{12} & = & 1087_{10} \end{array}$$

Therefore, $\text{t64}_{12} - \text{2e9}_{12} = 767_{12}$.

Duodecimal multiplication

The results of duodecimal multiplication are shown in Fig. 226-2, where $A \times B = C$. When, for example, 9_{12} is multiplied by 5_{12}, the product is 45_{10}, which is equal to $(3 \times 12^1) + (9 \times 12^0) = 39_{12}$. As another example, multiply 3e7_{12} by t4_{12}:

$$\begin{array}{rcr} \text{3e7}_{12} & = & 571_{10} \\ \times\ \ \text{t4}_{12} & = & 124_{10} \\ \hline \text{13t4} & & \\ \text{337t}\ & & \\ \hline \text{34e84}_{12} & = & 70804_{10} \end{array}$$

Therefore, $\text{3e7}_{12} \times \text{t4}_{12} = \text{34e84}_{12}$.

Duodecimal division

Using Fig. 226-2, duodecimal division will be defined by $A = C/B$. When, for example, 446_{12} is divided by 6_{12}:

$$\begin{array}{r} 89_{12} \\ 6_{12}\overline{)446_{12}} \\ \underline{40}\ \ \ \\ 46 \\ \underline{46} \\ 0 \end{array}$$

Therefore, $446_{12} \div 6_{12} = 89_{12}$.

As another example, divide 1te74_{12} by 5e_{12}:

×	0	1	2	3	4	5	6	7	8	9	t	e	} A
0	0	0	0	0	0	0	0	0	0	0	0	0	
1	0	1	2	3	4	5	6	7	8	9	t	e	
2	0	2	4	6	8	t	10	12	14	16	18	1t	
3	0	3	6	9	10	13	16	19	20	23	26	29	
4	0	4	8	10	14	18	20	24	28	30	34	38	
5	0	5	t	13	18	21	26	2e	34	39	42	47	
6	0	6	10	16	20	26	30	36	40	46	50	56	} C
7	0	7	12	19	24	2e	36	41	48	53	5t	65	
8	0	8	14	20	28	34	40	48	54	60	68	74	
9	0	9	16	23	30	39	46	53	60	69	76	83	
t	0	t	18	26	34	42	50	5t	68	76	84	92	
e	0	e	1t	29	38	47	56	65	74	83	92	t1	
B													

Figure 226-2

$$\begin{array}{r} \text{3t7}_{12} \\ \text{5e}_{12}\overline{)\text{1te75}_{12}} \\ \underline{159}\ \ \ \ \ \\ 527\ \ \ \\ \underline{\text{4e2}}\ \ \ \\ 355 \\ \underline{355} \\ 0 \end{array}$$

Therefore, $\text{1te75}_{12} \div \text{5e}_{12} = \text{3t7}_{12}$.

Example 226-1

Convert (a) 5t8_{12} into its equivalent decimal and binary numbers, and (b) 7834_{10} into its equivalent duodecimal number.

Solution

(a) $\text{5t8}_{12} = 5 \times 12^2 + 10 \times 12^1 + 8 \times 12^0$ (221-1)

$= 848_{10}$

But,

$$\begin{aligned} 848_{10} &= (1 \times 2^9) + (1 \times 2^8) + (0 \times 2^7) + (1 \times 2^6) \\ &\quad + (0 \times 2^5) + (1 \times 2^4) + (0 \times 2^3) \\ &\quad + (0 \times 2^2) + (0 \times 2^1) + (0 \times 2^0) \\ &= 1101010000_2 \end{aligned}$$

Therefore, $\text{5t8}_{12} = 1101010000_2$.

(b) By the repeated division method,

$$\frac{7834_{10}}{12} = 652, \text{ remainder } 10\ (t_{12})$$

$$\frac{652}{12} = 54, \text{ remainder } 4_{12}$$

$$\frac{54}{12} = 4, \text{ remainder } 6_{12}$$

$$\frac{4}{12} = 0, \text{ remainder } 4_{12}$$

Therefore, $7834_{10} = 464t_{12}$.

Example 226-2

What are the duodecimal values of (a) $8t3e_{12} + 748_{12}$, (b) $t4e6_{12} - 387t_{12}$, (c) $2e7_{12} \times 4t_{12}$, and (d) $19546_{12} \div 5e_{12}$?

Solution

(a)

$$\begin{array}{rcr} 8t3e_{12} & = & 15311_{10} \\ +\ 748_{12} & = & 1064_{10} \\ \hline 9587_{12} & = & 16375_{10} \end{array}$$

Therefore, $8t3e_{12} + 748_{12} = 9587_{12}$.

(b)

$$\begin{array}{rcr} t4e6_{12} & = & 17994_{10} \\ -\ 387t_{12} & = & 6430_{10} \\ \hline 6838_{12} & = & 11564_{10} \end{array}$$

Therefore, $t4e6_{12} - 387t_{12} = 6838_{12}$.

(c)

$$\begin{array}{rcr} 2e7_{12} & = & 427_{10} \\ \times\ \ 4t_{12} & = & 58_{10} \\ \hline 257t & & \\ et4\ & & \\ \hline 123et_{12} & = & 24766_{10} \end{array}$$

Therefore, $2e7_{12} \times 4t_{12} = 123et_{12}$.

(d)

$$\begin{array}{r} 376_{12} \\ 5e_{12}\overline{)19546_{12}} \\ \underline{159} \\ 384 \\ \underline{355} \\ 2e6 \\ \underline{2e6} \\ 0 \end{array}$$

Therefore, $19546_{12} \div 5e_{12} = 376_{12}$.

PRACTICE PROBLEMS

226-1. (a) Convert $7te4_{12}$ to a decimal number and (b) convert 38406_{10} to a duodecimal number.

226-2. (a) Convert $e6t3_{12}$ to a binary number and (b) 1110011000111_2 to a duodecimal number.

226-3. (a) Convert $45e7_{12}$ to an octal number and (b) 37451_8 to a duodecimal number.

226-4. (a) Convert $3t76e_{12}$ to a hexadecimal number and (b) convert $BACF_{16}$ to a duodecimal number.

226-5. What are the duodecimal values of (a) $t7e6_{12} + 3e5t_{12}$ and (b) $e7453_{12} - 874t_{12}$?

226-6. What are the duodecimal values of (a) $7t4_{12} \times ee_{12}$ and (b) $8044t_{12} \div e7_{12}$?

227 The arithmetic of complements

A computer might be designed to perform arithmetic operations by using addition only. Such a computer must use some method for identifying and manipulating both positive and negative numbers. This is normally accomplished by using complement arithmetic. When you wish to subtract B from A, the result, $A - B$, can be written $A + (-B)$. If you regard $(-B)$ as the complement of B, then you can carry out the subtraction process by adding A to the complement of B.

In any number system, the *complement* of a number is defined as the difference between the number and the next higher power of the base. Therefore,

$$C = r^D - N \tag{227-1}$$

where: N = the number
C = the complement of the number
D = the number of digits in the number
r = the base of the number system

For example, if the base is 10,

2_{10} is the complement of 8_{10}, $(2 = 10^1 - 8)$

26_{10} is the complement of 74_{10}, $(26 = 10^2 - 74)$

744_{10} is the complement of 256_{10}, $(744 = 10^3 - 256)$

Similarly in base 8,

2_8 is the complement of 6_8, $(10_8 - 6_8 = 2_8)$

4_8 is the complement of 74_8, $(100_8 - 74_8 = 4_8)$

522_8 is the complement of 256_8, $(1000_8 - 256_8 = 522_8)$

Consider the example of subtracting 26_{10} from 83_{10}. The simple subtraction is $83_{10} - 26_{10} = 57_{10}$. However, the complement of $26_{10} = 10^2 - 26_{10} = 74_{10}$; thus by addition,

$$\begin{array}{r} 83_{10} \\ +\ 74_{10} \\ \hline 157_{10} \end{array}$$

If the final carry-over of 1 is ignored, the addition process using complement-arithmetic produces the same answer as simple subtraction. This must be true because when you add 83_{10} and $10^2_{10} - 26_{10}$, the answer is $83_{10} + (10^2_{10} - 26_{10}) = 10^2_{10} + (83_{10} - 26_{10})$, or 100 plus the result of subtracting 26_{10} from 83_{10}.

The use of binary complements is important because it means that the same circuitry of a digital computer can be used for both addition and subtraction. Consequently, the total amount of circuitry required is reduced.

MATHEMATICAL DERIVATIONS

In chapters 222 through 226, the binary bits have been used to represent the magnitude of a number. However, because you are now considering positive and negative numbers, you need some method of indicating the sign of a number. The normal convention is to use 0 as the positive sign bit, which is positioned immediately to the left of the MSB. In a similar way, a negative number is preceded by a 1 bit. As examples,

$$0100111 = +39$$

$$1010010 = -18$$

If you wish to use addition in order to subtract 18 from 39, the +39 will remain in the binary form as shown, but the −18 must be converted into some form of complement.

2's-complement

From Equation 227-1, the complement of $1001_2 = 2^4 - 1001_2 = 16_{10} - 9_{10} = 7_{10} = 0111_2$. This is equivalent to interchanging 0 and 1 in the original number, then adding 1 to the LSB. The result is called the *2's-complement form.*

For example, the process of converting -37_{10} into its 2's-complement form is as follows:

$$\begin{array}{rrl} 37_{10} = & 100101_2 & \\ & 011010 & \text{(interchanging 0 and 1)} \\ + & 1 & \text{(Add 1 to the LSB)} \\ \hline & 011011 & \text{(2's-complement form)} \end{array}$$

When you include the sign bit, $-37_{10} = 1011011$ in its 2's-complement form.

To convert from the 2's-complement form back to the true binary value, repeat the 2's-complement process. Therefore, 011011 becomes 100100 + 1 or 100101, which is the original binary value. Because +37 = 0100101, and −37 in its 2's-complement form is 1011011, the complete operation when performed on an entire number, including the sign bit, changes a positive number to a negative number and vice-versa.

Combining two signed numbers

Now use the 2's-complement system to add together two signed numbers. If both numbers are positive, such as $+7_{10}$ and $+5_{10}$, neither number is expressed in its 2's-complement form, and the procedure is:

$$\begin{array}{rrl} & +7 \rightarrow \overset{+}{0} & 0111 \\ + & +5 \rightarrow \overset{+}{0} & 0101 \\ \hline \text{Sum} = & +12 \rightarrow \overset{+}{0} & 1100 \end{array}$$

When one number is positive (+7) and the other is a smaller negative number (−5), the −5 must be in its 2's-complement form, which is 11011. When you add +7 and −5, it is equivalent to subtracting 5 from 7, and you anticipate an answer of +2. The binary process is:

$$\begin{array}{rrl} & +7 \rightarrow \overset{+}{0} & 0111 \\ + & (-5) \rightarrow \overline{1} & 1011 \text{ (2's-complement form)} \\ \hline \text{Answer} = & +2 \quad \not{1}\overset{+}{0} & 0010 \end{array}$$

↑ discard

$$\text{Sum} = +2 \rightarrow \overset{+}{0} \ 0010$$

Although the sign bits are involved in the addition process, any carry-over from the sign bits is disregarded. Note that the same number of bits (4) is used for the magnitude of each number; this is essential, if the 2's-complement form is to be used.

Now add −7 and +5. This is equivalent to subtracting 7 from 5, so look for an answer of −2. The binary process is:

$$+5 \rightarrow \overset{+}{0}\ \ 0101$$

$$\underline{+(-7)} \rightarrow \bar{1}\ \ \underline{1001}\ \text{(2's-complement form)}$$

$$\text{Answer} \rightarrow \bar{1}\ \ 1110\ \text{(2's-complement form)}$$

$$\text{Sum} = -2 \rightarrow \bar{1}\ \ 0010\ \text{(true binary sum)}$$

Because the answer is negative, it is in its 2's-complement form and the true binary sum is $-(0010_2) = -2_{10}$.

Finally, add −7 and −5 for a sum of −12. Both numbers will be expressed in their 2's-complement form; therefore:

$$(-7) \rightarrow \bar{1}\ \ 1001\ \text{(2's-complement form)}$$

$$\underline{+(-5)} \rightarrow \bar{1}\ \ \underline{1011}\ \text{(2's-complement form)}$$

$$\text{Answer} \rightarrow \not{1}\ \bar{1}\ \ 0100\ \text{(2's-complement form)}$$

↑ discard

$$\text{Sum} = -12 \rightarrow \bar{1}\ \ 1100\ \text{(true binary sum)}$$

The carry-over from the sign bits is disregarded. Because the answer is negative, it is in its 2's-complement form, which must be converted to its true binary sum.

In each of the four combinations [(+7) + (+5) = +12, (+7) + (−5) = +2, (−7) + (+5) = −2 and (−7) + (−5) = −12], the magnitude of the sum was accommodated within the four bits available. However, if you add (+10) to (+8),

$$(+10) \rightarrow \overset{+}{0}\ \ 1010$$

$$(+8) \rightarrow \overset{+}{0}\ \ \underline{1000}$$

$$\text{Answer} \rightarrow \bar{1}\ \ 0010$$

the answer is wrong because it contains a negative sign bit, although the sign of the sum is clearly positive. Because the sum of +18 is not accommodated within the available four bits, there is an overflow into the sign bits and the answer is then wrong. Overflow is detected by the circuitry of the computer, and the necessary correction is made by using five bits.

Sometimes digital circuitry must multiply two signed numbers. Because digital circuitry uses the 2's-complement form, both numbers are first converted into their true binary forms, and then the multiplication is carried out. If the product of the two signed numbers is positive, the 0 sign bit is given to the true binary answer. However, if the product is negative, the true binary answer is converted to its 2's-complement form, and then given a 1 sign bit.

Example 227-1

What are the complements of (a) 213_{10}, (b) 61_8, and (c) 1011_2?

Solution

(a) Complement $= 10^3 - 213_{10} = 787_{10}$

(b) Complement $= 8^2 - 61_8 = 100_8 - 61_8 = 17_8$

(c) 2's-complement = 0100
$\underline{+\ 1}$
0101

Example 227-2

Carry out the following operations in binary, using the 2's-complement form when necessary: (a) (+8) + (+3), (b) (+8) + (−3), (c) (+3) + (−8), and (d) (−8) + (−3).

Solution

(a)

$$(+8) \rightarrow \overset{+}{0}\ \ 1000_2$$

$$\underline{(+3)} \rightarrow \overset{+}{0}\ \ \underline{0011_2}$$

$$\text{Sum} = +11 \rightarrow 0\ \ 1011_2$$

(b)

$$(+8) \rightarrow \overset{+}{0}\ \ 1000$$

$$\underline{+(-3)} \rightarrow \bar{1}\ \ \underline{1101}\ \text{(2's-complement form)}$$

$$\text{Answer} = +5\ \not{1}\ \overset{+}{0}\ \ 0101$$

↑ discard

$$\text{Sum} = +5 \rightarrow \overset{+}{0}\ \ 0101$$

(c)

$$(+3) \rightarrow \overset{+}{0}\ 0011$$

$$\underline{-(+8)} \rightarrow \bar{1}\ \underline{1000} \text{ (2's-complement form)}$$

$$\text{Answer} \rightarrow \bar{1}\ 1011 \text{ (2's-complement form)}$$

$$\text{Sum} = -5 \rightarrow \bar{1}\ 0101 \text{ (true binary form)}$$

(d)

$$(-8) \rightarrow \bar{1}\ 1000 \text{ (2's-complement form)}$$

$$\underline{+(-3)} \rightarrow \bar{1}\ \underline{1101} \text{ (2's-complement form)}$$

$$\text{Answer} \rightarrow \underset{\substack{\uparrow \\ \text{discard}}}{\not{1}}\ \bar{1}\ 0101 \text{ (2's-complement form)}$$

$$\text{Sum} = -11 \rightarrow \bar{1}\ 1011 \text{ (true binary form)}$$

PRACTICE PROBLEMS

227-1. What are the complements of (a) 217_{10}, (b) 324_8, (c) $6t7_{12}$, and (d) $BA8_{16}$?

227-2. Express -219_{10} in its 2's complement form.

227-3. Express $+83_{10}$ in its 2's complement form.

227-4. Express (a) $(+8) + (+7)$, (b) $(+8) + (-7)$, (c) $(+7) + (-8)$, and (d) $(-8) + (-7)$ in terms of their true binary sums.

228 Introduction to Boolean algebra

Boolean algebra was developed by the English logician and mathematician George Boole (1815–1864). In the spring of 1847, Boole wrote a pamphlet entitled *A Mathematical Analysis of Logic*. This was followed in 1854 by his more exhaustive treatise, *An Investigation of the Laws of Thought*. It is this later work that forms the basis for our present-day mathematical theories used for the analysis of logical processes.

Although conceived in the nineteenth century, little practical application was found for Boole's work until 1938, when it was discovered that his algebra could be adapted to the analysis of switching circuits. With the advent of modern computers, Boolean algebra has become an important subject in the understanding of complex digital circuitry.

CLASSES AND ELEMENTS

In our world, it is logical to visualize two divisions: all things of interest are in one division, and all things not of interest are in the other division. These two divisions make up a set (or a class) that is designated as the *universal class*. All things contained in the universal class are referred to as *elements* or *variables*. You can also visualize another class, that contains no elements; such a set is designated as the *null class*.

In a particular discussion, certain elements of the universal class can be grouped together to form combinations, which are known as *subclasses*. Each subclass of the universal class is dependent on its elements and the possible states (stable, unstable, or both) that these elements might have.

Boolean algebra is limited to the use of elements that only possess two possible logic states, both of which are stable. If you have two elements, A and B, their possible states can be designated in a number of ways, as shown in Table 228-1.

Table 228-1

Logic 1	Logic 0
True	False
Yes	No
High	Low
+10 V	0 V
ON	OFF
Closed Switch	Open Switch

Notice that in Boolean algebra, there are only two numbers: 0 and 1. Moreover, this form of algebra does not contain such concepts as squares, square roots, reciprocals, and logarithms.

If you have two elements, each of which has two states, there are 2^2, or 4, possible subclasses. If the states of A and B are true or false, and you use the connective word *AND*, the four subclasses are:

A true AND *B* false

A true AND *B* true

A false AND *B* true

A false AND *B* false

However, if the connective word *OR* is used, there are four additional subclasses:

A true OR *B* false

A true OR *B* true

A false OR *B* true

A false OR *B* false

MATHEMATICAL DERIVATIONS

Venn diagrams

A *Venn diagram* is a topographical picture of logic. Such a diagram is composed of the universal class, which is divided into subclasses. The number of these subclasses depends on the number of elements.

As an example, let A equal *cars* and B equal the color *red*. With the connective word AND, the four subclasses are:

Cars AND *Not red*

Cars AND *red*

Red AND *Not cars*

Not cars AND *Not red*

With the connective word OR, the four additional subclasses are:

Red OR *Not cars*

Not cars OR *Not red*

Cars OR *Not red*

Cars OR *red*

All these subclasses are shown in Fig. 228-1. The shaded area in each diagram represents the particular subclass.

Symbols in Boolean algebra

The symbols used in Boolean algebra are common in other branches of mathematics but in certain cases, their meaning is slightly different. The principal symbols are explained in Table 228-2.

It follows that logic circuits have only three basic operations: AND, OR, and NOT.

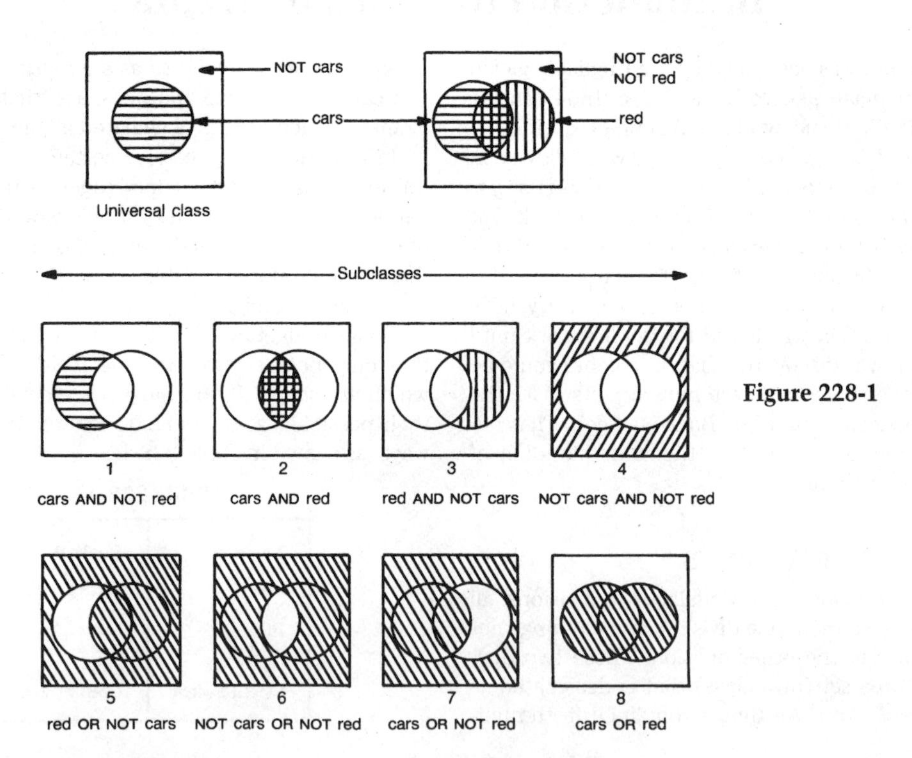

Figure 228-1

Table 228-2

Symbol	Meaning
=	As in conventional mathematics the equal sign represents the relationship of equivalence between the expressions which are so connected.
· or ×	These symbols indicate the logic product which is also known as the AND operation. Frequently this operation is indicated without the use of a symbol so that $AB = A \times B = A \cdot B$
+	The plus sign indicates the logic sum which is also known as the OR operation.
–	The overbar signifies logical complementation or inversion which is known as the NOT operation.
( ), [ ], { }	The grouping symbols mean that all the contained terms must be treated as a unit.

Truth tables

For any Boolean operation, there is a corresponding truth table, which shows the various outputs of the operation for each possible way the states of the input elements can be assigned. In a truth table, the states are designated as 0 and 1; if there are only two input elements, the corresponding truth table is as shown in Table 228-3.

Table 228-3

Input Elements		Output
A	B	$f(A, B)$
0	0	?
0	1	?
1	0	?
1	1	?

The operation's output will be a function of A and B and is, therefore, designated as $f(A,B)$. Notice that the top-to-bottom sequence of the input elements is the same as that of binary counting.

Example 228-1

Design a truth table for three input elements, designated A, B, and C.

Solution

Input Elements			Output
A	B	C	$f(A,B,C)$
0	0	0	?
0	0	1	?
0	1	0	?
0	1	1	?
1	0	0	?
1	0	1	?
1	1	0	?
1	1	1	?

PRACTICE PROBLEMS

228-1. Draw a Venn diagram to illustrate the subclass A AND B AND C.

228-2. Draw a Venn diagram to illustrate the subclass A OR B OR C.

228-3. Draw a Venn diagram to illustrate the subclass NOT A AND NOT B AND NOT C.

228-4. Design a truth table for four input elements designated A, B, C, and D.

229
The OR operation

The Venn diagram in Fig. 229-1A has two elements, or variables, which are designated A and B. The shaded area represents the subclass, which is $A + B$ in Boolean notation. The corresponding equation in Boolean algebra is expressed as:

$$f(A,B) = A + B \qquad (229\text{-}1)$$

This expression is called an OR *operation,* because it represents the last of the four OR subclasses illustrated in the Venn diagrams of Fig. 228-1. The equation $f(A,B) = A + B$ is read as either "$f(A,B)$ equals A plus B", or "$f(A,B)$ equals A OR B."

Figure 229-1B illustrates the truth table of the OR operation. When A and B are each 0, the output is also 0. If either A or B takes the value of 1, the output $f(A,B)$ equals 1. However, if both A and B are equal to 1, $f(A,B)$ is 1, not 2, as in binary addition because $A + B$ represents the *logic* sum; in Boolean algebra, only two numbers exist: 0 and 1.

Figure 229-1C shows the switching circuit that represents the OR operation. It consists of two switches in parallel. The circuit is ON (1) if either switch A or switch B is in the CLOSED (1) state. The circuit is also ON (1) if both switch A and switch B are CLOSED (1), but it is OFF (0) if, and only if, both switch A and switch B, are OPEN (0). Notice that the high and low values applied to the load are 5 V and 0 V, respectively.

Consider a situation in which Adam (A) and Brian (B) are members of a group which meets from time to time. The rule is:

Adam (A) present OR Brian (B) present, the meeting is held. It follows that:

Adam absent, Brian absent, the meeting is not held.
Adam absent, Brian present, the meeting is held.
Adam present, Brian absent, the meeting is held.
Adam present, Brian present, the meeting is held.

"Absent" and "present' correspond respectively to the logic states "0" and "1." The existence of the meeting depends on Adam and Brian, and can therefore be mathematically regarded as a function of A and B, ($f(A,B)$) with "not held" and "held" being represented

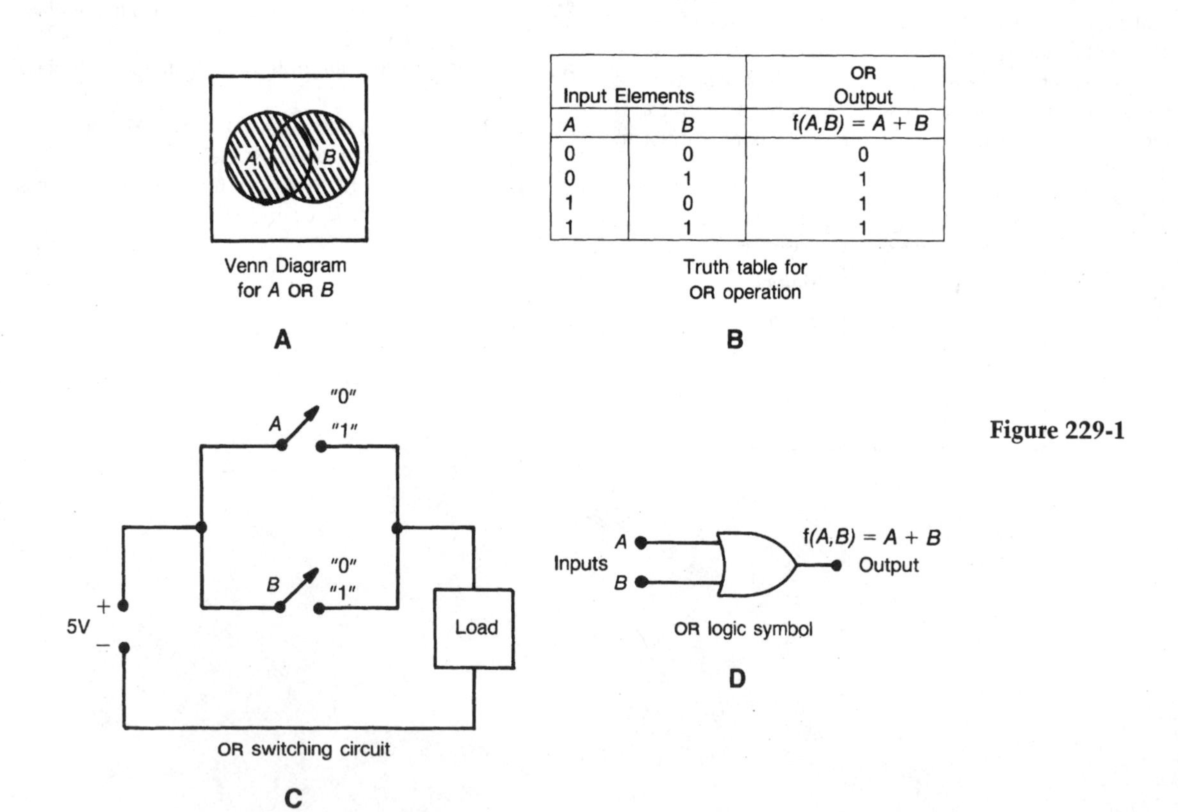

Input Elements		OR Output
A	B	$f(A,B) = A + B$
0	0	0
0	1	1
1	0	1
1	1	1

Figure 229-1

by the logic states "0" and "1." The following truth table will then apply to the "OR" operation:

Adam	Brian	Meeting
A	B	$f(A,B)$
0	0	0
0	1	1
1	0	1
1	1	1

The OR GATE

The OR gate is a digital circuit that has two or more inputs represented by the logic voltage levels 0 and 1. For example, the logic 1 level (high) might be +5 V, although the logic 0 level (low) is only +0.5 V. The output of the gate is the logic sum of the inputs, and has a voltage level (either 1 or 0), that is the result of carrying out the OR operation on the inputs. The logic symbol for the OR gate is shown in Fig. 229-1D.

If there are only two inputs, A and B, the OR gate operates in such a way that its output, A OR B, is low (0) if, and only if, both A and B are in the 0 state. By contrast, if either A or B or both are in the 1 state, the gate's output is high and in the 1 state.

With more than two inputs, the same principles apply. In all cases when one or more of the inputs is in the 1 state, the output is high. The outputs will be low only when all of the inputs are in the 0 state.

In practical digital circuits, the state ("1" or "0") at a particular point is commonly determined by a logic probe which is small in size and provides a direct readout. Alternatively, a voltmeter can be used, but such an instrument is more cumbersome and its reading has to be interpreted.

MATHEMATICAL DERIVATIONS

To summarize the results of the OR operation:

1. For the OR operation, the output is in the 0 state if, and only if, all of the inputs are in the 0 state. Therefore,

$$f(A,B,C\ldots) = A + B + C + \ldots = 0 \qquad (229\text{-}2)$$
$$\text{if } A = B = C = \ldots = 0$$

2. For the OR operation, the output is in the 1 state when any one of the inputs is in the 1 state. Therefore,

$$f(A,B,C\ldots) = A + B + C + \ldots = 1 \qquad (229\text{-}3)$$
$$\text{if } A \text{ or } B \text{ or } C \text{ or} \ldots = 1$$

3. For the OR operation, the output is in the 1 state if more than one of the inputs is in the 1 state. For example,

$$f(A,B,C\ldots) = 1 + 0 + 1 + 1 + 0 + 0 + 1\ldots$$
$$= 1.$$

Example 229-1

An OR gate has three inputs. Derive the corresponding Venn diagram, truth table, and switching circuit.

Solution

The solution is illustrated in Fig. 229-2.

Example 229-2

Figure 229-3 shows an OR gate with two inputs (A) and (B) whose logic voltage levels are given. Determine the output's logic voltage level as it varies with time.

Solution

The solution is illustrated in Fig. 229-3. The OR outputs are:

Between times T_1 and T_2: $A = 1, B = 0, A + B = 1$.

Between times T_2 and T_3: $A = 0, B = 0, A + B = 0$.

Between times T_3 and T_4: $A = 0, B = 1, A + B = 1$.

Between times T_4 and T_5: $A = 0, B = 0, A + B = 0$.

Between times T_5 and T_6: $A = 1, B = 0, A + B = 1$.

Between times T_6 and T_7: $A = 0, B = 0, A + B = 0$.

Between times T_7 and T_8: $A = 0, B = 1, A + B = 1$.

Between times T_8 and T_9: $A = 1, B = 1, A + B = 1$.

Between times T_9 and T_{10}: $A = 1, B = 0, A + B = 1$.

Time T_{10} and beyond: $A = 0, B = 0, A + B = 0$.

PRACTICE PROBLEMS

229-1. Write down the logic equation for the following statement:
"An alarm (A) is activated when: a window (W) is broken, OR a door (D) is forced, OR the combination lock of the safe (S) is touched between midnight and 7 am, OR a fire (F) occurs."

229-2. Design the truth table which corresponds to the logic equation of Practice Problem 229-1.

229-3. Illustrate the logic symbol for the equation of Practice Problem 229-1.

229-4. Figure 229-4 shows the two inputs (A and B) to an OR gate. Determine the output's logic voltage level as it varies with time.

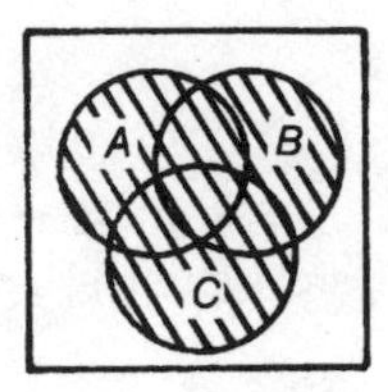

Venn diagram
for *A* OR *B* OR *C*

A

Input elements			OR Output
A	*B*	*C*	f(*A*,*B*,*C*) = *A* + *B* + *C*
0	0	0	0
0	0	1	1
0	1	0	1
0	1	1	1
1	0	0	1
1	0	1	1
1	1	0	1
1	1	1	1

Truth table for
OR operation
with three inputs

B

Figure 229-2

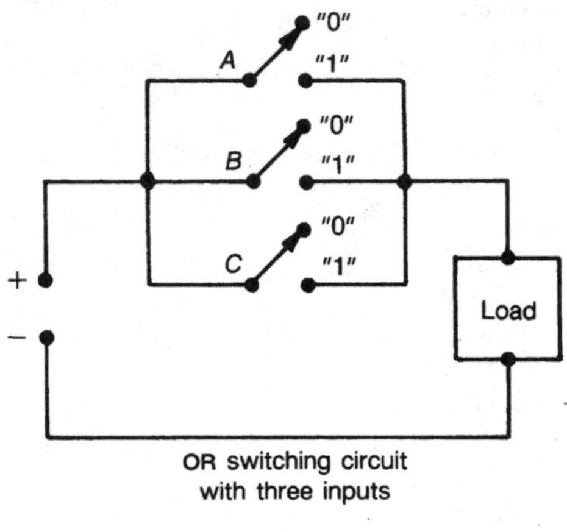

OR switching circuit
with three inputs

C

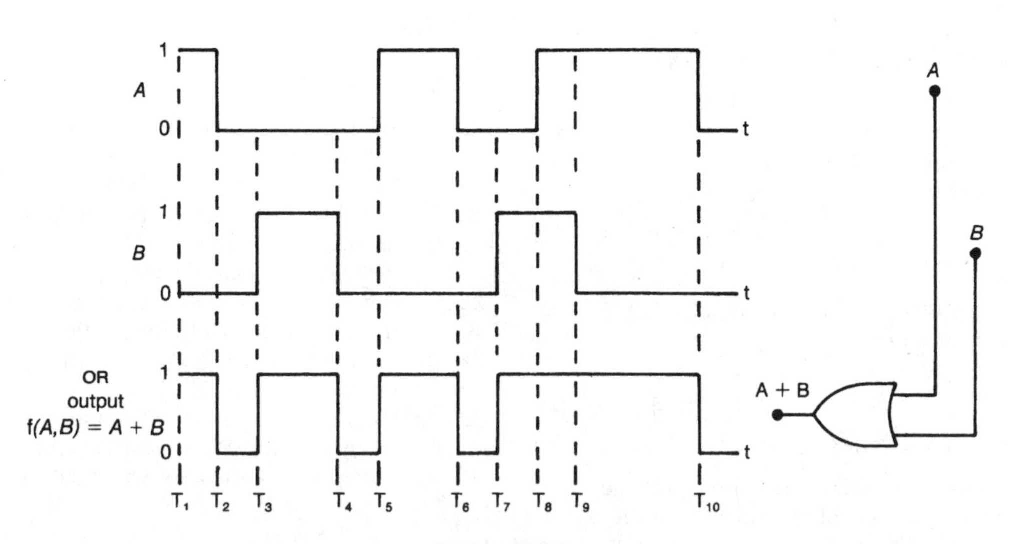

Figure 229-3

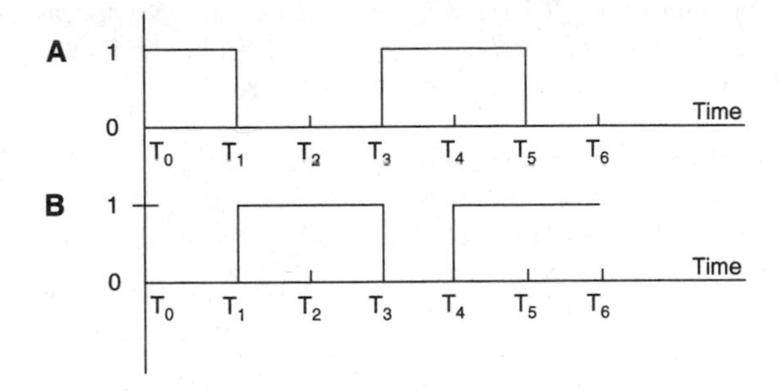

Figure 229-4

230
The AND operation

The Venn diagram in Fig. 230-1A has two elements, or variables, which are designated A and B. The shaded area represents the subclass, which is AB (or $A \cdot B$, or $A \times B$) in Boolean notation. The corresponding equation in Boolean algebra is expressed as:

$$f(A,B) = AB = A \cdot B = A \times B \qquad (230\text{-}1)$$

This expression is called an AND operation because it represents the second of the four AND subclasses illustrated in the Venn diagrams of Fig. 228-1. The equation $f(A,B) = AB$ is read as "$f(A,B)$ equals A AND B," which is the logic product of A and B.

Figure 230-1B depicts the truth table for the AND operation. When A and B are each 0, the output is also

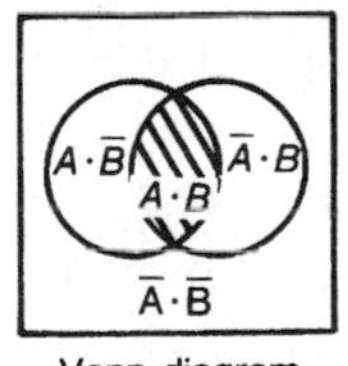

Venn diagram for A AND B

A

Input elements		AND Output
A	B	$f(A,B) = A \cdot B$
0	0	0
0	1	0
1	0	0
1	1	1

Truth table for AND operation

B

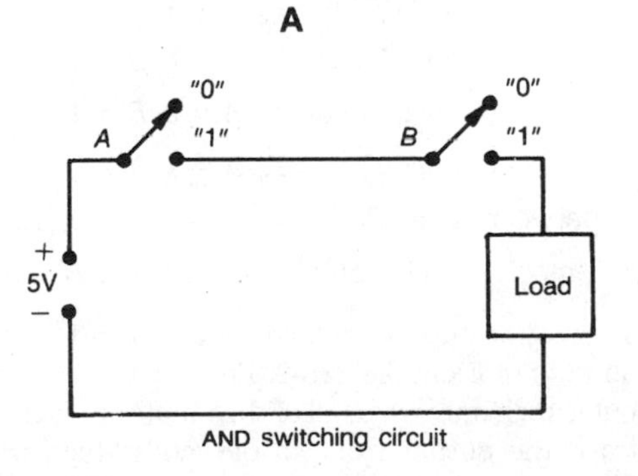

AND switching circuit

C

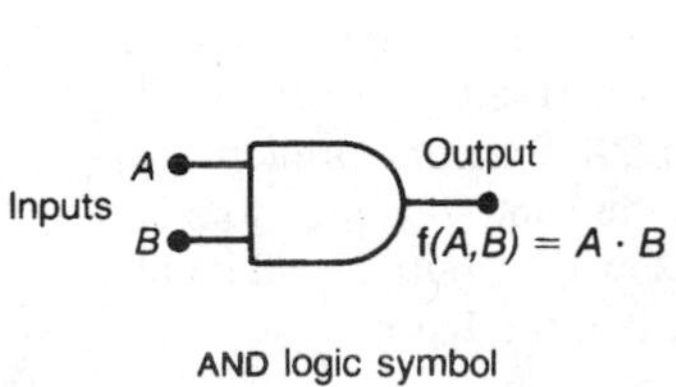

AND logic symbol

D

Figure 230-1

0. If either A or B takes the value of 0, $f(A,B)$ is 0. However, if both A and B are in the 1 state, $f(A,B)$ is 1. The rules for the AND operation are the same as those for binary multiplication.

Figure 230-1C shows the switching circuit that represents the AND operation. It consists of two switches in series so that the circuit is OFF (0) if either switch A or switch B is in the OPEN (0) state. The circuit is also OFF (0) if both switch A and switch B are OPEN (0). The circuit is ON (1) if, and only if, both switches A and B are CLOSED (1). Notice that the high and low logic values applied to the load are 5 V and 0 V, respectively.

THE AND GATE

The AND gate is a digital circuit that has two or more inputs represented by the logic voltage levels 0 and 1. For example, the logic 1 level (high) might be +5 V, while the logic 0 level (low) is only +0.5 V. The output of the gate is the logic product of the inputs and has a voltage level (either 1 or 0) that is the result of carrying out the AND operation on the inputs. The logic symbol for the AND gate is shown in Fig. 230-1D.

Refer to the meeting as described in chapter 229. The rule is changed to:
Adam (A) present AND Brian (B) present, the meeting is held. Under the new rule:
Adam absent, Brian absent, the meeting is not held.
Adam absent, Brian present, the meeting is not held.
Adam present, Brian absent, the meeting is not held.
Adam present, Brian present, the meeting is held.

The following truth table then applies to the "AND" operation:

Adam	**Brian**	**Meeting**
A	B	$f(A,B)$
0	0	0
0	1	0
1	0	0
1	1	1

If there are only two inputs (A and B) the AND gate operates in such a way that its output (A AND B) is high (1) if, and only if, A and B are in the 1 state. By contrast, if either A or B or both are in the 0 state, the gate's output is low and in the 0 state.

With more than two inputs, the same principles apply. In all cases, when one or more of the inputs is in the 0 state, the output is low. The output will be high only when all the inputs are in the 1 state.

MATHEMATICAL DERIVATIONS

To summarize the results of the AND operation:

1. For the AND operation, the output is in the 1 state if, and only if, all inputs are in the 1 state. Therefore,

$$f(A,B,C\ldots) = ABC\ldots \quad (230\text{-}2)$$
$$= 1$$
$$\text{if } A = B = C\ldots = 1$$

2. For the AND operation, the output is in the 0 state when any one or more of the inputs is in the 0 state. Therefore,

$$f(A,B,C\ldots) = ABC\ldots \quad (230\text{-}3)$$
$$= 0$$
$$\text{if } A \text{ or } B \text{ or } C \text{ or} \ldots = 0$$

Example 230-1

An AND gate has three inputs. Derive the corresponding Venn diagram, truth table, and switching circuit.

Solution

The solution is illustrated in Fig. 230-2.

Example 230-2

Figure 230-3 shows an AND gate with two inputs A and B whose logic voltage levels are given. Determine the output's logic voltage levels as they vary with time.

Solution

The solution is illustrated in Fig. 230-3. The AND outputs are:

Between times T_1 and T_2: $A = 1, B = 0, AB = 0$.
Between times T_2 and T_3: $A = 0, B = 0, AB = 0$.
Between times T_3 and T_4: $A = 0, B = 1, AB = 0$.
Between times T_4 and T_5: $A = 0, B = 0, AB = 0$.
Between times T_5 and T_6: $A = 1, B = 0, AB = 0$.
Between times T_6 and T_7: $A = 0, B = 0, AB = 0$.
Between times T_7 and T_8: $A = 0, B = 1, AB = 0$.
Between times T_8 and T_9: $A = 1, B = 1, AB = 1$.
Between times T_9 and T_{10}: $A = 1, B = 0, AB = 0$.
Time T_{10} and beyond: $A = 0, B = 0, AB = 0$.

These AND outputs should be compared with the OR outputs of Example 228-2. For the AND gate, the output is high only when all of the inputs are high. By contrast, the output from an OR gate is high when any input is high.

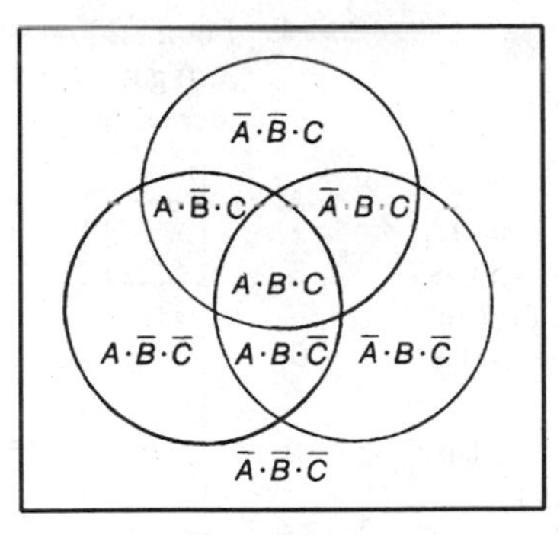

Venn diagram
for *A* AND *B* AND *C*

A

Figure 230-2

Input elements			AND Output
A	*B*	*C*	$f(A,B,C) = A \cdot B \cdot C$
0	0	0	0
0	0	1	0
0	1	0	0
0	1	1	0
1	0	0	0
1	0	1	0
1	1	0	0
1	1	1	1

Truth table for
AND operation
with three inputs

B

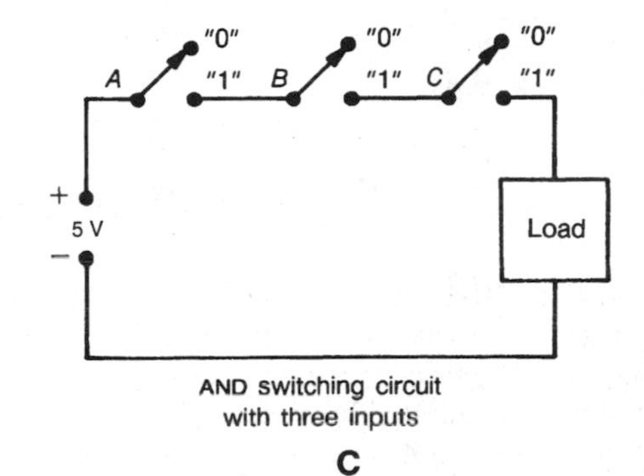

AND switching circuit
with three inputs

C

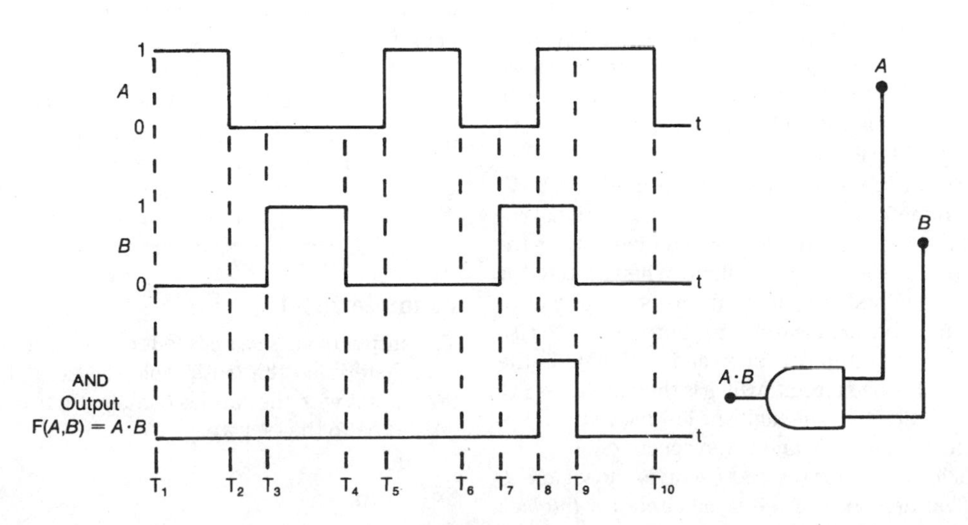

Figure 230-3

PRACTICE PROBLEMS

230-1. Write down the logic equation for the following statement:
A current (C), flowing in the dc series resistive circuit when a voltage source (V), is included in the circuit, AND the switch (S) is closed, AND the resistors (R) are operating within their normal limits, AND all the solder joints (J) meet acceptable standards.

230-2. Design the truth table that corresponds to the logic equation of Practice Problem 230-1.

230-3. Illustrate the logic symbol for the equation of Practice Problem 230-1.

230-4. Figure 230-4 shows two inputs (A and B) to an AND gate. Determine the output's logic voltage level as it varies with time.

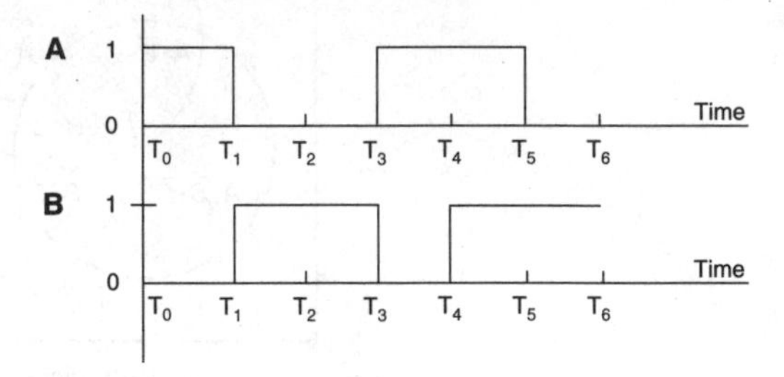

Figure 230-4

231 The NOT operation

The Venn diagram in Fig. 231-1A has one input element (or variable), which is designated as A and is represented by the clear area. The shaded area represents the Boolean output, which is indicated by $\overline{A}$ or A', and is the *complement* (or inverse) of A.

The corresponding equation is:

$$f(A) = \overline{A} = A' \tag{231-1}$$

This equation is read as "Output, $f(A)$ is NOT A." Notice that, unlike the AND and OR operations, the NOT operation can only be performed on a single input element.

Because the logic value of $\overline{A}$ is opposite to the logic value of A, it follows that: (a) if $A = 0, \overline{A} = 1$ because NOT 0 is 1 and (b) if $A = 1, \overline{A} = 0$ because NOT 1 is 0. The results of this reasoning are shown in the truth table of Fig. 231-1B.

The NOT switching circuit is shown in Fig. 231-1C. The requirement of a NOT circuit is that a signal injected at the input produces the complement, or *inversion*, of this signal at the output. When switch A is closed and is in state 1, the relay opens the circuit to the load and the load circuit is, therefore, OFF (0). However, when switch A is open and is in state 0, the relay closes the load circuit, which is then in the ON (1) condition. Notice that the high and low logic values applied to the load are 5 V and 0 V, respectively.

The logic symbol for the NOT circuit is shown in Fig. 231-1D. The presence of the small circle (or *bubble*) on a logic symbol is always an indication of inversion. The input and output symbols mean that the logic levels are interchanged; consequently, an input level of 1 corresponds to an output level of 0 and vice-versa.

MATHEMATICAL DERIVATIONS

The NOT operation is summarized as follows:

$$f(A) = \overline{A} = 1 \text{ when } A = 0.$$

$$f(A) = \overline{A} = 0 \text{ when } A = 1.$$

The OR, AND, and NOT circuits provide the three basic Boolean operations. Comparing the rules for these operations:

OR	AND	NOT
$0 + 0 = 0$	$0 \cdot 0 = 0$	$\overline{0} = 1$
$0 + 1 = 1$	$0 \cdot 1 = 0$	$\overline{1} = 0$
$1 + 0 = 1$	$1 \cdot 0 = 0$	
$1 + 1 = 1$	$1 \cdot 1 = 1$	

Example 231-1

The output of an OR gate is fed to the input of a NOT circuit. Determine the truth table for the output of the NOT circuit with the various combinations of the inputs (A and B) to the OR gate.

Solution

The solution is shown in Fig. 231-2.

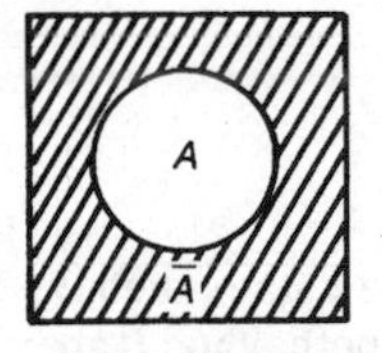

Venn diagram
for NOT operation

A

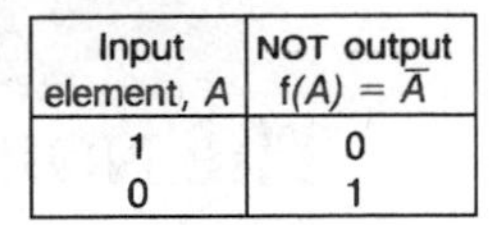

Input element, A	NOT output $f(A) = \overline{A}$
1	0
0	1

Truth table
for NOT operation

B

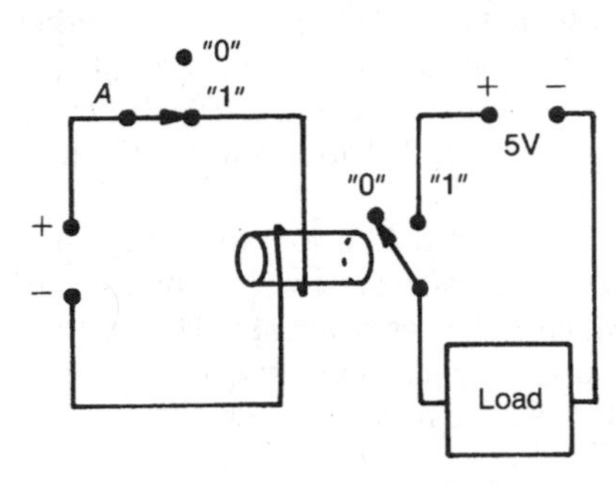

NOT switching circuit

C

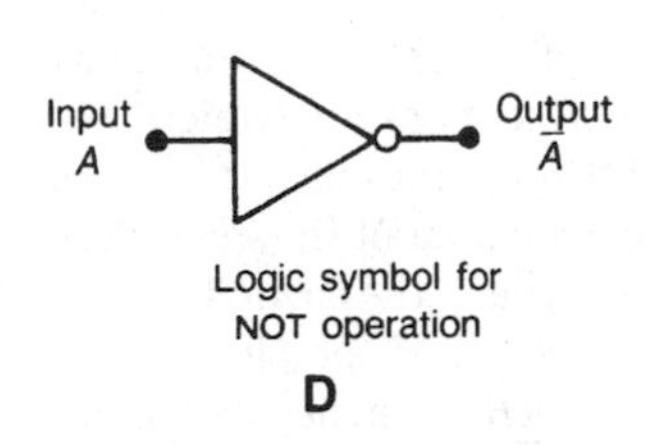

Logic symbol for
NOT operation

D

Figure 231-1

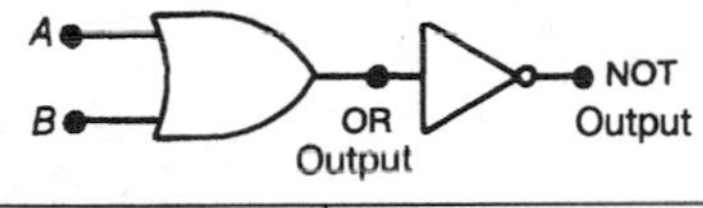

Input elements		Outputs	
A	*B*	OR	NOT
0	0	0	1
0	1	1	0
1	0	1	0
1	1	1	0

Figure 231-2

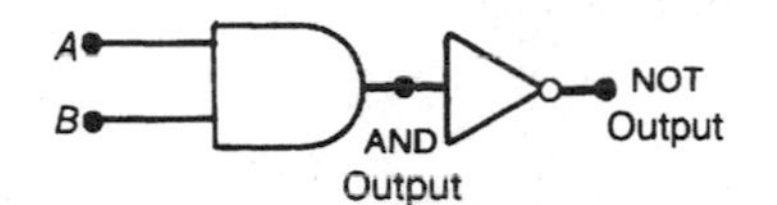

Input elements		Outputs	
A	*B*	AND	NOT
0	0	0	1
0	1	0	1
1	0	0	1
1	1	1	0

Figure 231-3

Example 231-2

The output of an AND gate is fed to the input of a NOT circuit. Determine the truth table for the output of the NOT circuit with the various combinations of the inputs (*A* and *B*) to the AND gate.

Solution

The solution is shown in Fig. 231-3.

PRACTICE PROBLEMS

231-1. Write down the logic equation for the following statement:
"In a traffic control system, turn on the red light (*R*) if the green light (*G*) is NOT on."

231-2. Design the truth table that corresponds to the logic equation of Practice Problem 231-1.

231-3. Illustrate the logic symbol for the equation of Practice Problem 231-1.

231-4. Figure 231-4 shows the input *A* to a NOT circuit. Determine the output's logic voltage level as it varies with time.

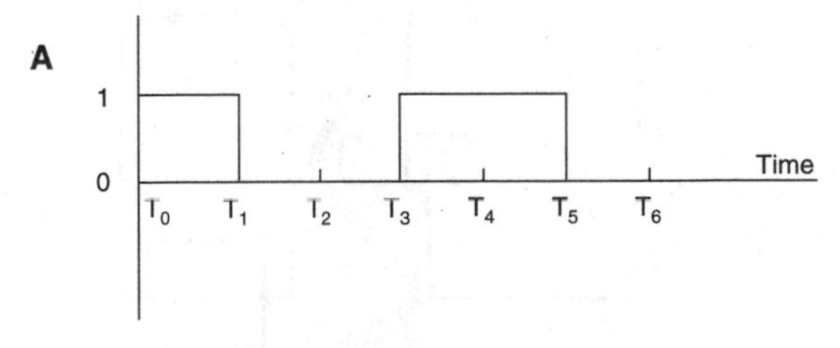

Figure 231-4

232
The NOR operation

If there are two input elements (A and B) each of which has two alternative stable states (0 and 1), there will be a total of 16 possible truth tables. Assuming that the 0,0,0,0 and 1,1,1,1 outputs have no use, there are 14 remaining tables that can be arranged in 7 pairs. In each pair, the outputs of one table will be the complements of the outputs from the other table. It follows that there must exist a truth table that is the complement (or inversion) of the table for the OR operation (chapter 229). This new truth table will refer to the NOR operation, which is a combination of the OR and NOT operations.

In Fig. 232-1A, the unshaded area of the Venn diagram represents the subclass A OR B; the shaded area is the subclass A OR B, when negated or complemented. The Boolean expression for A OR B, when negated or inverted is $f(A,B) = \overline{A \text{ OR } B} = \overline{A + B}$. DeMorgan's Theorems (chapter 228) show that $\overline{A \text{ OR } B} = \overline{A} \cdot \overline{B}$.

The truth table for the NOR operation is shown in Fig. 232-1B. This table shows that if either A or B or both A and B is in the 1 state, then the output $f(A,B)$ is in the 0 state. When both A and B are in the 0 state, the output $f(A,B)$ is in the 1 state.

The NOR equivalent switching arrangement is the result of combining the OR switching circuit and the NOT switching circuit (Fig. 232-1C). If either switch A or switch B, or both, is in the closed (1) position, the load circuit is inactive or OFF (0). However, if both of the switches A and B are open (0), the load circuit is active, or ON (1).

The logic symbol for the NOR gate is shown in Fig. 232-1D. It is a combination of the OR gate and the bubble, which indicates inversion. The NOR gate is commonly used in digital circuitry, and is equivalent to an OR gate followed by an inverter.

MATHEMATICAL DERIVATIONS

For the NOR operation, the Boolean equation is:

$$\text{Output}, f(A,B) = \overline{A + B} \qquad (232\text{-}1)$$

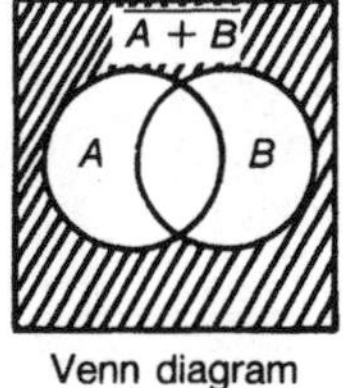

Venn diagram for NOR operation

A

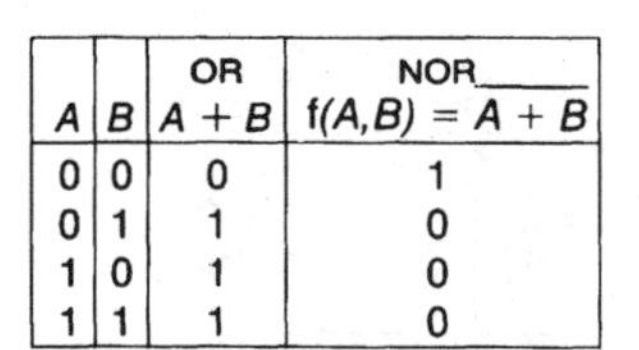

A	B	OR $A + B$	NOR $f(A,B) = \overline{A + B}$
0	0	0	1
0	1	1	0
1	0	1	0
1	1	1	0

Truth table for NOR operation

B

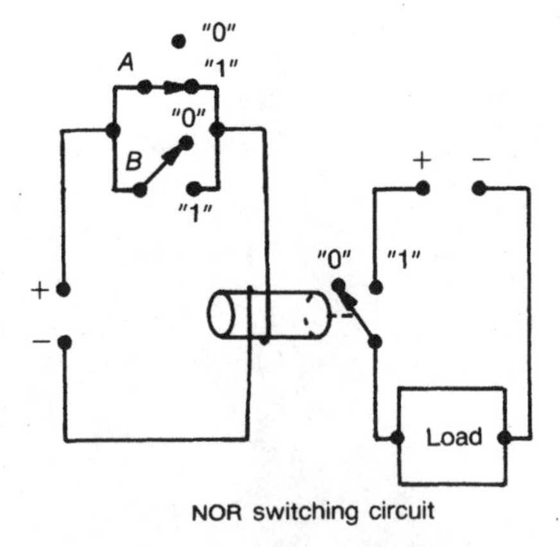

NOR switching circuit

C

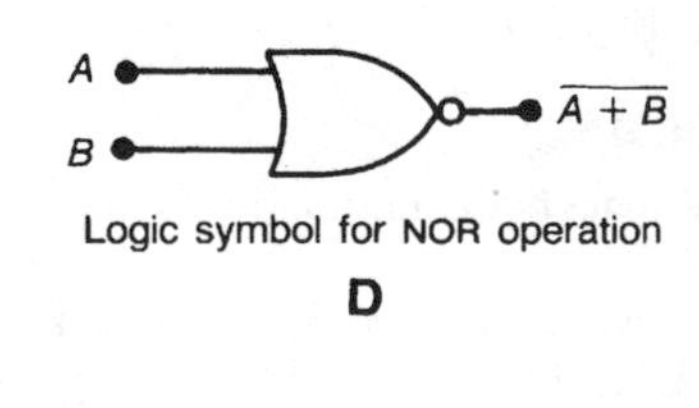

Logic symbol for NOR operation

D

Figure 232-1

Notice that $\overline{A + B}$ is not the same as $\overline{A} + \overline{B}$. For example, if $A = 1$ and $B = 0$, $\overline{A + B} = \overline{1 + 0} = \overline{1} = 0$, but $\overline{A} + \overline{B} = \overline{1} + \overline{0} = 0 + 1 = 1$.

Example 232-1

A NOR gate has three inputs, designated as A, B, and C. Determine the corresponding truth table. If the output of the NOR gate is passed to the input of a NOT circuit, obtain the Boolean expression for the inverter's output.

Solution

The truth table is as follows:

Inputs			NOR Output
A	*B*	*C*	*f*(*A*,*B*,*C*)
0	0	0	1
0	0	1	0
0	1	0	0
0	1	1	0
1	0	0	0
1	0	1	0
1	1	0	0
1	1	1	0

$$\text{NOR gate output} = \overline{A + B + C} \qquad (232\text{-}1)$$

$$\text{Inverter output} = \overline{\overline{A + B + C}} = A + B + C.$$

Example 232-2

Figure 232-2 shows the logic voltage levels of two inputs A and B to a NOR gate. Determine the logic voltage levels of the gate's output.

Solution

The output of the NOR gate is shown in Fig. 232-2. When the input elements are both in the 0 state, the output is in the 1 state. Under all other conditions, the output is in the 0 state.

PRACTICE PROBLEMS

232-1. Draw a Venn diagram to illustrate the subclass $\overline{A + B + C}$.

232-2. Write down the logic equation for the following statement:
"A group (G) contains three members whose names are Alan (A), Brian (B), and Charles (C). The group must NOT meet if A OR B OR C is present."

232.3 Illustrate the truth table and the logic symbol that correspond to the equation of Practice Problem 232-2.

232-4. Figure 232-3 shows the two inputs (A and B) to a NOR gate. Determine the output's logic voltage as it varies with time.

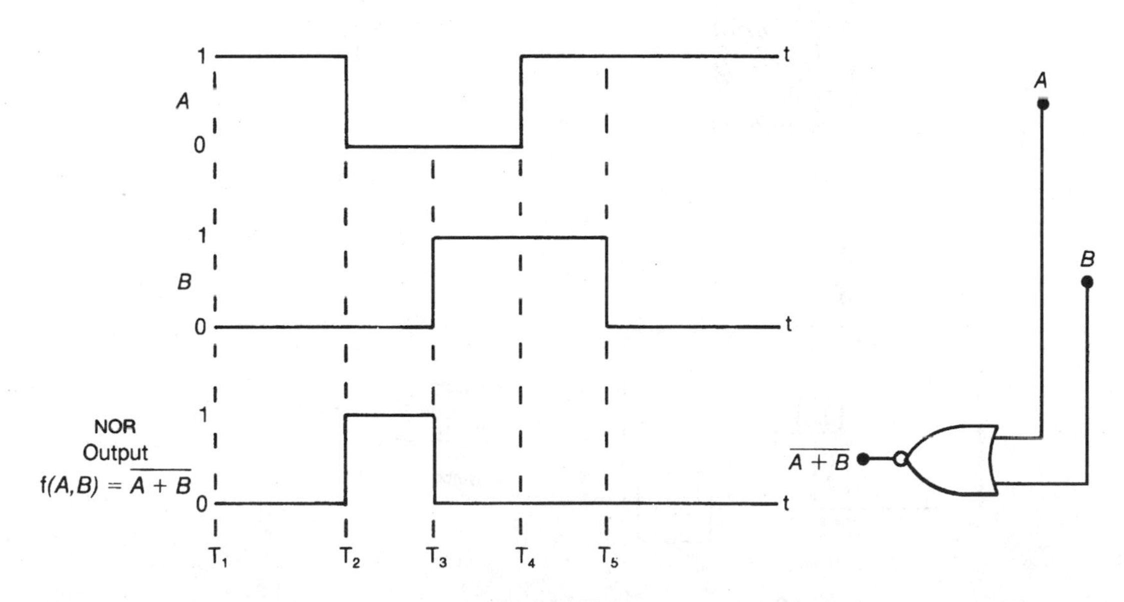

Figure 232-2

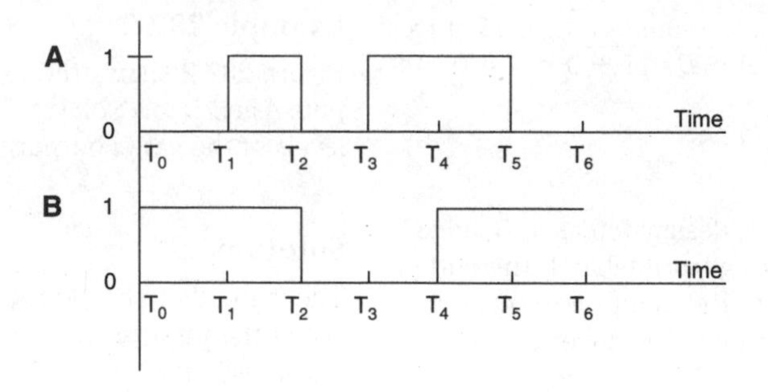

Figure 232-3

233
The NAND operation

In Fig. 233-1A the unshaded area of the Venn diagram represents the subclass *A* AND *B*. The shaded area is the subclass $\overline{A \text{ AND } B}$ when negated or complemented; this subclass, $\overline{A \text{ AND } B}$, is shown in the Venn diagrams of Fig. 228-1. The Boolean equation for *A* AND *B*, when negated or inverted, is $f(A,B) = \overline{A \cdot B}$, which is equal to $\overline{A} + \overline{B}$. (See DeMorgan's Theorem in chapter 238.)

The truth table for the NAND operation is shown in Fig. 233-1B. The table shows that if both *A* and *B* are in the 1 state, the output is in the 0 state. For all other possible combinations of the *A* and *B* states, the output is in the 1 state.

The equivalent NAND switching arrangement (Fig. 233-1C) is a combination of AND and NOT circuits. If

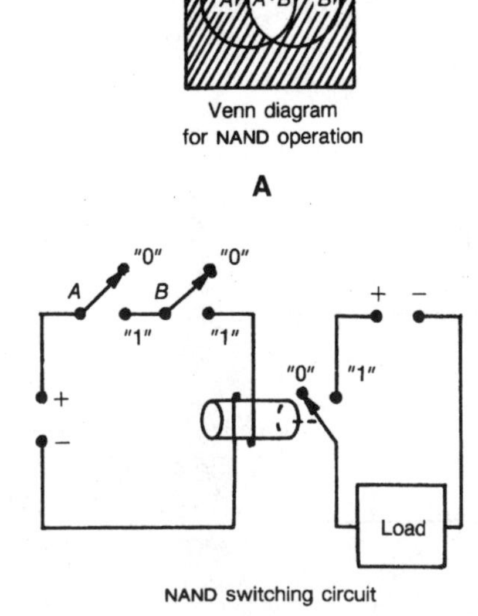

Venn diagram for NAND operation

A

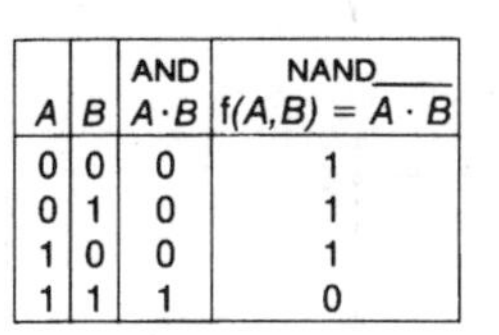

A	B	AND $A \cdot B$	NAND $f(A,B) = \overline{A \cdot B}$
0	0	0	1
0	1	0	1
1	0	0	1
1	1	1	0

Truth table for NAND operation

B

NAND switching circuit

C

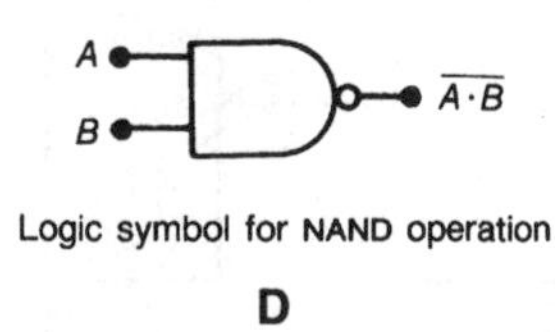

Logic symbol for NAND operation

D

Figure 233-1

either or both of the switches A and B are OPEN (0), the load circuit is ON (1). However, if both of the switches A and B are CLOSED (1), the load circuit is OFF (0).

The logic symbol for the NAND gate is shown in Fig. 233-1D. There are two input elements, A and B, into an AND gate whose output, $(A \cdot B)$ is fed into a NOT circuit; the final output, $f(A,B)$ is then $\overline{A \cdot B}$.

MATHEMATICAL DERIVATIONS

For the NAND operation, the Boolean equation is:

$$\text{Output}, f(A,B) = \overline{A \cdot B} \qquad (233\text{-}1)$$

Notice that the entire group $(A \cdot B)$ is complemented, and this is not the same as the logic product of the two elements when they are complemented separately $(\overline{A} \cdot \overline{B})$. For example, if $A = 1$ and $B = 0$, $\overline{A \cdot B} = \overline{0} = 1$, but $\overline{A} \cdot \overline{B} = 0 \cdot 1 = 0$.

Example 233-1

A NAND gate has three inputs, which are designated as A, B, and C. Determine the corresponding truth table. If the NAND gate is followed by an inverter, obtain the Boolean expression for the inverter's output.

Solution

The truth table is as follows:

Inputs			NAND Output
A	B	C	$f(A,B,C)$
0	0	0	1
0	0	1	1
0	1	0	1
0	1	1	1
1	0	0	1
1	0	1	1
1	1	0	1
1	1	1	0

$$\text{NAND gate output} = \overline{A \cdot B \cdot C} \qquad (233\text{-}1)$$

$$\text{Inverter output} = \overline{\overline{A \cdot B \cdot C}} = A \cdot B \cdot C$$

Example 233-2

Figure 233-2 shows the logic voltage levels of the input elements A and B to a NAND gate. Obtain the output waveform from the gate.

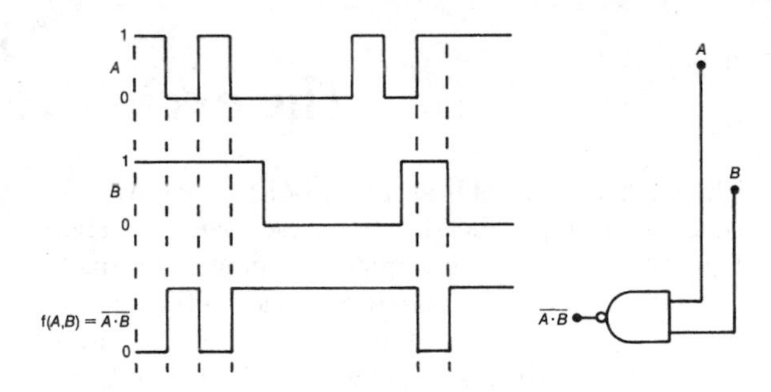

Figure 233-2

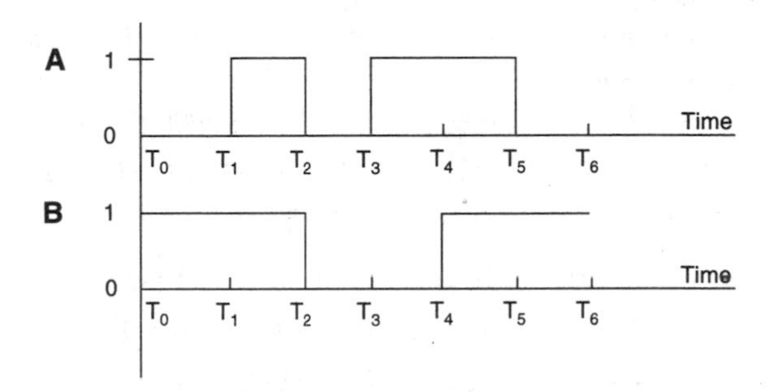

Figure 233-3

Solution

The solution is shown in Fig. 233-2. The NAND output is in the 0 state only when both of the inputs are in their 1 state. Under all other conditions, the NAND output is in the 1 state.

PRACTICE PROBLEMS

233-1. Draw a Venn diagram to illustrate the subclass $\overline{A \cdot B \cdot C}$.

233-2. Write down the logic equation for the following statement:
"A group (G) contains three members whose names are Alan (A), Brian (B), and Charles (C). The group must NOT meet if A AND B AND C are present."

233-3. Illustrate the truth table and the logic symbol that correspond to the equation of Practice Problem 233-2.

233-4. Figure 233-3 shows the two inputs to a NAND gate. Determine the output's logic voltage as it varies with time.

234
The exclusive-OR operation

The exclusive-OR (abbreviated EX-OR) operation is actually a special application of the OR operation (chapter 229) and appears quite frequently in digital circuitry. In the EX-OR operation, either of the inputs A or B must be in the 1 state in order for the output, $f(A,B)$ to be in the 1 state; however, if both A and B are in the 1 state or in the 0 state at the same time, the output is in the 0 state.

The Venn diagram for the EX-OR operation is shown in Fig. 234-1A. This diagram does not appear in any of those shown in Fig. 228-1; therefore, it is a new subclass.

The Boolean equation for the EX-OR operation is:

$$f(A,B) = \overline{A} \cdot B + A \cdot \overline{B} = A \oplus B \qquad (234\text{-}1)$$

If,

$$A = 0, B = 0,\ f(A,B) = 1 \cdot 0 + 0 \cdot 1 = 0$$
$$A = 0, B = 1,\ f(A,B) = 1 \cdot 1 + 0 \cdot 0 = 1$$
$$A = 1, B = 0,\ f(A,B) = 0 \cdot 0 + 1 \cdot 1 = 1$$
$$A = 1, B = 1,\ f(A,B) = 0 \cdot 1 + 1 \cdot 0 = 0$$

These results are shown in the truth table of Fig. 234-1B. Notice that the EX-OR gate always has only two inputs. There are no EX-OR gates with three or more inputs.

In the equivalent switching circuit (Fig. 234-1C) for an EX-OR gate, the two switches are mechanically linked together so that one or the other, but not both, can be closed at any particular time.

The EX-OR gate actually consists of inverters, AND gates, and an OR gate. The required combination of these logic circuits is shown in Fig. 234-2. However, the EX-OR gate is used sufficiently often to be given its own symbol (Fig. 234-1D).

MATHEMATICAL DERIVATIONS

For the EX-OR operation,

$$f(A,B) = \overline{A} \cdot B + A \cdot \overline{B} \qquad (234\text{-}1)$$
$$= A \oplus B$$

The $\oplus$ symbol is used to represent the EX-OR gate operation.

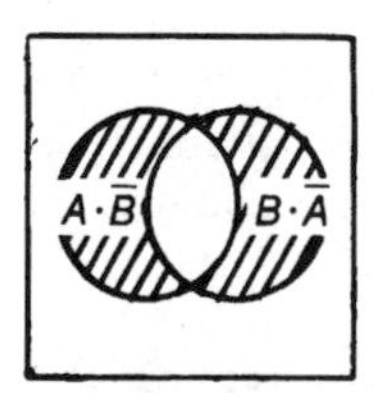

Venn diagram for EX-OR operation

A

Input elements		EX-OR output
A	B	$f(A,B) = A \cdot \overline{B} + B \cdot \overline{A}$
0	0	0
0	1	1
1	0	1
1	1	0

Truth table for EX-OR operation

B

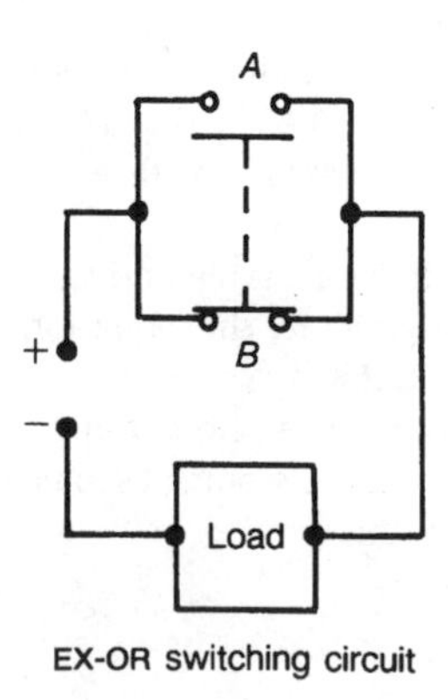

EX-OR switching circuit

C

A
B
$A \cdot \overline{B} + B \cdot \overline{A} = A \oplus B$

Logic symbol for EX-OR operation

D

Figure 234-1

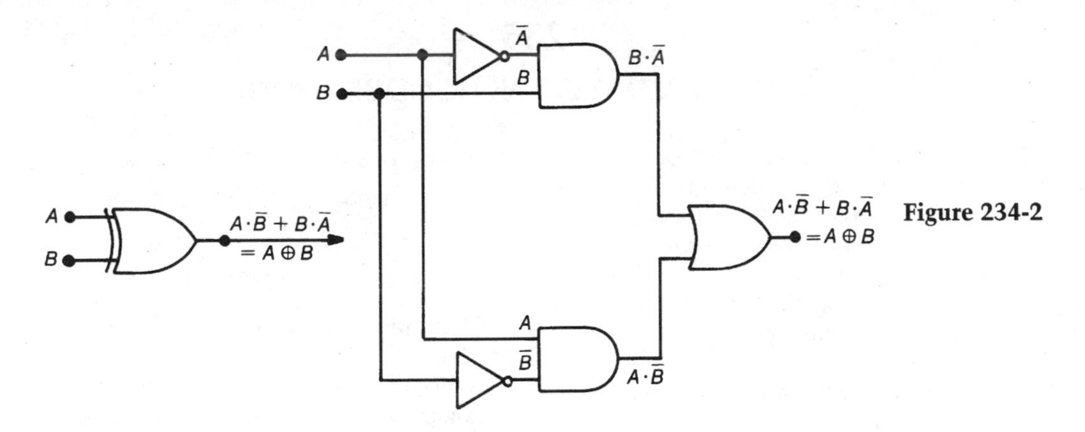

Figure 234-2

To summarize the EX-OR gate operation:

1. The output is only in the 1 state when the two input elements are in different states.
2. If the two input elements are in the same state, the output is in the 0 state.

Example 234-1

Figure 234-3 shows the logic voltage levels for the two inputs to an EX-OR gate. Obtain the waveform for the logic output.

Solution

The solution appears in Fig. 234-3. Notice that when B is in the 1 state, the output waveform is the inverse of the A waveform, but when B is in the 0 state, the A waveform and output waveform are the same.

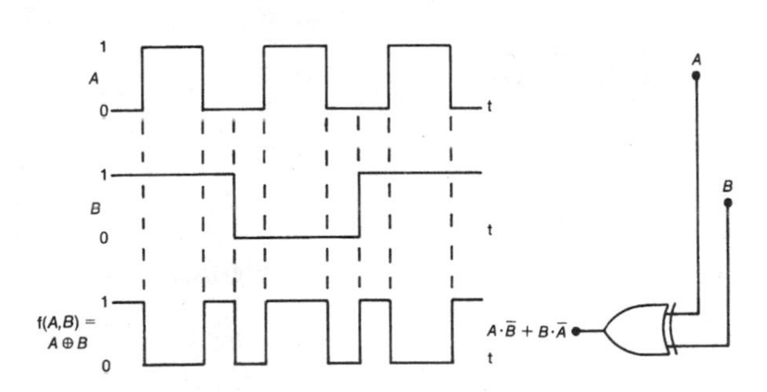

Figure 234-3

PRACTICE PROBLEMS

234-1. Write down the logic equation for the following statement:
"A group (G) contains two members whose names are Alan (A) and Brian (B). The group meets if A is present AND B is absent OR B is present AND A is absent."

234-2. Figure 234-4 shows the two inputs to an exclusive OR gate. Determine the output's logic voltage as it varies with time.

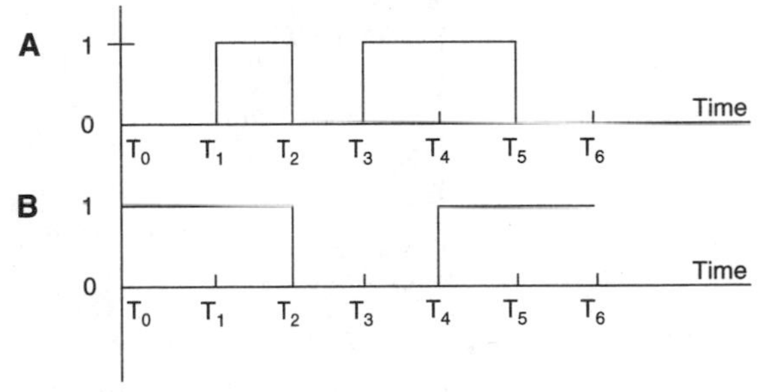

Figure 234-4

235
The exclusive-NOR operation

The exclusive-NOR (abbreviated EX-NOR) operation is the least common of all the operations discussed. It is merely the inverse of the EX-OR operation; consequently, an EX-NOR circuit can be formed by following an EX-OR gate with an inverter.

In the EX-NOR operation both of the input elements (A and B) must be in the same state (either 0 or 1) in order for the output $f(A,B)$ to be in the 1 state; however, if A and B are in opposite states at the same time, the output is in the 0 state.

The Venn diagram for the EX-NOR operation is shown in Fig. 235-1A. This diagram does not appear in any of those shown in Fig. 228-1, and is, therefore, a new subclass.

The Boolean equation for the EX-NOR operation is:

$$f(A,B) = A \cdot B + \overline{A} \cdot \overline{B} = \overline{A \oplus B} \quad (235\text{-}1)$$

Notice that this logic expression is not the same as $A \cdot B + \overline{A \cdot B}$, which must always equal 1.

If,

$$A = 0, B = 0,\ f(A,B) = 0 \cdot 0 + 1 \cdot 1 = 1$$

$$A = 0, B = 1,\ f(A,B) = 0 \cdot 1 + 1 \cdot 0 = 0$$

$$A = 1, B = 0,\ f(A,B) = 1 \cdot 0 + 0 \cdot 1 = 0$$

$$A = 1, B = 1,\ f(A,B) = 1 \cdot 1 + 0 \cdot 0 = 1$$

These results are shown in the truth table of Fig. 235-1B.

Notice that, like the EX-OR gate, the EX-NOR gate always has only two inputs. There are no EX-NOR gates with three or more inputs.

The equivalent switching circuit (Fig. 235-1C) for the EX-NOR gate is a combination of the switching circuit for the EX-OR gate and the inverter. The two switches (A and B) are mechanically linked so that one or the other, but not both, can be closed. If either A or B is closed (1), the switch in the load circuit is open so that the load circuit is OFF (0).

The EX-NOR gate is actually composed of inverters, AND gates, and an OR gate. The required combination

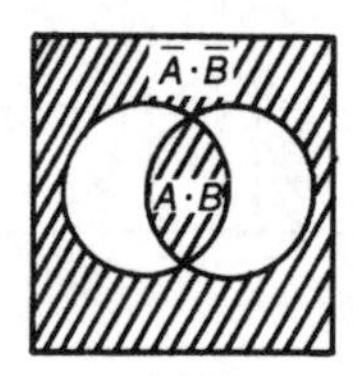

Venn diagram for EX-NOR operation

A

Input elements		EX-NOR output
A	B	$f(A,B) = A \cdot B + \overline{A} \cdot \overline{B}$
0	0	1
0	1	0
1	0	0
1	1	1

Truth table for EX-NOR operation

B

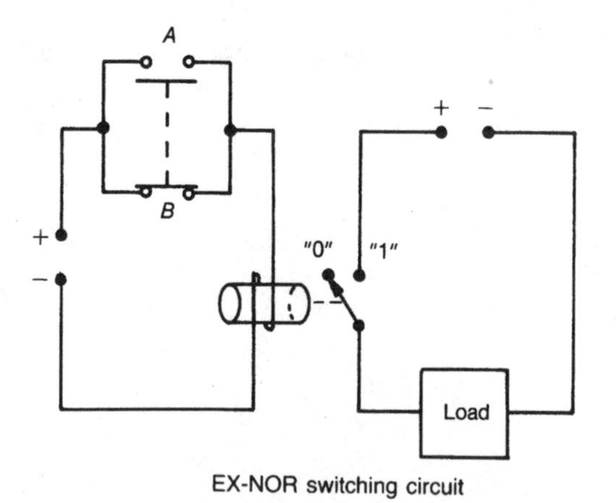

EX-NOR switching circuit

C

A, B → $A \cdot B + \overline{A} \cdot \overline{B} = \overline{A \oplus B}$

Logic symbol for EX-NOR operation

D

Figure 235-1

of these logic circuits is shown in Fig. 235-2. The corresponding symbol for the EX-NOR gate appears in Fig. 235-1D.

MATHEMATICAL DERIVATIONS

For the EX-NOR operation,

$$f(A,B) = A \cdot B + \overline{A} \cdot \overline{B}$$

This can be abbreviated to:

$$f(A,B) = \overline{A \oplus B} \qquad (235\text{-}1)$$

To summarize the EX-NOR operation,

1. The output is only in the 1 state when the two input elements are in the same state.
2. If the two input elements are in opposite states, the output is in the 0 state.

Example 235-1

Figure 235-3 shows the logic voltage levels for the two inputs to an EX-NOR gate. Obtain the waveform for the logic output.

Solution

The solution appears in Fig. 235-3. Notice that when B is in the 0 state, the output waveform is the inverse of the A waveform; however, when B is in the 1 state, the A waveform and the output waveform are the same.

PRACTICE PROBLEMS

235-1. Write down the logic equation for the following statement:
"A group (G) contains two members whose names are Alan (A) and Brian (B). The group meets if A is present AND B is present OR A is absent AND B is absent."

235-2. Figure 235-4 shows the two inputs to an exclusive NOR gate. Determine the output's logic voltage as it varies with time.

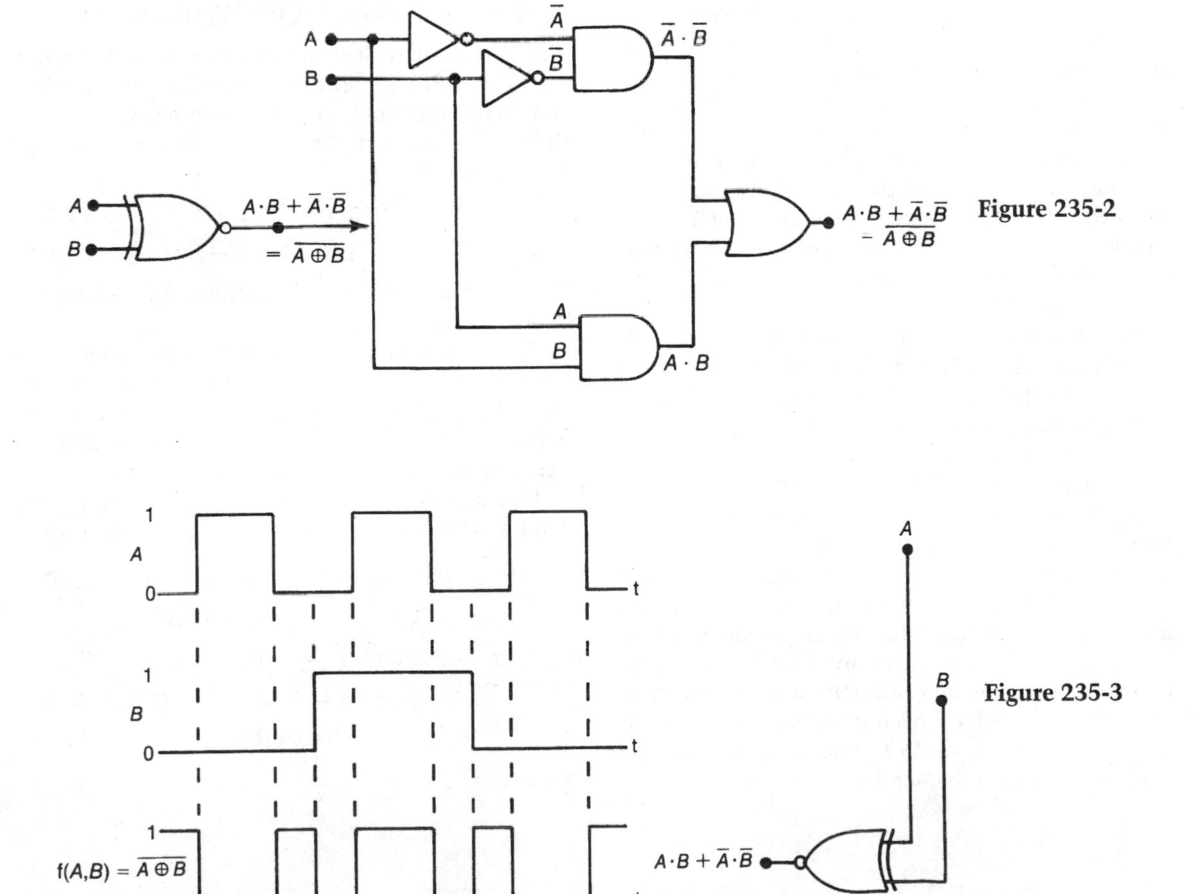

Figure 235-2

Figure 235-3

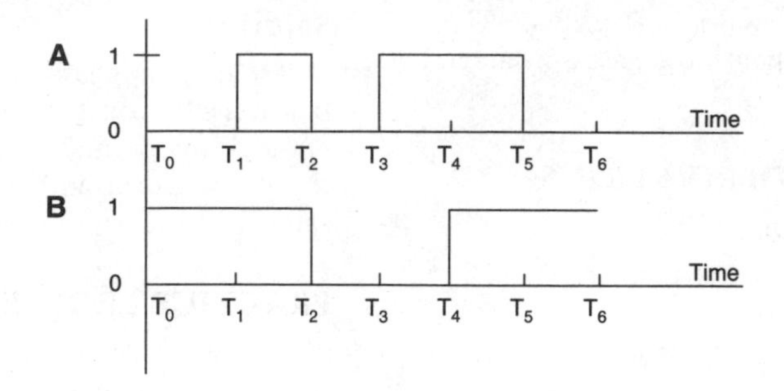

Figure 235-4

236 Combinatorial logic circuits

In chapters 229 through 235, you have looked at individual logic operations, each with its own Venn diagram, Boolean equation, and truth table. You can now investigate the result of combining inverters with the various gates to produce combinatorial logic circuits. In each such circuit, you will derive an expression for the Boolean output, and then obtain its truth table. However, the output expression might not be in its simplest form, so that you will need the axioms, laws, and theorems of Boolean algebra, as outlined in chapters 237 and 238.

Simplification is important because you might be given the required expression for the output, and then be asked to design the corresponding logic circuit. If the expression can be simplified, you can achieve the most economical solution by creating the necessary circuit with the minimum number of gates.

In the mathematical derivations, you will summarize the individual logic operations, and look at alternative symbols for these operations. In one set of alternative symbols, the input elements and the output are inverted, while the AND and OR operations are interchanged. The inversions are signified by adding bubbles to the input and output lines. Because both inputs and outputs have been inverted, you can regard these new symbols as DeMorgan equivalents of the standard symbols (chapter 238).

MATHEMATICAL DERIVATIONS

The most common logic operations are summarized in Fig. 236-1. For a particular operation, the standard and DeMorgan equivalent symbols show the same truth table. As a result, the following relationships must exist:

$$\text{NOR}\ \overline{A + B} = \overline{A} \cdot \overline{B} \qquad (236\text{-}1)$$

$$\text{NAND}\ \overline{A \cdot B} = \overline{A} + \overline{B} \qquad (236\text{-}2)$$

These results will be demonstrated by DeMorgan's Theorem (chapter 238).

The rectangular symbols were first approved in 1984 and are now used by some digital IC manufacturers and the military. The signs "1, &, ≥ 1" respectively indicate the NOT, AND, and OR operations, and the right-angled triangle "◺" represents inversion.

Consider the combinatorial logic circuit of Fig. 236-2A. The Boolean expression for the output is:

$$f(A,B,C) = \overline{(A \cdot B + \overline{C})} \cdot A \cdot C \qquad (236\text{-}1)$$

$$= A \cdot C \cdot \overline{\overline{C}} \cdot \overline{A \cdot B}$$

But $\overline{\overline{C}} = C$ and $C \cdot C = C$. Therefore,

$$f(A,B,C) = A \cdot C \cdot (\overline{A} + \overline{B}) \qquad (236\text{-}2)$$

$$= A \cdot \overline{B} \cdot C$$

because $A \cdot \overline{A} = 0$.

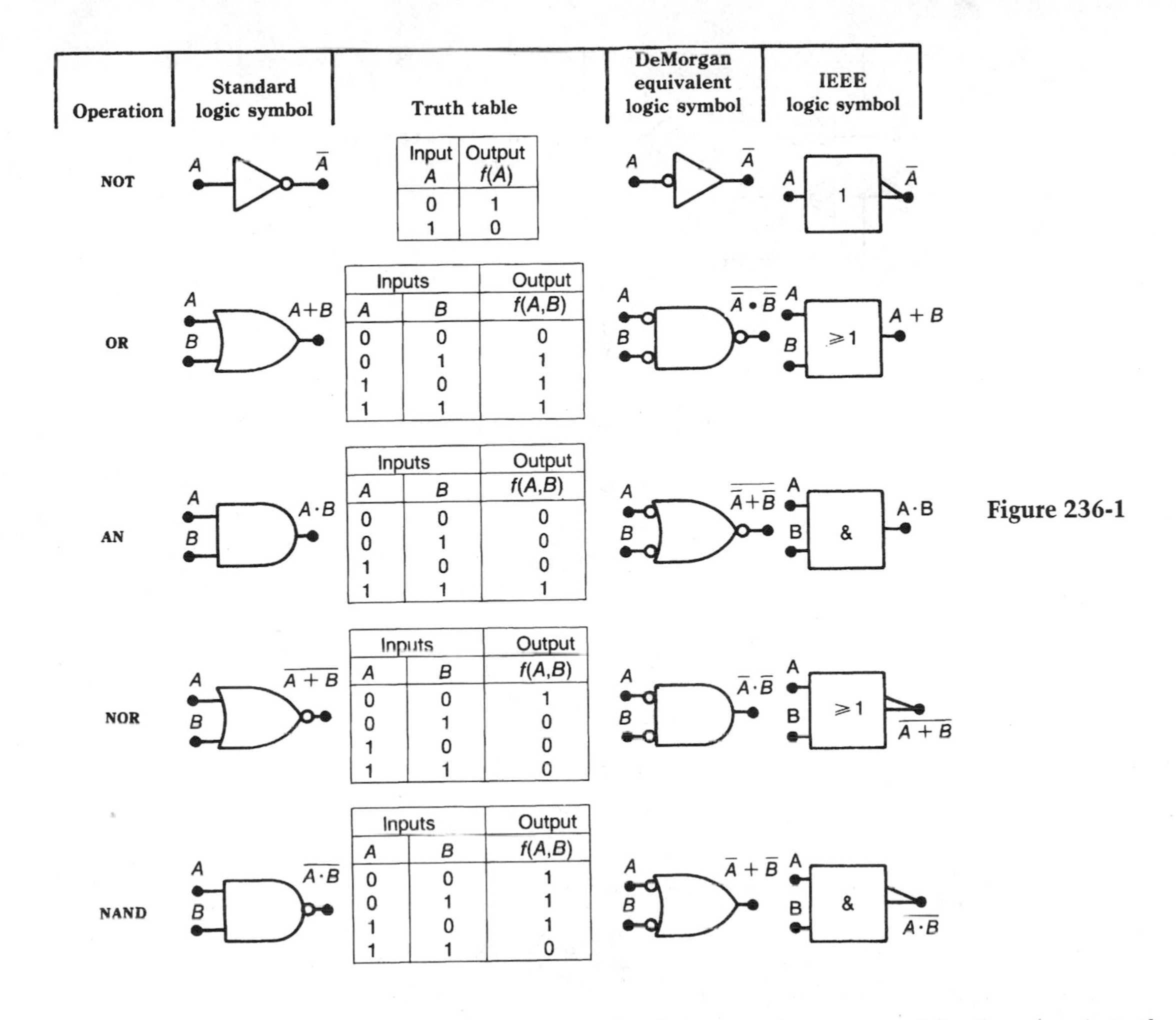

Figure 236-1

Consequently, the logic circuit with its three gates and one inverter can be simplified to the circuit of Fig. 236-2B. The corresponding truth table is shown in Fig. 236-2C.

Example 236-1

Obtain the Boolean expression for the output of the combinatorial logic circuit of Fig. 236-3A. Derive the corresponding truth table.

Solution

$$\text{Output, } f(A,B,C) = \overline{A}\cdot C\cdot(\overline{A+B}) \quad (236\text{-}1)$$
$$= \overline{A}\cdot C\cdot\overline{A}\cdot\overline{B}$$
$$= \overline{A}\cdot\overline{B}\cdot C$$

because $\overline{A}\cdot\overline{A}=\overline{A}$.

The simplified logic circuit and its corresponding truth table are respectively shown in Figs. 236-3B and C. Notice that the truth table can be derived by applying, in turn, each sequence of the three inputs to the original logic circuit. (See Fig. 236-3A for the inputs $A = 0, B = 0, C = 1$.)

Example 236-2

Develop a logic circuit for which the Boolean output expression is $f(A,B,C) = \overline{(\overline{A+B})\cdot C}$. Derive the corresponding truth table.

Solution

The NOR gate provides an output of $\overline{A+B}$ (see Fig. 236-1). The NOR gate output is then one input to a NAND gate, whose other input is C. The required logic circuit is shown in Fig. 236-4A.

Notice that:

$$f(A,B,C) = \overline{(\overline{A+B})\cdot C} \quad (236\text{-}1)$$
$$= \overline{\overline{A}\cdot\overline{B}\cdot C}$$
$$= \overline{\overline{A}} + \overline{\overline{B}} + \overline{C} = A + B + \overline{C} \quad (236\text{-}2)$$

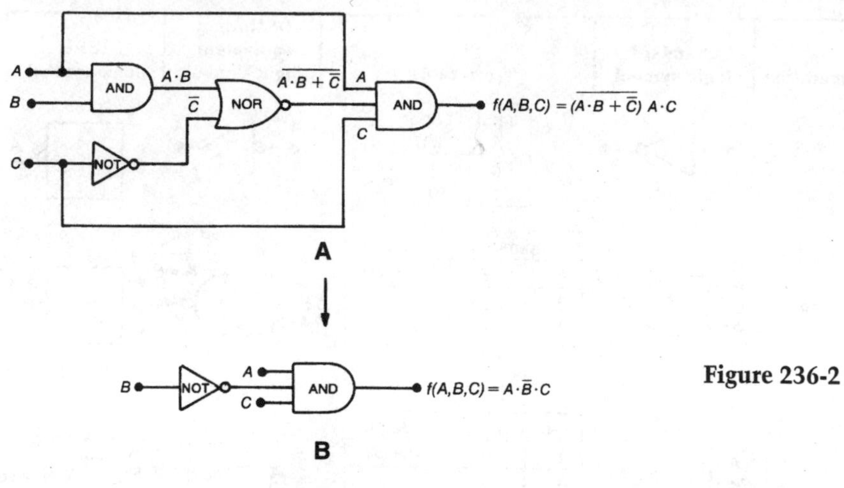

A

B NOT A C AND $f(A,B,C) = A \cdot \overline{B} \cdot C$

B

Figure 236-2

Inputs			Output
A	B	C	$f(A,B,C) = A \cdot \overline{B} \cdot C$
0	0	0	0
0	0	1	0
0	1	0	0
0	1	1	0
1	0	0	0
1	0	1	1
1	1	0	0
1	1	1	0

C

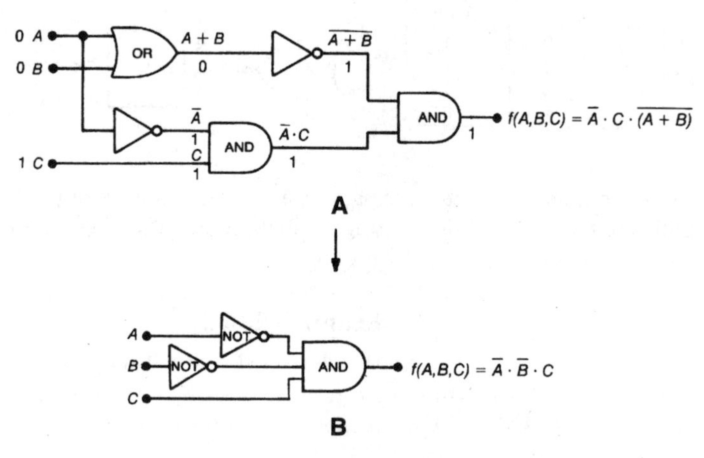

A

B

Figure 236-3

Inputs			Output
A	B	C	$f(A,B,C)$
0	0	0	0
0	0	1	1
0	1	0	0
0	1	1	0
1	0	0	0
1	0	1	0
1	1	0	0
1	1	1	0

C

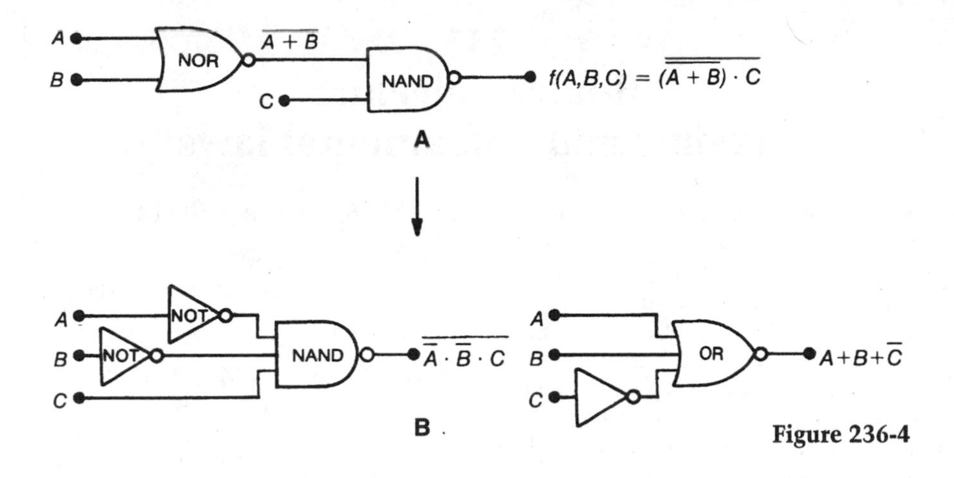

Figure 236-4

Inputs			Output
A	B	C	f(A,B,C)
0	0	0	1
0	0	1	0
0	1	0	1
0	1	1	1
1	0	0	1
1	0	1	1
1	1	0	1
1	1	1	1

C

The alternative logic circuits are shown in Fig. 236-4B and the corresponding truth table appears in Fig. 236-4C.

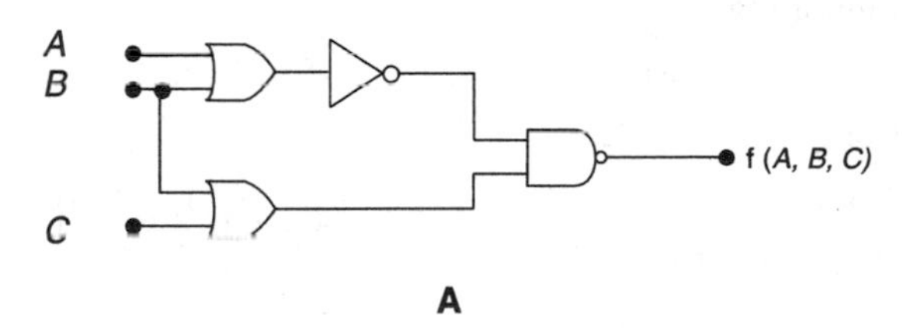

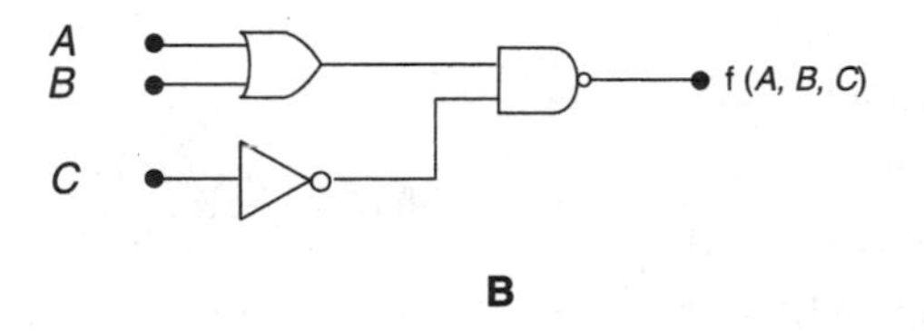

Figure 236-5

PRACTICE PROBLEMS

236-1. Draw a logic diagram to represent the equation $f(A,B,C) = \overline{B} + C + A \cdot B \cdot C$. Construct the corresponding truth table.

236-2. Simplify the equation of Practice Problem 236-1 and draw the new corresponding logic diagram.

236-3. Write down the Boolean equations for the logic circuits of Figs. 236-5A, B. Construct the corresponding truth tables.

236-4. Verify that the equation $f(A,B,C) = A + B + \overline{B} \cdot \overline{C}$ has the same truth table as that of Fig. 236-5A. Draw the corresponding logic circuit.

236-5. Verify that the equation $f(A,B,C) = \overline{A} \cdot \overline{B} + C$ has the same truth table as that of Fig. 236-5B. Draw the corresponding logic circuit.

237
Boolean algebra: axioms and fundamental laws

This chapter lists the fundamental laws and axioms of Boolean algebra. You should memorize these relationships because they are used to simplify logic expressions. Although these laws can change any expression, it is sometimes difficult to decide whether the result is in its simplest form. To overcome this difficulty, other methods of simplification—the Karnaugh map and the Harvard chart—are covered in chapters 239 and 240. For each axiom and law there is a corresponding switching circuit, logic diagram, and truth table.

MATHEMATICAL DERIVATIONS

Axioms

$$A + 0 = A \qquad (237\text{-}1)$$

$$A \cdot 0 = 0 \qquad (237\text{-}2)$$

$$A + 1 = 1 \qquad (237\text{-}3)$$

$$A \cdot 1 = A \qquad (237\text{-}4)$$

These axioms are true whether A is 0 or 1. They are illustrated in Figs. 237-1, 2, 3, and 4.

Laws

Identity Law

$$A = A \qquad \text{\{Fig. 237-5\}} \; (237\text{-}5)$$

Complementary Law

(a) $A \cdot \overline{A} = 0$ (237-6)

(b) $A + \overline{A} = 1$ (237-7)

{Fig. 237-6}

Idempotent Law

(a) $A \cdot A = A$ (237-8)

(b) $A + A = A$ (237-9)

{Fig. 237-7}

Commutative Law

(a) $A \cdot B = B \cdot A$ (237-10)

(b) $A + B = B + A$ (237-11)

{Fig. 237-8}

Associative Law

(a) $(A \cdot B) \cdot C = A \cdot (B \cdot C)$ (237-12)

(b) $(A + B) + C = A + (B + C)$ (237-13)

{Fig. 237-9}

Distributive Law

(a) $A \cdot (B + C) = (A \cdot B) + (A \cdot C)$ (237-14)

{Fig. 237-10}

(b) $A + (B \cdot C) = (A + B) \cdot (A + C)$ (237-15)

Double Negation Law

$$\overline{\overline{A}} = A \qquad \text{\{Fig. 237-11\}} \; (237\text{-}16)$$

Absorption Law

(a) $A \cdot (A + B) = A \cdot A + A \cdot B$ (237-17)

$= A + A \cdot B$

$= A \cdot (1 + B)$

$= A \cdot 1 = A$

{Fig. 237-12}

(b) $A + A \cdot B = A(1 + B)$ (237-18)

$= A \cdot 1$

$= A$

(c) $A + \overline{A} \cdot B = A + A \cdot B + \overline{A} \cdot B$ (237-19)

$= A + B \cdot (A + \overline{A})$

$= A + B$

Example 237-1

Simplify the Boolean equation $f(A,B,C,D) = A \cdot C + A \cdot D + B \cdot C + B \cdot D$. Derive the logic circuits for the original and the simplified expressions.

Solution

$$f(A, B, C, D) = A \cdot C + A \cdot D + B \cdot C + B \cdot D \qquad (237\text{-}14)$$

$$= A \cdot (C + D) + B \cdot (C + D)$$

$$= (A + B) \cdot (C + D)$$

The logic circuits for the original and simplified expressions are shown in Figs. 237-13A and B.

Example 237-2

Simplify the Boolean equation $f(A,B,C,D) = A \cdot B \cdot C + A \cdot B \cdot \overline{D} + A \cdot \overline{C} + \overline{A} \cdot \overline{B} \cdot \overline{C} \cdot \overline{D} + \overline{A} \cdot C$. Derive the logic circuits for the original and simplified expressions.

Solution

The logic circuit for the original equation appears in Fig. 237-14A.

$$f(A,B,C,D) = A \cdot B \cdot C + A \cdot B \cdot \overline{D}$$

$$+ A \cdot \overline{C} + \overline{A} \cdot \overline{B} \cdot \overline{C} \cdot \overline{D} + \overline{A} \cdot C$$

$$= A \cdot B \cdot C + A \cdot \overline{C} \qquad (237\text{-}11)$$

$$+\overline{A}\cdot\overline{B}\cdot\overline{C}\cdot\overline{D}+\overline{A}\cdot C+A\cdot B\cdot\overline{D}$$

$$=A\cdot(B\cdot C+\overline{C}) \quad (237\text{-}14)$$

$$+\overline{A}\cdot(\overline{B}\cdot\overline{C}\cdot\overline{D}+C)+A\cdot B\cdot\overline{D}$$

$$=A\cdot(B+\overline{C})+\overline{A}\cdot(\overline{B}\cdot\overline{D}+C) \quad (237\text{-}6)$$

$$+A\cdot B\cdot\overline{D}$$

$$=A\cdot B+A\cdot\overline{C}+\overline{A}\cdot\overline{B}\cdot\overline{D} \quad (237\text{-}14)$$

$$+\overline{A}\cdot C+A\cdot B\cdot\overline{D}$$

$$=(A\cdot B+A\cdot B\cdot\overline{D})+A\cdot\overline{C}$$

$$+\overline{A}\cdot C+\overline{A}\cdot\overline{B}\cdot\overline{D}$$

$$=A\cdot B+A\cdot\overline{C}+\overline{A}\cdot C \quad (237\text{-}18)$$

$$+\overline{A}\cdot\overline{B}\cdot\overline{D}$$

This form does, in fact, produce the simplest logic circuit (Fig. 237-14B). However, the simplification process for the equation can be carried one step further by factoring so that,

$$f(A,B,C,D)=A\cdot(B+\overline{C})+\overline{A}\cdot(C+\overline{B}\cdot\overline{D}) \quad (237\text{-}14)$$

The logic circuit for this equation requires six gates, as does the logic circuit for the original equation. The simplest logic circuit of Fig. 237-14B needs only five gates.

This example of simplification shows that the process is rather difficult at first, with no positive indication that the equation for the simplest logic circuit has been reached. Skill in making the correct choice can only be acquired by repeated use of the Boolean laws and axioms.

PRACTICE PROBLEMS

237-1. Simplify the Boolean equation: $f(A,B) = A\cdot A + A\cdot B + A\cdot\overline{B} + A\cdot\overline{A}$.

237-2. Simplify the Boolean equation: $f(A,B) = \overline{B}\cdot\overline{B} + \overline{B}\cdot B + \overline{B}\cdot\overline{A} + \overline{B}\cdot\overline{B}$.

237-3. Simplify the Boolean equation: $f(A,B) = A + B + A\cdot B + B\cdot C$. Draw the logic circuits for the original and simplified expressions.

237-4. Simplify the Boolean equation: $f(A,B) = A\cdot B + A\cdot C + A\cdot C + \overline{A\cdot C}$.

237-5. Simplify the Boolean equation: $f(A,B,C) = (A + B\cdot\overline{C})\cdot(\overline{A} + \overline{B} + \overline{C})$. Draw the logic circuits for the original and simplified equations.

237-6. Simplify the Boolean equation: $f(A,B,C) = (\overline{A} + B + \overline{B})\cdot(\overline{A} + \overline{B} + \overline{C})\cdot(A + \overline{B} + \overline{A})$.

237-7. Simplify the following Boolean equation: $f(A,B,C,D) = (\overline{A} + B + \overline{C})\cdot(\overline{A} + \overline{B} + \overline{C})\cdot(\overline{A} + D + \overline{C})\cdot(\overline{A} + \overline{D} + \overline{C})$.

237-8. Simplify the following Boolean equation: $f(A,B,C) = (\overline{A} + \overline{B})\cdot(\overline{A} + C)$.

237-9. Simplify the following Boolean equation: $f(A,B,C,D) = B\cdot C\cdot D + C\cdot D + C + A\cdot C\cdot D + A\cdot C + B\cdot C$.

237-10. Simplify the following Boolean equation: $f(A,B,C,D) = B + \overline{A}\cdot B + C\cdot D + A\cdot B + \overline{A}\cdot C\cdot D + D$.

237-11. Simplify the following Boolean equation: $f(A,B,C,D) = A\cdot B\cdot C\cdot(A + \overline{D})\cdot(C + \overline{D})$.

237-12. Simplify the following Boolean equation: $f(A,B,C,D) = (C + A)\cdot(C + A + \overline{B})\cdot(C + \overline{D})\cdot C$.

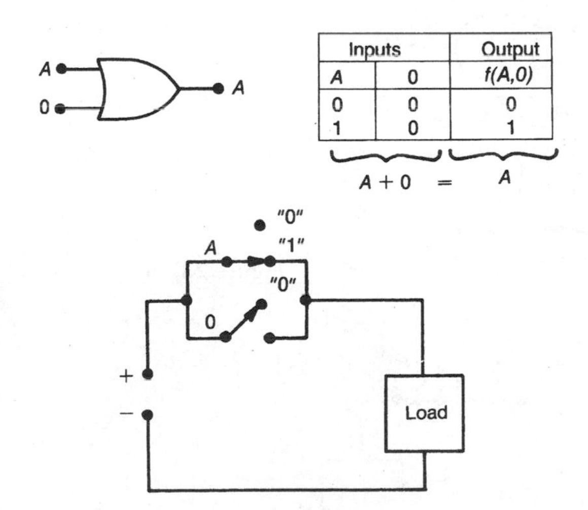

Figure 237-1

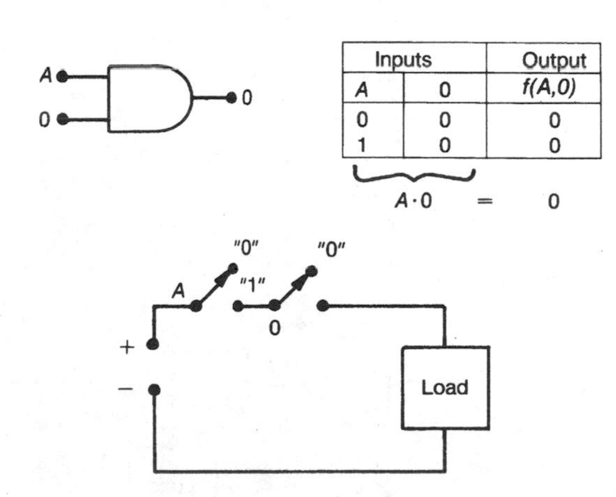

Figure 237-2

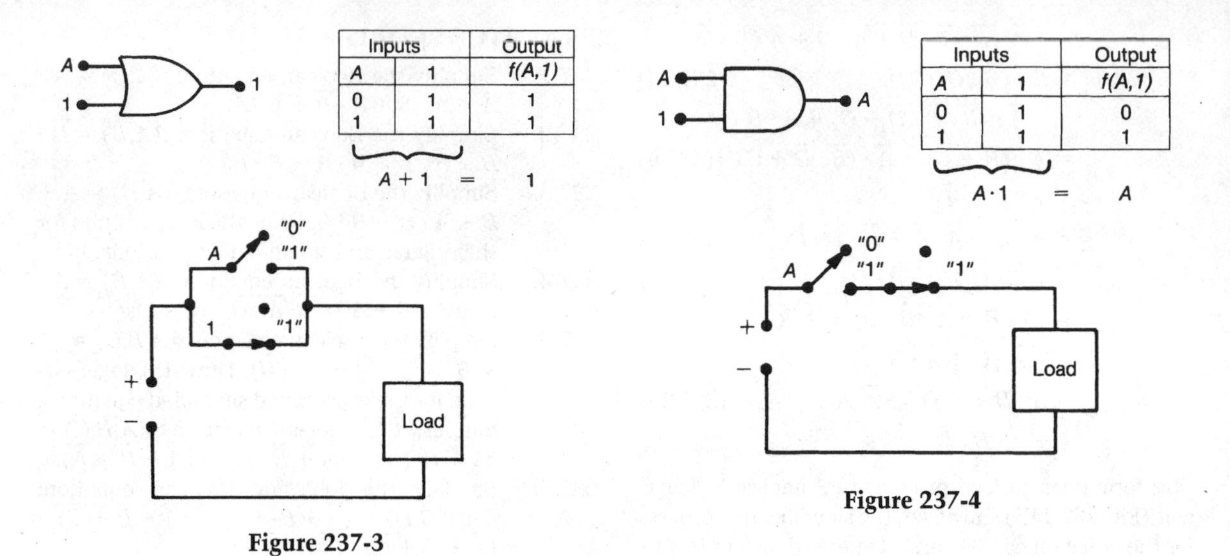

Figure 237-3

Figure 237-4

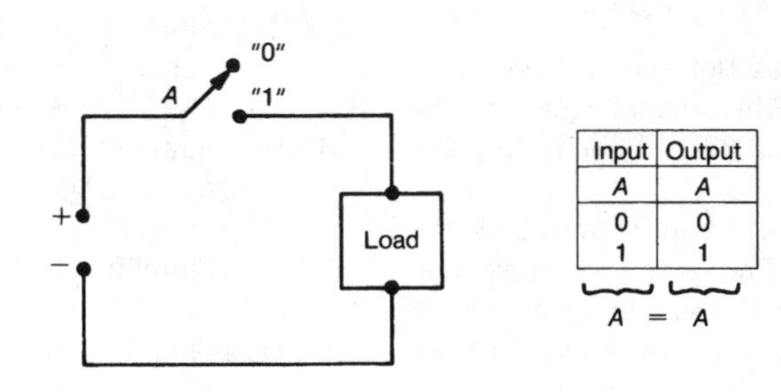

Figure 237-5

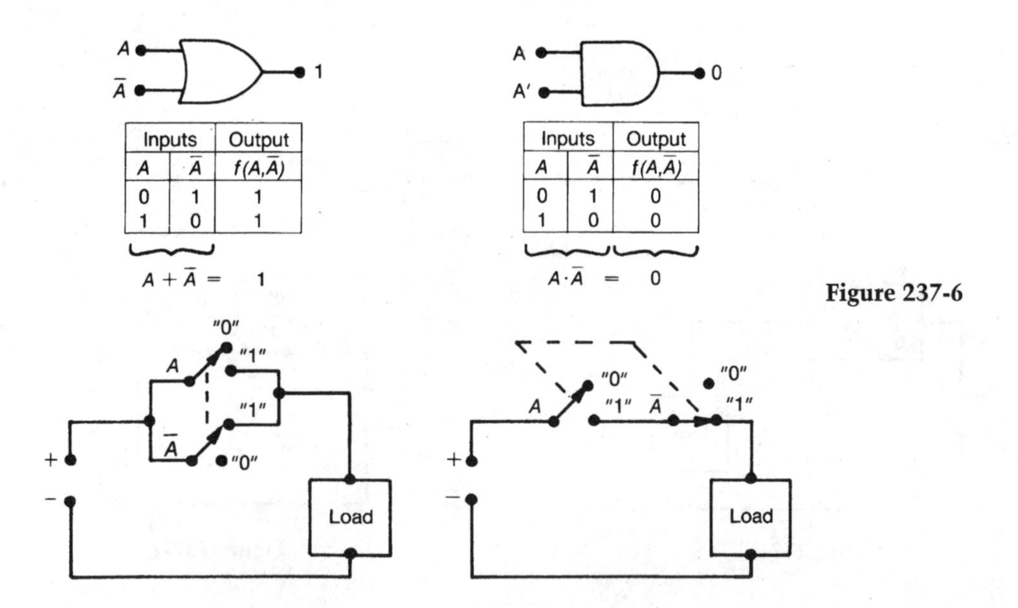

Figure 237-6

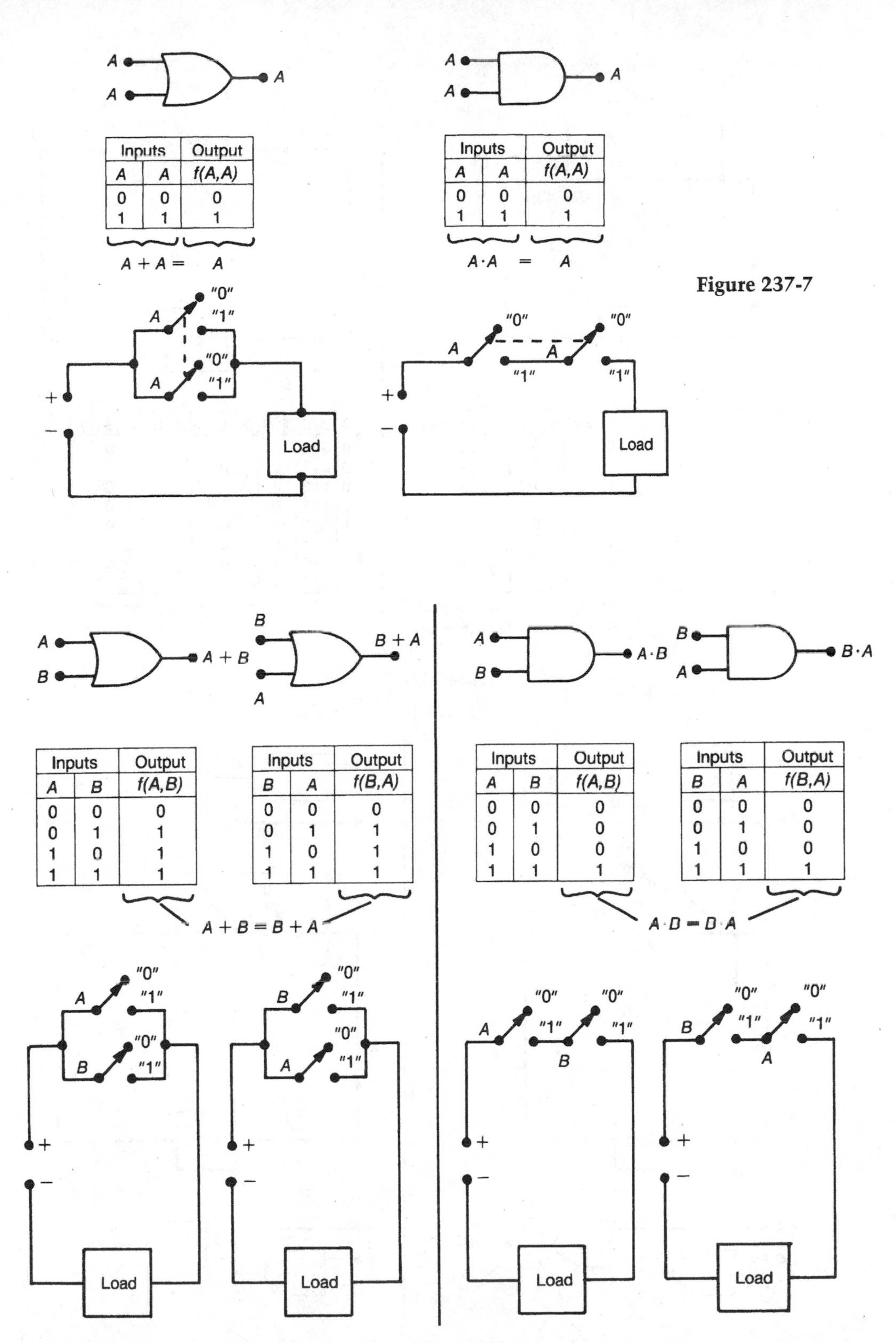

Figure 237-7

Figure 237-8

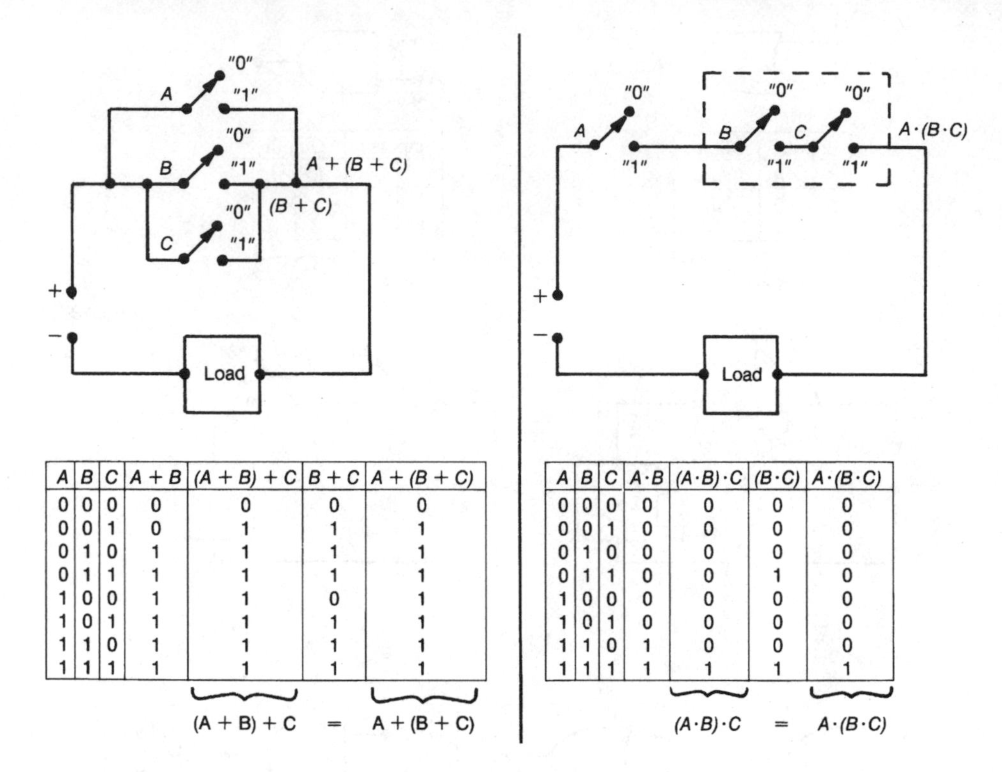

A	B	C	A + B	(A + B) + C	B + C	A + (B + C)
0	0	0	0	0	0	0
0	0	1	0	1	1	1
0	1	0	1	1	1	1
0	1	1	1	1	1	1
1	0	0	1	1	0	1
1	0	1	1	1	1	1
1	1	0	1	1	1	1
1	1	1	1	1	1	1

A	B	C	A·B	(A·B)·C	(B·C)	A·(B·C)
0	0	0	0	0	0	0
0	0	1	0	0	0	0
0	1	0	0	0	0	0
0	1	1	0	0	1	0
1	0	0	0	0	0	0
1	0	1	0	0	0	0
1	1	0	1	0	0	0
1	1	1	1	1	1	1

Figure 237-9A

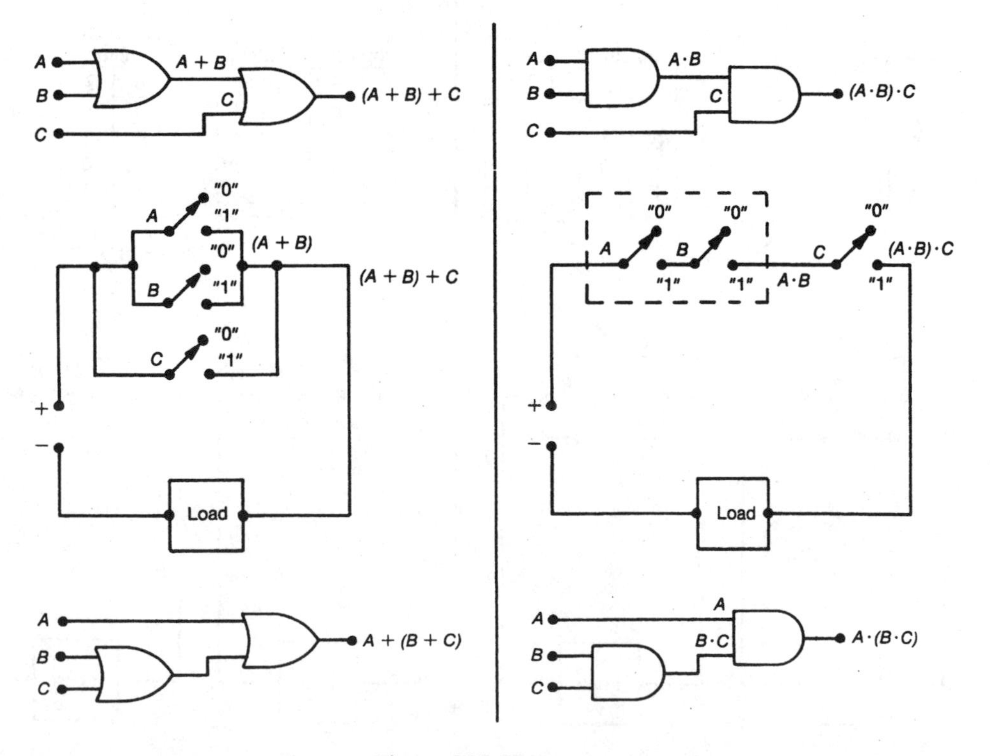

Figure 237-9B

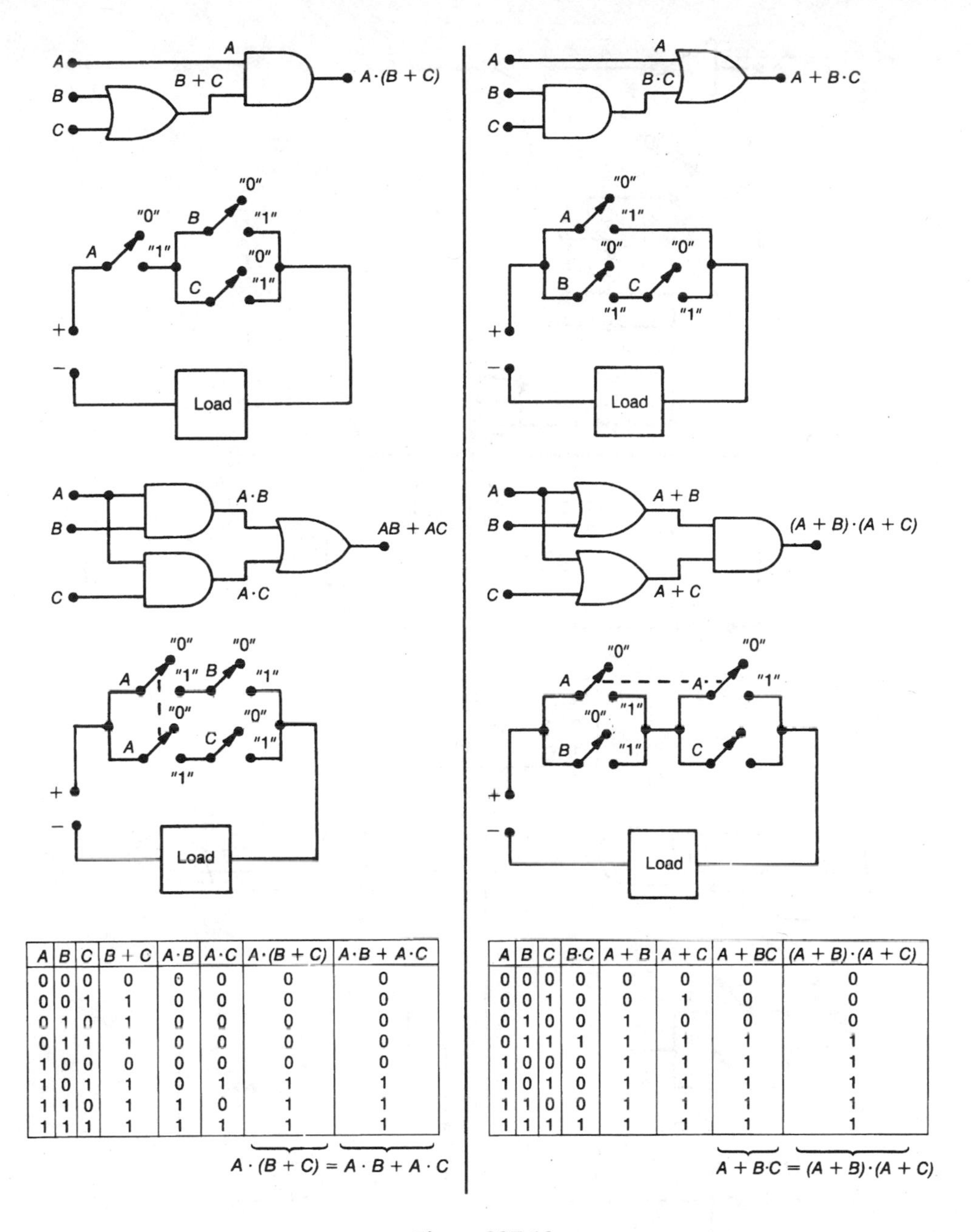

A	B	C	B + C	A·B	A·C	A·(B + C)	A·B + A·C
0	0	0	0	0	0	0	0
0	0	1	1	0	0	0	0
0	1	0	1	0	0	0	0
0	1	1	1	0	0	0	0
1	0	0	0	0	0	0	0
1	0	1	1	0	1	1	1
1	1	0	1	1	0	1	1
1	1	1	1	1	1	1	1

$A \cdot (B + C) = A \cdot B + A \cdot C$

A	B	C	B·C	A + B	A + C	A + BC	(A + B)·(A + C)
0	0	0	0	0	0	0	0
0	0	1	0	0	1	0	0
0	1	0	0	1	0	0	0
0	1	1	1	1	1	1	1
1	0	0	0	1	1	1	1
1	0	1	0	1	1	1	1
1	1	0	0	1	1	1	1
1	1	1	1	1	1	1	1

$A + B \cdot C = (A + B) \cdot (A + C)$

Figure 237-10

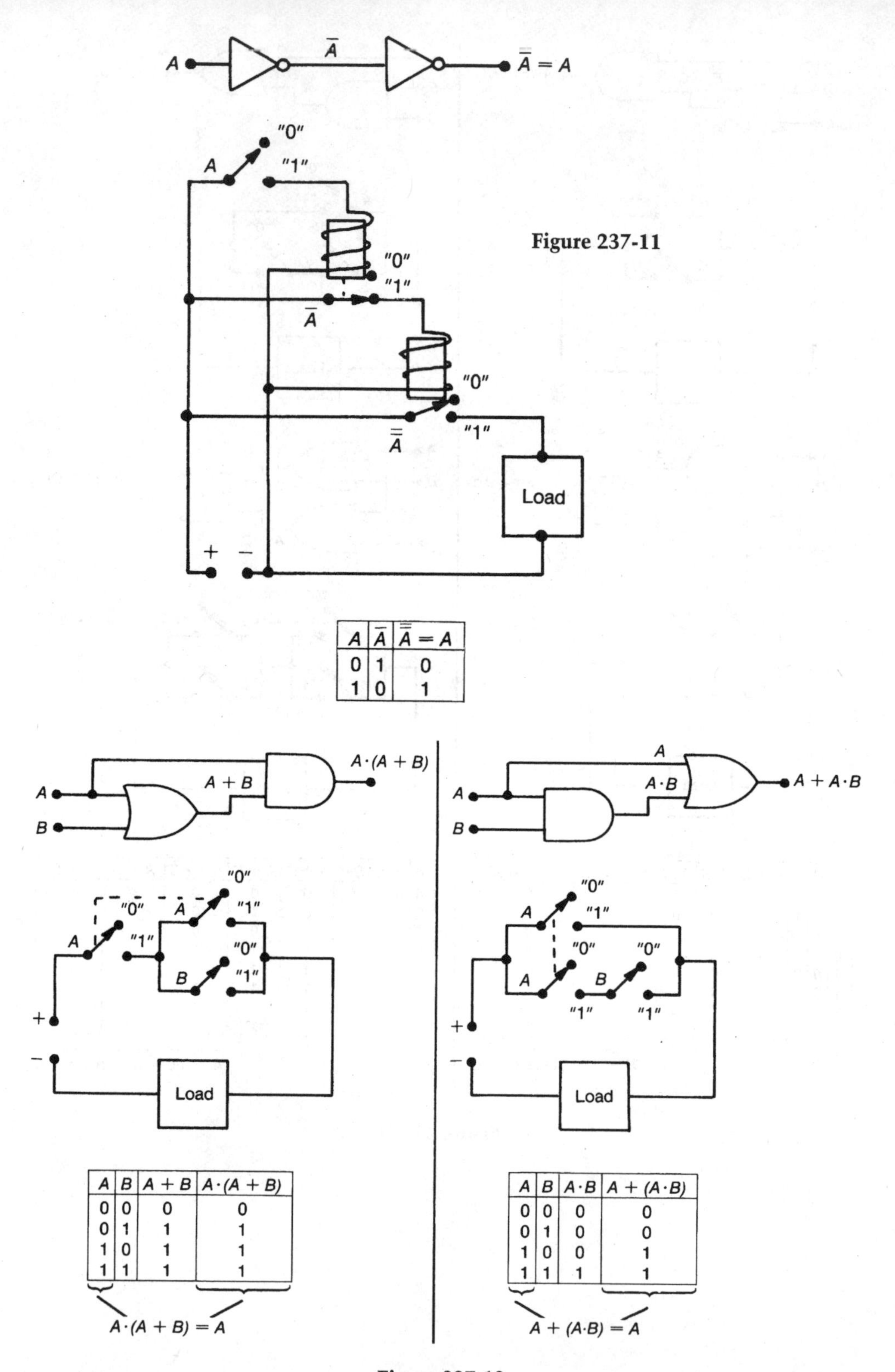

Figure 237-11

A	$\bar{A}$	$\bar{\bar{A}} = A$
0	1	0
1	0	1

A	B	$A + B$	$A \cdot (A + B)$
0	0	0	0
0	1	1	1
1	0	1	1
1	1	1	1

A	B	$A \cdot B$	$A + (A \cdot B)$
0	0	0	0
0	1	0	0
1	0	0	1
1	1	1	1

Figure 237-12

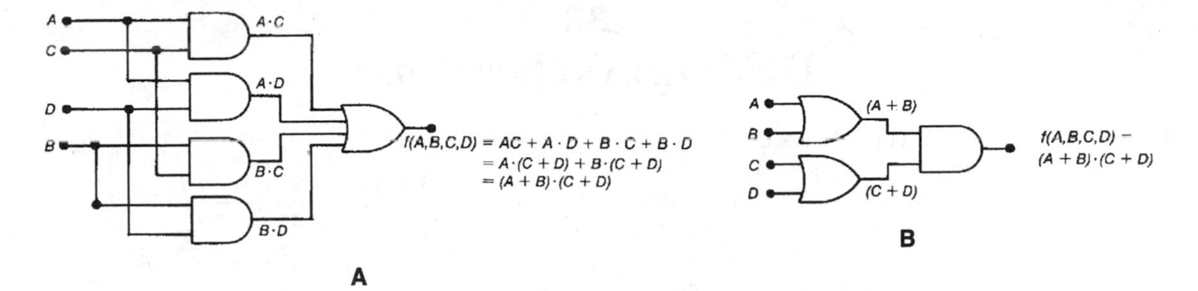

Figure 237-13

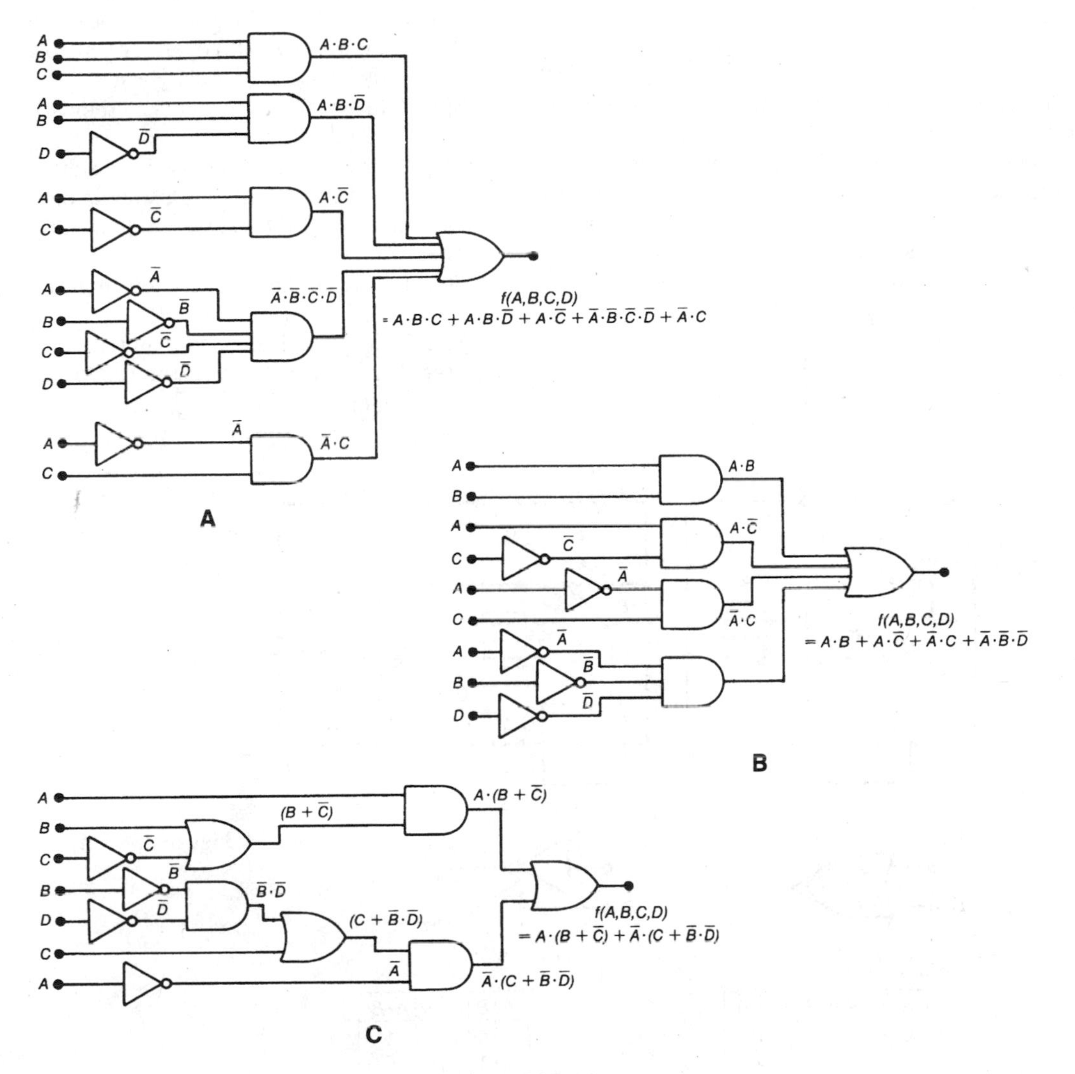

Figue 237-14

238
DeMorgan's theorems

A famous mathematician, Augustus DeMorgan (1806–1871), stated two theorems that relate to Boolean algebra. These theorems are very useful when you need to simplify expressions that contain the sum or product of two or more elements which are complemented (inverted). In equation form, the two theorems are:

$$\overline{A}\cdot\overline{B}\cdot\overline{C}\ldots = \overline{A + B + C + \ldots} \tag{238-1}$$

and,

$$\overline{A} + \overline{B} + \overline{C} + \ldots = \overline{A\cdot B\cdot C\ldots} \tag{238-2}$$

The equivalent switching circuits, logic diagrams, and truth tables for the DeMorgan theorems with two input elements are shown in Figs. 238-1A and B.

Equations 238-1 and 238-2 contain only single elements—A, B, C, and so on. However, each of these elements might be an expression containing a number of variables. As examples:

$$\begin{aligned} f(A,B,C) &= \overline{(A + B)\cdot\overline{C}} \\ &= \overline{A + B} + \overline{\overline{C}} \\ &= \overline{A}\cdot\overline{B} + C \end{aligned}$$

and,

$$\begin{aligned} f(A,B,C,D) &= \overline{A\cdot\overline{B}\cdot C + D} \\ &= \overline{A\cdot\overline{B}\cdot C}\cdot\overline{D} \\ &= (\overline{A} + \overline{\overline{B}} + \overline{C})\cdot\overline{D} \\ &= (\overline{A} + B + \overline{C})\cdot\overline{D} \end{aligned}$$

There is a step-by-step approach for obtaining the DeMorgan equivalent of a Boolean expression:

Step 1. Interchange AND operator symbols (·) with OR operator symbols (+).

Step 2. Invert all the elements so they are complemented.

Step 3. Invert (complement) the results of steps 1 and 2 to obtain the DeMorgan equivalent.

For example, the DeMorgan equivalent of $f(A,B) = A + B$ is:

Step 1. Change the OR operation to an AND operation. The result is $A\cdot B$.

Step 2. Invert each element to obtain $\overline{A}\cdot\overline{B}$.

Step 3. Invert $A\cdot B$ so that the DeMorgan equivalent of $A + B$ is $\overline{\overline{A}\cdot\overline{B}}$.

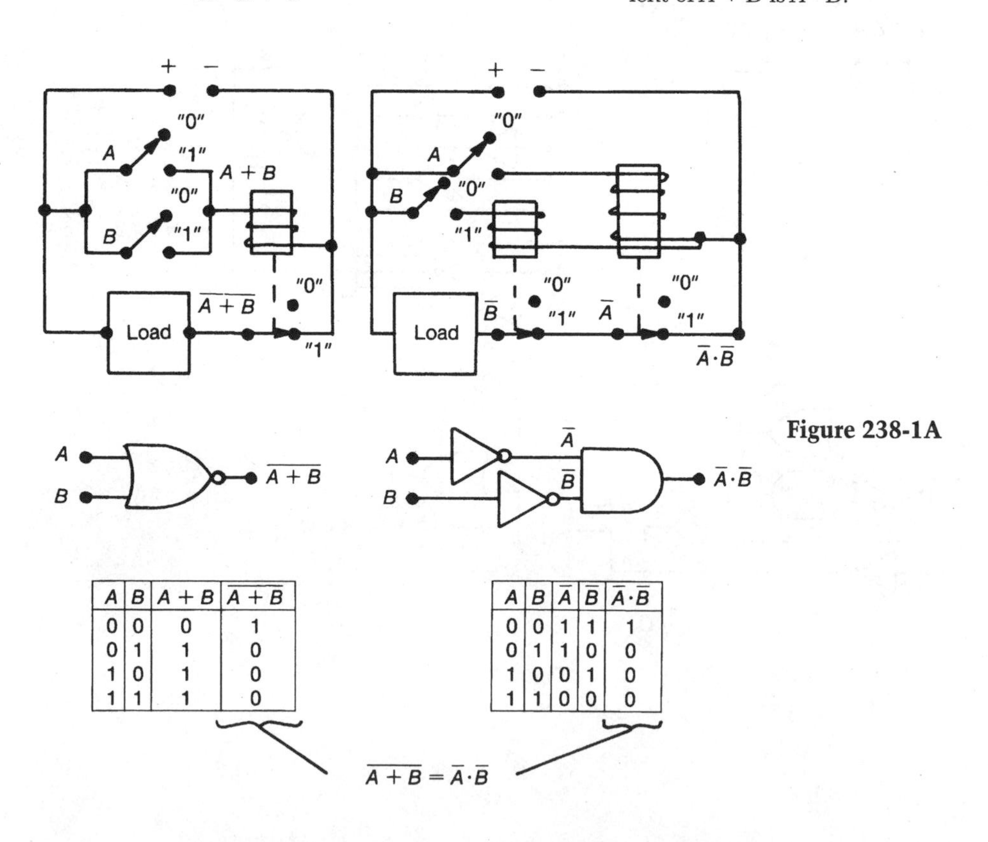

A	B	A + B	$\overline{A+B}$
0	0	0	1
0	1	1	0
1	0	1	0
1	1	1	0

A	B	$\overline{A}$	$\overline{B}$	$\overline{A}\cdot\overline{B}$
0	0	1	1	1
0	1	1	0	0
1	0	0	1	0
1	1	0	0	0

Figure 238-1A

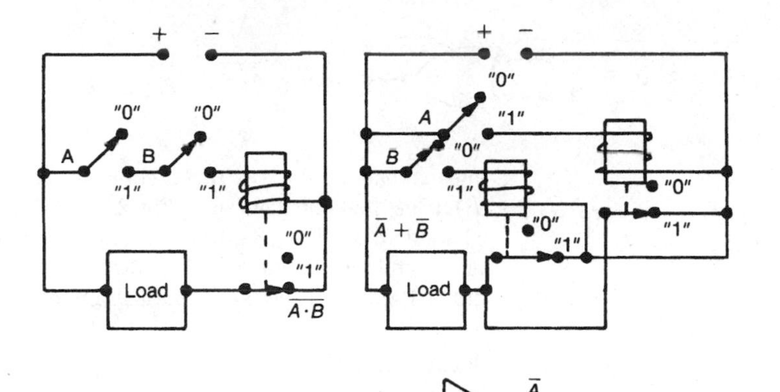

Figure 238-1B

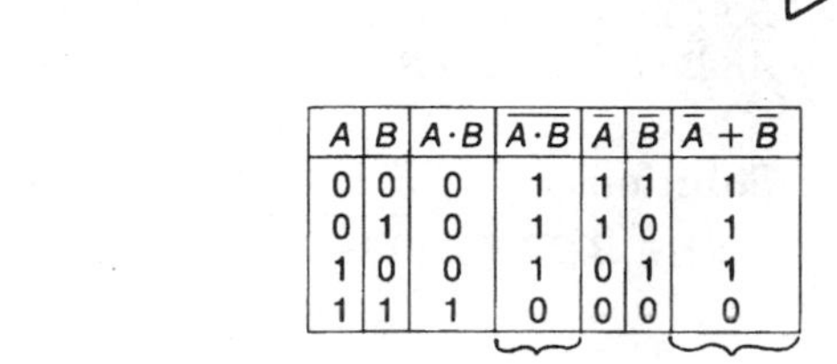

A	B	$A \cdot B$	$\overline{A \cdot B}$	$\overline{A}$	$\overline{B}$	$\overline{A} + \overline{B}$
0	0	0	1	1	1	1
0	1	0	1	1	0	1
1	0	0	1	0	1	1
1	1	1	0	0	0	0

$\overline{A \cdot B} = \overline{A} + \overline{B}$

Because,

$$A + B = \overline{\overline{A} \cdot \overline{B}} \quad (238\text{-}3)$$

$$\overline{A + B} = \overline{\overline{\overline{A} \cdot \overline{B}}} = \overline{A} \cdot \overline{B} \quad (238\text{-}1)$$

In terms of logic circuits, Equation 238-2 means that an OR gate is equivalent to an AND gate with both of the input elements and the output inverted. This relationship is illustrated in Fig. 238-2.

As a second example, consider the DeMorgan equivalent of $f(A,B) = A \cdot B$.

Step 1. Change the AND operation to an OR operation. The result is $A + B$.

Step 2. Invert each element so that you obtain $\overline{A} + \overline{B}$.

Step 3. Invert $\overline{A} + \overline{B}$ so that the DeMorgan equivalent of $A \cdot B$ is $\overline{\overline{A} + \overline{B}}$.

Because,

$$A \cdot B = \overline{\overline{A} + \overline{B}}, \quad (238\text{-}4)$$

$$\overline{A \cdot B} = \overline{\overline{\overline{A} + \overline{B}}} = \overline{A} + \overline{B} \quad (238\text{-}2)$$

Equation 238-3 can be interpreted to mean that an AND gate is equivalent to an OR gate in which both of the input elements, as well as the output, are inverted. This relationship is illustrated in Fig. 238-3.

MATHEMATICAL DERIVATIONS

Proofs of DeMorgan's theorems

Consider two elements C and D such that $C + D = 1$ and $C \cdot D = 0$. These relationships will only be satisfied by $C = 1$, $D = 0$ and $C = 0$, $D = 1$. Therefore, C and D are complementary, and $C = \overline{D}$, $D = \overline{C}$. Let $C = A + B$ and $D = \overline{A} \cdot \overline{B}$.

Then,

$$
\begin{aligned}
C + D &= A + B + \overline{A} \cdot \overline{B} \\
&= (A \cdot A + A \cdot B + A \cdot \overline{B} + A \cdot \overline{A}) \quad (237\text{-}7, 9) \\
&\quad + (A \cdot B + B \cdot \overline{B} + \overline{A} \cdot B) + \overline{A} \cdot \overline{B} \\
&= (A + B + \overline{A}) \cdot (A + B + \overline{B}) \quad (237\text{-}7) \\
&= (B + 1) \cdot (A + 1) \\
&= 1 \cdot 1 = 1 \quad (237\text{-}3)
\end{aligned}
$$

and,

$$C \cdot D = (A + B) \cdot \overline{A} \cdot \overline{B} = 0 \quad (237\text{-}6)$$

Therefore, C and D are complements so that:

$$\overline{C} = D \quad (238\text{-}1)$$

$$\overline{A + B} = \overline{A} \cdot \overline{B}.$$

A, B — $f(A,B) = A + B \equiv$ A, B — $f(A,B) = \overline{\overline{A} \cdot \overline{B}}$

Figure 238-2

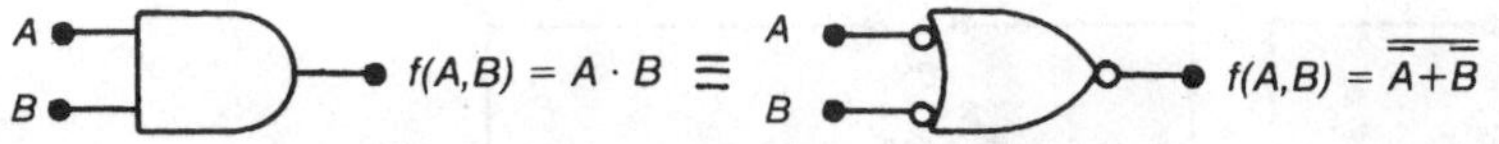

A, B → $f(A,B) = A \cdot B \equiv$ A, B → $f(A,B) = \overline{\overline{A}+\overline{B}}$

Figure 238-3

or,

$$A + B = \overline{\overline{A} \cdot \overline{B}} \qquad (238\text{-}3)$$

Now, let $C = \overline{A} + \overline{B}$ and $D = A \cdot B$.

Then,

$$\begin{aligned} C + D &= \overline{A} + \overline{B} + A \cdot B \\ &= \overline{A} \cdot \overline{A} + \overline{A} \cdot \overline{B} + \overline{A} \cdot B + \overline{A} \cdot \overline{B} + \overline{B} \cdot \overline{B} \\ &\quad + \overline{B} \cdot B + \overline{B} \cdot \overline{A} + \overline{B} \cdot \overline{B} + A \cdot B \\ &= (\overline{A} + \overline{B} + A) \cdot (\overline{A} + \overline{B} + B) && (237\text{-}7, 9) \\ &= (\overline{B} + 1) \cdot (\overline{A} + 1) && (237\text{-}7) \\ &= 1 \cdot 1 = 1 && (237\text{-}3) \end{aligned}$$

and,

$$C \cdot D = (\overline{A} + \overline{B}) \cdot (A \cdot B) = 0$$

Therefore, C and D are complements so that:

$$C = \overline{D} \qquad (238\text{-}3)$$

$$\overline{A} + \overline{B} = \overline{A \cdot B}$$

or,

$$A \cdot B = \overline{\overline{A} + \overline{B}} \qquad (238\text{-}4)$$

To summarize the equations for the DeMorgan theorems with two input elements:

$$\overline{A} \cdot \overline{B} = \overline{A + B} \qquad (238\text{-}5)$$

or,

$$A + B = \overline{\overline{A} \cdot \overline{B}} \qquad (238\text{-}6)$$

$$\overline{A} + \overline{B} = \overline{A \cdot B} \qquad (238\text{-}7)$$

or,

$$A \cdot B = \overline{\overline{A} + \overline{B}} \qquad (238\text{-}8)$$

Equations 238-5 and 238-7 are illustrated in the Venn diagrams of Figs. 238-4A and B. The shaded area of Fig. 238-4A represents $\overline{A} \cdot \overline{B}$ as well as $\overline{A + B}$ and means that NOT A AND NOT B is equivalent to NOT (A OR B). (See the Venn diagram in Fig. 228-1.) Similarly, the shaded area of Fig. 238-4B represents $\overline{A} + \overline{B}$ as well as $\overline{A \cdot B}$, and means that NOT A OR NOT B is equivalent to NOT (A AND B). (See the Venn diagram in Fig. 228-1.)

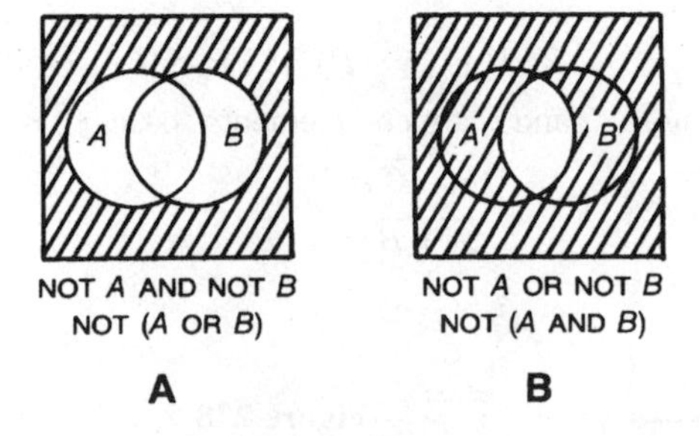

Figure 238-4

Example 238-1

Simplify the Boolean expressions (a) $f(A,B,C,D) = \overline{(\overline{A} + B) \cdot (C + \overline{D})}$, (b) $f(A,B,C) = A \cdot B \cdot C + A \cdot \overline{B} \cdot \overline{(\overline{A} \cdot \overline{C})}$, and (c) $f(A,B,C,D) = \overline{A} \cdot B \cdot \overline{(\overline{A} \cdot C \cdot D)} + \overline{A} \cdot \overline{B} \cdot C \cdot \overline{D} + A \cdot B \cdot \overline{C}$.

Solution

$$\begin{aligned} \text{(a) } f(A,B,C,D) &= \overline{(\overline{A} + B) \cdot (C + \overline{D})} \\ &= \overline{\overline{A} + B} + \overline{C + \overline{D}} \\ &= \overline{\overline{A}} \cdot \overline{B} + \overline{C} \cdot \overline{\overline{D}} \\ &= A \cdot \overline{B} + \overline{C} \cdot D \end{aligned}$$

This example illustrates one of the principles of simplification. The inversion overbar is broken by interchanging the AND and OR operations. This process is repeated until you are left with only single inverted elements.

$$\begin{aligned} \text{(b) } f(A,B,C) &= A \cdot B \cdot C + A \cdot \overline{B}\overline{(\overline{A} \cdot \overline{C})} \\ &= A \cdot B \cdot C + A \cdot \overline{B} \cdot (\overline{\overline{A}} + \overline{\overline{C}}) && (238\text{-}7) \\ &= A \cdot B \cdot C + A \cdot A \cdot \overline{B} && (237\text{-}16) \\ &\quad + A \cdot \overline{B} \cdot C \\ &= A \cdot B \cdot C + A \cdot \overline{B} + A \cdot \overline{B} \cdot C && (237\text{-}8) \\ &= A \cdot C \cdot (B + \overline{B}) + A \cdot \overline{B} \\ &= A \cdot C + A \cdot \overline{B} && (237\text{-}7) \\ &= A \cdot (\overline{B} + C) \end{aligned}$$

The logic circuit for the original expression requires four gates and three inverters, but the simplified expression only needs two gates and one inverter.

$$\begin{aligned} \text{(c) } f(A,B,C,D) &= \overline{A} \cdot B \cdot \overline{(\overline{A} \cdot C \cdot D)} + \overline{A} \cdot \overline{B} \cdot C \cdot \overline{D} \\ &\quad + A \cdot B \cdot \overline{C} \\ &= \overline{A} \cdot B \cdot (\overline{\overline{A}} + \overline{C} + \overline{D}) && (238\text{-}7) \\ &\quad + \overline{A} \cdot \overline{B} \cdot C \cdot \overline{D} + A \cdot B \cdot \overline{C} \\ &= A \cdot \overline{A} \cdot B + \overline{A} \cdot B \cdot \overline{C} + \overline{A} \cdot B \cdot \overline{D} \\ &\quad + \overline{A} \cdot \overline{B} \cdot C \cdot \overline{D} + A \cdot B \cdot \overline{C} \\ &= \overline{A} \cdot B \cdot \overline{C} + \overline{A} \cdot B \cdot \overline{D} && (237\text{-}6) \\ &\quad + \overline{A} \cdot \overline{B} \cdot C \cdot \overline{D} + A \cdot B \cdot \overline{C} \end{aligned}$$

$$= B \cdot \overline{C}(A + \overline{A}) + \overline{A} \cdot \overline{D}(B + \overline{B} \cdot C)$$
$$= B \cdot \overline{C} + \overline{A} \cdot \overline{D} \cdot (B + C) \qquad (237\text{-}19)$$

The original expression requires five inverters and five gates, but the simplified result only requires three inverters and four gates.

PRACTICE PROBLEMS

238-1. Use DeMorgan's theorems to simplify the Boolean equation: $f(A,B) = (A + B) \cdot \overline{A \cdot B}$.

238-2. Use DeMorgan's theorems to simplify the Boolean equation: $f(A,B,C,D,E,F) = A \cdot \overline{[\overline{B} + (C \cdot \overline{D} + \overline{E} \cdot F)]}$.

238-3. Use DeMorgan's theorems to simplify the Boolean equation: $f(A,B) = \overline{(A + B) \cdot \overline{C}}$.

238-4. Use DeMorgan's theorems to simplify the Boolean equation: $f(A,B,C) = \overline{A \cdot \overline{B} \cdot C + \overline{D}}$.

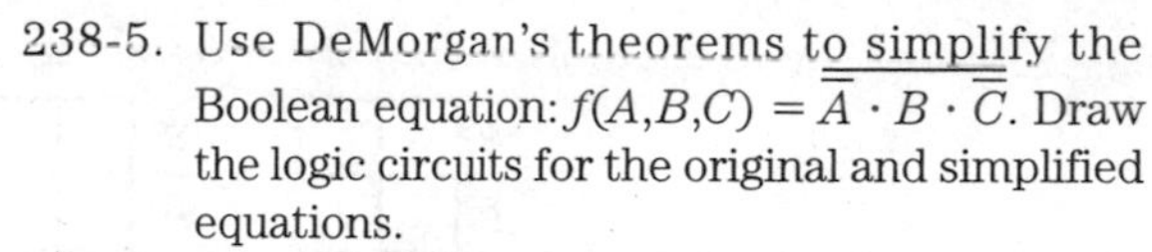

238-5. Use DeMorgan's theorems to simplify the Boolean equation: $f(A,B,C) = \overline{\overline{A} \cdot B \cdot \overline{C}}$. Draw the logic circuits for the original and simplified equations.

238-6. Use DeMorgan's theorems to simplify the Boolean equation: $f(A,B,C) = \overline{\overline{A} + \overline{B} \cdot C}$. Draw the logic circuits for the original and simplified equations.

238-7. Use DeMorgan's theorems to simplify the Boolean equation: $f(A,B,C,D) = \overline{A \cdot B \cdot \overline{C} \cdot D}$. Draw the logic circuits for the original and simplified equations.

238-8. Use DeMorgan's theorems to simplify the logic equation: $f(A,B,C,D) = \overline{A \cdot \overline{(B + \overline{C})} \cdot D}$. Draw the logic circuits for the original and simplified equations.

239
Karnaugh maps

Rather than use the Boolean laws and axioms to simplify logic expressions, you can use a Karnaugh map. Such a map provides a very quick and easy way to find the simplest form of a logic equation. As shown in Fig. 239-1, Karnaugh maps can be readily constructed for two, three, or four elements; but they become more unwieldly if five or six elements are involved.

Because each element has two possible states (such as A and B), the number of squares required is 2^N, where N is the number of variables. Consequently, for four, five, six, seven, and eight elements, the corresponding Karnaugh maps respectively contain 16, 32, 64, 128, and 256 squares. If the number of elements exceed six, the Karnaugh maps become too unwieldy and other methods of simplification should be used.

An exploded view of a four-element Karnaugh map is shown in Fig. 239-2. Notice the division of the map into labeled columns and rows. Entries into the map are placed in these columns and rows in accordance with the logic terms contained in the given Boolean expression.

In Fig. 239-2, the square in the upper left corner of the Karnaugh map contains the elements A, B, $\overline{C}$, and D; the adjacent lower square represents the elements A, B, $\overline{C}$, and $\overline{D}$; the next lower square contains A, $\overline{B}$, $\overline{C}$, and $\overline{D}$; and the lowest left-hand square has the elements A, $\overline{B}$, $\overline{C}$, and D. The other 12 squares in the map are similarly identified. Notice that the elements $A \cdot \overline{C}$ are contained in each of the four terms just identified. Because the elements $B \cdot \overline{B}$, and $D \cdot \overline{D}$ are also contained in the same four squares, these will disappear when the four squares are added together. In other words, the left vertical column represents the term $A \cdot \overline{C}$. This can be proven as follows:

$$A \cdot \overline{C} = A \cdot \overline{C} \cdot 1$$
$$= A \cdot \overline{C} \cdot (B + \overline{B}) \text{ since } B + \overline{B} = 1 \qquad (237\text{-}7)$$
$$= A \cdot B \cdot \overline{C} \cdot (D + \overline{D}) + A \cdot \overline{B} \cdot \overline{C} \cdot (D + \overline{D})$$
$$= A \cdot B \cdot \overline{C} \cdot D + A \cdot B \cdot \overline{C} \cdot \overline{D} \qquad (239\text{-}1)$$
$$+ A \cdot \overline{B} \cdot \overline{C} \cdot D + A \cdot \overline{B} \cdot \overline{C} \cdot \overline{D}$$

These terms have the same sequence as that of the four squares.

Notice that a two-element term is represented in Fig. 239-2 by four squares. Further study of the diagram reveals that a one-element term is represented by eight squares, but a three-element term is formed from two squares.

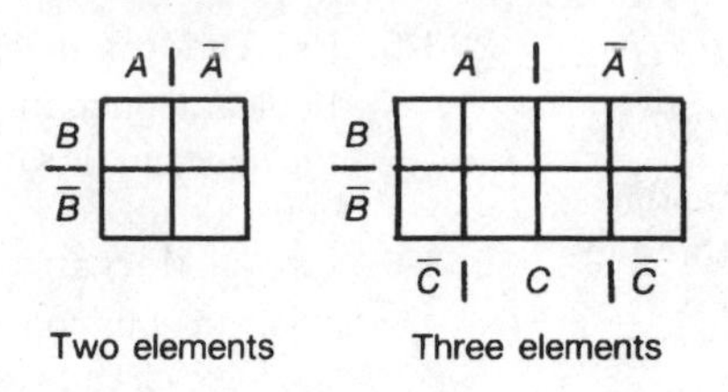

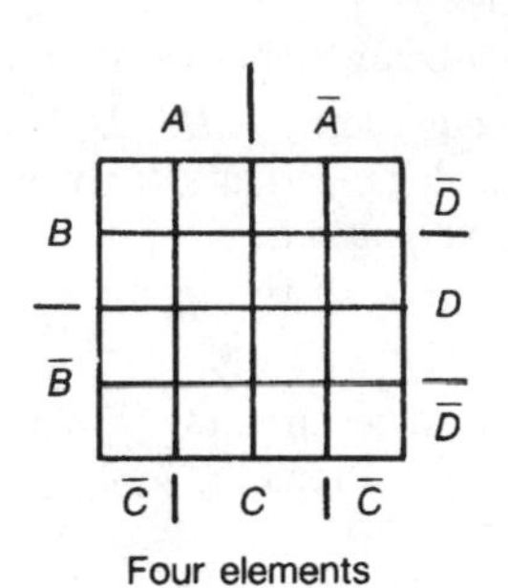

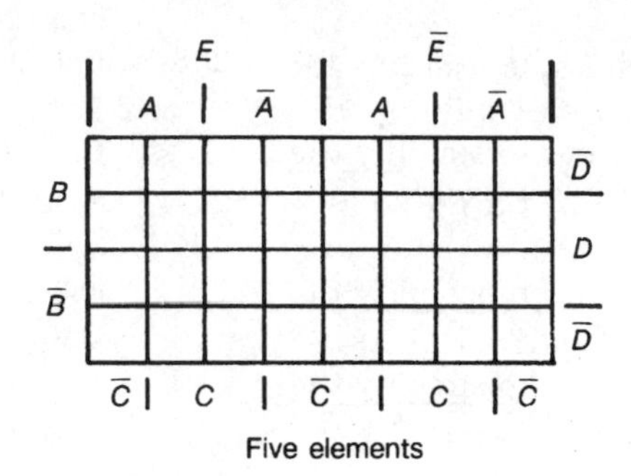

Figure 239-1

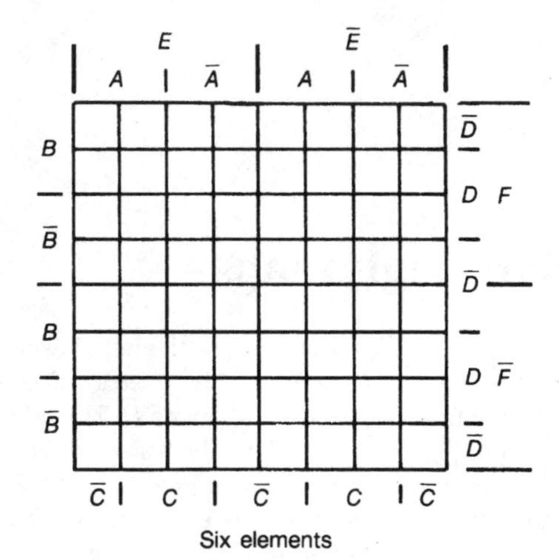

MATHEMATICAL DERIVATIONS

Now, use the Karnaugh map to simplify the logic equation:

$$f(A,B,C,D) = A \cdot B \cdot C + A \cdot B \cdot \overline{D} + A \cdot \overline{C} + \overline{A} \cdot \overline{B} \cdot \overline{C} \cdot \overline{D} + \overline{A} \cdot C.$$

Because there are four elements, you will need the Karnaugh map that contains 16 squares (Fig. 239-3). The step-by-step procedure is then:

Step 1. For the purpose of identification, label the squares from 1 to 16.

Step 2. On a term-by-term basis, plot the logic expression on the map. In other words, place a 1 in each of the squares that are required to represent a particular term.

The term $A \cdot B \cdot C$ is represented by squares 2 and 6, because $A \cdot B \cdot C \cdot D + A \cdot B \cdot C \cdot \overline{D} = A \cdot B \cdot C \cdot (D + \overline{D}) = A \cdot B \cdot C$.

The term $A \cdot B \cdot \overline{D}$ is identified by squares 1 and 2, because $A \cdot B \cdot \overline{C} \cdot \overline{D} + A \cdot B \cdot C \cdot \overline{D} = A \cdot B \cdot \overline{D} \cdot (C + \overline{C}) = A \cdot B \cdot \overline{D}$. Because a 1 already exists in square 2, it is only necessary to insert a 1 in square 1.

The term $A \cdot \overline{C}$ requires the squares 1, 5, 9, and 13 because Equation 239-1 has already shown that $A \cdot \overline{C} = A \cdot B \cdot \overline{C} \cdot D + A \cdot B \cdot \overline{C} \cdot \overline{D} + A \cdot \overline{B} \cdot \overline{C} \cdot D + A \cdot \overline{B} \cdot \overline{C} \cdot \overline{D}$.

The term $\overline{A} \cdot \overline{B} \cdot \overline{C} \cdot \overline{D}$ corresponds to square 16.

The term $\overline{A} \cdot C$ is the result of combining the squares 3, 7, 11, and 15 as follows:

$$\overline{A} \cdot B \cdot C \cdot \overline{D} + \overline{A} \cdot B \cdot C \cdot D$$

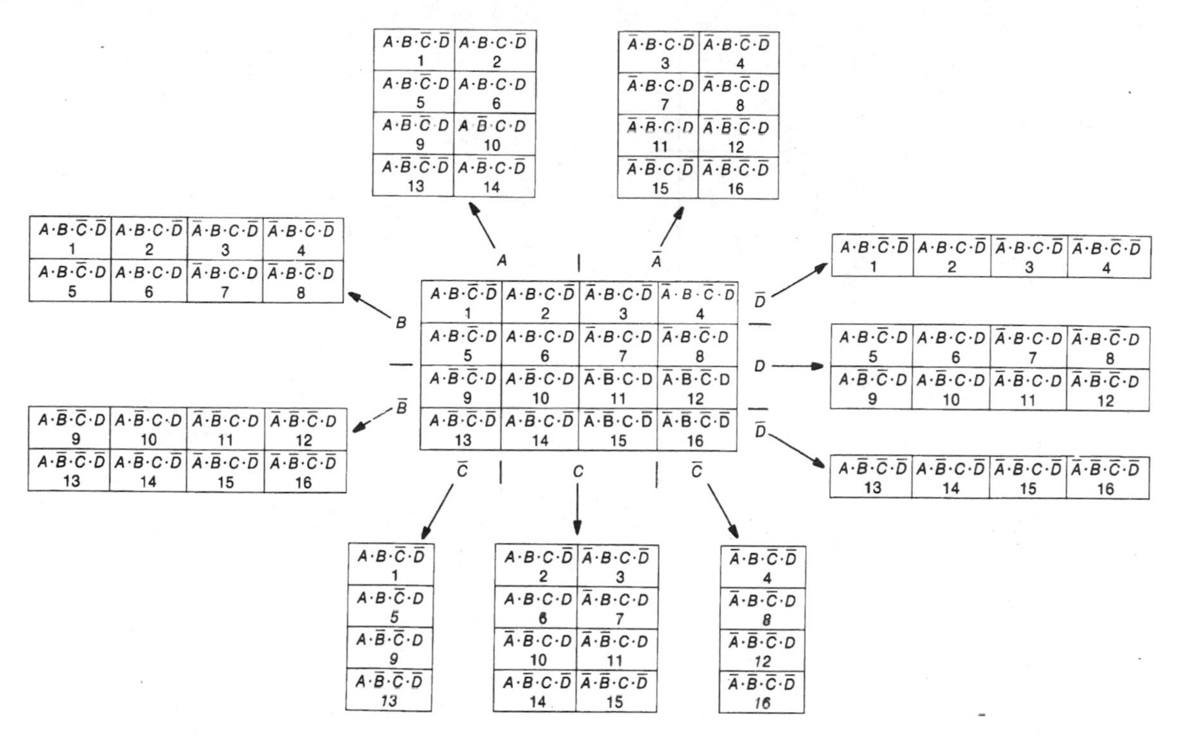

Figure 239-2

$$+\overline{A}\cdot B\cdot C\cdot D+\overline{A}\cdot\overline{B}\cdot C\cdot\overline{D}$$
$$=\overline{A}\cdot B\cdot C\cdot(\overline{D}+D)+\overline{A}\cdot\overline{B}\cdot C\cdot(D+\overline{D})$$
$$=\overline{A}\cdot B\cdot C+\overline{A}\cdot\overline{B}\cdot C$$
$$=\overline{A}\cdot C\cdot(B+\overline{B})$$
$$=\overline{A}\cdot C$$

Step 3. Obtain the simplified logic equation by using Fig. 239-3 and observing the following rules:

(a) If 1's are located in adjacent squares, or at opposite ends of any row or column, one of the elements can be dropped.

(b) Two of the elements can be dropped if: (1) any row or column of squares is filled with 1's; (2) any block of four squares is filled with 1's; (3) the four end squares of any adjacent rows or columns are filled with 1's; (4) the four squares of a corner are filled with 1's.

(c) Three of the elements can be dropped if: (1) two adjacent rows or columns are filled with 1's or if (2) the top and bottom rows or the right and left columns are completely filled with 1's.

(d) To reduce the original logic equation to its simplest form, all of the 1's must be included in the final expression. A particular 1 can be used more than once, but look for the largest possible combinations to form groups of eight, four, and two squares.

Notice that, although the Karnaugh map is shown as a table, it is considered to be a *cylinder*, so squares 13 and 16 are contiguous. In addition, the top of the table can be folded back and down, while the bottom of the table is folded back and up, so the "$\overline{D}$" squares become adjacent.

Using Rule (b) in Step 3, squares 1, 5, 9, and 13 are combined to yield $A\cdot\overline{C}$. Squares 3, 7, 11, and 15 are grouped to yield $\overline{A}\cdot C$. Squares 1, 2, 5, and 6 when taken together, are equivalent to $A\cdot B$. Finally, Rule (a) in Step 3 is used to group squares 15 and 16 and yields $\overline{A}\cdot\overline{B}\cdot\overline{D}$. All of these groups are shown in Fig. 239-3.

The Karnaugh map has produced the following simplification:

$$f(A,B,C,D)=A\cdot B\cdot C+A\cdot B\cdot\overline{D}+A\cdot\overline{C}$$
$$+\overline{A}\cdot\overline{B}\cdot\overline{C}\cdot\overline{D}+\overline{A}\cdot C$$
$$=A\cdot B+A\cdot\overline{C}+\overline{A}\cdot C+\overline{A}\cdot\overline{B}\cdot\overline{D}$$

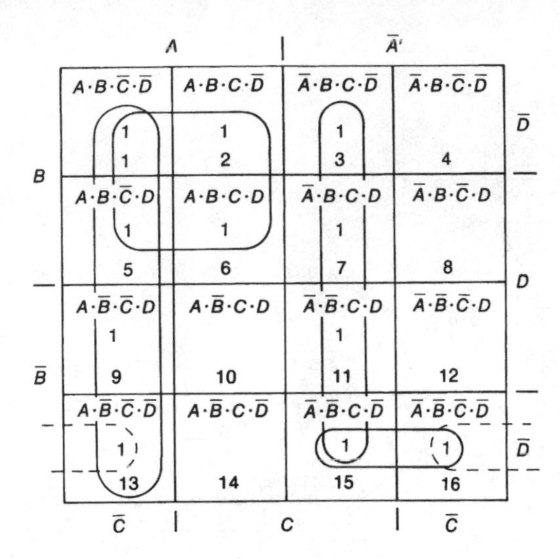

Figure 239-3

The original logic equation requires six gates, but the simplified expression requires five gates.

Alternatively, if squares 13 and 16 are combined by Rule (a),

$$f(A,B,C,D) = A \cdot B + A \cdot \overline{C} + \overline{A} \cdot C + \overline{B} \cdot \overline{C} \cdot \overline{D}$$

A single Boolean expression can, therefore, be represented by more than one simplified form.

A Karnaugh map also provides a convenient means of finding the complement of a logic expression. This is done by plotting the logic terms on a map; then, on a second map, placing 1's in all the squares that did not contain 1's in the original map.

As an example,

$$f(A,B,C) = A \cdot B \cdot C$$

simplify,

$$\overline{f(A,B,C)} = \overline{A \cdot B \cdot C}$$

Fig. 239-4A shows the original map, while the complement map appears in Fig. 239-4B. As illustrated, squares 3, 4, 7, and 8 are grouped to produce $\overline{A}$. Similarly, squares 5, 6, 7, and 8 are combined to form $\overline{B}$, although squares 1, 5, 4, and 8 are equivalent to $\overline{C}$. Therefore,

$$\begin{aligned} \overline{f(A,B,C)} &= \overline{A \cdot B \cdot C} \\ &= \overline{A} + \overline{B} + \overline{C} \end{aligned}$$

This result is DeMorgan's theorem, which appeared in Equation 238-2.

Example 239-1

Use a Karnaugh map to simplify the logic equation:

$$\begin{aligned} f(A,B,C,D) = {} & A \cdot \overline{B} \cdot \overline{C} + \overline{A} \cdot \overline{B} \cdot C \cdot D \\ & + A \cdot C \cdot \overline{D} + \overline{A} \cdot C \cdot \overline{D} + \overline{C} \cdot \overline{D} \end{aligned}$$

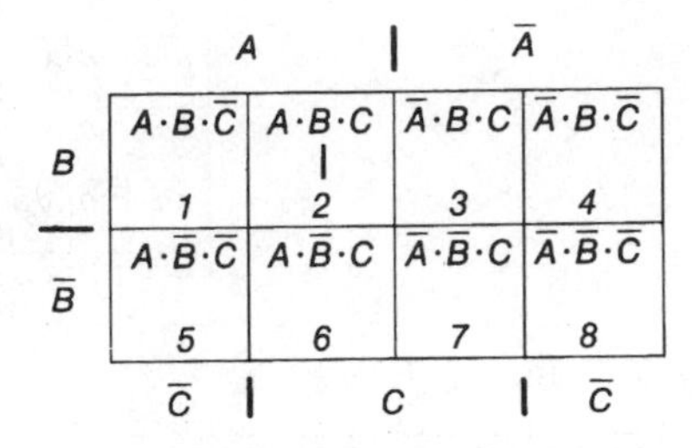

Karnaugh map for $f(A,B,C) = A \cdot B \cdot C$

A

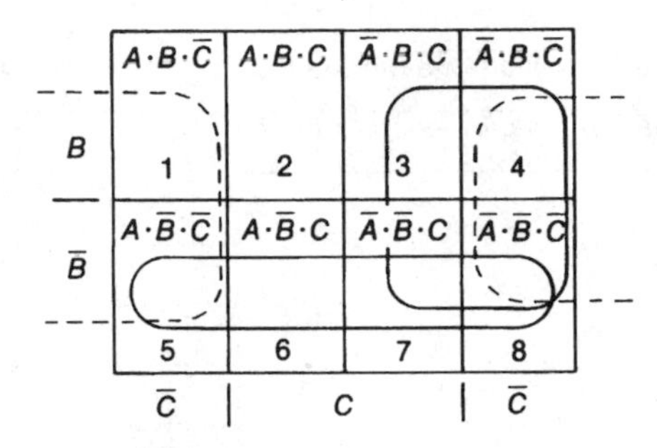

Karnaugh map for $\overline{f(A,B,C)} = \overline{A \cdot B \cdot C}$

B

Figure 239-4

Solution

$$\begin{aligned} f(A,B,C,D) &= A \cdot \overline{B} \cdot \overline{C} \cdot (D + \overline{D}) + \overline{A} \cdot \overline{B} \cdot C \cdot D \\ &\quad + A \cdot C \cdot \overline{D} \cdot (B + \overline{B}) \\ &\quad + \overline{A} \cdot C \cdot \overline{D} \cdot (B + \overline{B}) \\ &\quad + \overline{C} \cdot \overline{D} \cdot (\overline{A} + \overline{A}) \cdot (B + \overline{B}) \\ &= A \cdot \overline{B} \cdot \overline{C} \cdot D + A \cdot \overline{B} \cdot \overline{C} \cdot \overline{D} \\ &\quad + \overline{A} \cdot \overline{B} \cdot C \cdot D + A \cdot B \cdot C \cdot \overline{D} \\ &\quad + A \cdot \overline{B} \cdot C \cdot \overline{D} + \overline{A} \cdot B \cdot C \cdot \overline{D} \\ &\quad + \overline{A} \cdot \overline{B} \cdot C \cdot \overline{D} + A \cdot B \cdot \overline{C} \cdot \overline{D} \\ &\quad + \overline{A} \cdot B \cdot \overline{C} \cdot \overline{D} + A \cdot \overline{B} \cdot \overline{C} \cdot \overline{D} \\ &\quad + \overline{A} \cdot \overline{B} \cdot \overline{C} \cdot \overline{D} \end{aligned}$$

Notice that the term $A \cdot \overline{B} \cdot \overline{C} \cdot \overline{D}$ is repeated so that only ten 1's will be plotted in the four-element Karnaugh map of Fig. 239-5.

Rule (c). Squares 1, 2, 3, 4, 13, 14, 15, 16 are combined to yield $\overline{D}$.

Rule (a). Squares 9 and 13 are combined to yield $A \cdot \overline{B} \cdot \overline{C}$. Squares 11 and 15 are combined to yield $\overline{A} \cdot \overline{B} \cdot C$.

Therefore:

$$f(A,B,C,D) = \overline{D} + A \cdot \overline{B} \cdot \overline{C} + \overline{A} \cdot \overline{B} \cdot C.$$

PRACTICE PROBLEMS

239-1. Use Karnaugh mapping to simplify the following Boolean equation: $f(A,B) = A \cdot B + \overline{A} \cdot B + \overline{B} \cdot A + \overline{A} \cdot \overline{B}$.

239-2. Use Karnaugh mapping to simplify the following Boolean equation: $f(A,B,C) = A \cdot \overline{B} \cdot C + A \cdot B \cdot \overline{C} + \overline{A} \cdot B \cdot \overline{C}$.

239-3. Use Karnaugh mapping to simplify the following Boolean equation: $f(A,B,C) = A \cdot B \cdot \overline{C} + \overline{A} \cdot \overline{B} \cdot C + \overline{A} \cdot \overline{B} \cdot \overline{C} + A \cdot B \cdot C$.

239-4. Use Karnaugh mapping to simplify the following Boolean equation: $f(A,B,C,D) = A \cdot B \cdot \overline{C} \cdot \overline{D} + A \cdot \overline{B} \cdot \overline{C} \cdot D + \overline{A} \cdot \overline{B} \cdot \overline{C} \cdot D + A \cdot \overline{B} \cdot \overline{C} \cdot \overline{D} + A \cdot \overline{B} \cdot C \cdot D + A \cdot \overline{B} \cdot C \cdot \overline{D}$.

239-5. Use Karnaugh mapping to simplify the following Boolean equation: $f(A,B,C) = \overline{A} \cdot B + \overline{B} \cdot C + \overline{A} \cdot \overline{B} \cdot \overline{C}$. Draw the logic circuits for the original and simplified equations.

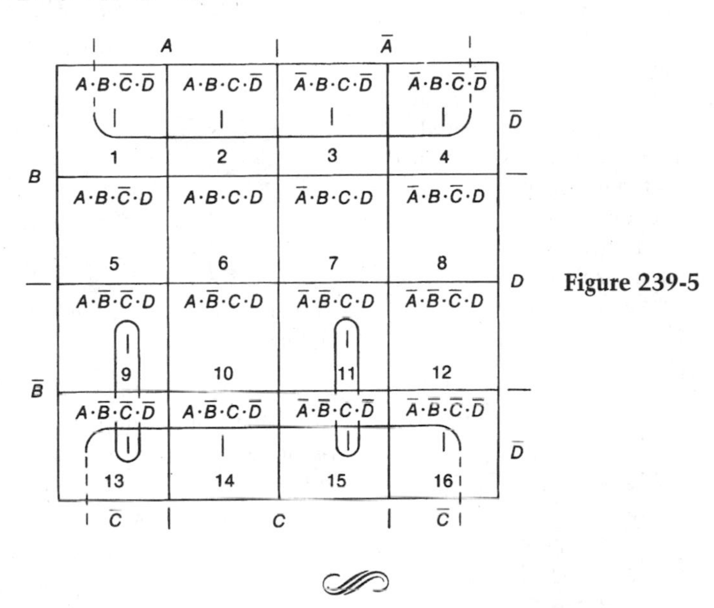

Figure 239-5

240 Harvard charts

The Harvard chart is another technique for simplifying Boolean expressions. With expressions containing four elements or less, the Karnaugh map is superior to the Harvard chart. When five or more elements are involved, however, the Harvard chart has distinct advantages.

As an example, consider the logic equation:

$$f(A,B,C) = A \cdot B \cdot \overline{C} + A \cdot B \cdot C + \overline{A} \cdot B \cdot C + \overline{A} \cdot B \cdot \overline{C} + A \cdot \overline{B} \cdot C$$

$$= A \cdot B \cdot (C + \overline{C}) + \overline{A} \cdot B \cdot (C + \overline{C}) + A \cdot \overline{B} \cdot C \qquad (237\text{-}14)$$

$$= A \cdot B + \overline{A} \cdot B + A \cdot \overline{B} \cdot C \qquad (237\text{-}7)$$

$$= B \cdot (A + \overline{A}) + A \cdot \overline{B} \cdot C \qquad (237\text{-}14)$$

$$= B + \overline{B} \cdot A \cdot C \qquad (237\text{-}7)$$

$$= B + A \cdot C \qquad (237\text{-}19)$$

Use the Harvard chart to obtain the same simplified result in the mathematical derivations.

MATHEMATICAL DERIVATIONS

A Harvard chart for three elements is shown in Fig. 240-1. The steps for obtaining the simplified result are:

Step 1. Draw lines through all the rows whose terms are not contained in the expression being simplified. In the example, Step 1 will apply to rows 1, 2, and 5.

Step 2. Starting with the leftmost column (column 1), cross out all terms that are the same as those already lined out in accordance with Step 1. Because A was lined out in rows 1, and 2, and A was lined out in row 5, all terms must be lined out in column 1.

Step 3. In column 2, only $\overline{B}$ was lined out in accordance with step 1. Therefore, all B's are circled because B is part of the final simplified result.

Step 4. By moving to the right of the chart, line out those items containing B in all of the rows that possess a circled B.

Step 5. Repeat Steps 3 and 4 for columns 3 and 4; no circled terms are discovered in these columns. However, in column 5, the term $A \cdot C$ is circled, and step 4 is repeated. All terms in columns 6 and 7 have now been lined out so that the simplification procedure has been concluded.

From the Harvard chart, $f(A,B,C) = B + A \cdot C$, which is the same simplified result that you obtained from using the Boolean laws and axioms.

Columns							
1	2	3	4	5	6	7	Rows
A	B	C	$A \cdot B$	$A \cdot C$	$B \cdot C$	$A \cdot B \cdot C$	
~~$\overline{A}$~~	~~$\overline{B}$~~	~~$\overline{C}$~~	~~$\overline{A} \cdot \overline{B}$~~	~~$\overline{A} \cdot \overline{C}$~~	~~$\overline{B} \cdot \overline{C}$~~	~~$\overline{A} \cdot \overline{B} \cdot \overline{C}$~~	1
~~$\overline{A}$~~	~~$\overline{B}$~~	~~C~~	~~$\overline{A} \cdot \overline{B}$~~	~~$\overline{A} \cdot C$~~	~~$\overline{B} \cdot C$~~	~~$\overline{A} \cdot \overline{B} \cdot C$~~	2
~~$\overline{A}$~~	(B)	~~$\overline{C}$~~	~~$\overline{A} \cdot B$~~	~~$\overline{A} \cdot \overline{C}$~~	~~$B \cdot \overline{C}$~~	~~$\overline{A} \cdot B \cdot \overline{C}$~~	3
~~$\overline{A}$~~	(B)	~~C~~	~~$\overline{A} \cdot B$~~	~~$\overline{A} \cdot C$~~	~~$B \cdot C$~~	~~$\overline{A} \cdot B \cdot C$~~	4
~~A~~	~~$\overline{B}$~~	~~$\overline{C}$~~	~~$A \cdot \overline{B}$~~	~~$A \cdot \overline{C}$~~	~~$\overline{B} \cdot \overline{C}$~~	~~$A \cdot \overline{B} \cdot \overline{C}$~~	5
~~A~~	~~$\overline{B}$~~	~~C~~	~~$A \cdot \overline{B}$~~	($A \cdot C$)	~~$\overline{B} \cdot C$~~	~~$A \cdot \overline{B} \cdot C$~~	6
~~A~~	~~B~~	~~$\overline{C}$~~	~~$A \cdot B$~~	~~$A \cdot \overline{C}$~~	~~$B \cdot \overline{C}$~~	~~$A \cdot B \cdot \overline{C}$~~	7
~~A~~	~~B~~	~~C~~	~~$A \cdot B$~~	($A \cdot C$)	~~$B \cdot C$~~	~~$A \cdot B \cdot C$~~	8

Figure 240-1

Example 240-1

By using a Harvard chart, simplify the logic equation:

$$f(A,B,C) = \overline{A} \cdot \overline{B} \cdot \overline{C} + \overline{A} \cdot B \cdot C + A \cdot B \cdot C + A \cdot \overline{B} \cdot \overline{C} + A \cdot \overline{B} \cdot C.$$

Solution

Start by simplifying the equation through the use of the Boolean laws and axioms.

$$f(A,B,C) = \overline{A} \cdot \overline{B} \cdot \overline{C} + \overline{A} \cdot B \cdot C + A \cdot B \cdot C + A \cdot \overline{B} \cdot \overline{C} + A \cdot \overline{B} \cdot C$$

$$= B \cdot C \cdot (\overline{A} + A) + A \cdot \overline{B} \cdot (\overline{C} + C) + \overline{A} \cdot \overline{B} \cdot \overline{C}$$

$$= B \cdot C \cdot 1 + A \cdot \overline{B} \cdot 1 + \overline{A} \cdot \overline{B} \cdot \overline{C} \quad (237\text{-}7)$$

$$= B \cdot C + \overline{B} \cdot (A + \overline{A} \cdot \overline{C})$$

$$= B \cdot C + \overline{B} \cdot (A + \overline{C}) \quad (237\text{-}19)$$

Columns							
1	2	3	4	5	6	7	Rows
A	B	C	$A \cdot B$	$A \cdot C$	$B \cdot C$	$A \cdot B \cdot C$	
~~$\overline{A}$~~	~~$\overline{B}$~~	~~$\overline{C}$~~	~~$\overline{A} \cdot \overline{B}$~~	~~$\overline{A} \cdot \overline{C}$~~	($\overline{B} \cdot \overline{C}$)	~~$\overline{A} \cdot \overline{B} \cdot \overline{C}$~~	1
~~$\overline{A}$~~	~~$\overline{B}$~~	~~C~~	~~$\overline{A} \cdot \overline{B}$~~	~~$\overline{A} \cdot C$~~	~~$\overline{B} \cdot C$~~	~~$\overline{A} \cdot \overline{B} \cdot C$~~	2
~~$\overline{A}$~~	~~B~~	~~$\overline{C}$~~	~~$\overline{A} \cdot B$~~	~~$\overline{A} \cdot \overline{C}$~~	~~$B \cdot \overline{C}$~~	~~$\overline{A} \cdot B \cdot \overline{C}$~~	3
~~$\overline{A}$~~	~~B~~	~~C~~	~~$\overline{A} \cdot B$~~	~~$\overline{A} \cdot C$~~	($B \cdot C$)	~~$\overline{A} \cdot B \cdot C$~~	4
~~A~~	~~$\overline{B}$~~	~~$\overline{C}$~~	($A \cdot \overline{B}$)	~~$A \cdot \overline{C}$~~	($\overline{B} \cdot \overline{C}$)	~~$A \cdot \overline{B} \cdot \overline{C}$~~	5
~~A~~	~~$\overline{B}$~~	~~C~~	($A \cdot \overline{B}$)	($A \cdot C$)	~~$\overline{B} \cdot C$~~	~~$A \cdot \overline{B} \cdot C$~~	6
~~A~~	~~B~~	~~$\overline{C}$~~	~~$A \cdot B$~~	~~$A \cdot \overline{C}$~~	~~$B \cdot \overline{C}$~~	~~$A \cdot B \cdot \overline{C}$~~	7
~~A~~	~~B~~	~~C~~	~~$A \cdot B$~~	($A \cdot C$)	($B \cdot C$)	~~$A \cdot B \cdot C$~~	8

Figure 240-2

From the Harvard chart of Fig. 240-2, the simplified result is:

$$f(A,B,C) = A \cdot \overline{B} + A \cdot C + \overline{B} \cdot \overline{C} + B \cdot C$$

At first glance, this does not appear to be the same result because there is an additional $A \cdot C$ term. However,

$$f(A,B,C) = A \cdot \overline{B} + A \cdot C + \overline{B} \cdot \overline{C} + B \cdot C$$

$$= A \cdot \overline{B} + A \cdot \overline{B} \cdot (C + \overline{C}) \quad (237\text{-}3, 7)$$

$$+ A \cdot B \cdot C + \overline{B} \cdot \overline{C} + B \cdot C$$

$$= A \cdot \overline{B} + A \cdot \overline{B} + A \cdot B \cdot C \quad (237\text{-}9)$$

$$+ B \cdot C + \overline{B} \cdot \overline{C}$$

$$= A \cdot \overline{B} + B \cdot C \cdot (1 + A) + \overline{B} \cdot \overline{C}$$

$$= A \cdot \overline{B} + B \cdot C + \overline{B} \cdot \overline{C} \quad (237\text{-}3)$$

$$= B \cdot C + \overline{B} \cdot (A + \overline{C})$$

The $A \cdot C$ term has been absorbed and the two results are now the same.

PRACTICE PROBLEMS

240-1. Use the Harvard chart to simplify the Boolean equation: $f(A,B,C,D) = A \cdot \overline{B} \cdot \overline{C} + A \cdot C \cdot \overline{D} + \overline{C} + \overline{D} + A \cdot C \cdot \overline{D} + \overline{A} \cdot \overline{B} \cdot C \cdot D$.

240-2. Use the Harvard chart to simplify the Boolean equation: $f(A,B,C) = A \cdot \overline{B} \cdot C + (B + \overline{C}) \cdot (\overline{B} + C)$.

9
PART
Satellite Communications

241
Introduction to satellite communications—the low-earth orbit

The basic elements of a satellite communications system are a spacecraft and two earth stations. The spacecraft must be capable of intercepting a signal from one earth station. After amplification, frequency translation, and filtering, the signal is then passed on to the second earth station. This is obviously an active process as opposed to a passive one. The first satellite to be positioned in a low-earth orbit was the passive Echo I. This radio reflector was an aluminum-coated Mylar balloon with a diameter of 30 m. Echo I was put into operation during August, 1960, some 3 years after the October 4, 1957, launch of Sputnik I, which contained only a transmitter and therefore cannot be categorized as a communications satellite.

The first active communications satellite was Courier I, which was launched on October 4, 1960, by the Department of Defense. It only operated for 17 days but was capable of storing up to 360,000 teletype words, which were transmitted from one earth station and then retransmitted by the spacecraft as it passed over another earth station.

The next milestone was the launching of the satellite Telstar I on July 10, 1962. This spacecraft was designed and built by Bell Telephone Laboratories. Approximately spherical in shape, its diameter was about 0.9 m with a mass of 77 kg. The satellite required 15 W for its operation, and this power was supplied from 19 nickel-cadmium cells which were recharged by the output from 3600 solar cells mounted on the spacecraft's spherical shell.

Telstar I was the first experimental satellite capable of simultaneous transmission and reception across the Atlantic. It was used to transmit black and white TV, color TV, multichannel telephony, telegraphy, and facsimile between the United States and Europe. The required electronic circuitry consisted of 1464 diodes, 1066 transistors, and one traveling-wave tube (TWT), which acted as the satellite's high-power amplifier (HPA).

To communicate with Telstar, a U.S. earth station was constructed at Andover, Maine. Its antenna used a very large reflector, which was 54 m long and 29 m high. This antenna system weighed 380 tons and was mounted on a circular track having a diameter of 55 m. The antenna system could then be rotated to follow the orbiting satellite. As part of the Telstar system, comparable land stations were built in England and France.

Echo, Courier, and Telstar were all placed in low-earth orbits (LEOs). Such satellites have relatively short orbital periods and circle the earth in less than 2 h (Fig. 241-1). Consequently they are only in sight of a particular land station for a limited period of time and must be tracked continuously if communications are to be maintained.

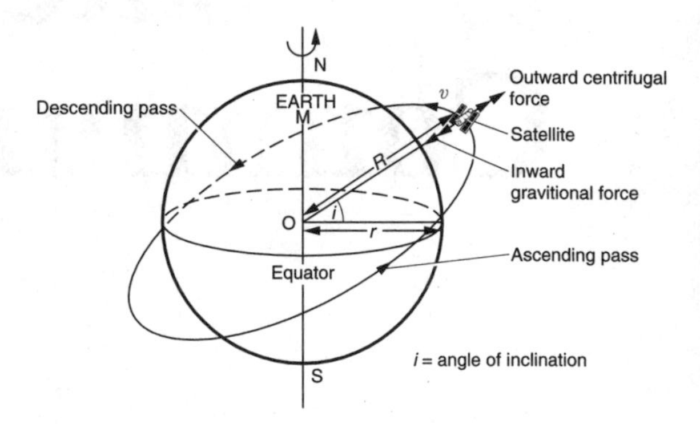

Figure 241-1

MATHEMATICAL DERIVATIONS

When a communications satellite is orbiting around the earth, it is pursuing a curved path so that an outward centrifugal force is exerted on the satellite. To illustrate the phenomenon of centrifugal force, attach a rock to a piece of string and then whirl the rock over your head. The faster the rock is whirled, the greater is the tension in the string. When the string eventually breaks, the rock will fly off due to the centrifugal force. The magnitude of the centrifugal force depends on the mass of the rock, its velocity, and the length of the string.

In equation form,

$$\text{Centrifugal force, } F_C = \frac{mv^2}{R} \qquad (241\text{-}1)$$

where: F_C = centrifugal force (N)
m = mass (kg)
R = radius of curved path (m)
v = velocity (m/s).

For a stable orbit the outward centrifugal force must be balanced by the inward gravitational force of attraction between the satellite and the earth (Fig. 241-1). By Newton's law,

$$\text{Gravitational force, } F_G = \frac{GMm}{R^2}. \qquad (241\text{-}2)$$

where: F_G = gravitational force (N)
G = universal gravitational constant (6.67×10^{-11} N-m^2/kg^2)
M = mass of the earth (5.975×10^{24} kg)
m = mass of the satellite (kg)
R = distance between the satellite and the center of the earth (m).

For a stable orbit,

$$F_C = F_G$$

$$\frac{mv^2}{R} = \frac{GMm}{R^2}$$

This yields

$$v = \sqrt{\frac{GM}{R}} \text{ m/s} \qquad (241\text{-}3)$$

or

$$R = \frac{GM}{v^2} \text{ m.} \qquad (241\text{-}4)$$

Equation 241-3 is true for any satellite traveling in a circular orbit around the earth. Consequently the satellite could be in an equatorial orbit or a polar orbit, or its orbit could have any angle of inclination between 0° and 90°. The angle of inclination is the angle between the plane of the satellite's orbit and the equatorial plane (Fig. 241-1).

Example 241-1

A satellite is traveling with a velocity of 12,000 mi/h. How far above the surface of the earth is the satellite's circular orbit?

Solution

Distance from the satellite to the center of the earth,

$$R = \frac{GM}{v^2} \qquad (241\text{-}4)$$

12,000 mi/h = 5364 m/s, radius of the earth, r = 3963 mi = 6.38×10^6 m.

Therefore,

$$R = \frac{6.67 \times 10^{-11} \times 5.975 \times 10^{24}}{5364^2}$$

$$= 13.85 \times 10^6 \text{ m.}$$

The distance of the satellite's orbit above the earth's surface is $13.85 \times 10^6 - 6.38 \times 10^6 = 7.47\times10^6$ m = 4640 mi.

Example 241-2

A satellite is moving in a circular orbit, which is 500 mi above the earth's surface. Calculate the velocity of the satellite in miles per hour.

Solution

Distance from the satellite to the center of the earth,

$R = 3963 + 500 = 4463 \text{ mi} = 7.185 \times 10^6$ m.

Velocity of the satellite,

$$v = \sqrt{\frac{GM}{R}} \qquad (241\text{-}3)$$

$$= \sqrt{\frac{6.67 \times 10^{-11} \times 5.975 \times 10^{24}}{7.185 \times 10^6}}$$

= 7477 m/s

= 16730 mi/h, rounded off.

PRACTICE PROBLEMS

241-1. A space shuttle is traveling in a circular orbit that is 400 mi above the earth's surface. Calculate the velocity of the shuttle.

241-2. A satellite is traveling in a circular orbit with a velocity of 9500 mi/h. As measured from the center of the earth, what is the radius of the orbit?

242
The geostationary satellite—angles of elevation and azimuth

As we learned in chapter 241, LEO satellites have relatively short orbital periods and circle the earth in only a few hours. By contrast, the journey of the moon (another earth satellite) around the earth is completed in approximately 28 days. Between these two extremes there must exist an equatorial circular orbit which lasts for one day. Such an orbit allows a satellite's position to be fixed *relative* to the earth, although the satellite's velocity in space is about 11,000 km/h. The satellite is then in a geostationary earth orbit (GEO) and is approximately 35,800 km above the earth's surface at the equator (Fig. 242-1). The main advantage of such a satellite is the use of virtually fixed antennas at the land stations since minimal tracking is required. However the extreme height of the satellite demands that the land stations have a transmitter power on the order of kilowatts, large directional antenna systems, and a highly sensitive low-noise receiver.

In 1960 Hughes Aircraft Company of Los Angeles started to develop an experimental geostationary satellite called Syncom. In order to use existing launch vehicles to carry the spacecraft to the required altitude, it was necessary to design a new type of satellite which was both lightweight and spin-stabilized. Syncom was finally launched on August 19, 1964, and was the first satellite to be placed in the geostationary orbit. It was subsequently used to transmit TV pictures from the Tokyo Olympic games. This success clearly

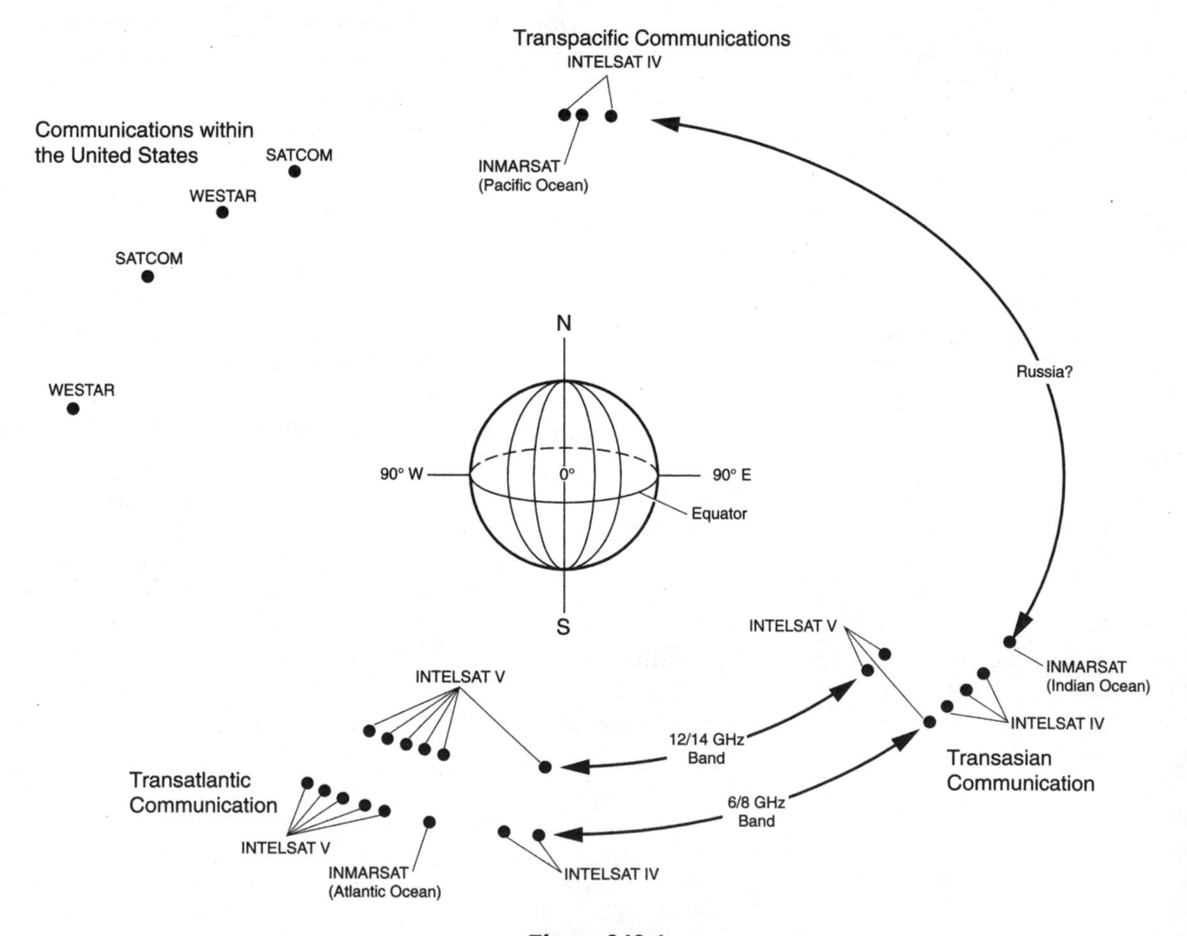

Figure 242-1

demonstrated the commercial possibility of satellite communications. As a result the U.S. Congress established the Communications Satellite Corporation (COMSAT) to oversee the development of satellite communications in the United States.

As a first step COMSAT contracted with Hughes Aircraft to design and construct the first *commercial* communications satellite for the International Telecommunications Satellite consortium (INTELSAT). Although the official name of this satellite was INTELSAT I, it was more commonly known as Early Bird, whose commercial service began on June 28, 1965. It provided 240 two-way translantic voice circuits which communicated with five land stations. Many times during its life Early Bird showed its worth by taking over from a failed transatlantic submarine cable. Early Bird was the first of a continuing series of evolving satellites which are now in their tenth generation.

By the end of 1984 there were more than 40 active satellites in the geostationary orbit. The 14 INTELSAT satellites were stretched over the Atlantic, Pacific, and Indian oceans and operated in conjunction with more than 500 land stations. Each of these satellites had a capacity of over 12,000 two-way voice circuits and a number of TV channels. Another international organization, INMARSAT, provided communications for ships at sea, and its satellites were positioned over the Atlantic, Pacific, and Indian oceans. Regional organizations such as SATCOM (RCA), GSTAR (GTE), and WESTAR (Western Union) have launched additional satellites into the increasingly crowded geostationary orbit (Fig. 242-1). The present U.S. standard for orbital spacing in the C band is 4° with a tolerance of ±0.1°, so that the maximum number of active geostationary satellites operating on the same frequencies is 360°/4° = 90. Although most of the satellites presently in orbit are controlled by the United States and the Russian republics, the remainder are owned and operated by such nations as Great Britain, Japan, China, Brazil, and India. Clearly the field of satellite communications is now a mature technology.

MATHEMATICAL DERIVATIONS

To an observer positioned on the sun the period of the earth's rotation on its axis is 24 h, or one solar day. This time interval takes into account the earth's orbit around the sun. However, the rotation of a satellite around the earth is independent of the sun's position, so that instead we must consider the period of the earth's rotation in relation to the fixed stars. The value of this period is 23.94 h, which is referred to as a sidereal day.

For a satellite to remain in a geostationary orbit, its direction of rotation must be the same as that of the earth and the orbit must be equatorial. In addition the period, T, of the satellite's orbit must be 23.94 h, so that the angular velocities of the satellite and the earth are the same.

Figure 242-2 shows a satellite in a geostationary orbit. The distance traveled by the satellite in one com-

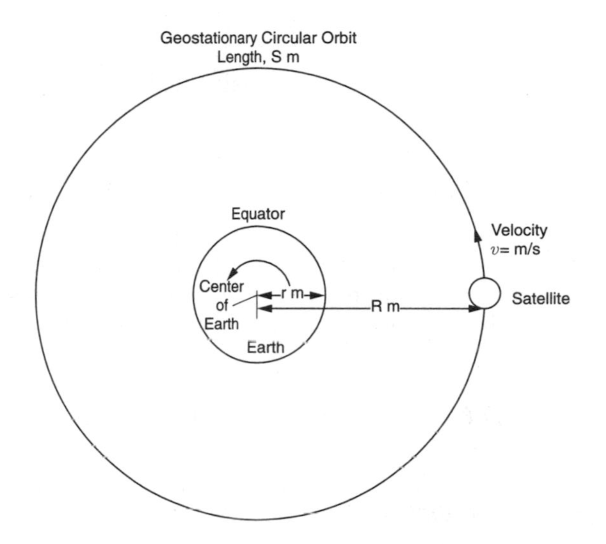

Figure 242-2

plete circular orbit is $S = 2\pi R$, and therefore the satellite's velocity is given by

$$v = \frac{S}{T} \qquad (242\text{-}1)$$

$$= \frac{2\pi R}{T}$$

$$= \frac{2\pi R}{23.94 \times 3600} \text{ m/s}$$

where v = satellite's velocity (m/s)
R = distance from the satellite to the center of the earth (m)
T = period of orbit (s).

Combining Equations 241-3 and 242-1,

$$v = \sqrt{\frac{GM}{R}} = \frac{2\pi R}{23.94 \times 3600} \text{ m/s}.$$

This yields

$$R = \sqrt[3]{\frac{GM \times 23.94^2 \times 3600^2}{4\pi^2}}$$

$$= \sqrt[3]{\frac{6.67 \times 10^{-11} \times 5.975 \times 10^{24} \times 23.94^2 \times 3600^2}{4\pi^2}}$$

$$= 4.217 \times 10^7 \text{ m}$$

$$= 26{,}203 \text{ mi.}$$

Substituting this result in Equation 241-3, the satellite velocity is

$$v = \sqrt{\frac{GM}{R}}$$

$$= \sqrt{\frac{6.67 \times 10^{-11} \times 5.975 \times 10^{24}}{4.217 \times 10^7}}$$

$$= 3047 \text{ m/s}$$

$$= 6877 \text{ mi/h.}$$

Therefore the velocity of a geostationary satellite is 6877 mi/h, and the distance between the satellite and the center of the earth is 26,203 mi. The distance from the satellite to the earth's surface is 26,203−3963 = 22,240 mi.

Angles of elevation and azimuth

Figure 242-3 is a three-dimensional diagram in which a geostationary satellite, *G*, is viewed from a land station, *S*. *TS* is a tangent to the earth at the position of the land station. The angle *GST* is in the vertical plane

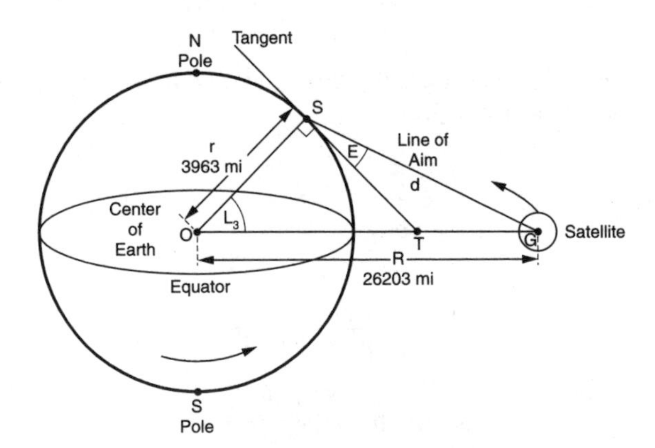

Figure 242-3

and measures the elevation, *E*, of the satellite. The value of *E* will depend on the latitude and longitude of the earth station and on the longitude of the geostationary satellite.

The value of *E* is given by

$$E = \tan^{-1}\left[\frac{\cos(L_1 - L_2)\cos L_3 - 3963/26{,}203}{\sqrt{(1 - \cos 2(L_1 - L_2)\cos 2L_3)/2}}\right] \quad (242\text{-}2)$$

where: E = angle of elevation (°)
L_1 = longitude of satellite (°)
L_2 = longitude of land station (°)
L_3 = latitude of land station (°).

Theoretically the value of *E* ranges from 0° if the land station is situated at the poles, to 90° if the land station is on the equator and directly beneath the satellite. An inclinometer can be used to obtain a direct measurement of the elevation angle.

For shallow angles of elevation (less than 10°), there is an increase in the signal's path length from the satellite through the earth's atmosphere to the receiver. This has two effects:

1. The signal suffers a higher degree of attenuation.
2. The signal is modulated by the atmospheric noise to a greater extent.

As a result the value of the signal-to-noise ratio at the receiver is lowered and is generally unacceptable if the angle of elevation is less than 5°. If the longitude of the satellite and the land station are the same, Fig. 242-3 becomes a two-dimensional diagram.

Using the sine rule in Δ *GSO*,

$$\frac{\sin(90° - E - L_3)}{\sin(90° + E)} = \frac{3963}{26{,}203} = 0.1512$$

$$\frac{\cos(E + L_3)}{\cos E} = \cos L_3 - \sin L_3 \tan E = 0.1512$$

$$E = \tan^{-1}\left[\frac{\cos L_3 - 0.1512}{\sin L_3}\right] \quad (242\text{-}3)$$

This result can be derived from Equation 242-2 by substituting $L_1 - L_2 = 0$.

If $L_3 = 75°$, $E = 6.38°$. It follows that in the polar regions, where L_3 is greater than 75°, it is impossible to receive adequate signals from geostationary satellites.

In order to aim a land station's antenna toward a geostationary satellite, it is required to know:

1. The angle of elevation, *E*, which is measured in the vertical plane.
2. The angle of azimuth, *A*.

The azimuth is measured in the horizontal plane and is the angle between the bearing of the satellite and true North. To an observer in the northern latitudes the geostationary orbit is an arc, which starts at the horizon in the SE direction. The arc reaches its maximum height in the direction that is due South of the observer and then ends at the SW horizon. Since the arc is not moving, the antenna of a land station may be aimed (Fig. 242-4) at any satellite whose position lies within the arc.

The angle of azimuth is given by

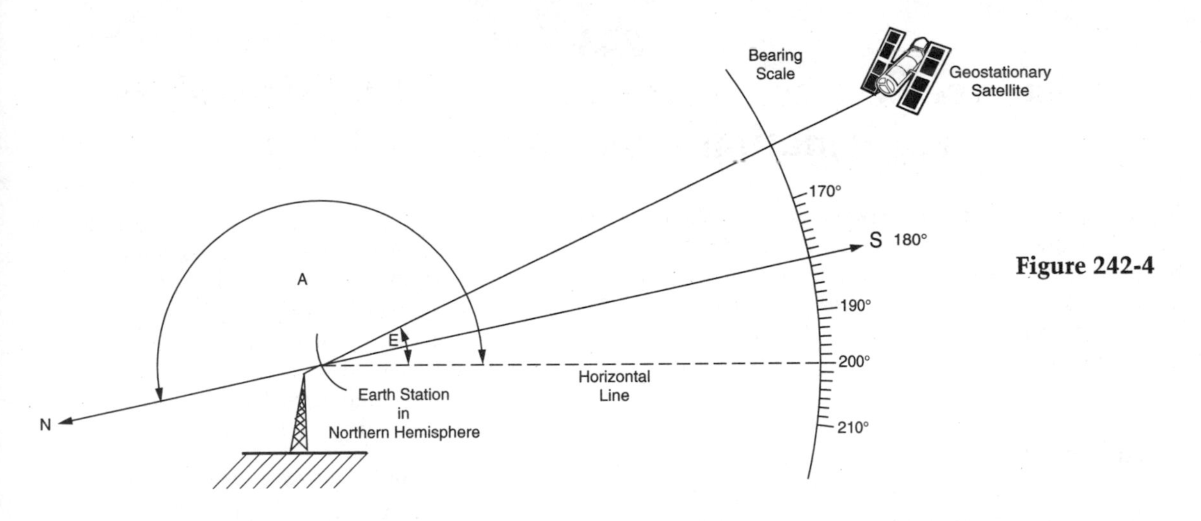

Figure 242-4

$$A = 180° + \tan^{-1}\left[\frac{\tan(L_1 - L_2)}{\sin L_3}\right] \qquad (242\text{-}4)$$

where: A = azimuth angle (°)
L_1 = longitude of satellite (°)
L_2 = longitude of land station (°)
L_3 = latitude of land station (°).

If required, the true heading can be converted to a magnetic compass bearing.

Example 242-1

An earth station is located at a position of latitude 42° N, longitude 98° W. If a geostationary satellite is positioned in the longitude of 112° W, calculate the angles of elevation and azimuth as the earth station's antenna is aimed at the satellite.

Solution

$$L_1 = 112°\text{ W}, L_2 = 98°\text{ W}, L_3 = 42°\text{ N}.$$

Angle of elevation,

$$E = \tan^{-1}\left[\frac{\cos(112° - 98°)\cos 42° - 0.1512}{(1 - \cos 2(112° - 98°)\cos 2 \times 42°)/2}\right] \qquad (242\text{-}2)$$

$$= \tan^{-1}\left[\frac{\cos 14° \cos 42° - 0.1512}{(1 - \cos 28° \cos 84°)/2}\right]$$

$$= 40.4°$$

Angle of azimuth,

$$A = 180° + \tan^{-1}\left[\frac{\tan(112° - 98°)}{\sin 42°}\right] \qquad (242\text{-}4)$$

$$= 180° + \tan^{-1}\left[\frac{\tan 14°}{\sin 42°}\right]$$

$$= 200.4°.$$

From the practical point of view the axis of the receiver's antenna can be moved in elevation and azimuth by means of motors which are remotely controlled by a computer. When a particular satellite is located, the necessary information can be fed into the computer; that same satellite can then be readily accessed in the future.

PRACTICE PROBLEMS

242-1. An earth station is located at a position of latitude 42° N, longitude 73° W. It is communicating with an INMARSAT geostationary satellite, which has a longitude of 15° W. Calculate the angles of elevation and azimuth as the earth station's antenna is aimed at the satellite.

242-2. Regarding the moon as a natural satellite which is 384,403 km from the center of the earth, calculate the moon's velocity in kilometers per hour.

243
Distance between a geostationary satellite and its earth station—spacing of satellites

The signal loss in decibels between an earth station's transmitter and a geostationary satellite (and between the satellite and another earth station's receiver) is primarily determined by the distance between the earth station and the satellite. This distance is dependent on the latitude and the longitude of the earth station as well as the longitude of the satellite.

Spacing of satellites

In its simplest form a commercial communications satellite system consists of an earth station transmitter, a spacecraft which behaves as a transponder, and an earth station receiver. To avoid interference the uplink frequencies for communicating between the transmitter and the satellite are different from those used on the downlink to the receiver. The 4/6-GHz (*C*) band is the most commonly used for commercial satellite communications. The uplink 6-GHz band covers the range of frequencies between 5925 MHz and 6425 MHz, while the downlink 4-GHz band extends from 3700 MHz to 4200 MHz. The bandwidth for either link is $6425-5925 = 4200-3700 = 500$ MHz.

The minimum allowed spacing in the geostationary orbit is determined by the earth station's ability to distinguish between the signals from adjacent satellites operating in the same band. The U.S. standard for the orbital spacing of *C*- band satellites is an assigned minimum of 4° with a tolerance of ±0.1°. Consequently two adjacent *C*-band satellites must not be closer than $2\times\pi\times 26{,}203$ mi$\times 4°/360° = 182.9$ mi (±48 mi). It also follows that the entire geostationary orbit must not contain more than 360°/4° = 90 *C*-band satellites. However, some of the more modern satellites operate in the 14-GHz/11-GHz *Ku* band (uplink, 14 GHz to 14.5 GHz; downlink, 11.7 GHz to 12.2 GHz). A *C*-band satellite and a *Ku*-band satellite can be placed in the same longitude without any interference problems. Since the minimum spacing for the *Ku*-band satellites is 3°, it is possible to place a further 360°/3° = 120 *Ku* satellites in the geostationary orbit. Between the *C* and *Ku* bands is the 7/8-GHz band that is used by military satellites.

MATHEMATICAL DERIVATIONS

Referring to Fig. 242-3,

$$d = \sqrt{R^2+r^2-2\,Rr\cos(L_1-L_2)\cos L_3} \quad (243\text{-}1)$$

where: d = distance between the geostationary satellite and the earth station (m)
R = distance from the satellite to the center of the earth (4.217×10^7 m)
r = radius of the earth (6.38×10^6 m)
L_1 = longitude of satellite (°)
L_2 = longitude of earth station (°)
L_3 = latitude of earth station (°).

Since the values of R and r are fixed, the maximum value of d occurs when $L_1 - L_2 = 90°$ and/or $L_3 = 90°$. This maximum value is $\sqrt{R^2+r^2} = \sqrt{(4.217\times10^7)^2+(6.38\times10^6)^2} = 4.625\times10^7$ m = 26,501 mi.

When $L_1 = L_2$ and/or $L_3 = 0°$, d has its minimum value of $d = \sqrt{R^2+r^2-2Rr} = R-r = 4.217\times10^7 - 6.38\times10^6 = 3.579\times10^7$ m = 22,240 mi.

Example 243-1

An earth station is located at the position of 42° N, 98° W. If a geostationary satellite is placed in the longitude of 112° W, determine the distance between the satellite and the earth station.

Solution

The distance between the satellite and the earth station is

$$d = \sqrt{(4.217\times10^7)^2+(6.38\times10^6)^2-2\times4.217\times10^7\times 6.38\times10^6\times\cos(112°-98°)\cos 42°} \quad (243\text{-}1)$$

$$= 3.78\times10^7 \text{ m}$$

$$= 23{,}507 \text{ mi.}$$

PRACTICE PROBLEMS

243-1. An earth station is located at a position of latitude 37° N, longitude 120° W. It is communicating with a geostationary satellite in a longitude of 176.5° E. Determine the distance between the satellite and the earth station.

243-2. In the geostationary orbit one satellite is operating in the *C* band whereas a second one uses the *Ku* band. If the distance between the two satellites is 2208 km, what is their angular separation in degrees?

244
The satellite signal—frequency-division multiplex

So far we know that the majority of the geostationary satellites operate in the 4/6-GHz *C* band and that the bandwidth for each satellite is 500 MHz. But what is the nature of the signals to be transmitted and how do we make efficient use of the large bandwidth available?

The signals fall into three main categories:

1. *Television signals* The video, chroma (color), and aural information is normally transmitted and received as analog signals since the use of digital signals would require an excessive bandwidth. In order to achieve an improvement in the signal-to-noise ratio at the receiver, the microwave carrier is frequency modulated by the TV information. This differs from TV broadcast, in which the picture signal information amplitude modulates the video carrier.
2. *Telecommunications (telephone) signals* The audio information can either be directly relayed in its analog form or be converted into a digital signal before modulation occurs. The use of digital transmission will improve the quality of the signal at the receiver.
3. *Computer signals* (digital data).

In addition satellites may be used to transmit and receive facsimile and telegraphy signals.

In order to make efficient use of the 500-MHz bandwidth a number of signals are combined to produce a composite signal. This process is made possible through the use of either frequency-division multiplex (FDM) or time-division multiplex (TDM). For example, the 500-MHz bandwidth is commonly divided into 24 channels so that it is possible to transmit and receive 24 separate signals simultaneously. Each channel is allocated a bandwidth of 36 MHz and can be used either for 1500 telephone signals, or for one color television signal, or for 6×10^7 bits per second (bps) of computer data.

Frequency-division multiplex

Let us consider the transmission and reception of a TV color signal via satellite. The bandwidth for the composite video and chroma information is 4.5 MHz. This frequency modulates the microwave carrier in the range of 5.925 GHz to 6.425 GHz. The bandwidth includes the video and chroma information together with the synchronizing, blanking, and equalizing pulses. The aural (sound) signal has a bandwidth of 50 Hz to 15 kHz and is frequency modulated on to a 6.8-MHz subcarrier.

The maximum frequency deviation of the composite video signal is 10.5 MHz, and the peak deviation of the aural subcarrier is 2 MHz. Consequently the total peak deviation is $10.5 + 2.0 = 12.5$ MHz, and the required bandwidth is $2(12.5 + 6.8) = 38.6$ MHz (Carson's rule). This can be reasonably accommodated by the allocated bandwidth of 36 MHz. Notice that 36 MHz is a large bandwidth when compared with the 6 MHz required by a terrestrial TV broadcast.

If a bandwidth of 36 MHz is assigned to each transponder, 12 such transponders have a total bandwidth of $36 \times 12 = 432$ MHz. Allowing for guard bands, the total assigned bandwidth of 500 MHz is adequate.

It is possible to double the use of the same frequency range and provide 24 channels by operating 12 channels with vertical transmitting and receiving antennas while the other 12 channels use horizontal transmitting and receiving antennas. For example, all the even-numbered channels (2, 4, 6,...24) use vertically polarized transmissions so that the odd-numbered channels (1, 3, 5,...23) are horizontally polarized. As a result there would be minimum interference between two channels with adjacent numbers such as channel 8 and channel 9.

To further our understanding of FDM we will assume that each of the 24 channels is simultaneously modulated by one color TV signal. Tables 244-1 and 244-2 list the carrier frequencies for the odd-numbered and the even-numbered channels, respectively. You will observe that in each table the channel width is 40 MHz, which accommodates the required bandwidth of 38.6 MHz and allows for a guard band of 1.4 MHz.

At the earth station transmitter the carrier frequency for channel 1 is 5945 MHz. After this carrier is frequency modulated by the information of the first TV signal, the sidebands will be accommodated within the frequency range of 5925 MHz to 5965 MHz. Using horizontal polarization this FM signal is transmitted on the uplink to the satellite. On the downlink the carrier frequency has been converted to 3720 MHz and the sidebands lie within the range of 3700 MHz to 3740 MHz. At the receiver a filter selects this signal, which is subsequently demodulated so that the first TV signal is reproduced (Fig. 244-1). The same process is applied to each of the odd-numbered channels. To summarize, the principle of FDM is to divide the available wide bandwidth into narrow bands, each of which is used to transmit a separate channel. A number of channels can then be transmitted as a composite signal over a common path by operating each channel at a different fre-

Table 244-1. Carrier frequencies for horizontally polarized odd-numbered channels.

Channel number	Uplink *C* band, MHz	Downlink *C* band, MHz	Uplink *Ku* band, MHz	Downlink *Ku* band, MHz
1	5945	3720	14,029	11,729
3	5985	3760	14,088	11,788
5	6025	3800	14,147	11,847
7	6065	3840	14,206	11,906
9	6105	3880	14,265	11,965
11	6145	3920	14,324	12,024
13	6185	3960	14,383	12,083
15	6225	4000	14,442	12,142
17	6265	4040		
19	6305	4080		
21	6345	4120		
23	6385	4160		

Table 244-2. Carrier frequencies for vertically polarized even-numbered channels.

Channel number	Uplink *C* band, MHz	Downlink *C* band, MHz	Uplink *Ku* band, MHz	Downlink *Ku* band, MHz
2	5965	3740	14,058.5	11,758.5
4	6005	3780	14,117.5	11,817.5
6	6045	3820	14,176.5	11,876.5
8	6085	3860	14,235.5	11,935.5
10	6125	3900	14,294.5	11,994.5
12	6165	3940	14,353.5	12,053.5
14	6205	3980	14,412.5	12,112.5
16	6245	4020	14,471.5	12,171.5
18	6285	4060		
20	6325	4100		
22	6365	4140		
24	6405	4180		

quency. Figure 244-2 illustrates the principle of FDM for the even-numbered channels.

Figure 244-3 shows another use of FDM in which 12 independent telephone voice signals are multiplexed into a composite signal. Each of the analog voice signals is allowed a bandwidth of 4 kHz so that we can expect that the bandwidth of the composite signals is $12 \times 4 = 48$ kHz. Voice signal 1 is fed to a balanced modulator which suppresses the 108-kHz carrier. The upper sidebands are filtered out so that the remaining lower sidebands extend from 104 kHz to 108 kHz and form part of the composite signal. Similarly voice signal 2 modulates a 104-kHz carrier. The lower sidebands range from 100 kHz to 104 kHz and appear in the composite signal. Finally voice signal 12 modulates a 64-kHz carrier and the output sidebands range from 60 kHz to 64 kHz. The composite signal therefore has sidebands whose frequencies extend from 60 kHz to 108 kHz, with the anticipated bandwidth of 48 kHz.

To provide more telephone signals for the 5945-MHz carrier of channel 1, the frequency range of one whole 12-voice-signal group is translated from 60–180 kHz up to 312–360 kHz. This is combined with four other similar groups whose frequency bands are 360–408 kHz, 408–456 kHz, 456–504 kHz, and 504–552 kHz. Then the composite signal for the new super-

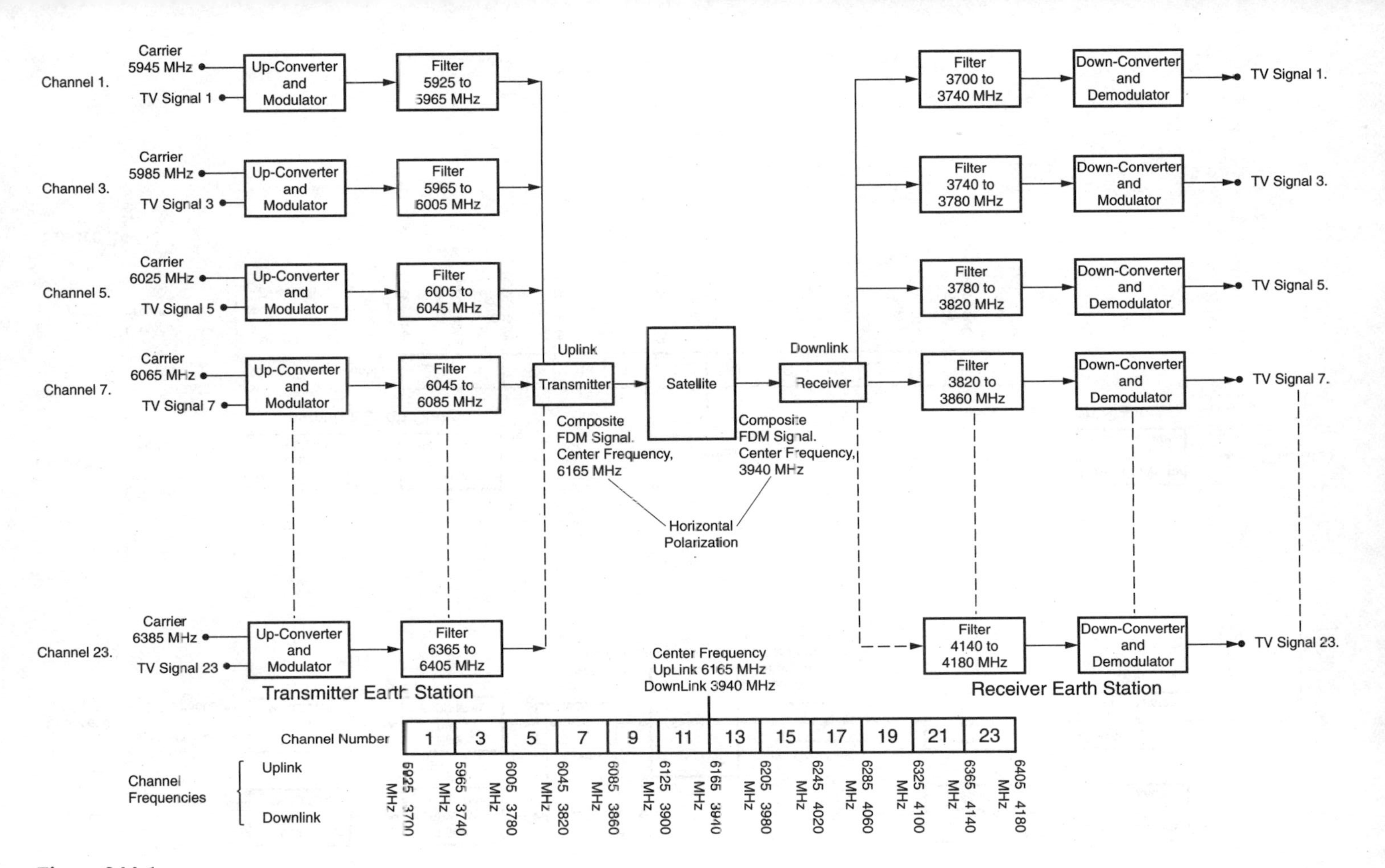

Channel 1.
Carrier 5945 MHz
TV Signal 1
Up-Converter and Modulator
Filter 5925 to 5965 MHz
Channel 3.
Carrier 5985 MHz
TV Signal 3
Up-Converter and Modulator
Filter 5965 to 6005 MHz
Channel 5.
Carrier 6025 MHz
TV Signal 5
Up-Converter and Modulator
Filter 6005 to 6045 MHz
Channel 7.
Carrier 6065 MHz
TV Signal 7
Up-Converter and Modulator
Filter 6045 to 6085 MHz
Channel 23.
Carrier 6385 MHz
TV Signal 23
Up-Converter and Modulator
Filter 6365 to 6405 MHz
Transmitter Earth Station
Uplink
Transmitter
Satellite
Downlink
Receiver
Composite FDM Signal. Center Frequency, 6165 MHz
Composite FDM Signal. Center Frequency, 3940 MHz
Horizontal Polarization
Filter 3700 to 3740 MHz
Down-Converter and Demodulator
TV Signal 1.
Filter 3740 to 3780 MHz
Down-Converter and Modulator
TV Signal 3.
Filter 3780 to 3820 MHz
Down-Converter and Demodulator
TV Signal 5.
Filter 3820 to 3860 MHz
Down-Converter and Demodulator
TV Signal 7.
Filter 4140 to 4180 MHz
Down-Converter and Demodulator
TV Signal 23.
Receiver Earth Station
Center Frequency UpLink 6165 MHz DownLink 3940 MHz
Channel Number
1 3 5 7 9 11 13 15 17 19 21 23
Channel Frequencies
Uplink
Downlink
5925 3700 MHz
5965 3740 MHz
6005 3780 MHz
6045 3820 MHz
6085 3860 MHz
6125 3900 MHz
6165 3940 MHz
6205 3980 MHz
6245 4020 MHz
6285 4060 MHz
6325 4100 MHz
6365 4140 MHz
6405 4180 MHz

Figure 244-1

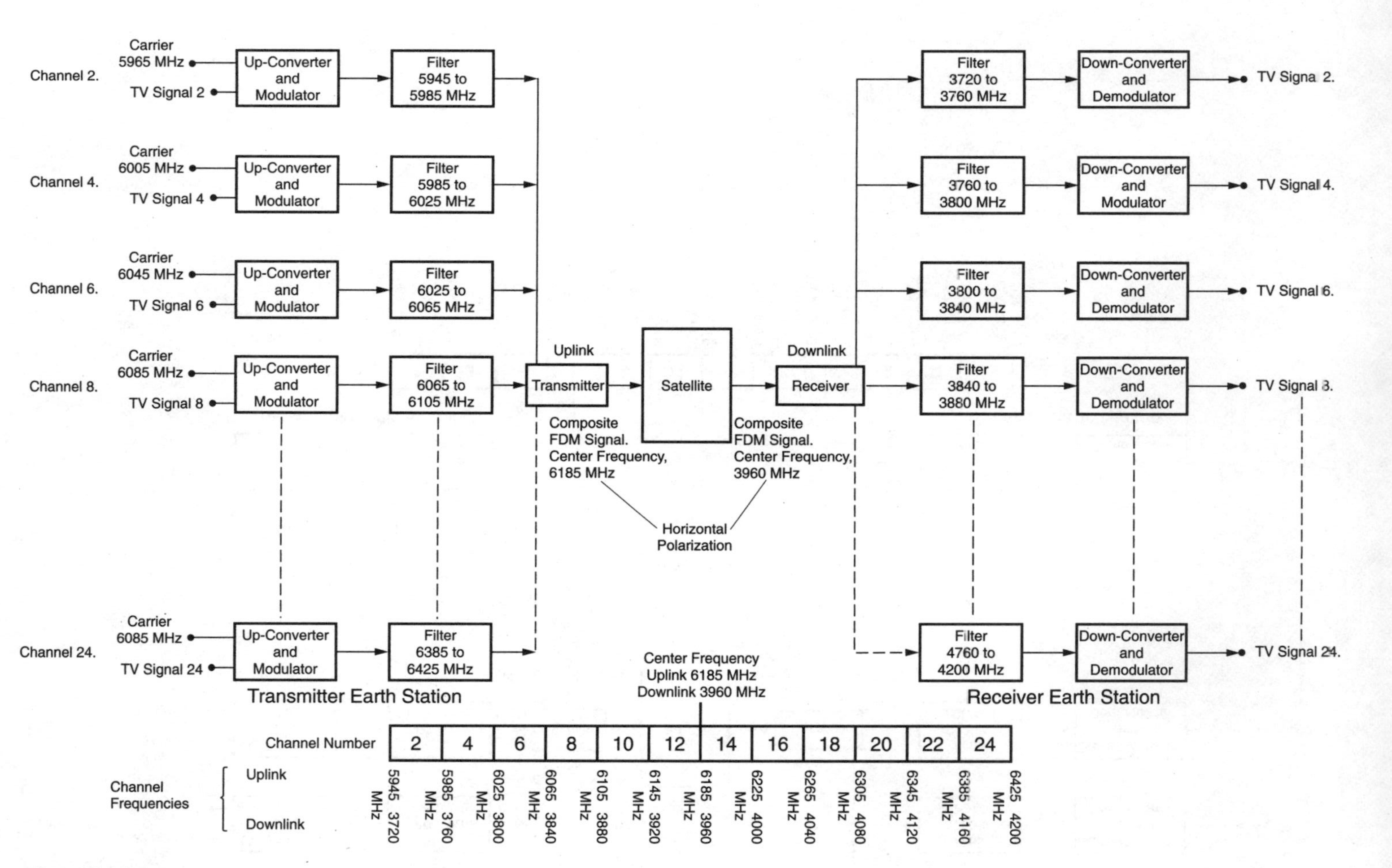

Channel 2.
Carrier 5965 MHz
TV Signal 2
Up-Converter and Modulator
Filter 5945 to 5985 MHz
Channel 4.
Carrier 6005 MHz
TV Signal 4
Up-Converter and Modulator
Filter 5985 to 6025 MHz
Channel 6.
Carrier 6045 MHz
TV Signal 6
Up-Converter and Modulator
Filter 6025 to 6065 MHz
Channel 8.
Carrier 6085 MHz
TV Signal 8
Up-Converter and Modulator
Filter 6065 to 6105 MHz
Channel 24.
Carrier 6085 MHz
TV Signal 24
Up-Converter and Modulator
Filter 6385 to 6425 MHz
Transmitter Earth Station
Uplink
Transmitter
Composite FDM Signal. Center Frequency, 6185 MHz
Satellite
Horizontal Polarization
Downlink
Receiver
Composite FDM Signal. Center Frequency, 3960 MHz
Filter 3720 to 3760 MHz
Down-Converter and Demodulator
TV Signa 2.
Filter 3760 to 3800 MHz
Down-Converter and Modulator
TV Signal 4.
Filter 3800 to 3840 MHz
Down-Converter and Demodulator
TV Signal 6.
Filter 3840 to 3880 MHz
Down-Converter and Demodulator
TV Signal 3.
Filter 4760 to 4200 MHz
Down-Converter and Demodulator
TV Signal 24.
Receiver Earth Station
Center Frequency Uplink 6185 MHz Downlink 3960 MHz
Channel Number
2 4 6 8 10 12 14 16 18 20 22 24
Channel Frequencies
Uplink
Downlink
5945 3720 MHz
5985 3760 MHz
6025 3800 MHz
6065 3840 MHz
6105 3880 MHz
6145 3920 MHz
6185 3960 MHz
6225 4000 MHz
6265 4040 MHz
6305 4080 MHz
6345 4120 MHz
6385 4160 MHz
6425 4200 MHz

Figure 244-2

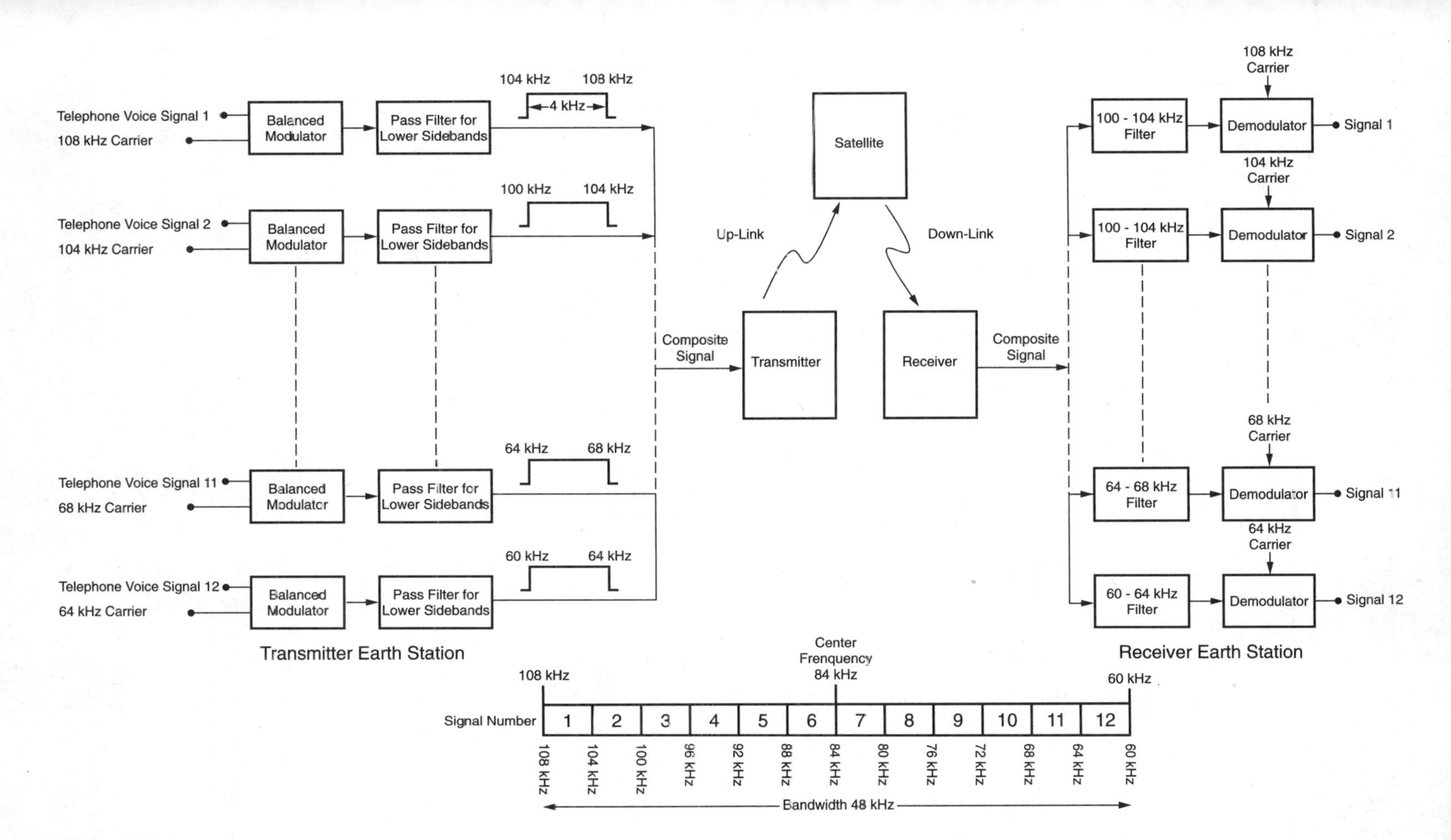

Telephone Voice Signal 1
108 kHz Carrier
Balanced Modulator
Pass Filter for Lower Sidebands
104 kHz
108 kHz
4 kHz
Telephone Voice Signal 2
104 kHz Carrier
100 kHz
104 kHz
Telephone Voice Signal 11
68 kHz Carrier
64 kHz
68 kHz
Telephone Voice Signal 12
64 kHz Carrier
60 kHz
64 kHz
Transmitter Earth Station
Composite Signal
Transmitter
Up-Link
Satellite
Down-Link
Receiver
Composite Signal
100 - 104 kHz Filter
64 - 68 kHz Filter
60 - 64 kHz Filter
108 kHz Carrier
104 kHz Carrier
68 kHz Carrier
64 kHz Carrier
Demodulator
Signal 1
Signal 2
Signal 11
Signal 12
Receiver Earth Station
108 kHz
Center Frenquency 84 kHz
60 kHz
Signal Number
1 2 3 4 5 6 7 8 9 10 11 12
108 kHz
104 kHz
100 kHz
96 kHz
92 kHz
88 kHz
84 kHz
80 kHz
76 kHz
72 kHz
68 kHz
64 kHz
60 kHz
Bandwidth 48 kHz

Figure 244-3

group will contain 5×12 = 60 voice signals whose sidebands range from 312 kHz to 552 kHz (60 × 4 = 240 kHz). In a like manner 10 supergroups can be multiplexed into a mastergroup of 600 voice signals. Six master-groups are then combined into a supermaster or jumbo group so that the final composite signal has a total of 3600 voice signals. This exceeds the requirement of 1500 telephone signals for a single channel of the satellite system.

On the receiver side the composite signal is first demodulated. Filters then separate out the various voice signals and each individual signal is passed on to the party being called.

So far we have discussed the application of FDM to the 4/6-GHz *C* band. However, the 12/14-GHz *Ku* band is also used for satellite communications (Fig. 242-1). The uplink band covers 14.0 GHz to 14.5 GHz, whereas the downlink band ranges from 11.7 GHz to 12.2 GHz. There are 16 channels, each with a bandwidth of 54 MHz and a guard band of 5 MHz. The carrier frequencies of these channels are shown in Tables 244-1 and 244-2. The even-numbered channels use vertical polarized transmissions whereas the odd-numbered channels are horizontally polarized. Advantages of operating in the *Ku* band, as compared with the *C* band, include:

1. There is less interference from terrestrial communications, which do not normally use the *Ku* band.
2. The effective radiated power from the satellite's transmitter can be increased.
3. The receiving antennas of earth stations do not have to be located in remote areas.
4. The required gain and beamwidth can be achieved with smaller antennas.

The main disadvantage of the *Ku* band is the greater degree of attenuation due to moisture in the atmosphere.

MATHEMATICAL DERIVATIONS

At the earth station transmitter, a composite video signal frequency modulates a 70-MHz subcarrier with a maximum frequency deviation of 18 MHz. This FM signal is passed to an up-converter (Fig. 244-4) that shifts the center frequency from 70 MHz to the carrier frequency, f_c, of the required uplink channel.

Referring to the principles of frequency conversion described in chapter 159, the frequency, f_o, of the microwave oscillator is given by

$$f_o = f_c - 70 \text{ MHz.} \qquad (244\text{-}1)$$

For example, channel 5 has an uplink carrier frequency of 6025 MHz. The frequency output of the microwave oscillator must therefore be 6025 − 70 = 5955 MHz. The mixer output also contains an FM signal with a carrier frequency of 5885 MHz, but this component is eliminated by the filter.

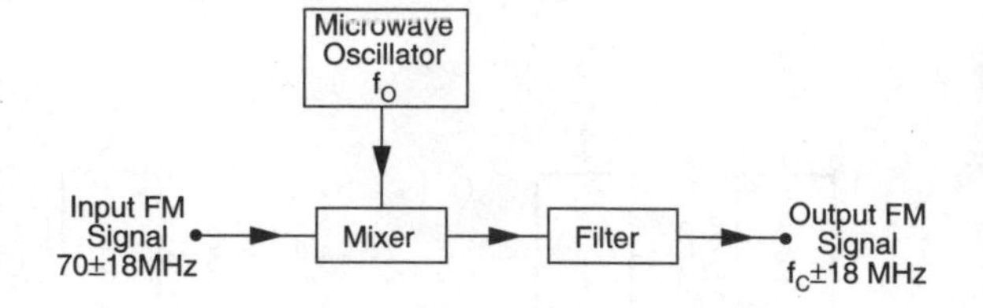

Figure 244-4

Similar frequency conversions appear in the earth station receiver.

Example 244-1

For the *C*-band channel 16, calculate the required frequency of the microwave oscillator in the up-converter circuit at the earth station transmitter.

Solution

Channel 16 carrier frequency, f_c = 6245 MHz.

Microwave oscillator frequency,

$$f_o = 6245 - 70 = 6175 \text{ MHz} \qquad (244\text{-}1)$$

Example 244-2

A satellite channel has a bandwidth of 36 MHz and is frequency modulated by a composite signal. If the modulation index is 2.0, determine the frequency deviation, f_d, and the bandwidth, f_m, of the composite signal. What are the possible composite signals that can occupy this bandwidth?

Solution

By Carson's rule,

$$\text{Channel bandwidth} = 2(f_d + f_m) = 36 \text{ MHz}$$

$$\text{Modulation index, } m = \frac{f_d}{f_m} = 2.$$

Therefore,

$$3f_m = 18 \text{ MHz.}$$

The signal bandwidth, f_m = 6 MHz and the frequency deviation is 12 MHz.

Alternative composite signals for a bandwidth of 6 MHz:

1. One color TV signal with sound
2. 1500 (= 6 MHz/4 kHz) telephone channels which are multiplexed by either FDM or TDM.

As an example, INTELSAT V satellites use the techniques of FM, FDM, and TDM.

PRACTICE PROBLEMS

244-1. A satellite channel has a bandwidth of 54 MHz and is frequency modulated by a composite signal. If the modulation index is 2.0, calculate the values of the frequency deviation, f_d, and the bandwidth, f_m, of the composite signal.

244-2. In an up-converter of a C-band earth station transmitter, the microwave oscillator generates an output frequency of 6015 MHz. Which channel is being frequency modulated by a composite signal?

245 Time-division multiplex

TDM is a method of combining a number of signals into a single composite signal without the use of filters. This is achieved by dividing a particular time frame into individual time slots, with each being assigned to one of the input signals. Consequently TDM operates by division of the time domain whereas FDM was involved with the division of the frequency domain. In the TDM system the input signals are transmitted in sequence. If the sequence of transmission is synchronized with the corresponding sequence of reception at the receiver, the individual signals will have maintained their separation.

TDM systems are used to transmit:

1. Digital input signals such as the binary messages of a computer
2. Analog voice communications, which are encoded into digital signals by pulse code modulation (PCM).

With PCM the instantaneous voltage of an analog voice signal is periodically sampled. For example, if the sampling rate is 8000 periods/s, the time interval between samples is 1/8000 s = 125 μs. The total voltage range in which the analog signal lies is divided into a number of standard levels. Since each of these levels is converted into a binary form, the number of levels must be equal to 2 raised to a certain positive exponent. A practical system may use 64 or even 128 levels, but for simplicity, Fig. 245-1 shows only 16 levels. At point A, for example, the instantaneous voltage is 10.2 V, but this would be quantized as level 10, which is equivalent to 1010 in binary code. In addition it is necessary to add one extra bit to each coded group for synchronization purposes. Consequently each sample is represented by 5 bits.

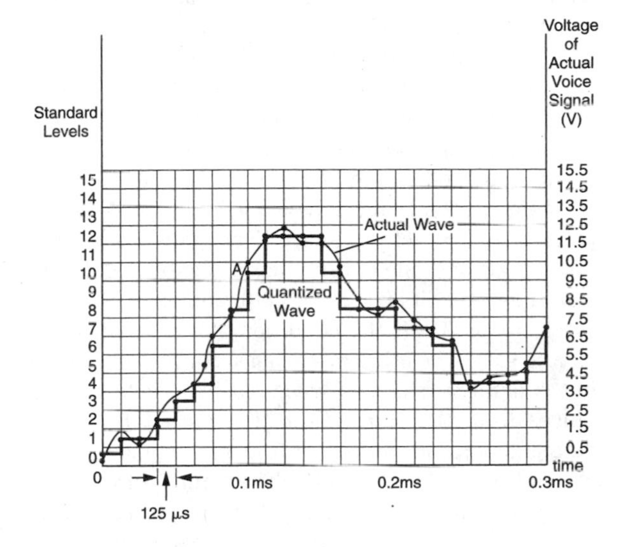

Figure 245-1

It is clear that there is a difference between the digitized signal and the analog waveform. The difference is reduced if the number of standard levels is increased. However, more levels equate to more bits and therefore a larger bandwidth. As a compromise 128 levels for voice communication is considered to be sufficient.

Now let us consider a practical TDM system which employs PCM. There are 24 separate 4-kHz voice signals with a sampling rate of 8000 periods/s. This satisfies the Nyquist criterion, which states: "In order for the voice signal at the receiver to be almost the same as the original analog signal at the transmitter, the rate of sampling (8000 times per second) must be twice the highest frequency (4 kHz) contained in the voice signal."

After passing through the PCM encoder circuit, each sample consists of eight pulses, seven of which represent the quantized value, so that there is a total of 2^7, or 128, standard levels. The eighth pulse is used for synchronization between the transmitter and the receiver. The width allocated to a single pulse is 0.625 μs, so that the time required for the group of eight pulses is $8 \times 0.625 = 5$ μs.

The time interval between sampling is 1/8000 s = 125 μs, of which only 1/25, or 5 μs, is used for one channel. Consequently TDM is used to divide the 125-μs periods into 25 channels, each of 5-μs duration. As shown in Fig. 245-2, the first 24 channels are allocated to separate voice signals, whereas channel 25 contains a single 5-μs synchronizing pulse.

At the transmitter each channel has a sampling circuit. All the sampling circuits are triggered simultaneously and the output sample for channel 1 then forms part of the composite signal. However, the output sample from channel 2 is passed through an artificial line which provides a delay time of 5 μs. Channels 3, 4, 5,..., 24 are delayed by 10, 15, 20,..., 115 μs, respectively, so that in the final composite signal, each successive interval of 5 μs is occupied by the digitized transmission of a different channel.

After demodulation at the receiver the whole of the composite signal is fed as one input to 24 digital gates (one for each channel). Every gate has two input signals and one output signal. The output signal is in the 1 state if and only if both input signals are in the 1 state. This is the condition that is satisfied by an AND gate. The other input signal comes from a gating generator which is triggered by synchronizing pulses from the transmitter. The output from the gating generator consists of 5-μs rectangular pulses with a frequency of 8000 Hz. The gating pulse to the channel 1 gate has no delay, but the pulses to the gates of channels 2, 3, 4,..., 24 are delayed by 5, 10, 15,..., 115 μs, respectively. Consequently the output from each gate only occurs during the appropriate time interval and the separation between the channels is maintained. Finally the outputs from the gates are fed to decoders which reproduce the original analog signals.

TDM does not require filters, and there is no theoretical limit to the number of channels that may be multiplexed. The main disadvantage of TDM is the requirement to transmit digital pulses which need a large bandwidth. However, it is possible to create a TDM system of 1200 channels by combining 50 PCM groups, each of 24 channels.

MATHEMATICAL DERIVATIONS

$$\text{Number of analog voice channels} = 24$$

$$\text{Highest audio frequency} = 4 \text{ kHz}$$

$$\begin{aligned}\text{Minimum sampling rate} &= 2 \times 4 \text{ kHz}\\ &= 8000 \text{ times per second}\end{aligned}$$

$$\text{Number of bits for each sample} = 8.$$

In channel 25 there is one synchronizing (framing) bit. A frame contains one complete set of 24 samples together with the additional synchronizing bit of channel 25.

$$\text{Total number of bits in one frame} = 8 \times 24 + 1 = 193$$

$$\text{Frame rate} = 8000 \text{ frames per second}$$

$$\begin{aligned}\text{Bit rate} &= \text{bits per frame} \times \text{frame rate} \quad (245\text{-}1)\\ &= 193 \times 8000\\ &= 1.544 \times 10^6 \text{ bps.}\end{aligned}$$

Example 245-1

TDM is used to sample 10 channels of analog signals each of which contains frequencies from 50 Hz to 25 kHz. Each sample is converted into 8 bits of digital information, and each frame consists of one sample from each channel together with one synchronizing bit. Calculate the values of the minimum sampling rate, the total number of bits per frame, and the bit rate.

Solution

$$\text{Minimum sampling rate} = 2 \times 25 \text{ kHz} = 50{,}000 \text{ times per second}$$

$$\text{Total number of bits per frame} = 10 \times 8 + 1 = 81$$

$$\begin{aligned}\text{Bit rate} &= 81 \times 50{,}000 \quad (245\text{-}1)\\ &= 4.05 \times 10^6 \text{ bps.}\end{aligned}$$

PRACTICE PROBLEMS

245-1. A time-division multiplexer is used to sample an analog signal which contains frequencies from 25 Hz to 5 kHz. Calculate the value of the minimum sampling rate. If each sample takes 10 μs, how many different TDM signals can be transmitted?

245-2. In Practice Problem 245-1, each sample is converted into 8 bits of digital information and each frame consists of one sample from each channel together with one synchronizing bit. Calculate the value of the bit rate.

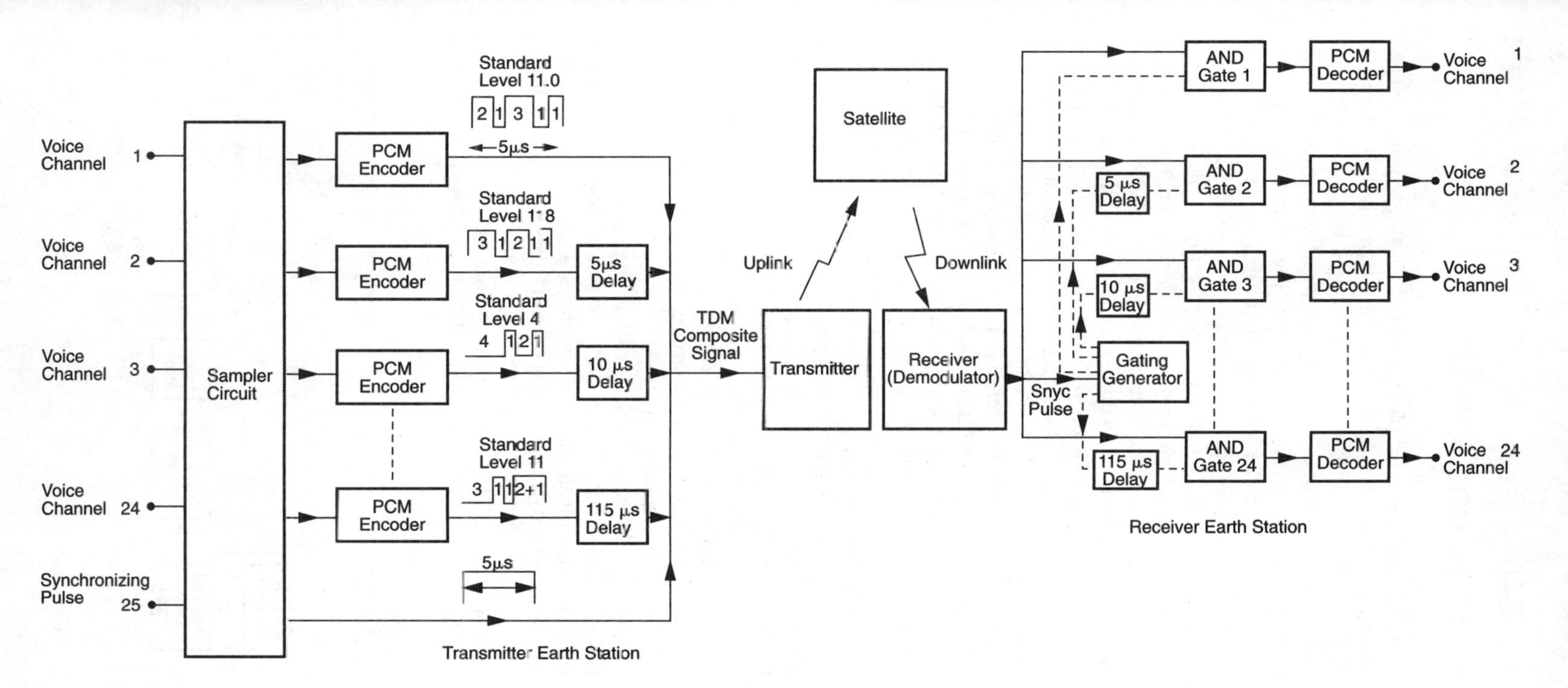

Voice Channel 1
Voice Channel 2
Voice Channel 3
Voice Channel 24
Synchronizing Pulse 25
Sampler Circuit
PCM Encoder
PCM Encoder
PCM Encoder
PCM Encoder
Standard Level 11.0
2 1 3 1 1
5μs
Standard Level 1·8
3 1 2 1 1
Standard Level 4
4 1 2 1
Standard Level 11
3 1 1 2+1
5μs
5μs Delay
10 μs Delay
115 μs Delay
Transmitter Earth Station
TDM Composite Signal
Transmitter
Uplink
Satellite
Downlink
Receiver (Demodulator)
Snyc Pulse
Gating Generator
5 μs Delay
10 μs Delay
115 μs Delay
AND Gate 1
AND Gate 2
AND Gate 3
AND Gate 24
PCM Decoder
PCM Decoder
PCM Decoder
PCM Decoder
Voice Channel 1
Voice Channel 2
Voice Channel 3
Voice Channel 24
Receiver Earth Station

Figure 245-2

246 The communications satellite

A communications satellite is based on the same principles as a terrestrial microwave link. When operating in the *C* band, the uplink 6-GHz signal is beamed skyward from a transmitter earth station. The satellite not only amplifies the received microwave signal, but also carries out frequency translation before retransmission. An electronic system which performs the functions of reception, amplification, frequency translation, and retransmission is called a repeater or a transponder. The downlink frequency is about 2 GHz lower than the uplink frequency. This frequency translation is essential to avoid interference between the uplink and downlink signals. Including the antenna gains, a satellite transponder provides a typical overall amplification of 135 dB. The signal received by the transponder is on the order of nanowatts and the retransmitted signal is between 5 and 10 watts.

As we discussed in chapter 244, a satellite may use FDM to obtain a total of 24 channels, each with a bandwidth, including guard bands, of 40 MHz. The even-numbered channels use a vertically polarized retransmitted signal while the odd-numbered channels employ horizontal polarization. Referring to Fig. 244-1, the uplink center frequency for the odd-numbered channels is 6165 MHz, and the corresponding downlink center frequency is 3940 MHz. Consequently the required amount of frequency translation is $6165 - 3940 = 2225$ MHz. Such a satellite would have 24 active transponders with 8 spares for back-up purposes (maintenance calls are difficult to schedule!). For the same reason a satellite normally has one or two additional low-noise amplifiers (LNAs) with translator circuits.

The block diagram of a transponder is shown in Fig. 246-1. The receiving antenna has a gain of 25 dB and feeds a diplexer and bandpass filters which separate the uplink signal into 12 individual channels. Since the signal level is so low, it is necessary to pass each channel to an LNA in order to obtain an adequate signal-to-noise ratio for the remainder of the transponder. As shown in the exploded view of the 20-ft-high INTELSAT satellite (Fig. 246-2), the LNA consists of a number of cascaded tunnel diode amplifier stages which must be operated at very low temperatures to reduce the amount of their internal noise. In addition tunnel diodes are immune to the background radiation encountered in the solar system and are therefore suitable for satellite communications. In more modern satellites low-noise amplification may be provided by four bipolar transistor stages.

In the translator the mixer device is a low-noise Schottky diode and the RF oscillator (Fig. 246-1) generates a low-level output at the frequency of 2225 MHz. In Fig. 244-1 the uplink frequencies of channel 5 extend from 6005 MHz to 6045 MHz (6025 MHz$\pm$20 MHz). After frequency translation the downlink frequencies range from $6005-2225 = 3780$ GHz to $6045-2225 = 3820$ GHz (3760 GHz$\pm$20 GHz). Since the mixing process produces both the sum and the difference of the mixer inputs, it is necessary to follow the mixer stage by a filter to eliminate the sum frequency component.

In many of the existing satellites the high-power amplifier (HPA) is a traveling wave tube (TWT) which is normally driven by four cascaded bipolar transistor stages. However, it is difficult to design such stages to produce the necessary output power over a lifetime of some 10 years. In more modern satellites which operate in the *C* band, the TWT has been replaced by cas-

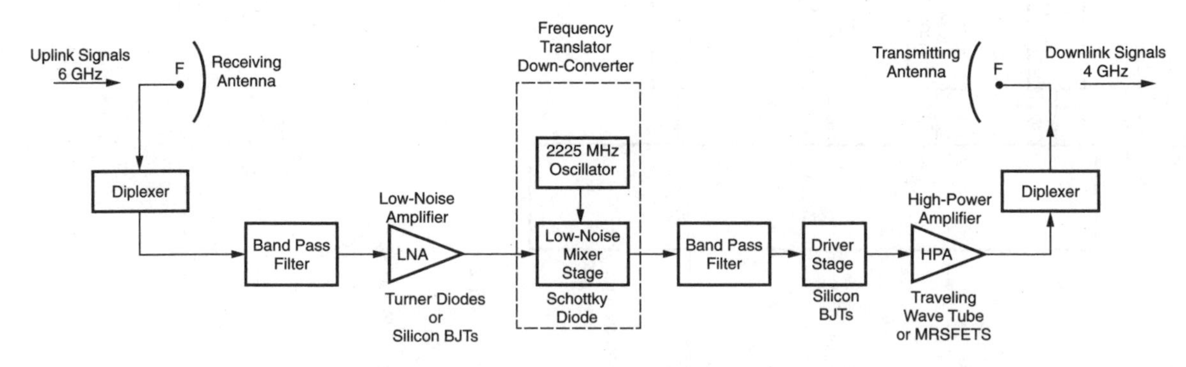

Figure 246-1

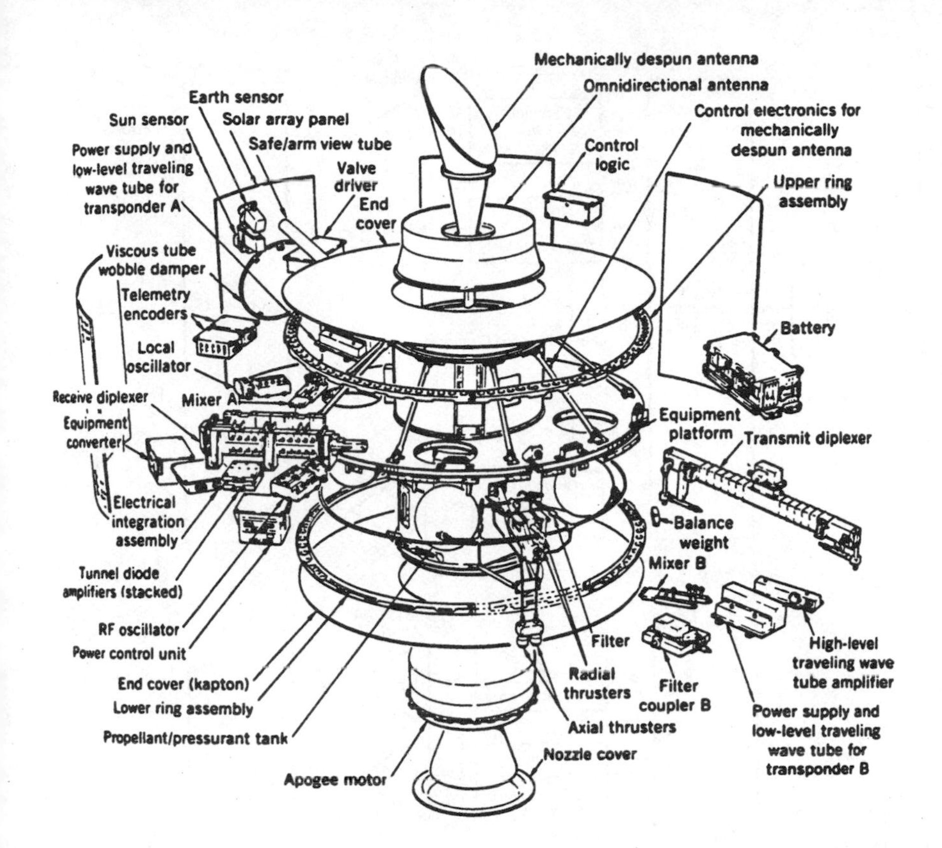

Figure 246-2

caded metal-semiconductor field-effect transistors (MESFETs), although TWTs are still used for the 12/14-GHz and 20/30-GHz bands.

The output from each HPA is passed to the transmitting diplexer, which recombines the various channels into the composite downlink signal. The final output power is between 5 W and 10 W, which is passed to the transmitting antenna whose gain is typically 20 dB to 25 dB.

If the satellite is used to accommodate 24 channels, there are two receiving antennas and two retransmitting antennas. The *even*-numbered channels are intercepted by the *vertical* receiving antenna and are then passed to one combination of the LNA and translator circuits. Since the same frequencies are being reused as described in chapter 244, a second LNA/translator combination will be needed for the odd-numbered channels which employ horizontal polarization. The block diagrams of the complete transponders are shown in Figs. 246-3A and B.

MATHEMATICAL DERIVATIONS

Satellite gain

Overall satellite gain in decibels,

$$G_T = G_{AR} + G_T + G_{AT} \qquad (246\text{-}1)$$

where: G_{AR} = receiving antenna's gain (dB)
G_T = transponder's gain (dB)
G_{AT} = retransmitting antenna's gain (dB).

Down-conversion

Center frequency of the uplink channel $= f_{cu}$

Frequency of the down-converter oscillator = 2225 MHz

Center frequency of the corresponding downlink channel,

$$f_{cd} = f_{cu} - 2225 \text{ MHz.} \qquad (246\text{-}2)$$

Example 246-1

For a particular geostationary satellite the receiving antenna's gain is 25 dB, the transponder's gain is 88 dB, and the retransmitting antenna's gain is 23 dB. What is the overall power gain ratio?

Solution

$$\text{Overall gain} = 25 + 88 + 23 = 136 \text{ dB} \quad (246\text{-}1)$$

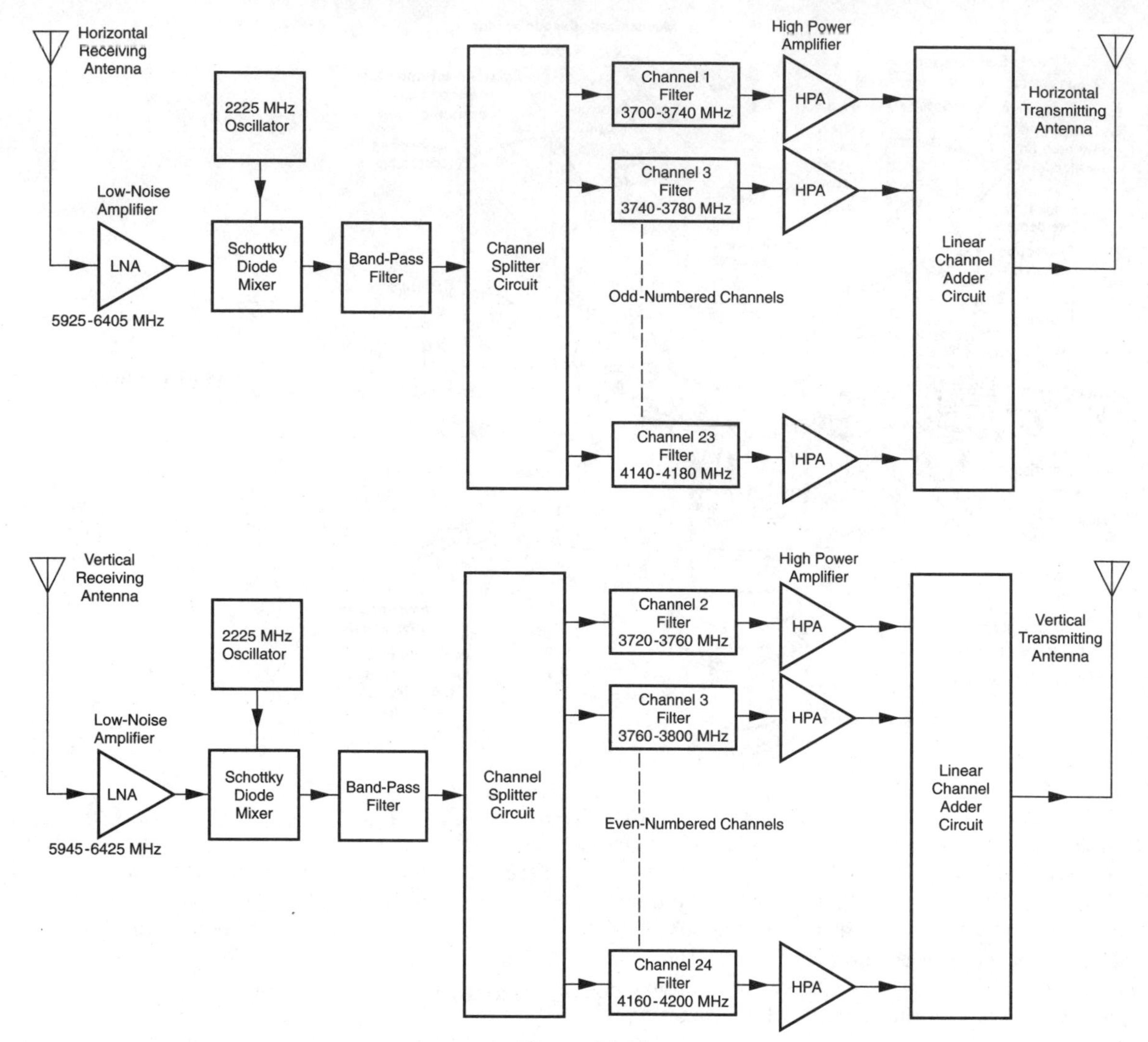

Figure 246-3

Overall power gain ratio $= 10^{13.6} = 4.0 \times 10^{13}$.

Example 246-2

A C-band satellite transponder receives, down-converts, and retransmits a 6125-MHz ± 18-MHz uplink signal. Calculate the value of the downlink carrier frequency and the channel's bandwidth.

Solution

Downlink carrier frequency,

$$f_{cd} = 6125 - 2225 = 3900 \text{ MHz} \qquad (246\text{-}2)$$

Bandwidth $= 2 \times 18 = 36$ MHz.

PRACTICE PROBLEMS

246-1. In Example 246-1, the power of the uplink signal arriving at the geostationary satellite is 40 pW. What is the effective radiated power of the corresponding downlink signal?

246-2. In the Ku-band, what is the output frequency of the oscillator in the satellite transponder's down-converter?

247
Launching of geostationary satellites

The two methods which enable a communications satellite to reach a geostationary orbit, employ:

1. *An unmanned expendable rocket* This places the satellite into an initial elliptical orbit whose plane is inclined to the equatorial plane of the earth. This angle of inclination depends on the site from which the rocket launch takes place. The main features of an elliptical orbit are shown in Fig. 247-1.

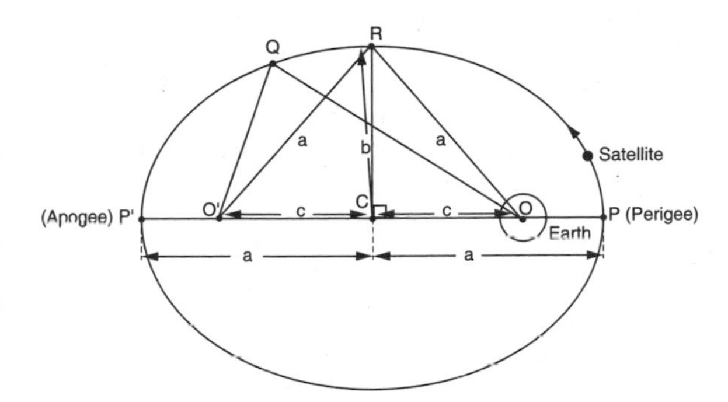

Figure 247-1

The point, O, represents the center of the earth and is also one focus of the ellipse. The point, O', is the other focus of the ellipse and is positioned such that $OP = O'P'$. If the satellite is instantaneously at the *perigee* point, P, it is closest to the center of the earth. By contrast if the satellite is at the *apogee* point, P', it is furthest from the earth's center. If the distances OP and OP' are known, the other parameters of the ellipse may be determined (see Mathematical Derivations)

The path of the ellipse is such that at any point, Q, which lies on the ellipse, $OQ+O'Q$ is equal to a constant. The value of this constant is the sum of the perigee and apogee distances $(OP+OP')$.

The rocket launch is arranged so that the apogee of the initial elliptical orbit lies on the circular geostationary orbit (Fig. 247-2). When the satellite reaches the P' position, its apogee motor (Fig. 246-2) fires and provides an acceleration which "kicks" the satellite from the inclined elliptical orbit into the geostationary circular orbit with its zero angle of inclination.

2. *The manned space shuttle* The shuttle is positioned in a low circular orbit, a few hundred miles above the earth's surface. After the satellite has been unloaded, its orbital attitude is increased by using a Hohmann transfer.

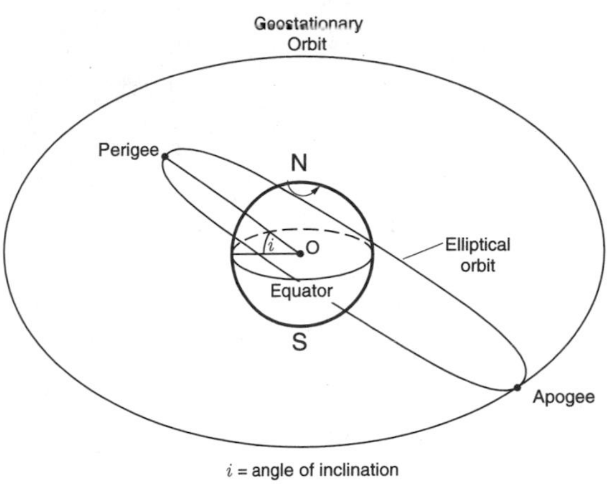

i = angle of inclination

Figure 247-2

The Hohmann transfer is a two-impulse maneuver between two circular coplanar orbits and requires two applications of thrust. To achieve the desired result, the direction and magnitude of the velocity increase must be accurately controlled. The path between the initial low circular orbit and the high geostationary orbit is called the transfer ellipse. The first application of thrust occurs at the point of departure, which is the perigee of the transfer ellipse (Fig. 247-3). Its apogee is at the point of injection where the second application of thrust causes the path of the satellite to change from the transfer ellipse to the geostationary orbit.

Launching satellites from the space shuttle has the obvious advantage that the space shuttle is reusable. Moreover launching from rockets tends to be less reliable.

Perturbations

Assume that a geostationary satellite has been placed in its correct longitude with a zero angle of inclination. Unfortunately it will not stay in this position because of additional forces which were not considered in the two-body (earth-satellite) analysis of chapter 242. These additional forces include the following:

1. The earth is not the only source of gravitational attraction on the satellite since there are other

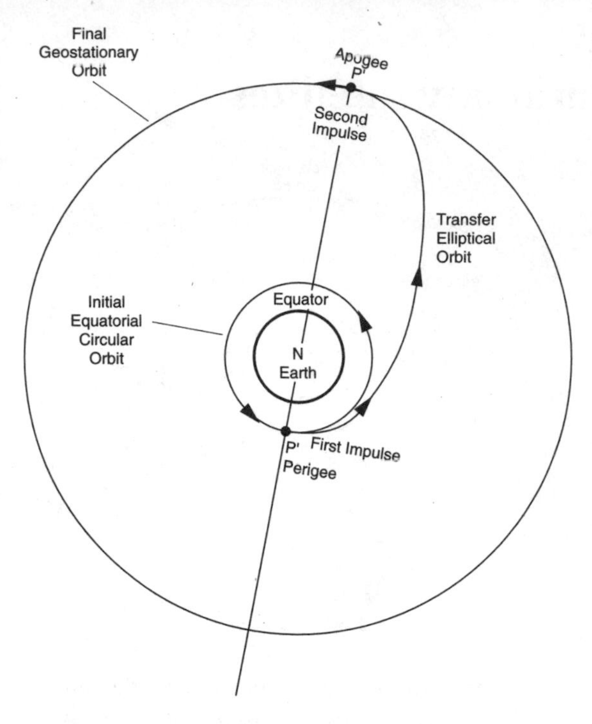

Figure 247-3

gravitational fields (mainly related to the sun and the moon) in space. This effect is greatest at higher altitudes [above 20,000 nautical miles (nm)], where the geostationary satellites reside. These forces cause a change in the satellite's angle of inclination.

2. The earth is not a spherically homogeneous mass, but an oblate spheroid with a bulge around the equatorial region. This additional mass causes the inward gravitational pull on the satellite to be directed away from the center of the earth. This is the major perturbation effect at medium altitudes (between 300 nm and 20,000 nm). Consequently satellites are not totally geostationary but change their positions by as much as 20 km over one day. Therefore the antenna systems of a land station must be automatically steerable to a limited degree.
3. The earth has an atmosphere that causes drag. This force is most significant at low altitudes (below 300 nm) and has little effect on geostationary satellites.

Any unexpected drift in a satellite's position is corrected by the use of radial and axial thruster rockets (Fig. 246-2). These eject hydrazine gas which is stored in tanks aboard the satellite. The amount of gas available is the principal factor in determining the useful lifetime of a communications satellite (typically about 10 years). Once the gas is exhausted, the satellite will drift out of orbit and is of no further use.

The attitude of a satellite is its position relative to the center of the earth. Once established it is essential that the attitude is stabilized in order that the satellite's antennas are always directed toward particular areas on the earth's surface. The radiation pattern of the satellite's transmitting antenna is called its "footprint," which is displayed on a map of the region so that the power received at any location can be obtained.

There are two methods of stabilizing the attitude of a satellite:

1. Any mass which is spinning on an axis possesses angular momentum and the property of gyroscopic inertia. This is sometimes spoken of as rigidity in space, which allows the axis of the mass to maintain its direction. The INTELSAT satellite of Fig. 246-2 uses this principle. The cylindrical body of the satellite is spun in one direction at a few hundred revolutions per minute. This means that the satellite's antennas must be spun at the same speed in the opposite direction (despun) so that these antennas are always pointing to the same positions on the earth's surface. As illustrated in Fig. 246-2, any wobble in the spin axis is corrected by the viscous-tube wobble damper.
2. The attitude of a satellite in space is determined in three dimensions. The attitude can therefore be stabilized through the use of three rotating cylinders whose axes are mutually perpendicular. Each of the cylinders will provide gyroscopic inertia to maintain the necessary rigidity in space. Any wobble can be corrected by varying the angular velocities of the cylinders. Satellites which use this form of stablization are referred to as 3-D. Since a 3-D satellite is not spinning, its outer body need not be cylindrical but may have a cross section that is rectangular or square.

Power supplies

On board a satellite dc electrical power is needed for a variety of tasks:

1. To operate the transponders
2. To spin the outer body of the satellite and despin the antenna assembly; alternatively, to rotate the stabilizing cylinders of a 3-D satellite
3. To operate the antenna subsystems which receive command signals from earth stations and transmit back telemetry signals
4. To operate the apogee motor
5. To operate the radial and axial gas thrusters

6. To operate the stepper motors, which rotate the solar sails.

The total power consumption ranges from a few watts to over 100 watts and is obtained from the output of several thousand solar cells. For a satellite that is spin-stabilized, the cells are mounted on the outer cylindrical body (Fig. 247-4). During the satellite's rotation only a fraction of the total number of cells is exposed to sunlight, and therefore the efficiency of the system is limited. By contrast, a 3-D satellite is attached to large solar sails (Fig. 247-5), which are flat panels on which the arrays of solar cells are mounted. The advantage of this method is that all the solar cells are exposed to direct sunlight on a continuous basis. Since the satellite is in a geostationary orbit, the solar sails must be rotated by a stepper motor once every 24 h so that the sails continue to point toward the sun.

With both types of satellite the output from the solar cells is fed to dc-to-dc converters, which then supply the necessary power to operate the satellite. Unfortunately solar cells deteriorate with time so that their output falls by about 80% after 8 years. The transponders represent the major load in the satellite system so that eventually there is insufficient power to operate all of the transponders. However, the satellite may continue to function for a few more years if command signals enable one or more transponders to be turned off.

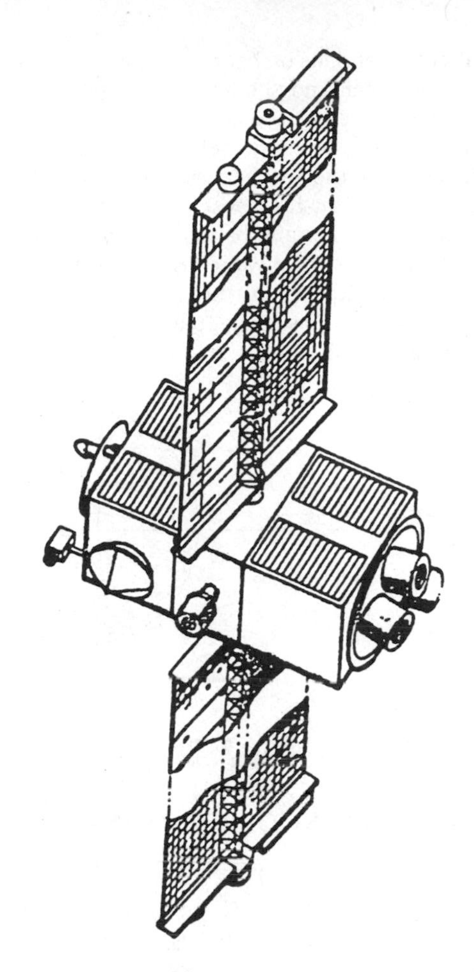

Figure 247-5

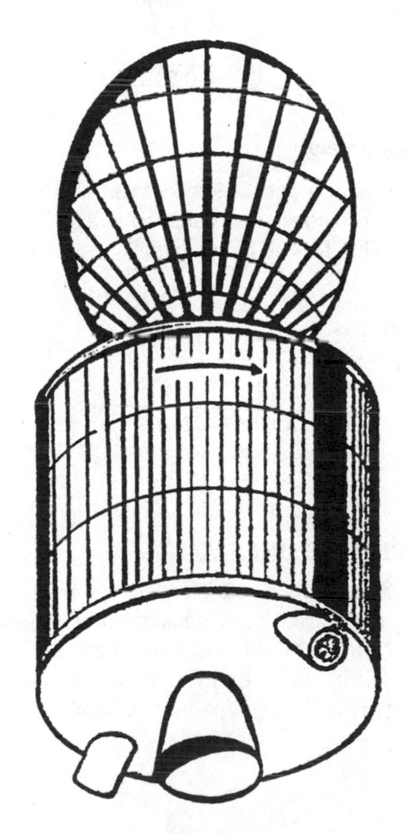

Figure 247-4

A satellite eclipse occurs when the earth is positioned between the sun and a satellite. The solar cells are no longer exposed to sunlight and the satellite must therefore rely on batteries to maintain its operation. Between eclipses, these batteries are recharged by the output from the solar cells.

Subsidiary antenna system

So far we have only considered the main directional antennas for receiving the uplink signals and retransmitting the downlink signals. However, a satellite contains many subsystems which are operated by command from the earth station. The same station also receives telemetry signals which provide information about conditions on board the spacecraft. During the initial launch and the subsequent positioning of the satellite, it is not possible to use the directional antennas to receive commands and transmit telemetry signals since the antennas will probably not be directed toward earth. Consequently the required antenna subsystem must be more omnidirectional in nature (Fig. 246-2). Such antennas are normally cone-shaped, with a radiation pattern (Fig. 247-6) that covers the entire earth's surface from which it is possible to aim at the satellite.

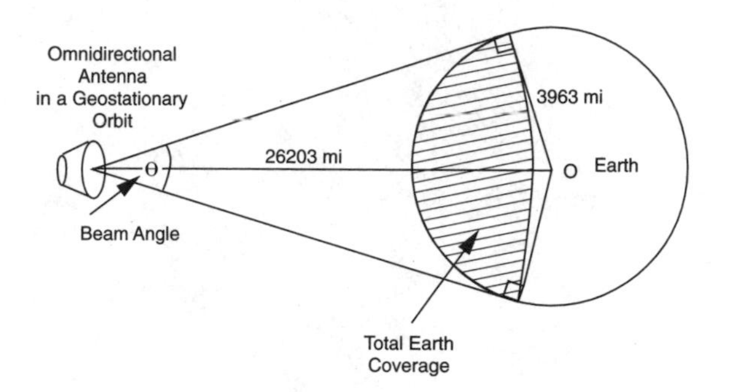

Figure 247-6

MATHEMATICAL DERIVATIONS

From Fig. 247-1,

$$OP + OP' = 2a \qquad (247\text{-}1)$$

$$OP' - OP = 2c. \qquad (247\text{-}2)$$

Then

$$b = \sqrt{a^2 - c^2} \qquad (247\text{-}3)$$

Eccentricity of the ellipse,

$$e = \frac{c}{a}. \qquad (247\text{-}4)$$

Example 247-1

A satellite in a transfer orbit has a perigee of 300 nm above the surface of the earth and an apogee of 19,360 nm. Referring to Fig. 247-1, determine the values of *a*, *b*, *c*, and the eccentricity of the ellipse.

Solution

The radius of the earth is 3440 nm. Therefore $OP = 300+3440 = 3740$ nm and $OP' = 19{,}360+3440 = 22{,}800$ nm.

Then,

$$a = \frac{OP + OP'}{2} \qquad (247\text{-}1)$$

$$= 13{,}270 \text{ nm}$$

and

$$c = \frac{OP' - OP}{2} \qquad (247\text{-}2)$$

$$= \frac{22{,}800 - 3740}{2}$$

$$= 9530 \text{ nm}.$$

Also,

$$b = \sqrt{a^2 - c^2} \qquad (247\text{-}3)$$

$$= \sqrt{13{,}270^2 - 9530^2}$$

$$= 9230 \text{ nm}.$$

Eccentricity of ellipse,

$$e = \frac{c}{a} \qquad (247\text{-}4)$$

$$= \frac{9530}{13{,}270}$$

$$= 0.718.$$

Example 247-2

A satellite is in a geostationary orbit and its antenna subsystem provides total earth coverage. What is the required value for the beam angle?

Solution

From Fig. 247-6,

$$\text{Beam angle} = 2\tan^{-1}\frac{3963}{26{,}203}$$

$$= 17.2°.$$

PRACTICE PROBLEM

247-1. A satellite is being transferred from a low circular orbit to a geostationary orbit. If the transfer ellipse has a perigee of 400 mi, refer to Fig. 247-1 and determine the values of *a*, *b*, *c*, and the ellipse's eccentricity.

248
The earth station transmitter

The block diagram of an earth station transmitter is shown in Fig. 248-1. It is assumed that this station is operated in the 4/6-GHz band and that the microwave carrier frequency is 6105 MHz, which corresponds to channel 9 (Table 244-1). Referring to chapters 244 and 245, either FDM or TDM has been used to provide a composite signal whose bandwidth matches that of the uplink carrier. This signal is first preemphasized to improve the signal-to-noise ratio at the receiver and then frequency modulates a 70-MHz subcarrier. The modulated 70-MHz signal is then passed to a frequency translator or up-converter which, in our example, changes the center frequency of the FM signal from 70 MHz to 6105 MHz. The microwave oscillator generates an output at a fixed frequency of 6035 MHz. The oscillator output and the 70-MHz FM signal are combined in a nonlinear mixer which produces new carriers at the sum (6035 + 70 = 6105 MHz) and difference (6035 − 70 = 5965 MHz) frequencies. The sum frequency is that of the desired carrier and is therefore passed by the filter following the mixer. At the same time the filter eliminates the difference component.

As an example it is assumed that the bandwidth of the composite signal is 6 MHz and that the modulation index is 2.0. The frequency deviation is 2 × 6 = 12 MHz and the bandwidth is 2(12 + 6) = 36 MHz. The uplink signal covers the range 6105 MHz ± 18 MHz, and therefore the available bandwidth is fully occupied.

So far the stages we have discussed would employ solid-state devices. However, the driver stage uses a low-power TWT, whereas the HPA contains either a water-cooled TWT or a multicavity klystron. Normally the parabolic reflector for the *C*-band transmitting antenna is up to 30 m in diameter.

The land stations fall into three categories:

1. *Standard A stations* These stations are allocated a maximum power of 8 kW for the total spectrum of their satellite communications. However, most stations use only a portion of their spectra, so that the normal power output is about 4 kW. This is usually achieved with two separate high-power output stages, which are either multicavity klystrons or water-cooled TWTs. These are driven by low-power TWTs, while all preceding stages employ solid-state devices. The high-gain antennas have parabolic reflectors with 30-m diameters and Cassegrain feeds.
2. *Standard B stations* These stations are capable of being transported so that their paraboloid antenna systems have diameters of only 11 m. This means that these stations operate only with high-gain satellites.
3. *Standard C stations* These fixed stations are designed to operate in the range of 14/11 GHz and have paraboloid antennas with diameters of about 16 m.

MATHEMATICAL DERIVATIONS

For a microwave antenna employing a parabolic dish reflector,

$$\text{Beamwidth angle, } \theta = \frac{70\lambda}{d} \text{ degrees} \quad (248\text{-}1)$$

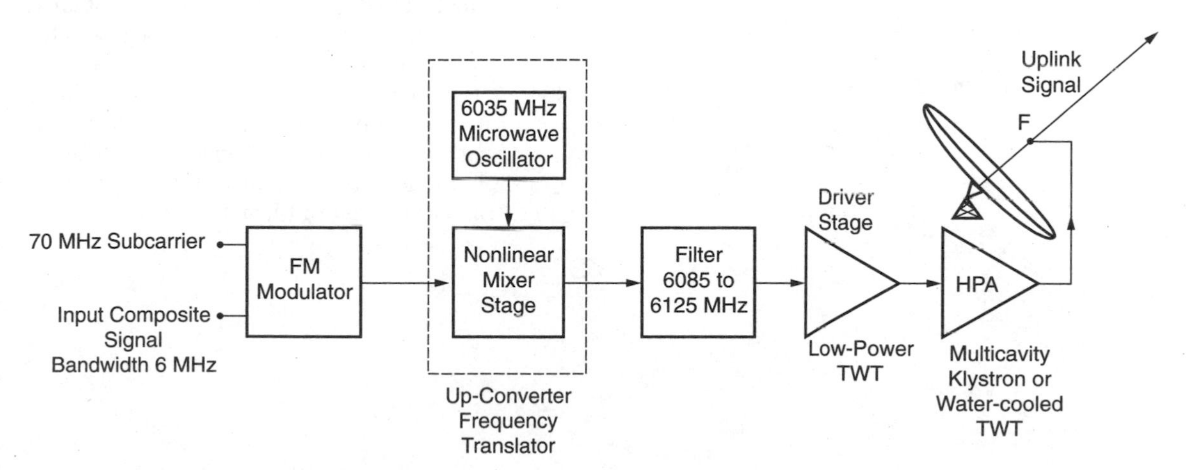

Figure 248-1

where: λ = wavelength (cm)
d = diameter of the parabolic reflector (cm).

Antenna power gain ratio,

$$G_p = 6\left(\frac{d}{\lambda}\right)^2. \qquad (248\text{-}2)$$

Antenna power gain in decibels,

$$G_p' = 10 \log\left[\frac{6d^2}{\lambda^2}\right]$$

$$= 10 \log 6 + 20 \log\left(\frac{d}{\lambda}\right)$$

$$= 7.8 + 20 \log\left(\frac{d}{\lambda}\right) \text{ dB.} \qquad (248\text{-}3)$$

The formulas for G_p and G_p' use the half-wave dipole (not the isotropic radiator) as the reference.

Example 248-1

A *C*-band transmitter is operating on a frequency of 6 GHz. Its antenna is a parabolic dish with a diameter of 25 m. Calculate the values of the beamwidth and the antenna gain (both as a power ratio and in decibels).

Solution

Wavelength of the transmission,

$$\lambda = \frac{30}{f}$$

$$= \frac{30}{6}$$

$$= 5 \text{ cm}$$

Diameter, d = 25 m = 2500 cm

$$\text{Beamwidth, } \theta = 70\frac{\lambda}{d} \qquad (248\text{-}1)$$

$$= 70 \times \frac{5}{2500}$$

$$= 0.14°$$

Antenna power gain ratio,

$$G_p = 6\left(\frac{d}{\lambda}\right)^2 \qquad (248\text{-}2)$$

$$= 6 \times \left(\frac{2500}{5}\right)^2$$

$$= 1.5 \times 10^6$$

Antenna power gain in decibels,

$$G_p' = 10 \log(1.5 \times 10^6)$$

$$\approx 62 \text{ dB.}$$

Example 248-2

In Example 248-1, the output power of the transmitter's final stage is 4 kW. Calculate in dBW the effective radiated power from the antenna.

Solution

Output power = 4 kW
= 4000 W
= 10 log 4000 dBW
= 36 dBW

Effective radiated power,

ERP = 36 + 62
= 98 dBW.

PRACTICE PROBLEMS

248-1. The transmitting antenna of a satellite's transponder has a parabolic reflector with a diameter of 1 m. If the satellite is operating in the 12/14-GHz band, determine the antenna's beamwidth and power gain for the center frequency of channel 9.

248-2. The transmitter of an earth station has an antenna whose parabolic reflector is 20 m in diameter. At the center frequency of channel 5 in the *Ku* band, calculate the values of the antenna's beamwidth and power gain.

249
The earth station receiver

Figure 249-1 is a block diagram of an earth station receiver. We will start by discussing the antenna, which must satisfy the following conditions:

1. In the geostationary orbit there is a 4° ± 0.1° separation between adjacent satellites which operate in the same frequency band. Consequently the beamwidth of the antenna must be sufficiently narrow to prevent any interference from neighboring satellites.
2. In chapter 244 we learned that FDM required the 12 odd-numbered channels in the 4/6-GHz band to be horizontally polarized while vertical polarization was used for the 12 even-numbered channels. Consequently the antenna must be capable of receiving both vertically and horizontally polarized signals.
3. The received signal level is on the order of 1 pW or less. To achieve an appropriate signal-to-noise ratio, it is essential that low-noise amplification begins at the focus of the antenna.

These conditions are met by the Hoghorn and Casshorn antennas. The selection of either the horizontally or the vertically polarized channel is achieved by rotating the feedhorn through 90°. The signals received by the horn antenna are directly fed to cooled low-noise preamplifiers which may either be masers or parametric amplifiers with varactor diode pumps. The preamplifier is immediately followed by a tunnel-diode LNA which drives a low-power TWT.

In more modern receivers the pump of the parametric amplifier is a transistor oscillator whose frequency is stabilized by an AFC circuit containing a crystal. The signal is then amplified by a series of cascaded FET stages which can be incorporated in a microwave integrated circuit.

The amplification so far achieved is on the order of 45 dB. Provided the power level exceeds 0.1 μW, the signal is sufficiently strong to be transferred through a waveguide which runs from the antenna to the main stages of the receiver. After more amplification the channels are passed to a power divider which feeds a number of filters.

The output of each filter is a particular channel whose center frequency is down-converted to 70 MHz. After further amplification at this intermediate frequency, the signal passes through an FM limiter stage and then to a discriminator. The output of the discriminator is fed to a deemphasis circuit which compensates for the degree of preemphasis applied at the transmitter. The channel which was once part of the transmitter's composite signal can now be transferred by microwave link or cable to its required destination.

We have described the operation of a single-conversion receiver. However, some receivers employ dual conversion with a first i-f of 880 MHz and a 70-MHz second i-f. As an example channel 14 in the 4/6-GHz band is vertically polarized and has a center downlink frequency of 3980 MHz with a bandwidth of 36 MHz. The first microwave oscillator generates 3100 MHz so that the output of the frequency translator is 3980 ± 18 − 3100 = 880 ± 18 MHz. After amplification the signal is passed to the second frequency translator whose oscillator generates 950 MHz. The output of the nonlinear mixer is then 950 − 880 ± 18 = 70 ± 18 MHz.

MATHEMATICAL DERIVATIONS

One of the merit factors for the complete receiver system is its equivalent internal noise temperature with all external noise eliminated by shorting the receiver input terminal to ground.

The internal noise, P_i, at the input terminals is also referred to as the noise floor, and is given by

$$P_i = \text{kTB watts} \qquad (249\text{-}1)$$

where: k = Boltzmann's constant (1.37×10^{-23} W · s/K)
T = equivalent noise temperature of receiver system (K)
B = bandwidth (Hz).

Typical values of the temperature, T, range from 100 kelvins to 200 kelvins (K) and depend on the design of the LNAs. Zero kelvin (0 K) corresponds to −273.2°C so that 200 K = −73.2°C.

For an earth station receiver the second important factor is the antenna's power gain, which depends on the dimensions of the parabolic reflector and the wavelength of the received signal.

The factors of antenna gain and noise temperature are combined into a figure-of-merit defined by

$$G/T \text{ figure-of-merit} = 10 \log \frac{G_p}{T}$$

$$= 10(\log G_p - \log T) \text{ dB/K} \qquad (249\text{-}2)$$

where: G_p = antenna power gain ratio
T = system noise temperature (K).

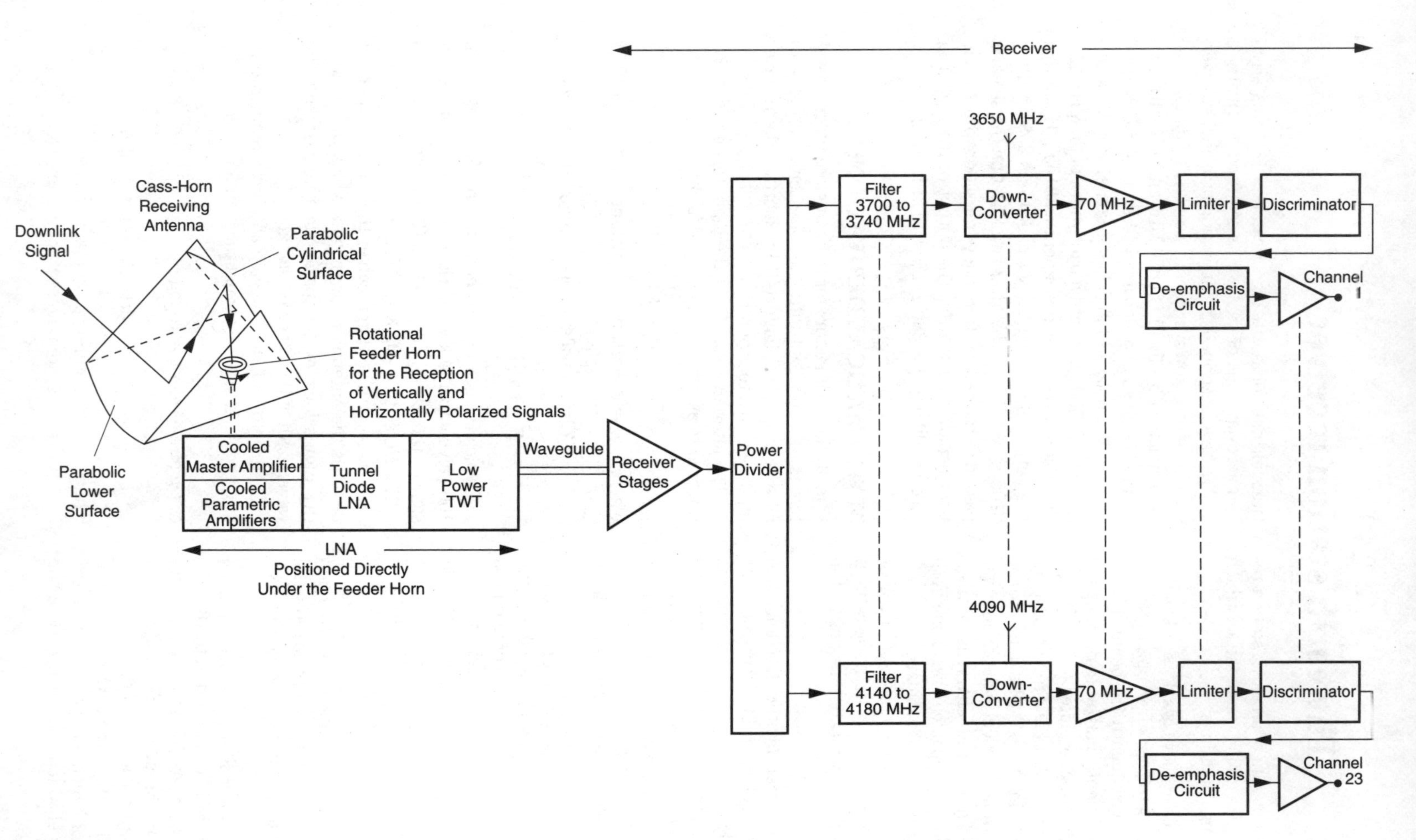
Receiver
Downlink Signal
Cass-Horn Receiving Antenna
Parabolic Cylindrical Surface
Rotational Feeder Horn for the Reception of Vertically and Horizontally Polarized Signals
Parabolic Lower Surface
Cooled Master Amplifier
Cooled Parametric Amplifiers
Tunnel Diode LNA
Low Power TWT
Waveguide
Receiver Stages
LNA
Positioned Directly Under the Feeder Horn
Power Divider
3650 MHz
Filter 3700 to 3740 MHz
Down-Converter
70 MHz
Limiter
Discriminator
De-emphasis Circuit
Channel 1
4090 MHz
Filter 4140 to 4180 MHz
Down-Converter
70 MHz
Limiter
Discriminator
De-emphasis Circuit
Channel 23

Figure 249-1

Example 249-1

An earth station receiver has an antenna gain of 48 dB with an equivalent noise temperature of 162 K. Determine the value of the system's G/T figure-of-merit.

Solution

$$G_p/T \text{ figure-of-merit} = 10(\log G_p - \log T) \quad (249\text{-}2)$$
$$= 48 - 10 \log 162$$
$$= 48 - 22.1$$
$$= 25.9 \text{ dB/K}.$$

Example 249-2

An earth station receiver operates in the 4/6-GHz band and has a total equivalent noise temperature of 125 K. If the receiving system's gain is 52 dB, find the internal receiver-noise level at the output terminals.

Solution

$$\text{Bandwidth of the downlink signal} = 4.2 - 3.7 \text{ GHz}$$
$$= 500 \text{ MHz}$$

Internal noise at the input terminals,

$$P_i = kTB \quad (249\text{-}1)$$
$$= 1.37 \times 10^{-23} \times 125 \times 500 \times 10^6$$
$$= 8.56 \times 10^{-13} \text{ W}$$

Internal noise level at the output terminals,

$$P_o = 8.56 \times 10^{-13} \times \text{antilog}(52/10)$$
$$= 1.36 \times 10^{-7} \text{ W}.$$

Example 249-3

The input signal power to a satellite's transmitting antenna is 7 W and this antenna has a gain of 25 dB. The space loss on the downlink is −190 dB while the receiving earth station's antenna has a gain of 45 dB. Calculate the minimum required gain for the receiving LNA.

Solution

The power level of 7 W is equivalent to 10 log 7 = 8.45 dBW. The power input to the receiving LNA is 8.45 + 25 − 190 + 45 = −111.55 dBW. The minimum required output from the LNA is 0.1 μW, or −70 dBW. The minimum required LNA gain is 111.55 − 70 = 41.55 dB. This is acceptable since the gain of the LNA is typically 45 dB.

PRACTICE PROBLEMS

249-1. The receiver of an earth station has an antenna whose gain is 50 dB with an equivalent noise temperature of 25 K. The equivalent noise temperatures of the LNA and the main receiver represent a total of 126 K. Obtain the value of the G/T figure-of-merit.

249-2. An earth station receiver operates in the 12/14-GHz band and has a total equivalent noise temperature of 130 K. If the receiving system's gain is 55 dB, find the internal receiver noise at the output terminals.

250
Analysis of the satellite communications system

Figure 250-1 shows the block diagram of the complete communications system using a C-band geostationary satellite. By using values already quoted in previous chapters, it is possible to determine the power level at various positions in the system.

Let the total power of the earth station transmitter be 4 kW, which is equivalent to 10 log 4000 = 36 dBW. The ERP of the transmitting antenna is 36 + 60 = 96 dBW. Allowing for the 200-dB uplink space loss, the power arriving at the satellite's receiving antenna is 96 − 200 = −104 dBW, or approximately 40 pW. If the power gain of the receiving antenna is 25 dB, the power input to the transponder is −104 + 25 = −79 dBW, or 12.5 nW. Assuming that the gain of the transponder alone is 88 dB, the output power to the transmitting antenna is −79 + 88 = 9 dBW, or 8 watts.

The gain of the satellite's transmitting antenna is 23 dB and the downlink space loss is 196 dB. If the antenna of the earth station's receiver has a gain of 50 dB, the power input to the LNA is 9 + 23 − 196 + 50

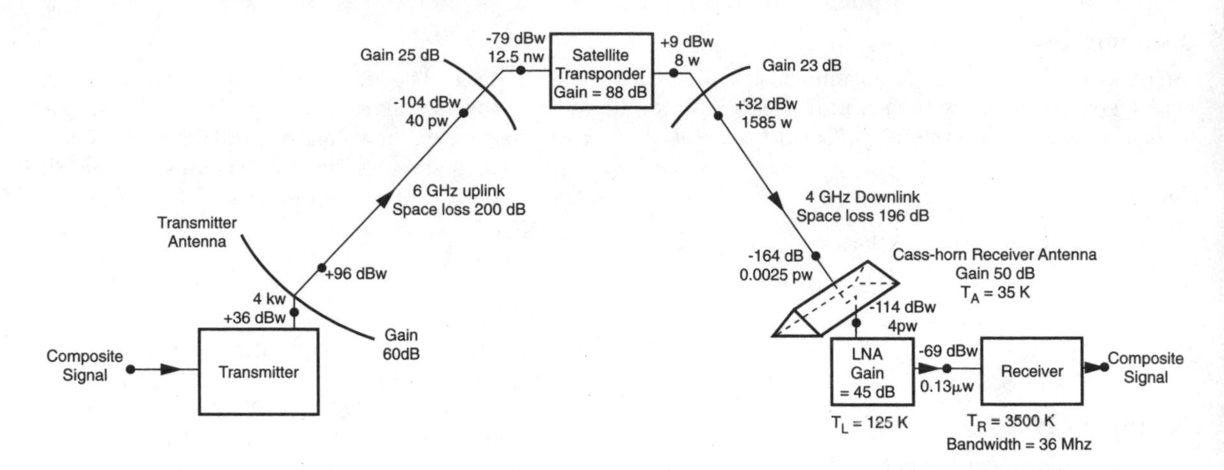

Figure 250-1

$= -114$ dBW, or 4 pW. If the LNA has a gain of 45 dB, the power level to the waveguide which joins the LNA to the main receiver is $-114 + 45 = -69$ dBW, or 0.13 μW. This exceeds the minimum requirement of -70 dBW.

MATHEMATICAL DERIVATIONS

Space loss

In the preceding analysis it was quoted that the uplink and downlink space losses were, respectively, 200 dB and 196 dB. The determination of these losses is as follows.

If a power, P_T, is transmitted through a free-space distance of d meters by direct-wave propagation, the received power, P_R, is given by

$$P_R = P_T\left(\frac{\lambda}{4\pi d}\right)^2 \qquad (250\text{-}1)$$

where: λ = wavelength of transmission (m)
P_R = received power (W)
P_T = transmitted power (W)
d = distance between transmitter and receiver (m).

Substituting $\lambda = c/f$ in Equation 250-1,

$$\frac{P_R}{P_T} = \left(\frac{c}{4\pi \times fd}\right)^2 \qquad (250\text{-}2)$$

where: c = velocity of light (3×10^8 m/s)
f = frequency (Hz).

Substituting the values of π and c in Equation 250-2,

$$\frac{P_R}{P_T} = 5.69 \times \frac{10^{14}}{f^2d^2}. \qquad (250\text{-}3)$$

If the units of f and d are changed to MHz and km, respectively,

$$\frac{P_R}{P_T} = \frac{5.69 \times 10^{14}}{(f \times 10^6 \times d \times 10^3)^2}$$

$$= \frac{5.69 \times 10^{-4}}{f^2d^2}. \qquad (250\text{-}4)$$

In terms of decibels, the space gain is

$$N = 10 \log(P_R/P_T)$$
$$= 10[\log(5.69 \times 10^{-4}) - 2 \log f - 2 \log d]$$
$$= -[32.4 + 20 \log f + 20 \log d] \text{ dB}. \qquad (250\text{-}5)$$

Therefore,

Space loss or attenuation = $32.4 + 20 \log f + 20 \log d$ dB.

As an example, the height of the geostationary orbit is 35.7×10^3 km and the uplink frequency in the C band is approximately 6000 MHz. Therefore the uplink space loss is $32.4 + 20 \log 6000 + 20 \log (35.7 \times 10^3) = 32.4 + 75.6 + 91.1 = 199.1 \approx 200$ dB, rounded off. By contrast the downlink frequency is approximately 4000 MHz and the space loss is $32.4 + 20 \log 4000 + 91.1 = 32.4 + 72.0 + 91.1 \approx 196$ dB, rounded off. Naturally these large free-space losses do not take into account any atmospheric absorption, which tends to be greater in the Ku band than in the C band.

Signal-to-noise ratio (SNR) analysis

To carry out an SNR analysis for the earth station receiver (Fig. 250-1), the following must be known:

1. Equivalent noise temperature of the receiving antenna—for example, 35 K

2. Equivalent noise temperature of the LNA—for example, 125 K. LNA gain is 45 dB, or 3.16×10^4
3. Equivalent noise temperature of the main receiver—for example, 3500 K
4. Receiver bandwidth, 36 MHz
5. FM SNR improvement factor due to limiting and deemphasis—for example, 35 dB.

The system containing the antenna, LNA, and main receiver has a total equivalent noise temperature, T_e, given by

$$T_e = 35 + 125 + \frac{3500}{3.16 \times 10^4}$$

$$= 160.1 \text{ K}$$

Input noise power,

$$P_N = kT_eB \qquad (249\text{-}1)$$

$$= 1.37 \times 10^{-23} \times 160.1 \times 36 \times 10^6$$

$$= 7.9 \times 10^{-14} \text{ W}$$

$$= -131 \text{ dBW}$$

Input SNR value = input power to LNA − noise power

$$= -114 - (-131)$$

$$= 17 \text{ dB}$$

Output SNR value $= 17 + 35$

$$= 52 \text{ dB.}$$

This exceeds the accepted minimum value of 48 dB.

Example 250-1

Determine the value of the uplink space loss for a geostationary satellite operating in the *Ku* band.

Solution

The uplink frequency in the *Ku* band is 14 GHz. Therefore,

$$\text{Uplink space loss} = 32.4 + 20 \log 14{,}000 + 20 \log(35.7 \times 10^3)$$

$$= 32.4 + 82.9 + 91.1 \qquad (250\text{-}5)$$

$$= 206.4 \text{ dB.}$$

Example 250-2

The output of a 12-GHz geostationary transponder is 5 W and its transmitting antenna has a parabolic reflector with a diameter of 1 m. The receiving antenna of the earth station has a gain of 50 dB and a noise temperature of 25 K. The LNA has a gain of 45 dB and a noise temperature of 130 K. The receiver has a bandwidth of 54 MHz, a noise temperature of 3500 K, and requires a minimum input signal level of 0.1 μW. Determine the actual signal power and signal-to-noise ratio at the input to the receiver.

Solution

Wavelength of satellite's transmission,

$$\lambda = \frac{30}{f}$$

$$= \frac{30}{12}$$

$$= 2.5 \text{ cm}$$

Gain of satellite's transmitting antenna,

$$G_p = 6\left(\frac{d}{\lambda}\right)^2 \qquad (248\text{-}2)$$

$$= 6\left(\frac{1 \times 10^2}{2.5}\right)^2$$

$$= 9.6 \times 10^3$$

$$= 39 \text{ dB.}$$

Since the output of 5 W is equivalent to 7 dBW, the ERP of the satellite's transmitting antenna is $7+39 = 46$ dBW.

$$\text{Downlink space loss} = 32.4 + 20 \log 12{,}000 + 20 \log(35.7 \times 10^3) \qquad (250\text{-}5)$$

$$= 32.4 + 81.6 + 91.1$$

$$= 205.1 \text{ dB}$$

$$\text{Input to the receiver} = 46 - 205.1 + 50 + 45$$

$$= -64.1 \text{ dBW}$$

$$= 0.4\ \mu\text{W.}$$

This exceeds the minimum required input level of 0.1 μW.

The LNA gain is 45 dB, or 3.16×10^4.

The total equivalent noise temperature, T_e, for the receiver system is

$$T_e = 25 + 130 + \frac{3500}{3.16 \times 10^4}$$

$$= 155.1 \text{ K}$$

$$\text{Input noise power} = kT_eB \qquad (249\text{-}1)$$

$$= 1.37 \times 10^{-23} \times 155.1 \times 54 \times 10^6$$

$$= 1.15 \times 10^{-13} \text{ W}$$

$$= -129.4 \text{ dBW.}$$

The signal-to-noise ratio at the input to the receiver is given by

$$\begin{aligned} \text{SNR} &= (46 - 205.1 + 50) - (-129.4) \\ &= +20.3 \text{ dB}. \end{aligned}$$

Advantages and disadvantages of satellite communications

Since the essential element of a satellite system is the spacecraft's transponder or repeater, this system may be compared with other systems which also employ repeaters. These include

1. Terrestrial microwave links
2. Coaxial copper cables
3. Optical fibers
4. Circular waveguides operated at frequencies above 10 GHz in the TE_{01} mode.

These four systems convey information from one position to another. By contrast the radiation pattern from the satellite's transmitting antenna may cover an area ranging from a single small city to the entire United States. Moreover this one satellite may be accessed by a large number of earth station receivers.

Multiple access of a particular transponder by a number of uplink and downlink land stations is achieved by either time-division multiplex access (TDMA) or frequency-division multiplex access (FDMA).

When comparing submarine coaxial copper or fiber-optic cables with satellite communications, the two systems are not really rivals but tend to complement one another. A satellite may be accessed by any earth station within its region while use of a cable is normally confined to the terminals to which it is connected.

In terms of reliability satellite transponders are excellent, but severe weather conditions may affect earth stations and the associated terrestrial links. By contrast submarine cables suffer more from damage, but their terminals are very reliable.

One disadvantage of satellites is the time delay between transmission and reception. The total distance between a satellite and back to earth is about 72,000 km, which corresponds to a time interval of approximately 0.24 s. Consequently on an international telephone call there will be a one-half-second interval between the end of a conversation and the beginning of the reply. Moreover, communications between computer systems involve microsecond timing, and these systems will have to be designed to take account of this delay. For a submarine cable the time delay naturally depends on its length and ranges from as little as 20 ms up to a maximum of 150 ms.

We have already discussed the effect that the signal received at the earth station from a satellite's transponder is extremely small (less than 1 pW). Such a small signal is very prone to interference from terrestrial microwave links which operate on the same 4/6-GHz band.

Finally, a geostationary satellite is incapable of communicating with an earth station which is located in the polar regions (chapter 242).

Recent developments

Modern electronic navigation techniques first made their appearance in the 1940s. The United States produced the Loran system while Europe developed Decca.

The position of a receiver was then obtained by detecting the difference in the arrival times of synchronized signals from a minimum of three land-based transmitters. Although both systems operated on frequencies close to 100 kHz, the Decca method measured the phase differences between the received signals, whereas Loran relied on the reception of short pulses, which were less susceptible to interference from sky waves.

The maximum operating range of both systems was limited by the changes in the propagation velocity as the ground wave traveled across different land and sea routes. To overcome this difficulty, the U.S. Navy developed the Omega system in 1957. The operating frequency was lowered to 10.2 kHz and eight transmitters were used to obtain a positional fix. At this low frequency the propagation loss was very small so that long-range operation was possible with an accuracy within 3 nautical miles. This is more than adequate for navigating on the open sea.

In 1964 the U.S. Navy developed a satellite system which enabled nuclear submarines to take intermittent fixes with a high degree of accuracy. The satellites were launched into low polar orbits with an altitude of only 1000 km and an orbital velocity of over 7 km/s. As one of these satellites passed overhead, a submarine could obtain its position by measuring the rate of change of the Doppler shift for the received signal. When this rate of change reached its maximum value, the satellite's position allowed the latitude of the submarine to be determined. The actual value of the Doppler change was a measure of the difference between the longitudes of the satellite and the receiver.

Since a fix was only possible when a satellite passed overhead, the system clearly did not provide continuously available positions. However, in 1973 the U.S. Armed Services proposed the Global Positioning System (GPS), which would allow an unlimited number of receivers whose positions could be determined at any time and any place on the earth's surface.

Successful operation of a GPS requires 18 satellites, which follow inclined orbits so that at least four satellites are available to each receiver. There are six orbits with three 3-D satellites tracking on each orbit (Fig. 250-2). All of these satellites must operate at an altitude of approximately 20,000 km, which corresponds to an orbital period of 11 h 58 min, or one-half of a sidereal day. On alternate passes each satellite travels over the same earth position while its track is stable against the background of the stars. All 18 satellites have now been launched and the system is therefore complete.

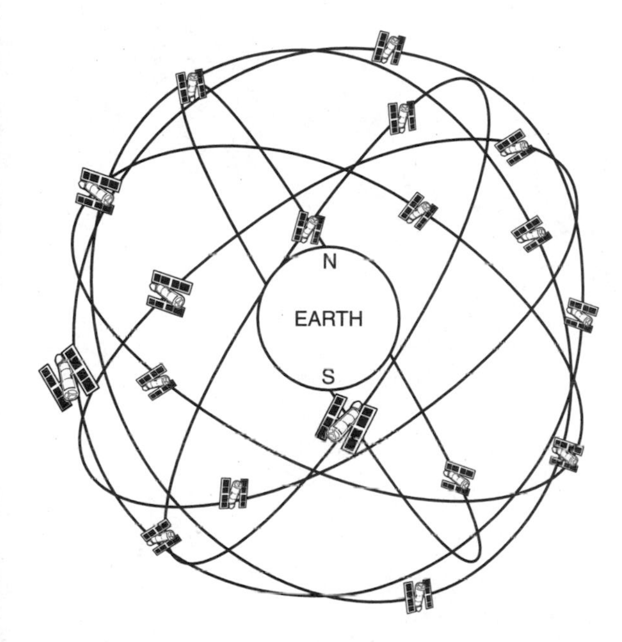

Figure 250-2

The position of a GPS receiver is calculated by computing the time delays of signals arriving from at least four satellites. Since the velocity of a radio signal in space is a constant, each time delay corresponds to the receiver's distance from a particular satellite. If the locations of the satellites are known, the receiver's position can be computed.

The timing of the satellite's position must be determined to a very high degree of accuracy. Each satellite therefore carries two rubidium clocks for short-term stability and two caesium clocks for long-term stability. The resultant accuracy of a receiver's position is within 20 m. Moreover if two GPS receivers remain stationary for more than 30 min, their distance apart can be determined to an accuracy of 5 mm+1 mm for each kilometer of separation. For example, if the receivers are approximately 2 km apart, their separation can be measured to an accuracy within $5 + 1 \times 2 = 7$ mm! This degree of accuracy is required by teams of surveyors who increasingly make use of GPS receivers. Surveying is already one of the main civilian markets for the GPS.

Over the years the GPS receiver's circuitry and its associated computer requirements have shrunk in size and are therefore cheaper to produce. Handheld receivers are already on the market, and these are general consumer products, which do not require the user to have any detailed knowledge of the GPS system.

Marine navigation is another market for GPS receivers. Most ships are already equipped with either Decca or Loran C systems which each cost about $800.00. As the demand for GPS receivers increases, their prices at the low end could probably drop to $500.00. The purchase of a GPS receiver would then be an attractive addition to a ship's navigation equipment, especially since the accuracy of the GPS system is far superior to that of Decca or Loran and the coverage is worldwide. Navigation on land is, of course, much simpler than at sea, but a hiker can often get lost in severe fog, even if he or she carries a compass and a map. However, in the future GPS receivers strapped to the wrist will enable hikers to determine their positions accurately under all weather conditions.

The largest future market for GPS receivers will probably be in cars. The position of a car equipped with such a receiver would be located on a precise road, even in large urban areas. This location would be displayed to the driver on some form of digitized map. In more sophisticated systems drivers could be given directions to their destinations or even given instructions on how to avoid traffic congestion, accidents, or other problems. In this connection Japanese car manufacturers have already announced that most of their cars in the future will carry some form of a GPS system.

Up to recent times navigation on the open seas was carried out by observing the stars. Now we can gaze up at the stars in wonder and leave navigation to consumer products.

PRACTICE PROBLEMS

250-1. Calculate the values of the uplink and downlink space loss at the center frequencies of channel 3 in the 12/14-GHz band.

250-2. The power output of the final stage in an earth station's transmitter is 1.5 kW. The operating frequency is 5.985 GHz and the transmitting antenna has a parabolic reflector whose diameter is 50 m. The receiving antenna of the geostationary satellite has a parabolic reflector whose diameter is 2 m. Calculate the signal power input to the satellite's transponder.

250-3. The power output from the final stage of an earth station's transmitter is 2 kW and the antenna gain is 55 dB. The space loss to a geostationary satellite is 200 dB and the satellite's transponder's antenna has a gain of 25 dB. If the transponder's output is 8 dBW, calculate the transponder's gain.

250-4. In Practice Problem 250-3 the gain of the transponder's transmitting antenna is 22 dB. The space loss on the downlink is 196 dB and the earth station's receiving antenna has a gain of 50 dB. If the input to the receiver is −68 dBW, calculate the required gain for the LNA.

250-5. In Practice Problem 250-4 the equivalent receiving antenna noise temperature is 30 K, whereas the LNA's equivalent noise temperature is 125 K. The receiver's equivalent noise temperature is 2500 K while the bandwidth is 36 MHz. Calculate the value of the signal-to-noise ratio at the output of the LNA.

10
PART
Fiber-optics technology

251
Introduction to fiber-optics technology

In part 9 we discussed the use of geostationary satellites for long-distance telephone communications. But how are telephone communications established over comparatively short distances, for example, within a city or between a network of cities? Up to 1970 the only answer was the coaxial cable, as described in chapter 167. A single cable system is in the form of a tube which contains a number of coaxial cables with copper conductors (Fig. 251-1). Transmission in one direction requires a single cable while another cable is needed for the reverse direction. Two additional cables may also be included as a backup, should a failure occur in any of the other coaxial lines. As an example, in a major system a tube may carry 22 cables—10 to transmit in one direction, 10 for the reverse direction, and two spares.

As we learned in chapters 244 and 245, the number of telephone channels carried by a single cable depends on the allocated bandwidth. If each telephone channel has a bandwidth of 4 kHz, we can use frequency-division multiplex to form a group of 12 channels in the frequency range of 60 kHz to 108 kHz. Five groups can then be stacked to form a 60-channel supergroup within the frequency range of 302 kHz to 542 kHz. Ten supergroups are similarly combined to produce a 600-channel mastergroup which requires a bandwidth of 3 MHz. Six stacked mastergroups becomes a supermastergroup with 3600 channels and a corresponding bandwidth of 18 MHz. If three supermastergroups are then combined, each cable will carry 10,800 channels with a bandwidth of approximately 54 MHz. The previously mentioned tube containing 22 coaxial cables would then be capable of carrying $10 \times 10{,}800$ = 108,000 simultaneous two-way conversations.

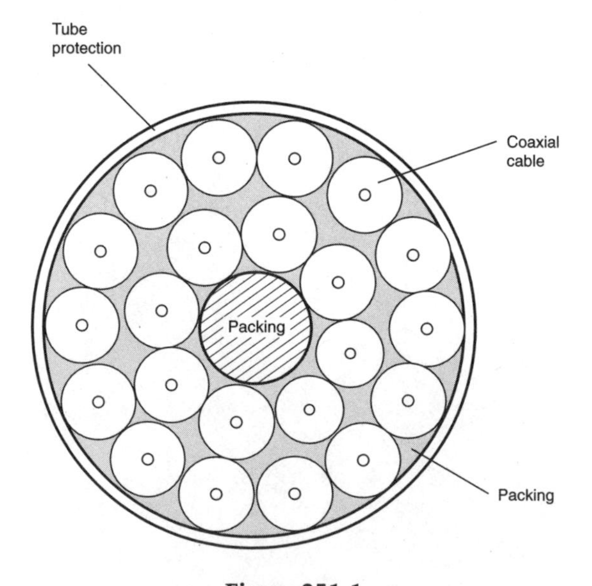

Figure 251-1

In chapter 167 we learned that a coaxial cable provides a certain degree of attenuation, which is typically 25 dB/km. To prevent severe attenuation of the various channels, solid-state amplifying repeaters must be included along the length of the tube. With a wide bandwidth such as 54 MHz, the spacing between adjacent repeaters is on the order of 1 mi. To prevent any change in the total gain of the repeaters, the individual gains of some of the repeaters are adjusted by control signals, which are sent down the tube from the terminals. Each repeater may require a 15-V dc power source for its operation, while the dc stations can be 80 mi apart. Since there will be 80 repeaters between the stations, each station must supply a high dc voltage to the series string of repeaters.

In addition to attenuation a coaxial cable provides a time delay as well as behaving as a low-pass filter. To counteract frequency and phase distortion, equalizers must be included along the cable system. With a 54-MHz bandwidth, adjacent fixed equalizers are separated by about 40 mi and counteract the anticipated amounts of distortion. At the end terminals of the cable system, adjustable equalizers may be used to eliminate any unexpected frequency and phase variations.

Submarine cables employ the same principles as coaxial cables for the transmission of telephone conversations. However, there are two differences:

1. To increase the number of channels carried by a particular bandwidth, submarine cables use 3-kHz voice circuits (as opposed to 4 kHz for coaxial cables).
2. In submarine cables a single coaxial cable allows for transmission to occur in both directions. This is achieved by using a different frequency band for each direction.

Over a 20-year period starting in 1970, a new fiber-optics technology emerged, which became the latest breakthrough in communications. A single optical fiber is a thin tube of transparent glass or plastic which has been manufactured to a high degree of purity in order to reduce attenuation as far as possible.

Historically the development of the fiber dates back to the early work of a Scottish scientist, John L. Baird who, as a pioneer of television, attempted in 1930 to transmit video information through an uncoated fiber. However, it was not until 1934 that research in Germany resulted in the successful transfer of signal through a glass fiber.

In the 1950s experimental attempts were made to transmit light through a cluster of fibers, and these led to the adoption of the term "fiber-optics." However, the development was slowed by the lack of a suitable light source. This difficulty was solved by the invention of the laser-light source in 1960. A laser's coherent output is monochromatic (single frequency) and has a low divergence. These are ideal properties for driving a fiber-optic system. Of all the lasers available, the heterojunction GaAs injection laser is one of the more commonly used.

Although the laser was a major breakthrough in fiber-optic communications, it was not until 1967 that experiments with cladded fiber guides were carried out at the Standard Telecommunications Company in England. These early optical fibers suffered from severe attenuation so that communication was only possible over short distances. However, by 1970 fibers became available with an attenuation of only 20 dB/km, and from that time on, the development of fiber-optic technology has been dramatic. By the mid-1980s there appeared a variety of efficient fiber-optic systems which offered superior quality and capacity to earlier technologies. We have now reached the point where submarine fiber-optic cables have been laid.

There are three types of fiber guide currently available.

1. Glass core with fused quartz cladding
2. Glass core with plastic cladding
3. Plastic core with plastic cladding.

In terms of attenuation glass fibers are far superior to plastic fibers. However, glass fibers are more expensive, heavier, and less flexible. Moreover plastic fibers are easier to install and can be subjected to greater stress. It follows that plastic fibers are used extensively over short runs of less than 1 km.

It is common practice to place a number of plastic or glass fibers in a single cable, which may include a steel core, buffer materials, and protective jackets. An example of such a cable is illustrated in Fig. 251-2; this cable should be compared with the tube of Fig. 251-1.

The advantages of an optical fiber over a comparable coaxial cable may be summarized as follows:

1. Improved physical factors. Copper is heavy, expensive, and there is a limited supply available. By contrast glass is much lighter and is manufactured from sand, of which there is a plentiful supply. Moreover glass is chemically inert and does not corrode or oxidize.

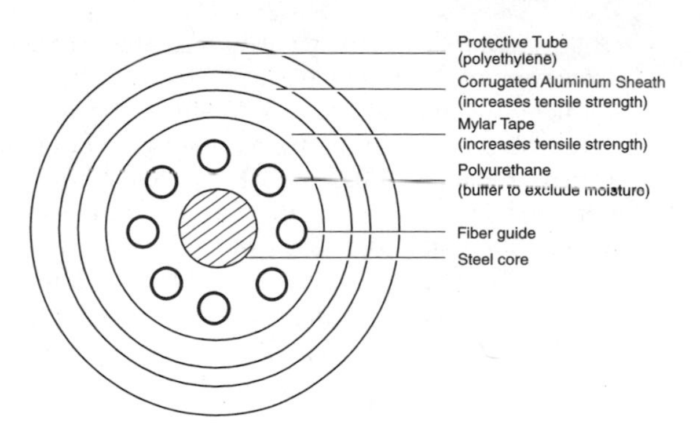

Figure 251-2

2. Greater bandwidth. The transmitted information (analog or digital) is used to modulate an infrared (black light) carrier which can easily be generated by an injection laser (chapter 258). Three commonly used wavelengths with low fiber losses are 1.55 μm, 1.30 μm, and 0.82 μm (micrometer or micron). These wavelengths correspond to frequencies which are on the order of 250 THz (see Mathematical Derivations). Such a high carrier frequency will easily permit a bandwidth that can extend up to 12 GHz. Clearly an enormous amount of information may be multiplexed on such a wide bandwidth. The figure of 12 GHz should be compared with the 54 MHz for a coaxial cable. Moreover a typical fiber with a capability of approximately 6000 channels has a diameter of 0.5 mm or less as opposed to 10 mm or more (1:20 ratio) for a comparable coaxial cable. Consequently when coaxial cables fail or have reached their maximum channel capacity, it is economically more attractive to replace a coaxial cable system with a far larger number of fibers, which can be accommodated within the same duct beneath a city's streets.

   Since fiber-optic cables are smaller than copper cables, it follows that they are also considerably lighter. In aircraft, for example, weight is a very important factor, and therefore it may be beneficial to replace some aviation circuitry using copper lines with optical-fiber systems.

3. Lower attenuation. As mentioned previously, the attenuation of a coaxial cable is on the order of 25 dB/km. In their initial development fibers suffered from severe attenuation (more than

1000 dB/km) due to impurities in the glass. However, new manufacturing techniques have reduced the attenuation to 3 dB/km or less, and research seems to indicate that the attenuation can be reduced to below 1 dB/km. Since adjacent coaxial cable repeaters were separated by 1 mi, comparable optical repeaters would only need to be 1 mi × 25 dB/km/1 dB/km = 25 mi apart. Systems operating over short distances would not need any repeaters, and over long distances the reduction in the number of repeaters would certainly lower the cost of the system.

4. Less interference. Since optical fibers are not conductors, they are not subject to electromagnetic interference. In addition there is very little crosstalk between two adjacent fibers. The same is, of course, true for two circular waveguides lying side by side.
5. Safety improvement. Since glass and plastic are insulators, such phenomena as sparks and short circuits cannot occur. This means that optical fibers can be replaced in the vicinity of volatile chemicals with no danger of explosion.

It would appear that the many advantages of fiber guides would lead to the immediate elimination of coaxial copper cables. However, there is considerable financial investment in the older technology so that only gradual replacement is possible. Moreover fiber-optic systems are so new that we have no idea how they will behave over an extended period of time. Finally the injection laser (light source) has an operating life of between 10^5 and 10^6 h, and this is therefore a weak link in an otherwise reliable system.

MATHEMATICAL DERIVATIONS

Because black light is an electromagnetic wave, its wavelength, λ, in free space is given by

$$\lambda = \frac{c}{f} \qquad (251\text{-}1)$$

where: c = velocity of EM waves (m/s)
f = frequency (Hz)
λ = wavelength (m).

Because $c = 3 \times 10^8$ m/s $= 300 \times 10^{12}$ μm/s and the frequencies of black light are on the order of terahertz (10^{12} Hz), Equation 251-1 becomes

$$\lambda = \frac{300}{f} \qquad (251\text{-}2)$$

where: f = frequency (THz)
λ = wavelength (μm).

Example 251-1

For fiber-optic communications, a commonly used wavelength with the black light source is 1.30 μm. Calculate the value of the corresponding frequency.

Solution

$$\text{Frequency}, f = \frac{300}{1.30} \qquad (251\text{-}2)$$

$$= 231 \text{ THz}$$

$$= 2.31 \times 10^{14} \text{ Hz.}$$

This frequency should be compared with the frequencies of the visible-light spectrum, which range from 375 THz (wavelength 0.8 μm) to 790 THz (0.38 μm).

PRACTICE PROBLEMS

251-1. A black-light source has a frequency of 366 THz. Calculate the value of the corresponding wavelength.

251-2. The black light from an injection laser has a wavelength of 1.55 μm. Calculate the value of the corresponding frequency.

252
Elements of a fiber-optic communications system

Figure 252-1 is a simplified block diagram of a fiber-optic communications system. The information to be transmitted is in either an analog or a digital form and is contained in a large number of channels, which have been either time-division multiplexed (TDM) or frequency-division multiplexed (FDM), as described in chapters 244 and 245. The multiplexing is necessary in order to make use of the wide bandwidth available.

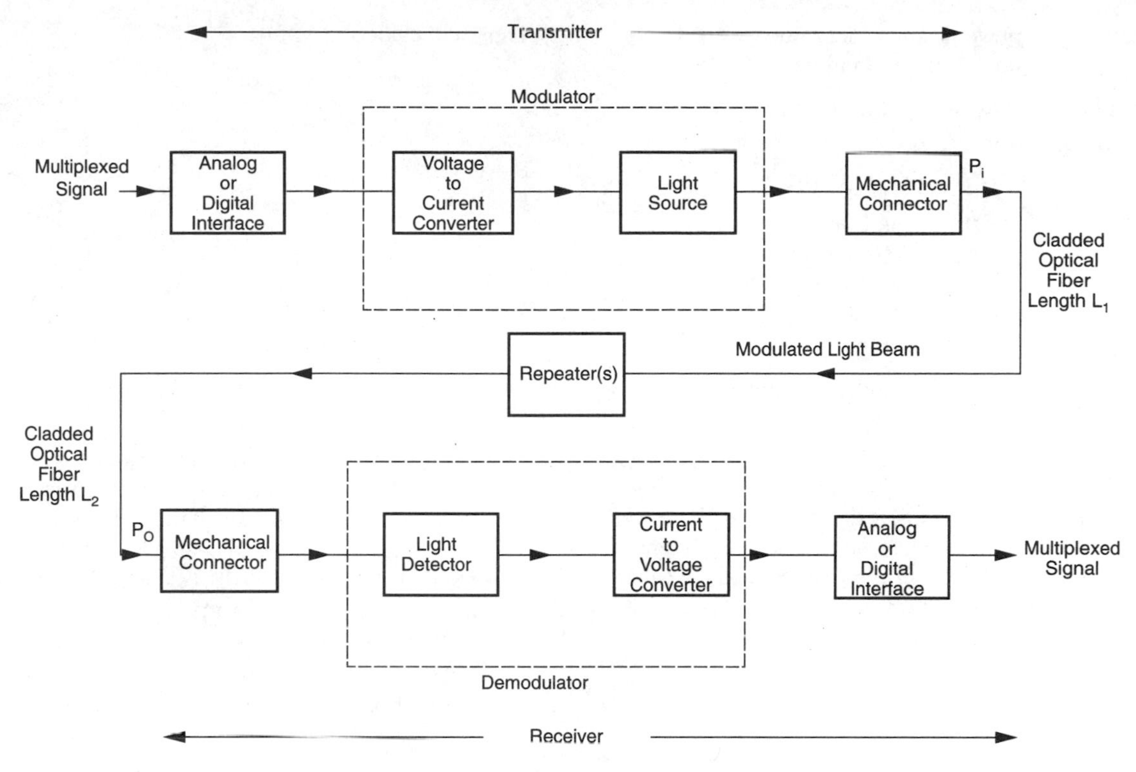

Figure 252-1

The analog or digital interface is used to match the output impedance of the information channels to the input impedance of the voltage-to-current converter. The required modulation relies on the fact that the intensity of an injection laser's output beam depends on the driving current in the bias circuit. When the level of this current is controlled by the analog or digital information, the narrow laser beam is amplitude modulated.

Similar results can be achieved with certain types of light-emitting diode (LED), but these devices tend to have larger active areas so that it is more difficult to provide efficient coupling between the light source and the optical fiber.

In making this mechanical connection it is essential that there be correct alignment between the light source and the fiber. For example, an axial misalignment of only 10% can cause a loss of more than 0.5 dB. A full description of the methods of connecting the light source to the fiber, one fiber to another, and the fiber to the light detector is given in chapter 257.

The light detector is basically the demodulator of the system. This objective may commonly be achieved by using either a PIN or an avalanche photodiode. In either case the modulated beam is allowed to enter the device, which then conducts with an output current that is proportional to the intensity of the light. This result is exactly opposite to the action of the laser source in which the original modulation took place.

The current output of the light detector is passed to a current-to-voltage converter whose output is the multiplex signal, which originated in the transmitter. Finally another analog or digital interface is required to match the output impedance of the light detector to the input impedance of the following circuitry. It should be emphasized that conventional communications electronics precede the light source and follow the light detector.

MATHEMATICAL DERIVATIONS

The decibel loss in a fiber guide is given by

$$\text{Loss} = \alpha \times l \text{ dB} \qquad (252\text{-}1)$$

where: α = guide's attenuation coefficient (dB/km)
l = length of the guide (km).

$$\text{Power loss ratio} = \frac{P_o}{P_i}$$

$$= 10^{(-\alpha l/10)} \qquad (252\text{-}2)$$

where: P_i = power input to the guide (mW)
P_o = power output from the guide (mW).

In Fig. 252-1, the power loss, P_L, from the transmitter to the receiver is given by

$$P_L = G_R - \alpha(l_1 + l_2) \text{ dB} \quad (252\text{-}3)$$

where: l_1 = length of the guide from the transmitter to the repeater (km)
l_2 = length of the guide from the repeater to the receiver (km)
G_R = gain of the repeater (dB).

The typical gain of the repeater is on the order of 45 dB to 50 dB.

Example 252-1

A light-emitting diode (LED) delivers a 0.5-mW power input to a fiber guide with an attenuation coefficient of 1.2 dB/km. If the length of the guide is 2 mi, calculate the value of the output power.

Solution

Because 1 mi = 1.61 km,

$$\text{Length of guide} = 2 \times 1.61 = 3.22 \text{ km}$$

$$\text{Guide loss} = 1.2 \text{ dB/km} \times 3.22 \text{ km} \quad (252\text{-}1)$$
$$= 3.864 \text{ dB}$$

$$\text{Power loss ratio} = 10^{(-3.864/10)} \quad (252\text{-}2)$$
$$= 10^{-0.3864}$$
$$= 0.41$$

$$\text{Output power, } P_o = 0.5 \times 0.41$$
$$= 0.21 \text{ mW.}$$

PRACTICE PROBLEMS

252-1. In Fig. 252-1, an injection laser delivers a 4-mW power input to a fiber guide with an attenuation constant of 1.1 dB/km. If the distance from the transmitter to the repeater is 44 km and the repeater's gain is 48 dB, calculate the value of the repeater's power output.

252-2. In Practice Problem 252-1, the distance from the repeater to the receiver is 35 km. Calculate the value of the power input to the receiver.

253 The physics of light

As we learned in chapter 251, the lower (red) end of the visible light spectrum commences at a frequency of about 375 THz, which corresponds to a wavelength of 0.8 μm. Since the wavelengths commonly used for fiber-optic communications are 0.82 μm, 1.30 μm, and 1.55 μm, it follows that their corresponding frequencies lie in the upper portion of the infrared region. At these frequencies the electromagnetic waves are commonly referred to as "black" light, whose behavior is similar to that of visible light.

MATHEMATICAL DERIVATIONS

The laws of refraction

The velocity of an EM wave depends on the medium through which the wave is traveling. Figure 253-1 shows a monochromatic light beam, which is traveling in a vacuum and then encounters an optically denser medium. The beam is approaching the medium at the angle of incidence, *i*, which is measured with respect to the normal. As the wave enters the medium, its velocity decreases and the wavefront is slewed *toward* the normal. One edge of the wavefront travels through the vacuum from *B* to *Q* in the same time as the other edge of the wavefront travels in the medium from *A* to *P*. This phenomenon is known as *refraction*, or bending, of the light beam. The direction in which the wavefront is moving into the medium is inclined to the normal at the angle of the refraction, *r*. Since distance = velocity×time,

$$\frac{BQ}{AP} = \frac{\text{speed of light beam in vacuum}}{\text{speed of light beam in medium}}.$$

This ratio is known as the medium's refractive index, whose letter symbol is μ. Naturally the value of μ has no units.

From the geometry of Fig. 253-1,

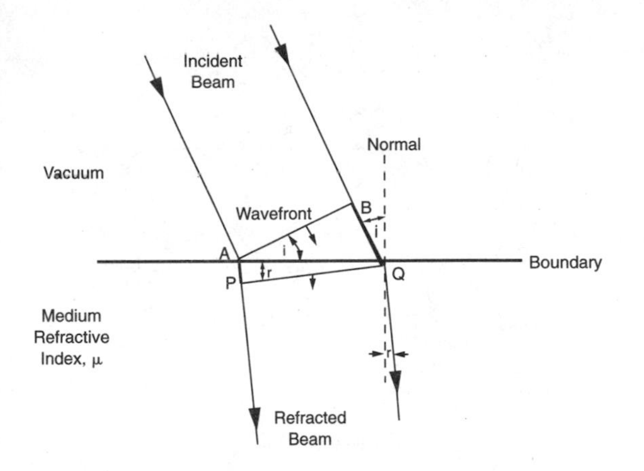

Figure 253-1

$$\sin i = \frac{BQ}{AQ} \quad \text{and} \quad \sin r = \frac{AP}{AQ}.$$

Therefore,

$$\frac{\sin i}{\sin r} = \frac{BQ/AQ}{AP/AQ} = \frac{BQ}{AP} = \mu$$

or

$$\text{Refractive index, } \mu = \frac{\sin i}{\sin r}. \qquad (253\text{-}1)$$

The laws of refraction may be summarized as follows:

First law The incident and refracted rays are in the same plane as the normal at the point of incidence and are on opposite sides of the normal.

Second law When light passes from one medium to another, the ratio of the sine of the angle of incidence to the sine of the angle of refraction is a constant whose value depends on the two media concerned and on the frequency of the light. The second law is often referred to as Snell's law.

In Fig. 253-2, a light ray is passing from medium A to medium B, which is optically denser. The refractive indices of the two media, A and B, are μ_A and μ_B, respectively. The velocity of the ray in medium A is the velocity of the ray in a vacuum divided by μ_A, whereas the velocity in medium B is the velocity in a vacuum divided by μ_B.

From Snell's law,

$$\frac{\sin A}{\sin B} = \frac{\text{velocity of light in medium } A}{\text{velocity of light in medium } B} = \frac{\text{velocity of light in a vacuum}/\mu_A}{\text{velocity of light in a vacuum}/\mu_B} = \frac{\mu_B}{\mu_A}.$$

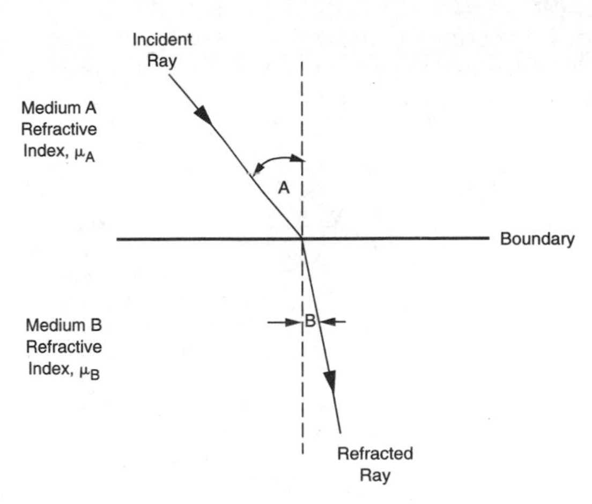

Figure 253-2

This yields

$$\mu_A \sin A = \mu_B \sin B. \qquad (253\text{-}2)$$

Table 253-1 shows the average values of the refractive indices for various media. These values do not indicate the small changes in the refractive indices for different frequencies. Table 253-2 illustrates these changes for distilled water and flint glass at opposite ends of the visible light spectrum. The fact that the

Table 253-1 Values of the refractive index for various media.

Medium	Refractive index
Vacuum	1.0
Air	1.0003
Water	1.33
Crown glass	1.52
Flint glass	1.65
Glass fiber	1.70
Fused quartz cladding	1.46
Diamond	2.42

Note: The values given are average values and do not indicate the small change in the refractive index with frequency.

Table 253-2 Variation in the value of the refractive index with frequency.

	Refractive index	
Medium	Red light	Violet light
Distilled water	1.331	1.343
Flint glass	1.643	1.685

value of the refractive index depends on frequency can be demonstrated experimentally by using a prism (Fig. 253-3) to split white light into its constituent colors.

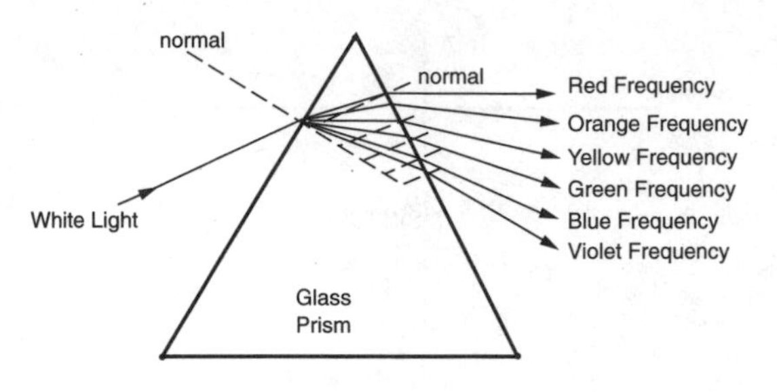

Figure 253-3

In Fig. 253-4, a light ray is traveling from a medium whose refractive index is μ_1 to another medium with a refractive index of μ_2. There are three possibilities:

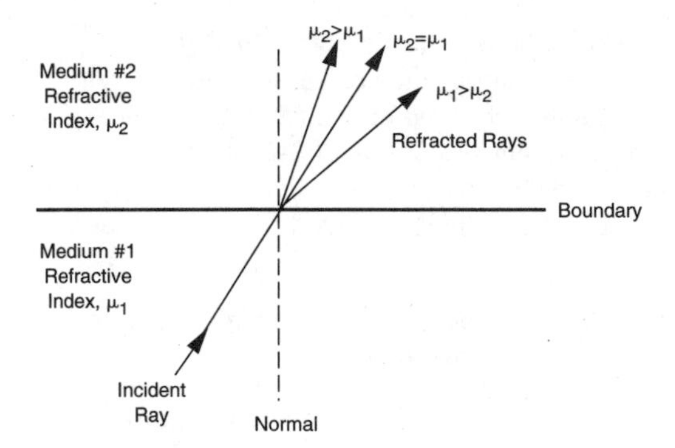

Figure 253-4

1. $\mu_2 > \mu_1$. The refracted ray is bent toward the normal.
2. $\mu_1 = \mu_2$. There is no refraction.
3. $\mu_1 > \mu_2$. The refracted ray is bent away from the normal.

As far as fiber-optic communications are concerned, the third possibility is the most important since the refractive index of the outer fused quartz cladding is less than that of the glass core.

Critical angle

In Fig. 253-5, an incident ray is traveling through a medium and then approaches a second medium with a lower refractive index. The refracted ray is therefore bent away from the normal. If the angle of incidence, i_1, is increased to a value of i_c there will be a critical condition in which the angle of refraction is 90°. At the

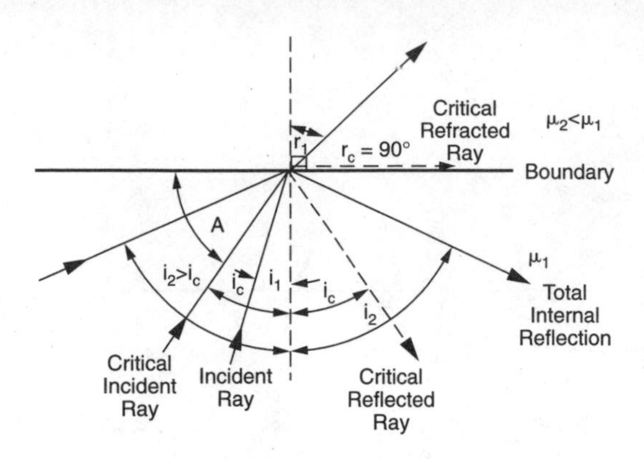

Figure 253-5

boundary between the two media, the ray is partially refracted and travels parallel to the boundary line. However, the ray is also partially reflected with the angle of reflection equal to the angle of incidence.

At the critical condition,

$$\mu_1 \sin i_c = \mu_2 \sin 90° = \mu_2.$$

This yields,

$$\text{Critical angle, } i_c = \sin^{-1} \frac{\mu_2}{\mu_1}. \tag{253-3}$$

If the angle of incidence exceeds its critical value ($i_2 > i_c$), no refracted wave exists and there is total internal reflection with equal angles of incidence and reflection. Referring to Table 253-1, the refractive index of diamond is much greater than that of glass. Assuming that a diamond and a piece of glass with the same physical dimensions are each surrounded by air for which $\mu \approx 1$, the critical angle for diamond is $\sin^{-1}$ (1/2.42) = 24.4°, whereas that of crown glass is $\sin^{-1}$ (1/1.652) = 41.1°. It follows that refracted rays can easily emerge from the glass whereas in the diamond there is a considerable amount of total internal reflection so that the escaping light will tend to consist of a concentration of rays; for this reason a diamond is said to "sparkle."

In Fig. 253-5 all rays that lie within the (internal) acceptance angle, A, will undergo total internal reflection.

$$\text{Acceptance angle, } A = 90° - i_c. \tag{253-4}$$

It follows that

$$\sin A = \cos i_c$$

$$= \sqrt{1 - \sin^2 i_c}$$

$$= \sqrt{1 - \left(\frac{\mu_2}{\mu_1}\right)^2}$$

$$= \sqrt{\frac{\mu_1^2 - \mu_2^2}{\mu_1^2}}$$

$$\text{Acceptance angle, A} = \sin^{-1}\sqrt{\frac{\mu_1^2 - \mu_2^2}{\mu_1^2}}. \qquad (253\text{-}5)$$

Example 253-1

In Fig. 253-2, let medium A be water and let medium B be flint glass. If the angle of incidence, $A = 41°$, determine the value of the angle B.

Solution

From Table 253-1, $\mu_A = 1.33$ and $\mu_B = 1.65$.

Angle of refraction,

$$B = \sin^{-1}\left(\frac{\mu_A \sin A}{\mu_B}\right) \qquad (253\text{-}2)$$

$$= \sin^{-1}\left(\frac{1.33 \sin 41°}{1.66}\right)$$

$$= 31.9°.$$

Since $\mu_B > \mu_A$, $B < A$ and the refracted ray is bent by $41° - 31.9° = 9.1°$ toward the normal.

Example 253-2

A glass fiber whose refractive index is 1.7 is placed in air. For a ray traveling in the fiber, what is the value of the critical angle of incidence?

Solution

In Equation 253-3, $\mu_1 \approx 1$ and $\mu_2 = 1.7$.

Critical angle of incidence,

$$i_c = \sin^{-1}\left(\frac{\mu_2}{\mu_1}\right)$$

$$= \sin^{-1}\left(\frac{1}{1.7}\right)$$

$$= 36.0°.$$

PRACTICE PROBLEMS

253-1. A light ray is traveling through a glass fiber whose refractive index is 1.65. If a ray strikes a fused quartz cladding (refractive index 1.40) at an angle of incidence equal to 25°, what is the angle of refraction on the cladding?

253-2. In Practice Problem 253-1, determine the critical angle of incidence at the glass-clad interface for which total internal reflection occurs. What is the value of the acceptance angle for the fiber?

254
Propagation of light through a cladded optical fiber

Figure 254-1 shows the source end of a homogeneous cylindrical glass fiber which is surrounded by a cladding such as fused quartz. Let us assume that the fiber and the cladding have respective refractive indices of μ_1 and μ_2. A ray from the source enters the fiber at an angle of incidence, i_1, to the first normal. Since the fiber's refractive index is greater than that of air, the ray is refracted toward the normal so that the angle of refraction, r_1, is less than the angle, i_1. The ray then travels through the fiber and approaches the second normal at the new angle of incidence, i_2, which equals $90° - r_1$. Unless i_2 is greater than the critical angle for the fiber–fused quartz interface, the ray will be refracted through the cladding and will not be propagated down the fiber.

MATHEMATICAL DERIVATIONS

At the first normal (Fig. 254-1),

$$\mu_0 \sin i_1 = \mu_1 \sin r_1. \qquad (253\text{-}2)$$

Assuming that i_2 is at its critical value,

$$\mu_1 \sin i_2 = \mu_2 \qquad (254\text{-}1)$$

or

$$\sin i_2 = \frac{\mu_2}{\mu_1}.$$

Because $\sin^2 i_2 + \cos^2 i_2 = 1$,

$$\cos i_2 = \sqrt{1 - \sin^2 i_2}$$

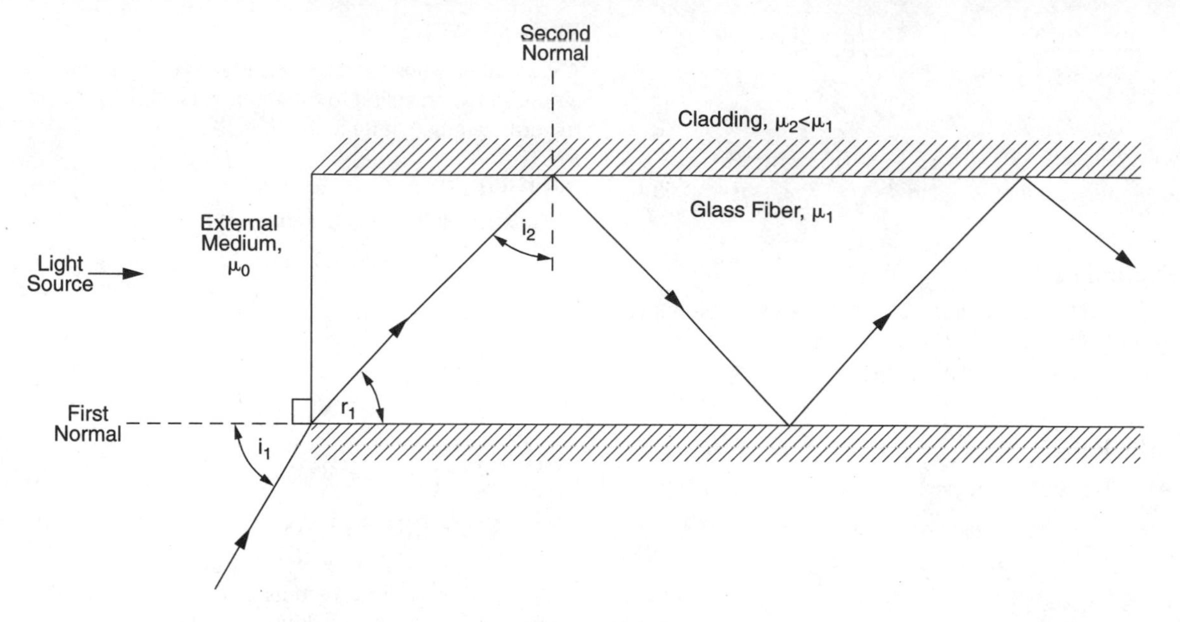

Figure 254-1

$$= \sqrt{1 - \frac{\mu_2^2}{\mu_1^2}}$$

$$= \sqrt{\frac{\mu_1^2 - \mu_2^2}{\mu_1^2}}. \qquad (254\text{-}2)$$

Since $i_2 = 90° - r_1$,

$$\cos i_2 = \sin r_1 = \sqrt{\frac{\mu_1^2 - \mu_2^2}{\mu_1^2}} \qquad (254\text{-}3)$$

Substituting for $\sin r_1$ from Equation 253-2,

$$\mu_0 \sin i_1 = \mu_1 \sqrt{\frac{\mu_1^2 - \mu_2^2}{\mu_1^2}} \qquad (254\text{-}4)$$

$$= \sqrt{\mu_1^2 - \mu_2^2}.$$

Since the external medium from the source to the fiber is probably air, $\mu_0 \approx 1$, and therefore

$$i_{1max} = \sin^{-1}\left(\sqrt{\mu_1^2 - \mu_2^2}\right). \qquad (254\text{-}5)$$

The value of i_{1max} is called the (external) acceptance cone's half-angle for which the ray is propagated down the fiber with a loss of 10 dB when compared with a ray that travels down the fiber's axis. If the initial angle of incidence exceeds i_{1max}, the ray will be refracted through the cladding and will be lost. Since the fiber and the cladding are three-dimensional, the direction of the ray from the source must be rotated about the axis of the fiber. The result is an (external) *acceptance cone* (Fig. 254-2) whose half-angle is equal to i_{1max}.

If the value of i_{1max} is increased, the greater the amount of light that the fiber can collect from the source. This merit factor is measured by the numerical aperture (NA), which is defined by

Numerical aperture,

$$\text{NA} = \sin i_{1max}$$

$$= \sqrt{\mu_1^2 - \mu_2^2}. \qquad (254\text{-}6)$$

Theoretically the value of the numerical aperture can range from zero to unity.

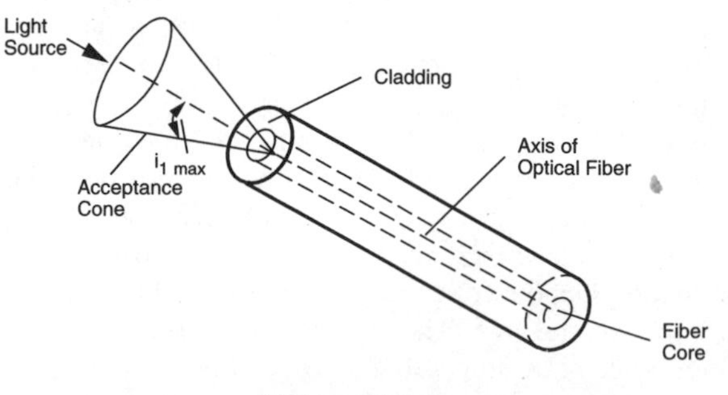

Figure 254-2

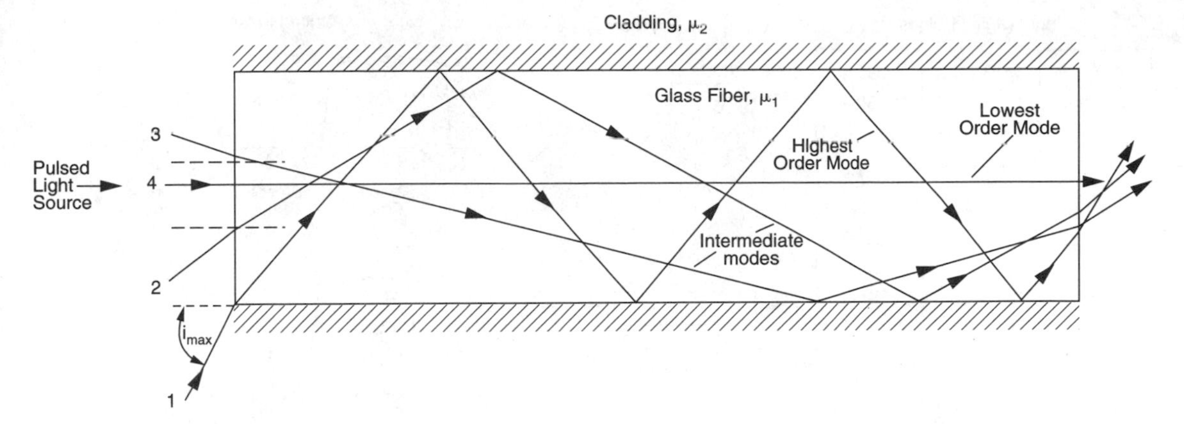

Figure 254-3

Pulse-spreading (modal dispersion)

We have examined the critical case in which the incident ray was located on the surface of the acceptance cone. However, within the cone all rays will approach the fiber surface at incident angles that are smaller than the critical value. Three of these rays are illustrated in Fig. 254-3.

Path 1 is followed by the critical ray that has four reflections over the length of the fiber. By contrast path 2 and path 3 require two reflections and one reflection, respectively. Path 4 represents the direct ray, which travels along the axis of the fiber with no reflection. Clearly path 4 is the shortest while path 1 is the longest. If the fiber has a constant refractive index, the rays will all travel at the same velocity, but since the paths have different lengths, the rays will not arrive simultaneously at the end of the fiber.

If the light from the laser source is modulated by a digital signal, the resultant pulses will contain rays with different modes. The term "mode" is used to refer to a particular path. When an individual pulse travels down a fiber, the various modes will cause the pulse to spread out in time, and consequently the amplitude of the pulse must decrease. This phenomenon is referred to as *pulse-width dispersion, modal dispersion,* or *pulse spreading.*

Figure 254-4 illustrates how a series of square-wave pulses become progressively more distorted as they are propagated down an optical fiber. Pulses A and C, which represent logic 1 states, are gradually lengthened in time and reduced in amplitude; in addition the pulse waveforms become more "rounded." In our example the A pulse spreads forward into the B time region, which is in the 0 state. If the fiber is sufficiently long, the A pulse will extend to the point where the B region is sampled so that this region will change from 0 to 1. This will occur if the amount of time spread for the pulse is equal to half of the interval, T, for 1 bit. The result will be an error in the transmission so that the amount of the pulse spread will limit the highest permitted rate (bits per second) of digital information.

At a particular position on an optical fiber each ray will have a different delay time. The difference in the delay times between the slowest and fastest rays is equal to the total pulse spread, $\Delta\tau$, which is the product of the fiber's length, L, and the pulse-spreading constant, δt. In equation form,

$$\Delta\tau = L \times \delta t \qquad (254\text{-}7)$$

where: $\Delta\tau$ = total pulse spread (ns)
δt = pulse-spreading constant (ns/km)
L = fiber length (km).

The highest rate of data transmission ($f_{b\max}$ bps) will occur when

$$\frac{T}{2} = \Delta\tau = L \times \delta t. \qquad (254\text{-}8)$$

Since $f_{b\max} = 1/T$, the highest rate of data transmission is

$$f_{b\max} = \frac{1}{2 \times L \times \delta t} \text{ bps} \qquad (254\text{-}9)$$

Apart from the modal dispersion there are two other phenomena which can cause pulse spreading:

1. *Material dispersion* In Table 253-2 we observed that the value of the refractive index for glass increases slightly as the frequency is raised. When a light beam is digitally modulated, Fourier analysis shows that a large number of harmonic components are introduced into the signal. Since a higher-frequency component will travel with a

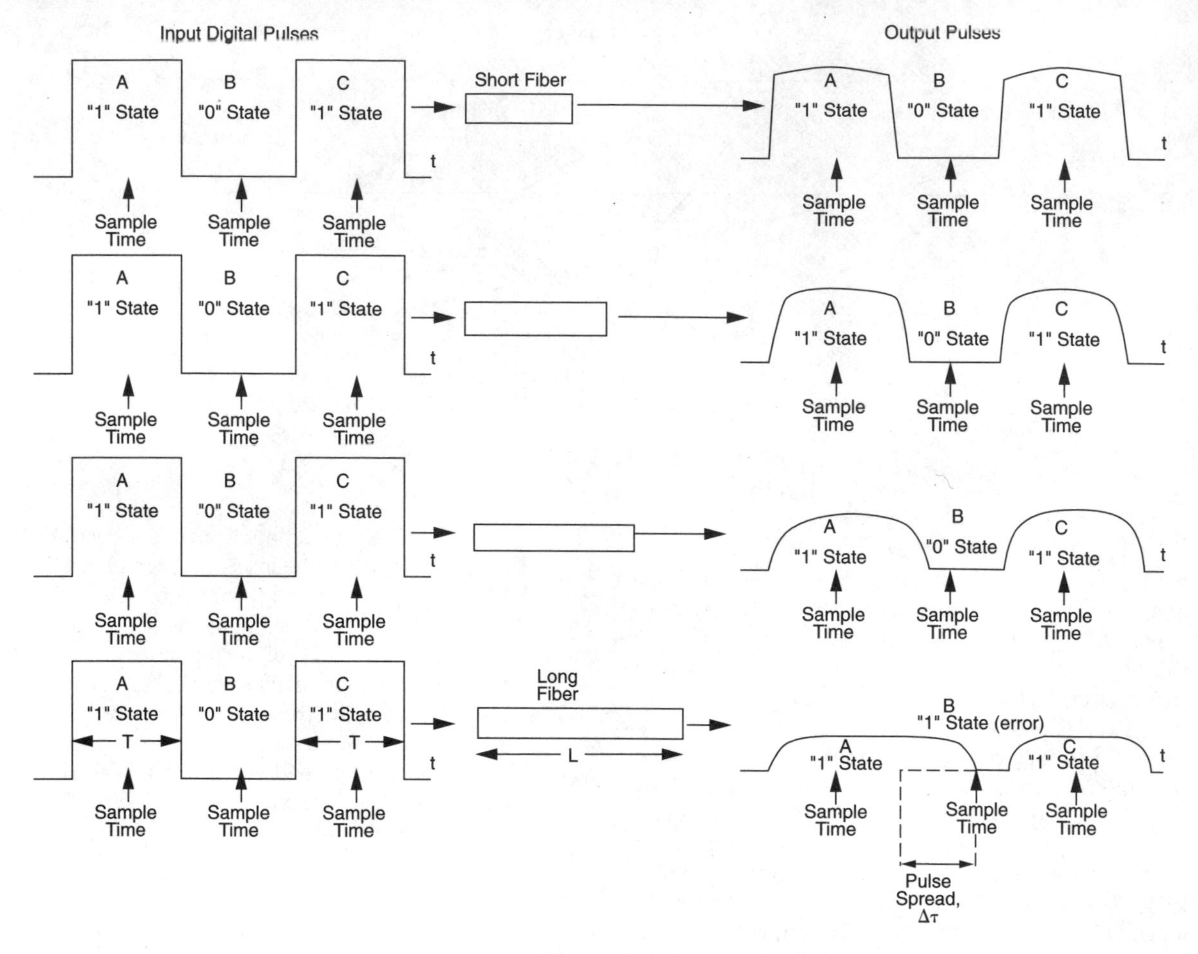

Figure 254-4

slower velocity, the amount of pulse spreading will be increased.

2. *Waveguide dispersion* Depending on the wavelength, between 15% and 25% of the light traveling down a fiber guide is trapped in the fused quartz cladding, which has a lower refractive index than the core. The light in the cladding therefore travels *faster* than that of the core, and further pulse spreading is the result. However, the effects of material dispersion and waveguide dispersion oppose one another and tend to cancel at a wavelength of 1.30 μm, which is commonly used in fiber-optic communications.

To take into account the effects of material and waveguide dispersion and to allow for a safety factor, Equation 254-9 is normally modified to

$$L \approx \frac{1}{5 \times f_{bmax} \times \delta t} \text{ km.} \qquad (254\text{-}10)$$

Example 254-1

In Fig. 254-2, $\mu_0 = 1$, μ_1 (optical fiber) $= 1.70$, and μ_2 (cladding) $= 1.46$. Determine the values of the acceptance cone's half-angle and the numerical aperture.

Solution

Half-angle of the acceptance cone,

$$i_{1max} = \sin^{-1}\sqrt{\mu_1^{\,2} - \mu_2^{\,2}} \qquad (254\text{-}5)$$

$$= \sin^{-1}\left(\sqrt{1.70^2 - 1.46^2}\right)$$

$$= 35.1^\circ$$

$$\text{Numerical aperture, NA} = \sin i_{1max} \qquad (254\text{-}6)$$

$$= \sin 35.1^\circ$$

$$= 0.575.$$

Example 254-2

The pulse-spreading constant of an optical fiber is 4 ns/km and the length of the fiber is 12 km. Calculate the value of the highest rate of data transmission.

Solution

Highest rate of data transmission,

$$f_{bmax} = \frac{1}{2 \times L \times \delta t} \qquad (254\text{-}9)$$

$$= \frac{1}{2 \times 12 \text{ km} \times 4 \text{ ns/km}}$$

$$= \frac{10^9}{96} \text{ bps}$$

$$= 10.4 \text{ Mbps.}$$

PRACTICE PROBLEMS

254-1. A glass fiber whose refractive index is 1.30 is placed in air. Calculate the value of the fiber's numerical aperture.

254-2. An optical fiber has a pulse-spreading constant of 1.8 ns/km. If the information rate is 6×10^7 bps, calculate the fiber's maximum length (without allowing for any safety factor).

255 Types of optical fiber

In chapter 254 we discussed the manner in which a light pulse is propagated down a *multimode* fiber. Since the light is able to take a number of possible paths with different lengths, the result is pulse spreading. This disadvantage could clearly be prevented if there was only one possible path for the light to follow. Such is the case with the *single-mode step-index* fiber.

Single-mode step-index fiber

In its simplest form this type of fiber is cylindrical in shape and is surrounded by air. The refractive index of the glass is uniform throughout the fiber and has a value of approximately 1.6. The term "step index" refers to the fiber's μ profile (Fig. 255-1A). There is an abrupt discontinuity (or step) between the refractive indices of the glass and the air. For this type of fiber the acceptance angle is

$$A = \sin^{-1}\sqrt{\frac{\mu_1^2 - \mu_2^2}{\mu_1^2}} \qquad (254\text{-}3)$$

$$= \sin^{-1}\sqrt{\frac{1.6^2 - 1^2}{1.6^2}}$$

$$= 51.3°$$

$$\text{Numerical aperture, NA} = \sin A \qquad (254\text{-}6)$$

$$= \sin 51.3°$$

$$= 0.78.$$

These high values of acceptance angle and numerical aperture indicate that it is relatively easy to couple the light source to the fiber. Provided the diameter of the glass is less than 10 μm (as opposed to 50 μm for multimode fibers), most of the light will travel along the fiber's axis while the remainder will only be involved with one or two reflections. The possible paths will nearly be equal in length so that there is a minimum amount of pulse spreading. When compared with the multimode type, the single-mode fibers will have a far greater potential bandwidth but are more difficult and more expensive to manufacture as well as being more fragile. Moreover there are more problems with joining together single-mode fibers. Consequently such fibers, which are surrounded by air, are only practical over short distances.

A single-mode fiber with a fused quartz cladding (Fig. 255-1B) is less fragile and is therefore more commonly used. However, since the refractive index (1.46) of fused quartz is high compared to that of air, the values of the acceptance angle and the numerical aperture are appreciably lower.

$$\text{Acceptance angle, } A = \sin^{-1}\sqrt{\frac{\mu_1^2 - \mu_2^2}{\mu_1^2}} \qquad (254\text{-}3)$$

$$= \sin^{-1}\sqrt{\frac{1.6^2 - 1.46^2}{1.6^2}}$$

$$= 24.1°$$

$$\text{Numerical aperture, NA} = \sin A \qquad (254\text{-}6)$$

$$= \sin 24.1°$$

$$= 0.41.$$

The values of the acceptance angle and the numerical aperture indicate that it is more difficult to provide adequate coupling between the light source and the fiber. It is therefore necessary to use a directive light source such as the injection laser as opposed to a light-emitting diode (chapter 258).

Multimode step-index fiber

This type of fiber has already been discussed in chapter 254. To summarize, the center core is much larger (Fig. 255-1C), with a diameter of approximately 50 μm. Consequently more light is accepted by the fiber but the numerous paths available cause pulse spreading, which reduces the possible bandwidth and the maximum value of the digital information rate.

Multimode graded-index fiber

The term "graded index" indicates that the refractive index of the glass is no longer uniform. The μ profile of Fig. 255-1D shows that the value of the refractive index is maximum at the center of the core and decreases with a parabolic distribution toward the perimeter. The ray that travels along the axis of the fiber has the lowest velocity. Other rays which deviate from the axis are propagated in regions that possess a lower value for the refractive index. This has two effects:

1. These rays are refracted back toward the axis, and their paths are longer than the shortest path along the fiber's axis.
2. Since the velocity of a ray is inversely proportional to the μ value, the refracted rays travel faster than the ray that is propagated along the fiber's axis. Because the effect of the longer path is offset by an accompanying higher velocity, the various rays tend to arrive simultaneously at the end of the fiber (Fig. 255-1D). The amount of pulse-spreading will then be greatly reduced.

To summarize, the features of the various types of fiber are:

1. *Single-mode step-index fiber*
   Disadvantages: Expensive and difficult to manufacture. Low values of acceptance angle and numerical aperture. Difficulty in joining or splicing two fibers together. Requirement for a directive light source such as an injection laser. Difficulty in coupling from the light source to the fiber.
   Advantages: Large bandwidth. High rate of data transmission. Minimum amount of pulse spreading.
2. *Multimode step-index fiber*
   Disadvantages: Less bandwidth. Lower rate of data transmission. Higher degree of pulse spreading.
   Advantages: Inexpensive and easy to manufacture. High values of acceptance angle and numerical aperture. Easy to join or splice two fibers together. Light-emitting diode may be used as a light source. Easy to couple from the light source to the fiber.
3. *Graded-index fiber*
   In all respects this type of fiber is a compromise between the single-mode and the multimode step-index fibers.

MATHEMATICAL DERIVATIONS

Single-mode fiber

In our study of circular waveguides (chapter 180) we learned that there was a certain maximum frequency (or minimum wavelength) at which only the dominant mode was propagated. If this frequency is exceeded, higher-order modes will exist as well as the dominant mode. The same effect occurs in single-mode fibers in which the minimum wavelength, λ_m, is given by

$$\lambda_m = \pi d \frac{\sqrt{\mu_1 (\mu_1 - \mu_2)}}{1.701} \qquad (255\text{-}1)$$

where: λ_m = minimum wavelength (μm)
d = diameter of core (μm)
μ_1 = refractive index of core
μ_2 = refractive index of cladding.

Single-mode operation will occur at frequencies for which the wavelengths are greater than the value of λ_m.

Graded-index and step-index fibers

For graded-index and step-index fibers the number of possible modes depends on the wavelength of the light, the diameter of the fiber's core, and the numerical aperture. The equations are:

Step-index fiber,

$$N = \left[\frac{2.22d \times (\text{numerical aperture})}{\lambda}\right]^2$$

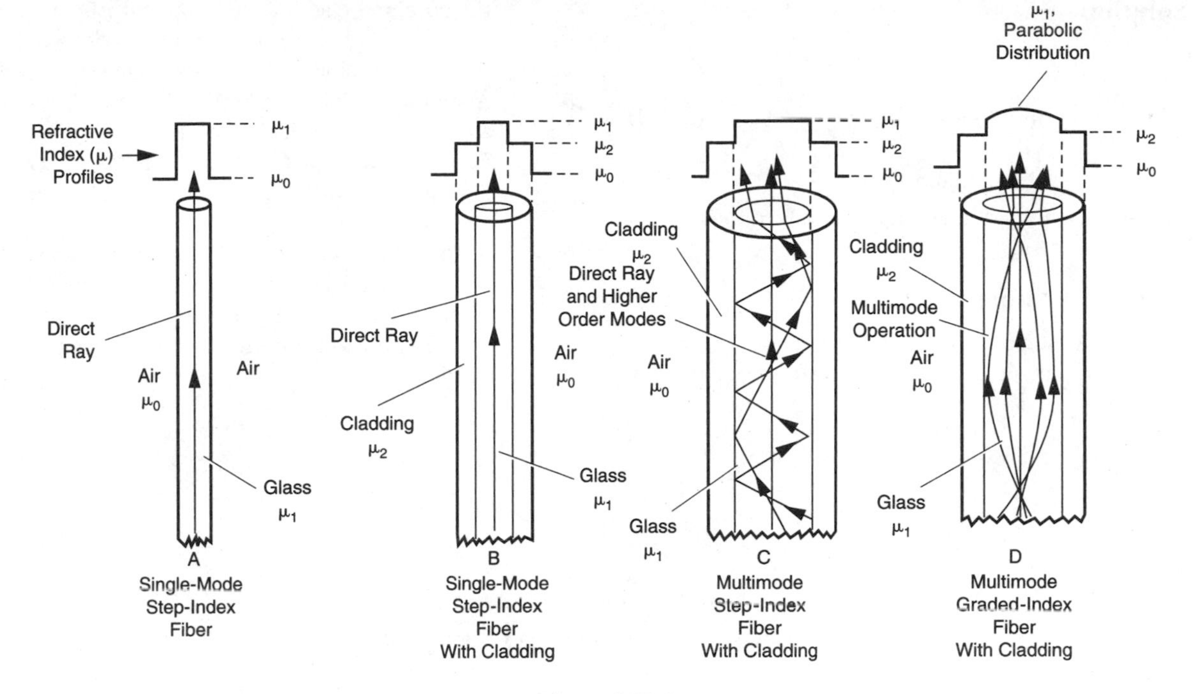

Figure 255-1

$$= \left[\frac{2.22d \times \sin(\text{acceptance angle})}{\lambda}\right]^2 \qquad (255\text{-}2)$$

Graded-index fiber,

$$N = \left[\frac{1.57d \times (\text{numerical aperture})}{\lambda}\right]^2$$

$$= \left[\frac{1.57d \times \sin(\text{acceptance angle})}{\lambda}\right]^2 \qquad (255\text{-}3)$$

where: N = number of possible modes
d = diameter of fiber's core (μm)
λ = wavelength of light (μm).

Example 255-1

A single-mode fiber has a glass core with a refractive index of 1.51 and a diameter of 3.2 μm. If the fused quartz cladding has a refractive index of 1.48, calculate the value of the minimum wavelength for a single-mode operation.

Solution

Minimum wavelength,

$$\lambda_m = \frac{\pi d \sqrt{\mu_1 (\mu_1 - \mu_2)}}{1.701} \qquad (255\text{-}1)$$

$$= \frac{\pi \times 3.2 \times \sqrt{1.51(1.51 - 1.48)}}{1.701}$$

$$= 1.26 \ \mu\text{m}.$$

Such a fiber would be suitable for the commonly used wavelengths of 1.30 μm and 1.55 μm.

Example 255-2

A step-index fiber has a core diameter of 50 μm. If the (internal) acceptance angle is 43° and the wavelength of the light is 1.30 μm, calculate the number of possible modes.

Solution

Number of possible modes,

$$N = \left[\frac{2.22 \times 50 \times \sin 43^\circ}{1.30}\right]^2 \qquad (255\text{-}2)$$

$$= 3391.$$

Example 255-3

The step-index fiber of Example 255-2 is replaced by a graded-index fiber with the same diameter. If the acceptance angle and the light's wavelength are unchanged, recalculate the number of possible modes.

Solution

Number of possible modes,

$$N = 3391 \times \left[\frac{1.57}{2.22}\right]^2 \qquad (255\text{-}3)$$

$$= 1696.$$

The use of a graded-index fiber has halved the number of possible modes.

PRACTICE PROBLEMS

255-1. A single-mode fiber has a glass core whose diameter is 2.0 μm. If the refractive indices of the core and cladding are 1.65 and 1.52, respectively, calculate the cut-off wavelength. What is the value of the numerical aperture?

255-2. A step-index multimode fiber has a core diameter of 25 μm. If the acceptance angle is 52° and the wavelength of the light is 1.55 μm, calculate the number of possible modes.

255-3. A graded-index multimode fiber has a core diameter of 36 μm. If the acceptance angle is 47° and the wavelength of the transmitted light is 1.55 μm, determine the number of possible modes.

256
Losses in optical fibers

A single fiber-optic cable is manufactured from glass which has a purity of 99.999%. It takes about 1 ton, or over 4900 kg, of dried silica sand to produce some 20 kg of purified glass, but this amount of glass will produce several miles of optical cable.

During manufacture a hollow glass tube with an outer cross-sectional area of about 20 cm^2 is cut to a length of approximately 1 m. This tube is rotated and vaporized doping compounds (which control the refractive index) are fed into its hollow center. A heating flame is then made to travel down the length of the tube. This causes some of the doping compounds to be deposited on the tube's inner surface. Any of the dopants which are not deposited, are then removed from the far end of the tube. The flame may travel a number of times down the tube until the required changes in the refractive index are achieved. This procedure is referred to as the modified chemical vapor deposition (MCVD) process.

After the deposition phase is completed, the tube is heated to a higher temperature. The tube then collapses owing to atmospheric pressure, and the result is a solid glass rod, which is known as a preform. The outer part of this preform will then behave as the cladding while the inner region is the fiber-optic cable.

In the last stage a vertical preform is passed slowly through a furnace, which is designed not to contaminate the glass in its plastic condition. This heating procedure allows the preform to be drawn into the final thin optical cable. A laser system monitors the cable's diameter by controlling the speed of the draw. As a final step the fiber is provided with a liquid protective coating, which is passed through a curing process.

Scattering losses

Although great care is taken in the manufacturing process, it is impossible to avoid minute irregularities, which occur primarily in the production of the preform and the subsequent drawing of the fiber. The dimensions of an individual irregularity are much less than the wavelength of the light traveling down the fiber. As a result the irregularities act as a scattering medium so that when a light ray encounters an irregularity, the light is diffracted in a number of directions. Some of the diffracted light penetrates through the cladding and causes a loss in the light power that reaches the end of the fiber cable. This form of loss is unavoidable and is referred to as Rayleigh scattering (Fig. 256-1). The degree of scattering is inversely proportional to the fourth power of the light's wavelength and determines the minimum loss of a fiber-optic cable.

Additional scattering losses occur in the manufacturing process due to mechanical defects. When the glass tube collapses in the preform process, small bubbles and minute cracks can appear in the core or the cladding. In the draw process the core diameter can change and over a short length the center of the core may no longer coincide with the axis of the cable. All these defects will increase the fiber's attenuation loss.

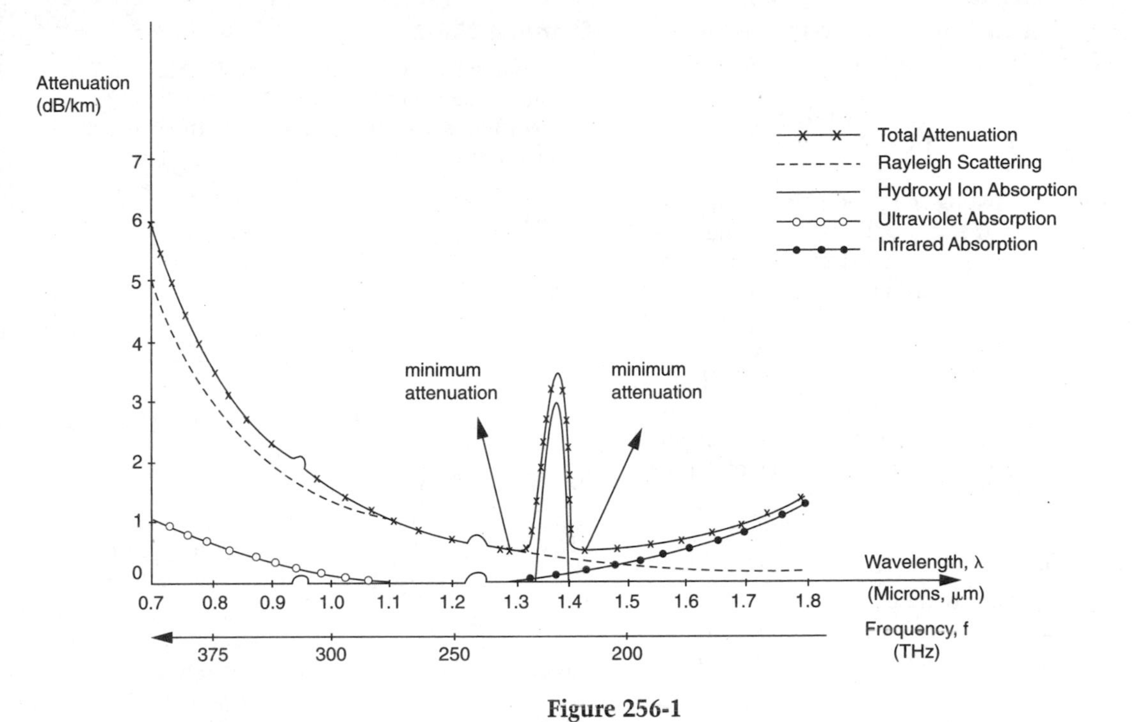

Figure 256-1

Absorption losses

The compounds from which a fiber guide is manufactured contains ions whose valence electrons can be lifted into the conduction band provided the light photons have sufficient energy. These absorption losses can be subdivided as follows:

Infrared absorption

Photons of light in the infrared band excite valence electrons associated with the dopants introduced into the core of the fiber guide. This absorption begins at a wavelength of approximately 1.25 μm and then increases as the wavelength is raised.

Ultraviolet absorption

The original pure silica contains oxygen ions whose electrons are excited by photons in the ultraviolet region. The effect is least at a wavelength of 1.1 μm and then increases as the wavelength is lowered (Fig. 256-1).

Hydroxyl ion absorption

As shown in Fig. 256-1, these absorption losses have peaks which occur over certain narrow wavelength bands. In the doping process the hydrogen atom of one additive combines with the oxygen atom of another additive to produce water with its hydroxyl (OH^-) ions. Such ions have a fundamental vibration wavelength of 2.7 μm, with smaller vibrations occurring at 1.38 μm, 1.24 μm, and 0.95 μm. Only the smaller vibrations are of significance in determining the wavelength to be used in fiber-optic communications.

Referring to Fig. 256-1, we can observe the following:

1. Minimum infrared and ultraviolet absorption occurs at the 1.2-μm wavelength.
2. The loss due to hydroxyl ions has its strongest peak of 3 dB/km at the 1.38-μm wavelength.
3. The overall loss curve has its minima at wavelengths of 1.29 μm and 1.44 μm. These values compare well with the commonly used wavelengths of 1.30 μm and 1.55 μm.

Experiments have shown that decreases in the surrounding temperature and increases in stress will raise a fiber guide's attenuation. Moreover attenuation will be greater if a fiber guide is subjected to a bend whose radius is less than 1000 times the guide's diameter. If the bend's radius is reduced below 100 times the diameter, the fiber can break.

MATHEMATICAL DERIVATIONS

The various losses are combined in the guide's attenuation constant, α, which is normally measured in decibels per kilometer. It follows that

$$\alpha = \frac{10 \log(P_1/P_o)}{l} \qquad (256\text{-}1)$$

where: P_i = input light power (W)

P_o – output light power (W)
l = length of fiber (km).

Equation 256-1 yields

$$P_o = P_i\, 10^{-\alpha l/10}. \qquad (256\text{-}2)$$

It is worth comparing the attenuation of an optical fiber with that of the transmission lines and waveguides of earlier chapters. A twisted pair using #22 AWG wire operating at 1 MHz has an attenuation of 30 dB/km. A similar degree of attenuation applies to a 2-cm coaxial cable at a frequency of 500 MHz. The waveguide is more efficient. An X-band guide has an attenuation of 3 dB/km, but even this pales by comparison with a single-mode fiber guide whose attenuation can be as low as 0.3 dB/km at a frequency of approximately 200 THz.

Example 256-1

The input light power to a fiber guide is 25 mW. If the length of the guide is 800 m with an attenuation constant of 0.4 dB/km, calculate the value of the output power.

Solution

$$\text{Output power, } P_o = P_i\, 10^{-\alpha l/10} \qquad (256\text{-}2)$$

$$= 25 \times 10^{-0.4\times 0.8/10}$$

$$= 23.2 \text{ mW.}$$

Example 256-2

The input light power to an optical fiber is 18 mW and the output light power is 16 mW. If the length of the fiber is 1300 m, calculate the value of the fiber's attenuation constant.

Solution

$$\text{Attenuation constant, } \alpha = \frac{10\log(P_i/P_o)}{l} \qquad (256\text{-}1)$$

$$= \frac{10\log(18/16)}{1.3}$$

$$= 0.4 \text{ dB/km, rounded off.}$$

PRACTICE PROBLEMS

256-1. The input light power to an optical fiber is 1.75 mW. If the attenuation constant of the fiber is 0.4 dB/km and the fiber is 18 km long with no repeaters, calculate the value of the output power to the photodetector circuit.

256-2. The input light power to an optical fiber is 25 mW and the output light power is 80 μW. If the length of the fiber is 17 km, obtain the value of the fiber's attenuation constant.

256-3. A photodetector circuit requires a minimum power output of 5 μW from an optical fiber. If the fiber's attenuation constant is 0.8 dB/km and the input power is 70 μW, calculate the maximum length of the fiber.

257 Fiber joins

In a fiber-optic system there must be at least three types of junction.

1. A connection from the light source to the fiber.
2. Connections between one fiber and another.
3. A connection from the fiber to the photodetector circuit.

As with the waveguides discussed previously, the overall fiber loss will be considerably increased if the joins are not made with a great deal of precision. Although a small radiation loss is inevitable with any join, it can be kept to a minimum if the core of one fiber is accurately aligned with the core of the other fiber. Misalignments fall into four categories, which are illustrated in Fig. 257-1.

1. Axial or lateral displacement is shown in Fig. 257-1A. The cores of the two fibers are displaced by a distance, d, which we will assume to be 0.5 μm. If the radius, r, of the core is 25 μm, the relative amount of displacement is d/r = 0.5 μm/25 μm = 0.02, which is experimentally found to introduce a radiation loss of about 0.5 dB.
2. Figure 257-1B shows an angular displacement of θ degrees between the axes of the two fibers. If θ = 5°, the amount of the relative misalignment is

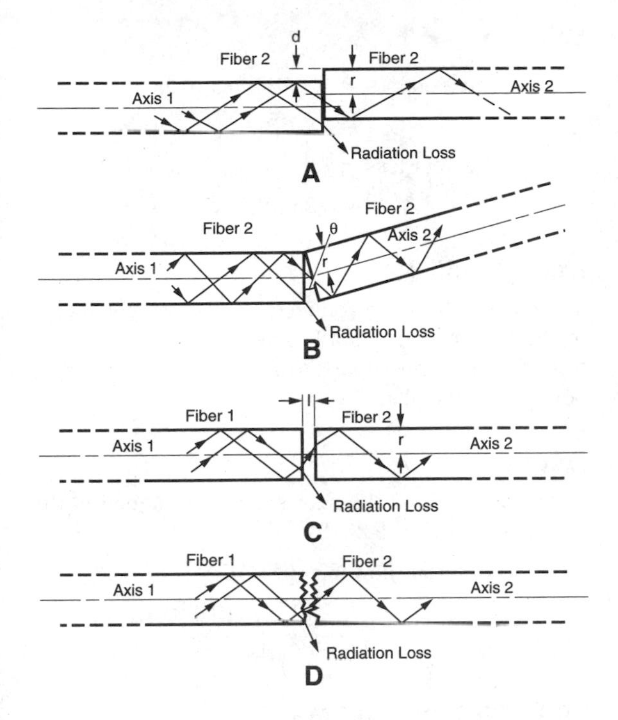

Figure 257-1

measured by sin θ = 0.087, which corresponds to an approximate loss of 0.3 dB.

3. When connectors are used, one fiber may be too short so that there is an air gap between the two fibers. Since the light from the transmitting source will tend to spread out, the length of the gap and the radius of the core must determine the amount of the loss. If the gap, l, is assumed to be 7.5 μm while the radius of the core is 25 μm, the value of the relative loss is l/r = 7.5 μm/25 μm = 0.3. This causes a loss of about 0.4 dB.
4. There are no mathematical calculations that determine the radiation loss due to rough surfaces. After placing a fiber in a connector, the rough end may be ground and polished to the required standard. A loss of 0.2 dB occurs if the end surfaces deviate by 2° from the perpendicular (Fig. 257-1D).

Of the four losses we have described, the loss due to axial displacement is generally the most significant. To reduce all losses to a minimum, the two fibers must be accurately spliced together. In one method the two fibers are first cut and cleaned and then accurately aligned. Finally the fibers are fused together with the aid of an electric arc. The loss from this type of splice is as little as 0.25 dB, but its use is normally confined to the laboratory. Under more practical conditions four alignment pins are used to enclose the two fibers being connected (Fig. 257-2). After the fibers have been carefully aligned by inspection through an opening, a containing tube is subjected to heat shrinking, so the pins apply pressure to the fiber and the alignment is maintained. Through the opening an epoxy is applied to the join and the whole assembly is then protected by a second heat-shrunk tube.

Connectors must be used in joining a fiber guide to the transmitting light source or to the photodetector circuit. In Fig. 257-3 the output light from the emitting source passes through a fiber guide. This guide is enclosed in a resilient ferrule, which is part of the source's housing package. The ferrule is then bolted into a bushing which has tapered openings at both ends. The output fiber guide is similarly enclosed in another resilient ferrule, which is placed in the threaded end of the bushing and held in position by a retaining cap. Since the ferrules are resilient, they are compressed by the bushing's tapers and the cores can then maintain their correct alignment.

The type of connector in Fig. 257-3 is also capable of joining the fiber guide to the photodetector. If both ends of the bushing are threaded, the same principle can be used to align the cores of two separate guides.

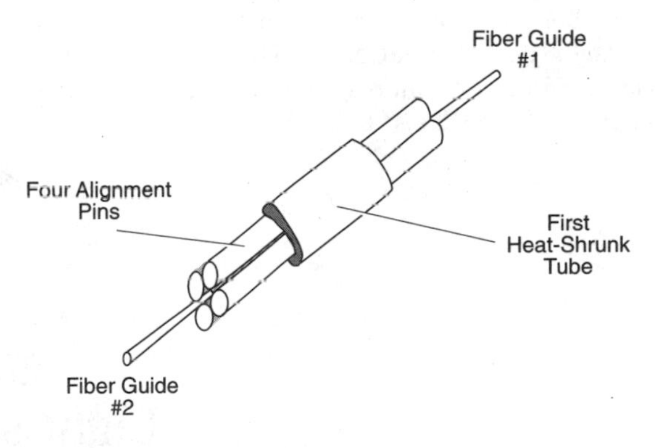

Figure 257-2

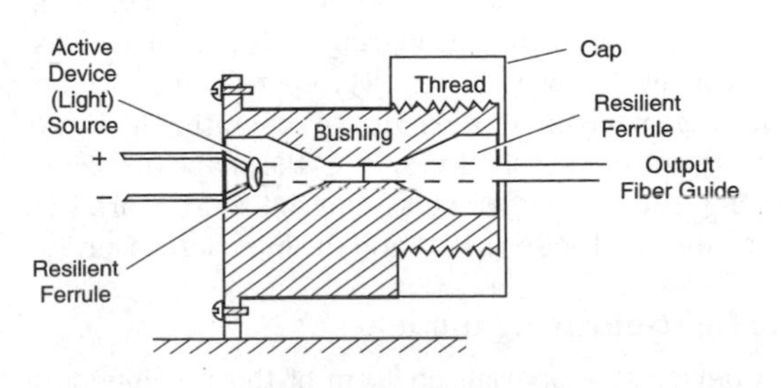

Figure 257-3

Such connectors produce a loss which is typically on the order of 1 dB.

MATHEMATICAL DERIVATIONS

Axial or lateral displacement

$$\text{Relative amount of displacement} = \frac{d}{r} \qquad (257\text{-}1)$$

where: d = displacement distance (μm)
r = core radius (μm).

Angular displacement

$$\text{Relative misalignment} = \sin\theta \qquad (257\text{-}2)$$

where: θ = angular displacement (°).

Air gap between fibers

$$\text{Relative loss} = \frac{l}{r}$$

where: l = length of gap (μm)
r = radius of core (μm).

All of these losses are normally measured in decibels.

Example 257-1

Three fiber joins have individual losses of 0.5 dB, 0.25 dB, and 0.3 dB. The total length of the fiber is 6 km and its attenuation constant is 1.2 dB/km. Calculate the total fiber loss in decibels.

Solution

$$\begin{aligned}\text{Attenuation loss} &= 1.2\ \text{dB/km} \times 6\ \text{km}\\ &= 7.2\ \text{dB}\\ \text{Total fiber loss} &= 7.2 + 0.5 + 0.25 + 0.3\\ &= 8.25\ \text{dB.}\end{aligned}$$

Example 257-2

In Example 257-1 the connector from the light source to the fiber introduces a loss of 1.0 dB. If the power from the light source is 1.8 mW, calculate the output power from the fiber.

Solution

The total loss from the light source to the end of the fiber is 1.0+8.25 = 9.25 dB.

$$\begin{aligned}\text{Output power} &= 1.8\times10^{(-9.25/10)}\\ &= 1.8 \times 10^{-0.925}\\ &= 0.21\ \text{mW.}\end{aligned}$$

PRACTICE PROBLEMS

257-1. A fiber is 11 km in length and has an attenuation constant of 0.8 dB/km. The power input to the fiber is 1.25 mW and the output power from the fiber is 110 μW. Calculate the decibel loss due to the joins in the fiber.

257-2. In Example 257-1, calculate the loss in microwatts owing to the joins in the fiber.

258 Light sources

The light sources used for fiber-optic communications may be compared by considering such parameters as (1) the level of power output, (2) response time, (3) the frequency spectrum of the light, (4) operating lifetime, (5) temperature sensitivity, and (6) the shape of the radiation pattern. The two most common sources are the Light Emitting Diode (LED) and the Injection Laser.

The light-emitting diode

This device is a specialized form of the p-n junction diode in its forward-biased condition. The forward bias lowers the energy "hill" so that conduction band electrons are able to move across the junction into the p-region, where they drop into holes and become valence electrons. This action must cause energy to be released since there is a gap between the energy level of the conduction-band electron and that of the valence electron. For particular materials this gap is such that the released energy takes the form of light photons. For example, a homojunction epitaxial LED may be manufactured from gallium arsenide, which is doped with silicon (Fig. 258-1A). The corresponding energy or band gap is 1.43 eV.

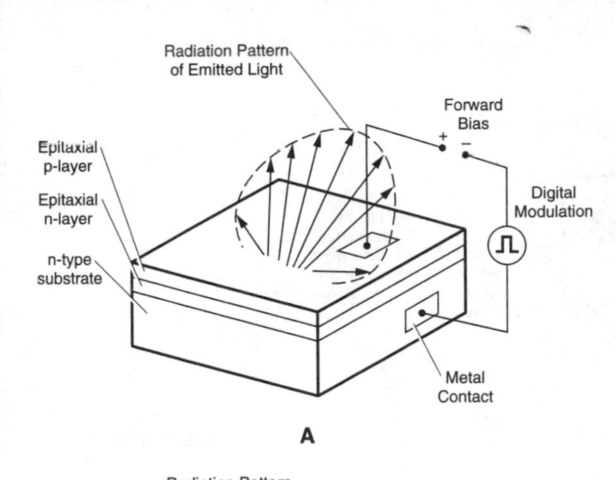

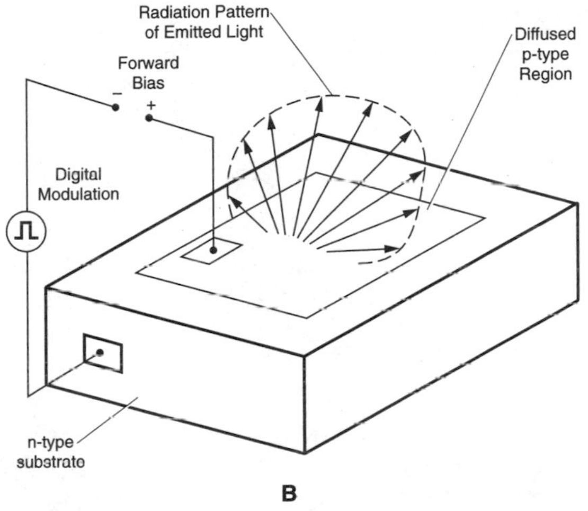

Figure 258-1

For a forward bias current of 120 mA the emitted light has a peak power of approximately 3.5 mW at the wavelength of 0.94 μm. However, the light has a wide bandwidth, which may extend from 0.7 μm to 1.1 μm. This is a major problem since each individual frequency travels with a different velocity down the fiber guide and the result is pulse dispersion, as already discussed in chapter 254. Moreover the turn-on/turn-off time for this type of LED can exceed 125 ns so that the rate of data transmission is limited to about 1×10^7 bps.

Better results are available from a homojunction planar diffused LED (Fig. 258-1B) in which zinc is diffused into a gallium arsenide n-type wafer which is doped with tellurium. For a forward bias current of 80 mA, a peak power of 0.5 mW occurs at a wavelength of 0.9 μm. The main improvement lies in the turn-on/turn-off time of less than 20 ns, which corresponds to a data transmission rate of about 6×10^7 bps.

The main disadvantage of homojunction LEDs is the tendency for the light to spread out in all directions so that there is a wide radiation lobe (Fig. 258-1) from a large emitting surface. It is therefore difficult to provide efficient coupling between the source and the short length of fiber guide which is joined to the connector (see Fig. 257-3).

Better results are available from a planar heterojunction LED, which is manufactured from a number of layers of different semiconductor materials (Fig. 258-2A). The wafer is 100 μm thick, but the light is confined to a small surface area at the center of the active region. A Burrus well is etched into the surface to a depth of 25 μm so that the narrow radiated beam can be coupled into a fiber guide with a ball end to assist in concentrating the light. However, this fiber is too wide to allow single-mode operation (Fig. 258-2B).

We have examined two types of the surface-emitting LEDs. It is possible to use the construction of Fig. 258-2A (without the Burrus well) and allow the light to be emitted from the edge of an active stripe (Fig. 258-3). This edge-emitting LED has a more directional beam, which is easier to couple to a fiber guide. However, the total amount of radiated light is less than that of the surface-emitting LED.

To summarize, LEDs are cheap and have a long life expectancy, on the order of 10^6 h. However, due to their large emitting surfaces, they cannot be coupled efficiently to the fiber guides, and single-mode operation is not possible. Moreover their emitted light has a wide bandwidth, which leads to pulse spreading. The output power from LEDs is directly proportional to the device's forward current and is also a function of the operating temperature (the higher the temperature, the less is the value of the output power).

The injection laser

In 1962 W. P. Dumke demonstrated that laser action was possible in a p^+-n^+ junction diode which was manufactured from gallium arsenide. The action of this semiconductor laser depended on pumping electrons and holes into the p^+- n^+ junction. For this reason the device is sometimes referred to as an injection laser diode, whose structure is illustrated in Fig. 258-4. Basically, we have a single crystal of gallium arsenide with a p-n junction formed by the standard diffusion method using heavily doped zinc to create the p^+ region. The parallel end faces form a resonator; one face is totally reflective, while the other is semireflective to allow a passage for the laser's output. By contrast, the side faces are roughened to suppress all unwanted modes of propagation.

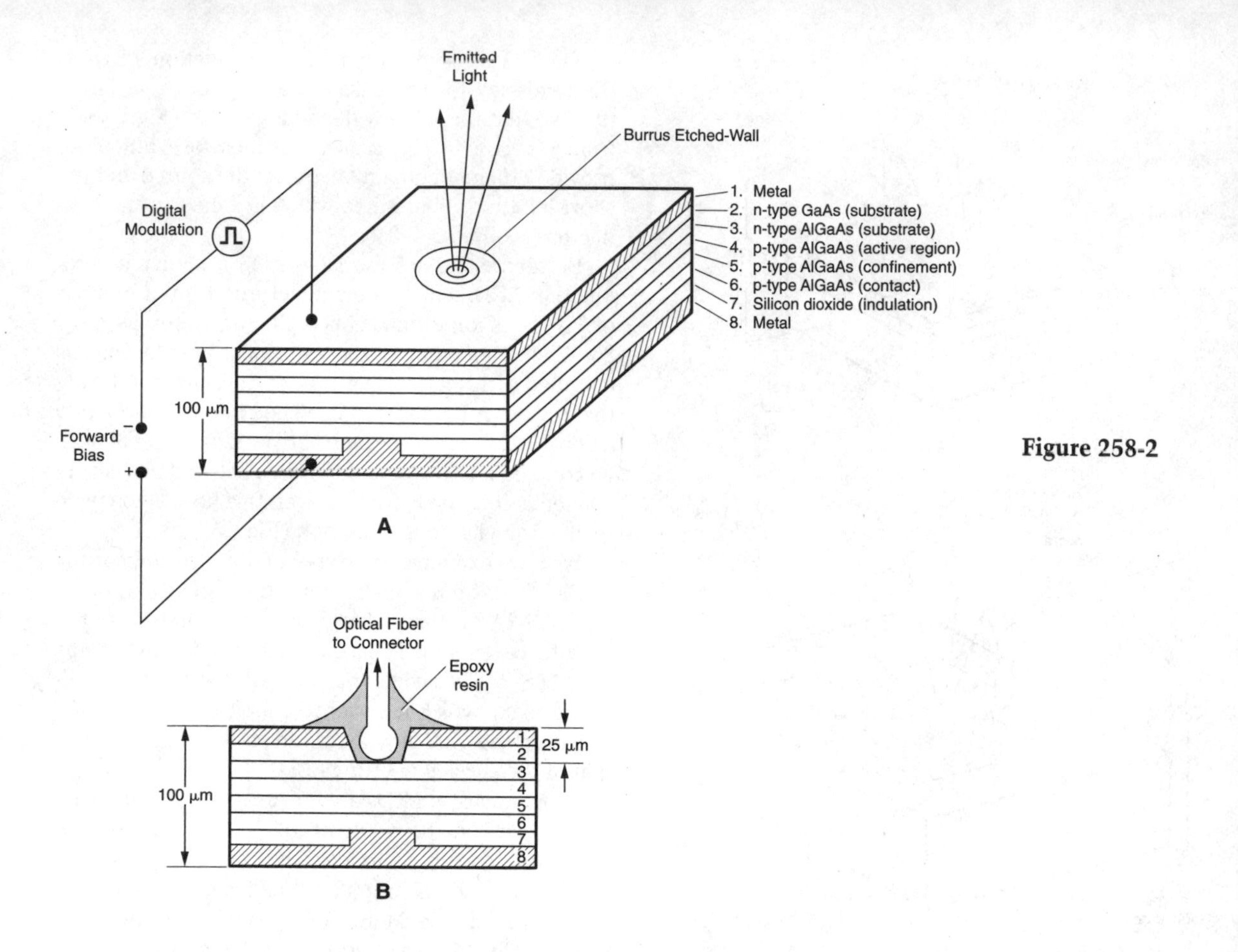
Emitted
Light
Burrus Etched-Wall
Digital
Modulation
1. Metal
2. n-type GaAs (substrate)
3. n-type AlGaAs (substrate)
4. p-type AlGaAs (active region)
5. p-type AlGaAs (confinement)
6. p-type AlGaAs (contact)
7. Silicon dioxide (indulation)
8. Metal
100 μm
Forward
Bias
A
Optical Fiber
to Connector
Epoxy
resin
25 μm
100 μm
1
2
3
4
5
6
7
8
B

Figure 258-2

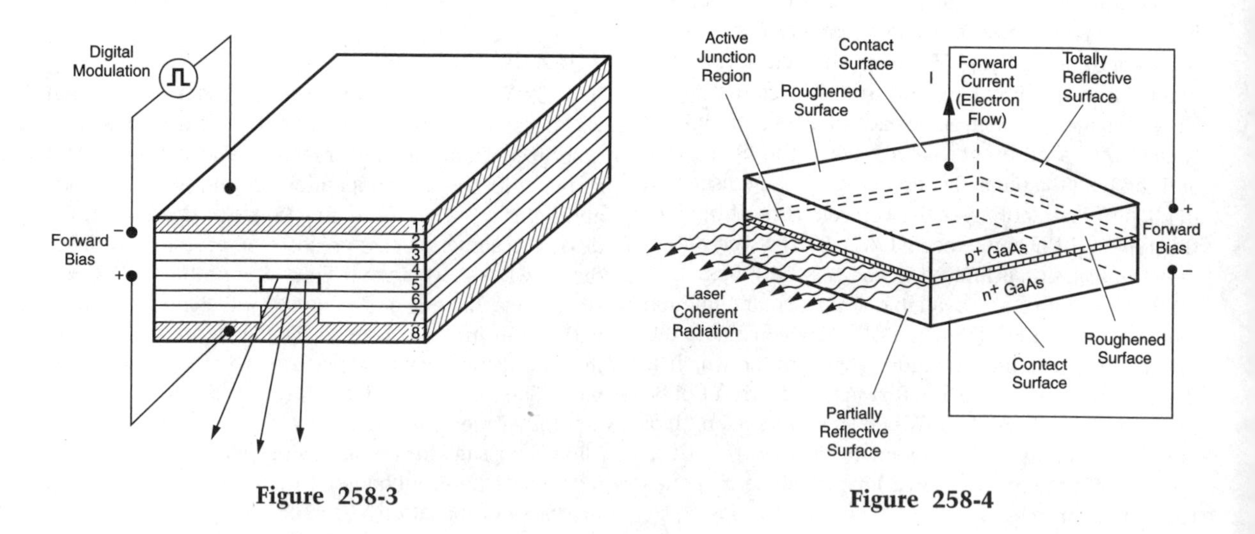
Digital
Modulation
Forward
Bias
1
2
3
4
5
6
7
8
Active
Junction
Region
Contact
Surface
Roughened
Surface
I
Forward
Current
(Electron
Flow)
Totally
Reflective
Surface
Forward
Bias
p+ GaAs
n+ GaAs
Laser
Coherent
Radiation
Roughened
Surface
Contact
Surface
Partially
Reflective
Surface

Figure 258-3

Figure 258-4

The operation of the laser diode depends on the action of the resonant cavity and on creating a population inversion in a very thin region at the junction. The energy levels involved are those of the conduction and valence bands, and suitable energy differences are only found in a semiconductor such as gallium arsenide, not in germanium or silicon.

The pumping action is provided by the forward dc bias across the junction. Under the condition of zero bias, the junction is depleted of charge carriers, but when a sufficiently high forward bias is applied, the potential barrier is overcome. Due to the heavy doping, large numbers of holes and electrons will be injected into the layer at the junction. As a result, this layer contains high concentrations of electrons and holes in the conduction and valence bands. These changes in the concentrations of the two energy bands represent the necessary population inversion and the layer at the junction is then called the inversion region. Because it is easier to inject electrons rather than holes, most of the inversion region lies on the p-side of the junction. Possible transitions can occur between the electrons at the bottom of the conduction band (where the electron concentration is greatest) and the top of the valence band (where the hole concentration is least).

Notice that the electron-hole recombination is taking place directly so that photons are emitted to create the laser action. The percentage of the original injected electron-hole pairs which result in emitted photons is about 70%, and the laser has an overall efficiency of 30%. By contrast, most electron-hole recombinations take place indirectly in silicon and germanium so that in their cases little radiation would result.

The transition that occurs in the GaAs laser corresponds to frequencies in the infrared region. The precise wavelength depends on the chemical composition of the gallium arsenide material and lies in the range of 0.75 μm to 0.90 μm (400 THz to 333 THz). However, if the principal wavelength of the light is 0.85 μm (353 THz), the source is not monochromatic and the laser output has a bandwidth of about 15 THz. The presence of the additional frequencies will lead to an appreciable amount of pulse spreading. In addition the large width of the emitting surface will only allow multimode operation.

Commercial homojunction injection lasers have a total volume on the order of a few cubic millimeters with emitting junction widths of between 75 μm and 1.5 mm. Such a laser can produce pulse power outputs between 2 and 70 W, with corresponding driving currents of 10 A to 250 A at room temperature. The pulses have durations on the order of hundreds of nanoseconds, with repetition rates between 10 and 20 kHz. The average output then lies in the milliwatt range. For CW operation it is possible to generate a power output of 1 W, which is far more than is necessary to drive a fiber-optic system.

Heterojunction lasers

The laser diode so far described contains only a single junction in one type of material. This limits the efficiency and makes it difficult to operate the homojunction device at room temperature. However, if multiple layers are used to form a heterojunction structure, the laser can then be used for continuous-wave (CW) operation with single-mode fibers. The secret is to confine the injected charge carriers to a very narrow region so that the population inversion can occur with lower levels of the forward current.

One successful method is to confine the very thin (less than 1 μm) laser active p-type GaAs region between two AlGaAs layers which are epitaxially grown on the p region (Fig. 258-5). Each AlGaAs layer then

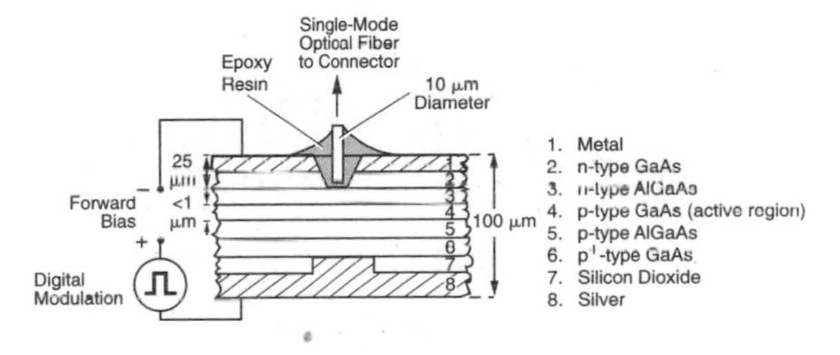

Figure 258-5

forms a boundary which injection electrons cannot cross because of the potential barriers existing at the GaAs/AlGaAs junctions. Not only does this double heterojunction restrict the injected carriers to the active region, but the changes in the refractive indices at the boundaries confine the laser output to a very narrow strip whose thickness is about 0.2 μm. The output of this laser can then be easily coupled to single-mode fiber-optic systems, which presently operate at a wavelength of 0.85 μm. However, the emitted light has a bandwidth of about 1.5 THz so that there will be a certain amount of pulse spreading.

Emissions in the 0.82-μm to 0.85-μm range for single-mode fibers can also be provided by ternary AlGaAs and InGaAs lasers. In each case a Burrus well is commonly used to couple the laser output to the fiber of a connector (Fig. 258-5). Typical life expectancies are on the order of 10^5 to 10^6 h. From Fig. 256-1, light emissions corresponding to wavelengths of 1.30 μm and 1.55 μm suffer appreciably less attenuation in optical fibers, and these frequencies can be generated by heterojunction InGaAsP lasers.

Monochromatic laser sources

A very close approximation to an ideal monochromatic (single-frequency) source is provided by modifying a gallium-arsenide-indium-phosphide (GaAsInP) injection laser (Fig. 258-6). Part of the manufacturing process involves slicing the laser into two parts with unequal lengths of 125 μm and 150 μm. Each part has the same substrate, but the slicing action provides a gap of about 5 μm between the two active sections, which are perfectly aligned. The result of this process is called a cleaved coupled cavity or C^3 laser whose output light has a wavelength of 1.3 μm with a bandwidth of less than 5 GHz.

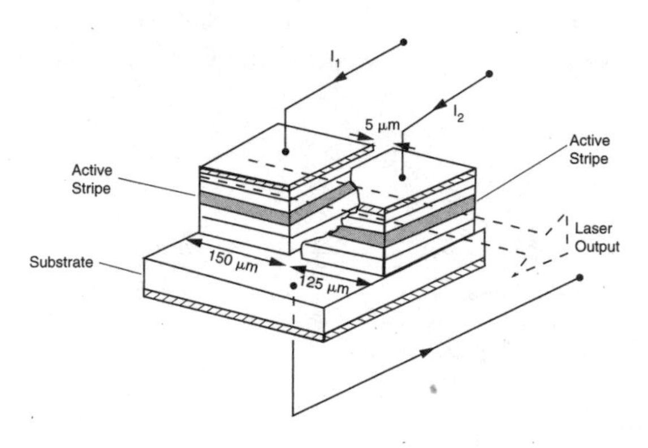

Figure 258-6

As shown in Fig. 258-6, the two active sections are separately driven by unequal currents I_1 and I_2. Consequently although each part behaves as an injection laser, the operating modes of one active section do not entirely match the modes of the other section. Owing to the coupling between the two sections, only those modes which are common to both sections will appear in the laser's output. All other modes will be suppressed. The two currents may then be externally adjusted to change the wavelengths of the various modes so that the two sections only possess one common mode, which can be efficiently coupled to a fiber with a diameter of about 5 μm.

Another monochromatic source for a single-mode fiber is the N^{3+}:YAG laser in which trivalent neodymium is introduced as an impurity into a crystal of yttrium-aluminum-garnet. Interaction between local electric fields within the crystal and the energy levels of the neodymium ions causes some of the energy levels to split. Consequently the number of possible transitions is increased. The result is a four-level laser, which is operated at a wavelength of 1.064 μm (282 THz) and has a very high power output. For example, an N^{3+}:YAG rod with a diameter of only 0.5 cm can emit an output of more than 100 W.

Similar results are obtainable if high-quality glass is used as the host for the neodymium impurity. Compared with the N^{3+}:YAG laser, the N^{3+}:glass laser is less expensive and has a greater power output at a wavelength of 1.06 μm. However the N^{3+}:glass laser has a wider spectrum and is therefore unsuitable as the source for single-mode fibers.

MATHEMATICAL DERIVATIONS

For an LED the principal wavelength of the emitted light is given by

$$\lambda = \frac{hc}{W} \tag{258-1}$$

where: λ = wavelength (m)
h = Planck's constant (6.626×10^{-34} J · s)
W = energy gap (J)
c = velocity of light (m/s).

Since $c = 3 \times 10^8$ m/s and 1 eV $= 1.619 \times 10^{-19}$ J,

$$\lambda = \frac{6.626 \times 10^{-34}\ \text{J} \cdot \text{s} \times 3 \times 10^8\ \text{m/s} \times 10^6\ \mu\text{m/m}}{W \times 1.619 \times 10^{-19}\ \text{J/eV}}$$

$$= \frac{1.242}{W} \tag{258-2}$$

where: λ = wavelength (μm)
W = energy gap (eV).

Example 258-1

A germanium LED has a band gap of 0.80 eV. What is the wavelength of the emitted light? In which band does the wavelength lie?

Solution

Wavelength of the emitted light,

$$\lambda = \frac{1.242}{0.80} \tag{258-2}$$

$$= 1.55\ \mu\text{m}.$$

This wavelength lies in the infrared region and provides a low attenuation in optical fibers (see Fig. 256-1).

Example 258-2

The emitted light of a gallium-indium-phosphide junction laser has a wavelength of 0.78 μm in the infrared region. What is the band gap of the material from which the laser is manufactured?

Solution

$$\text{Band gap, } W = \frac{1.242}{0.78} \qquad (258\text{-}2)$$

$$= 1.59 \text{ eV.}$$

PRACTICE PROBLEMS

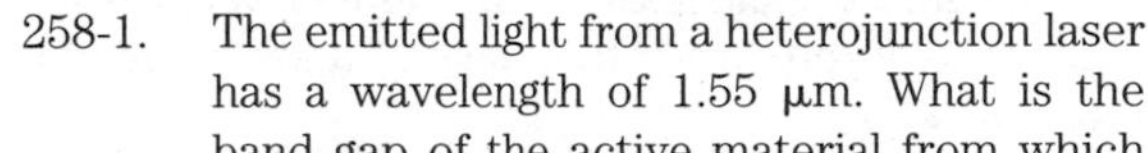

258-1. The emitted light from a heterojunction laser has a wavelength of 1.55 μm. What is the band gap of the active material from which the laser is manufactured?

258-2. A GaP light source has a band gap of 2.25 eV. Calculate the frequency of the emitted light.

259 Light detectors

For comparison purposes the principal parameters of photodetectors are:

1. *Dark current* This is a small reverse current which flows in the absence of any incident light. The presence of the dark current is primarily due to the existence of charge carriers created by thermal energy.
2. *Responsivity (radiation sensitivity)* Because the level of the output current from the photodetector is directly proportional to the power of the incident light, the detector's sensitivity can be expressed as the ratio of the output current to the light power. This is basically measured in amperes per watt, but the responsivity is normally on the order of microamperes per microwatt.
3. *Transit time* This is the time taken for the light-induced carriers to traverse the photodetector. Because this parameter will determine the rise and fall times, it will also control the value of the maximum information rate.
4. *Spectral response* This is the graph of the responsivity versus the wavelength of the input light.

The most commonly used photodetectors are discussed next.

The PIN diode

A PIN diode is basically a microwave device with three regions (Fig. 259-1). The center layer consists of n-type silicon, which is so lightly doped that it is regarded as intrinsic (pure) material. On one side of the intrinsic region is a narrow (heavily doped) p^+ region, while on the other side there is an equally narrow n^+ region. The combination of p^+-i-n^+ regions then gives the diode its name. Although gallium arsenide is used in the manufacture of PIN diodes, it presents more fabrication problems than silicon and has a lower power capability. In the construction process the intrinsic region is grown epitaxially on an n-type substrate and the p region is obtained by diffusion.

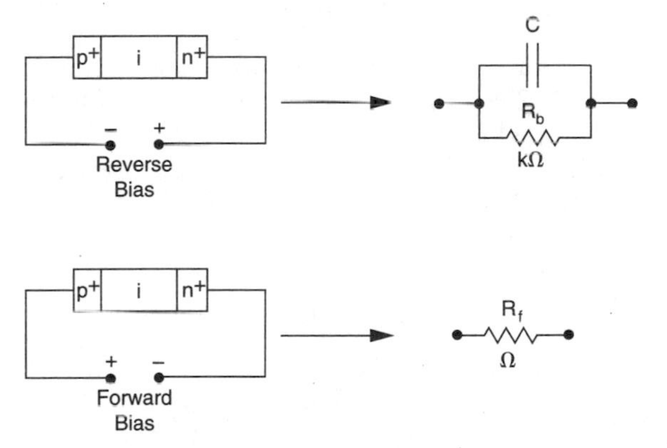

Figure 259-1

When the PIN diode is reverse biased, its equivalent circuit (Fig. 259-1) at microwave frequencies consists primarily of a small capacitance C, in parallel with a high reverse resistance R_b whose value is typically a few kilohms. However, under forward bias conditions, avalanche effects associated with the p^+ and

n^+ regions cause large numbers of holes and electrons to move into the intrinsic layer. The equivalent resistance, R_f, of the device then falls to a low value of a few ohms.

Because of the short transit time associated with the intrinsic region, the PIN diode's switching time is on the order of a few nanoseconds, with operating frequencies extending up to several terahertz.

When used as a photodetector the PIN diode is operated with a reverse bias level which is halfway between zero volts and the point of zener breakdown. It follows that a small "dark" current flows in the absence of any incident light. However, when photons are allowed to enter through a small window and fall onto the intrinsic region, the light energy is mainly absorbed and causes valence electrons to move into the conduction band. The result is to create a large number of hole-electron pairs. The existence of these additional charge carriers then allows a relatively large current to flow through the PIN diode and the external circuit, as shown in Fig. 259-2.

The band gap for silicon is 1.12 eV, so that from equation 258-2, the longest wavelength for the received light is

$$\lambda_{max} = 1.242/1.12 \qquad (258\text{-}2)$$
$$= 1.109\ \mu\text{m}.$$

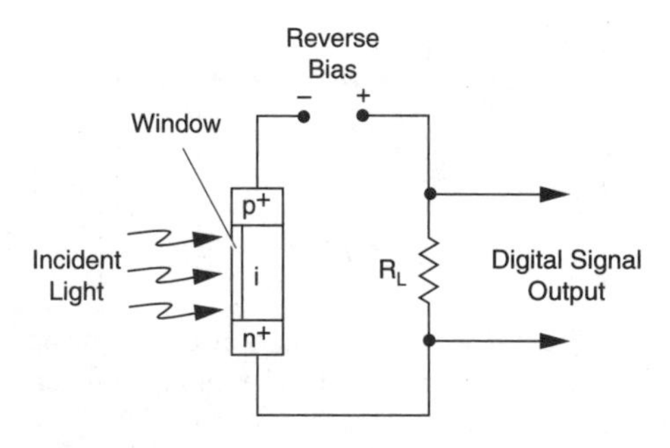

Figure 259-2

It follows that light with wavelengths of less than 1.109 μm (or frequencies greater than 271 THz) will cause valence electrons to move into the conduction band. In practice, the width of the intrinsic region controls the transit time so that the maximum response from the PIN diode occurs at a wavelength of 0.82 μm, which corresponds to a frequency of 366 THz. Figure 259-3 shows the spectral response of a PIN diode with a maximum responsivity of 0.85 μA/μW at the wavelength of 0.82 μm. In the 1.3-μm to 1.55-μm range it is

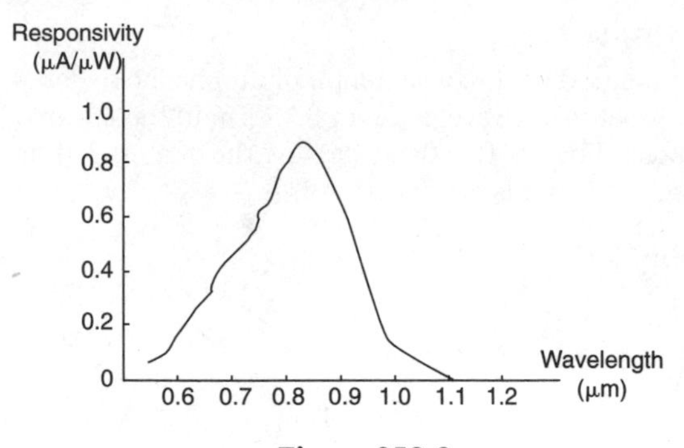

Figure 259-3

common practice to use either an InGaAs or an InGaAsP PIN diode.

Because the power of the incident light is normally a few microwatts, the output current is only a few microamperes or less. An adequate signal-to-noise ratio can be maintained by feeding the output of the PIN diode directly to a GaAs MESFET amplifier. Alternatively, the PIN diode can be incorporated into the circuitry of an IC comparator (Fig. 259-4).

With a reverse dc bias of 12 V, a silicon PIN diode has a rise and fall time on the order of 1 ns, which corresponds to a maximum information rate of 5×10^8 bps. Shorter rise and fall times can be obtained by higher levels of reverse bias.

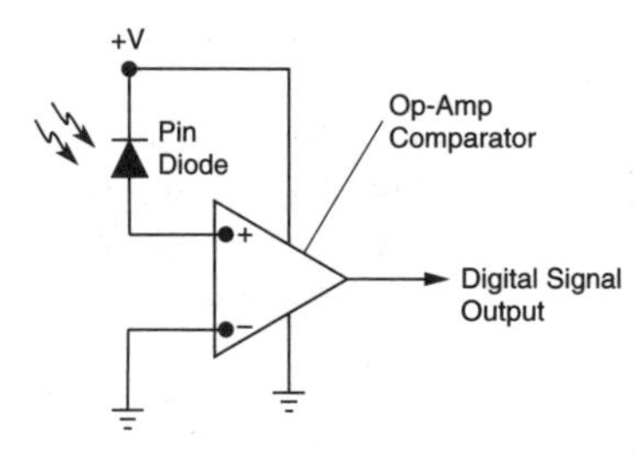

Figure 259-4

The Phototransistor

For lower information rates on the order of 3×10^4 bps it is possible to use a phototransistor, which is operated in the cut-off mode by applying zero volts between the emitter-base junction (Fig. 259-5). Under these conditions the dark current is only a few nanoamperes. Through a window the photons of light are allowed to strike the transistor's base region. Valence electrons then enter the conduction band, and there is an external current flow to create the output

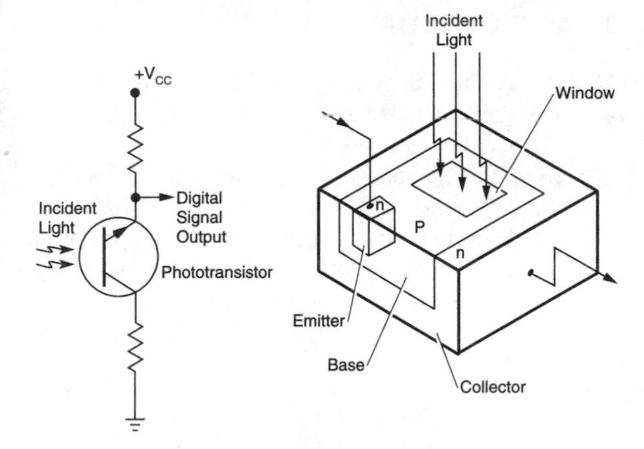

Figure 259-5

digital voltage from the collector. The responsivity of the circuit is on the order of 0.1 μA/μW, while the rise and fall times are about 2 μs.

The Avalanche Diode

This diode (Fig. 259-6) is manufactured from silicon and has an n^+-p-i-p^+ structure. The required reverse bias is normally more than 100 V. The incident photons pass through a window to the n^+ region and then penetrate the p layer to release a number of hole-electron pairs; ultimately the photons are absorbed in the intrinsic region. The hole-electron pairs then initiate an avalanche multiplication effect so that a large external current flows. In other words, the avalanche diode provides a current gain which is not available from the PIN diode.

Typical parameter values for an avalanche diode are as follows:

1. Spectral peak occurs at the commonly used wavelength of 1.3 μm.
2. Responsivity is on the order of 15 μA/μW, which is many times greater than that of the PIN diode.

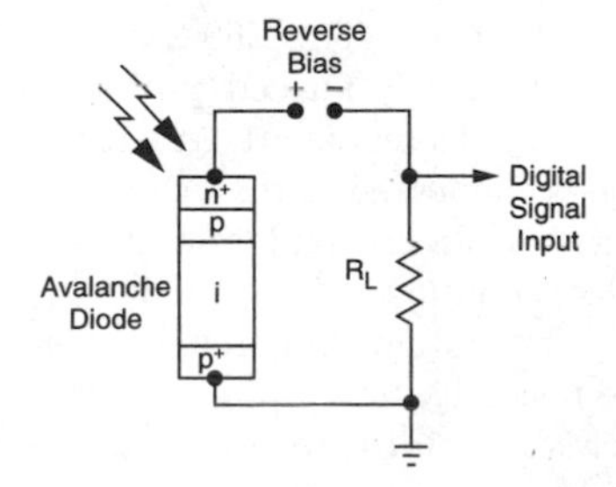

Figure 259-6

3. Dark current is 5 nA with a reverse bias of 100 V dc.
4. Life expectancy is about 10^8 h.
5. Rise and fall times are 0.25 ns, which corresponds to a maximum information rate of 2×10^9 bps. However, for higher information rates exceeding 3×10^9 bps, it is preferable to use a PIN diode with a high value of reverse bias.

To summarize, all photodiodes are operated with reverse bias and the output current for each photodiode is directly proportional to the power of the incident light. Photodiodes are compared in terms of their parameters such as spectral response, transit time, dark current, maximum bit rate, and responsivity.

MATHEMATICAL DERIVATIONS

The longest wavelength of the light received is

$$\lambda_{max} = \frac{1.242}{W} \quad (259\text{-}1)$$

where: λ_{max} = longest wavelength (μm)
W = band gap (eV).

Corresponding lowest frequency,

$$f_{min} = \frac{300}{\lambda_{max}} \quad (259\text{-}2)$$

where: f_{min} = lowest frequency (THz).

Example 259-1

A GaAs photodiode has an energy gap of 1.43 eV. Calculate the lowest frequency of the light that can be detected by the diode.

Solution

The longest wavelength that can be detected by the photodiode is given by

$$\lambda_{max} = \frac{1.242}{1.43} \quad (259\text{-}1)$$

$$= 0.87 \text{ μm}$$

$$\text{Lowest frequency, } f_{min} = \frac{300}{0.87} \quad (259\text{-}2)$$

$$= 345 \text{ THz.}$$

Example 259-2

A PIN photodiode has the spectral response of Fig. 259-3. The power of the incident light is 4 μW and its frequency is 400 THz. What is the value of the diode current?

Solution

The light's wavelength is

$$\lambda = \frac{300}{400} \qquad (259\text{-}2)$$

$$= 0.75\ \mu\text{m}.$$

From Fig. 259-3,

$$\text{Responsivity} = 0.57\ \mu\text{A}/\mu\text{W}$$

$$\text{Diode current} = 4\ \mu\text{W} \times 0.57\ \mu\text{A}/\mu\text{W}$$

$$= 2.3\ \mu\text{A, rounded off.}$$

PRACTICE PROBLEMS

259-1. A photodiode has a band gap of 0.68 eV. Calculate the lowest frequency of the light that can be detected by the diode.

259-2. The responsivity of a photodiode is specified as 1.5 μA/μW at a wavelength of 1.30 μm. Determine the output current if the power of the incident light is 75 μW at the frequency of 231 THz.

260 Analysis of fiber-optic networks and systems

In the 1960s the most common form of computer system was a large number of video terminals which were accessed to one mainframe computer. Access was normally achieved through modems connected into the telephone system. However, the advent of the microcomputer has led to dramatic changes. The cost of computer systems has fallen to less than $2000. Their use has therefore proliferated, with the result that a single company may now own a considerable number of microcomputers. Each of these computers was initially acquired for a single user, but there are obvious advantages to creating a local area network (LAN). A particular member of such a network is then able to communicate with all other members and may also have access to a mainframe computer. Moreover all expensive equipment, such as high-quality laser printers, can be shared.

In a typical LAN there may be over 100 members with a maximum separation of 3 km between individual members. It would be possible to tie the system together through modems and the telephone lines, but this solution would have two major disadvantages:

1. The data rate is limited to less than 3×10^3 bps.
2. One member cannot communicate simultaneously with several other members.

To overcome these disadvantages it is feasible to connect each network member to all other members by means of coaxial cables. This would increase the maximum data rate to more than 1×10^6 bps, but such a network is only suitable for a limited number of users. As shown in Fig. 260-1, the amount of connections required rises rapidly with the number of users in an arithmetic series. The equation is

$$C = \frac{N^2 - N}{2}$$

where: C = number of connections
N = number of users.

Five users only require $(5^2 - 5)/2 = 10$ ties, but 20 users need $(20^2 - 20)/2 = 190$ ties. Consequently for this type of network the number of users is normally less than 10.

The formation of a network is called its topology or architecture. If a central mainframe computer is not part of the network, the two most common topologies are the bus and the ring (Fig. 260-2). In each case only one more cable is required for each additional user. The cables themselves are normally 50-Ω coaxial lines, although the use of fiber-optic guides is becoming more common.

Referring to the bus network (Fig. 260-2A), the coaxial line is marked at equally spaced positions, which are normally less than 10 feet apart. At each of these positions a probe is connected to the inner conductor of the coaxial cable while a second probe makes contact with the outer conductor. This method is equivalent to operating the stations in parallel across the bus and avoids the necessity for cutting the coaxial line when a new user is added. To prevent reflection effects on the bus, it is normal practice to connect each end of the cable to a 50-Ω resistive termination.

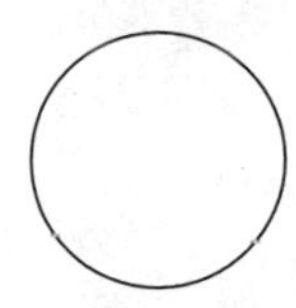

N = 1 Station
Number of Connections
$= \frac{1^2-1}{2} = 0$

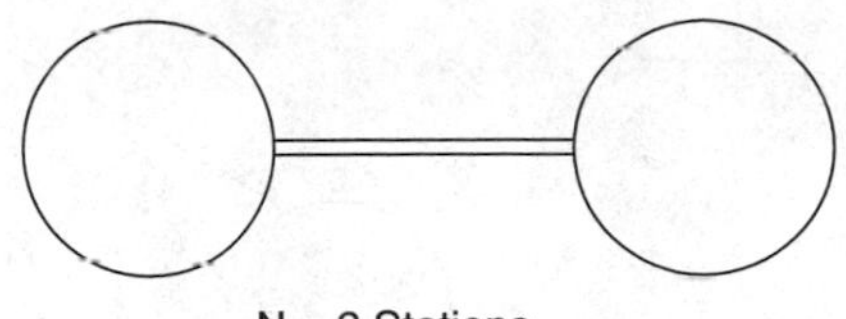

N = 2 Stations
Number of Connections
$= \frac{2^2-2}{2} = 1$

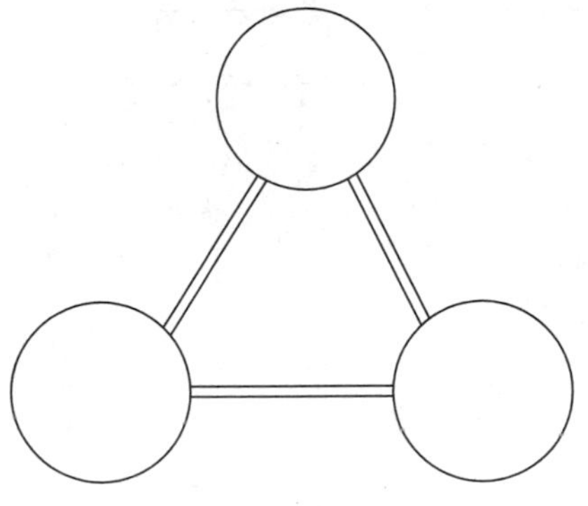

N = 3 Stations
Number of Connections
$= \frac{3^2-3}{2} = 3$

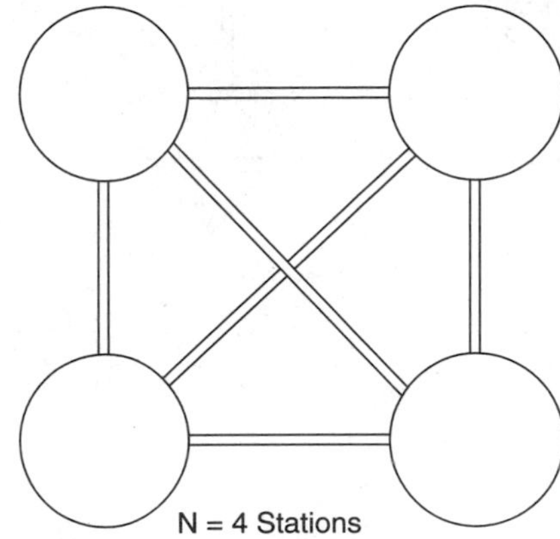

N = 4 Stations
Number of Connections
$= \frac{4^2-4}{2} = 6$

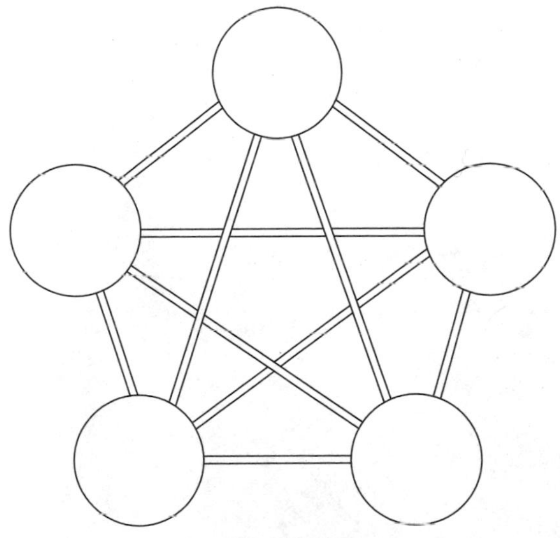

N = 5 Stations
Number of Connections
$= \frac{5^2-5}{2} = 10$

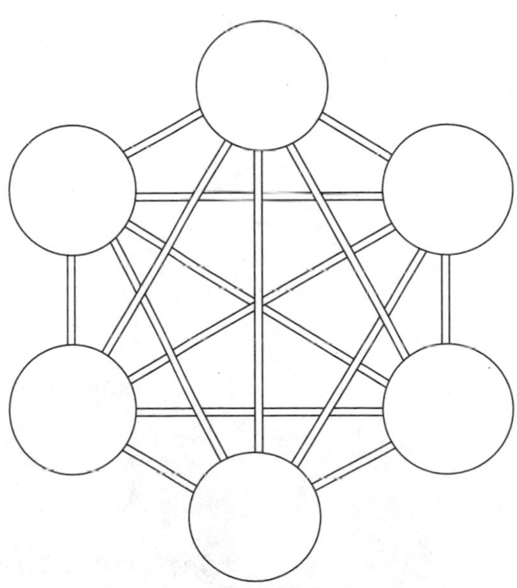

N = 6 Stations
Number of Connections
$= \frac{6^2-6}{2} = 15$

A

Figure 260-1

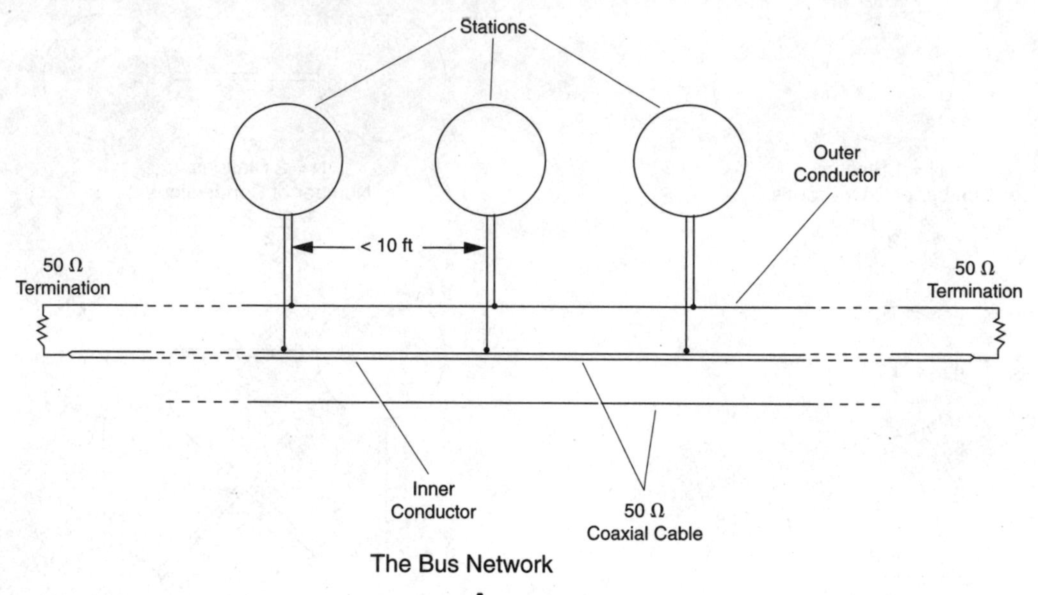

The Bus Network

A

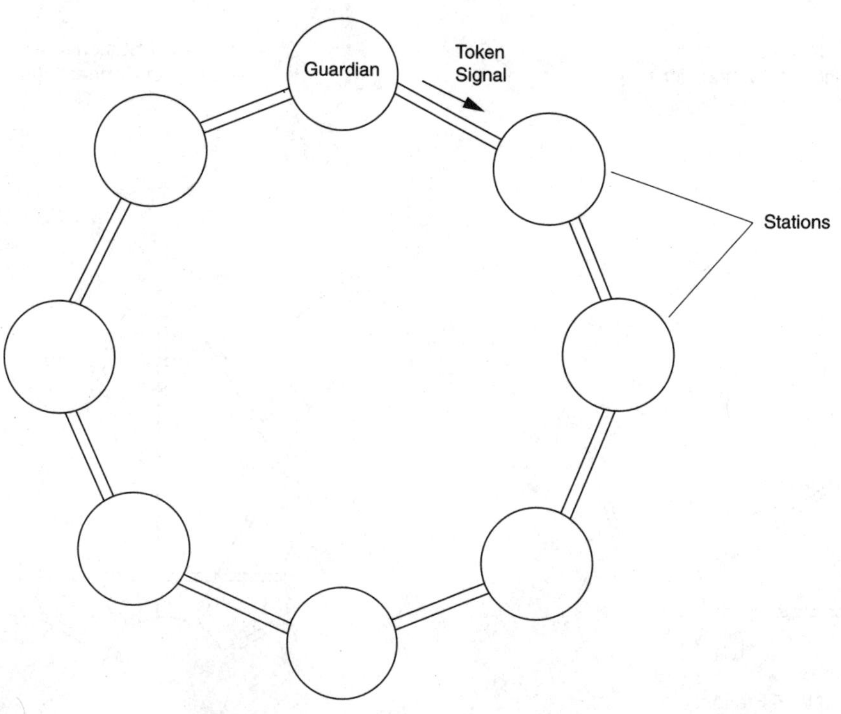

The Ring (Loop) Network

B

Figure 260-2

The ring or loop system (Fig. 260-2B) uses a circulating line so that all of the individual units are connected in series to form a continuous path. This method is inferior to the bus system because the processing time increases as more users are added to the system. Moreover if one of the stations fails, the whole network may cease to operate. Consequently the number of users is normally limited to 30 or less.

If a mainframe computer is part of the LAN, the topology of the network may be a star arrangement (Fig. 260-3), which uses modems and the telephone PBX system to provide the necessary connections. When compared with the bus and loop systems, the star method has serious disadvantages. These include:

1. Very slow processing time
2. Requirement for a large number of long interconnecting cables
3. Only a single member may transmit at any one time
4. The network ceases to function if the mainframe computer is "down."

It follows that the star topology is limited to less than 10 users.

To summarize, the bus topology possesses the highest data rate with a minimum number of cables. Moreover the data rate does not decrease appreciably if more users are added. It is therefore possible to create a bus LAN with more than 1000 users.

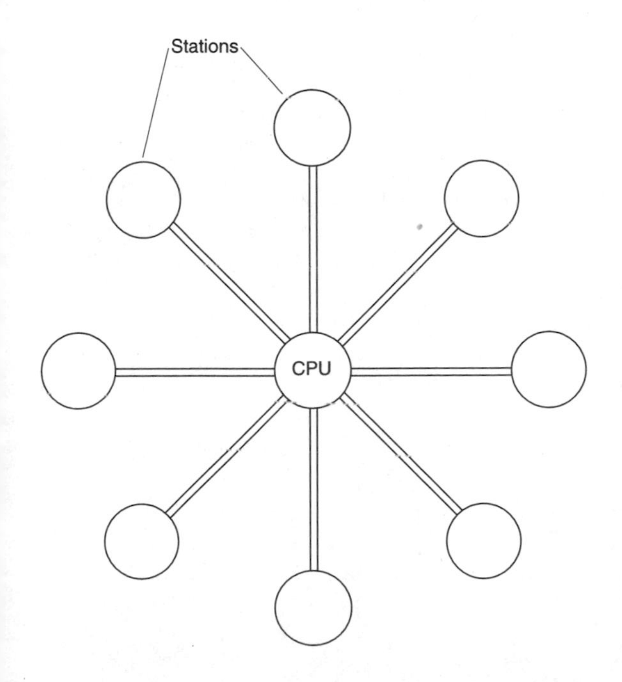

Figure 260-3

The Network Signal

LANs employ two alternative methods of transmitting information:

1. *Baseband* This is a system whereby the digital signal is fed directly to the coaxial cable bus. An example of a baseband LAN is the ETHERNET system, which first appeared in 1980 and was developed jointly by Xerox, Intel, and Digital Equipment Corporation. This is a high-data-rate network in which each bit is allocated a time interval of 100 ns. The corresponding data rate is therefore $1/(100 \text{ ns}) = 1 \times 10^7$ bps. However, to avoid interference from many units operating on the same baseband, the line can only carry one signal at a time. The cost of the connections can be reduced by using single-strand wire in ribbon form or in bundles.
2. *Broadband* In this system a number of information channels modulate a carrier by using frequency-division multiplexing (FDM) or time-division multiplexing (TDM). This allows a number of digital, video, and voice signals to be transmitted simultaneously. The broadband system is applicable to both bus and loop topologies and normally uses coaxial line connections to limit the levels of attenuation, interference, and crosstalk. These coaxial lines will be more expensive when compared with single-strand wire.

Methods of Access

In loop and star systems a commonly used access method is to designate one of the stations as the controller or central processing unit (CPU). All stations, including the controller, may be senders (talkers) which are involved with the transmission of data storage files, digital readings, and so on. Alternatively the stations may be receivers (listeners) which activate display units and printers. A station who wishes to transmit information identifies itself as a "talker" and requests permission from the controller. The system may be either in the "ready" or in the "busy" state. When it has been determined that the system is in the "ready" condition, the controller specifies the particular sender and the various receivers. As mentioned previously, this type of system has one major disadvantage. If the CPU fails, the entire network becomes inoperative although each station may continue to operate on an individual basis.

Clearly it is desirable to operate the ring system without a controller. In such an arrangement any sender would be able to transmit when necessary. However, there would be total confusion unless there is some way of indicating when the system is in the "ready" state. One of the stations is therefore designated as the guardian of the network. The guardian has special circuitry which allows it to generate a token signal. The token is a particular series of bits, which are fed into the ring. If a station wishes to transmit, it takes possession of the token. In this way each station is guaranteed access to the network. If the token circulates around the entire ring without any station taking possession, it is automatically removed by the guardian, which then inserts a new token into the ring. However, a failure in one of the stations may cause a ring network to become inoperative. As a result most networks use the bus topology.

With the bus system, token passing is sometimes used, but the most common method of access is called "carrier sense, multiple access with collision detection" (CSMA/CD). A station that desires to send a message initially inspects the data bus and determines whether any other station is transmitting. If the data bus is already being used, the potential sender senses the presence of the carrier and refrains from transmitting. When the carrier is no longer present, the station commences its transmission, and all other stations then sense the carrier and refrain from sending. This system appears to be foolproof, but unfortunately the bus topology is normally a large network so that two stations may be separated by several kilometers. Consequently a carrier may take an appreciable time to travel from one station to another. Two stations may therefore make a simultaneous determination that no carrier is present and start transmitting at the same time. As a result the two signals will collide, and this collision will be sensed by the stations who will immediately cease their transmissions. Subsequently the stations randomly inspect the data bus and make a second attempt to transmit when no carrier is sensed.

The following is a list of some of the more common networks:

- LOCALNET. A bus network which uses a broadband signal and the CSMA/CD method of access
- DOMAIN. A ring network which uses a broadband signal and the token method of access
- ETHERNET. A bus network which uses a baseband signal and the CSMA/CD method of access. As an example this type of network will be examined in greater detail.

The ETHERNET Network

As stated previously, the ETHERNET LAN system has a high data rate of 1×10^7 bps. Consequently the period for each bit is 10^{-7} s = 100 ns. In the middle of this time interval a positive transition is used to indicate the 1 state, while the 0 state is shown by a negative transition (Fig. 260-4). This has the advantage that every bit corresponds to a transition. At the end of each bit the voltage changes (if necessary) to the level required for the next transition. For example, if the next bit corresponds to a 1 state, the voltage must start at a low level to allow for the positive transition. However for a 0 state the voltage must start at a high level. This method of displaying bits is called Manchester encoding, which has two advantages when compared with the normal series of digital pulses:

1. Any 100-ns interval which contains no transition must be either an error or an indication that the signal has ended.
2. Each bit is contained within a precise period which can then be used for timing purposes.

The basic ETHERNET system is shown in Fig. 260-5. There are no more than three segments, each having a 50-Ω coaxial line with terminating resistors to avoid reflection effects. A normal line is about 1600 feet long, and if we assume that the tapping points are 8 feet

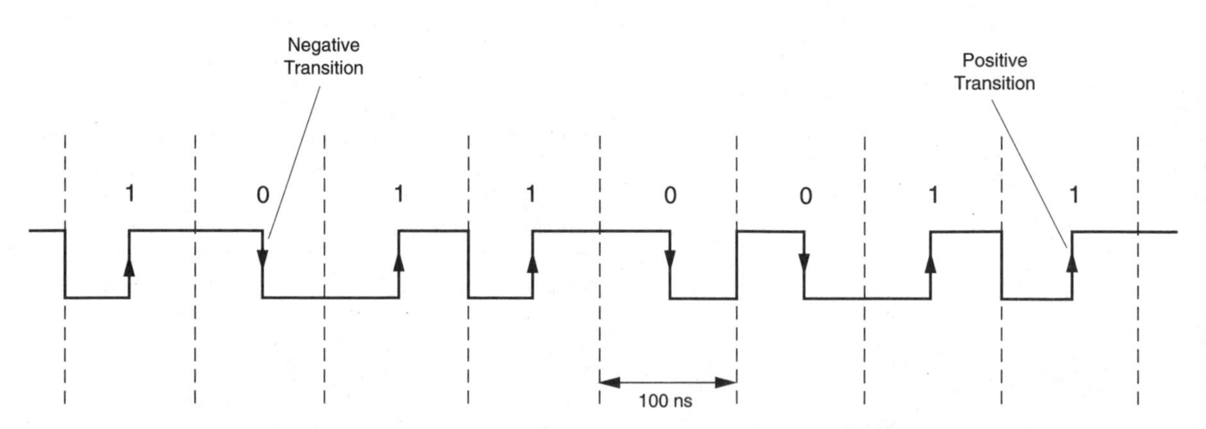

Figure 260-4

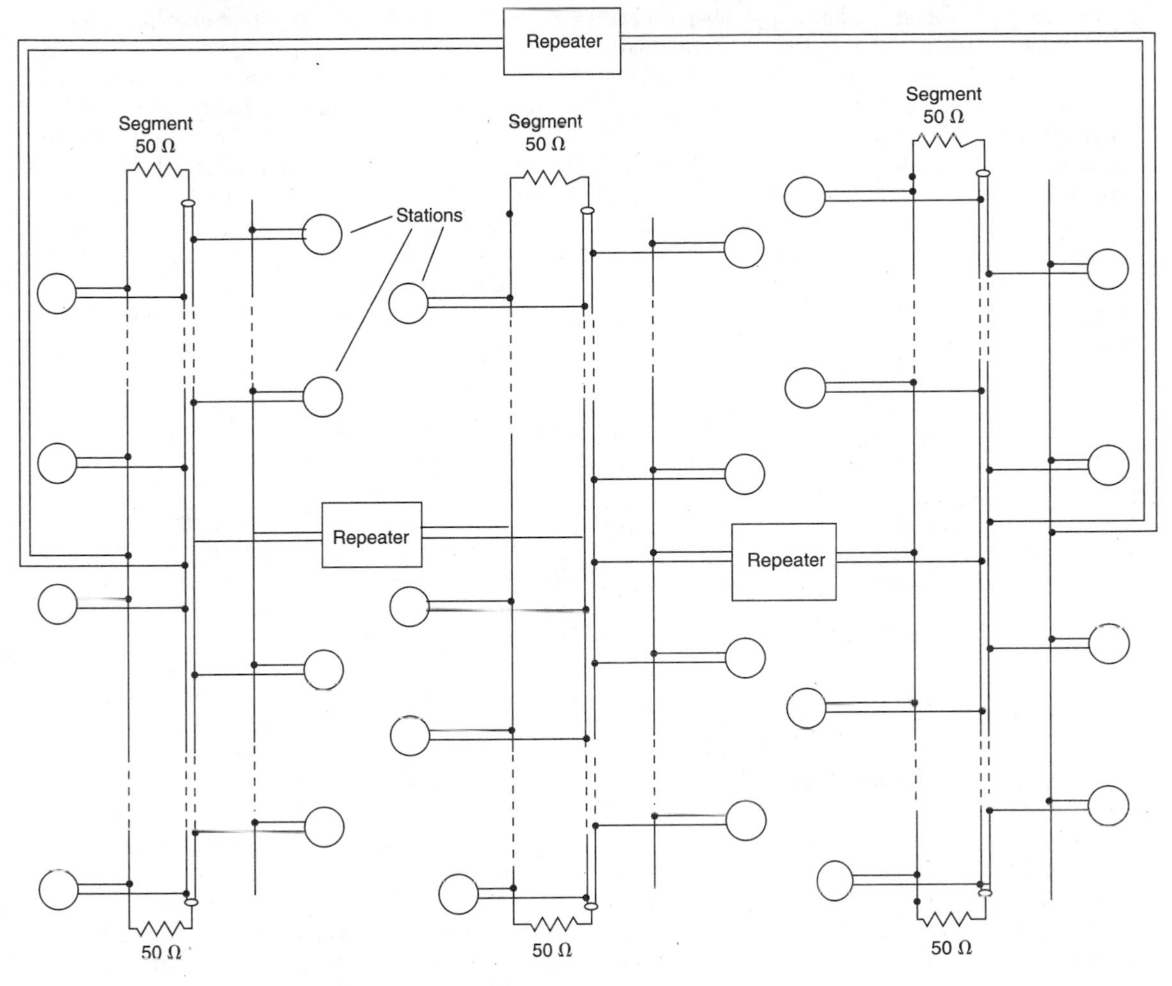

Figure 260-5

apart, the line could be used for a theoretical maximum of 1600/8 = 200 stations, or 3 × 200 = 600 stations for the entire system. In practice the total number would probably be less than 300 stations, with the limit being primarily determined by the line's attenuation factor. As a result the system is divided into three segments, which are joined together by amplifying repeaters. Access to the system is provided by the CSMA/CD method.

Fiber-optic LANS

The LANs described so far have used copper lines such as coaxial cables. These lines have two major disadvantages:

1. The data rate cannot normally exceed 2×10^7 bps.
2. The copper lines have an appreciable attenuation on the order of 20 dB/km, and consequently the network's span is generally less than 1 mi.

In the latest communications networks the data rate requirement has been increased to 1×10^9 bps with an associated span of more than 5 mi. These demands can only reasonably be met by a fiber-optic network with its wide bandwidth (GHz) and low attenuation (0.1 dB/km). Moreover optical-fiber systems can be easily expanded to meet future needs and have all of the advantages outlined in chapter 251.

Fiber-optic LANs use the same method of access and the same three topologies that we have already discussed.

1. *Star topology* The controlling station is located in the center and may be connected to 100 peripheral stations by means of a star coupler. For this type of coupler a signal to an input port causes equal signals from all output ports, but nothing appears at the input ports.

2. *Ring or loop topology* Since the fiber cables have so little attenuation, the distance between adjacent stations may be a few kilometers. With the stations effectively in series, a failure of one station may cause the whole system to be inoperative.
3. *Bus topology* As described previously, the stations are joined in parallel to the bus. With coaxial copper lines it was simple to tap the inner and outer conductors. No such method exists for fiber guides, which will require a four-port coupler for each station (Fig. 260-6). The losses caused by these couplers limit the number of possible stations.

These fiber-optic couplers divert some of the signal on the main fiber cable to each of the stations. This is commonly achieved by splicing two fibers to the main fiber. The division ratio is then determined by the degree of overlap between the cores of the output fibers and that of the input fiber. Alternatively the three fibers may be fused so that the modes that are propagating in the main fiber can penetrate into the other two fibers.

Metropolitan Area Networks (MANs)

As we have already discussed, LANs are used in the private domain to provide fast, cost-effective communications on a shared transmission medium. This is well suited to intercomputer traffic since, for example, ETHERNET has a data rate of 1×10^7 bps and the fiber distributed data interface (FDDI) can operate at more than 1×10^8 bps. The scale of some networks is already staggering. As an example, Citibank has a network spread over 94 countries. This system links over 800 mainframe and minicomputers with about 36,000 PCs.

When attempting to apply the same operating principles to both private and public domains, it is necessary to consider the fundamental differences in their communication environments. The LANs are primarily concerned with computer data buses and their need for wide bandwidths and universal access. By contrast public networks are mainly associated with telephony, which needs a much lower bandwidth as well as continuous use of the communication channels. Moreover, unlike the LAN, the public domain is only interested in networks that can operate over large modern cities. Consequently the distances involved may easily exceed 10 miles, whereas a single ETHERNET segment cannot cover even 1 mi. For long-distance systems the name is therefore changed from local area network to metropolitan area network (MAN).

A MAN operates on the same basic principles as the LAN. The shared transmission medium is composed of optical fibers, which allow a particular station in the network to communicate directly with any other station. The following are the main distinguishing features of these networks:

1. MANs can be used by many different groups while maintaining data security between the groups.
2. MANs are primarily designed for operation in the public domain.
3. MANs cover long distances and may be interconnected to provide high-speed communication services, both nationally and internationally.

With a shared transmission medium it is only possible to avoid chaos between competing users by establishing a method of access. In LANs the medium access control (MAC) was either by contention or by token. In the contention method, a station was free to use the medium at any time provided no other station was

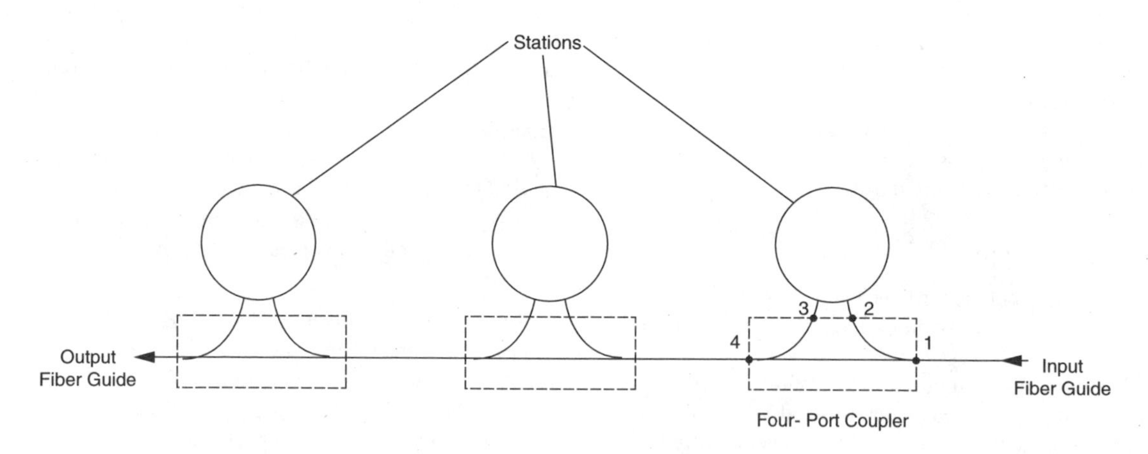

Figure 260-6

transmitting; the effects of collisions between simultaneous transmissions were avoided by means of CSMA/CD. The token method commonly occurs in ring networks. A token is circulated around the ring until it is held by a station during the transmission of its signal. No other station can transmit at the same time since no token is available.

Unfortunately neither method is suitable over distances of several kilometers. For example, a token will take an appreciable time to circulate around a long loop. It follows that there will be a "wasted" interval before a station can gain access. The larger the ring, the greater is the amount of wasted time. For a MAN the access method is based on a pair of buses. These run in opposite directions and allow for a distributed queue (line) of signal segments. This form of MAC is called "distributed-queue dual bus" (DQDB).

A number of MAN pilot projects are presently being studied in the United States, Europe, and Australia. The success of these projects may ultimately lead to a "coming together" of telecommunications and computing.

Recent Developments

New breakthroughs in fiber-optics technology are in the research and development stages. These include:

1. Halide glass fibers, which are ultraclear filaments for the transmission of infrared waves. The attenuation for these fibers is extremely low, so that the black light may literally travel hundreds of miles without the need for optical repeater stations.
2. In the future a single fiber may carry enough information to control a number of robots working on an automated assembly line. A fiber-optic cable buried deep in the ground will be used to determine whether the cable is passing through valuable mineral deposits or worthless mud. Finally, a fiber-optic thermometer may be inserted into human tissue as an aid to hyperthermia treatment for the destruction of cancer cells.

To summarize, an optic-fiber communications system has the advantages of large bandwidth, low cost, low attenuation, less weight, and minimum interference effects. It is not surprising that as existing coaxial cable links become overloaded, they are being replaced by fiber-optic tubes. Moreover in cable television systems, coaxial cables are rapidly being replaced by optical fibers in order to ensure better picture quality and greater reliability.

MATHEMATICAL DERIVATIONS

In previous chapters we have examined the individual elements of a fiber-optic system. These elements included the light source, the connector, the optical fiber, the light detector, and the receiver. If the system is considered as a whole and the lowest acceptable power input to the receiver is known, the minimum required power output from the light source can be determined.

The first system to be analyzed uses an LED source with a multimode fiber and has the following features:

1. The coupling loss from the light source to the connector is 18 dB.
2. The connector introduces a loss of 1 dB.
3. The multimode fiber has a diameter of 50 μm and a loss of 5 dB/km.
4. The length of the fiber is 3 km. This length is made up of six 0.5-km sections which are spliced together. The loss for each splice is 0.75 dB.
5. The coupling loss from the fiber to the light detector is 1.25 dB.
6. The lowest acceptable power input to the receiver is 0.1 μW, or −40 dBm.

The fiber loss is 5 dB/km × 3 km = 15 dB and the total splice loss is 5 × 0.75 = 3.75 dB; note that only five splices are required. The entire loss from the light source to the receiver is 18 dB + 1 dB + 15 dB + 3.75 dB + 1.25 dB = 39 dB. The maximum required output from the LED is −40 dBm + 39 dB = −1 dBm, or 0.79 mW. Practical LED light sources have output powers of up to 10 mW, so it is easy to obtain an adequate power margin at the receiver.

In the second system the source is an injection laser which is coupled to a single-mode fiber. The features of this system are:

1. The coupling loss from the source to the connector is 4 dB.
2. The connector introduces a loss of 0.75 dB.
3. The single-mode fiber has a 5-μm diameter and a loss of 0.4 dB/km.
4. The length of the cable is 20 km. This length is made up of twenty 1-km sections which are spliced together. The loss for each splice is 0.8 dB.
5. The coupling loss from the fiber to the light detector is 1 dB.
6. The lowest acceptable power input to the receiver is −40 dBm.

The fiber loss is 0.4 dB/km × 20 km = 8 dB and the total splice loss is 19 × 0.75 = 14.25 dB. The total loss from the light source to the receiver is 4.0 + 0.75 + 8.0 + 14.25 + 1.0 = 28 dB. The minimum required power output from the semiconductor laser is only −40 dBm + 28 = −12 dBm or $10^{-1.2}$ mW = 0.063 mW = 63 μW.

Bit error rate

One measure of a system's performance is the product of the bit rate and the length of the fiber between the light source and the photodetector. This parameter indicates the highest bps rate that can be sent over a 1-km length of optical fiber. For example, if the value of this parameter for a particular system is 8×10^7 bps-km, the same system can send 4×10^7 bps over a 2-km length of fiber or 2×10^7 bps over a 4-km length. In another case a bit rate–distance product of 2.5×10^{11} bps-km allows a maximum information rate of 1×10^9 bps over a distance of 250 km. The accompanying bit error rate must not be greater than 1 in 10^9. This means that noise can not cause more than one error in 10^9 bits. Analysis shows that this standard requires a minimum signal-to-noise ratio of 22 dB, or a peak signal voltage:rms noise voltage ratio of approximately 12.5. These ratios are quoted for a particular value of light input power.

Power budget

In designing a fiber-optic system it is necessary to balance such parameters as the power output of the source, the required power input to the photodetector, the losses in the system, the length of the fiber, the total amount of fiber dispersion, and the maximum bit rate. This balancing act is known as a power budget analysis, which is illustrated in Example 260-1.

Example 260-1

A fiber-optic system has the following parameters: LED power output 250 μW, minimum power input to the photodetector circuit 0.2 μW, total system attenuation 23 dB, length of fiber 5 km, fiber dispersion 5 ns/km, and maximum bit rate 6×10^6 bps. After allowing for a 6-dB safety factor, carry out a power budget analysis to determine whether the parameters are compatible.

Solution

To allow for a 6-dB safety factor, it is assumed that the total attenuation is equal to $23 + 6 = 29$ dB.

$$\text{Power input to the photodetector circuit} = 250 \times \text{antilog}(-29/10)$$

$$= 0.3\ \mu\text{W, rounded off.}$$

This result exceeds the minimum power requirement of 0.2 μW.

$$\text{Maximum fiber length} \approx \frac{1}{5 \times 6 \times 10^6 \times 5 \times 10^{-9}} \quad (254\text{-}10)$$

$$= 6.7\ \text{km.}$$

Since the length of the cable is only 5 km, the parameters are compatible.

Example 260-2

The parameters of a photodetector circuit (diode and preamplifier) are:

Wavelength of spectral response = 1.3 μm

Responsivity = 5 mV/μW

Required signal-to-noise ratio = 25 dB for 1.5 μW peak input power.

Calculate the minimum power level of the incident light to establish a bit error rate of 1×10^{-9}.

Solution

$$\text{Peak signal voltage} = 5\ \text{mV}/\mu\text{W} \times 1.5\ \mu\text{W} = 7.5\ \text{mV}$$

$$\text{Rms noise voltage} = 7.5/\text{antilog}(25/20) = 0.4\ \text{mV}$$

Peak signal voltage to establish a bit error rate of $1 \times 10^{-9} = 12.5 \times 0.4 = 5$ mV

Minimum power of the incident light

$$= \frac{5\ \text{mV}}{4.5\ \text{mV}/\mu\text{W}} = 1.1\ \mu\text{W} = -29.5\ \text{dBm.}$$

PRACTICE PROBLEMS

260-1. The parameters of an injection-laser fiber-optic system are: laser power output 1.5 mW, coupling loss from laser to connector 3.5 dB, single-mode fiber length 12 km, diameter 6 μm, loss 0.45 dB/km, number of splices 11 (each with a loss of 0.9 dB), and coupling loss from fiber to photodetector circuit 1.5 dB. Determine the level of the power input to the photodetector circuit.

260-2. The cable of a fiber-optic system has an attenuation constant of 2.8 dB/km and a length of 3.1 km. The loss between the light source and the fiber is 1.8 dB, and the fiber has a single splice with a loss of 1 dB. The loss between the fiber and the photodetector circuit is 1.5 dB. If the photodetector circuit requires a light power output of 7 μW, what is the necessary power input from the light source?

A
Elements of a remote-control AM station

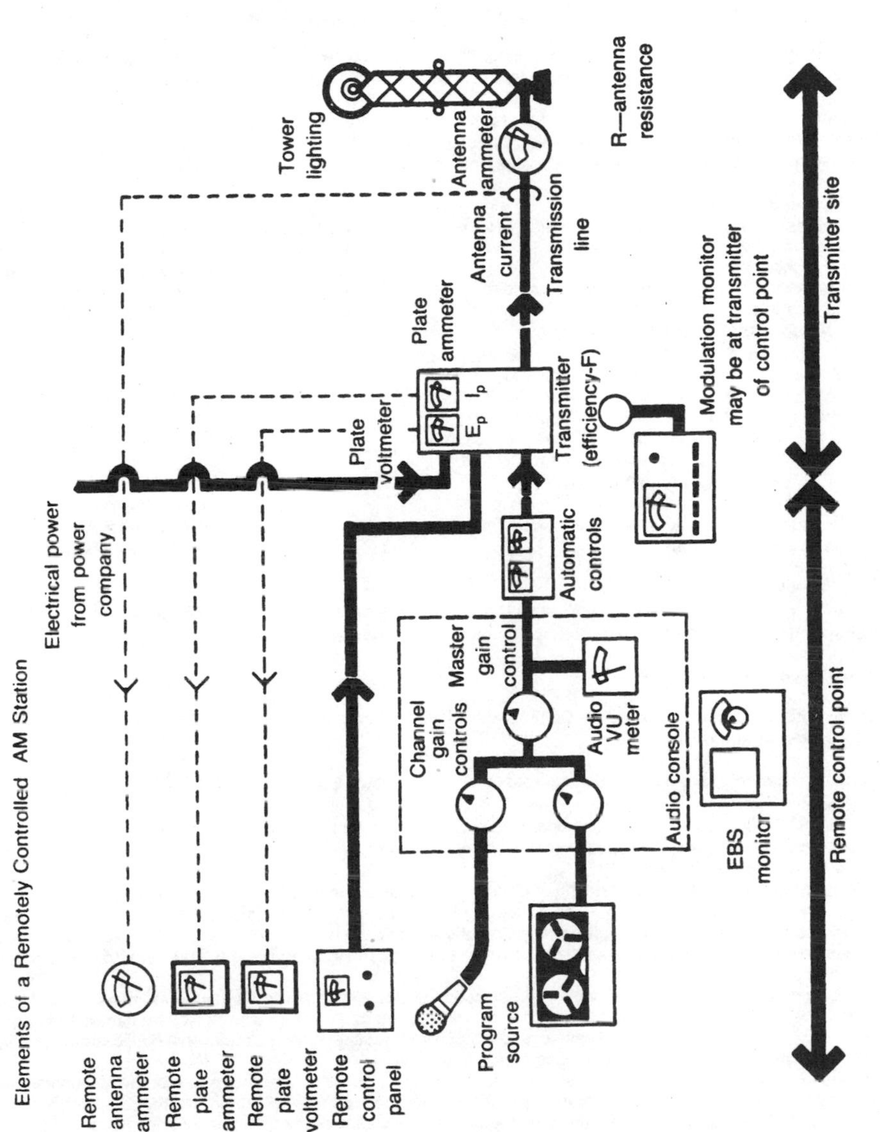

Figure A-1

B
AM station license

FCC Form 352

UNITED STATES OF AMERICA
FEDERAL COMMUNICATIONS COMMISSION

File No.: BL-13,040

Call Sign: W S L W

STANDARD BROADCAST STATION LICENSE

Subject to the provisions of the Communications Act of 1934, subsequent Acts, and Treaties, and Commission Rules made thereunder, and further subject to conditions set forth in this license, [1]the LICENSEE
REGIONAL RADIO, INC.
is hereby authorized to use and operate the radio transmitting apparatus hereinafter described for the purpose of broadcasting for the term ending 3 a.m. Local Time October 1, 1972
The licensee shall use and operate said apparatus only in accordance with the following terms:

1. On a frequency of 1310 kHz.
2. With nominal power of - watts nighttime and 5 kilo watts day time,
   with antenna input power of - watts - directional antenna nighttime .... [- current - amperes; - resistance - ohms]
   and antenna input power of 5 kilo watts non directional antenna daytime .... [antenna current 8.90 amperes; antenna resistance 63.0 ohms]
3. Hours of operation: Daytime as follows:

Jan.	7:30am	to	5:30pm;	Feb.	7.15am	to	6:00pm;
Mar.	6:30am	to	6:30pm;	Apr.	5:45am	to	7:00pm;
May	5:15am	to	7:30pm;	June	5:00am	to	7:45pm;
July	5:15am	to	7:45pm;	Aug.	5:30am	to	7:15pm;
Sep.	6:00am	to	6:30pm;	Oct.	6:30am	to	5:45pm;
Nov.	7:00am	to	5:15pm;	Dec.	7:30am	to	5:00pm;

   Eastern Standard Time (non-advanced)

4. With the station located at: White Sulphur Springs, West Virginia
5. With the main studio located at:
   73 East Main Street
   White Sulphur Springs, West Virginia

   Transmitter may be operated by remote control from 73 East Main Street, White Sulphur Springs, West Virginia

6. The apparatus herein authorized to be used and operated is located at: Rural area 0.75 mi. North of White Sulphur Springs, West Virginia
   North Latitude: 37° 48′ 34.5″
   West Longitude: 80° 17′ 59″

7. Transmitter(s): BAUER, FB-5V

(or other transmitter currently listed in the Commission's "Radio Equipment List, Part B, Aural Broadcast Equipment" for the power herein authorized).**

8. Obstruction marking specifications in accordance with the following paragraphs of FCC Form 715: 1, 3, 11, and 21
9. Conditions:
   **ANTENNA: 190′ (193′ overall height) uniform cross section, guyed, series excited vertical radiator.
   Ground system consists of 120 equally spaced, buried copper radials 106 to 190 feet in length plus 120 interspaced radials 50 to 106 feet in length.

The Commission reserves the right during said license period of terminating this license or making effective any changes or modification of this license which may be necessary to comply with any decision of the Commission rendered as a result of any hearing held under the rules of the Commission prior to the commencement of the license period or any decision rendered as a result of any such hearing which has been designated but not held, prior to the commencement of this license period.

This license is issued on the licensee's representation that the statements contained in licensee's application are true and that the undertakings therein contained so far as they are consistent herewith, will be carried out in good faith. The license shall, during the term of this license, render such broadcasting service as will serve public interest, convenience, or necessity to the full extent of the privileges herein conferred.

This license shall not vest in the licensee any right to operate nor any right in the use of the frequency designated in the license beyond the term hereof, nor in any other manner than authorized herein. Neither the license nor the right granted hereunder shall be assigned or otherwise transferred in violation of the Communications Act of 1934. This license is subject to the right of use or control by the Government of the United States conferred by Section 606 of the Communications Act of 1934.

[1]This license consists of this page and pages —

Dated: NOVEMBER 4, 1971

FEDERAL
COMMUNICATIONS
COMMISSION

Figure B-1

C
Elements of a directional AM station

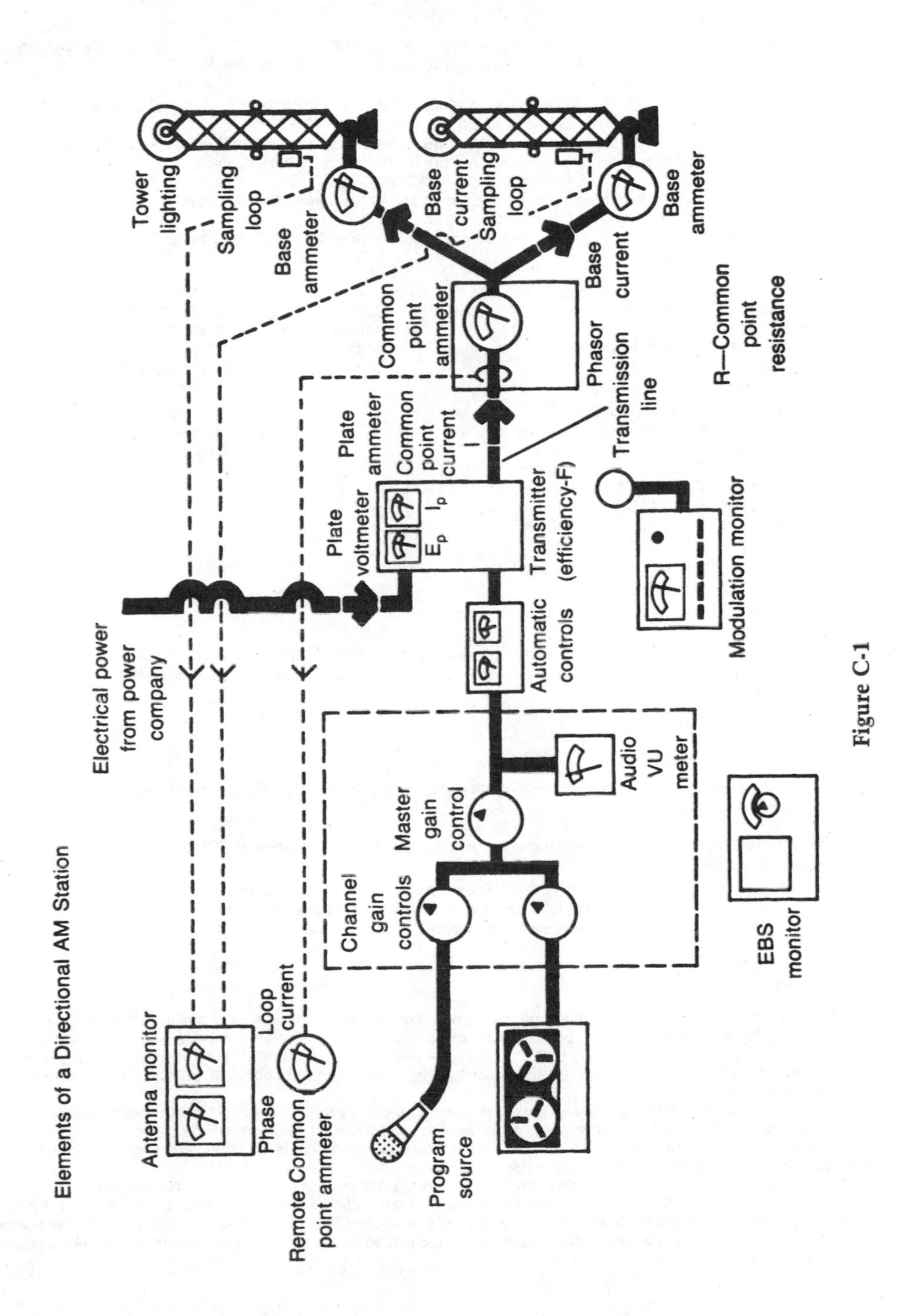

Figure C-1

D
Directional AM station license

FCC Form 352 — UNITED STATES OF AMERICA / FEDERAL COMMUNICATIONS COMMISSION — File No.: BR-989

STANDARD BROADCAST STATION LICENSE — Call Sign: K X X O
MAIN AND AUXILIARY TRANSMITTERS

Subject to the provisions of the Communications Act of 1934, subsequent Acts, and Treaties, and Commission Rules made thereunder, and further subject to conditions set forth in this license, [1]the LICENSEE

SAN ANTONIO BROADCASTING, INC.

is hereby authorized to use and operate the radio transmitting apparatus hereinafter described for the purpose of broadcasting for the term ending 3 a.m. Local Time JUNE 1, 1977

The licensee shall use and operate said apparatus only in accordance with the following terms:

1. On a frequency of 1300 kHz.
2. With nominal power of 1 kilo watts nighttime and 5 kilo watts daytime,

with antenna input power of 1.08 kilowatts - directional antenna nighttime	Common Point current	3.93	amperes
	Common Point resistance	70	ohms
and antenna input power of 5.4 kilowatts directional antenna daytime	Common Point current	8.79	amperes
	Common Point resistance	70	ohms

3. Hours of operation: Unlimited Time.

Average hours of sunrise and sunset:

Jan.	7:30 am	to	5:30 pm;	Feb.	7.15 am	to	6:00 pm;
Mar.	6:30 am	to	6:30 pm;	Apr.	6:00 am	to	7:00 pm;
May	5:15 am	to	7:30 pm;	June	5:00 am	to	7:45 pm;
July	5:15 am	to	7:45 pm;	Aug.	5:45 am	to	7:15 pm;
Sep.	6:00 am	to	6:30 pm;	Oct.	6:30 am	to	5:45 pm;
Nov.	7:00 am	to	5:15 pm;	Dec.	7:30 am	to	5:15 pm;

Central Standard Time (Non-Advanced).

Transmitters may be operated by remote control from 2805 East Skelly Drive, Tulsa, Oklahoma.

4. With the station located at: Tulsa, Oklahoma
5. With the main studio located at:
   2805 East Skelly Drive
   Tulsa, Oklahoma
6. The apparatus herein authorized to be used and operated is located at:
   8601 South Harvard
   Tulsa, Oklahoma

   North Latitude: 36° 02′ 19″
   West Longitude: 95° 56′ 07″

7. Transmitter(s): COLLINS, 820E-1 (Main)

   WESTERN ELECTRIC, 405-B2 (Auxiliary)

(or other transmitter currently listed in the Commission's "Radio Equipment List, Part B, Aural Broadcast Equipment" for the power herein authorized).

8. Obstruction marking specifications in accordance with the following paragraphs of FCC Form 715:**
9. Conditions: (See Page 1A.)

**TOWERS 1, 2, & 4: Paragraphs 1, 3, 12 & 21. Beacons and all obstruction lights shall be flashed, with flashing of towers synchronized so that at any instant two towers are lighted and one tower is not.

Tower 3: Paragraph 1.

The Commission reserves the right during said license period of terminating this license or making effective any changes or modification of this license which may be necessary to comply with any decision of the Commission rendered as a result of any hearing held under the rules of the Commission prior to the commencement of the license period or any decision rendered as a result of any such hearing which has been designated but not held, prior to the commencement of this license period.

This license is issued on the licensee's representation that the statements contained in licensee's application are true and that the undertakings therein contained so far as they are consistent herewith, will be carried out in good faith. The license shall, during the term of this license, render such broadcasting service as will serve public interest, convenience, or necessity to the full extent of the privileges herein conferred.

This license shall not vest in the licensee any right to operate nor any right in the use of the frequency designated in the license beyond the term hereof, nor in any other manner than authorized herein. Neither the license nor the right granted hereunder shall be assigned or otherwise transferred in violation of the Communications Act of 1934. This license is subject to the right of use or control by the Government of the United States conferred by Section 606 of the Communications Act of 1934.

File No. BR-989 | Call Sign: K X X O | Date: 12-17-75 DA- 2

1. DESCRIPTION OF DIRECTIONAL ANTENNA SYSTEM

No. and Type of Elements:	Four, triangular cross-section, guyed, series-excited vertical towers. Two towers used daytime, three used nighttime. A communications type antenna is side-mounted near the top of the W (No. 4) tower.
Height above Insulators:	284′ (135°)
Overall Height:	288′ (Towers 1 and 3); 290′ (Tower 4); 289′ (Tower 2)
Spacing Orientation:	West to West Center Tower, 273.5′ (130°) - Day; West to East Center Tower, 547′ (260°) East Center to East Tower, 547′ (260°) - Night Line of towers bears 72° true.
Non-Directional Antenna:	None used.

Ground System consists of 240 - 43′ buried copper wire radials equally spaced about each tower; 120 - 43′ to 284′ buried copper wire radials alternately spaced. All radials bonded together by a copper wire at a radius of 43′ from each tower. Copper strap between transmitter ground and bond straps at each tower.

2. THEORETICAL SPECIFICATIONS		E(No.1)	EC(No.2)	WC(No.3)	W(No.4)
Phasing:	Night	0°	−9.44°	-	0°
	Day	-	-	0°	−52°
Field Ratio:	Night	0.85	1.36	-	0.65
	Day	-	-	1.0	0.80

3. CREATING SPECIFICATIONS

		E(No.1)	EC(No.2)	WC(No.3)	W(No.4)
Phase Indication*	Night	17°	0°	-	8°
	Day	-	-	0°	57°
Antenna Base Current Ratio:	Night	0.608	1.0	-	0.473
	Day	-	-	1.0	0.818
Antenna Monitor Sample Current Ratio	Night	56	100	-	40
	Day	-	-	100	80

*As indicated by Potomac Instruments Antenna Monitor. PM-112

Figure D-1

E
Elements of an FM station

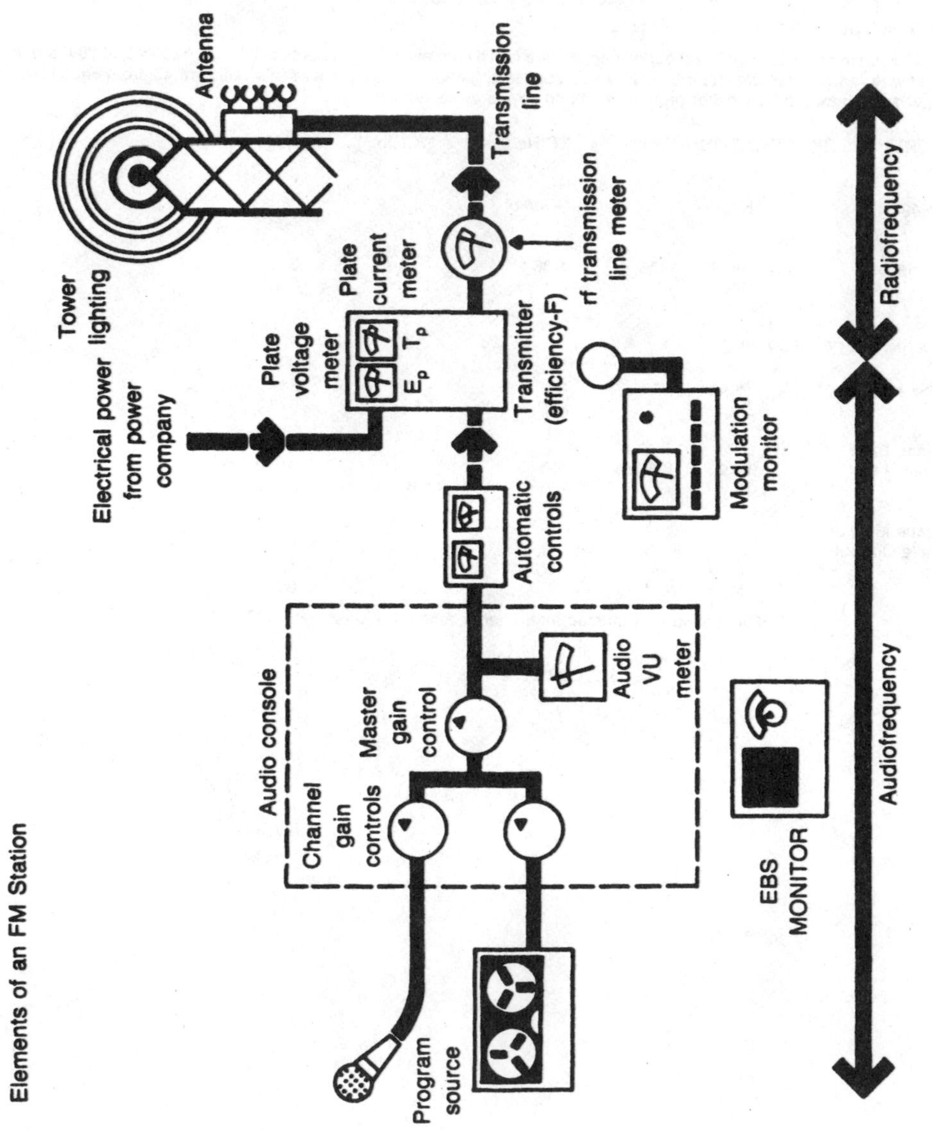

Figure E-1

F
FM station license

FCC Form 352-A

United States of America
FEDERAL COMMUNICATIONS COMMISSION

File No.: BRH-2019

Call Sign: W F Y N-FM

FM BROADCAST STATION LICENSE

Subject to the provisions of the Communications Act of 1934, as amended, treaties, and Commission Rules, and further subject to conditions set forth in this license, [1]the LICENSEE

FLORIDA KEYS BROADCASTING CORPORATION

is hereby authorized to use and operate the radio transmitting apparatus hereinafter described for the purpose of broadcasting for the term ending 3 a.m. Local Time FEBRUARY 1, 1979

The licensee shall use and operate said apparatus only in accordance with the following terms:

1. Frequency (MHz) . . . . : 92.5
2. Transmitter output power . . : 10 kilowatts
3. Effective radiated power . . : 25 kilowatts (Horiz.) & 23.5 kilowatts (Vert.)
4. Antenna height above average terrain (feet) . . . : 135′ (Horiz.) & 130′ (Vert.)
5. Hours of operation . . . . : Unlimited
6. Station location . . . . : Key West, Florida
7. Main studio location . . . . . :
   Fifth Avenue Stock Island
   Key West, Florida
8. Remote Control point . . . . . :

9. Antenna & supporting structure: North Latitude 24° 34′ 01″ West Longitude 81° 44′ 54″

   ANTENNA: COLLINS, 37M-5/300-C-5, Five-sections (Horiz. & Vert.), FM antenna side-mounted near the top of the north tower of WKIZ(AM) directional array. Overall height above ground 155 feet.

10. Transmitter location . . . . . . :

    Fifth Avenue Stock Island
    Key West, Florida

11. Transmitter(s) . . . . : COLLINS, 830-F-1A

12. Obstruction marking specifications in accordance with the following paragraphs of FCC Form 715: 1, 3, 11 & 21.
13. Conditions:

The Commission reserves the right during said license period of terminating this license or making effective any changes or modification of this license which may be necessary to comply with any decision of the Commission rendered as a result of any hearing held under the rules of the Commission prior to the commencement of the license period or any decision rendered as a result of any such hearing which has been designated but not held, prior to the commencement of this license period.

This license is issued on the licensee's representation that the statements contained in licensee's application are true and that the undertakings therein contained so far as they are consistent herewith, will be carried out in good faith. The license shall, during the term of this license, render such broadcasting service as will serve public interest, convenience, or necessity to the full extent of the privileges herein conferred.

This license shall not vest in the licensee any right to operate nor any right in the use of the frequency designated in the license beyond the term hereof, nor in any other manner than authorized herein. Neither the license nor the right granted hereunder shall be assigned or otherwise transferred in violation of the Communications Act of 1934. This license is subject to the right of use or control by the Government of the United States conferred by Section 606 of the Communications Act of 1934.

[1]This license consists of this page and pages —

Dated: January 28, 1976

Figure F-1

G
FCC emission designations

In 1987, the FCC introduced new emission designations into its rules and regulations. Each type of emission is now designated by three symbols of which the first and third symbols are letters, and the second or middle symbol is a number. The meanings of these symbols are:

First Symbol (letter)	Type of Modulation.
Second Symbol (number)	Nature of Signal.
Third Symbol (letter)	Type of Information.

FIRST SYMBOL

A	Amplitude Modulation. Double Sideband. Full Carrier.
C	Vestigial Sideband.
F	Frequency Modulation.
G	Phase Modulation.
H	Single Sideband. Full Carrier.
J	Single Sideband. Suppressed Carrier.
K	Pulse Amplitude Modulation.
L	Pulse Width (Duration) Modulation.
M	Pulse Position Modulation.
N	Unmodulated Carrier.
P	Unmodulated Pulse Sequence.
R	Single Sideband. Reduced Carrier.

SECOND SYMBOL

0	Absence of Any Modulation.
1	Telegraphy On-Off Keying without the use of a Modulating AF Tone.
2	Telegraphy by On-Off Keying of a Modulating AF Tone, or by the On-Off Keying of the Modulated Emission.
3	Analog Voice Communication.

THIRD SYMBOL

A	Telegraphy (aural reception).
B	Telegraphy (reception by automatic machine).
C	Facsimile.
D	Telemetry, Data Transmission.
E	Telephony.
F	Television (video signal).
N	No Information.

EXAMPLES OF EMISSIONS

A1A	Telegraphy by On-Off Keying. Previously Designated as A1.
A3C	AM Facsimile. Previously Designated as A4.
A3E	Amplitude Modulated, Double Sideband, Telephony. Previously Designated as A3.
C3F	Vestigial Sideband Transmission for Television's Video Signal. Previously Designated as A5C.
F1B	Frequency Shift Keying (FSK). Previously Designated as F1.
F3C	FM Facsimile. Previously Designated as F4.
F3E	FM Telephony. Previously Designated as F3.

G3E	Phase Modulated (PM) Telephony. Previously Designated as F3.
H3E	Single Sideband, Full Carrier. Previously Designated as A3H.
J3E	Single Sideband, Suppressed Carrier (SSSC). Previously Designated as A3J.
NON	Unmodulated Carrier. No Information. Previously Designated as A0.
R3E	Single Sideband, Reduced Carrier. Previously Designated as A3A.

Note that sometimes a designation is preceded by a number that represents the allowed bandwidth in kilohertz or hertz.

Examples are:

(1) 2k70A3E. The "*k*" indicates the position of the decimal point, and also that the bandwidth is measured in kilohertz. Therefore, the bandwidth is 2.70 kHz.

(2) 200HA1A The bandwidth is 200 Hz.

(3) 16k0F3E The bandwidth is 16.0 kHz.

H
FCC tolerances and standards

Carrier frequency

Standard AM broadcast stations	±20 Hz
Commercial FM broadcast stations	±2 kHz
Television broadcast stations—aural and visual transmitters	±1 kHz
Non-commercial educational FM broadcast stations:	
1. Licensed for power of more than 10 watts	±2 kHz
2. Licensed for power of 10 watts or less	±3 kHz
Studio transmitter link (STL)	0.005%
International broadcast stations	0.0015%

Public Safety Radio Services Frequency range	All fixed and base stations	All mobile stations	
		Over 3 W	3 W or less
MHz	Percent	Percent	Percent
Below 25	0.01	0.01	0.02
25 to 50	.002	.002	.005
50 to 450[1]	.0005	.0005	.005
450 to 470[2,3]	.00025	.0005	.0005
470 to 512	.00025	.0005	.0005
806 to 820	.00015	.00025	.00025
351 to 888	.00015	.00025	.00025
250 to 1,427[2]			
1,427 to 1,435[4]	.03	.03	.03
Above 1,435[2]			

[1]Stations authorized for operation on or before Dec. 1, 1961, in the frequency band 73.0–74.6 MHz may operate with a frequency tolerance of 0.005 percent.

[2]Radiolocation equipment using pulse modulation shall meet the following frequency tolerance: the frequency at which maximum emission occurs shall be within the authorized frequency band and shall not be closer than $1.5/T$ MHz to the upper and lower limits of the authorized frequency band where T is the pulse duration in microseconds. For other radiolocation equipment, tolerances will be specified in the station authorization.

[3]Operational fixed stations controlling mobile relay stations, through the use of the associated mobile frequency, may operate with a frequency tolerance of 0.0005 percent.

[4]For fixed stations with power above 200 watts, the frequency tolerance is 0.01 percent if the necessary bandwidth of the emission does not exceed 3 kHz. For fixed station transmitters with a power of 200 watts or less and using time division multiplex, the frequency tolerance can be increased to 0.05 percent.

Power

Transmitters of standard AM and FM commercial broadcast stations	10% below and 5% above
Aural and visual transmitters of TV broadcast stations	20% below and 10% above

Current

All currents	5%

Modulation—Standard AM Broadcast

Minimum modulation on *average modulation peaks*	85%
Maximum modulation on *positive modulation peaks*	125%
Maximum modulation on *negative modulation peaks*	100%
Maximum carrier shift allowed	5%

Temperature for Master-Oscillator Crystals

X-cut and Y-cut crystals	±0.1° C
Low temperature-coefficient crystals	±1.0° C

Final RF stage: Plate Voltage and Plate Current Meters

Accuracy at full-scale reading	2%
Maximum permissible full-scale reading: 5 times minimum normal reading	

Meters: Recording Antenna Current

Accuracy at full-scale reading	2%
Maximum permissible full-scale reading for the scale of current-squared meters	3 times minimum normal reading
Portion of scale used for accuracy with current-squared meters	Upper two-thirds

FCC standards

Standard AM Broadcast

Band	535–1605 kHz
Channel width	10 kHz

FM Commercial Broadcast

Band	88–108 MHz
Channel width	200 kHz
Transmitted AF range (main channel)	50 to 15000 Hz
100% modulation (deviation ratio = 5)	±75 kHz swing
Time constant for pre-emphasis and de-emphasis	75 microseconds

Television Broadcast

Bands

Channels 2 through 4	54 to 72 MHz
Channels 5 and 6	76 to 88 MHz
Channels 7 through 13	174 to 216 MHz
Channels 14 through 83	470 to 890 MHz
Channel width	6 MHz
Field frequency	60 Hz
Frame frequency	30 Hz
Lines per frame	525
Horizontal scanning frequency	15750 Hz
Aspect ratio	4 to 3
Visual bandwidth	4.9 MHz
Frequency separation between aural carrier frequency and channel upper limit	0.25 MHz
Frequency separation between visual carrier frequency (below) and aural carrier frequency	4.5 MHz
Frequency separation between visual carrier frequency (below) and chrominance sub-carrier frequency	3.579545 MHz ± 10 Hz

Reference white level	12.5% of peak carrier level (±2.5%)
Reference black level	70% of peak carrier level
Blanking level in a monochrome TV signal	75% of peak carrier level (±2.5%)
100% modulation for the aural FM transmission	±25 kHz swing
Transmitted AF range (main channel)	50 Hz to 15 kHz
Deviation ratio	1.667
Public Safety Radio Services	
Maximum audio frequency	3 kHz
A1A emission—maximum bandwidth	0.25 kHz
A3E emission—maximum bandwidth	8 kHz
Minimum modulation on average modulation peaks	70%
Maximum modulation on negative modulation peaks	100%

F3E emission Frequency band (MHz)	Authorized bandwidth (kHz)	Frequency deviation (kHz)
25 to 50	20	5
50 to 150	*20	*5
150 to 450	20	5
450 to 470	20	5
470 to 512	20	5
806 to 821	20	5
831 to 866	20	5

In each frequency band the deviation ratio is 1.667

*Stations authorized for operation on or before Dec. 1, 1961, in the frequency band 73.0–74.6 MHz may continue to operate with a bandwidth of 40 kHz and a deviation of 15 kHz.

Harmonic Attenuation

The mean power of emissions shall be attenuated below the mean power output of the transmitter in accordance with the following schedule:

1. On any frequency removed from the assigned frequency by more than 50% up to and including 100% of the authorized bandwidth: at least 25 decibels.
2. On any frequency removed from the assigned frequency by more than 250% of the authorized bandwidth: at least 35 decibels.
3. On any frequency removed from the assigned frequency by more than 250 percent of the authorized bandwidth: at least 43 plus 10 log (mean output power in watts) decibels or 80 decibels, whichever is the lesser attenuation.

I
The phonetic alphabet

A phonetic alphabet uses a particular standard word to substitute for each letter of the alphabet. When it is necessary to spell difficult proper names or other words that cannot easily be understood, the phonetic alphabet can be used to reduce the number of errors and, therefore, speed up communications. As an example, the use of the words “Bravo” and “Victor” will enable the receiving station to distinguish clearly between the letters “B” and “V,” which are similar in sound and can be easily confused. The international code of signals uses the following phonetic alphabet:

Letter	Code word	Letter	Code word
A	Alpha	N	November
B	Bravo	O	Oscar
C	Charlie	P	Papa
D	Delta	Q	Quebec
E	Echo	R	Romeo
F	Foxtrot	S	Sierra
G	Golf	T	Tango
H	Hotel	U	Uniform
I	India	V	Victor
J	Juliet	W	Whiskey
K	Kilo	X	X-Ray
L	Lima	Y	Yankee
M	Mike	Z	Zulu

J
Answers to practice problems

chapter 1

1-1. 785 J 1-2. 40 m 1-3. 866 J 1-4. 981 W 1-5. 94.2 m.

chapter 2

2-1. 1920 C 2-2. 0.75 A 2-3. 0.00949 kg 2-4. 15.6 C.

chapter 3

3-1. 48 J 3-2. 187.5 W.

chapter 4

4-1. 90 W 4-2. 270 Wh 4-3. 4.17 A 4-4. 1/4 or 0.25 4-5. 23.5 kWh.

chapter 5

5-1. 4.95 V, 7.5 mW 5-2. 0.68 A, 161 Ω 5-3. 2130 Ω 5-4. 2.7 J 5-5. 1.0 mS 5-6. 4 5-7. 2.82 mA, 31.0 mW 5-8. 9 5-9. 0.707 5-10. 60 V, 540 mW.

chapter 6

6-1. 120 Ω 6-2. 11.7 Ω 6-3. 0.53 Ω 6-4. 1.11×10^{-6} m^2 6-5. 0.0062 Ω 6-6. 142.8 m 6-7. 0.46 mV.

chapter 7

7-1. 3.42 Ω 7-2. 9.26 Ω 7-3. 0.0051 Ω/Ω/°C 7-4. +0.0036 Ω/Ω/°C 7-5. 6.4°C.

chapter 8

8-1. 1320 Ω 8-2. 3333 Ω, 3267 Ω 8-3. 13.6 mA 8-4. 17.3 V 8-5. 0.179 S 8-6. 2000 Ω 8-7. 1000 Ω, 10% 8-8. 10.5 Ω.

chapter 9

9-1. 1738 Ω 9-2. 2.26 mA 9-3. 11.1 V 9-4. 3500 V

chapter 10

10-1. 26.8 V 10-2. −98.2 V 10-3. −9.32 V.

chapter 11

11-1. 164 mW 11-2. +100 V, +100 V 11-3. 768 mW 11-4. +24 V, +18.1 V, +10.8 V 11-5. +24 V, +15.5 V, +15.5 V 11-6. +24 V, +24 V, +24 V.

chapter 12

12-1. 6 V 12-2. W + 5 V, X + 2.58 V, Y − 0.375 V, Z − 4 V 12-3. W + 5 V, X + 4.73 V, Y + 4.40 V, Z + 4 V 12-4. −0.8 V 12-5. +20.8 V.

chapter 13

13-1. 16 V, 57.6 mW 13-2. 33 kΩ 13-3. +66 V, +26.2 V 13-4. +50.8 V, 0 V 13-5. 3.8 W, 0.4 W.

chapter 14

14-1. 0.375 Ω 14-2. 95.6 W 14-3. 36 Ω, 25 W 14-4. 10.6 Ω.

chapter 15

15-1. 284 Ω, 3.5 mS 15-2. 166 Ω, 52.5 mA 15-3. 1 kΩ, 100 μA, 100 mA, 10 W 15-4. 25.0 mA 15-5. 4.7 kΩ 15-6. 56 kΩ 15-7. 96 15-8. 316 V.

chapter 16

16-1. Theoretically infinite 16-2. 128 mA 16-3. 0.24 W 16-4. 1 kΩ, zero 16-5. 773 Ω, 1.16 kΩ.

chapter 17

17-1. 8 V 17-2. 15 A 17-3. 10 A.

chapter 18

18-1. 5.15 mA 18-2. 7.38 mA 18-3. 23.8 mA.

chapter 19

19-1. 20.4 mA 19-2. 2.8 kΩ 19-3. 828 mW 19-4. 0.48 W 19-5. 40.2 mA 19-6.

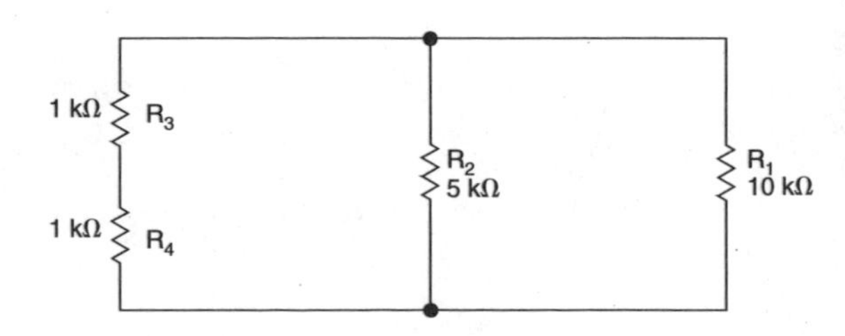

Figure J-19-6

chapter 20

20-1. 746 Ω 20-2. 0.38 mA 20-3. 2.72 mA 20-4. 10.9 mA 20-5. Zero. The bridge is virtually balanced.

chapter 21

21-1. 2.73 kΩ, 8.25 W   21-2. 208.5 V, 53.4 V, 7.5 V.

chapter 22

22-1. 0.5 Ω, 5.5 Ω, 96 A, 0.5 Ω   22-2. 150 A, 0.067 Ω, 33.3%   22-3. 48.6 V   22-4. 1.33 Ω, 101.3 V   22-5. 20%.

chapter 23

23-1. 2.25 Ω   23-2. 31 V, 4 Ω   23-3. 75 Ω   23-4. 625 W   23-5. 29.14 Ω, 85.4%, 0.858 Ω, 14.6%.

chapter 24

24-1. 8 A, 400 W   24-2. 10 A, 2.5 Ω.

chapter 25

25-1. 6.06 A   25-2. 1.6 A   25-3. 288 W   25-4. 4.8 Ω, 7.5 W   25-5. 0.133 Ω, 7.5 W.

chapter 26

26-1. 1.42 mA   26-2. 6 mA   26-3. + 24 V   26-4. I_1 = 11 mA, I_2 = 3 mA.

chapter 27

27-1. 10 V   27-2. 6 mA   27-3. i_1 = −6.78 mA, i_2 = −0.56 mA, i_3 = −3.10 mA   27-4. 14.54 V   27-5. 9.4 mA.

chapter 28

28-1. −5.24 V   28-2. +96 V   28-3. −2.6 V   28-4. +4 V   28-5. +10.16 V.

chapter 29

29-1. 120 mA   29-2. 15.88 μA   29-3. 12.8 mA   29-4. = −18 V   29-5. 3.33 mA   29-6. 0.94 mA.

chapter 30

30-1. Constant current, 346.6 mA; parallel resistance, 120 Ω.
30-2. Constant current, 16 mA; parallel resistance, 5 kΩ.
30-3. Constant current, 10 mA; parallel resistance, 2 kΩ.
30-4. Constant voltage, 8.98 V; series resistance, 1.38 kΩ.

chapter 31

31-1. E_{TH} = 80 V, R_{TH} = 10 kΩ   31-2. E_{TH} = 226 V, R_{TH} = 7.5 kΩ   31-3. E_{TH} = 13.76 V, R_{TH} = 15.5 kΩ   31-4. E_{TH} = 4.5 V, R_{TH} = 4.5 kΩ   31-5. E_{TH} = 14.02 V, R_{TH} = 1.79 kΩ.

chapter 32

32-1. I_N = 0, R_N = 45 kΩ   32-2. I_N = 1.22 mA, R_N = 1.052 kΩ   32-3. I_N = 1.22 mA, R_N = 2.03 kΩ   32-4. I_N = 4 mA, R_N = 176.5 Ω.

chapter 33

33-1. R_X = 4.44 kΩ, R_Y = 3.33 kΩ, R_Z = 6.67 kΩ   33-2. R_X = R_Y = 0.5 kΩ, R_Z = 1.5 kΩ   33-3. R_X = 3.33 kΩ, R_Y = 2.22 kΩ, R_Z = 6.67 kΩ   33-4. R_{XY} = 736 Ω, R_{YZ} = 1284 Ω, R_{ZX} = 2275 Ω.

chapter 34

34-1. 0.8 T   34-2. $7 \times 10^{-2} m^2$ 34-3. 975 μWb   34-4. 12.5 cm.

chapter 35

35-1. 0.42 N.m   35-2. 0.24 N.m   35-3. 8.6 A.

chapter 36

36-1. 255 AT/m, 306 AT   36-2. 3.94×10^5 AT/m, 1.6 mWb   36-3. 736 A/m, 8×10^{-4} T   36-4. 3185 A/m   36-5. 0.00476 T   36-6. 1515 A/m   36-7. 2.5 A   36-8. 83.5 A   36-9. 15.5×10^7 AT/m.

chapter 37

37-1. 2.79×10^5 SI units, 167 AT   37-2. 500 AT/m, 955, 1500 AT/m, 637   37-3. 5.9 A   37-4. 178 μWb   37-5. 1.714×10^6 SI units, 234 μWb.

chapter 38

38-1. 120 V   38-2. 100 m/s   38-3. 800 μWb.

chapter 39

39-1. 50 V   39-2. 7.5 A   39-3. 1 μs.

chapter 40

40-1. 9.0 mH   40-2. 11.25 mWb   40-3. 750 mH   40-4. 4116   40-5. 2408   40-6. 7.5 H.

chapter 41

41-1. 1.5 J   41-2. 30.2 J   41-3. 200 V, 40 J   41-4. 6 J.

chapter 42

42-1. 3 A/s, 15 V.

chapter 43

43-1. 20 μs, 5.0×10^4 A/s, 0.632 A, 0.95 A, 100 μs   43-2. 50 Ω, 16 ms, 0.8 H   43-3. 50 ms, 0.252 A, 0.316 A, 0.364

A, 0.392 A, 0.4 A 43-4. 4040 V, 808 A/s, 0.495 ms 43-5. 0.06 s 43-6. 0.275 s 43-7. 15.04 V, 23.5 V.

chapter 44

44-1. 100 μH 44-2. 6 V 44-3. 15 A/s 44-4. 16 V 44-5. 20 mH in series with a parallel combination of 40 mH and 60 mH 44-6. 40 H and 8 H 44-7. $I_1 = I_2 = I_T = 0$, $V_1 = V_2 = V_3 = 12$ V 44-8. $V_1 = V_2 = V_3 = 0$, $I_1 = 480$ mA, $I_2 = 120$ mA, $I_3 = 600$ mA.

chapter 45

45-1. 202 N, 12 μC.

chapter 46

46-1. 0.19824 μC, 8.0×10^5 V/m, 24.78×10^{-6} C/m^2 46-2. 7.07×10^{-6} C/m^2, 566×10^{-10} C.

chapter 47

47-1. 96 pF, 48 μC/m^2, 10^6V/m 47-2. 0.83 N 47-3. 500 pF 47-4. 0.1 μF 47-5. 0.27 mA, 175 μs.

chapter 48

48-1. 800 V, 16 μJ, 32 μJ. Difference = work done against the force of attraction between the capacitor plates. 48-2. 100 V 48-3. 3.75 μF, 1500 μC.

chapter 49

49-1. 5.76×10^{-2} J 49-2. 118.5 V, 39 V, 23.5 49-3. 180 μC, 2700 μJ, 5400 μJ, 30 V, 60 V 49-4. 10 μF 49-5. 345 V.

chapter 50

50-1. 8 μF, 192 V 50-2. 1.2 μF, 1600 μJ 50-3. 2300 μC 50-4. Zero 50-5. 144000 μJ 50-6. 68.6 V 50-7. 283 pF.

chapter 51

51-1. 207.6 V 51-2. −144 V, −144 V, −144 V, −240 V, −192 V, zero 51-3. +160 V, +240 V, 75 μs 51-4. +200 V, zero, +40 V, +40 V, 0.4 ms 51-5. +80 V, +192 V, 1.2 s 51-6. 30.46 mA, 12.56 mA, 17.9 mA, 17.38 mA, 17.38 mA, zero, 18.6 ms.

chapter 52

52-1. 250 kΩ 52-2. 100 kΩ 52-3. 1000 pF 52-4. 1000 Hz 52-5. 15 V.

chapter 53

53-1. 0.003 μF 53-2. 5×10^3 V/s 53-3. 2500 V/s.

chapter 54

54-1. 100000 Ω/V 54-2. 500 μA 54-3. 5 mA 54-4. 4.

chapter 55

55-1. 1000 Ω/V 55-2. 5.8 mA 55-3. 100 mA 55-4. 8 A 55-5. 533.33 Ω, 122.22 Ω, 11.11 Ω.

chapter 56

56-1. 57 V 56-2. 8.7 kΩ 56-3. 8.9 kΩ 56-4. 2160 Ω 56-5. 20000 Ω/V 56-6. 50 kΩ 56-7. 1.04 V, 4.8 V.

chapter 57

57-1. 6 kΩ 57-2. 70 kΩ 57-3. 60 μA 57-4. 19985 Ω, 40000 Ω, 20000 Ω, 10000 Ω, 60 Ω.

chapter 58

58-1. 15 58-2. 24% 58-3. 16%, 4.47 A.

chapter 59

59-1. 31.25% 59-2. 48%

chapter 60

60-1. 25 V 60-2. 3.6 cm.

chapter 61

61-1. 250 Hz 61-2. 25 μs 61-3. 60 Hz 61-4. 25 kHz, 75 kHz, 125 kHz 61-5. 50 kHz, 100 kHz, 150 kHz, 200 kHz 61-6. 1.1×10^6 rad/s 61-7. 22 V, 477 Hz 61-8. (a) 1.95^c (b) 207.4°.

chapter 62

62-1. 6.01 V 62-2. 52.3 V 62-3. Yes, 7.04 mA, 891 mW 62-4. 8 V, 10.95 V 62-5. 23.6 mA 62-6. 23.6 mA.

chapter 63

63-1. 29.9 V 63-2. 11.6 mA 63-3. 16.65 V 63-4. 7.03 V.

chapter 64

64-1. 597 Hz, 9.19 V 64-2. 2765 rad/s, 6.01 A, $i = 8.5 \sin 2765t$ 64-3. 1.09×10^5 rad/s.

chapter 65

65-1. 17.8 V, zero 65-2. 22.3 mA, 22.3 mA 65-3. 6.15 V, 10.65 V 65-4. zero, 25.2 V.

chapter 66

66-1. i_1 leads i_2 by 30° ($\pi^c/6$), 16.7 mA, +16.6°, 4.77 mA, +79.8° 66-2. v_1 leads v_2 by 90° ($\pi^c/2$), 6.79 V, 6.79 V, v_1 leads $v_1 + v_2$ by 35.7° 66-3. 23.8 V, 41.3 V, v_2 lags $v_1 + v_2$ by 60° ($\pi^c/3$).

chapter 67

67-1. 34.0 V, 8.55 mW, 428 mW, $v = 48.1 \sin(\omega t + \pi/6)$ V 67-2. $p = 4.28 [1 - \cos(2\omega t + \pi/3)]$ mW 67-3. 72 mA, 19.4 V, zero 67-4. 11.8 kΩ.

chapter 68

68-1. 118 mW 68-2. 2.66 V rms, 26.1 mW 68-3. 2.24 W 68-4. 36.4 mA, 436 mW.

chapter 69

69-1. 681 kHz 69-2. 9.31 kHz 69-3. 36.5 mA 69-4. $7.34 \sin(1.2 \times 10^6 t + \pi^c/2)$ V or $7.34 \cos(1.2 \times 10^6 t)$ V 69-5. 187 Ω, 374 Ω, 561 Ω 69-6. 53.5 mA, 26.7 mA, 17.8 mA.

chapter 70

70-1. 1.16 MHz 70-2. 0.0248 μF 70-3. 1.04 A 70-4. $8.44 \sin(7.5 \times 10^6 \times t - \pi^c/2)$ V 70-5. 2336 Ω, 1168 Ω, 779 Ω 70-6. 0.428 mA, 0.856 mA, 1.28 mA.

chapter 71

71-1. 704 VA, 562 W 71-2. 32.3° 71-3. 2.63 kΩ, 45°, 0.707, 71.2 mVA, 50.4 mW 71-4. 161 W.

chapter 72

72-1. 9.22 W 72-2. +45°, 0.707, lagging 72-3. 1927 Ω, 0.057 A, 68.4 V, 85.9 V, 3.90 W, 4.90 VAr, 6.27 VA, 0.62 lagging, +51.5° 72-4. 3.53 kΩ, 0.283 mA, 0.623 V, 0.782 V, 0.176 mW, 0.221 mVAr, 0.283 mVA, 0.62 lagging, +51.5° 72-5. 5.95 kΩ, 0.168 mA, 0.369 V, 0.929 V, 0.062 mW, 0.156 mVAr, 0.168 mVA, 0.369 lagging, +68.3° 72-6. 1100 Ω.

chapter 73

73-1. 0.707 leading, −45°, 0.894 leading, −26.6° 73-2. 1.04 kΩ, 1.92 mA, 1.31 V, 1.51 V, 2.52 mW, 2.90 mVAr, 3.84 mVA, 0.656 leading, −49.0° 73-3. 1.71 kΩ, 1.17 mA, 0.793 V, 1.85 V, 0.93 mW, 2.17 mVAr, 2.34 mVA, 0.40 leading, −66.6° 73-4. 240 Ω, 0.458 A, 91.7 V, 60.9 V, 41.9 W, 27.9 VAr, 50.4 mVA, 0.83 leading, −33.7° 73-5. 4.51 kΩ, 0.222 mA, 0.598 V, 0.802 V, 0.133 mW, 0.178 mVAr, 0.222 mVA, 0.60 leading, −53.2°.

chapter 74

74-1. 90 Ω, zero leading, −90° 74-2. 345 Ω, zero lagging, +90° 74-3. 220 Ω, 4.55 mA, 8.55 V, 7.55 V, 0 W, 4.55 mVAr, 4.55 mVA, zero lagging, +90° 74-4. 1.44 kΩ, 0.694 mA, 1.30 V, 2.30 V, 0 W, 0.694 mVAr, 0.694 mVA, zero leading, −90° 74-5. 11.25 kHz 74-6. 468 Ω, 0.235 A, 266 V, 156 V, 0 W, 63.0 VAr, 37.0 VAr, 25.85 VA, zero lagging, +90°.

chapter 75

75-1. 962 Ω, 0.707 leading, −45° 75-2. 732 Ω, 0.929 lagging, +21.7° 75-3. 404 Ω, 2.48 mA, 0.668 V, 7.48 V, 8.22 V, 1.65 mW, 1.84 mVAr, 2.48 mVA, 0.67 leading, −48.3° 75-4. 450 Ω, 4.44 mA, 1.20 V, 2.18 V, 3.78 V, 5.33 mW, 7.10 mVAr, 8.88 mVA, 0.60 leading, −53.1° 75-5. 618 Ω, 3.24 mA, 0.874 V, 3.18 V, 1.38 V, 2.83 mW, 5.83 mVAr, 6.28 mVA, 0.45 lagging, +63.2° 75-6. 895 kHz, 270 Ω.

chapter 76

76-1. 2.50 mA, 3.71 mA, 4.47 mA, 671 Ω, 7.50 mW, 11.13 mVAr, 13.41 mVA, 0.56 lagging, +56.0° 76-2. 2.50 mA, 1.86 mA, 3.12 mA, 962 Ω, 7.50 mW, 5.58 mVAr, 9.36 mVA, 0.80 lagging, +36.7° 76-3. 1.09 MHz 76-4. 91.7 mA, 72.9 mA, 117.1 mA, 939 Ω, 10.1 W, 8.02 VAr, 12.9 VA, 0.78 lagging, +38.5° 76-5. 0.833 mS, 0.663 mS, 1.06 mS 76-6. 48 Hz.

chapter 77

77-1. 3.70 mA, 2.76 mA, 4.62 mA, 2.16 kΩ, 37 mW, 27.6 mVAr, 46.2 mVA, 0.80 leading, −36.8° 77-2. 3.70 mA, 1.38 mA, 3.95 mA, 2.53 kΩ, 0.37 mS, 0.138 mS, 0.394 mS, 37 mW, 13.8 mVAr, 39.5 mVA, 0.94 leading, −20.5° 77-3. 14.7 kHz 77-4. 234 mA, 166 mA, 289 mA, 381 Ω, 25.7 W, 18.3 VAr, 31.8 VA, 0.81 leading, −36.1°.

chapter 78

78-1. 3.62 mA, 2.76 mA, 0.855 mA, 11.7 kΩ, 0 W, 8.55 mVAr, 8.55 mVA, zero lagging, +90° 78-2. 2.34 mA, 4.27 mA, 1.93 mA, 5.19 kΩ, 0 W, 19.3 mVAr, 19.3 mVA, zero leading, −90° 78-3. 12.6 kHz, theoretically infinite 78-4. 72.9 mA, 83.0 mA, 10.1 mA, 10.9 kΩ, 0 W, 1.11 VAr, 1.11 VA, zero leading, −90°.

chapter 79

79-1. 1.22 mA, 3.62 mA, 2.76 mA, 0.855 mA, 1.49 mA, 6.71 kΩ, 12.2 mW, 8.55 mVAr, 14.9 mVA, 0.82 lagging, +35.0° 79-2. 1.22 mA, 2.34 mA, 4.27 mA, 1.93 mA, 2.28 mA, 4.38 kΩ, 12.2 mW, 19.3 mVAr, 22.8 mVA, 0.555 leading, −57.6° 79-3. 12.6 kΩ, 8.2 kΩ 79-4. 13.4 mA, 72.9 mA, 83.0 mA, 16.8 mA, 6.55 kΩ, 1.47 W, 1.11 VAr, 1.84 VA, 0.80 leading, −37.0°.

chapter 80

80-1. 2.24 kΩ, 22.2 W 80-2. 1.414 kΩ, 29.8 W 80-3. 3 kΩ, 2 kΩ (capacitive reactance), 18.75 W 80-4. 3 kΩ, 4 kΩ (capacitive reactance), 18.75 W.

chapter 81

81-1. 412 pF 81-2. 539 Ω, 539 Ω, 15 Ω, 200 mA, 3 V, 108 V, 108 V, 600 mW, zero, 600 mVA, 1, zero 81-3. 561 μH 81-4. 364 pF.

chapter 82

82-1. 940 kHz, 65.5, 14.4 kHz 82-2. 193 mA, 947 kHz, 933 kHz 82-3. 27.7, 33.9 kHz, 81.6 mA 82-4. 1880 kHz, 131, 14.4 kHz.

chapter 83

83-1. 2.55 MHz, 6.07 μA, 45.2 μA, 45.2 μA, 6.07 μA, 0.206 μW 83-2. 7.44, 343 kHz, 7.92 kΩ 83-3. 2.55 MHz, 5.08, 502 kHz 83-4. 1.80 MHz, 5.26, 343 kHz.

chapter 84

84-1. 2.32 MHz, 0.97 μA, 80.7 μA, 80.7 μA 84-2. 103 kΩ, 82.6, 28.1 kHz 84-3. 51.5 kΩ, 41.3, 56.2 kHz 84-4. 1.64 MHz, 51.5 kΩ, 58.4, 28.1 kHz.

chapter 85

85-1. 950 kHz, 6.25×10^4 s^{-1} 85-2. 105 mA, $i = 105e^{-6.25 \times 10^4 t} \times \sin(2 \times \pi \times 950 \times 10^3 t)$ mA 85-3. 9.38×10^{-5} s.

chapter 86

86-1. 1.16 H 86-2. 12.5 A/s 86-3. 122 μH 86-4. 1.53 86-5. 0.58 H 86-6. 37.5 V 86-7. Zero.

chapter 87

87-1. 0.28 H 87-2. 0.5 87-3. 0.29 87-4. 0.75 87-5. 0.3 H 87-6. 1.0 H 87-7. 1.35 H.

chapter 88

88-1. 0.833 A, 144 Ω 88-2. 144 Ω 88-3. 30:1 step down 88-4. 92% 88-5. 55 W.

chapter 89

89-1. $11.4\angle 74.7°$, $12.04\angle -4.76°$, $7.28\angle 15.9°$, $8.06\angle -82.9°$, $5\angle 0°$, $6\angle +90°$, $5\angle -90°$, $7.28\angle 106°$, $9.84\angle -114°$ 89-2. $4.31 + j3.25$, $5.27 - j4.91$, 4, $+j8$, $-j11$, $-1.54 - j3.03$, $-3.94 + j2.56$ 89-3. (a) $5 + j6$, $7.81\angle 50.2°$ (b) $7 - j8 = 10.6\angle -48.8°$ (c) $+j11 - j3 = j8$, $8\angle 90°$.

chapter 90

90-1. $x = 1.02$, $y = 3.56$ 90-2. $16 + j1$, $16.03\angle 3.58°$, $-2 - j7$, $7.28\angle -106°$ 90-3. $-5.65 - j3.61$, $6.70\angle -147°$, $0.57 - j6.37$, $6.40\angle -84.9°$.

chapter 91

91-1. $4 - j1$, $4.12\angle -14.01°$ 91-2. $20\angle -40°$, $0.8\angle 100°$, $25\angle -140°$, $2\angle 15°$, $0.2\angle 70°$ 91-3. $-3 - j2$, $7\angle 130°$ 91-4. $15 - j2$, $15.1\angle -7.6°$, $-7 - j12$, $13.9\angle -120.3°$, $97.5\angle -35.9°$, $79.0 - j57.2$, $0.67\angle -84.7°$, $0.062 - j0.67$, $65.0\angle -120.6°$, $-33.1 - j55.9$, $3.48\angle 12.2°$, $3.40 + j0.74$, $0.124\angle 60.3°$, $0.0614 + j0.108$.

chapter 92

92-1. $5.84 - j2.88$ or $6.51\angle 26.3°$ 92-2. $1.54\angle -26.3°$ A 92-3. $6.16\angle -116.3°$ V, $9.13\angle 37.2°$ V 92-4. $1.57\angle -21.8°$, $1.27\angle 93.5°$ 92-5. 13.8 W. This may either be calculated from $10\text{ V} \times 1.54\text{ A} \times \cos 26.3°$ or $(1.57\text{ A})^2 \times 3\ \Omega + (1.27\text{ A})^2 \times 4\ \Omega$.

chapter 93

93-1. $2.99\angle 27.8°$ Ω 93-2. $0.435\angle -27.8°$ 93-3. $0.185\angle 21.9°$ A, $0.177\angle -45.0°$ A, $0.171\angle -59.1°$ A 93-4. 0.384 W.

chapter 94

94-1. $2.50\angle 86.7°$ mA, $5.60\angle -38.8°$ mA, $4.62\angle -12.7°$ mA 94-2. $9.025\angle -53°$ V, $11.2\angle -38.8°$ V, $19.0\angle -63.3°$ V 94-3. 102.8 mW 94-4. 62.5 mW, 40.5 mW.

chapter 95

95-1. $7.2\angle 146.3°$ mA, $10.0\angle 143.1°$ 95-2. $2.82\angle -45°$ mA 95-3. 112 mW 95-4. 99.7 mW, 12.4 mW.

chapter 96

96-1. $4.48\angle -26.6°$ mA 96-2. $6.32\angle 18.4°$ mA, $4.47\angle 63.4°$ mA 96-3. 140 mW 96-4. 80 mW, 60 mW.

chapter 97

97-1. $11.66\angle -120.95°$ V 97-2. $2.828\angle -45°$ mA 97-3. 62 mW.

chapter 98

98-1. $4.48\angle 26.6°$ mA 98-2. 368 mW 98-3. $13.44\angle 26.6°$ V.

chapter 99

99-1. $41.4 + j8.8$ V, $1.2 + j0.9$ kΩ 99-2. $11.3\angle 45°$ mA 99-3. 261 mW 99-4. $338\angle -87.7°$ V, $14 - j16.8$ kΩ.

chapter 100

100-1. $160\angle 110°$ mA, $-j2$ kΩ 100-2. $80\angle -70°$ mA 100-3. $5.16\angle 140.2°$ mA, $1.2 + j2.4$ kΩ 100-4. $0.344\angle 143.1°$ mA, $6.0 - j2.86$ kΩ.

chapter 101

101-1. 375 mW 101-2. $1.06\angle 4.5°$ kΩ.

chapter 102

102-1. $42.8\angle -130.1°$ V.

chapter 103

103-1. 0.064 103-2. 0.022 103-3. 9.9 Ω 103-4. 8 μH.

chapter 104

104-1. 165 V, 90 Hz 104-2. 6 104-3. 3300 rpm 104-4. 75 Hz.

chapter 105

105-1. 50 Hz, 64.5 Ω 105-2. 4.82 A 105-3. 3.41 A.

chapter 106

106-1. 60 Hz, 20.2 Ω 106-2. Zero 106-3. 5.45 A.

chapter 107

107-1. 220 V 107-2. 9.5 A 107-3. $109.5\angle 0°$ V, $109.5\angle 120°$ V, $109.5\angle -120°$ V 107-4. $10.95\angle 20°$ A, $10.95\angle 140°$ A, $10.95\angle -100°$ A 107-5. 3.38 kW.

chapter 108

108-1. 220 V 108-2. 19.1 A 108-3. $38.1\angle 25°$ A, $38.1\angle 145°$ A, $38.1\angle -95°$ A 108-4. 11.9 kW.

chapter 109

109-1. 26.8 dB, −30.8 dB 109-2. 602.6, 2.14×10^{-2} 109-3. +11.1 dB, 12.9 109-4. 501 mW, 17.4 μW 109-5. 65.8 dBm, −18.3 dBm 109-6. 21.6, 1.15×10^{-1} 109-7. 22.1 dB, −49.4 dB 109-8. 6.03 mW 109-9. 1.90 V 109-10. 34.6 dB.

chapter 110

110-1. −3.29, +2.00 110-2. +30.3, −49.8 110-3. 4.29 nepers, 37.3 dB.

chapter 111

111-1. 4 V, 5.1 V, 1.7 V, 1.02 V, 0.73 V 111-2. 18.8 V, 6 V, 4 V 111-3. 1000 Hz, 9.55 V, 2000 Hz, (−) 4.77 V, 3000 Hz, 3.18 V, 4000 Hz, (−) 2.39 V 111-4. 6.48 V, 0.72 V, 0.26 V 111-5. 31.8 V, 50 V, (−) 21.2 V, (−) 4.24 V, (−) 1.83 V 111-6. 63-7 V, 42.5 V, (−) 8.49 V, (−) 3.64 V.

chapter 112

112-1. 1.1025 A, 36.5 W 112-2. $1.56 \sin(\omega t + 14.04°) + 0.431 \sin(3\omega t + 41.05°) + 0.175 \sin(5\omega t + 120.1°)$ mA 112-3. 1.15 mA 112-4. 11.9 mW.

chapter 113

113-1. (a) 9.12 mA (b) 9.12 V 113-2. 8 mW 113-3. 5 MΩ 113-4. 4 V 113-5. 0.1 mA 113-6. 3.525 V.

chapter 114

114-1. +469 V, 937 V 114-2. 3100 V, 6200 V, 150 Hz 114-3. 80 mH.

chapter 115

115-1. 4:1 115-2. 8:1 115-3. 75 V 115-4. 1.74 H.

chapter 116

116-1. 110 V 116-2. 50 V.

chapter 117

117-1. 9.133 V 177-2. 9.183 V 117-3. 9.169 V.

chapter 118

118-1. 1.30 mA, 1.26 mA 118-2. 4.83 V, Yes 118-3. Saturation.

chapter 119

119-1. 8.7 V 119-2. 8.27 V, 6.35 V, 2.62 V 119-3. 9.4 V.

chapter 120

120-1. 5.1 V 120-2. 1.8 kΩ 120-3. 3.3 kΩ.

chapter 121

121-1. +7.3 V, −0.14 V   121-2. 3.3 kΩ   121-3. 3.3 kΩ   121-4. −12 V.

chapter 122

122-1. +10.5 V   122-2. 10.5 V, 9.8 V, 0.7 V   122-3. 4.54 mA, +8.2 V   122-4. 470 kΩ.

chapter 123

123-1. 10 Ω   123-2. 400, 1 kΩ   123-3. 0.28 V   123-4. 2, 201 kΩ   123-5. 8.6 Ω   123-6. 230, 860 Ω.

chapter 124

124-1. 18.5 Ω, 4.55 V   124-2. 216   124-3. 1.39 mV. In phase   124-4. 2.5   124-5. 18.5 Ω.

chapter 125

125-1. 24.5 Ω, 11.94 V   125-2. 0.988   125-3. 69.15 mV   125-4. 0.992 V   125-5. 202.45 kΩ.

chapter 126

126-1. 7 approx.   126-2. 60 approx.   126-3. 400 mW approx.   126-4. 6%.

chapter 127

127-1. 24.5 W   127-2. 78%. This is approximately equal to the maximum theoretical value of the efficiency   127-3. 100, 12.5 W.

chapter 128

128-1. 140 pF, 84, 63 kΩ   128-2. 134 kΩ   128-3. 32 mW, 0.64 mW   128-4. 0.65 mA.

chapter 129

129-1. 50 V, 9.3 V, 47 V   129-2. 6 W.

chapter 130

130-1. 20 mA, 6 V, 6666 μS   130-2. $I_D = 20\ (1 - V_{GS}/(-6))^2$ (a) 8.9 mA, 4444 μS (b) 2.2 mA, 2222 μS   130-3. − 1.5 V   130-4. 18 mA, 6 V, 6000 μS, 6 V < V_{DS} < 25 V   130-5. $I_D = 18\ (1 - V_{GS}/(-6))^2$ (a) 8 mA, 4000 μS (b) 2 mA, 2000 μS.

chapter 131

131-1. 2.8 mA, 9.3 V   131-2. 1.6 mA, 11.9 V   131-3. 1.33 mA, 20.5 V   131-4. 2.13 mA, 18 V   131-5. 0.93 mA, 18.7 V   131-6. 0.99 mA, 9.2 V.

chapter 132

132-1. 15.7   132-2. 31   132-3. 2000 μS   132-4. 0.89.

chapter 133

133-1. (a) 15 mA, 8 V (b) 8.4 mA, 2060 μS   133-2. 12 V   133-3. (a) 4 V, 7.1 mA (b) 3550 μS, 5330 μS.

chapter 134

134-1. 3.3   134-2. 0.23 V.

chapter 135

135-1. 12.5 mA   135-2. 4000 μS   135-3. 0.001   135-4. 5000 μS.

chapter 136

136-1. 4 V   136-2. 8000 μS, 32   136-3. 5.0   136-4. 0.92.

chapter 137

137-1. 10 kΩ   137-2. 260 kΩ   137-3. 75 kΩ   137-4. 150 kΩ   137-5. +14 V.

chapter 138

138-1. 16.7   138-2. 0.033   138-3. 21.

chapter 139

139-1. 0.1 μF   139-2. 6.6 kΩ.

chapter 140

140-1. 7.2 Hz, 18.4   140-2. 0.01 μF   140-3. 650 kΩ.

chapter 141

141-1. 1200 Hz   141-2. 0.006 μF   141-3. 65 kΩ.

chapter 142

142-1. 13.8 mA   142-2. Zero   142-3. Zero   142-4. −4.5 V.

chapter 143

143-1. 16.7 kΩ   143-2. 25 kΩ   143-3. 0.25 mA   143-4. 16.5 kΩ   143-5. 55 kΩ   143-6. 5.1 kΩ   143-7. The tube is cut off, and therefore, there is no value of r_p to be determined.

chapter 144

144-1. 2000 μS   144-2. 1.25 mA   144-3. 3000 μS   144-4. 5200 μS   144-5. 3000 μS.

chapter 145

145-1. 33 145-2. 237 V 145-3. −4.76 V 145-4. 47 145-5. 18.

chapter 146

146-1. +120 V, 960 mW 146-2. 20 146-3. 19.86 146-4. 60 V 146-5. 180 mW 146-6. 1.6 W, 11.25%.

chapter 147

147-1. 2 mA, 380 mW 147-2. 40 V, 1.5 mA 147-3. 10, 7.9% 147-4. 3 mA, 570 mW 147-5. 30 V, 1 mA 147-6. 10, 5.3%.

chapter 148

148-1. 820 Ω, 10%, 1/2 W 148-2. 20 μF, 10 WVdc 148-3. 0.005 μF 148-4. 23 V 148-5. 72.8 V, 55 V 148-6. Approximately 900 kHz 148-7. 7 V.

chapter 149

149-1. 250 kΩ 149-2. 820 Ω, 1/2 W, 25 μF, 10 WVdc, 0.1 μF, 200 WVdc 149-3. 33 kΩ, 82.5 149-4. 3300 μS 149-5. 1000 μS, 300 μS.

chapter 150

150-1. 15.3 150-2. 5.3% 150-3. 37.5 150-4. 8.3 150-5. 6%, 17.1, 8.4.

chapter 151

151-1. 23, 1.5 kΩ 151-2. 0.915, 11.8 MΩ, 642 Ω 151-3. 5.25 pF, 2 pF.

chapter 152

152-1. 75%, 62.5% 152-2. 1960 W, 640 W, 90 W 152-3. 83 1/3%, 0.27 cm 152-4. 133 1/3% 152-5. +4.3%.

chapter 153

153-1. 3 kW, 367.5 W, 367.5 W, 8.67 kW, 270 W 153-2. 1.797, 1.7985, 1.8015, 1.803 MHz, 6 kHz 153-3. 26.5% 153-4. 12.2% 153-5. 8% 153-6. 10 kHz.

chapter 154

154-1. 7.7%, 60.9% 154-2. 19.7% 154-3. 18.3% 154-4. 24.4%.

chapter 155

155-1. 396 W 155-2. 0.91 A 155-3. (a) 72 W, (b) 96 W, (c) 1280 W, (d) 592 W.

chapter 156

156-1. 250 Hz 156-2. ±0.00125% 156-3. ±1333 Hz.

chapter 157

157-1. 1.667 157-2. 1.667 157-3. 60, 23.2 kHz 157-4. 24.8 kHz 157-5. 5 kHz.

chapter 158

158-1. 34 pF 158-2. 4.0, 36 kHz 158-3. 24 kHz, 128.

chapter 159

159-1. 100.1 MHz 159-2. 1265 kHz.

chapter 160

160-1. 4.65 A 160-2. 0.8 160-3. 21.4 Ω.

chapter 161

161-1. 1.414 161-2. 9 mi 161-3. 18.75 A 161-4. 96% 161-5. 144 kW 161-6. 7.4 dB 161-7. 17.0 161-8. 2.55 kW

chapter 162

162-1. 3 kHz, 500 W 162-2. 3 kHz 162-3. 4:1, 6 dB

chapter 163

163-1. 25 kHz 163-2. 1.89 MHz, 10000, 63 kHz 163-3. +2400 Hz 163-4. 2500.003 kHz.

chapter 164

164-1. 4 Hz high 164-2. 11 Hz high.

chapter 165

165-1. 175.25, 179.75, 178.83, 221.0 MHz.

chapter 166

166-1. The surge impedance has the same 200-Ω value, but the line of this practice problem has a greater power capability.
166-2. The surge impedance has the same 81-Ω value, but the line of this practice problem has a lower power capability.
166-3. $487\angle -25.4°$ Ω.

chapter 167

167-1. 1575 V, 5.5 kW 167-2. 450 Ω, 2.025×10^4 167-3. 2.18 rad/m, 2.88 m 167-4. 0.6 m 167-5. (a) 437 Ω, (b) 5.12×10^{-4} dB/m, (c) 1.11 rad/m, (d) 5.7 m, (e) 0.43 A 167-6. 80.08 W.

chapter 168

168-1. 0.27, 185 W 168-2. 0.34 168-3. $0.39\angle 59°$ 168-4. 41 W, 229 W 168-5. 1.2 m.

chapter 169

169-1. 0.14, 375 Ω, 60 Ω 169-2. 1.6, 0.23, 13.3 W, 236.7 W, 120 Ω, 46.9 Ω 169-3. 1.4, 2.0 m 169-4. 1.32, 0.24, 265 Ω 169-5. 2.0.

chapter 170

170-1. 35 Ω 170-2. (a) short circuit (b) inductive reactance (c) open circuit 170-3. Theoretically infinite but, in practice, a high value of resistance 170-4. Parallel resonant L,C circuit.

chapter 171

171-1. 3.23 ft 171-2. 3.39 ft 171-3. 1.37, 85 Ω 171-4. Resistance and capacitive reactance.

chapter 172

172-1. 0.92 m, 0.76 m, 170 MHz 172-2. 5:1 172-3. 9.5 dB.

chapter 173

173-1. 13.04 THz 173-2. 103.3 mV/m, 2.83 $\mu W/m^2$.

chapter 174

174-1. 31.2 ft. 174-2. 32.8 ft. 174-3. Resistance and inductive reactance 174-4. 14.9% 174-5. 432 ft.

chapter 175

175-1. 1.52 and 1.68 175-2. Tower #4 175-3. Tower #5 175-4. +5.7%.

chapter 176

176-1. 0.96° 176-2. 670 μV.

chapter 177

177-1. 216 mi 177-2. 4370 ft. 177-3. 15 MHz.

chapter 178

178-1. 2.3° 178-2. 5577, 0.67 MW. 178-3. 9.1°.

chapter 179

179-1. (a) 14.428 cm, 2.078 GHz, (b) 7.214 cm, 4.156 GHz, (c) 6.808 cm, 4.407 GHz 179-2. (a) 9.09 cm, (b) 39.1°, (c) 11.7 cm, (d) 233 Mm/s, (e) 387 Mm/s, and (f) 54 rad/m 179-3. 485 Ω, 360 Ω 179-4. 2.08 GHz, 4.17 GHz 179-5. 5.43 cm 179-6. 12.7 GHz.

chapter 180

180-1. 3.33 GHz 180-2. 6 GHz 180-3. 2.15:1.

chapter 181

181-1. 0.667, 5.0, 25 Ω, 250 Ω 181-2. 35.6 W, 44.4 W 181-3. $76 - j105$ Ω 181-4. 31.3 W, 58.7 W, 1.5 m.

chapter 182

182-1. $0.48 - j0.97$ 182-2. $142 + j34$ Ω 182-3. 1250 Ω 182-4. $0.667 \angle -60.05°$, 5.0, 1500 Ω, 150 Ω.

chapter 183

183-1. $204 + j210$ Ω 183-2. $0.43 \angle +93°$, 0.13 λ 183-3. $80 - j60$ Ω 183-4. 200-Ω resistance 183-5. 20-Ω resistance.

chapter 184

184-1. 0.375 λ, 0.176 λ 184-2. 4.6 cm, 7.6 cm 184-3. 6.44 cm, 9.20 cm.

chapter 185

185-1. 2 dB, 3.0, $0.62 - j0.84$.

chapter 186

186-1. 6.56 cm 186-2. 3000 μs 186-3. 333 Hz 186-4. 750 W 186-5. 600 kW 186-6. 0.001 186-7. 0.0015 186-8. 800 yards.

chapter 187.

187-1. 1000 pF 187-2. 5 μH.

chapter 188

188-1. 15 188-2. 1 μs 188-3. 2 μs.

chapter 189

189-1. 94 μs 189-2. 8.6 nautical miles 189-3. 11.6 nautical miles.

chapter 190

190-1. 70 mph 190-2. 20 knots.

chapter 191

191-1. 3/7, 3/4, 2/7, 2/3 191-2. 10 11/18, 36 19/42, 46 7/18, 233 1/29 191-3. 87/16, 391/7, 1040/79, 534/35 191-4. (a) 4 1/2 (b) 269/312 (c) 2 13/15 191-5. (a) 2

11/12 (b) 11 (c) 380 35/48 191-6. (a) 4 (b) 5 7/11 (c) 9/46 191-7. 3 2/3, 3/8, 5/27.

chapter 192

192-1. 0.4, 0.1875, 0.075, 3.375 192-2. 5/32, 15/32, 5 5/16, 3 1/2000. 192-3. 7.255, 9.0012, 2.6255, –0.02, 6.20845 192-4. 7.70, 5, 0.00001225 192-5. 127.807, 10, 0.015, 0.256 192-6. 83%, 0.4%, 163%, 60%, 62.5% 192-7. 0.19, 0.973, 0.189, 0.07, 0.0025 192-8. 30, 70, 2.5, 0.75, 0.6 192-9. 7:3 or 2.33:1.

chapter 193

193-1. 5.28×10^3, 1.377×10^{-1}, 1.8628×10^5, 2.9809×10^{-5}, 1.745×10^{-2}, 3.21578×10^1, 3.67×10^{-3}, 3.65237×10^2, 2.99796×10^9, 9.6517×10^4 193-2. 3.7, 0.00473, 69430, 0.3857, 0.00000534 193-3. (a) 4.6563×10^5, (b) 8.9472×10^{-7}, (c) 6.9663×10^1 193-4. (a) 7.83×10^4, (b) 1.37×10^5, (c) 3.428×10^{-3}, (d) 8.441×10^{-6} 193-5. (a) 1.69×10^7, (b) 3.90×10^{11}, (c) 7.98×10^7, (d) 5.04×10^{-2}, (e) 4.8×10^4 193-6. (a) 1.52×10^4, (b) 7.5×10^{-4}, (c) 4×10^3, (d) 1.71×10^{13}, (e) 1.26×10^1 193-7. (a) 6.4×10^{10}, (b) 1.296×10^{27}, (c) 5.29×10^{24}, (d) 7.14×10^1, (e) 1.64×10^4, (f) 8.49×10^5 193-8. (a) 2.13×10^{-4}, (b) 1.49×10^4, (c) 4.08×10^{-1}, (d) 1.35×10^0, (e) 1.20×10^{-2}.

chapter 194

194-1. (a) 7, (b) 3, (c) −1.5, (d) −2 194-2. (a) 3, (b) 2, (c) 7, (d) −3

194-3. $C_3 = \dfrac{C_1C_2C_T}{C_1C_2 - C_2C_T - C_1C_T}$

194-4. $\sqrt{\dfrac{Ll}{A\mu}}$

194-5. $\dfrac{R_{T°C} - R_{20°C}}{R_{20°C}\,(T°C - 20°C)}$

194-6. $\dfrac{LlV_y}{2yV_x}$

194-7. $\dfrac{L}{R^2Q^2}$

194-8. $\dfrac{V_{GS}}{1 - \sqrt{\dfrac{I_D}{I_{DSS}}}}$.

chapter 195

195-1. (a) 7, 6, (b) 9, 3, (c) 2, 5, (d) 4, 5 195-2. (a) 2, 3, (b) 2, 12, (c) 5, 2, (d) −4/5, −1/10 195-3. $x = 2$, $y = 5$, $z = 5$ 195-4. 80000 Ω, 20000 Ω.

chapter 196

196-1. $i_1 = 3$, $i_2 = -2$, $i_3 = 1$ 196-2. −15 196-3. $i_1 = -2$, $i_2 = 3$, $i_3 = 1$, $i_4 = -1$.

chapter 197

197-1. +8.74 mA, +4.13 mA, +2.68 mA 197-2. +1.78 A, +0.68 A, +0.636 A.

chapter 198

198-1. 5.5, −2.33

198-2. $f = \dfrac{1}{2\pi}\sqrt{\dfrac{1}{LC} + \dfrac{R^2}{4L^2}} - \dfrac{R}{2\pi L}$

198-3. 2.92, 0.48 198-4. $1.25 + j1.2$, $1.25 - j1.2$.

chapter 199

199-1. $x = 2$, $y = 1$ 199-2. −0.56, 3.56 199-3. $4y + 3x - 25 = 0$ 199-4. (3, 12), (−3, −12) 199-5. (3, 0), (−3, 0) 199-6. $r^2(16 + 9\sin^2\theta) = 400$.

chapter 200

200-1. 129°, 45' 54", 2.26^c 200-2. 97.731°, 1.706^c 200-3. 214.744°, 214°, 44' 40.5" 200-4. 53°, 143°.

chapter 201

201-1. 70° 201-2. 3.47, 3.99 201-3. 5.76 201-4. 17.79 201-5. 22-1.

chapter 202

202-1. 4.08, 17.6°, 1.23 202-2. 47.2°, 42.8°, 16.1 202-3. 9.74, 73.6°, 33.1 202-4. 14.0, 15.7, −13.2, −11.9.

chapter 203

203-1. 10.22, 28.7°, 8.97 203-2. 57.9°, 32.1°, 29.48 203-3. 40.45, 44.4°, 21.06 203-4. 0.7732 203-5. 0.3472.

chapter 204

204-1. 1.81, 70.2°, 5.34 204-2. 23.1°, 66.9°, 49.7 204-3. 71.02, 51.4°, 91.2 204-4. 17.52°.

chapter 205

205-1. 1.055, 1.202, −1.798, −1.476 205-2. 1.6029, −2.6927, 249, 2.384 205-3. 3.230, −6.94, 2.55, −0.235 205-4. 31.1° 205-5. 34.3°.

chapter 206

206-1. 0.6691, 0.6018, −0.5446, −1.0000 206-2. −0.6157, −0.4695, 0.8988, −0.7455, −0.5640 206.3.

−6.772, 0.4040, −0.0349, 14.30, −0.5442 206-4. 2.904, 1.836, 0.2493 206-5. − 3.3 V

chapter 207

207-1. 63°, 4.21, 5.07 207-2. 21.6, 29.5°, 33.5° 207-3. 0.329, 46.4°, 96.6° 207-4. 33.1°, 43.3°, 103.6° 207-5. Two sides and a nonincluded angle do not define a triangle. There are two possible solutions: 2.25, 18.1°, 124.9°; 7.23, 87.9°, 55.1° 207-6. $e_1 + e_2 = 22.1 \sin(\omega t + 11.7°)$, $e_1 - e_2 = 12 \sin(\omega t + 113.8°)$.

chapter 208

208-1. $\dfrac{2 \tan A}{1 - \tan^2 A}$

208-2. 11°, 63°

208-3. (a) $\dfrac{1 - \cos 64°}{2}$,

(b) $\dfrac{1 + \cos 64°}{2}$

208-4. 64°, 22° (or vice-versa) 208-5. $1 - \cos^2 38°$ 208-6. $22 \sin 3000t \cos 1000t$ 208-7. 33°.

chapter 209

209-1. kf_o

209-2. $a^n + \dfrac{n}{1!} a^{n-1}b + \dfrac{n(n-1)}{2!} a^{n-2}b^2 + ---$

$+ \dfrac{n(n-1) --- (n-r+1)}{r!} a^{n-r} b^r$

$+ --- + b^n$

209-3. $\dfrac{L_p R_s^2}{\omega L_s^2}$

209-4. (a) $1 - \dfrac{x}{4} - \dfrac{x^2}{32} - \dfrac{x^3}{128} ---$,

(b) $1 + x - \dfrac{3}{2} x^2 + \dfrac{7}{2} x^3 ---$

209-5. (a) $1 + 6x + 24x^2 + 80x^3 ---$,

(b) $1 - \dfrac{x}{4} + \dfrac{3}{32} x^2 - \dfrac{5}{128} x^3 ---$,

(c) $1 + \dfrac{1}{3} x^2 - \dfrac{1}{9} x^4 + \dfrac{5}{54} x^6 ---$

chapter 210

210-1. 0.899 210-2. 0.817 210-3. 2.14 210-4. (a) 2.375 (b) 5.468 210-5. (a) 2 1/3 (b) 1 1/2 (c) zero (d) 1 1/2 (e) −4.

chapter 211

211-1. (a) 7, (b) $20x^3$, (c) $-18/x^4$, (d) $4/3\ x^{-1/3}$, (e) $(-33/8)x^{-7/4}$ 211-2. $21x^2 - 10x + 11$ 211-3. 1/2 211-4. 2/3 211-5. $2x + 1$.

chapter 212

212-1. $140(4x^3 - 5x)^{19}(12x^2 - 5)$ 212-2. $24x(3x^2 - 11)^{-2/3}$ 212-3. −4 212-4. $-3/2(4x^3 - x)^{-3/2}(12x^2 - 1)$ 212-5. $15(1 - 3\sqrt{3x^4 - 5x^3})^{-4}(3x^4 - 5x^3)^{-2/3}(12x^3 - 15x^2)$

212-6. $\dfrac{2\left(\omega L - \dfrac{1}{\omega C}\right)\left(L + \dfrac{1}{\omega^2 C}\right)}{R^2 + \left(\omega L - \dfrac{1}{\omega C}\right)^2}$.

chapter 213

213-1. $7\left[-3\sqrt[3]{4x^3 + 5x} \times (3x^4 - 8x^2)^{-4}(12x^3 - 16x) + \dfrac{(3x^4 - 8x^2)^{-3}(4x^3 + 5x)^{-2/3}(12x^2 + 5)}{3}\right]$

213-2. $6x^{-1/4}(3x^4 - 5x)^{-1/2} - \dfrac{32x^{3/4}(12x^3 - 10x)}{(3x^4 - 5x^2)^{3/2}}$

213-3. $\dfrac{j(R + j\omega C)C + (G + j\omega C)L}{2\sqrt{(R + j\omega L)(G + j\omega C)}}$

chapter 214

214-1. $\dfrac{-x}{(2x + 1)^3}$

214-2. $\dfrac{-x^2}{(x + 2)^4}$

214-3. $\dfrac{-x^2}{(x + 3)^5}$

214-4. $\dfrac{-\sqrt{1 + x^2}}{x^4}$

214-5. $j\left[\dfrac{L(G + j\omega C) - C(R + j\omega L)}{(G + j\omega C)^2}\right]$.

chapter 215

215-1. $x^2 \sin x$ 215-2. $5 \sin 2x$ 215-3. $x \sin 3x$ 215-4. $x^2 \cos x$ 215-5. $x \tan^{-1} x$ 215-6. $x \sin x \cos 3x$ 215-7. $x(2 \sin 4x + \sin 2x)/4$

chapter 216

216-1. (a) $3/x$, (b) $-1/x$, (c) $1/(2x)$ 216-2. (a) $\cot x$, (b) $-1/(1 - x)$, (c) $-2 \tan x$ 216-3. (a) $4/(4x + 3)$,

(b) $\sec x \csc x$, (c) $1 + \ln x$ 216-4. (a) $(1 - \ln x)/x^2$, (b) $1/(1 - x^2)$ 216-5. (a) $4e^{4x}$, (b) $-2e^{-2x}$, (c) $(1 + x)e^x$ 216-6. (a) $2xe^{x^2}$, (b) $e^{2x}(2 \sin 3x + 3 \cos 3x)$, (c) $(3x - 1 - x^2)e^{-x}$.

chapter 217

217-1. (a) $2 \cosh 2x$, (b) $3 \sinh 3x$, (c) $-4 \operatorname{csch}^2 x \coth x$ 217-2. (a) $-1/2 \cosh x \operatorname{csch}^2 x$, (b) $4 \tanh 2x \operatorname{sech}^2 2x$, (c) $-1/2 \operatorname{csch}^2 x/2$ 217-3. (a) $\coth^2 x$, (b) $\cosh 2x$, (c) $\tanh 2x$ 217-4. $\coth 3x$, (b) $\cosh^2 x$, (c) $\sinh^2 x/2$ 217-5. (a) $\coth^2 x$, (b) $\operatorname{sech} 2x$, (c) $\operatorname{csch} x$.

chapter 218

218-1. (a) $\frac{1}{y}, \frac{-x}{y^2}$ (b) $2ax + 2hy, 2hx + 2by$

(c) $\frac{-y}{x^2 + y^2}, \frac{x}{x^2 + y^2}$ (d) $\frac{2y}{(x + y)^2}, \frac{-2x}{(x + y)^2}$

(e) $\frac{1}{\sqrt{y^2 - x^2}}, \frac{-x}{y\sqrt{y^2 - x^2}}$

(f) $\frac{-x}{(x^2 + y^2)^{3/2}}, \frac{-y}{(x^2 + y^2)^{3/2}}$

218-2. (a) 0,1,1,0, (b) $6ax + 6by$, $6bx$, $6bx$, $6cy$, (c) $-1/x^2$, 0 0, $-1/y^2$, (d) All values are e^{x+y}, (e) $-y \cos x$, $-\sin x - \sin y$, $-\sin x - \sin y$, $-x \cos y$, (f) $\sinh x \cosh 2y$, $2 \cosh x \sinh 2y$, $2 \cosh x \sinh 2y$, $4 \sinh x \cosh 2y$ 218-3. 0 218-4. 0 218-5. $2(x^2 + y^2 + z^2)$.

chapter 219

219-1. Maximum $x = -2$, minimum $x = 3$ 219-2. There are no maxima or minima 219-3. $x = 1$ 219-4. $x = 3a/4$ 219-5. Maximum value 9/27, minimum value 0 219-6. Maximum value 3, minimum value 1/3 219-7. (1,2) 219-8. $(0, a\sqrt{3}), (0, -a\sqrt{3})$.

chapter 220

220-1. $(2/5)\sqrt{(x + 1)^5} - (2/3)\sqrt{(x + 1)^3} + K$ 220-2. $(1/2)[\sin^{-1} x + x\sqrt{1 - x^2}] + K$ 220-3. $\pi/4$ 220-4. $-x^2 \cos x + 2(x \sin x + \cos x) + K$ 220-5. 1 220-6. $(1/2)\ln(1 + x^2) + K$ 220-7. $2 - \ln 2$ 220-8. $e - 1$ 220-9. $-(1/5)e^{-x}(\sin 2x + 2\cos 2x) + K$

220-10. $\frac{e^4 - 4e^2 - 1}{8e^2}$

220-11. $x \tanh x - \ln(\cosh x) + K$ 220-12. $(1/3)\cosh^{-1}(3/4x) + K$ 220-13. $3\ln(x + 1) - 2\ln(x - 2) + K$

220-14. $\ln(x - 1) - \ln(x + 1) - \frac{2}{x + 1} + K$.

chapter 221

221-1. 65535_{10} 221-2. (a) 3781_{10} (b) 1499_{10}, (c) 581_{10}, (d) 455_{10} 221-3. 467.2421875_{10} 221-4. 2732.78125_{10} 221-5. 1654.38_{10} 221-6. 114.6875_{10}.

chapter 222

222-1. (a) 11011000100_2, (b) 1661_{10} 222-2. 111010100.111_2 222-3. 10101.001_2 222-4. 1000011101001_2 222-5. 10110111_2.

chapter 223

223-1. (a) 0001 0101 0111 0011, (b) 75539_{10} 223-2. (a) 0100 0101 1001, (b) 10110010011_2 223-3. 0001 0010 0001 1001.

chapter 224

224-1. (a) 892.375_{10}, (b) 2546_8 224-2. (a) 111000001001010111_2, (b) 1671_8 224-3. (a) 3361_8, (b) 17557_8 224-4. (a) 114315_8, (b) 600_8.

chapter 225

225-1. (a) 48343_{10}, (b) $120A4_{16}$ 225-2. (a) 1011011001000011_2, (b) 373_{16} 225-3. (a) 126070_8, (b) $9FD_{16}$ 225-4. (a) $EEF5_{16}$, (b) 16459_{16} 225-5. (a) $5CB64C_{16}$, (b) $31B_{16}$.

chapter 226

226-1. (a) 13672_{10}, (b) $1t286_{12}$ 226-2. (a) 100111000011011_2, (b) $431e_{12}$ 226-3. (a) 17133_8, (b) 9435_{12} 226-4. (a) $13AC3_{16}$, (b) 23813_{12} 226-5. (a) 12754_{12}, (b) $tt905_{12}$ 226-6. (a) 79818_{12}, (b) $83t_{12}$.

chapter 227

227-1. (a) 783_{10}, (b) 454_8, (c) 515_{12}, (d) 458_{16} 227-2. 100100111 227-3. 00101101 227-4. (a) 01111, (b) 00001, (c) 10001, (d) 11111.

chapter 228

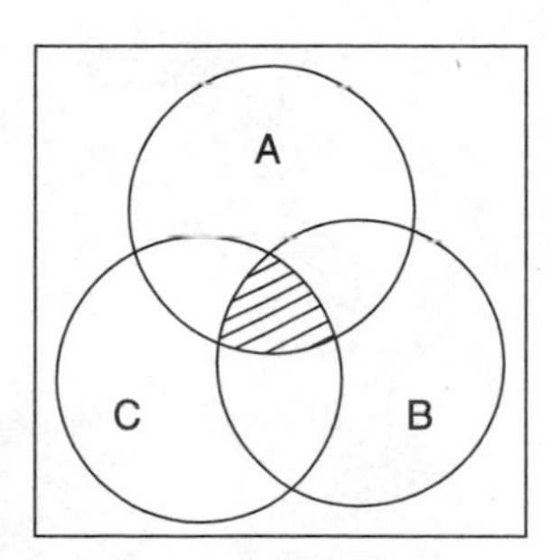

Figure J-228-1

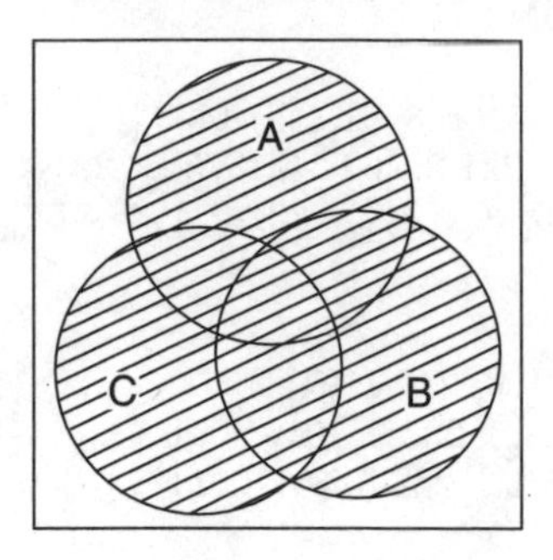

Figure J- 228-2

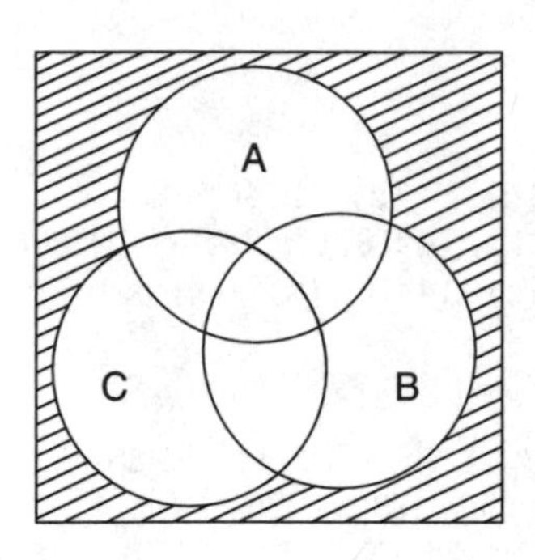

Figure J-228-3

A	B	C	D	f (A, B, C, D)
0	0	0	0	?
0	0	0	1	?
0	0	1	0	?
0	0	1	1	?
0	1	0	0	?
0	1	0	1	?
0	1	1	0	?
0	1	1	1	?
1	0	0	0	?
1	0	0	1	?
1	0	1	0	?
1	0	1	1	?
1	1	0	0	?
1	1	0	1	?
1	1	1	0	?
1	1	1	1	?

Figure J-228-4

chapter 229

229-1. $A = W + D + S + F$

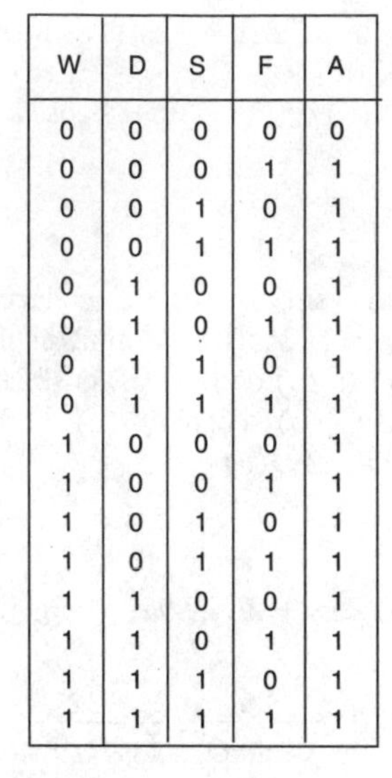

W	D	S	F	A
0	0	0	0	0
0	0	0	1	1
0	0	1	0	1
0	0	1	1	1
0	1	0	0	1
0	1	0	1	1
0	1	1	0	1
0	1	1	1	1
1	0	0	0	1
1	0	0	1	1
1	0	1	0	1
1	0	1	1	1
1	1	0	0	1
1	1	0	1	1
1	1	1	0	1
1	1	1	1	1

Figure J-229-2

229-3., 229-4.

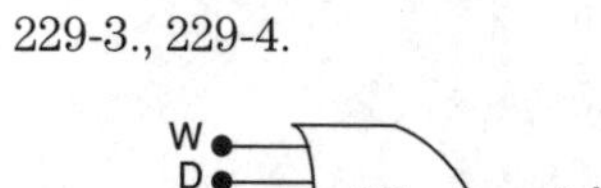

Fig J-229-3

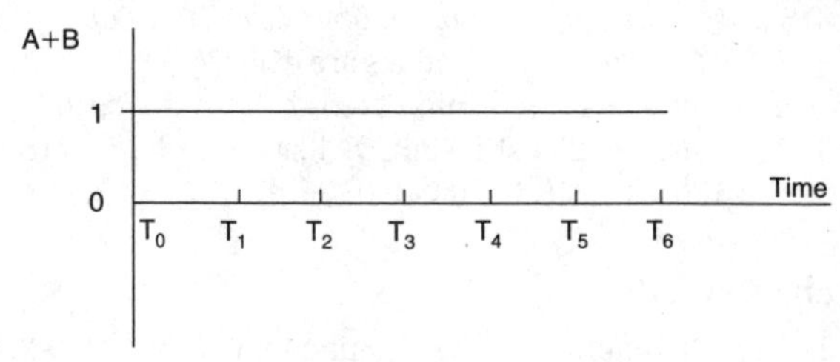

Figure J-229-4

chapter 230

230-1. $C = V \cdot S \cdot R \cdot T$

230-2

V	S	R	J	C
0	0	0	0	0
0	0	0	1	0
0	0	1	0	0
0	0	1	1	0
0	1	0	0	0
0	1	0	1	0
0	1	1	0	0
0	1	1	1	0
1	0	0	0	0
1	0	0	1	0
1	0	1	0	0
1	0	1	1	0

Figure J-230-2

V	S	R	J	C
1	1	0	0	0
1	1	0	1	0
1	1	1	0	0
1	1	1	1	1

Figure J-230-2 continued

230-3.

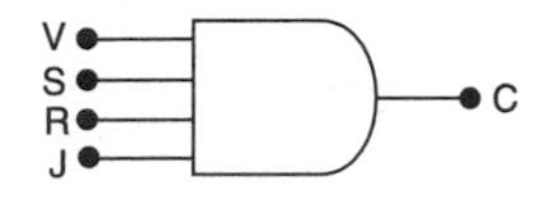

Figure J-230-3

230-4.

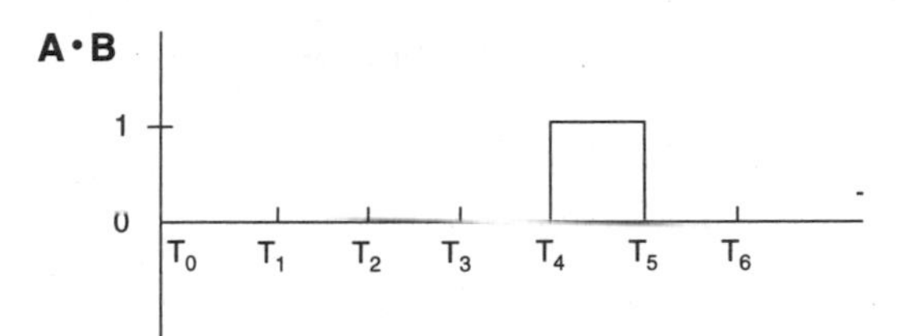

Figure J-230-4

chapter 231

231-1. $R = \overline{G}$

231-2.

G	$R=\overline{G}$
0	1
1	0

Figure J-231-2

231-3.

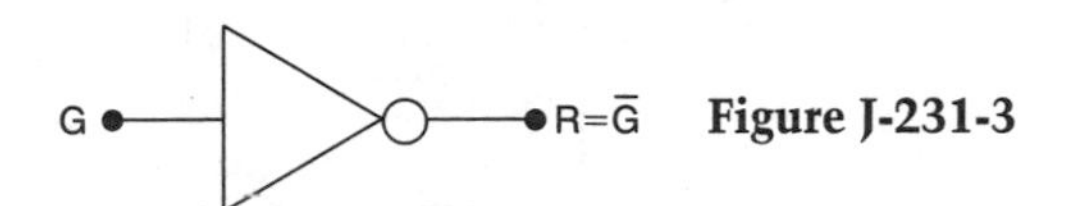

Figure J-231-3

231-4.

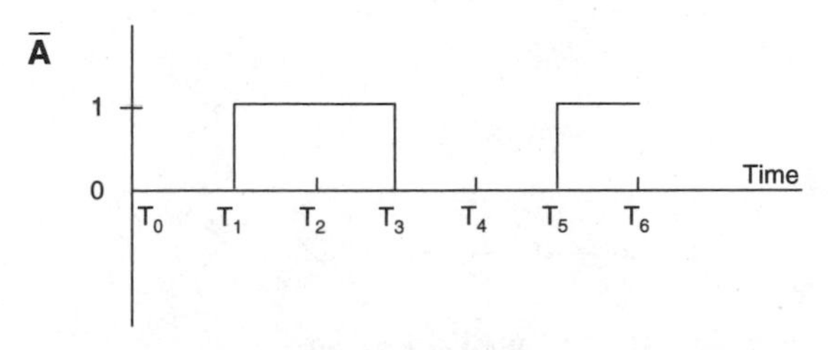

Figure J-231-4

chapter 232

232-1.

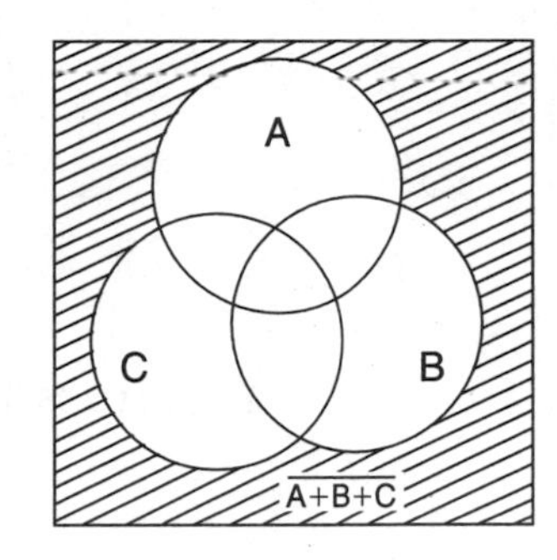

Figure J-232-1

232-2. $G = \overline{A + B + C}$

232-3

A	B	C	G
0	0	0	1
0	0	1	0
0	1	0	0
0	1	1	0
1	0	0	0
1	0	1	0
1	1	0	0
1	1	1	0

For A, B, C:
1=present
0=absent

For G:
1=meets
0=does not meet

Figure J-232-3

232-4.

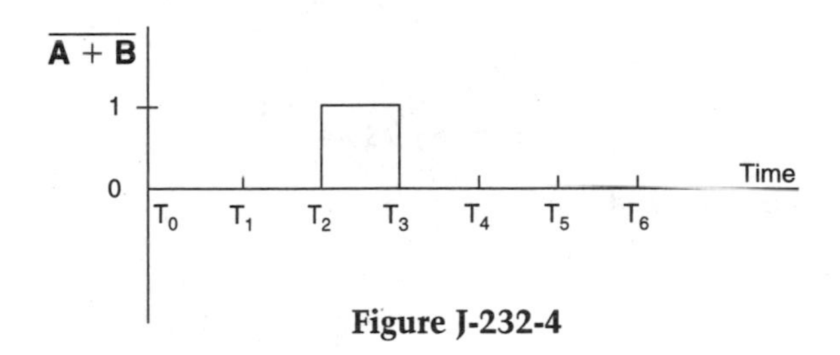

Figure J-232-4

chapter 233

233-1.

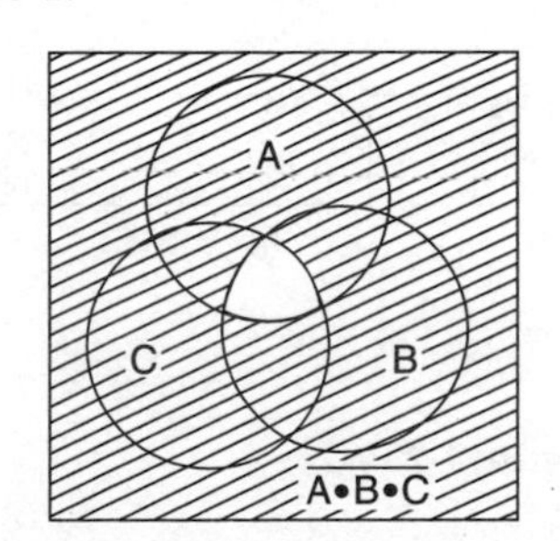

Figure J-233-1

233-2. $G = \overline{A \cdot B \cdot C}$

233-3

A	B	C	G
0	0	0	1
0	0	1	1
0	1	0	1
0	1	1	1
1	0	0	1
1	0	1	1
1	1	0	1
1	1	1	0

Figure J-233-3

For A, B, C:
1=present
0=absent

For G:
1=meets
0=does not meet

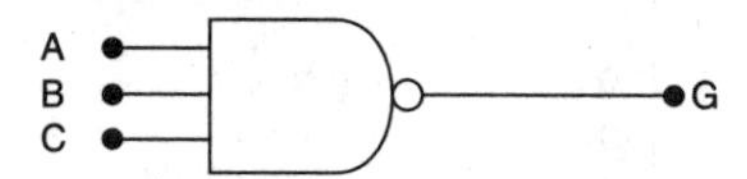

233-4.

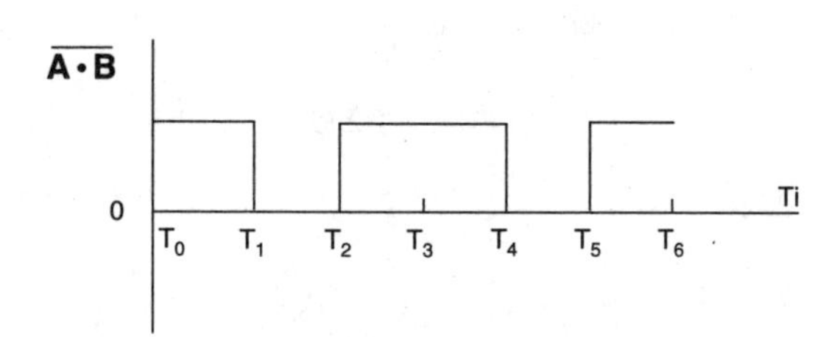

Figure J-233-4

chapter 234

234-1

$G = A \cdot \overline{B} + B \cdot \overline{A} = A \oplus B$

234-2.

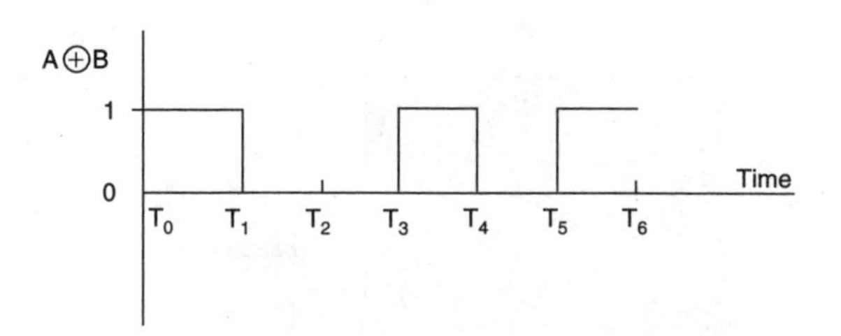

Figure J-234-2

chapter 235

235-1.

$G = A \cdot B + \overline{A} \cdot \overline{B} = \overline{A \oplus B}$

235-2.

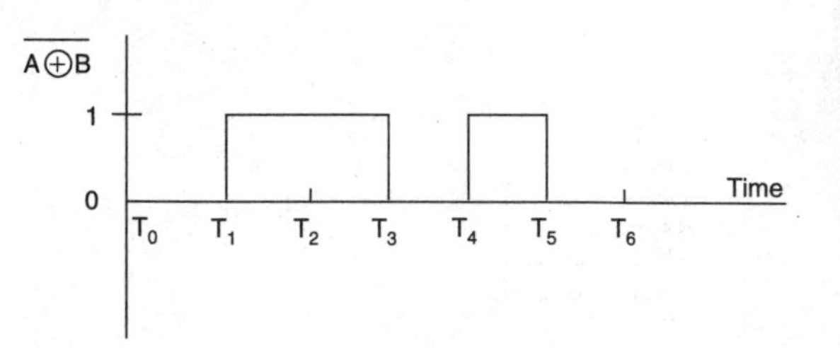

Figure J-235-2

chapter 236

236-1.

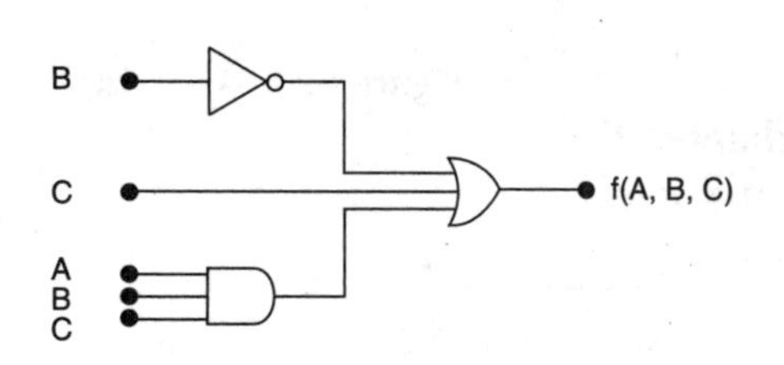

f(A, B, C)

A	B	C	f (A • B • C)
0	0	0	1
0	0	1	1
0	1	0	0
0	1	1	1
1	0	0	1
1	0	1	1
1	1	0	0
1	1	1	1

Figure J-236-1

236-2.

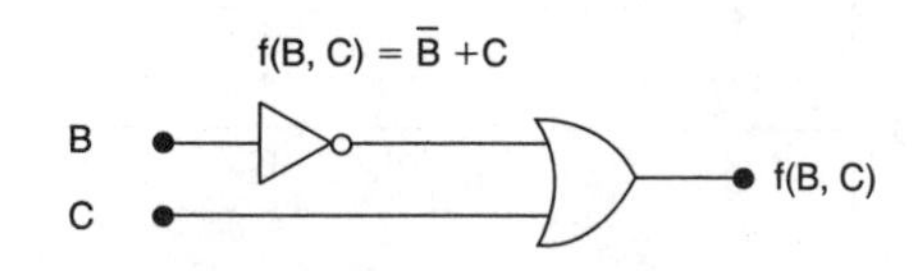

Figure J-236-2

236-3.

$f(A, B, C) = \overline{\overline{A + B} \bullet (B + C)}$

A	B	C	f (A,B,C,D)
0	0	0	1
0	0	1	0
0	1	0	1
0	1	1	1
1	0	0	1
1	0	1	1
1	1	0	1
1	1	1	1

$f(A, B, C) = \overline{(A + B) \bullet \overline{C}}$

A	B	C	f (A,B,C,D)
0	0	0	1
0	0	1	1
0	1	0	0
0	1	1	1
1	0	0	0
1	0	1	1
1	1	0	0
1	1	1	1

Figure J-236-3A, B

236-4.

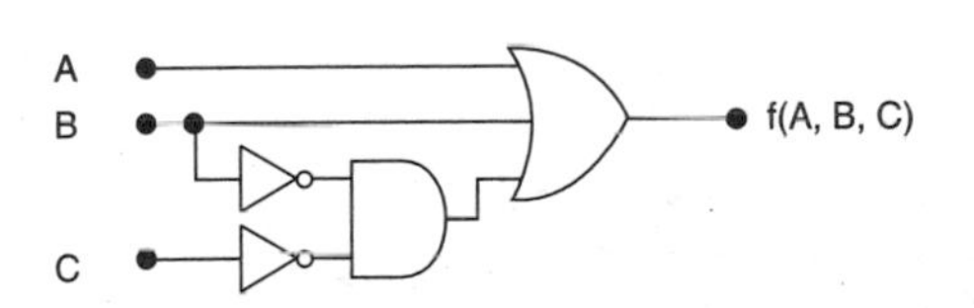

Figure J-236-4

236-5.

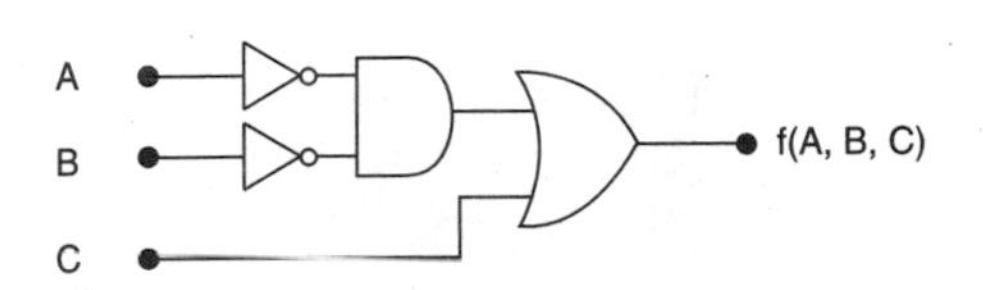

Figure J-236-5

chapter 237

237-1. $f(A) = A$ 237-2. $f(B) = \overline{B}$ 237-3. $f(A, B) = A + B$ 237-4. $f(A, B, C) = 1$ 237-5. $f(A, B, C) = A \cdot B + A \cdot \overline{C} + B \cdot \overline{C}$ 237-6. $f(A, B, C) = \overline{A} + \overline{B} + \overline{C}$ 237-7. $f(A, C) = \overline{A} + \overline{C}$ 237-8. $f(A, B, C) = \overline{A} + \overline{B} \cdot C$ 237-9. $f(C) = C$ 237-10. $f(B, D) = B + D$ 237-11. $f(A, B, C) = A \cdot B \cdot C$ 237-12. $f(C) = C$.

237-3.

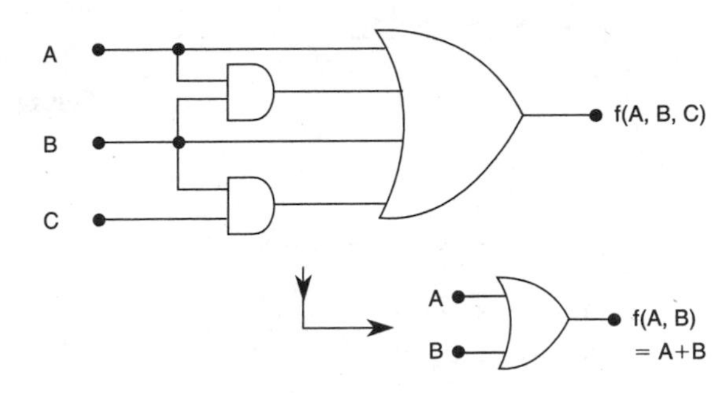

Figure J-237-3

237-5.

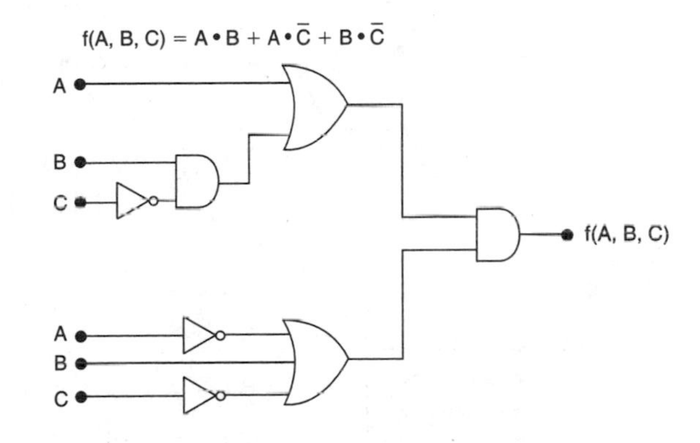

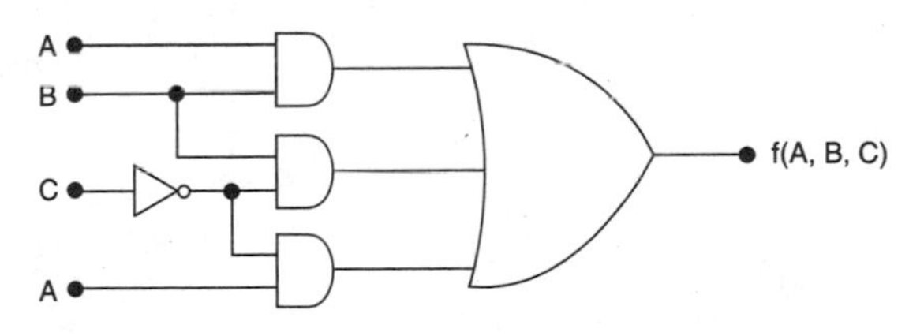

Figure J-237-5

chapter 238

238-1. $f(A, B) = A \cdot \overline{B} + \overline{A} \cdot B$ 238-2. $f(A, B, C, D, E, F) = \overline{A} + B \cdot [(\overline{C} + D) \cdot (E + F)]$ 238-3. $f(A, B, C) = \overline{A} \cdot \overline{B} + C$ 238-4. $f(A, B, C) = (\overline{A} + B + \overline{C}) \cdot C$ 238-5. $f(A, B, C) = \overline{A} + \overline{B} + C$ 238-6. $f(A, B, C) = A \cdot (B + \overline{C})$ 238-7. $f(A, B, C, D) = \overline{A} + \overline{B} + C \cdot D$ 238-8. $f(A, B, C, D) = \overline{A} + B + \overline{C} + \overline{D}$.

238-5.

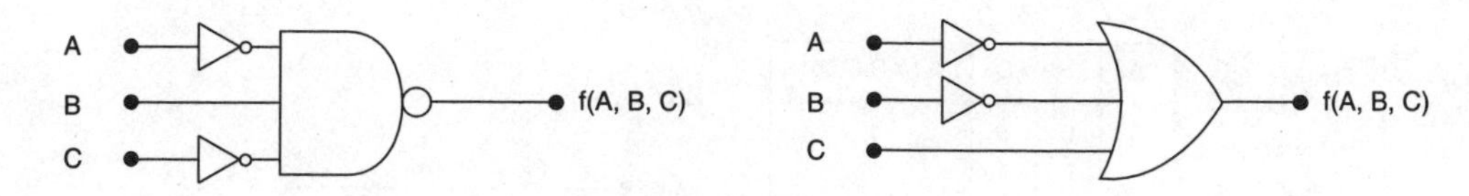

Figure J-238-1

238-6.

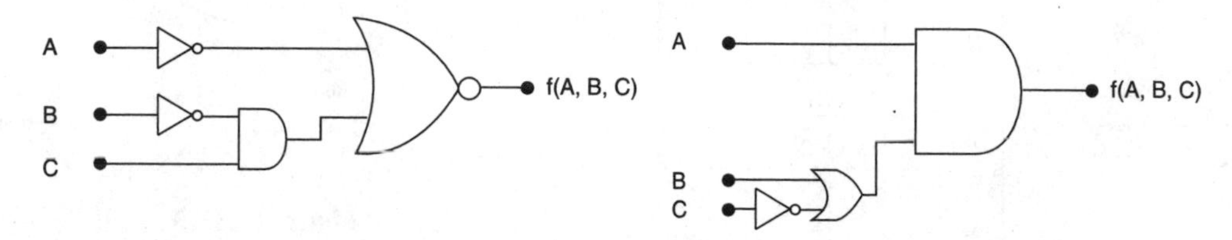

Figure J-238-6

238-7.

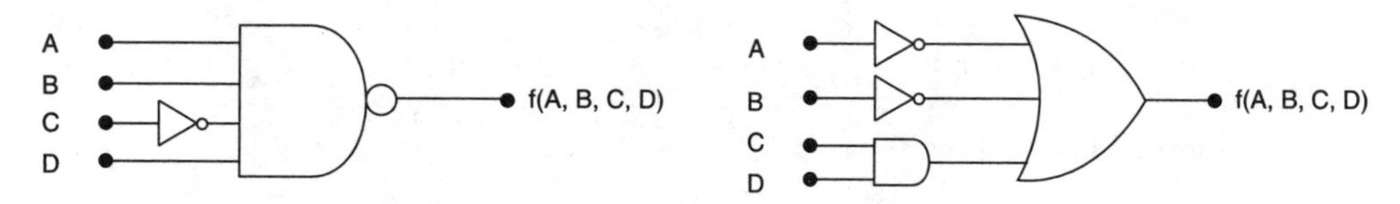

Figure J-238-7

238-8.

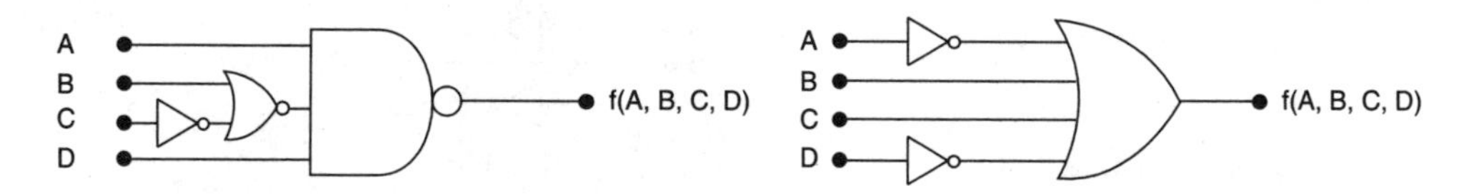

Figure J-238-8

chapter 239

239-1. $f(A, B) = 1$ 239-2. $f(A, B, C) = A \cdot \overline{B} \cdot C + B \cdot \overline{C}$ 239-3. $f(A, B) = A \cdot B + \overline{A} \cdot \overline{B}$ 239-4. $f(A, B, C) = A \cdot \overline{B} + \overline{A} \cdot \overline{B} \cdot \overline{C}$ 239-5. $f(A, B, C) = \overline{A} + \overline{B} \cdot C$.

239-5.

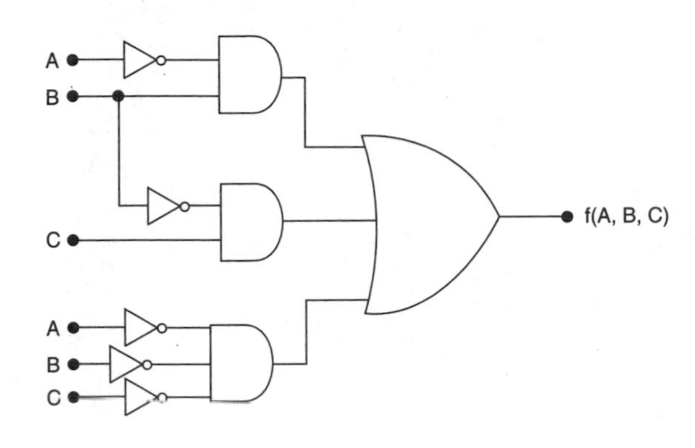

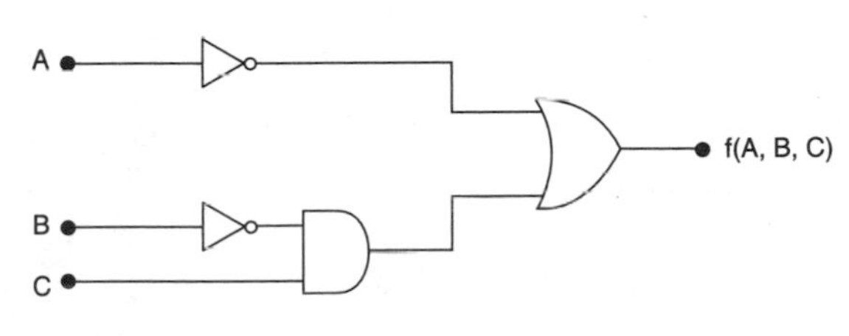

Figure J-239-5

chapter 240

240-1. $f(A, B, C, D) = A \cdot \overline{B} \cdot \overline{C} + A \cdot B \cdot C + D$ 240-2. $f(A, B, C) = A \cdot \overline{B} + B \cdot C + \overline{B} \cdot \overline{C}$.

chapter 241

241-1. 16,855 mph 241-2. 22.1×10^6 m.

chapter 242

242-1. 39.5°, 112.7° 242-2. 3666 km/h.

chapter 243

243-1. 24,937 mi. 243-2. 7.5°.

chapter 244

244-1. 18 MHz, 9 MHz. 244-2. Channel 8.

chapter 245

245-1. 10 channels. 245-2. 8.1×10^6 bps.

chapter 246

246-1. 1600 W. 246-2. 2300 MHz.

chapter 247

247-1. 15,102 mi, 7849 mi, 12,902 mi, 0.854.

chapter 248

248-1. 1.75°, 39.8 dB. 248-2. 0.075°, 67.3 dB.

chapter 249

249-1. 28.2 dB/K. 249-2. 2.81×10^{-7} W.

chapter 250

250-1. 206.5 dB, 205 dB. 250-2. 1.12 μW. 250-3. 95 dB. 250-4 48 dB. 250-5. 48 dB.

chapter 251

251-1. 0.82 μW. 251-2. 194 THz.

chapter 252

252-1. 3.65 μW. 252-2. 0.52 μW.

chapter 253

253-1. 29.9°. 253-2. 58°, 32°.

chapter 254

254-1. 0.83. 254-2. 4.6 km.

chapter 255

255-1. 1.71 μm, 0.64. 255-2. 796. 255-3. 14.3 km.

chapter 256

256-1. 0.33 mW. 256-2. 1.47 dB/km. 256-3. 14.3 km.

chapter 257

257-1. 1.6 dB. 257-2. 55 μW.

chapter 258

258-1. 0.85 eV. 258-2. 543 THz.

chapter 259

259-1. 164 THz. 259-2. 112.5 μW.

chapter 260

260-1. 14 μW. 260-2. 139 μW.

Index

A

A-band satellite stations, 703–704
absorption, 468, 729–730
absorption law, 662–669
acceleration, 2–3, 4
acceptance cones, 722–725
acceptor circuits, 202–205
acute angles, 557–559
Adcock antenna, 461–462
additive mixing, 411–413
admittance, 188–199
  Smith charts, 482–497
algebra, Boolean (see Boolean algebra)
algebra, complex, 228–230
alternating current (ac), 153–278
  acceptor circuits, 202–205
  admittance, 188–199
  algebra, complex, 228–230
  alternators, 156–158, 257–267
  analysis of circuit
  (See also sine wave)
    parallel branch, j operator, 236–237
    series-parallel, j operator, 234–235
  apparent power, 176, 178–180, 181–182
  attenuation, 271–272
  average value of sine wave, 161–162
  bandwidth, 214–216
  capacitive reactance, 174–176, 228
  coefficient of coupling, 219–221
  coils:
    mutually coupled, 218–221
    mutually coupled, series/parallel, 221–224
  conductance, 170–171
  core loss, 208
  current magnification, LCR circuit, 211
  cycles, 154–155
  damping, damped waves, 217–218
  decibels, 267–271
  delta-wye/wye delta transformations, 247–248
  dielectric loss, 208
  effective value, 158–161
  energy relationships, 207, 211, 215–216
  figure of merit, 206
  Fourier analysis, 154–155
  Fourier's theorem, 273–276
  frequency, 154–155
  fundamental of sine wave, 154–155, 273–276
  harmonics of sine wave, 154–155, 273–276
  hertz as unit of, 154–155
  impedance magnification, LCR circuit, 211
  impedance phasor, 181–182
  impedance, 176–177
  inductive reactance, 171–173, 228
  j operator, 228–230, 234–235, 236–237
  Kirchhoff's current law (KCL), 237–238
  Kirchhoff's voltage law (KVL), 237–238
  lagging power factor, 176, 193–196, 196–199
  leading power factor, 177, 180–182, 193–199
  load resistance, 200–202
  maximum power transfer to load (ac), 200–202
  mesh analysis, 239–240
  Millman's theorem, 243–244
  mutual inductance, 208, 218–221
  nepers, 271–272
  nodal analysis, 242–243
  nonsinusoidal voltages, currents, 276–278
  Norton's theorem, 245–247
  Ohm's law, 168
  open circuits, 194
  oscillation, free, in LC circuit, 210–218
  parallel branch circuit analysis, j operator, 236–237
  parallel resonant LCR circuit, 209–212
  parallel resonant tank circuit, 213–216
  peak values, 154–155
  peak-to-peak values, 154–155
  periods, 154–155
  phase relationships, 164–166
  phasor addition/subtraction, 166–167, 230–232
  phasor multiplication/division, 232–234
  phasor representation of voltage/current, 162–164
  polar notation, 228–230
  power factor, 176–182
  product-over-sum formula, 189, 234
  proximity effect, 208
  Pythagorean sum formula, 189
  Q factor, 205–209
  rationalization, 232–234
  reactance, 185–187
  real vs. imaginary values, 230–232
  reciprocity theorem, 249–250
  rectangular notation, 228–230
  rectangular/polar conversions, 228–230
  rectification, 275–276
  resistance and resistors, 168–169
  resonance, series LCR circuit, 202–205
  resonant frequent, 206
  root-mean-square (rms) value, 158–161
  sawtooth waves, 154–155
  selectivity, 205–209, 211, 214–216
  self inductance, 218–221
  series-parallel circuit analysis, j operator, 234–235
  sine waves, 154–155, 273–276
  sine-wave input voltages, series and parallel, 178–199
  skin effect, 207–209
  square waves, 154–155
  superposition theorem, 240–241
  susceptance, 188–199
  synthesis of waveform, 274
  Thevenin's theorem, 244–245
  transformers, 224–227
  transformers, radio frequency (RF) (see RF transformers)
  true power, 176–182
  voltage magnification factor (Q), 205–209
  waveform analysis, 273–276
alternators, 156–158, 257–267
  single-phase practical, 257–258
  three-phase, 260–267
  three-phase, delta connection, 265–267
  three-phase, wye connection, 263–265
  two-phase practical, 258–260
American Wire Gauge (AWG), 14
ampere, 2–3, 6
amplification factor, 355–356
amplifiers:
  amplification factor, 355–356
  audio, AM, 394
  Barkhausen criterion, 342

amplifiers (*cont.*):
beam power tube, 369–371
bias, 362–365
cathode-follower, 374–377
Class A, AB, B, C, 363
common-base, 309–310
common-emitter, 306–308
cross-over, 315
Darlington pairs, 310–312
DE MOSFET, 332–333
E MOSFET, 335–336
emission, primary and secondary, 367–371
emitter follower (common collector), 310–312
grounded-grid triode amp, 374–377
harmonic generators, 316–317
JFET, 327–329
negative feedback, 365, 371–374
Nyquist criterion, 342
operational amplifiers, 337–340
pentodes, 368–371
positive/regenerative feedback, 341–342
power, Class A, AB, B, 312–316
RF power amps, 391–396
RF power, Class C, 316–318
screen-grid tubes (tetrode), 367–371
tetrodes, 367–371
triode tube, 356–359
tubes, 365–371
variable mu-pentode, 369–371
video amplifier stage, 502–503
amplitude modulation (AM), 380–396, 417–418
antenna current, percentage changes, 385–387
audio amplifiers, 394
bandwidth, 382–385
carrier shift, 379
directional antenna for, 456–459
directional station, elements, 751
double-sideband system, 418
duplex communications, 394
emission designations, FCC, 756–757
intermodulation, 394
license, 750
license, directional AM, 752–753
microphones, 392–394
percentage of modulation, 380–396
plate modulation, 387–390
receivers, 406–413
sidebands, 382–387, 417–418
simplex communications, 394
single-sideband system, 417–418
tolerances, standards, FCC, 758–760
transmitters, 390–396, 749
amplitude modulation (AM) (*cont.*):
two-way communication systems, 394
AND operation, 640, 645–648
angle modulation, 397–399
angles, measuring, 556–559
angles, multiple, 582–584
angular frequency (velocity), 558–559
anodes, 5–6
anomalous propagation (anaprop), 465–472
antennas:
Adcock antenna, 461–462
antenna current, percentage changes, AM, 385–387
azimuth, angle of, 472
balanced/unbalanced antennas, 460–462
bearings, true and reciprocal, 460
Bellini-Tosi systems, 460
dipoles, 447–450
directional, for AM, 456–459
directors, 448–450
dish (see microwave antennas; parabolic reflectors)
effective radiated power, 415–417
efficiency, 455–456
electromagnetic (EM) waves in free space, 450–453
elevation, angle of, 472
field strength, 415–417
front-to-back ratio, 447–450
goniometers, 460
Hertz type, 444–447
image antenna, 454–456
induction fields, 450–453
loop antennas, 459–462
Marconi, 454–456
microwave, 472–473
operating power, direct/indirect method, 413–415
parabolic reflectors, 472–473
parasitic elements, 447–450
permeability, 453
permissivity, 453
polarization, 450–453
Poynting vector, 453
propagation of radio waves (see radio communications), 462–472
radar, 499–501
radiation, 450–453
reflectors, 447–450
rhombic, 468
satellites, 701–702
shunt feeding, 455
tower, reference tower, 458
transverse electric magnetic (TEM) waves, 450–453
antennas (*cont.*):
waveguides, 473–481
Yagi arrays, 448–450
antinodes, voltage, 433–441
apparent power, 176–182
associative law, 662–669
asymptote, 551
atmospheric conditions affecting, 465–472
atomic theory, 518
attenuation, 271–272
harmonic attenuation, 760
transmission lines, 426–428
automatic frequency control (AFC) circuit, 402
automatic gain control (AGC), 407–408
automatic volume control (AVC), 407–408
avalanche diodes, 739–740
average values, 161–162
azimuth, angle of, 472, 682–685

B

Baird, John L., 715
balanced/unbalanced antennas, 460–462
bandwidth, 205–209, 214–216
AM, 382–385
fiber-optics, 715
Barkhausen criterion, 342
base bias, transistors, 300–301
base, of number systems, 621–622
base, transistor, 297
baseband signals, fiber-optic, 743–748
batteries, 5–7
discharge/charge cycle, 55–57
voltage regulation, 54–57
B-band satellite stations, 703–704
beam power tube, 369–371
beat frequency, 406
Bellini-Tosi systems, 460
biasing:
amplifiers, 362–365
JFET, 324–327
transistors, 300–306
tubes, 362–365
binary numbers, 622–626
binary-coded decimal (BCD) system, 627–628
binomial series, 585–587
bipolar transistors, 297–306
bit error rate, 747–748
bleeder bias, 364
Boole, George, 639
Boolean algebra, 639–641
absorption law, 662–669
associative law, 662–669
axioms, fundamental laws, 662–669

Boolean algebra (*cont.*):
combinatorial logic circuits, 658–661
commutative law, 662–669
complementary law, 662–669
DeMorgan's theorems, 670–673
distributive law, 662–669
double negation law, 662–669
Harvard charts, 677–678
idempotent law, 662–669
identity law, 662–669
Karnaugh maps, 673–677
logic gate circuits, 642–652
branch circuits, analysis using j operator, 236–237
bridge circuits:
balanced bridge, 52
bridge rectifiers, 290–293
Wheatstone bridge circuit, 51–53
Wien bridge oscillator, 343–344
bridge rectifiers, 290–293
broadband signals, fiber-optic, 743–748
broadcast standards, 758–760
buffer stage, AM transmitter, 391–396
bus networks, 740–748

C

C,R integrator circuit, 135–138
C/R differentiator circuit, 131–132
calculus, differential, 590–593
calculus, integral, 613–618
capacitance and capacitors, 113–115
acceptor circuits, 202–205
C,R integrator circuit, 135–138
C/R differentiator circuit, 131–132
capacitive reactance, 174–176, 228
energy stored in, 115–116
parallel configuration, 120–123
RC time constants, 123–128
resonance, series LCR circuit, 202–205
series configuration, 116–119
sine-wave input voltage, L, C in parallel, 193–196
sine-wave input voltage, L, C in series, 182–185
sine-wave input voltage, R, C in series, 180–182
sine-wave input voltage, R,C in parallel, 190–193
sine-wave input voltage, R,L, C in parallel, 196–199
sine-wave input voltage, R,L,C in series, 185–187
Smith charts, 482–497
capacitive reactance, 174–176, 228
carbon microphones, 392–394
cardioid curves, in graphs, 555–556
carrier sense multiple access w/ collision detect (CSMA/CD), 744
carrier shift, 379
Cartesian coordinate system, 552–556
cathode-follower, 374–377
cathode-ray oscilloscopes (see oscilloscopes)
cathodes, 5–6
C-band satellite stations, 703–704
ceramic microphones, 393–394
cgs systems, 2
channels, JFET, 321
charge, 6
(See also batteries; capacitance and capacitors)
charge density, 111–113
choke filters, 283–288, 289–290
circular measurement, 557–559
circulators, waveguide, 480–481
cladding, fiber-optics, 715, 721–725
Class A power amplifier, 312–314, 363
Class AB power amps, 314–316, 363
Class B power amp, 314–316, 363
Class C RF power amps, 316–318, 363–364
classes, Boolean algebra, 639–641
coefficient of coupling, 219–221
coils, 92–94
coefficient of coupling, 219–221
electromagnetic induction, 94–95
Faraday's law, 94–95
left-hand (generator) rule, 95
Lenz's law, 94–95
mutual inductance, 218–221
mutually coupled, 218–221
mutually coupled, series/parallel, 221–224
parallel-aiding connections, 222–224
parallel-opposing connections, 222–224
polarity marks, 221–222
self inductance, 96–99, 218–221
sense dots, 221–222
series-aiding connection, 222–224
series-opposing connections, 222–224
collector feedback bias, bipolar transistor, 305–306
collector, transistor, 297
color code, resistors, 17–19
combinatorial logic circuits, 658–661
common collector (see emitter follower)
common-base amplifiers, 309–310
common-emitter amplifiers, 306–308
commutative law, 662–669
comparators, operational amplifiers, 337–340
complementary law, 662–669
complements, arithmetic of, 636–639
conductance, 10–12, 170–171
conjugates, of phasors, 233
constant current source, dc, 60–63
constant voltage source, dc, 60–63
contact bias, 364
continuity testing, 40–42
continuous-wave systems, 505–506
control grids, tubes, 350
coordinate systems, graphs, 547–556
copper loss, 251
core loss, 208
cosecant function, 573–576
cosine function, 567–570, 579
cotangent function, 573–576
coulomb, 6
Coulomb's law, 110–111
counting, 621–622
coupling:
coefficient of coupling, 219–221
coupling factor, transformers, 251, 253
critical angle, 720–721
critical frequency, 471–472
cross-modulation, 408, 412–413
cross-over, 315
crystal microphones, 393–394
crystals, piezoelectric, 418–423
frequency monitors, 423–424
Miller oscillator, 420–423
oscillators, 419, 420–423
Pierce oscillator, 421–423
response curves, 419–423
temperature effects, 419
current, 6, 8–9
ac, nonsinusoidal, 276–278
constant current source, dc, 60–63
current division rule (CDR), 44–45
decibel equivalents, 268–271
differentiator circuits, 128–135
displacement current, 116–119
dynamometer wattmeter, 149
effective value, 158–161
electrodynamometer movement, 148–149
electromagnetic induction, 94–95
electromotive force (EMF), 90–91
energy stored in electric field of capacitor, 115–116
energy stored in magnetic field of inductor, 99–100
equivalent generator, 63
Faraday's law, 94–95
flux density, 88
identical (practical) sources in series-parallel, 63–66
inductors in parallel, 107–110
inductors in series, 101–102

current (*cont.*):
Kirchhoff's current law (KCL), 36–39, 66–68, 237–238
L/R time constant, 103–106
left-hand (generator) rule, 95
Lenz's law, 94–95
magnetic field intensity, 90–91
magnetic flux, 87–88
magnetomotive force (MMF), 90–91
magnification of, LCR circuit, 211
mesh analysis, 69–70, 239–240
milliammeters, 141–143
Millman's theorem, 76–77, 243–244
motor effect, 88–89
moving-coil (D'Arsonval) meter
movement, 139–141
moving-iron meter, 147–148
nepers, 271–272
nodal analysis, 71–72, 242–243
Norton's theorem, 81–84, 245–247
phasor addition/subtraction, 166–167, 230–232
phasor multiplication/division, 232–234
phasor representation of voltage/current, 162–164
practical source in parallel, 63–66
practical source in series, 63–66
reciprocity theorem, 249–250
relative permeability, 92–94
reluctance, 92–94
rheostats, 32–34
right-hand rule, 89
ripple, 283–288
root-mean-square (rms) value, 158–161
Rowland's law, 92–94
sawtooth waveform differentiation, 133–134
self-inductance, 96–99
series-parallel resistors, 46–51
superposition theorem, 73–75, 240–241
Thevenin's theorem, 78–81, 244–245
wattmeters, 148–149
current division rule (CDR), 44–45
cut-off, transistors, 298
cut-off, tubes, 350
cycles, 154–155

D

D, E, F layers, 466–472
D'Arsonval meter movement, 139–141
dabble, 624–626
damping, damped waves, 217–218
dark current, 737
Darlington pairs, 310–312
decibels, 267–271
decimal numbers, 512–517, 620–622
de-emphasis circuits, 401–405
degrees, in angles, 556–559
Dellinger fade-outs 466–472
delta-wye transformations, 84–87, 247–248
DeMorgan's theorems, 670–673
depletion mode, MOSFET, 330
depletion-enhancement (DE) MOSFET, 331–333
derivatives, 603–609
detector stage, AM receiver, 407
detectors, light detectors, 737–740
determinants, 533–538
deviation, FM, 396
dielectric loss, 208
differential calculus, 590–593
differentiation, 590–593
exponential functions, 601–603
logarithmic functions, 601–603
of a function of a function, 593–595
product, 595–596
quotient, 597–598
trigonometric functions, 598–601
differentiator circuits, 128–135
C/R differentiator circuit, 131–132
L,R differentiator circuit, 132–133
operational amplifiers, 337–340
sawtooth waveform differentiation, 133–134
diffraction, 463–472
digital principles, 619–678
diodes:
avalanche diodes, 739–740
light detectors, 737–740
light-emitting diodes (LED), 732–737
phototransistors, 738–740
PIN diodes, 737–740
pn junction diodes, 280–281
varactor diode modulator, 401–405
zener diodes, 296–297
(See also solid-state; transistors)
dipole antennas, 447–450
direct current (dc), 1–151
acceleration, 2–3, 4
ampere, 2–3, 6
anodes, 5–6
batteries, 5–7
C,R integrator circuit, 135–138
C/R differentiator circuit, 131–132
capacitance and capacitors, 113–115
capacitors in parallel, 120–123
capacitors in series, 116–119
cathodes, 5–6
cgs systems, 2
charge, 6
charge density, 111–113
color code, resistors, 17
direct current (dc) (*cont.*):
conductance, 10–12
continuity testing, 40–42
coulomb, 6
Coulomb's law, 110–111
current, 6, 8–9
current division rule (CDR), 44–45
current source, constant, 60–63
delta-wye/wye-delta transformations, 84–87
differentiator circuits, 128–135
displacement current, 116–119
dissipation of power, 107
divider circuit, voltage, 26–29, 53–54
dynamometer wattmeter, 149
efficiency, percentage, 57–60
electric field intensity, 111–113
electric flux, 110–111
electrochemical equivalents, 6
electrodynamometer movement, 148–149
electrolysis, law of, 6
electrolytes, 6
electromagnetic induction, 94–95
electromotive force (EMF), 7–8, 90–91
electron-volt (eV), 8
electrostatic induction, 116
electrostatics, 110–111
energy or work, 4
energy stored in electric field of capacitor, 115–116
equivalent generator, 63
Faraday's law, 6, 94–95
flux density, 88
force, 4
force, unit of, 2–3
friction effect, 110–111
ground, 23–25
horsepower, 3–4
identical (practical) sources in series-parallel, 63–66
inductors in parallel, 107–110
inductors in series, 101–102
inductors, energy stored in, 99–100
integrator circuits, 135–139
international system (SI) units, 2
IR drop (see voltage drop), 19
joules, 3–4, 8–9
kilowatt-hours (kWh), 4, 8
Kirchhoff's current law (KCL), 36–39, 66–68
Kirchhoff's voltage law (KVL), 66–68, 66
L,R differentiator circuit, 132–133
L,R integrator circuit, 138
L/R time constant, 103–106
left-hand (generator) rule, 95

direct current (dc) (*cont.*):
Lenz's law, 94–95
magnetic flux, 87–88
magnetomotive force (MMF), 90–91
matched resistance, 58–60
maximum power transfer, 57–60
mesh analysis, 69–70
milliammeters, 141–143
Millman's theorem, 76–77
MKSA system, 2
motor effect, 88–89, 88
moving-coil (D'Arsonval) meter movement, 139–141
moving-iron meter, 147–148
newton-meters, 4
nodal analysis, 71–72
Norton's theorem, 81–84
Ohm's law, 10–12
ohmmeters, 145–147
open circuits, 40–42
oscilloscopes, 150–151
parallel circuits, 36–39
parallel voltage sources, 42–44
permeability of free space, 91
permittivity, absolute and relative, 111–113
potential voltages, 24–25
potentiometers, 32–34
power, 4, 8–9
power factor, 149
practical source in parallel, 63–66
practical source in series, 63–66
RC time constants, 123–128
relative permeability, 92–94
reluctance, 92–94
repulsion, 111
resistance, 10–12
resistors in parallel, 35–39
resistors in series resistors, 19–23
rheostats, 32–34
right-hand rule, 89
Rowland's law, 92–94
sawtooth waveform differentiation, 133–134
self-inductance, 96–99
series voltage-dropping resistor, 34–35
series-aiding voltage source, 30–32
series-opposing voltage source, 30–32
series-parallel resistors, 46–51
short circuits, 40–42
specific resistance, 13–14
static electricity, 110–111
superposition theorem, 73–75
temperature coefficient of resistance, 15–16
Thevenin's theorem, 78–81

direct current (dc) (*cont.*):
torque, 4
triboelectric effect, 110–111
velocity, 3, 4
voltage division rule (VDR), 26–29
voltage drop, 19–23
voltage reference level, 23–25
voltage regulation, 54–57
voltage source, constant, 60–63
voltmeters, loading effect, 143–145
volts, 2–3, 7–8
wattmeters, 148–149
watts, 2–4, 8–9
Wheatstone bridge circuit, 51–53
wire resistance, 14
work or energy, 4
directional antennas, 456–459
directors, antenna, 448–450
discriminators, 410–411
dish antennas (see microwave antennas; parabolic reflectors)
dispersion, fiber-optics, 723–725
displacement current, 116–119
dissipation of power, 107
distributed constants, transmission lines, 426–428
distributive law, 662–669
diversity reception, 465–472
dividers, voltage, 26–29, 53–54
voltage divider bias, bipolar transistors, 302–303
dominant modes, waveguide, 479–481
Doppler effect, 505–506
double dabble, 624–626
double negation law, 662–669
double-conversion system, FM receiver, 409
double-humping, RF transformers, 253–255
double-sideband system, 418
down-conversion, satellites, 697
ducting, 464–472
Dumke, W.P., 733
duodecimal number system, 633–636
duplex communications, 394
duplexers, waveguide, 481
dynamic microphones, 392–394
dynamometer wattmeter, 149

E

eddy current loss, 251
effective radiated power, 415–417
efficiency, 57–60
antenna, 455–456
transformer, 224–227
electric field intensity, 111–113
electric flux, 110–111
electrochemical equivalents, 6
electrodynamometer movement, 148–149
electrolysis, law of, 6
electrolytes, 6
electromagnetic (EM) waves in free space, 450–453
electromagnetic induction, 94–95
electromotive force (EMF), 7–8, 90–91
Faraday's law, 94–95
Lenz's law, 94–95
electron-volt (eV), 8
electrostatic induction, 116
electrostatics, 110–111
elevation, angle of, 472, 682–685
elimination, in math, 530–533
ellipse, in graphing, 551–556
emission designations, FCC, 756–757
emission, primary and secondary, 367–371
emitter bias, bipolar transistors, 303–305
emitter follower (common collector), 310–312
emitter, transistor, 297
energy, 4
motor effect, 88–89
relationships, ac circuit, 207, 211, 215–216
enhancement mode, MOSFET, 330
enhancement-only or E MOSFET, 333–336
equivalent generators, 63
ETHERNET, 744–748
exclusive-NOR operation, 656–658
exclusive-OR operation, 654–655
exponential functions, 587–590, 601–603
extra high tension (EHT) supplies, 286–287

F

farad, unit of capacitance, 113–115
Faraday's law, 6, 94–95
feedback:
negative feedback, 365, 371–374
positive/regenerative feedback, 341–342
ferrite isolators, waveguide, 479–481
fiber-optics, 713–748
absorption losses, 729–730
acceptance cones, 722–725
access methods, 743–748
analysis of networks, systems, 740–748
attenuation, 715–716
avalanche diodes, 739–740
bandwidth, 715

fiber-optics (*cont.*):
baseband signals, 743–748
bit error rate, 747–748
broadband signals, 743–748
bus networks, 740–748
carrier sense multiple access w/ collision detect (CSMA/CD), 744
cladding, 715, 721–725
components of optical fiber, 714–716
critical angle, 720–721
dark current, 737
elements of communication system, 716–718
ETHERNET, 744–748
frequency-division multiplex (FDM), 716–718
graded-index fiber, 726–728
infrared, ultraviolet, hydroxyl ion absorption, 729–730
joins, 730–732
lasers, 733–737
light detectors, 737–740
light-emitting diodes (LED), 732–737
local area networks (LAN), 740–748
LOCALNET, 744–748
losses, 728–730
material dispersion, 723–725
metropolitan area networks (MAN), 746–748
modified chemical vapor deposition (MCVD), 728
multimode fiber, 725–728
phototransistors, 738–740
physics of light, 718–721
PIN diodes, 737–740
power budget, 748
propagation of light through cladded fiber, 721–725
pulse spreading (modal dispersion), 723–725
radiation sensitivity, 737
Rayleigh scattering, 728–730
refraction, 718–721
refractive index, 719–721
responsivity, 737
ring networks, 740–748
scattering losses, 728–730
single-mode, step-index fiber, 725–728
sources of light, 732–737
spectral response, 737
star networks, 745–748
step-index fiber, 725–728
time-division multiplex (TDM), 716–718
topology of networks, 740–748
transit time, 737
waveguide dispersion, 724–725
field strength, 415–417
field-effect transistors (FET), 350
figure of merit, 206
filters:
choke filters, swinging, 290
choke, 280–290, 292
pi section, 284–288
ripple, 283–288
final stage, AM transmitter, 391–396
flux density, 88
force, unit of, 2–3, 4
formulas, 526
Foster-Seeley discriminators, 410–411
Fourier analysis, 154–155
Fourier's theorem, 273–276
fractional equations, 526
fractions, 508–512
frequency, 154–155
angular frequency (velocity), 558–559
bands of, 469–472
beat frequency, 406
critical frequency, 471–472
emission designations, FCC, 756–757
frequency bands, 469–472
frequency-division multiple access (FDMA), 710
frequency-division multiplex (FDM), 687–693, 716–718
heterodyning (beat frequency), 406
high frequency (HF) bands, 469–472
intermediate frequency, 406, 502–503
low frequency (LF) bands, 469–472
lowest usable frequency (LUF), 469–472
maximum usable frequency (MUF), 467–472
medium frequency (MF) bands, 469–472
monitor for, 423–424
optimum working frequency (OWF), 469–472
radar, 499–501
satellite communications, 687–693
Secant Law, 471–472
television broadcast, 424–426
ultra high frequency (UHF) bands, 469–472
very high frequency (VHF) bands, 469–472
very low frequency (VLF) bands, 469–472
(See also transmission lines)
frequency bands, 469–472
frequency diversity, 465–472
frequency modulation (FM), 396–412
automatic frequency control (AFC) circuit, 402
de-emphasis circuits, 401–405
circuit (*cont.*):
direct method of modulation, 400–405
emission designations, FCC, 756–757
frequency shift or deviation, 396
index of modulation, 397–399
indirect method of modulation, 402–405
JFET reactance modulator, 400–405
license, 755
phase modulator, 402–405
pre-emphasis circuits, 401–405
radar systems, 505–506
receivers, 406–413
television broadcast frequencies, 424–426
tolerances, standards, FCC, 758–760
transmitters, 400–405, 754
varactor diode modulator, 401–405
wide band FM, 397
frequency monitors, 423–424
frequency multipler circuits, 316–318
frequency shift or deviation, FM, 396
frequency-division multiple access (FDMA), 710
frequency-division multiplex (FDM), 687–693, 716–718
friction effect, 110–111
front-to-back ratio, 447–450
fundamental of sine wave, 154–155, 273–276

G

gain, satellite communications, 697
gates, logic, 642–658
AND, 645–648
exclusive-NOR (XNOR), 656–658
exclusive-OR (XOR), 654–655
NAND, 652–653
NOR, 650–652
NOT, 648–649
OR, 642–645
general numbers, 524
generators:
alternators, 156–158
equivalent generators, 63
harmonic generators, 316–317
left-hand (generator) rule, 95
geostationary satellites, 682–685, 686, 699–702
Giorgi, 2
goniometers, 460
graded-index fiber, 726–728
gradients, 557–559
graphs, 547–556
grid-leak bias, 364
ground, 23–25
ground or surface waves, 462–472
grounded-grid triode amp, 374–377

H

harmonic attenuation, 760
harmonic generators, 316–317
harmonics of sine wave, 154–155, 273–276
Harvard charts, 677–678
Hertz antennas, 444–447
hertz, 154–155
Hertz, Heinrich, 445
heterodyning (beat frequency), 406, 411–412
heterojunction lasers, 735–737
hexadecimal number system, 631–633
high frequency (HF) bands, 469–472
horsepower, 3–4
hump frequencies, RF transformers, 254–255
hyperbola, in graphing, 551–556
hyperbolic functions, derivatives, 603–607
hysteresis loss, 251

I

ideal transformers, 224–227
idempotent law, 662–669
identity law, 662–669
image antenna, 454–456
image channel, AM receiver, 407
imaginary vs. real values, 230–232
impedance, 176–177
  impedance phasor, 181–182
  load impedance, 489–491
  magnification of, LCR circuit, 211
  matched impedance, 429–432
  reflection coefficient, 432–441
  response curves, 206–207
  stubs, matching, 492–495
  transmission lines, 426–441, 489–491
  unmatched transmission lines, 432–441
index of modulation, 397–399
inductance and inductors:
  acceptor circuits, 202–205
  energy stored in magnetic field, 99–100
  figure of merit, 206
  inductive reactance, 171–173, 228
  L,R differentiator circuit, 132–133
  L,R integrator circuit, 138
  L/R time constant, 103–106
  mutual inductance, 208, 218–221
  parellel configuration, 107–110
  power factor, 206
  resonance, series LCR circuit, 202–205
  self inductance, 96–99,. 218–221
  series configuration, 101–102
inductance and inductors (*cont.*):
  sine-wave input voltages series and parallel, 178–199
induction:
  electromagnetic induction, 94–95
  electrostatic induction, 116
induction fields, antennas, 450–453
inductive capacitance, Smith charts, 482–497
inductive reactance, 171–173, 228
inflexion, 609–613
injection lasers, 733–737
instantaneous values, 129–130
insulated gate FET (IGFET), 321, 330
integral calculus, 613–618
integrator circuits, 135–139
  C,R integrator circuit, 135–138
  L,R integrator circuit, 138
  operational amplifiers, 337–340
intermediate frequency, 406, 502–503
intermediate stage, AM transmitter, 391–396
intermodulation, 394
international system (SI) units, 2
inverting/noninverting operational amplifiers, 337–340
ionospheric propagation, 463–472
IR drop (see voltage drop)
iron loss, 251

J

j operator, 228–230, 234–235, 236–237
joules, 3–4, 8–9
junction field-effect transistor (JFET), 321–329
  amplifier, 327–329
  biasing methods, 324–327
  channels, 321
  gate bias, 324–327
  insulated gate (IGFET), 321, 330
  metal-oxide (MOSFET), 321, 330–336
  pinch-off voltage, 322
  reactance modulator, 400–405
  self-bias, 324–327
  source bias, 325–327
  source follower, 327–329
  transconductance, 323

K

Karnaugh maps, 673–677
kilowatt-hours (kWh), 4, 8
Kirchhoff's current law (KCL), 36–39, 66–68, 237–238, 237
Kirchhoff's voltage law (KVL), 36–39, 66–68, 237–238
knife-edge diffraction, 470–472

L

L,R differentiator circuit, 132–133
L,R integrator circuit, 138
L/R time constant, 103–106
lagging power factor, 176, 193–199
lasers, 733–737
launching satellites, 699–702
layers of atmosphere, 466–472
leading power factor, 177, 180–182, 193–199
least significant digit (LSD), 620–622
left-hand (generator) rule, 95
license, AM station, 750, 752–753
license, FM station, 755
light detectors, 737–740
light, physical properties, 718–721
  absorption, 729–730
  critical angle, 720–721
  dark current, 737
  lasers, 733–737
  radiation sensitivity, 737
  Rayleigh scattering, 728–730
  refraction, 718–721
  refractive index, 719–721
  responsivity, 737
  spectral response, 737
  transit time, 737
light-emitting diodes (LED), 732–737
limiter, FM receiver, 409–410
linear equations, 524–530
line-of-sight transmissions, 464
literal numbers, 524
load impedance, 489–491
load lines, tubes, 360–362
load resistance, 200–202
loading effect, voltmeters, 143–145
local area networks (LAN), 740–748
LOCALNET, 744–748
logarithms, 587–590, 601–603
logic, 622–626
  Boolean algebra, 639–641
  combinatorial logic circuits, 658–661
  DeMorgan's theorems, 670–673
  gates, 642–658
  Harvard charts, 677–678
  Karnaugh maps, 673–677
loop antennas, 459–462
losses:
  fiber-optics, 728–730
  tranmission lines, 426–428, 430, 495–497
  transformer, 251
low frequency (LF) bands, 469–472
low-earth orbit (LEO) satellites, 680–681
lowest usable frequency (LUF), 469–472

M

magnetic field intensity, 90–91
magnetic flux, 87–88
magnetic storms, 467–472
magnetomotive force (MMF), 90–91
make-before-break switch, 142
Marconi antennas, 454–456
matched resistance, 58–60
mathematics, 507–618
matrices, 538–542
maxima, 609–613
maximum power transfer, 57–60
maximum usable frequency (MUF), 467–472
Maxwell, James C., 116
medium frequency (MF) bands, 469–472
mesh analysis, 69–70, 239–240, 540–542
metal-oxide semiconductor FET (MOSFET), 321, 330–336
  amplifiers
    DE MOSFET, 332–333
    E MOSFET, 335–336
  depletion mode, 330
  depletion-enhancement (DE) type, 331–333
  enhancement mode, 330
  enhancement-only or E MOSFET, 333–336
meters/measuring devices
  dynamometer wattmeter, 149
  electrodynamometer movement, 148–149
  frequency monitors, 423–424
  milliammeters, 141–143
  moving-coil (D'Arsonval) meter
movement, 139–141
  moving-iron meter, 147–148
  ohmmeters, 145–147
  oscilloscopes, CR, 150–151
  voltmeters, 143–145
  wattmeters, 148–149
metropolitan area networks (MAN), 746–748
microphones, 392–394
microwave antennas, 472–473
Miller effect, 337, 365
Miller oscillator, 420–423
milliammeters, 141–143
Millman's theorem, 76–77, 243–244
mils, 14
minima, 609–613
mixing, 411–413
MKSA system, 2
modal dispersion, 723–725
modified chemical vapor deposition (MCVD), 728
modulation:
  amplitude modulation (AM), 380–396, 417–418
  angle modulation, 397–399
  cross-modulation, 408, 412–413
  direct method of FM modulation, 400–405
  double-sideband system, 418
  emission designations, FCC, 756–757
  frequency modulation (FM), 396–412, 505–506
  frequency shift or deviation, 396
  high-level, 394
  index of modulation, 397–399
  indirect method of FM modulation, 402–405
  intermodulation, 394
  JFET reactance modulator, 400–405
  low-level, 394
  percentage of modulation, 380–396
  phase modulation (PM), 396–399
  phase modulator, 402–405
  plate modulation, 387–390
  sidebands, AM, 417–418
  single-sideband system, 417–418
  tolerances, standards, FCC, 758–760
  varactor diode modulator, 401–405
monochromatic lasers, 736–737
most significant digit (MSD), 620–622
motor effect, 88–89
moving-coil (D'Arsonval) meter
movement, 139–141
moving-iron meter, 147–148
multihop transmission, 467–472
multimode fiber, 725–728
multiplexing:
  frequency-division multiplex (FDM), 687–693, 716–718
  time-division multiplex (TDM), 693–695, 716–718
multiplicative mixing, 412–413
multipliers:
  frequency, 316–318
  voltage, 294–295
multivibrators, 347–348
mutual inductance, 208, 218–221

N

NAND operation, 652–653
negative feedback, 365, 371–374
negative logic, 622–626
nepers, 271–272
network topologies, fiber-optic, 740–748
Newton, Isaac, 2
newton-meters, 4
nodal analysis, 71–72, 242–243
NOR operation, 650–652
Norton's theorem, 81–84, 245–247
NOT operation, 648–649
null classes, Boolean algebra, 639–641
number systems, 620–622
Nyquist criterion, 342

O

obtuse angles, 557–559
octal number system, 628–631
Ohm's law, 10–12, 168
ohmmeters, 145–147
open circuits, 40–42, 194
operating power, transmitter, 413–415, 458
operational amplifiers, 337–340, 341–342
optimum working frequency (OWF), 469–472
OR operation, 640, 642–645
oscillators:
  AM transmitter, 390–396
  crystal, 419, 420–423
  damping, damped waves, 217–218
  free oscillation in LC circuit, 216–218
  Miller oscillator, 420–423
  Pierce oscillator, 421–423
  RC phase-shift, 345–347
  Wien bridge, 343–344
oscilloscopes, 150–151

P

parabola, in graphing, 549–556
parabolic reflectors, 472–473
parallel circuits, 36–39
parallel voltage sources, 42–44
parasitic elements, 447–450
peak values, 154–155
peak-to-peak values, 154–155
pentodes, 368–371
  variable mu-pentode, 369–371
percentage of modulation, 380–396
periods, 154–155
permeability, antennas, 453
permeability of free space, 91
  relative permeability, 92–94
permissivity, antennas, 453
permittivity, 111–113
phase modulation (PM), 396–399
  emission designations, FCC, 756–757
  phase modulator, 402–405
phase modulator, 402–405
phase velocity, 430–432
phasor representation of voltage/current, 162–164
  addition/subtraction, 166–167, 230–232

phasor representation of voltage/current (*cont.*):
conjugates, 233
impedance phasor, 181–182
multiplication/division, 232–234
rationalization, 232–234, 232
phonetic alphabet, 761
phototransistors, 738–740
pi section, 284–288
Pierce oscillator, 421–423
piezoelectric effect, 418–423
PIN diodes, 737–740
pinch-off voltage, JFET, 322
plate modulation, 387–390
pn junction diodes, 280–281
polar coordinate system, 552–556
polar notation, 228–230, 228
polar/rectangular conversions, 228–230
polarity marks, 221–222
polarization, antennas, 450–453
positional notation, 620–622
positive logic, 622–626
positive/regenerative feedback, 341–342
potential difference, resistors, 10–11
potential voltage, 24–25
potentiometers, 32–34
power, 4, 8–9
apparent power, 176, 178–182
current ratios, 271–272
decibels, 267–271
dissipation of power, 107
efficiency, percentage, 57–60
lagging power factor, 176, 193–199
leading power factor, 177, 180–182, 193–199
maximum power transfer to load (ac), 200–202
maximum power transfer, 57–60
nepers, 271–272
power factor, 149, 176–182
power ratio, 267–271
true power, 176, 178–182
voltage ratios, 271–272
power amplifiers (see amplifiers, power)
power factor, 149, 176–182
figure of merit, 206
inductor, 206
power supplies:
communication satellites, 700–702
constant current source, dc, 60–63
constant voltage source, dc, 60–63
extra high tension (EHT) supplies, 286–287
identical (practical) sources in series-parallel, 63–66
parallel voltage sources, 42–44
power supplies (*cont.*):
practical source in parallel, 63–66
practical source in series, 63–66
series regulator, 319–321
series-aiding voltage source, 30–32
series-opposing voltage source, 30–32
shunt regulator, 318–321
thermistor regulators, 319–321
varistor regulators, 319–321
voltage regulation (see voltage regulators)
(See also voltage regulators)
powers, of numbers, 518–524
Poynting vector, 453
practical transformer, 225–227
pre-emphasis circuits, 401–405
product-over-sum formula, 37, 189, 234
propagation of radio waves (see radio communications)
proximity effect, 208
pulse spreading (modal dispersion), 723–725
push-pull power amp, 314–316
Pythagorean sum formula, 189
Pythagorean theorem, 562–564

Q

Q factor, 205–209, 205
quadratic equations, 543–546
quarter-wave lines, 443–444

R

radar, 498–506
antenna rotation rate, 499–501
average power, 499–501
bandwidth of receiver, 502–503
continuous-wave systems, 505–506
discharge lines, 501–502
Doppler effect, 505–506
frequency modulated systems, 505–506
intermediate frequency, 502–503
parameters of pulsed radar set, 498–501
peak power, 499–501
power gain of antennas, 499–501
pulse duration, 498–501, 501–502
pulse repetition (period), 498–501
pulse repetition (rate), 499–501
pulse shape, 498–501
radio frequency, 499–501
range and intervals, 503–504
receivers, 502–503
video amplifier stage, 502–503
radiation sensitivity, 737
radiation, antennas, 450–453
radio communications, 379–506
absorption, 468
amplitude modulation (AM), 380–396, 417–418
angle modulation, 397–399
anomalous propagation (anaprop), 465–472
antenna current, percentage changes, AM, 385–387
antennas, 415–417, 444–450, 472–473
atmospheric conditions affecting, 465–472
audio amplifiers, 394
bandwidth, AM, 382–385
broadcast standards, 758–760
carrier shift, 379
continuous-wave systems, 505–506
critical frequency, 471–472
cross-modulation, 408, 412–413
D, E, F layers, 466–472
Dellinger fade–outs 466–472
diffraction, 463–472
direct method of FM modulation, 400–405
diurnal changes, 466–472
diversity reception, 465–472
Doppler effect, 505–506
double-sideband system, 418
ducting, 464–472
duplex communications, 394
electromagnetic (EM) waves in free space, 450–453
emission designations, FCC, 756–757
frequency bands, 469–472
frequency diversity, 465–472
frequency modulation (FM), 396–412
frequency monitors, 423–424
frequency shift or deviation, 396
ground or surface waves, 462–472
harmonic attenuation, 760
heterodyning (beat frequency), 406, 411–412
high frequency (HF) bands, 469–472
impedance matching, 432–441
index of modulation, 397–399
indirect method of FM modulation, 402–405
intermediate frequency, 406
intermodulation, 394
ionospheric propagation, 463–472
knife-edge diffraction, 470–472
layers of atmosphere, 466–472
line-of-sight transmissions, 464
load impedance, 489–491
low frequency (LF) bands, 469–472
lowest usable frequency (LUF), 469–472
magnetic storms, 467–472

radio communications (*cont.*):
maximum usable frequency (MUF), 467–472
medium frequency (MF) bands, 469–472
microphones, 392–394
mixing, 411–413
multihop transmission, 467–472
negative transmission, 424
operating power, direct/indirect
method, 413–415
optimum working frequency (OWF), 469–472
percentage of modulation, 380–396
phase modulation (PM), 396–399
piezoelectric crystals, 418–423
plate modulation, 387–390
propagation, 462–472
quarter-wave lines, 443–444
radar, 498–506
receivers, AM and FM, 406–413
rhombic antennas, 468
scatter, 465–472
seasonal changes, 466–472
sidebands, AM, 382–387, 417–418
simplex communications, 394
single-sideband system (SSSB), 417–418
skip distance/skip zone, 467–472
sky or indirect wave, 463–472
Smith charts, 482–497
space diversity, 465–472
space or direct waves, 462–472
sporadic E, 466–472
standing waves, 433–441
sudden ionospheric disturbances (SID), 466–472
sunspot cycles, 467–472
television broadcast frequencies, 424–426
tolerances, standards, FCC, 758–760
transmission lines, 426–444, 489–497
transmitters, AM, 390–396
transmitters, FM, 400–405
traveling waves, 430–432
tropospheric ducting, 464–472
tropospheric scatter, 465–472
two-way communication systems, 394
ultra high frequency (UHF) bands, 469–472
Universal Time Coordinated (UTC), 468
very high frequency (VHF) bands, 469–472
very low frequency (VLF) bands, 469–472
voltage standing wave ratio (VSWR), 441–443, 482–497

radio communications (*cont.*):
waveguides, 473–481
wide band FM, 397
(See also antennas; amplitude modulation; frequency modulation; radar; satellite communications)
range and interval, radar, 503–504
rationalization, 232–234
Rayleigh scattering, 728–730
RC phase–shift oscillator, 345–347
RC time constants, 123–128
reactance, 185–187, 185
capacitive reactance, 174–176, 228
inductive reactance, 171–173, 228
JFET reactance modulator, 400–405
Smith charts, 482–497
real vs. imaginary values, 230–232
receivers, AM, 406–413
receivers, FM, 406–413
receivers, satellite communications, 705–707
reciprocity theorem, 249–250
rectangular notation, 228–230
rectangular/polar conversions, 228–230
rectification and rectifiers, 275–276, 282–294
bridge rectifiers, 290–293
choke filter, swinging, 290
choke filters, 283–288, 289–290, 292
delta/wye connections, 285–286
extra high tension (EHT) supplies, 286–287
full-wave, 288–294
single phase, 288–292
three–phase, 292, 293
half-wave, 282–288
single-phase, 287
three-phase, 287, 285–286
pi section, 284–288
regulation, 285
ripple, 283–288
reference level, voltage, dc, 23–25
reflection coefficient, 432–441
reflectors, 447–450
refraction, 718–721
refractive index, 719–721
regenerative feedback, 341–342
regulators, voltage (see voltage regulators)
relative permeability, 92–94
reluctance, 92–94
repulsion, 111
resistance and resistors, 10–12, 13–14, 168–169
acceptor circuits, 202–205
C,R integrator circuit, 135–138
C/R differentiator circuit, 131–132
color code, 17–19

resistance and resistors (*cont.*):
composition resistors, 17–19
continuity testing, 40–42
current division rule (CDR), 44–45
delta-wye/wye-delta transformations, 84–87
efficiency, percentage, 57–60
identical (practical) sources in series-parallel, 63–66
IR drop (see voltage drop), 19
Kirchhoff's current law (KCL), 36–39, 66–68
Kirchhoff's voltage law (KVL), 66–68
L,R differentiator circuit, 132–133
L,R integrator circuit, 138
L/R time constant, 103–106
matching resistance, 58–60
maximum power transfer, 57–60
mesh analysis, 69–70
milliammeters, 141–143
Millman's theorem, 76–77
moving-coil (D'Arsonval) meter
movement, 139–141
nodal analysis, 71–72
Norton's theorem, 81–84
ohmmeters, 145–147
open circuits, 40–42
parallel circuits, 36–39
parallel configuration resistors, 35–39
potential difference, 10–11
practical source in parallel, 63–66
practical source in series, 63–66
product-over-sum formula, 37
RC time constants, 123–128
reflected resistance, transformers, 226–227
resonance, series LCR circuit, 202–205
series configuration resistors, 19–23
series voltage-dropping resistor, 34–35
series-parallel resistors, 46–51
short circuits, 40–42
sine-wave input voltages, series and
parallel, 178–199
Smith charts, 482–497
specific resistance, 13–14
superposition theorem, 73–75
temperature coefficient, 15–16
Thevenin's theorem, 78–81
tubes, ac/dc plate resistance, 351–353
voltage drop, 19–23, 34–35
Wheatstone bridge circuit, 51–53
wire, 14
resonance
parallel resonant LCR circuit, 209–212
parallel resonant tank circuit, 213–216
series LCR circuit, 202–205

resonant frequency, 206
responsivity, 737
RF power amps (see amplifiers, RF power)
RF transformers, 250–256
rheostats, 32–34
rhombic antennas, 468
ribbon microphone, 393–394
right angles, 557–559
right-hand rule, 89
ring networks, 740–748
ripple, 283–288
Rowland's law, 92–94

S

satellite communications, 679–712
  A-, B-, and C-band stations, 703–704
  analysis of satellite comm system, 707–712
  antennas, 701–702
  azimuth, angle of, 682–685
  commercial uses of, 682–685
  components of satellite, 696–698
  down-conversion, 697
  early satellite composition, 680–681
  elevation, angle of, 682–685
  frequencies, 687–693
  frequency-division multiple access (FDMA), 710
  frequency-division multiplex, 687–693
  gain, 697
  geostationary satellites, 682–685, 686, 699–702
  launching satellites, 699–702
  low-earth orbit (LEO), 680–681
  power supplies, 700–702
  receiver, earth station, 705–707
  signals of satellites, 693–695
  signals of, 687–693
  signal-to-noise (SNR) ratio, 708–712
  space loss, 708–712
  spacing of satellites, 686
  time-division multiple access (TDMA), 710
  time-division multiplex, 693–695
  transmitter, earth station, 703–704
saturation, transistors, 298
sawtooth waveform, 133–134, 154–155
scatter, 465–472
scattering losses, 728–730
scientific notation, 518–524
screen-grid tubes (tetrode), 367–371
seasonal changes 466–472
secant function, 573–576
Secant Law, 471–472
selectivity, 205–209, 211, 214–216
self-inductance, 96–99, 218–221
sense dots, 221–222
series regulator, 319–321
series voltage-dropping resistor, 34–35
series-aiding voltage source, 30–32
series-opposing voltage source, 30–32
series-parallel circuits, dc, 46–51
short circuits, 40–42
shunt regulator, 318–321
sidebands, AM, 382–387, 417–418, 756–757
signal bias, 364
signal-to-noise (SNR) ratio, 708–712
simplex communications, 394
simultaneous equations, 530–533
sine function, 564–567, 579–582
sine wave, 154–155
  analysis of waveforms, 273–276
  average value, 161–162
  cycles, 154–155
  effective value, 158–161
  Fourier analysis, 154–155
  Fourier's theorem, 273–276
  frequency, 154–155
  fundamental of sine wave, 154–155, 273–276
  harmonics of sine wave, 154–155, 273–276
  hertz as unit of, 154–155
  input voltages, series and parallel, 178–199
  nonsinusoidal ac voltage,current, 276–278
  peak values, 154–155
  peak-to-peak values, 154–155
  periods, 154–155
  phase relationships, 164–166
  rectification, 275–276
  root-mean-square (rms) value, 158–161
  sawtooth waves, 154–155
  sine waves, 154–155
  square waves, 154–155
  synthesis of waveform, 274
  (See also alternating current)
single-mode, step-index fiber, 725–728
single-sideband system (SSSB), 417–418
skin effect, 207–209
skip distance/skip zone, 467–472
sky or indirect wave, 463–472
Smith charts, 482–497
solid-state, 279–348
  amplifiers
    common-base, 309–310
    common-emitter, 306–308
    DE MOSFET, 332–333
    E MOSFET, 335–336
    JFET, 327–329
    operational amplifiers, 337–340
solid-state, amplifiers (*cont.*):
    positive/regenerative feedback, 341–342
    power, Class A, AB, C, 312–316
    RF power, Class C, 316–318
  Barkhausen criterion, 342
  base bias, bipolar transistor, 300–301
  bipolar transistors, 297–299
  bridge rectifiers, 290–293
  collector feedback bias, bipolar transistor, 305–306
  Darlington pairs, 310–312
  doubler, full-wave, 294–295
  doubler, half-wave, 294–295
  emitter bias, bipolar transistors, 303–305
  emitter follower (common collector), 310–312
  harmonic generators, 316–317
  insulated gate FET (IGFET), 321, 330
  junction field-effect transistor (JFET), 321–329
  metal-oxide semiconductor FET (MOSFET), 321, 330–336
  multipliers, frequency, 316–318
  multipliers, voltage, 294–295
  multivibrators, 347–348
  Nyquist criterion, 342
  operational amplifiers, 337–340
  oscillators:
    RC phase-shift, 345–347
    Wien bridge, 343–344
  pn junction diodes, 280–281
  positive/regenerative feedback, 341–342
  rectifier, half-wave, 282–288
  rectifiers, full-wave, 288–294
  source follower, JFET, 327–329
  voltage divider bias, bipolar transistors, 302–303
  voltage regulators, 318–321
  zener diodes, 296–297
source follower, JFET, 327–329
space diversity, 465–472
space loss, 708–712
space or direct waves, 462–472
spectral response, 737
split-tuning, RF transformers, 253–255
sporadic E, 466–472
square roots, 520–524
square waves, 154–155
squelch circuits, 408
standing waves, 433–441
star networks, 745–748
static electricity, 110–111
step-index fiber, 725–728
step-up/down transformers, 224–227

straight angles, 557–559
stubs, matching, 492–495
subclasses, Boolean algebra, 639–641
substitution, in math, 530–533
sudden ionospheric disturbances (SID), 466–472
sunspot cycles, 467–472
superposition theorem, 73–75, 240–241
supplementary angles, 557–559
susceptance, 188–199
switches, make-before-break, 142
synthesis of waveform, 274

T

tangent function, 570–572
tank circuits, 213–218
television frequencies for broadcast, 424–426
temperature coefficient of resistance, 15–16
tetrodes, 367–371
thermistor regulators, 319–321
Thevenin's theorem, 78–81, 244–245
time constants:
  C,R integrator circuit, 135–138
  C/R differentiator circuit, 131–132
  differentiator circuits, 128–135
  L,R differentiator circuit, 132–133
  L,R integrator circuit, 138
  L/R time constant, 103–106
  RC time constants, 123–128
time zones, 468
time-division multiple access (TDMA), 710
time-division multiplex, 693–695
topology of networks, 740–748
torque, 4
tower, reference tower, 458
transconductance, 323, 353–354
transformers, 224–227
  copper loss, 251
  coupling factor, 251, 253
  eddy current loss, 251
  efficiency of, 224–227
  hysteresis loss, 251
  ideal power transformer, 252
  ideal transformers, 224–227
  iron loss, 251
  losses, 251
  practical transformer, 225–227
  radio frequency (see RF transformers)
  reflected resistance, 226–227
  step-up/down transformers, 224–227
  (See also RF transformers)
transient state, 123
transistors:
  amplifiers, ???
  (See also diodes; solid-state)
    common-base, 309–310
    common-emitter, 306–308
    DE MOSFET, 332–333
    E MOSFET, 335–336
    JFET, 327–329
    power, Class A, B, C, 312–316
    RF power, Class C, 316–318
  base, 297
  base bias, 300–301
  bipolar transistors, 297–299
  collector, 297
  collector feedback bias, bipolar transistor, 305–306
  cross-over, 315
  cut-off point, 298
  Darlington pairs, 310–312
  emitter, 297
  emitter bias, bipolar transistors, 303–305
  emitter follower (common collector), 310–312
  field-effect (FET), 350
  insulated gate FET (IGFET), 321, 330
  junction field-effect transistor (JFET), 321–329
  metal-oxide semiconductor FET (MOSFET), 321, 330–336
  phototransistors, 738–740
  saturation point, 298
  source follower, JFET, 327–329
  transconductance, 323
  voltage divider bias, bipolar transistors, 302–303
transit time, 737
transmission lines, 426–444, 489–497
  antinodes, voltage, 433–441
  attenuation, 426–428
  distributed constants, 426–428
  heat effects, 426–428
  impedance matching, 429–432
  impedance, 426–441
  load impedance, 489–491
  losses, 426–428, 430, 495–497
  matched lines, 429–432
  phase velocity, 430–432
  quarter-wave lines, 443–444
  reflection coefficient, 432–441
  Smith charts, 482–497
  standing waves, 433–441
  stubs, matching, 492–495
  surge or characteristic impedance, 426–428
  traveling waves, 430–432
  tuned lines, 435
  unmatched lines, 432–441
transmission lines (*cont.*):
  velocity factor, 430–432
  voltage standing wave ratio (VSWR), 441–443, 482–497
  (See also fiber-optics; wire)
transmitters, AM, 390–396, 749
  audio amplifiers, 394
  buffer stage, 391–396
  frequency monitors, 423–424
  intermediate RF power amp, 391–396
  intermodulation, 394
  microphones, 392–394
  operating power, 413–415, 458
  oscillator stage, 390–396
  RF final stage, 391–396
  tolerances, standards, FCC, 758–760
  tower, reference tower, 458
  two-way communication systems, 394
transmitters, FM, 400–405, 754
  automatic frequency control (AFC) circuit, 402
  de-emphasis circuits, 401–405
  direct method of modulation, 400–405
  indirect method of modulation, 402–405
  JFET reactance modulator, 400–405
  operating power, direct/indirect method, 413–415
  phase modulator, 402–405
  pre-emphasis circuits, 401–405
  tolerances, standards, FCC, 758–760
  varactor diode modulator, 401–405
transmitters, satellite communications, 703–704
transverse electric magnetic (TEM) waves, 450–453
traveling waves, 430–432
triboelectric effect, 110–111
trigonometric functions, 576–579, 598–601
trigonometric transformations, 615
triodes, 350–351, 374–377
tropospheric ducting, 464–472
tropospheric scatter, 465–472
true power, 176, 178–182
truth tables, Boolean algebra, 641
tubes, 349–377
  amplification factor, 355–356
  amplifier tube, 365–371
  beam power tube, 369–371
  bias, 362–365
  bleeder bias, 364
  cathode-follower, 374–377
  Class A, AB, B, C amplifiers, 363–364
  contact bias, 364
  control grids, 350
  cut-off, 350

tubes (*cont.*):
dynamic characteristics, 359–362
emission, primary and secondary, 367–371
feedback voltages, 365
grid-leak bias, 364
grounded-grid triode amp, 374–377
load lines, 360–362
Miller effect, 365
negative feedback, 365, 371–374
pentodes, 368–371
plate resistances, ac/dc, 351–353
quiescent (dc) conditions, 357–359
screen-grid tubes (tetrode), 367–371
signal bias, 364
signal conditions, 357–359
tetrodes, 367–371
transconductance, 353–354
triode as amplifier, 356–359
triode tubes, 350–351, 374–377
variable mu-pentode, 369–371
voltage controlled devices, 350
tuned transmission lines, 435
two-way communication systems, 394

U

ultra high frequency (UHF) bands, 469–472
unipole (see Marconi antenna)
Universal Time Coordinated (UTC), 468

V

varactors diodes, modulator, 401–405
variable mu-pentode, 369–371
varistor regulators, 319–321
velocity, 3, 4
angular frequency, 558–559
differentiator circuits, 128–135
instantaneous, 129
velocity (ribbon) microphone, 393–394
velocity factor, 430–432
Venn diagrams, 640–641
vertex, 556
very high frequency (VHF) bands, 469–472
very low frequency (VLF) bands, 469–472
video amplifier stage, 502–503
voltage:
ac, nonsinusoidal, 276–278
antinodes, 433–441
constant voltage source, dc, 60–63
decibel equivalents, 268–271
voltage (*cont.*):
differentiator circuits, 128–135
divider circuit, 26–29, 53–54
drop, voltage drop, 19, 34–35
effective value, 158–161
efficiency, percentage, 57–60
energy stored in electric field of capacitor, 115–116
identical (practical) sources in series-parallel, 63–66
Kirchhoff's voltage law (KVL), 66–68, 237–238
maximum power transfer, 57–60
mesh analysis, 69–70
Millman's theorem, 76–77
moving-coil (D'Arsonval) meter
movement, 139–141
multiplier circuits, 294–295
nodal analysis, 71–72
Norton's theorem, 81–84, 245–247
parallel circuits, 36–39, 36
parallel voltage sources, 42–44
phase relationships, 164–166
phasor addition/subtraction, 166–167, 230–232
phasor multiplication/division, 232–234
phasor representation of voltage/current, 162–164
potential, 24–25
potentiometers, 32–34
practical source in parallel, 63–66
practical source in series, 63–66
reciprocity theorem, 249–250
reference level, 23–25
regulation (see voltage regulators)
root-mean-square (rms) value, 158–161
sawtooth waveform differentiation, 133–134
series voltage-dropping resistor, 34–35
series-aiding voltage source, 30–32
series-opposing voltage source, 30–32
series-parallel resistors, 46–51
superposition theorem, 73–75
Thevenin's theorem, 78–81, 244–245
voltage division rule (VDR), 26–29
voltmeters, loading effect, 143–145
voltage controlled devices, 350
voltage divider bias, bipolar transistors, 302–303
voltage division rule (VDR), 26–29
voltage drop, 19–23, 34–35
voltage magnification factor (Q), 205–209
voltage multipliers, doublers, half- and full-wave, 294–295
voltage regulators, 54–57, 318–321
rectifiers, 285
series regulator, 319–321
shunt regulator, 318–321
thermistor regulators, 319–321
varistor regulators, 319–321
voltage standing wave ratio (VSWR), 441–443, 482–497
voltmeters, loading effect, 143–145
volts, 2–3, 7–8

W

wattmeters, 148–149
watts, 2–4, 8–9
waveguides, 473–481
circular, 478–481
circulators, 480–481
dispersion, 724–725
dominant modes, 479–481
duplexers, 481
ferrite isolators, 479–481
rectangular, 473–477
weber, unit of magnetic flux, 87–88
Wheatstone bridge circuit, 51–53
wide band FM, 397
Wien bridge oscillator, 343–344
wire:
American Wire Gauge (AWG), 14
core loss, 208
dielectric loss, 208
mils, 14
mutual inductance, 208
proximity effect, 208
resistance, 14
skin effect, 207–209
(See also transmission lines)
work or energy, 4
wye-delta transformation, 84–87, 247–248

X

XNOR operation, 656–658
XOR operation, 654–655

Y

Yagi arrays, 448–450

Z

zener diodes, 296–297, 318–321

ABOUT THE AUTHOR

Victor F. C. Veley is Professor of Electronics Technology at the Los Angeles Trade-Technical College, where he served for many years as Dean of Science and High Technology. In other academic posts, he taught engineering electronics, microwave systems, and communications science. The author of Modern Electronics, published by Prentice Hall, he has also written numerous technical articles